Imaging in Neuroscience and Development

A LABORATORY MANUAL

EDITED BY

RAFAEL YUSTE
Columbia University

ARTHUR KONNERTH
Institut für Physiologie der Ludwig-Maximilians-Universität München

COLD SPRING HARBOR LABORATORY PRESS
Cold Spring Harbor, New York

Imaging in Neuroscience and Development
A LABORATORY MANUAL

Publisher	John Inglis
Acquisition Editor	David Crotty
Developmental Editor	Siân Curtis
Project Coordinator	Inez Sialiano
Permissions Coordinator	Maria Falasca
Production Editors	Pat Barker and Dotty Brown
Desktop Editors	Danny DeBruin and Stephanie Collura
Production Manager	Denise Weiss
Cover Designer	Ed Atkeson

Front cover artwork: Image showing a montage of a mouse embryo expressing the Runx3 transcription factor, a human brain, and functional map obtained by optical imaging. Courtesy of A. Grinvald and Y. Groner.

Back cover artwork: Two-photon images of living pyramidal neurons and their dendritic spines. Courtesy of A. Konnerth, R. Yuste, and Michael Noll-Hussong.

Library of Congress Cataloging-in-Publication Data

Imaging neuroscience and development : a laboratory manual / edited by Rafael Yuste, Arthur Konnerth.
p. cm.
Includes bibliographical references and index.
ISBN 0-87969-689-3 (hardcover : alk. paper) -- ISBN 0-87969-692-3 (pbk. : alk. paper)
1. Neurons--Laboratory manuals. 2. Microscopy--Laboratory manuals. 3. Developmental neurophysiology--Laboratory manuals. I. Yuste, Rafael. II. Konnerth, Arthur.
QP357.I48 2005
573.8'028'2--dc22

2004018170

10 9 8 7 6 5 4 3 2 1

With the exception of those suppliers listed in the text with their addresses, all suppliers mentioned in this manual can be found on the BioSupplyNet Web site at: http://www.biosupplynet.com.

All World Wide Web addresses are accurate to the best of our knowledge at the time of printing.

Procedures for the humane treatment of animals must be observed at all times. Check with the local animal facility for guidelines.

Certain experimental procedures in this manual may be the subject of national or local legislation or agency restrictions. Users of this manual are responsible for obtaining the relevant permissions, certificates, or licenses in these cases. Neither the authors of this manual nor Cold Spring Harbor Laboratory assumes any responsibility for failure of a user to do so.

All Cold Spring Harbor Laboratory Press publications may be ordered directly from Cold Spring Harbor Laboratory Press, 500 Sunnyside Blvd., Woodbury, NY 11797-2924. Phone: 1-800-843-4388 in Continental U.S. and Canada. All other locations: (516) 422-4100. FAX: (516) 422-4097. E-mail: cshpress@cshl.edu. For a complete catalog of all Cold Spring Harbor Laboratory Press publications, visit our World Wide Web Site http://www.cshlpress.com.

To our families

To train young people to grind lenses, and to found a sort of school for this purpose, I can't see there'd be much use ... because most students go there to make money out of science, or to get a reputation in the learned world. But in lens-grinding, and discovering things hidden from our sight, these count for nought.... And I am satisfied too that not one man in a thousand is capable of such study; because it needs much time, and spending much money; and you must always keep on thinking about these things, if you are to get any results. And over and above all, most men are not curious to know...

A.V. Leeuwenhoek, letter to Leibniz on 28 Sept 1715 in response to Leibniz's request that he should educate a school of younger men in his art.

Sometimes you can tell a lot about something just by looking.

YOGI BERRA

Contents

SECTION 3: CALCIUM IMAGING

SECTION 5: ADVANCED IMAGING AND SPECIAL APPLICATIONS

SECTION 8: PRINCIPLES AND INSTRUMENTATION

Foreword

We are witnessing a change in paradigms in biology. After a very successful century in which biological sciences have developed from a descriptive to an experimental science, biology is maturing into a full-blown quantitative science like its older sisters, physics and chemistry. In fact, a veritable merger between disciplines is taking place, to the point that biological problems are increasingly attacked by researchers with diverse scientific backgrounds. Even within the same research group, cross-fertilization of different approaches is becoming commonplace.

Essential to this new era is the development of new techniques. Indeed, it is clear that progress in research is tied inextricably to advances in techniques. This is particularly true for biology. For example, X-ray crystallography opened the way to solving the structure of DNA and to molecular biology. Technical advances in cloning and sequencing led to the recombinant DNA revolution that has given scientists the sequences of entire genomes. Advances in bioinformatics now allow researchers to access sequence data online in an instant, changing the way experiments are done. Moreover, quantitative methods and analysis tools, many imported from physics and engineering, are transforming traditional biological problems as diverse as developmental embryology, the study of biochemical pathways, or systems neuroscience. Finally, the synthesis of novel chemical tools, from indicator dyes to specific probes selected in combinatorial assays, is creating new fields and enabling experiments that, only a short time ago, were mere dreams. Overall, the traditional boundaries between different sciences are blurring, and future leaders in the research enterprise will increasingly have to equip themselves with an extraordinary array of technical approaches and analytical methods.

Imaging is playing a central role in this new era, as it represents the application of optical techniques to living samples. Recognizing the likely importance of imaging technologies, in 1991 Cold Spring Harbor Laboratory began a summer course solely devoted to imaging in neuroscience, in which novices could learn the often tricky art of applying these novel microscopy approaches to problems in cellular and developmental neurobiology. The great demand for this course resulted in the first edition of this book, *Imaging Neurons: A Laboratory Manual*, published in 2000, which was the first comprehensive review of the field. The success of this first manual and the commonality by which imaging techniques are being applied throughout biology have prompted us to prepare this second edition, in fact practically a new text altogether, in which neuroscience and other disciplines such as developmental biology, as well as many cell biological and biophysics applications, are also covered. This new manual fulfills a cherished goal of the Laboratory to provide professional technical education of the highest quality for the next generation of working scientists and, thus, to help usher in this new era.

Preface

One of the central themes that runs through the study of biology is the relation between form and function. Nature obviously does not discriminate between form and function; rather, this distinction has been artificially created by man to try to capture and comprehend the constant change and transformation of most biological systems. This dynamic aspect of biology is one of its most fascinating characteristics, and it draws generation after generation of students absorbed in understanding how an organism develops, how a cell functions, or how the brain works.

We define imaging as the application of optical techniques to study *living* organisms. It therefore constitutes a direct method to characterize the form and the function of cells and tissues. Imaging is a recent term of wide usage, and we use it here specifically to review imaging techniques that are carried out in the laboratory. Although it seems natural to use light to study the structure and function of cells or tissues, and although microscopists have been doing this with fixed preparations since Leeuwenhoek's time, it is only recently that imaging of living preparations has become standard practice. It is not an overstatement to say that imaging technologies have revolutionized research in many areas of biology. In addition to advances in microscopy such as differential interference contrast or the introduction of video technology and digital cameras, the development of methods to culture cells, to keep tissue slices alive, and to maintain living preparations on microscopes has opened new areas to biologists. The synthesis of fluorescent tracers and indicator dyes and the recent development of green fluorescent protein technology and quantum dots have made possible studies characterizing the form and function of cells with unprecedented detail, down to the single-molecule level. Until recently, confocal microscopy was the state-of-the-art imaging approach because of its superb spatial resolution and three-dimensional sectional capabilities; however, the development of two-photon excitation has enabled fluorescent imaging of small structures in the midst of highly scattered living media with reduced photodamage. Finally, other nonlinear optical techniques, such as second harmonic generation, appear well suited for the measurements of voltage and biochemical events in the plasma membrane and could therefore play a major role in the future.

This manual originated in the Cold Spring Harbor Laboratory course, "Imaging Structure and Function of the Nervous System," which has been taught since 1991. Every summer a dozen qualified applicants from all over the world are selected as students and trained in the latest technologies by a cadre of course directors, teaching assistants, and invited speakers, a teaching staff which outnumbers the students two, or even three, to one. The course runs for three weeks with a daily schedule of lectures in the morning, followed by intense laboratory work in the afternoon, evening, and well into the night. Since its inception, the course has quickly become a "watering hole" for the imaging community and especially for cellular and developmental neurobiologists, traditionally always open to microscopy and imaging approaches.

As course directors, we became motivated to pass the knowledge gathered every summer at the course to the entire community by publishing a manual, and it became quite natural to do this with CSHL Press. As experimentalists working in the laboratory, we also decided not to write a comprehensive textbook, but instead to make a practical manual using the combined expertise of the entire field. Our first manual, published in 2000, focused solely on neuroscience and was edited jointly with Fred Lanni. It was very well received by our colleagues and students.

The good reception of the first edition, together with rapid advances in imaging techniques, prompted us to prepare a second edition of the manual. At the same time, the increased blurring between neuroscience and developmental biology encouraged us to encompass both disciplines, so the original structure of the manual has been greatly revised, and many new chapters have been added. In addition, in this new version of our manual, we have decided to maintain a balance between comprehensive reviews of particular key topics and "practical guides," streamlined brief chapters with a more focused application or experiment in mind. The targeted audience of this new book includes students and researchers interested in imaging in neuroscience and developmental or cell biology. Our aim has been to publish a manual that, like other CSHLP manuals, investigators can have and consult at their setup or bench. We keep the theory to the fundamentals and concentrate instead on passing along the little tidbits of technical knowledge that make a particular technique or an experiment work and that are normally left out of the methods sections of scientific articles.

Section 1 is the equivalent of Imaging 101, where, starting with a practical chapter on how to keep cells on a microscope alive and healthy, we have collected a sample of chapters designed to get the experimentalist up and running without bogging him or her down with theory and unnecessary details. Section 2 covers developmental biology, with a combination of technique- and preparation-oriented chapters. Section 3 focuses on calcium imaging, reflecting both the importance of the study of calcium in today's biology and the quality and usefulness of the BAPTA-based calcium indicators. Section 4 covers photoactivation, concentrating on calcium and glutamate uncaging. The ability to optically manipulate the concentration of a substance in a small region of a cell or a tissue will turn imaging from a descriptive technique into an experimental technique. Section 5 encompasses a group of more specialized imaging needs, ranging from single-molecule imaging to nonlinear optical techniques such as second harmonic generation. In Section 6, we review applications of green fluorescent protein and luminescent imaging, areas that are cross-fertilizing all aspects of imaging and are bridging the gap between molecular biologists and physiologists. Section 7 is devoted solely to whole-animal imaging, something which is becoming increasingly more important in modern biology. There is an unavoidable overlap between this and the development section, and we have placed most in vivo development chapters in Section 2. Finally, the manual concludes with Section 8, which covers the basics of light microscopy, light sources, cameras, and image processing and novel technologies—liquid crystal, acousto-optical tunable filters, and grating systems.

Like its predecessor, this book is not a microscopy textbook. Although we do cover the basics, we refer readers interested in a comprehensive treatment of light microscopy to many of the excellent books published in the last decades. As the title implies, this manual also does not cover hospital-based imaging, like MRI, fMRI, PET, or MEG. Although they are, of course, imaging techniques in their own right, their nonoptical nature and their technical complexities preclude us from covering them in a laboratory manual. Finally, we do not cover electron microscopy or other forms of imaging of fixed specimens. Again, we refer the interested reader to the excellent monographs already published on these techniques.

We thank all the people and institutions that have made this manual possible. First, we salute the effort of all the other past and present instructors of the course (and similar imaging courses at other institutions). Running this course over the years has been a team effort that benefits the field as a whole, and this manual is the result of such cooperation and selflessness by all its instructors. Additionally, all the students of the course have kept us on our toes with their inquisitive questions, unparalleled enthusiasm, and late-night experiment sessions. Particular thanks go to the teaching assistants, who helped run the course smoothly and were instrumental in assembling and disassembling all the setups in record time. Special thanks to Sue Hockfield, Terri Grodzicker, Bruce Stillman, and Jim Watson, who conceived and supported the course over the years. The staff at CSHL Press has been exceptional in all respects, and our special gratitude goes to John Inglis, responsible for the first edition, and David Crotty, who generated the ideas and enthusiasm behind this second edition. In this second edition, Inez Sialiano, Liz Powers, Daniel deBruin, Pat Barker, Stephanie Collura, and Jan Argentine provided fuel to the fire to keep the book moving, and Siân Curtis edited the book with speed, precision, and intelligence and is the main person responsible for its timely publication.

A separate thanks goes to the institutions that have made the course possible. We were extremely lucky that practically all companies contacted generously lent equipment or donated supplies and reagents to the course. The imaging course, since its beginnings, has been financially supported by the generosity of the NINDS, the NIMH, and the Howard Hughes Medical Institute. In addition, we specially thank Fred Lanni for coediting the first edition of the manual and for all the wonderful late-night discussions of microscopy. Finally, we thank our families for sharing us with CSHL, and for supporting us in our careers as lens grinders.

RAFAEL YUSTE
ARTHUR KONNERTH

CHAPTER 1

Maintaining Live Cells and Tissue Slices in the Imaging Setup

Michael E. Dailey, Glen S. Marrs, and Dana Kurpius

Department of Biological Sciences, University of Iowa, Iowa City, Iowa 52242

The development of new fluorescent probes and imaging technologies for live-cell imaging has generated a good deal of excitement over the last decade. One of the challenges now facing the field is the application of these technologies to live biological samples that remain viable in the imaging setup. Although some preparations permit high-resolution imaging of live cells in vivo, in many cases (especially for cells of mammalian origin) it is still necessary to isolate the cells and tissues so that they may be stained and subsequently observed over time. Such procedures can compromise the health of the cells and tissues. The length of time needed to make the required observation will clearly dictate the kinds of measures taken to maintain the specimens. Experimental observations made over the course of many hours require a different level of specimen maintenance than those requiring only a few minutes. Under optimal conditions, it is possible to maintain viable tissues on the stage of a microscope for many hours—even days—while collecting valuable information on dynamic events in live cells and tissues.

How can biological samples be mounted and maintained to last long enough to make meaningful observations? This chapter offers some practical advice based on the authors' experiences imaging neurons and glia in neonatal and adult rodent brain tissue slices (O'Rourke et al. 1992; Dailey and Smith 1994, 1996; Dailey et al. 1994, 1999; Dailey and Waite 1999; Marrs et al. 2001; Stence et al. 2001; Dailey 2002; Grossmann et al. 2002; Petersen and Dailey 2004). Although the specific requirements of different samples will vary, there are some general principles to consider when attempting to maintain live specimens in the imaging setup. The emphasis here is on simple, low-cost approaches for maintaining live mammalian brain cells and tissue slices in imaging setups.

KEEPING THE SPECIMEN HAPPY IN THE IMAGING SETUP

There are several considerations for maintaining live cells and tissues in the imaging setup, including temperature, pH, gas exchange, and light exposure. The importance of these varies with the particular specimen, the spatial and temporal sampling requirements, and the expected length of observation.

Medium Considerations

The medium in which the specimen is mounted must support the cells and tissues for the duration of the observation. When working with cultured cells and tissues, it is necessary to consider the changes that may occur when the sample is taken from the environment of a culture incubator to ambient air. Most bicarbonate-based culture media are designed to buffer an air environment composed of ~5% CO_2, but in ambient air, these media rapidly (within ~5 minutes) become alkaline. Thus, by the time the sample is mounted, the gas content and pH of the culture media may have changed substantially. Such changes should be monitored. For determining pH changes, a pH indicator (phenol red) can be included in the chosen medium, or pH indicator strips (colorpHast; e.g., EMD Chemicals) may be used to test the media at different stages in the mounting and imaging procedure. In the authors' experience, a saline solution (e.g., phosphate-buffered saline) is sufficient to maintain neonatal mammalian brain tissue slices for a few hours. Longer observations usually require the more substantial nutritional support of culture media. In addition, long-term imaging of mammalian tissues derived from more mature animals (>postnatal day [PND]7) requires continuous perfusion of oxygenated media.

Mounting Live Specimens for Microscopic Observation

This section describes a simple mounting technique for rodent organotypic hippocampal slice cultures grown on filter culture membranes, a popular in vitro neural tissue slice preparation (Stoppini et al 1991; Gähwiler et al. 1997). These slice cultures are readily mounted and imaged because the culture membrane provides a stable platform that can be secured within a chamber without applying pressure directly to the tissues. The authors use translucent filter membrane inserts (Falcon 3090 or 3102) containing polyethylene terephthalate, track-etched porous membranes (1-µm pore size). Typically, two to five slices are cultured on a single membrane insert. Although it is possible to image samples through a translucent filter membrane using low-magnification or long-working-distance lenses, the highest resolution will be achieved by mounting samples in a way that allows direct imaging without an intervening filter membrane. The mounting technique described here (Fig. 1) is especially useful for imaging tissues through a coverslip on an inverted microscope system, but it could be adapted for an upright system in conjunction with saline or water-dipping lenses.

First, a membrane culture insert with attached organotypic tissue slices is removed from the culture well and the entire insert is placed upside down on a flat surface. A piece of membrane with attached slices is excised from the culture insert by making a rectangular cut through the membrane and around the slices using a razor blade or scalpel. The rectangular piece of membrane is then placed tissue side down in the center of a dry 22 x 50-mm glass coverslip. To minimize air pockets and bubbles, the membrane is lowered slowly onto the coverslip starting at one edge. A very small drop of cyanoacrylate glue is then applied to each of the shorter edges of the membrane, and is spread with the applicator tip across the membrane edge and onto the coverslip. Care is taken to avoid placing glue too near or on top of any tissue slices. The glue is allowed to dry for 10–20 sec. Then a custom- made Plexiglas slide (~3 mm thick) with an oblong hole (~1.5 x 3 cm) in the center is used to form a chamber around the tissue. First, a thin bead of silicone vacuum grease (e.g., Dow Corning) is laid down around the hole in the Plexiglas slide, and the slide is then inverted and carefully positioned around the membrane rectangle and sealed to the coverslip by application of firm pressure. The vacuum grease forms a stable, waterproof seal between the Plexiglas slide and the coverslip, thus forming an open chamber containing the specimen. It is important that the membrane rectangle and glue spots fit entirely inside the hole of the Plexiglas slide so that the grease ring is able to make a tight, leak-resistant seal. HEPES-buffered culture medium (~500 µl) is then added to the sealed well cavity directly on top of the membrane, with care to spread the medium across the entire floor of the well chamber. After transferring the chamber to the microscope stage, a second coverslip (22 x 40 mm or larger) is placed, without sealing, across the top of the well cavity to prevent evaporation of medium during imaging. Medium in the chamber cavity should not directly contact the top coverslip as this can increase focal drift during time-lapse experiments. It is important to mount the samples swiftly to prevent evaporation and thus maintain the salinity and pH of the medium.

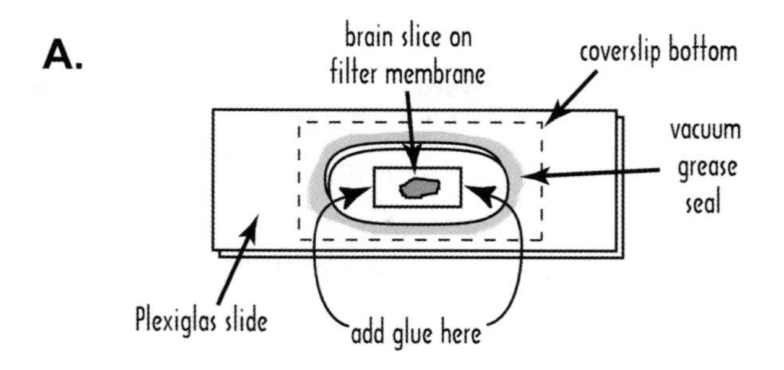

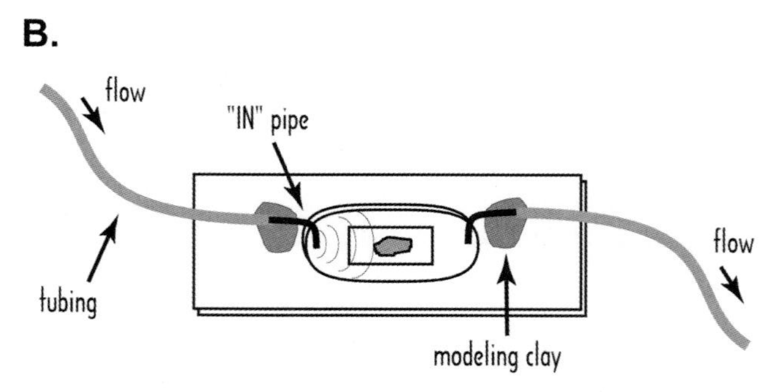

FIGURE 1. A simple specimen chamber for live-cell imaging. (*A*) The specimen chamber is made by drilling out an elongated hole (1.5 x 3 cm) through a thin (3 mm thick) piece of Plexiglas. A rectangular coverslip (22 x 50 mm) is attached to the bottom of the Plexiglas with a ring of silicone vacuum grease. The Plexiglas slide and coverslip are firmly pressed together. The specimen is secured to the coverslip using a plasma/thrombin clot, netting, tape, glue, or small weight, depending on the specimen. Here, a brain slice on a filter membrane is shown attached to a coverslip via cyanoacrylate glue. Then, the chamber is partially filled with medium and covered with a second glass coverslip or kept uncovered for perfusion. (*B*) Perfusate is delivered by gravity feed to one side of the open chamber and withdrawn by vacuum suction from the other side. The tips of two syringe needles are cut off, bent, and inserted into the ends of the tubing to serve as small "in" and "out" pipes. The pipes, which stick down into the chamber well, are held in place with tape or modeling clay. The "out" pipe works best if the tip of the needle is beveled, or if a small filter-paper wick is inserted into the end of the needle. This allows continuous withdrawal of perfusate, rather than a pulsed removal that can induce severe focus jumps. The position of the "out" pipe must be adjusted (raised) so as not to suck the chamber dry.

For some short-term (<2 hours) time-lapse experiments, it may be possible to maintain the specimen in a closed-chamber configuration for the duration of the experiment. Larger volumes of chamber media can support tissues for longer periods of time, although the pH of the chamber media should be monitored. When rather large pieces (or several pieces) of tissue are confined to small volumes of media, the pH of the media may significantly change (acidify) within a couple of hours as a consequence of tissue metabolism. This can induce changes in cell behavior resulting, for example, in greatly reduced cell motility or migration. Mounting the specimen in a larger volume of medium may help, or even periodic exchange of media may adequately support the sample for many hours (O'Rourke et al. 1992). A better solution for long-term imaging experiments is to set up a continuous perfusion system. The rate of perfusion need not be very high—1–2 ml of HEPES-buffered culture medium per hour can support developing rodent brain tissue slices and slice cultures on the stage of the microscope for many hours (Dailey and Smith 1996). The challenge is to set up a perfusion system that does not leak (thus minimizing risk of damage to microscope components) and that does not induce focus artifacts. Dropwise delivery of perfusate to the specimen chamber, as well as periodic sucking of medium from the chamber, can cause pressure changes that translate into focus jumps. A rather simple and inexpensive gravity-fed perfusion setup can be constructed mostly from common laboratory supplies (Fig. 2). This can be integrated into a microscope setup with an external stage heater, as shown in Figure 3. However, more elaborate chamber and perfusion systems can be constructed or purchased commercially.

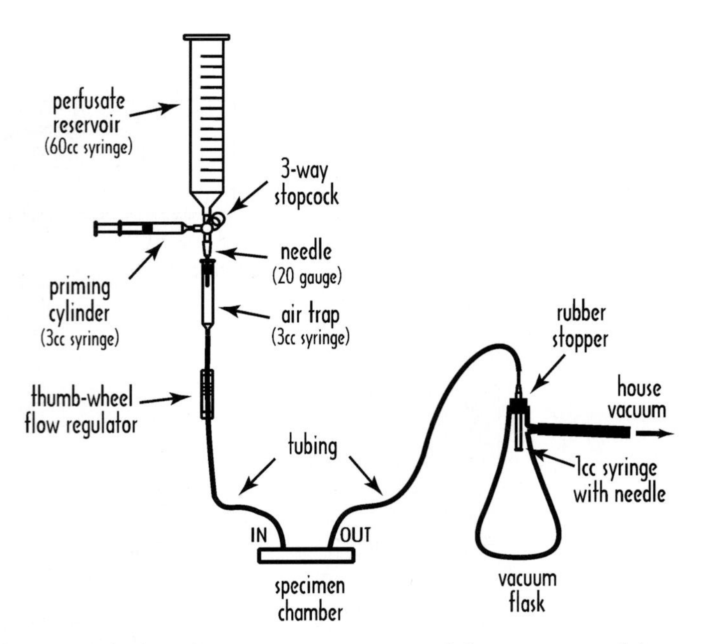

FIGURE 2. A simple gravity-feed perfusion system constructed from common lab supplies. Perfusate is added to the reservoir, and the system is primed using a 3-cc syringe. The flow of perfusate is monitored by observing drip rates in the air trap (3-cc syringe), and the rate is controlled by a thumb-wheel flow regulator taken from a standard medical IV set. A vacuum flask serves as the waste container.

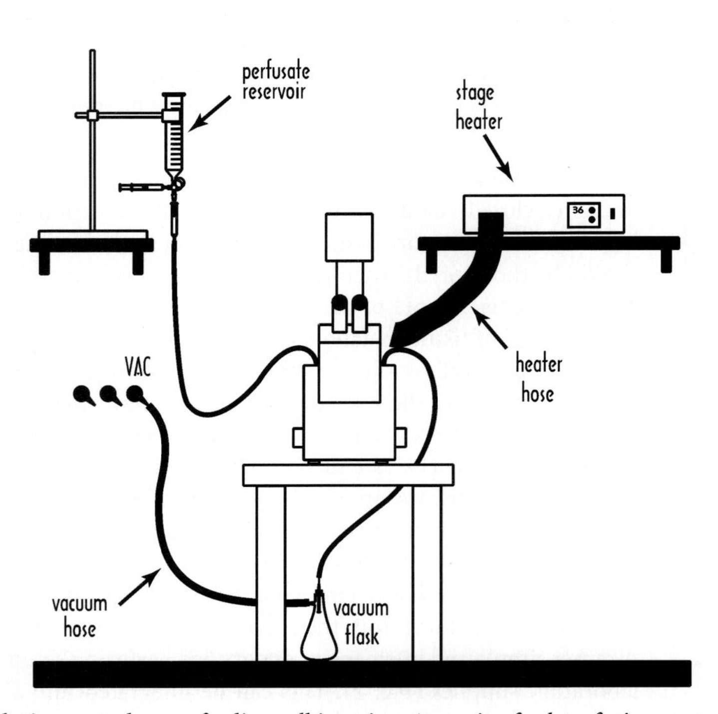

FIGURE 3. A simple, integrated setup for live-cell imaging. A gravity-feed perfusion system and forced-air stage heater are shown in the context of an inverted microscope.

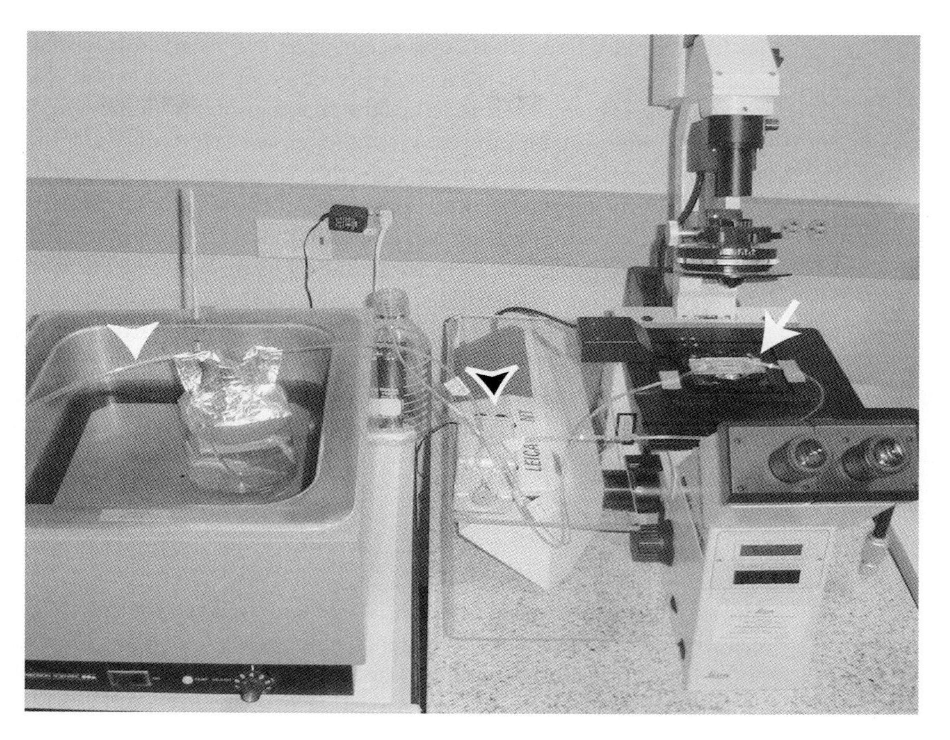

FIGURE 4. Perfusion setup for imaging adult rodent brain tissue slices on an inverted microscope platform. Warm perfusate is maintained in a covered reservoir (1-liter glass bottle) sitting within a warm water bath. For perfusates requiring oxygenation, a tube (*white arrowhead*) is run from the gas tank (95% O_2, not shown) into the reservoir perfusate, and the perfusate is continuously aerated with the gas. Perfusate is delivered to and removed from the specimen chamber using a two-channel, variable-speed pump (*black arrowhead*; e.g., Bioptechs) that maintains constant rates of perfusate delivery and removal (~1 ml/min). The specimen chamber (*arrow*) holds ~2 ml of perfusate. Tubes running to and from the specimen chamber are taped to the mobile portion of the microscope stage to minimize drag. Used perfusate is aspirated from the chamber and pumped into a waste container. Care must be taken to avoid dripping or spilling perfusate on any microscope or electronic components. Here, a custom-made Plexiglas shield (beneath the pump) was built to prevent perfusate from leaking onto electronic components in the confocal scan head.

For mature mammalian brain tissues, it is necessary to provide continuous perfusion of oxygenated medium. Figure 4 shows a simple perfusion arrangement that could be used to maintain mature brain tissue slices. Whatever the exact chamber configuration, for lengthy time-lapse experiments it appears important that the imaging setup provide sufficient tissue oxygenation and removal of metabolic by-products such as CO_2.

Temperature Considerations

Maintaining the specimen at a physiological temperature during the imaging session increases experimenter confidence that the observed phenomena truly reflect native behaviors and not artifacts of the experimental arrangement. Indeed, many dynamic cellular activities are temperature-dependent. In the case of mammalian cells and tissues, it is often desirable to maintain a constant temperature of 34–36°C. This can be accomplished by perfusing warmed medium through the specimen chamber or by blowing warm air onto the preparation. There are very elaborate ways to accomplish this, although it is also possible to put together a crude, inexpensive arrangement that will adequately maintain constant temperature for many hours. For example, a standard inexpensive hair dryer can be used to blow warm air onto the microscope stage. Make sure, however, that the sample is protected from evaporation, which may occur when blowing warm, nonhumidified air over the specimen chamber. Evaporation is reduced or prevented by using a closed chamber configuration, or by continuous replacement (perfusion) of the chamber medium. Blowing warm air from the objective side of the chamber (on an inverted microscope) will help reduce airflow over the chamber and will also serve to more efficiently heat the objective, which can be a significant heat sink when using oil-coupled objectives. On inverted platforms, use of (nondipping) water-immersion lenses in combination with forced-air stage heating can lead to rapid evaporation of the immersion water and loss of the image, but this problem can be overcome by using synthetic immersion medium that has a matched index of refraction (numerical aperature [NA]=1.335; R.P. Cargille Laboratories).

Unfortunately, heating the specimen can introduce further complications such as increased rate of fluorescence photobleaching, or focus drift due to thermal expansion across a stage that is heated nonuniformly. Preheating the stage and chamber can minimize temperature-induced focus drift. The medium should also be preheated to avoid the formation of bubbles within the chamber when cold medium is heated.

Collecting Images

How are images with sufficient spatial and temporal resolution collected without inducing changes in cell structure or physiology? A primary concern when imaging fluorescently labeled specimens is photodamage. The cellular consequences of photodamage can appear within seconds, and the effects of a single photodamaging event can persist for many minutes or even hours. A compromise between the desired spatial and temporal resolution, and the risk of photodamage to the sample, must be reached. This is a particularly important issue when constructing time-lapse image series. Increasing the incident illumination (until fluorophore saturation) can improve the signal-to-noise ratio (SNR) of the image and permit higher rates of image collection for a given SNR. However, most fluorescently stained samples cannot tolerate even modest doses of light for very long. The problem is particularly acute when attempting to sample a 3D tissue volume over time (4D imaging). It is often necessary to sacrifice some spatial and/or temporal resolution, but the experimental goals will dictate what must be sacrificed. For example, for fast physiological events (such as calcium spikes) it may be necessary to reduce exposure time and sacrifice some spatial resolution in order to collect images at the fastest rate possible. On the other hand, for relatively slow events (such as long-distance migration of cells over many hours), the required long imaging sessions can be achieved only at the cost of reduced temporal or spatial resolution.

These considerations raise the question of whether it is better to keep the intensity of incident illumination low, while averaging image frames over a longer period of time (low light, high sampling frequency), or whether it is preferable to turn up the light and reduce the number of images taken (high light, lower sampling frequency). In most scanning (confocal and two-photon) imaging systems, the amount of light delivered to the sample will be a primary determinant of image quality and temporal sampling. For a given incident illumination intensity, image quality (SNR) can be improved by increasing pixel dwell time or by frame or line averaging. Unfortunately, there is no clear consensus as to whether it is better (for the specimen) to absorb less intense light over a longer period of time or to allow the sample sufficient time to "recover" between events of higher illumination intensity. In live specimens, movement of organelles and cells can "smear" the image when several frames are averaged over the period of a few seconds. This is less of a problem when imaging relatively slow events such as cell migration or axon and dendrite growth, but it is acute when attempting to image fast physiological events such as calcium spikes, which tend to be averaged out. Some new commercially available scanning imaging systems are capable of "line averaging," which minimizes the smearing of fast events while maintaining high SNR. Suffice it to say that the optimal imaging parameters vary with different specimens and fluorescent tags, and they need to be determined empirically for each system.

Photodynamic damage of the sample may be reduced somewhat by the inclusion of antioxidants such as ascorbate (vitamin C, 25–30 mM) in the chamber medium. However, the pH of the medium should be monitored—addition of significant amounts of ascorbate will acidify the medium.

KNOWING WHEN THE SPECIMEN IS UNHAPPY

Deterioration of specimens in the imaging setup will limit the usefulness of observations. But what criteria might be used to determine the health of the specimen? The primary consideration would seem to be whether the essential features of cell morphology, physiology, and dynamic behavior change under "control" conditions during the imaging session. If abnormal morphological or physiological features are apparent in the first images, then the culturing, staining, or mounting conditions must be altered. If the changes occur only after the imaging session has begun, one would suspect a problem with the imaging protocol or possibly the mounting procedure.

What are the indicators of ill health? Different cells vary in their response to different conditions. Nerve cells are especially sensitive to imaging conditions. When stained with fluorescent membrane dyes, nerve cell processes can show abnormal features such as beading, blebbing, and triangulation of the branch points. The same cell types in different samples can show different patterns of beading and blebbing that may indicate different stress factors. Declining health can also affect the rates of cell growth and motility, usually resulting in a general decrease in activity. However, ill health is not always indicated by a decline in cellular activity. For example, compromised cells can exhibit a marked increase in Brownian-like organelle motions. Likewise, increases in physiological activity (calcium spikes) are sometimes seen when imaging cells that are loaded with fluorescent indicators (e.g., the calcium indicator fluo-3).

Whatever morphological or physiological criteria are used, a critical question is whether the specimen looks and behaves in a similar way through the course of the imaging session. For fluorescence imaging, it is important to remember that photodynamic damage can occur long before significant photobleaching is observed. The signs of ill health can be hard to discern, especially when experimental perturbations are involved. It can be helpful to compare the morphology of sample cells with control specimens that receive similar treatments (i.e., staining, mounting, drug application, etc.) but are imaged only at the end of the experiment. It can be instructive to sacrifice a few good-looking specimens to clearly define the sequence of changes that occur when the sample is subjected to phototoxic stress. Appropriate control experiments will help identify which cell behaviors are due to the experimental treatment and which are consequences of less-than-optimal imaging conditions.

CONCLUSION

Even as imaging technologies improve, the problems encountered in live-cell imaging can be daunting. Researchers are often limited by the photosensitivity of the biological sample, especially when fluorescent probes are used. Improving the performance of imaging systems will allow higher-quality images to be acquired with greater frequency, but the limits of the biological system will always be tested. Increasingly, imaging involves the simultaneous use of two or three different fluorophores in individual cells. It is necessary, therefore, to become thoroughly familiar with the limitations of both the imaging and biological systems so that sound conclusions about the observed biological phenomena can be made.

ACKNOWLEDGMENTS

Development of the techniques described herein was facilitated by grants from the National Institutes of Health (NS37159 and NS43468), the Whitehall Foundation (#S98-6), the Roy J. Carver Charitable Trust, and the American Cancer Society (#IN-122R) administered through the University of Iowa Cancer Center.

REFERENCES

Dailey M.E. 2002. Optical imaging of neural structure and physiology: Confocal fluorescence microscopy in live brain slices. In *Brain mapping: The methods,* 2nd edition (ed. A. Toga and J. Mazziotta), pp. 49–76. Elsevier, San Diego.

Dailey M.E. and Smith S.J. 1994. Spontaneous Ca^{2+} transients in developing hippocampal pyramidal cells. *J. Neurobiol.* **25:** 243–251.

———. 1996. The dynamics of dendritic structure in developing hippocampal slices. *J. Neurosci.* **16:** 2983–2994.

Dailey M.E. and Waite M. 1999. Confocal imaging of microglial cell dynamics in hippocampal slice cultures. *Methods* **18:** 222–230.

Dailey M.E., Buchanan J., Bergles D., and Smith S.J. 1994. Mossy fiber growth and synaptogenesis in rat hippocampal slices *in vitro*. *J. Neurosci.* **14:** 1060–1078.

Dailey M.E., Marrs G., Satz J., and Waite M. 1999. Concepts in imaging and microscopy: Exploring biological structure and function with confocal microscopy. *Biol. Bull.* **197:** 115–122.

Gähwiler B.H., Capogna M., Debanne D., McKinney R.A., and Thompson S.M. 1997. Organotypic slice cultures: A technique has come of age. *Trends Neurosci.* **20:** 471–477.

Grossmann R., Stence N., Carr J., Fuller L., Waite M., and Dailey M.E. 2002. Juxtavascular microglia migrate along brain capillaries following activation during early postnatal development. *Glia* **37:** 229–240.

Marrs G., Green S.H., and Dailey M.E. 2001. Rapid formation and remodeling of postsynaptic densities in developing dendrites. *Nat. Neurosci.* **4:** 1006–1013.

O'Rourke N.A., Dailey M.E., Smith S.J., and McConnell S.K. 1992. Diverse migratory pathways in the developing cerebral cortex. *Science* **258:** 299–302.

Petersen M.A. and Dailey M.E. 2004. Diverse microglial motility behaviors during clearance of dead cells in hippocampal slices. *Glia* **46:** 195–206.

Stence N., Waite M., and Dailey M.E. 2001. Dynamics of microglial activation: A confocal time-lapse analysis in hippocampal slices. *Glia* **33:** 256–266.

Stoppini L., Buchs P.-A., and Muller D. 1991. A simple method for organotypic cultures of nervous tissue. *J. Neurosci. Methods* **37:** 173–182.

CHAPTER 2

Video Microscopy, Video Cameras, and Image Enhancement

Masafumi Oshiro,[1] Lowell A. Moomaw,[1] and Ernst Keller[2]

[1]*System Division, Hamamatsu Photonics K.K., Hamamatsu City, Shizuoka 430-8587, Japan;*
[2]*Microscope Division, Carl Zeiss, Inc., Thornwood, New York*

Video microscopy is the combination of video technology and microscopy. This combination opens doors to two new fields of microscopy called video-enhanced contrast microscopy (VEC) and video-intensified microscopy (VIM). VEC involves the production of an image from a specimen that is invisible to the eye, either due to a lack of contrast or due to its spectral characteristics (UV or infrared). VIM involves imaging a specimen when the light levels are too low for standard cameras or, in some cases, even for the eye. Images are produced by VIM using image analysis computers (see p. 17).

The major components for VEC are a differential interference contrast (DIC) or phase-contrast microscope, a standard video rate high-resolution camera, a real-time image processor, and a high-resolution video monitor. The use of a video rate camera in conjunction with a real-time image processor greatly improves the contrast of the image and enables the use of the maximum aperture of the optical system. As a result, maximum resolution of the microscope is achieved with sufficient contrast to render the information perceivable.

The VIM method usually employs a fluorescence microscope, a low-light-level video camera, a real-time image processor, and a high-resolution video monitor. The low-light-level video camera and real-time image processor act together to integrate the image. As a result, detailed structure within the specimen is revealed and photo damage to the sample is minimized by reducing illumination intensity. Even in ultralow-light situations, such as bioluminescence and chemiluminescence, a photon-counting camera detects single photon events and an image processor integrates the image to generate gray-level information based on photon counting per pixel.

This rapidly growing field is well on its way to replacing conventional photomicroscopy. Technical advances in cameras, camera electronics, and image processors, as well as printers, have reached a point where the resolution of the video image becomes indistinguishable from that of photographic fine-grain film. Although it is substantially more expensive, the added technical benefits of video microscopy are considerable. These include:

- Direct display and optional recording of dynamic events, including time-lapse studies of live cells.
- Availability of a wide range of analog and digital cameras for every budget and to suit many specific applications.

Modified slightly, with permission, from Spector et al. 1998c. (Material provided by M. Oshiro, L.A. Moomaw, and E. Keller.)

- Electronic contrast enhancement via gain and black-level control, or digital image processing, reveals image information inaccessible to eye or film.
- Recording of exceptionally low light levels with time integrating or intensified (e.g., ICCD) cameras (see below). This is important in studies involving weak fluorescence emissions in vivo (e.g., following microinjection of fluorochrome-labeled proteins or transfection of cells with green fluorescent protein (GFP) constructs; see Spector et al. 1998a) or for minimizing the phototoxic effects of illuminating cells with high light intensities.
- Electronic color balancing with single, three-tube, or chip color cameras.
- Electronic image transfer to other locations.
- Optional image storage for further processing or image analysis.

To take full advantage of video-enhanced microscopy, it is important to know a few basic facts about video cameras and image processors. There are many versions of each, and each version has certain characteristics that must be considered for different applications.

DEFINITIONS

A number of terms used throughout this chapter that may be unfamiliar to the reader are defined below.

- **Analog and digital cameras:** Analog video signal is generated from the detector itself. This video signal is processed and converted to certain formats to interface to display monitors, image storage devices, and image processors. If this interface signal is analog, the camera is called an analog camera. If the interface signal is digital, the camera is called a digital camera. Analog cameras normally generate a standard video signal such as RS-170.
- **Integration camera:** The standard video camera's frame rate is 30 frames/sec. This means that incoming light is integrated in a detector for 1/30 sec and the signal is read out after the integration. To realize low-light sensitivity, integration cameras integrate incoming light for more than one frame, and the integrated signal is then read out. The sensitivity is increased in proportion with the number of integrating frames. However, maximum integration frames are limited by the dark noise of the camera itself.
- **Video rate camera:** A camera that generates a standard analog video signal, such as RS-170, is called a video rate camera.
- **Image processor:** An image processor modifies, analyzes, displays, and stores images.
- **Readout speed:** The speed at which each pixel is read out of a charge-coupled device (CCD) camera. The unit is MHz/pixel or kHz/pixel.
- **Linearity vs. lag:** Linearity is the relationship between incoming light to a video camera and the output signal from the video camera. If some signal generated in a detector remains in the detector after the reading, the leftover signal appears as the lag.
- **Frame buffer:** A frame buffer (or memory plane) stores one or more complete images for further image processing.
- **Horizontal vs. vertical resolution:** The horizontal and vertical spatial resolutions are not always the same. In the case of video rate cameras, the horizontal resolution is higher than the vertical. Normally, slow-scan cooled CCDs have square pixels, and the resolution for both directions is the same.
- **Geometric distortion:** Tube-type cameras and intensified cameras have geometric distortion. The distortion of tube cameras is due to the mechanism of the beam scanning. The distortion of intensified cameras is due to the intensifiers and relay optics, such as relay lens and tapered fiber plates.
- **Spectral range:** The spectral sensitivity range of detectors.
- **Temporal resolution:** The speed at which one complete image is acquired determines the temporal resolution. The temporal resolution of video rate is 33 msec. The temporal resolution of slow-scan cooled CCD cameras is low due to the slow readout speed and the time of signal integration.
- **Camera nonuniformity:** Detectors do not have uniform sensitivity over an entire area. Camera nonuniformity of tube cameras and intensified cameras is higher than that of CCD cameras.

VIDEO CAMERAS

The following are two basic classifications of cameras: high-light-level cameras and low-light-level cameras. A common component to the two classifications is a CCD camera. This is an electronic device that incorporates a two-dimensional photodetector, an amplifier, and a timing device that produces a signal that conforms to one of the industry standards. These standards are indicated by the terms RS-170, RS-330, RS-343, NTSC, RGB, and others. This signal transfers an image from the detector to an image processor, a storage device, or display device. The details of this operation can be found in a number of sources.

High-light-level Cameras

Although most low-light-level cameras work in high light levels, these cameras are expensive. If high-light-level images are to be recorded, the use of a high-light-level camera offers benefits in both cost and convenience. The following are characteristics of high-light/CCD cameras.

- Detectors available in 1/3″, 1/2″, 2/3″, and larger formats.
- The size of the detector does not determine the number of horizontal pixels.
- Horizontal resolution is a function of the number of horizontal pixels.
- Spectral sensitivity ranges from under 200 nm to 1000 nm.
- Linearity of the photometric response is very good.
- Virtually no lag.
- Will not be damaged by high light levels.
- No geometric distortion.
- Horizontal resolution is greater than vertical resolution.

The following special features of the high-light camera are essential for optimal VEC.

- Analog contrast enhancement to "stretch the gray levels" for low-contrast images.
- Analog shading correction used to compensate for illumination and camera nonuniformity.
- Video level indicator to adjust the proper light level for the camera.

Low-light-level Cameras

Low-light-level cameras use one of two methods to produce an image in situations where the number of photons is too low to create a meaningful image for the human eye. The camera must have an intensifier of some sort or the ability to integrate the photoelectrons on the surface of the detector. There are two approaches to intensification: One is the combination of image intensifiers and CCD cameras (referred to as intensified CCD camera [ICCD]), and the other is the inclusion of special detectors having electron multiplication mechanisms inside the detector (referred to as electron-multiplied CCD camera [EMCCD]). The integrated approach is almost always used in CCD cameras that have cooling capability and a mechanism to minimize the readout noise (referred to as integrating cooled CCD camera).

Intensified CCD Cameras

Advantages of ICCD Cameras

- Standard interface (RS-170, CCIR). The standard video format makes it easy to use video peripherals such as video recorders, image processors, and image processing software.
- The photocathode or the microchannel plate may be turned on and off at very high frequencies for recording very high speed events or for very fast time resolution.
- Fast frame rate (30 Hz for RS-170, 25 Hz for CCIR). The frame rate is fast enough for most real-time applications and focusing.

Disadvantages of ICCD Cameras

Usable only in low light levels. The camera can be used for high-light applications such as DIC and phase contrast only if the light source level is adjusted low enough for the detector. Under these conditions, the image quality is not as good as that of high-light detectors. The photocathode and the microchannel plate are at risk for burn-in when excessive light is applied. The damage can be minimized with a protection circuit; however, special caution is required.

Intensified CCD cameras are a combination of a Gen. II intensifier and a CCD camera. The two components can be coupled either by a fiber optic bundle or by a set of lenses. In either case, the intensifier uses a device called a microchannel plate (MCP) as the low-noise electron amplifier. The MCP provides more than one order higher gain compared to the SIT camera.

Characteristics of Intensified CCD Cameras

- Detector size is 2/3″ or 1″.
- Spectral range can be 200–950 nm.
- Horizontal resolution depends on the combination of coupling and the pixel dimensions and number of pixels in the attached CCD.
- Best sensitivity is achieved in low-light applications.
- Possibility of damage from too much light.
- Minimal problems with geometric distortion of images.
- Very good linearity.
- Very low lag characteristics.
- Horizontal resolution is greater than vertical resolution when a video camera is used, but it may be the same if a digital camera is used.
- Intensification and integration may be combined in one camera.

Photon-counting cameras are special versions of the ICCD cameras. The primary differences are the incorporation of a very high-gain MCP device to detect single photons, and a special photocathode to reduce the camera background noise, which limits detectability. This design produces tremendous gain but relatively low resolution when compared to a regular ICCD.

Characteristics of Photon-counting Cameras

- Detector size is 2/3″ or 1″.
- Spectral range is 200–950 nm.
- Horizontal resolution up to 400 TV lines.
- Very low camera background.
- Best sensitivity in low-light applications.
- Possibility of damage from more intense light.
- Very low geometric distortion of images.
- Very good linearity.
- Very low lag characteristics
- Real-time image processing is essential to eliminate camera noise.

The following special features for the low-light-intensified camera are essential for VIM.

- Intensifier protection circuit to avoid damage from excessive light.
- Manual and auto sensitivity adjustment for the intensifier.
- Signal level indicator to adjust the proper light level for the camera or the sensitivity of the camera.

Electron Multiplied CCD Camera

Electron multiplied CCD camera is a CCD that has an electron multiplication mechanism inside the detector. The mechanism of the electron multiplication is either electron bombardment (EB CCD) or electron impact ionization (EM CCD). Both provide adjustable electron gain. Because of the electron multiplication, the signal becomes relatively larger than the readout noise of the CCD, and it enables faster readout with "relatively" lower readout noise. However, the noise related to the dark current and signal itself cannot be reduced by this approach, and the camera has other noise sources such as fluctuation and variation of each pixel gain. If the electron multiplication gain is adjusted to minimum, the detector behaves like a regular CCD camera.

Advantages of EB and EM CCD Cameras

- All advantages of cooled CCDs apply to EB and EM CCD cameras.
- High-speed frame rate. Electron multiplication increases the signal compared to the noise caused by fast readout. With subarray and/or binning, the camera can achieve much faster frame rates than standard video rate cameras. This is useful for applications that require a higher frame rate.

Disadvantages of Electron Multiplied CCD Cameras

EM CCD cameras have no major disadvantages as compared to other cameras. A minor disadvantage, however, is that the variety of cameras available, in terms of number of pixels, chip size, spectrum responses, etc., is somewhat limited, as compared to cooled CCD cameras.

Electron bombardment CCD (EB CCD) cameras are a recent technological advance, providing a CCD, back-illuminated behind a photocathode within a vacuum tube. This arrangement provides high-gain signal without the limitations of a microchannel plate. Other benefits include high-spatial resolution and little chance of damage from high light exposure.

Characteristics of Intensified EB CCD Cameras

- Detector size is 2/3″ or 1″.
- Spectral range can be 200–900 nm, depending on the photocathode material.
- Resolution depends on the pixel size and number in the CCD.
- Best used in low-light applications, but moderate-light applications are also acceptable.
- Slight possibility of damage to photocathode from too much light.
- Minimal geometric distortion of images.
- Very good linearity.
- Very low lag characteristics.
- Good signal-to-noise ratio (SNR) characteristics in images.
- Intensification and integration may be combined in one camera.

Electron multiplication CCD (EM CCD) cameras are a new technology for signal amplification directly on the CCD without the requirement of a photocathode. Direct signal amplification provides high versatility, wide dynamic range, and no possibility of damage in high-light situations.

Characteristics of EM CCD Cameras

- Detector size is 2/3″.
- Spectral range depends only on the CCD itself (usually from 350 nm to 1000 nm).
- Resolution depends on the pixel size and number in the CCD.

- May be used in any light level.
- No possibility of damage to detector from too much light.
- No problems with geometric distortion of images.
- Very good linearity.
- No lag characteristics.
- Good S/N characteristics in images at low light, but high camera noise in high light.
- Intensification and integration may be combined in one camera.
- Camera noise and gain characteristics are extremely sensitive to camera temperature.

Integrating Cooled CCD Camera

Integrating cameras are almost always CCD cameras. In most instances, these cameras are cooled to keep the dark current low during integration and require a slow readout that can help to keep the readout noise low to prevent the loss of low-light-level signals. This camera is also excellent for intense light conditions. Traditionally, the major limitation was the low temporal resolution inherent in their operation. The temporal (time) resolution is limited by the necessity of waiting until enough photoelectrons accumulate on the detector to make an image and the time involved in reading out all the pixel rows at the degree of precision required. Recent improvements in readout speed with low noise realize faster frame rates, and temporal resolution is not a problem for most applications. A frame buffer or computer is required to hold an image with this type of camera, since the image can be produced only after integration (not continuously).

Advantages of Cooled CCD Cameras

- Low light to high light. Cooled CCDs can be used to obtain low-light and high-light images. At both levels, cooled CCDs generate a similar quality of image and there is no risk of detector burn in high light.
- Pixel manipulation (binning, subarray). Binning combines the pixels and handles multiple pixels as one. This increases sensitivity and reduces data size at the expense of spatial resolution, which is useful for very low-light imaging and for applications that require a higher frame rate. Subarray scan allows partial images to be read from the CCD. This maintains the original resolution and reduces data size, which is useful for applications that require a higher frame rate.

Disadvantages of Cooled CCD Cameras

- Nonstandard interface. A cooled CCD generates a digital video signal, and each cooled CCD has its own interface to a computer. This makes it difficult to use standard video peripherals such as video recorders, image processors, and image processing software. A movement to standardize interfaces, such as IEEE1394 and Camera Link, will make compatibility less problematic.
- Cooling required. CCDs are normally cooled by a thermoelectric cooler; however, secondary cooling may be required, such as water circulation for longer exposures.
- Slow frame rate. Because of slow readout and the existence of a mechanical shutter, the frame rate of a cooled CCD is slow. This limits the temporal resolution and makes focusing difficult under low light conditions. Frame transfer-type CCDs and interline CCDs are available that generate faster frame rates.

Characteristics of Cooled CCD Cameras

Since cooled CCD cameras have no standards for detector sizes or specifications, each device may have a variable number of pixels in both the horizontal and vertical directions. Resolution can be estimated by multiplying the number of pixels in each direction by a value between 0.7 and 0.9 (Kell Factor), with 0.7 being used most for microscopy applications. The number of pixels multiplied by their individual dimensions will provide the total detector area.

- Detector sizes are not standardized.
- The size of the detector does not determine the number of horizontal or vertical pixels.
- The resolution is related to the number of horizontal and vertical pixels.
- The size of the pixels is a factor in the sensitivity of the detector.
- The number of electrons that can be stored in each pixel (well depth) is a factor of the dynamic range and SNR at saturation.
- Temperature of the detector is a factor of the dynamic range and SNR.
- Readout speed is a factor of the dynamic range and SNR.
- Spectral range is 200–1000 nm.
- Very linear photometric response.
- Slow readout may limit applications due to low frame rate.
- No damage from high light levels.
- No geometric distortion.
- Integration time may limit applications.
- Groups of pixels may be summed (binned) to provide greater sensitivity and speed at the expense of spatial resolution.
- Discrete portions of the array may be read out (subarrayed) to increase speed at the expense of field of view without changing resolution or sensitivity.
- Horizontal and vertical resolution may be equal.

IMAGE PROCESSORS

Image processors are used to improve SNR and to change contrast on the images. Several types of image processors are available for video microscopy. They are classified into three groups:

- Hardware-based real-time image processor to enable real-time integration, background subtraction, frame averaging, and LUT (look up table). LUT converts input digital value to other digital values based on the content of the LUT. LUT is used to stretch the gray level and invert intensity (negative image). This is mainly used to improve image quality at up to 30 frames/sec and make the image perceivable. Video rate cameras generate images every 1/30 sec. If those images are processed synchronizing to the video rate, this is called real-time image processing.
- Software-based image analyzer to enable quantitative intensity measurement and/or morphological measurement.
- Combination of real-time image processor and image analyzer. If the original image quality from the camera is poor due to low light or low contrast, the result of the image analysis will also be poor. In this case, the combination of the image processor and the image analyzer will give the best result for the analysis.

VEC and VIM require an image processor with the following features.

- Real-time image averaging with background subtraction.
- Digital contrast enhancement using LUT.

VIDEO ADAPTERS

Several considerations are important in determining the mechanical and optical interface between TV or CCD camera and microscope, and sensible transfer factors to the video detector and monitor. First, the optical alignment of the microscope must be as close to perfect as one would require for critical photomicrography (see Spector et al. 1998b). Although some deficiencies in the microscope, such as shading across the field or poor contrast, can be electronically compensated, and although even simple image-processing techniques permit removal of dust and dirt on the optics, the best video images are still obtained by presenting the best possible optical image to the camera.

Most black and white or single-chip color cameras are equipped with so-called C-mounts, a female thread of 1″ diameter, 32 threads per inch, and a shoulder at a distance of 0.690″ from the detector surface. A corresponding C-mount adapter is provided by the microscope manufacturers, fits onto the camera port, and will, if all tolerances (also on the camera) are maintained, place the video detector in an image plane that is parfocal to other cameras or visual observation. Similarly, a so-called ENG-mount with bayonet will parfocalize 3-tube or 3-chip color cameras. Some special cameras utilize the Nikon F-F-mount, also a bayonet.

Most modern microscopes generate a fully color-corrected intermediate image (see Spector et al. 1998b) and make it directly available for video pickup. Only older microscopes require correcting optics in the video adapter, and most of the new C-mounts with 1× transfer factor contain no optics. Only objectives and tube lenses (in the infinity system) generate the video image. Potential internal reflections are kept to a minimum.

Detector Field of View Compared with Visual Field of View

Tube and the ever more popular chip cameras come in different sizes, with a clear trend toward smaller and smaller chips with more densely packed picture elements to retain similar resolutions. The following table lists some typical chip sizes and the actually utilized diagonal in millimeters:

Chip size	2/3″	1/2″	1/3″	1″
Diagonal	10.7 mm	8 mm	5.3 mm	15.9 mm

Selecting such small areas from a total field of 20–25-mm diameter relieves the microscope optical system from attaining a high degree of correction for off-axis aberrations, but also severely limits the area recorded. For this reason, video adapters, with C- or ENG-mounts have been developed with transfer factors of 0.5×, 0.63×, and 0.8×. Such reducing adapters can also be very useful to increase the image brightness in low-light-level conditions. Bear in mind, however, that the resolved detail in the intermediate image may no longer be recorded because the fixed pixel size of the detector may be larger than the point-to-point resolution in the image.

To assure that the optical resolution is fully transferred to the video image, 2.5× or 4× video adapters are often used. Magnification changers or zoom systems on the microscope, or all in combination, can be very useful. The following are just two examples:

- A 10×/0.3 objective has a point-to-point resolution of ~1 mm. At a transfer factor of 1x, the pixel size must be less than 10 μm, and for good sampling, 5 μm. Since the individual pixels for most arrays are 10–12 μm in size, information would be lost or a higher transfer factor would be required.
- A 100×/1.3 objective resolves 0.25 μm; in the intermediate image, the point-to-point resolution required for the detector is better than 25 μm, relatively easily accomplished by most modern chip cameras or CCDs.

The electronic magnification from video detector or target to final image on the monitor naturally depends on the size of the monitor. A 14′′ monitor and a 0.5′′ CCD camera would result in a 43× factor. Since the video monitor is always viewed from a distance of 1–2 m, the magnification to the eye becomes again 5–10×, and close to what one would see through the eyepieces.

VIDEO MICROSCOPY AND IMAGING PROCEDURES

Described here are the applications, system requirements, and methods for video-enhanced microscopy, fluorescence imaging, and luminescence and photon-counting imaging.

Video-enhanced Contrast Microscopy

VEC microscopy greatly improves the contrast of the image and allows maximal resolution of the microscope. This system is used in numerous applications of cell biology, including the following:

- Axonal transport studies.
- Monitoring neuronal growth cone activity.
- Cytoskeletal motion.
- Chromosomal motion.
- Near-infrared brain-section imaging.
- Checking electron microscope thin sections.

The system requirements for VEC include the following:

- High mechanical stability.
- Very high optical magnifications (2–40,000×).
- Maximum numerical aperture (NA) objectives.
- Maximum NA condensers.
- Minimum strain or strain-free optics.
- High-quality polarizers and analyzers (e.g., for DIC and polarized light microscopy).
- High-quality infrared blocking filters (except in the case of infrared brain-section imaging).
- Microscope stages with very fine gear ratios.
- Rotating stage mount.
- High intensity, stabilized, uniform illumination.
- High-resolution video camera.
- Analog gain and offset video controls.
- Real-time image processor with background subtraction and digital contrast enhancement.

PROCEDURE A

VEC Employing DIC Microscopy

1. Establish Köhler illumination (see Chapter 95).

   Notes: Use color LUTs in the image processor to help evaluate the evenness of the illuminator, because the eye is more sensitive to color changes than to gray-level changes.

   The use of fiber optic scramblers (see Spector et al. 1998b) is highly recommended.

2. Establish DIC (see Chapter 95); the condenser aperture diaphragm may be closed if needed at this point.

   Note: Closing the condenser aperture diaphragm will help increase not only the contrast, but also the depth of field to help find the sample.

3. Switch beamsplitter to video position.

   Note: A beamsplitter with 100% transmission to the video port is recommended for two reasons. It will increase the signal to the camera and eliminate possible external images from the eyepieces (due to room lights, etc.) from being superimposed on the image of the sample.

4. Open the aperture diaphragm to its maximum to obtain highest NA (see Spector et al. 1998b).
5. Adjust analog gain and offset knobs to increase video contrast as needed.
6. Recheck illuminator evenness and adjust as needed (see notes to step 1).
7. Increase the magnification with optical or mechanical means to final required magnification.

   Notes: It is sometimes advisable to initially set up at a lower magnification to aid in finding the sample because of the larger field of view.

Increasing optical magnification can be done by changing the relay eyepiece or an optical intermediate lens, called an optovar.

It is also possible to increase the magnification by moving the camera farther away from the microscope. This "drawtube" method has the advantage of continuous variable magnification, but at the expense of optical corrections, parfocality, and free working distance. Since monochromatic illumination is generally used, the loss of optical correction is not important.

The loss of parfocality is only of concern when changing objectives. The operator must take precautions to prevent the possible striking of objectives into the sample or coverslip.

The change in working distance at the front of the objective can be important when using thick samples or coverslips.

8. Carefully move the sample to an adjacent, but specimen-free, area and defocus slightly.

   Note: If the working distance of the objective permits, focus into the glass of the slide below the sample. This area is sure to be free of dirt, scratches, or other defects that will adversely affect the background image.
9. Acquire the background image with an image processor.
10. Return to live image, refocus specimen, and return to area of interest.
11. Start background subtraction function from live image. Any shading, dirt, or illumination defects inherent in the optical system should disappear from the final "background subtracted image."
12. Use the digital contrast enhancement mode of the image processor to "STRETCH" the gray levels of the image to maximize contrast.

    Notes: Use the arrows next to "HI" and "LOW" to change the values. Decreasing the value of "HI" will increase the brightness of light areas in the image. Increasing the value of "LOW" will decrease the brightness of dark areas. This combination results in the remaining gray levels within the image being separated by more shades, thus increasing contrast.

    Stretching is accomplished by adjusting the gray levels in the intensity histogram of the image. Reassign the brightest available gray level from the image—gray level 100, for example—to the brightest available gray level in the output histogram, gray level 255. Then reassign the darkest gray level in the image—gray level 50, for example—to the darkest gray level in the output histogram, gray level 0. The gray levels between 100 and 50 will now be automatically assigned new gray levels in a linear manner between 0 and 255 in the output. This change or stretching separates the previously similar gray levels, making them easier to perceive.

FLUORESCENCE IMAGING

The equipment employed for video-intensified microscopy (VIM) intensifies and integrates the image, even in low-light situations. The applications include the following:

- Immunofluorescence.
- Autofluorescence of plant tissues.
- Forensic medicine.
- Contaminant inspection.
- Bone-growth studies using tetracycline.
- Microcrack evaluation of materials.

The following are the system requirements:

- High mechanical stability.
- Maximum NA objectives.
- High-quality excitation filters, dichroic mirrors, and barrier filters. Quality is a function of spectral selectivity, plane parallel surfaces, and precise mounting angles (see Chapter 95).
- Good infrared blocking filters.
- High intensity, stabilized, and uniform illumination.
- High-sensitivity video camera with fast overload protection circuit *OR* a cooled CCD camera.

- Analog gain and offset video controls.
- Real-time image processor with image averaging and digital contrast enhancement.

PROCEDURE B

Epifluorescence Microscopy

1. Establish Köhler illumination in reflected light using a sample with a large area of fluorescing tissue or cells.

   Notes: A routine pathology slide with liver tissue stained with hematoxylin and eosin is a good choice, because this stain will fluoresce with almost any filter set, and liver tissue is very homogeneous.

   The adjustment of the illuminator for evenness is critical and can be best imaged using color LUTs to help the eye perceive subtle intensity differences (i.e., use color contrast rather than gray scales).
2. Change to the sample to be studied, select appropriate objective, and focus.

   Note: The ideal objective for fluorescence is the one that has the lowest ratio of magnification divided by NA, since brightness increases with the square of decrease in magnification and the square of increase in NA.
3. Make sure to adjust the field diaphragm of the epi-illumination system (see Chapter 95) to illuminate only the area visible with the video camera to prevent fading in the surrounding areas.
4. Adjust sensitivity control of intensified type cameras until just before video saturation *OR* adjust exposure time of cooled CCD until just before full-well capacity is reached.
5. Adjust analog gain and offset of the camera (if possible) to maximize contrast in the image before digitization in digital cameras.
6. Recheck the illuminator evenness and adjust it as needed (see note to step 2).
7. Using an image processor, pick the frame-averaging function and select the number of frames that are required to increase the SNR to an acceptable level *OR* apply binning, subarray, and/or gain as needed.

   Note: The SNR in the image can be seen as rapidly flashing bright points in each video frame when using intensified cameras. Although commonly referred to as "noise," these points actually reflect the photon variability within the image under very low light conditions.
8. Use the digital contrast enhancement of the image processor to "STRETCH" the gray levels of the image to maximize the contrast *OR* adjust the LUT of software to increase the contrast as needed.

   Note: Stretching is accomplished by adjusting the gray levels in the intensity histogram of the image. Reassign the brightest available gray level from the image—gray level 100, for example—to the brightest available gray level in the output histogram, gray level 255. Then reassign the darkest gray level in the image—gray level 50, for example—to the darkest gray level in the output histogram, gray level 0. The gray levels between 100 and 50 will now be automatically assigned new gray levels in a linear manner between 0 and 255 in the output. This change or stretching separates the previously similar gray levels, making them easier to perceive.

LUMINESCENCE AND PHOTON-COUNTING IMAGING

Photon-counting imaging is particularly useful for detecting single-photon events that occur, for example, in biological luminescence systems. Applications include the following:

- Monitoring gene expression with luciferase reporter genes.
- Calcium ion imaging using aequorin (see Chapter 79).
- ATP or glucose imaging in tissue.
- Real-time visualization of oxyradical burst activities.

The following are the requirements of the system:

- Microscope that has 100% transmission to video camera port.
- Maximum NA objectives with high transmission at the luminescent wavelength.
- Photon-counting video camera with single-photon detection capability and low camera background.
- Image processor with photon-counting and image-overlay capability.
- RGB color monitor.
- Light-tight room for the microscope to minimize stray light-noise problems.

The photon-counting camera generates images that have spots corresponding to the position of photon hits. An image processor uses the threshold method to extract the spots, and the spots are converted to the digital value of "1," whereas the other area is converted to the value of "0" (binary image). "SLICE" image is the accumulation of the extracted spots. The spots spread to more than one pixel, and the total counts of the "SLICE" image are more than the number of actual hits of the photons. If the center of gravity is calculated for each spot and the results are accumulated, the image is called a "GRAVITY" image. The total counts of the "GRAVITY" image and the number of actual hits are the same. The photon-counting camera generates some spots even in dark conditions. The accumulated image in the dark condition is called the "DARK" image. Pixel depth is the number of bits of the frame buffer or memory plane; 8 bits and 16 bits are standard. The number of bits determines the number of gray levels (8 bits, 256 gray levels; 16 bits, 65,536 gray levels).

PROCEDURE C

Photon-counting Imaging

1. Turn off the high voltage of the photon-counting video camera and focus on the specimen through the eyepieces using whichever illumination technique is appropriate for the sample. The photon-counting camera is very sensitive and fragile. Keep the high voltage OFF except when actually imaging.
2. Reduce to minimum the illuminating source in the microscope, and switch 100% of the light to the camera using the beamsplitter.
3. Make sure that there are no unnecessary optics (polarizer, fluorescence filter, beamsplitter, etc.) between the objective lens and the video camera port.
4. Set the sensitivity of the photon-counting video camera to zero and turn on the high voltage. If the image is too bright, the automatic protection circuit shuts off the high voltage. In this case, reduce the illumination light level using neutral density or color filters (see Fig. 94.1 in Spector et al. 1998b) and turn on the high voltage again. If the image is too dark, increase the sensitivity of the photon-counting camera.
5. Choose the resolution of the image.

   Note: The resolution selected for the image depends on the resolution required to see the details and the amount of intensity within the object. It is possible to use a smaller array of pixels if the object does not require higher resolution. This means that available memory can be subdivided into more individual memory planes, and less disk space will be needed for storage of each image. If the number of photons per pixel is less than 256, it is also possible to change the depth of each pixel to 8 bits from the default value of 16 bits. This has a similar effect on the memory by allowing more memory planes and requiring less disk space for storage.
6. Integrate 64 frames of the image and make contrast changes using histogram stretching. Then save the image as the reference or background image.
7. Turn off the illumination of the microscope and increase the sensitivity of the photon-counting camera to 10.
8. Adjust the light path of the microscope so that no light will reach the camera.
9. Set the discrimination level of photon-counting at 100 and select an integration time of 1 minute to create an image of the "dark noise" of the camera.

10. Select a memory number in which to store this "dark" image.
11. Integrate this "dark" image for 1 minute and make a note of the total count of photons or events.

    Note: The camera background is usually less than 10 counts/sec. More background is usually due to a light leak or illuminator afterglow. Check the box by draping it with a dark cloth and turning off all the room lights. If the count goes down, there is a light leak. If not, try putting an opaque object over the field diaphragm and a drape over the lamp housing vents, as a test to try to eliminate background light. DO NOT OPERATE THE ILLUMINATOR IN THIS CONDITION! The drape will burn.
12. Set the light path back to 100% to the camera.
13. Integrate the sample without the reporter (e.g., luciferin or aequorin) for 1 minute to check stray light conditions. Counts should be similar to the dark count reading.

    Note: The same checks should be made for light leaks and illuminator problems if the count is now much different from the dark count. If the illuminator is the problem, it will be necessary to provide an additional light baffle between the microscope and the illuminator.
14. Select the integration time for the sample and set memory output locations.

    Note: The integration time for a typical sample must usually be determined by trial and error. The simplest way is to integrate for periods of a minute or two at a time, recording the counts and continuing for another period until the area of the photon concentration can be distinguished from the background. If no object is distinguished after 30 minutes of integration, there is very little chance that luminescent objects will be detected with additional integration.
15. Select memory locations in which to store images of the samples.
16. After integration, use histogram stretch to enhance the image.
17. Superimpose the photon-counting image onto the reference image if necessary.

REFERENCES

Spector D.L., Goldman R.D., and Leinwand L.A., eds. 1998a. Heterologous expression of the green fluorescent protein. In *Cells: A laboratory manual,* vol. 2. *Light microscopy and cell structure*, pp. 78.1–78.21. Cold Spring Harbor Laboratory Press, Cold Spring Harbor, New York.

———. 1998b. Light microscopy. In *Cells: A laboratory manual,* vol. 2. *Light microscopy and cell structure*, pp. 94.1–94.53. Cold Spring Harbor Laboratory Press, Cold Spring Harbor, New York.

———. 1998c. Video microscopy and image enhancement. In *Cells: A laboratory manual,* vol. 2. *Light microscopy and cell structure*, pp. 95.1–95.15. Cold Spring Harbor Laboratory Press, Cold Spring Harbor, New York.

CHAPTER 3

The Application of Scientific-grade CCD Cameras to Biological Imaging

Mark Christenson

Photometrics, Tucson, Arizona 85706

The superior performance features of the CCD have made it the major sensor employed in scientific-grade imaging devices. These features include flatness of field, linearity of response over very large ranges, varying pixel sizes and formats, and low-noise on-chip amplifiers. When the CCD is cooled to minimize dark-charge accumulation, driven with low-noise, programmable electronics, and coupled to a high-quality analog-to-digital (A/D) converter, the resulting digital camera is both powerful and flexible for the scientific user. In this chapter, I define the performance capabilities of a scientific-grade digital camera and discuss the relevant camera features for several key application areas where digital cameras are currently being utilized.

ISSUES TO CONSIDER WHEN SELECTING CCD CAMERAS

Sensitivity

The sensitivity of a scientific-grade digital camera is based on its ability to maximize the measured signal, while simultaneously minimizing the noise components that can mask that signal. The fundamental signal measured by the CCD is in electrons, because the CCD utilizes the incoming photon's energy to promote an electron to the excited or conduction state. The electrons diffuse a small distance from the point of origin and become trapped in one of the potential wells that form the pixels on the CCD surface.

The signal level is directly proportional to several factors and can be described by the formula

$$S_e = \phi \times QE(\lambda) \times A \times \Delta t \qquad (1)$$

where S_e is the signal (electrons), ϕ is the photon-flux density (photons/μm^2/sec), $QE(\lambda)$ is the quantum efficiency, which is a function of the wavelength, A is the pixel area (μm^2), and Δt is the exposure time (sec). Most of these factors can be manipulated to maximize the signal and are discussed below.

Photon-Flux Density

The photon-flux density is determined primarily by the illumination conditions, the fluorescent sample, and the microscope optics, which are not discussed in this chapter. The one feature of the digital camera that can influence the photon-flux density is the type of window used and whether any coatings are applied to that window. For the best broadband transmission properties, a high-quality quartz window should be used with an excellent flatness and inclusion specification. In addition, custom antireflection coatings can be added to the windows to lower the 3.5% loss of light from each uncoated surface to less than 1%. If the camera is to be used for a wide variety of wavelengths, however, antireflection coatings can actually decrease the transmission outside of the optimized wavelength region. Specific information on window coatings should be obtained from each manufacturer, since the coating process depends on the coating methods and materials used.

Quantum Efficiency

The major determinant of signal intensity, and, hence, the sensitivity of the camera, is the quantum efficiency (QE) of the CCD. The QE is defined as the fraction of incoming photons at each wavelength of light that produces excited-state electrons in the detector. Each CCD has a unique QE profile that is dependent on the fabrication process and the overall structure of the device. Table 1 gives the peak QEs and the average QEs over the visible range for four types of CCDs: (1) the full-frame, front-illuminated Kodak 1317; (2) the interline, front-illuminated Sony 1300; (3) the frame-transfer, front-illuminated EEV 512 FT; and (4) the full-frame, back-illuminated Princeton Instruments 800 PB. The highest QEs are obtained with the back-illuminated sensors, designed to bring the light directly to the photosensitive region of the CCD. The front-illuminated devices are designed so that the light must pass through overlying gate structures that form the pixels on the CCD, thus lowering the light throughput. The full-frame, interline, and frame-transfer types of CCDs have similar peak and aver-

TABLE 1. CCD-sensor performance features

CCD type	Model[a]	Peak QE[b] average QE (400–700 nm)	Shuttering[c]/ Minimum exposure	Readout noise[d] (electrons)	Dynamic range[e] (bits)	Pixel size (µm)/format
Front-illuminated						
Full-frame	Kodak 1317	45% @ 690 nm 28% avg.	mechanical/20 msec	10	12	6.8 µm 1317 x 1035
Interline	Sony 1300	50% @ 420 nm 31% avg.	electronic/1 µsec	7	11–12	6.7 µm 1300 x 1030
Frame transfer	EEV 512 FT	45% @ 700 nm 25% avg.	electronic/10 msec	15	12	15 µm 512 x 512
Back-illuminated						
Full frame	PI 800 PB	80% @ 600 nm 70% avg.	mechanical/50 msec	5	14–16	15 µm 1000 x 800

[a]The Kodak 1317 is also called the KAF 1400, the Sony 1300 is also referred to as the ICX085, the EEV 512 FT is also called the EEV CCD-37. The PI 800 PB is a proprietary CCD produced for Princeton Instruments.

[b]QE is defined as the percentage of photons that converted into measurable electrons in the CCD. Peak and average QEs are for the visible portion of the spectrum (400–700 nm). The QE of the sensor determines the overall sensitivity of the device when the signals are above the total camera noise level.

[c]Full-frame-type detectors require a shutter to prevent smearing of the image. The minimum exposure time is dependent on the open and close time of the shutter and the physical size of the CCD. Since the interline only needs to move the stored charge underneath the storage mask in each pixel, the shuttering times can be much shorter than in full-frame devices, and the absence of moving parts means that there will be no vibrations during shuttering. The frame-transfer CCD needs to shift 512 rows under the permanent mask on the CCD, but this can be accomplished in 1.4 msec with no vibrations during shuttering.

[d]Readout noise is the fluctuation in the measured signal caused by the CCD and camera electronics. The number shown is the root mean square (rms) fluctuation in electrons, with the peak-to-peak noise being ~5 times the rms value. The noise values are typical numbers for slower readout rates of each of the CCDs. The noise values represent the limit of detectability of the device.

[e]Dynamic range is defined as the full-well capacity of the single pixel divided by the rms readout noise of that pixel. The full-well capacity is defined as the largest number of electrons that can be collected on a single pixel which still preserves the linearity of the measurement. Full-well capacities for the CCDs listed are 45,000, 20,000, 100,000, and 80,000 electrons, respectively.

age QEs, but the location of the peak is in the blue-green region of the spectrum for the interline and in the red for the others. This means that for imaging of blue-green fluorescence (e.g., using fura-2, BFP, GFP, or CFP) the Sony 1300 is often the best detector, whereas the other detectors have better performance for red and near-IR fluorescence (e.g., using fura-red, rhodamine, Cy5, or Cy7).

Pixel Area

Another contributor to overall signal strength is the cross-sectional area of each pixel. For maximum sensitivity, a large pixel area would seem to be optimal; however, the goal in fluorescence microscopy is often to observe small structural details within the cell, which requires smaller pixel sizes (see the discussion on Resolution, below). Fortunately, a scientific-grade CCD camera can be used in multiple modes using a capability called "binning." During readout of the CCD, adjacent lines and adjacent pixels can be grouped together to form square superpixels, increasing the total area and therefore the total number of electrons collected multiplicatively. Different binning ratios can be employed for each experiment, since this is all done under computer control, allowing the user to decide when to trade off resolution for sensitivity.

Exposure Time

One of the simplest ways to increase the signal is to increase the time of exposure of the CCD, with the signal intensity being linearly dependent on exposure time. Unfortunately, CCD devices exhibit a property of charge accumulation that is independent of the light signal; this is called the dark charge. The dark charge arises from irregularities in the CCD silicon layers where a fraction of the valence electrons have enough energy to become free-moving electrons and accumulate in the pixels. Although most pixels exhibit an average level of charge accumulation, some pixels build up charge at a much higher rate than others. These are called "hot pixels" and they characteristically appear as "stars in the night sky" on longer-time-period dark images (see Fig. 1).

Temperature

To minimize the dark-charge effect, the net energy of the silicon can be lowered by reducing the CCD temperature. Scientific-grade digital cameras employ sufficient thermoelectric cooling of the CCD to essentially eliminate both the average dark charge and the hot pixels from contributing to background signals under most operating conditions. For conditions where dark charge does accumulate to significant levels, the average dark charge can be measured with a dark image (exposure conditions identical to experimental conditions but with no light going to the CCD), and this dark image can be subtracted from the experimental image. The point-to-point fluctuation within the dark image, which is the dark noise, is permanently embedded in the image and degrades the overall signal-to-noise ratio (SNR). Therefore, the best solution is to prevent the dark charge from accumulating, rather than depending on dark-image subtraction to restore the image quality (see Fig. 1).

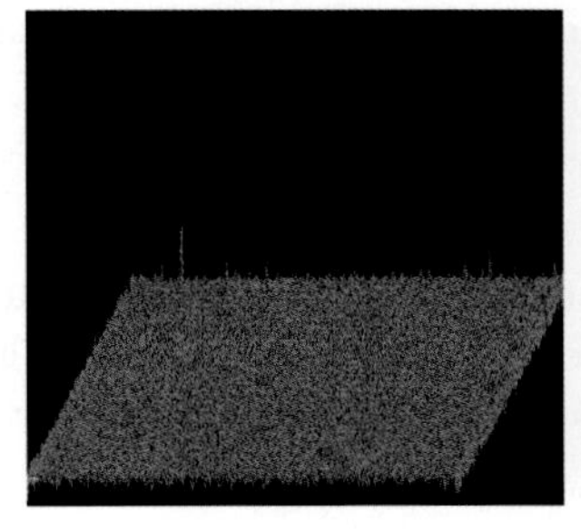
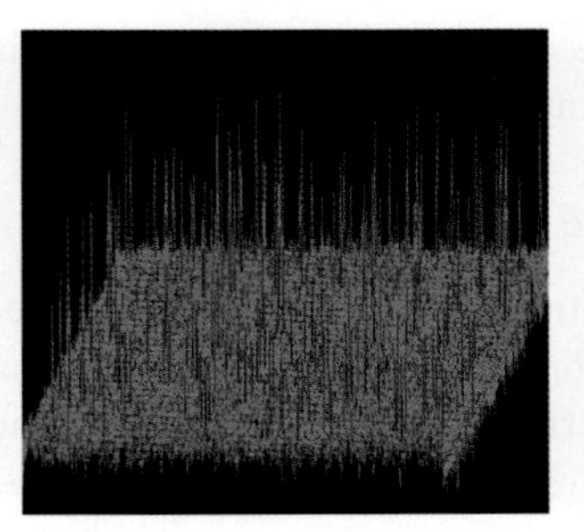
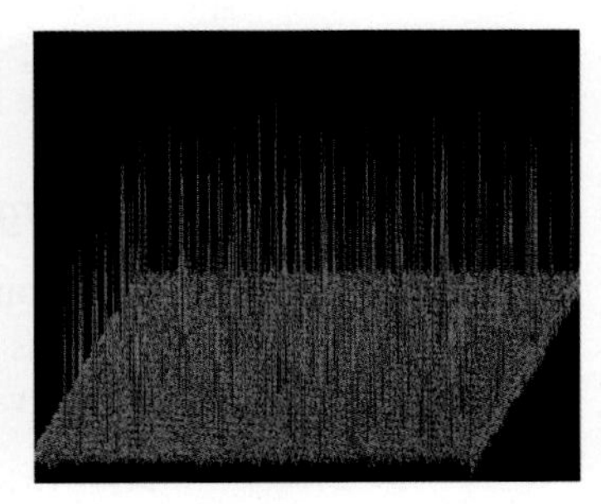

FIGURE 1. Effects of cooling on dark-charge accumulation. Dark-charge images of Sony Interline CCD cooled to –10°C (*left*), 0°C (*middle*), and +10°C (*right*) for 10-sec integration period. Cooling lowers the basal dark-charge level and also prevents the accumulation of dark charge in pixels with higher than average dark charge (hot pixels). These "hot pixels" lead to the "stars-at-night" pattern observed on images when inadequate cooling is used for a given exposure time.

Units of Measure

The signal can also be represented in the more familiar but arbitrary units of counts or gray levels:

$$S_c = S_e/G \tag{2}$$

where S_c is the signal (counts), S_e is the signal (electrons), and G is the gain (electrons per count). It is important to remember that a signal of 100 counts derived from one detector is not necessarily equivalent to 100 counts from another detector even after the no-light backgrounds have been subtracted from each. According to Equation 2, the relative gain levels of the detector must be known in order to compare the true signal in electrons. In addition, Equation 1 requires that the light-flux density arriving at the two detectors, the spectral characteristics of that light, and the exposure times be equivalent, and that the relative pixel area be accounted for. It is also apparent from Equation 2 that changing the gain of the CCD does not influence the magnitude of the actual measured signal and therefore does not influence the SNR appreciably.

Signal-to-Noise Ratio

To be accurately measured, the signal must be greater than the noise inherent in the measurement itself. This means that the SNR must be sufficiently large to guarantee the validity of the measurement, typically requiring a value of 2–5 as a minimum. Calculating the SNR for a detector allows a measure of the sensitivity limits of the device.

The noise elements in a scientific-grade CCD are added in a root-mean-square fashion due to the independent nature of the noise elements as follows:

$$N_t = (N_{sh}^2 + N_{ro}^2 + N_d^2)^{1/2} \tag{3}$$

where N_t is the total noise, N_{sh} is the photon shot noise, N_{ro} is the readout noise, and N_d is the dark-charge noise. These sources of noise are discussed below.

Photon Shot Noise

The photon shot noise is a fluctuation in point-to-point photon numbers that is inherent to the light signal itself and varies as the square root of the number of events measured:

$$N_{sh} = (S_e)^{1/2} \tag{4}$$

Readout Noise

The readout noise is a function of the CCD-amplifier structure, the method of readout, and the nature of the electronics in the camera that are used to read the CCD signal. This is a measured value, and most manufacturers will provide average numbers for general camera descriptions and individual camera measurements to users upon request. Generally, the readout noise increases with readout speed, which contributes to a decreased SNR. If the speed of readout is adjustable, the user can select the slower speed for conditions where the highest SNR is required and the faster speed for conditions where frame rates are critical.

Dark-Charge Noise

The noise contribution from the dark charge is exactly analogous to the photon shot noise, except that the electrons have been generated by thermal energy instead of photons. Therefore, the noise in the dark image varies with the square root of the number of events measured as well:

$$N_d = (S_d)^{1/2} \tag{5}$$

where S_d is the measured dark signal in electrons.

Putting the values for N_{sh} and N_d into Equation 3, we get

$$N_t = (S_e + N_{ro}^2 + S_d)^{1/2} \tag{6}$$

Robust cooling of the CCD lowers the dark noise to a negligible level, and careful electronic designs minimize the camera-readout-noise term. Given these conditions, we can calculate the minimal detectable light signal that can be measured with a given SNR as follows:

$$S/N = S_e/N_t = S_e/(S_e + N_{ro}^{\ 2})^{1/2} \tag{7}$$

Rearranging our terms, we can solve for S_e by using the quadratic equation and setting S/N = 2,

$$\text{Minimal detectable signal} = S_{e,md} = 2 + (1 + 4 \times N_{ro}^{\ 2})^{1/2} \tag{8}$$

Thus, for N_{ro} = 10 electrons, $S_{e,md}$ = 22 electrons, which is 2.2 times the readout noise. For a detector having a QE of 70%, this requires 32 photons per pixel, but for a detector with a QE of 30%, this would require even more; i.e., 73 photons per pixel. For detectors with robust cooling of the CCD, imaging under these conditions is limited by the readout noise of the detector. One way to increase the SNR under these light conditions is to use binning of the CCD to bring the net signal per superpixel to a higher value. Since the charges on multiple pixels are added before being digitized, the signal increases by the number of pixels binned, but only one unit of readout noise is added per superpixel. This gives an increased SNR, but at the expense of some spatial resolution.

As the signal increases significantly beyond the readout noise, the SNR becomes directly proportional to the photon shot noise:

$$S/N = S_e/N_t \_ S_e/(S_e)^{1/2} \cong (S_e)^{1/2} \tag{9}$$

To achieve images with maximal SNR, the exposure time can be increased until the signal approaches the single-pixel, full-well value. As the SNR increases, the visual quality of the image increases (see Fig. 2). For this reason, longer exposures are preferable when collecting publication-quality images and when better quantitative analysis is desired, although the user must verify that the imaging conditions do not significantly perturb the biological system being analyzed.

For applications that require an extremely high precision and hence, SNR (such as absorbance or intrinsic imaging), it is possible to increase the SNR even further by binning the CCD. The output node where the binned charge accumulates before digitization often has 2–3 times the single-pixel, full-well capacity. That means that the SNR can be increased by at least a factor of $\sqrt{2}$ or $\sqrt{3}$ by running at a full superpixel. Under full-well conditions, the readout noise does not significantly contribute to the total noise, so individual images can be combined to increase SNR even further, without a noise penalty.

Resolution

The resolving capability of a CCD detector is determined by the pixel size; therefore, the highest-resolution images are obtained by using a sensor with the smallest available pixels. Two adjacent structures are resolved by the CCD if their maxima fall on separate pixels with at least one pixel of lower signal intensity between them. This means that one feature requires at least two pixels to sample that feature adequately, and the smallest separation between objects that the CCD can accurately resolve is twice the single-pixel size. Therefore,

$$\text{CCD Resolution} = 2 \times X_p \tag{10}$$

where X_p is the pixel size (µm).

When imaging through the microscope, the resolution of the image is determined by the magnification of the objective and any tube lenses prior to the camera port. We can transpose the feature size in the sample to that on the CCD by this formula:

$$X_{im} = X_{ob} \times M_{obj} \times M_{tl} \tag{11}$$

where X_{im} is the image-feature size, X_{ob} is the object-feature size, M_{obj} is the magnification of the objective, and M_{tl} is the magnification of the tube lens.

We also know that the fundamental resolution limit of the microscope is determined by diffraction of light through the optics with the minimal detectable distance between two fluorescent point sources, as determined by the Rayleigh criterion:

$$d = 0.61 \times \lambda/\text{NA} \tag{12}$$

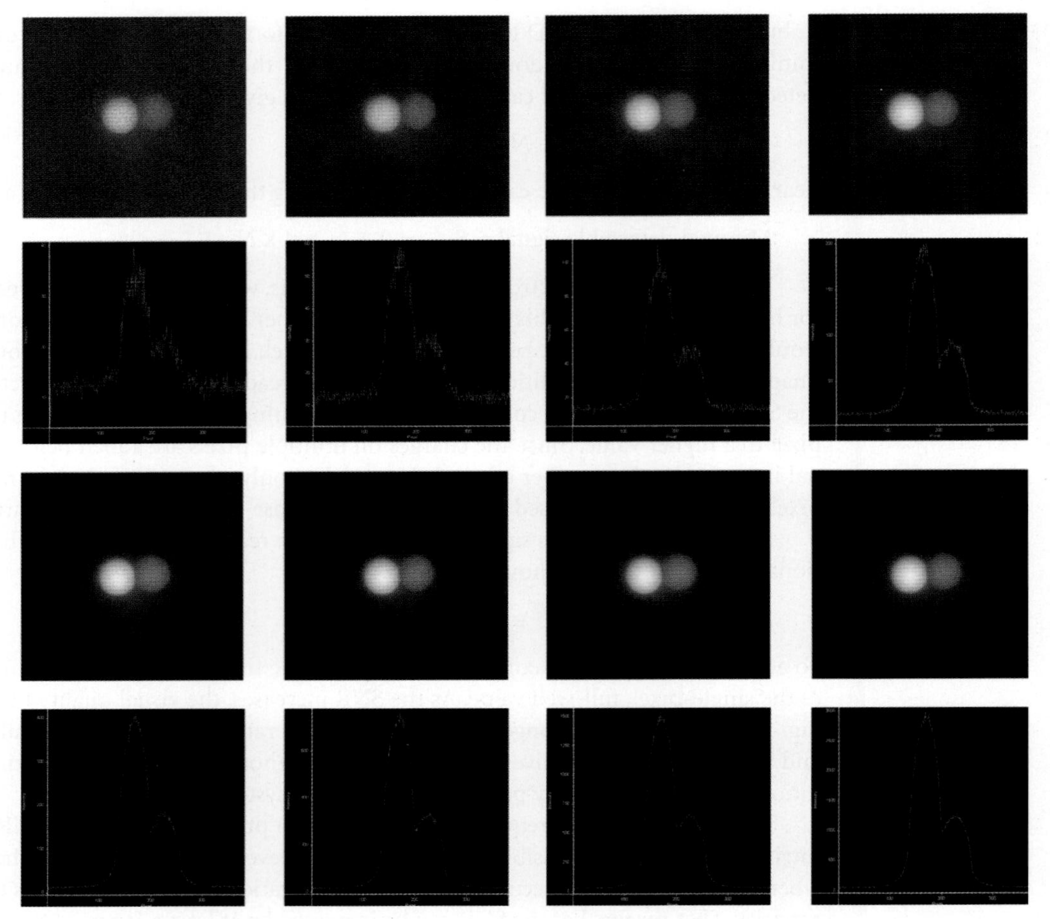

FIGURE 2. Increase in SNR with integration time. Images of fluorescently labeled beads for exposure times of 20 msec, 40 msec, 80 msec, 160 msec, 320 msec, 640 msec, 1280 msec, and 2560 msec, with each image scaled for contrast. Line profiles show the SNR inherent in the images. For each doubling of exposure time, the total number of photons measured per pixel doubles, while the total noise per pixel increases by √2, increasing the SNR by a factor of √2. This can be seen in the increasing quality of the image (notice the smoothing of the bead boundaries) and in the profile shown below each image.

where *d* is the minimal detectable distance between two fluorescent point sources, λ is the wavelength of light, and NA is the numerical aperture of the microscope objective.

Combining Equations 10, 11, and 12, we can calculate the ideal pixel size to collect a diffraction-limited image at a given magnification by setting the CCD resolution equal to the minimal image-feature size and solving for the pixel size:

$$X_P = (0.305 \times \lambda \times M_{obj} \times M_{tl})/NA \qquad (13)$$

Thus, using 520-nm light to image fluorescence with a 63x/1.4-NA objective with no tube lens (set $M_{tl} = 1$), the minimal pixel size required for diffraction-limited imaging would be 7.1 µm. Because the Sony 1300 Interline and the Kodak 1317 have pixels of 6.7 µm and 6.8 µm, respectively, both CCDs would meet this criterion (see Table 1). When using detectors with larger pixel sizes or working with lower-magnification, lower-NA lenses, a magnifying tube lens can be employed to reach the limiting resolution at the CCD, but with the penalty of adding more glass elements. Since the major goal of all live-cell imaging is to preserve the physiology being observed, any component that necessitates longer exposure conditions should usually be avoided even at the cost of some resolution.

Field of View

The field of view is the area of the object space being imaged by the CCD detector. The field of view (μm^2), which decreases with an increase in magnification, can be calculated by the following formula:

$$\text{Field of view} = N_x (N_y (X_P^2 / (M_{obj} \times M_{tl})^2 \quad (14)$$

where N_x is the number of pixels on the *x* axis, N_y is the number of pixels on the *y* axis, and X_p is the size of the pixels (assuming square pixels).

The larger the CCD, the greater the field of view that is available to the user. Currently, it is possible to obtain scientific-grade digital cameras with CCDs large enough to cover the whole microscope field of view; however, these systems tend to be used only for high-throughput survey work or for careful characterization of the microscope optics due to the higher costs of the large-format CCD sensors.

Dynamic Range

Dynamic range can be intrascenic or interscenic. Intrascenic dynamic range refers to the number of gray values that can be measured in a single frame, whereas interscenic refers to the number of gray values that can be measured by an imaging device with different exposure values. Typically, the intrascenic dynamic range is the more important, since it determines the utility of the device for individual scenes with dramatically varying intensity levels.

The dynamic range is calculated by taking the largest possible single-pixel signal, which is the full-well capacity, and dividing it by the smallest significant step size, which is the readout-noise value. If a step size is chosen that is smaller than the readout noise, the detector will be digitizing noise and will give the impression of more useful gray levels than are truly available. For example, the Kodak 1317 CCD has a full-well capacity of 45,000 electrons and a readout noise at 1 MHz of 10 electrons, meaning that the maximal dynamic range is 4500—large enough to digitize to 12 bits (4096 gray levels) but not to 16 bits (65,566 gray levels).

So how many bits are enough? Typically, the 8 bits (256 gray levels) of a video-type signal are good enough to determine that the intensity values in one part of the image are greater than those in another region. For truly quantitative imaging, however, a range of 10–12 bits (1024–4096 gray levels) is really required, and at times 16 bits (65,566 gray levels) can be useful to span the full range of the data in the sample.

Flexibility

The whole concept of flexibility implies the ability to tailor the readout mode of the digital camera to the experimental requirements. This can be done by exploiting the binning and subregion readout modes of a scientific digital-camera system (summarized in Table 2). Binning accumulates the charge from adjacent pixels during readout to make a superpixel, resulting in greatly enhanced signals but with a reduced pixel density over the same field of view. If the user can tolerate the loss in spatial resolution, the increased sensitivity can pay off in dramatically reduced exposure conditions, which may be critical for viability in time-resolved imaging experiments (Fig. 3).

Subregion readout is the ability to skip unwanted pixels during digitization, only retaining the data from the desired region of interest (ROI). Typically, this ROI is a rectangular region that is chosen from a reference image of the desired target. By selecting an ROI, the user reduces the readout time required and decreases the digital data load. The latter can be extremely important when collecting large numbers of images, since available RAM can become limiting during data collection and image processing or when displaying digital movies.

EXAMPLES OF APPLICATIONS

Imaging requirements and suggestions on CCD sensors and cameras are presented for the following three areas of imaging applications:

TABLE 2. Comparison of binning and subregion readout modes

Region size[a]		Readout time[b] Binning(msec)	Relative sensitivity[c]	Memory[d] (Mbytes)	Field of view[e] (%)
1000 x 1000	1 x 1	290	1	2.0×10^6	100
	2 x 2	150	4	0.5×10^6	100
	4 x 4	80	1	60.13×10^6	100
400 x 400	1 x 1	125	1	0.32×10^6	16
	2 x 2	68	4	0.08×10^6	16
	4 x 4	40	1	60.02×10^6	16
100 x 100	1 x 1	40	1	0.02×10^6	1
	2 x 2	26	4	0.005×10^6	1
	4 x 4	19	16	0.0013×10^6	1

[a]The pixel numbers shown are before binning; after binning the pixel numbers will be reduced by the binning ratio. For example, the 1000 x 1000 pixel region binned 4 x 4 becomes 250 x 250 superpixels.

[b]The readout time is calculated for the Sony Interline 1300 running at 5 MHz. Notice that the readout time does not vary linearly with the number of pixels that are digitized. This is due to the overhead involved in shifting all of the charge off the CCD, including the charge from pixels that are not digitized. For this megapixel sensor, a reduction in the number of pixels by 4 results in an approximate halving of the readout time. It can also be seen that the readout time of a 400 x 400 region binned 4 x 4 is equivalent to that of a 100 x 100 subregion binned 1 x 1.

[c]The relative sensitivity is calculated as the area of each superpixel relative to a single pixel.

[d]The memory for each single image is calculated from the total number of pixels multiplied by 2 bytes per pixel, since a minimum of 2 bytes is needed to encode each 12-bit value. To collect a time-lapse series of 100 images at 1000 x 1000 resolution with no binning would require 200 Mbytes of RAM; to collect the same series at 400 x 400 resolution binned 2 x 2 would require only 8 Mbytes of RAM.

[e]The field of view is calculated relative to the full 1000 x 1000 pixel image.

- High-resolution fluorescence imaging in fixed preparations
- High-resolution, static fluorescence imaging in living preparations
- Dynamic fluorescence imaging in living preparations

High-Resolution Fluorescence Imaging in Fixed Preparations

Some examples of this application area include direct or indirect immunofluorescence, fluorescence in situ hybridization (FISH), and neural tracing in fixed specimens.

Imaging Requirements

High-resolution imaging is the most important feature, since the main goals are to accurately image the localization of the fluorescence of one or multiple fluorescent probes. Intensities can vary from

FIGURE 3. Resolution and binning. Images of bodipy-phalloidin-stained actin filaments in fibroblasts taken at 40X magnification with the Sony Interline CCD (6.7 μm^2 pixels) at 1 x 1 binning (*left*), 2 x 2 binning (*middle*), and 3 x 3 binning (*right*). Most of the single microfilaments can be seen without distortion at both 1 x 1 and 2 x 2 binning, whereas 3 x 3 binning begins to degrade image quality. In regions where the filaments overlap, the 2 x 2 binned image begins to lose resolution, whereas the 3 x 3 binned image is at the limits of interpretation.

fairly strong to somewhat weak, so a moderate amount of integration on the CCD may be required. However, increasing the illumination intensity can also boost most signals. In the case of neural tracing, a large dynamic range is particularly useful since there will be important data in both the stronger and the weaker signal regions.

Camera Suggestions

These applications can be best accomplished by a digital camera with a high-resolution CCD. Since cooling, binning, and subregion readout are not critical, virtually any 10- to 12-bit camera system that utilizes the Sony 1300 or the Kodak 1317 will perform adequately in this relatively nondemanding application area.

High-Resolution, Static Fluorescence Imaging in Living Preparations

Some examples of this application area include imaging GFP-fusion proteins in vivo, fluorescence resonance energy transfer imaging, and vital fluorescent stains.

Imaging Requirements

The primary concern when imaging live cells should be to minimize the photodamage caused by short-wavelength illumination. This can be accomplished by using the most sensitive detector available for imaging. In addition, these experiments typically require high spatial resolution in order to identify the precise intracellular localization of the protein(s) being studied. A high dynamic range is also important here, because there is often a very large range of fluorescent intensities distributed throughout the cell that may encode important information. Another consideration is the digital data load that accumulates during time-lapse imaging experiments. To avoid the collection of unnecessary data, the subregion readout capability of a scientific digital camera can be exploited to collect only the pertinent data.

Camera Suggestions

The best solution in this category is a well-cooled digital camera with the back-illuminated type of CCD where the detection efficiency is maximal. The PI 800 PB is an ideal detector to use here, because it provides the smallest pixel sizes available in a back-illuminated CCD (15 x 15 μm) and has a peak QE of 80% (see Table 1). For applications where an even smaller effective pixel size is desired, the user can switch to a higher-power objective or add a relay lens to reach the desired final magnification on the CCD.

Since the readout noise of the back-illuminated CCD is strongly dependent on the rates of digitization, it is best to have dual A/D converters. By slowing the readout rate to 100 kHz (100,000 pixels per second), an approximately threefold increase in sensitivity can be obtained in the photon-limited regions of the image. The higher-speed A/D converter can be used for setup and focusing of the detector or for data collection when higher frame rates are required.

The second-best digital camera for this category is one based on the high-resolution Sony 1300 Interline or the Kodak 1317 CCDs. For GFP and its spectral derivatives BFP and CFP, the Sony 1300 Interline is superior due to its much higher QE in the blue-green portion of the spectrum, reaching 45% at 520 nm. Ideally, this detector should be cooled to eliminate overall dark current and to prevent any “hot pixels” from appearing.

Dynamic Fluorescence Imaging in Living Preparations

Some examples of this application area are calcium imaging with fura-2, fluorescent-vesicle or particle tracking in real time, fluorescence recovery after photobleaching (FRAP), and 4D imaging (X, Y, Z, t).

Imaging Requirements

All of these applications are dynamic and require at least a minimum speed to monitor the physiological events. The speeds required, however, vary greatly even in calcium imaging, where some calcium signals occur on half-minute timescales and others occur on half-second timescales. Sensitivity is important for all of these experiments, since they involve repeated exposure of the cells or tissues to potentially damaging excitatory light.

In the case of the fura-2 imaging and most FRAP experiments, the resolution of the imaging device is not as important, due to the diffuse distribution of the fluorescent signal. A moderately high dynamic range is advantageous for fura-2 experiments because the image ratios can be constructed from values with more significant digits. For the FRAP studies, a larger dynamic range allows a precise measurement of fluorescent probe values from the early recovery period until the final equilibration. Using the binning feature provides increased sensitivity for photon-limited conditions and increases imaging frame rates.

For vesicle or particle tracking and for 4D imaging, the targets usually require higher-resolution images, so a small pixel detector is advantageous. These applications can also exploit the subregion readout functions so that the imaging rates are higher and the data load is decreased.

Camera Suggestions

For moderate-speed imaging of fura-2, the Sony 1300 has a very good QE (45% at 520 nm) but perhaps too many pixels. This is not a fundamental problem for a scientific-grade camera, since the CCD can be run in binned mode to get even higher signals per superpixel. The EEV 512 FT can also be used for calcium imaging since the frame rates achievable with subregion readout and binning can be very high. Even though the QE response is lower for this chip (20% at 520 nm), the larger pixels contribute to a reasonable signal size and the smaller number of pixels reduces the overhead time in shifting charge at higher imaging rates. For extremely photon-starved imaging conditions, it is also possible to get the EEV 512 FT CCD with a fiber-optically coupled, green-sensitive, high-resolution image intensifier. This results in a detector with single-photon sensitivity and with high-imaging-rate capabilities.

Either the Sony 1300 or the EEV 512 FT can be used for FRAP experiments as well, with the choice depending on the wavelength of the fluorescent probe being used. For the tracking experiments or for 4D imaging, the higher-resolution Sony 1300 CCD may have the advantage, although the large number of pixels will give a large data overload if the subregion readout capabilities are not exploited. Both of the cameras have the dynamic range to meet the application needs.

FUTURE DIRECTIONS

The CCDs discussed here, and the scientific-grade digital cameras that use them, represent the current state of the art; however, both CCDs and the digital cameras that employ them are continuing to evolve rapidly. In the past, digital camera manufacturers were limited in the choice of CCDs that could be used for microscopy, but custom CCDs are now being designed that are more directly suited to the needs of the biological imaging community. With the development of better electronics, the CCDs and digital cameras that run them are yielding higher speeds with lower noise performance than was previously possible. This trend will continue, yielding high-performance detectors with better sensitivities and more pixels, and allowing greater frame rates than previously available, while maintaining the power of fully programmable, flexible control of camera-readout functionality.

CHAPTER 4

A Practical Guide: Differential Interference Contrast Imaging of Living Cells

Noam E. Ziv and Jackie Schiller

Rappaport Institute and the Technion Faculty of Medicine, Haifa 31096, Israel

This chapter describes the use of electronically enhanced, differential interference contrast (DIC) imaging techniques for the study of living cells. Protocols for DIC imaging of cells using an inverted microscope and for infrared (IR) imaging of brain slices using an upright microscope are presented. The basic principles of DIC microscopy are assumed. For an introduction to DIC microscopy see Chapter 95 or visit the wonderful "Molecular Expressions" Web site (http://www. microscopy.fsu.edu/primer/techniques/dic/dichome.html).

DIC microscopy (also known as Nomarski microscopy) is extremely useful for resolving individual cells and cellular organelles in live, unstained tissue. The data obtained are valuable in their own right, as well as for complementing fluorescence microscopy data. DIC microscopy can be applied to cell culture, brain slices, and even intact organisms (such as embryos). High-quality images of living cells in brain slices can also be obtained using a combination of DIC optics and IR video microscopy (Dodt and Zieglgänsberger 1994).

MATERIALS

Glass Substrates

Unlike bright-field transmission microscopy, in which contrast is obtained from differences in light absorbance, contrast in DIC microscopy is derived from differences in refractive index. The conversion of refractive index gradients to light intensity gradients relies on the use of polarized light. Thus, materials that affect light polarity cannot be used. In practice, this means that specimens cannot be grown or mounted on standard plastic substrates (such as polystyrene petri dishes or culture flasks) and must be grown or mounted on nonpolarizing materials, typically glass.

Cultured cells are usually grown on:

Glass coverslips (#1), usually maintained in 6-well plates or 35-mm petri dishes until used for experiments.

Glass-bottomed petri dishes from commercial sources (such as MatTek) or made "in house" by attaching coverslips with Sylgard 184 (Dow Corning) or a hot-melt glue gun, to the bottom of polystyrene petri dishes in which a hole has been bored.

Glass-bottomed chambers (such as Lab-Tek chamber slides and chambered cover glass, NUNC A/S).

Brain slices are usually mounted on glass-bottomed chambers, which can be custom-made or procured from commercial sources (e.g., Luigs & Neumann; Scientific Systems Design, Harvard Apparatus).

Microscopes

Microscopes, both inverted and upright, with suitable DIC optics can be purchased from all major brands. The major components are:

A polarizer in front of the light source (typically a halogen lamp).

An analyzer (another polarizer) in front of the imaging device.

Objectives suitable for DIC microscopy.

Modified Wollaston (Nomarski) prisms for each objective.

Matched Wollaston prisms within the condenser.

Imaging Device

The imaging devices used for DIC are usually high-quality monochrome video cameras (CCD or Newvicon), with high refresh rates (video rates or similar) with manual gain and offset adjustments. For IR DIC, the imaging device must have adequate sensor sensitivity in the IR range. Recently, high-speed CCD cameras have appeared on the market that can be used for both fluorescence and DIC microscopy (e.g., the SensiCam from Applied Scientific Instrumentation), reducing the alignment problems associated with the use of separate cameras for each imaging mode. Note, however, that on laser-scanning confocal microscopes, alternative sensors can be used (see application example below).

PROCEDURE A

DIC Imaging of Cells in Culture Using an Inverted Microscope

1. Mount the chosen cells (grown on a glass substrate) on the microscope stand.
2. Place a coverslip over the cells, avoiding air bubbles and liquid spillover. If necessary, use spacers to keep the coverslip away from the cells.
3. Verify that all DIC optical elements (polarizer, analyzer, Wollaston prisms) are introduced into the optical path.
4. Observe the cells through the oculars, focus, and adjust the condenser for Köhler illumination: Close the field diaphragm and adjust the condenser height until the field diaphragm is in focus. Center the diaphragm and reopen it.
5. Adjust the bias retardation using the objective Wollaston prism ("slider"), or the de Sénarmont compensator, if microscope is equipped with one.
6. Adjust the camera offset and gain settings until most of the camera's dynamic range is utilized.
7. Collect images.
8. At the end of the experiment, move the specimen to a relatively featureless region, defocus the image, and collect an image (a "mottle" image) at the same settings used throughout the experiment.
9. Use the image taken in step 8 to enhance those collected during the experiment, as described in the application example below.

PROCEDURE B

Infrared DIC Imaging of Brain Slices Using an Upright Microscope

1. Mount the tissue slice in the chamber and fix it in place using a metal U-shape grid made from platinum wire.
2. Adjust the condenser for Köhler illumination and open the field aperture to illuminate the whole field of view. This step is usually done while observing the specimen through the oculars.

3. Keep the aperture iris diaphragm open—contrast will be enhanced using the video camera contrast enhancement functions.
4. Place the IR filter in position and switch the light path to the video camera.
5. Increase the offset and gain of the camera to maximize the use of its dynamic range. Avoid saturation.
6. Adjust the camera's shading corrections to obtain an evenly illuminated field of view.
7. Collect images.

   Note: To obtain further magnification of the image, an extra magnification lens can be inserted into the light path (depending on the microscope configuration) or in front of the camera.

SHORT EXAMPLE OF APPLICATION

Although DIC microscopy can provide an extraordinary level of detail, the technique is more informative of cellular structures than of specific molecules. Fluorescence microscopy, on the other hand, provides detailed information on the distribution patterns of specific molecules and ions, but this information is often hard to interpret in the absence of additional structural information. Complementing fluorescence microscopy with DIC imaging is an excellent method to establish the structural context of fluorescence data. Unfortunately, combining DIC imaging with fluorescence microscopy using video cameras can be awkward. Light sources must be changed, various components must be switched in and out of the light path, and a separate camera is often used for each imaging mode.

On a laser-scanning confocal microscope, however, DIC and fluorescence imaging can be performed simultaneously at practically no cost in terms of exposure (Ryan et al. 1990, 1993). This can be done by placing a photodiode, or a photomultiplier, at the rear end of the condenser (at a conjugate location on the illumination lamp) and collecting the light used for epifluorescence excitation after it travels backward through the condenser (Fig. 1a). As most lasers used for confocal microscopy emit polarized light, the DIC analyzer is not required (which is fortunate, because this component would absorb much of the fluorescence emitted by the specimen). Because the objective Wollaston prism has only a marginal impact on the fluorescence signal, it can be left in the light path. The signal generated by the transmitted light detector is fed into the digitizing circuitry of the confocal microscope and displayed. As both the fluorescence and DIC images are generated by raster-scanning the focused laser beam over the specimen, the DIC and fluorescence images generated are in perfect register. Images obtained in this fashion are shown in Figure 1b and c. After further image enhancement by mottle subtraction (Fig. 1d and e) and contrast adjustment (Fig. 1f), the fluorescence and DIC images can be merged (Fig. 1g), revealing the structural context of fluorescence data.

ADVANTAGES AND LIMITATIONS

The major disadvantage of DIC microscopy is its requirement for glass substrates. More convenient substrates, such as polystyrene, can be used in phase contrast and Hoffman modulation contrast imaging techniques, but not in DIC. However, the objective and condenser apertures used in DIC are not obstructed as they are in other imaging techniques, enabling the microscope to be used at its full numerical aperture, resulting in a much higher resolution image.

Images produced in DIC microscopy have a distinctive shadow-cast appearance, as if they were illuminated from an angle and from above. It is important to be aware that the highlights and shadows do not necessarily represent actual topographical features.

Although DIC microscopy provides little information on the molecular level, it can provide extraordinarily detailed information on subcellular structures and their dynamics (see Forscher et al. 1987), and thus DIC microscopy is particularly suitable for studying cell motility and organelle dynamics.

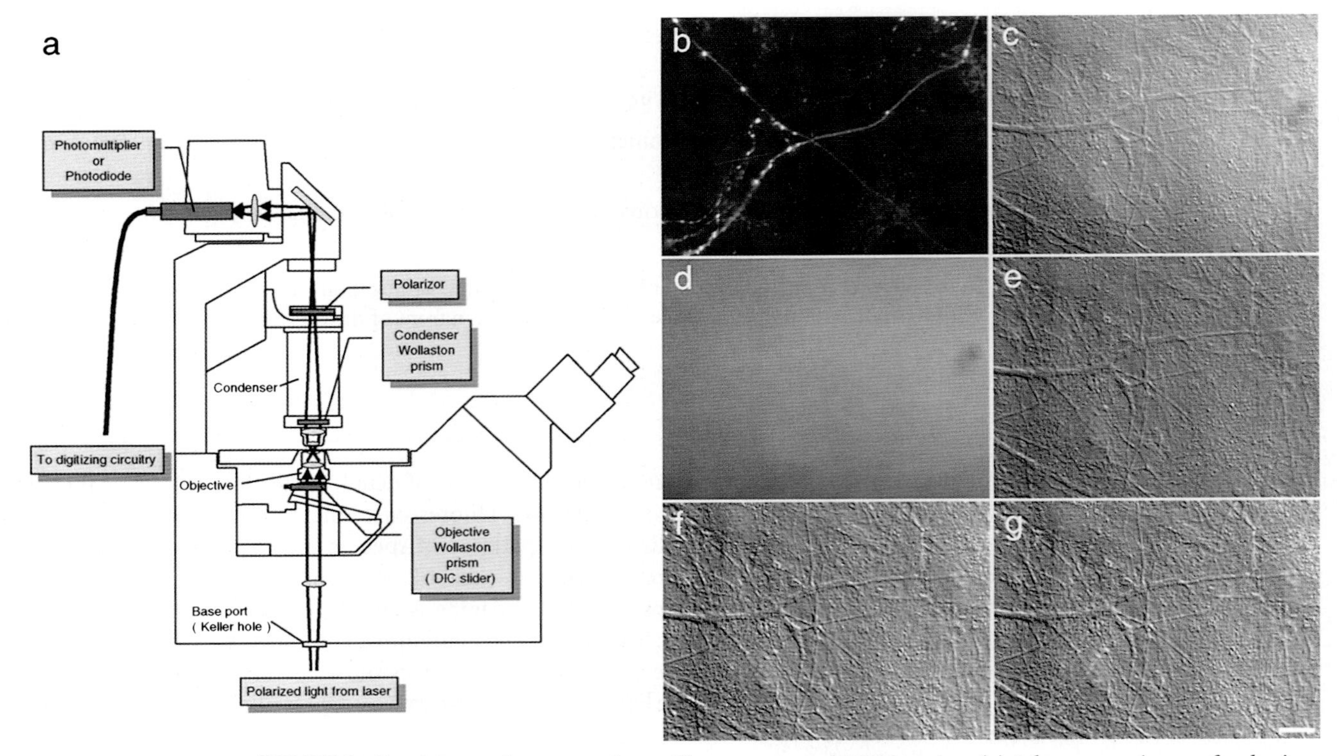

FIGURE 1. Simultaneous laser-scanning epifluorescence and DIC imaging. (*a*) A laser-scanning confocal microscope (LSCM) system equipped for DIC microscopy. As in most LSCM systems, the specimen is raster-scanned by a focused beam of laser light that serves to excite fluorescent substances in the specimen. The light emitted by these substances is collected by the objective and routed to photomultipliers that convert changes in light intensity to changes in an electrical signal (not shown). Here, the light passing through the specimen is also collected by the microscope condenser and focused onto a separate photodetector (a photodiode or, preferentially, a photomultiplier). The signal from this detector is then fed into the LSCM electronics, concomitantly with signals arising from the photomultipliers used for measuring the fluorescence. This results in the DIC and epifluorescence images being in perfect register. (*b*) An epifluorescence image of axons from a cultured hippocampal neuron expressing Synapsin I tagged with GFP. (*c*) A DIC image of the same field, collected concomitantly with the image shown in *b*. (*d*) An out-of-focus image of an empty region of the coverslip (a "mottle" image) collected at the end of the experiment. (*e*) The image of panel *c after* subtraction of the mottle image in panel *d*. (*f*) The image of panel *e after* digitally adjusting the gain and contrast. (*g*) The DIC image overlaid with the fluorescence image of the same field. Bar, 10 μm.

REFERENCES

Dodt H.U. and Zieglgänsberger W. 1994. Infrared videomicroscopy: A new look at neuronal structure and function. *Trends Neurosci.* **17:** 453–458.

Forscher P., Kaczmarek L.K., Buchanan J.A., and Smith S.J. 1987. Cyclic AMP induces changes in distribution and transport of organelles within growth cones of *Aplysia* bag cell neurons. *J. Neurosci.* **7:** 3600–3611.

Ryan T.A, Sandison D.R., and Webb W.W. 1990. Simultaneous DIC and fluorescence in laser scanning confocal microscopy. *Biophys. J.* **57:** A374.

Ryan T.A., Reuter H., Wendland B., Schweizer F.E., Tsien R.W., and Smith S.J. 1993. The kinetics of synaptic vesicle recycling measured at single presynaptic boutons. *Neuron* **11:** 713–724.

WWW RESOURCE

http://www.microscopy.fsu.edu/primer/techniques/dic/dichome.html Molecular Expressions Optical Microscopy Primer, Specialized Techniques

CHAPTER 5

A Practical Guide: Infrared Video Microscopy

Hans-Ulrich Dodt, Klaus Becker, and Walter Zieglgänsberger

Max-Planck-Institute of Psychiatry, 80804 Munich, Germany

This chapter describes how neurons and neuronal excitation can be visualized in brain slices. Infrared video microscopy can be used with all kinds of brain slices up to a thickness of 500 μm. To image the intrinsic optical signal (IOS), it is necessary to preserve long-range axonal projections, for example, in slices of the neocortex.

Infrared Video Microscopy Setup

Single neurons in thick brain slices cannot be seen with standard microscopy, because the neuronal network consists of a large number of neurons packed closely together. These cells act as birefringent-phase objects that scatter light very effectively. Any method for visualization of neurons in brain slices must, therefore, reduce this scattering of light. Light-scattering can be reduced by increasing the wavelength of illumination, by the optics used for contrast generation, and, indirectly, by electronic contrast-enhancement (Dodt 1992).

Infrared Illumination

The first reduction of light-scattering is achieved by the use of near-infrared (IR) radiation instead of visible light (Dodt and Zieglgänsberger 1990). IR radiation is scattered to a lesser extent than visible light, due to its longer wavelength. Standard halogen lamps serve adequately as light sources for IR radiation because their peak emission is in the near-IR range. The wavelength of illumination can be selected by placing a broadband interference filter in the filter holder of the microscope.

Microscope Setup

The microscope setup is built around an Axioskop FS microscope (Zeiss). For patch-clamping, water-immersion objectives with long working distances and high numerical apertures (NA) are required. The procedure discussed here uses an Olympus 60x (NA 0.9) water-immersion objective with a 2-mm working distance. This objective can be used without any noticeable reduction in image sharpness on Zeiss microscopes.

CONTRAST SYSTEM

Unstained neurons in brain slices are phase objects. To render them visible, their phase gradients have to be converted into amplitude gradients by the optics, which also have to provide optical sectioning.

The authors use the gradient-contrast system (Dodt et al. 1998, 2002). The aperture plane of the condenser is reimaged with a lens system between the rear of the microscope body and the lamp house to make it accessible for spatial filtering (Luigs and Neumann). A light-stop in the form of a quarter annulus is positioned in the illumination beam path. At a small distance downstream from the annulus, a diffuser is placed, generating a "gradient" of illumination across the condenser aperture plane. No spatial frequencies in the illuminating light are completely filtered out so that the image remains similar to the object. In addition, the curved form of the slit gives a gradient of illumination in two perpendicular directions in the image, left to right and up to down. This can be very helpful for visualization of dendritic branches running in different directions in the image. Because the light-stop blocks much of the illuminating light, only a part of the normal, illuminating light cone is used and, therefore, less stray light is generated in the slice. This is the same principle used in the slit lamps for ophthalmology. The contrast generated this way is so high that gradient contrast alone allows visualization of neurons in thick slices even with visible light by the naked eye.

ELECTRONIC CONTRAST ENHANCEMENT

Contrast enhancement in real time is most easily achieved by the use of video technology (Allen and Allen 1983) using cameras with IR-sensitive newvicon tubes (C2400-07, Hamamatsu). Cooled CCD cameras like the CoolSNAPro-cf (Mediacybernetics) can also be used, but these give a frame rate of only 10 Hz.

MICROMANIPULATORS

The experimental setup must allow simultaneous visualization and patch-clamping of neurons. After patch-clamping a neuron, it must be possible to move the microscope relative to the brain-slice chamber. This can be achieved by setting the microscope on a two-axis translation stage (Luigs and Neumann). The micromanipulator holding the patch pipette must also be motorized. Motorizing the translation stage of the brain-slice chamber, and the translation stage and focus drive of the microscope, is helpful.

INFRARED–DARK-FIELD MICROSCOPY SETUP

A further development of IR video microscopy allows the spread of neuronal excitation to be visualized directly at a macroscopic level. The brain slice is illuminated with a dark-field condenser, and the image of the slice in IR scattered light is projected by a low-power objective onto the target of the video camera (Dodt and Zieglgänsberger 1994). Because light-scattering of brain slices changes during neuronal excitation, the spread of neuronal excitation can be visualized by purely optical means, without the use of any dyes (MacVicar and Hochman 1991). To visualize the small (a few percent) light-scattering changes, a differential subtraction technique is employed. The image of the brain slice at rest is stored in the computer memory and subtracted, online, from the image of the slice during electrical stimulation. By strong digital contrast enhancement, areas of neuronal excitation in the brain slice become visible. These areas are then overlaid onto the black and white image of the brain slice by custom-made software.

PROCEDURE A

Patch-clamping in the Infrared

First, adjust the microscope for correct Köhler illumination: Bring a test object, e.g., a thread of the grid made for holding the brain slice, into focus with the 60x objective. Close the field diaphragm at

A

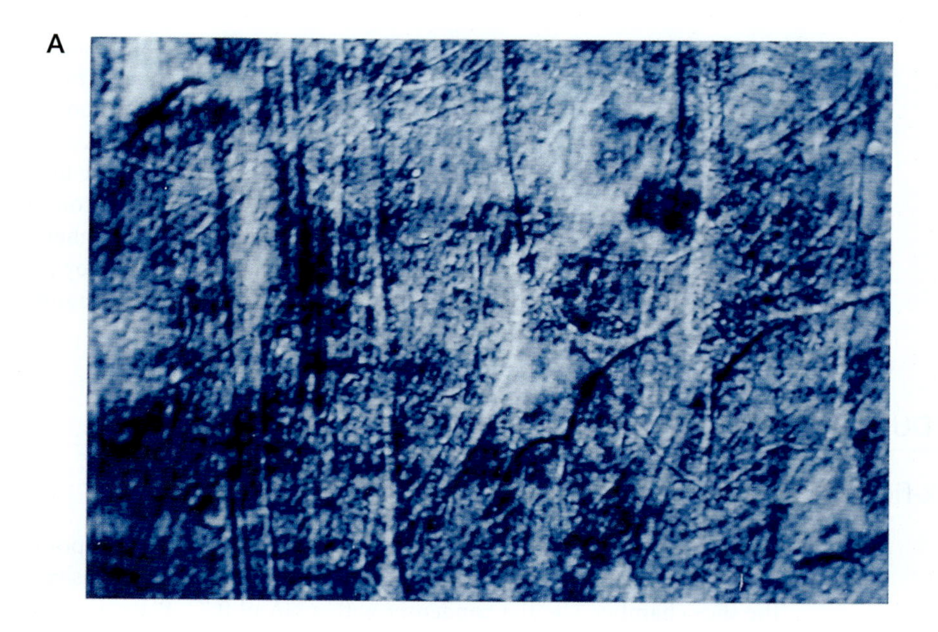

B

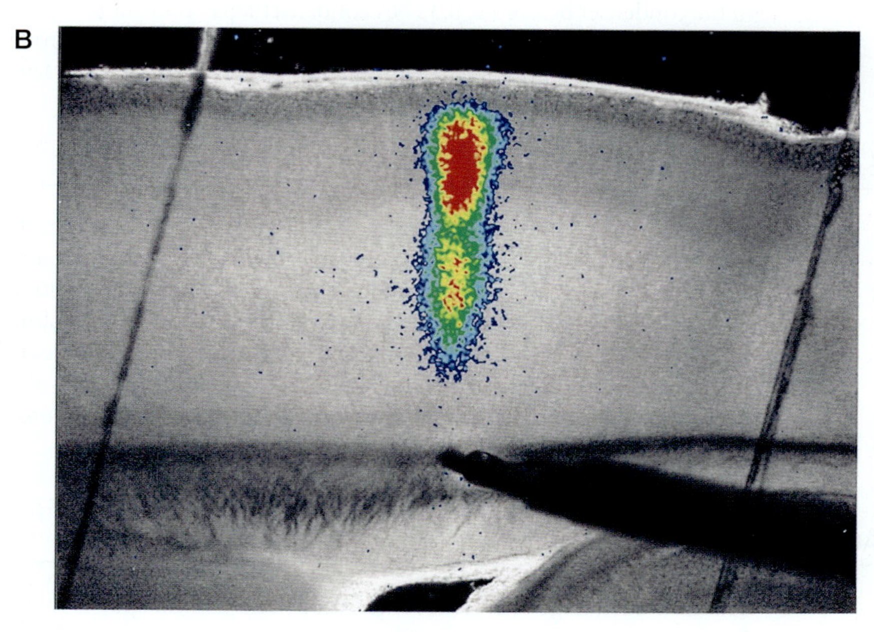

FIGURE 1. (*A*) Neuronal network in lamina 2/3 of the adult rat somatosensory neocortex, as seen with IR video microscopy using gradient contrast. Two pyramidal neurons, covered with small beaded structures, are visible on the right-hand side of the picture. A dendritic bundle on the left-hand side is overlaid with a dense fiber network, presumably axons. By focusing the microscope through different focal planes, a 3D impression of the neuronal network can be obtained. Image width, 100 µm. (*B*) Neuronal excitation in a neocortical slice visualized by IR dark-field video microscopy. Dark-field image of a sagittal neocortical slice overlaid with the IOS in the slice after tetanic stimulation (50 Hz for 2 sec) in layer VI. Image width, 5 mm.

the foot of the microscope, and adjust the position of the condenser with centering screws and focusing mechanisms. In the correct position, the diaphragm appears as a sharp round hole in the center of the field of view. Do this without the IR filter in visible light looking through the eyepieces of the microscope. Then rotate the IR broadband interference filter ($\lambda = 780 \pm 50$ nm, type KMZ 50-2, Schott) in the filter holder of the microscope into place. Place the brain slice in the recording chamber and bring the neurons into focus. For patch-clamping, it is convenient to search first at a lower magnification for "good" neurons in the slice and then to use higher magnification for approaching the selected neuron with the patch pipette. This can be achieved by placing an additional magnification changer in front of the video camera, allowing additional magnifications of 1x and 4x the magnification of the objective (Luigs and Neumann).

PROCEDURE B

IR Dark-field Microscopy to Visualize the Intrinsic Optical Signal

The Axiosop is also used for IR dark-field microscopy with a low-power objective. Put a dipping cone on a 2.5x objective (NA 0.12) to avoid fluctuations of light intensity by changes of the fluid level in the slice chamber. Use the condenser with a NA of 0.32. Place a circular light-stop (diameter 5 mm) in the condenser before the 0.32 NA condenser lens. After submerging the slice in the recording chamber in Krebs-Ringer solution, lower a concentric stimulation electrode (SNX-100, Fine Science Tools) onto the slice. Then use tetanic stimulation (50 Hz for 2 sec) of a few volts to elicit the IOS.

EXAMPLE OF APPLICATION

Figure 1A gives some idea of the complexity of structure that can be imaged with IR-gradient contrast. Because no light-consuming optical elements have to be placed in the beam path after the objective, gradient contrast can be combined with techniques such as fluorescence and photostimulation.

An example of an IOS, which, in the neocortex, exhibits a column-like shape, is given in Figure 1B. This technique allows investigation of the modulatory influences of many kinds of neuroactive substances on the spatial spread of neuronal excitation (Dodt et al. 1996; D'Arcangelo et al. 1997). Even the enhancement of inhibitory neurotransmission by a neuroactive steroid can be visualized (Dodt et al. 1996).

ADVANTAGES AND LIMITATIONS

The clear advantage of IR video microscopy and IOS imaging is that there is no risk of phototoxicity. Because no staining is necessary, all the neuronal elements that can be visualized by IR video microscopy can be seen at the same time. In contrast to imaging with voltage-sensitive dyes, IOS imaging can be performed over extended periods (hours) with no deleterious side effects of the dyes.

A disadvantage of IR video microscopy is the limitation set by the low inherent contrast of very fine neuronal structures like spines. To date, staining is the only way to reliably visualize spines. This may change as new techniques, such as transmission confocal microscopy, are developed.

REFERENCES

Allen R.D. and Allen N.S. 1983. Video-enhanced microscopy with a computer frame memory. *J. Microsc.* **129:** 3–17.

D'Arcangelo G., Dodt H.-U., and Zieglgänsberger W. 1997. Reduction of excitation by interleukin-1β in rat neocortical slices visualized using infrared-darkfield videomicroscopy. *Neuroreport* **8:** 2079–2083.

Dodt H.-U. 1992. Infrared videomicroscopy of living brain slices. In *Practical electrophysiological methods* (ed. H. Kettenmann and R. Grantyn), pp. 6–10. Wiley-Liss, New York.

Dodt H.-U. and Zieglgänsberger W. 1990. Visualizing unstained neurons in living brain slices by infrared DIC-videomicroscopy. *Brain Res.* **537:** 333–336.

———. 1994. Infrared videomicroscopy: A new look at neuronal structure and function. *Trends Neurosci.* **17:** 453–458.

Dodt H.-U., D'Arcangelo G., Pestel E., and Zieglgänsberger W. 1996. The spread of excitation in neocortical columns visualized with infrared-darkfield videomicroscopy. *Neuroreport* **7:** 1553–1558.

Dodt H.U., Eder M., Schierloh A., and Zieglgänsberger W. 2002. Infrared-guided laser stimulation of neurons in brain slices. *Sci. STKE* **120:** PL2.

Dodt H.-U., Frick A., Kampe K., and Zieglgänsberger W. 1998. NMDA and AMPA receptors on neocortical neurons are differentially distributed. *Eur. J. Neurosci.* **10:** 3351–3357.

MacVicar B.A. and Hochman D. 1991. Imaging of synaptically evoked intrinsic signals in hippocampal slices. *J. Neurosci.* **11:** 1458–1469.

CHAPTER 6

Confocal Microscopy: Principles and Practice

Alan Fine

Department of Physiology & Biophysics, Dalhousie University Faculty of Medicine, Halifax, Nova Scotia B3H 4H7, Canada

More than 15 years have passed since the introduction of the first commercial confocal laser-scanning microscope. The confocal microscope is now recognized as an invaluable tool for high-resolution fluorescence microscopy, and few biological research centers are without at least one such instrument. In this chapter, I outline the basic principles of confocal microscopy, relevant practical considerations, and various approaches to its implementation, principally from the perspective of visualizing rapid, small-scale phenomena in living tissue.

Much of the interest in confocal microscopy stems from the increasing reliance on fluorescent probes in contemporary biology. Fluorescent-labeled antibodies and ligands are essential tools for localizing specific molecules. Intracellular or membrane-bound fluorescent dyes are widely used to follow morphological changes in cells, and retrogradely transported fluorescent markers have been used to identify living neurons with particular projections for subsequent electrophysiologic or structural study. Voltage- and ion-sensitive indicator dyes have been used to observe patterns of electrical activity in large networks of neurons and in structures too small to be monitored with classic electrode techniques. The advent of green fluorescent protein (GFP) now permits the visualization of specific proteins, including engineered functional probes, in essentially unperturbed living tissue.

Unfortunately, fluorescence images are often severely degraded by light that is scattered or emitted by structures outside the plane of focus. This problem is particularly severe for thick specimens such as brain slices or whole embryos and is exacerbated by the poor depth discrimination of conventional (wide-field) light microscopy. These limitations have been only partly overcome by video image processing and deconvolution techniques (Inoué 1986), but have been greatly reduced by confocal optics.

THEORY OF CONFOCAL OPTICS

Principle

Confocal optics provides an alternative solution to the problem of image degradation by out-of-focus light, improving resolution by physical rather than electronic or computational means. The principle of confocal microscopy was described by Minsky more than 40 years ago (Minsky 1961). The crucial element of confocal optics is the projection of an image of a focally illuminated point in the specimen onto a small aperture in a conjugate focal plane (see Fig. 1). Light from the illuminated point can pass through the aperture and be detected; light from out-of-focus structures will be spread at, and blocked by, the aperture and will therefore be largely eliminated. By scanning the specimen across the point (or the point across the specimen; see below; see also Fig. 1), a 2D or 3D confocal image can be generated (Fine et al. 1988; Pawley 1995).

Factors Influencing Resolution

To appreciate the confocal improvement, it is useful to consider some of the factors influencing optical performance. A discussion of these factors is given below.

Point-spread Function

The resolution of a refractive system such as a microscope is set by its point-spread function (PSF), the 3D intensity distribution of an imaged point. For ideal optics (i.e., unaffected by spherical or other aberrations) and incoherent illumination, the PSF extends (to its first minimum) in the object focal plane to a radius of $r_A = 0.61\lambda/\text{NA}$, where λ is the wavelength of light and NA is the numerical aperture of the objective. This central zone of the PSF is referred to as the "Airy disk." The distance to the first minimum of the PSF along the optical axis is $2\lambda n/(\text{NA})^2$ (where n is the index of refraction for the object; see also Chapter 10). The PSF establishes the theoretical resolution limit of the system because the images of two points closer than these distances begin to overlap and, thus, cannot be distinguished. (*Note:* It is important here to distinguish resolution from detection. Subresolution particles can be detected if they have sufficient contrast against the background, although their dimensions

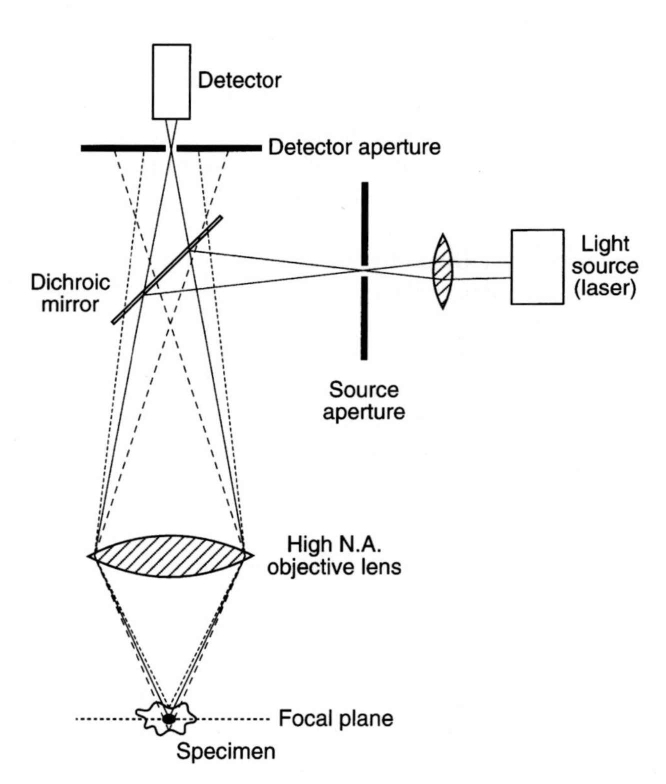

FIGURE 1. The confocal principle. Light from a point source is focused on a point in the specimen. Light (here, fluorescence) from the illuminated point is focused on, and passed through, a detector aperture; fluorescence from all other points is excluded.

will appear to be those of the PSF, regardless of their true size. Thus, even single fluorescent molecules can be detected if they are sufficiently bright. Similarly, for high-contrast structures larger than the resolution limit, it is possible to detect *changes* in size less than the resolution limit, if the same "edge" criterion [e.g., position at which intensity is half-maximum] is applied to the "before" and "after" images.)

Object Illumination and Detector Aperture

This classic resolution limit (the Rayleigh criterion) assumes wide-field illumination such that the PSFs are present and interfering at the same time. It is possible, therefore, to exceed this limit by various "superresolution" methods in which the instantaneous field of view is greatly reduced in both the object and image planes. In near-field fluorescence microscopy, for example, this is done by illuminating and detecting light through the submicron tip of a fiber optic placed close to the object. In confocal microscopy, a similar result is obtained by use of point illumination and conjugate-point detection. The actual improvement in resolution depends on the size and shape of the illumination spot and the detector aperture; a 1.4-fold improvement over the Rayleigh criterion has been obtained with circular apertures, and even greater improvements can be obtained at the expense of signal intensity by using annular apertures (Slater and Slayter 1992).

Deconvolution

The resolution of confocal images, as of ordinary wide-field images, can be further improved by deconvolution methods (Agard et al. 1989; Hosokawa et al. 1994). The image of a point is the PSF; thus, the image of a complex structure represents the convolution of the points in that structure with the PSF. Therefore, if the PSF of the imaging system can be estimated from known properties (e.g., NA), or empirically determined by imaging subresolution particles, the true structure of the object may be approximated by dividing the image point by point by the PSF. Various algorithms have been developed for efficient implementation of this computationally intensive procedure; through their use, resolution in raw images can be improved by a factor of 2 or more (Shaw 1995).

Specimen Thickness and Depth of Field

The axial dimension of the PSF ordinarily establishes the depth of field of the system. Although a large depth of field is useful in some circumstances, high-resolution applications generally require thin optical sectioning. For an infinitely thin specimen containing minute elements spaced more widely than r_A, the theoretical depth of field is defined as one-half of the axial extent of the PSF. Real specimens, on the other hand, may be relatively thick and contain numerous extensive fluorescent structures; for such specimens, the depth of field is greater because the PSFs generated by these structures or by light scattering can overlap, both within and beyond the focal plane. Under such circumstances, the wide-field fluorescence image is obscured by flare, decreasing the contrast between image features (and thus resolution) and increasing the apparent depth of field. The confocal aperture can eliminate this light from out-of-focus elements, reducing the depth of field toward its theoretical limit. *It is this elimination of out-of-focus light that is the most significant aspect of confocal optics* (see Fig. 2). The extent of this improvement depends on the size of the confocal aperture. Decreasing the aperture size improves resolution and rejection of out-of-focus light up to a limit, but it also decreases signal intensity; optimal signal-to-noise ratio (SNR) is generally achieved when the diameter of the aperture is ~75% that of the Airy disk (Wilson 1989, 1995).

By eliminating out-of-focus light and decreasing depth of field, confocal optics permits "optical sections" of less than 1 μm to be imaged even in relatively thick, scattering tissue. Serial optical sections can be generated simply by changing focal depth; because the specimen remains intact, these sections are necessarily in proper register, greatly simplifying subsequent computational reconstruction of 3D structure.

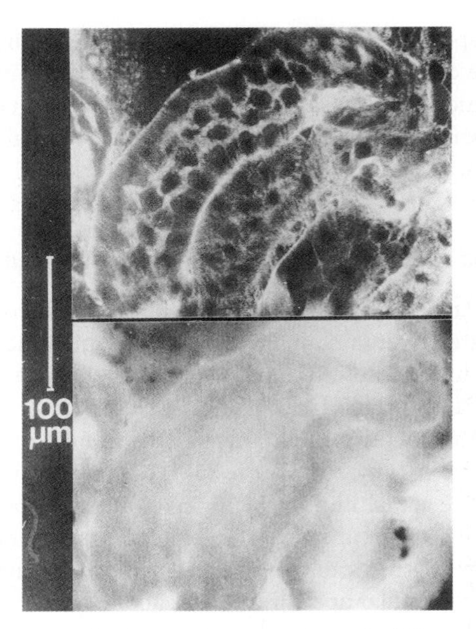

FIGURE 2. The confocal advantage. Cells close to the cut surface of a kidney glomerulus, stained with a fluorescent, lipophilic styryl dye, are clearly visible when imaged by confocal laser-scanning microscopy (*top*), but are obscured by out-of-focus light when viewed by ordinary wide-field fluorescence optics (*bottom*). Both images were obtained with the same 0.7-NA objective, focused at the same depth.

Other Factors

Other benefits of using confocal microscopes result from the nature of the electronic light detectors used, which facilitates digital image enhancement and can permit sensitive quantitative measurement of small, rapid changes in even low levels of fluorescence. In addition, computer control of many commercial instruments allows imaging to be easily integrated with other experimental manipulations, e.g., electrical stimulation.

SIGNAL OPTIMIZATION: PRACTICAL CONSIDERATIONS

To exploit the advantages offered by confocal optics, a number of important practical considerations must be addressed.

Signal Intensity and Signal-to-Noise Ratio

The foremost consideration is the problem of signal intensity, along with the associated issue of SNR. Confocal imaging depends on an aperture to exclude out-of-focus light, but, as a result of scattering and optical aberration, the aperture invariably also eliminates some light emanating from the focal point. Particularly for dim fluorescent objects, only a very small number of photons may reach the detector during the sampling interval; indeed, in many cases the number of photons detected per sample point (pixel) will be none or a few. Under these conditions, the inescapable random statistical fluctuations in photon flux can dominate the signal, eliminating contrast and resolution.

Because this intrinsic photon noise increases only with the square root of the mean light intensity, to increase the SNR it is essential to increase the detected light intensity. This can be done by decreasing the imaging rate and/or by averaging multiple images; however, for some applications, such as detection of fast transient events, this is not possible. The detector aperture cannot be opened too far without loss of confocality, so other means must be found to achieve this. Higher dye concentrations can be used, but beyond a certain concentration, self-quenching may reduce fluorescence, and high dye concentrations may perturb the behavior of the system under study (e.g., calcium indicator dyes can buffer the free-calcium concentration). Dyes with higher quantum yield or with excitation maxima more closely matched to the available laser lines can be substituted where available. Increasing illumination intensity is an obvious strategy, and most confocal microscopes incorporate lasers for this purpose. However, this approach is limited by several factors. At a certain illumination intensity, essentially all the fluorophore molecules in the illuminated spot will be excited even during brief exposure to the beam. At this point, the dye is saturated, and further increases in excitation-light intensity yield no additional fluorescence, although higher-power lines may permit the use of fluorophores whose excitation spectra are not well matched to the laser emission. With most commercial laser-scanning confocal microscopes, dye saturation occurs with single-line laser power of a few milliwatts; more powerful lasers confer little advantage. Furthermore, increasing illumination increases the rate at which fluorophores bleach, limiting the useful duration of investigation and generating toxic free radicals. (*Note:* Bleaching, which is an oxidative process, can be significantly reduced by the inclusion of antioxidants in the medium surrounding the preparation. For living cells, a particularly useful antioxidant is the water-soluble vitamin E derivative, Trolox [6-hydroxy-2, 5, 7, 8-tetramethyl-chroman-2-carbonate].) These considerations are usually dominant, so it is generally preferable to use the minimum possible illumination intensity.

A further important consideration concerning light sources is instability of output power, which can introduce additional noise and impair reproducibility. The best available lasers have output noise on the order of 0.1%. Some noise reduction can be obtained, at the expense of intensity, by feedback-regulated devices such as electro-optic modulators, but their frequency response may limit their usefulness to slow-speed scanning.

Optical-transfer Efficiency

Other routes to increasing detected light intensity involve increasing optical-transfer efficiency at all points of the optical path, and increasing detector sensitivity. Significant amounts of light are lost to scattering or absorption when light passes through lenses or is reflected from mirrors. Laser-scanning confocal microscopes may have eight or more mirrors for scanning and beam folding; thus, even 95% reflectance mirrors can lead to over 30% reduction in image intensity. Mirrors with ≥99% reflectance are available and should be used where possible. Dirt and misalignment can dramatically reduce optical-transfer efficiency and must be avoided. High-NA objectives not only improve resolution, but also increase light gathering: Epifluorescence intensity for spatially extended structures varies with the fourth power of NA. Thus, as a general rule, for any magnification, the highest-NA objective should be used; conversely, for any NA, the lowest-magnification objective should be used to provide maximal field of view and working distance. Objectives of similar nominal NA can vary substantially in optical-transfer efficiency at different wavelengths, however, and should be directly tested and compared.

Detector Quantum Efficiency

Most commercial confocal microscopes use photomultiplier tubes (PMTs) as light detectors because of their high gain and sensitivity. Simultaneous imaging at several wavelengths is easily achieved via dichroic mirrors and additional detectors. Quantum efficiency (QE; the percentage of incident photons that generate a signal) of all detectors is a function of wavelength; QE for PMTs with standard alkali photocathodes is seldom higher than 25% and can be much lower at longer (red) wavelengths. QE can be improved by using individually selected tubes, matching photocathode material to intended wavelength, with prismatic windows or similar arrangements. Very recently, GaAs and GaAsP photocathode PMTs have become available with substantially higher QE, particularly at red wavelengths. Silicon detectors such as charged-couple devices (CCDs) or avalanche photodiodes can have much higher QE (up to 90%), but generally have lower gain, higher background-noise levels (dark noise), and inferior high-frequency response (important when detecting several photons at each of a million pixels per second). Thermal electron emission contributes to dark noise, which can be reduced by cooling the detector. In addition, this noise is stochastic and uncorrelated from pixel to pixel, so can be reduced digitally, or by low-pass filtration of the detector output, prior to digitization.

Image Digitization

In most instruments, it is necessary (and useful) to digitize the detector output, generating discrete image-intensity samples corresponding to pixels. (Once digitized, images can be summed, averaged, and subtracted from or added to other images, and a wide range of image-enhancement and volume-reconstruction methods can be applied.) An important issue is the optimal spatial frequency of these samples. For nonperiodic structures, a sampling frequency at least 2.3 times the maximum spatial frequency in the image (the Nyquist criterion) is necessary to retain all image details. Thus, if r_A for the objective in use is 300 nm, pixel size should be smaller than 130 nm to obtain maximal detail in the *xy* plane. For laser-scanning confocal microscopes (see below), the temporal sampling rate (pixels/sec) may be constant, but the spatial sampling rate (pixels/μm) can be varied by changing the "zoom" factor. Sampling at spatial frequencies below the Nyquist criterion will lose spatial information, but can reduce bleaching and phototoxicity, considerations that may dominate for live-cell imaging; pixel sizes smaller than the Nyquist criterion may aid visualization, but provide no additional information, and needlessly increase bleaching and phototoxicity.

Scanning Mode

Temporal resolution becomes a serious consideration when the aim is to monitor fast phenomena. The Nyquist criterion applies equally in the time domain. Some scanning methods are too slow for these purposes. Even with fast-scanning methods (see below), it may not be possible to collect sufficient photons from small pixels over short intervals.

IMPLEMENTATION

Confocal imaging can be implemented in a variety of ways (see Fig. 3). The main designs used are described below.

Specimen-scanning Design

The light path can be fixed and the specimen scanned under the beam. This has the virtue of simplicity, permits scanning of large fields, and avoids lateral aberrations in the optics; however, it is slow, and impractical when the specimen is massive or attached to other (e.g., electrical recording) apparatus.

Nipkow Disk Design

Imaging can also be achieved by coordinated scanning of apertures, arranged in a spiral pattern as a Nipkow disk, across conjugate focal planes in the incident and image light paths (Fig. 3A). Disks with suitably spaced apertures permit confocal monitoring of many points at once, in turn allowing imaging at up to several hundred frames per second. In simple disk configurations, adequate simultaneous illumination of the multiple points is difficult to achieve, thus limiting sensitivity. This design has recently been substantially improved by incorporation in the incident light path of an array of microlenses, spinning together with the Nipkow disk, and aligned so as to focus virtually all the illumination onto the array of pinholes (Ichihara et al. 1996). This design is particularly useful for imaging rapidly changing specimens that are relatively bright or easily bleached.

Laser-scanning Design

Mirror-based

More intense illumination is possible by the use of a laser to scan a single focused spot across the specimen. A common strategy is illustrated in Figure 3B. The laser intensity is adjusted with a neutral-density filter, and the appropriate laser line is selected by a band-pass filter. A microscope, often equipped for ordinary epifluorescence, is used to focus the illumination spot on the specimen, with a dichroic mirror serving to introduce the illuminating beam into the microscope optical axis. A pair of computer-controlled galvanometer mirrors beyond the dichroic mirror steer the spot in a raster pattern over the specimen. Fluorescent light from the illuminated spot is returned and descanned by the same galvanometer mirrors, passed by the dichroic mirror, and focused onto an aperture in front of a detector. The digitized output of the detector is correlated in time with the position of the spot, permitting digital construction of an *xy* image. The true magnification of the image can be increased by compressing the raster into a smaller region about the optical axis ("zoom").

Acousto-optic Modulator-based

The mass of the galvanometer mirror (discussed above) makes it difficult to scan fast enough to generate images at the rates needed for observing very rapid phenomena (e.g., ion redistribution during action potentials). Higher scan speeds can be achieved with acousto-optic modulators (see Chapter 100), which use sound to induce refraction waves that behave like a diffraction grating. The degree of refraction is wavelength dependent, however, so that the (longer-wavelength) fluorescence cannot be descanned to a fixed point by the same device. A slit can therefore be used as the confocal aperture, with acceptable rejection of out-of-focus light, or a linear detector array such as a CCD can be used to select the moving point on the acousto-optically scanned line that corresponds to the stationary focal point (see Fig. 3C). Acousto-optic modulators or the related acousto-optic deflectors can also serve as ultrafast shutters or intensity regulators, useful for switching and balancing laser lines, minimizing unnecessary illumination and bleaching (e.g., during changes in beam scan direction and "flyback" during raster scanning), or reducing noise due to laser intensity fluctuations; these applications are compatible with all scanning designs.

Alternative Laser-scanning Designs

Simpler designs are possible. A single fiber optic can be used as both the point source and the detector aperture. The point can be scanned by mirrors in the light path as described above or, more simply, by vibrating the fiber in an image plane (see Fig. 3D). High-speed confocal imaging with no moving parts is also possible, using spatial light modulators to generate moving patterns of transmittance or reflectance (Liang et al. 1997; Verveer et al. 1998). All laser-scanning designs can be miniaturized, making possible head-mounted or endoscopic confocal imaging devices for in vivo applications.

Advances in Spectroscopy

Until recently, commercial instruments offered only a small number of detector channels for simultaneous imaging at different wavelengths, the latter usually determined by preselected interference filters with fixed spectral properties. Spectral dispersion by prism or diffraction grating offers greater flexibility. In one configuration (Leica TCS SP), an arbitrary spectral band can be selected by an adjustable slit formed between mirrored plates in front of the PMT; rejected portions of the spectrum are reflected to additional adjustable slits (Calloway 1999). In an alternative configuration (Zeiss META), the dispersed spectrum is projected onto a 32-channel PMT array, allowing simultaneous imaging within ~10-nm-wide emission bands across the visible spectrum. Signals derived from multiple fluorophores with known, overlapping emission spectra can be resolved from such data sets with suitable linear unmixing algorithms (Dickinson et al. 2001).

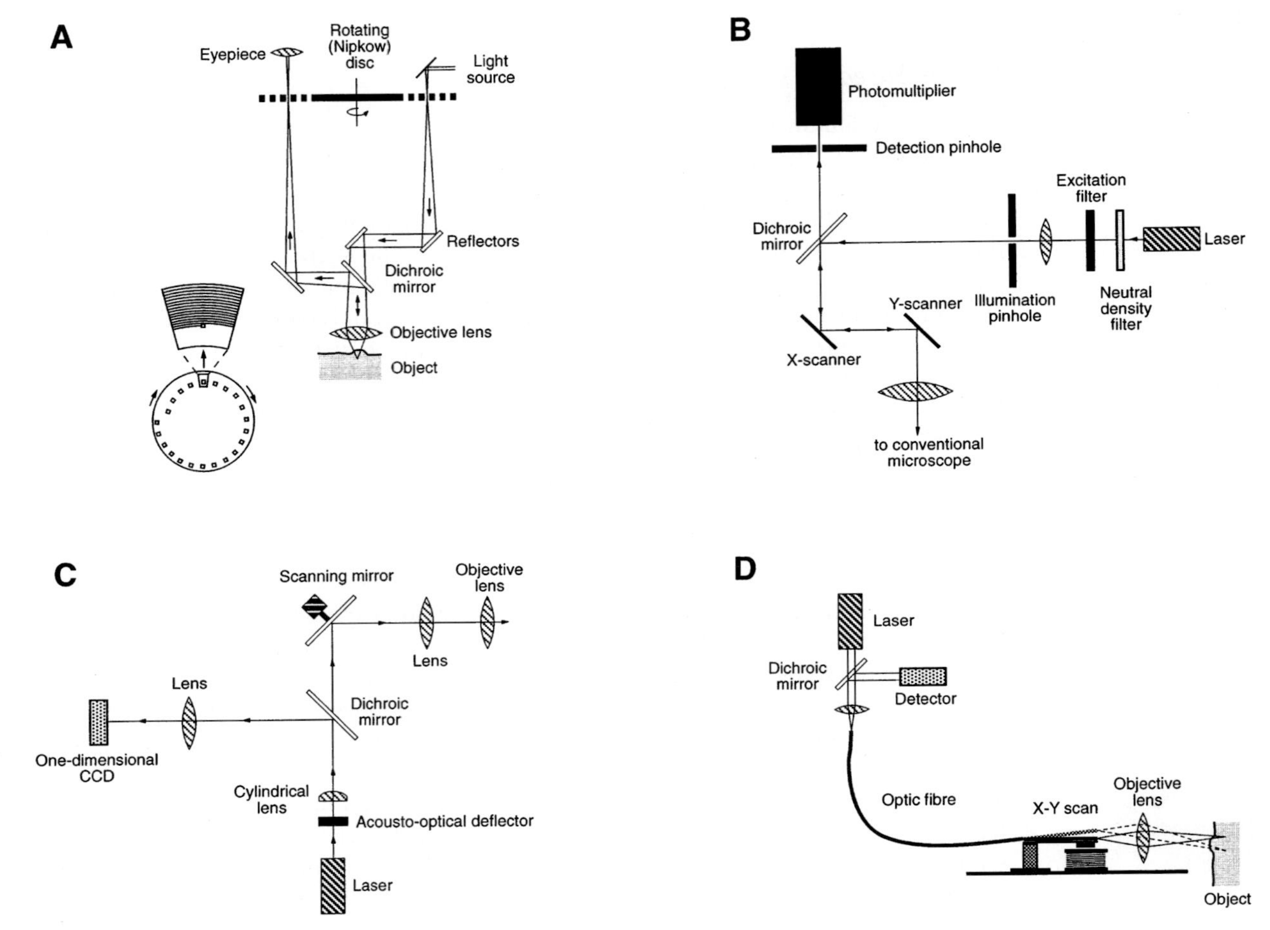

FIGURE 3. Confocal implementation. (*A*) A 2D image can be generated by spinning a spiral array of illumination apertures (a Nipkow disk) over a window on the specimen, thereby illuminating points on sequential lines (*insert*). A confocal image of the illuminated points is viewed through a similarly moving array in a conjugate image plane. (*B*) A spot of light from a fixed point source can be scanned across the specimen by mirrors; longer-wavelength fluorescence from the point is reflected back along the same light path through a dichroic reflector to a conjugate fixed-point aperture. The time-varying signal from the detector behind this aperture is converted by computer into a 2D image. (*C*) Scanning can also be performed by an acousto-optic deflector, but the wavelength dependency of these devices prevents their use for descanning the longer-wavelength fluorescence. Instead, a linear detector array can be used, sampling each element in sequence corresponding to the moving point. (*D*) A simpler configuration uses a fiber optic, vibrating in an image plane, as both point source and detector aperture.

CONCLUSION

Other methods for high-resolution imaging, such as atomic force microscopy, near-field optical microscopy, two-photon fluorescence microscopy, and digital deconvolution, are becoming progressively more powerful and affordable. Some of these can be applied to living tissue, and some can be used in conjunction with confocal microscopy. The power of confocal microscopy will continue to increase through instrumentation innovation and the development of new fluorescent probes.

REFERENCES

Agard D.A., Hiraoka Y., Shaw P.J., and Sedat J.W. 1989. Fluorescence microscopy in three dimensions. *Methods Cell Biol.* **30:** 353–378.

Calloway C.B. 1999. A confocal microscope with spectrophotometric detection. *Microsc. Microanal.* (suppl. 2) **5:** 460–461.

Dickinson M.E., Bearman G., Tille S., Lansford R., and Fraser S.E. 2001. Multi-spectral imaging and linear unmixing add a whole new dimension to laser scanning fluorescence microscopy. *BioTechniques* **31:** 1272–1278.

Fine A., Amos W.B., Durbin R.M., and McNaughton P.A. 1988. Confocal microscopy: Applications in neurobiology. *Trends Neurosci.* **11:** 346–351.

Hosokawa T., Bliss T.V.P., and Fine A. 1994. Quantitative three-dimensional confocal microscopy of synaptic structures in living brain tissue. *Microsc. Res. Tech.* **29:** 290–296.

Ichihara A., Tanaami T., Isozaki K., Sugiyama Y., Kosugi Y., Mikuriya K., Abe M., and Uemura I. 1996. High-speed confocal fluorescence microscopy using a Nipkow scanner with microlenses for 3-D imaging of siingle fluorescent molcule in real-time. *Bioimages* **4:** 57–62.

Inoué S. 1986. *Video microscopy.* Plenum Press, New York.

Liang M., Stehr R.L., and Krause A.W. 1997. Confocal pattern period in multiple-aperture confocal imaging systems with coherent illumination. *Opt. Lett.* **22:** 751–753.

Minsky M. 1961. Microscopy apparatus. U.S. Patent No. 3,013,467.

Pawley J.B., ed. 1995. *Handbook of biological confocal microscopy*, 2nd edition. Plenum Press, New York.

Shaw P.J. 1995. Comparison of wide-field/deconvolution and confocal microscopy for 3D imaging. In *Handbook of biological confocal microscopy*, 2nd edition (ed. J.B. Pawley), pp. 373–387. Plenum Press, New York.

Slater E.M. and Slayter H.S. 1992. *Light and electron microscopy.* Cambridge University Press, Cambridge, United Kingdom.

Verveer P.J., Hanley Q.S., Verbeek P.W., Van Vliet L.J., and Jovin T.M. 1998. Theory of confocal fluorescence imaging in the programmable array microscope (PAM). *J. Microsc.* **189:** 192–198.

Wilson T. 1989. Optical sectioning in confocal fluorescent microscopes. *J. Microsc.* **154:** 143–156.

———. 1995. The role of the pinhole in confocal imaging system. In *Handbook of biological confocal microscopy*, 2nd edition (ed. J.B. Pawley), pp. 167–182. Plenum Press, New York.

CHAPTER 7

Principles of Multiphoton-excitation Fluorescence Microscopy

Winfried Denk

Max-Planck Institute for Medical Research, Department of Biomed Optics, D-69120 Heidelberg, Germany

Fluorescence microscopy has been steadily gaining importance in quantitative biological research as dramatic improvements have been seen in fluorophores, optical systems, light sources, and detectors (see Chapter 2). In particular, the invention (Minsky 1961) and implementation of the confocal fluorescence microscope (for a useful collection of reprints, see Masters 1996; see also Chapter 2), usually by laser scanning, has for the first time allowed the observation of biological processes with high spatial resolution inside intact living tissue. This ability, often called "optical sectioning" (see Chapter 6), allows the spatial reconstruction of 3D specimens without the use of a microtome.

In this chapter, I review the physical mechanisms upon which the properties of multiphoton microscopy are based and discuss some practical aspects of its implementation. For further discussion of these issues, a number of review papers can be recommended: Denk et al. (1994, 1995a); Williams et al. (1994); Denk (1996); Potter (1996); Denk and Svoboda (1997) and original papers referenced therein.

THE THEORY OF MULTIPHOTON MICROSCOPY: A COMPARISON WITH CONFOCAL MICROSCOPY

Out-of-focus Light Rejection

For all its virtues, confocal fluorescence microscopy has a major flaw: It excites fluorophores with abandon, but detects only a small fraction of the generated fluorescence. This is unavoidable with one-photon absorption because the average intensity of the excitation light does not really vary with the distance from the focal plane (see below). In one-photon microscopy, optical sectioning is instead based on the rejection, by the confocal pinhole, of all fluorescence except that coming from the focus. For example, in a specimen several hundred micrometers thick, more than 99% of the generated fluorescence is wasted (even before taking into account the losses, by another factor of ~20, due to limited solid angle of detection and quantum efficiency [QE]). This loss of fluorescence is particularly unfortunate in the case of living specimens where certain chemical techniques for the avoidance of photobleaching and photodamage cannot be used because they involve the removal of oxygen.

Localized Excitation

Multiphoton microscopy (Denk et al. 1990) combines most of the advantages of confocal microscopy, including its optical sectioning properties and background rejection, with an almost complete utilization of the excited fluorescence (to the extent possible, given the limited acceptance angle and transmission of the collection optics, as well as the limited detector QE). The central idea—exciting only those fluorophores from which one wants to collect fluorescence—is simple enough, but impossible to achieve with one-photon (also called linear) absorption. The reason excitation confinement is impossible with one-photon absorption can be simply understood by considering the distribution of the fluorescence excitation in a focused laser beam, traversing a 3D object that is uniformly stained (for tissue, uniform staining is meant in a statistical sense). Now imagine the specimen divided into slices (perpendicular to the optical axis) of uniform thickness. For one-photon excitation, each slice produces the same amount of fluorescence, irrespective of the slice location relative to the focal point. In a thick specimen, extending along the optical axis for a large multiple (as high as several hundred times) of the focal depth (~1 μm for a high-NA lens), only a small fraction of slices, corresponding to the focal depth or imaging depth, will intersect the focal volume. The small fraction of the total fluorescence that comes from the focal volume (roughly equivalent to the observation volume of the confocal microscope) is all that is utilized for image generation. For one-photon excitation, this behavior is entirely unavoidable: The excitation light has to reach the focus and, on the way there, it intersects all the slices in front of the focal point and beyond, because most excitation light continues to propagate. Of course, the incident light has to get to the focus, even for multiphoton excitation (Goeppert-Mayer 1931; Kaiser and Garrett 1961), which, here, specifically means the simultaneous absorption of a number (≥2) of photons combining their quantum energies to bridge the gap between ground and excited molecular state. For one-photon excitation, the decay in excitation rate per fluorophore ($\propto z^{-2}$), as one moves away from the focus (by z), is balanced by the increase in beam cross-sectional area ($\propto z^2$); for two- (and greater than two) photon absorption, the excitation rate falls off much more quickly ($\propto z^{-2N}$) because it now depends on the light intensity to the Nth power, where N is the number of simultaneously absorbed photons. This decay of the excitation rate is so rapid that, for $N = 2$ and for all $N > 2$, even for an infinitely thick specimen, most of the fluorescence excitation occurs near the focus. It is important to appreciate that there is a qualitative difference in the excitation distribution as one moves from single to multiphoton excitation.

Detection

Once excitation has been successfully localized (to a volume roughly the size of the wavelength cubed, 1 femtoliter or less), there is no longer any great need for spatially selective detection. It follows that high-resolution objectives need not be used for collection of the fluorescence; a high-NA condenser, for example, will do just as well, if not better. There is no reason, however, to discard the fluorescence entering the objective, which, for a high-NA lens, can be a significant portion. This fluorescence is often sufficient to generate an image and is sometimes used exclusively, even in multiphoton laser-scanning microscopy. Threading the fluorescence light back through the scanning pathway (descanning), on the other hand, which is essential in a confocal microscope, can certainly be dispensed with (Denk et al. 1995a). Indeed, confocal detection (i.e., use of a confocal aperture) is often quite counterproductive because scattered fluorescence light (see below), which, in the case of localized excitation, carries perfectly good information (Denk et al. 1994), is lost.

Scattered Light

In scattering samples, confocal schemes are particularly wasteful (Denk et al. 1994; Denk 1996; Denk and Svoboda 1997). In a clear specimen, the photodynamic damage in the focal slice is comparable for confocal and multiphoton microscopy (assuming detection through the objective only) because a large fraction of the focal plane fluorescence is utilized, even in the confocal case. However, this is not the case for scattering tissue, where even focal-slice fluorescence is increasingly lost in the confocal microscope. This is unavoidable, simply because a scattered focal photon's provenience is tainted, as it has become indistinguishable from out-of-focus light; such a photon will and should be rejected by the confocal pinhole. For multiphoton excitation, incident light is also scattered, but such light can no longer excite because it is too dilute in time and space (Denk and Svoboda 1997). To make up for scattering losses, the excitation power can be increased (in the absence of significant one-photon absorption) without exacerbating damage. For whole-field detection, the detection efficiency does not decrease significantly with depth (Oheim et al. 2001); fluorescence photons eventually emerge from the tissue (quite possibly after being scattered many times) because most biological tissues, even if strongly scattering, usually do not absorb much light. Incidentally, but quite fortuitously, better depth penetration is seen in multiphoton excitation because light scattering is typically reduced (Svaasand and Ellingsen 1983) as the wavelength increases.

Resolution

The theoretical spatial resolution of the multiphoton microscope is reduced, as compared to an ideal confocal microscope using the same fluorophore. This is because the excitation wavelength in the multiphoton instrument is roughly double that in the confocal (Sheppard and Gu 1990; Gu 1996). In practice, however, the difference is much less, if present at all. This is because the finite pinhole size (which is needed for efficient detection; Gu and Sheppard 1991), chromatic aberration, and imperfect alignment of laser focus and detector pinhole all degrade resolution in the confocal microscope. The resolution of a multiphoton microscope can be improved by using confocal detection (Stelzer et al. 1994), but this comes at the cost of reduced collection efficiency, particularly for scattering samples, and with more severe chromatic aberration. Without confocal detection, chromatic aberration in the objective is only a minor concern when using multiphoton excitation. However, for highly chromatic systems, e.g., acousto-optical modulators, even the small spread of excitation wavelength that results from the shortness of the laser pulses needed to achieve efficient multiphoton excitation (Denk et al. 1990, 1995a) cannot be ignored.

Laser Pulse Width

Another potential problem in multiphoton microscopy is the increase in laser pulse width and resultant decrease in multiphoton excitation efficiency that is caused by group velocity dispersion (GVD) in conjunction with ultrashort pulses. GVD is the result of the difference in propagation speed for wave packets of different wavelengths in different optical materials (Denk et al. 1995a). The effect can be partially compensated for by using a "prism sequence" (Gordon and Fork 1984). In practice, GVD has not been a real problem because pulses with widths of around 100 fsec (as produced by commercial Ti:sapphire lasers, see Chapter 97), broaden only by a fraction of the original pulse width after passage through the microscope optics. However, if shorter excitation pulses (fractional broadening depends on the square of the inverse pulse width) and higher order ($\geq$3 photons) nonlinear processes are to be used, GVD compensation may become useful and even necessary.

INSTRUMENTATION

Practical Considerations

Excitation Light Source

The main difference between multiphoton excitation and confocal microscopy is the excitation light source. The average multiphoton excitation rate depends, like any nonlinear optical process, on the temporal structure of the excitation light and not just—as in the case of linear (one-photon) absorption, which is a special case in this respect—on the average intensity. In particular, the use of pulsed light with a very small-duty cycle enhances N-photon processes by the inverse duty cycle to the $(N-1)^{th}$ power. The use of "mode-locked" lasers, e.g., with pulse lengths of around 100 fsec and repetition rates of ~100 MHz, increases the multiphoton excitation rate enormously, as compared to a continuous-wave laser at the same average power: Rates are increased by factors of 10^5 and 10^{10} for two- and three-photon excitation, respectively. It is not only the average available laser power that dictates the use of mode-locked lasers. These lasers are often required to minimize residual one-photon absorption at the fundamental wavelength. The development of ultrashort-pulse mode-locked lasers was the breakthrough that made two-photon microscopy truly practical (Denk et al. 1990). (For a collection of useful reprints on these types of lasers, see Gosnell and Taylor 1991; see also Chapter 98.)

Wavelength Selection

Because the excitation wavelengths for two-photon microscopy and one-photon confocal laser-scanning microscopy are different (roughly on the order of 2:1, respectively), different dichroic beam splitters must be used; sometimes, the emission filter must also be changed to suppress the new excitation wavelength efficiently. Usually, however, elimination of the excitation light actually becomes easier, because the two-photon excitation wavelength is longer and separated from the fluorescence wavelengths by several hundred nanometers. Discrimination against excitation light can often be accomplished by using colored glass filters and is helped by the falloff in the sensitivity of photomultiplier tubes toward longer wavelengths.

SETTING UP A MULTIPHOTON MICROSCOPY SYSTEM

There are several ways of setting up a system for multiphoton excitation microscopy. Some of the options are discussed below.

- *Buy a complete commercial instrument* (e.g., Bio-Rad, Zeiss [femtosecond instruments], Leica [picosecond instrument]). The commercial solution will work right away, but it will be expensive and may not be optimal for the intended application.
- *Convert an existing laser-scanning microscope.* This will be the most sensible approach if a (underused) confocal microscope is already available. The main changes necessary are the replacement of dichroic beam splitters, filters, and those excitation path steering mirrors that are of the multilayer dielectric type. Detection efficiency can usually be gained by adding a "whole-field" (also called "external") detector (Denk et al. 1995a). This is essential if scattering samples are to be imaged. (For examples of this approach, see Chapter 8.)
- *Build an instrument from scratch.* This can be the most economical approach and, more importantly, allows the instrument to be tailored to achieve the best performance for a particular application. Technical expertise is required, however, and the resulting instrument may not be very user-friendly. An example of this approach can be found in Chapter 85.

EXAMPLES OF APPLICATIONS

Here, I mention some selected publications from my own and other laboratories that have used multiphoton microscopy in neuroscience and related applications. They include:

- Two-photon imaging and microspectrofluorometry of synaptic spines (Denk et al. 1995b, 1996; Yuste and Denk 1995; Koester and Sakmann 1998; Sabatini and Svoboda 2000; Sabatini et al. 2001; Oertner et al. 2002), calcium in hair-cell stereocilia (Denk et al. 1995c)
- Two-photon measurement of dendritic activity in vivo (Svoboda et al. 1997, 1999; Helmchen et al. 1999)

- Imaging in behaving rats using a miniaturized fiber-coupled two-photon microscope (Helmchen et al. 2001)
- Photochemical release of caged compounds (Denk 1994; Svoboda et al. 1996; Furuta et al. 1999; Matsuzaki et al. 2001)
- Imaging of GFP-labeled cells (Niswender et al. 1995; Potter et al. 1996; Kohler et al. 1997), NADH in tissue (Piston et al. 1995; Bennett et al. 1996), chromosomes in embryos (Summers et al. 1996)
- Three-photon imaging of intrinsic (Maiti et al. 1997) and extrinsic (Wokosin et al. 1996) fluorophores
- Observation of long-term changes in neurite morphology (Grutzendler et al. 2002; Trachtenberg et al. 2002)
- Measurement of visual stimulus-induced calcium signals in the retina (Denk and Detwiler 1999; Euler et al. 2002)

The application of multiphoton microscopy to other areas of imaging can be found in Chapters 70, 71, and 92.

REFERENCES

Bennett B.D., Jetton T.L., Ying G., Magnuson M.A., and Piston D.W. 1996. Quantitative subcellular imaging of glucose metabolism within intact pancreatic islets. *J. Biol. Chem.* **271:** 3647–3651.

Denk W. 1994. 2-Photon scanning photochemical microscopy: Mapping ligand-gated ion channel distributions. *Proc. Natl. Acad. Sci.* **91:** 6629–6633.

———. 1996. Two-photon excitation in functional biological imaging. *J. Biomed. Optics* **1:** 296–304.

Denk W. and Detwiler P.B. 1999. Optical recording of light-evoked calcium signals in the functionally intact retina. *Proc. Natl. Acad. Sci.* **96:** 7035–7040.

Denk W. and Svoboda K. 1997. Photon upmanship: Why multiphoton imaging is more than a gimmick. *Neuron* **18:** 351–357.

Denk W., Piston D.W., and Webb W.W. 1995a. Two-photon molecular excitation in laser scanning microscopy. In *The handbook of confocal microscopy* (ed. J. Pawley), pp. 445–458. Plenum Press, New York.

Denk W., Strickler J.H., and Webb W.W. 1990. Two-photon laser scanning fluorescence microscopy. *Science* **248:** 73–76.

Denk W., Sugimori M., and Llinas R. 1995b. Two types of calcium response limited to single spines in cerebellar Purkinje cells. *Proc. Natl. Acad. Sci.* **92:** 8279–8282.

Denk W., Holt J.R., Shepherd G.M.G., and Corey D.P. 1995c. Calcium imaging of single stereocilia in hair cells: Localization of transduction channels at both ends of tip links. *Neuron* **15:** 1311–1321.

Denk W., Yuste R., Svoboda K., and Tank D.W. 1996. Imaging calcium dynamics in dendritic spines. *Curr. Opin. Neurobiol.* **6:** 372–378.

Denk W., Delaney K.R., Gelperin A., Kleinfeld D., Strowbridge B.W., Tank D.W., and Yuste R. 1994. Anatomical and functional imaging of neurons using 2-photon laser scanning microscopy. *J. Neurosci. Methods* **54:** 151–162.

Euler T., Detwiler P.B., and Denk W. 2002. Directionally selective calcium signals in dendrites of starburst amacrine cells. *Nature* **418:** 845–852.

Furuta T., Wang S.S.H., Dantzker J.L., Dore T.M., Bybee W.J., Callaway E.M., Denk W., and Tsien R.Y. 1999. Brominated 7-hydroxycoumarin-4-ylmethyls: Photolabile protecting groups with biologically useful cross-sections for two photon photolysis. *Proc. Natl. Acad. Sci.* **96:** 1193–1200.

Goeppert-Mayer M. 1931. Ueber Elementarakte mit zwei Quantenspruengen. *Annal. Physik.* **9:** 273.

Gordon J.P. and Fork R.L. 1984. Optical resonator with negative dispersion. *Opt. Lett.* **9:** 153–155.

Gosnell T.R. and Taylor A.J., eds. 1991. *Selected papers on ultrafast laser technology*. SPIE Optical Engineering Press, Bellingham, Washington.

Grutzendler J., Kasthuri N., and Gan W.B. 2002. Long-term dendritic spine stability in the adult cortex. *Nature* **420:** 812–816.

Gu M. 1996. Resolution in 3-photon fluorescence scanning microscopy. *Opt. Lett.* **21:** 988–990.

Gu M. and Sheppard C.J.R. 1991. Effects of finite-sized detector on the OTF of confocal fluorescent microscopy. *Optik* **89:** 65–69.

Helmchen F., Fee M.S., Tank D.W., and Denk W. 2001. A miniature head-mounted two-photon microscope. High-resolution brain imaging in freely moving animals. *Neuron* **31:** 903–912.

Helmchen F., Svoboda K., Denk W., and Tank D.W. 1999. In vivo dendritic calcium dynamics in deep-layer cortical pyramidal neurons. *Nat. Neurosci.* **2:** 989–996.

Kaiser W. and Garrett C.B.G. 1961. Two-photon excitation in $CaF_2:Eu^{2+}$. *Phys. Rev. Lett.* **7:** 229–231.

Koester H.J. and Sakmann B. 1998. Calcium dynamics in single spines during coincident pre- and postsynaptic activity depend on relative timing of back-propagating action potentials and subthreshold excitatory postsynaptic potentials. *Proc. Natl. Acad. Sci.* **95:** 9596–9601.

Kohler R.H., Cao J., Zipfel W.R., Webb W.W., and Hanson M.R. 1997. Exchange of protein molecules through connections between higher plant plastids. *Science* **276:** 2039–2042.

Maiti S., Shear J.B., Williams R.M., Zipfel W.R., and Webb W.W. 1997. Measuring serotonin distribution in live cells with three-photon excitation. *Science* **275:** 530–532.

Masters B.R., ed. 1996. *Selected papers on confocal microscopy*. SPIE Milestone Series, SPIE Optical Engineering Press, Bellingham, Washington.

Matsuzaki M., Ellis-Davies G.C.R., Nemoto T., Miyashita Y., Iino M., and Kasai H. 2001. Dendritic spine geometry is critical for AMPA receptor expression in hippocampal CA1 pyramidal neurons. *Nat. Neurosci.* **4:** 1086–1092.

Minsky M. 1961. Microscopy apparatus. U.S. Patent No. 3013467.

Niswender K.D., Blackman S.M., Rohde L., Magnuson M.A., and Piston D.W. 1995. Quantitative imaging of green fluorescent protein in cultured cells: Comparison of microscopic techniques, use in fusion proteins and detection limits. *J. Microsc.* **180:** 109–116.

Oertner T.G., Sabatini B.L., Nimchinsky E.A., and Svoboda K. 2002. Facilitation at single synapses probed with optical quantal analysis. *Nat. Neurosci.* **10:** 10.

Oheim M., Beaurepaire E., Chaigneau E., Mertz J., and Charpak S. 2001. Two-photon microscopy in brain tissue: Parameters influencing the imaging depth. *J. Neurosci. Methods* **111:** 29–37.

Piston D.W., Masters B.R., and Webb W.W. 1995. 3-Dimensionally resolved NAD(P)H cellular metabolic redox imaging of the in-situ cornea with 2-photon excitation laser-scanning microscopy. *J. Microsc.* **178:** 20–27.

Potter S.M. 1996. Vital imaging: Two photons are better than one. *Curr. Biol.* **6:** 1595–1598.

Potter S.M., Wang C.M., Garrity P.A., and Fraser S.E. 1996. Intravital imaging of green fluorescent protein using two-photon laser-scanning microscopy. *Gene* **173:** 25–31.

Sabatini B.L. and Svoboda K. 2000. Analysis of calcium channels in single spines using optical fluctuation analysis. *Nature* **408:** 589–593.

Sabatini B.L., Maravall M., and Svoboda K. 2001. Ca(2+) signaling in dendritic spines. *Curr. Opin. Neurobiol.* **11:** 349–356.

Sheppard C.J.R. and Gu M. 1990. Image-formation in 2-photon fluorescence microscopy. *Optik* **86:** 104–106.

Stelzer E.H.K., Hell S., and Lindek S. 1994. Nonlinear absorption extends confocal fluorescence microscopy into the ultra-violet regime and confines the illumination volume. *Optics Commun.* **104:** 223–228.

Summers R.G., Piston D.W., Harris K.M., and Morrill J.B. 1996. The orientation of first cleavage in the sea urchin embryo, Lytechinus variegatus, does not specify the axes of bilateral symmetry. *Dev. Biol.* **175:** 177–183.

Svaasand L.O. and Ellingsen R. 1983. Optical properties of human brain. *Photochem. Photobiol.* **38:** 293–299.

Svoboda K., Tank D.W., and Denk W. 1996. Direct measurement of coupling between dendritic spines and shafts. *Science* **272:** 716–719.

Svoboda K., Denk W., Kleinfeld D., and Tank D.W. 1997. In vivo dendritic calcium dynamics in neocortical pyramidal neurons. *Nature* **385:** 161–165.

Svoboda K., Helmchen F., Denk W., and Tank D.W. 1999. Spread of excitation in layer 2/3 pyramidal neurons in rat barrel cortex *in vivo*. *Nat. Neurosci.* **2:** 65–73.

Trachtenberg J.T., Chen B.E., Knott G.W., Feng G., Sanes J.R., Welker E., and Svoboda K. 2002. Long-term in vivo imaging of experience-dependent synaptic plasticity in adult cortex. *Nature* **420:** 788–794.

Williams R.M., Piston D.W., and Webb W.W. 1994. 2-Photon molecular-excitation provides intrinsic 3-dimensional resolution for laser-based microscopy and microphotochemistry. *FASEB J.* **8:** 804–813.

Wokosin D.L., Centonze V.E., Crittenden S., and White J. 1996. Three-photon excitation fluorescence imaging of biological specimens using an all-solid-state laser. *Bioimaging* **4:** 1–7.

Yuste R. and Denk W. 1995. Dendritic spines as basic functional units of neuronal integration. *Nature* **375:** 682–684.

CHAPTER 8

Two-Photon Microscopy for 4D Imaging of Living Neurons

Steve M. Potter

Neuroengineering Laboratory, Coulter Department of Biomedical Engineering, Georgia Institute of Technology and Emory University, Atlanta, Georgia 30332

Two-photon laser-scanning fluorescence microscopy (Denk et al. 1990) has made it possible to image neurons over 600 μm deep within a living slice or organism, with submicrometer resolution and in three dimensions (3D) for many hours without photodamage (Potter et al. 1996a). With true 4D microscopy (i.e., 3D with time), it is now feasible to capture neural development and synaptic plasticity in the *act of happening*, eliminating many uncertainties associated with between-animal comparisons of fixed-tissue specimens. By using pulsed IR laser light to excite fluorescent labels (e.g., dyes, fluorescent proteins, or endogenous fluorophores) that are normally excited by visible light, excitation is restricted to the focal plane, greatly reducing photobleaching and phototoxicity (see Chapter 7). The IR illumination is scattered less than visible light, allowing imaging 2–3 times deeper than with standard confocal microscopy. In addition, because the fluorescent signal emanates only from the focus of the scanning IR laser beam within the specimen, no confocal aperture is necessary to remove the out-of-focus signal. This means that two-photon microscopy has an inherently higher signal-to-noise ratio (SNR) compared with confocal microscopy. Excellent references for two-photon microscopy, and for labeling and imaging living specimens, include Denk et al. (1995) and Terasaki and Dailey (1995).

In this chapter, I describe two-photon imaging hardware, pointing out potential pitfalls in microscope construction and operation. I also describe technical considerations for successful two-photon microscopy in two experimental systems: (1) 4D imaging of living neurons transplanted to cultured hippocampal slices from neonatal rats and (2) 4D imaging of dendritic spines within acute hippocampal slices from adult rats. Figure 1 shows the components of the imaging setup.

TWO-PHOTON HARDWARE

Although both confocal and two-photon microscopes scan a laser beam in the focal plane within the specimen, a two-photon microscope can actually be a much simpler device than a confocal microscope. There is no need to focus or even descan the emitted fluorescence to create an image. In most cases, however, it is probably easier to convert a confocal microscope to a two-photon microscope than to build a two-photon system from the ground up. I converted a Molecular Dynamics Sarastro 2000 confocal with an upright Nikon Optiphot II microscope, preserving its ability to be used in standard confocal mode (see Fig. 2) (Potter et al. 1996c). In every case where Fraser and I imaged the same

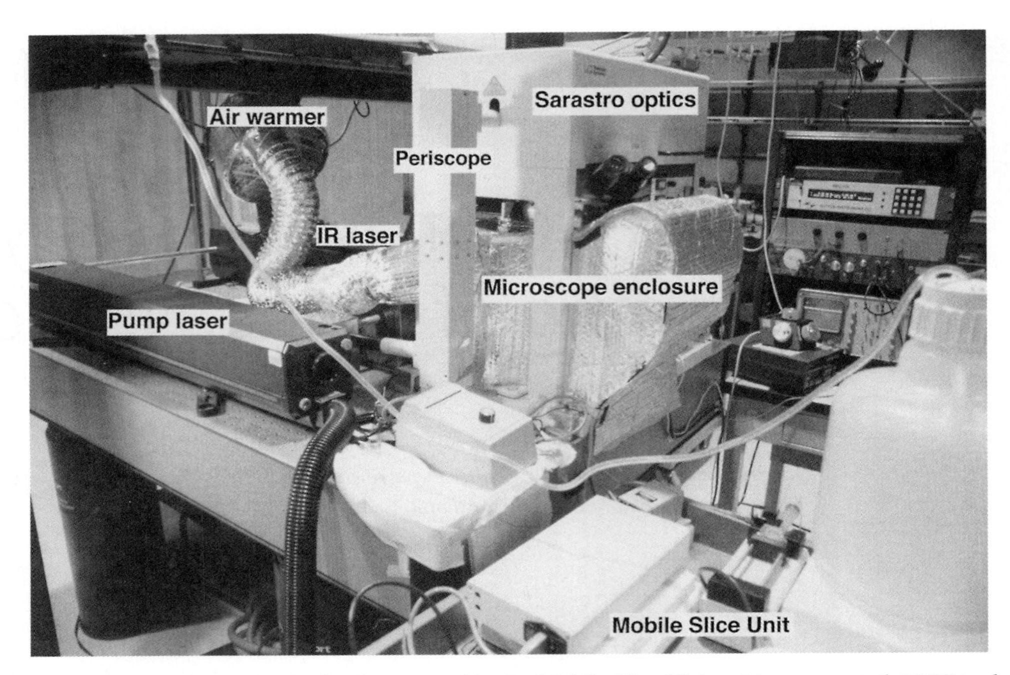

FIGURE 1. The imaging setup. In the foreground is the Mobile Slice Unit, with oxygenated ACSF tank and peristaltic and syringe pumps. In the center is the Molecular Dynamics Sarastro 2000 confocal microscope (converted for two-photon microscopy), enclosed in insulation and warmed by an egg incubator (*upper left*). Behind the microscope are the argon-ion pump laser and Ti:sapphire IR laser. The argon-ion laser has been replaced with a shoe-box-sized, all-solid-state Verdi pump laser (not shown).

specimen using confocal microscopy with visible excitation (either the Sarastro or the Bio-Rad 600) and two-photon microscopy (converted Sarastro), we got better resolution, deeper penetration, and less photodamage with two-photon microscopy. The confocal capabilities of the converted Sarastro are therefore seldom used.

Lasers

To produce adequate two-photon excitation, an IR laser that compresses all its output into very short (~100 fsec) pulses is necessary. Although the peak power of these pulses is enormous, the pulses are widely spaced, so the mean power is not enough to cause any heating of the specimen. (For a detailed discussion on lasers in multiphoton microscopy, see Chapter 79.)

We use the Coherent Mira900 titanium:sapphire laser, which is tunable from approximately 700 to 1000 nm. The Ti:sapphire laser converts green light from a pump laser into pulsed IR light.

The Coherent Verdi pump laser is superior to the Coherent Innova310 8-W argon-ion laser. The Verdi is an all-solid-state laser that can be powered from a normal 110-V outlet and requires no cooling water, unlike the Innova310. With only 5.5 W of green light, it produces more mode-locked (pulsed) IR output from the Mira900 compared with the 8-W argon laser, due to improved beam quality.

The IR beam is brought into the microscope with four dichroic mirrors, optimized for broadband IR reflection at a 45° angle (Newport BD.2). Between two of the mirrors is a beam expander (Newport T81-3X), which is used to focus the beam so that it is slightly larger than the back aperture of the objective lens. Without it, most of the large beam from the trinocular (scanner) eyepiece does not enter the lens, and two-photon excitation is greatly reduced. All beams are covered with tubes connected to sealed mirror boxes. This allows safe use, even by biologists, and protects the mirrors from dust buildup and accidental bumping.

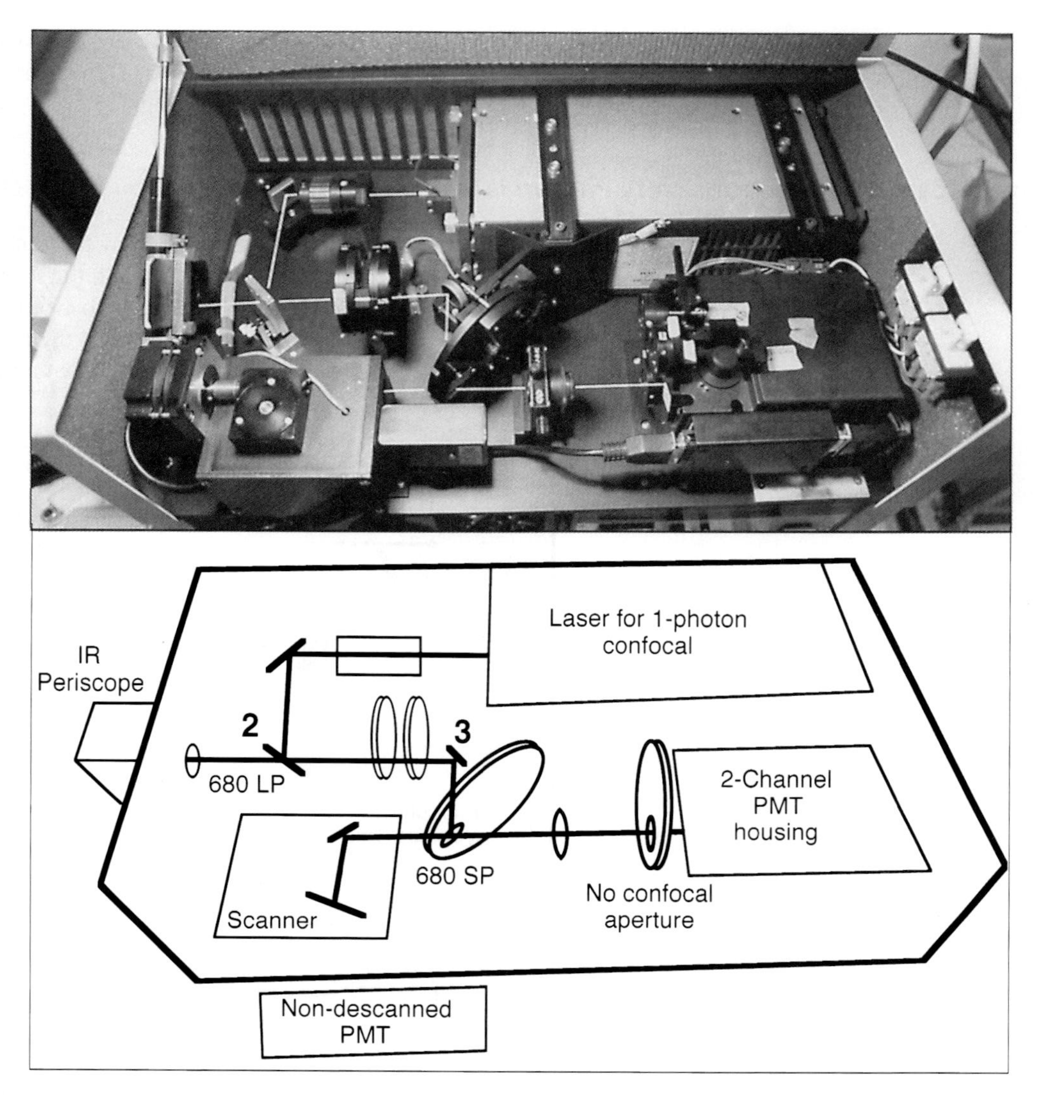

FIGURE 2. Optics of Molecular Dynamics Sarastro 2000, modified for two-photon microscopy. The pulsed IR laser beam is brought into the Sarastro by a periscope to the left. Before the periscope is a 3x "beam expander" (not shown) used to match the IR laser beam width to the back aperture of the objective lens. To switch from two-photon to confocal imaging, one need only block the IR beam, turn on the on-board laser (*upper right*), rotate the primary dichroic beam-splitter wheel to a long-pass mirror, and rotate the pinhole aperture wheel from the open hole to one of the three pinhole settings.

Conversion of the Sarastro 2000 Confocal to a Two-Photon System

The Sarastro 2000 has a very simple light path compared with most available confocals, which is desirable because valuable photons are inevitably lost at every mirror, lens, and filter. To convert it to a two-photon microscope, I replaced mirror 2 (see Fig. 2) with a 680-nm long-pass (680 LP) beam splitter that allows us to use either the IR laser or the on-board visible laser. We use a 680-nm short-pass (680 SP) dichroic mirror to separate excitation from emission in the primary beam-splitter wheel. Mirror 3 and the two scanning mirrors were coated with silver and a protective dielectric coating (Ventura Optical Industries). Silver reflects both visible and IR light well, unlike standard aluminum mirrors, which reflect IR poorly. The dielectric is crucial to prevent the silver from tarnishing. No changes were made to the OptiphotII, which has a motorized *z*-stage with 0.1-μm accuracy. The stage and scanner are controlled by an SGI Indigo workstation, running Molecular Dynamics ImageSpace software. No changes to the software were necessary for two-photon imaging.

Non-descanned Detection: Every Photon Is Sacred

To get images with the best SNR, it is important to collect as many photons as possible. Our microscope sends the emitted light through a trinocular eyepiece (beneath the scanner), two scanning mirrors, and an achromat lens before it gets to the on-board detectors. To avoid the losses associated with these optics, and to collect more of the photons that are scattered on their way out of the specimen, it is best to detect the emitted light as close to the specimen as possible (see Fig. 3). We replaced one

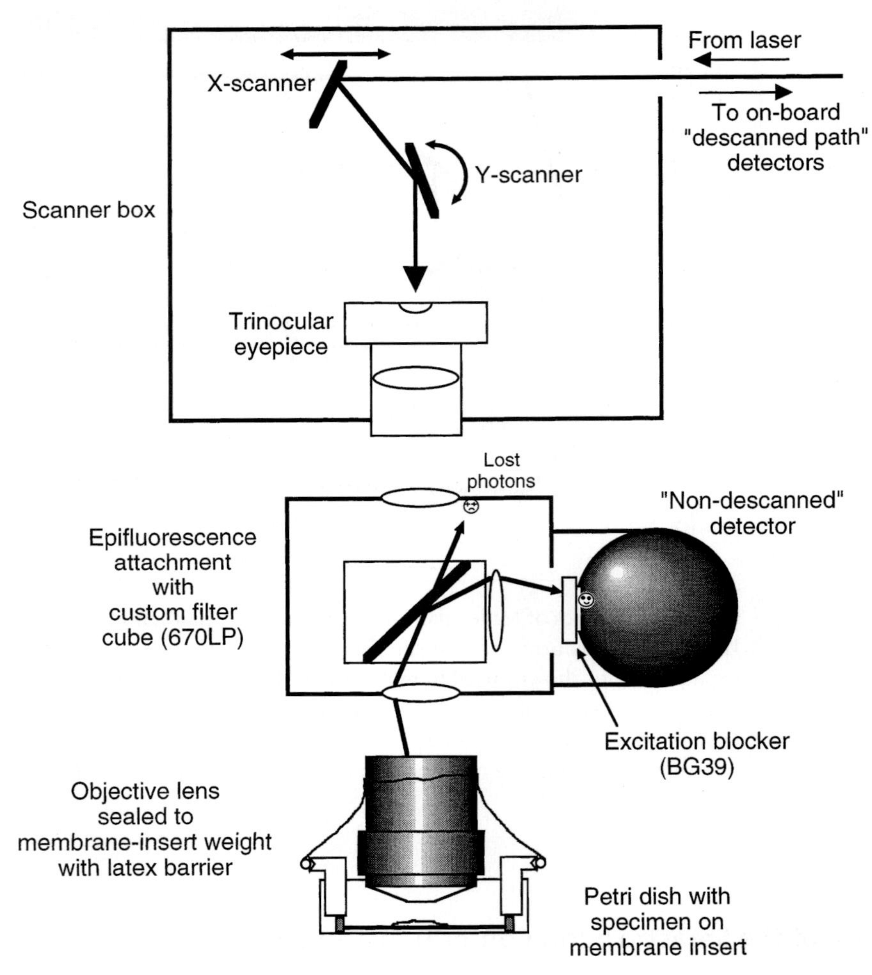

FIGURE 3. Non-descanned detection. By placing a detector close to the specimen, it is possible to collect more scattered photons and avoid losses associated with optics in the descanned path. For low background, this "non-descanned" detector must be well shielded from room lights and scattered IR excitation. To avoid damage to the sensitive photomultiplier tube, it should be automatically shut off when the transilluminator, the epi-illuminator, or the on-board laser is turned on. The Y-scanner reflects the IR beam down through a trinocular eyepiece and custom filter cube to the objective lens. Emitted fluorescence is reflected by the cube into a custom detector mounted directly in front of it, or if the cube is pulled out, it is descanned and sent through a 680-nm short-pass beam splitter and achromat lens to the on-board dual-channel detector (descanned path).

of the OptiphotII's epifluorescence filter cubes with a custom cube (Omega Optical) that transmits IR from above and reflects visible fluorescence forward to a photomultiplier tube (PMT; Hamamatsu R928) that was installed just below the binocular eyepiece. One antireflection-coated lens in the filter cube allows scattered photons that would not have made it to the on-board ("descanned path") detectors to enter this "non-descanned path" detector.

It is crucial to protect the PMT from scattered and reflected excitation light by sealing a BG39 colored-glass filter to the PMT housing. (A different filter may be more appropriate for red-emitting fluorophores.) I installed circuitry to turn off the PMT when any of the microscope's other light sources are turned on (substage tungsten lamp, epifluorescence lamp, or on-board confocal laser), to prevent accidental damage to the PMT. Although the original detectors are much less sensitive than the non-descanned detector, they are still used for double-labeling experiments.

Imaging Neurons Transplanted to Cultured Slices

Survival of transplanted neurons is often disappointingly low (Shetty and Turner 1995). Caltech professor Jerry Pine and I developed a model system for neuronal transplants that allows continuous imaging of transplant integration into the host tissue (Potter et al. 1996b; Fraser et al. 1997) to shed light on the dynamics of transplant migration and integration success or failure. The idea was to label a suspension of neurons from embryonic rat hippocampus with fluorescent membrane dye, wash away all free dye and labeled debris, and seed the cells onto cultured hippocampal slices from neonatal rats (see Fig. 4). 4D two-photon imaging showed that after one day in culture, the transplanted cells migrated throughout the slice and began to extend axons and dendrites. For an example of such a 4D movie, see the following WWW site: http://www.neuro.gatech.edu/groups/potter/movies.html.

At 17 days, transplanted neurons had developed pyramidal morphologies, with processes extending for several hundreds of microns (Potter et al. 1996b).

Cell Labeling

Lipophilic dyes, such as DiI or DiO (generic acronyms for dialkylindocarbocyanines and dialkyloxacarbocyanines, respectively), are often used to trace living neuronal processes. Small structures such as axons, filipodia, and dendritic spines contain little cytosol and are better visualized with a membrane

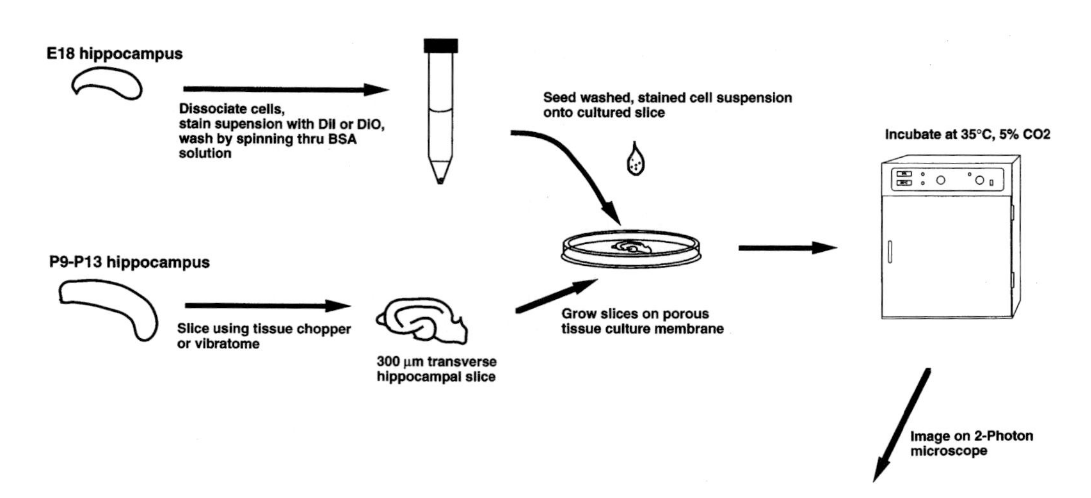

FIGURE 4. "In Slico" transplant model. Cells from an embryonic rat hippocampus are labeled and seeded onto cultured neonatal hippocampal slices. After some time in culture, transplant integration is observed in living slices using two-photon microscopy.

dye than a cytosolic dye. The problem is that these dyes tend to crystallize in aqueous media. Dye crystals are difficult or impossible to remove from a cell suspension, and will label host neurons if transplanted with the labeled cells. To surmount this problem, I used the more soluble 12-carbon form of DiI, along with a nonionic macromolecular surfactant Pluronic F127 (Lojewska and Loew 1986, 1987), to ensure that the dye remains completely dissolved and to aid the transfer of dye to the cell membrane. The 16-carbon form of DiO, which is slightly less soluble, was also used. Normally excited by blue light (485 nm), DiO is easier to excite with the Ti:sapphire laser than DiI (normally excited by green light around 548 nm). The following methods are from Potter et al. (1996b), which has further details.

PROCEDURE A

Preparation of Clean Cell Suspensions

1. Cut brain tissue into 1-mm pieces. Digest with 0.25% trypsin <!>, 0.02 mg/ml DNase (Sigma) in Hanks' balanced salt solution (no calcium or magnesium [GIBCO]) at 37°C for 15 min.
2. Gently wash pieces in plating medium. Repeat wash.

   Note: For hippocampus, use Neurobasal with B27 supplement (GIBCO) and 500 µM glutamine, 25 µM glutamate.
3. Gently triturate in 1 ml of plating medium with five passes through the 0.78-mm opening of a blue tip of a P-1000 pipetman.

   Note: This is much more reproducible than with the ubiquitous "fire-polished Pasteur pipette."
4. Decant suspended cells and triturate the remaining pieces once more (see step 3, above).
5. Gravity-filter the cell suspension through a 40-µm nylon mesh (Falcon) to remove large debris.
6. To remove smaller debris, layer 0.5 ml of 5% bovine serum albumin in 1x phosphate-buffered saline (BSA/PBS) under cell suspension in a 15-ml tube. Centrifuge the cell suspension through the BSA/PBS layer in a swinging-bucket rotor at 160*g* for 2 min.
7. Discard supernatant and resuspend cell pellet in plating medium by gentle trituration (three passes) as described in step 3 (above).

Caution: *See Appendix 3 for appropriate handling of materials marked with <!>.*

PROCEDURE B

DiI Labeling of Neurons in Suspension

1. Prepare a 40 mg/ml stock solution of $DiIC_{12}$ <!> (Molecular Probes) by dissolving the tar-like dye in DMF<!> containing 2.5% (w/v) Pluronic F127 (BASF)<!>.

   Note: This solution can be stored at –20°C and dissolved easily upon warming to room temperature.
2. Prepare staining solution: Add 2 µl of dye stock to 2 ml of plating medium in a 15-ml tube (final concentration: 40 µg/ml $DiIC_{12}$, 0.0025% Pluronic F127, 0.1% DMF).

   Note: This solution should easily pass through a 0.2-µm sterile filter with no diminution in color, indicating that it is completely dissolved.
3. Warm staining solution to 37°C in a water bath. Add approximately 0.5 ml of cell suspension (from step 7, above) with gentle mixing to the staining solution.

   Note: Labeling will be incomplete at room temperature due to decreased membrane turnover.
4. Incubate cells at 37°C for 15 min.
5. Layer 0.5 ml of 5% BSA/PBS under the cells in staining solution using a long Pasteur pipette.
6. Centrifuge in a swinging-bucket rotor at 160*g* for 6 min. Carefully decant the staining solution floating above the BSA/PBS, as well as most of the BSA/PBS.

7. To resuspend the stained cells, triturate (three passes) in 1 ml of plating medium using the P-1000 as described in step 3 of Procedure A.
8. Repeat steps 5–7 (above) to remove any remaining dissolved dye.

   Caution: *See Appendix 3 for appropriate handling of materials marked with <!>.*

PROCEDURE C

Labeled Transplants to Slice Cultures

These static (membrane) cultures are prepared according to the method developed by Stoppini and coworkers (Stoppini et al. 1991; Buchs et al. 1993). Routinely grown over 1 month in culture, they are several cell layers thick (~150 μm) and retain more of the original architechtonics than do roller-tube cultures.

1. Cut hippocampal (or other brain tissue) slices (300 μm) from postnatal day 7–11 rat pups using a tissue chopper or vibratome, keeping the slices submerged in Hanks' solution (GIBCO).
2. Add 1 ml of Organotypic Slice Culture Medium (Vanderklish et al. 1992) to a 35-mm petri dish and place a Millicell-CM membrane insert upon the medium. (Organotypic Slice Culture Medium: minimal essential medium [MEM] with Hanks' salts, no glutamine [GIBCO], 25% horse serum [Hyclone], 30 mM glucose, 5 mM $NaHCO_3$, 30 mM HEPES, 2.5 mM $MgSO_4$ <!>, 2 mM $CaCl_2$ <!>, 3 mM L-glutamine, 1 mg/liter insulin [Sigma], 10 ml/liter penicillin/streptomycin [Sigma P0781]. Adjust pH to 7.2 at 35°C.)

   Note: The membrane becomes clear when wet, facilitating imaging from below. The plastic ring on standard inserts gets in the way of petri dish lids, electrodes, and large objective lenses (upright scope), so this was initially cut off with a hot nichrome wire. I later persuaded Millipore to produce "Low height" inserts (#PICM0RG50) to avoid this hassle.
3. Transfer 1–3 slices to the membrane using a small spatula whose edges have been rounded with a grindstone.

   Note: The culture medium soaks through the membrane from below, and the moist slice is well oxygenated from above.
4. Incubate slices at 35°C in humidified 5% carbon dioxide atmosphere. Replace half the medium under the membrane with fresh medium weekly.

   Note: Slices will adhere well to the membranes in a day or two and flatten out to about 150 μm. For submerged imaging sooner than this, the slices must be glued to the membrane using a clot of fibrin (Brown et al. 1993).
5. Dilute freshly labeled cells (from step 6, p. 64) to 50,000 cells/ml in slice culture medium.
6. Place a drop (~5 μl) of cell suspension on each slice, at least one day after preparing slice cultures.

   Notes: In a few minutes after addition of the cell suspension, excess medium will soak through the membrane, and the cells will settle on and adhere to the slice. I have found that within a day, embryonic hippocampal neurons have migrated throughout the 150-μm thickness of a cultured slice. This method provides 20–50 labeled cells per slice.

   Caution: *See Appendix 3 for appropriate handling of materials marked with <!>.*

Environmental Control during Imaging

I enclose the microscope body within Reflectix Mylar bubble-plastic insulation (http://www.reflectix-inc.com/) (see Fig. 1), connected to a chicken egg incubator (Marsh Automatic Incubator, Lyon Electric) to warm the air to rat body temperature. Using this low-wattage heater (100 W), there are no thermostat-related temperature fluctuations. It takes about 12 hours for the microscope stage, objective, and other components of the setup to warm up to 35°C, but everything remains thermally stable during the imaging. This means there is no focus drift during protracted imaging sessions, even if the room temperature changes. The opaque enclosure also serves to block ambient light from entering the objective and causing unwanted background signal. This is especially important when using the sensitive non-descanned detector. For slices that have adhered to the membranes, I usually flood the slice with slice culture medium during imaging, and 40x/0.75-NA Nikon or 63x/1.2-NA Zeiss water-immersion objective is used. Nonadhered slices must be glued to the membrane using a fibrin clot (Brown et al. 1993). The membrane insert is weighted down by a custom-made, stainless steel ring. To prevent evaporation of medium and pH drift, a condom with the tip cut off (Trojan nonlubricated, Carter-Wallace) seals the ring to the objective lens (see Fig. 3). Because the objective lens is warm, it is also possible to image a moist nonflooded slice using an air objective, without condensation fogging the lens.

Data Acquisition

The two-photon absorption maximum for DiI is approximately 1030 nm (Xu et al. 1996), and the power of the Ti:sapphire laser falls off from 900 to 1000 nm; thus, an empirically determined excitation wavelength of 960 nm seems to give the best images of DiI-labeled specimens. At 900 nm, where the laser is more powerful, background autofluorescence increases more than DiI fluorescence. For imaging neurons labeled with DiO, two-photon excitation anywhere from 850 to 900 nm produces excellent signal-to-background ratios. Although photodamage is tens or hundreds of times less with two-photon microscopy than with standard wide-field fluorescence or confocal microscopy, it is possible to use too much laser power, excite already-excited fluorophores, and cause excessive dye bleaching. For this reason, I always set the PMT at 80–100% of its maximum voltage, and reduce the laser power to the minimum necessary to provide a good fluorescence signal at this gain, usually 10–30% transmission.

Each time point of a 4D time-lapse movie consists of a series of 20–50 *z*-sections, taken at 0.3- to 2-μm increments. With a 40x objective, a volume as large as 250 x 250 x 100 μm can be imaged every 5 minutes for over 8 hours. As long as the above guidelines are adhered to, there is no sign of photobleaching or phototoxicity of labeled neurons. To save on data storage and post-processing time, one can collect widely spaced (2 μm) z-sections and average two or three scans per section to reduce shot noise. However, better noise reduction is obtained by taking more single-scanned sections that are more closely spaced (0.5 μm or less), and then subjecting them to a 3 x 3 x 3 median filter using ImageSpace (or equivalent) software. Using 3D filters is only valid when each section is very similar to the two adjacent sections.

TWO-PHOTON, 4D TIME-LAPSE IMAGING OF DENDRITIC SPINES

In addition to its usefulness for revealing the dynamics of developing neural systems, two-photon microscopy is useful for revealing potential morphological correlates of synaptic plasticity in adult brain tissue. Dendritic spines have been imaged in acute hippocampal slices from juvenile rats, to study structural changes associated with synaptic facilitation (E. Schuman et al., unpubl.).

Maintenance of Healthy Slices

Maintaining acute slices poses a number of technical difficulties compared with organotypic cultured slices. The adult hippocampus is much more susceptible to excitotoxic injury from anoxia than the neonatal hippocampus. The Mobile Slice Unit (see Fig. 1), an instrument designed to keep acute slices healthy for up to 24 hours during imaging, was developed (A. Mamelak and S. Potter, unpubl.). This is a portable heart–lung machine that bubbles blood gas (95% oxygen/5% carbon dioxide) through artificial cerebrospinal fluid (ACSF) that is perfused continuously across the slice (150 ml/hour) by a peristaltic pump.

Medium Preparation

ACSF is equilibrated with blood gas several hours before use to prevent foaming on addition of B27 and then mixed with B27 nutrient supplement (GIBCO) using a syringe pump at 1:50 volume ratio (ACSF: 25 mM glucose [Edwards 1995], 126 mM NaCl, 2.5 mM KCl <!>, 1 mM $MgCl_2$ <!> , 2 mM $CaCl_2$ <!>, 1.25 mM sodium phosphate <!>, 26 mM $NaHCO_3$). The pH and osmolarity are adjusted to 7.3 and 295 mOsm, respectively.

Note: Phenol Red pH indicator is routinely included in the ACSF to ensure that the gassing is effective, because this is a carbonate-buffered solution. Ideally, an in-line pH sensor and oximeter would be included.

Temperature Control

As with cultured slices, we control the temperature of the microscope to prevent thermal focus drift, but imaging is usually conducted at 25°C, because acute slices kept at 37°C survive for only 6 hours or so.

Measurements

We routinely check the field potential of slices before and after imaging and can induce long-term potentiation by tetanic stimulation after 20 hours of imaging, as long as the osmolarity, pH, oxygenation, and temperature are carefully controlled.

Slice Movement during Imaging

Acute slices are not adhered to a membrane as the cultured slices are. To prevent movement during imaging, I use slice weights that are a modification of those described in Edwards et al. (1989). A 1-mm-diameter silver wire is shaped into a "U" and pressed in a vise until it is 50 μm thinner than the acute slice, which is usually cut at 500 μm. This weight is then glued to a Millicell-CM insert using cyanoacrylate glue. The weight is cut out of the insert, leaving membrane in the middle of the U, and a small hole is then cut in the membrane to allow access to the slice by electrodes or micropipettes. The porous membrane slice weights are transparent when wet and do not cut into the slice as the nylon fibers of the weights used by Edwards et al. (1989) do.

To allow several different sites to be imaged during one time-lapse session, we attach the slice chamber to a micromanipulator (e.g., Sutter MP-285). The manipulator controller can remember the location of several interesting regions and return to them in sequence under computer control. As long as the slice is well weighted down, the positional relocation is accurate to approximately 1 μm. This device also makes hunting around the slice for labeled cells much easier than using the microscope's mechanical stage translation knob, especially when imaging at high magnification.

Acute-Slice Labeling

A variety of approaches for labeling living slices have been developed. They include the following:

Biolistics and GFP

McAllister et al. (1995) had great success using biolistics to shoot gold particles coated with GFP-encoding DNA into slices of developing ferret cortex. However, it takes 12–24 hours for GFP to be expressed and oxidized, which is not acceptable for acute hippocampal slices.

Rusty-nail Approach

Slices can be labeled by drying a solution of DiI or DiO on a glass pipette to form small crystals and poking the slice with the pipette to deposit the crystals within it (Sorra and Harris 1997; S.M. Potter, unpubl.). Although quicker than the biolistics approach, this also takes several hours for maximal labeling of dendritic arbors, presumably due to the small contact between dye crystals and cell membrane. One can also combine the biolistics and rusty-nail approaches, using "DiOlistics," in which combinations of dyes are dried onto gold particles that are shot into brain tissue to provide rapid multicolor labeling of a random subset of neurons (Gan et al. 2000).

Oil-drop Approach

Hosokawa et al. (1992) and Dailey (Dailey and Smith 1993) successfully labeled acute and cultured slices by dissolving DiI or DiO in fish oil and applying a small drop of this solution to the surface of the slice. Unfortunately, the cells best labeled by this technique are the unhealthy ones near the cut surface. In addition, the oil drop must be removed by microaspiration before imaging, because it acts like a lens to defocus the excitation beam, reducing two-photon excitation.

Intracellular Dye Injection

We have produced some nice two-photon images of single neurons filled with fluorescein-dextran by intracellular iontophoretic injection. However, seldom do neurons survive for more than an hour after the trauma of impalement and electrode removal. Also, as mentioned on p. 63, membrane dyes produce clearer images of small structures such as dendritic spines.

Picospritzing a DMF Solution

We have had great success using a solution of DiO (16 mg/ml) in DMF. A very small amount can be applied by extracellular pressure injection into the middle of the slice. By varying pressure and pulse duration, any number from two to many neurons can be labeled. Because the carrier is miscible with water, the dye surrounds the cells and they become maximally labeled within 30 minutes. No toxicity has been observed from either the carrier or the dye.

Imaging and Post-processing

The demands and parameters for imaging spines in acute slices are similar to those used on cultured slices. However, to observe submicrometer changes in the structure of spines, which are typically 1–3 μm long and less than 1 μm wide, the smallest pixel size available should be used. Using a 63x/1.2-NA water-immersion objective lens, this is 0.08 μm. Thus, a 512 x 512 scan is only 40 μm wide. This exacerbates any drift problems and makes control of temperature and osmolarity (which affects swelling or shrinking of tissue) even more critical. Inevitably, some drift or movement must be removed during post-processing, for example, by using the "register" function in NIH Image. This Macintosh software is available free at the following WWW site: http://rsb.info.nih.gov/nih-image.

To virtually eliminate noise and greatly increase the *z*-resolution of the images, we have found it helpful to perform image deconvolution of the raw data (see Fig. 5). A model of how the microscope optics blurs the image is created by imaging subresolution fluorescent spheres (Molecular Probes MultiSpeck kit) under conditions as similar as possible to the imaging of the dendrites. The Huygens System is presently the only deconvolution software that deals appropriately with multiphoton images (Vankempen et al. 1997). It takes several hours on an SGI O2 workstation (Silicon Graphics) to deconvolve a single *z*-series, but the improvement is dramatic. As with the 3D median filter, slices must be collected at very close intervals (0.5 μm or less) for effective deconvolution.

IMPROVEMENTS FOR THE FUTURE

The benefits of two-photon microscopy for imaging living specimens are clear. However, the technology will not be widely implemented until the lasers needed become much cheaper. Many labs that are not mechanically or optically inclined are waiting for a true "turnkey" two-photon system, one that is designed from the ground up for optimum IR throughput and efficient, non-descanned detection.

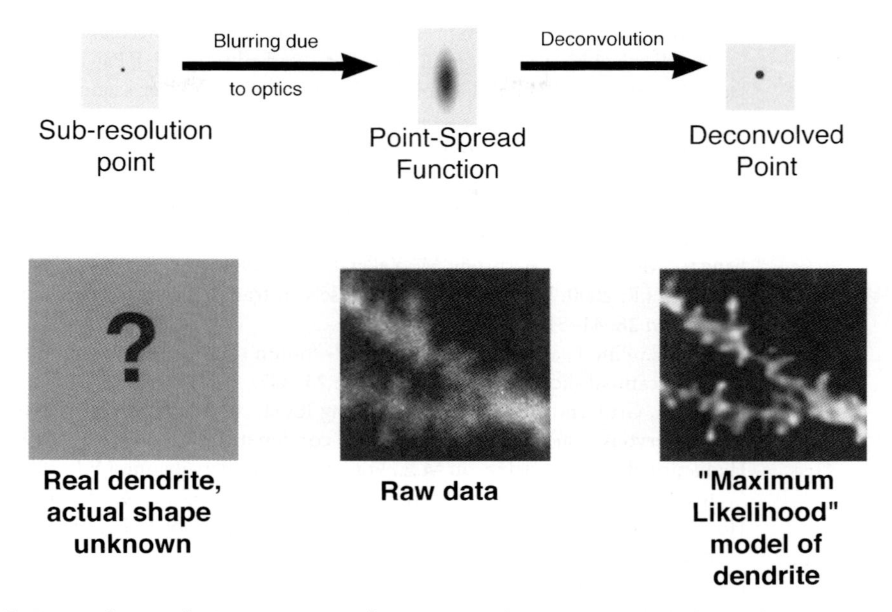

FIGURE 5. Image deconvolution. A 0.1-μm fluorescent sphere is imaged to obtain the microscope's point-spread function, which is then used to numerically deblur (deconvolve) a raw set of 3D-image data. We use the maximum likelihood estimation algorithm of the Huygens software suite.

Although most labels that are in current use with standard fluorescence microscopy also work with multiphoton microscopy, the potential exists for dyes optimized for multiphoton imaging (Albota et al. 1998).

Fluorescent proteins provide relatively noninvasive labeling for a variety of specimens. In collaborations with Paul Garrity and Peter Mombaerts, I have had great success using two-photon microscopy to image GFP-labeled neurons in living fruitfly optic lobes and mouse olfactory bulb (Potter et al. 1996c; Mombaerts et al. 1996, respectively). Transgenic mice expressing membrane-associated GFP in neuronal subpopulations would eliminate the need for tedious labeling procedures (Feng et al. 2000; Hadjantonakis et al. 2003).

ACKNOWLEDGMENTS

I thank Professors Jerry Pine and Scott Fraser for their continued support and guidance. I thank Arno Klein for technical assistance with the detector installation and a critical reading of this chapter. This work was supported by the Beckman Foundation and National Research Service Award F32NS10257-02 from the National Institute of Neurological Disorders and Stroke.

REFERENCES

Albota M., Beljonne D., Bredas J.L., Ehrlich J.E., Fu J.Y., Heikal A.A., Hess S.E., Kogej T., Levin M.D., Marder S.R., McCord-Maughon D., Perry J.W., Rockel H., Rumi M., Subramaniam G., Webb W.W., Wu X.L., and Xu C. 1998. Design of organic molecules with large two-photon absorption cross sections. *Science* **281:** 1653–1656.

Brown L.F., Lanir N., McDonagh J., Tognazzi K., Dvorak A.M., and Dvorak H.F. 1993. Fibroblast migration in fibrin gel matrices. *Am. J. Pathol.* **142:** 273–283.

Buchs P.A., Stoppini L., and Muller D. 1993. Structural modifications associated with synaptic development in area CA1 of rat hippocampal organotypic cultures. *Dev. Brain Res.* **71:** 81–91.

Dailey M.E. and Smith S.J. 1993. Confocal imaging of mossy fiber growth in live hippocampal slices. *Jpn. J. Physiol.* **43:** S183–S192.

Denk W., Piston D.W., and Webb W.W. 1995. Two-photon molecular excitation in laser-scanning microscopy. In *Handbook of biological confocal microscopy*, 2nd edition (ed. J. Pawley), pp. 445–458. Plenum Press, New York.

Denk W., Strickler J.H., and Webb W.W. 1990. 2-photon laser scanning fluorescence microscopy. *Science* **248:** 73–76.

Edwards F.A. 1995. Patch-clamp recording in brain slices. In *Brain slices in basic and clinical research* (ed. B.M. Rigor and A. Schurr), pp. 99–116. CRC Press, Boca Raton, Florida.

Edwards F.A., Konnerth A., Sakmann B., and Takahashi T. 1989. A thin slice preparation for patch clamp recordings from neurons of the mammalian central nervous-system. *Pflueg. Arch. Eur. J. Physiol.* **414:** 600–612.

Feng G., Mellor R.H., Bernstein M., Keller-Peck C., Nguyen Q.T., Wallace M., Nerbonne J.M., Lichtman J.W., and Sanes J.R. 2000. Imaging neuronal subsets in transgenic mice expressing multiple spectral variants of GFP. *Neuron* **28:** 41–51.

Fraser S.E., Pine J., and Potter S.M. 1997. 2-Photon time-lapse imaging of transplant integration in cultured rat hippocampal slices. *Soc. Neurosci. Abstr.* **23:** 347.

Gan W.B., Grutzendler J., Wong W.T., Wong R.O.L., and Lichtman J.W. 2000. Multicolor "DiOlistic" labeling of the nervous system using lipophilic dye combinations. *J. Neuron* **27:** 219–225.

Hadjantonakis A.K., Dickinson M.E., Fraser S.E., and Papaioannou V.E. 2003. Technicolour transgenics: Imaging tools for functional genomics in the mouse. *Nat. Rev. Genet.* **4:** 613–625.

Hosokawa T., Bliss T.V.P., and Fine A. 1992. Persistence of individual dendritic spines in living brain slices. *Neuroreport* **3:** 477–480.

Lojewska Z. and Loew L.M. 1986. Pluronic F127: An effective and benign vehicle for the insertion of hydrophobic molecules into membranes. *Biophys. J.* **49:** 521a.

———. 1987. Insertion of amphiphilic molecules into membranes is catalyzed by a high molecular weight nonionic surfactant. *Biochim. Biophys. Acta* **899:** 104–112.

McAllister A.K., Lo D.C., and Katz L.C. 1995. Neurotrophins regulate dendritic growth in developing visual-cortex. *Neuron* **15:** 791–803.

Potter S.M., Fraser S.E., and Pine J. 1996a. The greatly reduced photodamage of 2-photon microscopy enables extended 3-dimensional time-lapse imaging of living neurons. *Scanning* **18:** 147.

Potter S.M., Pine J., and Fraser S.E. 1996b. Neural transplant staining with DiI and vital imaging by 2-photon laser-scanning microscopy. *Scanning Microsc. Suppl.* **10:** 189–199.

Potter S.M., Wang C.M., Garrity P.A., and Fraser S.E. 1996c. Intravital imaging of green fluorescent protein using 2-photon laser-scanning microscopy. *Gene* **173:** 25–31.

Potter S.M., Zheng C., Koos D.S., Feinstein P., Fraser S.E., and Mombaerts P. 2001. Structure and emergence of specific olfactory glomeruli in the mouse. *J. Neurosci.* **21:** 9713–9723.

Shetty A.K. and Turner D.A. 1995. Enhanced cell-survival in fetal hippocampal suspension transplants grafted to adult-rat hippocampus following kainate lesions—A 3-dimensional graft reconstruction study. *Neuroscience* **67:** 561–582.

Sorra K. and Harris K. 1997. Stability in synapse number and size at 2 hr after long-term potentiation in hippocampal area CA1. *J. Neurosci.* **18:** 658–671.

Stoppini L., Buchs P.A., and Muller D. 1991. A simple method for organotypic cultures of nervous tissue. *J. Neurosci. Methods* **37:** 173–182.

Terasaki M. and Dailey M.E. 1995. Confocal microscopy of living cells. In *Handbook of biological confocal microscopy*, 2nd edition (ed. J. Pawley), pp. 327–346. Plenum Press, New York.

Vanderklish P., Neve R., Bahr B.A., Arai A., Hennegriff M., Larson J., and Lynch G. 1992. Translational suppression of a glutamate receptor subunit impairs long-term potentiation. *Synapse* **12:** 333–337.

Vankempen G.M.P., Vanvliet L.J., Verveer P.J., and Vandervoort H.T.M. 1997. A quantitative comparison of image-restoration methods for confocal microscopy. *J. Microsc.* **185:** 354–365.

Xu C., Williams R.M., Zipfel W., and Webb W.W. 1996. Multiphoton excitation cross-sections of molecular fluorophores. *Bioimaging* **4:** 198–207.

CHAPTER 9

A Practical Guide: Building a Two-Photon Laser-scanning Microscope

Jerome Mertz

Neurophysiologie et Nouvelles Microscopies, INSERM-CNRS, Ecole Supérieure de Physique et de Chimie Industrielles, Paris, France

The basic elements of a two-photon laser-scanning microscope (TPLSM) are shown in Figure 1 (Denk et al. 1990, 1995; Denk and Svoboda 1997; Mainen et al. 1999; Majewska et al. 2000; Williams et al. 2001; Tsai et al. 2002).

SYSTEM COMPONENTS

Laser

Two-photon excited fluorescence (TPEF) power is proportional to $P_0^2 \times r\tau$, where P_0 is the peak laser power at the focal point, r is the laser repetition rate, and τ is the laser pulse duration. Increasing any one of these parameters increases the power of TPEF to within the limits imposed by the application. If the sample cannot tolerate a high P_0 (e.g., if photobleaching is highly nonlinear [Hopt and Neher 2001]), then it is better to use a high rep-rate laser with long pulses. Equivalently, TPEF power is proportional to $\bar{P}^2/r\tau$, where $\bar{P}$ is the average laser power at the focal point. If the sample cannot tolerate a high $\bar{P}$ (e.g., because of excessive heating) or if the laser power is limited (e.g., because of attenuation in a thick tissue), then it is better to use a low rep-rate laser with short pulses (Beaurepaire et al. 2001). The most commonly used laser is a mode-locked Ti:sapphire ($\bar{P} \approx 1$ W; $r \approx 80$ MHz; $\tau \approx 100$ fsec).

Scanner

Laser XY scanning is typically performed with two small mirrors mounted on fast galvanometers (full angular swing [$\theta_1 \approx \pm 10°$] in less than 1 msec). Mirror coatings should be highly reflective at the laser wavelength (typically gold or protected silver, but not aluminum). Ideally, both mirrors should be at the same location. In practice, the mirrors are placed as close together as possible (this is simple, but not optically perfect), or the image of one mirror is projected onto the other (more optical elements are needed for this configuration). Video rate scanning can be performed with resonant galvanometers, polygonal mirrors, acousto-optic deflectors, or by multiple beam splitting (for references, see Williams et al. 2001). Scanning in the Z dimension is performed by translating the objective with a motor or piezoelectric transducer.

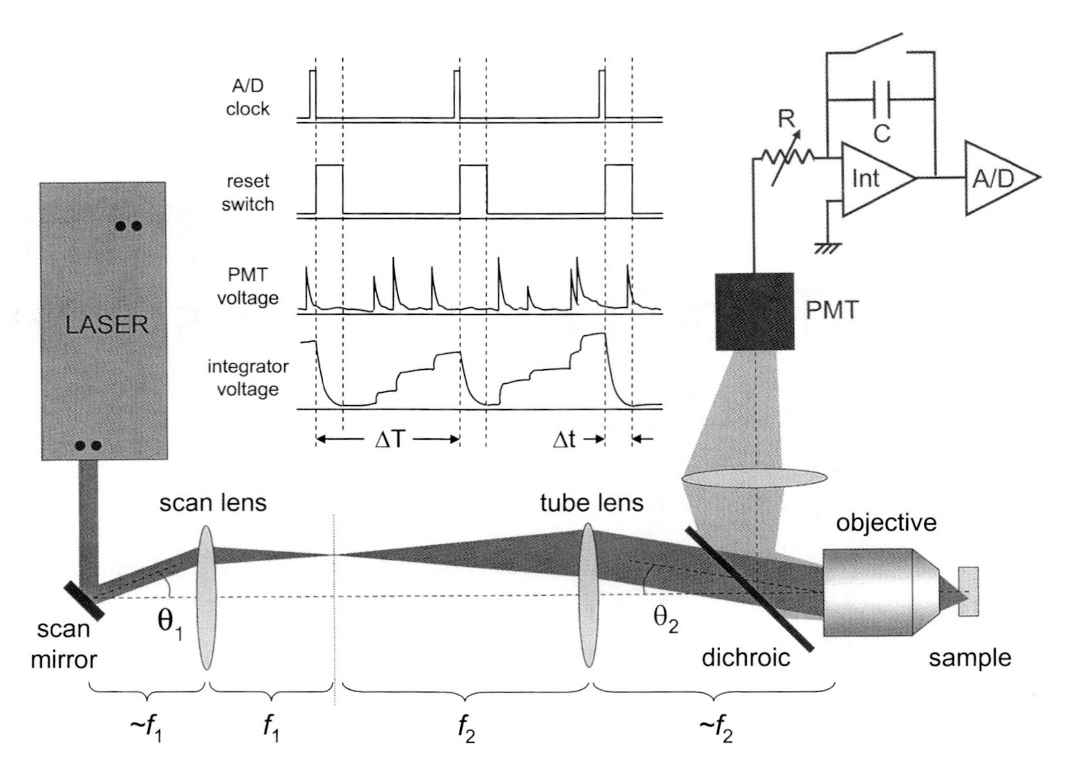

FIGURE 1. TPLSM with mirror scanning and analog electronics.

Excitation Optics

The roles of the scan and tube lenses are to expand the laser beam and to project the image of the scan mirrors onto the back aperture (BA) of the objective. For highest imaging resolution, the beam should overfill the BA to illuminate it uniformly. Alternatively, if power is a concern, as in deep-tissue imaging, the BA should be somewhat underfilled. The beam expansion factor is $M = f_2/f_1$, where f_1 and f_2 are the scan and tube lens focal lengths, respectively. Equivalently, $M = \theta_1/\theta_2$. The field-of-view (FOV) of a TPLSM, defined as the maximum focal-spot scan area, has a maximum radius $\theta_2 f_0$, where f_0 is the objective focal length. Hence, for a given galvanometer angular range, beam expansion unavoidably leads to a contraction of the FOV.

When designing a TPLSM, it is usually best to work backward, first deciding on a desired FOV and BA fill-factor for a specific objective. These prescribe the required M and galvanometer angular range (θ_1). Compromises have to be made, particularly if the TPLSM is designed for high speed, meaning that the scanning mirrors are small and hence M must be large.

To avoid clipping, the scan and tube lens radii should be at least $\theta_1 f_1$ and $\theta_2 f_2$, respectively. Because of its shorter focal length, the scan lens should be well corrected for spherical aberrations (a doublet can be acceptable). Typical focal lengths are $f_1 \approx 70$ mm and $f_2 \approx 200$ mm.

Resolution is entirely governed by the size of focal spot. The objective should be high numerical aperture (NA) (subject to working-distance constraints), and highly corrected for spherical aberrations (chromatic-aberration corrections are unimportant).

Collection Optics

Most TPLSMs operate in an epifluorescence configuration. The collection optics should project (image) the entire BA onto the detector's active area. The fluorescence collection efficiency is then proportional to NA^2. When the sample is thin or transparent, the collected fluorescence exits the BA as a parallel beam that can be descanned, permitting the use of a small-area high-sensitivity detector. When the sample is thick and turbid, although the fluorescence is still generated at the focal point, it *appears* to emerge from an extended area in the sample and no longer exits the BA as a parallel beam. Non-descanned detection is then prescribed, as shown in Figure 1, using an objective with as wide an FOV as possible. Optimal objectives for deep-tissue imaging should thus exhibit both high NA and low magnification (Oheim et al. 2001).

The dichroic shown in Figure 1 is long-pass (short-pass configurations are also possible). Color glass filters (e.g., BG22 or BG39, 1–3 mm) provide excellent laser rejection. Interference filters provide sharp cutoff wavelengths and can be long-pass, short-pass, or band-pass.

Detector

Detector response times should be shorter than the pixel duration. Advantages of photomultiplier tubes (PMTs) are (1) a very high internal gain, favoring shot noise over Johnson noise and (2) a large active area, convenient for non-descanned deep-tissue imaging; they are also relatively blind to laser wavelengths. Multialkali photocathodes are more sensitive to visible wavelengths than bialkali photocathodes, but they produce more dark current and usually require more laser rejection filters, possibly offsetting their higher sensitivity. Meshless cathodes are preferred (fluorescence should *not* be focused onto meshed cathodes since their gain is not uniform).

An advantage of avalanche photodiodes (APDs) is high quantum efficiency (Tan et al. 1999). The dark noise of APDs roughly scales with their active area. For ultra-low dark noise, the active area is typically $<0.5\ mm^2$, which is inconvenient for non-descanned applications.

New GaAsP photocathodes or hybrid detectors are available that are both highly sensitive and laser blind (and expensive).

Electronics

The duration T and extent X of a TPLSM image line are governed by the speed and range of the galvanometer sweep, which are user-defined. The number of pixels, N, into which the line is partitioned is arbitrary (in contrast to CCD cameras) and depends on the pixel acquisition rate, F, also user-defined. The duration, ΔT, and extent, ΔX, covered by each pixel are then $1/F$ and X/FT, respectively. Typically, ΔT is a few microseconds and ΔX should be no smaller than the focal spot size.

When using photon-counting detectors, a counter is reset to zero at the beginning of each pixel, incremented at every detector pulse, and read by a computer at the end of each pixel. The acquired counter values are color-coded and arranged into an image.

When using analog detectors, the signal must be integrated (Fig. 1). A feedback capacitor is short-circuited at the beginning of each pixel circuit, charged for a time, ΔT, and its voltage is read by a fast A/D converter at the end of each pixel. The integrator gain ($1/RC$) should be large, within the limitations that the amplifier should not saturate and $RC >> \Delta T$ (the latter condition ensures the photons are weighted equally). The reset duration should be long enough for the capacitor to fully discharge ($\Delta t > R_s C$, where R_s is the switch impedance when closed). Typically, $C \approx 1$ nF. The small dead time during reset can be eliminated with a backup integrator running in parallel.

REFERENCES

Beaurepaire E., Oheim M., and Mertz J. 2001. Ultra-deep two-photon fluorescence excitation in turbid media. *Opt. Commun.* **188:** 25–29.

Denk W. and Svoboda K. 1997. Photon upmanship: Why multiphoton imaging is more than a gimmick. *Neuron* **18:** 351–357.

Denk W., Piston D.W., and Webb W.W. 1995. Two-photon molecular excitation in laser-scanning microscopy. In *Handbook of confocal microscopy* (ed. J.B. Pawley), pp. 445–458. Plenum Press, New York.

Denk W., Strickler J.H., and Webb W.W. 1990. Two-photon laser scanning fluorescence microscopy. *Science* **248:** 73–76.

Hopt A. and Neher E. 2001. Highly nonlinear photodamage in two-photon fluorescence microscopy. *Biophys. J.* **80:** 2029–2036.

Kim K.H., Buehler C., and So P.T.C. 1999. High-speed, two-photon scanning microscope. *Appl. Opt.* **38:** 6004–6009.

Mainen Z.F., Maletic-Savatic M., Shi S.H., Hayashi Y., Malinow R., and Svoboda K. 1999. Two-photon imaging in living brain slices. *Methods* **18:** 151–155.

Majewska A., Yiu G., and Yuste R. 2000. A custom-made two-photon microscope and deconvolution system. *Pflugers Arch.* **441:** 398–408.

Oheim M., Beaurepaire E., Chaigneau E., Mertz J., and Charpak S. 2001. Two-photon microscopy in brain tissue: Parameters influencing the imaging depth (erratum *J. Neurosci. Methods* [2001] **112:** 205). *J. Neurosci. Methods* **111:** 29–37.

Tan Y.P., Llano I., Hopt A., Würriehausen F., and Neher E. 1999. Fast scanning and efficient photodetection in a simple two-photon microscope. *J. Neurosci. Methods* **92:** 123–135.

Tsai P.S., Nishimura N., Yoder E.J., White A., Dolnick E., and Kleinfeld D. 2002. Principles, design and construction of a two-photon scanning microscope for in vitro and in vivo studies. In *Methods for in-vivo optical imaging* (ed. R.D. Frostig), pp. 113–171. CRC Press, Boca Raton, Florida.

Williams R.M., Zipfel W., and Webb W.W. 2001. Multiphoton microscopy in biological research. *Curr. Opin. Chem. Biol.* **5:** 603–608.

CHAPTER 10

A Practical Guide: How to Build a Two-Photon Microscope Using a Confocal Scan Head

Volodymyr Nikolenko and Rafael Yuste

Department of Biological Sciences, Columbia University, New York, New York 10027

This chapter provides practical guidelines for the conversion of an Olympus Fluoview confocal microscope into a two-photon microscope. This enables the investigator to have access to two-photon microscopy without the large budget necessary to purchase a commercial instrument. Two-photon fluorescence microscopy allows deep-tissue imaging in highly scattering preparations and long-term imaging of live tissue without the photodamage that is caused by out-of-focus light (see Chapter 7). It is, therefore, an essential tool for imaging cells in physiologically relevant conditions such as acute or cultured brain slices or in vivo.

MATERIALS

Laser

The key component of the system is a femtosecond pulsed laser<!>, which can generate reasonable power in the near-infrared (IR) spectral region needed for convenient practical imaging (average power >50 mW). Our current setup uses the Chameleon laser from Coherent. This is a fully automated turnkey laser which can be tuned to any wavelength between 715 and 955 nm.

Optical Table

The laser beam is delivered to a modified Olympus Fluoview confocal laser-scanning system through the set of optical elements on the optical table (Fig. 1):

- Intermediate mirrors
- Spatial filter
- Retardation waveplate
- Pockels cell

BB1-E02 dielectric mirrors from Thorlabs are used as intermediate mirrors. These reflect >99% of light between 700 and 1150 nm at a 45° angle of incidence for all polarizations and do not introduce additional group velocity dispersion of ultrafast pulses.

An optical spatial filter, which has two planoconvex lenses and a pinhole in the focus of the first lens, acts as a simple telescope. It is used to restore a smooth Gaussian profile to the intensity of the laser beam cross-section and also to modify beam size to ensure proper overfilling of the back aperture of the microscope objective (Tsai et al. 2002). A retardation waveplate ($\lambda/2$ or $\lambda/4$) is used for complex experiments in which the polarization of the scanning laser beam must be controlled.

Pockels Cell

3030C from Quantum Technology, a nonlinear optical modulator, is included to allow the dynamic modulation of laser light intensity with excellent contrast ratio (>200:1) and submicrosecond temporal resolution. The temporal resolution of the Pockels cell is limited in practical terms only by the electronics of the high-voltage driver. The model in our current setup (302 from Quantum Technology) can work in DC-10 MHz modulation range (thus providing 100 nsec temporal resolution).

Scanning Microscope

A BX50WI Olympus upright scanning microscope is coupled to a modified Fluoview scanning unit. The Olympus Fluoview platform relieves the need for a separate, custom-built scanning system and software package (Majewska et al. 2000).

The scanning box contains the only essential optical component: a set of two galvanometer mirrors, which steer the laser beam and scan the image. The scanning box is optically linked to the infinity-corrected BX50WI microscope through a pupil transfer lens, which is part of the original Fluoview system. This lens, together with the tube lens of the microscope (original part of microscope) forms a telescope, which provides collimated light for the infinity-corrected objective lens (see optical scheme in Fig. 1). This telescope also approximately images the scanning mirrors onto the back aperture of the objective. This minimizes the movement of the laser beam at the back aperture, thus minimizing variation of laser power at the sample. The laser beam is reflected downward by a short-pass dichroic mirror (650DCSP from Chroma Technology), placed inside the standard trinocular tube of the Olympus microscope.

Fluorescence Detection

In the case of two-photon absorption, excitation of fluorescence is essentially limited to the diffraction-limited spot in the focal plane. This provides the 3D sectioning characteristic of two-photon microscopy. Since fluorescence from the excited region irradiates in all directions, it is important to use a high-numerical aperture objective in order to collect as many fluorescence photons as possible. Our system uses an external photomultiplier tube (PMT) as a detector. The tube is mounted on the camera port of the microscope's trinocular tube—along with a dichroic mirror, which transmits visible fluorescent light collected by the objective. An additional IR-blocking filter (BG39 from Chroma Technology) is placed in front of the PMT to filter out residual infrared fluorescent light reflected from the excitation path. The external PMT could also be positioned right next to the objective (see schematic in Fig. 1). In this case, the PMT would have to be mounted via a custom-made adapter with a long-pass dichroic mirror, which transmits excitation infrared light and reflects visible fluorescence to the detector. Positioning the PMT on top of the objective improves light collection efficiency but compromises the convenient positioning of micromanipulators.

By choosing appropriate dichroic mirrors, it is possible to use two channels of fluorescence imaging. Placing additional band-pass filters in front of detectors can also efficiently separate the emissions of two different fluorescent dyes. The Olympus Fluoview data acquisition board comes ready equipped with two independent input channels (see Section 8 in this manual for example applications and for advice on choosing filters and dichroic mirrors).

The PMT used in our instrument is the HC125-02 assembly from Hamamatsu, which comprises a bi-alkali, head-on PMT, a high-voltage power supply, and a wide bandwidth (8 MHz) signal preamplifier. This PMT provides a voltage signal that is proportional to the light intensity. The signal is fed directly into the signal input of the Olympus Fluoview data acquisition board. The standard Olympus Fluoview software is used in scanning mode for signal acquisition and reconstruction of a digital image.

To improve the signal-to-noise ratio, an additional intermediate amplifier is connected between the external PMT and the Fluoview signal input. Currently, our system uses a PE 5113 amplifier (AMETEK Advanced Measurement Technology; former EG&G and Perkin Elmer Instruments). This instrument has a variable gain and adjustable high/low-pass filters, and can be controlled through a user-friendly computer interface. The amplifier is very convenient for proper signal conditioning, since it allows proper filling of the dynamic range of the Fluoview signal analog-to-digital converter, something crucially important for imaging weak signals and quantitative measurements. For calibration curves of the available dynamic range of Olympus Fluoview signal inputs, see Nikolenko et al. (2003).

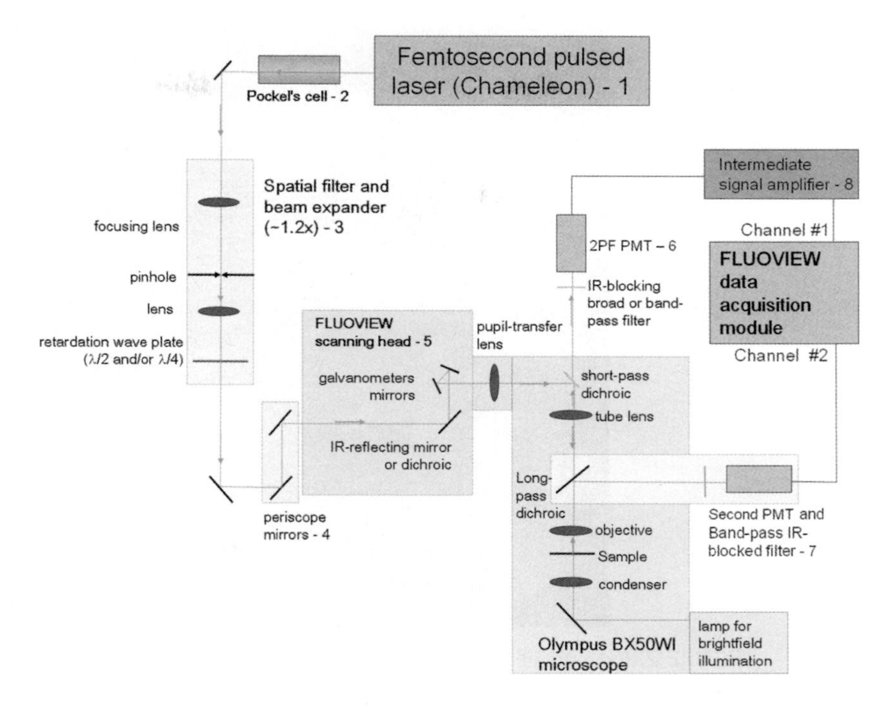

FIGURE 1. Optical design of the instrument. Key components: (1) NIR femtosecond pulsed laser; (2) Pockels cell modulator; (3) system of lenses, which works as spatial filter and beam expander (~1.2x in our case); (4) periscope mirrors (deliver laser beam from optical table level to scanning box of Fluoview, which is raised for upright microscope); (5) modified Olympus Fluoview scanning unit; (6) external PMT detector for 2P-fluorescence signal, attached to camera port of the microscope; (7)second external PMT attached to the microscope through custom-made adapter; (8) intermediate signal amplifier for matching dynamic range of signal source (PMT) and Fluoview data acquisition module.

PROCEDURE

Building a Custom-made Two-Photon Microscope

The full description of practical changes in the standard Olympus Fluoview confocal system can be found in Majewska et al. (2000) and Nikolenko et al. (2003). A brief summary of the necessary modification is presented here.

1. Install the pulsed femtosecond near-IR laser.
2. Build the external optical pathway from laser source to laser-scanning microscope with the optional spatial filter and Pockels cell.
3. Modify the scanning unit of a confocal laser-scanning microscope for scanning by the IR beam from external laser source. This essentially requires drilling a hole in the back of the scanning unit and replacing the internal dichroic mirror with an IR reflecting mirror (see full list of Olympus Fluoview modification in Majewska et al. 2000 and Nikolenko et al. 2003).
4. Place a short-pass dichroic mirror into the trinocular tube of the optical microscope and install the external detector (PMT) on the camera-imaging port of the trinocular tube.
5. Connect the detector signal output to the data acquisition input of the confocal system through the intermediate signal amplifier.

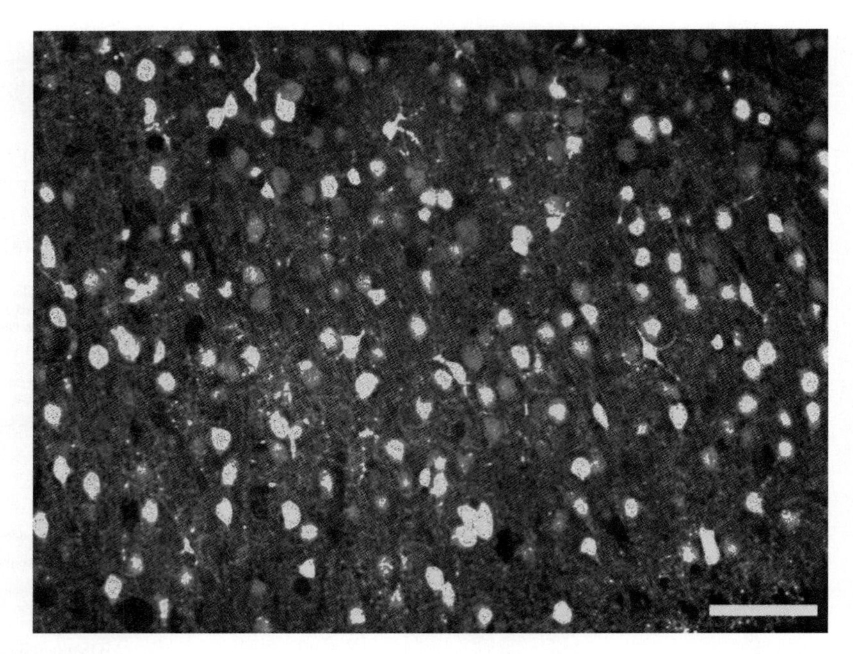

FIGURE 2. Example of imaging. A neocortical slice from a postnatal day-13 mouse, loaded with the Ca^{2+} fluorescence indicator indo-1AM. Two-photon fluorescence image acquired with ~730 nm excitation wavelength. Bar, 50 μm.

SHORT EXAMPLE OF APPLICATION

The custom-made two-photon microscope described here has been used successfully for long-term imaging of action potential activity in large (>1000) populations of neocortical neurons in acute brain slices, AM-loaded by Ca^{2+} fluorescent indicators such as fura-2 (Cossart et al. 2003) or indo-1 (Fig. 2).

ADVANTAGES AND LIMITATIONS

The two-photon system described here, based on the Olympus Fluoview confocal system, successfully combines a customized home-made system with a reliable commercial instrument. Although tailor-made for the chosen application, the individual elements of a home-made system can be difficult to maintain in proper working condition. However, recent advances are easing this problem. For example, the appearance on the market of turnkey femtosecond-pulsed tunable lasers, such as the Chameleon, eliminated the need for realignment of the optical path. However, potential users should be aware of the correct procedures for cleaning optical surfaces that are exposed to dust, changes in humidity, etc., and follow proper laser safety guidelines (see Appendix 3).

REFERENCES

Cossart R., Aronov D., and Yuste R. 2003. Attractor dynamics of network UP states in the neocortex. *Nature* **423:** 283–288.

Majewska A., Yiu G., and Yuste R. 2000. A custom-made two-photon microscope and deconvolution system. *Pflügers Arch.* **441:** 398–408.

Nikolenko V., Nemet B., and Yuste R. 2003. A two-photon and second-harmonic microscope. *Methods* **30:** 3–15.

Tsai P.S., Nishimure N., Yoder E.J., Dolnik E.M., White G.A., and Kleinfield D. 2002. Principles, design and construction of a two photon scanning microscope for in vitro and in vivo studies. In *Methods for in vivo optical imaging* (ed. R. Frostig), pp. 113–171. CRC Press, New York.

CHAPTER 11

In Vivo Electroporation during Embryogenesis

Catherine E. Krull,[1,2] Rebecca McLennan,[1] Sinead O'Connell,[1] Yaxiong Chen,[1] and Paul A. Trainor[3]

[1]Division of Biological Sciences, University of Missouri-Columbia, Columbia, Missouri 65202; [2]Cell and Developmental Biology, University of Michigan, Ann Arbor, Michigan 48109; [3]Stowers Institute of Medical Research, Kansas City, Missouri 64110

One essential approach to understanding embryogenesis is altering gene expression/function and analyzing the consequences on tissue organization and cell behavior. However, several model systems in which developmental processes have been extensively analyzed lack tools for genetic manipulation. In vivo electroporation, a relatively new technique (Muramatsu et al. 1996, 1997), provides several distinct advantages, including the opportunity to target gene manipulations to specific cell types while they are in their native environment. The goal of this chapter is to provide technical details about performing in vivo electroporation in various tissues in chicken and mouse embryos.

This approach allows investigators to manipulate gene expression and function in single cells and large numbers of cells, providing many possibilities in this era of functional genomics. Investigators can use this tool to quickly and inexpensively screen gene function prior to initiating more costly approaches, such as the construction of knockout and transgenic mice. Electroporation has been applied primarily to cells in whole embryos, but it can be used to alter gene expression in cells in slice preparations and explant cultures (Krull and Kulesa 1998; Fukuda et al. 2000; Haas et al. 2001). In addition, loss-of-function agents, including dsRNA or shRNA for RNAi experiments (Pekarik et al. 2003; Chesnutt and Niswander 2004) and morpholinos (Kos et al. 2001; Gerlach-Bank et al. 2004), can be successfully introduced into chickens via in vivo electroporation.

The approach also has broader applications, beyond chicken and mouse embryo manipulations. In vivo electroporation permits gene manipulation in species where gene transfer approaches have not been established, including ascidians and frogs (Corbo et al. 1997, 1998; Locascio et al. 1999; Eide et al. 2000). Promoter/enhancer screens could be accomplished using in vivo electroporation (Itasaki et al. 1999). In addition, investigators could transfect embryos with reporter constructs to follow cell lineage and normal cell migratory patterns. Coupled with time-lapse imaging, these types of studies could yield exciting data about fundamental events in development. Investigators are also referred to the following recent publications: Itasaki et al. 1999; Momose et al. 1999; Atkins et al. 2000; Yasuda et al. 2000; Osumi and Inoue 2001; Swartz et al. 2001a.

SINGLE-CELL ELECTROPORATION

Gene delivery into single cells offers the opportunity to manipulate gene expression and function and to examine subsequent effects on the behavior of a single cell. This approach is accomplished by the combination of electrical stimulation and DNA delivery via a single micropipette. It has been utilized successfully in *Xenopus* tadpoles and rat hippocampal slices. Single-cell electroporation was developed by Hollis Cline and her colleagues; the technique is beyond the scope of this chapter and readers are referred to Haas et al. 2001 for details (see Chapter 23).

IN VIVO ELECTROPORATION

The Basics

Electroporation involves the application of electrical field pulses that temporarily produce pores in the plasma membrane through which DNA is driven, due to its negative charge. For several years, electroporation has served as an effective tool for introducing DNA into bacteria, yeast, mammalian cell lines, and plant protoplasts (Neumann et al. 1982; Shillito et al. 1985; Potter 1988). In these scenarios, high voltage and short duration pulses are used, resulting in high percentages of cell death. The application of electroporation to chicken or mouse embryos was made feasible by modifications of voltage parameters and pulse duration. Typically, square wave pulses of low voltage and long duration are used to achieve DNA delivery and cell transfection in these embryos, with minimal cell death and high rates of survival. Of note, this method has also been used successfully to transfect *Drosophila* embryos and fish eggs/embryos (Buono and Linser 1992; Muller et al. 1993; Kamdar et al. 1995).

Important Considerations

Electroporation efficiency varies depending on the organization of the chosen tissue; highly compact tissues tend to limit DNA diffusion and, for a given set of electroporation conditions, transformation rates are reduced. Consequently, the voltage parameters (pulse size, pulse length, pulse number) and electrode placements must be determined empirically for each tissue. Parameters must also be modified according to the animal's age. Typically, younger embryos require lower voltages than older animals. Chicken embryos younger than stage 10 are particularly difficult to electroporate in ovo, without high mortality. One workable alternative is to transfect young embryos while growing them in modified New culture (Chapman et al. 2001).

The electroporator should be calibrated according to the efficiency of transfections and embryo survival rates. The voltage requirements of other experiments can be used as a general reference only. The electroporator should be connected to an oscilloscope, to acquire current readings.

The type of electrode (gold or platinum) must also be considered as well as their size and conductivity. Note that electrodes last for 3–4 months, depending on use, and should be replaced after this time.

It is important to understand how electric fields work. The Internet is an excellent source of information (e.g., www.jcphysics.com). It is also important to discuss experiments with colleagues who use this approach regularly.

Where possible, DNA should be microinjected into a lumenal surface. The lumen will then serve as a reservoir for the DNA.

MATERIALS

DNA

The authors have used two plasmids for in vivo electroporation in chicken and mouse embryos: (1) pCAX, which contains a chick β-actin promoter/CMV IE enhancer driving EGFP expression (Osumi and Inoue 2001); and (2) pMES, which has the same promoter/enhancer combination and an IRES-EGFP, generating a bicistronic message (Swartz et al. 2001a,b). Other variants of GFP are also useful, including RFP. DNA constructs should be sequenced, and expression of the chosen inserts and reporter genes should be tested in cell lines or primary cell cultures prior to use. DNA for electroporation should be prepared using a Qiagen Plasmid Maxi kit, eluting the DNA into water, *not* elution buffer. DNA can be quantified using a spectrophotometer; if the OD 260/280 = 1.8, then the DNA is sufficiently pure. A preparation with an OD of less than 1.6 is not suitable. Store the DNA in aliquots of 2–4 µl at 2–5 µg/µl in either PBS (1x final concentration) or water, at –20°C.

Equipment for Electroporation

Several companies make square wave electroporators that are suitable for these experimental manipulations. Investigators are advised to check with their colleagues who use this approach for their recommendations.

PROCEDURE A

Preparation of Chick Embryos and Electrodes

Embryos

1. Incubate fertilized chicken eggs horizontally, until the appropriate developmental stage, in a humidified egg incubator (e.g., GQF Manufacturing) at 38°C.

   Note: Use available resources to stage chicken embryos (Hamburger and Hamilton 1951; Bellairs and Osmond 1998).

2. After rinsing eggshells with 70% ethanol <!>, remove 2–3 ml of albumen by puncturing one end of the egg with an 18-gauge needle attached to a 3-cc syringe. Cover this hole immediately with Scotch Magic tape to prevent the embryo from drying out.
3. Apply two strips of tape to the top of the eggshell, place the egg on a piece of gauze in a bevel-edged watch glass (e.g., Fisher Scientific), and use scissors to cut a 1–2-cm diameter round window through the taped eggshell. Take care to avoid injuring the embryo and overlying membranes.
4. Locate the blastoderm—this will be somewhat opaque and white in color, compared to the yellow yolk. Enlarge the window using the scissors to enhance access to the embryo.
5. For embryos at stages 10–18, apply a solution of 10% india ink (Pelikan) in Howard Ringer's solution below the embryo to enhance contrast (Howard Ringer's solution [2 liters]: 14.4 g of NaCl, 0.45 g of $CaCl_2 \cdot 2H_2O$ <!>, 0.74 g of KCl <!>. Mix in distilled water to 2 liters. Adjust pH to 7.4. Sterile-filter and add to 125-ml sterile bottles.)
6. Mix india ink 1:10 in Howard Ringer's solution in a 5-ml plastic tube.
7. Use forceps to bend the tip of a 25-gauge needle to a 90° angle and attach it to a 1-cc syringe. Load the syringe with india ink solution.
8. Locate the interface between the blastoderm and the yolk. Carefully insert the needle at this point and slide it under and toward the embryo.
9. Expel a small amount of ink, without injuring the embryo or any overlying membranes.
10. Carefully remove the needle and gently shake the egg to disperse the ink solution. See *Avian Embryology* (Bronner-Fraser 1996) for other details about embryo preparation and dissection.

Electrodes (numbers in parentheses refer to the schematic in Fig. 1)

11. Select gold (0.5-mm diameter) or platinum (1-mm diameter) wire (e.g., Goldwire from Alfa Aesar or Platinum Rod from A-M Systems, respectively) and warm up a soldering iron.
12. Cut the chosen wire into 6–7-cm lengths (yielding three sets of electrodes) and use lead-free solder to attach a pin-stamped brass (Digikey, #82p-nd)(6) onto the gold or platinum wire (8).
13. Cut six pieces of wire insulation (heat-shrink tubing, Newark Electronics) into 5-cm lengths. Place the tubing over the metal wire (7), adjacent to the pin-stamped brass, and apply heat from a hair dryer or heat-shrink gun to seal the wire insulation to the wire.
14. Solder the gold-plated jack/socket (Newark Electronics)(5) onto the red and black stranded wires (both 22-gauge, Newark Electronics)(4).

    Note: The red and black stranded wires should be long enough to allow the electrodes to be held in the hand, or placed in a plastic electrode holder, without restriction.
15. Attach the red and black stranded wires to a banana connector (3), linked to a double banana BNC connector (2) with a male end.

    Note: This double banana BNC connector is then connected to the electroporation system output wiring (1). See the schematic diagram below. As an alternative, the BNC connector can be omitted and the red and black stranded wires can be connected directly to the electroporator.

    Caution: *See Appendix 3 for appropriate handling of materials marked with <!>.*

PROCEDURE B

In Vivo Electroporation of Chicken Neural Tube

Electroporation of embryos with DNA constructs, or other reagents, is now a very straightforward task, with the neural tube being the most straightforward application. However, experience in embryo manipulation, microinjection, and electroporation will enhance chances of future success in transfecting this and other tissues. The basic premise is to microinject DNA into the lumen of the neural tube, which will serve as a DNA reservoir during electoporation (see Fig. 2A). For microinjection novices, additional details can be obtained from Fraser (1996).

1. Prepare micropipettes using 1.5–1.8 x 100-mm glass capillaries (A-M systems) and a horizontal pipette puller (e.g., Sutter), or a microforge (e.g., Glassworks) with ~0.01-mm tip diameter.
2. Backfill a micropipette with the chosen DNA solution (see Materials section for advice on DNA preparation) containing a small amount of Fast Green <!> (solid or 1% solution) using a Hamilton syringe.
3. Connect the micropipette to a needle holder (e.g., Wright) and a micromanipulator (e.g., Narishige) that is linked to a picospritzer (General Valve).

    Note: This is used to direct the site of injection, the volume injected, and the injection rate.
4. Prepare the embryo for electroporation (as described above) and remove the vitelline membrane above the region to be microinjected with DNA. Add a few drops of sterile Howard Ringer's solution over the embryo.

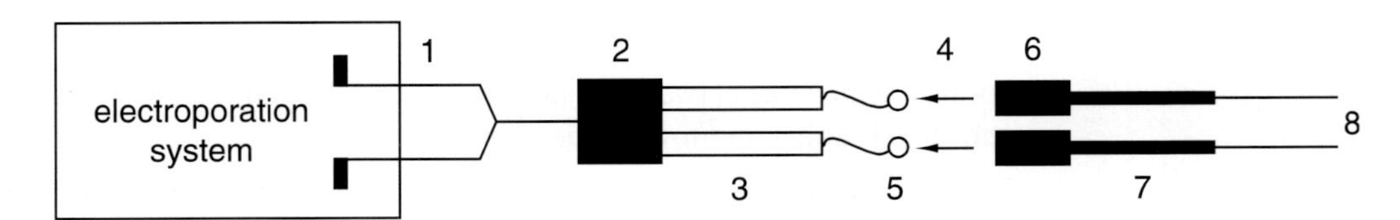

FIGURE 1. Schematic diagram showing the electrical setup for in vivo electroporation of chicken and mouse embryos.

Note: The Ringer's solution will protect the embryo from the heat generated by electroporation and will enhance the conduction of current.

5. Direct the micropipette to the lumen of the neural tube and carefully puncture the overlying ectoderm, allowing the pipette to penetrate the lumen, but no further.
6. Expel the DNA solution into the lumen.

   Note: Successful injection is indicated by a line of Fast Green dye that travels along the lumen of the neural tube. Failure is indicated by the presence of Fast Green solution outside the embryo.
7. Prepare electrodes as described above (see Procedure A), but bend the anode 1 mm from its tip, to an angle of 135°, and the cathode 2 mm from its tip, to an angle of 135°. Expose ~8 mm of wire (i.e., remove the insulation) at the tip of each electrode.
8. Transform each neural tube sector, as described in steps a, b, and c below.

   a. To transfect one half of the neural tube:

      i. Place the anode and cathode ~3–6 mm apart, on the area opaca, lateral to the embryo.

      Note: Do not touch the embryo proper or any future embryonic tissues with the electrodes.

      ii. Pass five 10–15-volt pulses, each of 50 msec duration.

      Note: The pulses will cause the DNA to move into neural tube cells adjacent to the anode.

      iii. After electroporation, apply a few drops of Ringer's solution to cool and hydrate the embryo, seal the eggshell with tape, and reincubate until the desired stage of development.

   b. To transfect the right ventral quadrant of the neural tube (see Fig. 2A):

      i. Place the cathode dorsally and slightly to the left of the midline of the embryo.

      ii. Place the anode ventral to the embryo, slightly to the right of the midline, inserting it at the interface between the blastoderm and the yolk.

      iii. Carefully pass the anode under or ventral to the embryo (~3 mm below), entering via the india ink injection site (see Procedure A).

      Note: Be careful not to puncture the embryo or its membranes.

      iv. Pass five 10–15-V pulses of 50 msec duration.

      v. After electroporation, apply a few drops of Ringer's, reseal the eggshell with tape, and reincubate until the desired stage of development (see Eberhart et al. 2002).

   c. To transfect a dorsal quadrant of the neural tube (to label neural crest or dorsal neural tube cells):

      i. Reverse the polarity of the electrodes before placing them as described in step b above.

      ii. Pass five 10–15-volt pulses of 50 msec duration.

      iii. Apply a few drops of sterile Ringer's solution on top of the embryo before sealing the eggshell with tape, and reincubate until the desired stage of development is achieved.

Caution: *See Appendix 3 for appropriate handling of materials marked with <!>.*

PROCEDURE C

In Vivo Electroporation of Chicken Somitic Mesoderm

The following procedure is for transforming the lateral half dermomyotome at forelimb level, to label muscle precursors that later migrate to the forelimb (see Fig. 2B). To transform the dorsomedial somite, place the anode (positive) electrode above the dorsomedial somite; to transfect the sclerotome, switch electrode polarity as shown in Figure 2, to drive DNA into sclerotomal cells.

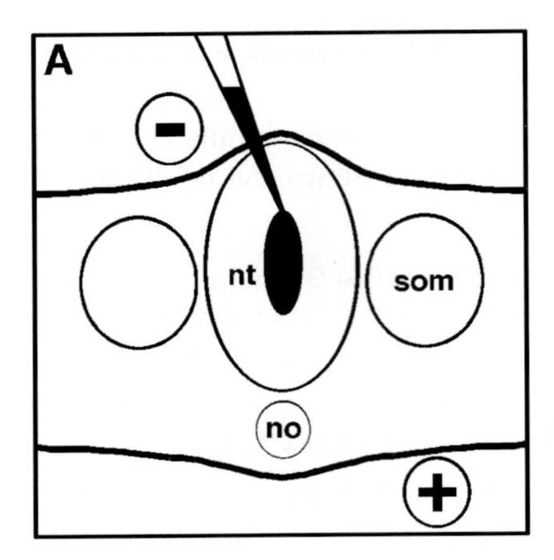

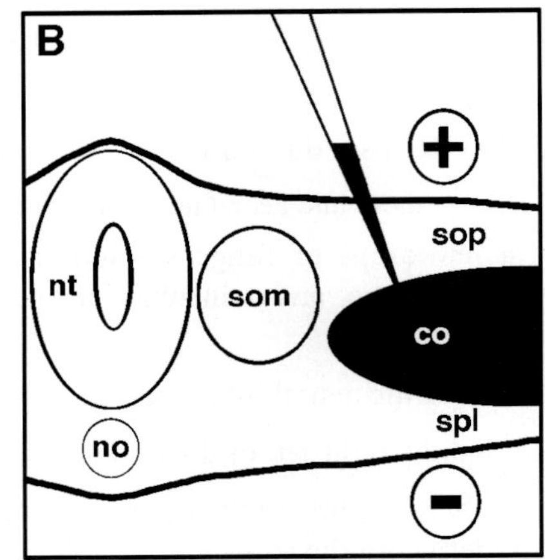

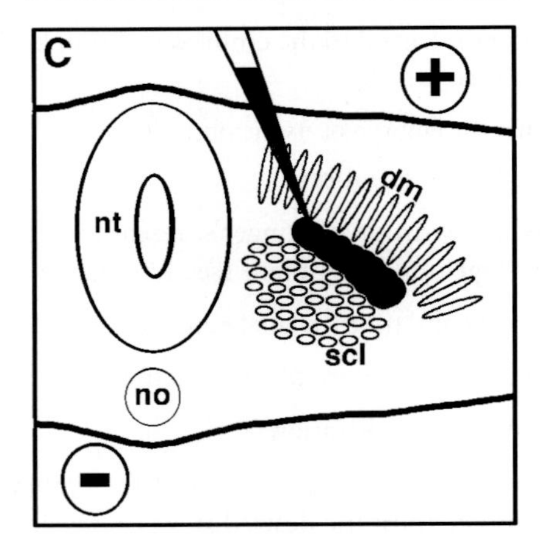

FIGURE 2. Schematic diagrams of in vivo electroporation in chicken embryos. (*A*) Chicken neural tube electroporation. DNA is microinjected into the lumen (*black*) of the neural tube (nt). Cathode (–) is placed dorsal to the neural tube; anode (+) is placed ventral to the embryo to achieve transfection of a ventral quadrant of the neural tube. Each electrode is placed lateral and parallel to the embryo to achieve half-neural tube transfections. (*B*) Chicken lateral plate mesoderm electroporation. DNA is microinjected into the coelom (co; *black*), which underlies the somatopleure (sop). Anode (+) is placed dorsal to the sop; cathode (–) is placed ventral to the splanchnic mesoderm (spl). (*C*) Chicken somitic dermomyotome electroporation. DNA (*black*) is microinjected into the extracellular space between the dermomyotome (dm) and sclerotome (scl). Cathode (–) is placed dorsal to the somite; anode (+) is placed ventral to the embryo, on the contralateral side.

1. Prepare stage-15 chicken embryos for electroporation, as described above. Carefully remove the vitelline membrane overlying the future forelimb area (somites #16–21; Christ and Ordahl 1995) using a sharpened tungsten needle.
2. Microinject plasmid DNA (2 μg/μl), containing a few crystals of Fast Green <!> , into the extracellular space between the dermomyotome and sclerotome of 1–6 somites at the forelimb level (see Fig. 2B).

   Note: These somites are undergoing compartmentalization into a dermomyotome and sclerotome at the time of microinjection.
3. Apply ~2 ml of Ringer's solution to the dorsal surface of the embryo, making a pool in which the anode will be placed.

4. Prepare platinum electrodes as described above (see Procedure A), but bend both ~3 mm from their tips, to an angle of 135°. Expose ~8 mm (i.e., remove insulation) at the tip of each electrode. Place the anode in the Ringer's solution dorsal to the embryo, above the lateral half dermomyotome. Insert the cathode via the india ink injection site (see Procedure A) and position it ventrally and contralaterally to the injected somites (see Fig. 2B). Neither electrode should touch the embryo.
5. Apply three 14-V pulses, each of 50 msec duration. Remove the electrodes, seal the eggshell with tape, and reincubate to desired stages.

Caution: *See Appendix 3 for appropriate handling of materials marked with <!>.*

PROCEDURE D

In Vivo Electroporation of Chicken Limb Mesoderm

See Figure 2C. *Note:* For electroporation of the hindlimb mesoderm, the same approach should be applied, using embryos at stages 15–16 of development and injecting DNA into the posterior coelom.

1. Prepare stage-13–14 chicken embryos for electroporation, as described above, and carefully remove the vitelline membrane overlying the forelimb lateral plate mesoderm.
2. Prepare platinum electrodes as described above (see Procedure A), but bend the anode ~3 mm from its tip to an angle of 135°. For the cathode, either use an unbent, point electrode or an electrode bent ~1 mm from its tip, again at an angle of 135°. Expose ~8 mm (i.e., remove insulation) at the tip of each electrode.
3. Microinject plasmid DNA, containing a few crystals of Fast Green, into the coelom, between the somatic and splanchnic mesoderm at forelimb levels.
4. Apply 0.5 ml of Ringer's solution to the dorsal surface of the embryo. Place the cathode in the Ringer's solution, but not in direct contact with embryonic tissue.
5. Insert the anode via the india ink injection site, between the blastoderm and yolk. Position the anode ventrally and parallel to, but not in direct contact with, the lateral plate mesoderm.
6. Apply three 9-V pulses of 50 msec duration.
7. Carefully remove the electrodes and apply 3–4 drops of Ringer's solution over the embryo, reseal the eggshell with tape, and reincubate until the desired stage of development.

PROCEDURE E

Preparation of Mouse Embryo Culture Medium

Chemically defined serum-free medium can be used to culture E10.5 embryos successfully for 24 hours, which negates the need for rat serum collection in some experiments (Moore-Scott et al. 2003). Unfortunately, this medium does not adequately support in vitro culture of embryos younger than E10.5. For these embryos, the following procedure for the preparation of mouse embryo culture medium containing rat serum must be followed.

Preparation of Rat Serum

1. Anesthetize large male rats with halothane, isoflurane, or similar anesthetic<!>, according to established animal-care procedures.

   Note: It is preferable to use a volatile gas that will evaporate quickly from the serum during later heat inactivation.

2. Place the rat supine and perform a laparotomy along the posterior midline or linea alba. Collect blood by inserting a 19-gauge hypodermic needle into the dorsal aorta and applying gentle negative pressure until blood flow ceases.

   Note: Excess pressure will cause the aorta to collapse.

3. Dispense the freshly collected blood <!> (usually 12–15 ml per rat) into 15-ml tubes and centrifuge at 1500*g* for 5–10 min.
4. Transfer the serum to a fresh tube and place it on ice. Squeeze the fibrin clot inside the 15-ml tube to extract the serum trapped inside.
5. Recentrifuge the collected serum at 3000 rpm for 5–10 min.
6. Pool the serum into 50-ml tubes and centrifuge again at 3000 rpm to remove the few remaining blood cells.
7. Dispense the serum into 5-ml aliquots and store at –20°C (up to 12 months) or –80°C (for longer periods).

Preparation of Mouse Embryo Culture Medium

8. Prepare 10 ml of penicillin/streptomycin/L-glutamine by mixing 10,000 units of penicillin, 10,000 µg of streptomycin <!> , and 200 ml of L-glutamine.
9. Dispense the mix into 10-ml aliquots. Add one aliquot to 500 ml of DMEM and store the remaining aliquots at –20°C.
10. Prepare 15-ml aliquots of DMEM+penicillin/streptomycin/L-glutamine. Store these at 4°C for a maximum of 4 weeks.
11. Heat-inactivate rat serum at 56°C for 30 min.
12. Complete the culture medium by mixing DMEM and rat serum in desired quantities, as defined by the age of the embryos being cultured (see Table 1).
13. Filter the medium through a 0.45-µm filter and dispense it into 2–3-ml aliquots in glass roller culture bottles (BTC Engineering).

    Note: Generally, 1 ml of culture medium is sufficient to support a single E9.5–10.5 embryo for up to 24 hours. For E7.5 and E8.5 embryos, two embryos can be cultured per 1 ml of culture medium.

14. Equilibrate culture medium in an appropriate oxygenated environment for at least 30 min before commencing embryo culture.

PROCEDURE F

General Preparation of Mouse Embryos

1. Sacrifice pregnant female mice by cervical dislocation or CO_2 inhalation.
2. Lay each mouse supine and wipe abdomen with 70% ethanol <!>. Make an incision along the dorsal midline from the level of the hindlimbs to the rib cage.
3. Pinch the uterus, but not the individual deciduas, and cut away the mesometrium (blood supply to the uterus and embryos).
4. Remove the two uterine horns and transfer into DMEM (containing penicillin/streptomycin/L-glutamine [see above for recipe]) at 37°C.
5. Using scissors, cut the uterus into sections, each containing a single deciduum.
6. With finely sharpened forceps, gently tear away the muscle tissue of the uterine wall to separate the decidua from the uterine tissue.

7. Transfer the decidua to a fresh dish containing DMEM ([see above for recipe]) and remove the maternal decidual tissue, leaving the yolk sac and amnion intact or simply attached, depending on the culture conditions required for a particular embryonic stage (see Table 1).

   Note: For additional details about embryo dissection, see *Manipulating the Mouse Embryo* (Nagy et al. 2003).

8. Stage mouse embryos according to *The Atlas of Mouse Development* (Kaufmann 1992).

   Caution: *See Appendix 3 for appropriate handling of materials marked with <!>.*

PROCEDURE G

In Vivo Electroporation of Mouse Tissue

Electroporation

1. To electroporate mouse embryos prior to culture, place them in a 6-cm petri dish filled with Tyrodes solution (Sigma).
2. Steady the embryo on the bottom of the dish using a pair of forceps, and inject DNA into the amniotic cavity or lumenal space using a pressure aspirator.
3. Withdraw the micropipette and place electrodes on either side of the embryo and pass current, as described in Table 1.
4. Transfer the embryos into equilibrated culture media and allow them to develop to the desired stage in a BTC roller culture chamber at 37°C.

Analysis

5. Establish the transformation efficiency of the electroporation procedure. GFP fluorescence can be observed in whole-mount embryos via a fluorescence dissecting microscope with GFP optics 8–12 hours post-electroporation.

TABLE 1. Mouse embryo staging and electroporation parameters

Age (E)	Culture med.	Gas/Oxygenation	Yolk sac	Electroporation	Distance
7.5[a]	DR50	5%O_2,5%CO_2,90%N_2	intact	26 V, 30 msec, 3P	0.5 cm
7.5–8.5[b]	DR50	5%O_2,5%CO_2,90%N_2	intact	10–15 V, 50 msec, 3-5P	0.5 cm
8.5–9.5[c]	DR50	5%O_2,5%CO_2,90%N_2	intact	15–20 V, 50 msec, 3-5P	0.5–1.0 cm
9.5–10.5[c]	DR75	20%O_2,5%CO_2,75%N_2	open	15–20 V, 50 msec, 5P	1.0–1.5 cm
10.5–11.5[c]	DR100	65%O_2,5%CO_2,35%N_2	open	15–20 V, 50 msec, 5P	1.5–2.0 cm
11.5[d]	Exo utero	—	intact	7–10 V, 100 msec, 3P	1 cm
11.5[e,f]	In utero	—	intact	20 V	5.8 cm
13.5[e,f]	In utero	—	intact	40 V	8.2 cm
15.5[e,f]	In utero	—	intact	60 V	9.3 cm

(D) DMEM supplemented with L-glutamine and penicillin/streptomycin. (R50, R75,R100) Rat serum percentage composition of culture media. (P) pulses.

[a]Mellitzer et al. (2002).
[b]Davidson et al. (2003).
[c]Osumi and Inoue (2001).
[d]Fukuchi-Shimogori and Grove (2001).
[e]Saito and Nakatsuji (2001).
[f]Takahashi et al. (2002).

Note: These observations can be extremely valuable for chicken embryos, as they save time in embryo collection.

6. Discard embryos with poor transformation rates and reincubate any embryos that require further growth before harvesting.
7. Assess the cell death rate in the target tissue. The mortality rate caused by the construct can be established easily using nile blue or acridine orange <!> staining or via tunel staining (Boehringer Mannheim, Promega; Trainor et al. 2002).
8. If desired, prepare embryos for sectioning; e.g., cryosectioning, paraffin sectioning, or vibratome sectioning (Nagy et al. 2003). Because fixation typically dampens or destroys GFP fluorescence, the addition of GFP antibody labeling (Boehringer Mannheim, Clontech) may be required.
9. Screen embryos by in situ hybridization or antibody labeling either in sections or whole mounts. This will verify transgene expression patterns in the embryo. Alternatively, perform western blot analysis on transfected tissues.
10. Use sectioned material for imaging using an upright or inverted fluorescence microscope or confocal microscopy.

Caution: *See Appendix 3 for appropriate handling of materials marked with <!>.*

The voltages listed in Table 1 for specific embryo stages are designed as a guide and provide a balance between high efficiency of electroporation (for DNA or RNA) and low levels of cell death. One of the principal determinants of electroporation voltage is the physical distance separating the electrodes, which equates to resistance. Normally this is in the 0.5–1-cm range. However, the electrodes can be spaced much farther apart, e.g., in a well-type system. In these instances, a correspondingly higher voltage is required for successful electroporation. Current is inversely proportional to distance, and a minimal electric field threshold is required for successful electroporation of DNA.

IN UTERO ELECTROPORATION

Until recently, one of the limitations of mouse whole-embryo culture was that embryos could only be cultured in vitro for 2–3 days maximum. When electroporation is performed in E7–10 embryos, in vitro culture does not permit subsequent analyses to be carried out during the later phases of fetal development or even postnatally, to assess long-term cell fate and differentiation. Recently, however, this obstacle has been overcome by both exo utero (Fukuchi-Shomogori and Grove 2001) and ultrasound-guided in utero (Saito and Nakatsuji 2001) electroporation. This in vivo approach has successfully and efficiently transformed embryos with DNA fragments up to 11 kb in length.

TROUBLESHOOTING

Problem

Poor electroporation efficiency; i.e., no or few cells are transformed and gene expression is transient.

Potential Solutions

1. Replace electrodes if they are more than 3 months old.
2. Adjust electrode placement to target a larger number of cells.
3. Test a different promoter.
4. Consider whether the normal regulatory controls of the embryo might influence the expression of your gene sequence of interest.

5. Check the concentration of the plasmid solution (it must be at least 2 μg/μl).
6. Test mammalian constructs for expression in chicken embryo or cell culture model.
7. Increase the voltage slightly and monitor cell to enhance transformation efficiency.
8. Increase the volume of DNA injected.
9. Apply Ringer's solution to enhance current conduction.
10. Decrease the time interval between DNA microinjection and electroporation.

Problem

Embryo survival is compromised.

Potential Solutions

1. Assess cell death using nile blue, acridine orange <!> , or tunel staining, to assay DNA fragmentation (Trainor et al. 2002). If elevated cell death is detected, reduce voltage accordingly.
2. Replace electrodes if they are more than 3–4 months old.
3. Determine whether electrodes are touching embryonic tissues during the electroporation and thereby damaging the embryo.
4. Confirm that DNA is aqueous solution or PBS, *not* TE. Prepare fresh DNA if impurities or contaminants are suspected.
5. Fast Green can interfere with survival of primitive-streak-stage embryos. Try Phenol Red<!> (0.1% final concentration) as an alternative tracer.
6. Reduce the voltage, pulse width, and/or number of pulses.

EXAMPLE OF APPLICATION

In vivo electroporation works well for targeting gene expression to particular cell types in mouse and chicken embryos (Fig. 3). EGFP expression can be observed in the anterior neural plate, migrating neural crest, and the anterior spinal cord in an E8.5 mouse embryo cultured for 24 hours in vitro after electroporation (Fig. 2A,B).

In chicken, approximately 60–85% of the cells in one half of the neural tube can be transfected to express the gene of interest and EGFP using in vivo electroporation (Fig. 3C). Transfections can be targeted more precisely to dorsal cell types, including neural crest (Fig. 3D). Smaller numbers of cells can be transfected by increasing the distance between electrodes and reducing the voltage.

ADVANTAGES AND LIMITATIONS

Advantages of In Vivo Electroporation

In vivo electroporation has, thus far, been utilized primarily for gain-of-function experiments for which it provides a systems approach, to test gene function within the native context of the embryo. It provides a simple and rapid assessment of gene function and facilitates targeted gene manipulation in specific tissues, leaving neighboring tissues unperturbed. Bicistronic plasmids can be used to generate GFP expression in transformed cells just 3 hours posttransformation. Antibody staining can then be used to determine the presence of the protein of interest. Typically, retroviral gene expression can be observed 18–24 hours posttransfection.

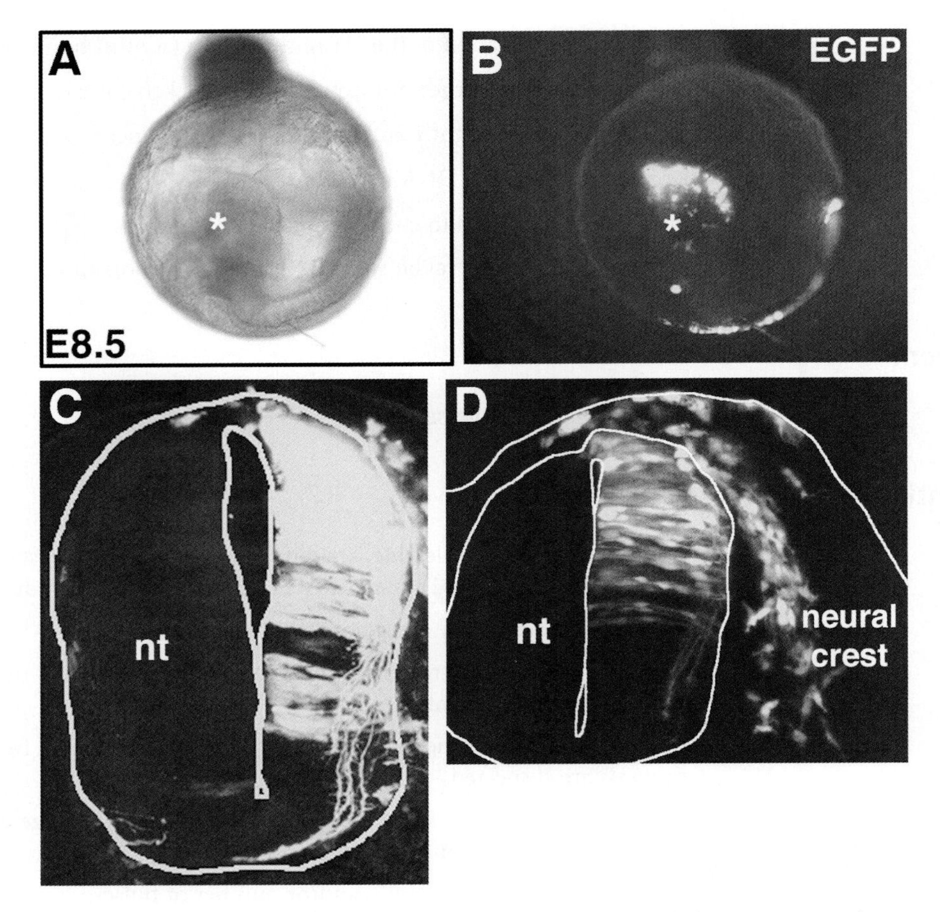

FIGURE 3. EGFP is strongly expressed in mouse and chicken embryos following in vivo electroporation. (*A*) Bright-field image of E8.5 mouse embryo in its yolk sac, cultured for 24 hours in vitro after electroporation. Asterisk indicates forming head. (*B*) Same embryo as in *A*; EGFP expression is apparent in the anterior neural plate, migrating neural crest, and the anterior spinal cord. (*C*) Transverse section showing that many cells in one half of the neural tube (nt) express EGFP, 24 hours after electroporation. (*D*) Transverse section showing that dorsal neural tube (nt) cells and migrating neural crest express EGFP, 24 hours after electroporation targeted to the dorsal quadrant of the neural tube.

The technique is applicable to a wide range of species (chicken, mouse, ascidians, plants, zebrafish, *Drosophila*) and tissue types, and although transgene expression is generally transient (DNA remains episomal), some transgene expression can last for 10 days or more.

In vivo electroporation offers a solution to the difficulties associated with embryonic-lethal mutations in null mutant mice, in the absence of conditional mutations. Loss-of-function experiments are now possible using RNAi, morpholinos, and constructs encoding dominant-negative receptors. These approaches are beginning to be more commonly used (Pekarik et al. 2003; Chesnutt and Niswander 2004; Eberhart et al. 2004; Gerlach-Bank et al. 2004).

Limitations of In Vivo Electroporation

There are several limitations associated with in vivo electroporation. Transformations are generally transient (see above), and results are not always reproducible from embryo to embryo. It is also more difficult to transform tissues that lack a lumenal surface, and targeting later developmental events can be problematic because of alterations in the structure, growth, and accessibility of tissues. One alternative to this later problem is to utilize explant culture methods.

In vivo electroporation of primitive streak/gastrulation-stage chicken embryos results in increased cell death. As an alternative, perform electroporation in modified New culture (Chapman et al. 2001).

ALTERNATIVES

One alternative to the general promoters/enhancers that have been used so far is to take advantage of cell-specific promoters. For example, the HB9 promoter targets gene expression to presumptive motor neuron progenitors and postmitotic motor neurons (Arber et al. 1999; Thaler et al. 1999) and has been used successfully to target gene expression to motor neurons using electroporation. Hox gene-specific enhancers/promoters have also been used successfully to target gene expression to particular rhombomeres (for review, see Itaskai et al. 1999). These agents have the potential to provide enhanced specificity.

OTHER APPLICATIONS

Explant/Slice Cultures

These hybrid paradigms that combine in vivo and in vitro environments should be ideal systems in which to utilize in vivo electroporation. They offer the opportunity to manipulate gene expression in particular cells at later stages of development, leaving other tissues unperturbed (Pu and Young 1990).

Invertebrates and Other Embryos

In vivo electroporation should prove to be a useful strategy to modify gene expression and function in embryos that lack genetics, including several invertebrate embryos. Investigators must modify the parameters discussed above, to enhance the success of this approach in their model system.

ACKNOWLEDGMENTS

This work is supported by grants from the National Institute of Mental Health and the Muscular Dystrophy Association to C.E.K. and by the Stowers Institute for Medical Research and a March of Dimes Basil O'Connor Scholar Research Award to P.A.T.

REFERENCES

Arber S., Han B., Mendelsohn M., Smith M., Jessell T.M., and Sockanathan S. 1999. Requirement for the homeobox gene Hb9 in the consolidation of motor neuron identity. *Neuron* **23:** 659–674.

Atkins R.L., Wang D., and Burke R.D. 2000. Localized electroporation: A method for targeting expression of genes in avian embryos. *BioTechniques* **28:** 94–96, 98, 100.

Bellairs R. and Osmond M. 1998. *The atlas of chick development*. Academic Press, San Diego, California.

Bronner-Fraser M., ed. 1996. *Methods in avian embryology*. Academic Press, San Diego, California.

Buono R.J. and Linser P.J. 1992. Transient expression of RSVCAT in transgenic zebrafish made by electroporation. *Mol. Mar. Biol. Biotechnol.* **1:** 271–275.

Chapman S.C., Collignon J., Schoenwolf G.C., and Lumsden A. 2001. Improved method for chick whole-embryo culture using a filter paper carrier. *Dev. Dyn.* **220:** 284–289.

Chesnutt C. and Niswander L. 2004. Plasmid-based short-hairpin RNA interference in the chicken embryo. *Genesis* **39:** 73–78.

Christ B. and Ordahl C.P. 1995. Early stages of chick somite development. *Anat. Embryol.* **191:** 381–396.

Corbo J.C., Levine M., and Zeller R.W. 1997. Characterization of a notochord-specific enhance from the *Brachyury* promoter region of the ascidian, *Ciona intestinalis. Development* **124:** 589–602.

Corbo J.C., Fujiwara S., Levine M., and Di Gregorio M. 1998. Suppressor of hairless activates *Brachyury* expression in the *Ciona* embryo. *Dev. Biol.* **203:** 358–368.

Davidson B., Tsang T., Khoo P.-L., Gad J., and Tam P. 2003. Introduction of cell markers into germ layer tissues of the mouse gastrula by whole embryo electroporation. *Genesis* **35:** 57–62.

Eberhart J., Swartz M.E., Koblar S.A., Pasquale E.B., and Krull C.E. 2002. EphA4 constitutes a population-specific guidance cue for motor neurons. *Dev. Biol.* **247:** 89–101.

Eberhart J., Barr J., O'Connell S., Flagg A., Swartz M.E., Cramer K., Tosney K.W., Pasquale E.B., and Krull C.E. 2004. Ephrin-A5 exerts positive or inhibitory effects on distinct subsets of EphA4-positive neurons. *J. Neurosci.* **24:** 1070–1078.

Eide F.F., Eisenberg S.R., and Sanders T.A. 2000. Electroporation-mediated gene transfer in free-swimming embryonic *Xenopus laevis. FEBS Lett.* **486:** 29–32.

Fraser S.E. 1996. Iontophoretic dye labeling of embryonic cells. In *Methods in avian embryology.* Academic Press, San Diego, California. pp. 147–160.

Fukuchi-Shimogori T. and Grove E.A. 2001. Neocortex patterning by the secreted signaling molecule FGF8. *Science* **294:** 1071–1074.

Fukuda K., Sakamoto N., Narita T., Saitoh K., Kameda T., Iba H., and Yasugi S. 2000. Application of efficient and specific gene transfer systems and organ culture techniques for the elucidation of mechanisms of epithelial-mesenchymal interaction in the developing gut. *Dev. Growth Differ.* **42:** 207–211.

Gerlach-Bank L.M., Cleveland A.R., and Barald K.F. 2004. DAN direct endolymphatic sac and duct outgrowth in the avian inner ear. *Dev. Dyn.* **229:** 219–230.

Haas K., Sin W.-C., Javaherian A., and Cline H.T. 2001. Single-cell electroporation for in vivo neuronal gene expression. *Neuron* **29:** 583–591.

Hamburger V. and Hamilton H.L. 1951. A series of normal stages in the development of the chick embryo. *J. Morphol.* **88:** 49–92.

Itasaki N., Bel-Vialar S., and Krumlauf R. 1999. Shocking developments in chick embryology: Electroporation and in ovo gene expression. *Nat. Cell Biol.* **1:** E203–207.

Kamdar K.P., Wagner T.N., and Finnerty V. 1995. Electroporation of *Drosophila* embryos. *Methods Mol. Biol.* **48:** 239–243.

Kaufman M.H. 1992. *The atlas of mouse development*, Elsevier, Amsterdam, The Netherlands.

Krull C.E. and Kulesa P. 1998. Embryonic explant and slice preparations for studies of cell migration and axon guidance. In *Cellular and molecular procedures in developmental biology* (ed. F. de Pablo et al.). Academic Press, San Diego, California. pp. 145–159.

Kos R., Reedy M.V., Johnson R.L., and Erickson C.A. 2001. The winged-helix transcription factor FoxD3 is important for establishing the neural crest lineage and repressing melanogenesis in avian embryos. *Development* **128:** 1467–1479.

Kos R., Tucker R.P., Hall R., Duong T.D., and Erickson C.A. 2003. Methods for introducing morpholinos into the chicken embryo. *Dev. Dyn.* **226:** 470–477.

Locascio A., Aniello F., Amoroso A., manzanares M., Krumlauf R., and Branno M. 1999. Patterning the ascidian nervous system: Structure, expression and transgenic analysis of the CiHox3 gene. *Development* **126:** 4737–4748.

Mellitzer G., Hallonet M., Chen L., and Ang S.-L. 2002. Spatial and temporal knockdown of gene expression by electroporation of double stranded RNA and morpholinos into early post implantation mouse embryos. *Mech. Dev.* **118:** 57–63.

Momose T., Tonegawa A., Takeuchi J., Ogawa H., Umesono K., and Yasuda K. 1999. Efficient targeting of gene expression in chick embryos by microelectroporation. *Dev. Growth Differ.* **41:** 335–344.

Moore-Scott B.A., Gordon J., Blackburn C.C., Condie B.G., and Manley N.R. 2003. New serum-free in vitro culture technique for midgestation mouse embryos. *Genesis* **35:** 164–168.

Muller F., Lele Z., Varadi L., Menczel L., and Orban L. 1993. Efficient transient expression system based on square pulse electroporation and in vivo luciferase assay of fertilized fish eggs. *FEBS Lett.* **324:** 27–32.

Muramatsu T., Mizutani Y., and Okumura J. 1996. Live detection of the firefly luciferase gene expression by bioluminescence in incubating chicken embryos. *Anim. Sci. Technol.* **67:** 906–909.

Muramatsu T., Mizutani Y., Ohmori Y., and Okumura J. 1997. Comparison of three nonviral transfection methods for foreign gene expression in early chicken embryos in ovo. *Biochem. Biophys. Res. Commun.* **230:** 376–380.

Nagy A., Gertsenstein M., Vintersten K., and Behringer R. 2003. *Manipulating the mouse embryo,* 3rd edition. Cold Spring Harbor Laboratory Press, Cold Spring Harbor, New York.

Neumann E., Schaefer-Ridder M., Wang Y., and Hofschneider P.H. 1982. Gene transfer into mouse lyoma cells by electroporation in high electric fields. *EMBO J.* **1:** 841–845.

Osumi N. and Inoue T. 2001. Gene transfer into cultured mammalian embryos by electroporation. *Methods* **24:** 35–42.

Pekarik V., Bourikas D., Miglino N., Joset P., Preiswerk S., and Stoeckli E.T. 2003. Screening for gene function in chicken embryo using RNAi and electroporation. *Nat. Biotechol.* **21:** 93–96.

Potter H. 1988. Electroporation in biology: Methods, applications, and instrumentation. *Anal. Biochem.* **174:** 361–373.

Pu H.F. and Young A.P. 1990. Glucocorticoid-inducible expression of a glutamine synthetase-CAT-encoding fusion plasmid after transfection of intact chicken retinal explant cultures. *Gene* **89:** 259–263.

Saito T. and Nakatsuji N. 2001. Efficient gene transfer into the embryonic mouse brain using in vivo electroporation. *Dev. Biol.* **240:** 237–246.

Shillito R., Saul M., Paszkowski J., Muller M., and Potrykus I. 1985. High efficiency direct gene transfer to plants. *BioTechnology* **3:** 1099–1103.

Swartz M., Eberhart J., Mastick G., and Krull C.E. 2001a. Sparking new frontiers: Using in vivo electroporation for genetic manipulations. *Dev. Biol.* **233:** 13–21.

Swartz M.E., Eberhart J., Pasquale E.B., and Krull C.E. 2001b. EphA4/ephrin-A5 interactions in muscle precursor cell migration in the avian forelimb. *Development* **128:** 4669–4680.

Takahashi M., Sato K., Nomura T., and Osumi N. 2002. Manipulating gene expressions by electroporation in the developing brain of mammalian embryos. *Differentiation* **70:** 155–162.

Thaler J., Harrison K., Sharma K., Lettieri K., Kehrl J., and Pfaff S.L. 1999. Active suppression of interneuron programs within developing motor neurons revealed by analysis of homeodomain factor HB9. *Neuron* **23:** 675–687.

Trainor P.A., Sobiezczuk D., Wilkinson D., and Krumlauf R. 2002. Signalling between the hindbrain and paraxial tissues dictates neural crest migration pathways. *Development* **129:** 433–442.

Yasuda K., Momose T., and Takahashi Y. 2000. Applications of microelectroporation for studies of chick embryogenesis. *Dev. Growth Differ.* **42:** 203–206.

CHAPTER 12

A Practical Guide: In Vivo Electroporation of Neurons

Stephen R. Price

Department of Anatomy and Developmental Biology, University College London, London, WC1E 6BT, United Kingdom

This chapter describes an approach to drive in vivo gene expression in neurons through the electroporation of DNA constructs during early chick development. Electroporation is a method of physically introducing DNA constructs into cells through the application of an electric field. This simple method is important as it allows the ectopic expression of transgenes with relative ease. Most neurons are accessible to ectopic gene expression by electroporation. Examples span the rostrocaudal extent of the central nervous system from olfactory epithelium (Renzi et al. 2000) to spinal cord neurons (Price et al. 2002). Furthermore, this technique has been used in both chick and mouse embryos (Osumi and Inoue 2001). Gene expression in restricted subclasses of neurons is also possible with the aid of specific promoter elements (Itasaki et al. 1999). For example, the mouse HB9 promoter drives expression selectively in motor neurons (Arber et al. 1999; William et al. 2003) and the math1, neurogenin1, and neurogenin2 promoters drive expression in distinct overlapping domains of the dorsoventral axis of the spinal cord (Timmer et al. 2001). It should be noted, however, that other methods exist for more focal electroporation of DNA constructs by the use of microcapillary electroporation into small numbers of cells (Haas et al. 2002) and for the delivery of DNA constructs by lipofection (Holt et al. 1990). These methods are, however, beyond the scope of this text.

MATERIALS

This technique requires reagents and equipment for incubating chicken embryos and for microinjection under a dissecting microscope.

The DNA used for transformation should be circular and dissolved at a concentration of 1–15 mg/ml in phosphate-buffered saline (PBS) (0.1 M sodium phosphate buffer at pH 7.4, 0.15 M NaCl), containing 0.1% Fast Green dye<!> (Sigma). The DNA encodes the chosen transgene under the control of an appropriate promoter element. The promoter may be one that drives pan-neuronal expression, for example, the β-actin promoter, or one that drives expression in a specific neuronal subtype.

The electroporator (e.g., Electrosquareporator ECM 830 from BTX, Genetronics) uses platinum/iridium (80/20) electrodes (250 μm diameter) (e.g., see Frederick Haer, UEPMGB-VNNNND). The electrodes are bent into an L-shape with the foot of the L about 1 cm in length. Any type of electrode holder may be used; this protocol was designed using a home-made apparatus with vertical displacement capability via a screw mechanism.

The borosilicate capillary tubes that will be used to introduce the DNA solution into the embryo should be tapered using a microcapillary pipette puller to a final diameter of ~200–500 µm.

Caution: *See Appendix 3 for appropriate handling of materials marked with <!>.*

PROCEDURE

In Vivo Electroporation of Neurons

1. Incubate fertilized chicken eggs on their sides (with the long axis of the egg horizontal) in a forced-draught incubator (39°C and 70% relative humidity) until the embryos are between stages 10 and 20 (Hamburger and Hamilton 1951).
2. Fill a tapered microcapillary tube with ~1 µl of the DNA solution. Prepare a second tapered capillary containing ~20 µl of Pelikan Fount black #17 india ink diluted 1:10 in PBS.
3. Remove an egg from the incubator and place it under a dissecting microscope.
4. Remove 3 ml of albumen from the blunt end of the egg using a 3-ml Luer-lok syringe fitted with a 21-gauge needle.
5. Use blunt-ended forceps to remove ~1–4 cm^2 of eggshell from the top of the egg to allow observation of the developing embryo. Tilt the egg until there is easy access to the rostrocaudal region of the embryo.
6. Pressure-inject the india ink into the albumen, directly under the embryo.

   Note: This facilitates observation of the developing embryo but is not necessary if the embryo is older than stage 12/13.
7. Pierce the membranes of the egg with the DNA-containing capillary tube, 1–2 cm away from the embryo, being careful not to interfere with the blood vessels.
8. Approaching either the rostral or caudal end of the embryo, insert the DNA-containing microcapillary assembly into the lumen of the CNS (central nervous system).

   Note: At early stages (stage 10–13), the capillary can be inserted directly into the forebrain vesicles. At later stages, when the head has turned, the capillary can be inserted into the cervical spinal cord just caudal to the hindbrain.
9. Push the microcapillary to the rostrocaudal level of the CNS in which gene expression is desired.

   Note: It is possible to push the capillary through the entire length of the spinal cord without causing significant damage to the embryo.
10. Slowly withdraw the microcapillary while pressure-injecting ~100 nl of the DNA solution into the lumen of the CNS.

    Note: The Fast Green dye in the DNA solution aids visualization of the DNA as it is forced into the lumen.
11. Move the egg to the electrode/electroporator assembly. Place the electrodes at the region of the CNS containing the DNA construct, parallel to the rostrocaudal axis of the CNS with each foot of the electrode as close to the embryo as possible but without actually touching it.

    Note: From stage 13 onward, the electrodes should not be placed adjacent to the heart—the electrical pulse will stop the heart and the embryo will die.
12. Transform the embryo using the electroporator (30-V pulse, 50-msec duration, 1-sec interval, 5 pulses).

Note: It may be possible to see the Fast Green dye move from the lumen to one side of the CNS.

13. Remove the electrodes immediately after the last pulse. Use water to clean the electrodes.

    Note: If the electrodes become coated in albumen, it is possible to clean them effectively by pulsing in a PBS solution.

14. Remove an additional 3 ml of albumen from the egg. Cover the hole in the egg with electrical tape and return it to the incubator until the desired stage.

    Note: This may be anything from an additional 6 hours to 7 days or more.

EXAMPLES OF APPLICATION

Figure 1 shows two examples of electroporation in neurons. Figure 1A shows the use of a chicken β-actin/rabbit β-globin hybrid promoter (pCAGGS; Niwa et al. 1991) to drive expression of nuclear-localized β-galactosidase. Figure 1B shows the use of the mouse HB9 promoter to drive expression of nuclear-localized β-galactosidase specifically in motor neurons of the ventral horn of the spinal cord (Arber et al. 1999; William et al. 2003).

ADVANTAGES AND LIMITATIONS

The main advantage of this technique is the speed with which the effects of ectopic gene expression can be assessed (Stern 2002). It has major time advantages over the generation of transgenic animals. It also allows the cell autonomy of the newly acquired trait to be assessed as the transgenes are misexpressed in a mosaic on one side of the CNS. Those neurons that did not acquire the transgene, either on the electroporated side of the CNS or on the contralateral side, provide ideal internal controls for the effect of the misexpression. The main disadvantage is that the technique does not provide a stable line of embryos with the given perturbation; each embryo generated is unique.

ACKNOWLEDGMENTS

The support and guidance of Thomas Jessell is gratefully acknowledged. Additionally, the panel in Figure 1B was modified from one kindly provided by C. William and T. Jessell.

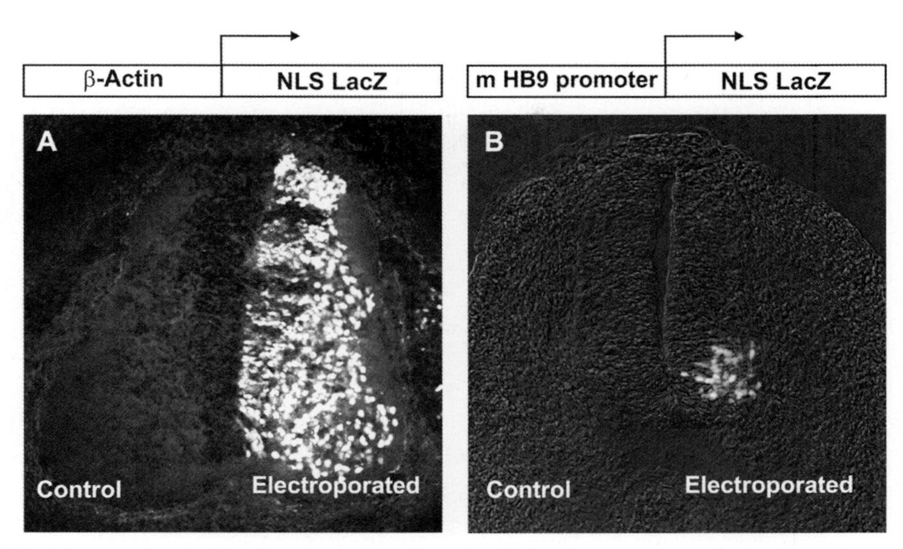

FIGURE 1. (*A*) Use of a chicken β-actin/rabbit β-globin hybrid promoter to drive expression of nuclear-localized β-galactosidase (NLS LacZ). (*B*) Use of the mouse HB9 promoter to drive expression of NLS LacZ. See text for details (see Acknowledgments).

REFERENCES

Arber S., Han B., Mendelsohn M., Smith M., Jessell T.M., and Sockanathan S. 1999. Requirement for the homeobox gene *Hb9* in the consolidation of motor neuron identity. *Neuron* **23:** 659–674.

Haas K., Jensen K., Sin W. C., Foa L., and Cline H. T. 2002. Targeted electroporation in *Xenopus* tadpoles in vivo—From single cells to the entire brain. *Differentiation* **70:** 148–154.

Hamburger V. and Hamilton H. 1951. A series of normal stages in the development of the chick embryo. *J. Morphol.* **160:** 535–546.

Holt C.E., Garlick N., and Cornel E. 1990. Lipofection of cDNAs in the embryonic vertebrate central nervous system. *Neuron* **4:** 203–214.

Itasaki N., Bel-Vialar S., and Krumlauf R. 1999. "Shocking" developments in chick embryology: Electroporation and in ovo gene expression. *Nat. Cell Biol.* **1:** E203–207.

Niwa H., Yamamura K., and Miyazaki J. 1991. Efficient selection for high-expression transfectants with a novel eukaryotic vector. *Gene* **108:** 193–199.

Osumi N. and Inoue T. 2001 Gene transfer into cultured mammalian embryos by electroporation. *Methods* **24:** 35–42.

Price S.R., DeMarco-Garcia N.V., Ranscht B., and Jessell T.M. 2002. Regulation of motor neuron pool sorting by differential expression of type II cadherins. *Cell* **109:** 205–216.

Renzi M.J., Wexler T.L., and Raper J.A. 2000. Olfactory sensory axons expressing a dominant-negative semaphorin receptor enter the CNS early and overshoot their target. *Neuron* **28:** 437–447.

Stern C.D. 2002. Induction and initial patterning of the nervous system—The chick embryo enters the scene. *Curr. Opin. Genet. Dev.* **12:** 447–451.

Timmer J., Johnson J., and Niswander L. 2001. The use of in ovo electroporation for the rapid analysis of neural-specific murine enhancers. *Genesis* **29:** 123–132.

William C., Tanabe Y., and Jessell T.M. 2003. Regulation of motor neuron subtype identity by repressor activity of Mnx class homeodomain proteins. *Development* **130:** 1523–1536.

CHAPTER 13

A Practical Guide: Single-Neuron Labeling Using Genetic Methods

Liqun Luo

Department of Biological Sciences, Neurosciences Program, Stanford University, Stanford, California 94305-5020

Our brain is composed of hundreds of billions of neurons, each of which has an elaborate shape and a complex pattern of connections. To untangle this complexity, it is often useful to visualize one neuron at a time.

There are many ways of visualizing individual neurons within their native environment. The classic method is Golgi staining, developed by Camillo Golgi (Golgi 1873), exploited to the fullest extent by Ramon y Cajal over a century ago (Cajal 1911), and still in wide use today. It is also possible to label single neurons in brain slices by filling them with a sharp or patch electrode, during physiology experiments (see Chapter 33), or by biolistic transfection, in which gold particles containing dye or expression constructs are "shot" into neurons in brain slices (see Chapter 14), or by single-cell electroporation (see Chapters 11, 12, and 23).

This chapter discusses single-neuron labeling using genetic methods. These methods can be divided into the following three broad categories: (1) using highly specific promoters (Fig. 1A) (see, e.g., Grueber et al. 2002), or chance insertions of transgenes (Fig. 1B) (e.g., Feng et al. 2000), to limit marker expression to a very small subset of isolated neurons; (2) using site-specific recombination within the same piece of DNA, such as "flip-out" in *Drosophila* (Fig.1C) (see, e.g., Basler and Struhl 1994; Wong et al. 2002), or Cre/loxP in mice (e.g., Huang et al. 2002), to limit marker expression in isolated neurons; and (3) using mitotic recombination to couple marker expression to cell division (Fig. 1D). This chapter focuses primarily on one specific example of this last category, the "MARCM system" (*M*osaic *A*nalysis with a *R*epressible *C*ell *M*arker) that was developed in *Drosophila* (Lee and Luo 1999), which, in principle, could also work in other genetic model organisms such as *Caenorhabditis elegans*, zebrafish, or mice. Examples of the other two categories are also given, and the pros and cons of these different methods are discussed.

Labeling single neurons in their native environment can be used for the following applications:

1. *To trace the axonal projection and dendritic elaboration patterns of individual neurons.* It is possible to characterize different neuronal types and their connection patterns within a specific brain region, or between different regions, to discern information flow and the logic of neural circuit organization. Although Ramon y Cajal has already done much of the work using the Golgi method (Cajal 1911), the complete axonal arbors are never recovered in Golgi stains. In addition, genetic tracing methods could allow those singly labeled neurons to be confined to specific neuronal types by using genetically defined expression of the label. Methods such as MARCM also allow the tracing of a neuron's lineage along with the visualization of its projections.

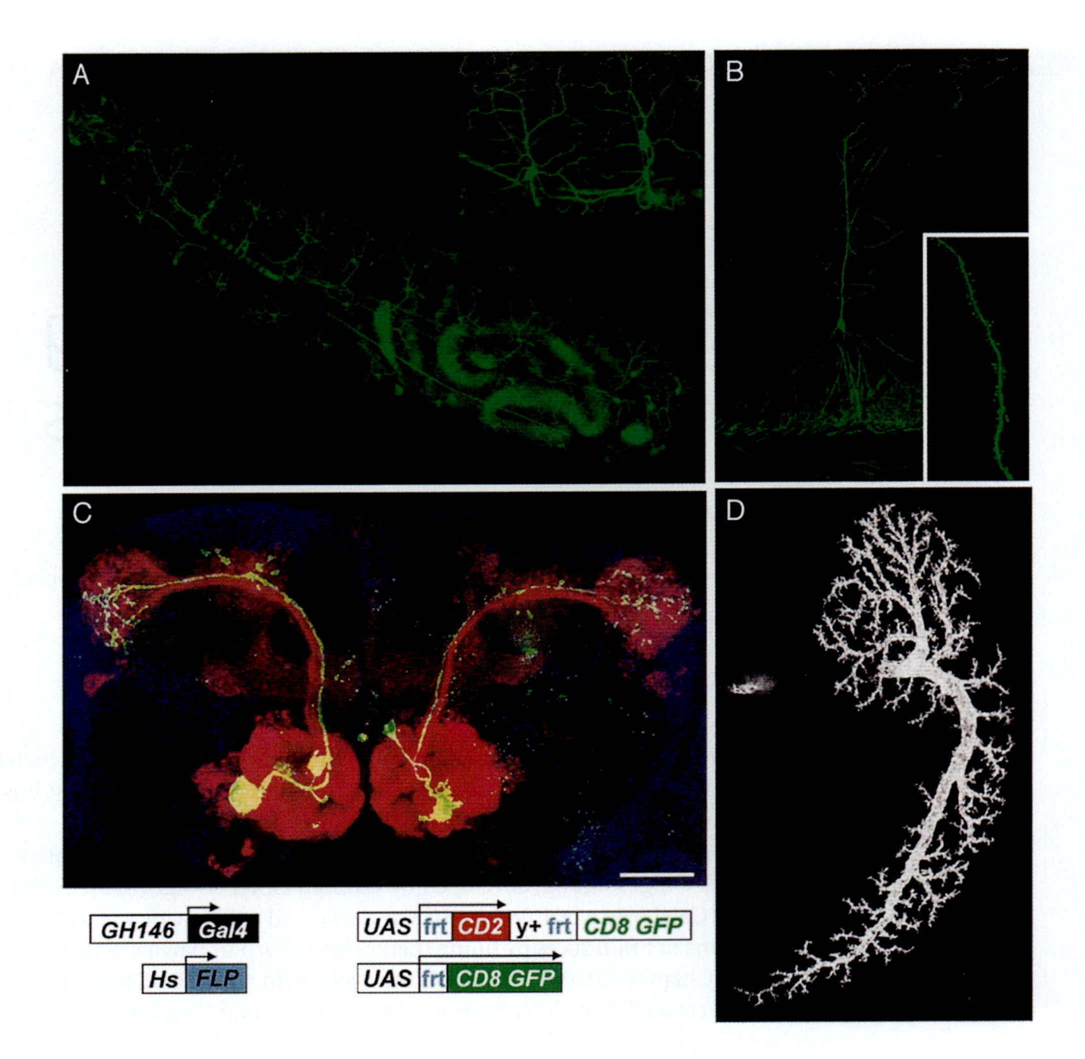

FIGURE 1. Examples of single-neuron labeling with different genetic methods. (*A*) A *pickpocket* promoter-EGFP transgene allows GFP expression in isolated type IV da sensory neurons throughout *Drosophila* larvae. Inset shows a higher magnification of two adjacent type IV da neurons. (*B*) A chance insertion of the Thy-1 promoter-GFP transgene in mouse (line GFP-M) allows high-level marker expression in isolated pyramidal neurons. Inset shows higher magnification of one dendritic segment; individual dendritic spines are clearly labeled. (*C*) In this example of the "flip-out" strategy, many olfactory projection neurons express a CD2 marker (*red*); heat shock induction of FLP recombinase results in deletion of intervening sequences between the two FRT sites, allowing expression of CD8-GFP (*green*) in isolated neurons. (*D*) The dendritic tree of a single VS1 giant cell in the *Drosophila* lobula plate, a high-order visual processing neuron, is visualized by the MARCM method. (*A*, Adapted, with permission, from Grueber et al. 2003 [©Elsevier]; *B*, adapted, with permission, from Feng et al. 2000 [©Elsevier]; *C*, adapted, with permission, from Wong et al. 2002 [©Elsevier]; *D*, adapted, with permission, from Scott et al. 2002.)

2. *To study molecular mechanisms of dendritic and axonal development and plasticity with high anatomical resolution.* Many neurons have complex dendritic branching patterns and axonal projections. To study the intrinsic and environmental regulation of neuronal morphogenesis, it is useful to study the consequences of genetic manipulation or extracellular stimulation with single-neuron resolution. Some of the genetic methods afford the possibility to disrupt endogenous genes or express transgenes only in labeled and isolated neurons, thus allowing analysis of specific functions of pleiotropic genes.

3. *To study the physiological functions of identified neurons in brain slices or in vivo.* Single neurons in wild-type or mutant animals labeled with vital dyes such as GFP could be used for electrophysiological recording or optical imaging in brain slices or in the intact animal. These will allow the characterization of neural circuit function in wild-type animals and genetic analysis of neural circuit formation and function using mutants. It is also possible to label single neurons with other markers, such as genetically encoded Ca^{2+} or voltage indicators, so that physiological properties of uniquely identifiable wild-type or mutant neurons can be assayed in behaving animals using state-of-the-art imaging techniques described elsewhere in this book.

PRINCIPLE OF THE MARCM SYSTEM IN *DROSOPHILA*

The principle of MARCM (Lee and Luo 1999, 2001) is schematically shown in Figure 2A. In brief, MARCM couples cell division with the possible expression of a marker. The marker is driven from a UAS (*U*pstream *A*ctivator *S*equence from GAL4)-promoter, and should be turned on when the yeast transcription factor GAL4 is present in the same cell. However, when the yeast GAL80 protein is also present, it effectively inhibits GAL4 activity in *Drosophila* cells. The *GAL80* transgene can be removed from one of the two daughter cells by a mitotic recombination event in the G_2 stage, prior to cell division; mitotic recombination is greatly facilitated by the action of site-specific FLP recombinase (FLP) activity on a pair of FRT (*F*LP *R*ecombinase *T*arget) sites present at exactly the same locus in a homologous chromosome pair. In the case of a fruitful segregation after mitotic recombination, one of the progeny cells is relieved of GAL80 inhibition and is consequently labeled with the marker under the control of UAS. Finally, if the *GAL80* transgene and a mutation of interest are placed distal to the paired FRT sites on homologous chromosomes, the loss of GAL80 is coupled with the generation of a homozygous mutant cell, and all its subsequent progeny. Thus, mutant cells can be selectively labeled in a mosaic animal.

It is possible to achieve single-cell labeling by activating FLP expression during the last cell division which generates the neuron, as has been done using an eye-specific promoter that targets FLP expression in the last division that generates three types of photoreceptors (Luo et al. 1999; Lee et al. 2001). A more broadly applicable way to label single neurons is to use transgenic flies in which FLP expression is driven by the heat-shock promoter (hs-FLP), allowing high-level ubiquitous expression of FLP in response to a brief heat shock (10 min to 1.5 hours at 37°C). Most *Drosophila* CNS neurons are generated in a typical asymmetric division scheme, as shown in Figure 2B; thus, a pulse of heat at a defined developmental time could label single-cell clones born at the time of heat shock (Fig. 2C, left). The same strategy can also generate a labeled neuroblast clone that contains all progeny born from the neuroblast (neural stem cell) following the heat shock (Fig. 2B; Fig. 2C, right), which is very useful for lineage analysis.

CONSTRUCTING FLIES FOR MARCM ANALYSIS

A minimum of six transgenes must be introduced into the same fly (MARCM fly) to generate labeled MARCM clones:

1. *A transgene encoding the FLP recombinase.* As discussed above, the heat-shock promoter or a tissue-specific promoter can be used to drive the expression of the FLP recombinase.

2, 3. *A pair of transgenes encoding the short FLP recognition target (FRT).* These are well established (Golic and Lindquist 1989; Xu and Rubin 1993; Chou and Perrimon 1996) and are widely used in *Drosophila.*

4. *A GAL4 line that allows marker expression in the tissue of interest.* Since the original description of the GAL4-UAS binary expression system (Brand and Perrimon 1993), thousands of well-characterized GAL4 lines have been generated, a true asset to the *Drosophila* community.

5. *A tubP-GAL80 transgene that ubiquitously expresses GAL80, the inhibitor of GAL4, using the tubulin-1α promoter.* Functional tubP-GAL80 insertions for each of the major chromosome arms (X, 2L, 2R, 3L, 3R) have been generated and recombined with the most widely used FRT insertions near the centromeres of these chromosome arms (Lee and Luo 1999; Luo et al. 1999).

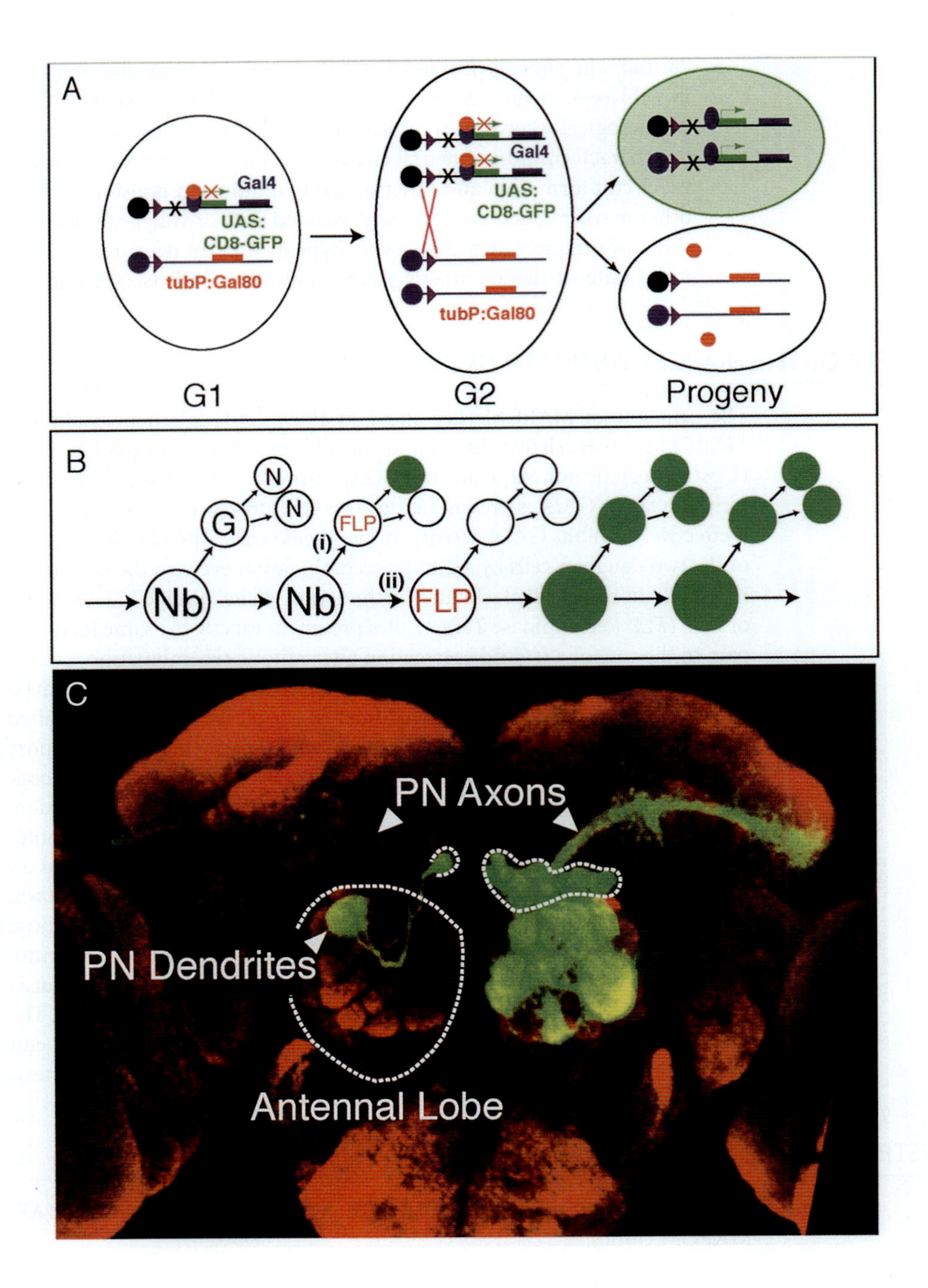

FIGURE 2. Principle of the MARCM system. (*A*) Schematic of the MARCM system. (*Black and blue circles*) Centromeres of homologous chromosomes; (*triangles*) FRT sites; (*red circles*) GAL80 protein; (*blue ovals*) GAL4 protein binding to the UAS sites; (*pink cross* at G_2) mitotic recombination; (*black x*) mutation of interest. See text for details. (*B*) A general schematic of neuroblast division in the *Drosophila* CNS illustrating the generation of single-cell (*i*) and neuroblast (*ii*) clones by inducing FLP activity during the appropriate cell division. (Nb) Neuroblast; (G) ganglion mother cell; (N) postmitotic neuron. (*C*) An example of single-cell (*left*) and neuroblast (*right*) clones of antennal lobe projection neurons (PNs) labeled in green, counterstained with a general neuropil marker (*red*) that allows visualization of glomeruli in the antennal lobe (outlined in the left hemisphere). The cell bodies of the single-cell and neuroblast clones are also outlined. (*A*, Adapted, with permission, from Lee and Luo 1999 [©Elsevier]; *B,C*, adapted, with permission, from Jefferis et al. 2001 [©Nature].)

6. *A UAS-marker transgene.* Extensive use has been made of the *UAS-mCD8-GFP* transgene that was developed by Lee and Luo (1999). It appears to efficiently label dendritic and axonal processes as demonstrated by the target of mouse lymphocyte marker, CD8, to the membrane. The transgene does not appear to have any observable deleterious effects (Williams et al. 2000). There are many other UAS markers available for labeling different subcellular structures including the cytoskeleton, presynaptic terminals, cytosol, and nucleus (see, e.g., Watts et al. 2003).

To generate MARCM clones homozygous for a mutation of interest (loss-of-function MARCM), the 7th component, a mutation, must be introduced into the MARCM fly. The mutation must be recombined onto the appropriate chromosome arm containing the FRT transgene.

To generate MARCM clones expressing a particular transgene only in labeled clones (gain-of-function MARCM), the 7th component would be a UAS transgene of interest that can be present on any chromosome arm of the MARCM fly; if it is present on the chromosome arm in trans to the tubP-GAL80 transgene, the transgene expression level will be doubled after mitotic recombination occurs.

It is also possible to generate MARCM clones that are homozygous for a mutation and, simultaneously, to express a second or even a third UAS transgene, in addition to the UAS marker, by incorporating the appropriate 8th or 9th genetic components into the same fly.

EXPERIMENTAL TIPS

The above requirement—introducing 6–9 genetic components into the same fly—may seem daunting. Yet in reality this "chromosome gymnastics" can be greatly simplified by constructing a set of "MARCM-ready" flies, which contain FLP, one FRT, GAL4, UAS marker, and *tubP-GAL80* transgenes all in one fly stock. To analyze a mutant phenotype, or misexpress a UAS transgene, all that is necessary is to recombine the mutant with the appropriate FRT transgene, or to introduce the transgene and the appropriate FRT transgene into the same fly. A simple cross with the MARCM-ready flies would readily generate MARCM clones homozygous for a given mutation, or clonally express a transgene, or both (see, e.g., Komiyama et al. 2003).

Many MARCM-related flies have been deposited into the Bloomington Stock Center MARCM stocks of general use. The Web page http://flystocks.bio.indiana.edu/gal80.htm contains a list of *tubP-GAL80* transgenes in combination with the common FRT transgenes. A FLP source, a UAS marker, and a GAL4 line must be included. Additional lines, containing the *hs-FLP*, *UAS-mCD8-GFP*, and pan-neural *elav-GAL4*, have also been deposited. One can use these additional transgenes, or replace them with preferred ones.

EXPERIMENTAL CAVEATS AND THEIR POSSIBLE SOLUTIONS

The following list discusses common problems that may be encountered when applying MARCM.

1. *Low-level marker expression.* This is most likely to be caused by GAL80 perdurance: the persistent activity of GAL80 protein, inherited from parental cells, in progeny cells that do not have the *tubP-GAL80* transgene. This problem can be especially significant in single-cell clones, in which there is no further dilution of inherited GAL80 protein. It usually takes at least 24 hours to visualize single-cell clones clearly, and 2–3 days to reach maximal labeling intensity. To ameliorate the problem, a stronger GAL4 can be used, or the dose of GAL4 can be doubled (for instance by using the FRT on the same chromosome arm as the GAL4, such that the MARCM clones will automatically have two copies of the GAL4 transgene). Increasing the UAS-marker dose can also improve labeling, although not as effectively as increasing the GAL4 dose. If it is essential to visualize single neurons shortly after birth, and no mutant generation/birth timing information is necessary, the flip-out method can be used instead (Fig. 1C), since marker expression is turned on immediately after the FLP event.

2. *Leaky expression of marker in non-clonal tissues.* This is most likely to be caused by the GAL4 line being too strong. At 25°C, *tubP-GAL80* can effectively repress UAS-marker expression in the

presence of many GAL4 lines, including *tubP-GAL4*. However, some GAL4 lines are particularly strong, surpassing *tubP-GAL80* repression. One way to overcome this problem is to reduce the experimental temperature, as GAL4 activity increases progressively from 18°C to 29°C. In fact, many of the GAL4 lines that can be effectively suppressed by *tubP-GAL80* at 25°C can no longer be effectively suppressed at 29°C. Another solution is to build a stronger GAL80. Since tubP-GAL80 can suppress *tubP-GAL4*, it is likely that the same strong promoter that drives GAL4 could be used to drive GAL80 and achieve effective repression.

3. *Low-efficiency clone generation.* This is most likely to be caused by the inherent bias of mitotic recombination. For instance, it is generally much more difficult to generate mitotic recombination during embryonic neuroblast division, as compared to postembryonic neuroblast division. One difference is the speed of the cell cycle. Efficiency of mitotic recombination would be reduced during a shorter G_2 phase. One potential way to overcome this difficulty is to allow *Drosophila* to develop at a lower temperature, slowing down the cell cycle and allowing more opportunity for mitotic recombination. Limited success in increasing MARCM clone efficiency has been achieved by growing the embryos at 16°C.
4. *Not all homozygous mutant cells are labeled.* A common misconception about the use of MARCM for analysis of loss-of-function mutations is that all homozygous mutant cells are always labeled, and therefore that the phenotypes seen are always caused by cell-autonomous effects. It is true that all labeled cells are homozygous mutant; however, the converse is only true if a ubiquitously expressing GAL4 line has been used (such as *tubP-GAL4*). In many cases, the GAL4 lines used to visualize MARCM clones express only in neurons, or in a subset of neurons. Homozygous mutant cells that are not labeled could be generated in a mosaic animal because they do not express GAL4. Hence, some phenotypes observed in labeled mutant neurons may be contributed by non-cell-autonomous effects of the mutation in unlabeled mutant cells. There are at least three ways to address this problem. First, specific promoters can be used to drive FLP, as has been done in *Drosophila* photoreceptors (Lee et al. 2001); as long as all cells expressing FLP also express GAL4, there is no chance of generating nonlabeled mutant cells. Second, clones can be induced using hs-FLP at specific stages of development, during which active cell divisions, and hence mitotic recombination, are restricted to a small subset of defined cell types. This has been done for inducing mushroom body enriched clones in newly hatched larvae (Lee and Luo 1999). Third, in the case that the above two conditions are not met, the cell autonomy of gene action can be verified by expressing a wild-type transgene under the control of UAS, only in cells homozygous mutant for the gene of interest, and assaying for rescue of mutant phenotypes (see, e.g., Komiyama et al. 2003).

EXAMPLES OF APPLICATIONS

Tracing Neural Circuits

Single-neuron labeling has contributed significantly to our understanding of the logic of olfactory information processing. From insects to mammals, a similar strategy has been used to represent olfactory information. Each olfactory receptor neuron (ORN) is likely to express only one type of odorant receptor; ORNs expressing a common receptor type converge their axons onto a common target region, called a glomerulus, in the olfactory bulb of the vertebrate brain or antennal lobe of the insect brain. Thus, there is a spatial "glomerular" map in the olfactory bulb or antennal lobe that represents the odor world. Olfactory information leaves the antennal lobe/olfactory bulb via projection neurons (PNs)/mitral cells, which send dendrites into glomeruli, where they synapse with ORN axons, and send axons to higher olfactory centers. How is the glomerular map represented in higher olfactory centers? Using systematic single-neuron labeling, two research groups, using two different methods, have found a highly stereotyped representation of the *Drosophila* glomerular map in the higher-order olfactory centers, with evidence for convergent and divergent projections. One of those methods is the MARCM system that is the focus of this chapter (Marin et al. 2002).

The second method utilizes the "flip-out" concept using the following transgene: *UAS-FRT-CD2-transcriptional stop-FRT-mCD8-GFP* (Fig. 1C). When this transgene is crossed to a PN GAL4 line, all PNs expressing the GAL4 will be labeled by CD2. If the flies also contain the hs-FLP transgene, a brief heat shock could induce postmitotic recombination between the two FRT sites, leading to the deletion of the intervening sequence and the consequent expression of mCD8-GFP in random individual PNs that express the GAL4 (Wong et al. 2002). Systematic correlation of PN axon projection patterns with their glomerular class was also achieved by Wong et al. (2002), with remarkably similar results compared to those obtained using the MARCM strategy.

The fact that MARCM couples the generation of labeled neurons with cell division gives it the added advantage of tracing a neuron's projections in conjunction with its lineage and birth timing. Systematic analysis of PNs born from different neuroblast lineages and at different times indicates that the dendritic projection pattern is determined by their lineage and birth order. PNs are therefore "prespecified" to the odors they will represent before they encounter their presynaptic partners, the ORN axons (Jefferis et al. 2001).

Live Imaging

Two examples are presented, illustrating how live imaging of isolated single neurons facilitates the study of dendritic morphogenesis and plasticity.

How do dendrites know when to stop growing? One mechanism, first described in the vertebrate retina, involves like-repels-like, or "dendritic tiling," such that neurons of the same type occupy nonoverlapping territories (Wassle and Riemann 1978). Using a promoter element highly specific for class IV da sensory neurons in *Drosophila*, it is possible to label all class IV neurons in the same animal. Because these neurons exhibit dendritic tiling, each individual dendritic tree can be clearly visualized (Fig. 1A), permitting live imaging of these dendrites as they tile. Avoidance of "iso-neuronal" branches was observed not only during initial development, but also after the initial tiling had been completed (Grueber et al. 2003).

How stable are synaptic connections in adult mammalian brain? Taking advantage of the chance insertion of transgenic lines expressing GFP analogs in isolated neurons (Fig. 1B) (Feng et al. 2000), repeated live imaging of dendritic segments and dendritic spines in live mouse brains was possible so that individual spines on the same dendritic segments could be followed for several months in the mouse's life (Grutzendler et al. 2002; Trachtenberg et al. 2002; see Chapters 22 and 85). Although different results were obtained regarding spine stability in two studies (concerning different areas of the brain), future research along the same lines is bound to shed light on the stability and dynamics of synaptic connections in the adult mammalian brain.

Genetic Analysis of Neuronal Morphogenesis

MARCM has been used to study gene function in axon growth, guidance, branching, pruning, and transport, as well as dendrite morphogenesis and targeting. In all of the four examples presented here, MARCM allows the analysis of pleiotropic genes in specific developmental events, and at the same time provides high-resolution phenotypic analysis.

Rho GTPases are key regulators of the actin cytoskeleton in response to extracellular signals; hence, it is not surprising that they are expressed quite ubiquitously and are thought to be required in all cells. MARCM allows the genetic manipulation of Rho GTPases, restricting their expression to a small subset of, or even single, identified neurons. This allows high-resolution phenotypic analysis (Fig. 3A, A′). For instance, it was found that progressive loss of combined activities of three Rac GTPases in *Drosophila* mushroom body neurons leads first to defects in axon branching, then in guidance, and finally in growth (Ng et al. 2002).

The ubiquitin-proteasome system (UPS) is essential for many cellular functions. However, MARCM analysis indicates that a small group of neurons, homozygous mutant for essential components for the entire UPS, such as the sole ubiquitin activating enzyme (E1) or proteasome subunits, have strikingly specific phenotypes. These neurons have largely normal axon growth and guidance,

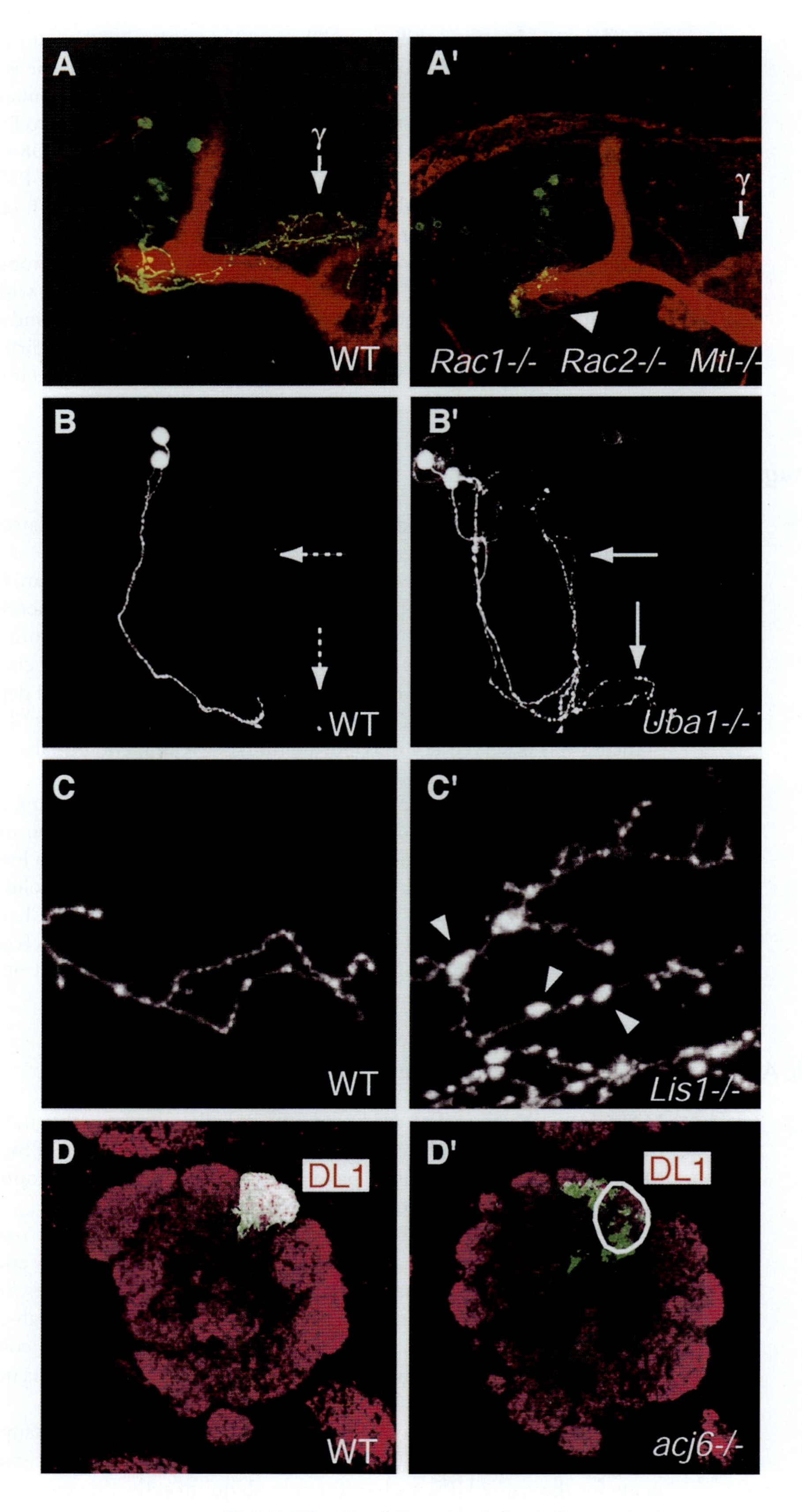

FIGURE 3. (*See facing page for legend.*)

but fail to prune their axons as their wild-type counterparts do (Fig. 3B, B′). These results indicate an essential role for the UPS in degenerative axon pruning during development (Watts et al. 2003).

Haploinsufficiency of human *Lis1* causes lissencephaly (smooth brain disorder). Reduced LIS1 activity, in both humans and mice, results in a neuronal migration defect, but homozygous *Lis1* mutant mice die in early embryonic development, before nervous system formation. The amino acid sequence of Lis1 is highly conserved from fungus to human. Using MARCM to analyze a small subset of labeled *Lis1* mutant neurons in *Drosophila*, it was possible to identify other functions of Lis1, including neuroblast proliferation, dendritic morphogenesis, and axonal transport (Fig. 3C, C′) (Liu et al. 2000). The *Lis1* mutant phenotypes are remarkably similar to mutant phenotypes of the dynein motor, consistent with the notion that they function together.

As a last example, the "prespecification" of the *Drosophila* olfactory projection neurons (PNs) according to lineage and birth order mentioned earlier begs the question of what molecular mechanisms are responsible for such precise wiring. Recent studies utilizing MARCM to generate loss- and gain-of-function neuroblast or single-cell clones revealed that two POU domain transcription factors play essential roles in translating lineage information into wiring specificity. For instance, single-cell PNs, homozygous mutant for the POU factor Acj6, fail to target their dendrites exclusively to the correct glomerulus (Fig. 3D, D′), indicating a role for Acj6 in regulating dendritic targeting precision (Komiyama et al. 2003).

ADVANTAGES AND LIMITATIONS

Comparing Genetic Methods with Golgi Stains

The genetic methods for single-neuron labeling described in this chapter have several advantages over traditional Golgi staining. The Golgi method labels random neurons, whereas genetic methods can use defined promoter elements to drive marker expression in neurons confined to a particular class(es). This can be quite important in tracing neural circuits and in studying genes that affect neuronal morphogenesis. (If a particular gene affects dendritic morphogenesis or axon projections, these processes can no longer be used as independent criteria in defining neuronal types.) In addition, genetic methods afford the possibility of labeling living neurons, allowing time-lapse studies of the dynamic behavior of neurons in their native environment. However, genetic methods are limited by the availability of defined promoter elements and existing technology (see below). Traditional methods, such as Golgi staining or intracellular dye fill, will still be very useful for many years to come. Finally, the more complete axonal filling achieved by genetic methods is also a major advantage over the Golgi method.

FIGURE 3. MARCM for analysis of gene function. (*A, A′*) Axonal projections (in *green*) from isolated mushroom body γ neuron single-cell clones that are wild type (*A*) or homozygous null for three *Rac* genes (*A′*), as visualized in adult brains. Axons from *Rac* mutant neurons stop short along their normal trajectories (*arrowhead in A′*). Brains are counterstained with anti-FasII antibody, revealing normal patterning of mushroom body axons in the heterozygous *Rac* brain (*A′*) as compared to wild type (*A*). (*B, B′*) Wild-type mushroom body γ neurons prune their larval-specific axonal branches (indicated by the *dashed arrows* in *B*) in early pupa; when the ubiquitin activating enzyme (encoded by *Uba1*) is mutated, axon pruning is blocked (*arrows* in *B′*). (*C, C′*) Compared with wild type (*C*), mushroom body γ neurons homozygous mutant for *Lis1* have numerous swellings along their axons (*arrowheads* in *C′*) in MARCM clones examined in adult. (*D, D′*) A wild-type DL1 projection neuron targets its dendrites to the DL1 glomerulus (*D*), whereas a DL1 projection neuron homozygous mutant for the POU factor *acj6* fails to restrict its dendritic projection to the DL1 glomerulus (*D′*). The antennal lobes are counterstained with a general neuropil marker that labels all glomeruli. (*A, A′*, Adapted, with permission, from Ng et al. 2002 [©Nature]; *B, B′*, adapted, with permission, from Watts et al. 2003 [©Elsevier]; *C, C′*, adapted, with permission, from Liu et al. 2000 [©Nature]; *D, D′*, adapted, with permission, from Komiyama et al. 2003 [©Elsevier].)

RELATIVE MERITS OF DIFFERENT GENETIC METHODS

Highly specific promoters allow robust and invariant labeling of the same neurons from animal to animal, and thus are excellent choices for studying the molecular mechanisms of neuronal morphogenesis. Because it is extremely rare to isolate promoter elements that allow single-neuron labeling in a defined brain area, this method is significantly limited by the availability of such special promoters. By the same token, chance insertions of transgenes that happen to be stable and label a very small subset of neurons could also be a very useful tool for studying that particular class of neurons. However, it may take a lot of trial and error, without any guarantees of success, to generate transgenic animals that sparsely label a particular class of neurons of interest.

"Flip-out" or "cre-out" marker gene expression, on the other hand, could theoretically be used to label any class of neurons. These methods require simple genetic manipulations, as compared to the MARCM method. The challenge here is to induce the recombination event at a frequency high enough to ease identification of labeled neurons, but low enough to permit single neurons to be traced. In *Drosophila*, this can be achieved by inducing heat shocks of different durations; in mice, the efficiency of the Cre recombinase is so high that, in most cases, all neurons expressing the target transgene (e.g., promoter-loxP-transcription stop-loxP-marker) and the Cre recombinase, even for a brief period, will result in all neurons being labeled, defeating the purpose of single-neuron labeling. Low-efficiency mutant versions of loxP or Cre recombinase deliberately reduce the frequency of recombination.

Two advantages of the MARCM system over the above genetic methods are that (1) MARCM allows labeling of only homozygous neurons; (2) since MARCM couples cell division with generation of labeled neurons, it is an excellent lineage-tracing tool (an example of determining that *Drosophila* PNs use lineage and birth order to determine their dendritic choice is given above). However, the MARCM method can be used only to knock out gene function at the time a neuron is born. In many cases it is useful to eliminate gene function later on; for example, in order to study the physiological function of the gene of interest in the mature brain. Another current limitation is that MARCM is used only in *Drosophila*. In theory, the same method could be established in other genetic model organisms such as zebrafish or mice; the rate of interchromosomal mitotic recombination determines how useful MARCM-like methods would be in such organisms.

ACKNOWLEDGMENTS

I thank all members of my laboratory over the last six years, whose work has contributed to many of the discussions in this chapter. I thank Takaki Komiyama for help with the figures, and Daniela Berdnik, Eric Hoopfer, Lisa Marin, Chris Potter, and Haitao Zhu for helpful comments. Research in my laboratory has been supported by grants from the National Institutes of Health, the McKnight Endowment Fund, the Human Frontiers Science Program, the Muscular Dystrophy Association, the Klingenstein Fund, the Sloan Program, and the Terman Fellowship.

REFERENCES

Basler K. and Struhl G. 1994. Compartment boundaries and the control of *Drosophila* limb pattern by hedgehog protein. *Nature* **368:** 208–214.

Brand A.H. and Perrimon N. 1993. Targeted gene expression as a means of altering cell fates and generating dominant phenotypes. *Development* **118:** 401–415.

Cajal S.R. 1911. *Histology of the nervous system of man and vertebrates.* Oxford University Press, Oxford, United Kingdom (English translation [1995]).

Chou T.B. and Perrimon N. 1996. The autosomal FLP-DFS technique for generating germline mosaics in *Drosophila melanogaster*. *Genetics* **144:** 1673–1679.

Feng G., Mellor R.H., Bernstein M., Keller-Peck C., Nguyen Q.T., Wallace M., Nerbonne J.M., Lichtman J.W., and Sanes J.R. 2000. Imaging neuronal subsets in transgenic mice expressing multiple spectral variants of GFP. *Neuron* **28:** 41–51.

Golgi C. 1873. Sulla struttura della sostanza grigia del cervello. *Gazetta medica lombarda* **IV**.

Golic K.G. and Lindquist S. 1989. The FLP recombinase of yeast catalyzes site-specific recombination in the *Drosophila* genome. *Cell* **59:** 499–509.

Grueber W.B., Jan L.Y., and Jan Y.N. 2002. Tiling of the *Drosophila* epidermis by multidendritic sensory neurons. *Dev. Biol.* **129:** 2867–2878.

Grueber W.B., Ye B., Moore A.W., Jan L.Y., and Jan Y.N. 2003. Dendrites of distinct classes of *Drosophila* sensory neurons show different capacities for homotypic repulsion. *Curr. Biol.* **13:** 618–626.

Grutzendler J., Kasthuri N., and Gan W.B. 2002. Long-term dendritic spine stability in the adult cortex. *Nature* **420:** 812–816.

Huang Z.J., Yu W., Lovett C., and Tonegawa S. 2002. Cre/loxP recombination-activated neuronal markers in mouse neocortex and hippocampus. *Genesis* **32:** 209–217.

Jefferis G.S., Marin E.C., Stocker R.F., and Luo L. 2001. Target neuron prespecification in the olfactory map of *Drosophila. Nature* **414:** 204–208.

Komiyama T., Johnson W.A., Luo L., and Jefferis G.S. 2003. From lineage to wiring specificity: POU domain transcription factors control precise connections of *Drosophila* olfactory projection neurons. *Cell* **112:** 157–167.

Lee C.H., Herman T., Clandinin T.R., Lee R., and Zipursky S.L. 2001. N-cadherin regulates target specificity in the *Drosophila* visual system. *Neuron* **30:** 437–450.

Lee T. and Luo L. 1999. Mosaic analysis with a repressible cell marker for studies of gene function in neuronal morphogenesis. *Neuron* **22:** 451–461.

———. 2001. Mosaic analysis with a repressible cell marker (MARCM) for *Drosophila* neural development. *Trends Neurosci.* **24:** 251–254.

Liu Z., Steward R., and Luo L. 2000. *Drosophila* Lis1 is required for neuroblast proliferation, dendritic elaboration and axonal transport. *Nat. Cell Biol.* **2:** 776–783.

Luo L., Lee T., Nardine T., Null B., and Reuter J. 1999. Using the MARCM system to positively mark mosaic clones in *Drosophila. Dros. Inform. Serv.* **82:** 102–105.

Marin E.C., Jefferis G. S. X. E., Komiyama, T., Zhu, H., and Luo, L. 2002. Representation of the glomerular olfactory map in the *Drosophila* brain. *Cell* **109:** 243–255.

Ng J., Nardine T., Harms M., Tzu J., Goldstein A., Sun Y., Dietzl G., Dickson, B. J., and Luo, L. 2002. Rac GTPases control axon growth, guidance and branching. *Nature* **416:** 442–447.

Scott E.K., Raabe T., and Luo L. 2002. Structure of the vertical system and horizontal system neurons of the lobula plate in *Drosophila. J. Comp. Neurol.* **454:** 470–481.

Trachtenberg J.T., Chen B.E., Knott G.W., Feng G., Sanes J.R., Welker E., and Svoboda K. 2002. Long-term in vivo imaging of experience-dependent synaptic plasticity in adult cortex. *Nature* **420:** 788–794.

Wassle H. and Riemann H.J. 1978. The mosaic of nerve cells in the mammalian retina. *Proc. R. Soc. Lond. B Biol. Sci.* **200:** 441–461.

Watts R.J., Hoopfer E.D., and Luo L. 2003. Axon pruning during *Drosophila* metamorphosis: Evidence for local degeneration and requirement for the ubiquitin-proteasome system. *Neuron* **38:** 871–885.

Williams D.W., Tyrer M., and Shepherd D. 2000. Tau and tau reporters disrupt central projections of sensory neurons in *Drosophila. J. Comp. Neurol.* **428:** 630–640.

Wong A.M., Wang J.W., and Axel R. 2002. Spatial representation of the glomerular map in the *Drosophila* protocerebrum. *Cell* **109:** 229–241.

Xu T. and Rubin G.M. 1993. Analysis of genetic mosaics in developing and adult *Drosophila* tissues. *Development* **117:** 1223–1237.

CHAPTER 14

A Practical Guide: Ballistic Delivery of Dyes for Structural and Functional Studies of the Nervous System

Wen-Biao Gan,[1] Jaime Grutzendler,[2] Rachel O. Wong,[3] and Jeff W. Lichtman[3]

[1]*Skirball Institute of Biomolecular Medicine, Department of Physiology and Neuroscience, New York University School of Medicine, New York, New York 10016;* [2]*Northwestern University, Chicago, Illinois 60611;* [3]*Department of Anatomy and Neurobiology, Washington University School of Medicine, St. Louis, Missouri 63110*

This chapter describes detailed protocols for rapid labeling of cells in a variety of preparations by means of particle-mediated ballistic (gene gun) delivery of fluorescent dyes. This method has been used for rapid labeling of cells with either lipid- or water-soluble dyes, in a variety of preparations at different ages. In particular, carbocyanine lipophilic dyes such as DiI have been used to obtain Golgi-like neuronal labeling in fixed and live cell cultures, brain slices, retinal explants, and fixed post-mortem human brain (Gan et al. 2000; Grutzendler et al. 2003). The ballistic technique can also be used for functional studies of living cells by loading cells with water-soluble compounds such as calcium indicators (Kettunen et al. 2002). This ballistic labeling technique is useful for studying neuronal connectivity, function, and pathology in the nervous system of living as well as fixed specimens.

MATERIALS

Biolistic Device

Helios Gene Gun System (Bio-Rad) or other custom-made ballistic devices (see http://thalamus.wustl.edu/nonetlab/)

Lipophilic Dyes<!>

DiI, DiO, or DiD (Molecular Probes, D-282, D-275, D-307)

Calcium Indicators

(e.g., Calcium Green-1 dextran, Molecular Probes)

Methylene chloride<!> (Sigma)
Tungsten particles (1.0–1.7-µm diameter, Bio-Rad)
Glass slides
Sonicator
Tefzel tubing (Bio-Rad)
Polyvinyl-pyrrolidone<!> (PVP, 10 mg/ml in distilled water, from Sigma or Bio-Rad)
Tubing prep station (Bio-Rad)
Isopore polycarbonate membrane filter with 3-µm pore size and 8.0×10^5 pores/cm^2 density (Millipore)

Caution: *See Appendix 3 for appropriate handling of materials marked with <!>.*

PROCEDURE

Ballistic Delivery of Dyes for Structural and Functional Studies of the Nervous System

Preparation of Stock Solutions/Lipophilic Dyes

1. For single-color labeling, dissolve 3 mg of lipophilic dye <!> such as DiI, DiO, or DiD in 100 μl of methylene chloride <!>(Sigma). For multicolor labeling, dissolve 7 mg of DiI, DiO, and DiD in 70 μl of methylene chloride in three separate microfuge tubes. Mix various proportions of the different dyes to obtain, for example, seven different dye combinations in seven tubes as follows: (a) 30 μl of DiI; (b) 30 μl of DiO; (c) 30 μl of DiD; (d) 15 μl of DiI: 15 μl of DiO; (e) 15 μl of DiI: 15 μl of DiD; (f) 15 μl of DiO: 15 μl of DiD; (g) 10 μl of DiI: 10 μl of DiO: 10 μl of DiD. Last, add 70 μl of methylene chloride to each tube. The final concentration for each dye combination is 3 mg in 100 μl of methylene chloride.

Coating Tungsten Beads

2. Spread tungsten particles (50–100 mg) evenly on a glass slide by mixing with a few drops of methylene chloride. Then add the dye solution (100 μl) to the particles on the glass slides. As methylene chloride evaporates quickly (within a minute), the lipophilic dye precipitates onto the tungsten particles. Scrape the dye-coated particles off the glass slide using a razor blade and collect them in a 10-ml test tube. For multicolor labeling, coat particles separately with different combinations of dyes and collect them together in a 10-ml test tube. Resuspend the particles in 3–10 ml of distilled water and sonicate <!> for 5–10 minutes to prevent formation of large clusters.

Bullet Preparation

3. Inject sonicated solutions of particles of a single color or mixtures of colors into a 30-inch-long Tefzel tube. To improve attachment of the beads to the tube walls, introduce a solution of polyvinyl-pyrrolidone <!> (PVP, 10 mg/ml in distilled water) and rapidly drain to pre-coat the tube prior to injecting the dye solutions. Then insert the tube into a "tubing prep station" (Bio-Rad), and allow the dye-coated particles to precipitate and settle onto the tube wall for ~3 min (depending on the desired labeling density) before slowly withdrawing the remaining liquid. To spread the beads evenly, rotate the tube on the prep station for ~5 min and dry with a constant air flow. Cut the air-dried tube into 13-mm pieces and store at room temperature, protected from light, for future use.

Particle Delivery

4. Deliver dye-coated particles to the preparation using a commercially available biolistic device ("gene gun," Bio-Rad). Such a particle delivery device can also be custom-made at much less expense (see http://thalamus.wustl.edu/nonetlab). To prevent large clusters of particles from landing on the tissue, causing nonuniform labeling and preventing single-cell resolution, it is useful to insert an Isopore polycarbonate membrane filter between the gene gun and the preparation (see below). A filter holder can be made by cutting a central hole (~20-mm diameter) on the bottom of a plastic petri dish and covering it with a metal mesh (~1-mm pore) glued to it. A membrane filter is then placed on top of the metal mesh and held immobile by a petri dish cover also with a central opening of ~20 mm diameter (Fig. 1). The filter holder is then placed on top of the target tissue and aligned with the gene-gun barrel prior to shooting (Fig. 1).

   Note: Density of labeling is controlled by using various gas pressures (60–200 psi helium gas), by changing the distance between the gun and the preparation (10–25 mm), or by shooting the tissue several times (see also below). Lower gas pressures and larger distance between the gun and the tissue lead to lower labeling densities. Higher gas pressures may cause membrane filters to break apart, and as a result, tissue injury and nonuniform labeling may occur.

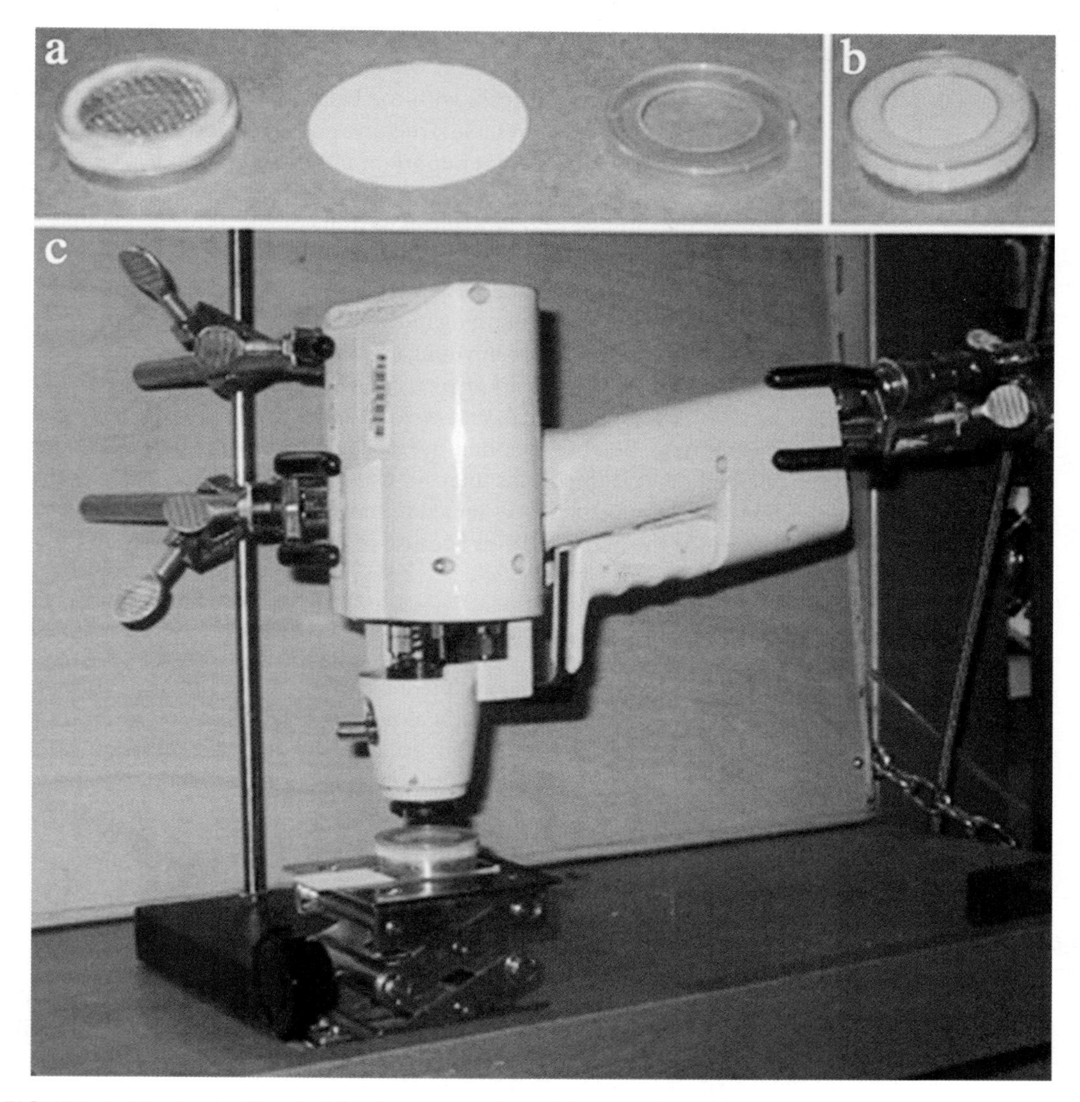

FIGURE 1. Membrane filter holder. (*a*) From left to right: metal mesh glued to a petri dish with 20-mm-diameter opening, membrane filter, and petri dish cover with ~20-mm-diameter opening. (*b*) Membrane filter is placed on top of metal mesh and held in place by the petri dish cover. (*c*) Membrane filter holder is placed on top of tissue and aligned with the gene gun barrel.

Labeling Density and Depth

5. The labeling efficiency is dependent on multiple factors such as bullet particle density, filter pore density, gun pressure, and distance from the tissue. To reduce the labeling density, a porous polycarbonate membrane is usually inserted between the gun and the target tissue. This prevents large clusters of dye-coated particles from landing on the preparation and causing diffuse dye labeling (Fig. 1c). The use of a membrane filter also protects the tissue from the air shock wave generated by the gun. Labeling efficiency can be increased either by producing bullets with more dye per particle or by shooting the preparation multiple times.

Tissue Fixation Conditions

6. The fixation condition is a critical variable in obtaining good-quality labeling of neuronal structures. For mouse brain slices, perfuse the animal with 10 ml of isosmotic PBS followed by rapid

perfusion with 40 ml of 4% paraformaldehyde<!> in PBS (20 ml/min). Remove the brain quickly from the skull and post-fix in 4% paraformaldehyde for 10 min. Label vibratome-cut brain slices (200–300 μm thickness with the ballistic method), and post-fix for 2–12 hours at room temperature to further preserve structures and allow dye diffusion. The duration of fixation prior to ballistic particle delivery is important for obtaining good-quality labeling. Over-fixation prior to shooting (>1 hour) often causes dye to spread between cells or nearby processes, and thus reduces the number of cells that can be individually resolved. On the other hand, the duration of post-fixation after ballistic labeling does not appear to cause dye spread between nearby cells, suggesting that dye spread and decreased labeling are probably due to failure of initial dye transfer from the particle to surrounding individual cells. The problem of dye spread between cells due to over-fixation is less severe in young animals, presumably because of greater extracellular space.

Fixation conditions are also critical when labeling postmortem human brain. It appears that labeling is best when small pieces of brain are fixed with 4% paraformaldehyde within 24 hours postmortem. Because lipophilic dye labeling requires the presence of an intact and continuous cell membrane, it is important to avoid using fixatives with substances that can dissolve membrane lipids, such as methanol contained in standard formalin. In addition, frozen specimens cannot be labeled by the lipophilic dyes because of the membrane disruption that occurs during the freezing and thawing process.

Dye Diffusion and Tissue Mounting

7. After delivery of particles, keep tissues such as brain slices in solution (e.g., 0.2 M PBS) for several hours to allow dye diffusion along neuronal processes. In fixed tissue, labeling takes longer than in living slices, presumably because of slower dye diffusion along the fixed plasma membrane. Following dye diffusion, mount tissues on glass slides using 80% glycerol in 0.2 M PBS or 100% glycerol.

Caution: *See Appendix 3 for appropriate handling of materials marked with <!>.*

LABELING WITH WATER-SOLUBLE DYES

The following modifications to the protocol are required for the delivery of water-soluble dyes: (1) Dissolve 1~2 mg of a dye, such as Calcium Green-1 dextran, in 12–25 μl of distilled water. (2) Spread 25~50 mg of tungsten particles (1.3 or 1.7 μm diameter) evenly onto a clean microscope glass slide. Use a razor blade to dissociate large clusters of particles until a fine powder is obtained. (3) Pipette the dissolved dye solution uniformly onto the tungsten particles and air-dry. Scrape the dye-coated particles off onto a piece of weighing paper with the razor blade. (4) Pour the powder of dye-coated particles (in a dry form, unlike the case of lipophilic dye loading) into a 15-cm-long Tefzel tubing, precoated with PVP (10 mg/ml in distilled water). Gently rotate and shake the tube until the particles spontaneously adhere onto the inner wall. Cut the particle-coated tube into 13-mm pieces and store in a dark, dry container at 4°C until use. (5) For particle delivery into the tissues, we use virtually the same protocol as the one described for lipophilic dyes.

When using water-soluble dyes, labeling density needs to be approximately fivefold higher than with lipophilic dyes to increase the probability of successful labeling, because dyes need to enter the cell soma and diffuse into the cytoplasm rather than diffuse along the cell membrane as with lipophilic dyes. Because of this, labeling with water-soluble dyes works best in preparations containing a high density of cell somata, such as the ganglion cell layer of the retina in which 15% of cells can be routinely labeled.

EXAMPLES OF APPLICATIONS

The ballistic technique has been used to label neurons in both live and fixed tissues, ranging from isolated neurons in culture to whole brain and brain slices (Fig. 2). When labeling at low densities, we typically observed only one dye-coated particle contacting either the soma or a branch of the dendritic

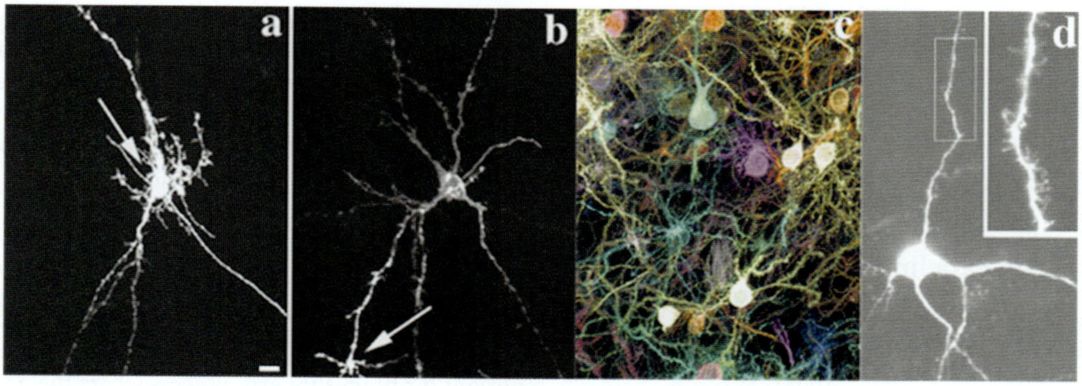

FIGURE 2. Ballistic labeling of cells in fixed and living tissue. An individual cell in a P10 mouse brain slice is labeled with lipophilic dye with a single dye-coated particle (*arrow*) landed on either the cell soma (*a*) or a dendritic arbor (*b*). (*c*) Labeling of many pyramidal neurons from a fixed brain slice from a P20 mouse that was shot with a combination of seven different lipophilic dyes. The image represents a collapsed view of 50 confocal planes covering ~25 μm of depth. Many of the cells had processes that extended beyond the upper or lower surface of the slice and hence have dendritic or axonal branches that end abruptly. (*d*) Labeling of a pyramidal neuron and spines (*inset*) with Calcium Green-1 dextran from an acute slice preparation of a P12 rat. Bar, 10 μm for *a–d*.

or axonal arbor of a labeled cell, indicating that a single particle contains enough dye to label one cell through retrograde and anterograde diffusion (Fig. 2a,b). Figure 2c shows multiple pyramidal neurons labeled in a fixed mouse cortical brain slice by delivering a mixture of particles coated with different combinations of dyes. Many neurons and their processes, including dendritic spines, appeared completely labeled and can be distinguished from each other due to the multicolor labeling. Labeled dendritic arbors and axons can often be followed for hundreds of micrometers. In living tissues, dendritic processes appeared labeled almost immediately (<5 min) after particles contacted the tissue (Gan et al. 2000; Kettunen et al. 2002; Grutzendler et al. 2003).

In addition to lipophilic dye labeling, the ballistic technique can be used to label cells with water-soluble compounds such as calcium indicators (Kettunen et al. 2002; Lohmann et al. 2002). Figure 2d shows a cortical pyramidal cell from an acute brain slice labeled successfully upon particle delivery of calcium indicators such as Calcium Green-1. Complete filling of dendritic arbors occurred within a minute. The intensity of labeling varied from cell to cell, most likely because particles carried differing amounts of indicator. In addition, these labeled cells demonstrated a rise in intracellular calcium levels upon stimulation with KCl, indicating that they remained viable after labeling (Kettunen et al. 2002; Lohmann et al. 2002; see Chapter 34).

ADVANTAGES AND LIMITATIONS

The ballistic method allows rapid delivery of fluorescent indicators into multiple cell types in both fixed and living preparations, ranging from embryonic tissue to adult postmortem human brain. This labeling technique provides several advantages over existing labeling methods.

Efficient Labeling of Multiple Cells

In comparison to intracellular injection of indicators using micropipettes, this approach allows labeling of many neurons within minutes rather than targeting cells individually. Dendrites in many cells are often labeled from a single indicator-coated particle, and the labeling process appears not to cause a significant perturbation of cellular structure or function (Gan et al. 2000; Kettunen et al. 2002; Lohmann et al. 2002). In addition, ballistic delivery can be very useful in immature cells in which dye injection through micropipettes is extremely difficult because of cell fragility and susceptibility to damage. Ballistic labeling of various cell types in fixed and live retina explants at various ages has also been successfully demonstrated (Lohmann et al. 2002; Rockhill et al. 2002; Strettoi et al. 2002; Sun et al. 2002). Finally, using various combinations of carbocyanine dyes, adjacent neurons can be labeled in different colors. Combined with high-density labeling, this multicolor feature is potentially useful for studying complicated neuronal connections (Gan et al. 2000).

Rapid Labeling

The ballistic technique permits immediate imaging of neuronal tissues as soon as the preparation is made, unlike GFP transfection techniques that require many hours for gene expression (Lo et al. 1994). This is particularly important for brain slice experiments where the health of the preparation can be maintained only for a limited period of time. In addition, the technique allows extensive dendritic arbors to be labeled rapidly without dialyzing cellular contents as seen in patch pipette loading (Majewska et al. 2000).

Labeling a Variety of Cell Types in Live and Fixed Tissues

The labeling technique involves passive dye transfer and diffusion that is independent of gene transcription and protein synthesis. Thus, any type of cells in fixed or living tissues can be labeled nonselectively. Furthermore, more complete labeling of fine neuronal structures such as dendritic spines is achieved with lipophilic dyes compared to immunolabeling. Transgenic mice expressing spectral variants of GFP in different subsets of neurons are now available (Feng et al. 2000). Such color variants, however, are limited to neuronal populations that happen to express GFP under control of specific promoters at a given postnatal age.

Labeling with Various Indicators and Molecules

In addition to delivering lipophilic dyes and calcium indicators as described here, the ballistic technique is potentially useful for delivering other indicators and molecules, such as voltage-sensitive dyes, dextran-conjugated pH indicators, ion indicators, and pharmacological reagents. This potential was exploited in recent studies using a rhodamine analog to investigate cell migration in embryonic brains (Alifragis et al. 2002). It is also possible to simultaneously deliver more than one indicator by co-coating the particles with multiple substances (O'Brien and Lummis 2004).

Combining the Ballistic Labeling Technique with Other Labeling Approaches

The ballistic labeling technique, in combination with other fluorescent dyes such as Thioflavin-S, has been used to study dendritic abnormalities near amyloid plaques in a mouse model of Alzheimer's disease and in human brain (Grutzendler et al. 2003). In addition, if immunostaining does not require tissue permeabilization and lipid extraction, antibody labeling can also be used in conjunction with lipophilic dye labeling (e.g., colocalization of dendritic spines with anti-synaptic vesicle protein immunoreactive puncta). Because antibody labeling usually requires extraction of lipids with detergents to facilitate antibody access to antigens, lipophilic dye labeling generally cannot be used in combination with immunostaining. One alternative is to first label neuronal structures with lipophilic dyes, photoconvert the labeled structures in the presence of diaminobenzidine (Gan et al. 1999), and subsequently perform immunohistochemistry. Another option is to label cells with water-soluble dyes, which are well retained after fixation. Indeed, it has been reported that live cells can be labeled upon ballistic delivery of 4-chloromethyl benzoyl amino tetramethyl rhodamine (CMTMR) and that CMTMR labeling can be combined with immunohistochemistry and in situ hybridization (Alifragis et al. 2002).

Despite many advantages, ballistic delivery of indicators also has some drawbacks: First, this approach is highly variable because dye crystal size, density, and penetration are difficult to control.

Using the commercially available Bio-Rad gene gun, the depth of penetration of dye-coated particles is generally ~20–30 μm and up to 60 μm. Therefore, most of the labeled cells are usually located near the surface. However, because of retrograde transfer of dyes, cells located deeper can also be labeled, particularly with lipophilic dyes. A potentially useful modification to the gene gun that can increase the depth of penetration substantially was reported (O'Brien et al. 2001). Such a modification may allow penetration of particles up to ~300 μm in brain slices. Second, because water-soluble dye-coated particles have to be in or near the cell somata for labeling to occur, it is difficult to get high-density labeling without showering the tissue with many particles and potentially leading to some tissue damage. Third, whereas labeling of dendritic structures is very good, especially with lipophilic dyes, labeling of axonal arbors over extensive distances has not been successful. This is likely because each particle carries a limited amount of dye on its surface. In the future, it may be possible to improve labeling by designing special beads that can carry a larger amount of dye. Another limitation of the technique is the difficulty in targeting a particular cell or region. However, by using a mask to cover parts of the preparation during shooting, one can prevent labeling of undesired areas and target a smaller region of interest.

In summary, we describe here in detail a ballistic technique that delivers fluorescent dyes and indicators into multiple cell types, providing rapid labeling of neurons and their processes both in vivo and in vitro. This approach should be useful for studying neuronal connectivity, function, and pathology in the nervous system of experimental animals, as well as in postmortem human tissue specimens.

REFERENCES

Alifragis P., Parnavelas J.G., and Nadarajah B. 2002. A novel method of labeling and characterizing migrating neurons in the developing central nervous system. *Exp. Neurol.* **174:** 259–265.

Feng G., Mellor R.H., Bernstein M., Keller-Peck C., Nguyen Q.T., Wallace M., Nerbonne J.M., Lichtman J.W., and Sanes J.R. 2000. Imaging neuronal subsets in transgenic mice expressing multiple spectral variants of GFP. *Neuron* **28:** 41–51.

Gan W.B., Bishop D., Turney S.G., and Lichtman J.W. 1999. Vital imaging and ultrastructural analysis of individual axon terminals labeled by iontophoretic application of lipophilic dye. *J. Neurosci. Methods* **93:** 13–20.

Gan W.B., Grutzendler J., Wong W.T., Wong R.O.L., and Lichtman J.W. 2000. Multicolor "DiOlistic" labeling of neuronal circuits using lipophilic dye combinations. *Neuron* **27:** 219–225.

Grutzendler J., Tsai J., and Gan W.B. 2003. Rapid labeling of neuronal populations through ballistic delivery of fluorescent indicators. *Methods* **30:** 79–85.

Kettunen P., Demas J., Lohmann C., Kasthuri N., Gong Y.D., Wong R.O.L., and Gan W.B. 2002. Imaging calcium dynamics in the nervous system by means of ballistic delivery of indicators. *J. Neurosci. Methods* **119:** 37–43.

Lo D.C., McAllister A.K., and Katz L.C. 1994. Neuronal transfection in brain slices using particle-mediated gene transfer. *Neuron* **13:** 1263–1268.

Lohmann C., Myhr K.L., and Wong R.O.L. 2002. Transmitter-evoked local calcium release stabilizes developing dendrites. *Nature* **418:** 177–181.

Majewska A., Tashiro A., and Yuste R. 2000. Regulation of spine calcium dynamics by rapid spine motility *J. Neurosci.* **20:** 8262–8268.

O'Brien J. and Lummis S.C. 2004. Biolistic and diolistic transfection: Using the gene gun to deliver DNA and lipophilic dyes into mammalian cells. *Methods.* **33:** 121–125.

O'Brien J.A., Holt M., Whiteside G., Lummis S.C., and Hastings M.H. 2001. Modifications to the hand-held Gene Gun: Improvements for in vitro biolistic transfection of organotypic neuronal tissue. *J. Neurosci. Methods* **112:** 57–64.

Rockhill R.L., Daly F.J., MacNeil M.A., Brown S.P., and Masland R.H. 2002. The diversity of ganglion cells in a mammalian retina. *J. Neurosci.* **22:** 3831–3843.

Strettoi E., Porciatti V., Falsini B., Pignatelli V., and Rossi C. 2002. Morphological and functional abnormalities in the inner retina of the rd/rd mouse. *J. Neurosci.* **22:** 5492–5504.

Sun W.Z., Li N., and He S.G. 2002. Large-scale morphological survey of mouse retinal ganglion cells. *J. Comp. Neurol.* **451:** 115–126.

CHAPTER 15

A Practical Guide: Imaging Embryonic Development in *Caenorhabditis elegans*

William A. Mohler and Ariel B. Isaacson

Department of Genetics and Developmental Biology, University of Connecticut Health Center, Farmington, Connecticut 06030-3301

Embryos are remarkable for their combination of pluripotency, three-dimensionality, and swiftness of subcellular and developmental rearrangements. Embryogenesis in the nematode *Caenorhabditis elegans* is uniquely suited among model systems to high-resolution dynamic imaging. Within a single high-magnification, high-NA microscope field, at submicrometer resolution, it is possible to observe several entire animals taking form. The full ~14-hour course of embryonic cleavage and morphogenesis of this transparent, free-living worm is essentially invariant (Sulston et al. 1983). Observing specific fluorescently labeled components during embryonic development promises to reveal the roles of organelles and molecules in an extremely diverse and reproducible set of contexts. Methods for imaging embryos labeled with membrane-specific vital dyes have been published elsewhere (Mohler and Squirrell 2000). Over the past few years, however, the *C. elegans* community has created a fast-growing collection of hundreds of transgenic strains expressing green fluorescent protein (GFP)-labeled versions of distinct endogenously expressed genes. The task of correlating the resulting expression and localization patterns in space and time is simultaneously alluring and technically demanding. This chapter describes the use of four-dimensional (4D) laser-scanning microscopy and subsequent data processing to record, portray, analyze, and compare the expression of fluorescently tagged gene products during development of the nematode embryo.

CHOICE OF IMAGING SYSTEM

Because of the simplicity with which *C. elegans* can be handled and transported, we have been able to acquire and compare 4D embryonic recordings of GFP fluorescence on a variety of optical-sectioning fluorescence microscopes. Laser-scanning microscopes are most practical, as optical sections are "real-time" output—unlike wide-field deconvolution systems, which require lengthy calculations to produce high-quality optical sections, and which do not tolerate rapid motion within the three-dimensional (3D) specimen. Fortunately, the variants of GFP typically used by *C. elegans* researchers are excited efficiently by widely used visible and ultrafast infrared lasers.

Both one-photon excitation with 488-nm (argon laser) light and two-photon excitation with 900-nm (Ti:sapphire laser) light can yield a high GFP signal, with tolerable rates of photobleaching and phototoxicity, and with low autofluorescence from unlabeled cells within the specimen. Two-photon imaging is preferable in order to record the full extent of embryogenesis continuously with high sensitivity and high resolution in space and time (Denk et al. 1990; Mohler and White 1998; Mohler and Squirrell 2000). Bleaching and cell damage occur only at the focus, and confocal fluorescence emission can be collected efficiently without the use of a pinhole. We have experience using two-photon microscopes based on the Zeiss LSM510, Bio-Rad 1024, and Olympus FV300 laser-scanning systems. In each case, however, a different output file format is used, which can complicate the creation of universally applicable custom-written image processing software.

An additional consideration arises when imaging late stages of embryogenesis. Point-scanning galvanometer-based systems like those mentioned above are typically limited in the rate at which individual optical sections can be acquired. Thus, a high-resolution 3D stack of sections through an embryo (i.e., a timepoint in a 4D recording) can take several minutes to acquire. This is sufficient time resolution to capture the developmental dynamics of cleavage and cell migrations in the first half of embryogenesis. However, after the onset of muscular contraction, the worm begins to writhe within the eggshell. Thus, the typical point-scanning rate of 3D stack acquisition results in distortions in the embryo reconstructed at each time point: Some features move and appear in more than one optical section, whereas other features may escape capture completely. Limited experimentation with a spinning microlens-scanning confocal microscope, marketed by Perkin-Elmer (UltraView), indicates that this system collects high-resolution stacks rapidly enough to "freeze" the 3D writhing embryo at each imaged time point. Thus, the position and intensity of GFP fluorescence can be readily localized in reconstructed images of late embryos, although the spatial correlation between signal sources may be lost between consecutive time-point stacks. Our preliminary observation is that vital imaging with the UltraView allows viability, sensitivity, and resolution comparable to that obtained with point-scanning two-photon imaging. Although several groups have constructed microlens-scanning two-photon systems, this promising technology has not yet been evaluated for vital imaging of *C. elegans* (Straub et al. 2000).

When considering approaches to producing rich 4D recordings of development, it can be useful to optimize the 4D resolution index = (time-point stacks/viable embryo) x (sections/stack) x (pixels/optical section) x (detected photons/pixel). For extended recordings of development, via any mode of laser-scanning fluorescence microscopy, it is necessary to minimize excitation power (to maintain viability of the specimen) and to maximize detector sensitivity (to compensate for low excitation power), even though the individual output images have lower signal-to-noise ratios than optimal "still" images. In practice, the photomultiplier or CCD detector is tuned to its highest usable gain, and the laser power is attenuated such that the output pixel intensity of a fluorescent specimen may not even approach the dynamic range limit of the detector. Noise pixels within the images can often be removed in postacquisition processing with a median or gaussian spatial filter. However, even unfiltered speckle noise often becomes visually negligible when a recording is viewed in animated playback.

CHOICE OF OBJECTIVE LENS

900-nm light is preferable to shorter wavelengths (down to 800 nm) in two-photon imaging of GFP and YFP (yellow fluorescence protein) (Mohler and White 1998). However, many microscope objectives do not efficiently transmit longer-wave infrared (IR) and therefore dramatically attenuate the excitation power of the beam. It is advisable to consult the manufacturer's specifications and to test available objectives before assuming that 900-nm imaging will be possible on any given system.

Our imaging has, so far, typically involved oil-immersion lenses, although sensitivity and resolution for regions of the embryo far (30–50 μm) from the coverslip may actually suffer as a result of this choice. Use of a water-immersion objective is theoretically optimal for matching refractive indices of the immersion medium and specimen, since live *C. elegans* embryos are mounted in aqueous media (see below). However, a water-immersion meniscus tends to evaporate rapidly, and time-lapse imaging over several hours becomes impractical. As an alternative, at least one organic immersion liquid (laser liquid 3421, Cargille Laboratories) is available with a refractive index of 1.33, matching the index of water, while being much less prone to evaporation. The theoretical predictions have yet to be tested in practice by comparing the quality of long-term 4D recordings using water- and oil-immersion.

SPECIMEN PREPARATION

Nematode stocks are maintained using standard methods (Brenner 1974). Two typical options exist for mounting embryos for imaging: agar mounts and suspended mounts. Agar mounting involves deposition of embryos on a gelled pad of 3% agarose in Egg Buffer (see below) and light compression of the embryos under an overlying coverslip. The slight pressure tends to make the embryos adopt either of two standard orientations. This trick was a critical factor allowing the determination of embryonic lineage (Sulston et al. 1983), and such reproducibility would be very desirable in making large numbers of comparable recordings from individual embryos. However, evaporation from the agar pad tends to cause such mounts to shift gradually over time, making consistency in long 4D recordings difficult. In practice, the suspended mount, in which uncompressed (more randomly oriented) embryos are adhered directly to the coverslip, is preferable. 4D spatial registration with these mounts is quite stable, and it is possible to digitally reorient embryos within the recorded volume of the 3D stack during postprocessing of the data.

PROCEDURE

Suspended Embryo Mount

1. Prepare a 1 mg/ml stock solution of poly-L-lysine (Sigma) in distilled water. Store in 50-ml aliquots at –20°C indefinitely. Coat the required number of #1.5 coverslips by spreading 1–2 μl of stock solution onto the coverslip surface. Allow to air-dry for 1 hour or more.
2. Excise embryos from gravid hermaphrodites in Egg Buffer and allow embryos to settle onto the surface of a poly-L-lysine–coated coverslip. (Egg Buffer [Hird and White 1993]): 118 mM NaCl, 48 mM KCl<!>, 3 mM $CaCl_2$<!>, 3 mm $MgCl_2$<!>, 5 mM HEPES at pH 7.2.)
3. Prepare a culture mount as follows:
   a. Invert the coverslip atop vacuum-grease feet on a slide, to create a chamber with a capacity of 100 μl. Fill the chamber with Egg Buffer.
   b. Seal the coverslip edges with 20 μl of melting-point-temperature bath oil (Sigma).

   Note: The bath oil prevents evaporation of medium from the chamber while permitting sufficient gas exchange for live culture.

   The sealed slides are viewed on the multiphoton microscope at room temperature (~20°C) without the need for special environmental controls.

   Caution: *See Appendix 3 for appropriate handling of materials marked with <!>.*

RAW DATA ACQUISITION AND STORAGE

In addition to the fluorescence confocal stack, a single transmitted light image is typically collected at each time point to allow identification and staging of embryos within the imaged field; this is helpful in setting up batch data processing, especially when the embryos are nonfluorescent for significant periods during development. The software controlling each commercial confocal scanning system mentioned has a different strategy for setting up automated acquisition of 4D data sets. Fortunately, however, most scanning software packages currently run under the Windows operating system. This simplifies the process of writing data files directly onto a networked file server during automated acquisition, with the resulting benefits of expandable terabyte-scale storage capacities to accommodate collection of readily accessible multi-gigabyte 4D raw data sets and automated data archival onto permanent media. Storage directly onto a networked server also permits monitoring of the progress of long recordings from remote computers, as well as commencement of automated data processing of the raw data even before the recording is complete.

PROCESSED DATA OUTPUT

Raw image stacks comprising a field of several randomly oriented embryos are typically processed to produce three useful standardized data types for each embryo: sliced-4D QuickTimeVR (QTVR) movies, stereo-4D QTVR movies, and embryonic developmental expression chronograms (EDECs). A finished sliced-4D QTVR allows free animation through focus and over time for an individual embryo, realigned into the canonical anterior (left), posterior (right), dorsal (up), ventral (down) orientation for display of *C. elegans* embryos (Fig. 1). A stereo-4D QTVR allows free rotational animation of 3D

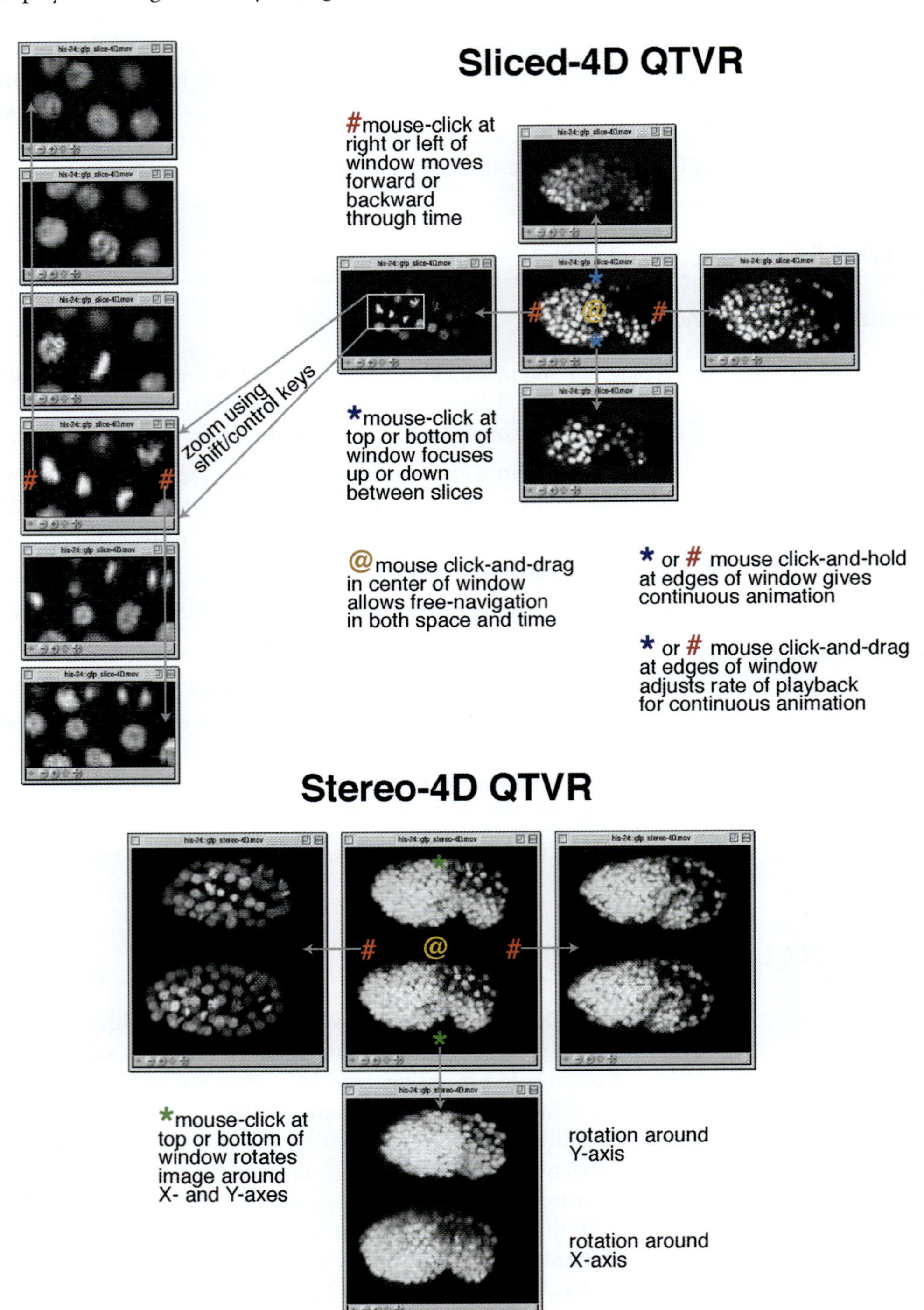

FIGURE 1. (*See facing page for legend.*)

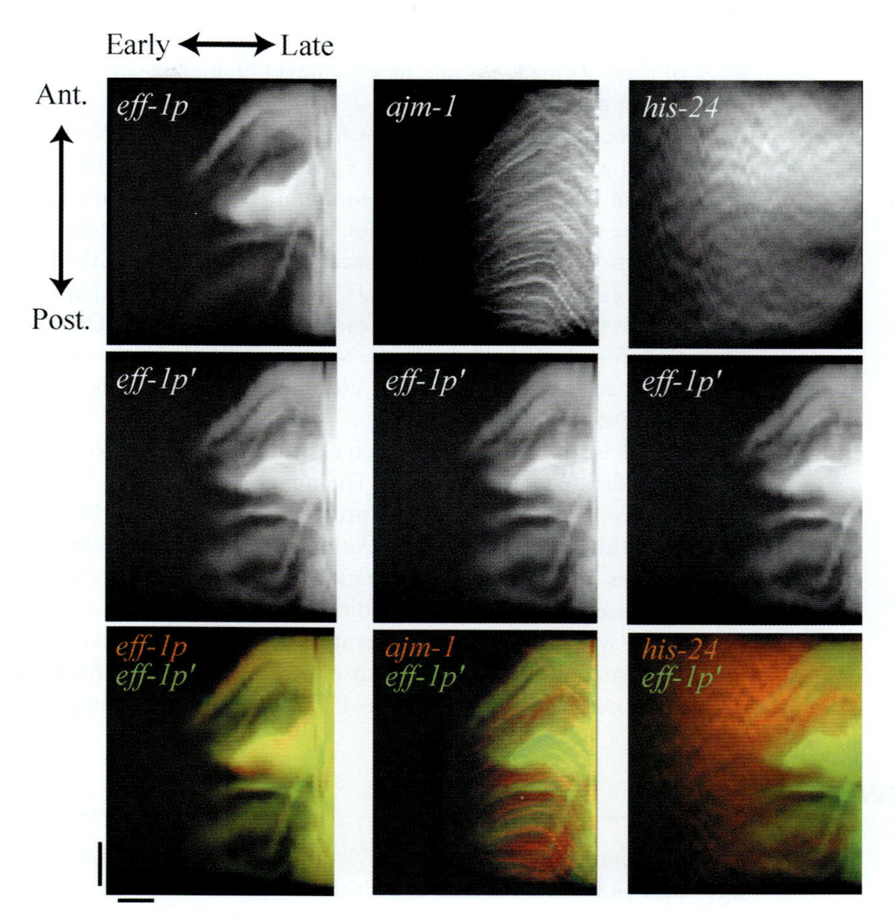

FIGURE 2. Embryonic developmental expression chronograms (EDECs) compiled from recordings of three different fluorescent strains. (*Top row*) EDECs from *eff-1p::gfp* (cytoplasmic expression in subset of differentiating epidermal precursors), *ajm-1::gfp* (junction-localized expression in all differentiating epidermal precursors), *his-24::gfp* (nucleus-localized expression in essentially all cell types). (*Middle row*) EDEC of a second *eff-1p::gfp* embryo (*eff-1p´*), for comparison with top row images. (*Bottom row*) Overlaid pseudocolor display for comparison of EDEC patterns shows reproducibility of the profiles for a given genotype (*eff-1p vs eff-1p´*) and allows correlation of time of expression, migration patterns of expressing cells, and subcellular localization for three genotypes with overlapping spatiotemporal distributions. Vertical bar, 10 µm. Horizontal bar, 60 min.

volume reconstructions (maximum point projection) around the *x* and *y* axes, as well as animation through time. Both sliced-4D and stereo-4D QTVRs benefit from several aspects of the established QuickTime data format: facile navigation of the movie using the computer mouse or keyboard, availability of free viewing software and Web browser plug-ins, streaming download in Web access to large movies, and flexible choices for data compression. Examples of both sliced-4D and stereo-4D QTVR movies may be viewed interactively at http://genetics.uchc.edu/MohlerLab/Movies.html. The third data type, an EDEC, represents the full 4D data set in a two-dimensional summary image—averaged fluo-

---

FIGURE 1. QTVR format movies of *his-24::gfp* fluorescence in a developing *C. elegans* embryo. (*Upper panel, right side*) Frames from a sliced-4D QTVR are arranged and annotated to demonstrate the interaction of the viewer with the movie on the computer screen. (*Upper panel, left side*) Image series shows zoomed display of ~3-min time steps in the mid-cleavage stage. DNA decorated by fluorescent histone protein reveals transition through the cell cycle for several nuclei. (*Lower panel*) Stereo-4D QTVR display of the same embryo at the same time points shown in the upper right panel. All animation control features shown are standard within the free QuickTime Player and browser plug-in software.

rescence intensity along the one dimension of the anteroposterior axis is plotted against time (Fig. 2). Each of these formats has unique advantages in analysis and comparison among expression patterns for different gene products.

The procedure for creating these finished data types currently involves two separate pieces of software running on the Macintosh OS 9 platform. First, a custom-written macro script, running within the program Object Image (http://simon.bio.uva.nl/object-image.html), creates the images that will make up the distinct finished movies. The macro filters detector noise and performs cropping, rotation, and projection operations on each selected embryo in the field. It also produces the 2D EDEC images for each embryo. Second, QTVR Authoring Studio (Apple Computer) collates, compresses, and formats the sets of images for each movie into sliced-4D and stereo-4D QTVR movies for each embryo. Efforts are now being made to combine both of these stages of processing into a single cross-platform package, by writing a comprehensive Java-based plug-in for the program ImageJ (http://rsb.info.nih.gov/ij/).

A database of data sets from each of the available fluorescent genotypes available in *C. elegans* will be of considerable scientific value. The imaging methodologies described here and the strategies for data processing and display should provide a good basis for the construction of such a public archive. Crucial in this effort will be the creation of standardized records of genetically informative fluorescence expression patterns, in formats that allow easy access via the Internet and both optical and machine-based comparisons. In addition to the major task of acquiring and processing 4D data sets, the project will demand development of new tools for making such correlations. These will likely include a synchronized interactive display of QTVR movies from different genotypes and adaptation of biometric algorithms to automated quantitative correlation between EDEC profiles, among others.

ACKNOWLEDGMENTS

We thank the following for their advice, reagents, and technical assistance in developing this technology: Kris Gunsalus, Zbigniew Iwinski, Ion Moraru, and John White. This work was supported by grants from the Muscular Dystrophy Association and the National Institutes of Health (HD43156) to W.A.M.

REFERENCES

Brenner S. 1974. The genetics of *Caenorhabditis elegans*. *Genetics* **77:** 71–94.

Denk W., Strickler J.H., and Webb W.W. 1990. Two-photon laser scanning fluorescence microscopy. *Science* **248:** 73–76.

Hird S.N. and White J.G. 1993. Cortical and cytoplasmic flow polarity in early embryonic cells of *Caenorhabditis elegans*. *J. Cell Biol.* **121:** 1343–1355.

Mohler W.A. and Squirrell J.M. 2000. Multiphoton imaging of embryonic development. In *Imaging neurons: A laboratory manual* (ed. R. Yuste et al.), pp. 21.1–21.11. Cold Spring Harbor Laboratory Press, Cold Spring Harbor, New York.

Mohler W.A. and White J.G. 1998. Multiphoton laser scanning microscopy for four-dimensional analysis of *C. elegans* embryonic development. *Opt. Expr.* **3:** 325–331.

Straub M., Lodemann P., Holroyd P., Jahn R., and Hell S.W. 2000. Live cell imaging by multifocal multiphoton microscopy. *Eur. J. Cell Biol.* **79:** 726–734.

Sulston J.E., Schiernberg E., White J.G., and Thomson J.N. 1983. The embryonic cell lineage of the nematode *Caenorhabditis elegans*. *Dev. Biol.* **100:** 64–119.

CHAPTER 16

Visualizing Morphogenesis in Frog Embryos

Lance A. Davidson[1] and John B. Wallingford[2]

[1]Department of Biology, University of Virginia, Charlottesville, Virginia 22904 and Department of Cell Biology, School of Medicine, University of Virginia Health System, Charlottesville, Virginia 22908; [2]Section of Molecular Cell and Developmental Biology & Institute for Cellular and Molecular Biology, Patterson Laboratories, University of Texas, Austin, Texas 78712

A comprehensive understanding of development requires detailed knowledge of both the tissue movements that shape the embryo and the individual cell behaviors underlying those movements. The *Xenopus laevis* embryo provides an ideal model system: It is particularly amenable to time-lapse study of both whole embryos and individual cells. Indeed, digital microscopy and high-resolution confocal techniques can be combined to reveal the cell biology that accompanies each of the distinct types of cell movements that drive gastrulation and neurulation in *Xenopus laevis* embryos. Furthermore, cells in *Xenopus* embryos are larger than those found in most other vertebrate model systems, making them excellent candidates for cell behavior studies. Moreover, the large cell size facilitates study of protein function and subcellular localization, where the dynamics of green fluorescent protein (GFP)-fusion proteins and fluorescently tagged proteins can be followed within cells that are actively undergoing morphogenetic movements. In addition, combining explant techniques with molecular manipulations allows the dissection of the mechanisms that drive these morphogenetic movements. By complementing live explant techniques with protein localization in fixed tissues, we both validate GFP overexpression studies and extend the relevance of the live studies in the more complex three-dimensional context of cells and tissues in the intact embryo. In this chapter, several useful techniques are presented for imaging cell and tissue movements in *Xenopus* embryos, including methods for enhanced visualization of gene expression and localization of protein distribution in whole embryos. Basic time-lapse imaging techniques to track the gross tissue movements in whole embryos are described, as well as confocal time-lapse techniques for resolving cell behaviors and protein dynamics within cells in tissue explants.

The techniques described here are applicable to the cell biology of morphogenesis: gastrulation and neurulation, whole-embryo imaging, whole-mount confocal microscopy, embryonic cell motility, and embryonic wound healing.

MATERIALS

In addition to the equipment and reagents mentioned below, the following materials are required: frog embryos, polystyrene petri dishes, modeling clay, coverslips (various sizes), diamond pencil, fix vials (scintillation vials), rocking platform (Clay Adams Nutator), equipment for embryo dissection, including forceps, hair loops, hair knives (Keller et al. 1999), and scalpels (no. 15). Unless otherwise noted, materials can be found at Fisher Scientific and reagents can be obtained from Sigma-Aldrich.

Caution: *See Appendix 3 for appropriate handling of materials marked with <!>.*

PROCEDURES

Embryonic tissues can be examined either in live samples where exposed cells can be visualized or in fixed whole-mount samples where cell shape, gene expression, and protein localization can be visualized within the full three-dimensional context of "cleared" embryos. Visualization of whole-mount, fixed embryos is important to validate both the use of GFP-fusion proteins for endogenous proteins and the use of the tissue explant as a model for in vivo movements. Conventional techniques for imaging continue to be useful and are cited, but more recent innovations are described in detail.

Fixing Tissues

A variety of fixatives can be used on whole embryos. These can be formaldehyde-based (Minimum Essential Medium with formaldehyde [MEMFA] fix; Sive et al. 2000) and methanol-based (Dent's fix; Fagotto and Gumbiner 1994), as well as fixes intended for specific targets such as microtubules (Gard's fix; Gard 1999) and extracellular matrix (TCA fix; Marsden and DeSimone 2001). Samples fixed with MEMFA can be processed for gene expression or cell shape, whereas other fixes work better for immunohistochemistry. This section presents a survey of the various fixatives and their uses.

MEMFA (4% Formaldehyde<!>, 0.1 M MOPS<!>, 2 mM EDTA, 1 mM $MgSO_4$<!>, pH 7.6)

Transfer 5–25 embryos to 5 ml of fixative. Fix overnight at 4°C or for 3 hours at room temperature. For long-term storage, dehydrate embryos and store them at –20°C in 100% ethanol<!> or methanol<!>.

Dent's Fix (80% Methanol<!>, 20% DMSO<!>)

This fixative was modified from Fagotto and Gumbiner (1994). Transfer 5–25 embryos to 5 ml of fixative and quickly rinse at least two times over 5 min. Fix embryos overnight at –20°C, transfer to 100% methanol for long-term storage. F-actin can be visualized with phalloidin if acetone is substituted for methanol.

Gard's Fix (Aldehyde<!>-based)

This fixative is from Lane and Keller (1997) and Gard (1999). Transfer 5–25 embryos to 5 ml of fixative. Fix overnight at 4°C. Process embryos immediately for best preservation of microtubule structures. Glutaraldehyde produces high fluorescence background that should be quenched with overnight incubation in $NaBH_4$<!> in PBS (phosphate-buffered saline).

TCA Fix (3% Trichloroacetic acid<!> in PBS)

Transfer 5–25 embryos to 5 ml of fixative. Fix overnight at 4°C. Store embryos for up to 1 week in TCA or dehydrate in methanol for long-term storage. TCA does not fix lineage tracers such as fluorescently conjugated dextrans.

Immunostaining

For this procedure, see Kay and Peng (1991); Selchow and Winklbauer (1997); and Sive et al. (2000).

Visualizing Gene Expression

RNA In Situ Hybridization

For a conventional RNA in situ protocol, see Harland (1991), and for modifications for a "fluorescent RNA in situ" protocol, see Davidson and Keller (1999). Fluorescent RNA in situs simply substitute peroxidase-conjugated Fab fragments for alkaline phosphatase(AP)-conjugated Fab fragments and substitute fluorescent tyramide substrates (Molecular Probes) for the AP substrates such as NBT/BCIP (Pharmacia) in the color reaction. Combined with confocal microscopy, this technique is particularly useful for discerning the cell-by-cell details of gene expression patterns (see Davidson and Keller 1999).

Visualizing Internal Morphology and Anatomy in *Xenopus*

Fixed samples are useful for examining the overall architecture of embryos and, in particular, patterns of gene or protein expression within the embryo. Once samples have been processed for in situ or antibody localization, they can be viewed and documented using a variety of techniques. Paraffin sectioning is commonly used, but it is time-consuming and labor-intensive (Sive et al. 2000). Here, three additional techniques are presented for examining deep tissues in *Xenopus* embryos: bisection, vibratome sectioning, and confocal optical sectioning. Vibratome sectioning produces samples of regular thickness, whereas hand bisection can produce nearly intact "half-mounts" where contiguous tissue architecture can be visualized. Sagittal and transverse sections can be achieved with conventional sectioning techniques, whereas more complex or unique preparations, such as coronal or *en face* sections, are more easily prepared by hand dissection of fixed embryos with scalpels. Confocal optical sectioning can be applied to both vibratome slices and bisected embryos, as well as intact whole embryos.

Confocal Microscopy

Confocal images of live and fixed samples shown here were collected with an entry-level laser-scanning confocal (PCM2000, Nikon) mounted on an inverted compound microscope (Eclipse TE200, Nikon). High-resolution images of samples that could be prepared in close proximity to the coverslip, such as bisected preparations or explants, were collected using high-magnification oil-immersion objectives (Nikon Plan Apo 60x, NA 1.4, working distance 0.21 mm; Nikon Plan Apo 40x, NA 1.0, working distance 0.16 mm). Lower-resolution images, or stacks of thick tissues, were collected using a 20x air objective (Nikon Plan Apo 20x, NA 0.75, working distance 1.0 mm). Images of fluorescein and GFP fluorophores were collected using 488-nm excitation from an Argon laser, whereas images of Alexa 568 or rhodamine were collected using 543-nm excitation from a Green HeNe laser. High spatial resolution was achieved for fixed specimens using the 20-μm pinhole, whereas the pinhole was frequently opened to 50 μm, reducing the ultimate resolution, to increase signal intensity and reduce phototoxicity in live preparations.

Bisection

Once embryos have been fixed and prepared by immunostaining or a fluorescence color reaction, they can be prepared for confocal sectioning. Carry out bisection by hand in PBS with 0.1% Tween-20 (PBST) in agarose-coated petri dishes. Use a hair loop or forceps (used only with fixed samples) to roll the embryo for dissection with a scalpel or with a fragment of razor blade (usually held with a hemostat or another set of forceps). Wash manually dissected samples twice in PBST to remove debris.

In preparation for clearing, dehydrate samples in 100% methanol<!> and rinse them further in 100% methanol (twice) over 20 minutes. These dehydrated samples can be stored at –20° C until immediately before clearing and confocal sectioning.

When performing in situ hybridization with markers of deep tissue, poor penetration of the reagents can prevent accurate labeling of deep cells. This problem can be avoided by routinely fixing embryos for 2 hours in MEMFA<!>, bisecting them, and then fixing them for an additional 30 min. The normal in situ protocol can then be performed on the "half-mount" embryos (see, e.g., Haigo et al. 2003).

Simple Vibratome Sectioning for Xenopus *Embryos*

For simple cross sections, embed embryos in low-melt agarose and section on a vibratome. The protocol is easy, inexpensive, and very fast. Consistently good results can be obtained with 150-µm sections. Reliable, although less consistent, results can be obtained with 50–100 µm sections. Fix embryos in MEMFA<!> for 3 hours (or longer) at room temperature, wash with PBST, and then transfer to 0.5x PBS. Place an embryo in a drop of PBS in the bottom of a standard weigh boat, and use a fine pipette to remove PBS. Place a drop of molten 4% low-melt agarose (in 0.5x PBS) on top of the embryo, and orient the embryo appropriately with forceps. This is best done under a stereoscope. Allow the agarose to cool; watch the embryo during this time to make sure that it remains in place. Once it has cooled, cut the drop of agarose containing the embryo into a cube with a razor blade, and secure this small block with cyanoacrylate adhesive onto a small metal block (custom-made) that will fit into the sample holder of the vibratome.

Section the block in a reservoir of 0.5x PBS. Remove the tissue slices from the reservoir with a paintbrush or a pipette. In some cases, the tissue will remain in the agarose, and in other cases the tissue slice will come free upon sectioning. Regardless, mount the samples in PBS in depression slides, under coverslips, for viewing on the compound scope.

Whole-Mount Confocal Optical Sectioning

Bleaching of *Xenopus* embryos is a commonly used technique for visualizing deep tissues in pigmented embryos (Sive et al. 2000). Our preferred method involves a brief incubation for as little as 60 min in 0.5x SSC with 5% formamide<!> and 1% H_2O_2<!> under a fluorescent lamp. However, the use of albino embryos is preferable to pigmented embryos that have been bleached; the bleaching can reduce the fluorescence signal and visual markers such as those deposited by alkaline phosphatase or horseradish peroxidase.

Clearing of *Xenopus* embryos is another commonly used technique for visualizing in situ probes or antibodies in deeper tissues, and the technique is well described elsewhere (Kay and Peng 1991; Sive et al. 2000). The "clearing" process renders the normally opaque *Xenopus* embryo nearly transparent and allows either stereoscopic imaging or high-resolution confocal imaging of deep tissues in intact whole embryos, thick vibratome sections, or bisected half-mounts.

The embryo is then oriented in the chamber (see below) such that the proper sectional plane is presented. Confocal microscopy allows serial sectioning by focusing up or down through the sample. Adjusting the pinhole allows thicker or thinner optical sections to be collected. With a well-bleached and well-cleared *Xenopus* embryo, confocal optical sections can be collected through the entire embryo.

Clearing requires immersion of samples in 2BB:BA (2:1 benzyl benzoate<!> to benzyl alcohol<!>), potent organic solvents, so mounting these embryos requires some care. For confocal microscopy, cleared embryos can be mounted in simple home-made chambers (see below).

Simple Chambers for Mounting and Imaging Cleared Embryos

2BB:BA-resistant mounting chambers should be made at least 24 hours prior to mounting samples (Fig. 1A). To make a single chamber, apply a thin coat of clear nail polish (Sally Hansen "Hard as Nails") to a nylon washer (3/8-inch diameter; e.g., Small Parts) and immediately place the washer, polish side down, onto a large coverslip (e.g., 45 x 50 mm). Allow the nail polish to dry and check for a complete seal. Chambers made this way will dry overnight and can be stored in a dust-free container until needed.

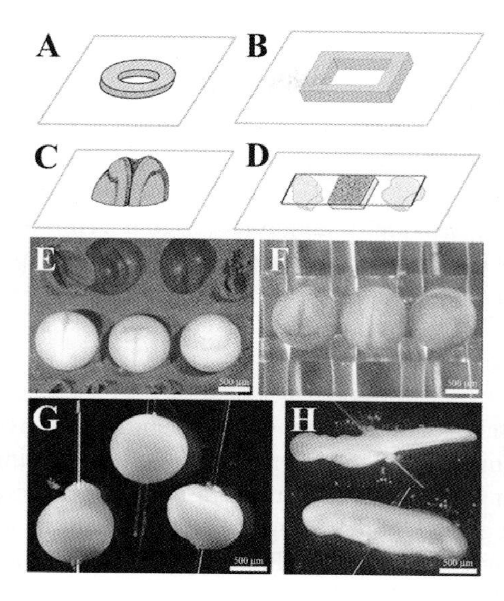

FIGURE 1. Chambers and methods for immobilizing samples, explants, and whole embryos. (*A*) Nylon washer for fixed and cleared samples; (*B*) acrylic chambers for live samples; (*C*) transverse orientation for a bisected cleared sample; (*D*) coverslip and silicon grease bridge for immobilizing explants; (*E*) clay wells for holding embryos; (*F*) Nitex grids for holding embryos; (*G*) plastic "spears" for immobilizing embryos in a particular orientation (late neurula stage albino embryos); (*H*) two of the same embryos in *G* showing normal development even while speared.

Clearing and Mounting

Samples dehydrated in methanol should be transferred with a disposable pipette (e.g., Fisher Scientific) to the well formed by the nylon washer prior to clearing. Transferring samples after clearing is not recommended as cleared samples tend to be completely transparent and are easily lost. Clear the samples by replacing methanol with 2BB:BA under a stereoscope using an automated pipettor to iteratively remove methanol/2BB:BA and replace with fresh 2BB:BA. Care should be exercised to prevent methanol or 2BB:BA from spilling out of the center of the washer since samples can easily "surf" out of the nylon washer well. Once samples have "settled" to the bottom of the well, orient them (Fig. 1C) under the stereoscope (a fluorescence-equipped stereoscope is very handy for this) with a hair-knife rinsed with methanol. Add more 2BB:BA to the well and place a small square coverslip over the washer. Small bubbles can be tolerated if they do not interfere with imaging or physically disturb the samples. Samples that do not clear in 2BB:BA hold residual water and must be removed, fully dehydrated, recleared, and remounted. Recover the samples by flushing the contents of the nylon washer well into a petri dish with 100% methanol<!>. Mounting chambers with cleared samples are best imaged immediately, but they can remain stable and clear for several days.

Samples can be photographed at any time, as whole mounts or as cleared samples, using a fluorescence-equipped stereoscope, an epifluorescence-equipped inverted compound microscope, or a confocal mounted on an inverted microscope. When using fluorescein-based detection, use 2BB:BA (pH 8) because the quantum efficiency of fluorescein is significantly reduced at low pH (Molecular Probes). 2BB:BA can take days or even weeks to reach pH 8 (check with pH paper) with NaOH<!> chips at the bottom of the bottle.

Visualizing Development in Live Tissues

Frog embryos have a long history of microsurgical manipulation. Development can be followed either in whole embryos or in tissue fragments dissected from embryos. Although frog embryos are not optically clear like mouse or zebrafish embryos, cells and subcellular structures within the embryo can be visualized through microsurgery and subsequent culture of tissue explants. Moreover, cells in developing frog embryos are five times (mesendoderm) to three times larger (neural plate) than analogous cells in mouse, avian, or zebrafish embryos. The large size of cells makes possible the identification of cellular protrusions (e.g., lamellipodia and filopodia) and subcellular protein localization studies on a par with those carried out on single cultured cells.

Targeted Expression

Stereotypical development of the 32-cell stage embryo has been extensively characterized (Dale and Slack 1987; Lane and Sheets 2002). This means that injections of mRNA, or lineage label, can be targeted to specific blastomeres that are fated to become specific tissues at specific locations with a high degree of accuracy. Furthermore, injected embryos can be screened at gastrula stages to ensure expression in only those tissues that were targeted (see Wallingford et al. 2002). Homotypic and homochronic grafts can also be used to express RNA or proteins in specific patterns or specific locations in the gastrula or neurula (see Wallingford and Harland 2001).

Whole Embryos

Development of whole embryos can be followed either at low magnification using a camera mounted on a stereoscope or at high magnification with an embryo mounted in a simple chamber. At low magnification, a field of embryos in various random orientations can be followed, but more control over the desired perspective can be gained by removing the embryo from its vitelline envelope and positioning it in a clay well, shaped by a flame-rounded Pasteur pipette. Embryos cultured in clay wells (Fig. 1E) or Nitex mesh (Small Parts) (Fig. 1F) will develop fewer abnormalities if the culture medium is augmented with both 0.1% BSA (bovine serum albumin) and antibiotic-antimycotic (Sigma). Typically, embryos can be filmed for up to 5 hours in simple clay wells. Longer-term time lapses (more than 24 hours) or time lapses through particularly violent morphogenetic movements such as neural tube formation have been collected from embryos pinned in place by short pieces of plastic, hair, or eyelash (Fig. 1G,H). A "speared" embryo can be held in place by cementing the ends of the spear using silicone grease (high vacuum, Dow Corning) and glass coverslip fragments.

Explants

With practice, tissue can be microsurgically isolated from any region of the embryo (Keller et al. 1999). Tissues can be isolated as cores or plugs across all cell layers or as finely as a single cell layer. Explanted tissue can be held gently under a coverslip fragment secured in place with silicone grease (high vacuum, Dow Corning) (Fig. 1D). When cultured in DFA (Davidson et al. 2002)—a specially formulated culture medium that matches the low-chloride and high-pH conditions of the interstitial fluid of the early *Xenopus* embryo (Gillespie 1983)—cells within explants continue in vivo programs of cell motility and cell rearrangement (Keller et al. 1985). In general, cells within physically isolated explants autonomously recapitulate movements within the embryo.

Simple Acrylic Chambers

We have developed a simple acrylic chamber that holds a small volume of medium; it can be sealed, cleaned, and reused easily with *Xenopus* tissue explants. Any reasonably equipped machine shop can cut cover-slide-sized pieces of 1/4-inch-thick acrylic (25 x 50 mm) and mill multiple wells in each (Fig. 1B). These acrylic chambers can be "glued" to large coverslips with silicone grease (high vacuum, Dow Corning). Multiple explants can be transferred into each of these wells and sandwiched under coverslip fragments by silicon grease. Silicon grease along the rim of the acrylic well can then be used to seal the well with a smaller square of coverslip (25 x 25 mm). Furthermore, a 200:1 dilution of an oxygen scavenger (e.g., Oxyrase, Oxyrase Corp.) can be added to the well before it is sealed without air bubbles. Inclusion of an oxygen scavenger does not alter early *Xenopus* development but reduces photobleaching and phototoxicity (Tanaka and Kirschner 1991; Adams et al. 2003).

Glass-bottomed Petri Dishes

Glass-bottomed petri dishes can be purchased (MatTek) or made in the lab from small polystyrene petri dishes. Using pliers, hold a penny in the flame of a bunsen burner until it is hot. Immediately place the hot penny in the bottom of the dish; it will melt a near-perfect circular hole in the bottom of the dish (it will stink, so do this in a chemical fume hood). Use a razor and sandpaper to trim any leftover plastic from the edge. Next, attach a large rectangular coverslip with silicone grease (high vacuum, Dow Corning) to the bottom of the dish, producing a petri dish with a coverslip bottom, perfect for imaging on an inverted compound microscope. Home-made chambers occasionally leak culture medium down into the microscope (which is very bad), so check the seal between the coverslip and the petri dish before use.

EXAMPLE OF APPLICATIONS

Live Probes

Examining Cellular Morphology with Membrane-targeted GFP

Expression of cell-membrane-targeted GFP provides an excellent means for examining cell morphology in living tissues. The GAP43-GFP fusion protein (Moriyoshi et al. 1996) and GFP fused to the CAAX (farnesylation) domain of Ras (memGFP; Wallingford et al. 2000) are both good reagents. Even at very high levels of expression, almost no GFP is detected in the cytoplasm. We typically inject 60–500 pg of mRNA encoding either GAP43-GFP or memGFP for visualization at gastrula stages. These reagents provide very bright labeling of cell membranes, allowing visualization not only of the cell body but also membrane protrusions such as filopodia and lamellipodia (Fig. 2A).

An alternative reagent for labeling cell membranes in *Xenopus* explants is FM4-64. This fluorescent lipophilic dye is good for visualizing cell bodies but fails to highlight protrusions. It has the additional advantage of not requiring injection; explants are cut and simply bathed in FM4-64 over about 10 min and then mounted for imaging. Its red channel fluorescence also makes it a nice complement to GFP fusions, although this reagent photobleaches rather rapidly and is not ideal for time-lapse.

Examining Subcellular Localization of GFP-fusion Proteins

By mRNA expression of GFP fusions to a gene of interest, it is possible to examine the subcellular localization of the fusion protein in explanted *Xenopus* tissues (Wallingford et al. 2000). Because of the difficulties associated with expressing too much of a fluorescent protein (see below), this technique requires a low concentration of the GFP fusion and optimal light collection. The large size of the *Xenopus* cell allows resolution to be sacrificed for photons. Inject approximately 50 pg of the mRNA of interest and carry out imaging using a high-NA objective.

Hazards of Overexpression

Fluorescent probes, both fluorophore-coupled proteins and GFP-fusion proteins, can be easily visualized in *Xenopus* cells. However, the degree to which the overexpressed probe reflects endogenous localization must be assessed cautiously. Plasma membrane (e.g., transmembrane or palmitoylated) proteins can quickly saturate available binding partners and become simple markers of the plasma membrane, losing their capacity to illustrate specific patterns of subcellular localization. Erroneous conclusions can also be drawn from overexpressed cytoplasmic proteins. For example, a nuclear-localizing GFP (Kroll and Amaya 1996) can be seen localized to the yolk-free regions of mesodermal cells in confocal sections from a marginal zone explant (Fig. 2C). *Xenopus* embryonic cells are filled with yolk granules, whose negative image is recorded by cytoplasm-filling fluorophores, coupled to dextran lineage tracers. Yolk granules and other vesicles are excluded from the subapical cortex in eggs and embryonic cells such that dextran lineage tracers appear to be localized to the cell periphery. In fact, it is simply more cytoplasm and thus more fluorophore per pixel. Both of these conditions can be controlled for. It is best to test plasma-membrane-localizing proteins at the lowest possible level needed for visualization. Localization should not change appreciably as expression levels are increased. Movement of proteins to the cell periphery can be checked by coinjection with a dextran lineage marker. Ratioing of the GFP-protein levels to the fluorescent dextran should provide an accurate image of

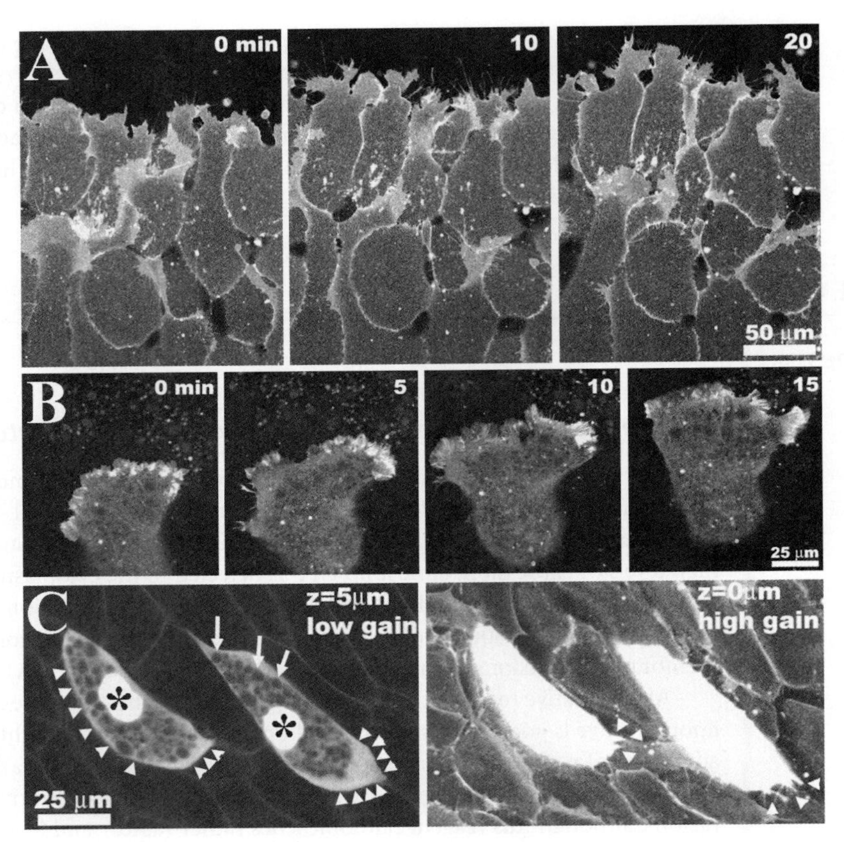

FIGURE 2. Live sample applications. (*A*) Sequence of time-lapse confocal frames of GAP43-GFP expressed in the leading edge of a sheet of migrating mesendoderm, collected with a 40x 1.0 NA lens. Cells direct monopolar lamellipodia away from the marginal zone. (*B*) Sequence of a single Alexa-568 actin-labeled mesendoderm cell against a background of unlabeled cells at the leading edge of a migratory sheet. (*C*) Hazards of overexpression. Two cells expressing nucGFP are seen within a deep confocal section at low gain (*left*) and in a confocal section collected at the surface of the coverslip with high gain (*right*) against a background of cells labeled with GAP43-GFP (only visible with high gain). Individual yolk granules (*arrows*) can be seen in the low-gain image where nucGFP appears "enriched" at the nucleus (*asterisk*), cortex, and mediolateral ends of the cells (*arrowheads*) where protrusions are visible in the high-gain image (*arrowheads*).

true localization. Of course, whenever possible, the best validation of live GFP-fusion protein localization is to determine whether GFP colocalizes with antibody staining of the endogenous protein.

Examining Actin Dynamics with Alexa-568 Actin

Blastomeres up to stage 7 (128–256 cells) can be injected with 1–2 ng equivalent per embryo of bovine-actin conjugated with Alexa-568 (Molecular Probes). Marginal zone explants were cut at stage 10 from injected embryos and plated onto a fibronectin-coated coverslip (in a well, as described above). A single labeled mesendoderm cell at the leading edge of the explant was tracked over time using a 1.4 NA 60x oil-immersion objective (Fig. 2B).

Fixed Probes

Fixed Alexa-568 Actin

Embryos were prepared as described above for live imaging. Whole embryos were then fixed at early gastrula stages in Dent's fix, cleared, and "half-mounted." 200 sagittal confocal sections (single section shown in Fig. 3A) were collected at 0.2-μm intervals and projected to a single plane (Fig. 3B). Actin-rich lamellipodia are seen extending from one end of these post-involution mesoderm cells.

Fluorescent RNA In Situ Hybridization

Whole embryos were cultured to late gastrula stages and then fixed and processed for fluorescent RNA in situ to visualize neuron-specific tubulin gene expression (n-tubulin; Chitnis et al. 1995). Samples were cleared and "half-mounted," and an image was collected with a cooled-CCD camera mounted on an epifluorescence-equipped stereoscope (Fig. 3C). Transverse confocal sections were collected and showed the restriction of n-tubulin RNA to prospective neurons (Fig. 3D).

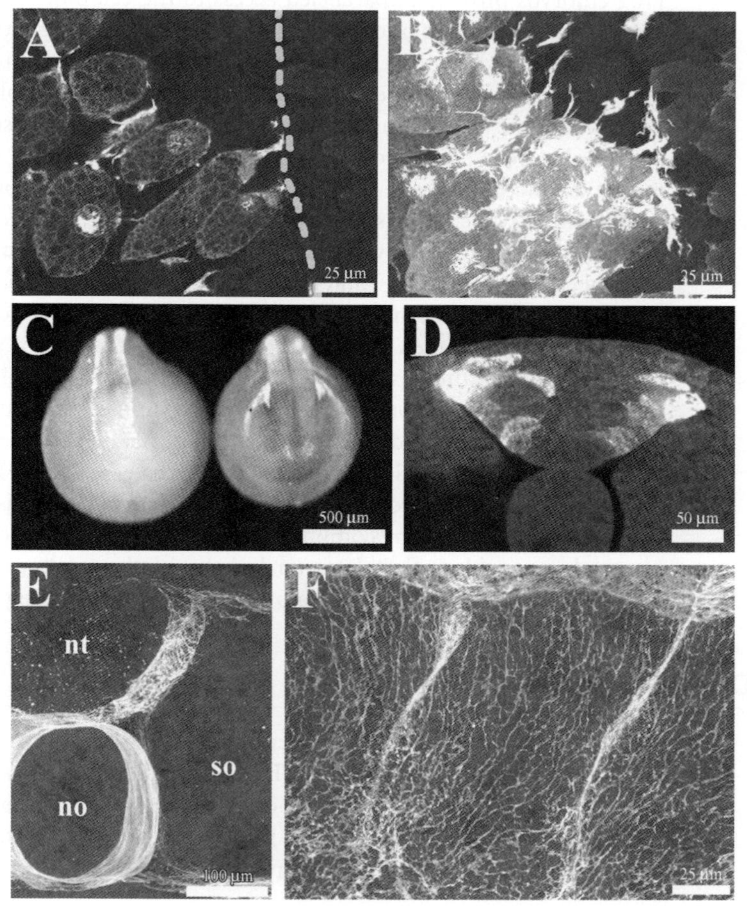

FIGURE 3. Fixed sample applications. (*A*) Single confocal section through the post-involution mesoderm shows polymerized actin localization in a patch of cells labeled with Alexa-568 actin. The embryo was injected with actin at the 32-cell stage and fixed in early gastrulation, and a sagittal half-mount was bisected and mounted in 2BB:BA. (*B*) An entire stack of confocal sections (one of which is shown in *A*) was projected onto a single plane using a "brightest point" algorithm. (*C*) Fluorescence detection of RNA in situ hybridization of neuron-specific tubulin expression in a transversely bisected whole embryo. After the color reaction, the embryo was bisected and cleared, and an image was collected with a cooled-CCD camera mounted on a stereoscope. (*D*) Single confocal section of a sample prepared as in *C* using a 20x objective. (*E*) Laminin localization in the projected image of an entire stack of confocal sections collected from a transversely bisected neurula stage embryo. Laminin takes on a lamellar form around the notochord (no) but is more fibrillar around the neural tube (nt) and surrounding the somites (so). (*F*) Image showing fibrillar organization and orientation on the dorsal surface of a single somite projected from a stack of images collected through an *en face* dissected preparation from a slightly later neurula stage embryo.

Immunofluorescence Localization of Proteins

Whole embryos were cultured to early tailbud stage, fixed with TCA, immunostained with a primary antibody to laminin (rabbit polyclonal; Sigma Aldrich) with a rhodamine-conjugated secondary antibody (Jackson ImmunoLaboratory), and "half-mounted." 80 transverse sections were collected at 1-μm intervals and projected to a slight offset (Fig. 3E). By positioning a dorsal fragment such that the somitic mesoderm can be seen *en face*, subtle differences in the fibrillar array of laminin, from fibrillar along the dorsal and ventral surface of the somite to laminar around the notochord and medial-laterally oriented over the somite, can be seen in the context of the intact tissue (Fig. 3F).

Advantages and Limitations

Frog embryos provide unparalleled access to the cell biology of morphogenesis. The elucidation of molecular mechanisms of cell biology has depended on the fortuitous conditions provided by lines of primary cells such as yeast or fish epidermal keratocytes, cultured cells such as 3T3 cells, or intracellular parasites such as *Listeria*. Particular problems in cell biology can be addressed in these systems because of the relative ease of using biochemistry, high-resolution imaging, and genetics. Rapid progress has been made because these techniques complement each other. Similarly, elucidation of the molecular mechanisms of morphogenesis, i.e., the cell biology of morphogenesis, will depend on complementary techniques. Cell biology studies using *Xenopus* embryos complement genetic studies of mouse and zebrafish development. Stable cell lines, biochemistry, high-resolution imaging, microsurgical explant techniques, and the rapid assessment of phenotypes in whole embryos make *Xenopus* a critical model system in understanding vertebrate morphogenesis.

One drawback often attributed to *X. laevis* is the lack of genetic data and a sequenced genome. However, both of these issues are being addressed presently. Model systems for vertebrate morphogenesis are genetically complex, zebrafish and mouse are pseudo-octoploid, and *X. laevis* is pseudotetraploid. A true diploid relative of *X. laevis*, *Xenopus tropicalis*, is currently being developed as a genetic model system (Hirsch et al. 2002; Khokha et al. 2002). In-bred and transgenic lines of *X. tropicalis* are being established, and a genome sequencing project is under way (Klein et al. 2002).

It is our belief that by combining cell biological analysis of morphogenesis—in systems that are genuinely amenable to time-lapse analysis—with the results of mutational analysis in genetic systems, progress can be made toward a comprehensive understanding of vertebrate development.

ACKNOWLEDGMENTS

This work was supported by an American Cancer Society postdoctoral fellowship (L.D.), the National Institutes of Health (USPHS/NICHHD R01-HD26402 to Douglas DeSimone and R01-HD25594 to Raymond Keller in whose labs L.D. carried out this work), and a Burroughs Wellcome Fund Career Award in the Biomedical Sciences to J.B.W. Confocal facilities were provided by the W. M. Keck Center for Cellular Imaging at the University of Virginia.

REFERENCES

Adams M.C., Salmon W.C., Gupton S.L., Cohan C.S., Wittmann T., Prigozhina N., and Waterman-Storer C.M. 2003. A high-speed multispectral spinning-disk confocal microscope system for fluorescent speckle microscopy of living cells. *Methods* **29:** 29–41.

Chitnis A., Henrique D., Lewis J., Ish-Horowicz D., and Kintner C. 1995. Primary neurogenesis in *Xenopus* embryos regulated by a homologue of the *Drosophila neurogenic gene Delta* (comments). *Nature* **375:** 761–766.

Dale L. and Slack J.M. 1987. Fate map for the 32-cell stage of *Xenopus laevis*. *Development* **99:** 527–551.

Davidson L.A. and Keller R.E. 1999. Neural tube closure in *Xenopus laevis* involves medial migration, directed protrusive activity, cell intercalation and convergent extension. *Development* **126:** 4547–4556.

Davidson L.A., Hoffstrom B.G., Keller R., and DeSimone D.W. 2002. Mesendoderm extension and mantle closure in *Xenopus laevis* gastrulation: Combined roles for integrin $\alpha_5\beta_1$, fibronectin, and tissue geometry. *Dev. Biol.* **242:** 109–129.

Fagotto F. and Gumbiner B.M. 1994. β-catenin localization during *Xenopus* embryogenesis: Accumulation at tissue and somite boundaries. *Development* **120**: 3667–3679.
Gard D.L. 1999. Confocal microscopy and 3-D reconstruction of the cytoskeleton of *Xenopus* oocytes. *Microsc. Res. Tech.* **44:** 388–414.
Gillespie J.I. 1983. The distribution of small ions during the early development of *Xenopus laevis* and *Ambystoma mexicanum* embryos. *J. Physiol.* **344:** 359–377.
Haigo S.L., Harland R.M., and Wallingford J.B. 2003. A family of *Xenopus* BTB/Kelch repeat proteins related to ENC-1: New markers for early events in floorplate and placode development. *Gene Expr. Patterns* **3:** 669–674.
Harland R.M. 1991. In situ hybridization: An improved whole-mount method for *Xenopus* embryos. *Methods Cell Biol.* **36:** 685–695.
Hirsch N., Zimmerman L.B., Gray J., Chae J., Curran K.L., Fisher M., Ogino H., and Grainger R.M. 2002. *Xenopus tropicalis* transgenic lines and their use in the study of embryonic induction. *Dev. Dyn.* **225:** 522–535.
Kay B.K. and Peng H.B. 1991. Xenopus laevis*: Practical uses in cell and molecular biology*. Academic Press, New York.
Keller R., Poznanski A., and Elul T. 1999. Experimental embryological methods for analysis of neural induction in the amphibian. *Methods Mol. Biol.* **97:** 351–392.
Keller R.E., Danilchik M., Gimlich R., and Shih J. 1985. The function and mechanism of convergent extension during gastrulation in *Xenopus laevis*. *J. Embyol. Exp. Morphol.* (suppl.) **89:** 185–209.
Khokha M.K., Chung C., Bustamante E.L., Gaw L.W., Trott K.A., Yeh J., Lim N., Lin J.C., Taverner N., Amaya E., Papalopulu N., Smith J.C., Zorn A.M., Harland R.M., and Grammer T.C. 2002. Techniques and probes for the study of *Xenopus tropicalis* development. *Dev. Dyn.* **225:** 499–510.
Klein S.L., Strausberg R.L., Wagner L., Pontius J., Clifton S.W., and Richardson P. 2002. Genetic and genomic tools for *Xenopus* research: The NIH *Xenopus* initiative. *Dev. Dyn.* **225:** 384–391.
Kroll K.L. and Amaya E. 1996. Transgenic *Xenopus* embryos from sperm nuclear transplantations reveal FGF signaling requirements during gastrulation. *Development* **122:** 3173–3183.
Lane M.C. and Keller R. 1997. Microtubule disruption reveals that Spemann's Organizer is subdivided into two domains by the vegetal alignment zone. *Development* **124:** 895–906.
Lane M.C. and Sheets M.D. 2002. Rethinking axial patterning in amphibians. *Dev. Dyn.* **225:** 434–447.
Marsden M. and DeSimone D.W. 2001. Regulation of cell polarity, radial intercalation and epiboly in *Xenopus*: Novel roles for integrin and fibronectin. *Development* **128:** 3635–3647.
Moriyoshi K., Richards L.J., Akazawa C., O'Leary D.D., and Nakanishi S. 1996. Labeling neural cells using adenoviral gene transfer of membrane-targeted GFP. *Neuron* **16:** 255–260.
Selchow A. and Winklbauer R. 1997. Structure and cytoskeletal organization of migratory mesoderm cells from the *Xenopus* gastrula. *Cell Motil. Cytoskelet.* **36:** 12–29.
Sive H.L., Grainger R.M., and Harland R.M., eds. 2000. *Early development of* Xenopus laevis: *A laboratory manual.* Cold Spring Harbor Laboratory Press, Cold Spring Harbor, New York.
Tanaka E.M. and Kirschner M.W. 1991. Microtubule behavior in the growth cones of living neurons during axon elongation. *J. Cell Biol.* **115:** 345–363.
Wallingford J.B. and Harland R.M. 2001. *Xenopus* Dishevelled signaling regulates both neural and mesodermal convergent extension: Parallel forces elongating the body axis. *Development* **128:** 2581–2592.
Wallingford J.B., Fraser S.E., and Harland R.M. 2002. Convergent extension: The molecular control of polarized cell movement during embryonic development. *Dev. Cell* **2:** 695–706.
Wallingford J.B., Rowning B.A., Vogeli K.M., Rothbacher U., Fraser S.E., and Harland R.M. 2000. Dishevelled controls cell polarity during *Xenopus* gastrulation. *Nature* **405:** 81–85.

CHAPTER 17

In Vivo Imaging of Synaptogenesis in the Embryonic Zebrafish

James D. Jontes and Stephen J Smith

Department of Molecular and Cellular Physiology, Stanford University School of Medicine, Stanford, California 94305

Imaging synaptogenesis in vivo is feasible. With the use of a strong neuronal promoter and a good synaptic marker–GFP fusion, it is possible to image synaptogenesis in the developing zebrafish embryo. With care and a reasonably efficient laser-scanning microscope (either two-photon or confocal), collecting data is not likely to be a serious obstacle. The zebrafish is a particularly good model system for studying neurodevelopmental events, such as synaptogenesis. The embryos are small and transparent and develop rapidly (events such as axon growth and synaptogenesis are well under way by 24 hours postfertilization; see Fig. 1). Moreover, the embryonic nervous system is highly stereotyped, with the hindbrain and spinal cord, in particular, being composed of a manageable number of identified neurons. Thus, it is possible to both observe and perturb normal developmental events and processes as they happen in their native environment of an intact, living embryo.

This chapter outlines the steps and techniques required to image the process of synaptogenesis in the developing zebrafish embryo, although most of the information will be generally useful and applicable to a wide range of biological questions. In addition to the information presented here, a number of excellent reference sources for zebrafish methods and protocols are recommended (Westerfield 1995; Detrich et al. 1999a,b; Nüsslein-Volhard and Dahm 2002).

LABELING SYNAPTIC PROTEINS

Transgenic Fish or Transient Expression?

One of the attractive features of the zebrafish is the relative ease with which transgenic lines expressing GFP or a GFP-fusion protein can be generated. It is possible to inject a linearized expression plasmid into a few hundred embryos, grow them to adulthood, and select those that integrated the transgene into their germ line. The primary advantage of such a strategy is that it reduces mosaicism of expression in the transgenic progeny (discussed below); i.e., the detailed pattern of expression is more

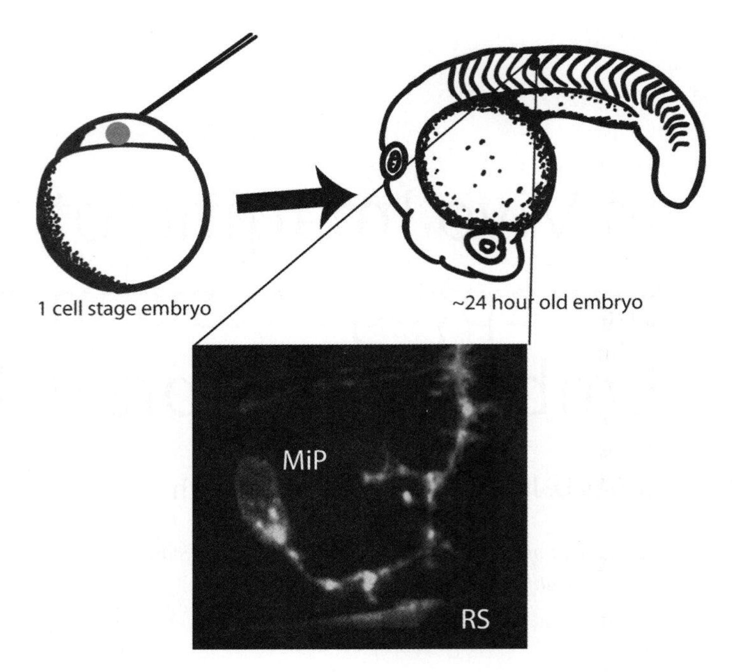

FIGURE 1. Transient gene expression in zebrafish embryos. This figure illustrates the general approach described in this chapter: DNA for a GFP-fusion protein is injected into a 1-cell-stage embryo. At about 24 hours postfertilization, the embryo is immobilized and imaged on a two-photon microscope. In the example shown, the embryo was injected with two expression plasmids: an *α1-tubulin* promoter/enhancer construct driving Gal4, and a UAS effector cassette driving zebrafish VAMP-GFP. The labeled neuron is a MiP (Middle Primary) motoneuron. Ventral to the motoneuron is the passing growth cone of a reticulospinal neuron.

consistent and reproducible from fish to fish. Moreover, it bypasses the need for the daily DNA injections required for transient gene expression; all that is necessary is to collect the embryos from the transgenic line, all of which will express the transgene, and to perform the imaging experiment. However, there are several potential drawbacks to using transgenic fish. First, it takes several months to generate a sexually mature F_1 generation in order to collect expressing F_2 embryos, a period that could be spent collecting data from transiently expressing embryos. Additionally, a dense field of GFP-labeled neurons may not be the most suitable situation in which to clearly image processes occurring in a single, identifiable cell. Finally, there is a significant cost, both in space and money, associated with maintaining a large number of transgenic fish lines.

For many purposes, injection of DNA into 1- to 4-cell-stage embryos may provide adequate results. It is rapid, generating GFP-expressing neurons, which last for several days, in less than 24 hours. In addition, the sparse labeling of neurons, akin to that seen by Golgi staining of neurons, often makes transient expression much more suitable for imaging experiments.

Promoters/Enhancers

Although it is possible to use ubiquitous promoters, such as CMV and EF1α, for expressing synaptic markers in zebrafish embryos, this is not advisable for imaging purposes because of their widespread expression patterns. The discussion here is restricted to pan-neural and neuron-specific promoters.

Pan-neural

Pan-neural promoters are active in a broad distribution of neuronal cell types. One such promoter is the *α1-tubulin* promoter from goldfish (Hieber et al. 1998; Goldman et al. 2001), which has been used to drive ectopic expression of GFP in several types of zebrafish neurons, including retinal ganglion cells, tectal neurons, reticulospinal neurons, spinal motoneurons, Rohon-Beard neurons, and spinal interneurons. The *HuC* zebrafish promoter has also been used to drive pan-neural GFP expression (Park et al. 2000; Gleason et al. 2003). It shows a similar expression pattern, but has the advantage of becoming active earlier and staying on longer than the *α1-tubulin* promoter.

Although pan-neural promoters/enhancers have a variety of uses, their broad expression patterns can cause problems. In cases where labeling a single, specific cell type is desired, it may be necessary to screen a large number of embryos to identify an individual suitable for imaging. This can be both tedious and time-consuming. The broad expression pattern may also complicate the field of view with additional labeled cells and neuronal processes. Finally, the mosaicism may make the imaging of specific neurons, such as the Mauthner cells (of which there are only two per embryo), totally impractical, since the probability of expressing a gene in any given cell is extremely low. Despite these potential limitations, pan-neural promoters can be extremely useful. In some cases, the broad expression pattern can be a tremendous boon, demonstrating whether a given observation is specific to a given neuron or subset of neurons, or whether it is present across a broad set of cell types. Similarly, it may be possible to observe and characterize differential effects of perturbations in different cell types. Such variation would not be visible using a more specific promoter.

Cell-specific Promoters

Although they are extremely attractive for use in some imaging experiments, the present limitation of cell-type-specific promoters is their relative paucity in zebrafish. Only a handful of good, cell-type-specific promoters have been published to date. A portion of the *islet-1* promoter (Higashijima et al. 2000), CMICP, is capable of driving expression in cranial motoneurons, as well as in other subsets of neurons in the developing brain. This promoter fragment also drives expression in the secondary motoneurons of the spinal cord. An alternative fragment of the *islet-1* promoter/enhancer, SSICP, drives expression specifically in primary sensory neurons: Rohon-Beard neurons in the spinal cord and trigeminal ganglion neurons. A related expression pattern is generated by the *nAchRβ3* promoter, which also expresses in retinal ganglion cells (Tokuoka et al. 2002). Currently, the least promiscuous neural promoter is the promoter for the olfactory marker protein (OMP) (Celik et al. 2002; Yoshida et al. 2002), which drives expression only in olfactory sensory neurons.

Use of the Gal4-UAS System

It is common that even a functional promoter will drive gene expression only weakly and inconsistently. Both the mosaicism inherent in transient expression from plasmid injections, and the low and variable levels of gene expression, can make some experiments impractical. To circumvent these problems, Köster and Fraser (2001) adapted the Gal4-UAS system, widely used in *Drosophila*, for transient gene expression in the zebrafish. Gal4 is a powerful yeast transcription factor, which binds to a specific 11-base-pair DNA sequence (the upstream activating sequence, UAS). It is possible to amplify transgene expression by placing Gal4 under the control of a tissue-specific promoter (the driver or activator construct), and by placing the transgene under the control of a minimal promoter and multiple UAS sites (the effector construct). This amplification results in increased gene expression in the cells of interest, as well as a greater number of expressing cells. An additional advantage of Gal4-UAS is that a single driver construct can activate multiple effector cassettes, simultaneously driving expression of multiple transgenes. However, it should be noted that, in the authors' experience, the Gal4-UAS system can be a bit leaky, with spillover expression in skin, muscle, and notochord, particularly during early development.

Troubleshooting New Promoters

With the progress of the zebrafish genome project (http://www.sanger.ac.uk/Projects/D_rerio) and the accessibility of zebrafish BAC libraries (e.g., CHORI-211, http://bacpac.chori.org), it is becoming increasingly easy to identify and obtain potentially useful promoter/enhancers. However, due to the complexities of gene regulation, no fragment of a regulatory region is guaranteed to work or to generate the expected expression pattern. In testing a newly identified promoter/enhancer, it is important to separate the promoter activity from other complicating issues. It is wise to test new promoters either by driving GFP directly or by driving expression of GFP through the Gal4-UAS system. Note, however, that there can be a dramatic falloff in expression, both in numbers of cells and in expression levels, when GFP is replaced with a fusion of interest (FOI). In some cases, it is important to linearize the expression plasmid (see Linearizing DNA, below), in others there is little or no advantage. However, anecdotally, linearization of the plasmid can mean the difference between seeing and not seeing GFP expression and, in at least one instance, linearization results in an apparent shift in the cell-type distribution.

Synaptic Proteins

A range of synaptically localized proteins, both pre- and postsynaptic, have been tagged with GFP and used as synaptic markers, both in vivo and in vitro. The most common presynaptic protein fusions are VAMP/synaptobrevin (Ahmari et al. 2000) and synaptophysin (Nakata et al. 1998), both of which are transmembrane proteins present in synaptic vesicles. Of the postsynaptic proteins, PSD-95 (Okabe et al. 1999, 2001; Marrs et al. 2001) and CaMKII (Shen and Meyer 1999; Gleason et al. 2003) have been commonly used in fusions. The choice of synaptic marker depends on the goals of the experiment, as well as the available promoter(s).

Since very few synaptic genes have been cloned from the zebrafish, a plausible first step in initiating a synaptogenesis project could be obtaining a gene of interest to fuse to GFP. This could be achieved, for example, by a fragment of a homologous gene from another species to screen one of the several zebrafish cDNA libraries that are available. This is a relatively good option if a high degree of sequence identity is expected at the DNA level: Library screening is rapid, reliable, and can yield a full-length clone. Potentially faster still is an iterative BLAST search of the zebrafish EST database (http://www.ncbi.nlm.nih.gov/genome/seq/DrBlast.html). If a partial sequence of a gene can be obtained, then a partial cDNA fragment can be obtained using RT-PCR. This fragment can then be used to probe a cDNA library for larger fragments, or the cDNA ends can be obtained using 5′- and 3′-RACE (rapid amplification of cDNA ends). It is often possible to piece together a full-length sequence from multiple ESTs. PCR primers can then be designed and the entire coding region obtained by RT-PCR. The authors have successfully employed each of these approaches to clone zebrafish genes for synaptic proteins.

The above discussion has ignored the easy alternative of expressing a homologous gene, which has already been cloned from another organism, in the zebrafish. This might be an attractive time-saving proposition, but there are pitfalls. A rat gene, for example, may appear to behave perfectly well in the zebrafish, but it is possible that, by expressing the gene in a heterologous species, the trait of interest may be subtly perturbed. The perturbation may be equivalent to expression of a mutant gene. In some cases, sequence conservation may be so high, and the protein so well characterized, that the possibility of generating artifactual results is low. Even in these cases, however, for the sake of a few weeks of work, it hardly seems worth performing a potentially flawed experiment.

Once the synaptic gene of interest has been cloned, a few decisions concerning design of the GFP fusion must be made. First, should the reporter gene be fused to the amino or carboxyl terminus? If a precedent has been established in the literature, this should be used as a guide. In the case of a novel GFP fusion, it may be necessary to generate and test both amino and carboxyl fusions. Of course, for novel fusion proteins there is an added burden of demonstrating that the marker is localized appropriately and, in some cases, that it is functional. The best, and perhaps the only reliable, way of verifying proper localization of a GFP fusion is with immunoelectron microscopy.

Because of the inherent inaccuracies of PCR, it is important that all PCR-generated fragments for use in fusion constructs be sequenced as a matter of routine. It is also useful to consider the reading frame of the FOI. Because multicolor experiments are increasingly important, the authors use the N1 or C1 reading frames, with respect to Clontech vectors, since DsRed, CFP, and YFP are commercially available in only the C1 and N1 reading frames. Generating all fusions in N1 or C1 will allow easy color-swapping in the future. The authors have also found that the addition of a polyglycine linker of 3–4 residues between GFP and the fusion may help expression, as does engineering a consensus Kozak sequence at the amino terminus of carboxy-terminal fusions to improve translation efficiency.

DNA INJECTION

Preparation of the Micropipette

Proper preparation of the micropipette is critical for rapid, efficient, and reproducible DNA injections. A pipette with too fine a point or too long and narrow a shaft will have difficulty piercing the chorion. On the other hand, a tip that is too blunt will tend to damage the embryo. Pulling appropriate micropipettes is an empirical endeavor and will involve a fair amount of initial trial and error. After pulling, the micropipette must be broken near the tip by pinching the end with a pair of fine forceps. In general, tips should have an external diameter of ~5–10 µm.

Linearizing DNA

Many vectors contain universal primer sites flanking the multiple cloning site; for example, T7, T3, Sp6, and M13 forward and reverse primer sites. For these cloning vectors, it is possible to generate a large quantity of linear expression cassette with minimal extraneous vector sequence by using pairs of universal primers in a PCR. The authors recommend High Fidelity (Taq) DNA Polymerase (Invitrogen) and 30 cycles of PCR under appropriate conditions (see Sambrook and Russell 2000).

For other plasmids, it may be more convenient to linearize the DNA by restriction digestion. Digestion should be performed under standard conditions at the appropriate temperature. The products of both PCR and restriction digestion purified using a commercial kit (e.g., Qiagen PCR Purification kit; Qiaquick Gel Extraction Kit [Qiagen]) prior to storage in aqueous solution at –20 to –80°C. It may also be useful to perform a transient expression assay to determine whether linearized plasmid is significantly more effective than circular DNA.

Preparation of DNA Microinjection Mix

In general, plasmid DNA purified by a standard commercial miniprep kit should be suitable for microinjection of zebrafish embryos. At its simplest, the DNA from a miniprep (assumed to be ~500 ng/µl) could be diluted ~1:10 with 0.1 M KCl, and then injected. For a strong promoter driving the expression of a nontoxic FOI, this should give reasonably good results. However, conditions must be determined empirically for each FOI. In general, a good starting point is ~50–100 µg/ml DNA (or 25–50 µg/ml each plasmid, if using Gal4:UAS) in a volume equivalent to 1/5 of the cell volume. As mentioned above, it may also be advisable to linearize the DNA for injections. The following procedure describes how to prepare the injection mix.

1. Thaw aliquots of chosen plasmids (linearized or not, as appropriate).
2. To a 0.5-ml microfuge tube, add the following:

   500 ng–1 µg of linearized plasmid DNA

   1 µl Phenol Red <!> , 0.5% (Sigma)

   0.1 M KCl <!> (or Danieau buffer*) to 10 µl

   (*Danieau buffer: 5 mM NaCl, 0.17 mM KCl <!> , 0.33 mM $CaCl_2$, 0.33 mM $MgSO_4$ <!>, 0.00001% methylene blue <!>)

3. Mix gently and load 1–5 µl of DNA solution into a micropipette using Eppendorf microloader tips.

Caution: *See Appendix 3 for appropriate handling of materials marked with <!>.*

Embryo Preparation and DNA Injection

Zebrafish are kept on a strict 14-hour-light/10-hour-dark cycle, and begin breeding shortly after the dawn of their light cycle. If the fish are properly maintained, a large number of embryos can be collected 20–30 minutes after the lights go on. Adult zebrafish will eat their eggs, so embryos must be removed promptly. The authors use plastic mesh inserts that fit in the bottom of 2-liter fish tanks. The embryos are then siphoned out through the mesh bottoms. To avoid fungal growth, which is a major cause of embryo death, embryos should be washed several times by serial transfer to dishes containing clean embryo medium (5 mM NaCl, 0.17 mM KCl <!> , 0.33 mM $CaCl_2$<!>, 0.33 mM $MgSO_4$<!>), supplemented with the antifungal agent, methylene blue (1:1000 dilution of 0.01% stock). In addition, any debris stuck to the chorions should be removed manually before transfer to the injection tray. An agar injection tray, similar to that described previously (Westerfield 1995; Meng et al. 1999; Gilmour et al. 2002), and shown in Figure 2a, is recommended. Embryos should be transferred to the tray and arrayed in the troughs, orienting the blastomere toward the micropipette (Fig. 2b), so that the micropipette enters the embryo directly. Injections into the yolk yield poor results, and the yolk causes extremely high background fluorescence. The dividing blastula is suitable for DNA injection through the 4-cell stage, during which time the individual cells remain connected by large cytoplasmic bridges. After penetrating the chorion and embryo, the DNA is pressure-injected, using a commercially available picospritzer, with 2–4 pulses of 20-msec duration at a pressure of 40–50 psi. An injection volume of ~1/5 a 1-cell-stage blastomere gives good results (Gilmour et al. 2002). After injection, embryos are moved to petri dishes containing fresh Embryo Medium (5 mM NaCl, 0.17 mM KCl <!>, 0.33 mM $CaCl_2$<!>, 0.33 mM $MgSO_4$<!>).

Caution: *See Appendix 3 for appropriate handling of materials marked with <!>.*

IMAGING

Preparation

In general, fairly large numbers of embryos must be screened in order to find the few that are most suitable for imaging experiments. First, under a dissecting microscope, all malformed embryos and any individuals that exhibit a noticeable degree of necrosis must be discarded. Then, ideally using a

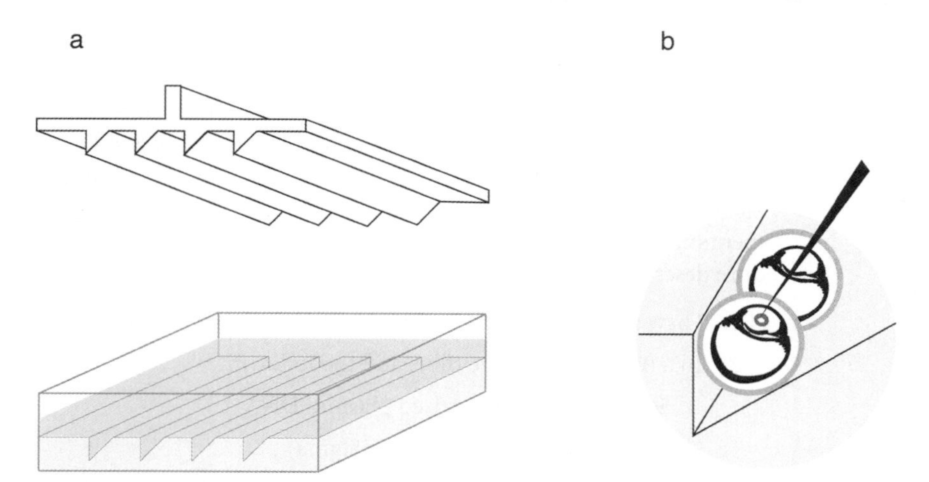

FIGURE 2. DNA injections. (*a*) Schematic of the plastic mold used to prepare for injection trays. This is essentially as described in *The Zebrafish Book* (Westerfield 1995), although similar molds are described by Meng et al. (1999) and by Gilmour et al. (2002). Preparation of injection trays is described in Protocol 3. (*b*) Illustration of DNA microinjection. A small volume of a dilute DNA solution is microinjected into the blastomere of a 1-cell-stage embryo. Typically, 100–200 embryos can be injected per sitting, since embryos up to the 4-cell stage are suitable for microinjections. Post-injection, the embryos are removed to petri dishes containing fresh Embryo Medium and allowed to develop.

fluorescence dissecting microscope, discard any embryos that are suspiciously bright. Although initial inclinations might make these individuals attractive, there are several practical and philosophical reasons to avoid them. First, very well labeled cells can sometimes exhibit abnormal morphologies or behaviors. They also tend to be more susceptible to phototoxic effects, even at reduced laser powers. It is likely that these cells are massively overexpressing the FOI and, unless overexpression is the objective of the experiment, the data may be misleading. The third phase of screening involves mounting dechorionated embryos in agarose and imaging them on the confocal or two-photon microscope.

Preparation and mounting of embryos is a crucial step in the imaging experiment. The authors use custom-made simple "imaging chambers," consisting of a glass slide with a border of either Sylgard or surgical wax to form a trough. Both simple and inexpensive, these chambers can also be saved and reused. An array of embryos can be placed on each slide, improving screening efficiency. The repeated mounting, imaging, and discarding of embryos until an appropriate individual is found can be tedious. However, each embryo should be mounted very carefully, since any mishandling can have a profound influence on the quality of the subsequent experiment. Each embryo should be mounted individually in a small drop of ~1.5% low-melting-point agarose, as described on page 144 and in Figure 3. If embryos need to be saved for later use or remounted, they can easily be freed from the agarose without damage. The chambers described above are for use with an upright microscope and water-immersion objectives and are not suitable for use with inverted microscope stands.

For embryos less than ~30 hours postfertilization, it is wise to include the anesthetic tricaine in the embryo medium, in order to prevent spontaneous muscle contractions. A 25x stock of 0.4% Tricaine <!> (MESAB or MS-222, Sigma) should be made up in 20 mM Tris, and adjusted to pH 7. Tricaine is not necessary for older embryos or larvae, which should instead be maintained in Embryo Medium containing 0.2 mM PTU <!> (1-phenyl-2-thiourea or phenylthiocarbamide, Sigma) to inhibit pigmentation, which can obscure microscopic observations. A 10 mM PTU stock solution should be made up in Embryo Medium.

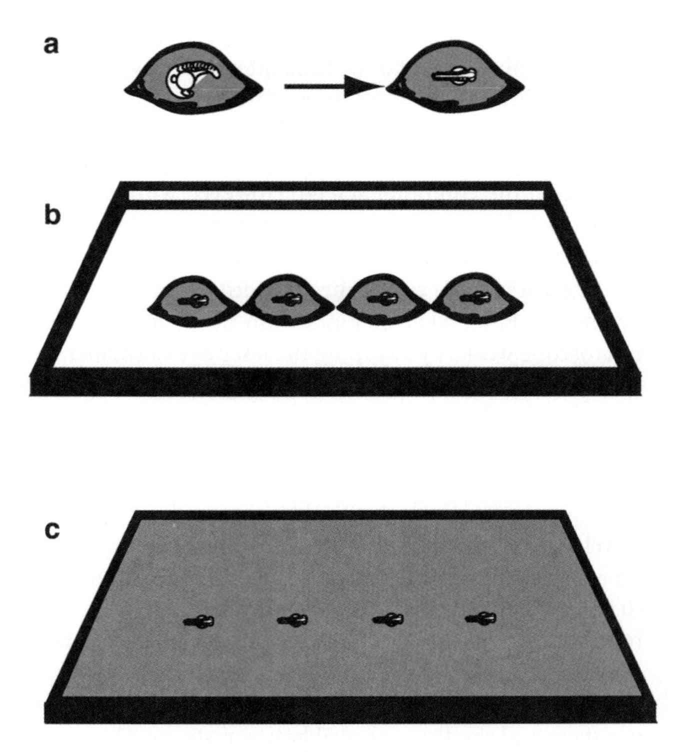

FIGURE 3. Mounting of embryos for imaging. (*a*) Individual embryos are placed on the glass surface of an imaging chamber in a small drop of molten agarose. They are then oriented so that the area of interest faces the microscope objective. (*b*) An array of embryos can be sequentially mounted on a single glass slide chamber. (*c*) Once the agarose has set for each of the individual embryos, the chamber is flooded with agarose to barely cover the embryos. This cements the embryos firmly to the glass and provides important stability for time-lapse imaging.

The following procedure describes in detail how to mount embryos for imaging.

1. In a clean borosilicate test tube (e.g., 16 x 150 mm), melt 0.15 g of low-melting-point agarose (Invitrogen) in 10 ml of Embryo Medium (1.5% final agarose concentration).
2. Keep agarose in 42–45°C water bath.
3. Add 0.4 ml of tricaine stock <!> to the agarose, and 0.8 ml to 20 ml of Embryo Medium (5 mM NaCl, 0.17 mM KCl <!>, 0.33 mM $CaCl_2$<!>, 0.33 mM $MgSO_4$<!>).
4. Using a Pasteur pipette, transfer an individual embryo to the test tube containing the agarose. Be careful to transfer as little excess liquid as possible. Remove the embryo in a small volume of agarose and place the embryo, in a small drop, in an imaging chamber.
5. Before the agarose sets, orient the embryo so that it is suitable for imaging (see Fig. 3). For orienting embryos we use short lengths of thin fishing line (~1 cm) that have been glued to the end of wooden handles (e.g., the wooden stick from a Q-tip applicator).
6. Repeat this process, carefully mounting each embryo individually on the glass slide.
7. Cement the mounted embryos to the slide by filling the bottom of the chamber with a layer of agarose. Let the agarose set.
8. Overlay the agarose with Embryo Medium containing 0.016% tricaine or 0.2 mM PTU <!> (see text above for details).

Caution: *See Appendix 3 for appropriate handling of materials marked with <!>.*

Imaging

In general terms, a good imaging experiment balances image contrast with biological reliability. It is important to minimize damaging the specimen, limiting exposure to the laser by controlling the laser power, the pixel size, and the time intervals between image stacks, while at the same time maintaining image quality and the usefulness of the data obtained. Balancing these issues is critical. The first place one can run into trouble is in admiring one's own work. After going through the prodigious amount of work to get to this point, an observer might be tempted to gaze ecstatically at the beautiful glowing cell on the computer screen. However, with each frame, the cell's health is being compromised, particularly if embryos are being screened at a high laser power. Early on, it is worth screening at minimum laser power and at the fastest scan rate.

Once a suitable cell has been identified for a time-lapse experiment, the imaging parameters must be chosen. First, what magnification needs to be used? The "electronic zoom" on a laser-scanning microscope works by sweeping the laser over a smaller area and distributing the same number of pixels across this smaller area, thereby magnifying the image. Since the pixel dwell times remain the same, more light goes into and is collected from a given volume. This improves the image quality but also increases the opportunity for phototoxicity and photobleaching. The image resolution also increases, to a limited extent, with electronic zoom. As the zoom increases and the pixel size decreases, there will come a point at which the pixel size is smaller than the resolution limit of the microscope, given the wavelength of the laser and the NA of the objective being used. For a digital image, like that collected on a laser-scanning microscope, the pixel size should be ~1/3 the resolution limit of the microscope ($0.61\lambda/NA$) to optimally reconstruct the object being imaged. Reducing the pixel size beyond this will result only in increased phototoxicity with no corresponding gain in image detail. For an experiment as sensitive as time-lapse imaging in vivo, this becomes an important consideration.

The number of frames to be collected also needs careful consideration, in terms of both images per stack and the time interval between stacks. The number of images in a stack will be experiment-dependent: depending on the z-dimensions of the object of interest, as well as the z-axis resolution of the microscope. In simple cases, like tracing an axonal or dendritic arbor, the spacing of the sections needs only to approximate the z-resolution of the microscope (~3x the x-y resolution for a high-NA objective, but usually greater than this), so that there are no discontinuities of arbor segments between sections. However, in cases where high-resolution data are needed and deconvolution of the image stack is to be performed, the volume needs to be over-sampled in z.

The time interval of the time lapse will depend on the timescale of the process being studied. For the observation of slower events like the overall growth of axonal and dendritic arbors, time intervals of 20–30 min or more may be satisfactory. For faster cellular dynamics like growth cone motility, filopodia growth/retraction, or protein trafficking in axons, time intervals of 1–5 min are more appropriate. For processes like vesicle recycling or calcium dynamics, even shorter timescales will be necessary. The appropriate collection parameters will balance the needs of the experiment with the risks of photodamage. For instance, imaging a growing axon labeled with VAMP-GFP on a two-photon microscope, a typical experiment might involve collecting 10–20 sections per time point at ~1-μm spacing and intervals of 1–2 min over a period of 1–4 hours. However, a DiI-labeled growth cone, imaged on a confocal microscope, may only be able to endure 5–6 sections per time point at 2-μm spacing at 1–2-min intervals for 1–2 hours before showing signs of deterioration.

Once a suitable cell has been identified and the imaging parameters have been chosen, it is now important to wait. Specimen stability is critical to the success of a time-lapse experiment. An ever-present danger in imaging is specimen drift. In some cases, drift may be due to an unstable microscope stage, but it is usually specimen-dependent. This is, in general, preventable. First, if there is insufficient agarose around the embryo to cement it to the glass of the imaging chamber, the drops of agarose will tend to slowly detach from the glass and float away. This is why the entire floor of the imaging chamber must be flooded with agarose. In addition, the microscope stage and the imaging chamber need to be shielded from drafts and thermal gradients. Any sort of air movement or temperature instability (including that due to buffer evaporation) can cause specimen drift. The microscope stage should be covered with a small box or some other protective covering, and a large reservoir of buffer should overlay the specimen, to minimize evaporation effects. Once the embryo has been mounted, covered with buffer, and set up for a time lapse, it is good practice to wait 15–20 min while the agarose gel settles and the imaging chamber microscope system comes to equilibrium. It is also a good idea to monitor the experiment for the first 5–10 frames to make any necessary focus adjustments. Finally, it may be worth collecting extra sections above and below the object of interest to catch any drift along the *z*-axis. An example of a successful time-lapse experiment is shown in Figure 4.

Image Analysis

Once data have been collected, the next steps are to generate a digital movie and to analyze the data (convert the salient information to numerical form). There are several options at this point. Commercial software packages can deal with image data, but they are quite expensive and are often cumbersome to use. Among the attractive alternatives is ImageJ, a *free* JAVA-based image analysis program originating from NIH-Image (http://rsb.info.nih.gov/ij/). This software is available for a range of platforms, including Apple, Unix, Linux, and Windows. ImageJ can also be easily extended through the use of custom plugins, which can be written by the user or can be downloaded from a free, public archive. Alternatively, Matlab is a science and engineering software package that can be supplemented with an extensive image-processing toolbox (http://www.mathworks.com/). The Matlab environment offers a high degree of flexibility, as the high-level language can be used to process images at the command line, by writing scripts to automate a series of steps or process a series of images (like those generated in a time-lapse experiment), or by generating GUI-based programs.

ADVANTAGES AND LIMITATIONS

The methods described here are easy to implement and master, and result in a relatively rapid and reliable way to obtain in vivo time-lapse data. Thus, it is possible to extract information on molecular and cellular dynamics in a near-perfect native environment in the developing nervous system of a vertebrate embryo. The zebrafish is an excellent vertebrate model system for forward genetics, and the advent of antisense morpholinos allows limited reverse-genetic approaches, as well. The availability of these tools greatly amplifies the power of in vivo time-lapse approaches, and vice versa. However, this method does have some limitations. In contrast to in vitro systems, such as hippocampal cell culture, this in vivo system currently lacks an array of antibodies directed against synaptic proteins that work

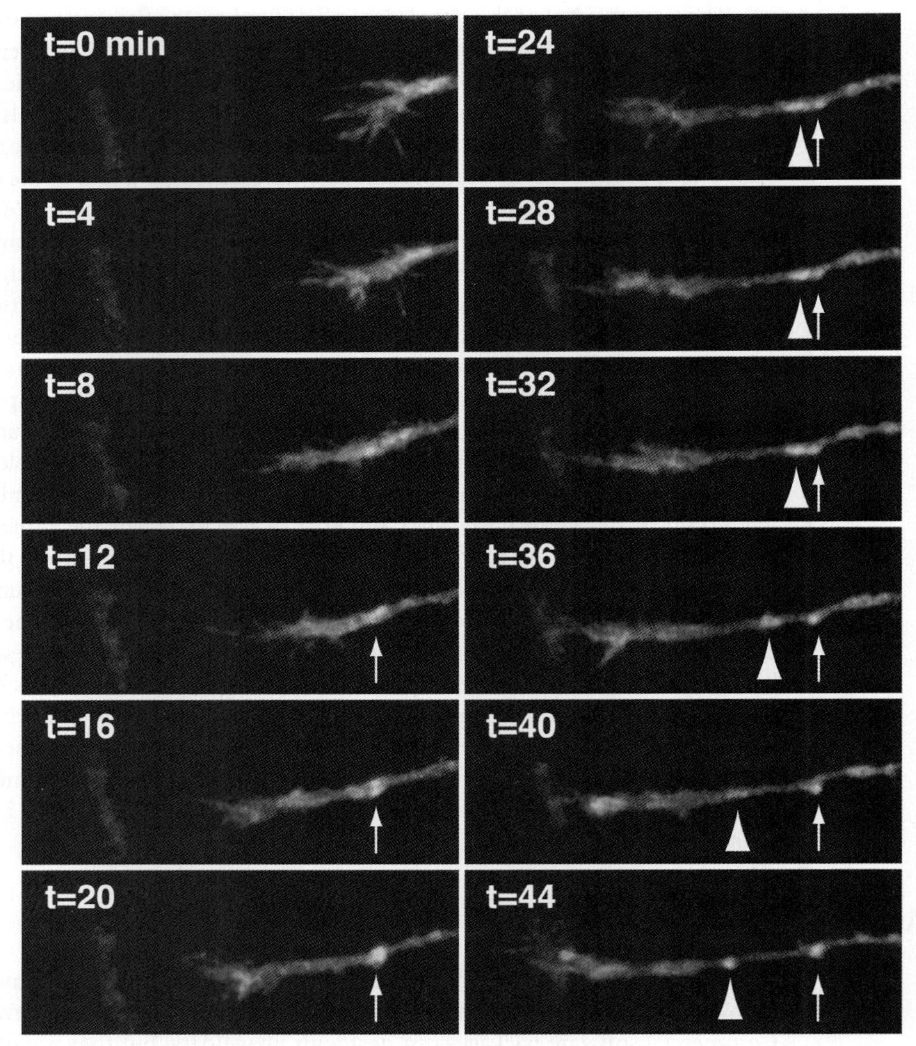

FIGURE 4. Example of a time-lapse imaging experiment. This time-lapse sequence is from an embryo that had been injected with the goldfish α*1-tubulin* promoter/enhancer::Gal4 and UAS::VAMP-GFP. The zebrafish VAMP was cloned by RT-PCR with primers designed against a coding sequence pieced together from multiple ESTs. VAMP/synaptobrevin is a single-pass transmembrane protein present on synaptic vesicles and important for docking and membrane fusion. It has been widely used as a marker of presynaptic terminals. Like most putative synaptic markers, VAMP has advantages and disadvantages. A mixed blessing of VAMP is that it has a tendency to label the plasma membrane early in development. This is good because it avoids the need for a second "counterstain" such as DsRed to provide cell morphology, but it also partially masks some of the axonal trafficking of "transport packets" (Ahmari et al. 2000), as well as the punctate labeling of forming synapses. In this time-lapse sequence, the growth cone of a Rohon-Beard sensory neuron is migrating rostrally in the dorsal spinal cord of a ~ 24-hour postfertilization embryo. VAMP-GFP is enriched in the growth cone and leaves behind a cluster of vesicles or vesicle precursors that first become distinct at ~12 minutes (*arrow*). This cluster stabilizes, then splits into two clusters of VAMP-GFP at ~24 minutes. The first cluster remains stable in its original spot, while the second cluster (*arrowhead*) migrates rostrally and stabilizes several micrometers away. These two VAMP-GFP clusters then remain for the duration of the time-lapse sequence. This sequence reveals two common mechanisms for stable VAMP-GFP punctum formation: formation directly by the migrating growth cone and splitting of clusters. These stable vesicle clusters are likely markers of nascent synapses, since more mature spinal axons are characterized by regularly spaced vesicle clusters that give the axons the appearance of beads on a string. This time-lapse experiment consisted of 120 time points spaced 2 minutes apart for a total duration of 4 hours. Each time point consisted of a 20-image stack spaced 1 μm apart, with a pixel size of 0.19 μm. The panels are maximum intensity projections of 10 images from their respective stacks. Prior to each projection, a median filter (radius = 1 pixel) was applied to each individual image. Otherwise, no contrast enhancement was applied to the images.

for whole-mount immunostaining. Moreover, it is not yet possible to use FM1-43 or other styryl dyes for the routine assay of presynaptic function. In addition, the images that can be obtained in principle (although not always in practice) from cultured neurons should be far superior to those obtained by imaging in vivo: High-NA objectives can be used at or near their optima, in contrast to the lower-NA, long-working-distance objectives that are used to image through layers of agarose and tissue. However, the lack of molecular reagents is likely to be remedied soon, and the technical limitations are not major obstacles to obtaining important biological information. In fact, much of the detail of neural development, including synaptogenesis, is likely to be revealed only by using in vivo systems such as the zebrafish.

ACKNOWLEDGMENTS

The authors thank members of the Smith lab for discussions and comments on the manuscript. J.D.J. is the recipient of a Burroughs Wellcome Fund Career Award in the Biological Sciences.

REFERENCES

Ahmari S.E., Buchanan J., and Smith S.J. 2000. Assembly of presynaptic active zones from cytoplasmic transport packets. *Nat. Neurosci.* **3:** 445–451.

Celik A., Fuss S.H., and Korsching S.I. 2002. Selective targeting of zebrafish olfactory receptor neurons by the endogenous OMP promoter. *Eur. J. Neurosci.* **15:** 798–806.

Detrich H.W., III, Westerfield M., and Zon L.I. 1999a. *The zebrafish: Biology. Methods Cell Biol.*, vol. 59. Academic Press, San Diego, California.

———. 1999b. *The zebrafish: Genetics and genomics. Methods Cell Biol.*, vol. 60. Academic Press, San Diego, California.

Dynes J.L. and Ngai J. 1998. Pathfinding of olfactory neuron axons to stereotyped glomerular targets revealed by dynamic imaging in living zebrafish embryos. *Neuron* **20:** 1081–1091.

Gilmour D.T., Jessen J.R., and Lin S. 2002. Manipulating gene expression in the zebrafish. *Pract. Approach Ser.* **261:** 121–144.

Gleason M.R., Higashijima S., Dallman J., Liu K., Mandel G., and Fetcho J.R. 2003. Translocation of CaM kinase II to synaptic sites in vivo. *Nat. Neurosci.* **6:** 217–218.

Goldman D., Hankin M., Li Z., Dai X., and Ding J. 2001. Transgenic zebrafish for studying nervous system development and regeneration. *Transgenic Res.* **10:** 21–33.

Hieber V., Dai X., Foreman M., and Goldman D. 1998. Induction of α1-tubulin gene expression during development and regeneration of the fish central nervous system. *J. Neurobiol.* **37:** 429–440.

Higashijima S., Hotta Y., and Okamoto H. 2000. Visualization of cranial motor neurons in live transgenic zebrafish expressing green fluorescent protein under the control of the islet-1 promoter/enhancer. *J. Neurosci.* **20:** 206–218.

Köster R.W. and Fraser S.E. 2001. Tracing transgene expression in living zebrafish embryos. *Dev. Biol.* **233:** 329–346.

Marrs G.S., Green S.H., and Dailey M.E. 2001. Rapid formation and remodeling of postsynaptic densities in developing dendrites. *Nat. Neurosci.* **4:** 1006–1013.

Meng A., Jessen J.R., and Lin S. 1999. Transgenesis. *Methods Cell Biol.* **60:** 133–148.

Nakata T., Terada S., and Hirokawa N. 1998. Visualization of the dynamics of synaptic vesicle and plasma membrane proteins in living axons. *J. Cell Biol.* **140:** 659–674.

Nüsslein-Volhard C. and Dahm R., eds. 2002. *Zebrafish: A practical approach.* Oxford University Press, New York.

Okabe S., Miwa A., and Okado H. 2001. Spine formation and correlated assembly of presynaptic and postsynaptic molecules. *J. Neurosci.* **21:** 6105–6114.

Okabe S., Kim H.D., Miwa A., Kuriu T., and Okado H. 1999. Continual remodeling of postsynaptic density and its regulation by synaptic activity. *Nat. Neurosci.* **2:** 804–811.

Park H.C., Kim C.H., Bae Y.K., Yeo S.Y., Kim S.H., Hong S.K., Shin J., Yoo K.W., Hibi M., Hirano T., Miki N., Chitnis A.B., and Huh T.L. 2000. Analysis of upstream elements in the HuC promoter leads to the establishment of transgenic zebrafish with fluorescent neurons. *Dev. Biol.* **227:** 279–293.

Sambrook J. and Russell D. 2001. *Molecular cloning: A laboratory manual*, 3rd edition. Cold Spring Harbor Laboratory Press, Cold Spring Harbor, New York.

Shen K. and Meyer T. 1999. Dynamic control of CaMKII translocation and localization in hippocampal neurons by NMDA receptor stimulation. *Science* **284:** 162–166.

Tokuoka H., Yoshida T., Matsuda N., and Mishina M. 2002. Regulation by glycogen synthase kinase-3β of the arborization field and maturation of retinotectal projection in zebrafish. *J. Neurosci.* **22:** 10324–10332.

Westerfield M. 1995. *The zebrafish book: A guide for the laboratory use of zebrafish* (*Brachydanio rerio*), 3rd edition. University of Oregon Press, Eugene.

Yoshida T., Ito A., Matsuda N., and Mishina M. 2002. Regulation by protein kinase A switching of axonal pathfinding of zebrafish olfactory sensory neurons through the olfactory placode-olfactory bulb boundary. *J. Neurosci.* **22:** 4964–4972.

CHAPTER 18

A Practical Guide: In Ovo Imaging of Avian Embryogenesis

Paul M. Kulesa[1] and Scott E. Fraser[2]

[1]Stowers Institute for Medical Research, Kansas City, Missouri 64110; [2]Beckman Institute 139-74, California Institute of Technology, Pasadena, California 91125

This chapter describes a method for following individual, fluorescently labeled cells and tissue in living chick embryos, using video and confocal time-lapse microscopy. This technique has been used to investigate two areas of chick morphogenesis: (1) early nervous system development, including neural crest cell patterning, and (2) somite formation. These phenomena have been imaged within the temporal window of chick development, starting from stage-8 embryos (Hamburger and Hamilton 1951) and continuing for up to 5 days. The technique can be extended to visualize earlier events such as gastrulation and later events such as organ and limb development; however, there are challenges caused by the fragility of the embryo and by motion of the heartbeat, respectively. The technique has been adapted to accommodate quail embryo culture and imaging: The quail egg is about one-quarter the size of the chicken egg, and its imaging availability is useful for assaying quail to chick (or chick to quail) tissue transplants, which are commonly used in cell lineage studies. With minor adjustments, the technique can be applied to visualize development in embryos of various egg shapes and sizes.

MATERIALS

Microscope

This technique uses an upright confocal laser scanning microscope (Bio-Rad MRC-500; see Chapter 5) equipped with an easy-to-assemble custom-made cardboard box and incubator (see Procedure B). The microscope must have a detachable stage that can accommodate the egg and holder beneath the objective.

Eggs

Fertile White Leghorn chicken eggs, 26–29 hours after laying, incubated at 38°C in an egg incubator (VWR).

Teflon Window

Teflon membrane (high-sensitivity, oxygen-permeable from Fisher; 3.8 cm x 7.5 cm x 15 μm), acrylic ring (~2.2 cm inner diameter x 2.6 cm outer diameter x 0.5 cm high), rubber O-ring (2.4 cm internal diameter x 2.1 cm external diameter).

PROCEDURE A

Preparation of Teflon Membrane Assembly

1. Use a heating block set to 37°C to melt a small piece of beeswax (Eastman Kodak) in a glass petri dish.
2. Hold the acrylic ring (as described in Materials section above) with forceps and dip it into the warmed beeswax for ~10 sec.
3. Lay the ring on a rectangular piece of Teflon membrane so that the beeswax is in contact with the Teflon.
4. Flip the Teflon and the ring over so that the ring is lying underneath the membrane.
5. Place a rubber O-ring (2.4 cm internal diameter x 2.1 cm external diameter) on top of the Teflon, directly above the acrylic ring, and push down. This will trap the membrane between the two rings like the surface of a drum.
6. After 5 minutes, remove the O-ring and cut the excess Teflon membrane away.
7. There may be parts of the Teflon membrane that do not adhere to the sides of the ring. In this case, roll the ring vertically along the heating block so that the beeswax between the ring and the Teflon melts. Pull on the Teflon membrane so that it is taut across the ring.
8. Allow the ring assembly to cool for at least 5 min.

PROCEDURE B

Microscope/Incubator Preparation

1. Construct a cardboard box (see Fig. 1B) around the microscope stage, using Velcro strips to hold the box sides together. Create small holes in the front and base of the box to allow microscope access.
2. Wrap Reflectix insulation (5/16-inch thick; Reflectix) around the outside of the box and secure it with clear packaging tape.
3. Place the chick incubator heater (Lyon Electric, TX7) either inside the box or at the side of the box into which a hole is cut to accommodate the face of the heater, away from the microscope.

PROCEDURE C

Egg Preparation

1. Rinse egg(s) with 70% ethanol and remove 3 ml of albumin from the caudal part of the egg using a 5-ml syringe and a 25-gauge needle.
2. Place the egg against an egg candler (Lyon Electric) and locate the position of the blastoderm/embryo.
3. Place an acrylic ring (~2.2 cm inner diameter x 2.6 cm outer diameter x 0.5 cm high, as used in the preparation of Teflon windows) on top of the eggshell over the center of the blastoderm and draw a circle around it.
4. Use Scotch tape to cover the portion of the eggshell (circle) that will be cut out for the window.
5. With scissors, cut a hole in the eggshell by following along the line drawn around the acrylic ring, taking care not to touch the embryo.

6. Inject the india ink solution (Pelikan 10% in Howard Ringer's solution [7.2 NaCl, 0.23 g of $CaCl_2$ <!>, 0.37 g of KCl <!> in 1 liter of water, pH 7.2–7.3]) under the blastoderm with a 25-gauge needle to visualize and stage the embryo.
7. Perform any necessary cell labeling and manipulation of the embryo.

   Caution: *See Appendix 3 for appropriate handling of materials marked with <!>.*

Inserting the Window into the Egg

8. After finishing all manipulations/labeling of the embryo, place the ring assembly into the hole cut in the eggshell such that it lies nicely on the surface of the vitelline membrane and yolk, above the embryo.
9. Seal the ring assembly into the eggshell, using a metal spatula to smear warm beeswax into the gap between the eggshell and the ring assembly. Make sure that there are no leaks of the egg contents.
10. Position the egg on the microscope stage, within the incubator, such that the embryo is visible with a low-magnification 5x or 10x objective through the Teflon membrane.

EXAMPLE OF APPLICATION

Figure 1 shows the setup and results of time-lapse imaging sessions in different regions of chick embryos. As shown in Figure 1A, the egg lies on the microscope stage, and events can be imaged directly through the Teflon membrane assembly, while the area is warmed to 37°C. Neuronal precursor cells were labeled with a fluorescent dye and visualized during the formation of the neural crest patterning of the branchial arches (Fig. 1C). The individual cells were followed after emergence from the neural tube to the destination sites (Fig. 1D) and displayed a much richer variety of individual and collective cell movements than previously assumed (Fig. 1E). These behaviors ranged from following a lead cell after filopodial contact to forming chain-like arrays and moving as a collective. In ovo imaging provides a more comprehensive picture of the migration pattern of neural crest cells than imaging in explanted embryos. This is due to the ability to follow cell movements near the neural tube, along the migratory routes, and near the branchial arches after embryo rotation (Kulesa et al. 2000; Kulesa and Fraser 2000).

ADVANTAGES AND LIMITATIONS

One of the main advantages of the in ovo technique is that the chick embryo remains intact in the natural setting of the egg. This provides nutrients for the embryo, obviating the need for culture media or other additives. Morphogenesis of the in ovo embryo resembles the natural process more closely than embryos cultured outside the egg, either as whole embryos or in tissue culture. The natural rotation of the chick embryo within the egg allows visualization and investigation of cell migration and tissue shaping, as well as organ and limb development.

Because the technique was developed to study embryos within a certain temporal window, its applications outside this window are limited by the embryos' physical states. The observation of earlier events, such as gastrulation, requires extremely gentle handling of the egg contents, due to the fragility of the embryo at this stage. For later events, after embryonic day 4 or so, natural movements of the embryo force constant refocusing of the microscope to maintain the region of interest in the focal plane. As the embryo continues to grow, it uses up more of the yolk and begins to drop below the surface of the Teflon window. This increases the distance across which the oxygen must diffuse through the Teflon membrane, thus limiting embryonic growth. Thus, extensions of the in ovo technique to study later events of chick morphogenesis should take care to maintain the level of the embryo close to the Teflon window.

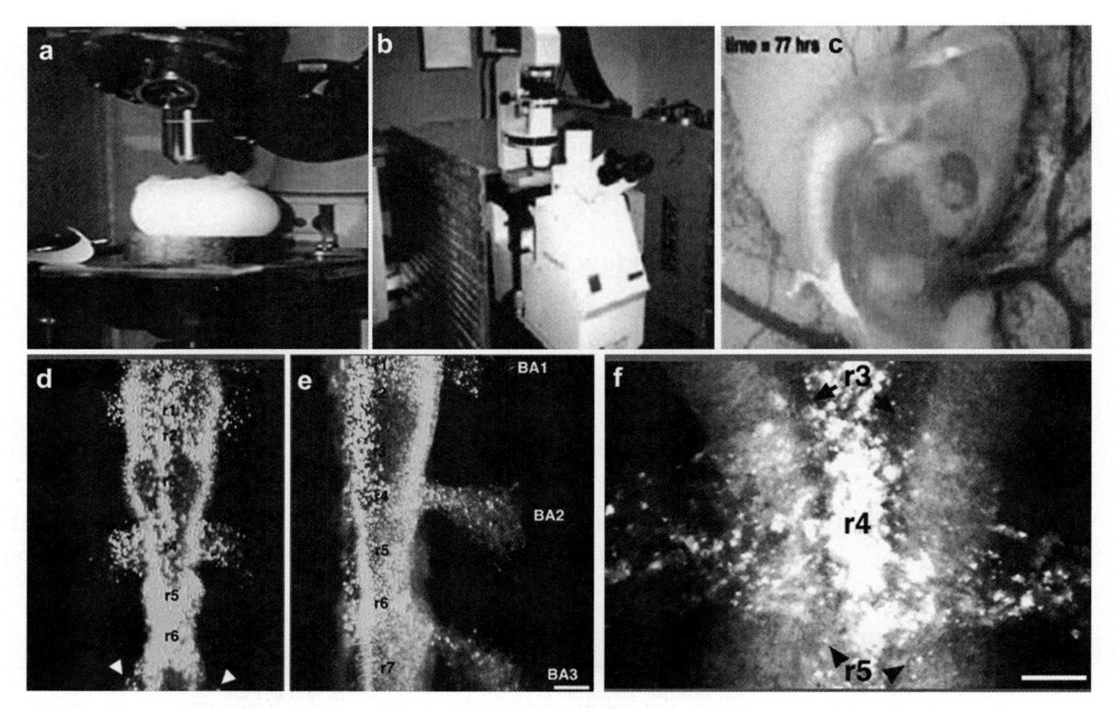

FIGURE 1. In ovo culture and imaging of chick embryogenesis. (*a*) The egg is placed on the microscope stage of an upright microscope. (*b*) A cardboard box is constructed to surround the microscope stage and includes a chick incubator heater to maintain the temperature at 37°C. (*c*) A typical chick embryo after insertion of a Teflon window and a confocal imaging session of 77 hr. (*d*) After labeling the neural tube with a fluorescent vital dye, the patterning of the cell movements, in this case the cranial neural crest cells, were followed in time-lapse confocal imaging of (*e*) the global pattern and (*f*) individual cell movements. Bar: (*e,f*) 50 μm.

ACKNOWLEDGMENTS

We thank Professor Dan Kiehart (Duke University) for the idea of the Teflon window, which he has used extensively for *Drosophila* imaging.

REFERENCES

Hamburger V. and Hamilton H.L. 1951. A Series of normal stages in the development of the chick enbryo. *J. Morphol.* **88:** 49–92.

Kulesa P.M. and Fraser S.E. 2000. In ovo time-lapse analysis of chick hindbrain neural crest cell migration shows interactions during migration to the branchial arches. *Development* **127:** 1161–1172.

Kulesa P.M., Bronner-Fraser M., and Fraser S.E. 2000. In ovo time-lapse analysis after dorsal neural tube ablation shows rerouting of chick hindbrain neural crest. *Development* **127:** 2843–2852.

CHAPTER 19

A Practical Guide: Time-lapse Imaging of the Formation of the Chick Peripheral Nervous System

Jennifer C. Kasemeier,[1] Frances Lefcort,[1] Scott E. Fraser,[2] and Paul M. Kulesa[3]

[1]Department of Cell Biology and Neuroscience, Montana State University, Bozeman, Montana 59717; [2]Beckman Institute 139-74, California Institute of Technology, Pasadena, California 91125; [3]Stowers Institute for Medical Research, Kansas City, Missouri 64110

This chapter describes a new culture and imaging technique for following individual, fluorescently labeled neural crest cells in sagittal slice explants of the chick neural tube, trunk region, with video and confocal time-lapse microscopy. This technique has been used to study chick neurogenesis, i.e., the patterning and cell migratory behaviors of neural crest and neural progenitor cells that form the dorsal root ganglia (DRG) and sympathetic ganglia (SG). Fluorescently labeled cells have been followed in time-lapse microscopy for an average of 24 hours and up to 36 hours. Embryos as young as 45 hours and up to 84 hours postfertilization have been imaged (Hamburger and Hamilton [HH] [1951] stage16–24). This technique may be used in both younger and older embryos, for example, to study somite formation and limb development.

MATERIALS

Eggs

This procedure uses fertile White Leghorn chicken eggs that have been stored in the cold for up to 1 week.

Microscope

This technique uses a laser-scanning confocal microscope (Zeiss LSM Pascal) on an inverted fluorescent compound microscope (Zeiss Axiovert) equipped with a home-made cardboard incubator (see Procedure B, Chapter 18). The box is heated by a tabletop incubator (e.g., Lyon Electric). A digital thermometer (Fisher) must be positioned on the microscope stage.

PROCEDURE

Time-lapse Imaging of the Formation of the Chick Peripheral Nervous System

Egg Preparation

1. After incubation at 38°C in an egg incubator (VWR), rinse egg(s) with 70% ethanol and remove 3 ml of albumin from the caudal part of the egg using a 5-ml syringe and a 25-gauge needle.
2. Perform any cell and tissue labeling or embryo manipulations prior to generating sagittal explants.
3. Use scissors (e.g., Fine Science Tools) to cut a hole in the eggshell large enough to manipulate the embryo, and discard the cut-out shell.
4. Place a few drops of Ringer's solution around the embryo. (Ringer's solution: 122 mM NaCl, 4.9 mM KCl<!>, 1.53 mM $CaCl_2$<!>, 0.81 mM Na_2HPO_4<!>, pH 7.4, also commercially available from Fisher Scientific.)
5. Cut a circle out of filter paper (Whatman) with a hole in the middle, large enough to fit around embryo, and place this filter ring on top of embryo so that the embryo is in the middle of the hole.

   Note: This filter paper "ring" eases removal of the embryo.
6. With the embryo positioned inside the filter ring, use scissors to cut around the outer perimeter of the circle, through the surrounding membranes and blood supply.
7. With a pair of tweezers (e.g., Fine Science Tools), gently remove the embryo and transfer it to a 100-mm petri dish with Ringer's solution.
8. Use a tungsten needle (e.g., A-M Systems) to cut out the region of the embryo from which the sagittal slice is to be taken, and transfer it to a new petri dish with fresh Ringer's solution (Fig. 1a). Lay the explant dorsal side up and hold in place with forceps.
9. Again using a tungsten needle, lightly cut along the midline of the spinal column, down the entire length of the explant (Fig. 1b).

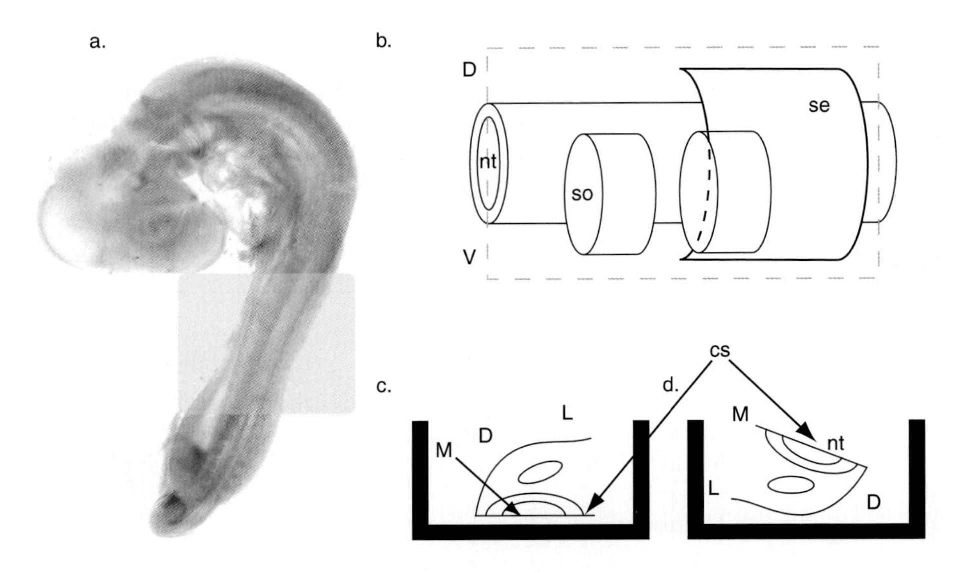

FIGURE 1. Schematic of sagittal explant technique. (*a*) 72-hr-old chick embryo, trunk explant taken from highlighted area. (*b*) Sagittal slice of trunk explant. Dotted lines indicate the plane of the cut. (*c*) Lateral surface side up explant; (*d*) lateral surface side down explant. (D) Dorsal side of embryo; (V) ventral side of embryo; (L) lateral surface of explant; (M) medial surface of explant; (nt) neural tube; (so) somite; (se) surface ectoderm.

Note: Cutting through the ectoderm will allow the spinal cord to start splaying open.

10. Still holding the embryo in place with the forceps, in one motion, insert a razor blade into the incision made by the tungsten needle. Slice down through the rest of the embryo.

    Note: This incision will slice the embryo in half and produce two sagittal explants.

Millicell Culture Insert and Petri Dish Assembly

11. Remove a filter (Millicell culture plate filter insert from Millipore, PICMORG50) from its package and place it in a large petri dish. Coat the surface of the filter with ~100 µl of fibronectin (20 µg/ml in phosphate buffer [GIBCO-BRL]) and incubate the filter in a covered petri dish for 15 min.
12. Remove any excess fibronectin by tilting the filter and pipetting the excess from the wall of the filter.
13. Coat the membrane with Ringer's solution and set aside.
14. Prepare a glass-bottomed petri dish using a 10 x 15-mm dish and a 25-mm circular coverslip (for detailed instructions, see Chapter 16).

Microscope/Heater Box Assembly

15. Assemble the cardboard box as described in Chapter 18 and ensure that the box fits snugly around the microscope.
16. Cut a hole in the box, large enough to accommodate the diameter of the incubator heater.
17. Feed the heater through the inside of the hole and pull out until the box catches on the outer lip of the incubator to hold it in place. Set the incubator temperature to 38°C.
18. Place the probe of the digital thermometer on the microscope stage such that it is an accurate readout of the temperature on the stage.

Culture Preparation and Observation

19. Prepare neural basal medium (GIBCO-BRL) containing B27 supplement (GIBCO-BRL) at 1:50.
20. Place 1 ml of B27-supplemented neural basal medium into the bottom of the adapted petri dish.
21. Place the fibronectin-coated Millipore filter on top of coverslip inside the petri dish.
22. Rinse the explant a few times in fresh Ringer's solution by carefully pipetting up and down.
23. Gently transfer the explant to the surface of the Millipore filter.
24. Use forceps to orient the explant medial or lateral side down, depending on area that is to be visualized.
25. Pipette excess liquid from the Millipore filter and from inside the petri dish so that the filter is sitting against the coverslip.
26. Place the lid on the petri dish and seal it with a piece of Parafilm.
27. When the microscope incubator temperature has stabilized, transfer the sample to the stage and observe.

EXAMPLE OF APPLICATION

The sagittal slice explant technique was used to study neural crest cell migration and genesis of the chick peripheral nervous system (PNS). Sagittal slice explants were taken from 72-hour (HH stage 20) chick embryos that had been injected with a green fluorescent protein (GFP) encoding vector into the neural tube at HH stage10 (Fig. 1a). Sagittal slice explants were prepared by cutting the tissue along the vertebrate axis (Fig. 1b) and laid either lateral side up (Fig. 1c) or down (Fig. 1d) on a filter. The sagittal explant is taken from the highlighted area (Fig. 1a).

To investigate the formation of the sympathetic ganglia (SG; Fig. 1c) fluorescently labeled trunk neural crest cells were visualized migrating out of the neural tube and coalescing to form the SG using time-lapse confocal microscopy. Images taken from a typical time-lapse imaging session show the time evolution of the SG (Fig. 2). Neural crest cells in the more anterior region of the explant emerge and migrate through the rostral half of the somites (Fig. 2b–d) and stop in the periphery to form the SG (Fig. 2b–e).

ADVANTAGES AND LIMITATIONS

The sagittal slice culture technique allows the visualization of exposed interior (medial) surfaces from a sagittal perspective. With the cut made down the midline of the embryo, events can be imaged which occur along the anteroposterior and dorsoventral axes of the embryo. With this technique, much of the normal embryo morphology is maintained since more of the intact embryo is retained in the explant. Previously, it had been difficult to study morphogenetic events that occur deep within the embryo.

The sagittal slice explant technique allows visualization of morphogenesis in later-stage embryos and in embryos where cell migratory paths occur in the dorsoventral plane. Imaging cellular events at later embryonic stages has been technically difficult due to complexities in embryonic configuration (i.e., flexures and twists) and because of movements associated with the beating heart. This technique obviates some of these problems by allowing tracking of fluorescently labeled cells that migrate along the surface of the exposed medial structures and those into deeper tissue because the

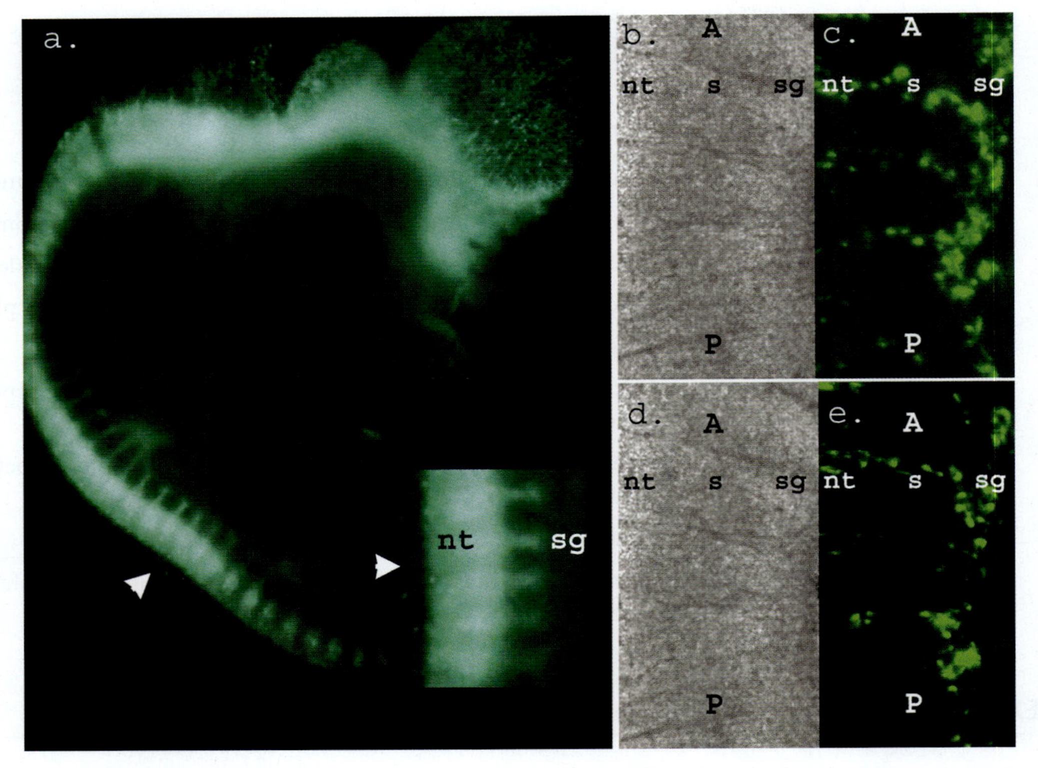

FIGURE 2. Before and after images taken from a typical time-lapse imaging session using the sagittal explant technique. (*a*) 84-hr-old chick embryo fluorescently labeled at $t = 36$ hr. Sagittal explant taken from highlighted trunk region. (*b–c*) Lateral side down image at $t = 0$, bright-field, fluorescence. (*d–e*) Lateral side down image at $t = 15$ hr, bright-field, fluorescence. (A) Anterior; (P) posterior; (nt) neural tube; (s) somite; (sg) sympathetic ganglia.

heart is separated from the slice explant. In addition, it has been difficult to follow cell trajectories along dorsoventral paths in typical cross-section slices. However, the sagittal slice technique allows these cell migratory pathways to be visualized. Hence, our sagittal slice explant method has proven to be ideal for imaging morphogenetic events which occur in later-stage chick embryos in the dorsoventral plane, such as formation of the DRG and SG.

With this preparation, two imaging approaches are possible: (1) Cells that migrate and remain superficially along the dorsolateral surface of the explant can be imaged and/or (2) the medial surface of the explant can be imaged to track cells that migrate within or adjacent to the spinal cord. Furthermore, because of the rostrocaudal temporal developmental gradient, one can image migrating cells at sequential stages of development within a single explant, depending on its length and the extent of cell labeling along the rostrocaudal axis.

The limitations to the sagittal slice culture technique include typical difficulties associated with slice culture preparations. First, there is a loss of tissue morphology over time. Second, outside of the egg, the embryo is subjected to a decrease in nutrition which induces loss of structure and degradation after 40 hours of incubation. Third, a sliced explant of an embryo has regions of its internal tissue exposed to the external environment. These excisions will induce a wound-healing effect on the neighboring tissue and slightly alter the morphology as compared to an in vivo embryo. However, the last aspect can be partially corrected by placing the cut slightly lateral to the midline of the neural tube to include the entire neural tube.

The chick sagittal slice technique has several advantages over other slice culture or whole-embryo techniques. These include the access to morphogenetic events within a three-dimensional embryo at early and late stages (HH stage13–26) of development and an opportunity to visualize phenomena along the vertebrate anterioposterior axis. This is important since many critical patterning events occur in a rostral-to-caudal manner in the embryo. Thus, the sagittal slice culture offers an ability to study neural development in living tissue and provides a viable option for imaging events which occur in the dorsoventral plane of the embryo and thereby obviate difficulties of imaging deep within whole embryos.

REFERENCES

Hamburger V. and Hamilton H.L. 1951. A series of normal stages in the development of the chick embryo. *J. Morphol.* **88:** 49–92.

CHAPTER 20

Imaging Mouse Embryonic Development

Elizabeth A.V. Jones,[1,2] Anna-Katerina Hadjantonakis,[3] and Mary E. Dickinson[2]

[1]*Department of Chemical Engineering and* [2]*Biological Imaging Center, California Institute of Technology, Pasadena, California 91125;* [3]*Developmental Biology Program, Sloan-Kettering Institute, New York, New York 10021*

Vertebrate development is a dynamic process that requires the careful orchestration of many events. In a relatively short amount of time, cells are added, deleted, transformed, and assigned roles that can be vital to the survival of the organism. Through sophisticated genetic manipulation and examination of naturally occurring mutations, hundreds of genes have been identified that have key roles in early mouse development. Understanding these roles fundamentally relies on understanding the phenotypic consequences of the mutations.

In many organisms that develop external to the mother, live imaging has been used to determine the relationships between cells that are involved in rapid morphogenetic processes (for review, see Lichtman and Fraser 2001). Since development is a continuous process, much more information about cellular phenotypes can be gained through continuous observation. Because mouse embryos develop in utero, most of what is known about complex phenotypes stems from embryos collected and fixed at several time points. Although culture and imaging of preimplantation embryos have been possible for some time now, postimplantation embryos have traditionally been cultured in roller culture, which does not permit time-lapse analysis of embryonic development using a microscope. To circumvent these limitations, we have adapted known culture methods for static culture on a microscope stage that allows dynamic imaging of embryos from gastrulation until early organogenesis (Jones et al. 2002; A.K. Hadkantonakis and V.E. Papaioannou, unpubl.). Whole-embryo culture, combined with stable fluorescent markers, such as fluorescent proteins, provides a powerful tool for understanding the cellular consequences of genetic manipulation in mice.

IMAGING MOUSE DEVELOPMENT

The development of the mouse embryo has been described in many volumes (including Theiler 1989; Rugh 1990; Kaufman 1992), and there are currently strong efforts to digitize atlases of fixed stages of development, making them more accessible and interactive (Davidson et al. 1997; Kaufman et al. 1998; Dhenain et al. 2001; MacKenzie-Graham et al. 2004). Atlases and descriptions of mouse development have been invaluable to understanding normal ontogeny as well as mutant phenotypes, but information about dynamic processes such as cell lineage or cell migration cannot be fully revealed from static images. For these reasons, methods for manipulating embryos ex vivo have been developed.

Protocols for culturing preimplantation embryos have been available for some time, and the procedure not only has provided great insight into early cleavage events, but has also been the cornerstone of transgenic and gene knockout technologies (Nagy et al. 2003). Imaging early postimplantation development is more difficult. Embryos at these stages require a rich supply of freshly prepared serum and have traditionally been grown in roller culture (Tam 1998). The constant movement of the media in roller cultures promotes gas exchange, and since the flow rate of gas is minimal, evaporation is limited. The protocol below shows how similar conditions can be used to culture mouse embryos on the stage of the microscope for fluorescence imaging. This protocol is recommended for mouse embryos from 6.5 to 9.5 days postcoitus (dpc) for 18–24 hours. After 10.0 dpc, embryos become far more difficult to maintain using this method, although a serum-free protocol for growing midgestation-stage

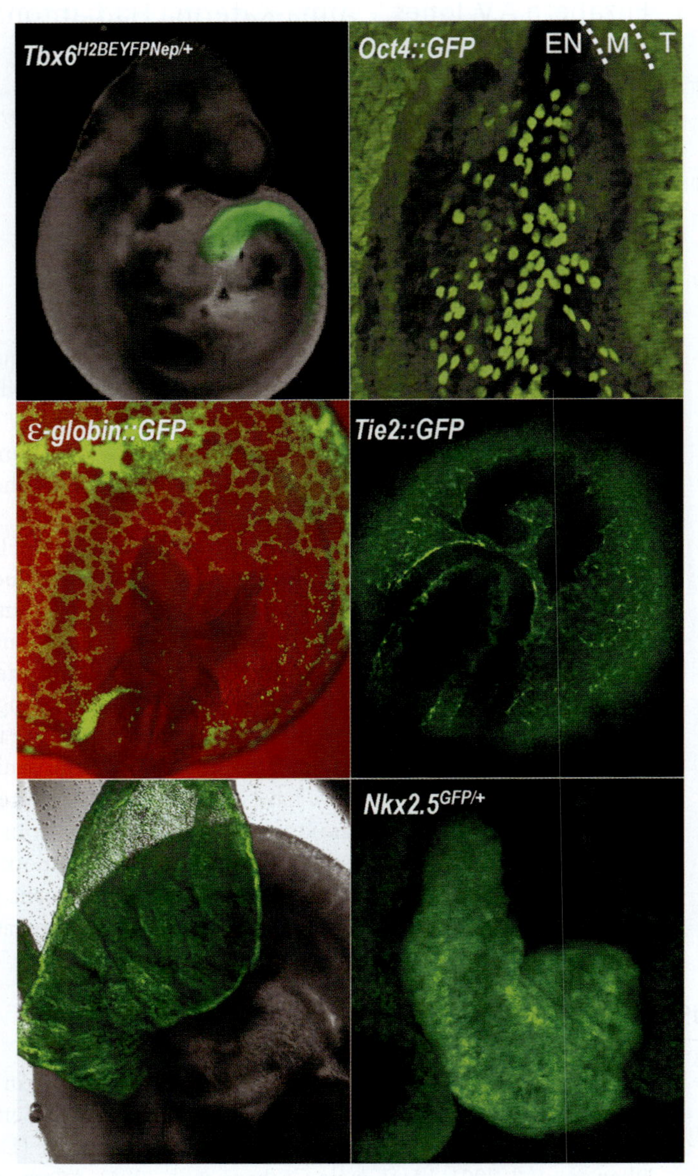

FIGURE 1. GFP expression (*green*) driven by various promoters. Tbx6 is expressed in the presomitic mesoderm and newly formed somites. Oct4 expression can be seen in the primordial germ cells. ε-globin marks the primitive erythroblasts. Tie2 is a marker of endothelial cells. Alpha-fetoprotein (AFP) expression is seen in extraembryonic tissues. Nkx2.5 expression is seen throughout the developing heart region.

embryos in culture has been described (Moore-Scott et al. 2003). These stages have also been studied successfully by labeling cells in embryos that are allowed to develop exo utero (Muneoka et al. 1986; Turner et al. 1990; Serbedzija et al. 1992; Saito and Nakatsuji 2001).

Many of the labeling techniques developed for other developmental models are applicable to mouse embryos grown in culture, including dye injection, tissue grafting (Quinlan et al. 1999), and retroviral labeling. The strength of the mouse model, however, is the ability to manipulate the genome in such a way that genetically encoded fluorescent proteins can be expressed in specific cells or tissues (Fig. 1). In fact, several genetic strains have been engineered to express fluorescent proteins in defined regions of the embryo. For example, the *Oct4* promoter has been used to drive fluorescence in primordial germ cells (Anderson et al. 1999). The *ε-globin* promoter has been used to express green fluorescent protein (GFP) only in the primitive erythroblasts (Dyer et al. 2001), and the *Hex* promoter can be used for specific labeling of the anterior visceral endoderm (AVE) (Rodriguez et al. 2001). In addition, many knockins, where fluorescent protein sequences are inserted into a locus to disrupt gene function, combine fluorescence imaging with mutant analysis; for example, the Nkx2.5$^{GFP/GFP}$ (Biben et al. 2000), or the Tbx6 mutant, which labels the presomitic mesoderm and newly formed somites (A.K. Hadjantonakis and V.E. Papaioannou, unpubl.). Additionally, internal ribosome entry sites (IRES) can be used to achieve single-copy protein expression (Rodriguez et al. 1999).

Although most current transgenics express fluorescent proteins throughout the cytoplasm of the cells of interest, fluorescent proteins that are specifically associated with distinct subcellular regions have also been produced (Fig. 2). In some cases, localization of the protein is conferred by engineering fusion proteins or by tagging with sequences recognized by protein trafficking machinery. Multiple-color transgenics can be made to help understand whether signal and putative responsive cells interact. With the advent of convenient tools for multispectral analysis and linear unmixing, such as those available on the Zeiss LSM 510 META, it is now possible to image labels with multiple colors without spectral bleed-through, even if the emission spectra are highly overlapping (for more information, see Dickinson et al. 2001, 2003; Lansford et al. 2001). Multiplexing allows more complex interactions between cells in the developing embryo to be understood.

MATERIALS

Media

Two types of media are used in mammalian embryo culture: a dissection medium and a culture medium.

Dissection Medium

The dissection medium is composed of 90% (v/v) DMEM/F-12 (GIBCO-BRL), 8% (v/v) heat-inactivated fetal bovine serum (GIBCO-BRL), 10 mM HEPES (Irvine Scientific), and 10 mM penicillin–streptomycin solution <!> (Irvine Scientific). This medium must be heated to 37°C before use in a water bath.

Culture Medium

The appropriate culture medium depends on the stage of the embryo. For younger embryos (embryonic day 5.5 [E5.5] to E7.5), the medium consists of 50% DMEM and 50% rat serum. For embryos between E7.5 and E9.5, DMEM/F12 is used instead of DMEM and the culture medium is supplemented with 10 mM HEPES and 10 mM penicillin–streptomycin solution (Irvine Scientific). The medium is filtered through a 0.2-μm filter to sterilize it. At all stages, it is necessary to heat and equilibrate the pH of the medium by placing it in a tissue culture incubator (5% CO_2, 37°C) for at least 1 hour.

The most important component to proper mammalian embryo culture is the use of high-quality rat serum. Although some commercial sources exist, home-made preparations consistently give superior results.

Caution: *See Appendix 3 for appropriate handling of materials marked with <!>.*

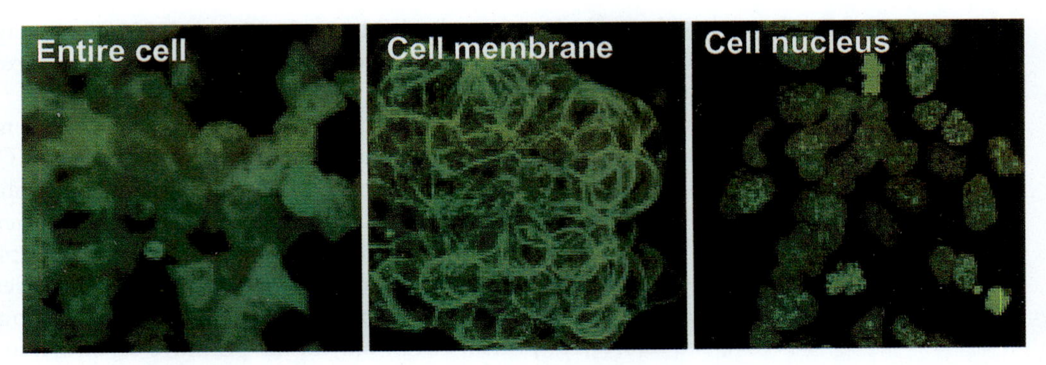

FIGURE 2. Fluorescent protein expression can be localized throughout the cytoplasm simply by expressing the protein. To localize proteins to the cell membrane, the fluorescent protein can be myristoylated. For localization to the nucleus, a fusion protein can be made of the fluorescent protein and histone 2B.

PROCEDURE A

Preparation of Rat Serum

1. Anesthetize adult male rats with ether <!>.
2. Spray the abdomen with 70% ethanol<!> and make a v-shaped incision into the lower abdomen.
3. Expose the dorsal aorta, which is located next to the larger vena cava. It is the smaller (~ 1-mm diameter) pulsing blood vessel (Fig. 3).
4. Puncture the aorta using a butterfly needle.

   Note: It is simplest to use a butterfly needle connected to a syringe needle (Vaculock system, BD Biosciences).
5. Once the dorsal aorta has been punctured with the butterfly needle, press the syringe needle into a Vacutainer tube (BD Biosciences), creating a mild suction to collect the blood.
6. Invert the Vacutainer tubes a few times during collection.

   Note: Vacutainer tubes must be used. The suction and the anticlotting agent provided by these tubes are essential for good results.
7. Once the rat has been exsanguinated, place the collected blood on ice.
8. Euthanize the rat and allow the ether to evaporate from the carcass.

   Note: Each rat gives approximately 2–3 ml of rat serum: therefore, process as many rats as needed for experiments.
9. Centrifuge the blood at 1300*g* for 20 min.
10. Remove and pool the supernatant; discard the pellet.
11. Centrifuge the serum again at 1300*g* for 10 min to remove any remaining cells and debris. Collect the supernatant.
12. Heat-inactivate the serum in a water bath for 30 min at 56°C with the lid partially unscrewed to allow the ether to evaporate.
13. Filter the serum using a 0.45-μm filter and aliquot.
14. Freeze aliquots and store at –80°C for up to 1 year.

    Note: Blood should be collected under low suction. At higher suction, the red blood cells can lyse, reducing the quality of the serum. It is possible to use a syringe for blood collection, rather than Vacutainer tubes; however, too much force will cause hemolysis, making the serum unusable.

    The use of ether is also essential to the procedure. Other anesthetics remain in the blood and can affect embryonic growth. Ether, however, will easily evaporate from the serum during heat-inactivation.

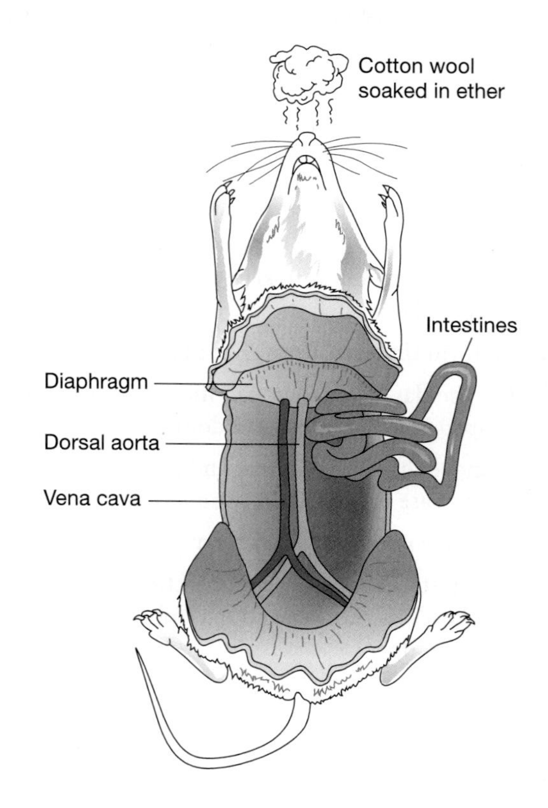

FIGURE 3. Preparation of rat serum, locating the dorsal aorta.

MICROSCOPE SETUP AND ENVIRONMENTAL CONTROL

Proper environmental control of the microscope stage is probably the single most important factor for successful culture. Mammalian embryos are especially sensitive to small changes in temperature (±3°C) and small amounts of evaporation. To keep the embryos at 37°C, construct a heater box that will fit around the microscope stage (Fig. 4), or purchase one from a commercial manufacturer (e.g., Zeiss). A heater box can also be constructed as described by Kulesa and Fraser in Chapter 18.

CULTURE CHAMBERS

For imaging on an inverted microscope, the embryos are routinely cultured in Lab-tek chambers with coverslip bottoms (Nalge Nunc) or 30-mm dishes with glass coverslip bottoms (MatTek). Gas is fed into the environment using 1/8-inch tubing. A hole is soldered into the lid of the chamber, and the gas is connected via a barbed polypropylene fitting (1/16 x 1/8 inch Cole-Parmer Instrument). For more detailed description of this setup, see the Procedures section.

The gas requirements for mammalian embryos change depending on the stage of embryonic development. The gas consists of 5% CO_2, variable oxygen concentration, and balanced nitrogen. For embryos between E4.5 and E6.5, the oxygen should be kept at 95%; embryos between E7.5 and E9.5 develop well with 20% oxygen. Since mammalian embryos are sensitive to even the slightest amount of evaporation, several components are required to prevent evaporation. First, the gas is bubbled through a gas-washing bottle (Fisher Scientific). Second, a thin layer of mineral oil (Sigma) is placed over the medium. Finally, the environment is sealed using silicon grease (Dow Corning) or Teflon tape. A description of how these components are assembled is provided in the Procedures section.

PROCEDURE B

Preparation of Embryos

Dissection

Timed matings and dissections are performed essentially as described by Nagy et al. (2003), except that we make every effort to keep the embryos at 37°C during the dissection. We use a warm hood and preheated media.

1. Perform timed matings. The presence of a vaginal plug is taken as 0.5 dpc.
2. On the day of interest, sacrifice the pregnant mouse according to local guidelines. Make a v-shaped incision into the abdominal cavity through the skin and body wall.
3. Remove the uterine horns and place in the warmed dissection medium (see Materials section). Then tease the muscle apart and remove the individual embryos, surrounded by decidua.
4. Carefully remove the deciduum using watchmakers tweezers (Dumont) to expose the embryo.
5. Using standard embryo dissection technique (as described by Nagy et al. 2003), remove Reichert's membrane and the trophoblast layer. Leave all other extraembryonic tissues intact up until 9.5 dpc.

   Note: For very young embryos, the removal of the Reichert's membrane requires the use of a micromanipulator. The ectoplacental cone is usually left on the embryo. The yolk sac can be removed from 8.5-dpc embryos and must be removed from 9.5-dpc embryos.
6. After dissection, use a Pasteur pipette to transfer the embryos to the culture chambers containing prewarmed culture medium (see Materials section). Be careful to transfer as little dissection medium as possible.

   Note: Do not culture embryos individually or in overcrowded conditions. They will not develop well. Each embryo requires 0.5–1 ml of heated medium. Limit older embryos (>7.5 dpc) to three embryos per chamber.
7. Place the culture chambers in a tissue culture incubator for at least 1 hour to allow the embryos to recover from dissection.

Embryo Immobilization

8. Immobilize the embryos using one of the following techniques to prevent them drifting out of focus during culture.

   Note: Embryo drift is especially significant at higher magnifications. At stage-8.5 dpc, the yolk sac has expanded and is quite buoyant. This makes the embryos susceptible to small currents in the medium, and the ectoplacental cone, which is heavier than the yolk sac, sinks toward the bottom of the dish, reorienting the embryo. At stage-9.5 dpc, the yolk sac can be removed, allowing the embryo to sit nicely on the bottom of the Lab-tek chamber.

   a. Orient embryos of all stages as required using a suction-holding pipette attached to a micromanipulator.

   b. If the culture is modified to use an upright microscope, immobilize the younger embryos (<7.5 dpc) in scratches in the bottom of the dish.

   c. Tie a piece of human hair or fine platinum wire in a knot around the ectoplacental cone and immobilize the embryo by "propping" it up.

   d. Make a wire hook and place it around the ectoplacental cone. Before medium is added, anchor the hook into wax or agarose that has been fixed to the bottom of the chamber.

   Note: If the yolk sac has expanded, immobilize the embryos in a way that avoids any pressure on the yolk sac or the growing embryo. Even slight pressure from a thin layer of agar or Nitex grating overlying the embryo will prevent circulation in the yolk sac.

PROCEDURE C

Time-lapse Imaging

1. Assemble the heater box and turn it on at least half an hour (although a full hour is preferable) before imaging starts to allow the equipment to warm up. Make sure that the gas-washing bottle (fitted with a fine diffuser) is also placed within the heater box (Fig. 4).
2. Place the embryos on the microscope stage and orient them as desired. Place a thin layer of embryo-tested mineral oil (Sigma) on top of the medium to reduce evaporation. When using Lab-tek chambers, fill the bottom of the space between the two chambers with silicon grease. This region will be used for gas inlet to the chamber environment.
3. Feed the gas outlet from the pressurized cylinder to the washing bottle through appropriate tubing and set the regulator on the pressurized cylinder as low as possible—too little gas is not a problem; too much gas will compromise the seal of the chamber and cause excess evaporation.
4. Use a soldering iron or a hot syringe needle to make a small hole in the chamber lid. This allows gas to be introduced into the culture chamber from the washing bottle.
5. Place a small male-male barbed polypropylene fitting in this hole and attach the external end of the fitting to the gas-washing bottle via appropriate tubing. Use silicon grease to ensure that the gas inlet is airtight.
6. Then place the embryos on the microscope stage and check their orientation. Seal the inner edges of the lid with silicon grease.

   Note: Alternatively, the chamber can be sealed with Teflon tape around the outer edges of the lid. If this method is used, seal the chamber before the embryos are placed on the microscope stage—no reorientation of the embryos is possible once they are on the stage.

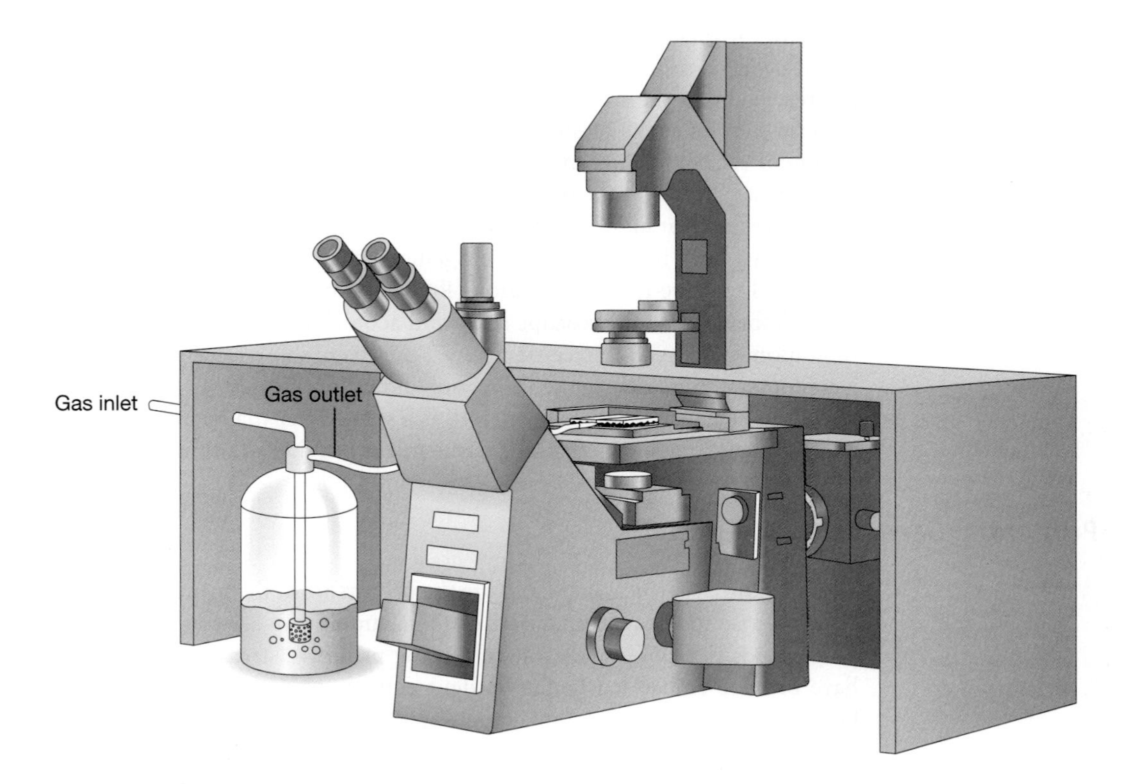

FIGURE 4. Heater box diagram. Static culture system for time-lapse imaging of postimplantation mouse embryos. The microscope and static culture are housed in a 37°C heater box. The gas mixture for the culture is humidified using a bubbler.

7. Collect images according to the needs of the chosen application (see below), taking care to minimize exposure to excitation light.

For many applications, it is sufficient to image once every 5–10 min. Images can be single frame or, if confocal imaging is being used, several sections can be taken along the *z* axis. Image acquisition is simpler at lower magnifications (e.g., 5X), since small shifts in the embryo position do not affect the image field of view as drastically. At higher magnifications (up to ~20X), the focus on the embryos must to be adjusted every hour or two. Using this technique, 5.5–8.5-dpc embryos can be cultured for up to 24 hours and 9.5-dpc embryos for 12–18 hours.

During the culture period, it is important to prevent any fluctuations in gas flow and temperature. At stage 8.0 dpc and older, embryos should also be observed to ensure the heart rate remains normal. Once culture is complete, the final morphology (as compared to normal in vivo development) of the embryos should be evaluated in order to assess whether development has occurred normally. Most embryos are then fixed in 4% paraformaldehyde for later analysis.

EXAMPLE OF APPLICATIONS

Anteroposterior Polarity

The anteroposterior polarity of the mouse embryo is not morphologically distinguishable until gastrulation at E6.5, when the site of primitive streak formation unequivocally defines the posterior of the embryo. Recent work has revealed that well before the onset of gastrulation, the mouse embryo initiates dynamic and asymmetric patterns of gene expression in the visceral endoderm, an extraembryonic cell layer that comprises a squamous epithelium surrounding the epiblast. Even though this extraembryonic tissue does not contribute to the body itself, it is involved in axis formation, with the anterior visceral endoderm (AVE) being essential for anterior patterning of the embryo.

Work involving the labeling of individual cells of the blastocyst with enhanced GFP (EGFP), followed by in vitro culture and/or reimplantation and subsequent time-lapse imaging at gastrula stages, has revealed that the spatial distribution of cells within the visceral endoderm can be traced back to polarity present at the blastocyst stage (for review, see Beddington and Robertson 1998). The distribution of visceral endoderm cells reflects characteristic cell movements. At pregastrula stages, cells of the AVE are located at the distal tip of the embryo, but they then are believed to move unilaterally to the future anterior and provide the first overt break in the symmetry of the embryo. This migration has recently been investigated in transgenic mice, where the regulatory elements of the *Hex* gene have been used to drive EGFP expression (Rodriguez et al. 2001; Srinivas et al. 2004). By culturing transgenic embryos on a microscope stage and acquiring time-lapse images, it was possible to follow the migration of the AVE. Interestingly, AVE cells were shown to change shape continuously and to project filopodial processes in their direction of motion, suggesting that they are migrating actively. It was also established that AVE cells migrate as a single layer that is in continuous contact with the underlying epiblast, suggesting that this tissue might provide additional directional cues.

Primordial Germ Cell Migration

The primordial germ cells (PGCs) are the precursors of the gametes in mice. They arise from cells that also give rise to the proximal part of the allantois and are routinely visualized as an alkaline phosphatase-positive cell population within the primitive streak and developing allantois. By E8.5, they have become embedded in the hindgut epithelium and thereafter migrate to the genital ridges to become incorporated into the developing gonads (for review, see Bendel-Stenzel et al. 1998).

In light of the uncertainty concerning the origin and migratory route taken by PGCs, Anderson et al. (2000) used regulatory elements for the *Oct4* gene to drive EGFP in primordial germ cells during their genesis and subsequent migration. Time-lapse microscopy was used to selectively follow the trajectory taken by primordial germ cells. From this work, Anderson et al. proposed a revised model PGC formation and behavior, suggesting that only the cells in the ventral posterior primitive streak contribute to the germ line and that the cells contained within the proximal allantois remain outside the embryo proper and do not contribute.

The same group of investigators (Bendel-Stenzel et al. 2000) went on to use Oct4::EGFP transgenic mice to image the later stages of PGC migration from the hindgut to the genital ridges, where the PGCs were observed to coalesce, both with each other and with somatic cells, to form the primary sex cords. The change in migratory path and behavioral properties of these cells has been linked to the differential expression of several members of the cadherin class of adhesion molecules.

Imaging Erythroblast Differentiation and Vascular Development

The culture of 8.5-dpc embryos has been used to visualize the formation of the early cardiovascular system (E. Jones et al., unpubl.). At approximately 7.5 dpc, isolated clusters of blood, surrounded by endothelial cells, form in the yolk sac of the embryo. These clusters expand and interconnect to form a random network known as the capillary plexus. This plexus connects to the heart and, as the heart begins to beat, the plexus is remodeled into a more mature, tree-like vascular system (for review, see Flamme et al. 1997). Using a mouse that expresses GFP driven by the *ε-globin* promoter (Dyer et al. 2001), it is possible to image the formation of primitive erythroblasts and visualize changes in the vasculature. Endothelial cells can also be observed directly using a mouse that expresses GFP driven by the *Tie2* promoter (Motoike et al. 2000). Both markers are being used to image erythroblast differentiation, the formation of the capillary plexus, as well as the subsequent remodeling of the capillary plexus (Fig. 1, panel 3). In addition to observing changes in morphology, the ε-globin::GFP mice have also been used to quantify flow velocities in the yolk sac during development (Jones et al. 2004).

ADVANTAGES AND LIMITATIONS

A clear advantage of optical imaging in mouse embryos comes from the ability to resolve the cellular and subcellular events that occur during normal development or as a consequence of genetic alterations. Many mouse models are used to understand the etiology and progression of human diseases and birth defects, and additional tools that provide greater mechanistic insight are always welcome. Although other imaging methods, such as ultrasound, optical coherence tomography, and magnetic resonance imaging, can provide insights into in utero development, these methods lack the cellular and even subcellular resolution offered by optical imaging. In addition, the number of fluorescent probes that can be used to identify specific cells or assay cell functions now extends into the thousands. Although specific contrast agents are still being developed for these other methods, improvements in both resolution and contrast are being made at a rapid pace.

The culture of mammalian embryos is, however, technically more difficult than more traditional models for imaging development. Mammalian embryos are exceedingly sensitive to temperature fluctuations, medium content, and physical pressure. A constant temperature must be maintained at 38°C to ensure proper growth. High-quality rat serum is essential for growth, and, at 8.5 dpc, even slight pressure on the yolk sac impairs circulation to the yolk sac, preventing the embryos from turning and, hence, advancing at a normal rate. The sensitivity of mammalian embryos to culture conditions can provide varying degrees of success, but time-lapse analysis provides a wealth of dynamic data, so multiple trials are often worth the data generated by a few good movies.

Movement during time-lapse imaging can be problematic in any organism, making it difficult to track cells and take quantitative measurements. Of the immobilization techniques described here, only the holding pipette offers significant control over the orientation of the embryo. In addition, the significant growth of the mammalian embryo between 7.5 and 9.5 dpc makes it difficult to keep all areas of interest in focus; it is important to choose the best objective lens for the resolution and magnification required. Furthermore, many algorithms have been used for the realignment of imaging sequences and can improve the information gained from such movies (Megason and Fraser 2003; Rupp et al. 2004).

The cup-shaped configuration and turning of the mouse embryo also increase the complexity of imaging and culturing techniques. The curvature of the embryo increases the number of optical sections needed to image a given region. Cell movements are also more complex to understand since they tend to occur across different optical sections. As the embryo increases in size, the radius of curvature is also in constant flux. These problems are not isolated to mouse embryos, however, and can be overcome using three-dimensional software for reconstructing and projecting the data and/or multiphoton imaging for improved depth penetration.

Dynamic imaging of mouse development provides yet another tool for a very robust genetic and developmental system. Whole-embryo culture of mammalian embryos for imaging combines molecular biology, genomics, and biological imaging, offering methods for understanding the interplay of intercellular communication and the signaling components important for orchestrating complex developmental events.

ACKNOWLEDGMENTS

We thank Chris Waters for technical support and Joaquin Gutierrez for assistance in animal husbandry. We also thank the National Institutes of Health and the Human Frontiers Science Program for funding.

REFERENCES

Anderson R., Copeland T.K., Scholer H., Heasman J., and Wylie C. 2000. The onset of germ cell migration in the mouse embryo. *Mech. Dev.* **91:** 61–68.

Anderson R., Fassler R., Georges-Labouesse E., Hynes R.O., Bader B.L., Kreidberg J.A., Schaible K., Heasman J., and Wylie C. 1999. Mouse primordial germ cells lacking β1 integrins enter the germline but fail to migrate normally to the gonads. *Development* **126:** 1655–1664.

Beddington R.S. and Robertson E.J. 1998. Anterior patterning in mouse. *Trends Genet.* **14:** 277–284.

Bendel-Stenzel M., Anderson R., Heasman J., and Wylie C. 1998. The origin and migration of primordial germ cells in the mouse. *Semin. Cell Dev. Biol.* **9:** 393–400.

Bendel-Stenzel M.R., Gomperts M., Anderson R., Heasman J., and Wylie C. 2000. The role of cadherins during primordial germ cell migration and early gonad formation in the mouse. *Mech. Dev.* **91:** 143–152.

Biben C., Weber R., Kesteven S., Stanley E., McDonald L., Elliott D.A., Barnett L., Koentgen F., Robb L., Feneley M., and Harvey R.P. 2000. Cardiac septal and valvular dysmorphogenesis in mice heterozygous for mutations in the homeobox gene *Nkx2-5*. *Circ. Res.* **87:** 888–895.

Davidson D., Bard J., Brune R., Burger A., Dubreuil C., Hill W., Kaufman M., Quinn J., Stark M., and Baldock R. 1997. The mouse atlas and graphical gene-expression database. *Semin. Cell Dev. Biol.* **8:** 509–517.

Dhenain M., Ruffins S.W., and Jacobs R.E. 2001. Three-dimensional digital mouse atlas using high-resolution MRI. *Dev. Biol.* **232:** 458–470.

Dickinson M.E., Bearman G., Tilie S., Lansford R., and Fraser S.E. 2001. Multispectral imaging and linear unmixing add a whole new dimension to laser scanning fluorescence microscopy. *BioTechniques* **31:** 1272.

Dickinson M.E., Simbuerger E., Zimmermann B., Waters C.W., and Fraser S.E. 2003. Multiphoton excitation spectra in biological samples. *J. Biomed. Opt.* **8:** 329–338.

Dyer M.A., Farrington S.M., Mohn D., Munday J.R., and Baron M.H. 2001. Indian hedgehog activates hematopoiesis and vasculogenesis and can respecify prospective neurectodermal cell fate in the mouse embryo. *Development* **128:** 1717–1730.

Flamme I., Frolich T., and Risau W. 1997. Molecular mechanisms of vasculogenesis and embryonic angiogenesis. *J. Cell Physiol.* **173:** 206–210.

Jones E.A., Crotty D., Kulesa P.M., Waters C.W., Baron M.H., Fraser S.E., and Dickinson M.E. 2002. Dynamic in vivo imaging of postimplantation mammalian embryos using whole embryo culture. *Genesis* **34:** 228–235.

Jones E.A., Baron M.H., Fraser S.E., and Dickinson M.E. 2004. Hemodynamic changes during the development of the mammalian vasculature. *Am. J. Physiol.: Heart Circ.* (in press).

Kaufman M.H. 1992. *The atlas of mouse development.* Academic Press, London.

Kaufman M.H., Brune R.M., Davidson D. R., and Baldock R.A. 1998. Computer-generated three-dimensional reconstructions of serially sectioned mouse embryos. *J. Anat.* **193:** 323–336.

Lansford R., Bearman G., and Fraser S.E. 2001. Resolution of multiple green fluorescent protein color variants and dyes using two-photon microscopy and imaging spectroscopy. *J. Biomed. Opt.* **6:** 311–318.

Lichtman J.W. and Fraser S.E. 2001. The neuronal naturalist: Watching neurons in their native habitat. *Nat.*

Neurosci. (suppl.) **4:** 1215–1220.

MacKenzie-Graham A., Lee E. F., Dinov I. D., Bota M., Shattuck D. W., Ruffins S., Yuan H., Konstantinidis F., Pitiot A., Ding Y. et al. 2004. A multimodal, multidimensional atlas of the C57BL/6J mouse brain. *J. Anat.* **204:** 93–102.

Megason S.G. and Fraser S.E. 2003. Digitizing life at the level of the cell: High-performance laser-scanning microscopy and image analysis for in toto imaging of development. *Mech. Dev.* **120:** 1407–1420.

Moore-Scott B.A., Gordon J., Blackburn C.C., Condie B.G., and Manley N.R. 2003. New serum-free in vitro culture technique for midgestation mouse embryos. *Genesis* **35:** 164–168.

Motoike T., Loughna S., Perens E., Roman B.L., Liao W., Chau T.C., Richardson C.D., Kawate T., Kuno J., Weinstein B.M., Stainier D.Y., and Sato T.N. 2000. Universal GFP reporter for the study of vascular development. *Genesis* **28:** 75–81.

Muneoka K., Wanek N., and Bryant S.V. 1986. Mouse embryos develop normally exo utero. *J. Exp. Zool.* **239:** 289–293.

Nagy A., Gertsenstein M., Vintersten K., and Behringer R. 2003. *Manipulating the mouse embryo: A laboratory manual*, 3rd edition. Cold Spring Harbor Laboratory Press, Cold Spring Harbor, New York.

Quinlan G.A., Trainor P.A., and Tam P.P. 1999. Cell grafting and labeling in postimplantation mouse embryos. *Methods Mol. Biol.* **97:** 41–59.

Rodriguez I., Feinstein P., and Mombaerts P. 1999. Variable patterns of axonal projections of sensory neurons in the mouse vomeronasal system. *Cell* **97:** 199–208.

Rodriguez T.A., Casey E.S., Harland R.M., Smith J.C., and Beddington R.S. 2001. Distinct enhancer elements control Hex expression during gastrulation and early organogenesis. *Dev. Biol.* **234:** 304–316.

Rugh R. 1990. *The mouse: Its reproduction and development.* Oxford University Press, New York.

Rupp P.A., Czirók A., and Little C.D. 2004. αvβ3 integrin-dependent endothelial cell dynamics in vivo. *Development* **131:** 2887–2897.

Saito T. and Nakatsuji N. 2001. Efficient gene transfer into the embryonic mouse brain using in vivo electroporation. *Dev. Biol.* **240:** 237–246.

Serbedzija G.N., Bronner-Fraser M., and Fraser S.E. 1992. Vital dye analysis of cranial neural crest cell migration in the mouse embryo. *Development* **116:** 297–307.

Srinivas S., Rodriguez T., Clements M., Smith J.C., and Beddington R.S. 2004. Active cell migration drives the unilateral movements of the anterioir visceral endoderm. *Development* **131:** 1157–1164.

Tam P.P. 1998. Postimplantation mouse development: Whole embryo culture and micro-manipulation. *Int. J. Dev. Biol.* **42:** 895–902.

Theiler K. 1989. *The house mouse: Atlas of embryonic development.* Springer-Verlag, New York.

Turner T., Bern H.A., Young P., and Cunha G.R. 1990. Serum-free culture of enriched mouse anterior and ventral prostatic epithelial cells in collagen gel. *In Vitro Cell. Dev. Biol.* **26:** 722–730.

CHAPTER 21

Imaging the Developing Retina

Christian Lohmann,[1,2] Jeff S. Mumm,[1] Josh Morgan,[1]
Leanne Godinho,[1] Eric Schroeter,[1] Rebecca Stacy,[1]
Wai Thong Wong,[1,3] Dennis Oakley,[1] and Rachel O.L. Wong[1]

[1]*Department of Anatomy and Neurobiology, Washington University School of Medicine, St. Louis, Missouri 63110;* [2]*Cellular and Systems Neurobiology, Max-Planck Institute for Neurobiology, 81377 München, Germany;* [3]*Scheie Eye Institute, Department of Ophthalmology, University of Pennsylvania, Philadelphia, Pennsylvania 19104*

The retina is one of the most optically accessible structures of the vertebrate central nervous system. Investigating neuronal function and development of the retina is facilitated not only by its transparency, but also by the stereotypic laminar organization of its cell bodies and connections (see Fig. 1a) (Cajal 1893; Wässle and Boycott 1991). Furthermore, unlike other regions of the brain, which must be sliced thinly for in vitro recordings, the live retina can be studied in vitro as a whole-mount preparation that comprises the entire intrinsic network. In vivo recordings of the developing retina can also be performed in animals, such as the zebrafish, that are largely transparent during embryonic stages. This chapter presents methods that have been adapted or developed to visualize retinal neurons for optical recordings both in vitro and in vivo.

During the past decade, live imaging methods have been used to investigate the cellular processes underlying the structural and functional development of the retina. The basic requirement for live imaging studies of the developing retina is to label cells with a nontoxic method that permits repeated imaging over time. This chapter discusses labeling techniques using fluorescent proteins and dyes that are routinely used for in vitro and in vivo imaging studies of the developing retina in our lab. Calcium imaging methods are presented in Chapter 34.

The first part of this chapter describes how to label and image retinal neurons in vitro using retinal whole-mount preparations. Retinal structure and circuitry remain intact in whole-mount preparations, and thus present an ideal model for in vitro studies of circuit structure and function (for review, see Sernagor et al. 2001). Protocols for ballistic delivery of plasmids and dyes to label retinal cells using a gene gun are provided in detail. The second section describes methods for imaging retinal cells of live zebrafish embryos. This preparation is ideally suited for investigating early events in the development of the retina, including cell division and migration (Das et al. 2003), as well as for elucidating mechanisms underlying circuit formation in vivo (Kay et al. 2004).

Ganglion cells that are closest to the surface in retinal whole mounts can be imaged using standard epifluorescence microscopes (typically, images are acquired with cooled CCD cameras). However, confocal or multiphoton imaging is necessary to obtain clearer views of cells deeper within the retina and to image cells in vivo. Examples of images acquired using the techniques described in this chapter are provided in Figures 2–4.

IN VITRO RECORDINGS OF RETINAL WHOLE MOUNTS

MATERIALS

Physiological Media

Ames medium (Sigma) is suitable for maintaining rabbit and ferret retinas (Ames and Nesbett 1981). Components for this medium are commercially available as a premixed powder. Pour the contents of one vial into a 1-liter flask, and add 800 ml of double-distilled water. Add either 4.76 g/l HEPES or 1.9 g/l sodium bicarbonate. If using HEPES, titrate the solution to a pH of 7.3–7.4 using 5 M NaOH<!>. Bring the final volume to 1 liter with double-distilled water. Bubble the solution with oxygen to use for retinal dissection and recording. If using bicarbonate buffer, bubble with carbogen (5% CO_2, 95% oxygen).

Artificial cerebral spinal fluid (ACSF) is used for mouse and chick retina. Add 800 ml of double-distilled water to a 1-liter flask, and then add the compounds in the order listed in Table 1. If using HEPES, titrate pH to 7.3–7.4 using 5 M NaOH. Finally, bring the volume of the ACSF up to 1 liter with double-distilled water.

Recording Chamber

Recording chambers typically used for electrophysiology are suitable. Recordings are performed on upright fixed-stage microscopes, using long-working-distance water objectives. The retina can be maintained for many hours (up to 36 hours) in a temperature-controlled and oxygenated environment within a simple home-made superfusion chamber. The configurations of such a chamber and the superfusion setup are shown in Figure 1c. Uniform heating is provided by a Cell Microtemp temperature controller through the ITO glass base (Cell Microtemp, HI-25 or HI-55) of the chamber. The black filter paper, onto which the retina adheres, can be held down with vacuum grease and/or with a piece of flattened platinum wire (diameter: 0.5–1 mm; Hauser and Miller) bent into a C-shaped ring (see Fig. 1c). Recordings are carried out between 32°C and 35°C. The superfusion medium is oxygenated prior to entering the chamber. A gravity feed system (50-ml syringe elevated above the microscope stage) is used for the inflow via polythene tubing (e.g., PE-10, PE-15 pr −20; A-M Systems).

A steady flow of fluid across the retina is achieved using a vacuum suction glass pipette at the outlet. A conical flask with side arm is used as a sump for the vacuum. The suction pipette is beveled by hand using fine-grade sandpaper; the beveled end is lowered to just below the surface of the medium in the central chamber (see Fig. 1c). Superfusion rates (typically around 1 ml/min) can be controlled using flow valves (Abbot Labs) on both the inflow and outflow.

TABLE 1. Composition of physiological solutions

	Mouse ACSF		Chick ACSF		Zebrafish Danieau's 30× stock	
Chemical components	g/l	mM	g/l	mM	g/l	mM
NaCl	6.95	119	7.29	125	101.7	1740
KCl	0.18	2.5	0.37	5	1.56	21
$MgCl_2 \cdot 6H_2O$	0.26	1.3	0.41	2	–	–
$CaCl_2 \cdot 2H_2O$	0.37	2.5	0.30	2	–	–
NaH_2PO_4	0.12	1.0	–	–	–	–
KH_2PO_4	–	–	0.17	1.25	–	–
Glucose	1.98	11	3.6	20	–	–
HEPES (free acid)	4.76	20	5.48	23	35.75	150
$MgSO_4 \cdot 7H_2O$	–	–	–	–	2.96	12
$Ca(NO_3)_2$	–	–	–	–	4.25	18

Caution: *See Appendix 3 for appropriate handling of materials marked with <!>.*

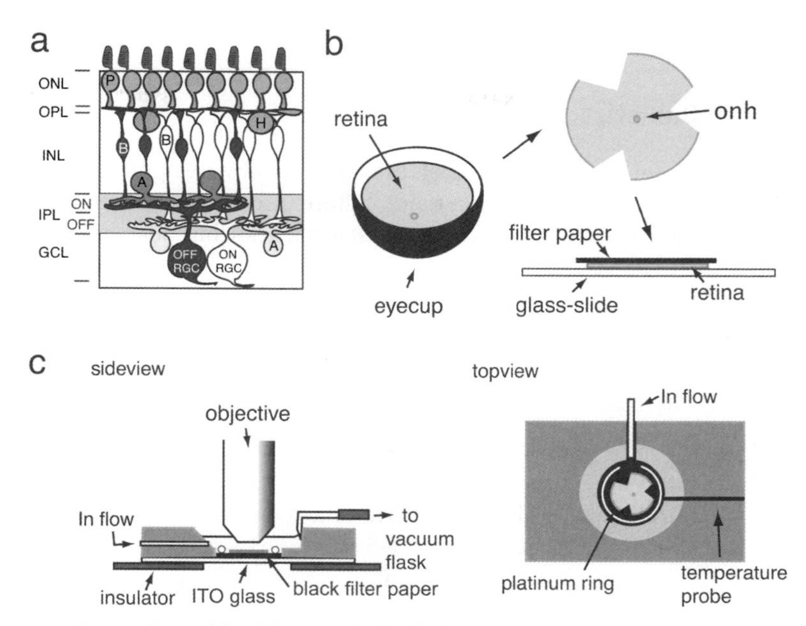

FIGURE 1. Preparation and recording from retinal whole mounts. (*a*) Schematic of a cross-section of the vertebrate retina, showing the laminar organization of cell bodies and their connectivity. (GCL) Ganglion cell layer; (INL) inner nuclear layer; (ONL) outer nuclear layer; (IPL and OPL) inner and outer plexiform layers, respectively. The IPL is further divided into two major functionally distinct sublaminae, which comprise connections from cells that are depolarized (ON) or hyperpolarized (OFF) by increased illumination. (*b*) The retina can be dissected from the eyecup, flattened on a glass slide by making pie-cuts and held flat by filter paper. (onh) Optic nerve head. (*c*) The retinal whole mount is then easily transferred, and held down, in a recording chamber by lifting the filter paper from the glass slide.

PROCEDURE A

Preparation of Retinal Whole Mounts

1. Enucleate the eyes from euthanized animals.
2. Remove the anterior portion of the eye by cutting around the cornea. Use fine forceps (#55 or 5) to remove the vitreous portion.
3. Dissect the retina quickly from the eyecup while it is immersed in cold (4°C), oxygenated medium. To do this, insert the tip of a pair of fine spring-scissors and make a few cuts between the retinal margin and the sclera. Pull the sclera away from the retina and detach the retina by cutting around the optic disk.
4. Float the retinal tissue onto a clean glass slide, scleral side down. Drain excess fluid and make pie cuts with a #10 scalpel blade to flatten the retina (Fig. 1b).
5. If imaging the ganglion cell layer, float the retina into the dissecting dish, flip it over using a fine paintbrush, and remount it on the glass slide with the photoreceptor/scleral surface upward. Otherwise, do not flip the retina over.
6. Drain excess fluid. Gently place a piece of dry black filter paper (Millipore, HABG01300) onto the surface of the flattened retina.
7. Moisten the filter paper with a few drops of medium (Ames [Sigma] or ACFS [see Materials section above]).
8. Transfer the filter paper and the adhering retina to an oxygenated and temperature-controlled recording chamber (see Fig. 1c).

PROCEDURE B

Cell Labeling with Ballistics

The three ballistics protocols presented below were optimized using the Bio-Rad Helios gene gun system. Filter insets (e.g., Falcon tissue culture inserts, 3.0 µm pore size, high-density, 6-well format) are used when delivering dye-coated tungsten particles, but not when delivering DNA-coated gold particles.

Method 1: DNA (Biolistics)

In this protocol (see Lo et al. 1994), 50 µg of DNA is used to coat 25 mg of gold particles to make ~50 sections of tubing ("bullets").

1. Weigh 25 mg of gold particles (1.0- or 1.3-µm gold particles; Bio-Rad) in a 1.5-ml microcentrifuge tube.
2. Add 100 µl of 0.05 M spermidine<!> (Sigma) to the gold particles and vortex for 3–5 sec. Then sonicate the mixture for 3–5 sec.
3. Add 50 µl of plasmid DNA (1 µg/µl) and vortex for 5 sec.
4. Add 100 µl of 1 M $CaCl_2 \cdot 2H_2O$ dropwise to mixture, vortexing between drops.
5. Allow the particles to settle for 10 min.
6. In the meantime, dry out ~75 cm of Bio-Rad Teflon tubing for ~15 min in the Bio-Rad-coating station. Insert new tubing into coating station, push against O-ring on left. Turn on the flow of ultrapure nitrogen<!> at 0.3–0.4 lpm (liters per min).

   Note: This step removes moisture and debris from the tubing, which could otherwise cause problems in step 12.
7. Spin down the now DNA-coated gold particles and remove the supernatant.
8. Wash the pellet three times with 1 ml of fresh 100% ethanol<!>, centrifuging and vortexing between each wash.
9. Carry out this step if particles do not stick well to the tubing wall when performing step 12. Prepare a 50 µg/ml solution of polyvinyl pyrrolidone (PVP)<!> in ethanol (a 1:400 dilution of a 20 mg/ml stock in 100% ethanol). Resuspend the particles in 3 ml of dilute PVP. Vortex and sonicate this suspension for 5–10 sec each just before drawing into the Teflon tubing.
10. Load the 3-ml suspension into a section of dried tubing, using a 5–10-ml syringe fitted with a small piece of silicon tubing that fits snugly over the Teflon tubing.
11. Lay the tubing horizontally on a tabletop, so that the gold will settle, and then carefully remove supernatant with a syringe, leaving the gold behind.
12. Load the tubing onto the coating station, and rotate for 20–30 sec. Then turn the nitrogen <!> on, gradually increasing the flow to 0.3–0.4 lpm, for 3–5 min to dry out the gold.
13. Cut the tubing into sections using a guillotine (provided in Bio-Rad kit), and store at 4°C in a capped vial with desiccant capsules (Ted Pella). The bullets can be stored under these conditions for ~30 days.

Method 2: Carbocyanine Dyes

1. Make up a solution of PVP <!> (0.01 mg/ml) in 100% ethanol<!>.

   Note: This will be used to precoat the Teflon tubing.
2. Using a 5–10-ml syringe fitted with a small piece of silicon tubing that fits snugly over the Bio-Rad tubing, draw the PVP solution into ~75 cm of Bio-Rad Teflon tubing.
3. Allow the tubing to stand for 2 min on the benchtop before withdrawing the solution from the tubing.
4. Dry the tubing while completing the remaining portion of the dye preparation (excess PVP may be purged with ultrapure nitrogen gas on the Bio-Rad coating station).

5. To coat the particles, place the chosen dye <!> in a 1.5-ml microfuge tube. For DiI, use ~2 mg; for DiO, ~4 mg; and for DiD, ~2.5 mg. Dissolve each dye in 200 μl of methylene chloride by vortexing. For dye combinations to make multicolored (7-color) bullets, see Gan et al. (2000).
6. Place 100–200 mg of tungsten particles (1.0–1.7 μm; Bio-Rad) on a clean glass slide. Use a different slide for each dye used.
7. Add dissolved dye to the tungsten particles and rapidly spread the particles across the surface of the glass slide to form a thin film.
8. Allow the particles to dry (several minutes) until they turn gray.
9. After drying, use a clean razor blade to gently scrape the dye-tungsten particles from the slide onto wax weighing paper. A fine powder should result.
10. Pour dye-tungsten particles into a 15-ml conical tube (Falcon).

    Note: To generate bullets containing more than one dye, add together the powders coated with the various dyes (see Chapter 14).
11. Add 3–5 ml of Milli-Q water to the powders, vortex, and sonicate for 1–2 min.
12. Draw the slurry into the precoated PVP tubing using a 5–10-ml syringe, again fitted with a small piece of silicon tubing that fits snugly over the PVP-coated tubing.
13. Allow the particles to settle for ~2–3 min, and then carefully draw off the supernatant using the syringe.
14. Dry the tubing in a gentle stream of nitrogen <!> using the coating station (0.2–0.4 lpm).
15. Cut the dried tubing with the guillotine (Bio-Rad) and collect the "bullets" on a large Kimwipe.
16. Gently shake the "bullets" in the Kimwipe (holding the ends of the Kimwipe together). This is important to ensure an even dispersion of the particles in the tubing.
17. Store the bullets in a capped vial with a desiccant pellet. They can be stored at room temperature for several months.

Method 3: Dextran-conjugated Fluorescent Dyes and Other Compounds

Even though dextran-conjugated fluorophores (Molecular Probes) are fixable, it is often desirable to codeliver a more "permanent" cellular label such as biotinylated dextran (see below). Co-delivery of FluoroRuby or Oregon Green 488-dextran with biotinylated dextrans is recommended. This enables immediate visualization of labeled cells, prior to processing for biotinylated dextrans. Tungsten particles can be coated with two or more dyes/markers, as described in the following method. All dextran-conjugated dyes are obtained from Molecular Probes.

1. Precoat tubing with PVP, as described in Method (2) above.
2. Place 1.5 mg of the chosen fluorescent dextran in a 1.5-ml microfuge tube and dissolve it in 15 μl of Milli-Q water.
3. Place 20–25 mg of tungsten particles (1.3 or 1.7 μm diameter; Bio-Rad) on a clean glass slide.
4. Add the dissolved dye to the tungsten particles and rapidly spread the mixture across the surface of the glass slide, forming a thin film.
5. Allow the particles to dry (several minutes) until they turn light gray.
6. After drying, use a clean razor blade to gently scrape the particles from the slide onto wax weighing paper. A fine powder should result.
7. Seal one end of the coated tubing with Parafilm and pour in the dried particles (use a 1-ml pipette tip cut off at the end to act as a funnel). Seal the other end of the tubing with Parafilm and gently shake, distributing the particles from one end of the tubing to the other until a dark film of particles is observed on the inner surface of the tubing.
8. When this has been done, pour excess particles back onto weighing paper and use these to coat additional lengths of tubing. The quantities specified in this recipe should be sufficient to coat 20 cm of tubing.
9. Cut the dried tubing with the guillotine (Bio-Rad) and collect the "bullets" onto a large Kimwipe.

10. Gently shake the bullets in the Kimwipe (holding the ends of the Kimwipe together). Store in capped vial with desiccant at 4°C.

Processing of the Retina for Biotinylated Dextran Labeling

1. After live imaging, fix the retina in 4% paraformaldehyde<!> in ACSF (see Table 1).
2. Rinse briefly in phosphate-buffered saline (PBS) and permeabilize with 0.5% Triton-X and 5% normal donkey serum (NDS in PBS; Jackson Immunochemicals) overnight at 4°C. (PBS comprises 0.01 M phosphate buffer and 9 g/l sodium chloride, pH 7.4).

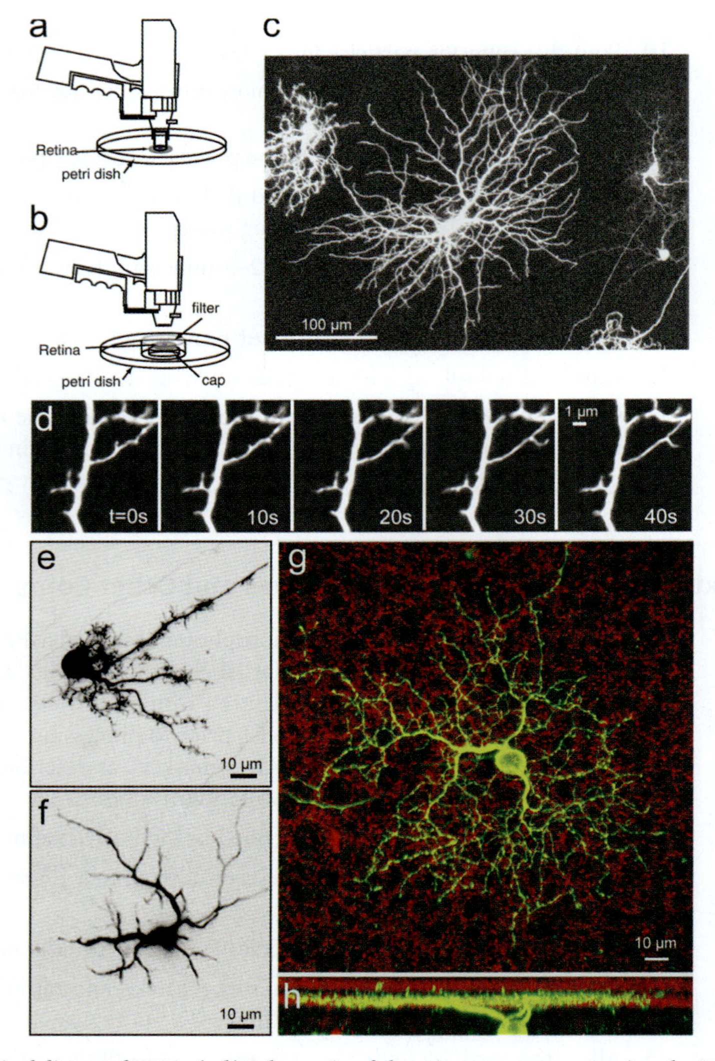

FIGURE 2. Ballistic delivery of DNA. (*a,b*) Schematic of shooting arrangement using the Bio-Rad Helios gene gun for delivering DNA (*a*) or dyes (*b*). For delivering DNA, the gun is used with its barrel. For delivering dyes, the outer guard of the barrel is removed and a filter is inserted between the retina and the gun (see text). (*c*) Confocal reconstruction of embryonic day-16 chick (E16) retinal ganglion cells expressing yellow fluorescent protein (YFP) after DNA delivery by ballistics 10 hr earlier. (*d*) Time-lapse images showing subtle changes in the morphology of small terminal processes of the dendrites in an E14 chick ganglion cell expressing YFP. (*e,f*) Expression of constitutively active small GTPases, Rac (*e*) and Rho (*f*). (*g*) Multiphoton reconstruction of immunolabeling in a week-old mouse retina for the vesicular acetylcholine transporter (VAChT; red) in the field of a GFP-expressing ganglion cell (*green*). (*h*) 90° rotation of the optical stack (*g*) provides a side view of the arbor of the ganglion cell, showing its stratification within one of the two VAChT immunopositive plexuses (see Stacy and Wong 2003). (*e,f,* Modified, with permission, from Wong et al. 2000.)

3. Rinse three times for 10 min each in PBS and block the tissue by submerging in 5% NDS for 30 min.
4. Incubate the retina with 1:1000 Alexa-488-conjugated streptavidin (or any other fluorophore conjugated to streptavidin; Molecular Probes) in 5% NDS for 70 min.
5. Rinse three times for 7 min each in PBS before mounting the tissue in Gelmount (Fisher).

USING THE GENE GUN

A schematic of the shooting arrangement is provided in Figure 2a and b. For delivering plasmid DNA, no filter is interposed between the gun and the retina (Fig. 2a). The optimal distance between the gun and the retina is 2.0–2.5 cm, using a shooting (helium) pressure of 35–40 psi. For delivering dyes, the outer guard of the barrel liner should be cut off with a pair of pliers. Membrane filters are interposed between the gun and the retina, as shown in Figure 2b (this catches large clumps of tungsten and reduces the pressure from the shock wave). The retina is placed on the smooth cap of a 15-ml Falcon plastic centrifuge tube and positioned about 1 cm from the membrane filter surface. Ideal helium pressures range between 80 and 100 psi. Change the filters every fourth shot as the filter becomes damaged.

After both plasmid and dye shooting, rinse the retina quickly in physiological medium and transfer to medium in an oxygenated chamber. If retinas are to be maintained overnight, include penicillin and streptomycin (1:1000 of 10^5 units/ml penicillin and 10^5 μg/ml streptomycin) in the medium. Retinas shot with plasmid should be incubated for 6–12 hours at 30°C, starting 3–6 hours after shooting.

IN VIVO IMAGING OF ZEBRAFISH RETINA

Injection of DNA plasmids into one-cell-stage zebrafish embryos (Fig. 4a) is used to create stable transgenic lines (see Table 2) (Stuart et al. 1988) and to achieve transient transgenic expression in embryonic and larval fish. Injection of cellular reporters that can be easily visualized, such as green fluorescent protein (GFP) and its spectral variants, is useful for imaging cellular dynamics in living zebrafish (Jontes et al. 2000; Köster and Fraser 2001; Das et al. 2003). Furthermore, fusions between fluorescent proteins and target proteins/domains allow protein localization and the study of subcellular structures in real time. A membrane-targeted fluorescent protein fusion is particularly useful for imaging cellular behavior, as saturation from the cell body of cytoplasmic fluorescent proteins can occlude fine detail around somata.

TABLE 2. Transgenic fish lines with expression in retina

Promoter	Reporter	Expression (eye)	References
Xenopus opsin (1.3 kb)	GFP·rhodopsin	rod photoreceptors (subsets of rods)	Perkins et al. (2002); Fadool (2003)
Zebrafish rod-opsin (1.1 and 3.7 kb)	GFP	rod photoreceptors (all rods, lines 1.1A, 1.1B, 3.7B; subsets of rods, line 3.7A)	Hamaoka et al. (2002)
Zebrafish rod-opsin (1.2 kb)	GFP	rod photoreceptors (subsets of rods)	Kennedy et al. (2001)
Zebrafish UV cone opsin	GFP	UV cone photoreceptors	Takechi et al. (2003)
Zebrafish nAChRβ3[a] (3.8 kb)	GFP	retinal ganglion cells (most?)	Tokuoka et al. (2002)
Zebrafish Tiggy-winkle Hh[a] (5.2 kb)	GFP	"retina," presumably ganglion cells	Du and Dienhart (2001)
Zebrafish Sonic Hh[a]	GFP	retinal ganglion cells (subsets)	Neumann and Nüsslein-Volhard (2000)
Rat GAP43	GFP	retinal ganglion cells	Udvadia et al. (2001)
Xenopus EF1α and Pax6 enhancer (quail)	M·GFP	neuroblastic early, then restricted to amacrine cell subsets; also lens cells	Kay et al. (2004)
Xenopus EF1α and Pax6 enhancer (quail)	M·CFP	approximately all retinal cells and lens cells	see Figure 4d,e,f (R.O.L. Wong, unpubl.)

[a]Abbreviations: (EF1α) Elongation factor 1-α; (Hh) hedgehog; (nAChRβ3) nicotinic acetylcholine receptor β-3; (M·FP) membrane-tagged XFP.

MATERIALS

Microscope

Stereomicroscope: Total magnification range of 5× to 40× is sufficient.

DNA

Highly purified DNA is critical for this procedure. After plasmid kit preparation, further purification with chaotropic salt treatment and silica binding is recommended. DNA is diluted in 1× Danieau's solution (Table 1) to a final concentration ranging from picograms to nanograms per microliter, determined empirically by the expression level desired and toxicity incurred. Dyes such as phenol red can be added at low concentrations to visualize the injection volume.

Microinjection Equipment

For injections, filament-containing micropipettes are "pulled" using a micropipette puller (e.g., Sutter Instruments, Model P-87) to a fine tip and trimmed under 40× magnification on Parafilm to a 60° angle using a razor blade. The final width of the micropipette tip should be close to 5 μm (i.e., just wide enough to penetrate the egg chorion without deflecting too much). Micropipettes with a filament 1.2-mm diameter, 4 inches long can be purchased from World Precision Instr. A micromanipulator (e.g., Narashige, models MN-151 or MN-153) fitted with a micropipette holder (e.g., Intracel, P/N 50-00XX-130-1) and a pressure injection system with foot switch (e.g., Intracel, Picospritzer II) are used for injection.

PROCEDURE C

Cell Labeling by Transgenic Expression of Fluorescent Proteins

1. Backfill the pipette with the DNA solution and mount in holder. Adjust the volume of the injection according to desired expression levels.

   Note: Generally, large-volume injections (1–2 nl) give more widespread expression than punctate injections (10–100 pl) and also increase the likelihood of germ-line transmission when making stable transgenic lines.

2. Collect eggs at light onset and array them one per channel in an egg-holding injection chamber (Sylgard 184, Dow Corning).

   Note: An SCSI computer ribbon cable glued to the bottom of a petri dish makes an excellent relief mold for a "channeled" egg holder. Pour Sylgard (Dow Corning) into the dish and let it cure. These silicone chambers are very durable and can be stored in 70% ethanol at room temperature for repeated use.

3. Orient the eggs against the wall of the injection chamber with the cell side up (Fig. 4a). Collect eggs every 15 min to maximize the numbers of single-cell-stage embryos.

4. Align the injection pipette at a 45° angle and penetrate first the chorion and then the cell membrane, while avoiding the yolk.

   Note: Initially, the injection volume should appear as a discrete bolus within the cytosol. If it diffuses away immediately, the cell membrane may not have been penetrated.

5. When a successful injection has been achieved, move to the next egg in the chamber and repeat.

6. When all eggs have been injected, rinse the embryos into a labeled petri dish containing Embryo Medium (0.3× Danieau's solution containing penicillin [100 units/ml] and streptomycin [100 μg/ml]) and incubate at 28.5°C.

IMAGING

The protocol presented here has been optimized for long-term time-lapse imaging of multiple zebrafish embryos.

1. At 16 hours postfertilization (hpf), transfer embryos into Embryo Medium (see step 6 above and Table 1) containing 1× PTU <!> and maintain them in this medium throughout the imaging

period. (50× PTU stock [1-phenyl-2-thiourea (Sigma)]: 10 mM PTU in 0.3× Danieau's [see Table 1], aliquot, and store at −20°C]).

Note: Addition of PTU inhibits melanin formation in the embryo, extending optical transparency past 24 hpf.

2. Screen embryos under a fluorescence-detecting stereoscope or microscope for the presence of the reporter and/or for an expression pattern of interest at various stages of development. To facilitate screening, array eggs in a microwell plate (e.g., 60 × 30-μl conical well plates from Nunc). Use a trimmed 200-μl pipette tip (large enough to accommodate a single egg) to pick up single eggs in 20–30 μl of medium and place them in individual wells.
3. Transfer the embryos of interest to petri dishes and, if necessary, manually remove the chorions with forceps under a stereoscope.
4. Just prior to imaging, anesthetize the embryos by transferring them to Embryo Medium containing 1× PTU (see above) and 1× tricaine (20× tricaine stock: 15 mM 3-aminobenzoic acid ethyl ester<!> [Sigma] in 20 mM Tris, pH 7). Aliquot and store at −20°C.
5. For high-resolution confocal imaging, further immobilize anesthetized embryos by transferring them to 0.5–1.0% low-melting-point agarose (Type VII: low-gelling temperature, Sigma). Maintain the agarose in liquid phase at 40°C. Add 1× PTU and 1× tricaine just before use. Transfer as little liquid as possible with the embryos.
6. Before the agarose sets, transfer the embryos to a warmed imaging dish (e.g., Center-Well Organ Culture Dish, Falcon) and array them as desired under stereoscope optics. Use fine forceps (#5) to nudge the embryos quickly, but very carefully, into position (Fig. 4b). Allow the agarose to gel (~30 min) and then immerse the embedded embryos in 0.3× Danieau's solution containing 1× PTU and 1× tricaine.
7. To maintain embryos at 28.5°C throughout the imaging session, use an ITO glass heating system connected to a temperature controller (e.g., Cell Micro Controls, TC^2-bip system). Place the imaging dish on an ITO glass, fitted with plastic holding chamber for the dish. The ITO glass must be fixed to the confocal microscope stage with dental wax or modeling clay.
8. Adjust temperature controller to 28.5°C and allow the system to equilibrate with objective in position before beginning imaging.
9. Image the embryos using long-working-distance water-immersion objectives with high NA (≥0.8) to enable resolution of fine subcellular structures such as neurites.

EXAMPLES OF APPLICATION

DNA-encoding GFP and its spectral variants are efficiently delivered to retinal ganglion cells at the vitreal surface of the retina using the biolistic method described above. Dendritic arbors are well labeled 8–12 hours after shooting (Fig. 2c). Changes to the dendritic structure can be followed easily over time using standard epifluorescence, confocal, or two-photon microscopy. For example, the dynamism of dendritic filopodia, fine structures abundant during development, is demonstrated in Figure 2d (Wong et al. 2000). Cocoating gold particles with DNA that encodes fluorescent proteins and constitutively active or dominant negative forms of other proteins is also possible (Fig. 2e,f) (Wong et al. 2000). After live imaging, the retina can be fixed with 4% paraformaldehyde and further immunolabeled for proteins of interest. For example, it has been possible to examine the spatial relationship between the arbors of biolistically labeled developing retinal ganglion cells and the plexuses of cholinergic amacrine cells, revealed by antibodies against the vesicular acetylcholine transporter (VAChT) (Fig. 2g,h) (see Stacy and Wong 2003).

Rapid visualization of the neurites of retinal neurons, in both whole-mount preparations (Fig. 3a,b) and slices of living tissue (Fig. 3c,d), can be obtained using DiOlistics (see also Sun et al. 2002). In addition, random labeling of neighboring cells with different colors enables a high density of cell labeling to be achieved while their arbors can still be distinguished. In some cases, it is also possible to visualize potential regions of contact between retinal neurons (Fig. 3d). Complete labeling of the arbors of

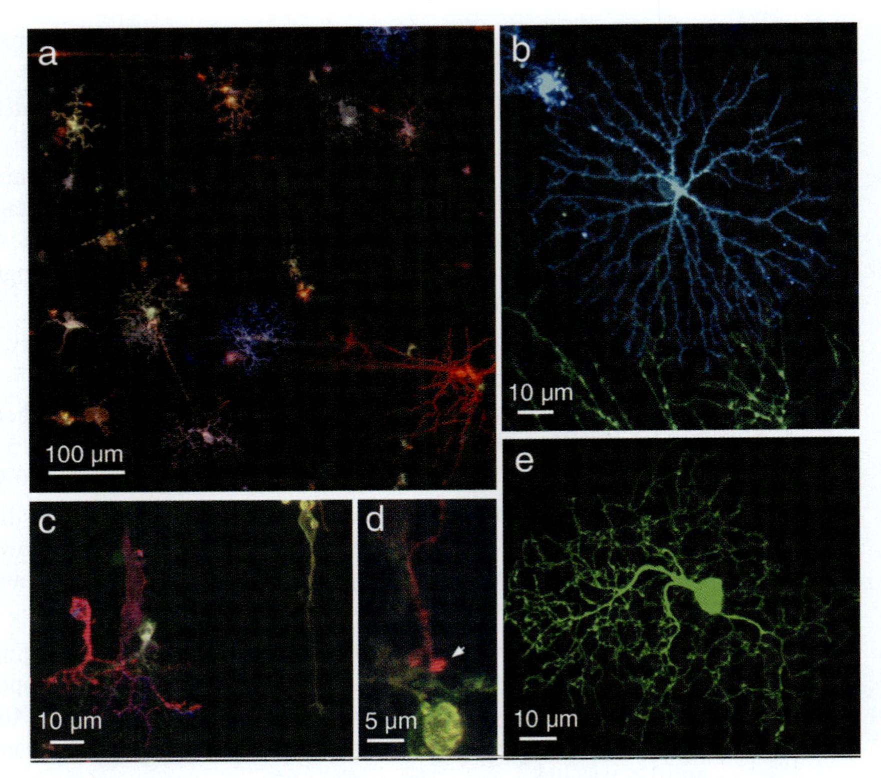

FIGURE 3. Retinal neurons labeled by ballistic delivery of dyes. (*a,b*) Multicolor carbocyanine dye labeling of an embryonic day-16 (E16) chick retina. Shown here are the dendritic arbors of various subtypes of retinal ganglion cells (*a*) and the arbor of a displaced amacrine cell (*b*). (*c*) Amacrine cells (*left*) and a bipolar cell (*right*) in a retinal slice. (*d*) Photoreceptor terminal (*red*) possibly contacting a horizontal cell (*green-yellow*) in a slice from a 3-week-old mouse retina. (*e*) Dendritic arbor of an E15 chick retinal ganglion cell.

retinal neurons can also be achieved within several minutes by shooting tissue with dextran-coupled dyes, such as Oregon Green 488-dextran (Fig. 3e).

Random subpopulations of retinal neurons in the zebrafish embryo can be labeled by injecting plasmids at the one-cell stage (Fig. 4c). Although expression is transient, it is often present long enough to facilitate time-lapse studies of neuronal development, such as that shown in Figure 4g. To obtain more stable and reproducible expression patterns, transgenic zebrafish can be generated (e.g., Fig. 4d–f). A summary of currently available transgenic lines in which retinal neurons express fluorescent proteins is provided in Table 2. Crossing transgenic lines that express different spectral variants of fluorescent protein, or injecting plasmids into transgenic lines, will be a powerful approach for simultaneous imaging of different types of retinal cells.

ADVANTAGES AND LIMITATIONS

All the methods presented here are suitable for time-lapse imaging of the living retina. Each method has strengths and weaknesses, and therefore, the choice of approach depends on the goal of the particular experiment.

In Vivo Versus In Vitro Imaging

In vivo imaging is the ideal approach for studying retinal development under natural conditions. This has been possible in the transparent zebrafish, a vertebrate whose retinal structure and development are similar to those of mammals. The challenge in performing in vivo imaging is finding ways to label cells

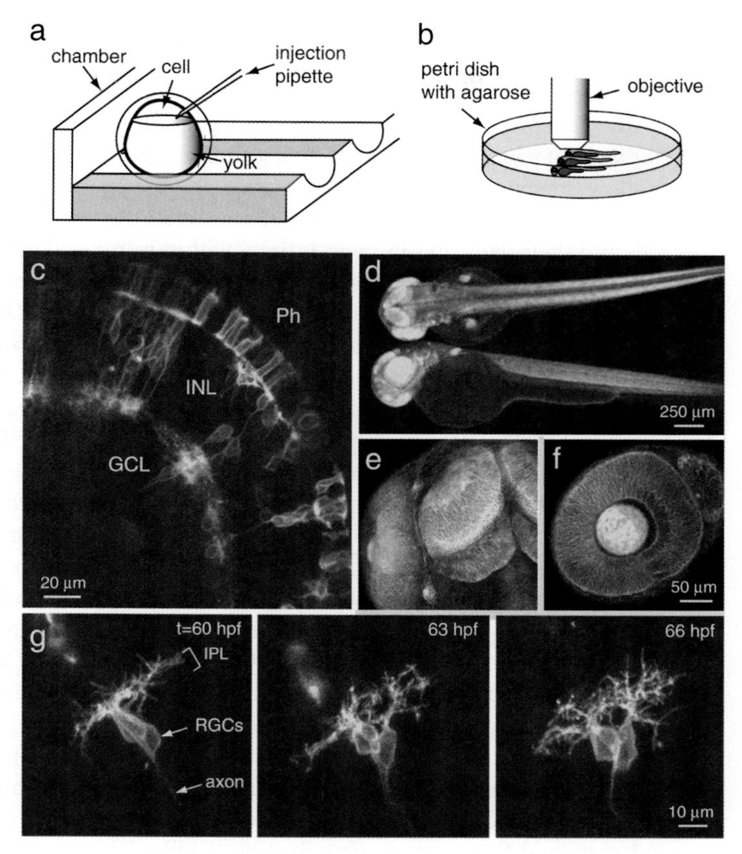

FIGURE 4. Imaging fluorescent retinal cells in zebrafish. (*a*) Injection of plasmid DNA into an embryo at the one-cell stage using glass micropipettes for transient or stable transgene expression. (*b*) Expression of fluorescent protein can be monitored in several embryos, lined up, and held immobilized in low-melting-point agarose. (*c*) Membrane-targeted YFP expression in the retina driven by the α-tubulin promoter (Köster and Fraser 2001) at 65 hr postfertilization. (Ph) Photoreceptors; (INL) inner nuclear layer; (GCL) ganglion cell layer. (*d*) Membrane-targeted CFP (cyan fluorescent protein)-expressing transgenic zebrafish line TG(Pax6::M-CFP^{Q01}); unpublished line from Wong Lab. (*e,f*) Higher-magnification views of the eye (*e,f*) and brain (*e*). (*g*) Time-lapse images of membrane-targeted YFP-expressing retinal ganglion cells showing changes in their dendrites in the inner plexiform layer (IPL) over time. (hpf) Hours postfertilization; (IPL) inner plexiform layer; (RGCs) retinal ganglion cells.

of interest in the living organism. Visualization of cells in the zebrafish is attained either by transient expression of fluorescent proteins or by the generation of stable transgenic lines. Transient expression of fluorescent proteins most often results in mosaic expression allowing imaging of individual or small clusters of cells. Stable lines can also be created in which specific subtypes of retinal neurons express different fluorescent proteins. Crossing lines in which retinal cells express different color fluorescent proteins potentially enables simultaneous visualization of pre- and postsynaptic cells during development.

Much can also be gained from isolated retinal whole mounts, or explants, in which retinal neurons can be viewed without distortion by the optics of the eye. *En face* views of the retina are also useful because subtypes of retinal neurons, such as various ganglion cell subtypes, are more readily identified (by their branching pattern) from this angle. In vitro imaging also enables easy manipulation of the composition of the extracellular milieu during pharmacological experiments. In addition, it is more straightforward to combine imaging studies with electrophysiology in the isolated retina (Lohmann et al. 2002) than in in vivo experiments. However, an obvious disadvantage of in vitro imaging is that the duration of recording is limited and the conditions are not the same as those in the animal, even though the intrinsic connectivity of the retina is maintained in whole-mount preparations.

Ballistic Methods of Cell Labeling

This chapter has described three different techniques for labeling retinal neurons by ballistic means for in vitro imaging. Of these techniques, delivery of DNA-encoding fluorescent proteins provides a superior signal-to-background ratio. This technique, however, requires a relatively long incubation period to allow for gene expression, which reduces the overall available duration of time-lapse imaging before the tissue degenerates. Ballistically administered dextran-coupled or carbocyanine dyes label many cells and their dendrites in their entirety within minutes. Thus, if imaging is to be carried out on the tissue shortly after it is removed, these methods are more suitable than plasmid delivery. Phototoxicity and bleaching must also be considered. GFP expression is by far the least phototoxic, as it is subject to minimal bleaching. In contrast, carbocyanine dyes are the most toxic for live imaging.

Labeling density is another important factor to consider. With a single shot, DiOlistics produces the highest number of labeled cells, in comparison to plasmid expression or shooting with dextrans. Carbocyanine dyes are lipophilic, and thus complete cellular labeling requires the particles to touch only the cell membrane. With plasmids, it is necessary for the particle to enter the cell. For dextrans, labeling of the cell is relatively rapid when the particles enter the cell body (Kettunen et al. 2002). With all particle types, the density of labeling can be altered by varying the coating procedure (by adding more or less dye) and the pressure during shooting (higher pressures deliver more particles), as well as how often the same tissue is shot (more than twice would tend to destroy the tissue).

Because neighboring cells tend to be labeled with different colors when multicolor carbocyanine bullets are used, visualization of overlapping neurites is best achieved using DiOlistics (see Chapter 14). However, labeling continues with time, and eventually even multiple colors are insufficient to help resolve neurites comprising the dense neuropil. Cells in fixed retinas can also be labeled by DiOlistics, but in general, cell labeling is better confined to individual cells in living tissue. In fixed material, carbocyanine dyes leach out over time, and, after a few days, cell membranes are no longer crisply labeled.

Like fluorescent protein expression, dextran-coupled dyes can provide relatively low background and sustained fluorescence after fixation. Unlike DiOlistics, dextran-conjugated dyes from particles that fail to enter the cell do not spread far in the extracellular matrix. A disadvantage of using dextrans is that cell filling is quite variable. Nevertheless, the density of labeling is often sufficient to select individual or small populations of cells for live imaging. We use dextran-coupled dyes primarily with calcium indicators (see Chapter 34) and when rapid labeling is necessary.

Finally, although this chapter has focused largely on methods that enable visualization of the morphology and dynamic behavior of retinal neurons during development, it is also possible to deliver DNA-encoding-activity indicators or mutant proteins into live zebrafish, or into retinal explants of other vertebrates using ballistic methods. For example, genetically encoded indicators for calcium (see Chapter 78) or chloride (Kuner and Augustine 2000) can be delivered using biolistics. One future development for increasing the usefulness of ballistic methods is to combine delivery of plasmid with dye. This would be helpful, for example, if calcium imaging with conventional chemical indicators were used to assay the activity of cells in response to the expression of a specific protein of interest.

ACKNOWLEDGMENTS

This work was supported by the National Institutes of Health and by Deutsche Forschungsgemeinschaft.

REFERENCES

Ames A. and Nesbett F.B. 1981. *In vitro* retina as an experimental model of the central nervous system. *J. Neurochem.* **37:** 867–877.

Cajal S.R. 1893. La rétine des vertébrés. *La Cellule* **9:** 121–225.

Das T., Payer B., Cayouette M., and Harris W.A. 2003. In vivo time-lapse imaging of cell divisions during neurogenesis in the developing zebrafish retina. *Neuron* **37:** 597–609.

Du S.J. and Dienhart M. 2001. Zebrafish *tiggy-winkle hedgehog* promoter directs notochord and floor plate green fluorescence protein expression in transgenic zebrafish embryos. *Dev. Dyn.* **222:** 655–666.

Fadool J.M. 2003. Development of a rod photoreceptor mosaic revealed in transgenic zebrafish. *Dev. Biol.* **258:** 227–290.

Gan W.-B., Grutzendler J., Wong W.T., Wong R.O.L., and Lichtman J.W. 2000. Multicolor "DiOlistic" labeling of the nervous system using lipophilic dye combinations. *Neuron* **27:** 219–225.

Hamaoka T., Takechi M., Chinen A., Nishiwaki Y., and Kawamura S. 2002. Visualization of rod photoreceptor development using GFP-transgenic zebrafish. *Genesis* **34:** 215–220.

Jontes J.D., Buchanan J., and Smith S.J. 2000. Growth cone and dendrite dynamics in zebrafish embryos: Early events in synaptogenesis imaged in vivo. *Nat. Neurosci.* **3:** 231–237.

Kay J.N., Roeser T., Mumm J.S., Godinho L., Mrejeru A., Wong R.O.L, and Baier H. 2004. Transient requirement for ganglion cells during assembly of retinal synaptic layers. *Development* **131:** 1331–1342.

Kennedy B.N., Vihtelic T.S., Checkley L., Vaughan K.T., and Hyde D.R. 2001. Isolation of a zebrafish rod opsin promoter to generate a transgenic zebrafish line expressing enhanced green fluorescent protein in rod photoreceptors. *J. Biol. Chem.* **276:** 14037–14043.

Kettunen P., Demas J., Lohmann C., Kasthuri N., Gong Y., Wong R.O.L., and Gan W.B. 2002. Imaging calcium dynamics in the nervous system by means of ballistic delivery of indicators. *J. Neurosci. Methods* **119:** 37–43.

Köster R.W. and Fraser S.E. 2001. Direct imaging of *in vivo* neuronal migration in the developing cerebellum. *Curr. Biol.* **11:** 1858–1863.

Kuner T. and Augustine G.J. 2000. A genetically encoded ratiometric indicator for chloride: Capturing chloride transients in cultured hippocampal neurons. *Neuron* **27:** 447–459.

Lo D.C., McAllister A.K., and Katz L.C. 1994. Neuronal transfection in brain slices using particle-mediated gene transfer. *Neuron* **13:** 1263–1268.

Lohmann C., Myhr K.L., and Wong R.O.L. 2002. Transmitter-evoked local calcium release stabilizes developing dendrites. *Nature* **418:** 177–181.

Miyawaki A., Llopis J., Heim R., McCaffery J.M., Adams J.A., Ikura M., and Tsien R.Y. 1997. Fluorescent indicators for Ca^{2+} based on green fluorescent proteins and calmodulin. *Nature* **388:** 882–887.

Neumann C.J. and Nüsslein-Volhard C. 2000. Patterning of the zebrafish retina by a wave of sonic hedgehog activity. *Science* **289:** 2137–2139.

Perkins B.D., Kainz P.M., O'Malley D.M., and Dowling J.E. 2002. Transgenic expression of a GFP-rhodopsin COOH-terminal fusion protein in zebrafish rod photoreceptors. *Vis. Neurosci.* **19:** 257R–264R.

Sernagor E., Eglen S.J., and Wong R.O.L. 2001. Development of retinal ganglion cell structure and function. *Prog. Retin. Eye Res.* **20:** 139–174.

Stacy R.C. and Wong R.O.L. 2003. Developmental relationship between cholinergic amacrine cell processes and ganglion cell dendrites of the mouse retina. *J. Comp. Neurol.* **456:** 154–166.

Stuart G.W., McMurray J.V., and Westerfield M. 1988. Replication, integration and stable germ-line transmission of foreign sequences injected into early zebrafish embryos. *Development* **103:** 403–412.

Sun W., Li N., and He S. 2002. Large-scale morophological survey of rat retinal ganglion cells. *Vis. Neurosci.* **19:** 483–493.

Takechi M., Hamaoka T., and Kawamua D. 2003. Fluorescence visualization of ultraviolet-sensitive cone photoreceptor development in living zebrafish. *FEBS Lett.* **553:** 90–94.

Tokuoka H., Yoshida T., Matsuda N., and Mishina M. 2002. Regulation by glycogen synthase kinase-3β of the arborization field and maturation of retinotectal projection in zebrafish. *J. Neurosci.* **22:** 10324–10332.

Udvadia A.J., Koster R.W., and Skene J.H. 2001. GAP-43 promoter elements in transgenic zebrafish reveal a difference in signals for axon growth during CNS development and regeneration. *Development* **128:** 1175–1183.

Wässle H. and Boycott B.B. 1991. Functional architecture of the mammalian retina. *Physiol. Rev.* **71:** 447–480.

Wong W.T., Faulkner-Jones B.E., Sanes J.R., and Wong R.O.L. 2000. Rapid dendritic remodeling in the developing retina: Dependence on neurotransmission and reciprocal regulation by Rac and Rho. *J. Neurosci.* **20:** 5024–5036.

CHAPTER 22

A Practical Guide: Long-term Two-Photon Transcranial Imaging of Synaptic Structures in the Living Brain

Jaime Grutzendler[1] and Wen-Biao Gan

Molecular Neurobiology Program, Skirball Institute, Department of Physiology and Neuroscience, New York University School of Medicine, New York, New York 10016

This chapter presents a detailed protocol for long-term transcranial imaging of neuronal structures in the brains of living mice, using two-photon microscopy. The method has been used to image individual dendritic spines and axonal varicosities in various mouse brain areas, such as visual, somatosensory, motor, and frontal cortices, over intervals of up to 4 months (Grutzendler et al. 2002). This long-term transcranial imaging approach allows detailed structural and functional changes of synapses to be monitored during learning and memory processes, as well as in neurological disease models. It also provides a sensitive tool to detect the effects of various pharmacological and therapeutic interventions on cells in the living brain.

MATERIALS

Transgenic Mice

The animals used for this protocol expressed fluorescent proteins (e.g., YFP) in the cytoplasm of cortical neurons (Feng et al. 2000).

Two-Photon Laser-scanning Microscope

This protocol can be used with commercially available multiphoton microscopes (e.g., Bio-Rad Radiance 2001) or custom-built multiphoton systems equipped with a mode-locked laser coupled to an upright fluorescence microscope (Olympus, BX50WI) (Majewska et al. 2000). The laser system is tunable from 690 nm to 1000 nm wavelength, with an 80-MHz pulse repeat and <100 fsec pulse width (Tsunami and Millenia Xs, Spectra Physics).

Caution: *See Appendix 3 for appropriate handling of materials marked with <!>.*

[1]*Present address:* Department of Neurology and Physiology, Northwestern University School of Medicine, 303 East Chicago Avenue, Chicago, Illinois 60611.

PROCEDURE

Long-term Two-Photon Transcranial Imaging of Synaptic Structures in the Living Brain

Skull-thinning Procedure

1. Anesthetize transgenic mice by intraperitoneal injection (0.2 ml/20 g body weight) of 20 mg/ml ketamine <!> and 3 mg/ml xylazine <!> in 0.9% NaCl.
2. Thoroughly shave the hair over most of the scalp using a conventional razor blade. Perform a midline scalp incision using standard microsurgical tools.

   Note: The incision usually extends approximately from the neck region to the frontal area.
3. Localize the brain area to be imaged using stereotactic coordinates (Paxinos and Franklin 1997). Use a fine forceps to remove the soft tissue attached to the skull over the area to be imaged.
4. Then, under a dissecting microscope, use a high-speed micro-drill (Fine Science Tools) to thin a circular area of skull (typically 1–1.5 mm in diameter), directly above the region of interest.

   Note: Drilling should be done intermittently and the drill bit can be immersed periodically in a cold artificial cerebrospinal fluid (ACSF) solution to minimize heat-induced tissue injury.
5. After removing the external compact bone (periostium), carefully thin the middle spongy bone layer to about 75% of its original thickness.

   Note: Some bleeding from bony canaliculi may occur during the thinning process. This bleeding will usually stop spontaneously or by carefully blotting the blood with an absorbing tissue.
6. After removing the majority of the spongy tissue, look for concentric cavities within the bone. These can usually be seen under the dissecting microscope and are an indication that the drill is close to the internal compact bone layer. At this stage, continue to thin the bone using a microsurgical blade (Surgistar #6900).
7. Hold the blade at a ~45° angle, taking great care not to push the skull downward against the brain surface, and not to break through the bone—minor brain trauma or bleeding may cause inflammation and disruption of neuronal structures.
8. Continue scraping the bone until a very thin (<30 μm), smooth preparation is achieved (Fig. 1a).

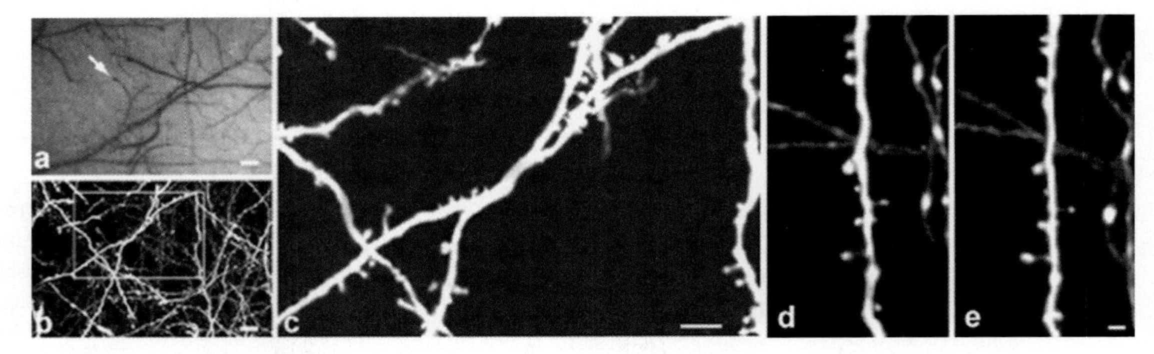

FIGURE 1. Long-term transcranial two-photon imaging of fine neuronal structures. (*a*) CCD camera view of the thinned skull. The cortical vasculature can be seen clearly through the thinned bone. This vasculature pattern remains stable over several months and can be used as a landmark to relocate the imaged region at subsequent time points. Arrow indicates the region imaged in subsequent imaging sessions. Bar, 50 μm. (*b*) Two-dimensional projection of a 3D stack of dendritic branches and axons in the primary visual cortex (60x, digital zoom = 1). The stack was 50 μm deep (2 μm step size). The boxed region was then imaged at higher digital zoom (as shown in *c*). Bar, 5 μm. (*c*) High-power 2D projection of a 3D stack (60x, digital zoom = 3) reveals clear neuronal structures including dendritic shafts, axonal varicosities, and dendritic filopodia spines (10 μm reconstructed, 0.70 μm step size). Bar, 2 μm. (*d, e*) Axonal and dendritic branches from two animals imaged 3 days apart show the same spines and boutons at the same locations. (Adapted from Grutzendler et al. 2002). Bar, 1 μm.

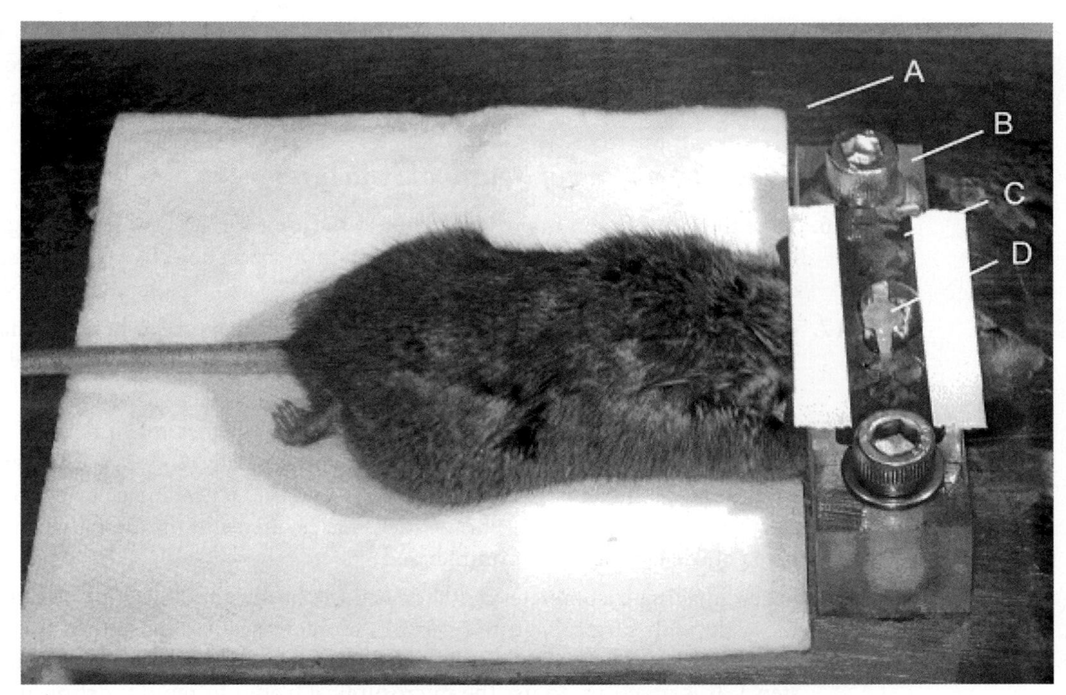

FIGURE 2. A head immobilization device for reducing movement artifacts during imaging. (A) 25 x 15 x 0.3-cm stainless steel plate; (B) 15 x 15 x 15-mm blocks of acrylic material; (C) three double-edge razor blades glued together for stabilization of the mouse skull; (D) the thinned skull area to be imaged is exposed and visible through the hole in the middle of the blades.

Note: The optimal degree of skull thinning can be determined by looking at the preparation under a conventional fluorescence microscope. Dendrites and spines in the area of interest should be clearly visualized at this stage (Fig. 1b).

Construction of Head Restraint

9. Glue two 15 x 15 x 15-mm acrylic blocks to a 25 x 15 x 0.3-cm stainless steel plate (see Fig. 2).

   Note: The blocks should be placed about 8 cm from one of the short sides of the plate and 4 cm from each other. Drill a hole into each block, large enough to accommodate a ¼-inch screw.

10. Take three conventional razor blades and cut off their sharp edges using a pair of scissors. Glue the blades together in a stack using cyanoacrylate glue.
11. Immobilize the animal's skull by gluing it to the triple razor blade. Place a small amount of cyanoacrylate glue around the hole in the blade and press it against the skull, leaving the area to be imaged visible through the hole (Fig. 2). Be very careful to avoid contaminating the thinned skull area with glue.
12. Allow the glue to set fast (10–15 min) and then place the mouse face down with the blades supported by the acrylic blocks. Tighten the screws to immobilize the blades completely.
13. To remove traces of unpolymerized glue, thoroughly wash the area to be imaged with artificial cerebrospinal fluid. If the preparation is not washed thoroughly, the microscope objectives can pick up trace amounts of glue. To avoid problems, the objectives should be kept immersed in water after imaging and cleaned periodically with 100% ethanol.

Mapping the Imaging Area for Future Relocation

14. To ensure relocation of the imaged area at a later date, use a stereomicroscope with a CCD camera to obtain a high-quality picture of the meningeal blood vessels (Fig. 1a).

15. Place the mouse under an epi-fluorescence microscope and select a specific area for two-photon imaging. Using the meningeal vasculature pattern seen under fluorescence, identify and mark the same area on the CCD camera map (Fig. 1a).

Two-Photon Imaging of Neuronal Structures

16. Tune the two-photon microscope laser to the appropriate wavelength (e.g., 920 nm for YFP) and image the area of interest using a 60X water-immersion objective (Olympus, NA 0.9). Use ACSF for objective immersion.
17. Obtain an image stack of fluorescently labeled neuronal processes at low magnification (1X zoom). These images are used in conjunction with the CCD camera map, obtained in step 14, to relocate the imaged area at a later date (Fig. 1b).
18. Without changing the position of the stage, take a higher magnification image of the same area (e.g., 3X zoom, Fig. 1c). The stack depth is typically 100–200 μm beneath the pial surface. Additional zoomed images can be taken around the central image by moving the imaging position electronically. To minimize phototoxicity, laser intensities in the range of 10–20 milliwatts (measured at the sample) are recommended.
19. When imaging is complete, gently detach the head immobilization device from the skull. Suture the scalp with surgical suture (6-0 silk) to allow healing and return the mouse to its cage until required for the next viewing (Fig. 1d,e). For subsequent viewing sessions, repeat protocol from step 1. It is necessary to use the microsurgical blades to remove debris and re-thin the skull window to obtain optimal image quality.
20. View and analyze the image stacks with NIH image software—a plug-in for reading Bio-Rad PIC files or Fluoview Tiff files can be downloaded at http://rsb.info.nih.gov/ij/.

EXAMPLE OF APPLICATION

The transcranial imaging approach has been used to examine dendritic spine stability in living YFP-expressing mice over intervals up to 4 months (Grutzendler et al. 2002). Figure 1 shows examples of images obtained over a 3-day period in mice at 4 months of age. Note the remarkable stability of the number and location of adult spines and axonal varicosities between the two views (Fig. 1d,e).

ADVANTAGES AND LIMITATIONS

Two-photon microscopy is an increasingly important tool for imaging the structure and function of brain cells in vivo (Denk et al. 1990). Transcranial two-photon imaging of neuronal structures is a relatively straightforward and minimally invasive method for high-resolution study of structural changes in dendrites, spines, and axons, by repeated imaging over periods ranging from minutes to months. This approach avoids the need for cranial window preparations (Svoboda et al. 1997; Trachtenberg et al. 2002; see Chapter 85). This may be important, because skull removal can induce a significant local, inflammatory reaction with microglial activation (Bacskai et al. 2001). The long-term impact of such an inflammatory reaction on the neuronal structures remains to be investigated. Furthermore, the thin-skull approach allows long-term imaging of neuronal structures immediately after the initial surgery, as opposed to the open-skull method in which optimal imaging quality appears to be achieved many days after the craniotomy has been performed (Trachtenberg et al. 2002).

The limitations of the thin-skull technique are mainly related to the fact that skull thickness is critical for image quality, and it takes a lot of practice to consistently obtain a preparation of the optimal thickness for spine imaging. The thin-skull approach has been used in various studies, including intrinsic optical imaging of cortical maps (Frostig et al. 1990; Masino et al. 1993), two-photon imaging of the cerebral vasculature (Yoder and Kleinfeld 2002) and amyloid plaques (Christie et al. 2001). These studies, however, investigated structures much larger than dendritic spines and did not require the skull to be very thin (Christie et al. 2001; Yoder and Kleinfeld 2002). In the authors' experience, to obtain high-resolution images of synapses, the skull thickness must be less than 30 μm. On the other hand, over-thinning the skull can lead to cortical injury, presumably due to deformation of the skull under the pressure of the blade. In such cases, the thinning process generally leads to mild neuronal injury, mainly manifested by axonal and dendritic blebbing, and eventual disappearance of fluorescent structures. Injury can be prevented by using the blade at an angle and by avoiding pushing the skull against the cortical surface. Another limitation stems from the need to re-thin the skull for each new imaging session. This can limit the frequency of imaging sessions, as the optical properties of the preparation may gradually deteriorate with repeated shaving of the skull.

In summary, the transcranial two-photon microscopy technique allows repeated imaging of fine neuronal structures (spines, filopodia, axonal boutons) in living mice. This technique is likely to be very useful for addressing many interesting questions related to long-term structural and functional changes of cells in the living mouse brain.

REFERENCES

Bacskai B.J., Kajdasz S.T., Christie R.H., Carter C., Games D., Seubert P., Schenk D., and Hyman B.T. 2001. Imaging of amyloid-beta deposits in brains of living mice permits direct observation of clearance of plaques with immunotherapy. *Nat Med.* **7:** 369–372.

Christie R.H., Bacskai B.J., Zipfel W.R., Williams R.M., Kajdasz S.T., Webb W.W., and Hyman B.T. 2001. Growth arrest of individual senile plaques in a model of Alzheimer's disease observed by in vivo multiphoton microscopy. *J. Neurosci.* **21:** 858–864.

Denk W., Strickler J.H., and Webb W.W. 1990. Two-photon laser scanning fluorescence microscopy. *Science* **248:** 73–76.

Feng G., Mellor R.H., Bernstein M., Keller-Peck C., Nguyen Q.T., Wallace M., Nerbonne J.M., Lichtman J.W., and Sanes J.R. 2000. Imaging neuronal subsets in transgenic mice expressing multiple spectral variants of GFP. *Neuron* **28:** 41–51.

Frostig RD., Lieke E.E., Ts'o D.Y., and Grinvald A. 1990. Cortical functional architecture and local coupling between neuronal activity and the microcirculation revealed by *in vivo* high-resolution optical imaging of intrinsic signals. *Proc. Natl. Acad. Sci.* **87:** 6082–6086.

Grutzendler J., Kasthuri N., and Gan W.B. 2002. Long-term dendritic spine stability in the adult cortex. *Nature* **420:** 812–816.

Majewska A., Yiu G., and Yuste R. 2000. A custom-made two-photon microscope and deconvolution system. *Pflügers Arch. Eur. J. Physiol.* **441:** 398–408.

Masino S.A., Kwon M.C., Dory Y., and Frostig R.D. 1993. Characterization of functional organization within rat barrel cortex using intrinsic signal optical imaging through a thinned skull. *Proc. Natl. Acad. Sci.* **90:** 9998–10002.

Paxinos G. and Franklin K.B.J. 1997. *Mouse brain in stereotaxic coordinates.* Academic Press, San Diego, California.

Svoboda K., Denk W., Kleinfeld D., and Tank D.W. 1997. In vivo dendritic calcium dynamics in neocortical pyramidal neurons. *Nature* **385:** 161–165.

Trachtenberg J.T., Chen B.E., Knott G.W., Feng G., Sanes J.R., Welker E., and Svoboda K. 2002. Long-term in vivo imaging of experience-dependent synaptic plasticity in adult cortex. *Nature* **420:** 788–794.

Yoder E.J. and Kleinfeld D. 2002. Cortical imaging through the intact mouse skull using two-photon excitation laser scanning microscopy. *Microsc. Res. Tech.* **56:** 304–305.

CHAPTER 23

In Vivo Time-lapse Imaging of Neuronal Development

Edward S. Ruthazer, Kurt Haas, Ashkan Javaherian, Kendall Jensen, Wun Chey Sin, and Hollis T. Cline

Cold Spring Harbor Laboratory, Cold Spring Harbor, New York 11724

The realization in recent decades that developing and even adult neurons exhibit remarkable structural dynamics has encouraged an increasing number of neuroscientists to study the structural remodeling of neurons that takes place in the intact animal. This "neuronal ethology," the study of neurons in their natural habitat, makes sense for many reasons. The complex landscape of the developing brain is rich in signals that regulate cell proliferation and fate, guidance cues for migration and process outgrowth, as well as opportunities to interact with neighboring neurons and glia and to form synapses. Often, the relevant cues are not the absolute levels of a signaling ligand in the environment, but rather changes in local concentration or distribution of one or more ligands, which can be difficult to predict and replicate in vitro. Furthermore, other factors that contribute to development and plasticity, such as patterned neural activity or sensory experience, are most meaningful in an intact nervous system. In vivo time-lapse imaging experiments have revealed mechanisms which modulate branch formation, growth, and stabilization that could never have been surmised from studies in fixed tissue (Kaethner and Stuermer 1992; Balice-Gordon and Lichtman 1993; O'Rourke et al. 1994; Feng et al. 2000; Lendvai et al. 2000; Gan et al. 2003).

Collecting accurate time-lapse imaging data in vivo, however, is impeded by considerations that do not severely affect experiments in cultured cells. First and foremost is the scattering of light by the overlying tissue. This problem can be considerable when imaging deep brain structures, even in largely transparent animals, such as *Xenopus* tadpoles or zebrafish, and becomes more severe in adult brain, due to increases in myelination and changes in lipid components of neuronal membranes. In an effort to improve signal-to-noise ratios, more intense excitation light can be used, but this greatly increases the risk of severe photodynamic cell damage. Another challenge to imaging cells in vivo is the ability to discretely label individual cells, or specific populations of cells, in the brain structure of interest. Technological advances in microscopy, and fluorophore design and delivery, combined with clever strategies for the use of transparent and transgenic animals for imaging studies, have helped overcome many of these problems, bringing in vivo neuronal imaging to prominence in recent years. This chapter reviews recent technological progress in in vivo time-lapse imaging and describes the protocols used by our laboratory, where applicable.

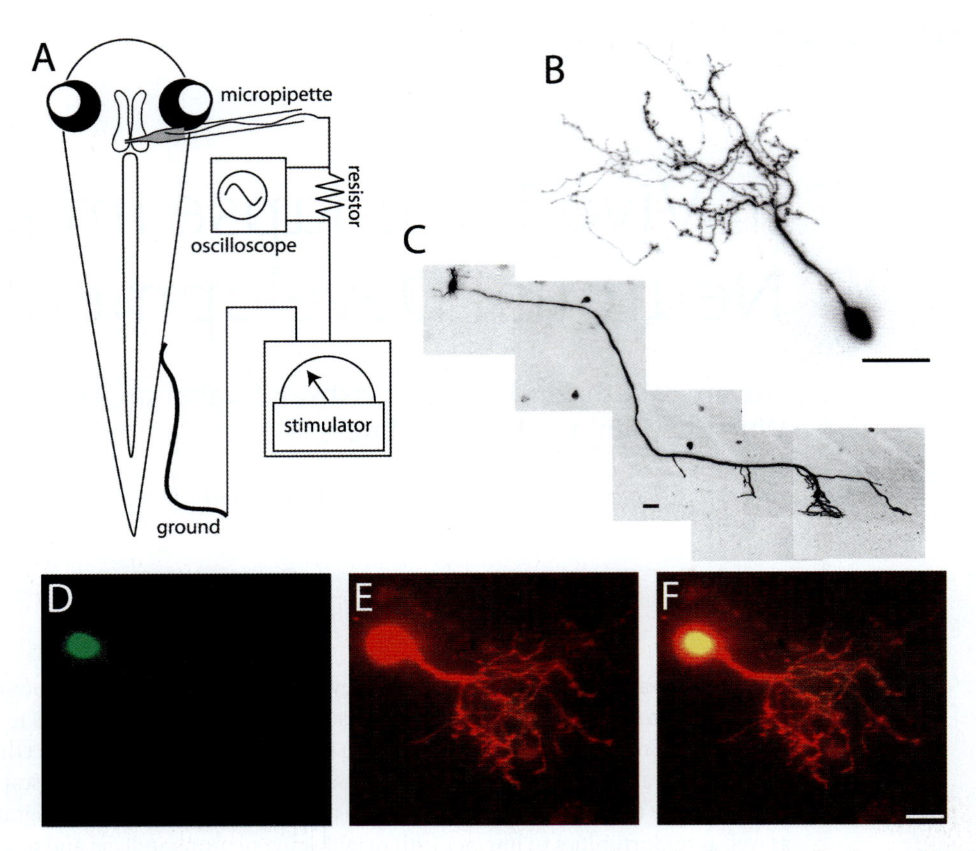

FIGURE 1. Single-cell electroporation. (*A*) Circuit diagram and electrode placement for dye or DNA electroporation of single tectal neurons in tadpoles. (*B*) Tadpole tectal neuron transfected by electroporation with plasmid encoding enhanced GFP. (*C*) Example of electroporated motor neuron in spinal cord. (*D–F*) Example of coelectroporation of fluorescently tagged Morpholino oligos (*D*) with Texas Red dextran (*E*). (*F*) Merged image. (*D, E, F,* Modified, with permission, from Haas et al. 2002 [© Blackwell].) Bars, 20 µm.

LABELING OF NEURONS FOR IN VIVO IMAGING

Fluorescent Dextrans

Fluorescent dye-conjugated dextran (3K molecular weight [m.w.] for retrograde labeling and either 3K or 10K m.w. for anterograde labeling) is a water-soluble neuronal tracer, available in a near-limitless variety of fluorophore conjugates, ranging from relatively photostable Alexa dyes to calcium reporters such as Calcium Green. When used in conjunction with fixable biotinylated dextran amine, which can be visualized postmortem in histological sections by incubation with streptavidin-HRP (horseradish peroxidase), labeled material can be taken all the way from in vivo fluorescence imaging to the electron microscope (Reiner et al. 2000). The mechanism by which fluorescent dextrans are taken up by neurons is not well understood, but either pressure injection or iontophoretic delivery of a 5% dextran solution readily labels many cells (Glover et al. 1986; Schmued et al. 1990). It has been demonstrated that fibers severed during the injection procedure are preferentially labeled (Gahtan and O'Malley 2001). Iontophoresis may therefore increase labeling efficiency by electroporation of the plasma membranes of cells and processes in the vicinity of the ejection micropipette. Electroporation occurs when micropores in the plasma membrane are induced in an electric field, thereby permitting charged molecules to be driven directly into the nucleus. Images of dextran-labeled neurons can be collected for many days to weeks after labeling, without obvious clumping or degradation of the fluorophore, as long as measures are taken to reduce phototoxicity. Thus, fluorescent dextrans may be the most versatile and easily delivered neuronal tracers available. Despite their ease of delivery, or perhaps because of it, it can be difficult to restrict fluorescent dextran labeling to single neurons within a brain structure. However, single-cell electroporation is a relatively simple means to label single cells with dextrans (see below).

Lipophilic Vital Dyes

Lipophilic fluorescent dyes, which include carbocyanine dyes like DiI (1,1'-dioctadecyl-3,3,3',3'-tetramethylindocarbocyanine perchlorate), and styryl dyes such as DiA (4-[4-(dihexadecylamino)styryl]-*N*-methylpyridinium iodide), are well suited to single-cell labeling (Honig and Hume 1986). Because lipophilic dyes are insoluble in the aqueous environment of the brain or eye, a solution of the dye, dissolved in either DMSO <!>, DMF <!>, or ethanol <!> and delivered by pressure or iontophoresis through a glass micropipette, will precipitate instantaneously, resulting in little or no uptake of the dye outside of the small injection site. Only cells in contact with the dye crystal will be labeled as the dye diffuses throughout the lipid-rich plasma membrane of the cell, resulting in the sometimes complete labeling of the dendritic and axonal processes of neurons within minutes. Over several days, lipophilic dyes label intracellular membrane compartments in cells, probably as a result of membrane cycling. Considerable success has also been achieved with targeted delivery of the dye, by juxtacellular iontophoresis under microscopic visualization, to label single neurons in the brains of living tadpoles using a method adapted from that of Paul Myers (Myers and Bastiani 1993).

Lipophilic dyes with a fairly wide range of spectral properties are available. For in vivo confocal imaging, both DiI and DiO have been found to give satisfactory results, but the long-wavelength carbocyanine dye, DiD, offers superior labeling with relatively little phototoxicity and tissue autofluorescence at its excitation wavelength of 644 nm and emission peak at 665 nm. Unfortunately, DiI, and DiD in particular, have properties that reduce their usefulness for two-photon imaging. DiO and DiA, on the other hand, efficiently undergo two-photon excitation at 880 nm, resulting in bright images of neurons and their processes (Ruthazer and Cline 2002).

Genetic Labeling by Fluorescent Protein Expression

The use of green fluorescent protein (GFP), from the jellyfish *Aequorea victoria*, represents a revolutionary step forward in in vivo imaging (see Section 6). GFP has finally made it practical for anatomists to take advantage of the myriad tools and techniques of molecular biology, at the same time bringing in vivo imaging into the mainstream of cell biology (Chalfie et al. 1994). The discovery of useful fluorescent proteins from corals, including a red variant, and the use of mutagenesis to develop enhanced GFPs and GFP-like proteins with a broad range of spectral properties, has made the GFP family of proteins as versatile as extrinsic dyes for most applications (Tsien 1998; Baird et al. 2000; Fradkov et al. 2000; Bevis and Glick 2002; Karasawa et al. 2003). A number of calcium- and membrane-voltage-sensitive GFPs also have been introduced, further extending their practical applications (Miyawaki et al. 1999; Nakai et al. 2001; Sakai et al. 2001; see Section 6). The following two properties of GFP make it a near-ideal fluorophore for in vivo imaging: The protected fluorophore-in-a-β-barrel structure of GFP renders it relatively resistant to photobleaching and free-radical generation, and GFP is efficiently excited by two-photon excitation (Ormo et al. 1996).

What truly sets GFP apart from other tracers is the fact that it can be introduced into cells by genetic manipulation. This opens the door to a huge number of cellular imaging possibilities. For example, genes encoding proteins fused to GFP have been used extensively to study the trafficking of molecules within the cell (Wang and Hazelrigg 1994). Transgenic animals from nematodes to mammals have been created that express GFP in defined subtypes of neurons (Chalfie et al. 1994; Amsterdam et al. 1995; Yeh et al. 1995; Feng et al. 2000). These approaches have been especially useful for in vivo imaging under limited circumstances, where GFP-expressing neurons are present at low enough density to permit discrimination of individual cells.

In other cases, such as for anatomical tracing, it is necessary to label a small number of cells at a defined site in the brain. In frogs, DNA can be injected or transfected by lipofection into early embryos, resulting in a mosaic of crudely targeted GFP-expressing cells (Ohnuma et al. 2002). Alternatively, this can be accomplished by making stereotaxic injections of virus for targeted delivery of DNA encoding GFP to cells at one site in the brain (Wu et al. 1995; Hermens and Verhaagen 1998). One particularly clever use of a GFP-expressing virus is to simultaneously target Cre recombinase expression to a restricted set of cells in transgenic animals in which a gene of interest has been flanked with Cre-substrate *loxP* sites. Cre recombinase will then excise the *loxP*-flanked ("floxed") gene, effectively creating a highly localized gene knockout in cells that can be marked by GFP expression (Anton and Graham 1995; Kaspar et al. 2002). The usefulness of virus-mediated gene transfer, however, can be limited by toxicity of the virus, slow or weak gene expression, and limitations on insert size.

Electroporation of DNA is an alternative method that is becoming increasingly popular. Commonly used to transfect avian embryos, the method now has been applied, with impressive results, in mammalian and amphibian systems (see Chapter 11). Electroporation can be used in situ to transfect neurons and glia in the central nervous system (CNS) of the tadpole (Fig. 1). For single-cell electroporation, the tip of a glass micropipette, filled with a DNA solution, is placed next to the cell to be electroporated under visual guidance provided via an upright microscope, and current is passed through the pipette (Haas et al. 2001). For transfecting larger numbers of cells, DNA solution is pressure-injected into the brain ventricle, or directly into the tissue, and current is passed through the target tissue using a pair of closely spaced platinum plate electrodes (Haas et al. 2002). These methods are both effective and versatile for labeling cells, including neurons, glia, proliferating cells in the CNS, and muscle cells in the *Xenopus* tadpole, as well a variety of cell types in several preparations (see Chapter 11). In contrast to the viral and transgenic methods outlined above, DNA purified from bacteria, by a simple miniprep, can be used to electroporate neurons directly. Moreover, multiple genes can be used to cotransfect the same cells with high efficiency, either by simply electroporating the cells with a cocktail of different plasmids mixed together or by using single plasmids carrying multiple independent promoters. In principle, any two molecules that migrate together in an electric field can be used to cotransform cells by electroporation. For example, electroporation may be used to deliver fluorescent dextrans to reveal the morphology of a single cell, together with morpholino oligonucleotides to inhibit the expression of specific gene products in that cell (Fig. 1D–F) (Haas et al. 2002). Expression levels under the control of the cytomegalovirus promoter are extremely high, with GFP typically filling tectal dendrites at 12 hours, and reaching the distal tips of retinal ganglion cell axons within about 24 hours of electroporation. Use of constructs driven by stronger or weaker promoters, such as the neuron-specific enolase promoter, provides added versatility to this method.

MATERIALS FOR PROCEDURE A

This protocol uses a micromanipulator mounted on an upright microscope with a long working distance air objective, ~20x. Electric pulses are provided by an electrical stimulator (e.g., Grass SD9) via silver wire electrodes (0.25 mm diameter) and moist Kimwipe tissues.

Glass Micropipettes

These should resemble patch pipettes with 0.5-µm to 1-µm tips to give about 10 MΩ resistance when filled with standard recording internal solution. If the tips are too long, they will bend and break on the skin of the tadpole. Borosilicate standard wall with filament, outer diameter = 1.5 mm, inner diameter = 0.86 mm from Warner Instrument are recommended, pulled on a Sutter P-87 puller using a box filament. For DiI iontophoresis and either dextran or DNA electroporation, tip shape is critical to avoid labeling multiple cells and to avoid clogging. A good tip can be reused many times in the same labeling session.

Fluorescent Dye or Plasmid DNA Labeling Solution

DiI is 0.01–0.05% in absolute ethanol; fluorescent dextran is 5% (w/v) in 0.1 M phosphate buffer or water; plasmid DNA is 0.1–5.0 µg/µl in water or elution buffer. An endotoxin-free plasmid purification kit is recommended for plasmid preparation. Dye solutions can all be stored at –20° C for 1 month, although fluorescent dyes are generally best made fresh from powder. Care should be taken to avoid exposure of fluorescent dyes to light at all times.

PROCEDURE A

Labeling Individual Neurons in the Brains of Live *Xenopus* Tadpoles

1. Backfill the pipette with labeling solution and mount it on the micromanipulator set up on an upright microscope
2. Connect the circuit as follows: Positive lead of stimulator > silver wire > labeling solution in micropipette > [gap] > tadpole > moist Kimwipe > silver ground wire > negative lead of stimulator.

   Note: For electroporation with DNA, the polarity of the stimulator must be **REVERSED**—DNA is negatively charged and will move toward the positive electrode. To monitor for blockade of the pipette tip, a 10-KΩ resistor can be inserted (optionally), in series, anywhere along the circuit (except between the tadpole and the pipette, of course), and an oscilloscope can be used in parallel to measure the potential drop across the resistor (Fig. 1A).
3. Anesthetize the tadpole in MS-222<!>solution for about 1 minute. (MS-222 anesthetic solution: 0.02% MS-222 [Sigma] in rearing solution. The solution can be stored for 1 month at 4°C. Rearing solution: 58.2 mM NaCl, 0.67 mM KCl <!> , 0.34 mM $CaNO_3$<!> , 0.78 mM $MgSO_4$<!>, 50 mM HEPES. Adjust to pH 7.4.)
4. Transfer the tadpole onto the stage of the microscope with a large-bore dropper pipette, carefully orienting it dorsal-side-up on the moist Kimwipe by gently sliding it with a fine brush.
5. Advance the pipette into the brain of the tadpole, positioning the tip directly at the area to be labeled.
6. Apply pulses from the stimulator. Settings for amplitude and pulse duration must be set empirically for each tissue and pipette shape.

   Note: A good starting point is 3–5 pulses with an amplitude setting of about 30–80 V on the Grass stimulator. For DiI, pulse durations of 1–10 msec are recommended. For dextrans, pulses of 10–100 msec can be used. For DNA electroporation, delivery of 3–5 brief (~1 sec) bursts of 1-msec pulses at a frequency of 200 Hz gives high transfection efficiency. Note that although dextran solution almost never clogs the pipette tip, both DiI and DNA solutions can accumulate and block the tip. When this happens, it is often possible to rescue the pipette by temporarily reversing the polarity of current through the pipette and delivering one or two brief pulses.
7. Withdraw the pipette, and use a dropper or fine brush to transfer the tadpole to fresh rearing solution.

   Note: Sometimes the dye solution sticks to the outside of the pipette and leaves a track of dye along the pipette path. The best way to avoid this is to apply negative current prior to retracting the pipette from the injection site. This dilutes the dye in the pipette tip and prevents leakage and clogging.

Caution: *See Appendix 3 for appropriate handling of materials marked with <!>.*

The fluorescence microscope allows the dye injection to be monitored during and after labeling. Although it is very beautiful to watch the dye spread throughout the cell, it is critical to remember that looking at the fluorescent dye in the cell during the labeling process is one of the easiest ways to kill the cell. It is recommended that the labeled cell be observed very briefly if at all. For DNA electroporation, there is no instant feedback regarding labeling success, as GFP expression takes many hours to develop. For this reason, it can be useful to practice single-cell electroporation with a dextran solution, prior to attempting DNA electroporation.

The success rate in labeling single isolated neurons is slightly lower for DNA electroporation than for dye iontophoresis, but cells imaged with GFP typically remain brighter and healthier over many days of imaging, ultimately resulting in a higher yield of cells that can be used for analysis. If multiple injection sites are used for each animal, the success rate for labeling single neurons in each animal can approach 80–90%. This method has been used to label cells in *Xenopus* tadpoles between stages 39 and 50 (Nieuwkoop and Faber 1956). In principle, it should be useful in older and younger animals. As the animal gets older, the thicker skin becomes more difficult to penetrate without breaking the electrode tip. This problem can be addressed by making a tiny hole in the skin through which the electrode is inserted.

PROCEDURE B

Dye Labeling Retinal Ganglion Cell Axons in Live *Xenopus* Tadpoles

The same basic materials and procedures are suitable for labeling RGC axons as for tectal cells. However, it is possible to use a high-magnification dissecting microscope rather than an upright transmitted light microscope for RGC labeling, which may provide a greater range of approach angles for the pipette. The objective is to introduce a bolus of dye by iontophoresis into the neural retina where the ganglion cell somata are located and the axons course, with the expectation that it will be taken up by a small number of the axons.

1. Anesthetize and transfer the tadpole to the microscope stage as described in Procedure A above.
2. Orient the tadpole so that the dye-filled pipette, when advanced, will enter the eye, perpendicular to its surface at the border between the lens and retina.
3. Advance the pipette, under visual guidance, to the part of the neural retina directly behind the lens, entering the eye at the margin of the lens (Fig. 2A). Be careful to avoid the large blood vessel surrounding the lens of the eye. If this is damaged, there is a high probability that the tadpole will not survive due to blood loss.
4. Apply current as described in Procedure A above. The dye should visibly eject to form a dense deposit in the eye. If the dye does not come out, try reversing polarity to unblock the tip. In addition, the tip can sometimes be withdrawn and broken back very slightly against the moist Kimwipe; although this will eventually reduce the effectiveness of the pipette, it will relieve the blockage.
5. Transfer the tadpole to clean rearing solution.

MATERIALS FOR PROCEDURE C

This procedure requires the use of a dissecting microscope, a pressure injection system (Picospritzer II, General Valve) and two micromanipulators, one equipped with a glass micropipette holder and one to position platinum plate electrodes. Electric pulses are provided by an electrical stimulator (e.g., Grass SD9) and a high-voltage 3 μF capacitor, via platinum plate electrodes, which can be made from pipette puller filaments (each 1 x 2 mm, mounted about 1 mm apart at the end of a rod held by the second micromanipulator).

The glass micropipettes and plasmid DNA solutions described for Procedure A are also suitable for this procedure, although it is useful to add a tiny amount of Fast Green dye to help visualize the solution as it is injected into the eye. The pipette tip should be broken back very slightly to prevent clogging.

PROCEDURE C

Electroporation of Retinal Ganglion Cell Axons in Live *Xenopus* Tadpoles

1. Connect the circuit as a direct loop across the plate electrodes, with the capacitor in parallel as follows: Positive lead of stimulator > one lead of capacitor > one plate electrode > tadpole > second plate electrode > other lead of capacitor > negative lead of stimulator. DNA is negatively charged and will transfect cells on the side nearest to the positive electrode (Fig. 2B).
2. Back-fill a glass micropipette with the DNA solution and mount it onto the pipette holder held by the micromanipulator. Break back the tip until solution can first be seen with the dissecting microscope coming from the tip when pressure is applied.
3. Anesthetize the tadpole in MS-222 solution <!> (see Procedure A for recipe) and place it on a moist Kimwipe to keep it from moving while the pipette is inserted into its eye. Position the pipette at the border between the lens and the retina, as described in the previous section.

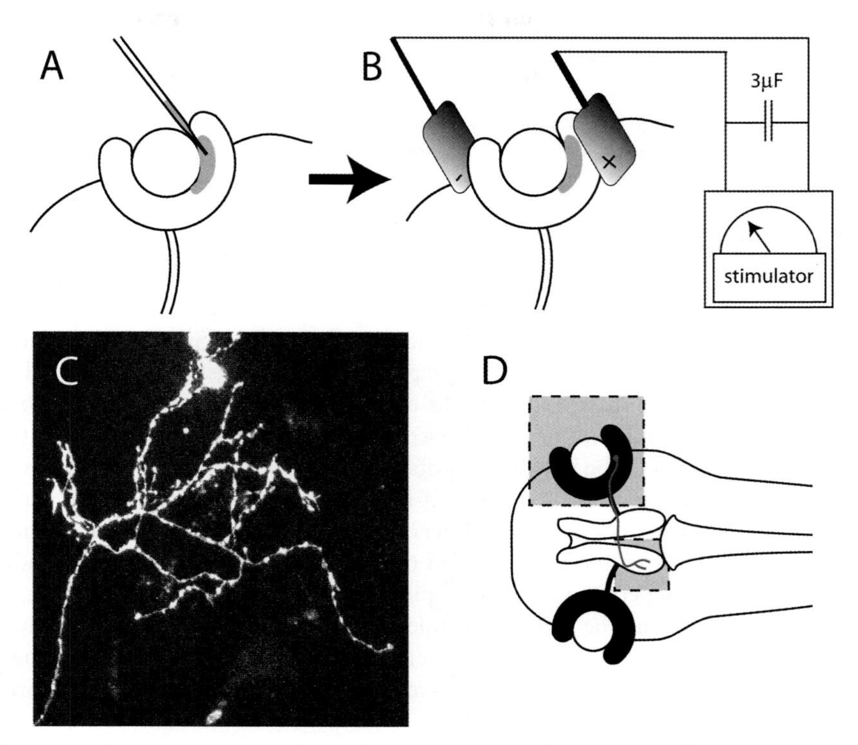

FIGURE 2. Retinal electroporation of plasmid DNA. (*A*) Plasmid solution, with Fast Green to provide contrast, is pressure-injected into the vitreous humor by insertion of a micropipette at the margin of the lens and sclera. (*B*) Platinum plate electrodes are placed on the surface of the tadpole just across the eye. Negatively charged DNA will move toward the positive electrode. A capacitor, placed in parallel, creates the exponential decay pulse shape for optimal transfection efficiency. (*C*) Within 24 hr, labeled axons are visible in the contralateral optic tectum. Example of a retinal ganglion cell axon expressing enhanced yellow fluorescence protein in a stage-48 tadpole. (*D*) Diagram of the tadpole head with the eye and tectum boxed.

4. Pressure-inject just enough of the DNA solution into the vitreous humor to see the Fast Green dye in the eye (Fig. 2A).
5. Withdraw the pipette and lower the plate electrodes so that they barely touch opposite sides of the tadpole's eye. The part of the retina closest to the positive electrode will be electroporated (Fig. 2B).
6. Apply about 5 pulses of 1.6-msec duration at 30–60 V across the electrodes.

   Note: The capacitor in parallel will convert this pulse from a square wave into a lower-intensity pulse with an exponential decay. This waveform is thought to be ideal for first creating micropores in the plasma membranes of cells and then driving DNA into the electroporated cells.

   Small bubbles should appear at the electrode, indicating that current is flowing. A large number (>10) of small bubbles is a better indication of successful electroporation than fewer larger bubbles. To increase the number of cells across the retina that are labeled, reverse the polarity of the stimulator and repeat the electroporation.
7. Retract the electrodes, and use a dropper or brush to transfer the tadpole to a bowl containing fresh rearing solution.

GFP-expressing ganglion cell axons should be visible in the optic tectum within 24 hours of electroporation using this method (Fig. 2C). The number of labeled axons is highly variable and appears to decrease with the age of the tadpole at the time of electroporation. Electroporation of retinal cells in tadpoles at stage 45–46 generally gives rise to one to five well-labeled axons, although this number is influenced by volume and concentration of the DNA solution.

It is also possible to use this approach for transfecting large groups of cells in the brain. To electroporate groups of cells in the optic tectum, the DNA solution is injected into the brain ventricle, and current is passed across the midbrain by placing the electrodes directly on the overlying skin. By changing the size, contact area, and positioning of the plate electrodes, this method can be used to electroporate cells along the rostrocaudal extent of the brain and in the spinal cord.

EXAMPLE OF APPLICATION: IMAGING LABELED NEURONS IN THE LIVING TADPOLE BRAIN

Anesthetized animals, hours to days after labeling, are mounted in a small pool of 0.02% MS-222 solution under a coverslip in a silicone polymer (e.g., Sylgard) chamber, custom-made to fit the animal in the same orientation with repeated imaging sessions. The animal should sit snugly in the chamber with the dorsal aspect of the animal immediately under, and in contact with, the coverslip. The animal should not be compressed by the coverslip or by the walls of the chamber. If the animal is compressed in the imaging chamber, blood flow in the brain will slow significantly. If animals of different stages are to be imaged, or if images are to be collected over a period of days during which the animals grow, separate chambers must be made for the animals as they grow. Test the chamber using a tadpole with a single labeled neuron. Record two images of the tadpole, being sure to reposition the animal between images. Despite the repositioning, the overall three-dimensional structure of the neuron should be superimposable at the two short-interval time points. Although some changes in dendritic arbor structure are observed within a 5-min interval, these changes are usually confined to short branch tips and do not prevent superimposition of the major branches of the arbor with a good chamber.

ADVANTAGES AND LIMITATIONS

Time-lapse Imaging by Confocal Microscopy

The confocal microscope offers an advantage over conventional fluorescence microscopy in that it can acquire information about the full three-dimensional structure of neurons in vivo. To collect a high-quality *z*-series of neurons over time in the living animal requires a balance between image quality, the number of optical sections and step size in the *z*-dimension, laser intensity and illumination time, and the number of time points and imaging intervals in the time series. Image quality itself is also a balance of many parameters that have a bearing on cell and animal survival: scan rate, dwell time, laser intensity, excitation wavelength, and number of scans averaged for each optical section. Prolonged laser illumination is not only toxic to dye-labeled cells, but it can also kill animals that contain no dye.

Optimal parameters for time-lapse image collection depend on the number of time points to be collected, and the total period over which the images are to be collected. If images are to be collected over a period of several days, it is critical that the first images be collected under the most conservative conditions possible, to minimize laser exposure, preserve cell health, and minimize dye fading. Collecting repeated images over short intervals is more taxing on the viability of a neuron, and the animal, than collecting the same number of images over longer intervals. Imaging dyes excited by longer wavelengths (e.g., DiD) is less toxic to the animals than imaging those excited by shorter wavelengths (e.g., FITC-dextran). No apparent advantage is gained by using antioxidants with DiI imaging, although these chemicals may improve viability when imaging with shorter-wavelength dyes. Antioxidants can reportedly alter *N*-methyl-D-aspartate (NMDA) receptor activity (R.Y. Tsien, pers. comm.) and may alter function of other proteins as well. GFP exhibits relatively low phototoxicity, compared with any of the extrinsic dyes that we have used. Because GFP is continually replenished by the cell, fading of fluorescence is also less of a problem, particularly for daily imaging protocols.

When imaging a live three-dimensional neuron embedded in scattering brain tissue, fluorescence from parts of the cell outside the plane of focus poses a serious problem. First of all, out-of-focus light degrades the sharpness of the image in the focal plane. This problem has traditionally been reduced by either image deconvolution or confocal microscopy (Chapters 6 and 95). Image deconvolution software actually makes use of the information content in the out-of-focus light to create a sharper three-dimensional image, but it is computationally demanding and cannot correct for scattering of light in the specimen. Laser-scanning confocal imaging essentially takes the opposite approach by optically discarding both the out-of-focus light and the photons scattered by the specimen. This is achieved by focusing the returning emission light through a pinhole aperture placed in front of the photomultiplier tube in the optical path. The major problem with laser-scanning confocal microscopy for in vivo imaging is that substantial fluorophore excitation takes place outside the plane of focus, but contributes little to the final image.

The defocusing of the scanning excitation laser beam outside the focal plane results in lower-intensity exposure to excitation light, but also broadens the profile of the scanning beam from a diffraction-limited point to a circular section of the light cone (Fig. 3A). As the beam scans, any one point in the specimen is actually exposed to excitation light for a longer period of time outside the plane of focus than within the focal plane. Consequently, each optical section added to a *z*-series, captured on a laser-scanning confocal microscope, is roughly equivalent to scanning all the other optical sections in the *z*-series one additional time. For complex neuronal dendritic arbors of tectal neurons in the tadpole, it is not unusual to collect more than 40 optical sections.

Time-lapse Imaging by Two-Photon Microscopy

The laser-scanning two-photon microscope overcomes many of the difficulties associated with in vivo imaging. Two-photon excitation occurs when a fluorophore simultaneously absorbs two photons, each of approximately half the energy (double the wavelength) that induces single-photon fluorescence. Two-photon excitation is normally an infinitesimally rare event, but ultrafast pulsed lasers can deliver intense bursts of photons to the sample that converge in the focal plane, achieving the photon density required for two-photon excitation, but only in the plane of focus. Consequently, any fluorophore above or below the plane of focus will not be excited, eliminating nearly all phototoxic damage to the specimen outside the focal plane (Fig. 3B). Furthermore, because fluorescence excitation occurs only at the diffraction-limited spot being scanned across the focal plane, all emitted photons can be assumed to originate at that site. This means that, instead of discarding scattered photons, all the emitted photons can be collected, resulting in brighter images, greatly reducing photodamage, and allowing images to be collected deep inside scattering tissue like brain. In live *Xenopus* tadpoles, laser-

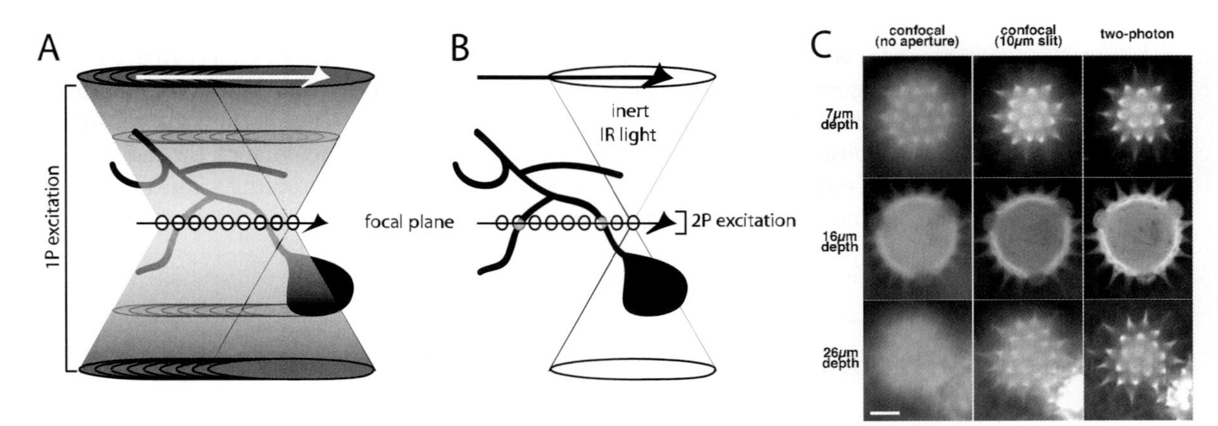

FIGURE 3. Two-photon microscopy causes less photodamage outside the focal plane. (*A*) With laser-scanning confocal microscopy, the scanning beam of excitation light excites the labeled cell outside the focal plane. Planes exposed to less intense excitation are exposed longer as the beam scans. (*B*) Two-photon excitation minimizes photodamage outside the plane of focus, as infrared light is relatively harmless to living tissue at the levels of exposure used. (*C*) Examples of laser-scanning microscopic images of a pollen grain. The confocal aperture improves one-photon excitation image quality by reducing out-of-focus emission light, but, with increasing depth, it eventually overwhelms the signal from the image plane. Two-photon excitation provides sharper, brighter images at these depths because there is no excitation of the overlying fluorescent material. Bar, 10 µm. (Reprinted, with permission, from Ruthazer and Cline 2002 [©Elsevier].)

scanning two-photon microscopy can be used to collect 1-μm-spaced *z*-series stacks of labeled axons and dendrites, at 10-min intervals, for over 2 hours, with no evidence of photodynamic damage. Finally, two-photon excitation also uses infrared light, which is less scattered by living tissue, so the depth penetration of the imaging is greatly increased.

IMAGING DYES OF DIFFERENT WAVELENGTHS

Because many common fluorophores have broad two-photon cross sections, the same two-photon excitation wavelength can be used to excite multiple fluorophores in the specimen. In this case, the fluorophores are distinguished only by their emitted-light wavelengths. This approach is excellent for colocalization studies, as it eliminates the misalignment of different-color excitation beams that may occur with confocal microscopy due to chromatic aberration. On the other hand, the ease in conventional epifluorescence microscopy with which different fluorophores can be excited independently is useful for quantitative measurements of relative fluorescence intensity.

IMAGE ANALYSIS AND MORPHOMETRY

One of the most critical and time-consuming steps in three-dimensional time-lapse imaging is data analysis. Three fundamental morphological properties of neurons can be extracted from images: total arbor length, individual branch-tip number and length, and arbor complexity (e.g., branch density and branch order). Time-lapse imaging adds further useful information about growth rates and branch-tip dynamics, which would be impossible to extract from static images in fixed material. To take full advantage of these data, it is important to use a strategy for morphometric analysis that incorporates both structural and temporal information.

The ideal software would perform automatic tracing of dendritic and axonal arbors and be able to identify the same branch tip at each time point. Because some errors are inevitable, a user-friendly editor should allow drawings of branches to be compared with the raw data and modified easily. Automated neuron reconstruction is still in its infancy, although a few top-of-the-line professional products do offer some degree of automated drawing. In our experience to date, identifying and correcting errors produced by these programs remain too time-consuming to justify their use, although we remain optimistic about future advances.

Manual reconstructions can be performed using a freely distributed version of the popular and powerful NIH Image analysis software for the Macintosh, called Object-Image (Vischer et al. 1999). This program permits a nondestructive overlay to be traced manually, in three dimensions, on a confocal *z*-series stack (Fig. 4). By simultaneously running a set of custom macros for time-lapse analysis, it is possible to assign an identity number to each unique branch tip and thus follow it through the time series (Ruthazer and Cline 2002). The following procedure illustrates the basic process of morphometric analysis using custom Object-Image macros.

PROCEDURE D

Morphometric Analysis Using Custom Object-Image Macros

1. Open all the *z*-series stacks that make up the time series and rename them in temporal order as 1, 2, 3, etc. (Fig. 4A, B).
2. Calibrate the *x-y* scale by selecting the menu item *Analyze:SetScale* and entering scale values of pixels per micron derived from an image of a slide reticle, then for the *z* scale by selecting menu item *Stacks:Stack Info* and entering the *z*-axis step distance in microns that was used to capture the *z*-series stack.
3. Create an "object file" which will contain the overlay data. To do this, select menu item *Objects:New Object File*... and choose a name for the file. Drag one "SegLine" from the Object Type window into the Sequence window. Activate *3D Objects* and set Collect Mode to *Single*. Click *OK*.

4. Run the *Initialize for Counting* macro to create columns in the Object-Image mini-spreadsheet for branch-tip reference number and time-point information.
5. Trace the cell by pointing and clicking in the appropriate optical sections, using the object tool from the tools window. Each branch up to its first branch point should be drawn as an independent object (Fig. 4A, B). Upon reaching the end of each branch, close the object by hitting the TAB key. Then activate the *mark branchtips* macro item from the *Special* menu, which will request a reference number. The number entered, as well as the name of the window that was active at that moment, will be recorded. Each occurrence of a particular branch at different time points (i.e., windows) should be assigned the same reference number.
6. To export the measurement data for analysis in a spreadsheet, select the menu item *Objects:Export Object Results* (Fig. 4C). For display purposes, the menu item *Objects:Export XYZ Data* will create a file that can be read by Rotater (http://raru.adelaide.edu.au/rotater/) (Fig. 4 D, E). Rotater is a three-dimensional dynamic ray-tracing display program that provides a very nice way to display reconstructions.

In the end, a list of branch tips, their three-dimensional coordinates, and their lengths at each time point are generated. This information can then be analyzed easily in spreadsheet programs to investi-

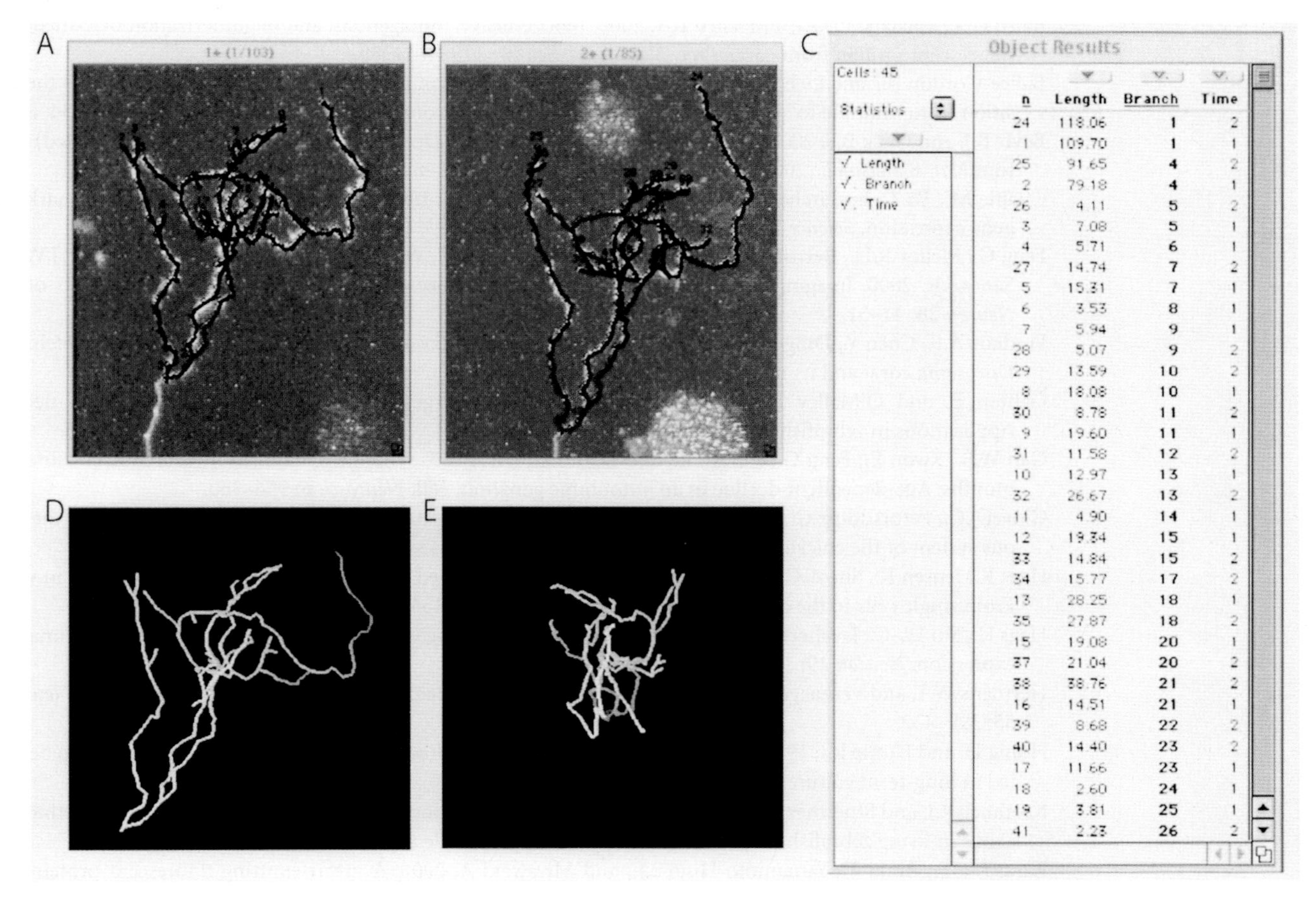

n	Length	Branch	Time
24	118.06	1	2
1	109.70	1	1
25	91.65	4	2
2	79.18	4	1
26	4.11	5	2
3	7.08	5	1
4	5.71	6	1
27	14.74	7	2
5	15.31	7	1
6	3.53	8	1
7	5.94	9	1
28	5.07	9	2
29	13.59	10	2
8	18.08	10	1
30	8.78	11	2
9	19.60	11	1
31	11.58	12	2
10	12.97	13	1
32	26.67	13	2
11	4.90	14	1
12	19.34	15	1
33	14.84	15	2
34	15.77	17	2
13	28.25	18	1
35	27.87	18	2
15	19.08	20	1
37	21.04	20	2
38	38.76	21	2
16	14.51	21	1
39	8.68	22	2
40	14.40	23	2
17	11.66	23	1
18	2.60	24	1
19	3.81	25	1
41	2.23	26	2

FIGURE 4. Time-lapse axon reconstruction using Object-Image software. (*A*) Projection of *z*-series captured at first time point. The manually traced overlay (*black lines and numbers*) shows the object identifier number at the tip of each branch during tracing. (*B*) Projection of *z*-series from second time point. Normally, tracing would be performed in three dimensions on the stack, rather than on a projected image. Tracing in three dimensions permits true branch growth to be distinguished from apparent changes caused by shifts in the orientation of the animal. (*C*) Object Results list generated during tracing that identifies each individual branch with a branch number and time information permitting its growth and dynamics to be followed over time. (*D*) Rotater ray-trace image of axon from *A*, which can be viewed from all angles. (*E*) Same axon rotated perpendicularly as it would appear looking down the main axon shaft.

gate aspects such as branch growth rates, branch-tip dynamics, and lifetimes. We also have created macros to perform more elaborate analyses, such as three-dimensional Sholl analysis on the reconstructed arbors (Ruthazer and Cline 2002). All of these macros are available, with detailed instructions, on our Web page (http://clinelab.cshl.edu/methods.html).

ACKNOWLEDGMENTS

This work was supported by the National Institutes of Health, National Science Foundation, Hoffritz Fund, NARSAD, and an endowment from the Charles Robertson Family.

REFERENCES

Amsterdam A., Lin S., and Hopkins N. 1995. The *Aequorea victoria* green fluorescent protein can be used as a reporter in live zebrafish embryos. *Dev. Biol.* **171:** 123–129.

Anton M. and Graham F.L. 1995. Site-specific recombination mediated by an adenovirus vector expressing the Cre recombinase protein: A molecular switch for control of gene expression. *J. Virol.* **69:** 4600–4606.

Baird G.S., Zacharias D.A., and Tsien R.Y. 2000. Biochemistry, mutagenesis, and oligomerization of DsRed, a red fluorescent protein from coral. *Proc. Natl. Acad. Sci.* **97:** 11984–11989.

Balice-Gordon R.J. and Lichtman J.W. 1993. In vivo observations of pre- and postsynaptic changes during the transition from multiple to single innervation at developing neuromuscular junctions. *J. Neurosci.* **13:** 834–855.

Bevis B.J. and Glick B.S. 2002. Rapidly maturing variants of the *Discosoma* red fluorescent protein (DsRed) (erratum *Nat. Biotechnol.* [2002] **20:** 1159). *Nat. Biotechnol.* **20:** 83–87.

Chalfie M., Tu T., Euskirchen G., Ward W.W., and Prasher D.C. 1994. Green fluorescent protein as a marker for gene expression. *Science* **263:** 802–805.

Feng G., Mellor R.H., Bernstein M., Keller-Peck C., Nguyen Q.T., Wallace M., Nerbonne J.M., Lichtman J.W., and Sanes J.R. 2000. Imaging neuronal subsets in transgenic mice expressing multiple spectral variants of GFP. *Neuron* **28:** 41–51.

Fradkov A.F., Chen Y., Ding L., Barsova E.V., Matz M.V., and Lukyanov S.A. 2000. Novel fluorescent protein from *Discosoma* coral and its mutants possesses a unique far-red fluorescence. *FEBS Lett.* **479:** 127–130.

Gahtan E. and O'Malley D.M. 2001. Rapid lesioning of large numbers of identified vertebrate neurons: Applications in zebrafish. *J. Neurosci. Methods* **108:** 97–110.

Gan W.B., Kwon E., Feng G., Sanes J.R., and Lichtman J.W. 2003. Synaptic dynamism measured over minutes to months: Age-dependent decline in an autonomic ganglion. *Nat. Neurosci.* **6:** 956–960.

Glover J.C., Petursdottir G., and Jansen J.K. 1986. Fluorescent dextran-amines used as axonal tracers in the nervous system of the chicken embryo. *J. Neurosci. Methods* **18:** 243–254.

Haas K., Jensen K., Sin W.C., Foa L., and Cline H.T. 2002. Targeted electroporation in *Xenopus* tadpoles in vivo—From single cells to the entire brain. *Differentiation* **70:** 148–154.

Haas H., Sin W.-C., Javaherian A., Li Z., and Cline H.T. 2001. Single-cell electroporation for in vivo neuronal gene expression. *Neuron* **29:** 1–9.

Hermens W.T. and Verhaagen J. 1998. Viral vectors, tools for gene transfer in the nervous system. *Prog. Neurobiol.* **55:** 399–432.

Honig H. and Hume R.I. 1986. Fluorescent carbocyanine dyes allow living neurons of identified origin to be studied in long-term cultures. *J. Cell Biol.* **103:** 171–187.

Kaethner R.J. and Stuermer C.A. 1992. Dynamics of terminal arbor formation and target approach of retinotectal axons in living zebrafish embryos: A time-lapse study of single axons. *J. Neurosci.* **12:** 3257–3271.

Karasawa S., Araki T., Yamamoto-Hino M., and Miyawaki A. 2003. A green-emitting fluorescent protein from *Galaxeidae* coral and its monomeric version for use in fluorescent labeling. *J. Biol. Chem.* **278:** 34167–34171.

Kaspar B.K., Vissel B., Bengoechea T., Crone S., Randolph-Moore L., Muller R., Brandon E.P., Schaffer D., Verma I.M., Lee K.F., Heinemann S.F., and Gage F.H. 2002. Adeno-associated virus effectively mediates conditional gene modification in the brain. *Proc. Natl. Acad. Sci.* **99:** 2320–2325.

Lendvai B., Stern E.A., Chen B., and Svoboda K. 2000. Experience-dependent plasticity of dendritic spines in the developing rat barrel cortex in vivo. *Nature* **404:** 876–881.

Miyawaki A., Griesbeck O., Heim R., and Tsien R.Y. 1999. Dynamic and quantitative Ca^{2+} measurements using improved cameleons. *Proc. Natl. Acad. Sci.* **96:** 2135–2140.

Myers P.Z. and Bastiani M.J. 1993. Cell-cell interactions during the migration of an identified commissural growth cone in the embryonic grasshopper. *J. Neurosci.* **13:** 115–126.

Nakai J., Ohkura M., and Imoto K. 2001. A high signal-to-noise $Ca^{(2+)}$ probe composed of a single green fluorescent protein. *Nat. Biotechnol.* **19:** 137–141.

Nieuwkoop P.D. and Faber J. 1956. *Normal table of* Xenopus laevis (Daudin). Elsevier, Amsterdam.

O'Rourke N.A., Cline H.T., and Fraser S.E. 1994. Rapid remodeling of retinal arbors in the tectum with and without blockade of synaptic transmission. *Neuron* **12:** 921–934.

Ohnuma S., Mann F., Boy S., Perron M., and Harris W.A. 2002. Lipofection strategy for the study of *Xenopus* retinal development. *Methods* **28:** 411–419.

Ormo M., Cubitt A.B., Kallio K., Gross L.A., Tsien R.Y., and Remington S.J. 1996. Crystal structure of the *Aequorea victoria* green fluorescent protein. *Science* **273:** 1392–1395.

Reiner A., Veenman C.L., Medina L., Jiao Y., Del Mar N., and Honig M.G. 2000. Pathway tracing using biotinylated dextran amines. *J. Neurosci. Methods* **103:** 23–37.

Ruthazer E.S. and Cline H.T. 2002. Multiphoton imaging of neurons in living tissue: Acquisition and analysis of time-lapse morphological data. *Real-Time Imaging* **8:** 175–188.

Sakai R., Repunte-Canonigo V., Raj C.D., and Knopfel T. 2001. Design and characterization of a DNA-encoded, voltage-sensitive fluorescent protein. *Eur. J. Neurosci.* **13:** 2314–2318.

Schmued L., Kyriakidis K., and Heimer L. 1990. In vivo anterograde and retrograde axonal transport of the fluorescent rhodamine-dextran-amine, Fluoro-Ruby, within the CNS. *Brain Res.* **526:** 127–134.

Tsien R.Y. 1998. The green fluorescent protein. *Annu. Rev. Biochem.* **67:** 509–544.

Vischer N.O., Huls P.G., Ghauharali R.I., Brakenhoff G.J., Nanninga N., and Woldringh C.L. 1999. Image cytometric method for quantifying the relative amount of DNA in bacterial nucleoids using *Escherichia coli. J. Microsc.* **196:** 61–68.

Wang S. and Hazelrigg T. 1994. Implications for bcd mRNA localization from spatial distribution of exu protein in *Drosophila* oogenesis. *Nature* **369:** 400–403.

Wu G.-Y., Zou D.-J., Koothan T. and Cline H.T. 1995. Infection of frog neurons with vaccinia virus permits in vivo expression of foreign proteins. *Neuron* **14:** 681–684.

Yeh E., Gustafson K., and Boulianne G.L. 1995. Green fluorescent protein as a vital marker and reporter of gene expression in *Drosophila. Proc. Natl. Acad. Sci.* **92:** 7036–7040.

CHAPTER 24

Optical Projection Tomography: Imaging 3D Organ Shapes and Gene Expression Patterns in Whole Vertebrate Embryos

James Sharpe

MRC Human Genetics Unit, Western General Hospital, Edinburgh EH4 2XU, United Kingdom

This review is a brief introduction to optical projection tomography (OPT) as a method for imaging whole vertebrate embryos. To place OPT in its proper context within the field of three-dimensional (3D) embryo imaging, this chapter starts with a brief overview of the previously existing techniques, both destructive and nondestructive. Next is a description of how OPT works and its advantages over the previously described approaches. Finally, two practical sections are presented on what OPT is particularly good for, and how to get the most out of OPT in terms of specimen preparation and imaging modes.

PREEXISTING TECHNIQUES FOR IMAGING WHOLE EMBRYOS

Physical Sectioning

Until recently, the most common technique for recreating a 3D representation of a small biological specimen has been to physically cut it into hundreds of sections, photograph each one, and then recombine the images in a computer (Davidson and Baldock 2001). Although time-consuming, this approach has the advantage that any type of optical staining may be used, because all that is needed is a series of digital images. Unfortunately however, when a specimen is cut into thin sections, the data become essentially two-dimensional (2D). This means that information about the alignment of each section to its neighbor (the third dimension) is lost. In addition, for many common types of sectioning (microtome, cryosections), the embedding material (paraffin wax, Optical Cutting Temperature [OCT] compound) is not rigid enough to prevent the tissue from stretching and deforming when it is mounted onto glass slides. These two factors prevent reconstruction of the true 3D shape of the specimen from sections alone. For some experiments, this does not matter, but, for tasks such as phenotyping, it can be important to use a technique, such as OPT, that preserves true shape.

New sectioning techniques have been developed to speed up the process and overcome the "true-shape" problem, e.g., block-face imaging (e.g., EFIC [episcopic fluorescence image capture], see Weninger and Mohun 2002). Instead of photographing each section after it has been cut and mounted on a glass slide, the section is imaged just *before* it is cut—at the surface of the block. If the positions of the camera and block are carefully controlled such that the focal plane always coincides with the cut surface of the block, the problems of both section deformation and alignment are simultaneously overcome. This approach can generate very high resolution reconstructions of large specimens with excellent shape preservation. However, so far, the common forms of EFIC cannot be used to image a gene expression pattern as a positive signal—they image a colored stain by its ability to quench the autofluorescence of the tissue. This means that within a single specimen, a strong signal cannot formally be distinguished from an absence of tissue. Very recently, a new version of this approach has been reported, which uses high-energy laser ablation of tissue instead of a knife to remove the tissue sections (http://www.nature.com/nsu/030609/030609-2.html). This allows it to be used on soft tissues, which are difficult to embed. However, at the time of writing, this has not yet been formally published.

In another new technique, section deformation is reduced by the use of a strong embedding material, and the problem of alignment is overcome by the use of fiducial markers. This approach is the basis of the external marker-based automatic congruencing (EMAC) technique (Streicher et al. 2000). This method uses vertically drilled holes around the specimen as fiducial markers and resin as the embedding material. Cut sections are digitized, and the drilled holes are recognized by computer algorithms, thereby allowing fast automatic registration of the tissue. Although this approach does take longer than EFIC, and the registration of sections is not as accurate, it does allow colored stains to be analyzed (as with normal paraffin wax sections). This is very important for gene expression analysis because the most widespread protocol for detection of mRNAs involves the enzymatic conversion of BCIP/NBT (5-bromo-4-chloro-3-indolyl phosphate/nitroblue tetrazolium) to a purple precipitate (Hammond et al. 1998). Although immunohistochemistry does work routinely using fluorescence-based detection, the number of proteins for which specific antibodies exist represents only a fraction of the genes in the genome. It therefore appears that bright-field imaging will remain an important method for detecting gene expression patterns for some time (until alternative fluorescence-based techniques, such as tyramide amplification, become routinely reliable).

Nondestructive Imaging

The following nondestructive imaging techniques can be used to generate images from living embryos (OPT has yet to be optimized for this purpose). However, none of the techniques is ideal for analyzing 3D gene-expression-fixed-embryos.

X-ray MicroCT (Computed Tomography)

This method is well suited to specimens around the 1-cm size and displays a resolution that is in principle ideal for embryos (<10 μm). However, because X-rays are poorly absorbed by embryonic tissues, and they are not significantly absorbed by colored stains, the technique cannot be used to image a BCIP/NBT in situ pattern, nor can it be used to capture the histology of the tissue. (Histological data can be obtained using this technique, but the samples must be washed in a solution containing metal atoms, such as barium; see unpublished results on the Web site of the Mouse Imaging Centre in Toronto: http://mice1.ocgc.ca/mouseimage/Micro.htm.)

Microscopic Magnetic Resonance Imaging (MRI)

This technique has a big advantage over microCT in that it can detect useful contrast between the tissues of untreated embryos (Dhenain et al. 2001). However, the resolution is significantly lower than that achieved by microCT (~25 μm in the studies by Schneider et al. 2003). Another drawback is the paucity of assays for molecule-specific tissue labeling compared to optical techniques. Arguably, the most important application for microMRI will be its ability to image dynamically changing 3D expression patterns in living embryos, using high-contrast agents (Louie et al. 2000).

Confocal Microscopy and Multiphoton Microscopy

3D optical sectioning approaches have many advantages over nonoptical techniques (Potter et al. 1996). In particular, there is now available a veritable battery of fluorescent dyes which can be used in conjunction with a wealth of techniques to label biologically significant molecular distributions. At the cellular and cell-cluster levels, confocal microscopy generates impressive 3D results and is invaluable for cell biology. However, when it comes to analyzing tissues more than a few hundred micrometers in depth, confocal microscopy fails to image the entire intact specimen. In addition, another drawback to confocal and multiphoton approaches that is sometimes overlooked is their inability to generate 3D reconstructions of colored (absorption-based) staining patterns (as opposed to fluorescence patterns). There are many common assays for which the most reliable signal-detection method is still the production of a dark colored precipitate, for example, the use of BCIP/NBT to visualize whole-mount in situ hybridization. Although such precipitates can be detected easily on physically cut sections, by bright-field microscopy, this particular assay for gene expression patterns cannot be imaged in 3D by fluorescence microscopy.

Optical Coherence Tomography (OCT)

This technique has been used in a few developmental studies (Yelbuz et al. 2002), and it is the optical equivalent of ultrasound tomography—photons are sent into the specimen in one direction, and those reflected by the tissue are detected as they emerge. Their time of flight is calculated using interferometry. OCT can be used to image living tissue, and can do so down to a depth of 2–3 mm. However, due to its reliance on interferometry, this approach is more limited than confocal microscopy in terms of signal detection. Since the light reflected from the specimen must be coherent with a reference beam, it is not possible to use fluorescence-based assays to detect gene expression patterns.

Microscopic Ultrasound Imaging

Real microscopic ultrasound imaging is actually better-suited to whole-embryo imaging than OCT, as it displays far greater depth penetration (it can image an entire mouse embryo, alive, within the uterus of a pregnant female; see Foster et al. 2003). Again, its strength lies in its ability to image living embryos, and in this case, more convenient access to embryos within a pregnant mouse compared to MRI. However, for fixed specimens, this approach does not provide high resolution (achieving voxels ~50 μm across), nor does it feature convenient gene-expression assays.

HOW OPT WORKS

The optical 3D microscopy techniques mentioned above (confocal microscopy, multiphoton microscopy, and optical coherence tomography) use a geometric arrangement that can be called *section tomographly*. This essentially means that they sample the properties of a specimen at discrete points in space (confocal microscopy samples the fluorescence of a specimen, whereas OCT samples reflectiveness). In these cases, each measurement has explicit *x,y,z* coordinates and can be mapped directly to a voxel in the resulting reconstruction. In contrast, imaging techniques that use the principle of *projection tomography* (often called computed tomography) collect data, which sum a measured property along a linear projection that traverses the entire specimen (see Kak and Slaney 1988). These projection data, which are gathered from a series of orientations through the specimen, do not directly represent voxels and must be transformed in order to recover the original structure of the specimen (using a filtered back-projection algorithm). Projection tomography is widely employed to generate high-resolution reconstructions in techniques that use X-rays, γ-rays, or electron beams, as these rays are not deflected as they pass through their respective specimens (Massoud and Gambhir 2003; Midgley and Weyland 2003). A related form of tomography has also been applied to optical systems in the medical technique known as diffuse optical tomography (DOT) (for review, see Ntziachristos and

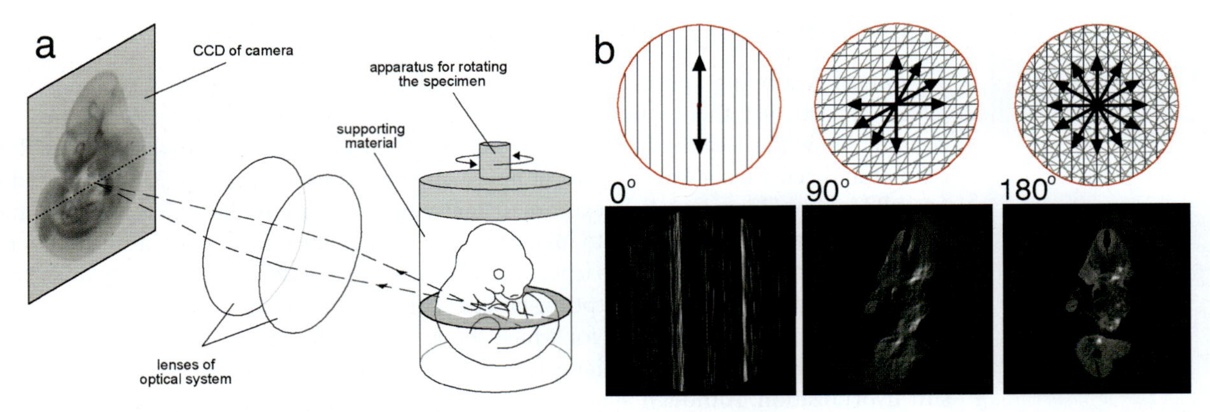

FIGURE 1. How OPT works. (*a*) Apparatus for obtaining OPT images. Light from the specimen (*dashed lines*) is focused to create an image on the CCD of the camera. The apparatus is aligned such that projections through a section perpendicular to the axis of rotation (*gray oval*) are focused onto a single row of pixels (*dotted line*). (*b*) As projections from successive angles (0°, 90°, 180°) are accumulated in the back-projection algorithm, a representation of the section is gradually reconstructed. (*Top row*) Accumulated orientations of projection; (*bottom row*) emerging reconstruction during the process. Although techniques such as X-ray CT require only 180° of information, OPT benefits from a full 360° rotation because the limited depth of focus means that views from opposite orientations do not contain the same information.

Chance 2001). However, light rays are very prone to scattering as they pass through biological tissue, so DOT employs algorithms different from those of OPT, which attempt to compensate for these distortions and can produce only low-resolution reconstructions.

The approach adopted in OPT (Sharpe et al. 2002) is to suspend the specimen in an index-matching liquid to (1) reduce the scattering of light and (2) reduce heterogeneities of refractive index throughout the specimen. This means that light passes through the specimen in more or less straight lines and the back-projection algorithm can therefore generate relatively high resolution images. The liquid most often used is BA:BB or Murray's Clear (a mixture of benzyl alcohol and benzyl benzoate). Inside the OPT scanning device, the specimen is maintained within the liquid and rotated through a series of angular positions (<1° apart in our current design), and an image is captured at each orientation. The apparatus is carefully aligned to ensure that the axis of rotation is perpendicular to the optical axis, so that projection data pertaining to each plane are collected by a linear row of pixels on the CCD of the camera (Fig. 1).

There are two imaging modes for OPT. The first is transmission imaging, or *bright-field OPT*. This mode corresponds closely to the method used by an X-ray CT scanner; i.e., the specimen is illuminated from the side opposite the detector and light rays pass through the specimen and are absorbed by varying amounts. The CCD camera records an image similar to a quantitative shadow of the specimen (the major difference being that OPT uses lenses to focus an image, whereas CT captures a genuine shadow).

The second mode is emission tomography, or *fluorescence OPT*. As before, due to the use of light and the consequent ability to employ lenses to create focused images, this form of emission tomography is very different from a technique like positron emission tomography (PET). In fluorescence OPT, the entire specimen is illuminated with a specific wavelength, and fluorescently emitted light is then focused into an image on the CCD. Usually, the excitation is provided on the same side as the detection.

In both modes of OPT, the fact that lenses are used to focus an image means that OPT suffers one technical drawback compared to its X-ray equivalent: As with all optical imaging systems, there is only a limited depth of focus, and this usually cannot encompass the entire specimen. OPT therefore generally takes advantage of a compromise, which yields good results—the focal plane is positioned halfway between the axis of rotation and the edge of the specimen closest to the lenses. This maximizes the focused information obtained from the specimen, while bypassing the need to image at multiple depths and keeping the imaging time to a minimum.

The resolution of OPT depends on the size and type of specimen. However, for a 10.5 dpc (days postcoitum) mouse embryo, it is typical to reconstruct the data into 5-µm voxels, obtaining a resolution good enough to pinpoint individual cells if they are labeled and surrounded by unlabeled cells. In unlabeled specimens, single-cell membranes can be clearly seen.

Many of OPT's advantages derive from its being an optical technique. The fact that optical staining techniques (such as immunohistochemistry) are sophisticated enough to pinpoint tissues expressing one specific gene underlies why OPT is so suited to morphology analysis: The precise shape of an organ can be discovered without having to dissect it away from its neighboring tissue. At a more specific level, OPT also displays advantages over fluorescent optical techniques such as confocal microscopy because, in addition to handling fluorescent signals, it can also image the colored stains used in mRNA gene expression analysis.

WHAT IS OPT PARTICULARLY GOOD FOR?

OPT is an evolving technology, so it would be premature to define a complete list of its applications; however, it is useful to list the major areas in which OPT is proving to be effective. One of OPT's strengths is its ability to provide an overview of the whole specimen. From a practical point of view, two kinds of imaging role can be distinguished: (1) providing the 3D shape of a structure or collection of structures and (2) recording the complete distribution of a signal throughout the whole specimen.

Phenotyping

One of the most important applications for OPT is to help analyze mutant phenotypes. Although the scientific community is producing mouse mutants at an increasing rate, is widely recognized that our ability to define what has "gone wrong" at a morphological level is often inadequate. A common conclusion is that many of these mutants are in fact phenotypically normal or that many genes do not have critical roles in the development of all the tissues they are expressed in. This may be true in many cases; however, we have found that the ability to generate 3D reconstructions of a particular organ relatively fast for a number of specimens (e.g., by immunohistochemistry) often highlights morphological defects that were never spotted before. A particular advantage of examining organ shapes interactively on a computer screen is that differences are much easier to spot than when examining histological sections. Features that appear insignificant on a 2D section may become obviously important once seen in the context of the surrounding 3D tissue. The result is that differences between wild-type and mutant phenotypes are easy to detect, even for researchers with little anatomical training.

A Deeper Understanding of Normal Development

Vertebrate embryo development is a complex, dynamic process involving the growth and transformation of 3D shapes over time. Although embryological studies now span more than a century of research, and we know a lot about the topological connectivity of developing organs, the true 3D shape of these structures is often still not well understood. The ability to create 3D reconstructions of labeled organs relatively easily should allow various areas of morphology to be more fully studied, including embryological time courses with a high temporal resolution, comparative embryology to explore phylogenetic relationships, and the variability inherent in "normal" development.

Analyzing Gene Expression Patterns at the mRNA or Protein Level

OPT is particularly useful for exploring the expression patterns of genes. This can be done at the RNA level (using whole-mount in situ hybridization) or at the protein level (using immunohistochemistry). At its current speed, it allows a given project to analyze the complete patterns (i.e., of the whole embryo) of many more genes than is possible using physical sectioning approaches. Although it does

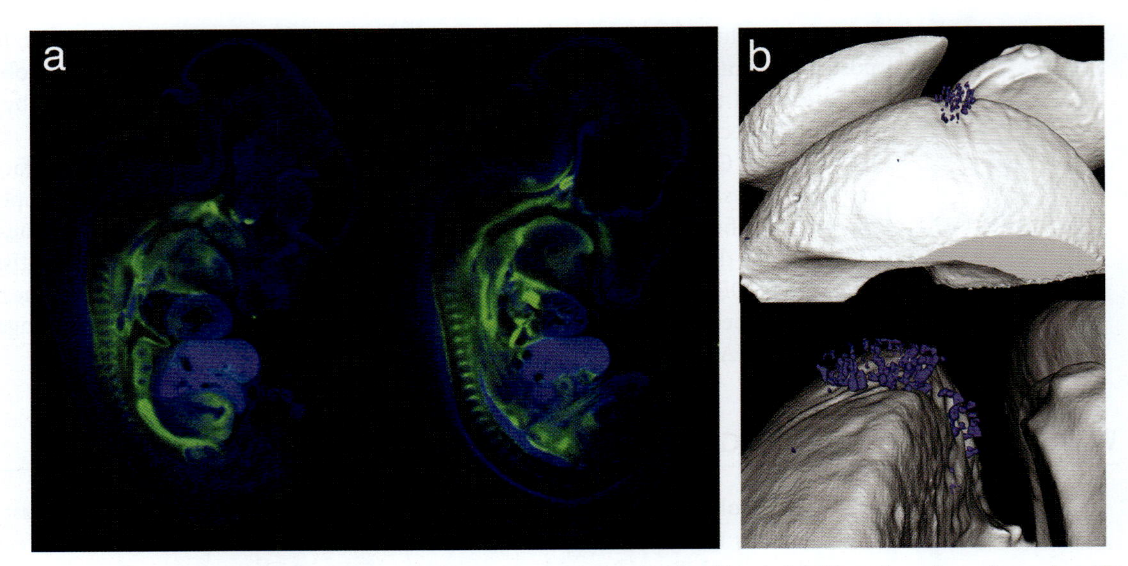

FIGURE 2. OPT imaging of LacZ-expressing tissue, stained using X-gal. (*a*) Virtual sections through a 3D OPT image of a transgenic 12.5-dpc mouse embryo. Expression of the *lacZ* reporter gene (*green*) can clearly be localized within many internal organs of the embryo. Since *lacZ* was detected using X-gal (producing a blue precipitate), this signal could not be imaged by confocal. (Reproduced with kind permission of Sharon Mulroy and Richard Sanford.) (*b*) 3D rendering of a small clone of LacZ-expressing cells in the developing brain of a 12.5-dpc mouse embryo (see Wilkie et al. 2004).

not provide subcellular resolution, a project designed to assess the patterns of 50 specimens (e.g., 10 genes at 9.5 dpc, 10.0 dpc, 10.5 dpc, 11.0 dpc, and 11.5 dpc) would be quite feasible using OPT. This would require the cutting of ~10,000 sections using more traditional approaches.

As with endogenous gene expression patterns, transgenic reporter constructs can also be analyzed with OPT. Figure 2 shows the result of OPT imaging a LacZ-expressing embryo.

Exploring the Distribution of Labeled Subpopulations of Cells

Another useful function of OPT is the ability to localize labeled cells within the embryo. Figure 2b shows an example of a small clone of LacZ-expressing cells in the developing brain of a mouse embryo (Wilkie et al. 2004). The high resolution clearly shows the distribution of the cells in the medial part of the developing brain. Seeing such images in their true 3D shape is important—the results from these experiments were initially analyzed in 2D sections, and some of the early conclusions had to be modified upon viewing the OPT results. In general, if a colored stain is used (such as X-gal staining of the LacZ-expressing cells shown here), small clusters of cells and, in some cases, single cells will be detectable. If a fluorescent dye is used (such as by fluorescent immunohistochemistry), single cells can be clearly identified.

PRACTICAL ISSUES OF USING OPT FOR VARIOUS TYPES OF EXPERIMENTS

There are two general points to make about preparing specimens for OPT imaging. The first is that trying to image a specimen that has both a colored stain (e.g., the blue precipitate from X-gal staining) and a fluorescent signal (e.g., fluorescent immunohistochemistry) may not work well. Although it is indeed useful in many cases to combine fluorescent and bright-field *imaging modes* for a single specimen (as described below), it is not recommended to combine the two types of *staining*. Since the colored stain will absorb some of the light emitted from the fluorescent dye, the results produced may contain artifacts. This problem may be improved algorithmically in the future; information from the bright-field images should make it possible to estimate how much of the fluorescent light was absorbed by the colored stain.

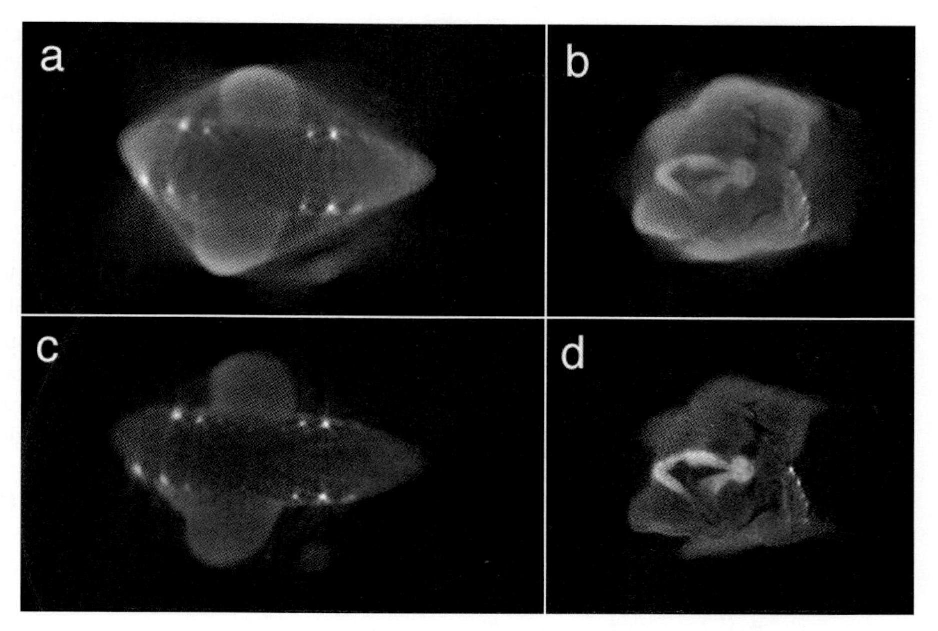

FIGURE 3. Reconstruction artifacts caused by an overstained specimen. Two virtual sections from a specimen imaged by bright-field OPT. (*a,b*) The overstaining of this specimen means that both the signal and the background levels are too high. The appearance of the reconstruction is fuzzy and the edges of the specimen are not clear. (*c,d*) The best way to improve the results is to stain the specimen more weakly. However, a way to reduce the problem for specimens that have already been overstained is to use a longer wavelength. (*Top panels*) Imaged with white light; (*bottom panels*) imaged with light above 700 nm. This improves both the clarity of the tissue (*c*) and the detection of the gene expression pattern (compare *b* with *d*).

The second point is that the optimal intensity for a good OPT scan of a colored stain is usually weaker than the best staining for visual inspection. This means that OPT is more sensitive to low-level expression, which can be an advantage. However, it also means that, when staining a specimen such as a LacZ transgenic embryo with X-gal, when the staining looks good "by eye," it will often be too dark for OPT. If not enough light passes through all orientations of the embryo, the algorithm is unable to produce an accurate reconstruction. Consequently, the best results for bright-field OPT tend to be from specimens that were specifically prepared with this technique in mind.

Histology/Anatomy of Unstained Specimens

For mouse embryos, 13.5 dpc or younger, the best results are obtained using "autofluorescence." This may be due as much to the fluorescence of the fixative as to fluorescence of the tissue itself. Fluorescence images work better for small specimens because (1) fluorescence OPT generates higher-resolution images than bright-field OPT and (2) in small, unstained embryos, there tends to be insufficient optical density to absorb much light during bright-field imaging. However, as specimens become larger, or more differentiated, more light is absorbed by the tissue, and optimal technique shifts away from autofluorescence imaging and toward the bright-field approach. As specimens become larger still, the choice of wavelength for bright-field illumination becomes important. Shifting toward the infrared part of the spectrum (e.g., with a wavelength up to 800 nm) can significantly improve the imaging depth (see Fig. 3).

Immunohistochemistry: Protein Patterns

The optimal way to explore protein expression patterns is to use fluorescently tagged antibodies. As mentioned above, fluorescence imaging generates higher resolution than bright field, and, in addition, it appears to be impossible to "overstain" the embryo. Increasing the intensity of a colored stain, by definition, increases the opacity of the tissues, whereas increasing the strength of a fluorescent dye does not. Whole-mount immunohistochemistry can be performed on large embryos and even adult tissues. However, working with larger or more differentiated tissues tends to increase problems of background fluorescence. In some cases, this can be overcome by bleaching the tissue. Penetration of antibody staining may also be particularly problematic in certain tissues (e.g., skin) and appears to vary from antibody to antibody.

In Situ Hybridization: "Gene Expression Patterns"

The most widely used assay for whole-mount in situ hybridization (WMISH) relies on an alkaline phosphatase enzyme to catalyze the conversion of BCIP/NBT to a purple precipitate (Hammond et al. 1998). In this case, it is not the OPT imaging that limits the size of the specimen, it is the staining technique itself. Apparently, WMISH cannot penetrate all regions of an intact 12.5-dpc embryo. Since the technique is limited to embryos of 11.5 dpc or younger, the best OPT results will come from combining bright-field imaging (for the in situ pattern) with fluorescent imaging (for the histology/anatomy of the specimen). This is similar to the LacZ results shown in Figure 2. The main complication for WMISH/OPT imaging is achieving the optimal staining level. For best results, multiple staining assays should be carried out in parallel, varying the concentration of the final substrate (or temperature/incubation time) and the staining (see Fig. 3).

Transgenic Reporter Constructs

Because LacZ staining encounters fewer penetration problems than WMISH (described above), it can be employed for larger specimens, including adult organs. For smaller embryos, the best results are obtained by treating the specimen as for immunohistochemistry, using a fluorescently labeled anti-β-galactosidase antibody to benefit from the advantages of fluorescent imaging mentioned above. However, if the X-gal method (blue precipitate) is required, good results can also be obtained, especially if the staining is not too strong. Again, a number of different staining intensities should be explored, and the fact that weak staining may wash out of the specimen while in benzyl benzoate:benzyl alcohol (BA:BB) should be borne in mind. (In rare cases, the specimen must be imaged as soon as the BA:BB has equilibrated within it.) As in the case of WMISH, the best results may be achieved by combining bright-field imaging (to detect the X-gal pattern) with fluorescent imaging (to capture the histology of the specimen). However, since X-gal can work on quite large specimens (e.g., some adult mouse organs), there may be cases where only a bright-field scan is required as the higher opacity of the tissue will give poor results with a fluorescent scan.

Green fluorescent protein (GFP) is another reporter gene of interest to many researchers. Unfortunately, it appears that the standard GFP is not stable enough for its fluorescence to persist in alcohols. As a result, good-resolution OPT images of the GFP signal in transgenic tissues have not been obtained. It is possible that, in the future, newly engineered versions of these proteins may be more stable in alcohols.

Lypophilic Dyes: Tracing of Axons and Cell Movements

Currently, tracing of axons and cell movements using lypophilic dyes is a problematic area for OPT. The achievement of high-resolution images requires a good index-matching solution such as BA:BB. Since these solvents disrupt lipid membranes, lypophilic dyes tend to rapidly dissolve out of the specimen. The discovery of alternative solvents for use with OPT, or cross-linking techniques for the dyes, could possibly overcome this limitation in the future.

FUTURE PROSPECTS

So far, OPT has proven useful in the major areas described above: analysis of morphology, especially for phenotyping, and the recording of complete 3D gene expression patterns at the level of both mRNA and protein. These will probably remain the key applications in the short term; however, technological improvements may open up some new areas as well.

In particular, through improvements to the imaging mode (e.g., laser scanning instead of wide-field imaging), it may be possible to record the development of embryos or organs in culture. This would allow us to follow the dynamics of a 3D gene expression pattern as it changes during development in, for example, a mammalian embryo (using transgenic fluorescent protein reporter genes). Another development, which could contribute to this goal, is multispectral imaging, in which the intensity of fluorescently emitted light is recorded for many different wavelengths. This approach, which has been applied to other fluorescent imaging techniques, can allow multiple fluorochromes to be imaged within the same specimen.

Finally, projects such as the Edinburgh Mouse Atlas Gene Expression Database are creating a framework within which information about the spatial distribution of thousands of genes can be stored (http://genex.hgu.mrc.ac.uk). This will be an invaluable resource for developmental biologists. Since OPT imaging is far more amenable to automation than traditional section-cutting approaches, it suggests for the first time that generating 3D expression data for all genes in the genome is conceivable. Such a resource would certainly be an exciting prospect for our ongoing attempts to understand developmental biology.

ACKNOWLEDGMENTS

I thank Harris Morrison for invaluable help in the on-going development of this imaging technique, Leonard Hay for continued help in designing the hardware, and Duncan Davidson for useful comments on this chapter. Development of OPT was funded by the MRC.

REFERENCES

Davidson D. and Baldock R. 2001. Bioinformatics beyond sequence: Mapping gene function in the embryo. *Nat. Rev. Genet.* **2:** 409–417.

Dhenain D., Ruffins S., and Jacobs R.E. 2001. Three-dimensional digital mouse atlas using high-resolution MRI. *Dev. Biol.* **232:** 458–470.

Foster F.S., Zhang M., Duckett A.S., Cucevic V., and Pavlin C.J. 2003. In vivo imaging of embryonic development in the mouse eye by ultrasound biomicroscopy. *Invest. Ophthalmol. Vis. Sci.* **44:** 2361–2366.

Hammond K.L., Hanson I.M., Brown A.G., Lettice L.A., and Hill R.E. 1998. Mammalian and *Drosophila* dachsund genes are related to the Ski proto-oncogene and are expressed in eye and limb. *Mech. Dev.* **74:** 121–131.

Kak A.C. and Slaney M. 1988. *Principles of computerized tomographic imaging.* IEEE Press, New York.

Louie A.Y., Huber M.M., Ahrens E.T., Rothbacher U., Moats R., Jacobs R.E., Fraser S.E., and Meade T.J. 2000. In vivo visualization of gene expression using magnetic resonance imaging. *Nat. Biotechnol.* **18:** 321–325.

Massoud T.F. and Gambhir S.S. 2003. Molecular imaging in living subjects: Seeing fundamental biological processes in a new light. *Genes Dev.* **17:** 545–580.

Midgley P.A. and Weyland M. 2003. 3D electron microscopy in the physical sciences: The development of Z-contrast and EFTEM tomography. *Ultramicroscopy* **96:** 413–431.

Ntziachristos V. and Chance B. 2001. Probing physiology and molecular function using optical imaging: Applications to breast cancer. *Breast Cancer Res.* **3:** 41–46.

Potter S.M., Fraser S.E., and Pine J. 1996. The greatly reduced photodamage of 2-photon microscopy enables extended 3-dimensional time-lapse imaging of living neurons. *Scanning* **18:** 147.

Schneider J.E., Bamforth S.D., Grieve S.M., Clarke K., Bhattacharya S., and Neubauer S. 2003. High-resolution, high-throughput magnetic paragraph sign resonance imaging of mouse embryonic paragraph sign anatomy using a fast gradient-echo sequence. *MAGMA* **16:** 43–51.

Sharpe J., Ahlgren U., Perry P., Hill B., Ross A., Hecksher-Sorensen J., Baldock R., and Davidson D. 2002. Optical projection tomography as a tool for 3D microscopy and gene expression studies. *Science* **296:** 541–545.

Streicher J., Donat M.A., Strauss B., Sporle R., Schughart K., and Muller G.B. 2000. Computer-based three-dimensional visualization of developmental gene expression. *Nat. Genet.* **25:** 147–152.

Weninger W.J. and Mohun T. 2002. Phenotyping transgenic embryos: A rapid 3-D screening method based on episcopic fluorescence image capturing. *Nat. Genet.* **30:** 59–65.

Wilkie A.L., Jordan S.A., Sharpe J.A., Price D.J., and Jackson I.J. 2004. Widespread tangential dispertion and extensive cell death during early neurogenesis in the mouse neocortex. *Dev. Biol.* **267:** 109–118.

Yelbuz T.M., Choma M.A., Thrane L., Kirby M.L., and Izatt J.A. 2002. Optical coherence tomography: A new high-resolution imaging technology to study cardiac development in chick embryos. *Circulation* **106:** 2771–2774.

CHAPTER 25

Imaging the Development of the Neuromuscular Junction

Mark K. Walsh[1] and Jeff W. Lichtman[2]

[1]*Wilmer Eye Institute, Johns Hopkins Medicine, Baltimore, Maryland 21287;* [2]*Department of Molecular and Cell Biology, Harvard University, Cambridge, Massachusetts 02138*

Although fixed tissues are often easier to analyze, many biological questions cannot be answered by experimentation on dead tissue. The main problem is that the single-time-point data that one obtains from fixed material do not provide insights into dynamic events (Lichtman and Fraser 2001). This is a particular problem in studies of the development of synaptic circuits in mammals. Because individual neurons make unique connections and these are thought to be guided by the particular experience of an animal, it is likely that populations of neurons are not developing in temporal lockstep. Thus, at any single time point there is a range of states. Deciding how to order these, how long a cell spends in any particular state, and whether some of these states are reversible cannot simply be approached without time-lapse imaging. The authors' interest has centered on the developing neuromuscular junction, where several axons from different neurons temporarily co-occupy the same neuromuscular junction (Sanes and Lichtman 1999; Lichtman and Colman 2000). The aim has been to understand the dynamic changes that lead to the removal of all but one axon from each junction. This chapter describes in vivo time-lapse methods to study this phenomenon with epifluorescence microscopy.

TECHNOLOGICAL ADVANCES

XFP Mice

Due to the close proximity of the synaptic terminals of different motor neurons at each neuromuscular junction during the period of multiple innervation, motor neuron inputs must be differentially labeled in a way that permits them to be distinguished from one another. This is an interesting example of resolving details that are below the resolution of light microscopy by using more than one color. Labeling two axons that innervate an individual neuromuscular junction with different colors has been accomplished with activity-dependent dye uptake (Lichtman et al. 1985; Barry and Ribchester 1995) and selective lipophilic dye labeling of different axons with different colored dyes (Balice-Gordon et al. 1993; Gan and Lichtman 1998). Unfortunately, these techniques do not work well when the aim is to follow structures over relatively long periods (days, weeks) as the dyes move into other compartments and spread to other cells. This technical problem is solved by transgenic expression of cytoplasmic green fluorescent protein (GFP) or its spectral variants yellow fluorescent protein (YFP) and cyan fluorescent protein (CFP) in subsets of motor neurons under the regulatory control of the *thy-1* promoter (Feng et al. 2000).

For reasons that are not entirely clear (see discussion in Feng et al. 2000), some of the XFP transgenic lines of mice express fluorescent protein in only a subset of neurons serving a single function. For example, only several of the α motor neurons projecting to any particular muscle express YFP in the “YFP-H” line. Due to the small numbers of neurons that express in some of these subset lines, the entire axonal projections of single neurons can be analyzed. Unlike a Golgi stain, however, this approach has two important advantages: First, it is a vital label so the same axon can be studied over time and, second, this labeling can be combined with other fluorescent dyes, providing a degree of spatial resolution not possible with one dye (see below).

Imaging

Vital fluorescence imaging places demands on detectors. Tissue movement and phototoxicity (at least in principle) associated with long exposures require that the camera have the sensitivity to retrieve an acceptable image with short exposure times and moderate to very low fluorescence emission. These constraints can partially be overcome by choosing objectives with high numerical aperture (NA) and carefully matching the camera pixel size and the resolution limit so there is no empty magnification in the image. Last, when dealing with living animals, time is of the essence, because the duration of anesthesia has a direct effect on the postoperative health of the subject. For this reason, automation of as much of the acquisition as possible is advisable. The methods described here aim to allow imaging of multiple fluorophores at multiple z-depths to be carried out as quickly as possible.

Several descriptions of in vivo techniques for the neuromuscular junction have already been written (see, e.g., van Mier and Lichtman 1994; Balice-Gordon 1997). This chapter concentrates on those issues that pertain to very young animals and newer acquisition techniques.

METHODS

Double Transgenic Mice and Multiple-Color Neuromuscular Junctions

Transgenic mice that express spectral variants of cytoplasmic GFP in motor neurons can be used to separately label different axons with different colors in the same mouse (Feng et al. 2000). For example, crossing a transgenic line that expresses one color in all motor neurons with another line that expresses a spectrally distinguishable second color in only a small subset of motor neurons produces “double” transgenic mice in which individual neuromuscular junctions can contain two competing and differentially labeled neuronal inputs (Walsh and Lichtman 2003). Alternatively, two mice, each expressing a different fluorophore in subsets of neurons, can be crossed. This latter approach has the advantage that the deployment of two axonal arbors can be compared; the disadvantage is that many animals must be screened to find an appropriately labeled muscle (Kasthuri and Lichtman 2003). The former approach is more efficient for living animals, although even with this technique many animals must be screened to find individuals in which the stochastic subset expression was useful. Various transgenic lines that express either CFP (lines CFP-D or CFP-5 or CFP-23) or YFP (YFP-F or YFP-16) in all motor neurons (100% lines) can be crossbred with mice that express GFP or YFP (GFP-S, YFP-H) or CFP (CFP-S), respectively, in subsets of motor axons (subset lines). In this way, double transgenic mice in which all motor axons express one fluorescent protein and a subset of motor axons additionally express a second spectrally distinct color are generated. The aim was to find individual multiply innervated neuromuscular junctions during early postnatal life, in which two axonal inputs are labeled different colors. In addition, a red-shifted fluorescently conjugated α-bungarotoxin (e.g., Alexa Fluor 594 α-btx, Molecular Probes) was used to label the high-density acetylcholine receptors (AChRs) in the postsynaptic membrane with a third color (Fig. 1). Several results indicate that the time course, stages, and outcome of synapse elimination are not affected by the expression of fluorescent proteins in neurons in these transgenic mice (Walsh and Lichtman 2003).

Note: Three of the above mentioned transgenic lines are commercially available through The Jackson Laboratory (http://jaxmice.jax.org):

YFP-16: 100% motor neuron line; strain name = B6.Cg-TgN(Thy1-YFP)16Jrs

YFP-H: subset line; strain name = B6.Cg-TgN(Thy1-YFPH)2Jrs

CFP-23: 100% motor neuron lines; strain name = B6.Cg-TgN(Thy1-CFP)23Jrs

In Vivo Time-lapse Imaging

An accessible neck muscle, the sternomastoid, in living anesthetized double transgenic animals was chosen for this method. This muscle has unusually large junctions, and it is accessible with a midline (bloodless) incision in the ventral neck and requires retraction of only the overlying submandibular and sublingual salivary glands to obtain unfettered access to the muscle surface. In addition, this muscle is not covered with a thick connective tissue sheath as are some limb and back muscles, thus making optical access quite good. The studies carried out in the Lichtman lab begin with screening muscles in living mice at the beginning of the second postnatal week for two-color neuromuscular junctions. It is necessary to wait until the second week because, in the subset transgenic lines, fluorescent protein intensity increases with postnatal age. In addition, perioperative mortality rates decrease with increasing age.

Note: To effectively assess synaptic morphology, neuromuscular junctions must be viewed en face. This is because optical resolution in the light microscope is severalfold worse in the depth dimension. In addition, scatter from overlying tissue reduces the resolution of deep structures, so superficial junctions should be imaged preferentially. Last, for the authors' studies, the two inputs must be differentially labeled. These constraints impose a low yield on this experiment. The most serious limitation comes from the requirement for multiple colors at one junction. In the authors' experience, the subset mouse lines typically have, throughout the body, only a few labeled motor axons in the second postnatal week. For example, at postnatal day 8 (PND 8), in the YFP-H subset line, only 10% of sternomastoid muscles have axons that intensely express fluorescent protein (Keller-Peck et al. 2001). Often those axons are too deep for effective imaging using standard epifluorescence. Therefore, many (sometimes dozens) animals may need to be screened to find multi-innervated, multicolored, superficial en face neuromuscular junctions.

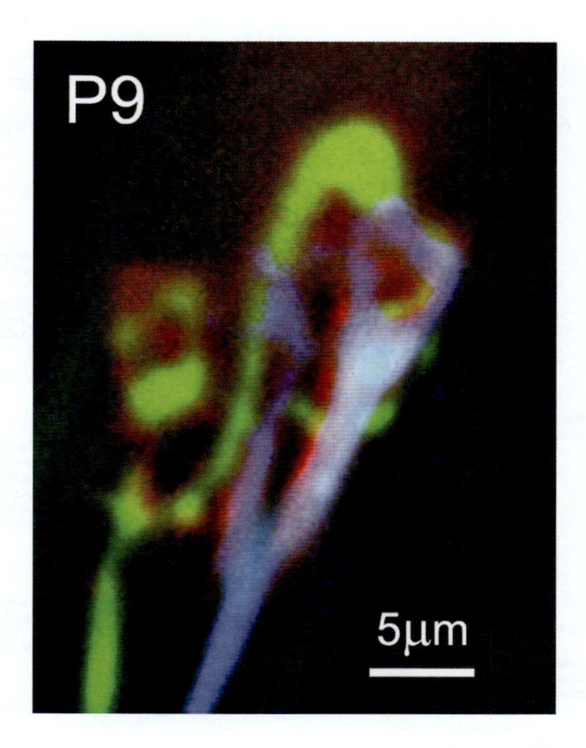

FIGURE 1. Multi-innervated, multicolored neuromuscular junction. This is from a living PND 9 double transgenic mouse pup in which all of the axons express YFP (*green*) and only a small subset of axons express CFP (*blue*). The postsynaptic acetylcholine receptors are labeled with Alexa Fluor 594 α-bungarotoxin (*red*).

PROCEDURE

Imaging the Development of the Neuromuscular Junction

1. For in vivo studies, handle the mouse pups with latex gloves to avoid rejection/cannibalization when returned to parents at the end of the imaging session.
2. Anesthetize neonatal mouse pups, between PND 7 and 10; inject individuals (intraperitoneal or subcutaneous above the lower back) with ketamine/medetomidine pup cocktail <!> (dose is ~0.1–0.2 ml /5 g of body weight) using a 1-ml 26G 3/8 Tuberculin syringe.

   Note: Ketamine-medetomidine pup cocktail is a 1:5 dilution of adult cocktail in saline (1 ml adult cocktail:4 ml saline). To prepare adult cocktail, mix 0.75 ml of ketamine (100 mg/ml) with 1 ml of medetomidine (1 mg/ml) and then add 7 ml of saline. This mixture contains 8.57 mg/ml of ketamine and 0.114 mg/ml of medetomidine.
3. Once anesthetized (usually by 2–3 min postinjection), place each pup supine on a piece of gauze on a magnetized stainless steel platform (~10 cm x 20 cm x 1–2 mm). See Figure 2 for details of setup.

   Note: The platform should have two holes drilled into one end so that it may be fastened with screws to the microscope stage. It should also be fitted with a thin heating element (~5 cm x 10 cm x 4 mm; e.g., custom-sized heating tape from Cole-Parmer Instrument) secured with a small magnet to the middle of the stage. Underneath the heating element, between the heating tape and the steel platform, a temperature probe (e.g., Cole-Parmer Instrument, probe must be compatible with the below-mentioned temperature controller) is secured.
4. Secure the gauze at the proximal end (closest to experimenter) by wrapping a thin rubber band around the entire steel platform. Secure the mouse pup's head by passing the rubber band through its mouth.
5. Secure the pup's paws using rubber bands with small plastic cuffs that can be cinched down around the ankles of the animal. Anchor the rubber bands with small magnets. Connect the heating element and temperature probe to a temperature controller (e.g., Digi-Sense Temperature Controller, Cole-Parmer Instrument) and set the thermostat to 35°C.
6. Place a thermometer between the body of the animal and the gauze pad to ensure that overheating (>37°C) does not occur.

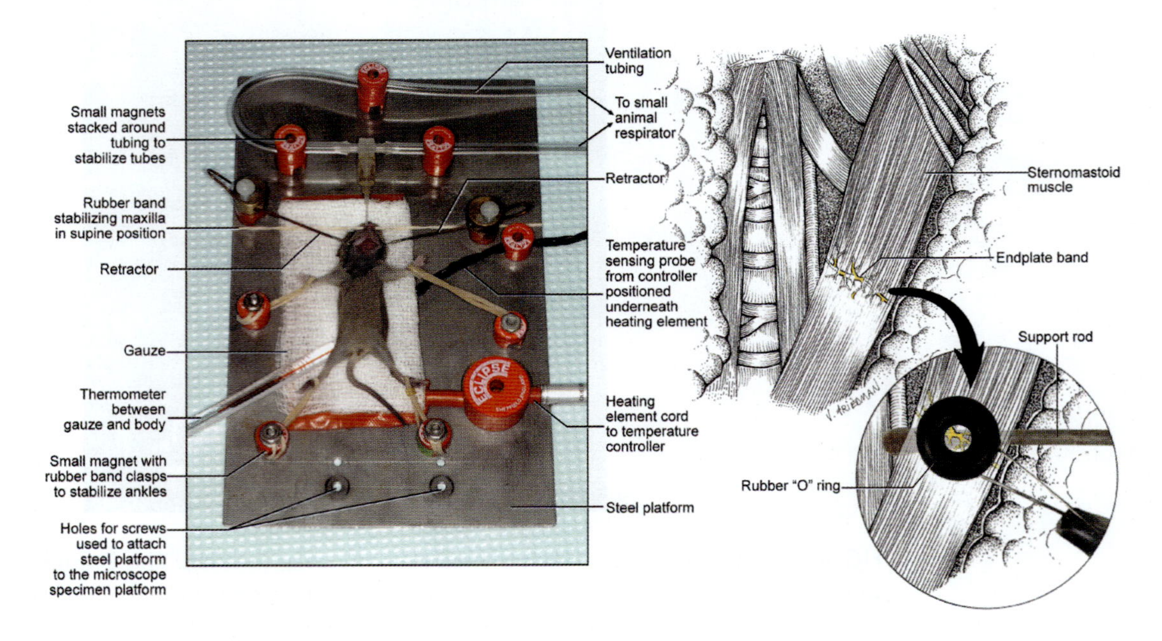

FIGURE 2. Setup for in vivo imaging. See the Procedure for details.

Intubation

7. Because ketamine-medetomidine anesthesia depresses respiration, it is necessary to mechanically ventilate the unconscious animal. Intubate the animal with a small, slightly curved, polyethylene tube (~600 μm in diameter and about 1 cm long).

   Note: The tube must be of sufficient firmness both to hold a curve upon molding and not to kink easily.

8. Connect the tube via a series of small connecting tubes of increasing diameters to the T-connector of a small animal respirator (e.g., MiniVent Type 845, Hugo Sachs Electronik).

   Note: The entire intubation tube, with connecting tubes of increasing diameter, is ~2 cm in length.

9. Set the respirator at ~80–100/min with a respiratory volume of ~200–300 μl.

10. Once the mouse has been successfully set up on the respirator, remove the hair on the ventral region of the neck using a commercial depilatory cream.

11. Disinfect the region by swabbing with Betadine and then 70% ethanol.

12. Using aseptic technique (dissecting tools should be sterilized using a heated bead sterilizer, make a superficial 1–1.5-cm midline incision in the ventral neck.

    Note: The midline incision is bloodless because few vessels cross the midline.

Intubation Troubleshooting

Routine intubation of mouse pups requires practice. The intubation tube should be checked prior to use for patency and lubricated by dipping it in a drop of water and making sure it can blow appropriately sized bubbles (200–300 μl). If necessary, the pup's tongue can be held to the left side of its mouth with a pair of blunt serrated forceps. With the tip of the intubation tube curving up toward the ceiling, insert it into the mouth of the animal (ventral side up). Insert the tube into the trachea while the ventilator is working. A slight scooping motion that follows the curvature of the intubation tube may facilitate intubation. Some slight resistance to insertion is possible, but do not force the tube. If blood is drawn, too much force has been exerted. Successful intubation is demonstrated by symmetric chest expansion synchronized to the ventilator pump. If, after several attempts, there is no success, re-wet and recheck the intubation tube for patency and reposition the mouse pup's head and neck. If the tube becomes clogged (it usually does become clogged over the course of weeks), unclog it with a 26G 3/8 Tuberculin syringe or similar fine needle. Once the animal is sufficiently anesthetized, it should no longer attempt to breathe spontaneously.

13. Using blunt dissection (i.e., pulling rather than cutting), separate the salivary glands in the midline. Retract the overlying salivary glands laterally to expose the sternomastoid muscle (see Fig. 3).

    Note: If necessary, the salivary glands may be retracted using wire retraction hooks attached to small magnets.

14. Label the postsynaptic sites at neuromuscular junctions by applying a 0.5 μg/ml solution of Alexa Fluor 594 α-btx<!> (a red-shifted fluorescently conjugated α-bungarotoxin available from Molecular Probes). After 2 min, rinse the area with 30 ml of physiological lactated Ringer's solution (Baxter Healthcare).

    Note: Because the toxin competes with acetylcholine, it is important to use a low dose. Approximately 3% of the AChRs (determined by quantitative fluorescence) are labeled using this approach.

15. After rinsing, fill the neck with Ringer's solution and screen both the right and left sternomastoids for superficial neuromuscular junctions of interest using an epifluorescence microscope (e.g., Nikon Eclipse E800); the screening is done using highly filtered light and CCD imaging. The authors use a cooled CCD camera (e.g., MicroMAX:512BFT, Roper Scientific Princeton Instruments) that has a thinned, back-illuminated chip, and long working distance, dipping cone, water-immersion objectives (e.g., Nikon 10x, 0.3 NA and 60x, 1.0 NA water objectives).

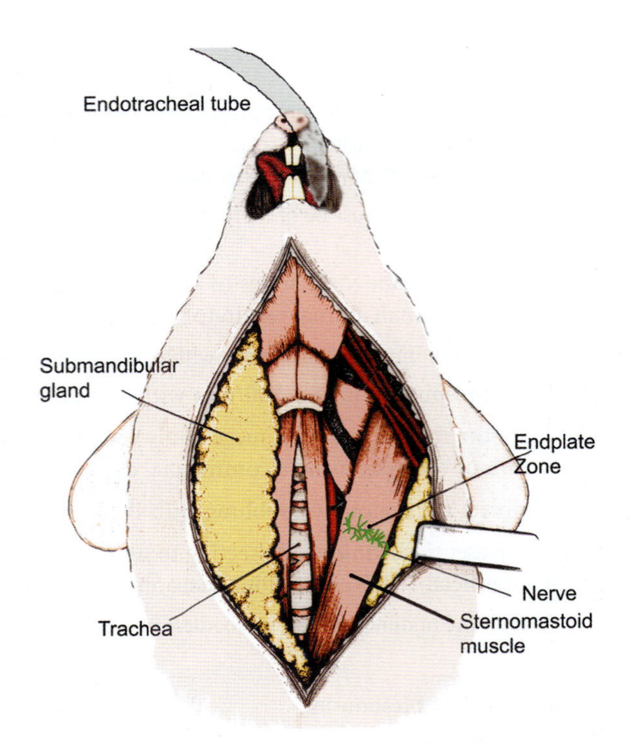

FIGURE 3. Schematic of the exposed sternomastoid muscle of an intubated mouse pup. Retraction of the overlying glands reveals the superficial sternomastoid muscle. The neuromuscular junctions are clustered toward the middle of the muscle in the endplate band, near where the nerve enters the lateral part of the muscle. (Adapted, with permission, from Lichtman et al. 1987 [©Society for Neuroscience].)

16. After imaging, prepare mouse pups for high-resolution imaging by slightly elevating the sternomastoid. This can be achieved by placing a small polished support rod (~1 mm x 3 mm; made by flattening the end of a small stainless steel rod of ~700 µm diameter) under the muscle.

    Note: The support can be attached to a small manipulator on a magnetic base to make adjustment of the platform easier. The authors also place a small rubber O-ring (~4 mm external diameter with ~1.5–2 mm internal diameter attached to a small manipulator) gently on the surface of the muscle around the endplate band (in the middle of the muscle). This helps to stabilize the muscle and provide a water well for the immersion long-working-distance objectives.

17. Excite the label by illuminating the tissue with a xenon arc lamp and rapidly switching between exciting filters (Lambda DG-4, Sutter Instrument).

18. Acquire low-power images (10x, 0.3 NA water objective) of the endplate band (axons and receptors using single dichroic filters, e.g., Chroma Technology) using Metamorph imaging software (Universal Imaging).

    Note: These images will be used as maps for relocating individual junctions at the next imaging session.

19. Then image individual junctions at higher power (60x, 1.0 NA water objective) utilizing a piezoelectric objective Z-axis stepper (Physik Instrumente) to obtain stacks (usually 1-µm steps) of images. See below for details of imaging technique.

20. After the imaging session is complete, suture the wound using simple interrupted stitches (e.g., 9-0 Ethilon monofilament nylon, Ethicon).

21. Give the animal an intraperitoneal injection of anesthetic reversal agent, i.e., atipamezole (Antisedan, Orion) 5 µg per 5 g of body weight (or 0.02 ml of a 0.25 mg/ml solution in saline per 5 g of body weight).

    Note: This injection not only reverses the medetomidine, it also decreases the chances of anesthetic overdose.

22. To prevent dehydration, administer a subcutaneous injection of 0.1 ml of sterile lactated Ringer's solution.
23. Place the animal on a heated blanket in an oxygenated chamber until it has recovered sufficiently (usually less than 2 hr). Take care not to overheat the pup.
24. Before returning it to its parents, gently rub the back of the pup in the used bedding of its parents' cage; adding a bit of urine scent to the pup may help prevent its rejection.
25. Subsequent operations can be performed when the pup has returned to its regular feeding schedule (typically 1–2 days, possibly longer).
26. For subsequent operations, re-anesthetize the pup, locate the same neuromuscular junction (utilizing the low-power maps of the endplate band), and repeat the imaging procedure. Due to natural acetylcholine receptor turnover, relabeling of AChRs with new α-btx may be necessary after several days.

Caution: *See Appendix 3 for appropriate handling of materials marked with <!>.*

IMAGING

For three-color junctions, a triple dichroic (e.g., Chroma XF91) may be used to obtain all three colors at each image plane without having to switch the dichroic filter (the Lambda DG-4 was used to switch excitation light). This allows perfect alignment of the three channels. Separate stacks are sometimes obtained for each of the three colors using separate filter cubes (e.g., Chroma). Exposure times for most images are typically ~0.5 sec per plane. For these experiments, use the maximal intensity available with the fiber optic light source (Lambda DG-4) with minimal exposure times to reduce movement artifacts. The cubes are chosen to maximize separation of the various fluorophores: YFP and GFP (exciter-HQ500/20x, dichroic-Q515LP, emitter-HQ535/30m), CFP (exciter-D436/20x, dichroic-455DCLP, emitter-D480/40m), and red-shifted α-btx (exciter-D560/40x, dichroic-595DCLP, emitter-D630/60m). These dichroic filters easily separate fluorescent proteins from labeled AChRs. They also sufficiently separate CFP from YFP, and GFP from CFP, which is often problematic. The alignment of the images made with the single filter cubes is based on the triple-dichroic images, which have more bleed-through but are a useful reference. Two-dimensional images are then derived from stacks of images by extracting in-focus information from each image plane to create image mosaics (e.g., Adobe Photoshop). The Metamorph imaging software allows fully automated acquisition of multiple colored images at each image plane using various excitation filters and a single triple dichroic emission filter at multiple *z* planes, providing images through the entire depth of the neuromuscular junction. Exposure times, *z* step depths with beginning and ending *z* positions, and various excitation wavelengths are all preset in Metamorph before each acquisition.

Note: In most cases, movement from the respiration of the animal is transferred to the sternomastoid muscle. Such movement makes alignment of stacks difficult and can also cause blurring of individual frames. In these cases, we shut off the animal respirator during the acquisition of an image stack. If the animal is sufficiently anesthetized and respirated, one may have a window of nearly 30 sec before the animal begins to spontaneously breathe again. In some cases even with the respirator off, there is too much vibration/movement from the cardiac cycle to allow imaging of a particular junction. In these cases, manipulating the steel platform or O-ring sometimes helps. It is fair to say, however, that movement remains one of the most serious impediments to high-quality images in vivo.

IN VIVO IMAGING CONTROLS

This protocol is designed to avoid perturbation of normal development. For example, α-btx is known to block AChRs, so an extremely low dose (~3% of AChRs labeled) is used to label the postsynaptic AChRs. Given the large safety factor of the neuromuscular junction, even in neonates (>40% of AChRs must be blocked to disrupt transmission; Wareham et al. 1994), this dose is not expected to block postsynaptic activity. In addition, the anesthetics used are thought not to affect either motor neuron excitability (Kakinohana et al. 2000) or structural dynamics. These were chosen over volatile anesthetics, for example, which disrupt actin-based dendritic mobility (Kaech et al. 1999). Despite these precautions, it does remain possible that the observations may be artifacts, induced by labeling or imaging procedures. Therefore, in control experiments, the in vivo imaging system was disrupted in a number of ways in order to determine how the system reacts to both phototoxic and mechanical insults (Walsh and Lichtman 2003). Nerve crush lateral to the muscle resulted in discontinuous or beaded fluorescence in axons within 24 hours, and most junctions were no longer apposed by any fluorescence (see also Nguyen et al. 2002). Muscle damage (caused by crushing the center of the muscle, which may also cause direct nerve damage) can result in long terminal and/or nodal sprouts, some of which can be greater than 100 μm in length and arise within 2 days) (see also Rich and Lichtman 1989; van Mier and Lichtman 1994). In the authors' experience, signs of such damage (denervation, beaded fluorescence, or long sprouts) are rarely observed (<5% of repeatedly imaged animals; mostly during the early days of technique refinement). When injuries were sustained, they were usually due to accidental trauma that was obvious at the time of infliction. It is important to realize that practice with in vivo imaging techniques improves the quality and reliability of the data obtained. Phototoxicity is another important impediment to the collection of good data. Various fluorophores have been tested by illuminating age-matched neuromuscular junctions in vivo at full intensity, continuously, for one hour with the appropriate excitation wavelengths (60x, 1.0 NA water-immersion objective). No signs of damage or phototoxicity (not even photobleaching) were seen either immediately or after 24 hours. This is in spite of the fact that the excitation is, in some cases, hundreds of times greater than that normally experienced by junctions during imaging. In addition, junctions that were imaged repeatedly showed no pre- or postsynaptic structural features in addition to those normally present in control junctions of similar developmental ages, imaged for the first time (Walsh and Lichtman 2003).

DISCUSSION

Although in vivo imaging is a powerful technique for studying mammalian synaptic development, there are several interpretive pitfalls that require some care. For example, analysis of two axon colors is complicated when studying the progeny of two crossed transgenic lines in which all axons express one color and a small subset expresses a second color. Interpretation of what axons are doing in this situation depends on whether one is looking at an axon that has the distinct color or an axon which expresses only the common color. For example, it is possible to conclude that an input axon which expresses the subset fluorophore has advanced into the territory of an axon expressing the common color. However, it is not possible to conclude that an axon expressing the common color has advanced into the territory occupied by an axon expressing both colors. The obvious way to circumvent this interpretative problem is to generate mice in which competing axons each express one color fluorescent protein exclusively (Kasthuri and Lichtman 2003), for example, by crossing two different subset-expressing lines. Unfortunately, the probability of finding subset expression using this technique is low and independent between the lines, so the size of the screening population required to find junctions of interest is dramatically increased. Several novel approaches to labeling individual motor neurons with varying colors are currently being investigated.

The current technology also suffers from a lack of tools for alignment and removal of out-of-focus light. Better deconvolution software that is both faster and easier to use would facilitate the production of 2D projections from stacks of images. As an alternative, this kind of developmental study could be done using a confocal microscope (S.G. Turney and J.W. Lichtman, in prep.). In addition, the current work on imaging axonal arbors and synaptic terminals for other types of neurons represents only the first attempts that have been made (see, e.g., Gan et al. 2003). In conclusion, although each in vivo preparation presents its own challenges, three major hurdles have been cleared: (1) survivability of young animals after multiple surgeries, (2) imaging both the pre- and postsynaptic sides of a synapse, and (3) imaging the deployment of different axons innervating the same postsynaptic cell. Recent advances in our own in vivo imaging attempts augur well for the future of this kind of work.

REFERENCES

Balice-Gordon R.J. 1997. In vivo approaches to neuromuscular structure and function. *Methods Cell Biol.* **52:** 323–348.

Balice-Gordon R.J., Chua C.K., Nelson C.C., and Lichtman J.W. 1993. Gradual loss of synaptic cartels precedes axon withdrawal at developing neuromuscular junctions. *Neuron* **11:** 801–815.

Barry J.A. and Ribchester R.R. 1995. Persistent polyneuronal innervation in partially denervated rat muscle after reinnervation and recovery from prolonged nerve conduction block. *J. Neurosci.* **15:** 6327–6339.

Feng G., Mellor R.H., Bernstein M., Keller-Peck C., Nguyen Q.T., Wallace M., Nerbonne J.M., Lichtman J.W., and Sanes J.R. 2000. Imaging neuronal subsets in transgenic mice expressing multiple spectral variants of GFP. *Neuron* **28:** 41–51.

Gan W.B. and Lichtman J.W. 1998. Synaptic segregation at the developing neuromuscular junction. *Science* **282:** 1508–1511.

Gan W.B., Kwon E., Feng G., Sanes J.R., and Lichtman J.W. 2003. Synaptic dynamism measured over minutes to months: Age-dependent decline in an autonomic ganglion. *Nat. Neurosci.* **6:** 956–960.

Kaech S., Brinkhaus H., and Matus A. 1999. Volatile anesthetics block actin-based motility in dendritic spines. *Proc. Natl. Acad. Sci.* **96:** 10433–10437.

Kakinohana M., Motonaga E., Taira Y., and Okuda Y. 2000. The effects of intravenous anesthetics, propofol, fentanyl and ketamine on the excitability of spinal motoneuron in human: An F-wave study. *Masui* **49:** 596–601.

Kasthuri N. and Lichtman J.W. 2003. The role of neuronal identity in synaptic competition. *Nature* **424:** 426–430.

Keller-Peck C.R., Walsh M.K., Gan W.B., Feng G., Sanes J.R., and Lichtman J.W. 2001. Asynchronous synapse elimination in neonatal motor units: Studies using GFP transgenic mice. *Neuron* **31:** 381–394.

Lichtman J.W. and Colman H. 2000. Synapse elimination and indelible memory. *Neuron* **25:** 269–278.

Lichtman J.W. and Fraser S.E. 2001. The neuronal naturalist: Watching neurons in their native habitat. *Nat. Neurosci.* (suppl.) **4:** 1215–1220.

Lichtman J.W., Magrassi L., and Purves D. 1987. Visualization of neuromuscular junctions over periods of several months in living mice. *J. Neurosci.* **7:** 1215–1222.

Lichtman J.W., Wilkinson R.S., and Rich M.M. 1985. Multiple innervation of tonic endplates revealed by activity-dependent uptake of fluorescent probes. *Nature* **314:** 357–359.

Nguyen Q.T., Sanes J.R., and Lichtman J.W. 2002. Pre-existing pathways promote precise projection patterns. *Nat. Neurosci.* **5:** 861–867.

Rich M. and Lichtman J.W. 1989. Motor nerve terminal loss from degenerating muscle fibers. *Neuron* **3:** 677–688.

Sanes J.R. and Lichtman J.W. 1999. Development of the vertebrate neuromuscular junction. *Annu. Rev. Neurosci.* **22:** 389–442.

van Mier P. and Lichtman J.W. 1994. Regenerating muscle fibers induce directional sprouting from nearby nerve terminals: Studies in living mice. *J. Neurosci.* **14:** 5672–5686.

Walsh M.K. and Lichtman J.W. 2003. In vivo time-lapse imaging of synaptic takeover associated with naturally occurring synapse elimination. *Neuron* **37:** 67–73.

Wareham A.C., Morton R.H., and Meakin G.H. 1994. Low quantal content of the endplate potential reduces safety factor for neuromuscular transmission in the diaphragm of the newborn rat. *Br. J. Anaesth.* **72:** 205–209.

CHAPTER 26

A Practical Guide: Imaging Synaptogenesis by Measuring Accumulation of Synaptic Proteins

Camin Dean and Peter Scheiffele

Columbia University, Department of Physiology and Cellular Biophysics, Center for Neurobiology and Behavior, New York, New York 10032

Synapse formation and modification involve the successive recruitment of pre- and postsynaptic signaling molecules (Scheiffele 2003). The aim of the method presented here is to investigate the function of specific synaptic factors in gain- and loss-of-function experiments by measuring protein accumulation at synapses using quantitative immunohistochemistry, confocal microscopy, and image analysis. The procedure is described for dissociated cultures of hippocampal neurons, but it can be extended to various neuronal cell types in culture.

MATERIALS

Endotoxin-free Preparations of Expression Vectors for Transfection Marker (e.g., EGFP), and cDNAs of Interest

In most primary neurons, more robust expression is observed with cellular promoters (e.g., β-actin or synapsin promoter), rather than viral promoters (such as cytomegalovirus or SV40).

Transfection Reagent

Lipofectamine 2000 (Invitrogen).

Markers

The following commercially available primary antibodies to synaptic proteins have proven to be useful markers: rabbit anti-synapsin I (Chemicon), mouse anti-VAMP2/synaptobrevin (clone 69.1, Synaptic Systems), mouse anti-bassoon (Stressgen), rabbit anti-GluR2/3 (Chemicon), mouse anti-PSD-95 (clone 7E3-1B8, Affinity Bioreagents), mouse anti-MAP2 (clone AP20, Roche).

Microscope

This procedure requires the use of a confocal microscope.

Image Analysis Software

Software must include thresholding and object analysis functions, e.g., Metamorph (Universal Imaging Corporation), Open Lab (Improvision), or Simple PCI (Compix).

Caution: *See Appendix 3 for appropriate handling of materials marked with <!>.*

PROCEDURE

Neuronal Cell Culture

1. Isolate dissociated hippocampal neurons from E18 rats according to standard protocols (Banker and Cowan 1977; Peacock et al. 1979).
2. Plate the cells at 25,000–100,000 cells/cm^2 on poly-L-lysine-coated glass coverslips (12-mm diameter, Carolina Biologicals) in 24-well plates. Maintain cells in 500 µl of Complete Medium (Neurobasal medium [Invitrogen], supplemented with 2 mM Glutamax, penicillin-streptomycin<!> [10 units/ml and 1µg/ml, respectively], and 2% B-27 supplement [Brewer et al. 1991]).

Transfection

3. Just prior to transfection, remove culture medium (save) and replace it with 400 µl of Complete Medium lacking antibiotics.
4. Prepare DNA–lipid complexes for transfection.
   a. Dilute 0.25 µl of Lipofectamine 2000 transfection reagent in 50 µl of Neurobasal Medium.
   b. Add 0.6 µg of DNA in 50 µl of Neurobasal Medium.
   c. Incubate the solutions separately for 5 min at room temperature. Then mix both solutions and incubate for 30 min.
5. Slowly add the lipid–DNA complexes to the culture dish. Incubate for 4 hours at 37°C, 5% CO_2. Replace medium with 50% conditioned medium (saved from cells prior to transfection) and 50% fresh Complete Medium.

   Note: Cells can be transfected at various stages in culture (1–24 days in vitro). A common transfection marker, such as enhanced green fluorescent protein (EGFP), should be used to cotransfect the cells along with the cDNA of interest. The marker will outline the morphology of the transfected cell and allow image collection and quantitation to be performed in a blinded fashion. There are batch-to-batch variations in the transfection reagent. The amount of Lipofectamine 2000 added per transfection should be titrated for optimal transfection efficiency and cell health.

Immunostaining and Image Acquisition

6. Fix and analyze cultures by immunohistochemistry 36–96 hours posttransfection.
7. Carry out immunostaining for all samples in parallel, under identical conditions.
8. Collect and analyze images blind with respect to the transfected molecule.

   Note: Pinhole settings should be kept constant at one Airey unit. Laser power and photomultiplier settings should be constant for all samples and should be set such that pixels representing the brightest structures of interest do not reach saturation. Synaptic puncta are generally 0.5–2 µm in diameter, depending on the protein immunostained and the fluorophore used.
9. For quantitation, generate a maximum projection image of eight to ten 0.5-µm *z*-sections for each of ten or more different cells, for the control and test conditions. Save images as tiff files (RGB mode). Remember to add a scale bar during image acquisition.

Image Analysis

10. For unbiased quantitation of synaptic protein accumulation, select synaptic puncta based on their size and staining intensity using the thresholding function of the image analysis software.
    a. Open tiff files in the analysis program.
    b. Select the color channel for the synaptic protein of interest. Set lower-intensity threshold such that diffuse nonsynaptic signals are excluded but synaptic protein concentrations are included (set the upper threshold to maximum brightness of 255; e.g., see Fig. 1). For all images analyzed, identical thresholding values must be used.

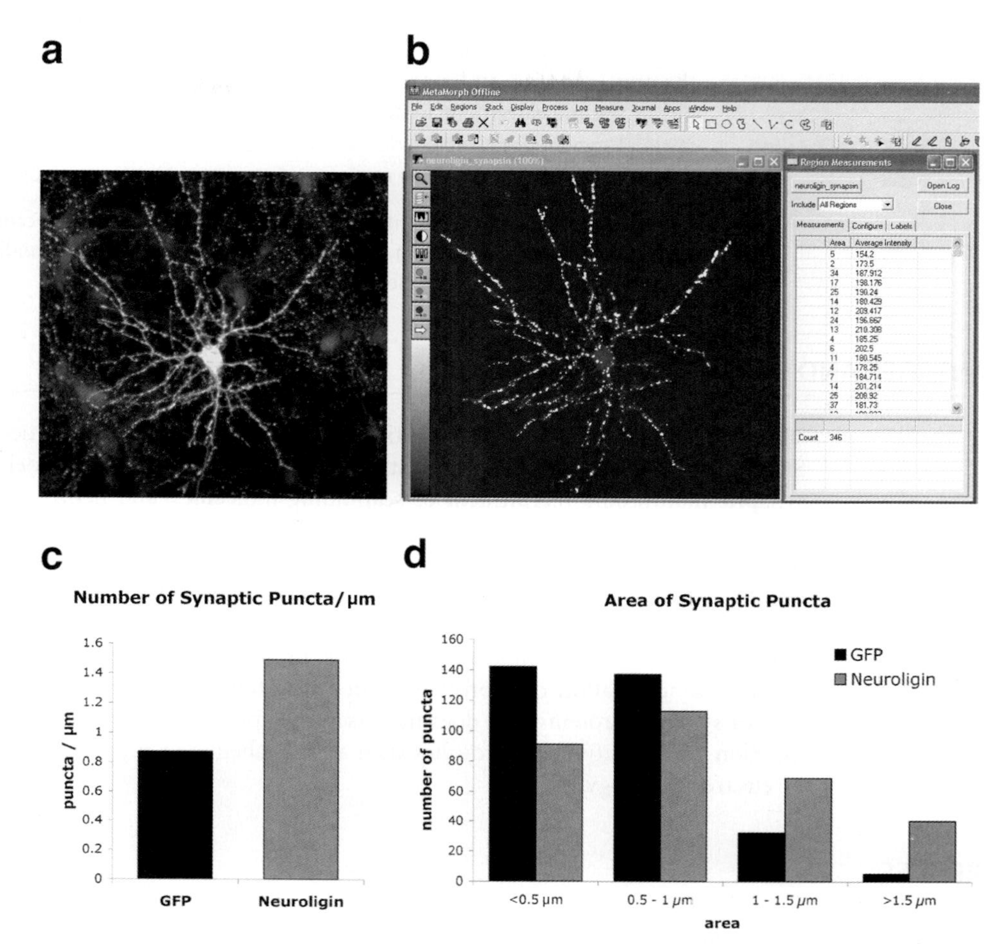

FIGURE 1. (*a*) An example of a confocal image of a hippocampal neuron transfected with the postsynaptic adhesion molecule neuroligin-1 and stained for the synaptic vesicle marker, synapsin. (*b*) The same image in Metamorph, showing the synaptic puncta on the transfected cell selected as regions, and the region measurements showing quantitation of number, area, and average intensity of puncta. (*c*) An example of a quantitation of the number of puncta per micrometer for a single EGFP and a single neuroligin-expressing cell. (*d*) Example of binning quantitative data on the area of synaptic puncta for two conditions, a neuroligin-1-transfected cell, and a EGFP-transfected control cell.

c. Create regions around the thresholded areas that denote synaptic puncta.

d. Measure the average intensity and the area of all regions. Exclude regions with an area smaller than 0.25 μm^2 to avoid scoring small bright areas or individual bright pixels that do not represent synaptic structures.

Note: Besides their concentration at synapses, some synaptic proteins show a brightly stained intracellular pool in the perinuclear region which must be excluded from analysis. This can be done either by excluding all thresholded regions larger than 5 μm^2 or by excluding all regions within a defined circular area above and around the nucleus of all transfected cells.

e. The raw data can be further analyzed and normalized in the following ways.

   i. Determination of number, area, and average intensity of synaptic puncta that are colocalized with the transfected cell: Threshold the signal of the transfection marker to select the entire area of the transfected cell. Subsequently select areas overlapping with the transfection marker using the colocalization function of the analysis software. The disadvantage of this approach is that because pre- and postsynaptic markers do not show complete overlap, parts of the area of synaptic puncta will be excluded and missed.

ii. Determination of density of synaptic puncta (areas) per length of axon or dendrite: This is the preferred readout for experiments. Highlight all areas above the size and intensity threshold in the measurement window. Subsequently, manually count the number of highlighted areas along the transfected neuronal processes and divide the number by the length of the process analyzed to obtain a measure of synaptic puncta per dendrite (or axon) length.

Note: Data obtained by any of the above paradigms can be grouped according to the number, size, and average intensity of puncta. Differences between experimental conditions can be most clearly detected by binning the data (see Fig. 1).

ADVANTAGES AND LIMITATIONS

This simple approach permits a quantitative analysis of protein recruitment to developing synapses. Simultaneous analysis of pre- and postsynaptic markers can be used to delineate sequential events in synaptic maturation, recruitment of scaffolding molecules, and neurotransmitter receptor subunits (Rao et al. 1998; Dean et al. 2003). The protocol described here can be combined with overexpression of wild-type and mutant proteins, application of pharmacological agents, and function-blocking antibodies, as well as suppression of gene expression by short interfering RNAs.

One limitation of this analysis is that the readout strongly depends on the sensitivity of the antibodies used for the immunostaining. Better immunological reagents will suggest higher apparent synaptic concentration of a protein. More importantly, this method assays the relative concentration of synaptic proteins, but does not assess synaptic function. It should therefore be used in conjunction with functional approaches such as FM-labeling or measurements of synaptic transmission by electrophysiology.

REFERENCES

Banker G.A. and Cowan W.M. 1977. Rat hippocampal neurons in dispersed cell culture. *Brain Res.* **126:** 337–342.

Brewer G.J., Torricelli J.R., Evege E.K., and Price P.J. 1993. Optimized survival of hippocampal neurons in B27-supplemented Neurobasal, a new serum-free medium combination. *J. Neurosci. Res.* **35:** 567–576.

Dean C., Scholl F.G., Choih J., DeMaria S., Berger J., Isacoff E., and Scheiffele P. 2003. Neurexin mediates the assembly of presynaptic terminals. *Nat. Neurosci.* **6:** 708–716.

Peacock J.H., Rush D.F., and Mathers L.H. 1979. Morphology of dissociated hippocampal cultures from fetal mice. *Brain Res.* **169:** 231–246.

Rao A., Kim E., Sheng M., and Craig A.M. 1998. Heterogeneity in the molecular composition of excitatory postsynaptic sites during development of hippocampal neurons in culture. *J. Neurosci.* **18:** 1217–1229.

Scheiffele P. 2003. Cell-cell signaling during synapse formation in the CNS. *Annu. Rev. Neurosci.* **26:** 485–508.

CHAPTER 27

A Practical Guide: Imaging Retinotectal Synaptic Connectivity

Susana Cohen-Cory

Department of Neurobiology and Behavior, University of California, Irvine, California 92697

This chapter describes a method for imaging synaptic sites in individual axon or dendritic terminals in live, developing embryos. By selectively targeting expression of green fluorescent protein (GFP)-tagged pre- and postsynaptic proteins to developing neurons in vivo, it is possible to visualize synaptic sites and to correlate their distribution and dynamics with changes in the morphology of axon or dendritic arbors labeled with a red fluorescent molecule.

The techniques described here may be applied to almost any neuronal circuit or neuron type, as long as the embryonic tissues are accessible for transfection and the neurons of interest are located at a depth that is within the working distance of the objective. These techniques have been optimized for *Xenopus* embryos, but they are applicable to other vertebrate systems such as zebrafish and chick embryos.

In the *Xenopus* tadpole, simultaneous visualization of neuronal morphology and synaptic specializations can be achieved by expressing fluorescently labeled proteins in individual neurons. Chimeric constructs coding for fluorescent and synaptic proteins are targeted to neurons just before they become postmitotic. Synaptobrevin II, a synaptic vesicle membrane protein, is a reliable marker for visualizing presynaptic sites, as GFP-tagged synaptobrevin is targeted to the axon terminal and is highly concentrated at synaptic terminals (Ahmari et al. 2000; Alsina et al. 2001). PSD-95, the major component of the postsynaptic specialization at glutamatergic synapses, can also serve as an accurate marker of developing synapses. GFP-tagged PSD-95 concentrates at postsynaptic sites and can be used to visualize postsynaptic specializations both in vivo (see Fig. 1 below) and in vitro (Marrs et al. 2001; Okabe et al. 2001).

MATERIALS

Plasmid Constructs

This protocol was optimized using chimeric genes coding for natural or enhanced GFP and the complete sequence of synaptobrevin II (gift from M.M. Poo) or PSD-95 (El-Husseini et al. 2000; gift from D. Bredt) as fluorescent reporters for pre- and postsynaptic sites in live *Xenopus* tadpoles. GFP-chimeric constructs were subcloned into pCS2+ expression plasmids (driven by the cytomegalovirus promoter) (Turner and Weintraub 1994). pDsRed2 (Clontech) is used for fluorescently labeling neurons and visualizing dendritic and axon arbor morphologies.

Tadpoles

Tadpoles are raised in rearing solution (60 mM NaCl, 0.67 mM KCl <!>, 0.34 mM $Ca(NO_3)_2$, 0.83 mM $MgSO_4$, 10 mM HEPES at pH 7.4, 40 mg/l gentamycin) and anesthetized with 0.05% tricane methanesulfonate <!> (Finquel, Argent Laboratories) during experimental manipulations. Transfections are carried out on the retinal or brain neuroepithelium of stage-20–24 embryos. Developmental stages were determined according to the method of Nieuwkoop and Faber (1956).

Laser-scanning Confocal Microscope

A laser-scanning confocal microscope equipped with argon (488 nm excitation; 10% neutral density filter) and HeNe (543-nm excitation) lasers <!> is suitable for these studies. 515/30-nm (barrier) and 605/32-nm (band-pass) filters allow separation of signals obtained at the two wavelengths.

Image Analysis Software

This protocol uses the Metamorph image analysis software (Universal Imaging) to analyze individual optical sections from confocal stacks. Image analysis programs with similar capabilities may also be appropriate.

Caution: *See Appendix 3 for appropriate handling of materials marked with <!>.*

PROCEDURE

Imaging Retinotectal Connectivity in *Xenopus laevis* Tadpoles

Plasmid Lipofection

1. Mix the GFP-tagged (1 μg/μl) plasmid DNA to be used for DNA lipofections with an equimolar amount (1:1) of DsRed2 plasmid DNA.
2. Mix 1 μg of plasmid DNA mix with 3 μl of the lipofecting agent DOTAP (Boheringer Mannheim).
3. Mount anesthetized *Xenopus* embryos over a bed of clay in a petri dish.
4. Use a picospritzer (Picospritzer III, General Valve) to microinject a small volume (5–10 nl) of the DNA:DOTAP mix directly into the retinal or brain neuroepithelium of the anesthetized tadpoles with a fine-pulled glass micropipette.
5. Allow the embryos to recover from anesthesia in fresh rearing solution containing 0.001% phenylthiocarbamide <!> (to prevent melanocyte pigmentation) until imaging.

Time-lapse Imaging

6. Select embryos with individual neurons labeled with DsRed2 showing specific, punctate GPF labeling in their terminals for time-lapse imaging.

   Note: Typically, tadpoles at stage 43–45 of development (2–3 days after lipofection) contain clearly distinguishable double-labeled arbors that can be imaged by confocal microscopy.
7. Mount the anesthetized tadpoles over a bed of 1–2% agar in a petri dish and secure them with a slice anchor (Warner Instruments).
8. Collect thin optical sections (0.5–1.0 μm) through the entire extent of the axon or dendritic arbor with a laser-scanning confocal microscope.

   Note: A 60–63x magnification, water-dipping objective is recommended. Confocal images are obtained simultaneously at the two wavelengths (488 and 543 nm), below saturation levels, with minimal gain and contrast enhancements.

Data Analysis

All analysis is performed from raw confocal images with no postacquisition manipulation or thresholding. To characterize the distribution of GFP-labeled puncta (green) to particular regions within the DsRed2 labeled arbors (red), pixel-by-pixel overlaps from individual optical sections obtained at the two wavelengths are analyzed. GFP-labeled puncta of 0.5–1.0 μm^2 in size (size of smallest puncta observed) and 150–255 pixel intensity values in 8-bit images are usually considered single synaptic clusters. To obtain a second measure of synaptic cluster number, overlapped images are digitized, selected for color (yellow; locations of complete red and green overlay with hue and pixel intensity values between 16–67 and 150–255, respectively), and binarized, and the number of pixels representing the GFP-labeled puncta is counted by the imaging software. Total arbor length can be measured from binarized images of digitally reconstructed axons. Counting the total number of pixels from the first branchpoint provides a relative measure of the cumulative length of all branches per axon terminal. Other parameters (i.e., branchpoint number and location) can be counted manually. Yellow regions of complete red and green overlap are identified, counted, and related to arbor morphology.

EXAMPLE OF APPLICATION

Visualizing Synapse Dynamics in Arborizing Axon and Dendritic Arbors

Figure 1 shows an example of a *Xenopus* retinal ganglion cell axon (A) and of a *Xenopus* tectal neuron (B) imaged in vivo by confocal microscopy. In Figure 1A, the reconstruction of the three-dimensional axon arbor, labeled with GFP-synaptobrevin (green) and DsRed2 (red), illustrates the relationship between synapse distribution (yellow, Overlay) and axon arbor morphology. Axon arbors of similar complexity span 60–90 µm in depth. In Figure 1B, image reconstruction of a PSD95-GFP (green), DsRed2 (red) double-labeled tectal neuron illustrates dendritic complexity and postsynaptic cluster distribution along the branching arbor. Punctate-specific localization of PSD-95-GFP (yellow, Overlay) can be observed along dendritic arbors, which were imaged 3 days after plasmid lipofection

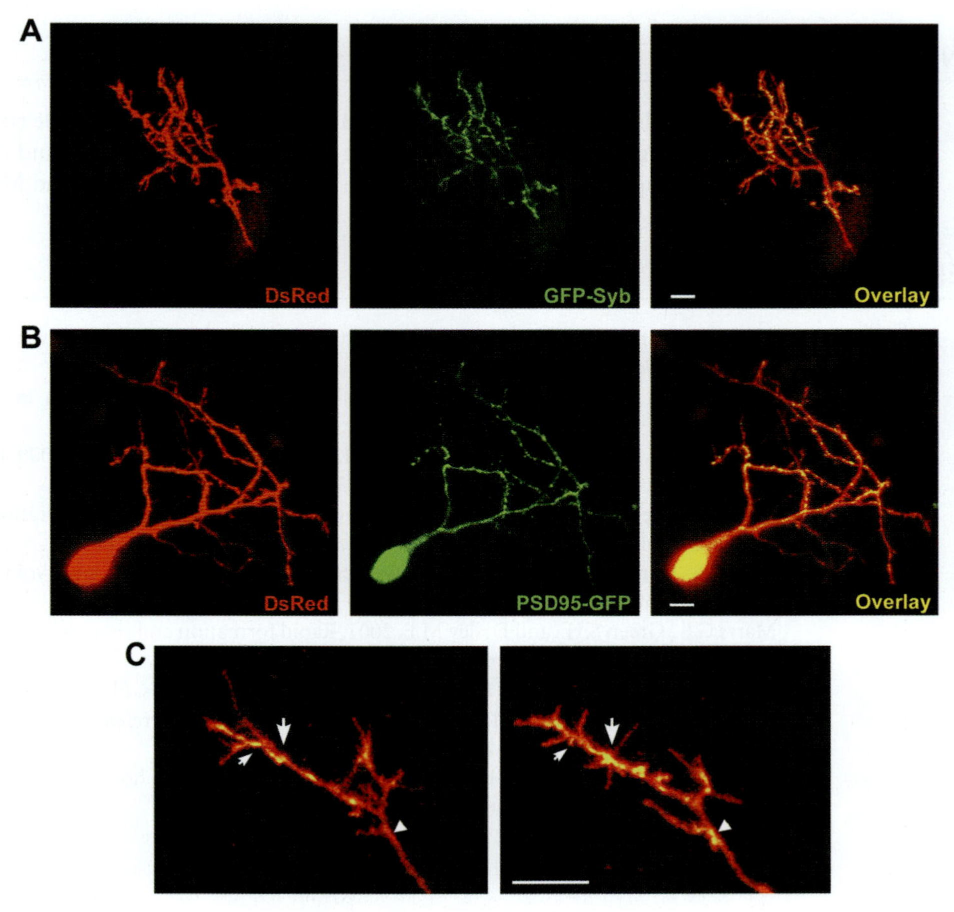

FIGURE 1. See text for details. Bars, 10 µm.

(stage 45). Dendritic arbors of similar complexity span 90–120 μm in depth. In Figure 1C, time-lapse images (0 and 6 hr) of portions of an individual axon arbor show that new branches originate at axon arbor sites rich in GFP-labeled synaptic clusters (white arrows) as well as the recruitment of new GFP-labeled synaptic clusters (arrowhead) along an axon branch.

ADVANTAGES AND LIMITATIONS

Simultaneous transfection of neuronal precursor cells with expression plasmids coding for several fluorescent proteins can be easily achieved by lipofection (Holt et al. 1990). A great advantage of using lipofection is that only a few neurons per embryo are transfected and that 80–90% of those neurons express the plasmids. This allows resolution of individual arbors labeled with two fluorescent probes. For example, targeting GFP-synaptobrevin expression directly to retinal neurons, in combination with DsRed2, allows simultaneous visualization of synaptic sites and arbor morphology in individual retinal axons branching in the optic tectum. Similarly, targeting GFP-tagged PSD-95 expression in combination with DsRed2 directly to brain neurons allows simultaneous visualization of postsynaptic sites and dendritic arbor morphology in individual postsynaptic neurons. These techniques can be used to determine dynamics and distribution of almost any pre- or postsynaptic protein, as long as appropriate controls are run to determine specific localization of the fluorescent reporters (Ahmari et al. 2000; Alsina et al. 2001).

This technique is limited in its application because it can only be used in accessible embryonic systems, and because of the need to transfect neurons early in development, prior to their final differentiation. The latter limitation, however, can be overcome by the use of novel electroporation techniques (Haas et al. 2002).

ACKNOWLEDGMENTS

I thank B. Alsina, B. Lom, A.L. Sanchez, B. Hu, and T. Vu for invaluable contributions to this work, and Drs. M.M. Poo and D. Bredt for their gift of the GFP-synaptobrevin and PSD-95-GFP constructs. This work was supported by the National Eye Institute and the Arnold and Mabel Beckman Foundation.

REFERENCES

Ahmari S.E., Buchanan J., and Smith S.J. 2000. Assembly of presynaptic active zones from cytoplasmic transport packets. *Nat. Neurosci.* **3:** 445–451.

Alsina B., Vu T., and Cohen-Cory S. 2001. Visualizing synapse formation in arborizing optic axons in vivo: Dynamics and modulation by BDNF. *Nat. Neurosci.* **4:** 1093–1101.

El-Husseini A.E., Schnell E., Chetkovich D.M., Nicoll R.A., and Bredt D.S. 2000. PSD-95 involvement in maturation of excitatory synapses. *Science* **290:** 1364–1368.

Haas K., Jensen K., Sin W.C., Foa L., and Cline H.T. 2002. Targeted electroporation in *Xenopus* tadpoles in vivo—From single cells to the entire brain. *Differentiation* **70:** 148–154.

Holt C.E., Garlick N., and Cornel E. 1990. Lipofection of cDNAs in the embryonic vertebrate central nervous system. *Neuron* **4:** 203–214.

Marrs G.S., Green S.H., and Dailey M.E. 2001. Rapid formation and remodeling of postsynaptic densities in developing dendrites. *Nat. Neurosci.* **4:** 1006–1013.

Nieuwkoop P.D. and Faber J. 1956. *Normal table of* Xenopus laevis. Elsevier, North Holland, The Netherlands.

Okabe S., Miwa A., and Okado H. 2001. Spine formation and correlated assembly of presynaptic and postsynaptic molecules. *J. Neurosci.* **21:** 6105–6114.

Turner D.L. and Weintraub H. 1994. Expression of achaete-scute homolog 3 in *Xenopus* embryos converts ectodermal cells to a neural fate. *Genes Dev.* **8:** 1434–1447.

CHAPTER 28

A Practical Guide: Intrinsic Optical Imaging of Functional Map Development in Mammalian Visual Cortex

Tobias Bonhoeffer and Mark Hübener

Max-Planck Institute of Neurobiology, Am Klopferspitz 18A, 82152 München-Martinsried, Germany

Optical imaging based on intrinsic signals is a technique, developed ~15 years ago (see Chapter 88), that takes advantage of the fact that neuronal activity is reflected in the intrinsic optical properties of brain tissue. It can be used to measure brain activity over large areas, with high spatial resolution (<100 μm), in a relatively noninvasive manner. These properties have made intrinsic imaging ideal either for experiments where different cortical maps are obtained simultaneously from the same cortical area or for studies where map development is followed over several weeks in the same animal. This chapter concentrates on the latter kind of experiments and shows how recent progress has facilitated chronic imaging not only in cats and ferrets, but also in the genetically amenable mouse.

The general principle of the technique is not new. It has been known for decades that brain activity can be observed directly, without the use of extrinsic probes, by taking advantage of optical signals which are intrinsic to the brain, and which reflect neuronal activity (Hill and Keynes 1949; Chance et al. 1962; Cohen 1973; Jobsis 1977). It was only much later, however, that hemoglobin-based intrinsic signals were used to monitor cortical activity in the living brain (Grinvald et al. 1986).

An advantage of intrinsic optical imaging is that, after an initial chamber implantation (see below), it is completely noninvasive. This is particularly helpful in developmental studies following cortical maps over time in individual animals. Until recently, conventional techniques allowed only inter-animal comparisons of functional maps at a given age (e.g., see Singer et al. 1981). These early studies could not address how cortical maps develop and change over time in a single animal. Consequently, it was not possible to determine, for example, whether orientation or ocular dominance maps change during early development or whether they remain stable once established. For these and similar questions, the development of intrinsic imaging presents a major advance. This chapter shows how intrinsic imaging can be used as a chronic technique, allowing the time course of the development of cortical maps to be studied in a single animal during a period of up to 1 year.

BASICS OF INTRINSIC IMAGING

Many publications deal extensively with the various sources of the different components of the intrinsic optical signal (for reviews on this topic, see Chapter 88 and Bonhoeffer and Grinvald 1996).

Understanding the origin of the different components of the signal is essential for optimizing the protocol for different applications. The basic procedures used in intrinsic imaging experiments are also described in detail elsewhere (see Chapter 88) (Bonhoeffer and Grinvald 1996; Bonhoeffer 1999); this chapter concentrates exclusively on issues of chronic imaging.

CHRONIC INTRINSIC IMAGING

Chronic intrinsic imaging benefits greatly from the noninvasiveness of the method and allows observation of the development of functional maps in single animals. The first chronic imaging experiments were performed in cats (Kim and Bonhoeffer 1994; Gödecke and Bonhoeffer 1996) and ferrets (Chapman et al. 1996; Chapman and Bonhoeffer 1998). More recently, the technique has also been successfully applied to monkeys (Shtoyerman et al. 2000; Chen et al. 2002), rats (Polley et al. 1999), and mice (Mrsic-Flogel et al. 2003).

Chronic recordings necessitate a number of changes to the procedures used for acute experiments. First, in young animals, the bone of the skull may be too soft for the chamber to be firmly attached. In these cases, agar can be used to stabilize the brain (Chapman et al. 1996), or a cranial glass window can be inserted flush into the bone (M. Vaz Afonso, pers. comm.). Alternatively, recordings can be made through the thinned (Frostig et al. 1990; Bosking et al. 1997) or intact skull (Schuett et al. 2002). In any case, for good spatial resolution, the brain *must* be sufficiently stable to provide a good optical interface.

Second—perhaps not unexpectedly for an experiment that relies strongly on the oxy/deoxy transition of hemoglobin—proper respiration of the animal is essential for successful imaging. Although it has been shown that good-quality intrinsic signal imaging maps can be obtained in a spontaneously breathing animal (Rubin and Katz 1999; Meister and Bonhoeffer 2001), artificial respiration has proven far more reliable. Despite the technical challenges, intubation is highly recommended for all intrinsic signal imaging experiments, even for animals weighing as little as 20 g.

Another concern is the anesthesia regime during chronic imaging experiments, particularly in those involving very young animals. A wide range of anesthetics, including halothane, isofluorane, ketamine, fentanyl, and urethane, are available. In our hands, gas anesthesia using isofluorane or halothane is most successful. It provides a high level of control over the animal's anesthetic state while not interfering noticeably with the source of the intrinsic signal, i.e., the coupling of neural activity to blood flow and oxygenation.

In chronic imaging experiments, using very young animals, the physiological state of the animal significantly affects the duration and success of the experiment. Sterility of the surgical environment is paramount, and routine implementation of additional (nutritional) measures is highly recommended. Offering the animals additional food (milk) and limiting the number of siblings being fed by the same mother have proven to be extremely beneficial to the animals' health and strength in mice, rats, ferrets, and cats.

Exemplary data from a chronic imaging experiment on the development of orientation domains in ferret visual cortex are shown in Figure 1. These data illustrate the entire time course of the development of orientation domains in ferret visual cortex, from the time that the domains are first observed (postnatal day [PND]32) to the time at which they are fully developed (PND43). In these experiments, four different orientation maps were acquired for each age studied (only two maps are shown for each age). To better display the overall layout of the iso-orientation domains and their development, "polar maps" are shown in the column on the right. In these maps, different colors indicate different orientations, and the brightness of the color indicates the degree of orientation tuning (for more details on the calculation of such maps, see Bonhoeffer and Grinvald 1993, 1996). Both the single orientation maps and the polar maps in Figure 1 show that, between PND32 and PND43, the overall structure of the orientation domains changes very little. Patch-by-patch comparison of the maps reveals that, when the first hint of an orientation map is detected at PND32, the structure of that map, although immature, is already remarkably similar to that seen in the mature map (PND43). The polar maps shown in Figure 1 illustrate particularly well how, over the period of 11 days, a functional map, with only very weak orientation preferences (dark colors), develops into a crisp, clear orientation preference map while maintaining its overall structure.

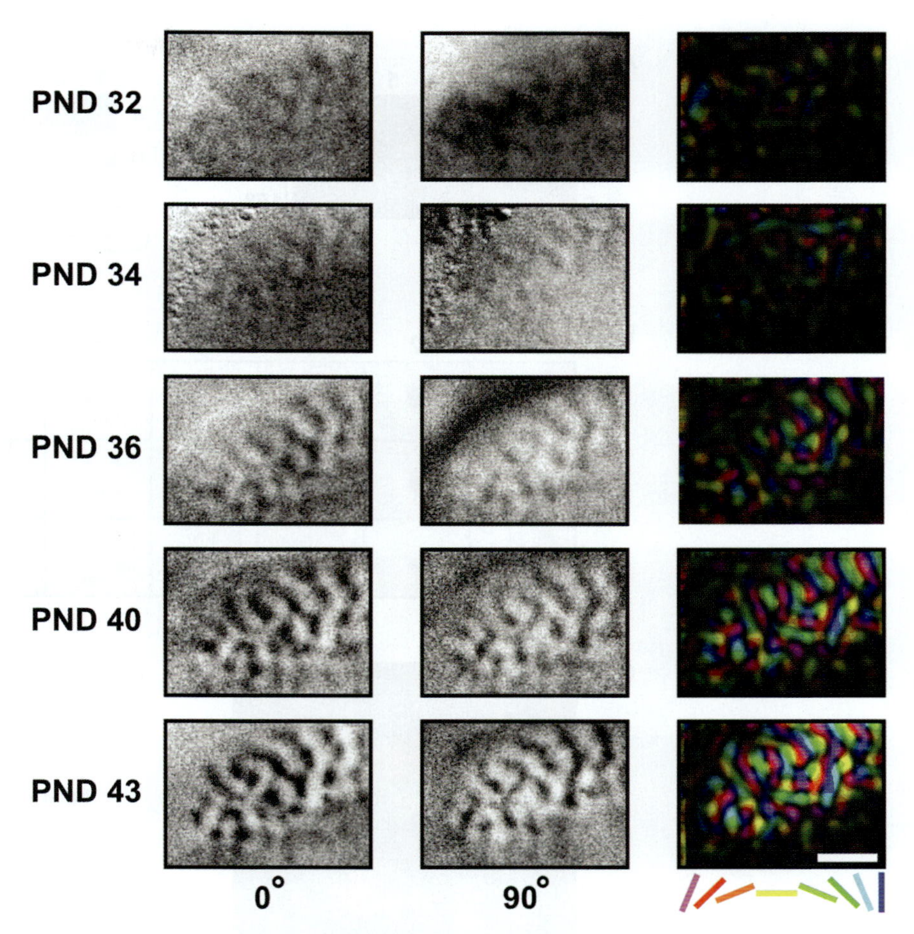

FIGURE 1. Chronic recording of developing orientation maps. Orientation maps obtained from ferret visual cortex (area 17) over a period of 11 days. Postnatal age of the ferret is shown at the left of each row. The two left-hand columns show single-condition maps in response to different orientations (0° and 90°). The right-hand column shows a "polar map" of orientation preference. In this map, color indicates the preferred angle for every cortical location. In addition, the brightness of the color indicates the strength of orientation preference. Bar, 2 mm.

During the past few years, a number of improvements have made chronic imaging an even more useful tool. First, intrinsic imaging can now be performed on rodents such as rats (e.g., see Masino et al. 1993; Dowling et al. 1996; Rubin and Katz 1999; Meister and Bonhoeffer 2001) and mice (Schuett et al. 2002; Kalatsky and Stryker 2003). The development of intrinsic imaging techniques in mice, in particular, promises to be very useful. In that respect, the recent achievement of imaging retinotopic maps in mouse visual cortex (Schuett et al. 2002) holds great promise. It will allow the use of genetics to scrutinize the array of molecules thought to be important for pathfinding in the visual system with respect to their influence on the formation of functional maps. The recent flurry of imaging studies in the olfactory system of mice and rats (Rubin and Katz 1999, 2001; Uchida et al. 2000; Meister and Bonhoeffer 2001) opens another interesting avenue: The olfactory system is particularly suitable for studying the intricate interplay between function and development in the rodent brain.

A further technical advance, which impinges directly onto the ability to do chronic recordings, has recently been reported by Kalatsky and Stryker (2003). These authors showed that a new Fourier-based data acquisition and analysis method can speed up the process of data acquisition considerably. Since data acquisition time is often the limiting factor when working on young and therefore fragile animals, this new method can greatly improve the quality and usefulness of the data.

Figure 2 shows an example where the visual cortex of a mouse was mapped in the traditional way, using 21 different stimulus locations in visual space. The figure shows the meticulous and precise

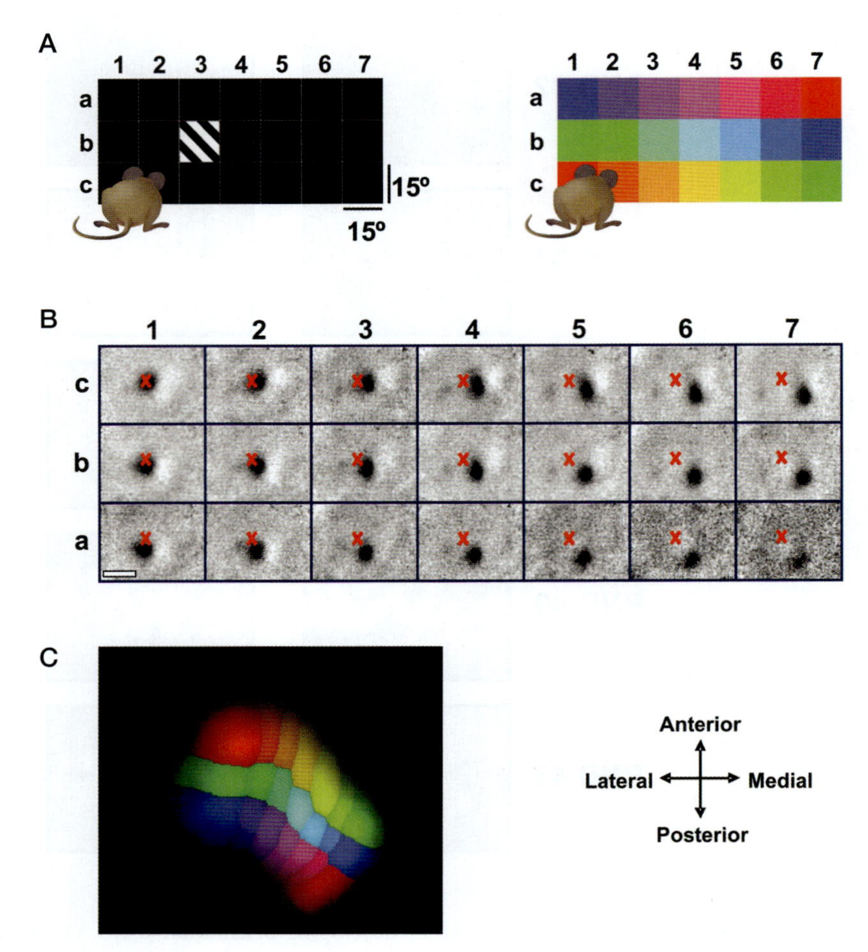

FIGURE 2. Mapping retinotopy in mouse visual cortex. (*A*) Experimental setup: Small visual stimuli are presented in random order at different locations in the visual field. (*B*) Activity maps are derived for each stimulus location. Note the systematic shift of the activated region relative to the red cross, which was placed at the same position in all maps. (*C*) A complete map of retinotopy is generated by assigning a color to each pixel, based on the stimulus location to which it best responded (see panel *A*). The response strength is encoded by pixel intensity, such that nonresponsive regions of the cortex appear dark. Bar, 2 mm.

mapping of the visual field onto the visual cortex. Up to now, the development of retinotopic maps has not been examined in the way that the development of orientation maps has been studied in the ferret visual system (Fig. 1). However, such studies should be very useful for better understanding the developmental events underlying the formation of retinotopic maps in the visual system.

Another area in which chronic optical imaging in mice should be useful is the adaptation of the cerebral cortex to changes in the sensory environment. Initial forages have been made in this direction using the retinal lesion model (Mrsic-Flogel et al. 2003). Such experiments should be very useful in elucidating the role of molecular factors involved in cortical plasticity and reorganization.

CONCLUSIONS

Recent technical advances in intrinsic optical imaging allow chronic experiments in which changes in cortical maps can be observed over time in single animals. These experiments have been extended to animals amenable to genetic manipulation. Chronic experiments like these are crucial for the understanding of cortical map development. The fact that these experiments can now be performed in the cerebral cortex of the mouse has the important implication that the influence of single genes on the formation (or reorganization) of maps can be studied in great detail. Together with new data acquisition tools, this has further enhanced the power of chronic intrinsic imaging. It is safe to predict that the method will continue to gain acceptance and that it might soon be a universal tool for the systems-oriented developmental neurobiologist.

REFERENCES

Bonhoeffer T. 1999. Intrinsic signal optical imaging as a tool to visualize the development of functional maps in the mammalian visual cortex. In *Imaging: A laboratory manual* (ed. R. Yuste et al.), pp. 46.1–46.10. Cold Spring Harbor Laboratory Press, Cold Spring Harbor, New York.

Bonhoeffer T. and Grinvald A. 1993. The layout of iso-orientation domains in area 18 of cat visual cortex. Optical imaging reveals a pinwheel-like organization. *J. Neurosci.* **13:** 4157–4180.

———. 1996. Optical imaging based on intrinsic signals: The methodology. In *Brain mapping: The methods* (ed. A.W. Toga and J.C. Mazziotta), pp. 55–97. Academic Press, San Diego, California.

Bosking W.H., Zhang Y., Schofield B., and Fitzpatrick D. 1997. Orientation selectivity and the arrangement of horizontal connections in tree shrew striate cortex. *J. Neurosci.* **17:** 2112–2127.

Chance B., Cohen P., Jobsis F., and Schoener B. 1962. Intracellular oxidation-reduction states in vivo. *Science* **137:** 499–508.

Chapman B. and Bonhoeffer T. 1998. Overrepresentation of horizontal and vertical orientation preferences in developing ferret area 17. *Proc. Natl. Acad. Sci.* **95:** 2609–2614.

Chapman B., Stryker M.P., and Bonhoeffer T. 1996. Development of orientation preference maps in ferret primary visual cortex. *J. Neurosci.* **16:** 6443–6453.

Chen L.M., Heider B., Williams G.V., Healy F.L., Ramsden B.M., and Roe A.W. 2002. A chamber and artificial dura method for long-term optical imaging in the monkey. *J. Neurosci. Methods* **113:** 41–49.

Cohen L.B. 1973. Changes in neuron structure during action potential propagation and synaptic transmission. *Physiol. Rev.* **53:** 373–418.

Dowling J.L., Henegar M.M., Liu D., Rovainen C.M., and Woolsey T.A. 1996. Rapid optical imaging of whisker responses in the rat barrel cortex. *J. Neurosci. Methods* **66:** 113–122.

Frostig R.D., Lieke E.E., Ts'o D.Y., and Grinvald A. 1990. Cortical functional architecture and local coupling between neuronal activity and the microcirculation revealed by *in vivo* high-resolution optical imaging of intrinsic signals. *Proc. Natl. Acad. Sci.* **87:** 6082–6086.

Gödecke I. and Bonhoeffer T. 1996. Development of identical orientation maps for two eyes without common visual experience. *Nature* **379:** 251–254.

Grinvald A., Lieke E.E., Frostig R.D., Gilbert C.D., and Wiesel T.N. 1986. Functional architecture of cortex revealed by optical imaging of intrinsic signals. *Nature* **324:** 361–364.

Hill D.K. and Keynes R.D. 1949. Opacity changes in stimulated nerve. *J. Physiol.* **108:** 278–281.

Jobsis F.F. 1977. Noninvasive, infrared monitoring of cerebral and myocardial oxygen sufficiency and circulatory parameters. *Science* **198:** 1264–1266.

Kalatsky V.A. and Stryker M.P. 2003. New paradigm for optical imaging: Temporally encoded maps of intrinsic signal. *Neuron* **38:** 529–545.

Kim D.-S. and Bonhoeffer T. 1994. Reverse occlusion leads to a precise restoration of orientation preference maps in visual cortex. *Nature* **370:** 370–372.

Malonek D. and Grinvald A. 1996. Interactions between electrical activity and cortical microcirculation revealed by imaging spectroscopy: Implications for functional brain mapping. *Science* **272:** 551–554.

Masino S.A., Kwon M.C., Dory Y., and Frostig R.D. 1993. Characterization of functional organization within rat barrel cortex using intrinsic signal optical imaging through a thinned skull. *Proc. Natl. Acad. Sci.* **90:** 9998–10002.

Meister M. and Bonhoeffer T. 2001. Tuning and topography in an odor map on the rat olfactory bulb. *J. Neurosci.* **21:** 1351–1360.

Mrsic-Flogel T.D., Vaz Afonso M., Eysel U.T., Bonhoeffer T., and Hübener M. 2003. Reorganization of mouse visual cortex after retinal lesions. *Soc. Neurosci. Abstr.* 567.4.

Polley D.B., Chen-Bee C.H., and Frostig R.D. 1999. Two directions of plasticity in the sensory-deprived adult cortex. *Neuron* **24:** 623–637.

Rubin B.D. and Katz L.C. 1999. Optical imaging of odorant representations in the mammalian olfactory bulb. *Neuron* **23:** 499–511.

———. 2001. Spatial coding of enantiomers in the rat olfactory bulb. *Nat. Neurosci.* **4:** 355–356.

Schuett S., Bonhoeffer T., and Hübener M. 2002. Mapping retinotopic structure in mouse visual cortex with optical imaging. *J. Neurosci.* **22:** 6549–6559.

Shtoyerman E., Arieli A., Slovin H., Vanzetta I., and Grinvald A. 2000. Long-term optical imaging and spectroscopy reveal mechanisms underlying the intrinsic signal and stability of cortical maps in V1 of behaving monkeys. *J. Neurosci.* **20:** 8111–8121.

Singer W., Freeman B., and Rauschecker J.P. 1981. Restriction of visual experience to a single orientation affects the organization of orientation columns in cat visual cortex. A study with deoxyglucose. *Exp. Brain Res.* **41:** 199–215.

Uchida N., Takahashi Y.K., Tanifuji M., and Mori K. 2000. Odor maps in the mammalian olfactory bulb: Domain organization and odorant structural features. *Nat. Neurosci.* **3:** 1035–1043.

CHAPTER 29

How Calcium Indicators Work

Stephen R. Adams

Department of Pharmacology, University of San Diego, La Jolla, California 92093

INTRODUCTION

In the last two decades, imaging of fluorescent indicators specific for Ca^{2+} has revealed its often spectacular spatial dynamics, such as rhythmic oscillations or standing gradients, within single groups or individual cells, in unprecedented detail. This short review describes how the more widely used indicators work and discusses some new indicators from the laboratory of R.Y. Tsien. More detailed information concerning the biological use of such indicators can be found in other chapters in this manual or in a number of reviews (Tsien 1993, 1999; Kao 1994).

The currently used Ca^{2+} indicators have a modular design consisting of a metal-binding site (or sensor) coupled in some way to a fluorescent dye. Combining different sensors with different dyes results in numerous indicators suited to a wide range of experiments and equipment.

Ca^{2+}-BINDING SITES

All chemical indicators are based on BAPTA (Fig. 1), a pH-insensitive homolog of EGTA. BAPTA retains its high selectivity for Ca^{2+} (K_d ~ 100 nM at pH 7.0) in the presence of competing millimolar concentrations of Mg^{2+} and has fast on–off rates for metal binding as a result of aromatizing the aliphatic nitrogens of EGTA (Tsien 1980). Indicators based on BAPTA can be incorporated into cells by temporarily masking the carboxylates as acetoxymethyl (AM) esters. These hydrophobic uncharged molecules passively diffuse across membranes but release the impermeant polycarboxylate indicator after cleavage of the AM esters by intracellular esterases (see Chapter 34).

BAPTA-based Chelators: Approaches for Modifying Affinity for Ca^{2+}

The binding site of BAPTA-based chelators can be further modified to change its affinity, allowing sensitivity to Ca^{2+} in a wider range (nanomolar to millimolar) by three general methods (see Table 1).

- *Addition of Electron-donating or Electron-withdrawing Groups to the Aromatic Ring(s) of BAPTA*

  Electron-withdrawing groups, such as nitro ($-NO_2$) or fluoro (–F) on the aromatic ring(s) in the para-position to the nitrogen, weaken the binding for Ca^{2+} to 42 µM (for nitro substitution on one ring) and 7.4 mM (nitro substitution on both rings; Pethig et al. 1989). Electron-donating groups, such as $-CH_3$, result in a modest increase in affinity and are used in many Ca^{2+} indicators, including fura-2 and fluo-3.

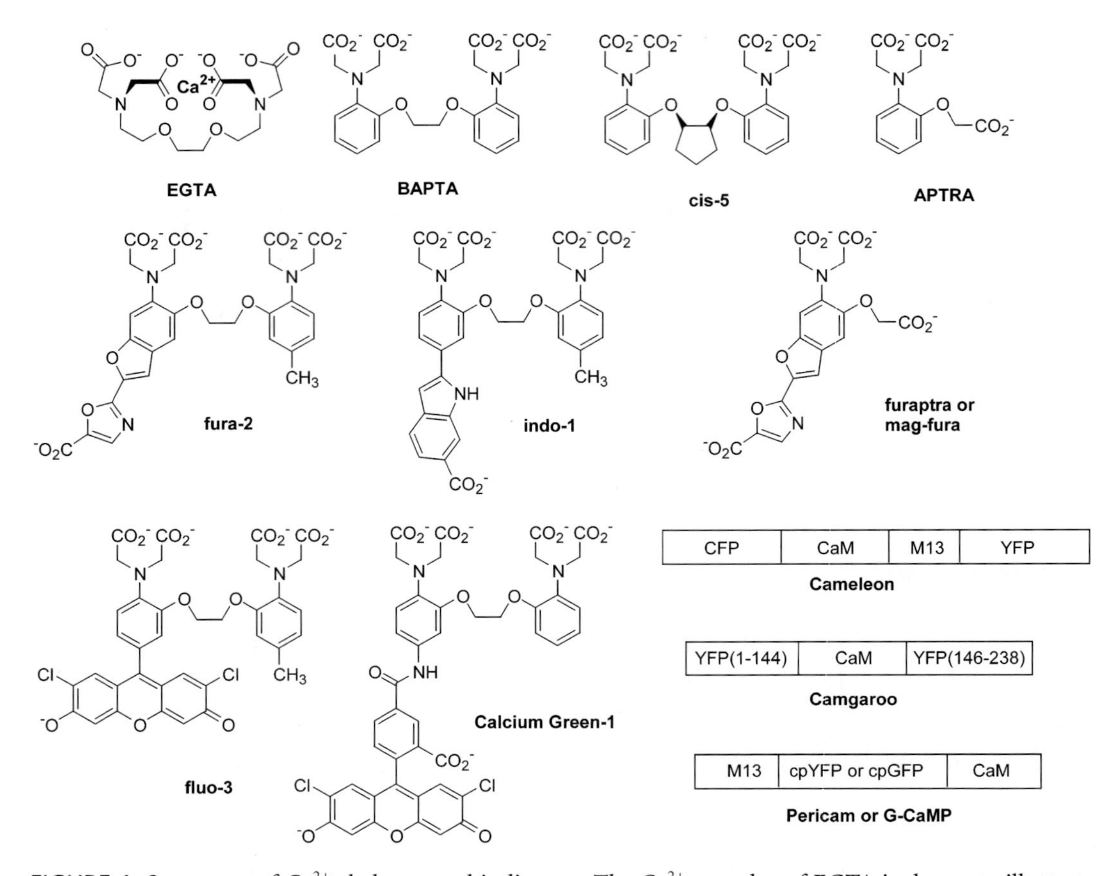

FIGURE 1. Structures of Ca^{2+} chelators and indicators. The Ca^{2+} complex of EGTA is shown to illustrate the complexation of the metal by the four carboxylates, two nitrogens, and two oxygens. The remaining chelators and indicators are shown in their Ca^{2+}-free form, but they bind Ca^{2+} in a similar manner, apart from APTRA and cameleon. Abbreviations: (APTRA) 2-amino phenol-*N,N,O*-triacetic acid; (BAPTA) 1,2-bis-(2-aminophenoxy)ethane-*N,N,N′*,N′-tetraacetic acid; (CFP) cyan-emitting mutant of green fluorescent protein (GFP); (CaM) calmodulin; (M13) calmodulin-binding peptide of myosin light-chain kinase; (YFP) yellow-emitting mutant of GFP; (cpYFP) circularly permuted YFP.

- *Addition of Modifying Groups That Sterically Alter Ca^{2+} Binding*

  Another approach is to sterically alter Ca^{2+} binding by incorporating the ethylene ($-CH_2CH_2-$) group of BAPTA into a carbo- or heterocycle or by substituting these hydrogen atoms for methyl groups (Adams et al. 1988). For example, trans-5 binds Ca^{2+} with an affinity of 6 µM, making it useful as a low-affinity Ca^{2+} buffer (Ranganathan et al. 1994). Cis-5 (Fig. 1) has a fourfold increased affinity for Ca^{2+} and is incorporated into nitr-7, a structurally related photolabile chelator that releases Ca^{2+} on photolysis (Adams et al. 1988). However, these modifications are hard to predict in advance, and the increased hydrophobicity decreases loading of cells by AM-ester derivatives.

- *Reduction in the Number of Coordinating Ligands for Ca^{2+} Chelation*

  The final and most successful approach for decreasing Ca^{2+} binding has been to decrease the number of coordinating ligands involved in metal chelation. The APTRA (Fig. 1) series of chelators have about half of the metal-binding site of BAPTA with the addition of a further carboxy ligand (Levy et al. 1988). Ca^{2+} binding is reduced to about 18 µM, but the affinity for Mg^{2+} is increased to 1.45 mM (S.R. Adams and R.Y. Tsien, unpubl.) within its physiological range. Ca^{2+} prefers eight ligands (APTRA has only five), unlike Mg^{2+}, which is hexavalent. APTRA and its fluorescent derivatives (furaptra or mag-fura, mag-indo, and Magnesium Green) were originally used as indicators of

TABLE 1. BAPTA- and APTRA-based chelators: Approaches for modifying affinity for Ca^{2+}

Ca^{2+} chelator[a]	Wavelength (nm)[b]		Ca^{2+} affinity (μM)[c]	Mg^{2+} affinity (mM)[c]	Comments/ Reference[d]
Electron-withdrawing substituents					
	excitation	emission			
Fura-FF	330/370	510	20	> 100	London et al. (1994)
Fluo-5N	491	516	90	> 100	orange and red versions available; Haugland (2002)
Calcium Green-5N	506	532	14	> 100	Oregon Green, orange and red versions available; Haugland (2002)
Sterically modifying groups					
Trans-5	NA	NA	6	>100	Ca^{2+} buffer
Cis-5	NA	NA	0.020	4.2	Ca^{2+} buffer
Decreased ligands					
Furaptra (Mag-fura)	330/369	510	25	1.9	
Mag-indo	350	390/480	35	2.7	
Magnesium Green	506	531	6	1.0	orange and red versions available; Haugland (2001)

[a]Ca^{2+} buffers and indicators with altered affinity for Ca^{2+} sorted by the method of modification.
[b]Excitation and emission maxima for indicators except fura-FF are from Haugland (2002). Two values for excitation or emission refer to the Ca^{2+}-bound and Ca^{2+}-unbound states for ratiometric indicators. (NA) Not applicable.
[c]Ca^{2+} and Mg^{2+} affinities measured in 0.1–0.15 M ionic strength, pH 7.0–7.4, 20–23°C, except those for fura-FF, are from Haugland (2002).
[d]References are provided only for indicators not described in the text.

cytosolic Mg^{2+} (Raju et al. 1989; Haugland 2002) and only later used to measure high Ca^{2+} transients or Ca^{2+} in the intracellular stores. Such measurements are always susceptible to misinterpretation if there are concurrent Mg^{2+} changes. A further advantage to using APTRA-based indicators is the decrease in molecular weight and loss of a hydrophobic aromatic ring that should aid loading of AM esters into cells. Recent modifications of APTRA and BAPTA have produced chelators and indicators sensitive to Ca^{2+} concentrations in the low-micromolar to low-millimolar range and are suitable for measuring the elevated Ca^{2+} found in stimulated cells, Ca^{2+} stores, or serum (S.R. Adams and R.Y. Tsien, unpubl.).

Changes in the Affinity of Cameleons for Ca^{2+}

The newest class of indicators, the cameleons (Miyawaki et al. 1997), utilize CaM fused to color variants of the GFP. The affinity of such indicators can be conveniently decreased by mutation of key chelating glutamate residues to glutamines in the first and third Ca^{2+}-binding loops of CaM. Cameleons responsive to Ca^{2+} concentrations ranging from 0.1 mM to low millimolar now exist. With other related genetically encoded indicators, such as camgaroos (Baird et al. 1999) and pericams (Nagai et al. 2001), a similar approach should be feasible.

COUPLING OF Ca^{2+} BINDING TO CHANGES IN FLUORESCENCE

Ca^{2+} binding to an indicator is coupled to changes in fluorescence by mechanisms leading to the following three outcomes:

- The binding of Ca^{2+} results in a shift in excitation, and sometimes emission peaks, allowing excitation or emission ratioing (e.g., quin-2, fura-2, indo-1, and "ratiometric" pericam).

- The binding of Ca^{2+} leads to a change in fluorescence intensity, but no wavelength shifts (e.g., fluo-3, rhod-2, the Calcium Green family of indicators, and pericams and camgaroos).
- The binding of Ca^{2+} results in changes in fluorescent resonance energy transfer (FRET; e.g., cameleons).

Quin-2, Fura-2, and Indo-1

The first indicators (e.g., quin-2, fura-2, and indo-1; see Fig. 1) integrated the Ca^{2+}-binding site with small fluorophores (such as benzofurans and indoles) that are excitable in the UV region of the spectrum (Tsien 1980; Grynkiewicz et al. 1985). Direct conjugation of the chelating aminodiacetate moiety, $-N(CH_2CO_2-)_2$ with the planar fluorophores results in shifted excitation peaks (and sometimes also emission) upon Ca^{2+} binding. Exciting these peaks alternately while collecting the emission allows excitation ratioing, whereas emission ratioing requires excitation at a single wavelength with collection at two separate emission peaks. Ratioing helps correct for variations in indicator concentration, cell shape, lamp and detector fluctuations, and photobleaching, and enables calibration of the fluorescent signal in terms of free Ca^{2+} concentration. Mechanistically, in the absence of Ca^{2+}, the aminodiacetate group lies planar to the aromatic ring system, thereby maximizing electron donation by the nitrogen. Ca^{2+} coordination twists this group out of planarity and causes a decrease in the electron density of the fluorophore, thereby decreasing the excitation wavelength. Rigidization of the aminodiacetate group by bound Ca^{2+} also probably decreases its radiationless deactivation of the excited state, thereby increasing the fluorescent quantum yield.

Fluo-3, Rhod-2, and the Calcium Green Family

Successive indicators such as fluo-3, rhod-2 (Minta et al. 1989), and the Calcium Green family (Kuhn 1993; Haugland 2002) incorporated the more fluorescent fluorescein and rhodamine fluors, which operate at visible wavelengths that cause less cellular autofluorescence than UV excitation. All these long-wavelength Ca^{2+} indicators use a different photochemical mechanism for coupling the Ca^{2+} chelation to changes in fluorescence, which results in no excitation or emission shifts, but only changes in emission intensity upon binding Ca^{2+}.

In these molecules, the BAPTA moiety is twisted out of plane with the fluorophore (through steric interaction between the juxtaposed aromatic rings), preventing direct electronic coupling. However, upon excitation by light, in the absence of Ca^{2+}, the electron-rich nitrogen of the aminodiacetate group can transfer an electron to the electron-poor fluorophore before a photon can be emitted as fluorescence. The products of electron transfer, a cation and anion radical, cannot fluoresce and rapidly recombine directly or indirectly to re-form the ground state. When Ca^{2+} binds to the aminodiacetate group, the nitrogen's electron density is decreased, electron transfer occurs less, and the fluorescence is not quenched. Binding Ca^{2+} therefore leads to a change in fluorescence intensity but no wavelength shifts. The degree of quenching and the extent of relief by Ca^{2+} reflect the distance and types of bonds between the Ca^{2+}-binding site and the fluorophore. Long-wavelength indicators can be readily synthesized by reacting simple BAPTA derivatives with fluorescent labeling reagents (Kuhn 1993). Provided the two components are sufficiently close or appropriately linked, the resulting molecule's fluorescence becomes Ca^{2+}-sensitive. Using this principle, an infrared Ca^{2+} indicator, cyan-4 (Ozmen and Akkaya 2000; S.R. Adams and R.Y. Tsien, unpubl.), has been synthesized that may be useful in certain cells and tissues containing molecules which strongly absorb or fluoresce in the visible part of the spectrum, such as hemoglobin (e.g., blood-containing tissues), chlorophyll (plants), and rhodopsin (photoreceptors).

Genetically Encoded Indicators: Cameleon, Camgaroo, G-CaMP, and Pericam

Cameleons (see Fig. 1) were the first genetically encoded Ca^{2+} indicators. They are protein-based, resulting from the fusion of CaM and M13 (a CaM-binding peptide from myosin light-chain kinase) with cyan and yellow mutants of GFP (CFP and YFP) at the amino and carboxyl termini, respectively (Miyawaki et al. 1997). They utilize a third mechanism for coupling Ca^{2+} binding to changes in fluorescence. When CFP is excited at 430 nm, FRET occurs from CFP to YFP, resulting in re-emission as yellow light. (For a detailed discussion on the principles of FRET and the photochemical mechanisms of various indicators utilizing FRET, see Chapter 74). FRET is exquisitely sensitive to changes in the distance (in the range of 20–100 nm) and the orientation of the two fluorophores involved, so the well-known conformation change of calmodulin upon binding Ca^{2+} results in a significant change in FRET and emission ratio. In low Ca^{2+}, the more extended conformation of CaM inhibits FRET and gives a low ratio of YFP to CFP emission, whereas high Ca^{2+} results in more FRET and a higher ratio. One limitation of cameleons, however, is their requirement for 430-nm excitation, not a convenient laser line. This excitation wavelength is now available on newer imaging systems, although two-photon IR-excitation appears to work well (Fan et al. 1999). Recent developments have improved the expression and folding of cameleons in cells and decreased their sensitivity to pH and chloride anions (Griesbeck et al. 2001).

New indicators in this class are camgaroo (Baird et al. 1999), G-CaMP (Nakai et al. 2001), and pericams (Nagai et al. 2001), all of which involve insertion of CaM into a single fluorescent protein (usually YFP) at a site close to the chromophore. Ca^{2+} binding alters the pH sensitivity of the YFP, resulting in large changes in fluorescence intensity. Different versions of pericam either increase or decrease fluorescence intensity at a single wavelength or are excitation ratiometric indicators.

Genetically encoded indicators have been targeted using molecular and cell biology techniques (transfection and transgenics) to different tissue and cell types and various subcellular locations such as the endoplasmic reticulum, mitochondria, and the nucleus. This contrasts greatly with conventional indicators that can often be easily loaded into cells as AM esters but are not restricted to specific cytoplasmic compartments or tissues. Genetically encoded indicators have been particularly successful in organisms that do not load AM esters, such as plants (Allen et al. 2001), worms (Kerr et al. 2000; Suzuki et al. 2003), flies (e.g., Reiff et al. 2002; Liu et al. 2003), and fish (Higashijima et al. 2003).

CONCLUSION

A wide variety of Ca^{2+} indicators with a range of affinities are now available that can be excited by UV or visible light and readily incorporated into many cells as AM esters (chemical indicators) or by molecular biology techniques (cameleons, pericams, and camgaroos).

REFERENCES

Adams S.R., Kao J.P.Y., Grynkiewciz G., Minta A., and Tsien R.Y. 1988. Biologically useful chelators that release Ca^{2+} upon illumination. *J. Am. Chem. Soc.* **110:** 3212–3220.

Allen G.J., Chu S.P., Harrington C.L., Schumacher K., Hoffmann T., Tang Y.Y., Grill E., and Schroeder J.I. A defined range of guard cell calcium oscillation parameters encodes stomatal movements. 2001. *Nature* **411:** 1053–1057.

Baird G.S., Zacharias D.A., and Tsien R.Y. 1999. Circular permutation and receptor insertion within green fluorescent proteins. *Proc. Natl. Acad. Sci.* **96:** 11241–11246.

Fan G.Y., Fujisaki H., Miyawaki A., Tsay R.K., Tsien R.Y., and Ellisman M.H. 1999. Video-rate scanning two-photon excitation fluorescence microscopy and ratio imaging with cameleons. *Biophys J.* **76:** 2412–2420.

Griesbeck O., Baird G.S., Campbell R.E., Zacharias D.A., and Tsien R.Y. 2001. Reducing the environmental sensitivity of yellow fluorescent protein. Mechanism and applications. *J. Biol. Chem.* **276:** 29188–29194.

Grynkiewicz G., Poenie M., and Tsien R.Y. 1985. A new generation of Ca^{2+} indicators with greatly improved fluorescence properties. *J. Biol. Chem.* **260:** 3440–3450.

Haugland R.P. 2002. *Handbook of fluorescent probes and research chemicals*, 9th edition. Molecular Probes, Eugene, Oregon.

Higashijima S.I., Masino M.A., Mandel G., and Fetcho J.R. 2003. Imaging neuronal activity during zebrafish behavior with a genetically encoded calcium indicator. *J Neurophysiol.* **90:** 3986–3997.

Kao J.P.Y. 1994. Practical aspects of measuring $[Ca^{2+}]$ with fluorescent indicators. *Methods Cell Biol.* **40:** 155–181.

Kerr R., Lev-Ram V., Baird G., Vincent P., Tsien R.Y., and Schafer W.R. 2000. Optical imaging of calcium transients in neurons and pharyngeal muscle of *C. elegans*. *Neuron* **26:** 583–594.

Kuhn M.A. 1993. 1,2-Bis(2-aminophenoxy)ethane-N,N,N′,N′-tetraacetic acid conjugates used to measure intracellular Ca^{2+} concentration. In *Fluorescent chemosensors for ions and molecule recognition* (ed. A.W. Czarnik), pp. 147–161. American Chemical Society, Washington, D.C.

Levy L.A., Murphy E., Raju B., and London R.E. 1988. Measurement of cytosolic free magnesium ion concentration by ^{19}F NMR. *Biochemistry* **27:** 4041–4048.

Liu L., Yermolaieva O., Johnson W.A., Abboud F.M., and Welsh M.J. 2003. Identification and function of thermosensory neurons in *Drosophila* larvae. *Nat. Neurosci.* **6:** 267–273.

London R.E., Rhee C.K., Murphy E., Gabel S., and Levy L.A. 1994. Nmr-sensitive fluorinated and fluorescent intracellular calcium ion indicators with high dissociation constants. *Am. J. Physiol.* **266:** C1313–C1322.

Minta A., Kao J.P.Y., and Tsien R.Y. 1989. Fluorescent indicators for cytosolic calcium based on rhodamine and fluorescein chromophores. *J. Biol. Chem.* **264:** 8171–8178.

Miyawaki A., Llopis J., Heim R., McCaffery J.M., Adams J. A., Ikura M., and Tsien R.Y. 1997. Fluorescent indicators for Ca^{2+} based on green fluorescent proteins and calmodulin. *Nature* **388:** 882–887.

Nagai T., Sawano A., Park E.S., and Miyawaki A. 2001. Circularly permuted green fluorescent proteins engineered to sense Ca^{2+}. *Proc. Natl. Acad. Sci.* **98:** 3197–3202.

Nakai J., Ohkura M., and Imoto K. 2001. A high signal-to-noise Ca^{2+} probe composed of a single green fluorescent protein. *Nat. Biotechnol.* **19:** 137–141.

Ozmen B. and Akkaya E.U. 2000. Infrared fluorescence sensing of submicromolar calcium: Pushing the limits of photoinduced electron transfer. *Tetrahedron Lett.* **41:** 9185–9188.

Pethig R., Kuhn M., Payne R., Adler E., Chen T.H., and Jaffe L.F. 1989. On the dissociation constants of BAPTA-type calcium buffers. *Cell Calcium* **10:** 491–498.

Raju B., Murphy E., Levy L.A., Hall R.D., and London R.E. 1989. A fluorescent indicator for measuring cytosolic free magnesium. *Am. J. Physiol.* **256:** C540–C548.

Ranganathan R., Bacskai B.J., Tsien R.Y., and Zuker C.S. 1994. Cytosolic calcium transients: Spatial localization and role in *Drosophila* photoreceptor cell function. *Neuron* **13:** 837–848.

Reiff D.F., Thiel P.R., and Schuster C.M. 2002. Differential regulation of active zone density during long-term strengthening of *Drosophila* neuromuscular junctions. *J. Neurosci.* **22:** 9399–9409.

Suzuki H., Kerr R., Bianchi L., Frokjaer-Jensen C., Slone D., Xue J., Gerstbrein B., Driscoll M., and Schafer W.R. 2003. In vivo imaging of *C. elegans* mechanosensory neurons demonstrates a specific role for the MEC-4 channel in the process of gentle touch sensation. *Neuron* **39:** 1005–1017.

Tsien R.Y. 1980. New calcium indicators and buffers with high selectivity against magnesium and protons: Design, synthesis, and properties of prototype structures. *Biochemistry* **19:** 2396–2404.

———. 1993. Fluorescent and photochemical probes of dynamic biochemical signals inside living cells. In *Fluorescent chemosensors for ions and molecule recognition* (ed. A.W. Czarnik), pp. 130–146. American Chemical Society, Washington, D.C.

———. 1999. Monitoring cell Ca^{2+}. In *Calcium as cellular regulator* (ed. E. Carafoli and C. Klee), pp. 28–54. Oxford University Press, Oxford, United Kingdom.

CHAPTER 30

Some Quantitative Aspects of Calcium Fluorimetry

Erwin Neher

Department of Membrane Biophysics, Max-Planck-Institute for Biophysical Chemistry, 37070 Göttingen, Germany

Ca^{2+} indicator dyes, by necessity, are Ca^{2+} chelators, because it is the binding of Ca^{2+} to dye molecules that induces the change in fluorescence on which the Ca^{2+} signal is based. As chelators, once introduced into a cell, they contribute to cellular Ca^{2+} buffering. It has been a question of much debate to what extent this "added Ca^{2+} buffer" (exogenous Ca^{2+} buffer) changes Ca^{2+} homeostasis and the signals of interest. In this chapter, this problem is discussed, emphasizing the distinction between the influence of the dyes on amplitudes (which may be not so severe) and on the dynamics of Ca^{2+} signals (which may be drastic). It is pointed out that once the Ca^{2+} buffering action of dyes relative to intrinsic Ca^{2+} buffers is understood for a given preparation, Ca^{2+} dyes can be used as very versatile tools for studying both Ca^{2+} concentrations and Ca^{2+} fluxes. Some of my own experiences in calibrating the indicator dye fura-2 are described in detail. These refer exclusively to experiments in which the dye is loaded into the cell via a patch pipette, because AM-ester loading introduces problems, which very often prohibit precise quantitative conclusions.

THE INFLUENCE OF Ca^{2+} DYES ON Ca^{2+} SIGNALS

Buffering Capacity of Ca^{2+} Dyes

Ca^{2+} indicators act as Ca^{2+} buffers because they bind Ca^{2+} in a reversible manner. In the following discussion, the term Ca^{2+} buffer is used in the strict sense to designate Ca^{2+}-binding molecules—either endogenous or added by the experimenter. For my arguments, it is important that such Ca^{2+} buffers are not confused with other mechanisms (Ca^{2+} pumps and exchangers) that sequester or pump Ca^{2+} into organelles or into other compartments. The Ca^{2+}-buffering power of such molecules is most conveniently characterized by the so-called Ca^{2+}-binding ratio, κ (Mathias et al. 1990; Neher 1995):

$$\kappa \equiv \frac{d[\mathrm{CaB}]}{d[\mathrm{Ca}^{+}]} = \frac{\mathrm{B_T}}{K_\mathrm{D}} \times \frac{1}{(1 + [\mathrm{Ca}^{2+}]/K_\mathrm{D})^2} \tag{1}$$

Here [CaB] is the concentration of Ca^{2+}-bound buffer, $[Ca^{2+}]$ is the concentration of free Ca ions, B_T is the total concentration of buffer, and K_D is its dissociation constant (for review, see Neher 1995, 1998). One minimum requirement for unperturbed $[Ca^{2+}]$ measurement can readily be written down, by postulating that the Ca^{2+}-binding ratio κ_B of the dye (the added Ca^{2+} buffer) should be smaller than the Ca^{2+}-binding ratio κ_s of the cytoplasm,

$$\kappa_B << \kappa_s \tag{2}$$

Here, the cellular Ca^{2+} buffers, for simplicity, are represented by one hypothetical Ca^{2+}-buffering species, S. κ_s can be measured by a number of techniques and turns out to be in the range of 40–200 for many neuronal cell types (for review, see Neher 1995). In many cases where it has been studied, its Ca^{2+} dependence has been found to be remarkably constant (Xu et al. 1997), indicating that endogenous Ca^{2+} buffers have low affinity (in that case, Eq. 1 reduces to $\kappa = B_T/K_D$). Some types of neurons, however, express specific Ca^{2+}-binding proteins at high concentration. In these cases, κ_s may have a value of 1000 or more (Roberts 1993; Fierro and Llano 1996).

How much indicator dye is tolerable to satisfy Equation 2? Using a κ_s value in the middle of the range indicated above (100), assuming that Ca^{2+} signals around 100 nM are of interest, and inserting an in vivo value for the K_D of fura-2 (see below) of 238 nM (Zhou and Neher 1993a), we find according to Equation 1 that κ_s is equal to κ_B for a total concentration of dye B_T of 48 μm. Thus, Equation 2 is quite restrictive, given the fact that many imaging studies use dye concentrations well above 100 μm. It is, however, borne out by experiments on several types of mammalian central neurons in which it was shown that dye concentrations in the range of 50–100 μM severely distort peak values and time courses of Ca^{2+} signals (Helmchen et al. 1996, 1997). Higher dye concentrations are tolerable for low-affinity dyes such as Calcium Green-5N, for which the Ca^{2+}-binding ratio is smaller, according to Equation 1.

Effects of Exogenous Ca^{2+} Buffers on Ca^{2+} Dynamics

In many cases, however, there is another aspect that may impose restrictions much more severe than those of Equation 2. It turns out that most of the endogenous Ca^{2+} buffer in cells with relatively low κ_s is immobile, probably being fixed to organelles, membranes, or cytoskeletal elements (Zhou and Neher 1993a). This means that endogenous buffer in the unperturbed cell retards diffusion of Ca^{2+}, localizing Ca^{2+} changes and slowing Ca^{2+} redistribution within a cell. Adding a mobile Ca^{2+} buffer, such as a dye, in this situation dramatically speeds Ca^{2+} diffusion, since the mobile dye competes with the fixed endogenous buffer for Ca^{2+} and will "shuttle" Ca^{2+} from subcellular regions of high concentration to those of low concentration. Ca^{2+} transport in the presence of both fixed and mobile buffers can be described by an effective diffusion coefficient D_{app} of the form (Wagner and Keizer 1994; Gabso et al. 1997):

$$D_{app} = (D_{Ca} + \sum_m \kappa_m D_m)/(1 + \sum_m \kappa_m + \sum_f \kappa_f) \tag{3}$$

where κ_m and D_m represent Ca^{2+}-binding ratios and diffusion coefficients of mobile buffer species, and κ_f represents Ca^{2+}-binding ratios of fixed buffer species. Unfortunately, κ_m values for some cell types are so small that they cannot be measured. Zhou and Neher (1993a) argued that for adrenal chromaffin cells, the value might be within $2 < \kappa_m < 7$, which, together with reasonable estimates for the other parameters of Equation 3, would give values for D_{app} between 2×10^{-7} and 4×10^{-7} cm^{2}sec^{-1}. Adding 50 μM fura-2 to this background would change D_{app} three- to fivefold. Fura-2 concentrations less than 5 μM would be required to maintain D_{app} close to its value in an unperturbed cell. A similar conclusion was reached by Gabso et al. (1997), who measured D_{app} in axons of cultured *Aplysia* cells. One must conclude from these considerations that the dynamics of Ca^{2+} signals, inasmuch as they are determined by Ca^{2+} diffusion, are not faithfully represented by Ca^{2+}-imaging studies, basically because no studies are performed at dye concentrations as low as required. The situation is relaxed, however, in cell types that express mobile Ca^{2+}-binding proteins, such as cochlear hair cells (Roberts 1993), cerebellar Purkinje cells (Fierro and Llano 1996), or muscle cells. When studying developmental or activity-induced changes in the expression of Ca^{2+}-binding proteins (for review, see Heizmann and Braun 1995), investigators should be aware that such changes will manifest themselves in Ca^{2+} dynamics only at extremely low dye concentrations (Gabso et al. 1997).

INDICATOR DYES AS TOOLS TO MEASURE Ca^{2+} FLUXES

It has been demonstrated that in many cell types, extremely low concentrations of indicator dyes are required, if amplitudes and time courses of Ca^{2+} signals are to be recorded faithfully (see above). The reason is that dyes compete with endogenous Ca^{2+} buffers for Ca^{2+} and that even at concentrations below 100 μM, dyes are quite successful in this competition. This situation can be turned into an advantage if Ca^{2+} fluxes rather than Ca^{2+}-concentration signals are of interest in a research project. Once the level of endogenous Ca^{2+} buffers is known in a given cell type (e.g., let it be equivalent to 50 μM fura-2), a moderate increase in dye concentration (such as to 500 μM) will secure that the majority of Ca^{2+} entering a cell during a short stimulus will bind to the dye. The fluorescence change measured at a Ca^{2+}-sensitive wavelength (such as at 380 nm for fura-2 or as measured with Calcium Green) will then be directly proportional to the amount of Ca^{2+} entering during the stimulus. It is then relatively straightforward to calibrate a fluorescence setup for Ca^{2+} flux and to measure the contribution of Ca^{2+}-inward current in nonspecific cationic currents (for review, see Neher 1995) or to estimate the amount of Ca^{2+} released from intracellular stores (Ganitkevich 1996).

CALIBRATING A FURA-2 SETUP FOR QUANTITATIVE FLUORIMETRY

Many of the aspects discussed above—particularly estimating endogenous Ca^{2+} buffer strength from the changes in Ca^{2+} signals induced by increasing dye concentration—require a very careful calibration of the fluorescence setup. A variety of procedures have been described previously for such calibrations (Grynkiewicz et al. 1985; Poenie 1990; Williams and Fay 1990). Incomplete hydrolysis of ester-loaded dye or compartmentalization of dye, however, very often prevents a precise determination of calibration constants (Tsien 1989; see Appendix in Zhou and Neher 1993a). Therefore, the following descriptions refer exclusively to experiments in which the indicator dye is loaded into the cell by diffusion from a patch pipette.

For such types of measurements an in vivo calibration, such as described by Almers and Neher (1985), is appropriate. Thereby, the fluorescence ratio is determined in whole-cell recordings, and three different pipette solutions are used, which set $[Ca^{2+}]$ to one of three values: zero (by including an excess of free Ca^{2+} buffer), very high (by including excess Ca^{2+}), and intermediate (by including appropriate amounts of Ca^{2+}-bound and free buffer). From these three ratios, the calibration constants of the equation by Grynkiewicz et al. (1985) (see Eq. 4, below) can be calculated. It is advisable to perform at least three to five measurements under each condition, thus the procedure is quite laborious. Additionally, recalibration is necessary from time to time because of aging (or after replacement) of the light source or when new batches of dyes are introduced. Thus, calibration may use up a substantial portion of measuring time on a given setup. It is my experience that the calibration procedure is safer and, on the whole, less time-consuming if in addition to the three types of measurements mentioned above, two additional measurements are performed during a calibration. These, together with the conventional calibration, are described below, following discussion of some of the intrinsic problems of calibration. For further guidelines on calibrating calcium indicators, see Chapter 31.

Problems Encountered in Ca^{2+}-indicator Calibration

The Dissociation Constant of the Calibrating Buffer and Its pH Dependence

The accuracy of the calibration ultimately depends on the correctness of the dissociation constant of the Ca^{2+} buffer used in the preparation of the calibrating solutions. Values for these parameters can be found in tables (see, e.g., Martell and Smith 1977) for standard conditions (usually 0.1 ionic strength and 20°C or 25°C). Corrections have to be applied to these values to account for ionic strength and temperature of the experiment. An additional complication arises from the fact that the apparent K_D of EGTA, the most commonly used calibration buffer, is strongly pH-dependent (Miller and Smith 1984). A pH error of 0.1 unit will cause an error in the apparent K_D close to 50%. This is a serious problem because the pH changes when Ca^{2+} binds to EGTA (and replaces H^+). It arises because calibration solutions are usually made in small quantities, which are difficult to titrate correctly. In any case, pH of the calibration solution should be measured immediately before or after a calibration. Above all, it is advisable to use BAPTA instead of EGTA for calibration purposes, because BAPTA is much less sensitive to pH (above 7.0) than EGTA. Tsien (1980) gives a K_D for BAPTA of 107 nM, as measured in 100 mM KCl at 22°C. However, this value is quite sensitive to ionic strength and temperature (Harrison and Bers 1987). Under conditions typical for pipette-filling solutions (~150 mM total salt concentration, pH 7.2, 1 mM free Mg^{2+}, 22°C) a value of about 225 nM is calculated from the parameters given by Harrison and Bers (1987). Oheim (1995) measured a value of 221 nM by potentiometric titration in a solution that was formulated to allow measurement of Ca^{2+} currents in the whole-cell configuration (120 mM CsCl, 20 mM HEPES, 2 mM BAPTA, 5 mM NaCl, 1 mM $MgCl_2$, pH 7.2). Thus, a value of 225 nM (Zhou and Neher 1993a) seems to be quite appropriate for an in vivo condition. It should be kept in mind that this value is defined in terms of concentration of Ca^{2+} (not activity), and that the concentration of calcium in the cytosol may be quite different from that in the pipette during the calibration measurement, even if the pipette solution is at equilibrium with the cytosol (for a discussion of this problem, see Neher 1995).

Autofluorescence

For typical measurements in adrenal chromaffin cells using 50–100 μm fura-2, autofluorescence of the cell may be as large as 5–10% of total fluorescence after dye loading. This value may be even larger in brain-slice preparations. Autofluorescence can be partially accounted for if its values at the wavelengths used are measured during the cell-attached configuration and subtracted from corresponding values after loading. (This then also corrects for some fluorescence originating from the pipette!) However, two problems arise with such subtraction:

- Shortly after obtaining the cell-attached configuration, fura-2-containing solution spilled from the pipette during the approach may still be present in the measuring field. Care must be taken to either wash away such solution by bath perfusion or to allow enough time for it to diffuse away.
- Some of the autofluorescence is mobile, probably representing NADH, and will be lost by washout during whole-cell recording. Both total fluorescence and the mobile contribution vary from cell to cell, and more so, from cell type to cell type, depending also on the metabolic state of cells (Chance et al. 1965). Thus, the procedure to subtract all autofluorescence overcompensates, leading to errors of variable degree in the calibration constants. This is particularly serious when the value R_{max} is determined at high $[Ca^{2+}]$, because fura-2 fluorescence at long excitation wavelength (385 nm) is very small under these conditions. For very accurate measurements at high $[Ca^{2+}]$, the mobile part of autofluorescence should be measured separately and allowed for. For normal work, it may be sufficient to perform calibration and test measurements under conditions that are as similar as possible, such that these errors cancel. Nevertheless, it is advisable to determine, in a few test measurements, the order of magnitude of autofluorescence, its variability, and its time course during a whole-cell recording without dye to be able to estimate the order of magnitude of the problem. It has also been found that low autofluorescence (in the excitation range 360–380 nm) is an indicator for healthy cells, and preselecting cells with low autofluorescence is quite helpful for obtaining stable whole-cell patch recordings.

Slow Changes in Fluorescence Properties

Fluorescence properties of cytosol change slowly during prolonged recordings (5–30 min). This may be due to cell swelling (Zhou and Neher 1993b) or to washout of cellular constituents, with the result that the cytosol becomes more similar with time to pipette-filling solution. Strictly speaking, a set of calibration parameters is only valid within a certain time window following establishment of whole-cell recording. Again, care should be taken to perform critical measurements as similar to the corresponding calibration measurements as possible; i.e., at about the same time after break-in (i.e., entry of the pipette into the cell).

Calibration Procedure

The relationship between the fluorescence ratio R and $[Ca^{2+}]$ was given by Grynkiewicz et al. (1985) as

$$[Ca^{2+}] = K_{eff}\,(R - R_0)/(R_1 - R) \tag{4}$$

where K_{eff}, R_0, and R_1 are three calibration constants, which have to be determined for a given setup. The three constants are related to K_D, the dissociation constant of the indicator dye, and to a fourth constant, the isocoefficient α (Zhou and Neher 1993a), through the equation

$$K_D = K_{eff}\,(R_0 + \alpha)/(R_1 + \alpha) \tag{5}$$

Therefore, Equation 4 can be written as:

$$[Ca^{2+}] = K_D \frac{(R_1 + \alpha)}{(R_0 + \alpha)} \times \frac{(R - R_0)}{(R_1 - R)} \tag{6}$$

Once K_D is known for a given dye and a given cell type, it is convenient to base a calibration on the measurements of R_0, R_1, and α, which are relatively easy to determine (see below). Alternatively, Equation 5 may be used to calculate K_D from R_0, R_1, K_{eff}, and α. Since K_D is a property of the indicator dye, which should not depend on the specific fluorimeter, such a calculation is a good check for the validity of the procedures employed. K_D of fura-2 was found to be 238 nM in adrenal chromaffin cells (Zhou and Neher 1993a) in calibrations based on BAPTA, when the K_D of BAPTA was assumed to be 225 nM. This value is close to the value of 259 nM determined by titration of fura-2 in a standard intracellular saline (Oheim 1995). However, this result is quite different from studies on skeletal muscle cells (Konishi et al. 1988) where the K_D of fura-2 was found to be higher by a factor of 3–4 in the presence of myoplasmic proteins as compared to the value in simple saline. Two explanations can be given for these discrepancies. First, problems of interaction of fura-2 with intracellular constituents seem to be much more severe in muscle cells than in other cell types. For instance, diffusion of fura-2 is retarded much less in neuronal axons (Gabso et al. 1997) than in muscle (Baylor and Hollingworth 1988). Second, the calibration buffer used by Zhou and Neher (1993a) is based on BAPTA, a compound much more similar to fura-2 than EGTA, as used in the studies on muscle. Therefore, the K_D of fura-2 in cytosol is expected to be quite close to the K_D of BAPTA in cytosol, although both values may be quite different from those in the calibration buffer. Given our calibration procedure, which interprets all concentration values as those in the calibration buffer (at equilibrium with cytosol), the resulting "cytosolic" K_D of fura-2 is expected to be also quite similar to the one in the calibration buffer.

As argued above, calibration measurements should be performed in a way as similar as possible to the corresponding test measurements. The fluorescence of a region of interest (such as an entire cell) should be measured at two appropriate wavelengths after loading cells with the same solution as is used during experiments, except that $[Ca^{2+}]$ is fixed to one of three values (see below) by adding excess buffer or excess $[Ca^{2+}]$. The fluorescence values should be corrected for background and the fluorescence ratios should be calculated. In all cases, solutions should be titrated to pH 7.2 immediately before use. More details on the procedures used in my laboratory can be found in Neher (1989) and Zhou and Neher (1993a).

Measurements of Calibration Constants

R_o

Measure the limiting fluorescence ratio (derived from two wavelengths, 1 and 2) at low $[Ca^{2+}]$ using a standard internal solution containing 0.1 mM fura-2 and 10 mM BAPTA-tetrapotassium salt (Sigma). (Standard Internal Solution: 145 mM cesium glutamate, 8 mM NaCl, 1 mM $MgCl_2$<!>, and 2 mM Mg-ATP, 10 mM HEPES, 0.3 mM GTP.)

MEASUREMENT OF R_o

Supplement standard internal solution with 0.1 mM fura-2 and 10 mM BAPTA-tetrapotassium salt (Sigma).

MEASUREMENT OF R_1

Supplement standard internal solution with 10 mM $CaCl_2$<!>.

Caution: *See Appendix 3 for appropriate handling of materials marked with <!>.*

MEASUREMENT OF K_{EFF}

Supplement standard internal solution with 0.1 mM fura-2, 6.6 mM Ca-BAPTA, and 3.3 mM BAPTA-tetrapotassium salt.

Note: A whole-cell recording is established, with care being taken that access resistance (the resistance between measuring pipette and cell) is smaller than 20 MΩ and that the fluorescence ratio is measured at a time when the fluorescence readings at both wavelengths have reached a steady state (after ~3–4 min in adrenal chromaffin cells).

This ratio can be used directly as R_0 of Equations 4 and 6.

R_1

Measure the limiting fluorescence at high $[Ca^{2+}]$ in the same way as for R_0, except that 10 mM $CaCl_2$ is added instead of BAPTA (see solution preparation above).

Note: Particular attention has to be given to the series resistance because pipettes tend to seal off at high $[Ca^{2+}]$, and powerful Ca^{2+}-sequestering mechanisms are at work to reduce cytosolic $[Ca^{2+}]$ (for a discussion of the problem of diffusion between a pipette and the cytosol, see Pusch and Neher 1988; Matthias et al. 1990).

It is advisable to provoke Ca^{2+} influx (e.g., by depolarization in the case of cells expressing Ca^{2+} channels) or Ca^{2+} release from stores to verify that cytosolic $[Ca^{2+}]$ has reached its maximum possible value. The fluorescence ratio under these conditions is the quantity R_1 of Equations 4 and 6.

K_{EFF}

Once R_0 and R_1 have been determined, obtain a further measurement at intermediate $[Ca^{2+}]$ to determine K_{eff} from Equation 4.

Note: Supplementing standard internal solution with 0.1 mM fura-2, 6.6 mM Ca-BAPTA, and 3.3 mM BAPTA-tetrapotassium salt is expected to result in a free $[Ca^{2+}]$ of $2K_D$, BAPTA or 450 nm (see above).

Measuring the cellular fluorescence ratio, R_2, allows the calculation of K_{eff} (in nM) from Equation 4 according to:

$$K_{eff} = 450 \times (R_1 - R_2)/(R_2 - R_0) \tag{7}$$

Auxiliary Measurements

It has been argued above that it is advisable to complement the calibration by two more auxiliary measurements, which provide checks for consistency. They also can be used to monitor instrumental changes. These are (1) measurement of the isocoefficient and (2) measurement of the fluorescence ratio of a fluorescence standard, such as fluorescent beads.

THE ISOCOEFFICIENT

If $F_1(t)$ and $F_2(t)$ denote the background-corrected fluorescence signals from a ratiometric dye measured at excitation wavelengths 1 and 2, respectively, it is possible to find a coefficient α, the so-called isocoefficient, for which the linear combination $F(t)$

$$F(t) = F_1(t) + \alpha\, F_2(t) \tag{8}$$

is independent of $[Ca^{2+}]$ (Zhou and Neher 1993a; Naraghi et al. 1998). Once α has been determined for the pair of wavelengths used, it is possible to calculate a Ca^{2+}-independent fluorescence signal (proportional to the total concentration of dye), even if neither of the two wavelengths is exactly at the isosbestic point. Such a signal is useful for monitoring dye loading and for calculating Ca^{2+} signals from single-wavelength measurements in cases when, during a certain time interval, the total dye concentration and its spatial distribution can be considered as constant (Naraghi et al. 1998). The isocoefficient is readily determined during any standard experiment in which a brief Ca^{2+} excursion can be induced, such as by a short depolarization. Such an excursion will show up in both $F_1(t)$ and $F_2(t)$, and it is straightforward to display F(t) on a computer screen according to Equation 8 with a range of values for the parameter α. The correct isocoefficient can then be chosen as that value of α which minimizes the excursion. A more complicated procedure, appropriate for Ca^{2+} imaging, is given by Naraghi et al. (1998).

Once the isocoefficient has been determined, it is advisable to calculate the K_D of the dye, according to Equation 5, and to verify that this value (which is a property of the dye) is reasonable. As mentioned above, the $K_{D,fura}$ was found to be 238 nM by Zhou and Neher (1993a) in adrenal chromaffin cells. Once this value has been determined for a batch of dye and for a given cell type, it is more convenient to calculate $[Ca^{2+}]$ according to Equation 6, which does not require the knowledge of K_{eff}.

The Bead Ratio

The most common cause for changes in calibration parameters on a given setup is aging of the lamp. This can be monitored by measuring the fluorescence ratio, R_B, of a suitable fluorescence standard. Fluoresbrite B/B beads (Polyscience) turned out to be quite useful for this purpose, although their fluorescence properties were found to depend on the ionic composition of the medium (Hoth 1995). If this "bead ratio," as measured at the time of calibration, is designated R_B^C, and a slightly different ratio R_B^t is measured at some later time, or after a lamp change, then the calibration can be readily updated by multiplying both R_0 and R_1 by the factor R_B^t/R_B^C. K_{eff} should be updated according to Equation 5, based on the knowledge of K_D (as calculated from the original calibration) and the new values of R_0, R_1, and α (the latter to be measured under the new conditions). R_B^C and R_B^t should be measured in the same Ringer-type solution. It is advisable to measure the bead ratio repeatedly, at least once every week of experimentation, and to compare both its value and the fluorescence values at the individual wavelengths to corresponding values during the last calibration. This comparison is particularly important following lamp changes and other major changes on a given setup. Unexpected variations will alert the experimenter to misalignments of the optical path, to deterioration of the light source or the detection device, and to many other problems, which readily can invalidate weeks of experimentation.

ACKNOWLEDGMENTS

Work in my laboratory related to the subject of this review was supported by the Behrens-Weise-Stiftung and by the Deutsche Forschungsgemeinschaft (SFB 406 and SFB 523), by a grant from the European Community (CHRX-CT94-0500), and by the Human Science Frontier Program (RG-4/95B). I thank my colleagues Ralf Schneggenburger and Uri Ashery for helpful suggestions on the manuscript.

REFERENCES

Almers W. and Neher E. 1985. The Ca signal from fura-2 loaded mast cells depends strongly on the method of dye-loading. *FEBS Lett.* **192:** 13–18.

Baylor S.M. and Hollingworth S. 1988. Fura-2 calcium transients in frog skeletal muscle fibres. *J. Physiol.* **403:** 151–192.

Chance B., Williamson J.R., Jamieson D., and Schoener B. 1965. Properties and kinetics of reduced pyridine nucleotide fluorescence of the isolated and in vivo rat heart. *Biochem. Z.* **341:** 357–377.

Fierro L. and Llano I. 1996. High endogenous calcium buffering in Purkinje cells from rat cerebellar slices. *J. Physiol.* **496:** 617–625.

Gabso M., Neher E., and Spira M. 1997. Low mobility of the Ca^{2+} buffers in axons of cultured *Aplysia* neurons. *Neuron* **18:** 473–481.

Ganitkevich V.Y. 1996. The amount of acetylcholine mobilisable Ca^{2+} in single smooth muscle cells measured with the exogenous cytoplasmic Ca^{2+} buffer, Indo-1. *Cell Calcium* **20:** 483–492.

Grynkiewicz G., Poenie M., and Tsien R.Y. 1985. A new generation of Ca^{2+} indicators with greatly improved fluorescence properties. *J. Biol. Chem.* **260:** 3440–3450.
Harrison S.M. and Bers D.M. 1987. The effect of temperature and ionic strength on the apparent Ca-affinity of EGTA and the analogous Ca-chelators BAPTA and dibromo-BAPTA. *Biochim. Biophys. Acta* **925:** 133–143.
Heizmann C.W. and Braun K. 1995. *Calcium regulation by calcium-binding proteins in neurodegenerative disorders.* Springer-Verlag, Heidelberg.
Helmchen F., Borst G.G., and Sakmann B. 1997. Calcium dynamics associated with a single action potential in a CNS presynaptic terminal. *Biophys. J.* **72:** 1458–1471.
Helmchen F., Imoto K., and Sakmann B. 1996. Ca^{2+} buffering and action potential-evoked Ca^{2+} signaling in dendrites of pyramidal neurons. *Biophys. J.* **70:** 1069–1081.
Hoth M. 1995. Calcium and barium permeation through calcium release-activated calcium (CRAC) channels. *Pflueg. Arch. Eur. J. Physiol.* **430:** 315–322.
Konishi M., Olson A., Hollingworth S., and Baylor S.M. 1988. Myoplasmic binding of fura-2 investigated by steady-state fluorescence and absorbance measurements. *Biophys. J.* **54:** 1089–1104.
Martell A.E. and Smith R.M. 1977. *Critical stability constants.* Plenum Press, New York.
Mathias R.T., Cohen I.S., and Oliva C. 1990. Limitations of the whole cell patch clamp technique in the control of intracellular concentrations. *Biophys. J.* **58:** 759–770.
Miller D.J. and Smith G.L. 1984. EGTA purity and the buffering of calcium ions in physiological solutions. *Am. J. Physiol.* **246:** C160–C166.
Naraghi M., Müller T.H., and Neher E. 1998. 2-Dimensional determination of the cellular Ca^{2+} binding in bovine chromaffin cells. *Biophys. J.* **75:** 1635–1647.
Neher E. 1989. Combined fura-2 and patch clamp measurements in rat peritoneal mast cells. In *Neuromuscular junction* (ed. L.C. Sellin et al.), pp. 65–76. Elsevier, Amsterdam.
———. 1995. The use of fura-2 for estimating Ca buffers and Ca fluxes. *Neuropharmacology* **34:** 1423–1442.
———. 1998. Usefulness and limitations of linear approximations to the understanding of Ca^{2+} signals. *Cell Calcium* **24:** 345–357.
Oheim M. 1995. "Methodische Voraussetzungen zur Untersuchung der Calcium Diffusion und Calcium Pufferung im Cytosol lebender Chromaffinzellen." Diploma thesis, University of Göttingen, Germany.
Poenie M. 1990. Alteration of intracellular fura-2 fluorescence by viscosity: Simple correction. *Cell Calcium* **11:** 85–91.
Pusch M. and Neher E. 1988. Rates of diffusional exchanges between small cells and a measuring patch pipette. *Pflueg. Arch. Eur. J. Physiol.* **411:** 204–211.
Roberts W.M. 1993. Spatial calcium buffering in saccular hair cells. *Nature* **363:** 74–76.
Tsien R. 1980. New calcium indicators and buffers with high selectivity against magnesium and protons: Design, synthesis, and properties of. *Biochemistry* **19:** 2396–2404.
———. 1989. Fluorescent probes of cell signaling. *Annu. Rev. Neurosci.* **12:** 227–253.
Wagner J. and Keizer J. 1994. Effects of rapid buffers on Ca^{2+} diffusion and Ca^{2+} oscillations. *Biophys. J.* **67:** 447–456.
Williams D. and Fay F. 1990. Intracellular calibration of the fluorescent calcium indicator fura-2. *Cell Calcium* **11:** 75–83.
Xu T., Naraghi M., Kang H., and Neher E. 1997. Kinetic studies of Ca^{2+} binding and Ca^{2+} clearance in the cytosol of adrenal chromaffin cells. *Biophys. J.* **73:** 532–545.
Zhou Z. and Neher E. 1993a. Mobile and immobile calcium buffers in bovine adrenal chromaffin cells. *J. Physiol.* **469:** 245–273.
———. 1993b. Calcium permeability of nicotinic acetylcholine receptor channels in bovine adrenal chromaffin cells. *Pflueg. Arch. Eur. J. Physiol.* **425:** 511–517.

CHAPTER 31

Calibration of Fluorescent Calcium Indicators

Fritjof Helmchen

Abteilung Zellphysiologie, Max-Planck-Institut für medizinische Forschung, 69120 Heidelberg, Germany

During the past decades, many different fluorescent indicators have been developed for measuring intracellular ion concentrations (Tsien 1989; Haugland 1993; see Chapter 29). Of particular interest has been the design of various fluorescent calcium indicators, motivated by the fundamental role of Ca^{2+} in cellular functions such as contraction, secretion, and gene activation. For a quantitative understanding of the physiological functions of Ca^{2+} (e.g., for determination of dose-response curves), the measured fluorescence values must be calibrated in terms of intracellular free-calcium concentration ($[Ca^{2+}]_i$). This chapter summarizes the basic methods, equations, and calibration procedures that have been used for this conversion. In addition, a different type of calibration is described, which is used to quantify total calcium fluxes.

Fluorescent calcium indicators are universal tools for studying intracellular signaling and are therefore employed in all fields of cell biology. For example, in neurobiology, they have been applied to study the role of calcium in neurotransmitter release and in postsynaptic signal integration. For each specific application, the most suitable indicator dye can be chosen from a large palette of indicators, which differ with respect to their affinity, mobility, solubility, and fluorescence properties. Two large groups of indicators exist: (1) synthetic organic molecules (mostly derived from the fast calcium buffer BAPTA; Tsien 1980), which can be loaded into cells and subcellular compartments using various techniques (Tsien 1989), and (2) genetically encoded calcium indicators based on fluorescent proteins (for reviews, see Guerrero and Isacoff 2001; Miyawaki 2003; see Chapter 74). A common feature of all indicators is that their fluorescence is sensitive to $[Ca^{2+}]_i$ in at least one respect; Ca^{2+} binding may cause changes in fluorescence intensity, it may result in spectral shifts or changes in FRET (fluorescence resonance energy transfer) efficiency, which enable fluorescence ratioing, and/or it may change the fluorescence lifetime (Fig. 1). Described here are the basic principles of converting these fluorescence changes to $[Ca^{2+}]_i$ changes. Note that equivalent considerations apply to measurements using fluorescent indicators of other ion species (e.g., H^+, Na^+, and Cl^-).

CHANGES IN FLUORESCENCE INTENSITY

The emission intensity F arising from an observation volume V, e.g., a single cell loaded with a fluorescent dye, depends on the number of dye molecules, the illumination intensity I_0, the dye absorption coefficient α, the quantum yield of the dye Q_F, the photon-collection efficiency Φ of the optical setup, and the quantum efficiency (QE) of the detector Q_D:

$$F = \Phi\, Q_D\, Q_F\, \alpha\, I_0\, n = S \cdot n \quad (1)$$

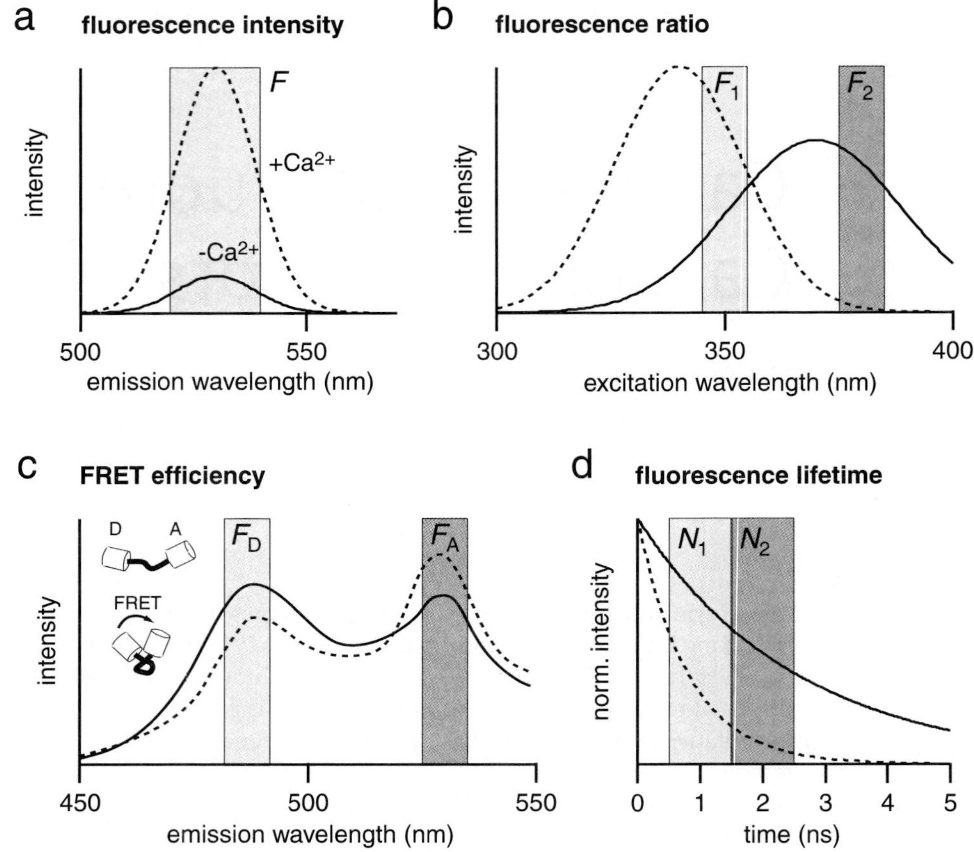

FIGURE 1. Changes in fluorescence properties of calcium indicators used for calibration. (*a*) Changes in fluorescence intensity result from changes of the quantum yield and the absorption of a dye upon Ca^{2+} binding. An emission spectrum, similar to that of Calcium Green-1, is shown schematically with higher intensities of the Ca^{2+}-bound form (*dashed line*) compared to the unbound form (*solid line*). Other indicators (e.g., fura-red [not shown]) show decreases in intensity. In single-wavelength measurements, the dye is excited at a single wavelength and the emission intensity *F* is collected in a spectral window around the peak of the emission spectrum. (*b*) Spectral shifts allow ratiometric measurements because the ratio of the fluorescence intensities F_1 and F_2, measured at two different wavelengths in this case, is sensitive to Ca^{2+} binding. The drawing schematically shows an excitation spectrum similar to that of fura-2 at zero Ca^{2+} concentration (*solid line*) and at saturating Ca^{2+} levels (*dashed line*). The peak of the spectrum shifts to shorter wavelength upon Ca^{2+} binding. Note that excitation at 360 nm results in $[Ca^{2+}]_i$-insensitive fluorescence emission. This is the so-called isosbestic wavelength. (*c*) Changes in FRET efficiency are used in the case of tandem GFP-based indicators, which use a donor (D) fluorescent protein linked via a Ca^{2+}-sensitive spacer to an acceptor (A) fluorescent protein. The emission spectrum of a yellow cameleon (with ECFP and EYFP as donor and acceptor, respectively) is shown schematically at low (*solid line*) and high (*dashed line*) $[Ca^{2+}]_i$ level (excitation ~430 nm). The distance between the two fluorescent proteins decreases upon Ca^{2+} binding, which, as a result of enhanced FRET efficiency, causes the donor emission (F_D) to decrease and the acceptor fluorescence (F_A) to increase, thus permitting ratiometric measurements. (*d*) Alternatively, changes in the fluorescence lifetime have been used for calibration. Fluo-3, for example, shows a shorter fluorescence lifetime in the Ca^{2+}-bound form (*dashed line*) compared to the unbound form (*solid line*) as illustrated. Other indicators show an increase of fluorescence lifetime upon Ca^{2+} binding. As a result, the ratio of the number of photons N_1 and N_2 that are detected in two time windows during the fluorescence decay is sensitive to $[Ca^{2+}]_i$. Note that not all indicators necessarily are $[Ca^{2+}]_i$-sensitive with respect to all the properties shown in (*a–d*).

where n is the molar amount of dye present in V. All factors that depend on dye properties or the experimental setup can be "lumped" together as a constant factor S. In the case of a calcium indicator, the molar amounts n_f and n_b of the free and the Ca^{2+}-bound form must be considered separately. They differ with respect to their quantum yield and their absorption, and therefore contribute to F with different factors, S_f and S_b, respectively:

$$\begin{aligned} F &= S_f\, n_f + S_b\, n_b \\ &= F_{min} + (S_b - S_f) n_b \\ &= F_{max} - (S_b - S_f) n_f \end{aligned} \tag{2}$$

This equation has been rewritten using the definitions for the fluorescence at zero Ca^{2+} concentration $F_{min} = S_f\, n_{tot}$ and at saturating $[Ca^{2+}]_i$ levels $F_{max} = S_b\, n_{tot}$ with $n_{tot} = n_f + n_b$, and assuming a fluorescence increase upon Ca^{2+} binding. The indicator fluorescence thus depends on the relative amounts of the free and bound forms. Assuming 1:1 binding ratio of Ca^{2+} with the dye, n_f and n_b vary according to the law of mass action

$$K_d = \frac{n_b\,[Ca^{2+}]_i}{n_f} \tag{3}$$

where K_d is the dissociation constant of the indicator. Equations 1–3 are the basis for all conversion equations that relate emission intensities to $[Ca^{2+}]_i$ values and are described in the following sections. First, however, some assumptions and restrictions concerning the dye concentration and the subtraction of background fluorescence are discussed.

Dye Concentration and Background Subtraction

A central assumption underlying Equation 1 is that all dye molecules sense the same illumination intensity, meaning that no inner filtering occurs. The intensity of light absorbed by a fluorescent layer of thickness l is given by the Beer-Lambert law

$$I_{abs} = I_0\,(1 - 10^{-\varepsilon lc}) \approx I_0 \ln(10)\, \varepsilon lc \tag{4}$$

where ε is the molar extinction coefficient and c the dye concentration. The approximation of a linear relationship between I_{abs} and c is valid only if $c < (\ln(10)\varepsilon l)^{-1}$. This sets an upper limit to the useful dye concentration range. Extinction coefficients of calcium indicators are typically in the range of 20,000–100,000 M^{-1} cm^{-1} (Molecular Probes). Thus, for measurements on small cells with 10-μm diameter, the indicator concentration should be well below 5–20 mM. In thick cuvettes of about 1-cm path length, which are sometimes used for an in vitro calibration, much smaller concentrations must be used (typically 1 μM). At very high concentration, dye fluorescence may also be reduced because of self-quenching. The most severe restriction on the concentration of high-affinity indicators, however, often arises from the buffering of Ca^{2+} by the indicator. Depending on the endogenous Ca^{2+}-buffering capacity, even a relatively low concentration of a high-affinity dye (100 μM) may reduce and prolong $[Ca^{2+}]_i$ changes severalfold (Neher 1995; Chapter 32). Low-affinity dyes can partly circumvent this problem, because they can be used at relatively high concentrations without significantly altering $[Ca^{2+}]_i$ dynamics. The resulting fluorescence signals, however, may be very small.

Optical components, the bathing solution, and endogenous fluorophores all add background to the indicator fluorescence given in Equation 2. Since background fluorescence increases the noise level, it should be minimized (Moore et al. 1990). Even so, background fluorescence must be subtracted from the observed fluorescence before applying equations for conversion to $[Ca^{2+}]_i$ ($F = F_{obs} - F_{back}$). In single-cell imaging experiments in tissue slices, a relatively large amount of background arises from endogenous fluorophores of the surrounding tissue, especially with UV excitation. In addition, the background may vary during an experiment due to bleaching (Eilers et al. 1995). Therefore, it is necessary to perform background measurements throughout the experiment in a region neighboring the cell, e.g., before and after each stimulation (Helmchen et al. 1996). A more severe problem occurs when tissue is loaded with AM-esters of calcium indicators (see Chapter 44). In this case, the background can still be estimated from a region near the cell under investigation; however, upon stimulation, it may vary due to $[Ca^{2+}]_i$-sensitive background indicator fluorescence. If it is variable, it must be monitored continuously and subtracted frame by frame.

Single-Wavelength Measurements

One class of calcium indicators responds to Ca^{2+} binding with an up- or downscaling of the fluorescence intensity without showing appreciable spectral shifts (see Fig. 1a). Almost all synthetic indicators excited in the visible-wavelength range belong to this group. As a result, all available information on Ca^{2+} concentration is obtained by exciting the dye at a single wavelength.

In principle, the fluorescence signal F, measured at a single excitation wavelength, can be converted to $[Ca^{2+}]_i$ by assuming equilibrium between Ca^{2+} and the indicator and combining Equations 2 and 3:

$$[Ca^{2+}]_i = K_d \frac{n_b (S_b - S_f)}{n_f (S_b - S_f)} = K_d \frac{F - F_{min}}{F_{max} - F} \qquad (5)$$

Although this equation is readily applicable to cuvette measurements (e.g., measurements of cell suspensions and also calibration measurements), it is difficult to apply to imaging experiments because F_{min} and F_{max} would have to be determined individually for each observation volume (corresponding to each pixel). If, however, ratios of intensities are used instead of absolute intensities, then variations in cell thickness and total dye concentration, as well as illumination heterogeneities, cancel out. For time-dependent measurements, this is achieved by expressing the signal as relative fluorescence change

$$\Delta F/F = (F - F_0)/F_0 \qquad (6)$$

where F_0 denotes the background-subtracted prestimulus fluorescence level. For $\Delta F/F$ ("delta F over F"), the following conversion equation can be derived (Lev-Ram et al. 1992):

$$[Ca^{2+}]_i = \frac{[Ca^{2+}]_{rest} + K_d \dfrac{(\Delta F/F)}{(\Delta F/F)_{max}}}{\left(1 - \dfrac{(\Delta F/F)}{(\Delta F/F)_{max}}\right)} \qquad (7)$$

where $[Ca^{2+}]_{rest}$ is the resting calcium concentration, and $(\Delta F/F)_{max}$ is the maximal change upon dye saturation, which can often be estimated using strong stimulation. The major drawback of this and other related single-wavelength equations (Vranesic and Knöpfel 1991; Neher and Augustine 1992; Wang et al. 1995) is that an a priori knowledge of $[Ca^{2+}]_{rest}$ is required, for example, from an initial ratiometric measurement (see below). In the case of combined electrical and fluorometric recordings, however, during which the health of the cell can be judged from electrophysiological parameters, a reasonable value of $[Ca^{2+}]_{rest}$ may be assumed (50–100 nM) to obtain a rough estimate of $[Ca^{2+}]_i$ using Equation 7.

An alternative single-wavelength approach that circumvents the necessity for an independent measurement of $[Ca^{2+}]_{rest}$ is based on a rearrangement of Equation 5 (Maravall et al. 2000):

$$[Ca^{2+}]_i = K_d \frac{F/F_{max} - 1/R_f}{1 - F/F_{max}} \qquad (8)$$

where R_f denotes the dynamic range (F_{max}/F_{min}) of the indicator. The idea is that the ratio of the actual fluorescence F to the saturating fluorescence F_{max} reflects the $[Ca^{2+}]_i$ level, given that the dynamic range R_f is known (R_f must be determined for a particular indicator initially). The resting $[Ca^{2+}]_i$ level is directly related to the inverse of $(\Delta F/F)_{max}$ (the closer $[Ca^{2+}]_{rest}$ already is to indicator saturation, the smaller is $[\Delta F/F]_{max}$). $[Ca^{2+}]_i$ calibration therefore requires intermittent measurements of F/F_{max} or $(\Delta F/F)_{max}$ during an experiment, which, for high-affinity indicators, can be achieved by inducing trains of high-frequency action potentials (Maravall et al. 2000).

If only small fluorescence changes are evoked and if the indicator is far from saturation (e.g., in the case of low-affinity calcium indicators), the single-wavelength equations can be linearized to provide an estimate of the change in $[Ca^{2+}]_i$

$$\Delta[Ca^{2+}]_i = \frac{K_d}{(\Delta F/F)_{max}} (\Delta F/F) \quad (\Delta F/F << (\Delta F/F)_{max}) \tag{9}$$

For low-affinity indicators it may, however, be difficult to induce a large enough calcium influx to determine the saturating fluorescence changes.

Dual-Wavelength Ratiometric Measurements

A second group of indicators undergoes shifts in the excitation or emission spectrum upon Ca^{2+} binding (Fig. 1b). These spectral shifts can be exploited for $[Ca^{2+}]_i$ calibration because the ratio $R = F_1 / F_2$ of the intensities measured at two wavelengths in this case depends on $[Ca^{2+}]_i$. The ratiometric method is the most often applied method of calibration, because R is independent of dye concentration, optical path length, and illumination intensity. The intensities F_1 and F_2 are given according to Equation 2:

$$\begin{aligned} F_1 &= S_{f1}\, n_f + S_{b1}\, n_b \\ F_2 &= S_{f2}\, n_f + S_{b2}\, n_b \end{aligned} \tag{10}$$

Note that for actual measurements, both intensities must be corrected for background fluorescence at the corresponding excitation or emission wavelengths before taking the ratio. From Equations 3 and 10, the standard equation for ratiometric measurements can be derived (Grynkiewicz et al. 1985):

$$[Ca^{2+}]_i = K_{eff} \frac{R - R_{min}}{R_{max} - R} \tag{11}$$

with the ratios at zero Ca^{2+} concentration, $R_{min} = (S_{f1}/S_{f2})$, at saturating Ca^{2+} concentrations, $R_{max} = (S_{b1}/S_{b2})$, and an effective binding constant $K_{eff} = K_d\,(S_{f2}/S_{b2})$.

The design and use of ratio imaging systems have been described in a number of reviews (see, e.g., Neher 1989; Tsien and Harootunian 1990). Dual-wavelength measurements with high time resolution require rapid switching of wavelengths. As an alternative, Equation 11 can be used to obtain an initial value of $[Ca^{2+}]_{rest}$ for subsequent single-wavelength measurement and application of Equation 7. For a more detailed description of how to use fura-2, the most popular ratiometric dye, see Chapter 30. The ratiometric method has also been extended to mixtures of nonratiometric dyes that result in Ca^{2+}-sensitive fluorescence ratios (Lipp and Niggli 1993; Oheim et al. 1998).

Changes in FRET Efficiency

Genetically encoded calcium indicators consist either of a single modified green fluorescent protein (GFP) exhibiting Ca^{2+}-sensitive fluorescence (Baird et al. 1999; Nakai et al. 2001) or they are based on changes in FRET between two spectral variants of GFP. For example, cameleons are fusion proteins of two fluorescent proteins linked via a spacer consisting of calmodulin and the calmodulin-binding peptide M13 (Miyawaki et al. 1997). Ca^{2+} binding to calmodulin causes a conformational change, bringing the two fluorescent proteins closer together and causing an increase in the FRET efficiency, which can be read out by measuring the change in the ratio of the donor and acceptor emission intensities, respectively (Fig. 1c). Since the emission spectra of donor and acceptor typically have some overlap, cross-talk may have to be taken into account (Gordon et al. 1998). For further information on genetically encoded calcium indicators, see Chapters 74 and 78.

CHANGES IN FLUORESCENCE LIFETIME

As an alternative to the intensity, the fluorescence lifetime can be used as an indicator of Ca^{2+} binding (Lakowicz et al. 1992; Draaijer et al. 1995). After the end of excitation, the fall-off in fluorescence intensity reflects the lifetime of the excited state of the dye molecules (Fig. 1d). In the simplest case, the decay is described by a single exponential curve with a fluorescence lifetime constant τ_L, typically in the nanoseconds range. The calcium indicators quin-2, fura-2, fluo-3, and Calcium Green-1 all respond to Ca^{2+} binding with increases or decreases in τ_L (Draaijer et al. 1995).

One method to measure the fluorescence lifetime is time-gated photon detection following a brief exciting laser pulse. The ratio of the numbers of photons N_1 and N_2 that are detected in two time windows following the excitation pulse provides an effective fluorescence lifetime $\tau_{eff} = \Delta t/\log(N_1/N_2)$, where Δt is the width of the windows (Fig. 1d). τ_{eff} can also be obtained from the phase shift and the change in amplitude when a modulated light source is used (Lakowicz et al. 1992). Both the free and the Ca^{2+}-bound form of the indicator contribute to the effective fluorescence lifetime, which relates to $[Ca^{2+}]_i$ via the equation:

$$[Ca^{2+}]_i = K_{app} \frac{\tau_{eff} - \tau_{min}}{\tau_{max} - \tau_{eff}} \tag{12}$$

where τ_{min} and τ_{max} denote the lifetime of the bound and free indicator form, respectively, assuming a lifetime decrease, and K_{app} denotes an apparent dissociation constant.

Since lifetimes are independent of dye concentration and illumination intensity, a calibration according to Equation 12 is readily applied to imaging experiments. K_{app} depends on the relative intensities of the free and bound indicator forms. Interestingly, the apparent affinity of an indicator showing a spectral shift can be tuned in a wide range simply by selecting the excitation wavelength (Szmacinski and Lakowicz 1995). Lifetime-based Ca^{2+} imaging has rarely been applied so far, probably because of the advanced technical requirements. Microscopic techniques that inherently use pulsed laser sources for excitation, such as two-photon laser-scanning microscopy (see Chapter 7), may be adapted easily for lifetime measurements. Notably, a fast fluorescence lifetime imaging (FLIM) microscope has recently been described, enabling $[Ca^{2+}]_i$ measurements with frame rates of more than 50 Hz (Agronskaia et al. 2003). Finally, measurement of the fluorescence lifetime of the combined donor/acceptor emission can also be used as an alternative means to determine changes in FRET efficiency (Harpur et al. 2001).

CALIBRATION METHODS

Calibration Solutions

All of the methods described above require the determination of calibration parameters. Therefore, the dependence of the relevant fluorescence property (intensity, ratio, or lifetime) on $[Ca^{2+}]_i$ must be measured using a set of solutions of known Ca^{2+} concentration. Calibration solutions are commercially available from Molecular Probes. Dissociation constants of Ca^{2+} buffers and indicator properties, however, depend on temperature, pH, and the ionic strength of the solution (Groden et al. 1991). Therefore, it may be necessary to prepare custom buffered solutions in order to mimic experimental conditions. For guidelines on the preparation of calibration solutions, see Tsien and Pozzan (1989).

In Vitro Calibration

For an in vitro calibration, glass capillaries or microslides (e.g., from Vitrodynamics) are filled with calibration solutions. Fluorescence is measured under conditions as close as possible to the experimental conditions; e.g., capillaries should be imaged under the microscope used for imaging experiments. If a water-immersion objective is used, the ends of the microslides must be sealed before mounting them on the microscope. As an example, the determination of the three ratiometric parameters R_{min}, R_{max}, and K_{eff}, for a dye such as fura-2, requires at least three calibration solutions with low, intermediate, and high $[Ca^{2+}]_i$. Typically, a highly buffered solution (e.g., with 10 mM BAPTA), a solution adjusted to a calcium concentration around the K_d of the indicator, and a solution with a saturating calcium concentration (>10 mM) are used.

Fluorescence is measured at both excitation wavelengths for each calibration solution. Background is also measured, using capillaries filled with dye-free solution, and subtracted. The ratios

R are calculated for each $[Ca^{2+}]_i$ level, and the three parameters are obtained from this set of measurements using Equation 11.

Unfortunately, the parameters obtained from in vitro calibrations may give unreliable results. This is because the behavior of fluorescent dyes is altered in the viscous cytosolic environment and by intracellular binding or uptake (Moore et al. 1990). Apparent negative $[Ca^{2+}]_i$ values are a clear indication of this problem when in vitro calibration parameters are applied to fluorescence data from cells. Poenie (1990) suggested that viscosity could be corrected for by multiplying R_{min} and R_{max} by a factor of 0.7–0.85.

In Vivo Calibration

Where possible, an in vivo calibration is preferable. In this case, the cells under investigation are filled directly with buffered $[Ca^{2+}]_i$ solutions. Patch-clamp experiments, for example, provide direct access to the cytosol (Neher 1989). For calibration in these experiments, whole-cell recordings with the pipette containing the intracellular solution buffered to different $[Ca^{2+}]_i$ levels are obtained. Small cells such as chromaffin cells are readily filled with the calibration solutions (within minutes), and a set of calibration measurements can be obtained as described for the in vitro calibration. For large cells, the loading of the cell may take significantly longer. In addition, strong extrusion mechanisms may prevent $[Ca^{2+}]_i$ from reaching and maintaining the known level of the pipette solution. In these cases, calibration measurements should be taken close to the pipette tip and with low access resistance (Eilers et al. 1995). In general, reliable values for R_{min} can be obtained in this way, but the determination of R_{max} and K_{eff} is more susceptible to the problems of obtaining a stable clamp of the $[Ca^{2+}]_i$ level. R_{max}, however, can also be estimated at the end of, or during, an experiment by applying a strong stimulation, e.g., a long high-frequency train of action potentials, to saturate the indicator.

Calcium-Flux Measurements

Calcium indicators have mainly been used for measuring the concentration of free calcium ions. However, as calcium chelators, they can significantly alter the magnitude and time course of $[Ca^{2+}]_i$ changes (see Chapter 32). This problem of indicator buffering is separate from the problem of an accurate calibration: Even if the calculated concentration values are correct, they may not reflect the $[Ca^{2+}]_i$ levels reached in the absence of the indicator. On the other hand, buffering by the indicator can be exploited to measure Ca^{2+} fluxes (Schneggenburger et al. 1993; Neher 1995). When loaded in excess into a cellular compartment, the indicator molecules out-compete the endogenous Ca^{2+} buffers and capture virtually all ions that enter the cytosol. Under these "overload" conditions, the change in absolute fluorescence intensity is proportional to the total calcium charge Q_{Ca} injected:

$$\Delta F = (S_b - S_f)\Delta n_b = \frac{(S_b - S_f)}{2F_c} Q_{Ca} = f_{max} Q_{Ca} \tag{13}$$

where F_c is Faraday's constant. Calibration in this case refers to the determination of the proportionality f_{max}, for example, by electrical recordings of pure calcium currents and the simultaneous measurement of the evoked fluorescence changes. Subsequently, changes in absolute fluorescence can be directly converted to calcium charge. The overload method has been used to determine fractional Ca^{2+} currents through ligand-gated ion channels (Schneggenburger et al. 1993; Neher 1995; Bollmann et al. 1998) and the total calcium influx during an action potential (Helmchen et al. 1997; Bollmann et al. 1998).

EXAMPLE OF APPLICATION

To illustrate several of the calibration techniques described above, Figure 2 shows examples of calcium measurements from large presynaptic terminals ("calyces of Held") in the medial nucleus of the trapezoid body (MNTB) (see also Helmchen et al. 1997). Using an acute brain slice preparation, nerve terminals were loaded with various indicators at different concentrations, via whole-cell patch

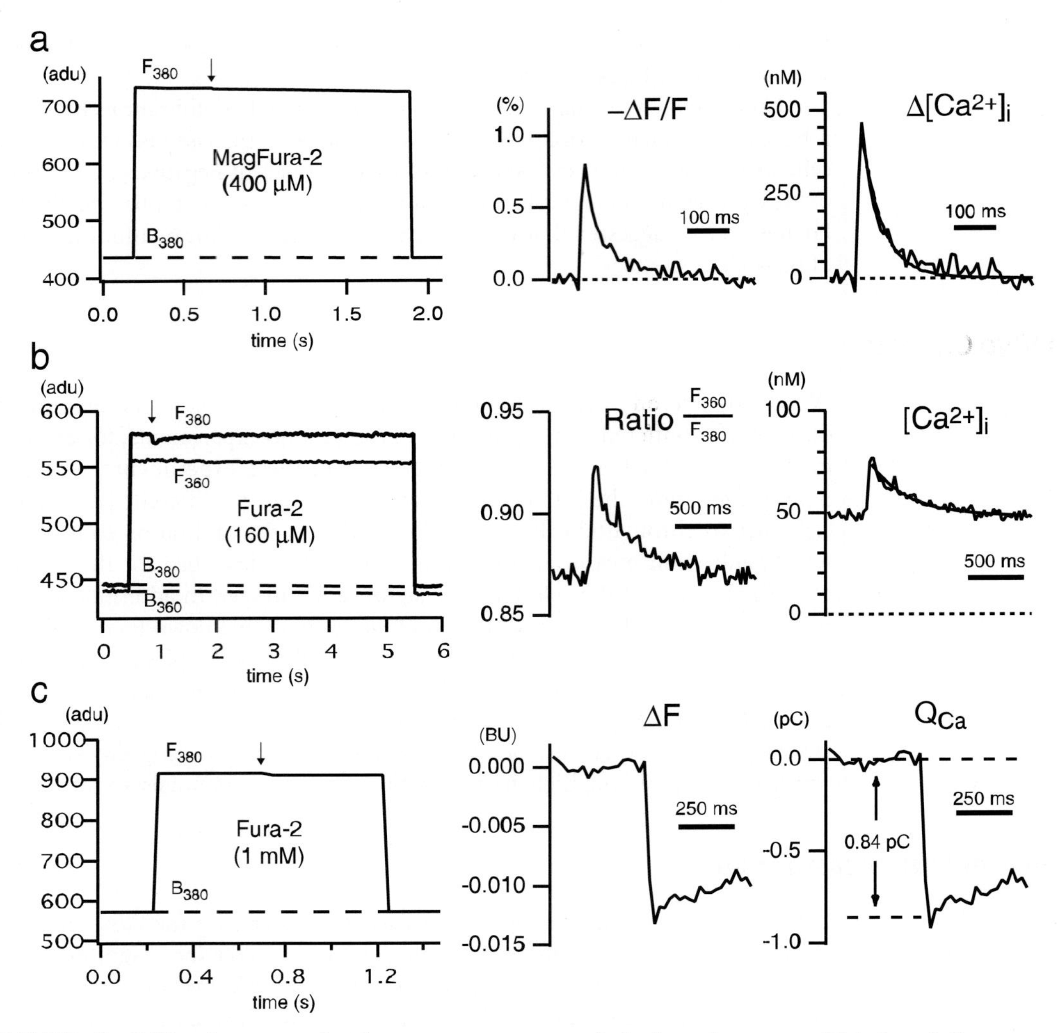

FIGURE 2. Calibration examples from measurements of single-action potential-induced fluorescence changes in calxy-type presynaptic terminals in the medial nucleus of the trapezoid body (MNTB). The fluorescence (F) averaged over the entire terminal was measured using a fast CCD camera (expressed in analog-to-digital units [adu]). Background fluorescence (B) from a nearby region was determined immediately before and after each measurement (interpolated by dashed lines). The timing of single action potentials is indicated by arrows. The temperature was 35°C. (*a*) Single-wavelength measurement using the low affinity indicator MagFura-2. A single action potential caused a small $\Delta F/F$ change of less than 1%, which was converted to $\Delta[Ca^{2+}]_i$ using Equation 9. Average of 50 traces. (*b*) Ratiometric measurement using fura-2. The ratio between the fluorescence intensities at the isosbestic wavelength (F_{360}) and at 380 nm (F_{380}) was evaluated and converted to $[Ca^{2+}]_i$ using Equation 11. Note that an estimate of $[Ca^{2+}]_{rest}$ is obtained, but that the $[Ca^{2+}]_i$ transient is profoundly reduced in amplitude and prolonged due to the added fura-2 Ca^{2+}-buffering capacity. No averaging. (*c*) Calcium flux measurement using fura-2 overload. At high fura-2 concentration, a single action potential induces only a small fluorescence decrement ΔF_{380} of about 5 adu ($\approx$1% $\Delta F/F$), which is expressed in bead units (BU) and then converted to the total calcium charge Q_{Ca} using Equation 13. Average of 20 traces. For further details, see text.

pipettes. In the panels on the left in Figure 2, the raw fluorescence data are shown, including pre- and poststimulus measurements of the background fluorescence. These data were then evaluated as either relative fluorescence change $\Delta F/F$, ratio, or absolute fluorescence change (Fig. 2, middle panels). Finally, fluorescence changes were converted to calcium concentration or charge by applying the set of predetermined calibration parameters.

Single-wavelength measurements using the low-affinity indicator MagFura-2 were performed to obtain an estimate of the presynaptic $[Ca^{2+}]_i$ dynamics with minimal distortion by indicator buffering (Fig. 2a). At 380-nm wavelength, MagFura-2 fluorescence decreases upon Ca^{2+} binding. For calibration, a dissociation constant of K_d = 45 μM was assumed and $(-\Delta F/F)_{max}$ was determined to be 78% using long trains of action potentials at 200 Hz. Because a single action potential induced only a very small fluorescence change, the linearized Equation 9 could be applied, revealing a large (several hundred nanomolars) and brief (decay time constant 50 msec) calcium transient.

In a separate set of experiments, fura-2 was used at a moderate concentration (160 μM), and ratiometric measurements were performed using fast 360/380-nm wavelength switching with a monochromator-based illumination system (see Chapter 99). Calibration parameters were determined using an in vivo calibration procedure by loading terminals with internal solutions clamped to zero, intermediate, and high $[Ca^{2+}]_i$, respectively (resulting in R_{min} = 0.77; R_{max} = 3.15; K_{eff} = 1117 nM). Note that in the case of using the isosbestic wavelength in the nominator, the K_d of the dye is given by $K_d = K_{eff} \cdot R_{min} / R_{max}$, yielding K_d = 273 nM for fura-2 in our case. Conversion using these parameters gave an estimate of about 50 nM for $[Ca^{2+}]_{rest}$. The amplitude (26 nM) and the decay time constant (364 msec) of the single action-potential-induced calcium transient were, however, clearly altered compared to the MagFura-2 measurement (Fig. 2b). This is attributed to the relatively large added Ca^{2+}-buffering capacity compared to the endogenous Ca^{2+}-buffering capacity (see Chapter 32).

Finally, fura-2 was used at high (1 mM) concentration to overload the presynaptic terminal with added buffer. In this case, the change in absolute fluorescence intensity is evaluated (Fig. 2c). To normalize for changes in the imaging system over time (e.g., aging of the arc lamp), it is necessary to express absolute fluorescence in units of a fluorescent standard, typically fluorescent beads ("bead units" or BU), which must be measured on each day of the experiment (Schneggenburger et al. 1993). The proportionality constant f_{max}, which is required for conversion to total calcium charge (Eq. 13), was determined in separate experiments using pure calcium currents as f_{max} = 0.0144 BU pC^{-1} (Helmchen et al. 1997).

Together, these examples demonstrate how different methods of calcium calibrations can be used to quantify various aspects of the calcium signaling system in small neuronal compartments.

ADVANTAGES AND LIMITATIONS

In summary, many different fluorescent calcium indicators are by now available and different methods exist for quantifying Ca^{2+}-signaling parameters from the observed fluorescence signals. Quantitative measurements require a careful understanding of the indicator properties under the experimental conditions and of possible interference of other binding partners (e.g., H^+, Mg^{2+}). As shown in the examples in Figure 2, employment of different types and/or concentrations of indicators can be highly informative. Single-wavelength measurements provide the highest possible time resolution and should be used to measure fast signals. Combined with initial ratiometric measurements, or following careful determination of the dissociation constant and the dynamic range of the indicator, single-wavelength measurements can also be used for calibrated $[Ca^{2+}]_i$ measurements.

A general problem with calcium measurements using imaging systems is, however, that they represent the $[Ca^{2+}]_i$ levels averaged over a certain cytosolic volume, in the best case over the diffraction-limited focal volume. This means that they are insensitive to highly localized $[Ca^{2+}]_i$ gradients and may underestimate the actual $[Ca^{2+}]_i$ level reached at the site of action, e.g., the binding to a Ca^{2+} sensor. Two approaches may overcome this problem: (1) Ca^{2+} uncaging by flash photolysis (see Chapter 46), causing spatially homogeneous $[Ca^{2+}]_i$ elevations, which can be quantified using the methods described here, and (2) application of indicator forms that are targeted to the intracellular sites of interest, e.g., by genetic means.

ACKNOWLEDGMENTS

I thank Samual S.-H. Wang for critical comments on an earlier version of this chapter.

REFERENCES

Agronskaia A.V., Tertoolen L., and Gerritsen H.C. 2003. High frame rate fluorescence lifetime imaging. *J. Phys. D Appl. Phys.* **36:** 1655–1662.

Baird G.S., Zacharias D.A., and Tsien R.Y. 1999. Circular permutation and receptor insertion within green fluorescent proteins. *Proc. Natl. Acad. Sci.* **96:** 11241–11246.

Bollmann J.H., Helmchen F., Borst J.G.G., and Sakmann B. 1998. Postsynaptic Ca^{2+} influx mediated by three different pathways during synaptic transmission at a calyx-type synapse. *J. Neurosci.* **18:** 10409–10419.

Draaijer A., Sanders R., and Gerritsen H.C. 1995. Fluorescence lifetime imaging, a new tool in confocal microscopy. In *Handbook of biological confocal microscopy*, 2nd edition (ed. J.B. Pawley), pp. 491–505. Plenum Press, New York.

Eilers J., Schneggenburger R., and Neher E. 1995. Patch clamp and calcium imaging in brain slices. In *Single-channel recording*, 2nd edition (ed. B. Sakmann and E. Neher), pp. 213–229. Plenum Press, New York.

Gordon G.W., Berry G., Liang X.H., Levine B., and Herman B. 1998. Quantitative fluorescence resonance energy transfer measurements using fluorescence microscopy. *Biophys. J.* **74:** 2702–2713.

Groden D.L., Guan Z., and Stokes B.T. 1991. Determination of Fura-2 dissociation constants following adjustment of the apparent Ca-EGTA association constant for temperature and ionic strength. *Cell Calcium* **12:** 279–287.

Grynkiewicz G., Poenie M., and Tsien R.Y. 1985. A new generation of Ca^{2+} indicators with greatly improved fluorescence properties. *J. Biol. Chem.* **260:** 3440–3450.

Guerrero G. and Isacoff E.Y. 2001. Genetically encoded optical sensors of neuronal activity and cellular function. *Curr. Opin. Neurobiol.* **11:** 601–607.

Harpur A.G., Wouters F., and Bastiaens P.I. 2001. Imaging FRET between spectrally similar GFP molecules in single cells. *Nat. Biotechnol.* **19:** 167–169.

Haugland R. 1993. Intracellular ion indicators. In *Fluorescent and luminescent probes for biological activity* (ed. W.T. Mason), pp. 34–43. Academic Press, London.

Helmchen F., Borst J.G.G., and Sakmann B. 1997. Calcium dynamics associated with a single action potential in a CNS presynaptic terminal. *Biophys. J.* **72:** 1458–1471.

Helmchen F., Imoto K., and Sakmann B. 1996. Ca^{2+} buffering and action potential-evoked Ca^{2+} signaling in dendrites of pyramidal neurons. *Biophys. J.* **70:** 1069–1081.

Lakowicz J.R., Szmacinski H., Nowaczyk K., and Johnson M.L. 1992. Fluorescence lifetime imaging of calcium using Quin-2. *Cell Calcium* **13:** 131–147.

Lev-Ram V., Miyakawa H., Lasser-Ross N., and Ross W.N. 1992. Calcium transients in cerebellar Purkinje neurons evoked by intracellular stimulation. *J. Neurophysiol.* **68:** 1167–1177.

Lipp P. and Niggli E. 1993. Ratiometric confocal Ca^{2+}-measurements with visible wavelength indicators in isolated cardiac myocytes. *Cell Calcium* **14:** 359–372.

Maravall M., Mainen Z.F., Sabatini B.L., and Svoboda K. 2000. Estimating intracellular calcium concentrations and buffering without wavelength ratioing. *Biophys. J.* **78:** 2655–2667.

Miyawaki A. 2003. Visualization of the spatial and temporal dynamics of intracellular signaling. *Dev. Cell* **4:** 295–305.

Miyawaki A., Llopis J., Heim R., McCaffery J.M., Adams J.A., Ikura M., and Tsien R.Y. 1997. Fluorescent indicators for Ca^{2+} based on green fluorescent proteins and calmodulin. *Nature* **388:** 882–887.

Moore E.D.W., Becker P.L., Fogarty K.E., Williams D.A., and Fay F.S. 1990. Ca^{2+} imaging in single living cells: Theoretical and practical issues. *Cell Calcium* **11:** 157–179.

Nakai J., Ohkura M., and Imoto K. 2001. A high signal-to-noise Ca^{2+} probe composed of a single green fluorescent protein. *Nat. Biotechnol.* **19:** 137–141.

Neher E. 1989. Combined fura-2 and patch clamp measurements in rat peritoneal mast cells. In *Neuromuscular junction* (ed. L. Sellin et al.), pp. 65–76. Elsevier, Amsterdam.

———. 1995. The use of fura-2 for estimating Ca buffers and Ca fluxes. *Neuropharmacology* **34:** 1423–1442.

Neher E. and Augustine G. 1992. Calcium gradients and buffers in bovine chromaffin cells. *J. Physiol.* **450:** 273–301.

Oheim M., Naraghi M., Müller T.H., and Neher E. 1998. Two dye two wavelength excitation calcium imaging: Results from bovine adrenal chromaffin cells. *Cell Calcium* **24:** 71–84.

Poenie M. 1990. Alteration of intracellular Fura-2 fluorescence by viscosity: A simple correction. *Cell Calcium* **11:** 85–91.

Schneggenburger R., Zhou Z., Konnerth A., and Neher E. 1993. Fractional contribution of calcium to the cation current through glutamate receptor channels. *Neuron* **11:** 133–143.

Szmacinski H. and Lakowicz J.R. 1995. Possibility of simultaneously measuring low and high calcium concentrations using Fura-2 and lifetime-based sensing. *Cell Calcium* **18:** 64–75.

Tsien R.Y. 1980. New calcium indicators and buffers with high selectivity against magnesium and protons: Design, synthesis, and properties of. *Biochemistry* **19:** 2396–2404.

———. 1989. Fluorescent probes of cell signaling. *Annu. Rev. Neurosci.* **12:** 227–253.

Tsien R.Y. and Harootunian A.T. 1990. Practical design criteria for a dynamic ratio imaging system. *Cell Calcium* **11:** 93–109.

Tsien R.Y. and Pozzan T. 1989. Measurement of cytosolic free Ca^{2+} with quin2. *Methods Enzymol.* (part S) **172:** 256–262.

Vranesic I. and Knöpfel T. 1991. Calculation of calcium dynamics from single wavelength fura-2 fluorescence recordings. *Pflügers Arch.* **418:** 184–189.

Wang S.S.-H., Alousi A.A., and Thompson S.H. 1995. The lifetime of inositol 1,4,5-triphosphate in single cells. *J. Gen. Physiol.* **105:** 149–171.

CHAPTER 32

A Single-Compartment Model of Calcium Dynamics in Nerve Terminals and Dendrites

Fritjof Helmchen[1] and David W. Tank[2]

[1]*Abteilung Zellphysiologie, Max-Planck-Institut für medizinische Forschung, 69120 Heidelberg, Germany;* [2]*Department of Molecular Biology, Princeton University, Carl Icahn Laboratory, Princeton, New Jersey 08544*

This chapter describes a single-compartment model of calcium dynamics that has been applied to fluorescence measurement of changes in intracellular free-calcium concentration ($[Ca^{2+}]_i$) in neurons. The model describes intracellular calcium handling under simplified conditions, for which analytical expressions for the amplitude and the time constants of $[Ca^{2+}]_i$ changes can be explicitly derived. In particular, it reveals the dependence of the measured $[Ca^{2+}]_i$ changes on the calcium indicator concentration. Applied to experimental data from small cells or subcellular compartments, the model equations have been extremely useful for obtaining quantitative information about essential parameters of Ca^{2+} influx, buffering, and clearance. The chapter also illustrates several situations in which the basic assumptions do not hold, e.g., when calcium diffusion, dye saturation, or kinetic effects become significant. Finally, we discuss how the changes in calcium dynamics that are explained by the model have been exploited for measuring properties of calcium-driven reactions, such as those regulating short-term synaptic enhancement, vesicle recycling, and adaptation.

The single-compartment model treats a cellular compartment as a homogeneous, well-mixed compartment, typically containing several pools of fast-acting nonsaturating buffers (including the fluorescent indicator), as well as a simple linear extrusion mechanism. The model has been particularly useful in describing the buildup and decay of $[Ca^{2+}]_i$ produced by sodium action potentials in nerve terminals, dendrites, and dendritic spines (Regehr and Tank 1994; Sabatini et al. 2001). Rapid (<1 msec) and local (<0.1 µm) actions of calcium ions near their entry site, such as the triggering of neurotransmitter release by calcium microdomains (see Chapter 40), are not addressed by this approach. Rather, the model describes the dynamics of longer-lasting (typically >10 msec) and more spatially homogeneous $[Ca^{2+}]_i$ elevations, sometimes referred to as "residual free calcium," that are commonly measured in dendrites and nerve terminals with high-affinity calcium indicators such as fura-2 and Calcium Green-1.

MODEL ASSUMPTIONS AND PARAMETERS

The single-compartment model is based on the assumption that Ca^{2+} gradients and diffusion can be neglected over the timescale of interest, making it possible to treat the cytosolic volume as a single, homogeneous compartment. But when is this assumption of spatial homogeneity justified? For a

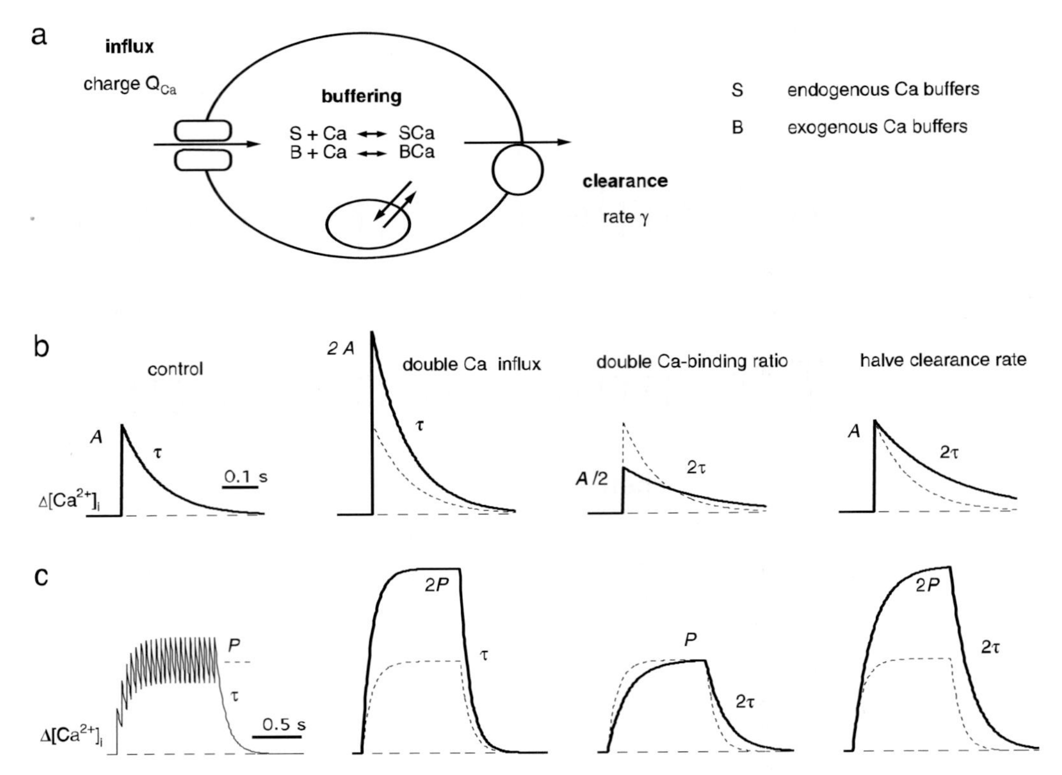

FIGURE 1. (*a*) Schematic drawing of the single-compartment model. Ca^{2+} clearance from the cytosol includes extrusion via the plasma membrane, uptake into intracellular organelles, and binding to slow endogenous buffers. Two pools of endogenous rapid Ca^{2+} buffers (*S*) and exogenously introduced buffers (*B*) are considered. (*b*) A single, brief calcium influx leads to an exponential $[Ca^{2+}]_i$ transient (*left*). Changes in influx, buffering, or clearance parameters alter the shape of the transient (*middle* and *right*). For comparison the control trace is drawn as dashed curves. (*c*) Summation of $[Ca^{2+}]_i$ transients during a train of stimuli for 1 sec at 20 Hz (*left*). The accumulation of the average $[Ca^{2+}]_i$ is described by an exponential rise to a steady-state plateau level *P* and an exponential decay. The dependence of the $[Ca^{2+}]_i$ accumulation on changes in influx, buffering, and clearance is shown as in *b*. Note that a change in the Ca^{2+}-binding ratio does not affect the plateau level.

compartment of radius *r*, a characteristic diffusion time is given by $t_d = r^2/(6D_{Ca})$, where D_{Ca} is an effective Ca^{2+} diffusion constant that takes into account the slowdown of Ca^{2+} diffusion caused by binding to cytosolic proteins (see Chapter 30). For $r = 1$ μm, the size of most mammalian nerve terminals and dendrites, and $D_{Ca} \approx 10–20\ \mu m^2\ sec^{-1}$ (Gabso et al. 1997; Murthy et al. 2000), t_d is about 10 msec. Following the cessation of influx, Ca^{2+} gradients dissipate within this time period. Changes in $[Ca^{2+}]_i$ that occur on the timescale of 0.1 sec (widely observed in small neural processes) can be well described using a single-compartment model that considers Ca^{2+} influx, buffering, and clearance as the relevant biophysical processes (Fig. 1a). The approximations used to describe these processes are discussed below.

Influx

We limit our attention to pulses of calcium influx that are brief compared to the time required for calcium clearance. This condition is quite reasonable for short depolarizing voltage steps and in particular for the calcium influx through voltage-gated calcium channels during the repolarizing phase of an action potential, which lasts less than a millisecond (Borst and Helmchen 1998). Ca^{2+} clearance times in presynaptic terminals and dendrites are typically at least one order of magnitude longer. A brief Ca^{2+} pulse at time t_p can be described by a delta function and the increase in total calcium concentration ($\Delta[Ca^{2+}]_T$)

$$j_{in} = \Delta[Ca^{2+}]_T \delta(t - t_p) = \frac{Q_{Ca}}{2FV}\,\delta(t - t_p) \quad (1)$$

F is Faraday's constant, V the volume of the cellular compartment, and Q_{Ca} the calcium charge injected.

Buffering

Most of the calcium ions entering the cytosol become bound by endogenous calcium-binding proteins and by exogenous calcium buffers introduced into the cell by the investigator. The initial changes in free-calcium concentration depend on the Ca^{2+}-buffering capacity of the rapid Ca^{2+} buffers. Usually, a pool S of rapid endogenous buffers and a pool B of exogenously introduced rapid buffers are considered (Fig. 1a). The Ca^{2+}-binding ratio of the exogenous buffer (κ_B), which usually is the fluorescent indicator itself, is defined as the ratio of the change in buffer-bound Ca^{2+} over the free-Ca^{2+} change (Neher and Augustine 1992). κ_B depends on the total buffer concentration $[B]_T$, its dissociation constant K_d, and $[Ca^{2+}]_i$

$$\kappa_B = \frac{\partial[BCa]}{\partial[Ca^{2+}]_i} = \frac{[B]_T K_d}{([Ca^{2+}]_i + K_d)^2} \approx \frac{[B]_T}{K_d} \qquad ([Ca^{2+}]_i << K_d) \tag{2}$$

An analogous expression exists for the Ca^{2+}-binding ratio of the endogenous buffer (κ_S). The maximal Ca^{2+}-binding ratio ($[B]_T/K_d$) occurs at $[Ca^{2+}]_i$ levels that are well below the K_d. In its simplest version, the single-compartment model assumes that κ_S and κ_B are $[Ca^{2+}]_i$-independent, corresponding to the idea that there are only small $[Ca^{2+}]_i$ changes from a resting level during the experiment. Under these conditions, changes in free-calcium concentration are proportional to changes in total calcium concentration (see Eq. 5). A value of $\kappa_B = 100$ implies that for each free-calcium ion, there are 100 ions bound to the buffer B. Note that Ca^{2+}-binding ratios are dimensionless and that experimentally determined values for the endogenous buffers range from 20 to 2000 (Table 1). Also note that Ca^{2+} indicators, which generally have fast kinetics, are potent exogenous Ca^{2+} buffers and therefore must be included in pool B.

Clearance

Calcium ions are either sequestered into intracellular organelles or extruded via the plasma membrane until $[Ca^{2+}]_i$ decays back to a resting level of typically 30–100 nM (Helmchen et al. 1996; Maravall et al. 2000). Most single-compartment models used for the analysis of experimental imaging data have used a calcium clearance mechanism linearly dependent on the deviation of $[Ca^{2+}]_i$ from the resting calcium level with a clearance rate γ

$$j_{out} = -\gamma\,\Delta[Ca^{2+}]_i = -\gamma([Ca^{2+}]_i - [Ca^{2+}]_{rest}) \tag{3}$$

This expression provides a good fit to experimental data in several systems (Neher and Augustine 1992; Tank et al. 1995; Helmchen et al. 1996, 1997). Nevertheless, Equation 3 is generally considered to be the low-$[Ca^{2+}]_i$ limit of a saturable enzyme-driven extrusion mechanism, e.g., one following Michaelis-Menten kinetics. Experimental data suggest that saturation effects might only be found under rather extreme stimulus conditions.

DYNAMICS OF SINGLE $[Ca^{2+}]_i$ TRANSIENTS

Combining these kinetic descriptions for influx, buffering, and clearance, the differential equation for the temporal dynamics of $[Ca^{2+}]_i$ following a single, brief calcium influx is

$$\begin{aligned} \frac{d[Ca^{2+}]_T}{dt} &= j_{in} + j_{out} \\ \frac{d[Ca^{2+}]_i}{dt}(1+\kappa_S+\kappa_B) &= \Delta[Ca^{2+}]_T\,\delta(t-t_p) - \gamma\,\Delta[Ca^{2+}]_i \end{aligned} \tag{4}$$

This equation simply states that the change in total calcium equals the increase in calcium minus the clearance. Assuming constant Ca^{2+}-binding ratios, the analytical solution of Equation 4 is an exponential function with amplitude A and decay time constant τ (Fig. 1b). The relationships between A and τ and the model parameters are as follows

$$A = \frac{Q_{Ca}/(2FV)}{(1+\kappa_S+\kappa_B)} \tag{5}$$

$$\tau = \frac{(1+\kappa_S+\kappa_B)}{\gamma} \tag{6}$$

Both A and τ depend on the Ca^{2+}-binding ratios. Note, however, that the product $A\,\tau$, which is the integral of the $[Ca^{2+}]_i$ transient (the area "underneath" the exponential curve), is independent of the Ca^{2+}-binding ratios. Figure 1b illustrates that changes in influx, total buffering capacity, and clearance rate lead to specific changes of the $[Ca^{2+}]_i$ transient shape (Sabatini and Regehr 1995). Although changes in influx or clearance affect only either A or τ, both magnitudes depend on the Ca^{2+}-buffering capacity. Because of this dependence, A and τ may be altered by introducing a calcium indicator dye. A description of how this can be exploited to obtain an estimate of the endogenous Ca^{2+}-binding ratio κ_S is given below.

SUMMATION OF $[Ca^{2+}]_i$ TRANSIENTS DURING REPETITIVE CALCIUM INFLUX

If multiple calcium influx pulses occur on a timescale that is short compared to the $[Ca^{2+}]_i$-decay time, the amplitude of the $[Ca^{2+}]_i$ transient simply increases with the greater total calcium charge Q_{Ca}. For example, the amplitude of $[Ca^{2+}]_i$ transients has been shown to scale approximately linearly with the number of high-frequency action potentials in presynaptic terminals (Regehr et al. 1994), in Purkinje cell axons (Callewaert et al. 1996), and in mammalian pyramidal cell dendrites in vivo (Svoboda et al. 1997). Such a linear increase would be expected to break down for more stimuli when calcium influx per pulse changes or saturation of buffers and pumps occurs. A linear sum of individual transients can also approximate the buildup of $[Ca^{2+}]_i$ during repetitive calcium influx as it occurs during firing of a neuron at a lower constant frequency. Consider a train of stimuli starting at time $t = 0$ and with a time interval of Δt (frequency $f = 1/\Delta t$). The $[Ca^{2+}]_i$ level above resting level immediately before the $(n + 1)$th stimulus is given by a geometric progression (Regehr et al. 1994)

$$\Delta[Ca^{2+}]_i(n\Delta t) = A\sum_{i=1}^{n} e^{-(i\Delta t)/\tau} = \frac{A}{(e^{\Delta t/\tau}-1)}(1-e^{-(n\Delta t)/\tau}) \tag{7}$$

where A is the amplitude and τ the time constant of each individual transient. For stimulation frequencies $f<1/(2\tau)$, there is little buildup, and individual, spaced transients result, but for higher frequencies, the transients add up and $[Ca^{2+}]_i$ exponentially reaches a steady state in which influx and clearance balance, and $[Ca^{2+}]_i$ fluctuates around a plateau level (Fig. 1c). The time constant of the rise and the decay following the end of stimulation are both given by τ. The mean plateau level P reached at steady state is proportional to the frequency (Tank et al. 1995; Helmchen et al. 1996)

$$P = A\tau f = \frac{Q_{Ca}}{2FV\gamma} f \tag{8}$$

with the integral of the single calcium transient ($A\tau$) as the proportionality constant. P is independent of the Ca^{2+}-binding ratios, indicating that Ca^{2+} buffers can affect the transient dynamics of $[Ca^{2+}]_i$, but not its steady-state levels. Figure 1c summarizes the effects of changes in influx, buffering, and clearance on the $[Ca^{2+}]_i$ accumulation during repetitive stimulation. As in the case of the single $[Ca^{2+}]_i$ transient, each change in these biophysical parameters leads to a characteristic alteration of the overall shape of the $[Ca^{2+}]_i$ change (Tank et al. 1995).

APPLICATIONS

One of the main advantages of the single-compartment model is that it takes into account the effect of calcium indicator buffering. This provides a framework within which to estimate how severely the indicator perturbs calcium signaling, and, in addition, it suggests systematic variation of the indicator concentration as a tool to determine endogenous calcium-handling parameters and to reveal unperturbed calcium dynamics.

Estimates of Endogenous Ca^{2+}-binding Ratio and Clearance Rate

The endogenous Ca^{2+}-binding ratio κ_s and the clearance rate γ can be estimated by measuring changes in the $[Ca^{2+}]_i$ transient decay time produced by increasing concentrations of an exogenous Ca^{2+} buffer such as BAPTA or fura-2 (Fig. 2a). This follows directly from Equation 6, which can be rearranged to give

$$\begin{aligned} \tau &= a_1\kappa_B + a_0 \\ a_1 &= 1/\gamma \\ a_0 &= (1 + \kappa_S)/\gamma \end{aligned} \tag{9}$$

The decay time constant τ should be linearly related to the exogenous buffer capacity κ_B(Tank et al. 1991). The inverse of the measured slope (a_1) provides the clearance rate, and the negative x-axis intercept ($-a_0/a_1$) provides an estimate of the endogenous buffer capacity κ_S (Neher and Augustine 1992). An example of this method, taken from a study of $[Ca^{2+}]_i$ transients in pyramidal cell dendrites, is shown in Figure 2c. Two methods have been employed to provide known exogenous buffer concentrations: dialysis with patch pipettes (Neher and Augustine 1992; Helmchen et al. 1996; Maravall et al. 2000) and intracellular sharp microelectrode injection of a fluorescent buffer, from which fluorescence intensity can be measured and used to calculate concentrations (Tank et al. 1995). An elegant and effective method that should be used whenever possible is to monitor the indicator buffering effect during the dye loading process in individual experiments, avoiding cell-to-cell variability (Fig. 2b–d) (Neher and Augustine 1992; Helmchen et al. 1996; Sabatini et al. 2002). The actual indicator concentrations in this case are back-calculated from the estimated final concentration.

Estimates of Amplitude A and Total Calcium Charge Q_{Ca}

Compared to the unaltered $[Ca^{2+}]_i$ transients, the amplitude A, which is obtained from a Ca^{2+} indicator measurement, is reduced by a factor $(1 + \kappa_S)/(1 + \kappa_S + \kappa_B)$ (see Eq. 5). Similar to the analysis of decay time constants, the inverse of the amplitude (A^{-1}) is expected to depend linearly on the exogenous Ca^{2+}-buffer capacity κ_B (Eq. 5). This relationship can be used to obtain another estimate of κ_S (Fig. 2d). The absolute values of A can either be measured using ratiometric or nonratiometric calibration techniques (see Chapter 31) or it can be estimated from the degree of saturation of the fluorescent dye (see below). If the total buffering capacity of the cytoplasm is known and with an estimate of the compartment volume from morphology, the total calcium charge Q_{Ca} can be calculated by rearranging Equation 5

$$Q_{Ca} = 2FVA(1 + \kappa_S + \kappa_B) \tag{10}$$

Alternatively, Q_{Ca} can be measured directly using excessive dye loading of the compartment to obtain fluorescence changes proportional to Q_{Ca} (Neher 1995; Borst and Helmchen 1998). When individual calcium transients cannot be easily resolved, Q_{Ca} can be determined from the frequency dependence of the initial slope of the calcium buildup produced by a stimulus train (Tank et al. 1995).

Estimates of Unperturbed Calcium Dynamics

Because many cells have a relatively low endogenous Ca^{2+}-buffering capacity (see below), even the lowest concentrations of indicator dyes that can be used for fluorescence measurements can often cause significant perturbation of calcium signals. Systematic variation of the exogenous Ca^{2+}-buffering

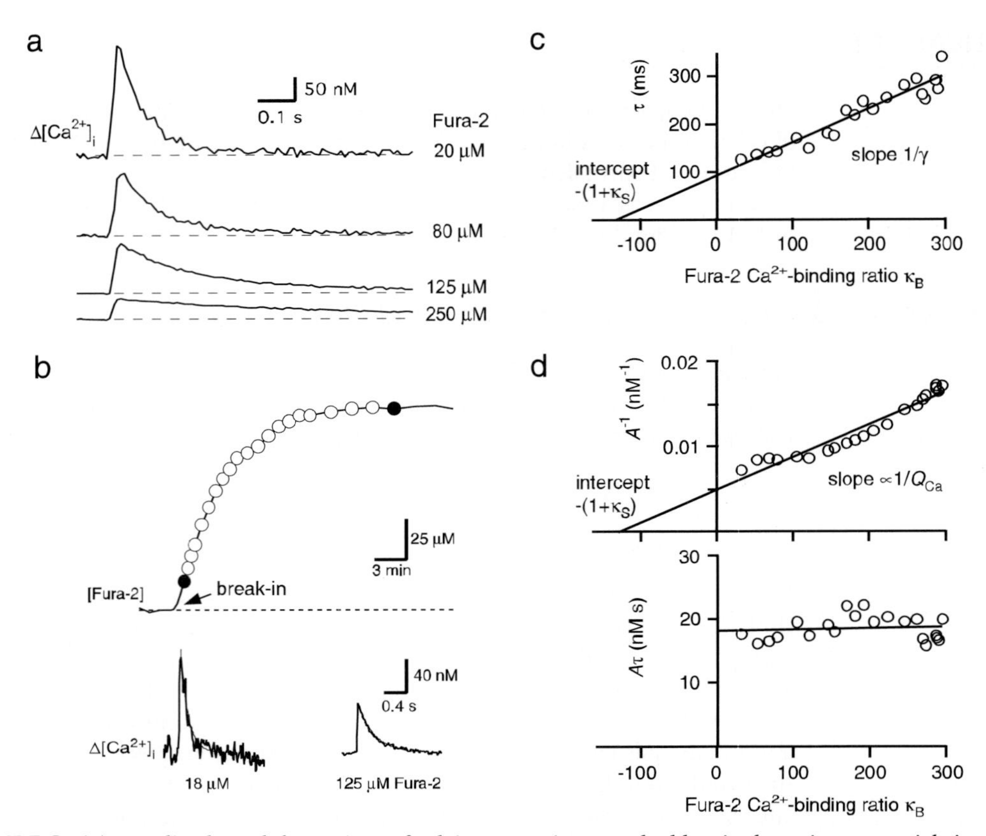

FIGURE 2. (*a*) Amplitude and decay time of calcium transients evoked by single-action potentials in proximal apical dendrites of neocortical pyramidal neurons depend on the fura-2 concentration (four different experiments). (*b*) This dependence can also be measured in single experiments by monitoring action-potential-induced calcium transients during the loading of the indicator. Following break-in with a whole-cell patch pipette, the fluorescence exponentially reaches a steady-state plateau level, which represents equilibration with the patch-pipette solution containing 125 μM fura-2. Calcium transients were measured at 20–60-sec time intervals (*circles*). Two example calcium transients at low and high fura-2 concentration, respectively (*solid circles*), are shown at the bottom. (*c*) Analysis of the time constant of calcium transients during fura-2 loading. The endogenous Ca^{2+}-binding ratio κ_S and the clearance rate γ can be determined by plotting the decay time constant as a function of the exogenously introduced Ca^{2+}-buffering capacity. The extrapolation to zero fura-2 Ca^{2+}-binding ratio yields an estimate of the time constant in the absence of indicator. (*d*) Similarly, κ_S and the extrapolated calcium transient amplitude can be determined from a plot of the inverse of the measured amplitudes versus κ_B. As a control, the product of amplitude and time constant is also plotted, which should not depend on κ_B. (Modified, with permission, from Helmchen et al. 1996.)

capacity κ_B in this case also permits the estimation of the amplitude and time course of unperturbed calcium transients, yielding the best possible approximation of the intracellular $[Ca^{2+}]_i$ changes as they occur in cells under physiological conditions. The unperturbed amplitude and decay time constant are obtained by extrapolation of the regression lines from the plots of τ and A^{-1} against κ_B to zero exogenous Ca^{2+}-binding ratio (Fig. 2c,d). Another method to approximate unperturbed dynamics is the use of low-affinity indicators, which introduce less exogenous Ca^{2+}-buffering capacity, but usually require averaging to resolve the small signals (Helmchen et al. 1997; Sabatini et al. 2002).

EXAMPLE OF APPLICATION

Changes in calcium dynamics similar to those illustrated in Figure 1 have been measured in small cell somata, presynaptic terminals, dendrites, and axons. The results of these experiments provide evidence for the applicability of the single-compartment model as well as quantitative estimates for the model

TABLE 1. Some representative experimental results for the basic parameters of calcium dynamics in dendrites and nerve terminals

Preparation	Amplitude A (nM)	Time constant τ (msec)	Ca^{2+}-binding ratio κ_S	Clearance rate γ (sec^{-1})	References
Dendrites					
Neocortical L5 pyramidal neurons	260	70	120	1700	Helmchen et al. (1996)
Neocortical L2/3 pyramidal neurons	190–240	60–90	110	1800	Koester and Sakmann (2000); Kaiser et al. (2001)
Hippocampal CA1 neurons	150–240	90	60–180	700–2000	Helmchen et al. (1996); Maravall et al. (2000)
Dendritic spines; hippocampal CA1 neurons	1500–1700	12–15	20	1200–1600	Sabatini et al. (2002)
Bitufted interneuron in neocortex	140	200	300	1500	Kaiser et al. (2001)
Cerebellar Purkinje cells	—	400	900–2000	—	Callewaert et al. (1996); Fierro and Llano (1996)
Nerve terminals					
Neocortical pyramidal cell nerve terminals	500–1000	60	140	2600	Cox et al. (2000); Koester and Sakmann (2000)
Dentate granule cell axonal boutons	900	40	20	500	Jackson and Redman (2003)
Hippocampal mossy fiber terminals	5–10	1000	—	—	Regehr et al. (1994)
Hippocampal CA3-CA1 synapse	—	40	—	—	Sinha et al. (1997)
Crayfish neuromuscular junction	5–10	4000	600	80–100	Tank et al. (1995)
Cerebellar granule cell to Purkinje cell synapse	300	150	—	—	Regehr and Atluri (1995)
Frog retinotectal synapse	140	100	—	—	Feller et al. (1996)
Calyciform terminals in brain stem	400–500	50	40	900	Helmchen et al. (1997)

The reported amplitudes and time constants refer to single-action potential-evoked calcium transients. They depend on the exogenous buffer conditions and temperature (see original publications for specific conditions). (–) indicates not determined.

parameters. Table 1 provides a comparison of observed parameters and lists the corresponding references.

In general, it was found that relatively large and rapid calcium signals are associated with action potentials. In central nervous system (CNS) pyramidal cells, the $[Ca^{2+}]_i$ changes in dendrites and nerve terminals that are caused by single action potentials were estimated to have several hundred nanomolar amplitude and to decay within 100 msec or less (Table 1). The reason for this fast signaling capacity is a relatively low endogenous Ca^{2+}-buffering capacity in the range of 20–200. GABAergic interneurons and, in particular, cerebellar Purkinje neurons appear to have a higher Ca^{2+}-buffering capacity, most likely because of their different cytosolic repertoire of calcium-binding proteins. Most recently, the single-compartment model has been applied to characterize calcium handling in dendritic spines (Sabatini et al. 2002), revealing a low value of κ_S of about 20, which consequently resulted in fast spineous calcium transients with a decay time constant of about 10–15 msec.

ADVANTAGES AND LIMITATIONS

This section describes several extensions that can be added to the single-compartment model if certain assumptions cannot be made. In addition, a description is given of how the model has helped to characterize calcium-dependent cellular processes.

Buffered Calcium Diffusion

In addition to Ca^{2+} buffering, fluorescent indicators may perturb intracellular calcium signals because of their mobility. This is relevant if Ca^{2+} influx occurs locally and subsequently spreads via diffusion (e.g., along a dendrite). The single-compartment model can be extended to such situations if the diffusional exchange occurs on a rather long spatial scale so that local chemical equilibrium can still be assumed. In this case, calcium diffusion is described by an apparent diffusion constant that depends on the diffusion constants D_m and the Ca^{2+}-buffering capacity of mobile buffers κ_m, as well as the Ca^{2+}-buffering capacity of fixed buffers κ_f (Gabso et al. 1997). In general, immobile buffers tend to hold Ca^{2+} in place and thereby slow down diffusion, whereas highly mobile buffers such as fura-2 or Calcium Green-1 facilitate the spread of Ca^{2+} (see Chapter 30).

Deviations from Linear Behavior

When $[Ca^{2+}]_i$ changes are appreciable compared to the dissociation constant of the dominant intracellular buffers, or when buffer kinetics are slow compared to the decay time constant, exponential decaying transients are not observed. The following sections describe expected deviations under these two circumstances.

Saturation of Buffers and Pumps

If buffer kinetics are still rapid but $[Ca^{2+}]_i$ changes become appreciable compared to the K_d values of the intracellular buffers, then Equation 4 is still appropriate, but κ_B and κ_S are not constant and the solution is not an exponential decay. Numerically calculated decay curves as well as measured transients show pronounced upward curvature (faster decays) at higher $[Ca^{2+}]_i$ levels on a semilog plot (Tank et al. 1995) because the effective buffering capacity is reduced at these levels, due to partial saturation of the buffers. An opposite effect would be expected from saturation of clearance mechanisms at very high $[Ca^{2+}]_i$ levels. Saturation in the clearance rate would also be expected to alter the linear frequency dependence of the plateau level during long stimulus trains (Eq. 8). A related buffer saturation effect can be exploited to provide an estimate of the amplitude A of individual $[Ca^{2+}]_i$ transients when ratiometric measurements of absolute calcium concentration are not feasible; for example, when fluorescence is measured from a large region of AM-loaded brain slices or intact brains (Regehr and Atluri 1995; Feller et al. 1996). Except when confocal or two-photon techniques for volume resolution are used, such measurements contain contaminating fluorescence from other structures. In these cases, A can be estimated from the degree of saturation of a high-affinity calcium indicator

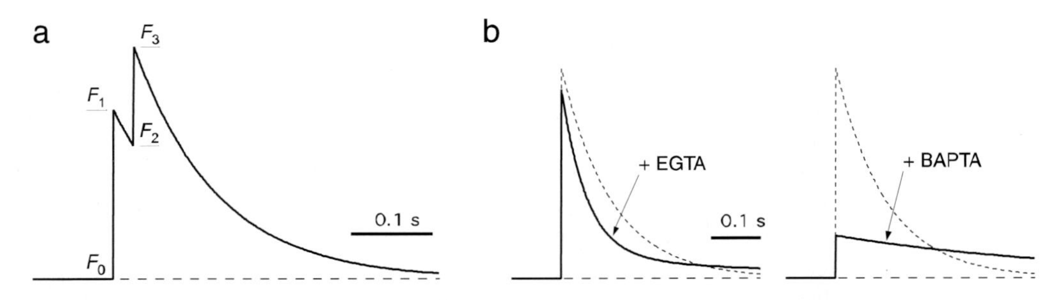

FIGURE 3. (*a*) The amplitude A of the $[Ca^{2+}]_i$ transient can be estimated from the degree of saturation of a fluorescent indicator with a known dissociation constant (see text). Schematic drawing of the indicator fluorescence change upon two brief calcium injections. The $[Ca^{2+}]_i$ change per stimulus is assumed to be equal, but the change in fluorescence F is smaller for the second stimulus. (*b*) Schematic illustration of the effect of adding Ca^{2+} buffers with different kinetic properties on individual control calcium transients (*dashed curves*). The fast buffer BAPTA changes the $[Ca^{2+}]_i$ transient shape according to equilibrium considerations. A slow buffer like EGTA, in contrast, leads to a faster initial decay and a subsequent slow phase, which are explained by its slow association rate.

(Regehr and Atluri 1995; Sabatini and Regehr 1995; Feller et al. 1996). Figure 3a shows the fluorescence change of an indicator evoked by two subsequent brief calcium injections. The second injection evokes a smaller fluorescence change than the first, due to partial saturation of the indicator. Assuming that the change in calcium concentration is the same for the two stimuli, its magnitude can be estimated from the ratio α of the two fluorescence changes

$$\Delta[Ca^{2+}]_i = \frac{([Ca^{2+}]_{rest} + K_d)(1-\alpha)}{2\alpha} \tag{11}$$

where $\alpha = (F_3 - F_2)/(F_1 - F_0)$ and assuming $F_2 \approx F_1$ (Feller et al. 1996). This method, however, relies on reasonable estimates of the resting $[Ca^{2+}]_i$ level and the dissociation constant of the indicator.

Slow Buffers

Radical departures in calcium dynamics from the expected results (Eqs. 4–6) have been observed when slow calcium buffers such as EDTA or EGTA were introduced as exogenous buffers into presynaptic nerve terminals (Atluri and Regehr 1996; Feller et al. 1996) and nerve cell dendrites (Markram et al. 1998). A comparison between the effects on a $[Ca^{2+}]_i$ transient of adding a fast buffer (BAPTA) versus a slow buffer (EGTA) is illustrated schematically in Figure 3b. Addition of BAPTA reduces the amplitude of the transient while prolonging its decay, consistent with expectations from Equations 5 and 6. Addition of EGTA, however, has only a small effect on the amplitude *A*, but changes the shape of the transient by providing a faster initial decay, while also producing a slower subsequent phase. As easily demonstrated in numerical simulations, the early faster decay is produced by Ca^{2+} binding to EGTA on a timescale longer than the response time of the Ca^{2+} indicator (Atluri and Regehr 1996; Feller et al. 1996; Markram et al. 1998). The fast time constant can be approximated by adding the association rate of EGTA Ca^{2+} binding to the clearance mechanisms (Atluri and Regehr 1996)

$$\tau_{fast} = \frac{(1+\kappa_S+\kappa_B)}{\gamma + k^{on}_{EGTA}[EGTA]} \tag{12}$$

The longer time constant of the second decay phase is produced by the same mechanism as for BAPTA: More bound calcium must be removed per unit change of free calcium ions. It is given by Equation 6 after addition of the Ca^{2+}-binding ratio κ_{EGTA}. Kinetic effects might be particularly relevant in small structures displaying rapid Ca^{2+} handling, such as dendritic spines, since the sharing of Ca^{2+} between proteins will be a complicated function of their relative kinetics and the time course of calcium influx. Proteins with slow binding kinetics could be bypassed following single action potentials and activated significantly only during prolonged or synaptic activation of Ca^{2+} influx (Sabatini et al. 2002).

Single-Compartment Models and the Measurement of Calcium-driven Reactions

The single-compartment models summarized here were developed to explain basic properties of calcium transients observed in imaging experiments on small structures such as presynaptic terminals and dendrites. The analysis also provides a set of tools, however, to alter calcium dynamics, which has proved to be important in the biophysical characterization of calcium-driven reactions. For example, the systematic prolongation of calcium transients with increasing exogenous buffer concentrations helped to demonstrate that the time constant of $[Ca^{2+}]_i$ decay was the rate-limiting step in the decay of calcium-driven short-term synaptic enhancement at an invertebrate synapse (Delaney and Tank 1994). In a mammalian synapse, addition of exogenous buffer and manipulation of the calcium plateau by changes in stimulus frequency were used to alter calcium dynamics to determine kinetic rate constants of a calcium-driven reaction involved in short-term synaptic enhancement (Regehr et al. 1994). Similar approaches have been used in characterization of vesicle mobilization in chromaffin cells (Heinemann et al. 1993), of mitochondrial Ca^{2+} uptake and release in bullfrog sympathetic neurons (Friel and Tsien 1994), of adaptation of cellular excitability (Sobel and Tank 1994), and of synaptic facilitation in the cerebellum (Atluri and Regehr 1996).

In summary, this chapter described the basic equations and parameters of a simple model of calcium dynamics that is applicable to action-potential-evoked $[Ca^{2+}]_i$ transients in nerve terminals and dendrites. Methods have been introduced to quantify the main parameters of calcium influx, buffering, and clearance. It can be concluded that the single-compartment model is a suitable starting point for a more detailed characterization of Ca^{2+} handling in small cellular compartments.

REFERENCES

Atluri P.P. and Regehr W.G. 1996. Determinants of the time course of facilitation at the granule cell to Purkinje cell synapse. *J. Neurosci.* **16:** 5661–5671.

Borst J.G.G. and Helmchen F. 1998. Calcium influx during an action potential. *Methods Enzymol.* **293:** 352–371.

Callewaert G., Eilers J., and Konnerth A. 1996. Axonal calcium entry during fast "sodium" action potentials in rat cerebellar Purkinje neurones. *J. Physiol.* **495:** 641–647.

Cox C.L., Denk W., Tank D.W., and Svoboda K. 2000. Action potentials reliably invade axonal arbors of rat neocortical neurons. *Proc. Natl. Acad. Sci.* **97:** 9724–9728.

Delaney K.R. and Tank D.W. 1994. A quantitative measurement of the dependence of short-term synaptic enhancement on presynaptic residual calcium. *J. Neurosci.* **14:** 5885–5902.

Feller M.B., Delaney K.R., and Tank D.W. 1996. Presynaptic calcium dynamics at the frog retinotectal synapse. *J. Neurophysiol.* **76:** 381–400.

Fierro L. and Llano I. 1996. High endogenous calcium buffering in Purkinje cells from rat cerebellar slices. *J. Physiol.* **496:** 617–625.

Friel D.D. and Tsien R.W. 1994. An FCCP-sensitive Ca^{2+} store in bullfrog sympathetic neurons and its participation in stimulus-evoked changes in $[Ca^{2+}]_i$. *J. Neurosci.* **14:** 4007–4024.

Gabso M., Neher E., and Spira M.E. 1997. Low mobility of the Ca^{2+} buffers in axons of cultured *Aplysia* neurons. *Neuron* **18:** 473–481.

Heinemann C., von Rüden L., Chow R.H., and Neher E. 1993. A two-step model of secretion control in neuroendocrine cells. *Pflügers Arch.* **424:** 105–112.

Helmchen F., Borst J.G.G., and Sakmann B. 1997. Calcium dynamics associated with a single action potential in a CNS presynaptic terminal. *Biophys. J.* **72:** 1458–1471.

Helmchen F., Imoto K., and Sakmann B. 1996. Ca^{2+} buffering and action potential-evoked Ca^{2+} signaling in dendrites of pyramidal neurons. *Biophys. J.* **70:** 1069–1081.

Jackson M.B. and Redman S.J. 2003. Calcium dynamics, buffering, and buffer saturation in the boutons of dentate granule-cell axons in the hilus. *J. Neurosci.* **23:** 1612–1621.

Kaiser M.M.K., Zilberter Y., and Sakmann B. 2001. Back-propagating action potentials mediate calcium signalling in dendrites of bitufted interneurons in layer 2/3 of rat somatosensory cortex. *J. Physiol.* **535:** 17–31.

Koester H.J. and Sakmann B. 2000. Calcium dynamics associated with action potentials in single nerve terminals of pyramidal cells in layer 2/3 of the young rat neocortex. *J. Physiol.* **529:** 625–646.

Maravall M., Mainen Z.F., Sabatini B.L., and Svoboda K. 2000. Estimating intracellular calcium concentrations and buffering without wavelength ratioing. *Biophys. J.* **78:** 2655–2667.

Markram H., Roth A., and Helmchen F. 1998. Competitive calcium binding: Implications for dendritic calcium signaling. *J. Comp. Neurosci.* **5:** 331–348.

Murthy V.N., Sejnowski T.J., and Stevens C.F. 2000. Dynamics of dendritic calcium transients evoked by quantal release at excitatory hippocampal synapses. *Proc. Natl. Acad. Sci.* **97:** 901–906.

Neher E. 1995. The use of fura-2 to estimate Ca buffers and Ca fluxes. *Neuropharmacology* **34:** 1423–1442.

Neher E. and Augustine G. 1992. Calcium gradients and buffers in bovine chromaffin cells. *J. Physiol.* **450:** 273–301.

Regehr W.G. and Atluri P.P. 1995. Calcium transients in cerebellar granule cell presynaptic terminals. *Biophys. J.* **68:** 2156–2170.

Regehr W.G. and Tank D.W. 1994. Dendritic calcium dynamics. *Curr. Opin. Neurobiol.* **4:** 373–382.

Regehr W.G., Delaney K.R., and Tank D.W. 1994. The role of presynaptic calcium in short-term enhancement at the hippocampal mossy fiber synapse. *J. Neurosci.* **14:** 523–537.

Sabatini B.L. and Regehr W.G. 1995. Detecting changes in calcium influx which contribute to synaptic modulation in mammalian brain slice. *Neuropharmacology* **34:** 1453–1467.

Sabatini B.L., Maravall M., and Svoboda K. 2001. Ca^{2+} signaling in dendritic spines. *Curr. Opin. Neurobiol.* **11:** 349–356.

Sabatini B.L., Oertner T.G., and Svoboda K. 2002. The life cycle of Ca^{2+} ions in dendritic spines. *Neuron* **33:** 439–452.

Sinha S.R., Wu L.-G., and Saggau P. 1997. Presynaptic calcium dynamics and transmitter release evoked by single action potentials at mammalian synapses. *Biophys. J.* **72:** 637–651.
Sobel E.C. and Tank D.W. 1994. In vivo Ca^{2+} dynamics in a cricket auditory neuron: An example of chemical computation. *Science* **263:** 823–826.
Svoboda K., Denk W., Kleinfeld D., and Tank D.W. 1997. *In vivo* dendritic calcium dynamics in neocortical pyramidal neurons. *Nature* **385:** 161–165.
Tank D.W., Regehr W.G., and Delaney K.R. 1991. Modeling a synaptic chemical computation: The buildup and decay of presynaptic calcium. *Soc. Neurosci. Abstr.* **17:** 578.
———. 1995. A quantitative analysis of presynaptic calcium dynamics that contribute to short-term enhancement. *J. Neurosci.* **15:** 7940–7952.

CHAPTER 33

A Practical Guide: Dye Loading with Patch Pipettes

Jens Eilers[1] and Arthur Konnerth[2]

[1]Department of Neurophysiology, Carl-Ludwig-Institute of Physiology, University of Leipzig, 04103 Leipzig, Germany; [2]Institute of Physiology, Ludwig-Maximilians University Munich, 80336 Munich, Germany

This chapter describes the loading of individual cells with fluorescent probes via patch pipettes. The patch-clamp methodology has been successfully used for single-cell dye labeling in cultured neurons, brain slices, and in vivo preparations. A broad range of dyes can be used with this loading technique. Markers for morphological reconstruction (e.g., Lucifer yellow), ion-sensitive indicator dyes for monitoring second-messenger cascades (e.g., fura-2), and dye-labeled proteins for FRET, FCS, and FRAP studies (fluorescence resonance energy transfer, fluorescence correlation spectroscopy, and fluorescence recovery after photobleaching, respectively) are all suitable for patch-clamp loading. The most widespread application of this technique has been for Ca^{2+} imaging.

This chapter assumes familiarity with the standard procedures of whole-cell patch-clamp recordings (see, e.g., Marty and Neher 1995).

MATERIALS

Equipment

This procedure requires the use of standard electrophysiological equipment (amplifier, manipulator, pipette holder, pipettes, etc.; for more details, see Penner 1995) and a fluorescence detection system (e.g., standard epifluorescence microscope).

Pipettes

It is important to choose the right kind of pipette for this technique. Dye should be loaded with borosilicate pipettes (e.g., from Hilgenberg) rather than soft, soda glass pipettes. The latter release trace amounts of cations that may interfere with voltage-gated channels (Cota and Armstrong 1988) and, potentially, ion-sensitive dyes (Grynkiewicz et al. 1985).

The electrical resistance of the patch pipette, R_P, is of special importance for fast and efficient dye-loading. R_P, which should not be confused with the "series resistance" or "access resistance" (R_s), see below, is determined by the conductivity of the pipette solution and the geometry of the pipette (i.e., by the length of the pipette taper and the width of the pipette opening). Thus, R_P can be finely tuned by changing the settings of the pipette puller (for details, see Penner 1995). Depending on the experimental requirements, and the cell type under study, the optimal R_P value may vary considerably. A low R_P permits a fast dialysis of remote cellular regions (Fig. 1B; see also Pusch and Neher 1988; Rexhausen 1992). However, it is more difficult to maintain long-lasting recordings with low-resistance pipettes. A higher R_P should be chosen if multiple measurements at a steadily increasing dye concentration are needed, as for measurements of the calcium-binding ratio with the calcium flux approach (Schneggenburger et al. 1993; Neher 1995; Chapter 30).

The cell size usually imposes limits on the diameter of the pipette tip and, consequently, on R_P. As a general rule, large cells tolerate larger tips than small cells or fine cellular processes (Sakmann and Stuart 1995). For instance, patch pipettes with resistances of 2.0–3.5 M Ω (measured with KCl-based pipette solution) are well suited for long-lasting recordings from the large cerebellar Purkinje cell bodies (Fig. 1B), which have a diameter of 25–30 µm.

Solutions

Important! Use highly purified water (e.g., W3500 from Sigma) for dissolving the dyes and for preparing the pipette solutions. Trace amounts of heavy metals or other contaminants may severely affect the fluorescence properties of the fluorophores (Grynkiewicz et al. 1985). For examples of dye-containing pipette solutions, see Table 1.

Biological Sample

Acutely dissociated cells, cultured cells, acute brain slices, or in vivo preparations are suitable. Note that in some preparations, white matter, glia cells, and/or the perineural net may hinder successful patch-clamp recordings. Extensive "cleaning" of the cell membrane, as described by Edwards et al. (1989), may be necessary in these cases.

TABLE 1. Examples of solutions

Solution	Composition	Notes
External saline	125 mM NaCl, 2.5 mM KCl<!>, 2 mM $CaCl_2$<!>, 1 mM $MgCl_2$<!>, 1.25 mM NaH_2PO_4<!>, 26 mM $NaHCO_3$, 20 mM glucose, aerated with 95% O_2 and 5% CO_2	~ pH 7.3 at 21°C
Pipette solution 1	140 mM potassium gluconate, 10 mM HEPES, 10 mM NaCl, 0.5 mM $MgCl_2$<!>, 4 mM Mg-ATP, 0.4 mM Na_3-GTP, 0.5 mM EGTA[a], dye of choice, pH 7.3 (adjusted with KOH<!>)	for current-clamp recordings
Pipette solution 2	140 mM CsCl<!>, 10 mM HEPES, 10 mM TEA, 0.5 mM $MgCl_2$<!>, 4 mM Mg-ATP, 0.4 Na_3-GTP, 0.5 mM EGTA[a], dye of choice, pH 7.3 (adjusted with CsOH<!>)	for voltage-clamp recordings

Caution: See Appendix 3 for appropriate handling of materials marked with <!>.

Once prepared, dye-containing pipette solutions should be stored at –20°C until use. Fresh solutions should be prepared every 1–2 weeks. For efficient use of expensive dyes with different pipette solutions, it is advisable to prepare appropriately concentrated (e.g., 125%) dye-free pipette solutions and a separate stock solution of the dye that will yield the desired concentration after dilution. Mix (and sonicate) stock solutions just prior to use. Dissolved in purified water, most dyes can be stored at –20°C for several months.

[a]When using ion-sensitive indicator dyes such as fura-2, no additional substances that buffer the ion under study (i.e., EGTA or BAPTA) should be included in the pipette solution. The competition between the indicator dye and the additional buffer would seriously complicate the interpretation of the measurements.

PROCEDURE

Dye Loading through Patch Pipettes

1. Using Microloaders (Eppendorf), fill the chosen pipettes with 5–10 µl of the appropriate dye-containing solution.

   Note: Do not use syringe needles since these may release heavy metals; see above.

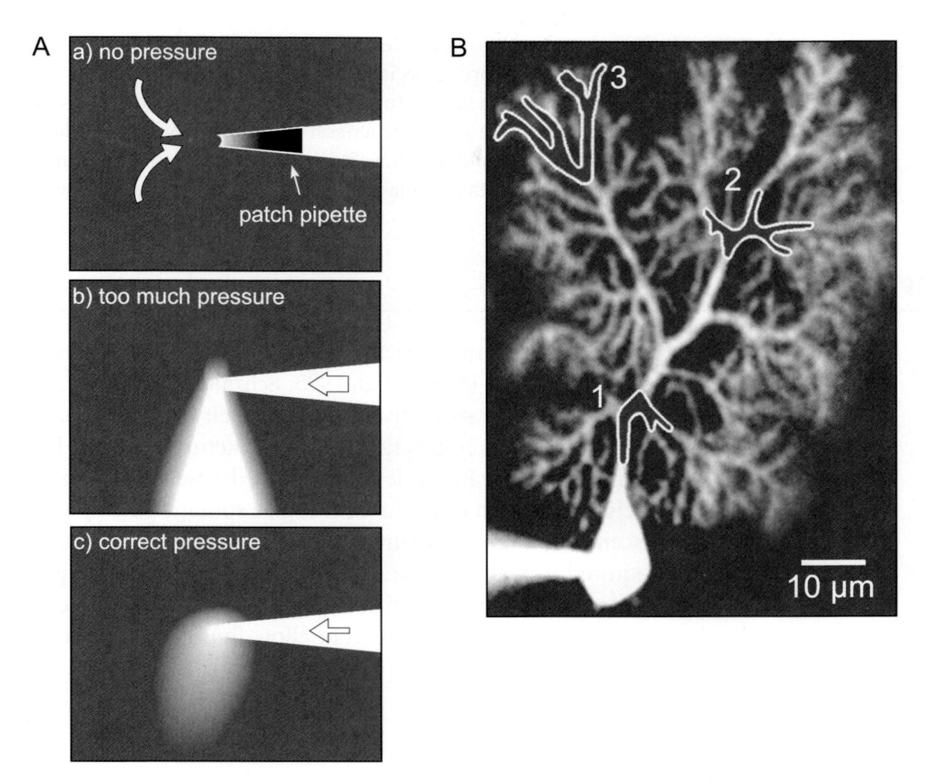

FIGURE 1. (*A*) Minimal pressure should be applied to the patch pipette while approaching the tissue. A schematic drawing showing the effect of different pressure settings, as viewed through fluorescence optics. (*a*) If no pressure is applied, bath solution enters the pipette due to capillary forces. Thus, the composition of the pipette solution is modified and successful seal formation may become impossible. (*b*) If the pressure applied to the pipette is too high, a strong jet of dye-containing solution is expelled from the pipette tip, and surrounding tissue may become heavily stained. (The bath solution flows from the top to the bottom.) (*c*) At the correct pressure setting, only a minimal amount of dye is ejected. (*B*) Dye-loading of distal dendrites. A Purkinje neuron in a cerebellar slice preparation was loaded with the calcium indicator dye fura-2 via a somatic patch pipette. During the whole-cell recording, repeated measurements of the dendritic fluorescence at the calcium-insensitive wavelength were performed (Rexhausen 1992). From these data, the times of half-maximal loading were estimated to be about 2.5 min, 16 min, and 30 min for regions 1, 2, and 3, respectively. (Modified, with permission, from Rexhausen 1992.)

2. Use standard techniques to lower the pipette into the recording chamber, containing the chosen slice, and apply positive pressure to prevent dust floating on the surface from entering the pipette. Once the pipette tip is in the bath, set the positive pressure to a level that prevents the ejection of too much dye (see Fig. 1A). Apply a brief pulse of stronger pressure immediately before seal formation to clean the surface of the cell.
3. Perform seal formation and establish the whole-cell configuration as per standard electrophysiological recordings. Immediately after the patch is ruptured, the dye will start to diffuse into the cell.

   The time required to reach a certain cytosolic dye concentration will depend on the following factors:

   - The dye concentration in the pipette
   - The series resistance (R_s) of the pipette (Pusch and Neher 1988)
   - The mobility of the dye molecule (Pusch and Neher 1988)
   - The cell geometry (Rexhausen 1992)

Note: A difficulty frequently encountered, especially in brain slices (Edwards et al. 1989), is that R_s tends gradually to increase during whole-cell recordings to levels that seriously hinder dye loading and proper voltage control. Therefore, especially for long-lasting experiments, the first minutes of the whole-cell recordings should be devoted entirely to the establishment of a low and stable series resistance by applying suction pulses to the pipette whenever R_s tends to increase. Although sometimes this approach may destroy the whole-cell configuration, it is preferable to lose a cell during the first minutes than to load it for a prolonged time with an unfavorable R_s and compromise on the degree of dye loading.

EXAMPLE OF APPLICATION

To load remote cellular compartments such as axons (Callewaert et al. 1996) or terminal dendrites and spines of central neurons (Eilers et al. 1995; Schmidt et al. 2003), the time that is required to reach a certain dye concentration strongly depends on their distance from the patch pipette and the geometrical characteristics of the connecting cellular structures. Figure 1B shows an example from a study of dye diffusion in cerebellar Purkinje neurons (Rexhausen 1992). In this experiment, the cell was loaded with the calcium-indicator dye fura-2 through a somatic patch pipette for 120 min (R_s = 5.5–6.7 MΩ). The fluorescence of distinct dendritic compartments was monitored during the loading phase, and the time required to reach half of the dye concentration in the pipette solution was estimated (Fig. 1B). Thus, for remote compartments, about 30–40 min of whole-cell recordings may be needed before the dye concentration reaches levels that allow fluorometric recordings with a reasonable signal-to-noise ratio.

ADVANTAGES AND LIMITATIONS

In comparison to other staining techniques such as dye loading with membrane-permeable dyes (e.g., of the so-called AM-type; Tsien 1981) or dye injection via sharp electrodes (von Blankenfeld and Kettenmann 1992), the patch-clamp dye loading technique offers important advantages. First, all cell types that can be studied with the whole-cell patch-clamp technique are suitable for staining with patch-clamp dye loading. This is of crucial importance, especially when studying cells that are too small to tolerate impalement with a sharp electrode, or neurons that are extensively covered by glia cells and, therefore, cannot be efficiently stained with AM dyes. A further advantage is that the intracellular dye concentration can be rather precisely predicted, either during the loading phase (Pusch and Neher 1988; Rexhausen 1992; Schneggenburger et al. 1993) or after an appropriate loading time, when the cytosolic dye concentration is close to that of the pipette solution. Thus, quantitative studies relying on accurate knowledge of cytosolic dye concentrations are possible (Neher 1995).

The drawbacks of dye loading via patch pipettes are associated with (1) the difficulties of the whole-cell patch-clamp technique (staining tissue with membrane-permeable dyes is obviously an easier technique), (2) the possible washout of cytosolic constituents during prolonged whole-cell recordings (Pusch and Neher 1988), and (3) the limitations in the size of dye-labeled proteins that can be loaded via pipettes (proteins of up to 150 kD have successfully been used; Pusch and Neher 1988).

In conclusion, whole-cell patch-clamp recordings represent a versatile loading technique that allows combined electrophysiological and optical measurements at a quantitative level.

REFERENCES

Callewaert G., Eilers J., and Konnerth A. 1996. Axonal calcium entry during fast ‘sodium’ action potentials in rat cerebellar Purkinje neurones. *J. Physiol.* **495:** 641–647.

Cota G. and Armstrong C.M. 1988. Potassium channel “inactivation” induced by soft-glass patch pipettes. *Biophys. J.* **53:** 107–109.

Edwards F., Konnerth A., Sakmann B., and Takahashi T. 1989. A thin slice preparation for patch clamp recordings from neurones of the mammalian central nervous system. *Pflügers Arch.* **414:** 600–612.

Eilers J., Augustine G.J., and Konnerth A. 1995. Subthreshold synaptic Ca^{2+} signalling in fine dendrites and spines of cerebellar Purkinje neurons. *Nature* **373:** 155–158.

Grynkiewicz G., Poenie M., and Tsien R.Y. 1985. A new generation of Ca^{2+} indicators with greatly improved fluorescence properties. *J. Biol. Chem.* **260:** 3440–3450.

Marty A. and Neher E. 1995. Tight-seal whole-cell recordings. In *Single-channel recordings*, 2nd edition (ed. B. Sakmann and E. Neher), pp. 31–52. Plenum Press, New York.

Neher E. 1995. The use of fura-2 for estimating Ca buffers and Ca fluxes. *Neuropharmacology* **34:** 1423–1442.

Penner R. 1995. A practical guide to patch-clamping. In *Single-channel recordings*, 2nd edition (ed. B. Sakmann and E. Neher), pp. 3–30. Plenum Press, New York.

Pusch M. and Neher E. 1988. Rates of diffusional exchange between small cells and a measuring patch pipette. *Pflügers Arch.* **411:** 204–211.

Rexhausen U. 1992. "Bestimmung der Diffusionseigenschaften von Fluoreszenzfarbstoffen in verzweigten Nervenzellen unter Verwendung eines rechnergesteuerten Bildverarbeitungssystems." Ph.D. thesis, University of Göttingen, Germany.

Sakmann B. and Stuart G.J. 1995. Patch-pipette recordings from the soma, dendrites, and axon of neurons in brain slices. In *Single-channel recordings*, 2nd edition (ed. B. Sakmann and E. Neher), pp. 199–211. Plenum Press, New York.

Schmidt H., Brown E., Schwaller B., and Eilers J. 2003. Diffusional mobility of parvalbumin in spiny dendrites of cerebellar Purkinje neurons quantified by fluorescence recovery after photobleaching. *Biophys. J.* **84:** 2599–2608.

Schneggenburger R., Zhou Z., Konnerth A., and Neher E. 1993. Fractional contribution of calcium to the cation current through glutamate receptor channels. *Neuron* **11:** 133–143.

Tsien R.Y. 1981. A non-disruptive technique for loading calcium buffers and indicators into cells. *Nature* **290:** 527–528.

von Blankenfeld G. and Kettenmann H. 1992. Application of Lucifer yellow as intracellular marker in electrophysiology. In *Practical electrophysiological methods* (ed. H. Kettenmann and R. Grantyn), pp. 309–315. Wiley-Liss, New York.

CHAPTER 34

A Practical Guide: Calcium Imaging of the Retina

Christian Lohmann,[1] Jay Demas, Josh L. Morgan, and Rachel O.L. Wong

Department of Anatomy and Neurobiology, Washington University School of Medicine, St. Louis, Missouri 63110

Calcium imaging has been used to simultaneously monitor the activity patterns of populations of retinal neurons, as well as to visualize spontaneous or light-driven calcium signals within individual neurons and their processes. As in other tissues, a major technical challenge in performing calcium imaging of the retina is the loading of cells of interest with calcium indicator dyes. Retinal neurons, in tissue or in dissociated culture, can be loaded by incubation with AM-ester forms of the indicators (e.g., Ratto et al. 1988; Wong et al. 1995), injection via intracellular pipettes (e.g., von Gersdorff and Matthews 1994; Euler et al. 2002) or by Calistics, a ballistic method using a gene gun (Kettunen et al. 2002; Lohmann et al. 2002). Indicators can also be injected into the optic nerve to retrogradely label ganglion cells (Sernagor et al. 2000), or directly into the retina itself to label clusters of retinal cells (Baldridge 1996). The choice of loading method depends primarily on the age of the tissue to be studied, the need to label a single cell or a population of cells, and the requirement for discriminating between labeled neuronal processes. Here, we provide protocols for AM-ester incubation and Calistics (ballistic delivery) that are used to image the developing retinae from a variety of species.

MATERIALS

Calcium Indicators

Molecular Probes offers a large assortment of AM-ester and dextran-coupled calcium indicators (different spectral characteristics, low or high affinity, ratiometric or non-ratiometric).

Biological Sample

Retinal tissue from the species of interest. For advice on dissection of retinal tissue and its preparation for imaging and image acquisition methods, see Chapter 21.

For Procedure A: AM-ester Loading

Calcium indicator as AM-ester derivate (e.g., fura-2, AM, special packaging, Molecular Probes)

Dimethyl sulfoxide (DMSO) <!>

Pluronic acid (Pluronic F-127<!>)

Physiological solution depending on the species (e.g., Ames medium or ACSF)

(See box continued on top of next page.)

[1] *Present address:* Cellular and Systems Neurobiology, Max-Planck Institute of Neurobiology, Am Klopferspitz 18a, 81377 München, Germany.

For Procedure B: Preparation of Cartridges (Calistics)

Calcium indicator as dextran conjugates, (e.g. Oregon Green 488 BAPTA-1 [MW 10,000] dextran, Molecular Probes)

Tungsten M-25 Microcarriers 1.7 μm (Bio-Rad)

Glass slides

Razor blades

Polyvinylpyrrolidone (PVP, Sigma)

GoldCoat tubing (Bio-Rad Cartridge Kit)

Nitrogen gas tank

Biorad "prep station"

Storage vials, desiccant pellets

For Procedure C: Ballistic Delivery (Calistics)

Gene Gun (Bio-Rad Helios)

Helium gas tank (ultra-high purity)

Culture plate insert (3 μm pore size, Millipore)

Caution: *See Appendix 3 for appropriate handling of materials marked with <!>.*

PROCEDURE A

AM-ester Loading

Preparation of Incubation Solution (~10 μM Indicator, 0.001% Pluronic Acid)

1. Add 50 μl of DMSO<!> to one vial (50 μg) of the chosen calcium indicator. Mix well.
2. In a separate tube, add 5 μl of pluronic acid <!> (2.5% in DMSO) to 5 ml of the appropriate physiological solution. Mix well and then combine the indicator solution with the physiological solution.

Incubation of Retinae

3. (a) When using retinae from young animals, transfer the retinae to incubation solution and keep at room temperature for 30–45 min.

   (b) When using retinae from older animals (e.g., mice or ferrets, more than about 2 weeks old), add a few drops of 1 μg/μl calcium indicator in DMSO/0.001% pluronic acid onto the surface of the retina. Leave the solution on the tissue for several seconds. If the solution is reluctant to penetrate the tissue, make small slits on retina and drop concentrated indicator solution (1 μg/ml) over the slits (as described by Wong and Oakley 1996).
4. After 30–45 min at room temperature, increase the temperature to 30–35°C and incubate for an additional 10–15 min. Wash retinae in the appropriate physiological solution (two washes, 5 min each) and then transfer them to a recording chamber for experimentation.

Caution: *See Appendix 3 for appropriate handling of materials marked with <!>.*

PROCEDURE B

Preparation of Cartridges (Calistics)

Preparation of Cartridges

1. Dissolve 1.5 mg of dextran-conjugated indicator in 15 μl of distilled water and place 25 mg of tungsten microcarriers on a microscopic slide.
2. Add the indicator solution to the microcarriers. Mix well and spread evenly across the slide. Allow the slide to dry.

3. Using a razor blade, scrape the now indicator-coated particles off the slide and onto a piece of weighing paper.
4. Pre-coat 25 cm of GoldCoat tubing by filling it briefly with 1% PVP<!> (w/v, in water). Remove the solution and dry the tubing by passing nitrogen through it (e.g., using a Bio-Rad "prep station").
5. Seal one end of the tubing with Parafilm and then pour indicator-coated microcarrier particles into it (use a 1-ml pipette tip as a funnel). When all the particles are inside the tubing, seal the other end.
6. Shake the tubing gently until the particles stick to the inside wall of the tubing and are evenly distributed.
7. Cut the tubing into 13-mm-long cartridges (Bio-Rad tubing cutter or a razor blade) and store in a desiccator at 4°C, protected from light for up to 4 weeks.

Caution: *See Appendix 3 for appropriate handling of materials marked with <!>.*

PROCEDURE C

Ballistic Delivery (Calistics)

1. Load a cartridge into the barrel of the gene gun and set the helium pressure to ~80–90 psi. Use a pressure gauge that allows fine adjustment between 0 and 250 psi (e.g., Puritan-Bennett model 60).
2. Remove excess physiological solution from retina(e) and place a culture plate insert over the tissue.

   Note: The insert prevents damage to the tissue resulting from the pressure wave created when the gun is fired, and it eliminates clumps of particles fired from the cartridge.
3. Hold the gene gun ~2 cm above the retina and fire.
4. Wash the retina(e) twice in physiological solution and then transfer to a recording chamber for experimentation.

   Note: See Chapter 21 for tips on shooting, controlling the density of labeling, and reducing tissue trauma during shooting.

SHORT EXAMPLE OF APPLICATION

Figure 1 provides examples both of indicator-loading techniques presented in the chapter and of calcium imaging in the retina. Spontaneous calcium waves that spread across the developing retina can be observed following AM-ester incubation (Fig. 1a,b). Calcium imaging at higher magnifications permits simultaneous imaging of populations of cells or recording from individual cells both in vitro (Fig. 1c) and in vivo (Fig. 1d). The arbors of retinal neurons at the retinal surface can be filled easily with indicator by intracellular injection (Fig. 1e) and by Calistics (Fig. 1f,g) on whole-mount preparations. Intracellular injection and Calistics are also suitable for labeling cells in retinal cross sections (Fig. 1h).

ADVANTAGES AND LIMITATIONS

The relative ease and effectiveness of calcium indicator loading of retinal tissue are summarized for AM-ester incubation, intracellular injection, and Calistics in Table 1. The AM-ester forms of indicators are particularly suitable for loading large populations of retinal neurons. However, because the retina has a very dense neuropil, it is often hard to differentiate axons and dendrites from neighbor-

TABLE 1. Comparison of effectiveness of different loading methods

Method of loading	AM-ester	Micropipette	Calistics
Effective loading ages	immature only mouse ≤P12 (Bansal et al. 2000) rabbit ≤P13 (Zhou and Zhao 2000) chick ≤P1 (Wong et al. 2000) ferret ≤P23 (Wong and Oakley 1996)	all ages	all ages
Loading time	~60 min	30 min	≤5 min
% of cells loaded	~90–100	<1	≤10
Control of indicator concentration	poor	excellent	limited

(P) Postnatal day.

ing cells in AM-ester-loaded tissue. It is easier to image individual axonal or dendritic arbors in sparsely labeled cells. Sparse labeling can be achieved by intracellular injection or by Calistics. Delivering indicators via micropipettes has the advantage that individual cells can be targeted, and electrophysiological recordings can be simultaneously obtained. However, Calistics has two major advantages over loading with pipettes, patch pipettes in particular: (1) Calistics is fast and does not require tedious maneuvers to impale neurons without damage; (2) the intracellular milieu of the neuron is not perturbed because the cell membrane is sealed immediately after the particle enters the cell. Thus, there is no washout of intracellular components, as can occur in the whole-cell patch configuration.

FIGURE 1. (*a*) Low-magnification image of fura-2 AM loading of a whole-mount preparation from a newborn mouse retina. The optic nerve head (ONH) is indicated. (*b*) Time-lapse "difference" images showing a spontaneously occurring calcium wave spreading across the retina shown in *a*. Difference images are generated by digitally subtracting each image frame from the previous frame. Light regions indicate areas of elevated intracellular calcium, averaged over 1 sec. Imaging was carried out using a Hamamatsu SIT camera. (*c*) Multiphoton image of fura-2-loaded cells in a retinal whole mount from an embryonic day (E) 13 chick. The differences in the fluorescence intensity of individual cells demonstrate variability in the intracellular concentration of free indicator, following incubation with AM-esters. (GCL) Ganglion cell layer; (FL) fiber layer; (IPL) inner plexiform layer. (Reprinted, with permission, from Wong et al. 1998.) (*d*) Fura-2 loading of cells in the retina (R) and lens (L) of a zebrafish embryo. Images were collected at 35 hr postfertilization. Fura-dextran (10,000 MW, Molecular Probes) was injected, using a micropipette, into the yolk at the one-cell stage. Three puffs of 50-mM fura-dextran (Molecular Probes) in sterile distilled water were delivered at 20 psi, 20 msec duration, using a picrospritzer (General Valve). Image obtained using a multiphoton microscope. (*e*) Two-photon image of a starburst amacrine cell from an adult rabbit retina, labeled by intracellular injection with Oregon Green BAPTA-1. (Reprinted, with permission, from Euler et al. 2002 [©2002 Macmillan].) (*f*) Three ganglion cells labeled with dextran-conjugated Oregon Green BAPTA-1 using Calistics. The difference in the fluorescence intensity of the cells likely reflects variability in the concentration of indicator per microparticle loaded using Calistics. To avoid effects of calcium buffering by the indicator, relatively dim cells (2 and 3) should be chosen for study (see Kettunen et al. 2002). Images were acquired using a confocal microscope. (*g*) An E15 chick retinal ganglion cell labeled with Oregon Green BAPTA-1 by Calistics. Traces below depict the percent change in fluorescence intensity as a function of time for the three dendritic regions indicated. Relatively large, global (Gb) increases in intracellular calcium levels throughout the cell are observed. Smaller calcium elevations (L) occurring at separate times in local segments of dendrites are also seen (see Lohmann et al. 2002). Recording obtained using a cooled CCD (Sensicam; Cooke). (*h*) An image of an E16 chick retinal ganglion cell (fluorescence represented by dark pixels; inverted contrast) with its dendritic arbor stratifying in the inner plexiform layer (IPL) where it receives synaptic connections from retinal interneurons. The ganglion cell was labeled by Calistics in a vertical slice of the retina. In the background is a Nomarski image of the retinal slice. The image was reconstructed using an Olympus confocal microscope (FV500).

However, unlike pipette injections and AM-ester incubation of immature tissue, the ballistic approach is currently limited to cells whose somata are located within 50 µm of the tissue surface. A recent modification of the gene gun barrel may make deeper penetration possible (O'Brien et al. 2001).

Finally, it is possible to combine calcium imaging with other methods of dye labeling. For example, AM-ester loading can be used in conjunction with carbocyanine dye labeling (Gan et al. 2000). Additionally, different fluorescent dyes, indicators, or compounds can be co-delivered by intracellular injection and by Calistics. For Calistics, one useful combination is to co-coat tungsten particles with calcium indicator dextrans and biotinylated dextran amine (Molecular Probes), a compound that can be processed to provide a permanent cellular marker after tissue fixation (see Chapter 21). Another important application is to co-coat tungsten particles with indicator and caged compounds that are activated upon ultraviolet excitation (Lohmann et al. 2002).

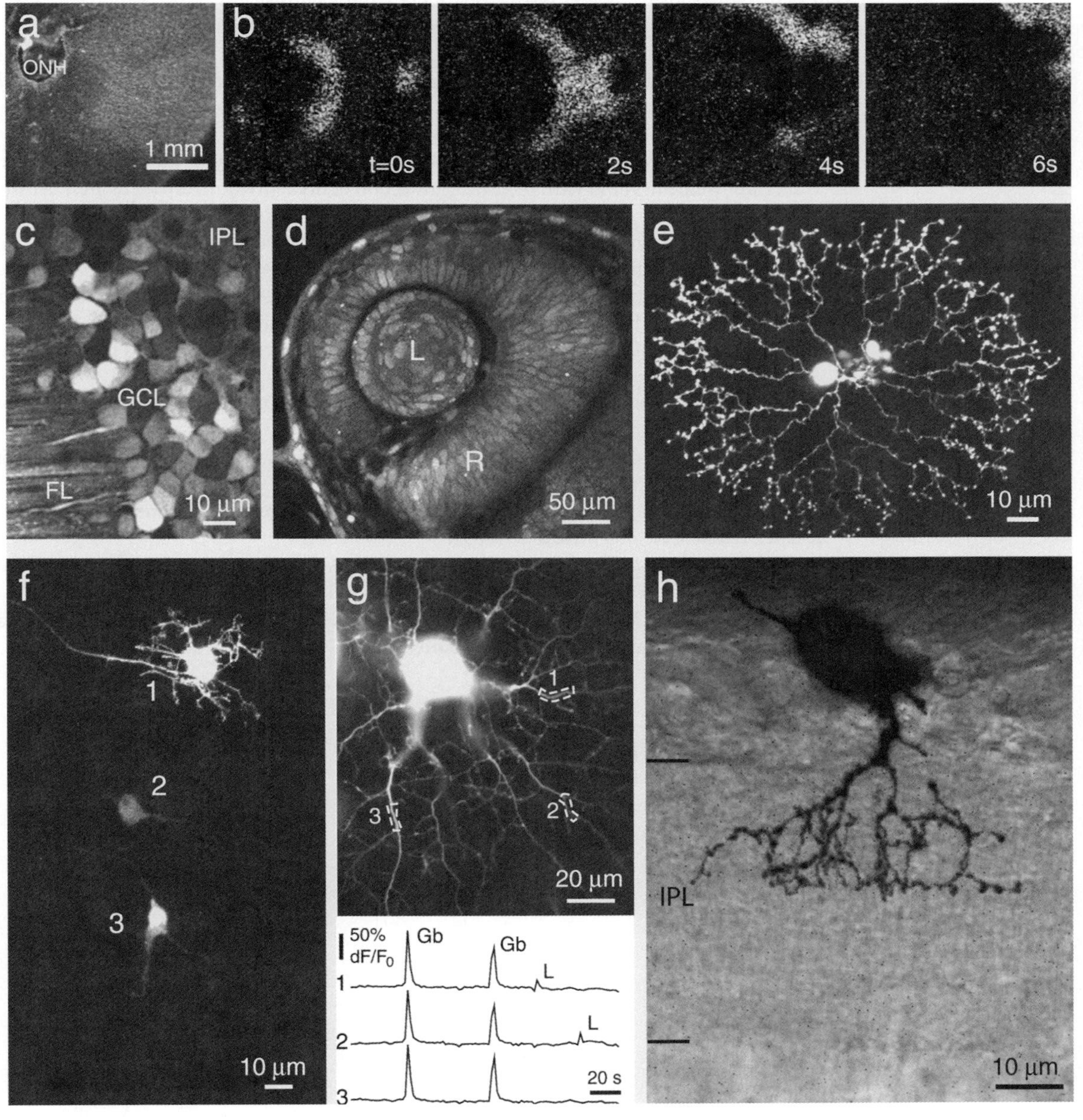

FIGURE 1. (*See facing page for legend.*)

ACKNOWLEDGMENTS

This work was supported by the Deutsche Forschungsgemeinschaft (C.L.), the National Science Foundation (J.D.), and the National Institutes of Health (R.O.L.W., J.L.M.).

REFERENCES

Baldridge W.H. 1996. Optical recordings of the effects of cholinergic ligands on neurons in the ganglion cell layer of mammalian retina. *J. Neurosci.* **16:** 5060–5072.

Bansal A., Singer J.H., Hwang B.J., Xu W., Beaudet A., and Feller M.B. 2000. Mice lacking specific nicotinic acetylcholine receptor subunits exhibit dramatically altered spontaneous activity patterns and reveal a limited role for retinal waves in forming ON and OFF circuits in the inner retina. *J. Neurosci.* **20:** 7672–7681.

Euler T., Detwiler P.B., and Denk W. 2002. Directionally selective calcium signals in dendrites of starburst amacrine cells. *Nature* **418:** 845–852.

Gan W.B., Grutzendler J., Wong W.T., Wong R.O.L., and Lichtman J.W. 2000. Multicolor "DiOlistic" labeling of the nervous system using lipophilic dye combinations. *Neuron* **27:** 219–225.

Kettunen P., Demas J., Lohmann C., Kasthuri N., Gong Y., Wong R.O.L., and Gan W.B. 2002. Imaging calcium dynamics in the nervous system by means of ballistic delivery of indicators. *J. Neurosci. Methods* **119:** 37–43.

Lohmann C., Myhr K.L., and Wong R.O.L. 2002. Transmitter-evoked local calcium release stabilizes developing dendrites. *Nature* **418:** 177–181.

O'Brien J.A., Holt M., Whiteside G., Lummis S.C., and Hastings M.H. 2001. Modifications to the hand-held Gene Gun: Improvements for in vitro biolistic transfection of organotypic neuronal tissue. *J. Neurosci. Methods* **112:** 57–64.

Ratto G.M., Payne R., Owen W.G., and Tsien R.Y. 1988. The concentration of cytosolic free calcium in vertebrate rod outer segments measured with fura-2. *J. Neurosci.* **8:** 3240–3246.

Sernagor E., Eglen S.J., and O'Donovan M.J. 2000. Differential effects of acetylcholine and glutamate blockade on the spatiotemporal dynamics of retinal waves. *J. Neurosci.* **20:** RC56 (1–6)

von Gersdorff H. and Matthews G. 1994. Dynamics of synaptic vesicle fusion and membrane retrieval in synaptic terminals. *Nature* **367:** 735–739.

Wong R.O.L. and Oakley D.M. 1996. Changing patterns of spontaneous bursting activity of on and off retinal ganglion cells during development. *Neuron* **16:** 1087–1095.

Wong R.O.L., Chernjavsky A., Smith S.J., and Shatz C.J. 1995. Early functional neural networks in the developing retina. *Nature* **374:** 716–718.

Wong W.T., Sanes J.R., and Wong R.O.L. 1998. Developmentally regulated spontaneous activity in the embryonic chick retina. *J. Neurosci.* **18:** 8839–8852.

Wong W.T., Myhr K.L., Miller E.D., and Wong R.O.L. 2000. Developmental changes in the neurotransmitter regulation of correlated spontaneous retinal activity. *J. Neurosci.* **20:** 351–360.

Zhou Z.J. and Zhao D. 2000. Coordinated transitions in neurotransmitter systems for the initiation and propagation of spontaneous retinal waves. *J. Neurosci.* **20:** 6570–6577.

CHAPTER 35

Calcium Imaging of Identified Astrocytes in Hippocampal Slices

Jian Kang,[1] Gregory Arcuino,[1] and Maiken Nedergaard[2]

[1]New York Medical College, Valhalla, New York 10595; [2]Center for Aging and Developmental Biology, University of Rochester, Rochester, New York 14615

Astrocytes account for one-third of the mass of the nervous system and outnumber neurons tenfold (Privat et al. 1995). They form an extensive network of line processes that is highly coupled through gap junctions and intimately associated with neuronal and vascular elements (Ransom 1995). The study of astrocytic function is critically dependent on imaging of fluorescence ion indicators because, for all practical purposes, astrocytes are electrically silent. Calcium imaging techniques are readily applied to astrocytes in cultures, but primary cultures do not replicate the intermingling of astrocytic and neuronal processes in the brain and are therefore poor models for defining the functional importance of these cells. Over the last few years, electrical recordings have been combined with imaging of calcium indicators in acutely prepared hippocampal slices. In this chapter, we discuss our experience studying astrocytes in situ.

IDENTIFYING ASTROCYTES AND NEURONS IN HIPPOCAMPAL SLICES

Using DIC Microscopy

During electrophysiological recordings, the standard approach is to identify cells with highly negative resting membrane potentials and the absence of action potentials upon depolarization as astrocytes (Walz and MacVicar 1988). However, in calcium imaging studies, many cells are visualized simultaneously, and it is not feasible to obtain recordings from them all. The use of more traditional approaches for identifying astrocytes in fixed tissues, including immunoreactivity against glia fibrillary acidic protein (GFAP) (Dani et al. 1992) or biocytin staining (Robinson et al. 1994), is not an option in live tissues. Our goal was, therefore, to establish whether astrocytic morphology under differential inference contrast (DIC) microscopy allows the investigator to identify astrocytes and thus discriminate between astrocytes and neurons during imaging experiments. We have primarily studied astrocytes in hippocampal slices, but similar observations have been made in cortical slices. For information on DIC, see Chapter 1. We used an Olympus RX50 upright microscope equipped with DIC optics. Slices in a thin 1.5-ml chamber mounted on the microscope stage were perfused with a slice solution. Cells were visualized with a 40× water-immersion lens. Figure 1 illustrates a characteristic field of the stratum radiatum, containing both astrocytes and neurons. Interneurons (Fig. 1a, red arrow) are characterized

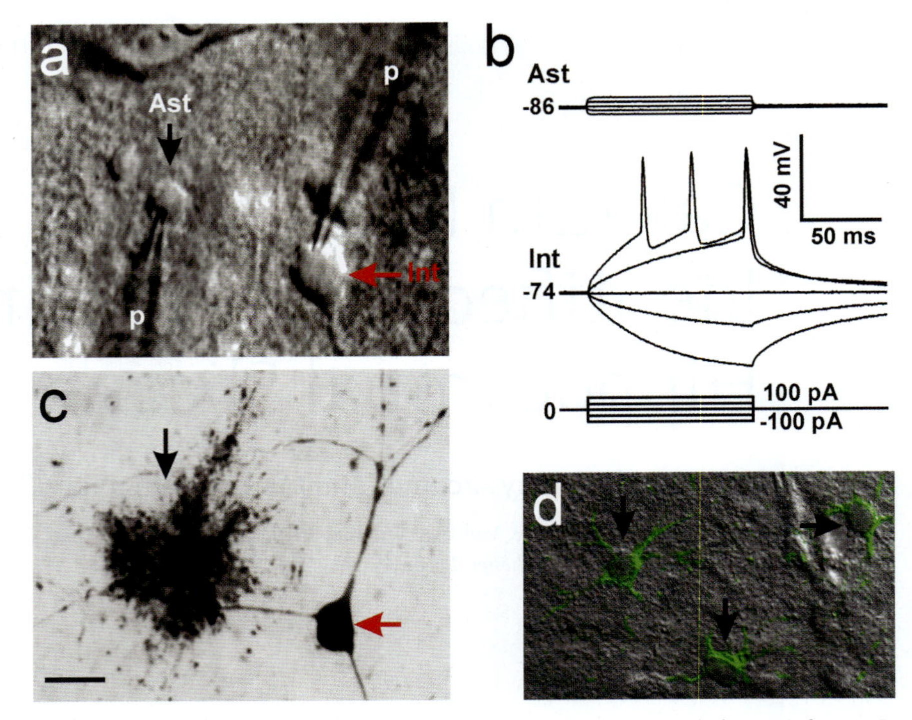

FIGURE 1. Identification of interneurons and astrocytes in hippocampal slices under DIC optics. (*a*) An astrocyte (*black arrow*) characterized by small, rounded cell bodies and poorly defined irregular processes, and an interneuron (*red arrow*) in the stratum radiatum having large soma with large dendrites that may project in all directions, were patched by two pipettes (p). (*b*) Electrophysiological properties of interneurons (Int) and astrocytes (Ast). The bottom lines indicate depolarization currents delivered to patch pipettes. Astrocytes do not fire upon depolarization and have a low input-resistance. (*c*) Biocytin staining of the interneuron and the astrocyte in *a*. Biocytin (0.5%) was injected into both of these cells to compare DIC with cellular morphology. The cell identified by DIC as an astrocyte (*black arrow*) was characterized by an extensive array of branched short processes, phenotypical for astrocytes. In contrast, the presumed interneuron (*red arrow*) had few, but evenly sized, less branched, processes, a morphology that is stereotypical for neurons. The size of biocytin-staining astrocytes seems larger than that of the same cell under DIC optics. This might result from the mixture of short processes with the cell body. Bar, 20 μm. (*d*) Overlapping of DIC optics (background) and GFAP staining (*green*) in a field confirms that three cells (*black arrows*) that were identified as astrocytes under DIC optics are GFAP-positive.

by large and bright somata (diameter, 16.1 ± 2.7 pm, mean ± S.D.) with large dendrites, whereas astrocytic somata (Fig. la, black arrow) are small and round (diameter, 8.1 ± 1.7 pm, mean ± S.D.) without large processes. The characteristics of the cells, as revealed by DIC morphology, were confirmed by electrophysiological recordings, biocytin staining, and GFAP staining. The use of slices prepared from transgenic mice with expression of green fluorescence protein (GFP) under control of the GFAP promoter is a new and excellent approach to identify astrocytes in live tissue (Zhuo et al. 1997; Nolte et al. 2001).

Confirming DIC Morphology

Recording Electrophysiological Properties

Electrophysiological properties confirmed the DIC morphological criteria (above) for identifying interneurons and astrocytes. Recordings were taken from a total of 205 cells with a DIC morphology typical for astrocytes, and none fired action potentials upon depolarization (Fig. 1b, Ast); and from 107 interneurons, all of which fired action potentials (Fig. 1b, Int). Astrocytes also have more negative resting membrane potentials (−86 ± 0.5 mV, mean ± S.E.M., $n = 87$) than interneurons (−74 ± 0.6 mV, $n = 90$) and the input resistance of astrocytes is low (15.0 ± 1.1 MΩ, $n = 10$), reflecting both high ion conductance and gap junctions between cells.

PROCEDURE

Staining with Biocytin

DIC morphology was further evaluated by biocytin staining. Brain slices were prepared from 8- to 20-day-old (PND8-20) Sprague-Dawley rats of either sex. The following procedure was used for biocytin staining.

1. Place slices in a recording chamber (1.5 ml) and perfuse with Slice Solution.

   (Slice Solution: 126 mM NaC1, 2.5 mM KC1<!>, 1.25 mM NaH_2PO_4<!>, 2 mM $MgC1_2$<!>, 2 mM $CaC1_2$<!>, 10 mM glucose, 26 mM NaH_2CO_3, pH 7.4, when the solution is gassed with 5% CO_2 and 95% O_2, at room temperature [23–24°C]).

   Note: Rats are anesthetized with pentobarbital sodium <!> (55 mg/kg) and decapitated. Coronal brain slices (300 μm) are cut with a vibratome (Technical Products International) in a cutting solution containing 2.5 mM KCl, 1.25 mM NaH_2PO_4, 10 mM $MgSO_4$<!>, 0.5 mM $CaCl_2$, 10 mM glucose, 26 mM $NaHCO_3$, and 230 mM sucrose. Slices containing the hippocampus are incubated in the slice solution gassed with 5% CO_2 and 95% O_2, for 1 hour, then transferred to the recording chamber.

2. Using two patch pipettes, load 0.5% biocytin into each cell type under study. First, patch an interneuron with a biocytin-containing pipette and then patch an astrocyte with another biocytin-containing pipette. (0.5% Biocytin: Dissolve biocytin hydrochloride (Sigma) in the pipette solution. Use freshly prepared solution.) (Pipette Solution:120 mM K-gluconate <!>, 10 mM KC1<!>, 1 mM $MgCl_2$<!>, 10 mM HEPES, 0.1 mM EGTA, 0.025 mM $CaCl_2$<!>, 1 mM ATP, 0.2 mM GTP, 4 mM glucose. Adjust pH to 7.2 with KOH <!> .)

   Note: For dye loading using patch pipettes, see Chapter 33.

3. Without moving the patch pipettes, perfuse the slices with fixative for 10 min. Carefully withdraw the pipettes.

   (8% Paraformaldehyde Solution: Add 40 g of paraformaldehyde<!> (Sigma) to 400 ml of distilled water, heat to 60°C, and add 2–3 drops of 1 N NaOH <!>. Adjust final volume to 500 ml with distilled water. Store this stock solution at 4°C.) (Fixative: Add 40 μl of saturated picric acid <!> (Sigma) and 80 μl of 25% glutaraldehyde <!> [Fisher] to 9.88 ml of 0.2 M phosphate-buffered saline [PBS, Sigma]. Dilute this solution 1:1 with the 8% paraformaldehyde <!> solution. Adjust pH to 7.4 with NaOH<!>. The final concentration is 4% paraformaldehyde <!>, 0.2% picric acid <!>, 0.1% glutaraldehyde <!>.)

4. Post-fix the slices for 10–12 hours. Wash four times in 0.1 M PBS.
5. Incubate in 0.3% I (in 0.1 M PBS) at room temperature for 15 min. Wash four times in PBS.
6. Incubate in 0.2% Triton X-100/0.2% albumin (in 0.1 M PBS) at room temperature for 45 min. Wash four times in PBS.
7. Stain the fixed slices with the ABC kit <!> (Vectastain Elite, Vector Laboratories).

   Note: Perform staining according to the manufacturer's directions.

8. Mount with Permount <!> (Fisher).

Caution: *See Appendix 3 for appropriate handling of materials marked with <!>.*

Procedural Notes

The cell identified by DIC as an astrocyte (Fig. 1c, black arrow) was characterized by an extensive array of branched short processes, phenotypical for astrocytes. In contrast, the presumed interneuron (Fig. 1c, red arrow) had few, but evenly sized, less branched processes—a morphology stereotypical for neurons. A total of eight pairs of interneurons and astrocytes were injected with biocytin, and in all cases, biocytin staining confirmed cell identification under DIC.

DETECTION OF ASTROCYTIC Ca^{2+} SIGNALS IN SLICES BY CCD CAMERA, CONFOCAL, AND TWO-PHOTON LASER SCANNING MICROSCOPY

Three imaging techniques for detecting astrocytic Ca^{2+} signals in slices are compared below. Two-photon laser scanning microscopy has a special advantage for detecting astrocytic Ca^{2+} signals in intact brain tissue due to the excellent penetration of long-wavelength excitation and the low level of phototoxic damage associated with infrared excitation. Of note, several groups, including our own, have found that astrocytes preferentially load with AM dyes in slices prepared from rats older than PND 9 days (Liu et al. 2004). The mechanism underlying the preferential loading of astrocytes is unknown, but it is specific to intact brain tissue because neurons, endothelial cells, and pericytes load well with AM dyes in cultures.

Detection of Ca^{2+} Signals from Multiple Astrocytes in a Large Field of Slices

Quantitative fluorescence microscopy of calcium signaling in astrocytes has been used extensively in the past. This approach offers high temporal and spatial resolution when working with cultured astrocytes, but fluorescence microscopy is associated with considerable problems when adapted to imaging hippocampal slices. UV excitation is not tolerated well by the slices and penetrates poorly (<40 μm). The result is that only astrocytes at the surface of the slice can be imaged. The membrane potential of surface astrocytes in hippocampal slices generally is reduced to −60 to −65 mV, suggesting damage during slice preparation. Detection of astrocytic calcium signaling by fluorescence microscopy is therefore not recommended (Fig. 2a). Use of single-photon confocal microscopy has gained increasing popularity over the past 5 years. Astrocytic processes are clearly viable using this approach, but the technique suffers from the same problems that arise with fluorescence microscopy. The most important problems are poor penetration of the laser light and photodamage (Fig. 2b). The first sign of photodamage is that injured astrocytes begin to display spontaneous Ca^{2+} oscillations. Typically, the cytosolic Ca^{2+} levels in healthy astrocytes are stable and Ca^{2+} oscillation is evident only following exposure to neurotransmitters, such as glutamate or ATP. In contrast, astrocytes suffering from photodamage exhibit frequent Ca^{2+} oscillations that are not blocked by a mixture of receptor antagonists, including CNQX, APV, and suramin. The mechanism by which photodamage triggers astrocytic Ca^{2+} oscillations is not known, but it is consistently observed after short-term exposure to 488-nm excitation in hippocampal slices. After prolonged imaging, the cells exhibit sustained increases in calcium, reflecting irreversible damage (Fig. 2b).

The best currently available approach to image astrocytes in intact tissue is two-photon laser scanning microscopy (Fig. 2c). The excitation typically utilized in two-photon imaging of, for example, fluo-4/am, is in the range of 800–900 nm. Long-wavelength excitation easily penetrates 150 μm into the slice and is associated with less phototoxicity than single-photon excitation. Using two-photon excitation, it is therefore possible to image healthy astrocytes, deep within the slice, over a prolonged period of time. However, it should be noted that two-photon imaging has a lower spatial resolution than confocal imaging, resulting in somewhat "grainy" images. Another disadvantage is that only a few astrocytes are imaged in each optical section, due to the fine focus of two-photon excitation. In most experiments, therefore, it is necessary to obtain multiple optical sections (1–5 μm steps) and project these images. Finally, it should be noted that photobleaching also occurs during two-photon imaging, but to a lesser extent than after 488-nm excitation.

Calcium Imaging

Acute hippocampal slices are cut from rat PND13–15. After recovering for 30 min at room temperature, slices are incubated at 30–32°C for 60–80 min in cutting solution containing a membrane-

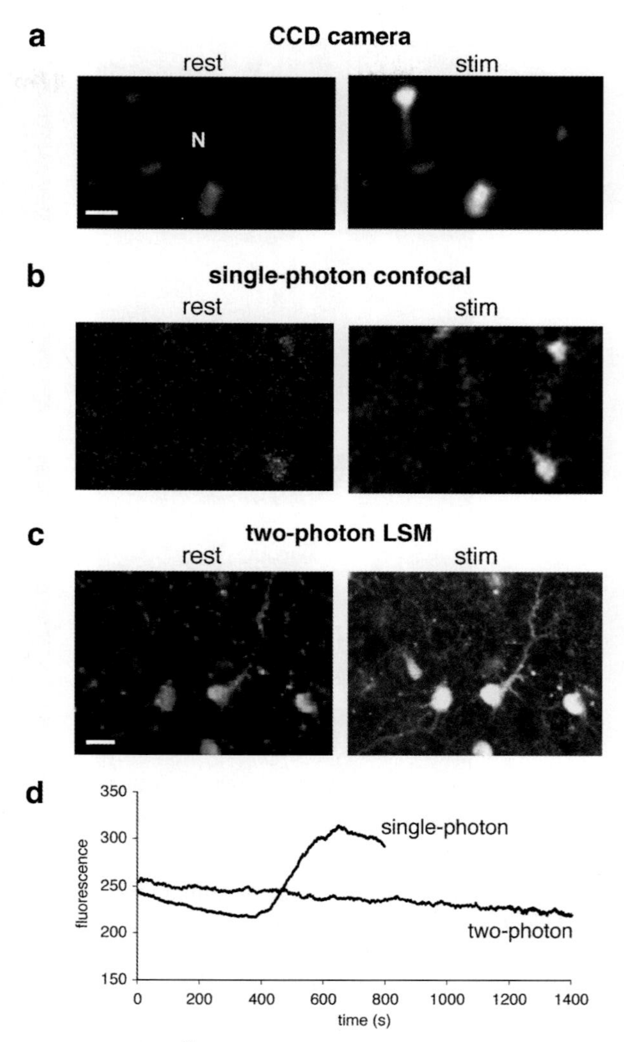

FIGURE 2. Comparison of astrocytic Ca^{2+} signals in hippocampal slices in response to external stimulation. (*a*) CCD camera (*b*) single-photon confocal microscopy, and (*c*) two-photon laser scanning microscopy. (*a*) Astrocytic Ca^{2+} signals were stimulated by interneuronal spikes that were evoked by delivering depolarizing pulses through the patch pipette in an interneuron (N), and detected by the fluorescence microscopy with a CCD camera. (*b,c*) Astrocytic Ca^{2+} signals were triggered by application of ATP (200 μM). Bar, 10 μm. (*d*) Fluo-4 emission as a function of time of imaging. Images of fluo-4/AM loaded astrocytes in hippocampal slices were collected every 5 sec with a 20× objective. Note that both single-photon and two-photon imaging is associated with photobleaching (emission decreasing as a function of time). Also, single-photon excitation is associated with an irreversible increase in the intensity of fluo-4 emission increases after 400 sec due to phototoxicity.

permeable Ca^{2+} indicator dye, acetoxymethyl (AM) ester of fluo-4 (5 μM) in a 5-ml incubation tube gassed by 95% O_2 and 5% CO_2. After loading, slices are washed in artificial cerebrospinal fluid for 20 min and then transferred into the recording chamber under the microscopy. ATP (100 μM) or other agonists is applied after a 5-min recording of baseline activity.

Detection of Ca^{2+} Signals from Fine Astrocytic Processes by Loading of Membrane-impermeable Ca^{2+} Indicators through Patch Pipettes

Ca^{2+} signals in a single astrocyte and its processes can be detected by two-photon laser scanning microscopy of astrocytes filled with membrane-impermeable fluorescence Ca^{2+} indicators, like

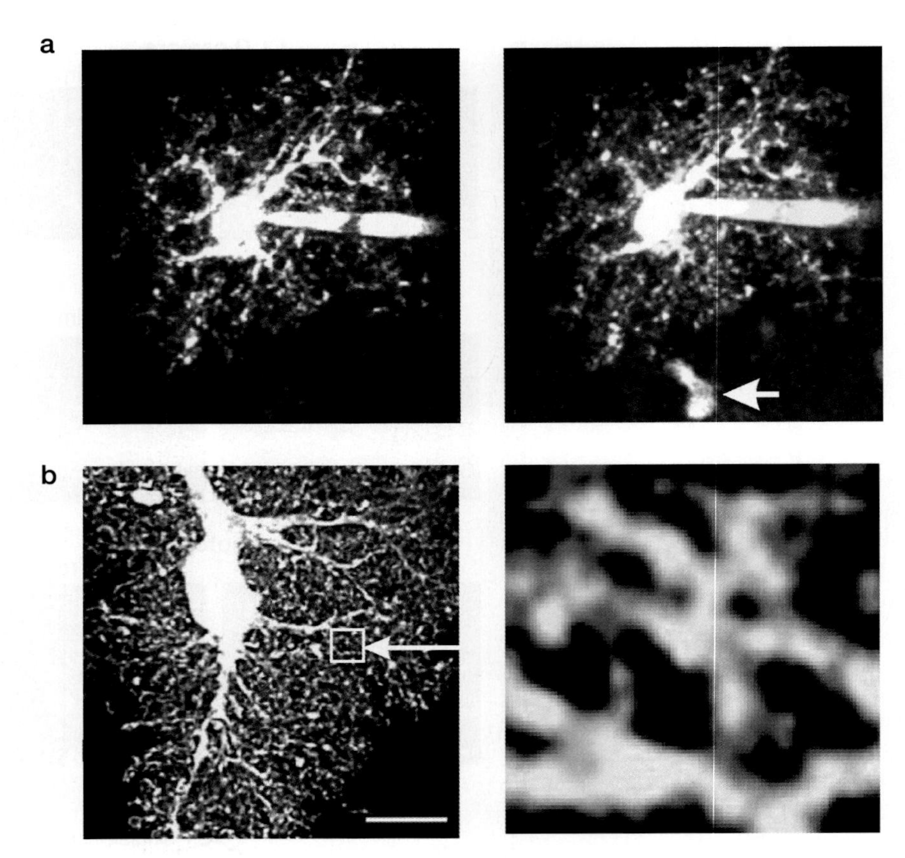

FIGURE 3. Single processes of an astrocyte can be visualized using two-photon laser scanning microscopy. (*a*) An astrocyte patched by a pipette containing 5 μM fluo-4. (*Left panel*) Five minutes after patching the cell. (*Right panel*) Thirty minutes after patching the cell. A neighboring cell was loaded through gap-junction diffusion (*arrow*). (*b*) An eGFP-expressing astrocyte. A recombinant adenovirus carrying eGFP was injected into the cortex of 10-day-old rat pups. Four days later, cortical slices were prepared and eGFP was visualized by two-photon laser-scanning microscopy. Single astrocytes expressed eGFP strongly (*left*). Their fine laminated processes were clearly identified in the enlarged scale (*right*). Bar, 10 μm.

fluo-4, which are readily loaded into astrocytes through patch pipettes. The advantage of the approach is that fine astrocytic processes are clearly outlined and subcellular changes in Ca^{2+} can, therefore, be visualized. The patch-clamp recording also provides information regarding cell membrane potential and ionic currents. Another advantage of pipette loading is that gap junction coupling can be detected by cell–cell diffusion (Fig. 3a). The disadvantage of loading single astrocytes with patch pipettes is that the patching often causes prolonged increases in cytosolic Ca^{2+}. Astrocytes are highly sensitive to mechanical stimulation, and many groups have noted that astrocytes failed to respond to external stimulation after patching. Another problem is that astrocytes are relatively small cells and that their cytosolic content is quickly dialyzed (within minutes of breaking the membrane).

Transgenic expression of Ca^{2+}-sensitive fluorescent proteins is expected to be achieved in the near future (Zhuo et al. 1997; O'Callaghan et al. 2003). Viral delivery of these probes, or generation of mice expressing such proteins under the GFAP promoter, may rapidly alter current protocols for imaging of astrocytes. Adenoviral delivery of eGFP (Ca^{2+}-insensitive) to astrocytes (Nolte et al. 2001) has allowed the astrocytes and their fine processes to be clearly observed (Fig. 3b).

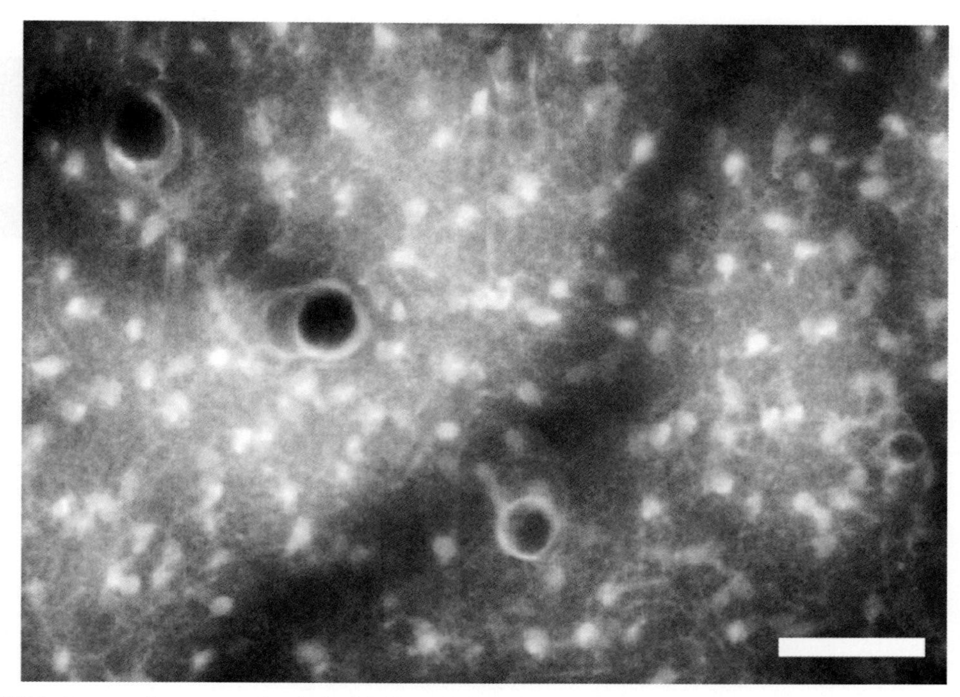

FIGURE 4. Image of astrocytes loaded with fluo-4/AM taken 50–150 μm below the pial surface in vivo. The image was constructed by projecting several serial sections. The astrocytes were loaded by adding fluo-4/AM to an ACSF solution applied to the exposed cortical surface. Penetrating arteries surrounded by astrocytic endfeet are clearly outlined (Modified from Hirase et al. 2004.)

Imaging of Astrocytes in Live Animals

Imaging of astrocytes in live animals is still in its infancy. Several groups have imaged neurons in intact brain of anesthetized rats or mice (Svoboda et al. 1997; Stosiek et al. 2003), but only one group has successfully applied Ca^{2+} imaging technique to astrocytes (Fig. 4) (Hirase et al. 2004). This has been achieved by visualizing astrocytes loaded with fluorescent Ca^{2+} indicators using two-photon laser scanning microscopy via a cranial window. This approach is certain to offer new insights into the role of astrocytes in normal physiological situations, as well as in acute pathophysiology, including stroke, head trauma, and spinal cord injury.

Detection of Ca^{2+} Signals from Photolysis of Caged Compound in a Single Astrocyte

Astrocytes can be loaded by incubating brain slices with NP-EGTA-AM or through a patch pipette filled with membrane-impermeable NP-EGTA. The Ca^{2+} signals in astrocytes are triggered by uncaging Ca^{2+} from NP-EGTA (Fig. 5a). The advantage of the technique is that Ca^{2+} levels in a single astrocyte can be enhanced by laser beam, and Ca^{2+}-driven release of transmitters can be tested (Lui et al. 2004). Ca^{2+} signals can also be triggered by uncaging of IP3, which evokes Ca^{2+} release from intracellular stores. Uncaging of IP3 triggers the release of Ca^{2+} from specific intracellular stores, whereas uncaging of Ca^{2+} may reflect Ca^{2+} signals from different sources. When focusing on functions of Ca^{2+} release from intracellular stores, uncaging IP3 is the preferred triggering method. If studying the overall effects of intracellular Ca^{2+} increase, NP-EGTA uncaging is the better choice.

Photolysis of the Caged Compound NP-EGTA

A photolysis system (Prairie Tech.) was used to uncage NP-EGTA and release calcium. A UV laser beam (337 nm) from a nitrogen-pulsed laser (VSL-337ND; Laser Science) was focused through a

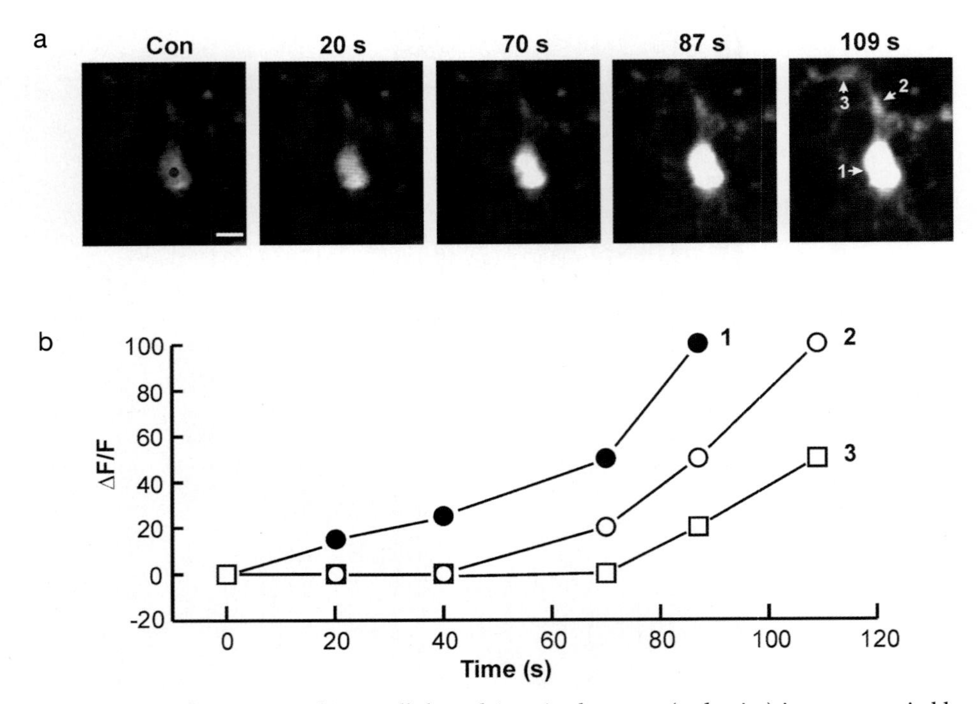

FIGURE 5. (*a*) Local uncaging of intracellular calcium in the soma (*red point*) is accompanied by an intracellular calcium wave that spreads to a distal process detected by a two-photon confocal microscopy. Bar, 10 μm. (*b*) Time course of calcium increases in the soma (1), a proximal process (2), and a distal process (3). Images were acquired with an Olympus 60× water-immersion lens, excited at 810 nm, every 0.5 sec.

DIC 40× water-immersion objective to an optical spot of 10 μm in diameter. A red spot from a weak 635-nm diode laser was coupled with the UV beam to guide the position of uncaging.

Fluorescent imaging was performed using confocal or two-photon confocal Olympus Fluoview confocal microscopes. The 488-nm line of the argon laser was used for excitation, and emission was low-pass filtered at 510 nm. Fluorescence intensities were analyzed by using ImageJ software. Changes in $[Ca^{2+}]_i$ are given as $\Delta F/F$, where F is the baseline fluorescence level and ΔF is determined by the relative change from the base level.

CONCLUSION

Using DIC microscopy, astrocytes in hippocampal slices were identified by their small, round somata that lack large processes. Electrophysiological properties, and biocytin and GFAP staining confirmed the DIC morphological criteria for identifying astrocytes. Astrocytes in slices prepared from PND13–15 rats load well with fluo-4 AM and can be detected by two-photon confocal microscopy. Two-photon confocal microscopy is a better tool for detecting Ca^{2+} signals in astrocytes in slices than fluorescence microscopy and single-photon microscopy. Uncaging of Ca^{2+} or IP3 can be used to obtain further information on the functions of astrocytic Ca^{2+} signals in neuronal circuits.

REFERENCES

Dani J.W., Chernjavsky A., and Smith S.J. 1992. Neuronal activity triggers calcium waves in hippocampal astrocyte networks. *Neuron* **8:** 429–440.

Hirase H., Qian L., Bartho P., and Buzsaki G. 2004. Calcium dynamics of cortical astrocytic network in vivo. *PLoS Biol.* **4:** 1–6.

Liu Q.S., Xu Q., Arcuino G., Kang J., and Nedergaard M. 2004. Astrocyte-mediated activation of neuronal kainate receptors. *Proc. Natl. Acad. Sci.* **101:** 3172–3177.

Nolte C., Matyash M., Pivneva T., Schipke C.G., Ohlemeyer C., Hanisch U.-K., Kirchhoff F., and Kettenmann H. 2001. GFAP promoter-controlled EGFP-expressing transgenic mice: A tool to visualize astrocytes and astrogliosis in living brain tissue. *Glia* **33:** 72–86.

O'Callaghan D.W., Tepikin A.V., and Burgoyne R.D. 2003. Dynamics and calcium sensitivity of the Ca^{2+}/myristoyl switch protein hippocalcin in living cells. *J. Cell Biol.* **163:** 715–721.

Privat A., Gimenez-Ribotta M., and Ridet J. 1995. Morphology of astrocytes. In *Neuroglia* (ed. H. Kettenmann and B.R. Ransom), pp. 3–22. Oxford University Press, New York.

Ransom B.R. 1995. Gap junctions. In *Neuroglia* (ed. H. Kettenmann and B.R. Ransom), pp. 299–318. Oxford University Press, New York.

Robinson S.R., Hampson E.G., Munro M.N., and Vaney D.I. 1994. Unidirectional coupling of gap junctions between neuroglia. *Science* **265:** 1018–1020.

Stosiek C., Garaschuk O., Holthoff K., and Konnerth A. 2003. In vivo two-photon calcium imaging of neuronal networks. *Proc. Natl. Acad. Sci.* **100:** 7319–7324.

Svoboda K., Denk W., Kleinfeld D., and Tank D.W. 1997. In vivo dendritic calcium dynamics in neocortical pyramidal neurons. *Nature* **385:** 161–165.

Walz W. and MacVicar B. 1988. Electrophysiological properties of glial cells: Comparison of brain slices with primary cultures. *Brain Res.* **443:** 321–324.

Zhuo L., Sun B., Zhang C.L., Fine A., Chiu S.Y., and Messing A. 1997. Live astrocytes visualized by green fluorescent protein in transgenic mice. *Dev. Biol.* **187:** 36–42.

CHAPTER 36

Imaging Microscopic Calcium Signals in Excitable Cells

Mark B. Cannell, Angus J.C. McMorland, and Christian Soeller

Department of Physiology, University of Auckland School of Medicine, Grafton, Auckland, New Zealand

The discovery of localized microscopic calcium release events has led to a major advance in our understanding of signal transduction (for review, see Berridge 1997) by establishing the idea that cell function is regulated at a *subcellular* level. As a generalization, cell behavior is probably regulated by the summation of subcellular transduction events, which are generally linked nonlinearly to subsequent stages of signal transduction. It therefore follows that it will be impossible to fully describe signal transduction by measuring average levels of cytoplasmic messengers. Instead, the spatial nonuniformity of second-messenger signals must be considered to better understand how stimulus leads to response. One area in which imaging of subcellular signaling has revolutionized our understanding of signal transduction is in muscle excitation–contraction coupling, where microscopic calcium release events called "calcium sparks" occur (Fig. 1A) (Cheng et al. 1993; see also López-López et al. 1994; Nelson et al. 1995; Tsugorka et al. 1995; Klein et al. 1996). There is evidence that calcium sparks (or spark-like events) may occur in at least two different neuronal preparations (Melamed-Book et al. 1999). This chapter discusses the factors that are important in experimentation on calcium sparks, although many of the comments are also applicable to live-cell microscopy and microfluorimetry.

FACTORS LIMITING SIGNAL DETECTION IN MICROFLUORIMETRY

Consider a fluorescent probe at a concentration of 10^{-5} M. In a confocal imaging volume of 0.2 x 0.2 x 0.6 μm (0.024 fl) there will then be ~145 probe molecules. The maximum applicable illumination intensity is limited by the risk of photodamage at high light levels. Thus, the highest acceptable excitation rate of the probe is likely to be <10% of maximum. At higher rates, a population inversion (where most of the probe molecules are in the excited state) will develop and increase the risk that a second excitation photon will arrive before the molecule decays to the ground state, which may promote photodamage. A typical probe molecule has a fluorescence lifetime of <10 nsec and a quantum efficiency of 0.3, emitting 3×10^6 photons/sec at 10% of maximum excitation rate. The entire confocal volume then will therefore emit ~4×10^8 photons/sec, and a standard confocal microscope, with a detection efficiency of ~1%, will report a photon flux of ~4×10^6 photons/sec. If a 1-msec time resolution is desired with 500 pixels per line for reasonable spatial sampling (resulting in a pixel dwell time of 2 μsec), then the signal in each pixel is ~9 photons, giving a signal-to-noise ratio (SNR) (which is proportional to the square root of the number of detected photons) of ~3. This figure is close to that achieved in experiments and largely explains the noise that is so evident in calcium spark recordings.

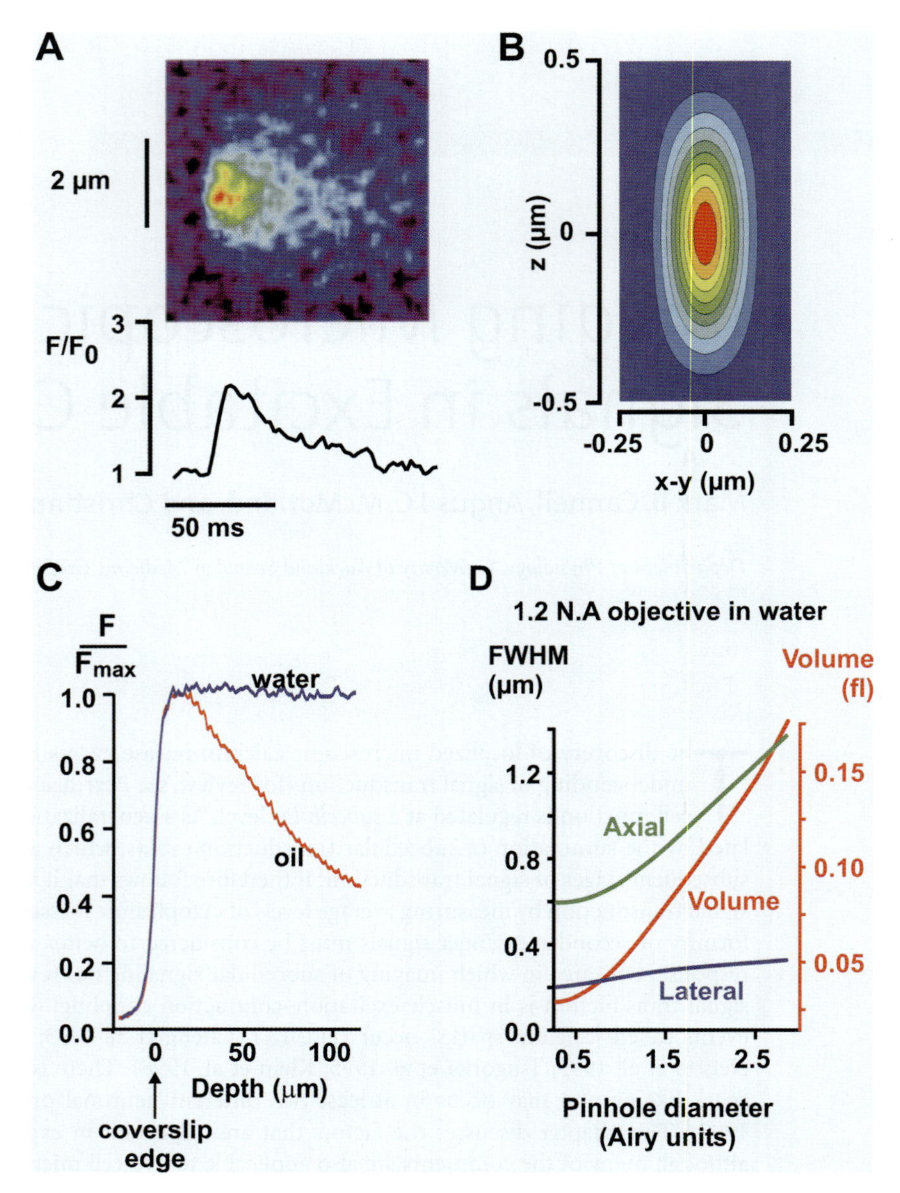

FIGURE 1. Confocal imaging of "calcium sparks." (*A*) Line scan image of a typical, spontaneous calcium spark from a rat ventricular myocyte. One line through the cell is scanned repetitively, and each scan line is displayed adjacent to the previous scan line (Cannell et al. 1994) so that the time course of the spark can be resolved (which is shown below as a line plot of the fluorescence through the center of the spark). Note the relatively small size, the rapidity of fluorescence changes, and the noise levels associated with the spark line scan image. This image has been low-pass-filtered with a 3 x 3 spatiotemporal filter to reduce photon noise. (*B*) Theoretical point spread function (PSF) computed for a 1.2-NA water-immersion objective and closed pinhole (for details of the computation, see Van der Voort and Brakenhoff 1990). The PSF characterizes the resolution of a confocal microscope, and optimal alignment should be routinely checked by measuring the PSF with subresolution fluorescent beads to ensure optimal detection efficiency. The contour lines indicate isointensity lines from 10% to 90% of peak signal. (*C*) Illustration of the loss of signal due to refractive index mismatch. The two curves show the measured dependence of fluorescence signal on distance from the coverslip edge (*arrow*) using oil- and water-immersion objectives in an aqueous sample. The refractive index mismatch due to use of an oil-immersion objective results in strong signal decay when imaging at depths >30 µm. (*D*) The relationship between pinhole size, resolution, and SNR. As pinhole diameter increases, lateral (*blue*) and axial (*green*) resolution (inversely proportional to their FWHMs) decreases. However, decreasing resolution also has the effect of increasing the confocal volume (*red*), which increases the signal collected and thus the SNR (proportional to the square root of the signal intensity) in the image. For example, opening the pinhole from 1 to 2 Airy units will decrease the lateral resolution by only 20% from ~0.25 µm to ~0.3 µm, but will increase the confocal volume and signal strength by 70%, resulting in a 30% improvement in SNR.

Furthermore, this signal must be detected on top of any detector noise and probe background signal (which is generally greater than zero). The best probe is therefore one that produces a minimal background signal in the absence of the ion of interest (e.g., fluo-4).

These rough calculations emphasize the importance of maximizing optical and detector efficiency, and the need to use a probe that is optimal for detecting the ion of interest. They also show the importance of a short fluorescence lifetime and resistance to photodamage (as well as a damage-resistant cell system). A short fluorescence lifetime may be associated with a small Stoke's shift, which may place extra demands on the design of excitation, emission, and dichroic filters.

SPATIAL AND TEMPORAL RESOLUTION

The SNR will fall with increasing spatial and temporal resolution (as sample volume and pixel dwell time decrease), so some trade-off between these quantities may be necessary. In more concrete terms, a doubling of linear resolution within the detection volume will require an eightfold reduction in temporal resolution to achieve the same SNR. This trade-off between temporal and spatial resolution is sometimes poorly appreciated and is one of the reasons that "real-time" confocal systems can be of limited use for detecting small, fast events (such as calcium sparks). At a practical level, one should consider the time taken to acquire a pixel in the image to decide the likely limit of resolution in the detection system. In most conventional (mechanical) scanning confocal microscopes, pixel dwell times are on the order of 10^{-6}sec. Thus, for a fluorochrome emitting 10^{8} photons/sec, one might expect to detect ~1 photon per pixel/excited fluorochrome molecule in a confocal resolution volume (represented by a pixel or voxel). This figure is a rough guide to likely signal levels. It therefore follows that improvements in SNR (having chosen a particular dye) can be achieved only by (1) sacrificing spatial resolution (to increase the number of fluorochromes in the detected volume), (2) increasing the fluorochrome concentration (which will increase the buffering power and therefore perturb the system more strongly), or (3) increasing the pixel dwell time. In our measurements of calcium sparks, we have generally reduced the spatial resolution of the confocal microscope (by slightly opening the pinhole) to give a resolution of about 0.4 x 0.4 x 0.8 μm (x,y,z, respectively), which does not seem to be too much of a problem since calcium sparks are larger than this (see Cheng et al. 1993; Cannell et al. 1994; and below). On the other hand, stationary point confocal measurements have been used to detect even smaller amplitude events ("calcium blips"; Parker and Yao 1996) and to achieve higher temporal resolution (see, e.g., Escobar et al. 1994) in exchange for a complete loss of spatial information. In our studies on calcium sparks, we have generally used line scanning to retain some spatial information (along the scan line), while preserving a millisecond time resolution (for a detailed description of line scanning, see Cannell et al. 1994).

To achieve good spark records, great care should be taken when choosing the method by which to introduce the indicator into the cell; most of the indicator must remain in the cytoplasm and not in intracellular organelles. When patch-clamping cells, this can be ensured by incorporating the probe in the pipette-filling solution, which then dialyzes the cell (see, e.g., Cannell et al. 1987). The indicator of choice was originally fluo-3, although this has now been superceded by the brighter fluo-4. When using membrane-permeant forms of the indicator, loading conditions must be carefully adjusted. Fortunately, cardiac muscle cells appear to load quite readily with fluo-4 and there is little difference between the amplitude and time course of calcium sparks after introducing the indicator by either method (see Cannell et al. 1995). Nevertheless, loading can be problematic and the cell's health should be monitored by recording its response to electrical stimuli. A healthy cardiac cell should shorten by >10% on electrical stimulation and the fluo-4 fluorescence should increase by a factor >3. In other words, if the cells look healthy and contract vigorously, there should always be a large fluorescence change from the indicator. This simple test can prevent a lot of wasted time during experimentation. Other cells (such as smooth muscle) can also be tested by electrical or pharmacological stimulation and should give similar amplitude fluorescence changes. Put another way, if you cannot detect a good fluorescence change during a large cell-wide increase in calcium levels, you will not be able to detect calcium sparks.

EVENT AVERAGING

Using electrical stimulation, in conjunction with Ca^{2+} channel antagonism, it is possible to image successive spark events at the same site and to average them to improve SNR (see Bridge et al. 1999). This approach relies on the assumption that sparks occurring at one site have a stereotypical time course and spatial profile.

For certain experiments, imaging sparks at a single site is not practical, or even possible. However, it is still possible to construct an "average" event by combining sparks from distinct sites, provided the individual images can be aligned. To achieve this, we require a reference point for each spark so that excessive spatiotemporal blurring associated with the averaging process does not occur. The center of mass (or centroid) of the fluorescence signal can provide such a reference, but it should be noted that the researcher cannot control the relationship of each spark to the scanned image. In this case, the resulting average combines sparks from slightly different optical planes, and any measures derived from the average must take this into account. For example, average spark amplitudes may become non-modal, even if the underlying events have a preferred amplitude (Izu et al. 1998).

MICROSCOPE PREPARATION

It is always useful to have a test specimen on hand to check that the microscope is working properly. We usually use fluorescent beads that are 0.2 μm in diameter (Molecular Probes) and easily detected at illumination rates that bleach them <1% per scan. This size of bead can also be scanned in the x–z direction to show whether the microscope alignment is correct. A test slide can be prepared by wiping a small quantity of such beads onto a coverslip (in the same way that a blood smear is prepared) and then embedding the beads in a drop of Sylgard (Dow Corning) with a microscope slide placed on top. It takes only about 30 min to prepare ten of these slides, and they will last for months if kept in a dark drawer. They therefore also provide a good reference to check the microscope sensitivity. Confocal microscope light detectors (photomultipliers) are easily damaged by accidental exposure to high light levels, and forgetting to place the correct barrier filter in the system can cause exposure to damaging light levels (although photomultiplier damage can be quite subtle and is often only manifest as a reduction in sensitivity and a slight increase in background signal rather than complete tube failure). Again, the test beads can provide a reference if there is any doubt about whether the detector efficiency has fallen (provided images of the samples of the same beads were stored for reference).

A common loss of sensitivity in the confocal microscope arises from misalignment. The excitation light should pass through the center of the objective rear aperture and all other components in the optical train. When reflected off a front surface mirror, via an objective lens (that is focused on the mirror), the excitation light will pass back along the optical train and should be centered on the pinhole (make sure you do not expose the photomultiplier during such tests). Degradation of the point spread function (see Fig. 1B) of the microscope (measured by imaging our subresolution beads, see Pawley 1995) will reveal serious errors in alignment, but the confocal pinhole may "hide" more subtle problems. Small alignment errors lead to a large drop in signal strength as the pinhole is closed (while imaging subresolution test beads). As a rough guide, the signal from 0.2-μm beads should fall by not more than ~50% when the pinhole diameter is closed from fully open to normal operating diameters (i.e., as used in experiments). A large loss in signal as the pinhole is closed is almost always due to misalignment. It is a good idea to become familiar with the alignment procedure of the microscope and *not* treat the microscope as an expensive black box or simply rely on a service engineer to align it correctly; most service engineers are more concerned with spatial resolution than detection efficiency.

The objective lens of the microscope is a critical component that must be selected for high throughput and numerical aperture (NA). The best lenses for detecting calcium sparks have NA of >1.2 and, if the cells are very close to the coverslip, oil-immersion lenses can be used. If the cells are some distance from the coverslip (>50 μm), a water-immersion objective must be used to prevent the loss of signal associated with the spherical aberration that develops (due to the refractive index mismatch; see Fig. 1C and Chapter 20 in Pawley 1995). Before buying an objective, try to negotiate a reasonable period for testing the objective; if this is not possible, try another manufacturer.

Unfortunately, the best water-immersion objectives are very expensive (~$6000) but this represents a modest investment for such a critical component of a confocal microscope.

Finally, a word of caution: Physiological saline destroys microscopes. If an accident occurs and saline comes in contact with the microscope, stop the experiment and wash the affected parts immediately with distilled water, then alcohol to displace the water. Do not delay cleaning up a saline spill, and never be afraid to seek help in such an event. Lenses must be cleaned carefully to remove all traces of grease and water. Mechanical parts can be given a light spray with a light oil such as WD40 to inhibit the corrosion that may develop after saline exposure (but be careful not to put any oil onto optical parts).

CALCIUM SPARKS

Calcium sparks were first observed in quiescent cardiac myocytes and are caused by local sarcoplasmic reticulum calcium release, which diffuses to occupy a region of the cell about 2 μm across (Cheng et al. 1993). Their discovery was quite unexpected and resulted from imaging fluorescence from the calcium indicator fluo-3 with a confocal fluorescence microscope. Although cardiac cells had been examined previously with wide-field imaging methods, calcium sparks had not been previously observed (see Wier et al. 1987) because out-of-focus fluorescence reduces the imaging contrast for small subcellular events. Since a calcium spark originates from a very small volume (<10 fl; Cheng et al. 1993), signal-to-noise levels are low and some care is needed to achieve good recordings. Furthermore, the background fluorescence at resting cytosolic calcium levels ($[Ca^{2+}]_{rest}$ ~ 100 nM) should be as small as possible to improve signal contrast, and this has led to fluo-3 (and now fluo-4) being the most suitable commercially available indicator. Using fluo-3, cardiac sparks have normalized fluorescence (F/F_0) ≈2 (see Cheng et al. 1996 and Fig. 1A). Although fluo-4 is not a ratiometric indicator, an estimate of the cytosolic calcium concentration ($[Ca^{2+}]_i$) can be made from the equation

$$[Ca^{2+}]_i = KR/[(K/[Ca^{2+}]_{rest} + 1) - R]$$

where K is the dissociation constant of the dye and R a pseudo-ratio image (F/F_0), obtained by dividing images by a control image (see also Cheng et al. 1993), which for a typical $R = 2$ and $[Ca^{2+}]_{rest}$ of 100 nM implies that the peak of the calcium spark is about 300 nM. Since the peak of the calcium spark is reached in 10 msec, a calcium flux of 3 pA can be estimated (Cheng et al. 1993; Blatter et al. 1997). More recent estimates based on detailed mathematical modeling of the Ca^{2+} spark suggest that it arises from a flux of >7 pA (Soeller and Cannell 2002), suggesting that a Ca^{2+} spark arises from >15 RyRs: a result that is consistent with a clear mode in spark amplitudes (Bridge et al. 1999).

It has been shown that spark amplitude is not regulated by membrane potential (Cannell et al. 1995; López-López et al. 1995) and, at the foot of the calcium current activation curve, the rate of spark production increases with the same voltage dependence as the calcium current (Cannell et al. 1995; Santana et al. 1996), which shows that a single L-type calcium channel can activate a spark. At more positive potentials, calcium channel antagonists can be used to reduce the numbers of calcium sparks evoked by depolarization, thus allowing them to be counted and the probability of their occurrence (P_s) to be determined. Using this approach, Santana et al. (1996) showed that P_s was a nonlinear function of the single-channel current, as proposed by Cannell et al. (1994) from consideration of the increase P_s needed to explain the whole-cell calcium transient. Furthermore, the data of Santana et al. (1996) showed that P_s was not proportional to the whole-cell current, but was instead determined by the amplitude of the single L-type channel current. This observation shows that control of sarcoplasmic reticulum (SR) release channel gating by the *local microenvironment* is fundamental to cardiac excitation–contraction (E-C) coupling and emphasizes the benefit/importance of imaging subcellular signal transduction.

By using the line scan mode of the confocal microscope, calcium sparks have been shown to decline with a half-time of ~25 msec (see Fig. 1A). The local $[Ca^{2+}]_i$ associated with a calcium spark is therefore quite different from the time course of the whole-cell $[Ca^{2+}]_i$ transient, which reaches a peak of about 1.5 μM and declines with a half-time of about 164 msec. The time course of decline of $[Ca^{2+}]_i$ during a spark is faster than the whole-cell transient, which can be explained by two factors (Cheng et al. 1993): (1) Increased calcium uptake by intracellular calcium buffers (since the calcium transient associated with a spark is smaller than the whole-cell transient buffers are less saturated) and (2) diffusion of calcium from its site of release will also help reduce $[Ca^{2+}]_i$ (during the electrically evoked $[Ca^{2+}]_i$ transient, calcium increases everywhere, so the contribution of diffusion to the decline of $[Ca^{2+}]_i$ would be reduced). By inhibiting SR calcium uptake, Gomez et al. (1996) confirmed that diffusion was the major contributor to the decline of the spark, although SR calcium uptake also made a significant contribution to the kinetics and spatial properties of the spark.

CONCLUSION

Under the right conditions and with good equipment, spark-like events about an order of magnitude smaller than calcium sparks may be detected (see Parker and Yao 1996; Shirokova and Rios 1997; Gordienko et al. 1998). This suggests fluxes of calcium on the order of 10^4 ions, which is close to the levels achieved during patch-clamp recording of single-channel currents! Since optical techniques can be noninvasive, they offer the unique ability to study signal transduction inside intact living cells. Therefore, application of optical probes and confocal/multiphoton microspectrofluorimetry can be expected to spark further new insights into biological signal transduction.

Checklist for Confocal Imaging of Calcium Sparks

1. Is your imaging setup optimal? Check using (bead) test sample.
   - Can the beads be seen clearly?
   - When viewing the beads, do they blur evenly as you focus above and below them? Asymmetric blurring indicates spherical aberration.
   - Is the spatial resolution appropriate? Maximum resolution is not required. 0.4 x 0.4 x 0.8 μm ought to be sufficient.
   - Is the microscope aligned correctly? Check signal loss as pinhole is closed.
   - Check to ensure that the excitation, dichroic, and emission filter wavelengths are correct.
2. Is the sample loaded correctly and alive?
   - Are the intracellular stores loaded with Ca^{2+}? Applying caffeine (or some other suitable Ca^{2+} release agonist) should reveal a strong Ca^{2+} response if the stores are well loaded and active.
   - Does the cell respond to electrical stimulation?
   - Can whole-cell Ca^{2+} transients be seen clearly in wide-field (non-confocal) mode?
3. Is your imaging regime appropriate?
   - Are you using the highest possible NA objective, with the appropriate immersion medium?
   - Is the temporal resolution sufficient? To resolve a 10-msec upstroke, an interline time of not more than 3 msec is required. Increasing zoom results in more pixels being acquired per unit distance. If these pixels are then combined to restore the original resolution, the SNR will increase, but so does the photodamage.
   - Does the pixel resolution meet Nyquist's criterion of half the spatial resolution? For example, a lateral spatial resolution (apparent size of beads) of 0.4 μm requires a sampling distance of 0.2 μm per pixel.
   - Always set the detector gain at maximum. Quite high illumination powers are needed for imaging sparks. If you do not see Ca^{2+} sparks, increase the laser power until you do, or until the cell dies.

REFERENCES

Berridge M.J. 1997. Elementary and global aspects of calcium signalling. *J. Exp. Biol.* **200:** 315–319.

Blatter L.A., Huser J., and Rios E. 1997. Sarcoplasmic reticulum Ca^{2+} release flux underlying Ca^{2+} sparks in cardiac muscle. *Proc. Natl. Acad. Sci.* **94:** 4176–4181.

Bridge J.H.B., Ershler P.R., and Cannell M.B. 1999. Properties of Ca sparks evoked by action potentials in mouse ventricular myocytes. *J. Physiol.* **518:** 469–478.

Cannell M.B., Berlin J.R., and Lederer W.J. 1987. Effect of membrane potential changes on the calcium transient in single rat cardiac muscle cells. *Science* **238:** 1419–1423.

Cannell M.B., Cheng H., and Lederer W.J. 1994. Spatial non-uniformities in $[Ca^{2+}]_i$ during excitation-contraction coupling in cardiac myocytes. *Biophys. J.* **67:** 1942–1956.

———. 1995. The control of calcium release in heart muscle. *Science* **268:** 1045–1050.

Cheng H., Lederer W.J., and Cannell M.B. 1993. Calcium sparks: Elementary events underlying excitation-contraction coupling in heart muscle. *Science* **262:** 740–744.

Cheng H., Lederer M.R., Lederer W.J., and Cannell M.B. 1996. Calcium sparks and $[Ca^{2+}]_i$ waves in cardiac myocytes. *Am. J. Physiol.* **270:** C148–C159.

Escobar A.L., Monck J.R., Fernandez J.M., and Vergara J.L. 1994. Localization of the site of Ca^{2+} release at the level of a single sarcomere in skeletal muscle fibres. *Nature* **367:** 739–741.

Gomez A.M., Cheng H., Lederer W.J., and Bers D.M. 1996. Ca^{2+} diffusion and sarcoplasmic reticulum transport both contribute to $[Ca^{2+}]_i$ decline during Ca^{2+} sparks in rat ventricular myocytes. *J. Physiol.* **496:** 575–581.

Gordienko D.V., Bolton T.B., and Cannell M.B. 1998. Variability in spontaneous subcellular calcium release in guinea-pig ileum smooth muscle cells. *J. Physiol.* **507:** 707–720.

Izu L.T., Wier W.G., and Balke C.W. 1998. Theoretical analysis of the Ca^{2+} spark amplitude distribution. *Biophys J.* **75:** 1144–1162.

Klein M.G., Cheng H., Santana L.F., Jiang Y.H., Lederer W.J., and Schneider M.F. 1996. Two mechanisms of quantized calcium release in skeletal muscle. *Nature* **379:** 455–458.

López-López J.R., Shacklock P.S., Balke C.W., and Wier W.G. 1994. Local stochastic release of Ca^{2+} in voltage-clamped rat heart cells: Visualization with confocal microscopy. *J. Physiol.* **480:** 21–29.

———. 1995. Local calcium transients triggered by single L-type calcium channel currents in cardiac cells. *Science* **268:** 1042–1045.

Melamed-Book N., Kachalsky S.G., Kaiserman I., and Rahamimoff R. 1999. Neuronal calcium sparks and intracellular calcium "noise". *Proc. Natl. Acad. Sci.* **96:** 15217–15221.

Nelson M.T., Cheng H., Rubart M., Santana L.F., Bonev A.D., Knot A.D., and Lederer W.J. 1995. Relaxation of arterial smooth muscle by calcium sparks. *Science* **270:** 633–637.

Parker I. and Yao Y. 1996. Ca^{2+} transients associated with openings of inositol trisphosphate-gated channels in *Xenopus* oocytes. *J. Physiol.* **491:** 663–668.

Pawley J., ed. 1995. *Handbook of biological confocal microscopy*, 2nd edition. Plenum Press, New York.

Santana L.F., Cheng H., Gómez A.M., Cannell M.B., and Lederer W.J. 1996. Relation between the sarcolemmal Ca^{2+} current and Ca^{2+} sparks and local control theories for cardiac excitation-contraction coupling. *Circ. Res.* **78:** 166–171.

Shirokova N. and Rios E. 1997. Small event Ca^{2+} release: a probable precursor of Ca^{2+} sparks in frog skeletal muscle. *J. Physiol.* **502:** 3–11.

Soeller C. and Cannell M.B. 2002. Estimation of the sarcoplasmic reticulum Ca^{2+} release flux underlying Ca^{2+} sparks. *Biophys J.* **82:** 2396–2414.

Tsugorka A., Rios E., and Blatter L.A. 1995. Imaging elementary events of calcium release in skeletal muscle cells. *Science* **269:** 1723–1726.

Van der Voort H.T. and Brakenhoff G.J. 1990. 3-D image formation in high-aperture fluorescence confocal microscopy: A numerical analysis. *J. Microsc.* **158:** 43–54.

Wier W.G., Cannell M.B., Berlin J.R., Marban E., and Lederer W.J. 1987. Cellular and subcellular heterogeneity of $[Ca^{2+}]_i$ in single heart cells revealed by fura-2. *Science* **235:** 325–328.

CHAPTER 37

Monitoring Presynaptic Calcium Dynamics with Membrane-permeant Indicators

Wade G. Regehr

Department of Neurobiology, Harvard Medical School, Boston, Massachusetts 02138

Classic studies have established that calcium is important in rapidly triggering the release of neurotransmitter (Katz and Miledi 1967; Llinás et al. 1982) and in use-dependent synaptic modulation, such as facilitation and post-tetanic potentiation (Katz and Miledi 1968; Charlton et al. 1982; Delaney et al. 1989). Until recently, most of what was known regarding the role of calcium in presynaptic terminals was based on the squid giant synapse and the neuromuscular junction. It has been difficult to extend such studies to the mammalian brain due to the small size of presynaptic terminals in the mammalian CNS. By way of comparison, the volumes of synaptic boutons at the crayfish neuromuscular junction and the squid giant synapse are, respectively, approximately 300 and one million times as large as that of typical boutons in the mammalian CNS, which are less than a micron in diameter. To directly examine the role of calcium in synaptic transmission in the mammalian brain, it is necessary to measure presynaptic calcium levels, to detect presynaptic calcium influx, and to determine the time course of presynaptic calcium entry.

In this chapter, I describe a simple and reliable procedure that can be used to study presynaptic calcium transients in rodent brain slices (Regehr and Tank 1991). Rather than attempting to measure calcium levels in individual terminals, the strategy is to make an aggregate measurement from many presynaptic terminals. Although my colleagues and I have concentrated our efforts on cerebellar granule cell presynaptic terminals and hippocampal mossy fibers, many other synapses are suited to this method. The chapter includes the procedure used to obtain such labeling and discusses labeling conditions and indicators required for studies of presynaptic signaling.

APPROACH FOR STUDYING PRESYNAPTIC CALCIUM TRANSIENTS

A fluorescent indicator is dissolved in external saline, and this labeling solution is applied to a small region of the brain slice. Acetoxymethyl (AM) ester forms of the indicators are used, which pass through the membrane and enter the cell where they are hydrolyzed, thereby becoming ion-sensitive and membrane-impermeant (Tsien 1981). These indicators diffuse within cells, and with time, the indicator concentration equilibrates throughout each cell. This method overcomes barriers to labeling that are apparent in adult brain slices.

This approach is best suited to the study of well-segregated, nonmyelinated fiber tracts that contain a homogeneous population of presynaptic terminals. For example, consider the labeling obtained by applying the dye Magnesium Green (Molecular Probes) to stratum lucidum in the hippocampal CA3 region (Regehr et al. 1994). Glia, inhibitory neurons, pyramidal cells, and axons (fibers of passage) are indiscriminately labeled near the application site. Among the structures loaded with dye are the mossy fibers, which are the axons, and associated presynaptic boutons of hippocampal granule cells. They traverse the cell-body layer of the CA3 region, where they synapse primarily on the proximal dendrites of the pyramidal cells (Claiborne et al. 1986; Chicurel and Harris 1992). As a consequence of the favorable anatomy of this fiber tract, away from the fill site the only structures labeled with Magnesium Green are the axons and their associated terminals. In Figure 1, mossy fibers and their periodic presynaptic boutons are labeled with Magnesium Green (Haugland 1996; Zhao et al. 1996), and staining of the CA3 pyramidal cell bodies is indicated in red. For mossy fibers, the volume of presynaptic boutons is four or five times that of the axons. Moreover, calcium channels are usually concentrated in terminals as opposed to axons (Delaney et al. 1991; Robitaille and Charlton 1992). Thus, fluorescence transients reported by calcium indicators well away from the fill site are dominated by signals from mossy-fiber boutons. This example illustrates the type of favorable anatomy that is required for this technique.

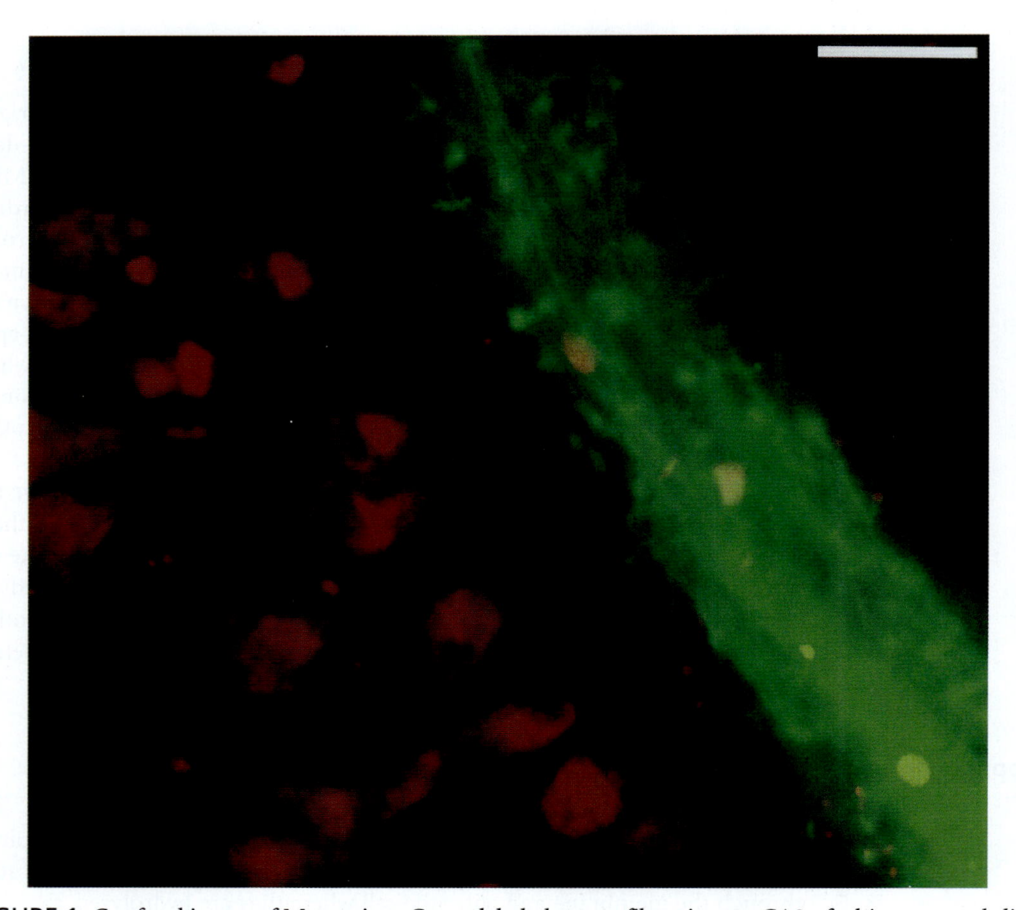

FIGURE 1. Confocal image of Magnesium Green-labeled mossy fibers in area CA3 of a hippocampal slice. The cell bodies of CA3 pyramidal cells are labeled with a Nissl staining dye and shown in red. Bar, 25 μM.

PROCEDURE

Loading Fiber Tracts with Membrane-permeant Indicators

The following protocol is used for loading indicator dye into rodent brain slices.

1. Prepare labeling solution as follows:
   a. Dissolve 50 μg of AM-ester indicator <!> (e.g., fura-2 AM; Molecular Probes) in 20 μl of a solution of 25% pluronic acid/75% DMSO<!> (Molecular Probes) in a vial.

      Note: Fast green is included in the labeling solution to allow visualization during loading.

   b. Mix by vortexing for 30 sec. Sonicate for 2 min.
   c. Add 400 μl of artificial cerebrospinal fluid (ACSF) and 20 μl of 0.1% fast green to the vial. (ACSF: 125 mM NaCl, 26 mM $NaHCO_3$, 1.25 mM NaH_2PO_4<!>, 2.5 mM KCl<!>, 1.0 mM $MgCl_2$<!>, 2.0 mM $CaCl_2$<!>, 25 mM glucose. The solution is bubbled with 95% O_2 and 5% CO_2.)
   d. Mix by vortexing for 30 sec, sonicate for 2 min, and filter (Z-spin plus 0.45-μm filter, Gelman Sciences). Use freshly prepared solution and keep on ice to minimize premature de-esterification.
2. Place a brain slice in the recording chamber and hold down with a nylon grid.
3. Position a suction electrode (20–30 μm in diameter) above the surface of the slice.
4. Fill a loading electrode with labeling solution and position near the suction electrode.

   Note: Presynaptic calcium transients have been measured with many different types of indicators (Molecular Probes), including BTC (coumarin benzothiazole-based calcium indicator), fura-2, Calcium Green-2, Calcium Orange, Magnesium Green, mag-fura-5, furaptra, fura-2-ff, Calcium Crimson, and Magnesium Orange (Grynkiewicz et al. 1985; Raju et al. 1989; Iatridou et al. 1994; Haugland 1996; Zhao et al. 1996). In the author's experience, the indicators Calcium Green-5N, Calcium Orange-5N, and Oregon Green-5N (Haugland 1996) are not suited to this loading method; although they exhibit low-affinity properties in vitro, their properties change within presynaptic terminals where they behave like high-affinity indicators. Although the primary focus of our studies has been on presynaptic calcium, this loading procedure also works with indicators for other ions, such as SBFI (sodium-binding benzofuran isophthalate) for detecting sodium (Minta and Tsien 1989; Regehr 1997).

5. Apply positive pressure on the loading electrode and suction on the other electrode. Adjust the flow rate to maintain a rapidly flowing, well-confined stream of dye.

   Note: We find aquarium pumps convenient for the application of positive pressure on the loading electrode.

6. Lower the electrodes together near the surface of the slice.
7. Carefully lower the loading electrode to the slice surface and place into the slice until the flow of solution from the pipette reduces or stops entirely.
8. Increase the pressure on the loading electrode until the flow from the labeling pipette resumes.

   Note: There should be sufficient suction during steps 6 and 7 to capture all of the labeling solution with the suction pipette. In addition, the slice is superfused with external solution (ACSF) at a high flow rate (4 ml/min). In this way, only a small region of the slice is exposed to the indicator, and undesirable background labeling is avoided.

Procedural Notes

Typically, labeling times range from 2 to 60 min, depending on the application, the dye used, and the desired degree of loading.

One to two hours is generally required for the indicator to equilibrate within the cell. Slices are usually loaded at 24°C and stored at this temperature during the equilibration period. Ionic transients can be measured for many hours with slices loaded in this manner. Although it is possible to load and store slices at elevated temperatures, we generally choose not to do so, because the indicator is removed from the presynaptic terminals at physiological temperatures. This rate of removal is not sufficiently fast to prevent high-temperature studies, but prolonged storage at such high temperatures results in a significant loss of dye.

Compartmentalization of the dye and the associated problems that have been observed with AM dyes do not affect the recordings. It appears that hydrolysis occurs near the fill site and that, although the indicator may load intracellular organelles near the fill site, more distant sites are not exposed to the AM-ester indicator; the only dye reaching them is the acidic membrane-impermeant form.

Fibers are stimulated with an extracellular electrode placed in the middle of the labeled tract. Fluorescence transients are recorded from a region several hundred microns from the stimulus electrode.

Illumination is provided by either a 150-W xenon bulb or a 100-W halogen bulb powered by a low-noise power supply. Fluorescence changes are detected with either a photodiode or a photomultiplier. The size of the region from which measurements are taken, the detector, and the light source are all tailored to the dye being used and to the application.

SELECTING THE APPROPRIATE INDICATOR

Although the same basic procedure is used for all experiments of this type, the choice of the indicator and the degree of loading depend on the specifics of the experiment. To illustrate this, consider the use of high- and low-affinity indicators for three different applications: (1) measurement of the time course of presynaptic calcium transients, (2) detection of changes in the amplitude of presynaptic calcium influx, and (3) measurement of the time course of presynaptic calcium entry (presynaptic calcium current).

- *Measurement of the time course of presynaptic calcium transients.* Calcium transients on the tens-of-milliseconds to tens-of-seconds timescale have been implicated in the use-dependent forms of plasticity: facilitation, augmentation, and post-tetanic potentiation. As shown in Figure 2A, accurate determination of the time course of calcium transients on these timescales relies on the correct choice of indicator (Regehr and Atluri 1995; Feller et al. 1996). A comparison of the time course of fluorescence transients measured with two different indicators is shown in Figure 2A. Magnesium Green, which has a dissociation constant for calcium of 7 μM, is effective at detecting large calcium transients (Zhao et al. 1996) and provides an accurate measure of the time course of the calcium transient. In contrast, the high-affinity indicator fura-2 (dissociation constant ~200 μM) is sufficiently sensitive to slow the transient as a consequence of saturation of the indicator (Regehr and Atluri 1995).
- *Detection of changes in the amplitude of presynaptic calcium influx.* Such measurements are required in determination of the relationship between calcium influx and transmitter release, as well as in studies of the contribution of presynaptic calcium-channel modulation to alterations in synaptic strength. Consider the measurement of fluorescence changes for stimulus trains in which fibers are activated twice in rapid succession (Fig. 2B). With a low-affinity indicator, such as Magnesium Green, the second stimulus produces the same increment in fluorescence, but with fura-2 the fluorescence transient produced by the second stimulus is smaller. We have previously shown that the difference between the fluorescence transients reported by these two indicators is a reflection of their different affinities. For low-affinity calcium-sensitive indicators, such as BTC, Magnesium Green, mag-fura-5, furaptra, fura-2-ff, and others, there is a roughly linear relationship between calcium entry and the resulting ΔF/F signals, and it is a simple matter to detect changes in calcium entry. For high-affinity calcium indicators, such as fura-2, Calcium Orange, and Calcium Crimson, there is a distortion of the calcium transient due to saturation of the indicator. Despite the saturation of high-affinity indicators such as fura-2, these indicators can be put to good use in the study of presynaptic terminals. Their sensitivity makes them well suited to detecting small elevations of calcium, and they also have the advantage that they are relatively insensitive to magnesium. Furthermore, their degree of saturation provides a way of quantitating calcium levels reached in

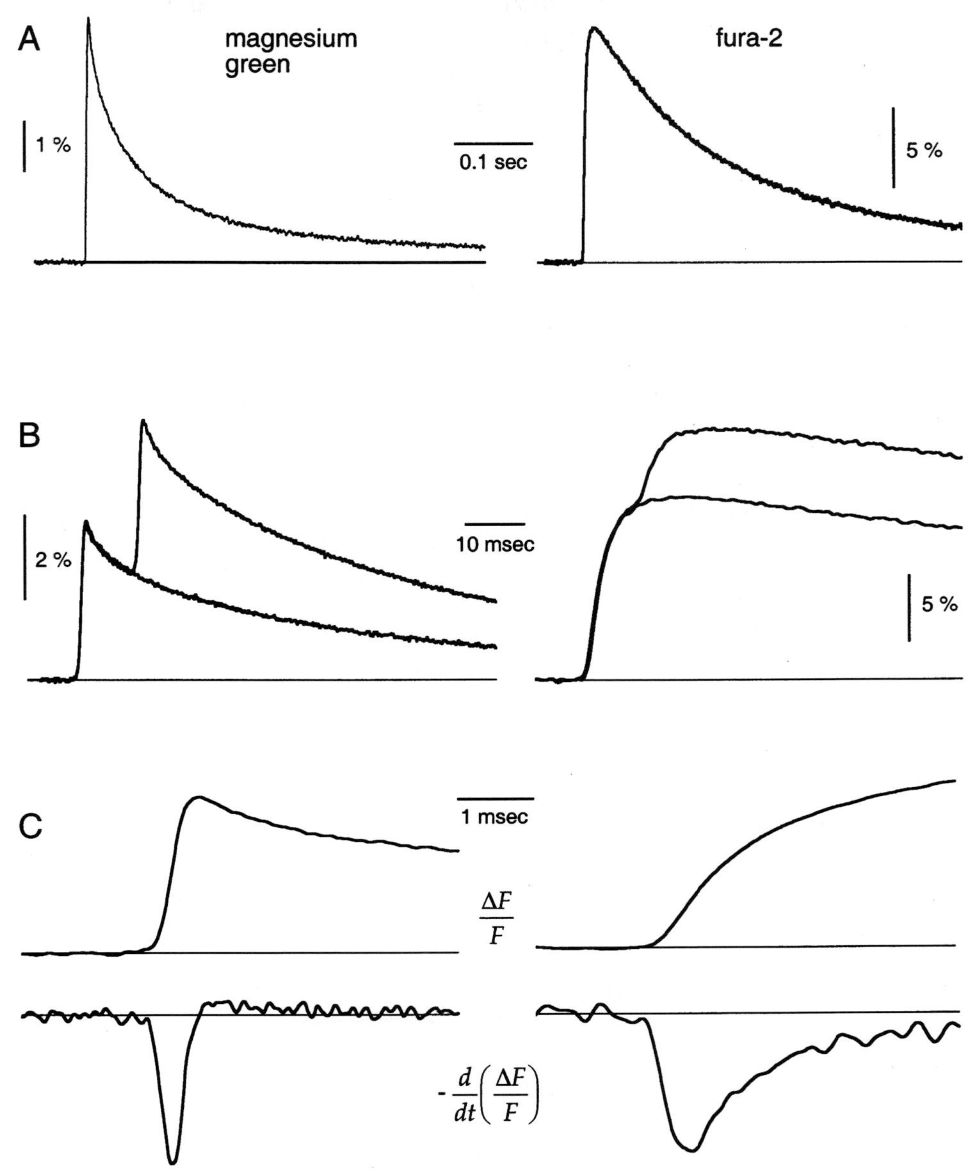

FIGURE 2. Detection of presynaptic calcium transients, changes in calcium influx, and presynaptic calcium current in cerebellar parallel fibers. The low-affinity calcium indicator Magnesium Green (for calcium, K_d = 7 μM) and the high-affinity calcium indicator fura-2 (K_d = 200 nM) were used to measure the time course of ΔF/F signals. Fluorescence changes produced by a single-stimulus pulse (*A*) and by stimulation with one and two pulses separated by 10 msec (*B*). (*C*) Measurement of the onset response. An increase in calcium corresponds to a decrease in fura-2 fluorescence for the wavelengths used in this experiment, and the ΔF/F signals for fura-2 are presented as positive signals for clarity. The bottom traces in *C* are the derivatives of the fluorescence transients, and they are presented as negative going signals for clarity and for ease of comparison with presynaptic calcium currents. Experiments were conducted in cerebellar slices from young rats. For *A* and *B*, T = 24°C; for *C*, T = 34°C.

presynaptic terminals and, with care, can be used to quantitate changes in calcium entry (for a detailed description, see Regehr and Atluri 1995; Sabatini and Regehr 1995).

- *Measurement of the time course of presynaptic calcium entry.* This application is, in essence, the determination of the presynaptic calcium current (Sabatini and Regehr 1996, 1998). In this application, the calcium indicator serves to integrate the calcium that enters the terminal. We have shown that for cerebellar parallel fibers, differentiating the fluorescence transient provides a good estimate of the time course of presynaptic calcium entry for low-affinity indicators, but that high-affinity indicators, such as fura-2, fail to capture the time course of presynaptic calcium entry. For this application, the choice of a suitable indicator depends on the loading conditions and on the properties of the endogenous buffer (for a detailed description, see Sabatini and Regehr 1998). Examples of the derivatives of the fluorescence transients for fura-2 and Magnesium Green are shown in Figure 2C.

DETERMINING THE APPROPRIATE LOADING CONDITIONS

It must be recognized that the introduction of an indicator into a cell always alters the presynaptic calcium dynamics to some extent. With a loading method, such as that described here, it is important to determine how much indicator has been introduced into the presynaptic terminal and what effect this might have on the parameter being measured. With high-affinity indicators, such as fura-2, it is straightforward to use the degree of saturation as a measure of the extent of calcium buffering. For example, compare the measurements of fluorescence transients produced by one and two stimuli in Figure 3A. For a fiber tract that has a low concentration of fura-2 (Fig. 3A, left), there is a much smaller incremental increase in fluorescence for the second stimulus than for the first, indicating that the calcium transients are sufficiently large to begin to significantly saturate the response of the indicator. In a fiber tract that has a much higher concentration of fura-2 (Fig. 3A, right), the size of the calcium transient is reduced by the increased buffer capacity, and there is much less saturation of the indicator. Thus, for a high-affinity indicator, the degree of saturation is highly sensitive to the amount of indicator introduced into the presynaptic terminal. Changes in the degree of saturation can even be used to quantitate the amount of buffer introduced into the presynaptic terminal, relative to the buffer capacity of the endogenous buffer, using an approach similar to that described previously (Sabatini and Regehr 1995).

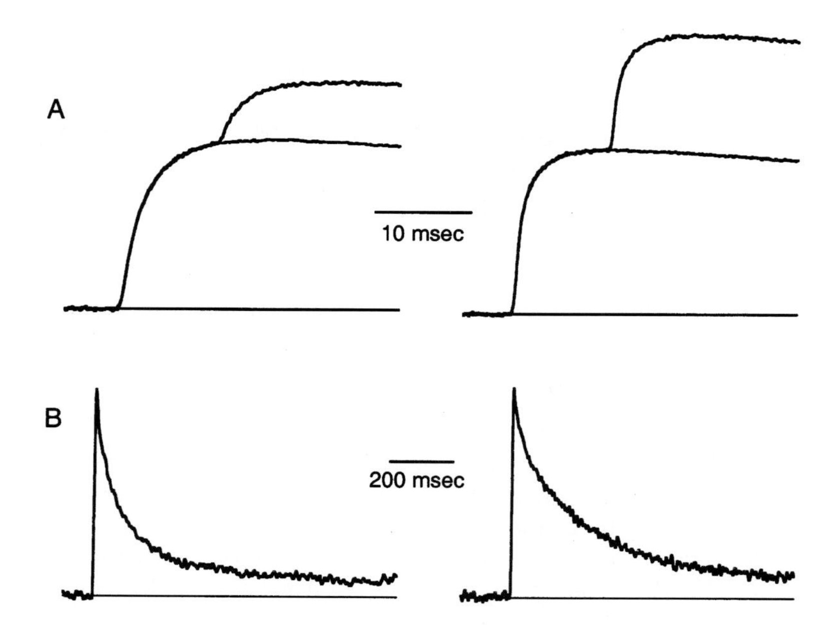

FIGURE 3. Assessing the degree of indicator loading. (*A*) The effect of loading time and indicator concentration on the saturation of fluorescence responses. Presynaptic terminals were loaded for 5 min (*left*) and 30 min (*right*). (*B*) The effect of loading on mag-fura-5-fluorescence transients. Experiments were conducted at room temperature in cerebellar parallel fibers from young rats. Presynaptic terminals were loaded for 5 min (*left*) and 30 min (*right*). For *A* and *B*, T = 24°C. Peak changes in ΔF/F produced by single stimuli are (*A*) 4.85% (*left*) and 5.9% (*right*), (*B*) 0.96% (*left*) and 0.97% (*right*).

For low-affinity indicators, the easiest way to determine the extent of loading and the degree to which calcium signaling is distorted is to measure the time course of the calcium transient, as in Figure 3B. Lightly labeled fiber tracts provide an accurate measure of the calcium transient, whereas transients in heavily labeled tracts are significantly slowed. The degree of slowing reflects the extent to which the buffer capacity of the presynaptic terminal has been increased by the indicator (Neher and Augustine 1992; Tank et al. 1995).

The choice of indicator and loading conditions depends on the type of measurement being made. Time-course measurements are best made with a lightly labeled, low-affinity indicator. Changes in the amplitude of calcium influx can also be made with a low-affinity indicator, but even an overloaded terminal provides an accurate means of quantitating changes in calcium influx. For detection of presynaptic calcium currents, low-affinity indicators provide a good measure of the time course of calcium entry, but the temporal fidelity of this measurement is actually improved by deliberately overloading the terminals.

CONCLUSION

The loading procedure described here has proven to be a useful tool in the study of calcium's role in synaptic transmission in the mammalian brain (Regehr et al. 1994; Wu and Saggau 1994; Mintz et al. 1995; Atluri and Regehr 1996; Dittman and Regehr 1996; Sabatini and Regehr 1996, 1997, 1998; Chen and Regehr 1997). Provided that care is taken in the choice of indicator and loading conditions, it is possible to measure the time course of presynaptic calcium transients, to quantify changes in calcium influx, and to measure the time course of presynaptic calcium entry.

ACKNOWLEDGMENTS

I thank P. Atluri, C. Chen, J. Dittman, B. Peters, B. Sabatini, and M. Xu-Friedman for helpful comments. This work was supported by National Institutes of Health grant RO1-NS-32405-01.

REFERENCES

Atluri P.P. and Regehr W.G. 1996. Determinants of the time course of facilitation at the granule cell to Purkinje cell synapse. *J. Neurosci.* **16:** 5661–5671.

Charlton M.P., Smith S.J., and Zucker R.S. 1982. Role of presynaptic calcium ions and channels in synaptic facilitation and depression at the squid giant synapse. *J. Physiol.* **323:** 173–193.

Chen C. and Regehr W.G. 1997. Consquences of presynaptic PKA activation at the granule cell to Purkinje cell synapse. *J. Neurosci.* **17:** 8687–8694.

Chicurel M. and Harris K.M. 1992. Three-dimensional analysis of the structure and composition of CA3 branched dendritic spines and their synaptic relationships with mossy-fiber boutons in the rat hippocampus. *J. Comp. Neurol.* **325:** 169–182.

Claiborne B.J., Amaral D.G., and Cowan W.M. 1986. A light and electron microscopic analysis of the mossy fibers of the rat dentate gyrus. *J. Comp. Neurol.* **246:** 435–458.

Delaney, K., Tank D.W., and Zucker R.S. 1991. Presynaptic calcium and serotonin-mediated enhancement of transmitter release at crayfish neuromuscular junction. *J. Neurosci.* **11:** 2631–2643.

Delaney K.R., Zucker R.S., and Tank D.W. 1989. Calcium in motor nerve terminals associated with posttetanic potentiation. *J. Neurosci.* **9:** 3558–3567.

Dittman J.S. and Regehr W.G. 1996. Contributions of calcium-dependent and calcium-independent mechanisms to presynaptic inhibition at a cerebellar synapse. *J. Neurosci.* **16:** 1623–1633.

Feller M.B., Delaney K.R., and Tank D.W. 1996. Presynaptic calcium dynamics at the frog retinotectal synapse. *J. Neurophysiol.* **76:** 381–400.

Grynkiewicz G., Poenie M., and Tsien R.Y. 1985. A new generation of Ca^{2+} indicators with greatly improved fluorescence properties. *J. Biol. Chem.* **260:** 3440–3450.

Haugland R.P. 1996. *Handbook of fluorescent probes and research chemicals.* Molecular Probes, Eugene, Oregon.

Iatridou H., Foukaraki E., Kuhn M.A., Marcus E.M., Haugland R.P., and Katerinopoulos H.E. 1994. The development of a new family of intracellular calcium probes. *Cell Calcium* **15:** 190–198.

Katz B. and Miledi R. 1967. The timing of calcium action during neuromuscular transmission. *J. Physiol.* **189:** 535–544.

———. 1968. The role of calcium in neuromuscular facilitation. *J. Physiol.* **195:** 481–492.

Llinás R., Sugimori M., and Simon S.M. 1982. Transmission by presynaptic spike-like depolarization at the squid giant synapse. *Proc. Natl. Acad. Sci.* **79:** 2415–2419.

Minta A. and Tsien R.Y. 1989. Fluorescent indicators for cytosolic sodium. *J. Biol. Chem.* **264:** 19449–19457.

Mintz I.M., Sabatini B.L., and Regehr W.G. 1995. Calcium control of transmitter release at a cerebellar synapse. *Neuron* **15:** 675–688.

Neher E. and Augustine G.J. 1992. Calcium gradients and buffers in bovine chromaffin cells. *J. Physiol.* **450:** 273–301.

Raju B., Murphy E., Levy L.A., Hall R.D., and London R.E. 1989. A fluorescent indicator for measuring cytosolic free magnesium. *Am. J. Physiol.* **256:** C540–C548.

Regehr W.G. 1997. Interplay between sodium and calcium dynamics in granule cell presynaptic terminals. *Biophys. J.* **72:** 2476–2488.

Regehr W.G. and Atluri P.P. 1995. Calcium transients in cerebellar granule cell presynaptic terminals. *Biophys. J.* **68:** 2156–2170.

Regehr W.G. and Tank D.W. 1991. Selective fura-2 loading of presynaptic terminals and nerve cell processes by local perfusion in mammalian brain slice. *J. Neurosci. Methods* **37:** 111–119.

Regehr W.G., Delaney K.R., and Tank D.W. 1994. The role of presynaptic calcium in short-term enhancement at the hippocampal mossy fiber synapse. J. *Neurosci.* **14:** 523–537.

Robitaille R. and Charlton M.P. 1992. Presynaptic calcium signals and transmitter release are modulated by calcium-activated potassium channels. *J. Neurosci.* **12:** 297–305.

Sabatini B.L. and Regehr W.G. 1995. Detecting changes in calcium influx which contribute to synaptic modulation in mammalian brain slice. *Neuropharmacology* **34:** 1453–1467.

———. 1996. Timing of neurotransmission at fast synapses in the mammalian brain. *Nature* **384:** 170–172.

———. 1997. Control of neurotransmitter release by presynaptic waveform at the granule cell to Purkinje cell synapse. *J. Neurosci.* **17:** 3425–3435.

———. 1998. Optical measurement of presynaptic calcium current. *Biophys. J.* **74:** 1549–1563.

Tank D.W., Regehr W.G., and Delaney K.R. 1995. A quantitative analysis of presynaptic calcium dynamics that contribute to short-term enhancement. *J. Neurosci.* **15:** 7940–7952.

Tsien R.Y. 1981. A non-disruptive technique for loading calcium buffers and indicators into cells. *Nature* **290:** 527–528.

Wu L.-G. and Saggau P. 1994. Adenosine inhibits evoked synaptic transmission primarily by reducing presynaptic calcium influx in area CA1 of hippocampus. *Neuron* **12:** 1139–1148.

Zhao M., Hollingworth S., and Baylor S.M. 1996. Properties of tri- and tetracarboxylate Ca2+ indicators in frog skeletal muscle fibers. *Biophys. J.* **70:** 896–916.

CHAPTER 38

A Practical Guide: Two-Photon Calcium Imaging of Spines and Dendrites

Jesse H. Goldberg and Rafael Yuste

Department of Biological Sciences, Columbia University, New York, New York 10027

This chapter describes an approach to two-photon calcium imaging in dendrites and spines of living neurons. The technique can be applied to acute slices from hippocampus, cerebellum, and neocortex, as well as to slice and neuronal cultures (Yuste and Denk 1995; Holthoff et al. 2002; Sabatini et al. 2002). It uses two fluorescence detection channels to provide quantitative estimates of calcium concentration and to minimize the required concentrations of calcium indicator. A method for estimating calcium concentrations from fractional changes in fluorescent light intensity ($\Delta F/F$) is presented, along with two methods for loading neurons with calcium indicator. First, individual cells can be loaded via patch pipette with calcium indicator and Alexa 594 in a recording chamber (see Chapter 33). Alternatively, if the dialysis that occurs during sustained whole-cell recording is a concern (e.g., during spine motility experiments), calcium indicator can be loaded using the bolus injection technique (Majewska et al. 2000).

MATERIALS

General

This protocol requires equipment and reagents for whole-cell patch-clamping or bolus injection (see Chapter 33).

Calcium Indicators

The potassium salt form of a calcium indicator in the green emission range—fluo-4, Calcium Green-1 (CG-1), or Oregon Green BAPTA (OGB) and Alexa Fluor 59 (Molecular Probes)

Microscopy

Two-photon microscope (See Chapter 10) equipped with a recording chamber and a dichroic mirror, which separates red and green light, (565 DCXR from Chroma Technology). At an angle of 45° the dichroic mirror transmits ~80% of light between ~580 nm and 880 nm (Red Channel—IR from laser to sample and red Alexa 594 fluorescence back to internal PMT) and reflects light below 530 nm (Green Channel—emission from calcium indicator) to external PMT. We place an additional bandpass 510/40 (transmits between 490 and 530) filter in front of the external PMT to reduce scattered red light.

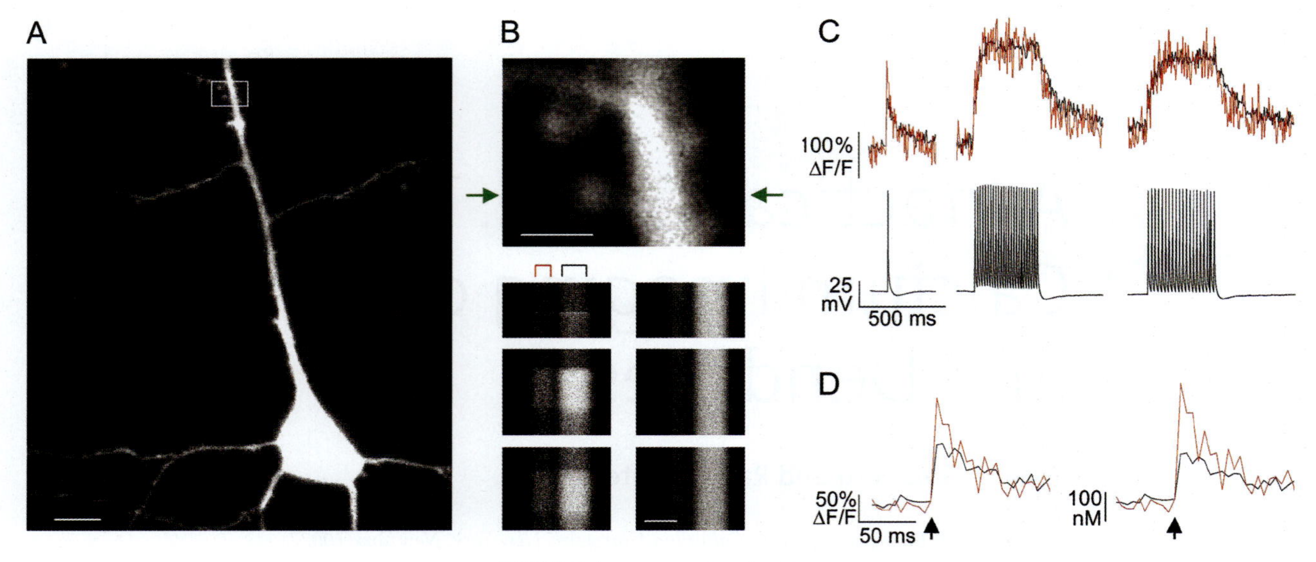

FIGURE 1. See text for details.

PROCEDURE

Loading Neurons with Calcium Indicator and Alexa 594

1. For whole-cell recording: Patch the neuron of choice with a pipette filled with 20–100 μM of green-emission calcium indicator and 20–100 μM Alexa Fluor 594<!> in standard internal recording solution, such as: 130 mM $KMeO_4$, 5 mM KCl <!> , 5 mM NaCl, 10 mM HEPES, 2.5 mM Mg-ATP, 0.3 mM GTP, titrated to pH 7.3. (See Chapter 33.)

2. After 20–30 min of recording, dendrites are well visualized on the red channel, and the green channel can be used to image calcium changes during stimulation (see Fig. 1).

   Note: Resting indicator fluorescence is necessary to visualize dendrites if calcium imaging is going to be performed with only one (green) channel. In this case, a higher concentration of indicator (> 100 μM) must be used.

   CAUTION: *See Appendix 3 for appropriate handling of materials marked with <!>.*

Alternatively, for the bolus injection technique, prepare millimolar concentrations (1–5 mM) of calcium indicator in internal recording solution and hold the cell in whole-cell mode for ~3 min. When patching the cell, target a region of the cell body that does not overlie the nucleus; this will ensure cell viability after electrode withdrawal. Twenty minutes after withdrawal, the dendritic tree should be loaded satisfactorily and will be ready for calcium imaging.

Estimating Changes in Calcium Concentration from ΔF/F

Maravall et al. (2000) developed a method for estimating changes in calcium concentrations that is ideal for use during two-photon excitation of calcium indicators with large dynamic ranges (such as fluo-4 and OGB). Briefly, the approach takes advantage of the ability to estimate $\Delta F/F_{max}$ during experiments by achieving near-indicator saturation with strong stimulation, such as during high-frequency action potential (AP) trains. An alternative method was pioneered by Feller and coworkers (Feller et al. 1996). By comparing the ΔF/F plateau at two frequencies, $v_2 > v_1$, the percentage, x, of true indicator saturation ($\Delta F/F_{max}$) can be estimated using the following equation:

$$x = 100 \cdot \frac{1 - Q\, v_1/v_2}{1 - v_1/v_2}$$

where Q equals the plateau ΔF/F reached during v_2 divided by that reached during v_1 (see Fig. 1). Equipped with a derived value of $\Delta F/F_{max}$, $\Delta[Ca^{2+}]$ can be solved by

$$\Delta[\mathrm{Ca}] = K_d\left[\left((\Delta F/F)_{max}+1\right)(1-R^{-1})\left(\frac{\Delta F/F}{\left((\Delta F/F)_{max}-\Delta F/F\right)(\Delta F/F)_{max}}\right)\right]$$

where R is the dynamic range of the indicator, and K its dissociation constant (Maravall et al. 2000; Sabatini et al. 2002).

SHORT EXAMPLE OF APPLICATION

Figure 1 shows a L2/3 pyramidal neuron from a slice from p11 mouse visual cortex filled with 25 μM Alexa Fluor 594 and 50 μM fluo-4. The morphology of the cell and region of interest, indicated in Figure 1A and B, were obtained from the calcium-independent Alexa 594 fluorescence recorded on the red channel. Line scans through spine and dendrite, at positions indicated by the green arrows in Figure 1B, were performed using the calcium-dependent green channel, (B, bottom left), during 1 AP, and during 66- and 50-Hz trains (B, from top to bottom). Note the that the Alexa Fluor 594 signal does not change in fluorescence, reflecting its insensitivity to calcium (B, bottom right). In Figure 1C, calcium fluorescence signals acquired from the green channel over spine and dendrite, at positions indicated by red and black brackets, respectively, in Figure 1B, are plotted above time-locked APs. $\Delta F/F_{max}$ was estimated from the $\Delta F/F_{plateau}$ reached during the 66-Hz (v_2) and 50-Hz (v_1) AP trains as indicated above. In Figure 1D, the calcium signal during a single AP is plotted as %ΔF/F, left, and as $\Delta[Ca^{2+}]$, right. Arrowheads indicate timing of AP generation. Note how the spine $\Delta[Ca^{2+}]$ signal appears "stretched" compared to the %ΔF/F signal, reflecting how ΔF/F sublinearly reflects $\Delta[Ca^{2+}]$ as indicator saturation is approached.

ADVANTAGES AND LIMITATIONS

Two-Channel Calcium Imaging

Calcium imaging in small structures such as dendrites and spines is complicated by the fact that the calcium indicators, by binding calcium, effectively perturb calcium efflux, and diffusion. To better approximate physiological conditions, exogenous buffer capacity must be reduced by using low-affinity indicators, such as Oregon Green BAPTA-6F, or, as described here, low indicator concentrations. One problem associated with low indicator concentrations is insufficient resting fluorescence and, hence, poor visualization of the dendritic tree. This problem can be solved by using a separate channel to visualize morphology by using a calcium-insensitive fluorophore with nonoverlapping emission (Fig. 1).

An important consideration when working with low indicator concentrations, however, is indicator saturation. This can be advantageous when examining small calcium influxes such as during single APs, since it allows estimation of $\Delta F/F_{max}$, and, therefore, $\Delta[Ca^{2+}]$ (see above). However, if large calcium signals, such as during N-methyl-D-aspartate (NMDA) receptor activation or release from intracellular stores, are of interest, the use of higher indicator concentrations or low-affinity indicators, such as Oregon Green BAPTA 6-F, is recommended.

Choice of Indicator

CG-1, OGB, and fluo-4 have all been used successfully in single-channel imaging, and all three indicators are excited well by a mode-locked laser at 800 nm. However, each indicator is suited to different conditions. When imaging with one channel (using the indicator's resting fluorescence to visualize the dendritic tree), CG-1 and OGB (150–200 μM) are recommended for their high fluorescence at resting calcium concentration. However, partly because of their high fluorescence, their increase in intensity on binding calcium is compromised. These indicators are, therefore, unsuitable for detecting small signals. Fluo-4 has a low fluorescence at resting calcium concentration. This means that higher concentrations (200–400 μM) are necessary for one-channel use, but fluo-4 has a large dynamic range and is ideal for imaging small or heavily buffered signals, such as single APs in cortical interneurons (Goldberg et al. 2003). Fluo-4 is also a good indicator to use in tandem with Alexa 594 during two-channel image acquisition (Fig. 1).

REFERENCES

Feller M.B., Delaney K.R., and Tank D.W. 1996. Presynaptic calcium dynamics at the frog retinotectal synapse. *J. Neurophysiol.* **76:** 381–400.

Goldberg J.H., Tamas G., and Yuste R. 2003. Ca^{2+} imaging of mouse neocortical interneurone dendrites: Ia-type K^{+} channels control action potential backpropagation. *J. Physiol.* **551:** 49–65.

Holthoff K., Tsay D., and Yuste R. 2002. Calcium dynamics of spines depend on their dendritic location. *Neuron* **33:** 425–437.

Majewska A., Tashiro A., and Yuste R. 2000. Regulation of spine calcium compartmentalization by rapid spine motility. *J. Neurosci.* **20:** 8262–8268.

Maravall M., Mainen Z.F., Sabatini B.L., and Svoboda K. 2000. Estimating intracellular calcium concentrations and buffering without wavelength ratioing. *Biophys. J.* **78:** 2655–2667.

Sabatini B.L., Oertner T.G., and Svoboda K. 2002. The life cycle of $Ca^{(2+)}$ ions in dendritic spines. *Neuron* **33:** 439–452.

Yuste R. and Denk W. 1995. Dendritic spines as basic units of synaptic integration. *Nature* **375:** 682–684.

CHAPTER 39

A Practical Guide: Measurement of Free Ca^{2+} Concentration in the Lumen of Neuronal Endoplasmic Reticulum

Natasha Solovyova and Alexei Verkhratsky

The University of Manchester, School of Biological Sciences, Manchester M13 9PT, United Kingdom

This chapter describes a technique for the simultaneous monitoring of free calcium concentrations in the cytosol ($[Ca^{2+}]_i$) and within the lumen of the endoplasmic reticulum (ER) ($[Ca^{2+}]_L$) of cultured/freshly isolated dorsal root ganglia (DRG) neurones. The method uses two synthetic fluorescent Ca^{2+} probes (fluo-3 and mag-fura-2), in combination with fluorescence microscopy and a whole-cell patch-clamp technique.

This approach has been used successfully in acutely isolated/cultured DRG neurons, in Purkinje neurones, acutely isolated from cerebellar slices, and in cultured astrocytes. Similar techniques, using low-affinity Ca^{2+} indicators (mag-fluo-4, mag-fura-2, fluo-3FF) have been used extensively for monitoring $[Ca^{2+}]_L$ dynamics in a variety of nonexcitable cells (e.g., in gastric epithelial cells [Hofer and Machen 1993]; in pancreatic acinar cells [Mogami et al. 1998]; in pancreatic β cells [Tengholm et al. 2001]; in endothelial cells [Dedkova and Blatter 2002]) and in muscle cells (Shmigol et al. 2001).

In this protocol, isolated neurons are first loaded with the membrane-permeant, low-affinity Ca^{2+} indicator, mag-fura-2, which preferentially, though not exclusively, accumulates in the ER (see Thomas et al. 2000). Cells are then loaded with the membrane-impermeant, high-affinity calcium indicator fluo-3 K_5 using the whole-cell patch-clamp configuration. This second loading removes the majority of cytosolic mag-fura-2, replacing it with fluo-3 K_5. Mag-fura-2 and fluo-3 K_5 signals can be separated by virtue of their distinct excitation properties (340 nm and 380 nm for mag-fura-2 and 488 nm for fluo-3). $[Ca^{2+}]_L$ values are calculated using the 340/380 nm ratio with the equation given in Procedure C, step 1.

MATERIALS

Membrane-permeant Low-Affinity Calcium-sensitive Fluorescent Indicator

Mag-fura-2 is a ratiometric probe (i.e., it demonstrates spectral shift upon Ca^{2+} binding) that was originally developed as a Mg^{2+} indicator (Raju et al. 1989). Notwithstanding this history, mag-fura-2 is ideally suited for $[Ca^{2+}]_L$ measurements (1) because of its propensity to accumulate in subcellular compartments and (2) because the probe affinity for Ca^{2+} (K_D ~ 50 μM) is much higher than that for Mg^{2+} (K_D ~ 5 mM) (see Hyrc et al. 2000) so that the probe can accurately report free Ca^{2+} values between tens to several hundreds of μM—i.e., within the characteristic range of $[Ca^{2+}]_L$.

Membrane-impermeant High-Affinity Calcium-sensitive Fluorescent Indicator

Fluo-3 K_5 (Molecular Probes) has a K_D of ~325–390 nM, which allows it to measure cytosolic Ca^{2+} concentrations within the range of ~20 to ~2000 nM.

Real-Time Video Imaging

This protocol was designed to use an Olympus IX70 inverted microscope (40XUV objective) equipped with a charge-coupled device (CCD)-cooled intensified camera (Pentamax Gene IV, Roper Scientific). The illumination was provided by a monochromator (Polychrom IV, TILL Photonics). The equipment was linked to a Windows 98 workstation. Data were collected, stored, and analyzed using the MetaFluor/MetaMorph software (Universal Imaging).

Whole-Cell Voltage-Clamp Recordings

These recordings were made using an EPC-9 amplifier run by the PC-based PULS software (both from HEKA) (see Penner 1995).

PROCEDURE A

Staining Neurons with Mag-Fura-2/AM and Fluo-3 K_5

1. Dilute an aliquot of mag-fura-2/AM stock solution<!> (1 mM in DMSO<!>) in standard extracellular saline to a final concentration of 2.5–5 μM (135 mM NaCl, 3 mM KCl<!>, 2 mM $CaCl_2$<!>, 20 mM glucose, 20 mM HEPES/NaOH<!>, pH 7.4).
2. Load cultured or freshly isolated cells with mag-fura-2 by bathing them in the mag-fura-2/AM extracellular saline solution and incubating for 30 min at 37°C.
3. Carefully drain off the mag-fura-fura-2/AM/saline solution and wash the cells with prewarmed *dye-free* intra-pipette solution (intra-pipette solution: 122 mM $CsCl_2$<!>, 20 mM TEA-Cl; 3 mM Na_2ATP, 0.05 mM fluo-3 K_5; 10 mM HEPES/CsOH<!>, pH 7.3).
4. Incubate the cells in standard extracellular saline for a further hour at 37°C.

   Note: This step ensures complete de-esterification of the probe.
5. Establish whole-cell patch-clamp configuration on the cells (that are now loaded with mag-fura-2) with a patch pipette, containing intra-pipette solution supplemented with 0.05 mM fluo-3 K_5 (from 1 mM stock in DMSO).

   Note: This step washes out the cytosolic portion of the mag-fura-2 loading and introduces fluo-3 K_5 in its place. For information on the kinetics of intracellular dialysis, see Solovyova and Verkhratsky (2002). Any remaining cytosolic fluorescence can be additionally quenched by adding Mn^{2+} to the dialyzing solution (see Tse et al. 1994).

Caution: *See Appendix 3 for appropriate handling of materials marked with <!>.*

PROCEDURE B

Real-Time Video Imaging

1. Illuminate the cells alternately at 340, 380, and 488 nm using a monochromator at a cycle frequency 0.5–5 Hz.
2. Capture data at 510 ± 15 nm.

PROCEDURE C

Intracellular Calibration of Mag-Fura-2 Signals

1. Calculate $[Ca^{2+}]_L$ using the 340/380 nm ratio with the following equation

   $$[Ca^{2+}]_L = K^* (R - R_{min})/(R_{max} - R) \cdot R_{max}$$

   where R is ratio obtained during measurement, R_{min} is the fluorescence ratio of Ca^{2+}-free dye, and R_{max} is the ratio for dye saturated by Ca^{2+}. The parameter K^* comprises the value of the affinity of dye to Ca^{2+} ions (K_d – dissociation constant of the dye) and the constant β is determined by the optical properties of the experimental setup. The K_d, though known for each dye, may vary depending on the physical properties of the solution (e.g., viscosity, temperature, pH). For practical reasons, K_d is usually determined empirically by measuring R for a given Ca^{2+} concentration and using it in the formula with known values for R_{min} and R_{max}.
2. Determine the R_{min} and K^* values by exposing the neurons to 10–20 μM ionomycin and 3–4 calibrating solutions with known Ca^{2+} concentrations (e.g., with $[Ca^{2+}]$ <10 nM; 100 μM; 400 μM and 10 mM).

   Calibrating solutions: 130 mM NaCl, 3 mM KCl<!>; 10 mM HEPES/KOH<!>, 10 mM glucose, 0.01 mM ionomycin, pH 7.3; the desirable Ca^{2+} concentration is achieved by adding Ca^{2+}/EGTA or Ca^{2+}/BAPTA buffer; appropriate concentrations of Ca^{2+} and EGTA/BAPTA can be calculated using widely available calculators (e.g., http://www.stanford.edu/~cpatton/webmaxcE.htm).

   Caution: *See Appendix 3 for appropriate handling of materials marked with <!>.*

EXAMPLE OF APPLICATION

Figure 1 shows an example of simultaneous $[Ca^{2+}]_L$ and $[Ca^{2+}]_i$ recordings from a single DRG neuron as it responds to a brief (5 sec) application of 20 mM caffeine or to the entry of Ca^{2+} into the cytoplasm via voltage-operated Ca^{2+} channels. The latter provides an example of genuine Ca^{2+}-induced Ca^{2+} release (CICR). Both treatments provoked the release of Ca^{2+} from intracellular stores, resulting in a transient increase of $[Ca^{2+}]_i$ and a transient decrease of $[Ca^{2+}]_L$.

ADVANTAGES AND LIMITATIONS

Intraluminal free Ca^{2+} regulates numerous signaling cascades that converge in the ER (Berridge 2002; Bootman et al. 2002; Verkhratsky and Petersen 2002), giving rise to fast signaling events (e.g., Ca^{2+} release or Ca^{2+} re-uptake) and long-lasting adaptive responses (e.g., posttranslational protein folding controlled by Ca^{2+}-dependent chaperones). Therefore, the ability to make direct measurements of $[Ca^{2+}]_L$ is important for further understanding of the intimate mechanisms of ER signaling processes.

The use of low-affinity Ca^{2+} indicators, such as mag-fura-2, provides a relatively simple and reliable means of monitoring $[Ca^{2+}]_L$. Because mag-fura-2 has a much higher affinity for Ca^{2+} than it does for Mg^{2+}, the indicator should respond to changes in intracellular free Ca^{2+} with little interference from changes in intracellular Mg^{2+}. The combination of mag-fura-2/AM ester indicator com-

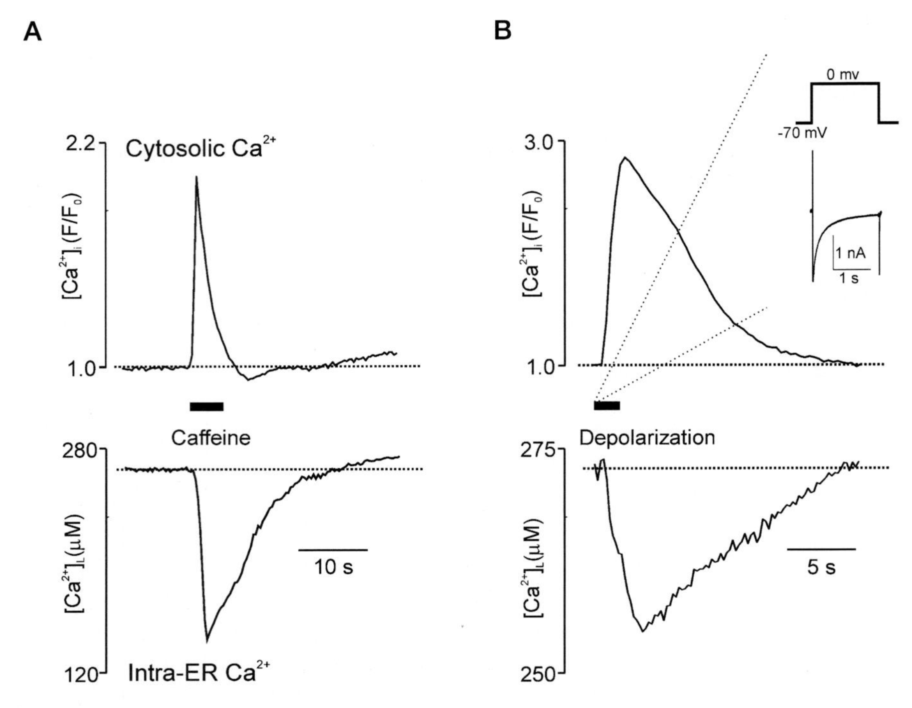

FIGURE 1. Simultaneous recordings of $[Ca^{2+}]_i$ (cytosolic Ca^{2+}) and $[Ca^{2+}]_L$ (intra-ER Ca^{2+}) in an isolated dorsal root ganglion neuron upon activation of Ca^{2+} release from ER stores. Calcium release was activated by a brief (5 sec) application of caffeine (*A*) or by Ca^{2+} entry through voltage-gated Ca^{2+} channels (*B*). In the latter case, the voltage-clamped neuron was depolarized from –70 mV to 0 mV for 2 sec. The insert shows a voltage protocol and calcium current recorded in the presence of 1 μM tetrodotoxin. Both caffeine and Ca^{2+} entry resulted in a transient decrease in $[Ca^{2+}]_L$, indicative of Ca^{2+} release from the ER.

partmentalized within the ER lumen and fluo-3 (loaded using a patch pipette), allows (1) simultaneous but independent monitoring of Ca^{2+} dynamics in the ER and in the cytosol (see Solovyova et al. 2002); and (2) the application of established electrophysiological protocols to the cell of interest. The use of a ratiometric probe for monitoring intra-ER Ca^{2+} permits accurate calibration and quantitative measurement of $[Ca^{2+}]_L$; it also reduces the effects of uneven dye loading, photobleaching, uneven cell thickness, and leakage of the dye. However, unlike ER-targeted cameleons (Yu and Hinkle 2000; Varadi and Rutter 2002) and aequorins (Montero et al. 1995), mag-fura-2 cannot be targeted exclusively to the ER. Indeed, mag-fura-2 is known to compartmentalize to a variety of organelles. The dye accumulated in the organelles, other than the ER, contributes to the overall mag-fura-2 signal and can interfere with estimation of $[Ca^{2+}]_L$ levels. Yet, the dynamic range of synthetic fluorescent Ca^{2+} probes significantly exceeds that of chameleons and, hitherto, low-affinity dyes remain the indicators of choice for measuring low-amplitude $[Ca^{2+}]_L$ signals in response to direct electrical stimulation of nerve cells under voltage/current clamp conditions.

ACKNOWLEDGMENTS

This research was supported by The Biotechnology and Biological Sciences Research Council (U.K.) and The Wellcome Trust.

REFERENCES

Berridge M.J. 2002. The endoplasmic reticulum: A multifunctional signalling organelle. *Cell Calcium* **32:** 235–249.

Bootman M.D., Petersen O.H., and Verkhratsky A. 2002. The endoplasmic reticulum is a focal point for co-ordination of cellular activity. *Cell Calcium* **32:** 231–234.

Dedkova E.N. and Blatter L.A. 2002. Nitric oxide inhibits capacitative Ca^{2+} entry and enhances endoplasmic reticulum Ca^{2+} uptake in bovine vascular endothelial cells. *J. Physiol. Lond.* **539:** 77–91.

Hofer A.M. and Machen T.E. 1993. Technique for in situ measurement of calcium in intracellular inositol 1,4,5-trisphosphate-sensitive stores using the fluorescent indicator mag-fura-2. *Proc. Natl. Acad. Sci.* **90:** 2598–2602.

Hyrc K.L., Bownik J.M., and Goldberg M.P. 2000. Ionic selectivity of low-affinity ratiometric calcium indicators: Mag-Fura-2, Fura-2FF and BTC. *Cell Calcium* **27:** 75–86.

Mogami H., Tepikin A.V., and Petersen O.H. 1998. Termination of cytosolic Ca^{2+} signals: Ca^{2+} reuptake into intracellular stores is regulated by the free Ca^{2+} concentration in the store lumen. *EMBO J.* **17:** 435–442.

Montero M., Brini M., Marsault R., Alvarez J., Sitia R., Pozzan T., and Rizzuto R. 1995. Monitoring dynamic changes in free Ca^{2+} concentration in the endoplasmic reticulum of intact cells. *EMBO J.* **14:** 5467–5475.

Penner R. 1995. A practical guide to patch-clamping. In *Single-channel recording* (ed. B. Shakmann and E. Neher), pp. 3–30. Plenum Press, New York.

Raju B., Murphy E., Levy L.A., Hall R.D., and London R.E. 1989. A fluorescent indicator for measuring cytosolic free magnesium. *Am. J. Physiol.* **256:** 540–548.

Shmigol A.V., Eisner D.A., and Wray S. 2001. Simultaneous measurements of changes in sarcoplasmic reticulum and cytosolic [Ca^{2+}] in rat uterine smooth muscle cells. *J. Physiol. Lond.* **531:** 707–713.

Solovyova N. and Verkhratsky A. 2002. Monitoring of free calcium in the neuronal endoplasmic reticulum: An overview of modern approaches. *J. Neurosci. Methods* **122:** 1–12.

Solovyova N., Veselovsky N., Toescu E.C., and Verkhratsky A. 2002. Ca^{2+} dynamics in the lumen of the endoplasmic reticulum in sensory neurones: Direct visualisation of Ca^{2+}-induced Ca^{2+} release triggered by physiological Ca^{2+} entry. *EMBO J.* **21:** 622–630.

Tengholm A., Hellman B., and Gylfe E. 2001. The endoplasmic reticulum is a glucose-modulated high-affinity sink for Ca^{2+} in mouse pancreatic β-cells. *J. Physiol. Lond.* **530:** 533–540.

Thomas D., Tovey S.C., Collins T.J., Bootman M.D., Berridge M.J., and Lipp P. 2000. A comparison of fluorescent Ca^{2+} indicator properties and their use in measuring elementary and global Ca^{2+} signals. *Cell Calcium* **28:** 213–223.

Tse F.W., Tse A., and Hille B. 1994. Cyclic Ca^{2+} changes in intracellular stores of gonadotropes during gonadotropin-releasing hormone-stimulated Ca^{2+} oscillations. *Proc. Natl. Acad. Sci.* **91:** 9750–9754.

Varadi A. and Rutter G.A. 2002. Dynamic imaging of endoplasmic reticulum Ca^{2+} concentration in insulin-secreting MIN6 Cells using recombinant targeted cameleons: Roles of sarco(endo)plasmic reticulum Ca^{2+}-ATPase (SERCA)-2 and ryanodine receptors. *Diabetes* **51:** S190–201.

Verkhratsky A. and Petersen O.H. 2002. The endoplasmic reticulum as an integrating signalling organelle: From neuronal signalling to neuronal death. *Eur. J. Pharmacol.* **447:** 141–154.

Yu R. and Hinkle P.M. 2000. Rapid turnover of calcium in the endoplasmic reticulum during signaling. Studies with cameleon calcium indicators. *J. Biol. Chem.* **275:** 23648–23653.

Imaging Intracellular Calcium-concentration Microdomains at a Chemical Synapse

Rodolfo Llinás and Mutsuyuki Sugimori

Department of Physiology and Neuroscience, New York University School of Medicine, New York, New York 10016

More than 30 years ago, it was hypothesized that calcium ion concentration might be responsible for the trigger of transmitter release from specialized active zones at the presynaptic terminal of chemical junctions (cf. Katz 1969). Presynaptic voltage-clamp determination of calcium inward current at the release site (Llinás et al. 1976, 1981a; cf. Augustine and Charlton 1986; Augustine et al. 1991) indicated a current flow on the order of several hundred nanoamperes. Considering that calcium entry was shown to be exclusively localized at the sites of transmitter release using the photoprotein aequorin (Llinás and Nicholson 1974), it was proposed that calcium concentration should be particularly high against the internal surface of the plasmalemma next to the voltage-gated calcium channel clusters (Llinás 1977). Calculation of calcium concentration next to the membrane, given the levels of calcium current measured, gave an estimate on the order of 600 μM (Llinás et al. 1981a). The actual solution for calcium concentration at the mouth of the calcium channels was given by Chad and Eckert (1984), Simon and Llinás (1985), and Zucker and Fogelson (1986). Those papers agreed on the fact that calcium concentration would be quite high near the mouth of the calcium channel. The small well-localized profiles of increased $[Ca^{2+}]$ were named "microdomains." Each calcium channel opening was thought to produce a rapid (1–2 μsec) increase in intracellular $[Ca^{2+}]$ ($[Ca^{2+}]_i$) that was maintained for the average open-time of the channel, after which $[Ca^{2+}]_i$ would rapidly return to its preopening level (Simon and Llinás 1985). Transmitter release would thus be triggered by a large, transient increase in $[Ca^{2+}]$ localized to the immediate vicinity of the synaptic vesicles. These concentration transients exist only as dissipative structures of short duration; the latency for calcium activation of the release process was on the order of 200 μsec, which suggested that calcium had to act very close to the release site (see Fig. 1) (Llinás 1977).

OBSERVING CALCIUM CONCENTRATION MICRODOMAINS

During the last decade, direct measurements of calcium concentration profiles in the presynaptic terminal of the squid have been performed (Llinás et al. 1992), using the photoprotein *n*-aequorin-J (Shimomura et al. 1990). This photoprotein was chosen because its low binding constant for calcium

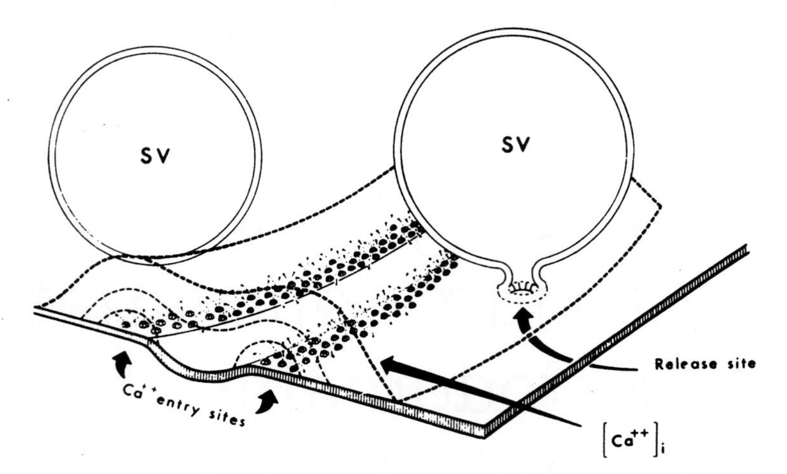

FIGURE 1. Diagram of the active zone, modeled after freeze-fracture micrographs of the active zone in the neuromuscular junction. The voltage-dependent calcium current is assumed to enter the terminal through channels believed to be the intramembranous particles seen in the freeze-fracture (calcium entry sites). Broken lines illustrate the hypothetical distribution of intracellular calcium, $[Ca^{2+}]_i$, and its relation to the synaptic release sites. Calcium current is assumed to generate a well-localized change in calcium concentration in the vicinity of the active zone, triggering vesicle (SV) opening at that location (see Llinás 1979).

triggers photon emission in ~750 µsec (Silver et al. 1994) and calcium concentration transients in the range of 300–400 µM (Llinás et al. 1992, 1994; Sugimori et al. 1994). In addition, it has the advantage of not modifying substantially the buffering properties of the intracellular milieu.

PROCEDURE

Injecting Aequorin

The hybrid synthetic *n*-aequorin-J, with a sensitivity to $[Ca^{2+}]_i$ of 10^{-4} M, developed by Shimomura et al. (1990), is so far the best indicator. This protein may be obtained from Dr. Shimomura, at the Marine Biological Laboratory (Woods Hole, Massachusetts). (Because the extraction and preparation procedure for *n*-aequorin-J is very complex, and because it is partly synthetic, Dr. Shimomura is likely to inquire about the research protocol and aims and may require an extended time to prepare the indicator.) The following protocol has been developed for injecting aequorin in isolated squid giant synapse presynaptic termini.

1. Fill a microelectrode with 2.5 mg/ml *n*-aequorin-J, 0.5 M potassium acetate, and 0.1 mM dextran tetramethylrhodamine (m.w. 70,000; Molecular Probes) or 2.5 mg/ml fluorescein-conjugated nonactive aequorin for injection marking.

   Note: Because aequorin is activated by metallic silver, the electrode connection must utilize a fine wire located as far from the microelectrode tip as possible. However, because of the very small amount of indicator available, a balance must be reached on how much material to use in each electrode. Backfilling the pipette to the top of the thin shaft (assuming 8–10 mm length) provides sufficient distance to protect the aequorin in the distant two thirds of the shaft.

2. Apply pressure to inject the indicator solution into the isolated squid giant synapse presynaptic terminal. Pressure pulses must be given continuously when the electrode is placed into the bathing artificial seawater to prevent injecting spent aequorin into the terminal.

   Note: Aequorin is pressure-injected intracellularly using 20-pound pressure pulses of 200–500 msec duration at 1 Hz.

3. Monitor the distribution of the dye using a 40x/0.8-NA water-immersion objective lens.

   Note: The distribution of the injected aequorin inside the terminal is best determined by a co-injection of a fluorophore (e.g., coinjection dyes used in step 1; see Fig. 2A). This allows a direct visualization of the diffusion profile using a fluorescence microscope with a 40x water-immersion objective lens.

Procedural Notes

Because aequorin does not regenerate once it has been triggered by calcium, it is necessary to protect the indicator from contacting calcium during the injection process. Thus, all instruments utilized in the handling of the protein must be soaked for at least 1 day in EGTA and then rinsed with very low calcium Ringer. The water to be utilized for this procedure must be calcium-free to 10^{-8} M. In particular, microelectrode tubing must be free of calcium.

Detecting Aequorin Light

Aequorin luminescence imaging requires a high-quality DMCP (dual microchannel plate) image intensifier (Hamamatsu Argus 100), or equivalent, operated in the photon-counting mode. Best results are obtained using a large-aperture objective (1.15 NA). The images can be stored on either a conventional videotape or a fast-tape system (NAC HSD 400) allowing 33-msec and 5-msec time resolution, respectively, or a digital system (Kodak HSC) allowing 250-µsec resolution (Silver et al. 1994). To ascertain whether the system is capable of detecting the signal, tetanic presynaptic stimulation (Fig. 2D,E) is the most useful approach (this procedure consists of 10–50 electrical stimuli of 0.1-msec duration at 50–100 Hz) as such stimulus evokes small points of light localized at the active zone. Once the site of calcium entry is localized, single stimuli can be triggered, and the light produced by each stimulus can be gathered over a certain number of trials. In a healthy preparation, the number of spontaneous quantum-emission domains (QEDs) observed at any time is very small. We define microdomains as the superposition of many (5–20) QEDs arising close enough to define a well-circumscribed spot with an average diameter of the order of 0.5 µM. Such microdomains are distributed in our preparation such that they coincide with the synaptic active zones.

As demonstrated in Figure 2, D and E, activation of calcium entry following tetanic activation of presynaptic action potentials (100 Hz) is characterized by the appearance of evoked calcium microdomain sites that repeat themselves during the tetanus stimulation. This is in contrast to a similar set of measurements obtained in the absence of stimulation and related, most probably, to the normal spontaneous release or to shot noise (Fig. 2B,C); the latter, however, is much shorter in duration than a calcium microdomain (Silver et al. 1994).

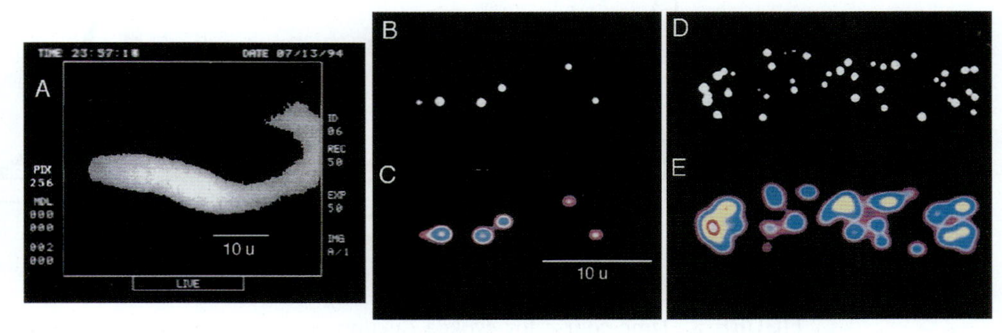

FIGURE 2. Fluorescence image (*A*) and light emission (*B–E*) from a squid synaptic preterminal injected with *n*-aequorin-J, which was used as a $[Ca^{2+}]_i$ indicator. *n*-Aequorin-J has a sensitivity to $[Ca^{2+}]$ on the order of 10^{-4} M and was developed by Shimomura et al. (1989). This photoprotein, which was mixed with a fluorescent dextran (*A*), was chosen because its low binding constant requires relatively large calcium concentrations to activate photon emission (≈300–400 µM), which can be triggered with a short delay (≈750 µsec; Silver et al. 1994). The dots in *B–E* are localized directly over the presynaptic terminal (*A*), but shown at a higher magnification (see scale bar). The leftmost point for light emission corresponds to the left end of the terminal as shown in *A*.

TIME COURSE OF THE MICRODOMAIN

The time course of the calcium current produced by a voltage step that simulates an action potential is shown in Fig. 3A, superimposed on the time course of a microdomain evoked by a similar spike. Its duration is on the order of 1.2 msec (Fig. 3B). Note that although the time course is very similar, light emission seems to outlast that of the transmembrane current. This is to be expected, however, because high-level light emission does leave a wake in the photosensors, as they have a relaxation time. If connected for such lag, the time course of the image becomes closer to I_{Ca} (see Fig. 3C) (Sugimori et al. 1994).

This duration of light emission is in general agreement with the time course for spike-evoked transynaptic current and model results (Simon and Llinás 1985) for the distribution and time course for transient calcium concentration profiles, given by the equation

$$C(r,t) = \frac{2.0 \times 10^3 \times j}{4.0 \times \pi \times \text{FAR} \times \text{Dca} \times r} \, \text{erfc} \left(\frac{r}{2.0 \times \sqrt{\text{Dca} \times t}} \right)$$

The analytical solution for three-dimensional (3D) calcium diffusion where $C(r,t)$ is the calcium concentration at time t and distance r from the channel; *Dca* is the diffusion coefficient for calcium; FAR is the Faraday constant; and *erfc* is the complementary error function. Using this formalism, it was calculated that the calcium concentration profile would reach a steady-state level nearly equal to 300 μM

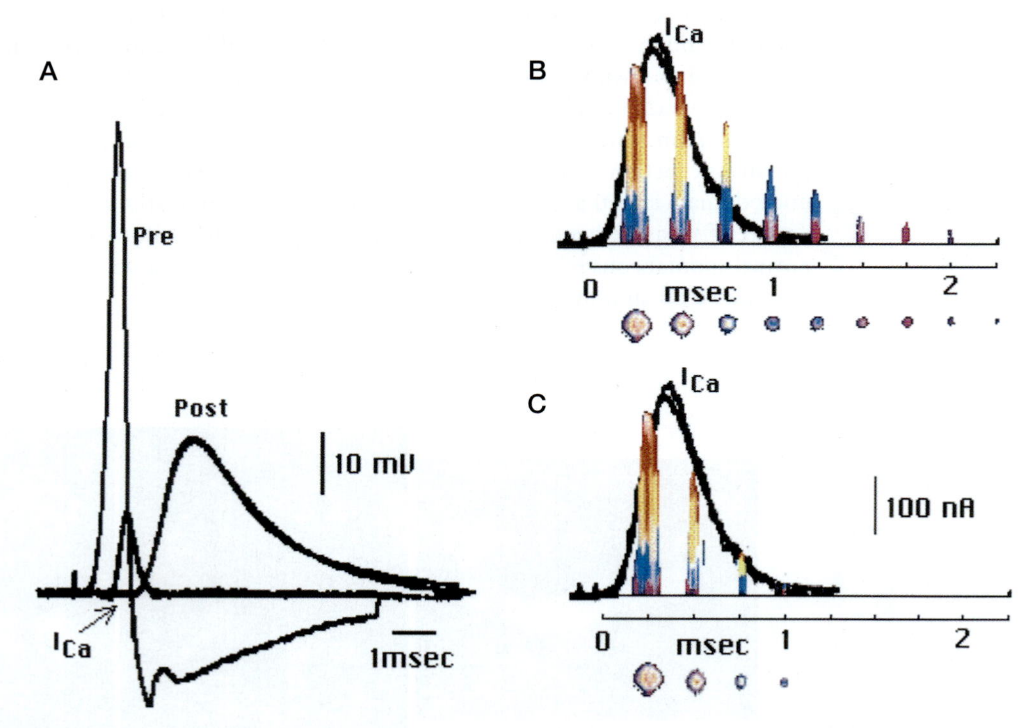

FIGURE 3. Time course for evoked calcium-concentration microdomains. (*A*) Illustration of the time course for a presynaptic action potential (Pre) that simulates the voltage profile of a presynaptic spike recorded from the same fiber, before block of sodium and potassium channels (modified from Sugimori et al. 1994); the calcium current (I_{Ca} [*arrow*]), which is basically a tail current with an overall duration nearly equal to 800 μsec; and the postsynaptic potential (Post) obtained in previous experiments (Silver et al. 1994). (*B,C*) Superimposition of the time course for the calcium microdomain (amplitude on a relative scale) and the time course for the calcium current. (*B*) Amplitudes of raw light measurements taken every 250 μsec. (*C*) Amplitudes after correction for afterglow. The area of the colored circles under each time point relates to the amplitude of the light emission during that time bin.

at 500 Å from the channel pore in 1 µsec and would decay to a level of 10 µM in a similar period for the same distance (Simon and Llinás 1985). Given the sensitivity of *n*-aequorin-J, the results are in keeping with modeling predictions and are similar to that observed for the calcium concentration profile seen in individual frog muscle sarcomeres following the induction of action potentials (Monck et al. 1994).

HOW MANY IONS IN A MICRODOMAIN?

The number of calcium ions detected in an evoked microdomain may be estimated as follows. Consider that a presynaptic action potential generates an I_{Ca} of about 300 nA (Llinás et al. 1982; Lin et al. 1990) and a current of 0.5 pA/channels, and that a single channel allows the flow of approximately 150–200 calcium ions (Pumplin et al. 1981; Stanley 1993), then approximately 6×10^5 channels are opened to release 5,000–10,000 vesicles (Miledi 1967; Llinás et al. 1981a,b), or approximately 15,000, calcium ions/vesicle (Llinás et al. 1982). (This is in contrast to the estimated minimum of 200 ions/vesicle [1 calcium channel] for the ciliary ganglion [Stanley 1993].) Given that the number of active zones in a presynaptic terminal is about 5,000–10,000 (Pumplin and Reese 1978), then the number of channels open per action potential for each active zone is approximately 100. If single channels allow the flow of 150–200 calcium ions (Pumplin et al. 1981; Stanley 1993), an evoked microdomain, using *n*-aequorin-J and the present imaging technique, may represent an influx of calcium of about 15,000 ions.

CONCLUSION

The issue of calcium concentration microdomains continues to develop as a concept, bringing together calcium entry via a transmembrane channel with its localized role as a second messenger. The next set of steps might relate to the molecular organization that builds around such sites. Indeed, calcium microdomain generation demands that this very precise, organized triggering event be equally precisely coupled with the molecular sensors that will take advantage of such exquisite signals.

ACKNOWLEDGMENT

Support for this work was received from the National Institutes of Health–National Institute of Neurological Disorders and Stroke (grant NS-13742).

REFERENCES

Augustine G.J. and Charlton M.P. 1986. Calcium dependence of presynaptic calcium current and post-synaptic response at the squid giant synapse. *J. Physiol.* **381:** 619–640.

Augustine G.J., Adler E.M., and Charlton M.P. 1991. The calcium signal for transmitter secretion from presynaptic nerve terminals. *Ann. N.Y. Acad. Sci.* **635:** 365–381.

Chad J.E. and Eckert R. 1984. Calcium domains associated with individual channels may account for anomalous voltage relation of Ca-dependent response. *Biophys. J.* **45:** 993–1000.

Katz B. 1969. The release of neural transmitter substances. In *The Sherrington Lectures*, vol. 10. C.C. Thomas, Springfield, Illinois.

Lin J.-W., Sugimori M., Llinás R., McGuinness T., and Greengard P. 1990. Effects of synapsin I and calcium/calmodulin-dependent protein kinase II on spontaneous neurotransmitter release in the squid giant synapse. *Proc. Natl. Acad. Sci.* **87:** 8257–8261.

Llinás R. 1977. Calcium and transmitter release in squid synapse. *Soc. Neurosci. Symp.* **2:** 139–160.

———. 1979. The role of calcium in neuronal function. In *The neurosciences: Fourth study program* (ed. F.O. Schmitt and F.G. Worden), pp. 555–571.

Llinás R. and Nicholson C. 1974. Aequorin study of the suppression potential in the squid giant synapse. *Biol. Bull.* **147:** 489.

Llinás R., Steinberg I.Z., and Walton K. 1976. Presynaptic calcium currents and their relation to synaptic transmission: A voltage-clamp study in squid giant synapse and a theoretical model for the calcium gate. *Proc. Natl. Acad. Sci.* **73:** 2520–2523.

———. 1981a. Presynaptic calcium currents in squid giant synapse. *Biophys. J.* **33:** 289–322.

———. 1981b. Relationship between presynaptic calcium current and postsynaptic potential in squid giant squid synapse. *Biophys. J.* **33:** 323–352.

Llinás R., Sugimori M., and Silver R.B. 1992. Microdomains of high calcium concentration in a presynaptic terminal. *Science* **256:** 677–679.

———. 1994. Localization of calcium concentration microdomains at the active zone in the squid giant synapse. In *Molecular and cellular mechanisms of neurotransmitter release* (ed. L. Stjärne et al.), pp. 133–136. Raven Press, New York.

Llinás R., Sugimori M., and Simon S.M. 1982. Transmission by presynaptic spike-like depolarization in the squid giant synapse. *Proc. Natl. Acad. Sci.* **79:** 2415–2419.

Miledi R. 1967. Spontaneous synaptic potentials and quantal release of transmitter in the stellate ganglion of the squid. *J. Physiol.* **192:** 379–406.

Monck J.R., Robinson I.M., Escobar A.L., Vergara J.L., and Fernandez J.M. 1994. Pulsed laser imaging of rapid Ca^{2+} gradients in excitable cells. *Biophys. J.* **67:** 505–514.

Pumplin D. and Reese T.S. 1978. Membrane ultrastructure of the giant synapse of the squid *Loligo pealei. Neuroscience* **3:** 685–696.

Pumplin D., Reese T.S., and Llinás R. 1981. Are the presynaptic membrane particles the calcium channels. *Proc. Natl. Acad. Sci.* **78:** 7210–7213.

Shimomura O., Musicki B., and Kishi V. 1989. Semi-synthetic aequorins with improved sensitivity to Ca^{2+} ions. *Biochem. J.* **261:** 913–920.

Shimomura O., Inoye S., Musicki B., and Kishi Y. 1990. Recombinant aequorin and recombinant semi-synthetic aequorins, cellular Ca^{2+} ion indicators. *Biochem. J.* **27:** 309–312.

Silver R.B., Sugimori M., Lang E.J., and Llinás R. 1994. Time-resolved imaging of Ca^{2+}-dependent aequorin luminescence of microdomains and QEDs in synaptic terminals. *Biol. Bull.* **187:** 293–299.

Simon S.M. and Llinás R. 1985. Compartmentalization of the submembrane calcium activity during calcium influx and its significance in transmitter release. *Biophys. J.* **48:** 485–498.

Stanley E.F. 1993. Single calcium channels and acetylcholine release at a presynaptic nerve terminal. *Neuron* **11:** 1007–1011.

Sugimori M., Lang E.J., Silver R.B., and Llinás R. 1994. High-resolution measurement of the time course of calcium-concentration microdomains at squid presynaptic terminals. *Biol. Bull.* **187:** 300–303.

Zucker R.S. and Fogelson A.L. 1986. Relationship between transmitter release and presynaptic calcium influx when calcium enters through discrete channels. *Proc. Natl. Acad. Sci.* **83:** 3032–3036.

CHAPTER 41

Monitoring Intramitochondrial Calcium with Rhod-2

Markus Hoth and Richard S. Lewis

Department of Molecular and Cell Biology, Stanford University School of Medicine, Stanford, California 94306-5426

The development of membrane-permeant fluorescent Ca^{2+} indicators by Tsien and colleagues (Grynkiewicz et al. 1985; Minta et al. 1989) revolutionized studies of Ca^{2+} signaling by enabling quantitative measurements of cytosolic free Ca^{2+} ($[Ca^{2+}]_i$) in single, intact cells. Despite their great utility for measuring cytosolic Ca^{2+}, however, it has long been recognized that several of these dyes tend to accumulate not only in the cytoplasm, but also in organelles such as the endoplasmic reticulum (ER) and mitochondria (Malgaroli et al. 1987; Steinberg et al. 1987). Sequestration by organelles is often viewed as a nuisance, because fluorescence from the organellar dye contaminates the cytoplasmic Ca^{2+} signal, and effort is often spent to minimize dye compartmentation (Malgaroli et al. 1987). However, the realization that Ca^{2+} uptake and release by organelles also have essential roles in Ca^{2+} signaling has fueled growing interest in developing methods that would allow measurements of intra-organelle $[Ca^{2+}]$ changes. Thus, a number of recent studies have exploited Ca^{2+} dye compartmentalization to permit such measurements, turning a drawback into an advantage.

This chapter focuses on the use of the fluorescent dye, rhod-2 (Minta et al. 1989), to measure intramitochondrial calcium concentrations ($[Ca^{2+}]_m$) in single cells. In recent years, a number of important functions for mitochondrial Ca^{2+} uptake and release have been brought to light, revealing new roles for mitochondria as signaling organelles in excitable and nonexcitable cells. These activities include shaping the time course of Ca^{2+} transients (Ali et al. 1994; Friel and Tsien 1994; Werth and Thayer 1994; Herrington et al. 1996), contributing to posttetanic potentiation at synapses (Tang and Zucker 1997), matching the rate of ATP production to metabolic demand (Hansford 1994; Hajnoczky et al. 1995; Drummond and Fay 1996), promoting Ca^{2+}-wave propagation (Jouaville et al. 1995; Simpson and Russell 1996; Ichas et al. 1997), and modulating the activity of Ca^{2+} channels in the ER (Jouaville et al. 1995; Simpson and Russell 1996) and the plasma membrane (Hoth et al. 1997). Although in many cases it is not possible to calibrate the response of intramitochondrial rhod-2 precisely, much useful qualitative information has been obtained using this dye. Some of the important practical issues for using rhod-2 to measure $[Ca^{2+}]_m$ are described below, along with our own experiences with rhod-2, and a brief survey of alternative methods that have been used to estimate $[Ca^{2+}]_m$.

LOADING RHOD-2 SELECTIVELY INTO MITOCHONDRIA

The general strategy for loading mitochondria with rhod-2 is based on diffusion of the acetoxymethyl ester of rhod-2 (rhod-2/AM) from the cytoplasm into the mitochondrial matrix, followed by esterolysis and exposure of the carboxylates necessary for binding Ca^{2+}. Of several Ca^{2+} indicators based on the structure of BAPTA (see Chapter 29), rhod-2 is generally the best at loading into mitochondria, probably because only rhod-2 has a net positive charge in its fully esterified membrane-permeant form, thereby promoting uptake by the strongly negative mitochondrial membrane potential. Mitochondrial uptake may be further optimized by use of the reduced form, dihydro-rhod-2/AM (see below). However, it is important to note that in contrast to the targeting of protein-based Ca^{2+} indicators to specific subcellular locations, the specificity of mitochondrial labeling by rhod-2 is considerably less stringently controlled and may vary widely according to cell type and loading conditions. For this reason, in each case, it is important to assay the degree of loading specificity, ideally with a combination of criteria described below.

Criteria for Selective Loading of Mitochondria

To determine the specificity of loading into mitochondria, it is advisable to perform a combination of assays based on the following criteria:

- *Dye compartmentation.* At low concentrations, digitonin preferentially solubilizes cholesterol-rich membranes (such as the plasma membrane), leaving intracellular membranes intact. Under these conditions, cytoplasmic fluorescence abruptly decreases as dye escapes the cytoplasm, leaving behind a particulate fluorescence due to compartmentalized and immobilized dye (Jou et al. 1996). Compartmentalized dye can be distinguished from dye bound to fixed cytoplasmic sites by its insensitivity to quenching by cytosolic Mn^{2+} (Jou et al. 1996). Quenching of the cytoplasmic dye in intact cells is achieved by activating Mn^{2+} influx through Ca^{2+} channels in the plasma membrane.
- *Morphology.* The pattern of staining is commonly used to assay for mitochondrial location, although this can be misleading in certain cells. The mitochondria can appear as bright punctate spots and/or a filamentous pattern, depending on the cell type and conditions (Jou et al. 1996; Simpson and Russell 1996; Babcock et al. 1997; Hoth et al. 1997). The mitochondrial network is generally less complex than the ER and more disperse than the Golgi of interphase cells.
- *Colocalization with a mitochondrial marker.* A more stringent test is colocalization of rhod-2 with a mitochondrial marker. Convenient markers for this purpose include rhodamine 123 and MitoTracker Green FM (both available from Molecular Probes), which have spectral properties that distinguish them from rhod-2 (Hoth et al. 1997). Likewise, mitochondrially targeted green fluorescent protein (GFP) (Rizzuto et al. 1995) should be useful in colabeling studies.
- *Sensitivity to mitochondrial inhibitors.* Depolarization of mitochondria or blockade of the Ca^{2+} uniporter prevents potential-driven uptake of Ca^{2+} by mitochondria, and hence should inhibit Ca^{2+}-dependent changes in the mitochondrial rhod-2 signal. Useful reagents for mitochondrial depolarization include membrane-permeant protonophores (e.g., CCCP [carbonyl cyanide *m*-chlorophenylhydrazone] and FCCP [carbonyl cyanide *p*-trifluoromethoxyphenylhydrazone]) and inhibitors of the electron transport chain (e.g., antimycin A1, 1799) in combination with the ATP synthase inhibitor, oligomycin (Hajnoczky et al. 1995; Jou et al. 1996; Simpson and Russell 1996; Babcock et al. 1997; Hoth et al. 1997). Ruthenium red applied internally through a patch pipette or at a high external concentration inhibits the uniporter (Jou et al. 1996; Babcock et al. 1997).

In most published attempts to measure $[Ca^{2+}]_m$ with rhod-2, a significant amount of the dye is cytoplasmic, setting certain limits on the quantitation of mitochondrial $[Ca^{2+}]$. In addition, several groups have reported staining of punctate structures within the nucleus that are most likely nucleoli (Rutter et al. 1996; Babcock et al. 1997; Hoth et al. 1997). In practice, these structures can be easily distinguished from mitochondrial staining, as they occur within the nucleus and tend to be dimmer and morphologically distinct from mitochondrial structures (see Fig. 1).

Practical Considerations

In general, the ability to load mitochondria with rhod-2 varies among cells, and conditions must be adapted for the particular cell type being studied. A number of approaches have been used in differ-

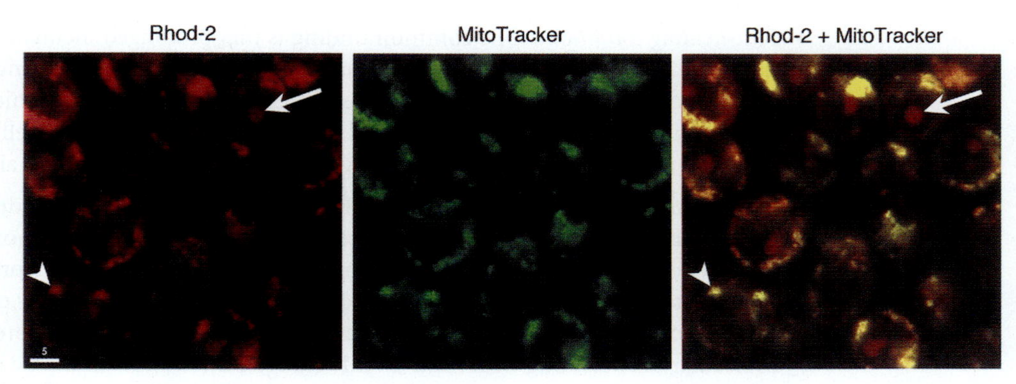

FIGURE 1. Rhod-2/AM labels mitochondria in Jurkat T cells. Confocal micrographs of Jurkat cells after incubation with rhod-2/AM (*left*) or MitoTracker Green FM, a mitochondrial marker (*center*). Rhod-2 labeling of mitochondria (*arrowhead*) is easily distinguished from labeling of the nucleolus (*arrow*). An overlay of the two images demonstrates colocalization of the two dyes in the mitochondria (*right*). Bar, 5 μm. (Reprinted, with permission, from Hoth et al. 1997 [©Rockefeller University Press].)

ent cells, as summarized in Table 1. For details, see the original papers; here, an overview of the general findings and pecularities that have been encountered is provided. A detailed description of the our experience of loading mitochondria in Jurkat cells, a human leukemic T-cell line, is given below.

- *Loading conditions.* Loading is most often done in serum-free media because some sera appear to contain significant esterase activity that cleaves the dye before it has a chance to cross the cell plasma membrane. Pluronic F-127 (Molecular Probes) is often used to increase the solubility of the esterified dye under these conditions. In some cells, loading specificity may be enhanced at low temperature (4°C), perhaps by reducing the chance that esterified cytoplasmic dye will be hydrolyzed before it can enter mitochondria (Trollinger et al. 1997).

TABLE 1. Survey of methods used to load mitochondria with rhod-2

Cell type	Indicator and concentration[a]	Time and temperature: loading	Time and temperature: post-loading[a]	Criteria for specificity[b]	Reference
Astrocyte	1 μM rhod-2/AM	5 min, 37°C	1 hr, 37°C	morphology photobleaching[c] digitonin Mn^{2+} quench CCCP, ruthenium red	Jou et al. (1996)
Oligodendrocyte	5 μM rhod-2/AM	30 min, room temp.	24 hr, 37°C	morphology FCCP, antimycin	Simpson and Russell (1996)
Chromaffin cell	1 μM rhod-2/AM	35–50 min, room temp.		morphology CCCP	Babcock et al. (1997)
Cardiac myocyte	10 μM rhod-2/AM + 10% FCS	60 min, 4°C	3–5 hr, 37°C, no FCS	morphology (compared to rhodamine 123, TMRM[d])	Trollinger et al. (1997)
CHO	2 μM rhod-2/AM	30 min, 37°C		morphology	Rutter et al. (1996)
Hepatocyte	1–2 μM dihydro-rhod-2/AM + 2 mM acetoacetate	60 min, 37°C	12–16 hr, 37°C	1799 + oligomycin	Hajnoczky et al. (1995)
Jurkat T cell	4–8 μM rhod-2/AM	30–40 min, room temp. or 37°C		MitoTracker rhodamine 123 CCCP	Hoth et al. (1997)

[a]Specific conditions may be indicated.
[b]Tests for the specificity of loading are described in the text (see Criteria for selective loading of mitochondria).
[c]Lack of recovery after photobleaching.
[d]Tetramethylrhodamine methyl ester (TMRM)

- *Post-loading conditions.* One common finding is that prolonged incubation of the cells after washing rhod-2/AM out of the medium enhances the specificity of mitochondrial staining. This is presumably due to the tendency of many cells to actively pump organic anions like fura-2 out across the plasma membrane (Di Virgilio et al. 1990). Alternatively, whole-cell dialysis through a patch pipette is an effective means of removing the cytosolic dye (Babcock et al. 1997).
- *Rhod-2/AM versus dihydro-rhod-2/AM.* It has been suggested that dihydro-rhod-2/AM, a reduced form of rhod-2/AM, may provide more specific signals from mitochondria because it exhibits Ca^{2+}-dependent fluorescence after it is oxidized, a process that occurs preferentially in mitochondria due to the high oxidizing potential of the inner matrix (Hajnoczky et al. 1995). A simple protocol for converting rhod-2/AM to dihydro-rhod-2/AM with sodium borohydride is distributed by Molecular Probes along with rhod-2/AM (Haugland 1996). It should be noted that dihydro-rhod-2/AM may not be advantageous for all cell types. In Jurkat leukemic T cells, no selective mitochondrial localization of the reduced dye was detected using a wide range of loading conditions (loading in culture medium or Ringer's solution; varying incubation time from 30 to 180 min, time after incubation from 0 to 16 hr, and temperature from 22°C to 37°C, ± Pluronic F-127). The main problem was that the dye fluorescence was weak and the background signal from cytosolic dye was too high to allow measurements of mitochondrial [Ca^{2+}].

PROCEDURE A

Mitochondrial Loading

The following protocol is used for loading mitochondria in Jurkat T cells. Actively dividing, log-phase cells grown in suspension are used for loading.

Loading

1. Prepare a stock solution of 2 mM rhod-2/AM<!> (Molecular Probes) in anhydrous DMSO <!> (Pierce). Store at –20°C. Dilute the stock solution in Ringer's solution to obtain a final concentration of 4–8 µM rhod-2/AM. Mix by vortexing at high speed for 15 sec and use immediately.
2. Incubate Jurkat T cells with 4–8 µM rhod-2/AM for 30–40 min at room temperature in the dark (Hoth et al. 1997). (Ringer's solution: 155 mM NaCl, 4.5 mM KCl<!>, 2 mM $CaCl_2$<!>, 1 mM $MgCl_2$<!>, 10 mM D-glucose, 5 mM HEPES; adjust pH to 7.4 with NaOH<!>.)

   Note: Ringer's solution is better suited for selective loading of rhod-2 into mitochondria than culture medium (RPMI 1640 supplemented with 10% fetal calf serum [FCS]). This difference may be due to esterase activity of serum. The loading temperature (room temperature or 37°C) is not critical.
3. Wash three times in Ringer's solution to remove excess dye.

Caution: *See Appendix 3 for appropriate handling of materials marked with <!>.*

PROCEDURE B

Assaying for Selective Loading of Mitochondria

1. Prepare a stock solution of 1 mM MitoTracker Green FM<!> in DMSO<!>. Store at –20°C. Dilute the stock solution in Ringer's solution (see above) to a final concentration of 100 nM and shake vigorously for 5–10 sec to disperse.
2. Double-label Jurkat T cells with 4–8 µM rhod-2/AM (for preparation, see above) and with 100 nM MitoTracker Green FM (Molecular Probes) for 30 min at 20–25°C.

Note: The cells are simultaneously labeled with the two solutions. To test whether mitochondria were selectively stained, the pattern of rhod-2 labeling was compared to that of the classic mitochondrial dye, rhodamine 123, in double-labeled cells using confocal microscopy. The two patterns appeared to overlap very well, but rapid bleaching of rhodamine 123 prevented a detailed comparison. Better results were obtained with MitoTracker Green FM, which was found to bleach much more slowly.

3. Wash three times in Ringer's solution to remove excess dye.

Imaging

Cells were imaged with a MultiProbe 2010 confocal laser-scanning microscope (Molecular Dynamics) employing a Nikon Diaphot 200 inverted microscope and 60X/1.4-NA Plan Apo objective. Cells were illuminated by the 488-nm and 568-nm emission lines of a krypton/argon laser<!> at 10% power, and fluorescence was collected simultaneously at 515–545 nm (bandpass filter, Chroma Technology) for MitoTracker Green FM and at wavelengths greater than 590 nm (longpass interference filter, Chroma Technology) for rhod-2.

Rhod-2 and MitoTracker Green FM were colocalized in most intracellular structures (Fig. 1). The only notable difference between the labeling patterns was due to rhod-2 labeling of some round intranuclear structures, most likely representing the nucleoli. Nucleolar and mitochondrial staining are easily distinguished by virtue of the nuclear location and lower fluorescence of the nucleolus (Fig. 1, left panel).

EXPERIMENTAL SETUP FOR TIME-LAPSE IMAGING OF $[Ca^{2+}]_m$

The imaging system described below was used to make time-lapse recordings of $[Ca^{2+}]_m$ in single Jurkat T cells (Hoth et al. 1997). In this particular experiment, $[Ca^{2+}]_m$ was monitored in response to Ca^{2+} entry through store-operated Ca^{2+} channels.

Instrumentation

Cells are imaged on a Zeiss Axiovert 35 microscope equipped with a 63X/1.25-NA Zeiss Neofluar objective. A 75-W xenon arc lamp provides illumination, and neutral-density filters attenuate the excitation-light intensity to reduce dye bleaching. An electronic shutter is used to block illumination between sampling periods, further minimizing dye bleaching and photodamage to the cell. Excitation (540 ± 12 nm) and emission (605 ± 28 nm) wavelengths are selected using bandpass filters (Chroma Technology) in the light path. The fluorescence emission is captured with an intensified CCD camera (Hamamatsu) and digitized and analyzed using a VideoProbe imaging system (ETM Systems).

PROCEDURE C

Preparation of Cells

1. Prepare a stock solution of 10 mg/ml phytohemagglutinin-P (Difco) in Ringer's solution. Store at –20°C. Dilute the stock solution in Ringer's solution (for preparation, see above) and use immediately.
2. Allow rhod-2-loaded cells (from above) to adhere to poly-L-ornithine-coated glass coverslip chambers on the stage of the microscope.

   Note: Chambers are constructed by gluing a No. 1 cover glass with Sylgard (Dow Corning) across a 10-mm hole cut in the bottom of a 35-mm plastic petri dish. Before plating cells, clean the glass with absolute ethanol, and place a drop of poly-L-ornithine (1 mg/ml) on the glass for 15 min, then wash it off with distilled water.

3. Incubate the cells with 10 μg/ml phytohemagglutinin (PHA) for 30 min at 22–25°C.

   Notes: Movement of mitochondria poses a serious problem for time-lapse imaging, particularly for relatively spherical cells such as T cells. To minimize movement artifacts, the attached cells are treated with PHA to flatten them. The flattening action of PHA may be specific for T cells, and other methods may therefore be more applicable to other cell types.

   In control experiments using fura-2-loaded cells, PHA did not appear to affect Ca^{2+} signals or responses to mitochondrial inhibitors, implying that it does not alter mitochondrial Ca^{2+} uptake.

4. Prepare a stock solution of 1 mM thapsigargin TG<!> (LC Laboratories) in anhydrous DMSO<!>. Store at –20°C. Dilute the stock solution in Ca^{2+}-free Ringer's solution to a final concentration of 1 μM and shake vigorously for 5–10 sec to disperse. Use immediately.

5. Focus the microscope on the mitochondria prior to collecting each data set.

6. Treat cells with 1 μM thapsigargin in Ca^{2+}-free Ringer's solution for 10 min.

7. Perfuse the chamber with Ringer's solution containing 2 mM Ca^{2+}.

   Notes: Cells were treated with TG to deplete intracellular Ca^{2+} stores and activate store-operated (CRAC) Ca^{2+} channels in the plasma membrane. Subsequent addition of Ca^{2+} and the ensuing influx of Ca^{2+} caused the mitochondrial rhod-2 fluorescence to increase by ~43% within 3 min (Fig. 2), whereas areas of diffuse rhod-2 fluorescence adjacent to the "hot spots" increased only ~10% under the same conditions. The increase in mitochondrial rhod-2 fluorescence was prevented by preincubation with 1 μM CCCP (Fig. 2), which rapidly depolarizes mitochondria and effectively inhibits their ability to take up Ca^{2+} (Thayer and Miller 1990; Babcock et al. 1997). These experiments demonstrate that mitochondria take up Ca^{2+} entering the T cells through Ca^{2+} channels in the plasma membrane.

   Cells are imaged within 2 hours of loading with rhod-2, usually allowing the collection of three or four data sets per preparation.

CALIBRATION PROCEDURE

To estimate $[Ca^{2+}]_m$ from rhod-2 fluorescence, it is necessary to perform an intramitochondrial calibration of the dye. Calibration is difficult, because rhod-2 is a single-wavelength dye, and loading is often not entirely specific. To our knowledge, this problem has so far only been approached by one group in a study of chromaffin cells (Babcock et al. 1997). Their protocol is described briefly below.

First, rhod-2 fluorescence was corrected for autofluorescence from nonloaded cells (<5% of total signal) and for the exponential decline (presumably due to photobleaching) before and after recovery from a stimulus. The corrected signals were converted to Ca^{2+} concentrations with the standard calibration equation

$$[Ca^{++}]_m = K_D \frac{F - F_{Ca\text{-}free}}{F_{Ca\text{-}sat} - F}$$

where F is the corrected fluorescence; K_D of rhod-2 for Ca^{2+} (500 nM) was measured in vitro and assumed to apply in the cell; and $F_{Ca\text{-}sat}$ and $F_{Ca\text{-}free}$ are the fluorescence intensities of Ca^{2+}-saturated or Ca^{2+}-free rhod-2, respectively. The dynamic range of the dye, $F_{Ca\text{-}sat}/F_{Ca\text{-}free}$, was originally reported to be quite low (~3; Minta et al. 1989), but purer rhod-2 preparations, now available from Molecular Probes, yield a ratio of approximately 15 (Babcock et al. 1997). $F_{Ca\text{-}free}$ and $F_{Ca\text{-}sat}$ were estimated from $F_{Mn\text{-}sat}$ measured in vivo by treatment of the cell with ionomycin and 2 mM Mn^{2+}, and from the $F_{Ca\text{-}free}/F_{Ca\text{-}sat}$ and $F_{Mn\text{-}sat}/F_{Ca\text{-}sat}$ ratios measured in vitro. This indirect procedure was chosen because the $F_{Mn\text{-}sat}/F_{Ca\text{-}sat}$ ratio measured in vivo was significantly greater than the ratio measured in vitro, possibly because ionomycin is better able to saturate mitochondrial rhod-2 with Mn^{2+} than with Ca^{2+}. The results give resting $[Ca^{2+}]_m$ in the range of 100–200 nM, with peak values in the micromolar range following Ca^{2+} influx across the plasma membrane. These values are consistent with those measured by other groups in isolated mitochondria or mitochondrial clusters in cells using ratiometric Ca^{2+} dyes (Miyata et al. 1991; Schreur et al. 1996; Griffiths et al. 1997).

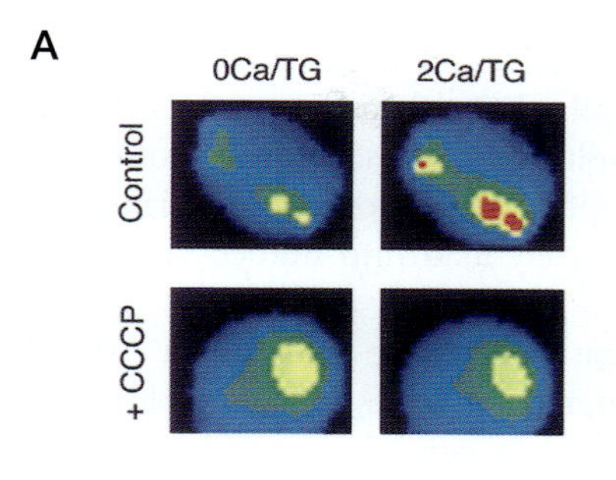

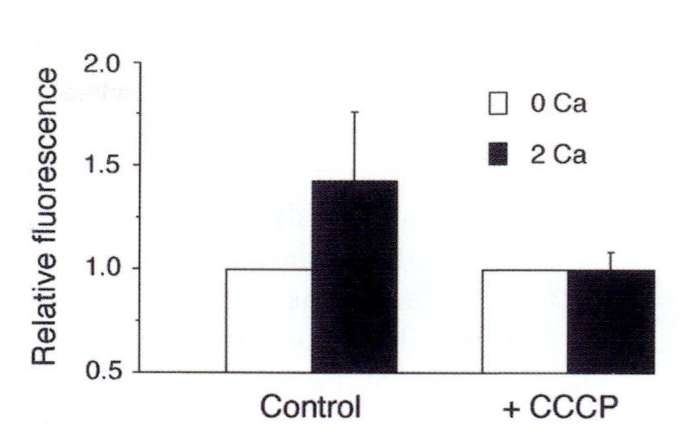

FIGURE 2. Mitochondrial Ca^{2+} imaging with rhod-2. (*A*) In the pseudocolored video-microscopic images of a single TG-treated cell, rhod-2 fluorescence in the mitochondria increases within 3 min after readdition of 2 mM Ca^{2+} (*top*). 1 μM CCCP prevents the fluorescence increase in another cell (*bottom*). Color scale: From blue to red indicates from lower to higher rhod-2 fluorescence. (*B*) Quantification of rhod-2 fluorescence in "hot spots" in cells of the experiment shown in *A*. Hot-spot fluorescence was measured before and 3 min after readdition of 2 mM Ca^{2+}, and the fluorescence intensity before readdition was normalized to 1. Because of the presence of rhod-2 in the cytoplasm, mitochondrial rhod-2 fluorescence was not calibrated in terms of $[Ca^{2+}]_m$, but instead was expressed relative to the initial fluorescence intensity. Thirty-five cells (Control) and 28 cells (+CCCP) were analyzed. Error bars reflect standard deviations. (Reprinted, with permission, from Hoth et al. 1997 [©Rockefeller University Press].)

OTHER APPROACHES TO MEASURE INTRAMITOCHONDRIAL CALCIUM

AM esters of other fluorescent Ca^{2+} dyes such as indo-1 or fluo-3 have been used successfully to measure $[Ca^{2+}]_m$ in cardiac myocytes (Miyata et al. 1991; Chacon et al. 1996; Schreur et al. 1996; Griffiths et al. 1997). In contrast to rhod-2, however, these dyes most often do not accumulate preferentially in mitochondria. Thus, the cytosolic dye must be either quenched with Mn^{2+} (Miyata et al. 1991; Schreur et al. 1996) or removed by prolonged incubation at 37°C to allow export from the cell (Griffiths et al. 1997), or additional markers must be used to identify mitochondria (Chacon et al. 1996). The advantage of indo-1 is that ratiometric measurements can be carried out which are insensitive to variations in excitation-light intensity, focus, and dye concentration. The applicability of indo-1 to cells other than cardiac myocytes will depend on the degree of mitochondrial loading and whether cytosolic dye can be removed or quenched effectively enough to give a mitochondrion-specific signal.

Targeted protein-based indicators of $[Ca^{2+}]$ provide alternative, highly specific mitochondrial Ca^{2+} probes. These include recombinant forms of the luminescent photoprotein aequorin (Rizzuto et al. 1992, 1994; see also Chapter 79) and "cameleons," Ca^{2+} indicators based on variants of GFP linked by calmodulin (CaM) and a CaM-binding peptide (Miyawaki et al. 1997; see also Chapter 29). An important advantage of these probes is that they can be directed by targeting sequences to highly specific locations within the cell, such as the ER, mitochondria, Golgi, or nucleus. It is difficult to measure aequorin emissions at the single-cell or subcellular level, however, because each aequorin molecule emits only a single photon upon Ca^{2+} binding and must be restored by a slow process involving exchange of a fresh lumiphore (coelenterazine) for the exhausted one. For this reason, time integration with specialized instrumentation (e.g., a two-stage, photon-counting, intensified CCD camera) is necessary (Rutter et al. 1996). Fluorescent probes in general offer a much higher signal, and thus cameleons may be easier to apply to measurements of $[Ca^{2+}]_m$ at the single-cell or subcellular level (Miyawaki et al. 1997).

FUTURE PERSPECTIVES

Because its spectral properties do not change upon binding Ca^{2+}, the rhod-2 signal is easily contaminated by bleaching, movement out of the focal plane, and changes in excitation-light intensity, dye concentration, or organelle shape. In addition, the loading is often not entirely specific, placing further limitations on quantitation of $[Ca^{2+}]_m$. Despite these disadvantages, rhod-2 has been rather useful in gaining more insight in mitochondrial Ca^{2+} handling and its dependence on cytosolic $[Ca^{2+}]$. When imaged using confocal or deconvolution microscopy, much of the cytosolic signal can be minimized and $[Ca^{2+}]_m$ more precisely quantified. Ideally, one would wish for a ratiometric indicator with a high dynamic range that would selectively load into mitochondria. Cameleons (Miyawaki et al. 1997) show great promise in this regard.

One of the most challenging current goals in the study of Ca^{2+} signaling is to understand how local domains of elevated Ca^{2+} are generated and what their physiological roles are. In this context, it will be important to study the cross-talk between $[Ca^{2+}]_m$ and Ca^{2+} microdomains in the ER and cytosol. Quantitative simultaneous measurements of $[Ca^{2+}]_m$, $[Ca^{2+}]_{ER}$, and $[Ca^{2+}]_i$ provide an approach to tackle these problems.

ACKNOWLEDGMENTS

We thank Dr. Barbara Niemeyer for criticial reading of the manuscript. Work in the Lewis laboratory has been supported by postdoctoral fellowships from the Boehringer Ingelheim Fonds and the Human Frontier Science Program (M.H.) and by National Institutes of Health grant GM-45374 (R.S.L.).

REFERENCES

Ali H., Maeyama K., Sagi-Eisenberg R., and Beaven M.A. 1994. Antigen and thapsigargin promote influx of Ca^{2+} in rat basophilic RBL-2H3 cells by ostensibly similar mechanisms that allow filling of inositol 1,4,5-trisphosphate-sensitive and mitochondrial Ca^{2+} stores. *Biochem. J.* **304:** 431–440.

Babcock D.F., Herrington J., Goodwin P.C., Park Y.B., and Hille B. 1997. Mitochondrial participation in the intracellular Ca^{2+} network. *J. Cell Biol.* **136:** 833–844.

Chacon E., Ohata H., Harper I.S., Trollinger D.R., Herman B., and Lemasters J.J. 1996. Mitochondrial free calcium transients during excitation-contraction coupling in rabbit cardiac myocytes. *FEBS Lett.* **382:** 31–36.

Di Virgilio F., Steinberg T.H., and Silverstein S.C. 1990. Inhibition of fura-2 sequestration and secretion with organic anion transport blockers. *Cell Calcium* **11:** 57–62.

Drummond R.M. and Fay F.S. 1996. Mitochondria contribute to Ca^{2+} removal in smooth muscle cells. *Pflueg. Arch. Eur. J. Physiol.* **431:** 473–482.

Friel D.D. and Tsien R.W. 1994. An FCCP-sensitive Ca^{2+} store in bullfrog sympathetic neurons and its participation in stimulus-evoked changes in $[Ca^{2+}]_i$. *J. Neurosci.* **14:** 4007–4024.

Griffiths E.J., Stern M.D., and Silverman H.S. 1997. Measurement of mitochondrial calcium in single living cardiomyocytes by selective removal of cytosolic indo 1. *Am. J. Physiol.* **273:** C37–C44.

Grynkiewicz G., Poenie M., and Tsien R.Y. 1985. A new generation of Ca^{2+} indicators with greatly improved fluorescence properties. *J. Biol. Chem.* **260:** 3440–3450.

Hajnoczky G., Robb-Gaspers L.D., Seitz M.B., and Thomas A.P. 1995. Decoding of cytosolic calcium oscillations in the mitochondria. *Cell* **82:** 415–424.

Hansford R.G. 1994. Physiological role of mitochondrial Ca^{2+} transport. *J. Bioenerg. Biomembr.* **26:** 495–508.

Haugland R.P. 1996. *Handbook of fluorescent probes and research chemicals,* 6th edition. Molecular Probes, Eugene, Oregon.

Herrington J., Park Y.B., Babcock D.F., and Hille B. 1996. Dominant role of mitochondria in clearance of large Ca^{2+} loads from rat adrenal chromaffin cells. *Neuron* **16:** 219–228.

Hoth M., Fanger C.M., and Lewis R.S. 1997. Mitochondrial regulation of store-operated calcium signaling in T lymphocytes. *J. Cell Biol.* **137:** 633–648.

Ichas F., Jouaville L.S., and Mazat J.P. 1997. Mitochondria are excitable organelles capable of generating and conveying electrical and calcium signals. *Cell* **89:** 1145–1153.

Jou M.J., Peng T.I., and Sheu S.S. 1996. Histamine induces oscillations of mitochondrial free Ca^{2+} concentration in single cultured rat brain astrocytes. *J. Physiol.* **497:** 299–308.

Jouaville L.S., Ichas F., Holmuhamedov E.L., Camacho P., and Lechleiter J.D. 1995. Synchronization of calcium waves by mitochondrial substrates in *Xenopus laevis* oocytes. *Nature* **377:** 438–441.

Malgaroli A., Milani D., Meldolesi J., and Pozzan T. 1987. Fura-2 measurement of cytosolic free Ca^{2+} in monolayers and suspensions of various types of animal cells. *J. Cell Biol.* **105:** 2145–2155.

Minta A., Kao J.P., and Tsien R.Y. 1989. Fluorescent indicators for cytosolic calcium based on rhodamine and fluorescein chromophores. *J. Biol. Chem.* **264:** 8171–8178.

Miyata H., Silverman H.S., Sollott S.J., Lakatta E.G., Stern M.D., and Hansford R.G. 1991. Measurement of mitochondrial free Ca^{2+} concentration in living single rat cardiac myocytes. *Am. J. Physiol.* **261:** H1123–1134.

Miyawaki A., Llopis J., Heim R., McCaffery J.M., Adams J.A., Ikura M., and Tsien R.Y. 1997. Fluorescent indicators for Ca^{2+} based on green fluorescent proteins and calmodulin. *Nature* **388:** 882–887.

Rizzuto R., Simpson A.W., Brini M., and Pozzan T. 1992. Rapid changes of mitochondrial Ca^{2+} revealed by specifically targeted recombinant aequorin. *Nature* **358:** 325–327.

Rizzuto R., Bastianutto C., Brini M., Murgia M., and Pozzan T. 1994. Mitochondrial Ca^{2+} homeostasis in intact cells. *J. Cell Biol.* **126:** 1183–1194.

Rizzuto R., Brini M., Pizzo P., Murgia M., and Pozzan T. 1995. Chimeric green fluorescent protein as a tool for visualizing subcellular organelles in living cells. *Curr. Biol.* **5:** 635–642.

Rutter G.A., Burnett P., Rizzuto R., Brini M., Murgia M., Pozzan T., Tavare J.M., and Denton R.M. 1996. Subcellular imaging of intramitochondrial Ca^{2+} with recombinant targeted aequorin: Significance for the regulation of pyruvate dehydrogenase activity. *Proc. Natl. Acad. Sci.* **93:** 5489–5494.

Schreur J.H., Figueredo V.M., Miyamae M., Shames D.M., Baker A.J., and Camacho S.A. 1996. Cytosolic and mitochondrial $[Ca^{2+}]$ in whole hearts using indo-1 acetoxymethyl ester: Effects of high extracellular Ca^{2+}. *Biophys. J.* **70:** 2571–2580.

Simpson P.B. and Russell J.T. 1996. Mitochondria support inositol 1,4,5-trisphosphate-mediated Ca^{2+} waves in cultured oligodendrocytes. *J. Biol. Chem.* **271:** 33493–33501.

Steinberg S.F., Bilezikian J.P., and Al-Awqati Q. 1987. Fura-2 fluorescence is localized to mitochondria in endothelial cells. *Am. J. Physiol.* **253:** C744–C747.

Tang Y. and Zucker R.S. 1997. Mitochondrial involvement in post-tetanic potentiation of synaptic transmission. *Neuron* **18:** 483–491.

Thayer S.A., and Miller R.J. 1990. Regulation of the intracellular free calcium concentration in single rat dorsal root ganglion neurones in vitro. *J. Physiol.* **425:** 85–115.

Trollinger D.R., Cascio W.E., and Lemasters J.J. 1997. Selective loading of Rhod 2 into mitochondria shows mitochondrial Ca^{2+} transients during the contractile cycle in adult rabbit cardiac myocytes. *Biochem. Biophys. Res. Commun.* **236:** 738–742.

Werth J.L. and Thayer S.A. 1994. Mitochondria buffer physiological calcium loads in cultured rat dorsal root ganglion neurons. *J. Neurosci.* **14:** 348–356.

CHAPTER 42

Generation of Controlled Calcium Oscillations in Nonexcitable Cells

Ricardo E. Dolmetsch and Richard S. Lewis

Department of Molecular and Cellular Physiology, Stanford University School of Medicine, Stanford, California 94305

Oscillations of cytosolic free calcium ($[Ca^{2+}]_i$) occur in a wide variety of cells following stimulation of surface receptors, particularly those linked to the production of inositol 1,4,5-trisphosphate (IP3) (Berridge and Galione 1988; Fewtrell 1993). Oscillations have been proposed to offer important advantages for cell signaling, including protection against the toxic effects of tonic $[Ca^{2+}]_i$ elevation, avoidance of desensitization, and improvement of signaling fidelity (Berridge and Galione 1988; Tsien and Tsien 1990). However, several serious obstacles have thwarted attempts to test for these and other functions. $[Ca^{2+}]_i$ oscillations triggered by receptor stimulation are often heterogeneous in amplitude and frequency, both within populations of cells and in single cells over time. In addition, the surface receptors involved often activate additional signaling molecules, such as heterotrimeric G proteins or protein kinases. These factors make it difficult or impossible to determine the contribution of particular patterns of Ca^{2+} signaling to specific cell responses.

To circumvent these problems, we developed a method for generating synchronous $[Ca^{2+}]_i$ oscillations in a population of cells independently of cell-surface receptors. With this "calcium-clamp" method, defined $[Ca^{2+}]_i$ oscillations are generated by controlling the timing and magnitude of Ca^{2+} influx through constitutively activated Ca^{2+} channels in the plasma membrane. We have used this method to explore the role of $[Ca^{2+}]_i$ oscillations in controlling gene expression in T lymphocytes (Dolmetsch et al. 1998). This approach may be generally useful for studying the cellular consequences of oscillations and the mechanisms of Ca^{2+} regulation in excitable and nonexcitable cells. This chapter describes the system and discusses some of the important factors involved in its operation.

THEORY OF OPERATION OF THE CALCIUM CLAMP

The Ca^{2+} clamp is based on the idea that if the plasma membrane of a cell is rendered sufficiently permeable to Ca^{2+}, then changes in $[Ca^{2+}]_i$ can be driven by changes in the concentration of extracellular Ca^{2+} ($[Ca^{2+}]_o$). Under such conditions, raising $[Ca^{2+}]_o$ causes cytoplasmic $[Ca^{2+}]$ to rise because the rate of Ca^{2+} influx through channels transiently exceeds the rate of clearance by pumps and other transporters in the plasma membrane and organelles. Subsequent removal of $[Ca^{2+}]_o$ causes $[Ca^{2+}]_i$ to decline, because the rate of Ca^{2+} clearance now exceeds the rate of influx. Thus, by simply changing $[Ca^{2+}]_o$ in a repetitive fashion, it is possible to alter the balance of fluxes to favor either Ca^{2+} influx or efflux and thereby generate controlled oscillations. After a suitable period of stimulation, cells can be analyzed for a particular Ca^{2+}-dependent response.

CONSTRUCTION OF A CALCIUM CLAMP

The three essential parts of the Ca^{2+} clamp include a method to increase the Ca^{2+} permeability of the plasma membrane, a means of rapidly changing $[Ca^{2+}]_o$, and a system for continuously measuring $[Ca^{2+}]_i$. Below is a description of the Ca^{2+} clamp, followed by a description of how it can be used to generate $[Ca^{2+}]_i$ oscillations with defined frequency and amplitude.

Increasing the Ca^{2+} Permeability of the Plasma Membrane

The predominant Ca^{2+} influx pathway in many nonexcitable cells consists of store-operated Ca^{2+} channels in the plasma membrane that open in response to the depletion of the IP3-sensitive Ca^{2+} store (Putney and Bird 1993; Parekh and Penner 1997; Lewis 1999). Thapsigargin (TG), an irreversible inhibitor of endoplasmic reticulum (ER)-Ca^{2+} ATPases, can effectively deplete the store by preventing the re-uptake of Ca^{2+}, and thereby activate store-operated Ca^{2+} channels. The irreversible action of TG offers the advantage of economy over other methods of elevating $[Ca^{2+}]_i$ such as calcium ionophores. Because ionophores act reversibly to increase Ca^{2+} permeability, they must be applied continuously; the cost of continuous perfusion can become significant with experiments lasting several hours, such as those involving gene expression as an endpoint. A possible disadvantage that must be considered is that store depletion may alter the signaling pathways being studied. This does not appear to be a major problem with regard to Ca^{2+}-dependent gene expression in lymphocytes, as strong expression occurs in cells treated with maximal concentrations of TG for several hours (Negulescu et al. 1994; Dolmetsch et al. 1998).

Rapid Solution Exchange

To generate controlled oscillations, the Ca^{2+} clamp must be able to change the concentration of Ca^{2+} surrounding a large population of cells rapidly and uniformly. Because of cellular feedback processes that regulate $[Ca^{2+}]_i$ on a timescale of seconds, the size of the signal produced in each cell is highly influenced by the rate, as well the magnitude, of changes in $[Ca^{2+}]_o$. We have observed that if the perfusion rate is too slow or the solution exchange is inefficient, the peak $[Ca^{2+}]_i$ reached during each oscillation is signficantly higher in cells near the perfusion inlet than in cells at the opposite end of the chamber. Obviously, this problem greatly complicates the interpretation of the experiment, and therefore it is essential to assay the $[Ca^{2+}]_i$ responses of cells located at the extremes of the chamber to confirm that they are homogeneous.

Below, we describe a perfusion system that rapidly and uniformly exchanges the solution surrounding 3×10^5 to 4×10^5 cells (Fig. 1A). It consists of a solenoid-controlled valve that rapidly selects one of two perfusion solutions containing different calcium concentrations and a laminar flow chamber that permits the gentle and uniform exchange of solutions surrounding the cells.

Perfusion System

Two 500-ml culture bottles containing high- or low-Ca^{2+} medium (for preparation, see below) are used. The culture bottles are immersed in a 42°C water bath and pressurized to 2 psi with 95% air and 5% CO_2 to maintain proper pH and to control the perfusion rate. The temperature of the water bath and additional forced-air heating of the underside of the chamber are adjusted to keep the cells at 37°C, given the perfusion flow rate.

The two culture bottles are connected with silicone tubing to a solenoid-activated Y valve (valve 3-132-900 with driver 90-1-100; General Valve) mounted on the microscope stage within 4 cm of the

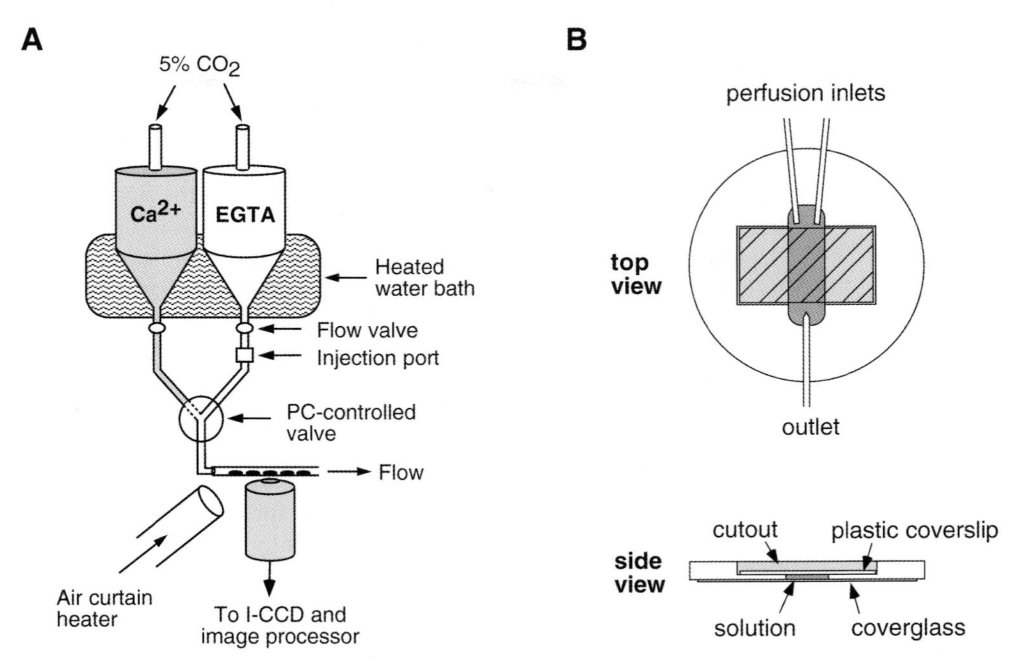

FIGURE 1. Schematic diagram of the calcium clamp. (*A*) Cells preloaded with fura-PE3/AM are attached to a laminar-flow coverslip chamber. Thapsigargin is applied through the injection port to deplete intracellular Ca^{2+} stores and activate CRAC channels. Thereafter, a solenoid-operated valve under computer control switches the perfusion solution between Ca^{2+}-containing and Ca^{2+}-free media. (Reprinted, with permission, from Dolmetsch et al. 1998 [©Macmillan].) (*B*) Top and side views of the laminar-flow chamber, drawn to scale. Dimensions are given in the text.

chamber. The outlet port of the valve is connected to the chamber as described below. Signals (5v) from a digital I/O board (PIO-12, Keithley Instruments) controlled by the image-processing program (VideoProbe; ETM Systems) trigger the valve, determining which of the two solutions flows through the chamber.

Preparation of High- or Low-Ca^{2+} Medium

Use deficient RPMI 1640 (Irvine Scientific), 25 mM HEPES, 0.2% BSA. For high-Ca^{2+} medium, supplement with 1 mM Ca^{2+} (total [Ca^{2+}] = 1.5 mM). For low-Ca^{2+} medium, supplement with 0.5 mM EGTA (calculated free [Ca^{2+}] = 50 μM). Deficient RPMI 1640 without phenol red, biotin, or riboflavin is used, as these contribute to autofluorescence.

Laminar-flow Chamber

The chamber consists of a #1 coverglass permanently mounted across a rectangular hole (6 x 20 mm) in a Plexiglas disk (Fig. 1B). The glass is pretreated with poly-L-lysine, and preloaded cells in suspension (see below) are allowed to settle onto and adhere to it. A plastic coverslip (13 x 23 mm) is then mounted across the fluid-filled well with silicone grease, creating a flow chamber with dimensions of 13 x 6 x 0.8 mm (total volume of 60 μl). Two stainless steel 18-gauge tubes placed at one end of the chamber direct solution inflow directly into the space between the two coverslips. At the opposite end of the chamber, a sharp 18-gauge needle connected to a suction line collects the solution flowing out of the chamber. We have confirmed that flow through the chamber is essentially laminar and uniform by measuring the exchange rate of a solution containing fluorescein. At a flow rate of 3.5 ml/min, the rate of exchange is rapid and only varies slightly throughout the chamber; in 15 locations spanning the length and width of the chamber, 99% exchange occurred within 3 seconds with a standard deviation of 5%.

TIME-LAPSE IMAGING OF CYTOSOLIC CALCIUM

Practical Considerations

To ensure the proper operation of the calcium clamp, it is critical to monitor $[Ca^{2+}]_i$ throughout the duration of the experiment. Unexpected variations in the size of $[Ca^{2+}]_i$ responses among different batches of cells can be misleading if $[Ca^{2+}]_i$ is not measured in the same cells being assayed for the downstream response. A common approach is to measure $[Ca^{2+}]_i$ using standard video-microscopic techniques in cells loaded with a ratiometric calcium-indicator dye like fura-2. One problem we have encountered with fura-2 is that most of the dye is lost from T cells after a 1-hour incubation at 37°C, precluding $[Ca^{2+}]_i$ measurements during the minimum time needed to stimulate detectable levels of gene expression (3 hours). This problem can be overcome by using fura-PE3 (TefLabs), a derivative of fura-2 that is more resistant to export or leakage from cells. fura-PE3/AM (AM, acetoxymethyl ester) does not load Jurkat T cells as easily as fura-2/AM, making it necessary to increase the concentration and incubation time and to reduce the level of fetal calf serum (FCS) in the loading medium. Optimal loading was achieved with 2 μM fura-PE3/AM for 1 hr at 37°C in medium containing 2% FCS. Cells also need a longer time after loading (1 hr) to fully de-esterify the dye. We can routinely record $[Ca^{2+}]_i$ in fura-PE3-loaded T cells for up to 3 hr at 37°C.

Instrumentation

Our imaging system is based on an inverted microscope (Zeiss Axiovert 35) equipped for epifluorescence and connected to an intensified CCD camera (C2400-87; Hamamatsu) and a Videoprobe image processor (ETM Systems). A computer-controlled filter wheel (Lambda-10; Sutter Instruments) switches between the 350- and 380-nm excitation wavelengths to collect a pair of images for calculating $[Ca^{2+}]_i$ every 5–10 sec. The system and details of calibration are described in other publications (Dolmetsch and Lewis 1994).

CONTROLLING OSCILLATION AMPLITUDE AND FREQUENCY WITH THE CALCIUM CLAMP

After being loaded with fura-PE3 and adhering to the chamber, Jurkat T cells were treated with 1 μM TG for 5 minutes in the absence of extracellular Ca^{2+} to irreversibly deplete IP3-sensitive stores and maximally activate Ca^{2+} release-activated calcium (CRAC) channels. Following removal of TG, Ca^{2+}-containing medium was applied for 10, 15, or 30 sec every 100 sec, alternating with Ca^{2+}-free medium. As shown in Figure 2A, this protocol elicits average oscillation amplitudes of 358, 857, or 1217 nM (means of 200–300 cells). Similar results were obtained by varying the $[Ca^{2+}]_o$ instead of the length of application, in which case the shape of the oscillations varied with $[Ca^{2+}]_o$. Oscillations can be generated in T cells at different frequencies (Fig. 2B) up to a maximum of ~1/60 sec^{-1}; beyond this frequency the rate of clearance is not high enough to bring $[Ca^{2+}]_i$ back to baseline between spikes. Within the population of cells, the amplitudes of $[Ca^{2+}]_i$ oscillations are relatively uniform. Figure 2C shows the distribution of average $[Ca^{2+}]_i$ values in a population of oscillating cells and in cells exposed to constant $[Ca^{2+}]_o$. The two distributions overlap closely, suggesting that cell-to-cell variation in $[Ca^{2+}]_i$ arises from intrinsic differences in each cell's Ca^{2+} signaling machinery (e.g., the number of CRAC channels or Ca^{2+} pumps), rather than from a solution-switching artifact of the calcium clamp.

SUMMARY AND FUTURE PERSPECTIVES

The calcium clamp described above can be used to generate relatively uniform $[Ca^{2+}]_i$ oscillations in a large population of cells. Because it relies on rapid changes in Ca^{2+} influx through endogenous channels, this method may also recreate the naturally occurring $[Ca^{2+}]_i$ gradients within the cell. This is likely to be true for T cells, in which naturally occurring $[Ca^{2+}]_i$ oscillations appear to result from peri-

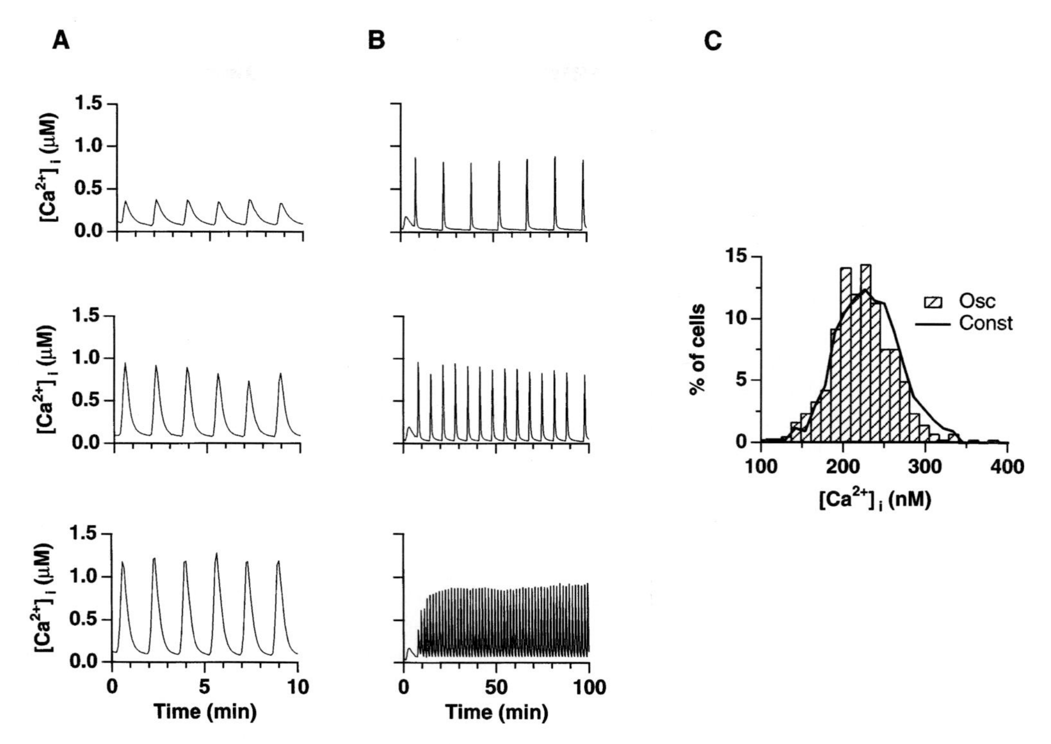

FIGURE 2. Creating defined patterns of $[Ca^{2+}]_i$ elevation. (*A*) Control of oscillation amplitude. Average $[Ca^{2+}]_i$ is plotted for 200–300 cells exposed to 1.5 mM Ca^{2+} for 10 sec (*top*), 15 sec (*middle*), or 30 sec (*bottom*) every 100 sec. (*B*) Control of frequency. 1.5 mM Ca^{2+} was applied for 30 sec every 900 sec (*top*), 400 sec (*middle*), or 100 sec (*bottom*). (*C*) Histogram of the average $[Ca^{2+}]_i$ (average = 225 nM) in single cells stimulated with oscillating (*bars*) or constant (*solid line*) extracellular $[Ca^{2+}]$. (*A,B,* Reprinted, with permission, from Dolmetsch et al. 1998 [©Macmillan].)

odic activation of CRAC channels; however, it may not be true for other nonexcitable cells in which oscillations arise directly from repetitive release from Ca^{2+} stores. The main strength of the approach lies in its ability to vary the amplitude and frequency of oscillations and to maintain these parameters constant over hours. These features allowed us to examine how $[Ca^{2+}]_i$ oscillations increase the ability of small Ca^{2+} signals to activate gene expression, and how the frequency of the oscillations contributes to the selectivity of gene activation in T cells (Dolmetsch et al. 1998). The ability to create controlled patterns of $[Ca^{2+}]_i$ elevation should be useful in studying Ca^{2+}-signaling mechanisms in a large number of different cell types.

Variations of the technique described in this chapter may be better suited to particular cells. Although we adjusted $[Ca^{2+}]_o$ to control the amount of Ca^{2+} entry, membrane depolarization could be varied instead (Negulescu et al. 1994). In nonexcitable cells, periodic exposure to high-K^+ medium could elicit $[Ca^{2+}]_i$ oscillations by transiently reducing the driving force for Ca^{2+} entry through tonically activated channels. In excitable cells, repetitive application of high K^+ could be used to generate oscillations by periodically activating voltage-gated Ca^{2+} channels. In cells that do not express suitable endogenous Ca^{2+} channels, it may be possible to transfect with a heterologous Ca^{2+} channel that is directly activated by depolarization or by extracellular ligands such as glutamate or acetylcholine.

It is worthwhile to note that the system described here is not a true clamp, in the sense that the Ca^{2+} flux is not controlled by feedback to achieve the desired value of $[Ca^{2+}]_i$. We found that feedback control was not necessary to generate periodic $[Ca^{2+}]_i$ oscillations of relatively uniform amplitude in T cells. However, the Ca^{2+} permeability of the plasma membrane and the activity of endogenous clearance mechanisms do place certain limitations on the shape, amplitude, and frequency of oscillations that can be produced in individual cells. These limitations may be overcome to some extent by allowing the imaging software to control $[Ca^{2+}]_o$ in a graded fashion in response to measured changes in $[Ca^{2+}]_i$. Alternatively, patch-clamp methods could be adapted to control $[Ca^{2+}]_i$ in a single cell using feedback to control the membrane potential. Such modifications will increase the versatility and precision of the system and its applicability to studies of Ca^{2+} regulation.

REFERENCES

Berridge M.J. and Galione A. 1988. Cytosolic calcium oscillators. *FASEB J.* **2:** 3074–3082.

Dolmetsch R.E. and Lewis R.S. 1994. Signaling between intracellular Ca^{2+} stores and depletion-activated Ca^{2+} channels generates $[Ca^{2+}]_i$ oscillations in T lymphocytes. *J. Gen. Physiol.* **103:** 365–388.

Dolmetsch R.E., Xu K., and Lewis R.S. 1998. Calcium oscillations increase the efficiency and specificity of gene expression. *Nature* **392:** 933–936.

Fewtrell C. 1993. Ca^{2+} oscillations in non-excitable cells. *Annu. Rev. Physiol.* **55:** 427–454.

Lewis R.S. 1999. Store-operated calcium channels. *Adv. Second Messenger Phosphoprotein Res.* **33:** 279–307.

Negulescu P.A., Shastri N., and Cahalan M.D. 1994. Intracellular calcium dependence of gene expression in single T lymphocytes. *Proc. Natl. Acad. Sci.* **91:** 2873–2877.

Parekh A.B. and Penner R. 1997. Store depletion and calcium influx. *Physiol. Rev.* **77:** 901–930.

Putney J.W., Jr. and Bird G.S. 1993. The inositol phosphate-calcium signaling system in nonexcitable cells. *Endocr. Rev.* **14:** 610–631.

Tsien R.W. and Tsien R.Y. 1990. Calcium channels, stores, and oscillations. *Annu. Rev. Cell Biol.* **6:** 715–760.

CHAPTER 43

A Practical Guide: High-Speed Imaging of Calcium Waves in Neurons in Brain Slices

William N. Ross,[1] Takeshi Nakamura,[2] Shigeo Watanabe,[1] Matthew E. Larkum,[3] and Nechama Lasser-Ross[1]

[1]*Department of Physiology, New York Medical College, Valhalla, New York 10595;* [2]*Laboratory of Cellular Neurobiology, Tokyo University of Pharmacy and Life Science, Hachioji-shi, Tokyo 192-0392, Japan;* [3]*Abteilung Zellphysiologie, Max-Planck-Institüt für Medizinische Forschung, D-69120 Heidelberg, Germany*

This chapter describes a method for measuring and displaying spatially extended Ca^{2+} release waves in dendrites, with sufficient temporal resolution to determine the sites of wave initiation and the characteristics of wave propagation.

This technique has been used to study Ca^{2+} waves in hippocampal CA1 pyramidal neurons (see, e.g., Nakamura et al. 1999, 2002) and neocortical pyramidal neurons (Larkum et al. 2003). Similar technology has been used to study Ca^{2+} waves in hippocampal CA3 pyramidal neurons (Kapur et al. 2001). The method presented here can be used in many kinds of experiments. It can be used (1) where it is necessary to examine $[Ca^{2+}]_i$ changes over spatially extended regions, (2) where high temporal resolution is needed to follow the time course of $[Ca^{2+}]_i$ changes accurately in different locations, (3) where the *z*-sectioning accuracy of confocal recording is not needed (or when it is a hindrance), and (4) where simultaneous electrical recording is useful to correlate $[Ca^{2+}]_i$ changes with specific electrical events.

A membrane-impermeant Ca^{2+} indicator is used to stain the cytosol of the neurons of interest. Many different indicators can be used. There is usually a tradeoff between an indicator with high sensitivity (high affinity), which may buffer and distort transient changes in $[Ca^{2+}]_i$, and an indicator with low sensitivity (low affinity), which follows changes in $[Ca^{2+}]_i$ more accurately. Ca^{2+} waves are particularly sensitive to the buffering effects of indicators because they involve the regenerative release of Ca^{2+} from internal stores. This protocol uses 100–250 μM bis-fura-2 (high affinity) and 500 μM furaptra (low affinity), both available from Molecular Probes.

Imaging data are collected using a cooled CCD camera operated in the frame transfer mode. The current generation of cameras (e.g., Roper Scientific Quantix 57 or RedShirt Imaging, NeuroCCD-SM256) has high quantum efficiency for detecting the fluorescence emission of most current indicators and has adequate spatial and temporal resolution for these kinds of measurements. For example, images are typically recorded at 50 Hz using 3 x 3 pixel binning (170 x 170 effective pixels) using the Quantix camera. This binning still gives a resolution of 0.7 μm using a 60X water-immersion lens on an Olympus BX50WI upright microscope. Custom-designed software is used to synchronize the optical data with electrical measurements recorded with an analog-to-digital (A/D) converter. The same program controls the pixel binning and frame interval (for details, see Lasser-Ross et al. 1991). A second program generates the displays.

MATERIALS

Reagents

High-affinity Ca^{+} indicator, e.g., bis-fura-2, 100–250 μM

Low-affinity Ca^{2+} indicator, e.g., furaptra, 500 μM

Equipment

Cooled CCD camera, capable of operating in frame transfer mode (e.g., Quantix 57 from Roper Scientific, or NeuroCCD-SM256 from RedShirt Imaging) linked to a PC workstation, running software appropriate for synchronizing optical data with electrical measurements, and displaying data

Equipment for dye loading with patch pipettes

Microscope, equipped with 40x and/or 60x water-immersion lens, e.g., Olympus BX50WI

Stimulating electrode

Note: A sharpened tungsten electrode (e.g., model TM33B01KT, WPI, Sarasota, FL) or a blunt patch pipette filled with artificial cerebrospinal fluid (ACSF) works well.

Biological Sample

Brain slice, cut from the region of interest, usually from rats or mice

PROCEDURE

Indicator Loading

1. Load the neurons with the chosen Ca^{2+} indicators using patch pipettes.

   Note: The use of patch pipettes for loading cells with Ca^{2+} indicators is now a standard technique, and no detailed explanation is given here. Patch pipettes are preferred to other loading methods because the concentration of indicator can be controlled. Usually, a stock solution of indicator is prepared and an aliquot of the stock is added to the appropriate standard intracellular solution. The loading electrode usually remains in place and is used to record the electrical response from the neuron. See Nakamura et al. (1999).

Stimulation of Ca^{2+} Waves

2. Place an extracellular stimulating electrode in the dendritic field near the cell and within 100 μm of the soma. Evoke Ca^{2+} waves by applying repetitive high-frequency stimulation.

   Note: See Zhou and Ross (2002) for an analysis of the effective stimulation parameters.

Measurement of Ca^{2+} Waves

3. Ca^{2+} waves are detected as a propagating change in fluorescence corresponding to changes in $[Ca^{2+}]_i$. There is nothing special about the calibration of the magnitude of the $[Ca^{2+}]_i$ change.

Single-wavelength measurements are necessary if high time resolution and maximum signal-to-noise ratios (SNRs) are desired. Standard corrections for indicator bleaching and background fluorescence should be applied. See Nakamura et al. (1999).

EXAMPLE OF APPLICATION

Figure 1 shows a Ca^{2+} wave evoked in the dendrites of a neocortical pyramidal neuron by repetitive stimulation (100 Hz for 0.5 sec). The synaptic response generated an initial action potential that led to a near-synchronous $[Ca^{2+}]_i$ increase as the spike back-propagated over the dendrites. This change in $[Ca^{2+}]_i$ was followed by a delayed, larger-amplitude response that was not synchronous and which extended over a limited range of the dendritic field. The high temporal and spatial resolution of the measurements clearly separate the two components of the response and show that the waves were initiated at several locations near branch points. The "line scan" display is particularly useful for following wave propagation.

ADVANTAGES AND LIMITATIONS

High-speed cooled CCD cameras are good detectors for imaging spatially extended dendritic Ca^{2+} waves. Single detectors (photodiodes or photomultipliers) have high temporal resolution but cannot follow the propagation of the wave. Confocal microscopes, operated in the line-scan mode, also have high temporal resolution, but the "line" cannot track the complex geometry of the dendrites. Even video-rate confocals, which have adequate temporal and spatial resolution for these measurements, are not ideal because the dendrites do not lie in the plane of a single *z*-section. Therefore, these instruments can only follow a Ca^{2+} release wave in a restricted spatial region. CCD cameras can detect

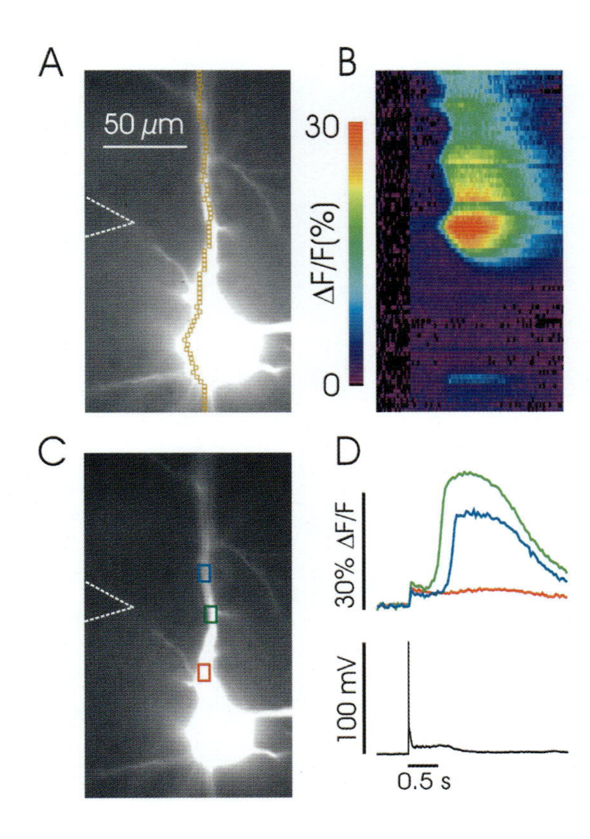

FIGURE 1. A synaptically activated Ca^{2+} wave in a neocortical pyramidal neuron. The cell was patched under DIC optics in a slice from the cortex of a 3-week-old rat. It was loaded with 200 μM bis-fura-2 via a patch electrode on the soma and stimulated via a tungsten electrode (*dotted arrow*) at 100 Hz for 0.5 sec. Two views of the resulting changes in $[Ca^{2+}]_i$ are shown in *B* and *D*. *D* shows the time course of the fluorescence changes at the regions of interest (ROIs) (indicated by colored rectangles in *C*). There was an initial response caused by the back-propagating spike, which was, within the temporal resolution of the measurements, synchronous at all three locations. The time courses and amplitudes of the delayed responses differ at the three ROIs. *B* shows a "line scan" of the same data. A selected series of pixels along the somato-dendritic axis, as shown on the cell image (*A*), was chosen for analysis. The pseudocolor image (*B*) shows the time course of the $[Ca^{2+}]_i$ changes at each of these locations. The timescale is the same as for the optical and electrical traces shown in *D*. With this view, it is clear that several waves initiated at different locations in the dendrites (close to branch points) and propagated along the main dendrite in both directions. In this cell, the waves did not propagate into the soma. (Adapted from Larkum et al. 2003.)

fluorescence changes in a larger volume of the tissue slice. Therefore, they can follow wave propagation even if the dendrites are only approximately planar. On the down side, this increased detection volume does lead to a loss of spatial resolution and, consequently, the possible misassignment of the signal to the wrong dendritic branch.

High speeds with CCD cameras are achieved with frame transfer chips and pixel binning. The shutter is never closed with this technique, maximizing the total amount of light detected, improving the SNR of the measurement. The low noise, linearity, and high digital accuracy of scientific-grade cameras are preserved. Developments in CCD camera technology, made since this technique was originally described (Lasser-Ross et al. 1991), have greatly improved the spatial and temporal resolution of the data.

REFERENCES

Kapur A., Yeckel M., and Johnston D. 2001. Hippocampal mossy fiber activity evokes Ca^{2+} release in CA3 pyramidal neurons via a metabotropic glutamate receptor pathway. *Neuroscience* **107:** 59–69.

Larkum M.E., Watanabe S., Nakamura T., Lasser-Ross N., and Ross W.N. 2003. Synaptically activated Ca^{2+} waves in layer 2/3 and layer 5 rat neocortical pyramidal neurons. *J. Physiol.* **549:** 471–488.

Lasser-Ross N., Miyakawa H., Lev-Ram V., Young S.R., and Ross W.N. 1991. High time resolution fluorescence imaging with a CCD camera. *J. Neurosci. Methods* **36:** 253–261.

Nakamura T., Barbara J.-G., Nakamura K., and Ross W.N. 1999. Synergistic release of Ca^{2+} from IP_3-sensitive stores evoked by synaptic activation of mGluRs paired with backpropagating action potentials. *Neuron* **24:** 727–737.

Nakamura T., Lasser-Ross N., Nakamura K. , and Ross W.N. 2002. Spatial segregation and interaction of calcium signalling mechanisms in rat hippocampal CA1 pyramidal neurons. *J. Physiol.* **543:** 465–480.

Zhou S. and Ross W.N. 2002. Threshold conditions for synaptically evoking Ca^{2+} waves in hippocampal pyramidal neurons. *J. Neurophysiol.* **87:** 1799–1804.

CHAPTER 44

A Practical Guide: Imaging Action Potentials with Calcium Indicators

Jason N. MacLean and Rafael Yuste

Department of Biological Sciences, Columbia University, New York, New York 10027

The understanding of neuronal circuits has been, and will continue to be, greatly advanced by the simultaneous imaging of action potentials in neuronal ensembles. This chapter describes "bulk" loading of brain slices with acetoxymethyl (AM) ester calcium indicators in order to monitor action potential activity in large populations of neurons simultaneously. The imaging of calcium influx into neurons provides an indirect, but accurate, measure of action potential generation in individual neurons. Single-cell resolution and thus the easy identification of every active cell is the key advantage of the technique.

Starting with Yuste and Katz (1991), this method has been successfully applied throughout the central nervous system including neocortex (e.g., Mao et al. 2001; Cossart et al. 2003), hippocampus (e.g., Tanaka et al. 2002), the cerebellum (e.g., Ghozland et al. 2002), striatum (e.g., Mao and Wang 2003), and spinal cord (e.g., Voitenko et al. 1999), utilizing either acute slices, cultured slices, or cultured dissociated neurons. Recently, the technique has also been applied in vivo to mouse neocortex (Stosiek et al. 2003).

MATERIALS

Cell-permeant Calcium Indicators

Different AM ester calcium indicators under similar experimental conditions have different loading efficacies. Ranked from best loading to worst: mag-fura-2>fura-2>indo-1>Calcium Green-1>fluo-4>Oregon-Green BAPTA-1>Calcium Orange>Calcium Crimson. Several of the indicators are also ranked below by their efficacy as detectors of action potentials: fluo-4>fura-2>mag-fura-2. The majority of our experimental procedures use fura-2. However, the methods described below can be used for any of the listed indicators.

Microscopy

Both single-photon and two-photon excitation of calcium indicators can be utilized to monitor calcium changes in large populations of neurons.

PROCEDURE A

Loading Embryonic and Neonatal Acute Cortical Slices

1. Dissolve 50 µg of fura-2AM (Molecular Probes) in 48 µl of DMSO<!> and 2 µl of Pluronic F-127 (Molecular Probes) for a final concentration of 1 mM.

   Note: Because the solubility of calcium AM indicators is rather poor, vortex the solution for 10–15 min prior to use.

2. Load the slices in oxygenated artificial cerebral spinal fluid (ACSF) and add enough fura-2AM solution for a final concentration of 10 µM fura-2AM, for 30–60 min.
3. Remove slices from loading chamber and place into incubation chamber containing oxygenated ACSF. Imaging may begin after a 30-min recovery period.

Caution: *See Appendix 3 for appropriate handling of materials marked with <!>.*

PROCEDURE B

Loading Juvenile and Adult Acute Cortical Slices

1. Prepare 1 mM fura-2 AM as indicated above (A1).
2. Place slices in loading chamber containing 2.5 ml of oxygenated ACSF.
3. Pipette 5–10 µl of fura-2AM solution on top of each slice.

   Note: This results in a high initial concentration of fura-2AM. The concentration decreases as the fura-2AM diffuses away from the site of application, resulting in a final concentration in the entire chamber of ~10–20 µM.

4. Load slices in the dark for 20–30 min at 35–37°C with 95% O_2/5% CO_2 lightly ventilated into the chamber.

   Note: As a rule of thumb, the loading time should be 10 min, plus as many minutes as the age of the animal in postnatal days.

5. Remove slices from the loading chamber and place into incubation chamber containing oxygenated ACSF for wash. The wash occurs through simple diffusion. Imaging may begin after a 30-min recovery period.

PROCEDURE C

Focal Loading-Slice Painting

Although local neural connections are maintained in the procedure described above (see Kozloski et al. 2001), long axonal projections, such as thalamocortical axons in thalamocortical slice preparations (Agmon and Connors 1991), may be compromised. A modification of Procedure A can be used to circumvent this problem and can also be used in vivo (Stosiek et al. 2003).

1. Dissolve 50 µg of fura-2AM (Molecular Probes) in 13 µl of DMSO<!> and 2 µl of Pluronic F-127<!> (Molecular Probes), achieving a final concentration of 3.3 mM. Vortex the solution for 10–15 min.
2. Deposit the slices into the first loading chamber containing 2 ml of ACSF, ventilated with 95% O_2/5% CO_2, and place onto microscope stage.
3. Fill a fire-polished pipette (tip diameter ~30 µm) with 7.5 µl of fura-2AM.
4. Insert the filled pipette into a standard patch-clamp electrode holder, with tubing attached, and, using a micromanipulator, place the pipette tip 100–200 µm above the surface of the slice. Apply 5–10 psi positive pressure to the pipette. Slowly (1–2 µl in 1 min) move the pipette across the surface of the slice using the manipulator, covering the area of interest with the dissolved fura-2AM.

5. Move the slices to the second loading chamber containing 2.5 ml of ACSF and the remaining 7.5 μl of fura-2AM.
6. Load the slices in the dark for 20–30 min (as a rule of thumb, 10 min plus the age of the animal in days) at 35–37°C with 95% O_2/5% CO_2 lightly ventilated into the chamber.
7. Remove the slices from the loading chamber and transfer them to an incubation chamber containing oxygenated ACSF for wash. Imaging may begin after a 30-min recovery period.

PROCEDURE D

Imaging of Calcium AM Indicators

Two-photon imaging of calcium indicators allows highly sensitive detection of changes in neuronal calcium concentration with relatively little bleaching and photodamage. However, the simultaneous imaging of large populations of neurons is necessarily slow. Epifluorescent imaging of calcium indicators is sufficient to detect changes in neuronal calcium concentration and has the advantage that, using a fast camera, the detection of changes can be of higher temporal resolution than in the two-photon system. However, fluorescent imaging of bulk-loaded slices is subject to rapid bleaching and is also prone to high background fluorescence. Finally, in our experience, spinning disk confocals, together with fast cameras, can be used to image thousands of neurons simultaneously, without significant photobleaching and with good signal-to-noise ratios, over long periods of time.

EXAMPLE OF APPLICATION

Figure 1 illustrates the two-photon imaging of fura-2 and the utilization of fura-2 as an indicator of neuronal activity. Figure 1A illustrates thousands of loaded neurons in a neocortical slice imaged using a two-photon microscope. Figure 1B reveals the correlation between neuronal activity and a change in fluorescence as detected by epifluorescence. The fluorescence of fura-2 decreases in the

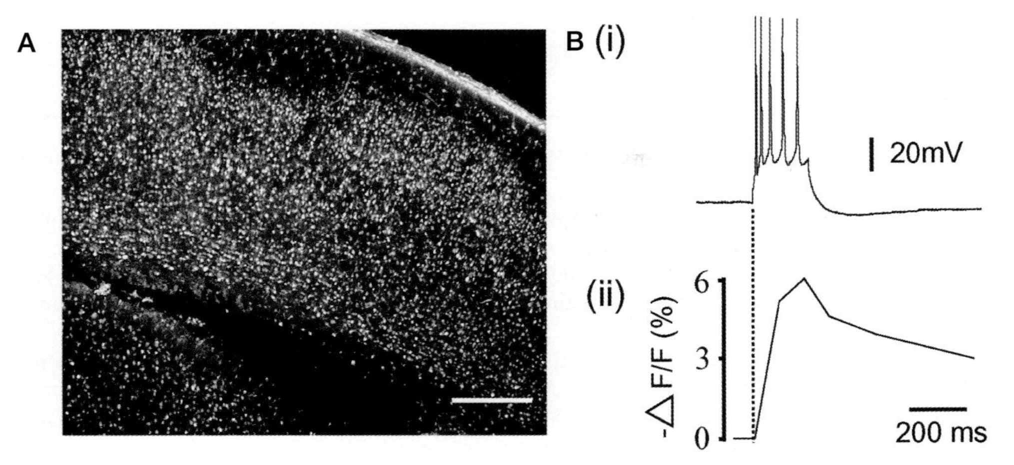

FIGURE 1. High-resolution calcium imaging of neuronal activity. (*A*) Two-photon image of transverse section of mouse neocortex loaded with fura-2AM. Bar, 250 μm. Note how thousands of neurons can be adequately visualized. Courtesy of Dr. Hajime Hirase. (*B*) Correspondence between action potentials and somatic fluorescence change. (i) Whole-cell recording of a burst of action potentials in response to a single intracellular depolarizing current step (150 pA, 200 msec). (ii) Normalized fluorescence change for the recorded neuron imaged with a cooled CCD camera (Micromax, Princeton Instruments). Recording pipette contained 50 μm of fura pentapotassium salt (comparable to the intracellular concentration of fura following bulk loading). Although individual action potentials cannot be resolved, onset of neuronal activity is accurately imaged.

presence of calcium, thus the negative fluorescent change in Figure 1B(ii). As illustrated, even one-photon imaging of fura-2 is sufficient to detect and resolve action potential generation in a single neuron within a field of hundreds or thousands of loaded cells. Because of the rapid time course for the onset of its response, fura-2 lends itself to the elucidation of network dynamics when simultaneously imaging large populations of neurons.

ADVANTAGES AND LIMITATIONS

The use of calcium AM ester indicators provides a very robust method of imaging activity in large populations of neurons simultaneously. In our experience, voltage-sensitive dyes are insufficient for the detection of activity in an individual neuron within a population of neurons such as a cortical slice. Their nonspecific staining pattern and poor signal strength currently make optical recording with single-cell resolution unfeasible in cortical slices (Yuste et al. 1997). Calcium indicators (Tsien 1989) that can be bulk-loaded into brain slices using their AM ester derivatives (Yuste and Katz 1991) act as very good, albeit indirect, measures of action potential generation (Smetters et al. 1999). These still provide the best means of imaging activity in large populations of neurons when single-cell resolution is desirable.

The limitations of the technique are due to the properties of the dyes themselves. Calcium indicators, being charged molecules, do not easily cross the cell membrane and need, therefore, to be microinjected. To circumvent this problem, AM ester derivatives of the indicators were synthesized (Tsien 1981). The AM esters mask negative charges, making the indicator molecules more lipophilic and membrane-permeant, thus allowing them to enter the cell. Once inside the cell, cytoplasmic esterases hydrolyze the acetyl ester linkage, releasing formaldehyde and free indicator, which then accumulates intracellularly as it is once again charged. However, the dependence on intracellular enzymatic cleavage makes this process cell-dependent. This can result in differential loading efficiency in different neurons. In addition, the increased hydrophobicity of the AM ester derivatives of the indicators can cause problems in delivering sufficient amounts to their targets. This problem becomes significant in adult preparations, where the slice painting method appears to be the best loading strategy. Finally, although the time constant for the onset of quenching indicator fluorescence in neurons is rapid, the offset is proportionally slow due to saturation of the dye. Thus, although the calcium indicators are excellent measures for the onset of activity, they do not provide adequate temporal resolution to detect single action potentials during a burst of action potentials.

REFERENCES

Agmon A. and Connors B.W. 1991. Thalamocortical responses of mouse somatosensory (barrel) cortex in vitro. *Neuroscience* **41:** 365–379.

Cossart R., Aronov D., and Yuste R. 2003. Attractor dynamics of network UP states in the neocortex. *Nature* **423:** 283–288.

Ghozland S., Aguado F., Espinosa-Parrilla J.F., Soriano E., and Maldonado R. 2002. Spontaneous network activity of cerebellar granule neurons: Impairment by in vivo chronic cannabinoid administration. *Eur. J. Neurosci.* **16:** 641–651.

Kozloski J., Hamzei-Sichani F., and Yuste R. 2001. Stereotyped position of local synaptic targets in neocortex. *Science* **293:** 868–872.

Mao L. and Wang J.Q. 2003. Group I metabotropic glutamate receptor-mediated calcium signalling and immediate early gene expression in cultured rat striatal neurons. *Eur. J. Neurosci.* **17:** 741–750.

Mao B.Q., Hamzei-Sichani F., Aronov D., Froemke R.C., and Yuste R. 2001. Dynamics of spontaneous activity in neocortical slices. *Neuron* **32:** 883–898.

Smetters D.K., Majewska A., and Yuste R. 1999. Detecting action potentials in neuronal populations with calcium imaging. *Methods* **18:** 215–221.

Stosiek C., Garaschuk O., Holthoff K., and Konnerth A. 2003. In vivo two-photon calcium imaging of neuronal networks. *Proc. Natl. Acad. Sci.* **100:** 7319–7324.

Tanaka E., Uchikado H., Niiyama S., Uematsu K., and Higashi H. 2002. Extrusion of intracellular calcium ion after in vitro ischemia in the rat hippocampal CA1 region. *J. Neurophysiol.* **88:** 879–887.

Tsien R.Y. 1981. A non-disruptive technique for loading calcium buffers and indicators into cells. *Nature* **290:** 527–528.

———. 1989. Fluorescent probes of cell signaling. *Annu. Rev. Neurosci.* **12:** 227–253.

Voitenko N.V., Kostyuk E.P., Kruglikov I.A., and Kostyuk P.G. 1999. Changes in calcium signalling in dorsal horn neurons in rats with streptozotocin-induced diabetes. *Neuroscience* **94:** 887–890.

Yuste R. and Katz L.C. 1991. Control of postsynaptic Ca^{2+} influx in developing neocortex by excitatory and inhibitory neurotransmitters. *Neuron* **6:** 333–344.

Yuste R., Kleinfeld D., and Tank D.W. 1997. Functional characterization of the cortical microcircuit with voltage-sensitive dye imaging of neocortical slices. *Cereb. Cortex* **7:** 546–558.

CHAPTER 45

Imaging Calcium Transients in Developing *Xenopus* Spinal Neurons

Nicholas C. Spitzer, Laura N. Borodinsky, and Cory M. Root

Neurobiology Section and Center for Molecular Genetics, Division of Biological Sciences, University of California at San Diego, La Jolla, California 92093

The roles of electrical activity in regulating neuronal development are well documented, and the use of tetrodotoxin (TTX) to block voltage-gated sodium channels has demonstrated the critical functions of sodium-dependent action potentials that generate an influx of calcium ions. Studies of the ontogeny of electrical excitability, however, have revealed calcium-dependent action potentials at early stages of development, motivating investigation of various forms and functions of calcium-dependent excitability that are insensitive to TTX. This chapter describes methods for imaging different classes of calcium transients in the spinal neurons of *Xenopus,* both in vitro and in vivo.

The procedures described here were developed to image intracellular Ca^{2+} in embryonic *Xenopus* (amphibian) spinal neurons grown in dissociated cell culture and in the intact neural tube (the developing spinal cord), focusing on early stages of neuronal differentiation, around the time of neural tube closure. The methods can also be applied to explant and organotypic cultures. The procedures are sufficiently simple that they could be further adapted to dissociated neuronal cell cultures from other developing embryos, and to embryonic spinal cords of vertebrates such as zebrafish, as well as ganglia in the developing nervous systems of invertebrates.

Ca^{2+} SPIKES, GROWTH CONE Ca^{2+} TRANSIENTS, AND FILOPODIAL Ca^{2+} TRANSIENTS

Recent studies have led to the discovery of three classes of spontaneous transient elevations of intracellular Ca^{2+} and have shown that they regulate distinct aspects of differentiation in a frequency-dependent manner, prior to synapse formation: (1) Ca^{2+} spikes are generated by developmentally transient Ca^{2+}-dependent action potentials and regulate expression of neurotransmitters and maturation of voltage-dependent potassium current (Gu et al. 1994; Gu and Spitzer 1995; Watt et al. 2000; Borodinsky et al. 2004). (2) Growth cone calcium transients are generated locally in the growth cone and regulate the rate of axon extension (Gu and Spitzer 1995; Gomez and Spitzer 1999). (3) Filopodial calcium transients are produced at the tips of filopodia and regulate growth cone turning (Gomez et al. 2001).

PROCEDURE A

Preparation of Dissociated Cell Cultures

Prepare cultures from embryos of *Xenopus laevis* at the neural plate stage (stage 15; Nieuwkoop and Faber 1967), several hours after the final DNA synthesis (birthday) of the first neurons to be generated in the neural tube (Spitzer and Lamborghini 1976; Lamborghini 1980). Dissect tissue from the posterior presumptive spinal cord region with jewelers' forceps and dissociate in divalent cation-free medium (58.8 mM NaCl, 0.67 mM KCl<!>, 0.4 mM EDTA, and 4.6 mM Tris; pH adjusted to 7.8 with HCl<!>) and plate on 35-mm tissue-culture dishes (e.g., Costar) in minimal medium containing four salts and a buffer (58.8 mM NaCl, 0.67 mM KCl, 1.31 mM $MgSO_2$<!>, 2 mM or 10 mM $CaCl_2$<!>, and 4.6 mM Tris; pH adjusted to 7.8 with HCl) (Ribera and Spitzer 1989). These cultures contain both neurons and myocytes, and neuromuscular junctions form readily.

For preparation of neuron-enriched and myocyte-free cultures, excise tissue from the posterior presumptive spinal cord region and treat with collagenase B (1 mg/ml; e.g., Sigma) in minimal medium for 5–15 min. Mesoderm, endoderm, and notochord begin to separate and can then be peeled away from the ectodermal layer including neural folds. Dissociate cells of the ectodermal layer in divalent cation-free medium and plate on 35-mm tissue-culture dishes (Gu et al. 1994). These neuron-enriched cultures contain neurons and nonneuronal cells but are free of myocytes; synapses between neurons are rare.

In both types of cell culture, neurons are present at low density, with 50–100 per culture dish. Cells in these cultures survive for several days, despite the minimal medium, by metabolizing intracellular inclusions of yolk and lipid. Substantial differentiation of neurons occurs during this period and parallels differentiation in vivo, perhaps as a result of complete provision of natural nutrients.

For studies in which larger numbers of neurons and growth cones are needed, prepare cultures of dissociated spinal neurons and neural tube explants from older embryos (stages 20–22). Grow cells in medium containing 2 mM Ca^{2+} on untreated tissue-culture plastic or on various biological substrates (e.g., laminin, fibronectin, vitronectin, tenascin, and poly-D-lysine) coated onto tissue-culture plastic or acid-washed glass coverslips, usually at saturating concentrations.

Caution: *See Appendix 3 for appropriate handling of materials marked with <!>.*

PROCEDURE B

Preparation of Spinal Cords

Prepare embryonic spinal cords from embryos at neural tube stages (stages 18–30), overlapping the period studied in culture. To image the dorsal spinal cord, pin embryos dorsal-side uppermost in small wells cut in Sylgard (Dow Corning; other materials would suit as well)-coated 35-mm tissue-culture dishes and incubate in collagenase B for 5–15 min (Gu et al. 1994; Borodinsky et al. 2004). Remove skin and surrounding myotomes using jewelers' forceps and electrolytically sharpened tungsten needles to expose the dorsal surface of the spinal cord. Access the ventral spinal cord in a similar manner; pin embryos ventral-side uppermost and remove mesoderm, endoderm, and the notochord to expose the ventral surface of the spinal cord. The medium for dissociated cell cultures is used for spinal cords.

Imaging

For imaging Ca^{2+} spikes and growth cone Ca^{2+} transients in vitro and in vivo, incubate cells in culture or freshly dissected whole spinal cords for 30 min with the fluorescent Ca^{2+} indicator fluo-4AM (Molecular Probes; 5 μM dye dissolved in 0.01% pluronic acid<!>/0.1% DMSO<!>) for most experiments. Wash out free dye with three rinses of culture medium over a 30-min period. This dye is useful for measurement of transient signals at relatively low levels of illumination, allowing signal acquisition over long periods without photodynamic damage to cells (Gu et al. 1994; Gu and Spitzer 1995).

Image dye-loaded cultured cells or spinal cords for periods of 1 hr with a Bio-Rad MRC1024 15-mW Kr/Ar ion laser<!> confocal system with an Olympus microscope and 20x, 40x, or 60x water-immersion objectives or equivalent. During imaging experiments, cells are usually superfused at a rate of several ml min^{-1} with saline containing 2 mM Ca^{2+}. To minimize photodynamic damage, operate the laser under low power, and attenuate light to no greater than 1% using neutral density filters. Collect images at 5- or 10-sec intervals; illuminate individual cells for <0.2 sec. Use confocal line scans at 4-msec sweep to measure rapid changes in fluorescence along a single axis, displaying sweeps horizontally in a raster array (Gu et al. 1994).

Caution: *See Appendix 3 for appropriate handling of materials marked with <!>.*

Imaging for Extended Periods

Ca^{2+} dynamics can be imaged for as long as 5 hr in culture (Gu and Spitzer 1995). A low rate of image capture (0.1 Hz) coupled with continuous superfusion of culture medium facilitates examination of Ca^{2+} transients for these long periods. The Ca^{2+} indicator slowly becomes sequestered in vesicles, but cytosolic levels remain sufficiently high to enable recording of Ca^{2+} transients (Fig. 1). Neurons continue to differentiate after imaging, suggesting that there is little photodynamic damage. In contrast, imaging single cells in the embryonic neural tube becomes difficult after periods of ~1 hr, as cells move in the neural tube, and it is difficult to track individual neurons with confidence.

Imaging Growth Cone Transients In Vivo

Because growth cones have a small volume and contain low levels of Ca^{2+} indicator, selectively load local regions of neural tube with fluo-4AM to permit visualization of labeled growth cones extending into unlabeled regions. Dissect stage 21–25 embryos in a modified Ringer's solution (MR; 100 mM NaCl, 2 mM KCl<!>, 2 mM $CaCl_2$<!>, 1 mM $MgCl_2$<!>, 5 mM HEPES; pH adjusted to 7.4 with HCl<!>) at 20°C. Remove anterior or posterior portions of skin and somites from one side of the neural tube of stage 21–25 embryos using an electrolytically sharpened tungsten needle and collagenase B (0.1 mg/ml) in MR. Bathe partially exposed neural tubes in 10 μM fluo-4AM (with 0.01% pluronic acid<!>/0.1% DMSO<!>) for 1 hour. Remove the remaining skin and somites after washing in MR, revealing unlabeled portions of the neural tube, and pin down embryos laterally on a Sylgard dish and perfuse with MR at 1 ml/min during confocal imaging (Fig. 2) (Gomez and Spitzer 1999).

Imaging Filopodial Transients In Vitro

Load neurons with 2–5 μM fluo-4AM (with 0.01% pluronic acid<!>/0.1% DMSO<!>) in MR, for 1 hour at 20°C, rinse and image at frequencies of .06–8 Hz on Bio-Rad MRC 600 argon laser or MRC 1024 krypton/argon laser confocal systems mounted on Olympus upright microscopes or equivalent. Perform 8-Hz imaging with the MRC 1024 using either a 60x water (NA 0.9) or 100x oil (NA 1.4) objective (Gomez et al. 2001). Use an acquisition box of 100 x 40 pixels to increase collection speed, and zoom to achieve pixel sizes between 0.2 and 0.4 μm (Fig. 3). Perform reduced-speed Ca^{2+} imaging (.06–1 Hz), used for imaging of global growth cone transients, on the MRC 600 using a 764 x 512-pixel collection box and a 40x Fluor (NA 1.2) oil objective. Invert loose coverslips with adherent neurons for use with oil objectives by first sealing them on a coverslide with high-vacuum grease (e.g., Dow Corning), leaving inlet and outlet ports when necessary for exchange of solutions.

Use ratiometric imaging of fluo-4 and fura-red signals to assess the possibility of artifactual fluorescence elevations due to volume changes or movement. For combined fluo-4/fura-red imaging, excite neurons loaded with 5 μM fluo-4AM and 5 μM fura-red AM with the 488-nm laser line of the MRC 1024 and collect images at both 522 ± 35 and ≥ 585 nm.

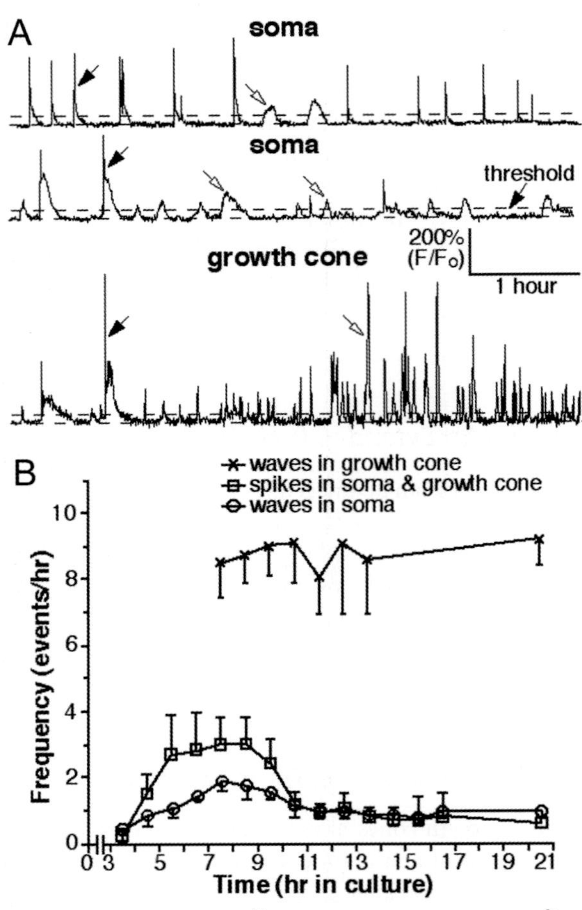

FIGURE 1. Patterns of spontaneous neuronal Ca^{2+} spikes and growth cone Ca^{2+} transients in cultured neurons over the period in which differentiation is altered by removal of extracellular Ca^{2+}. Sequences of fluorescence images of cells loaded with fluo-3AM were acquired by confocal microscopy. (*A*) (*Top*) Spikes (*filled arrows*) and longer transients (*open arrows*) recorded from somata of two neurons during a 5-hr period from 7 to 12 hr in culture. (*Bottom*) Spikes and growth cone transients recorded from a growth cone of the second neuron during the same period. Growth cone Ca^{2+} transients are generated at higher frequencies and have longer durations than spikes (33 ± 4 sec vs. 9 ± 1 sec; mean ± S.E.M., $n > 20$). Spikes are propagated throughout neurons, associated with Ca^{2+} action potentials and Ca^{2+}-induced Ca^{2+} release, and eliminated by Ca^{2+}-channel blockers; growth cone Ca^{2+} transients propagate decrementally, and only large events in the growth cone are detected at the soma. (*B*) The incidence of spikes is temporally regulated and is highest between 5 and 10 hr in culture. In contrast, the incidence of growth cone transients is spatially regulated, is higher in growth cones than in cell bodies, and is constant in growth cones during the period examined. Values are means ± S.E.M. for $n = 9–24$ neurons (≥3 cultures). (Adapted, with permission, from Gu and Spitzer 1995 [© Macmillan].)

Imaging Calcium Quantitatively

Use fura-2AM (Molecular Probes) in some experiments to ratiometrically estimate intracellular Ca^{2+} concentrations (Gu et al. 1994; Gu and Spitzer 1995). Dissolve dye in DMSO<!> and add to cultures to achieve a final concentration of 2 μM. Elicit fluorescence emission at 500 nm by excitation at 340 and 380 nm and image with a Zeiss Photoscope and SIT camera (or equivalent). Use identical equipment and optical settings for calibrations and for experiments. Estimate the intracellular Ca^{2+} concentration from the ratio image, using the equation: $[Ca^{2+}] = K_D\,((R-R_{min})/(R_{max}-R))F_o/F_s$. F_o and F_s are fluorescence intensities at 380 nm in 0 mM and saturating $[Ca^{2+}]$. R_{min} and R_{max} are ratios of

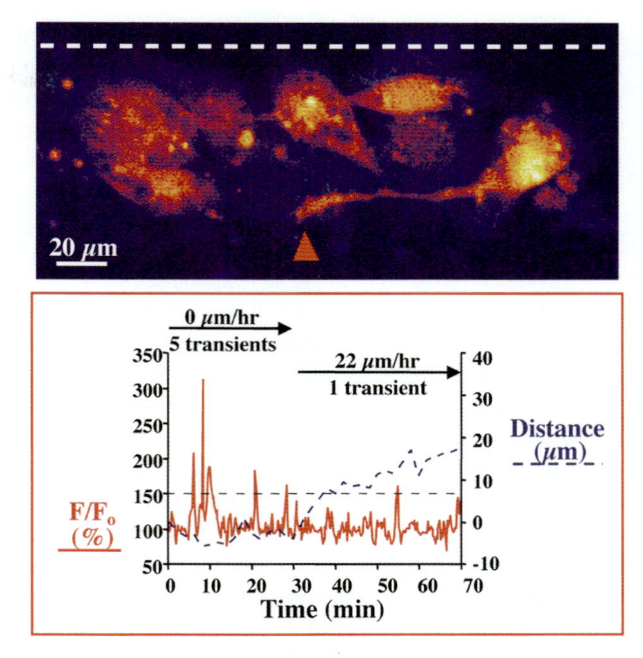

FIGURE 2. Growth cone Ca^{2+} transients of a dorsolateral ascending interneuron (DLA) imaged in the intact neural tube. (*Top*) The DLA growth cone (*arrowhead*) ascends near the sulcus limitans in a stage-22 spinal cord. (*Bottom*) Fluo-3 fluorescence and rate of outgrowth of this DLA growth cone show that it does not grow during the first 30 min of imaging when five Ca^{2+} transients exceeding 150% of baseline (threshold, *dashed line*) are produced, but accelerates to 32 μm/hr after transients subside. (Adapted, with permission, from Gomez and Spitzer 1999 [© Macmillan].)

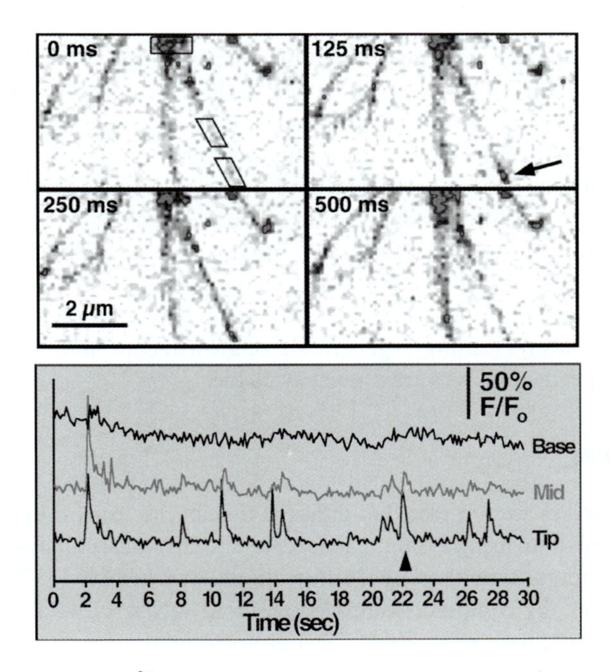

FIGURE 3. Spontaneous filopodial Ca^{2+} transients visualized by high-speed imaging of fluo-4 fluorescence. (*Top*) Time-lapse confocal images of fluo-4-loaded filopodia. Regions selected for quantification at the tip, middle, and base of one filopodium are shown at 0 msec. A local Ca^{2+} transient at the tip occurs at 125 msec (*arrow*). (*Bottom*) Normalized fluorescence changes (scale represents percent over baseline) within regions of this filopodium captured at 125-msec intervals. Ca^{2+} transients occur at high frequency at the tip and move toward the growth cone. Arrowhead indicates transient visualized at top. (Adapted, with permission, from Gomez et al. 2001.)

fluorescence excited at 340 and 380 nm under experimental conditions of 0 mM and saturating [Ca^{2+}]. K_D is 135 nM (Grynkiewicz et al. 1985). Calibration entails examination of fura-2-loaded cells, permeabilized with 5 μM ionomycin for 30 min and incubated in 10 mM Ca^{2+} to yield R_{max} and F_s, or incubated in 0 mM Ca^{2+} plus 2 mM EGTA and 2 mM Mn^{2+} to yield R_{min} and F_o (Grynkiewicz et al. 1985; Holliday and Spitzer 1990).

DATA ANALYSIS

Assess the impact of imaging on cell survival by comparing counts of cells in imaged fields at the time of imaging and one day later; adjacent fields in the same cultures serve as controls. Score four cultures, each with ≥50 neurons. Evaluate the effect of loading cells with Ca^{2+} indicators in a similar manner.

Analyze fluorescent pixel intensities of regions of neurons in each image with the freely available NIH Image program (W. Rasband, NIH; other programs may be used). Normalize changes in fluorescence intensity of each neuron to its baseline fluorescence intensity. Score propagated Ca^{2+} spikes and localized growth cone Ca^{2+} transients as events greater than twice the standard deviation of the noise, and distinguish them on the basis of their kinetics. Reconfirm each spike and growth cone transient by visual examination of time-lapse movies, to avoid analysis of spurious signals arising from cell movements or neighboring cells. Quantify filopodial fluorescence and normalize changes to baseline using MetaMorph (Universal Imaging) or NIH Image software; events exceeding 10% of baseline fluorescence are scored as filopodial transients. The relative fluorescence change of fluo-4 over the range of interest is proportional to the actual concentration of intracellular Ca^{2+} (Kao et al. 1989; Cornell-Bell et al. 1990; Holliday and Spitzer 1990).

PERTURBATION EXPERIMENTS

Suppression of Ca^{2+} Transients

Suppress Ca^{2+} transients by introduction of 0-Ca^{2+} medium containing 1 mM EGTA or by exposure to 10 mM Ni^{2+} to block influx, or 1 μM BAPTA-AM <!> to sequester intracellular Ca^{2+} (Gu et al. 1994; Gu and Spitzer 1995; Gomez and Spitzer 1999; Gomez et al. 2001; Borodinsky et al. 2004). Effects of BAPTA are unlikely to be due to generation of toxic metabolites, since longer exposure to the AM (acetoxymethyl) ester of fluo-4 does not produce similar effects. Assess the involvement of intracellular Ca^{2+} stores by application of 1 μM thapsigargin <!> (e.g., Sigma) to inhibit the Ca^{2+} ATPase of the endoplasmic reticulum. Ca^{2+} spikes are also blocked by a cocktail containing 20 nM calcicludine <!> (e.g., Calbiochem), 1 μM GVIA ω-conotoxin<!>, 1 μM flunarizine, and 1 μg/ml TTX <!> (e.g., Sigma) (Gorbunova and Spitzer 2002), to block voltage-dependent Ca^{2+} and Na^+ channels. The Ca^{2+} entry pathways leading to generation of growth cone and filopodial transients have not been identified, and specific blockers are not yet available.

Imposition of Spikes

Spikes are stimulated at different frequencies in vitro by culturing neurons in 250 μl of Ca^{2+}-free saline medium using a volume reducer and continuously superfusing them with this medium at 2.5 ml/min. The composition of saline is automatically switched for 15–20 sec to one containing 100 mM KCl plus 2 mM Ca^{2+} by computer-controlled solenoid valves (General Valve) (Gu and Spitzer 1995; Watt et al. 2000; Borodinsky et al. 2004). This protocol prevents neurons from generating spontaneous Ca^{2+} spikes (in Ca^{2+}-free saline) and depolarizes them in the presence of Ca^{2+} (KCl plus Ca^{2+}) to simulate spikes (Fig. 4A).

Manipulating Growth Cone Ca^{2+} Dynamics with Caged Compounds

To manipulate Ca^{2+} dynamics in growth cones, coload neurons with diazo-2AM (caged BAPTA<!>, 100 nM; Molecular Probes) or NP-EGTA-AM (caged Ca^{2+}, 5 μM; Molecular Probes) in conjunction

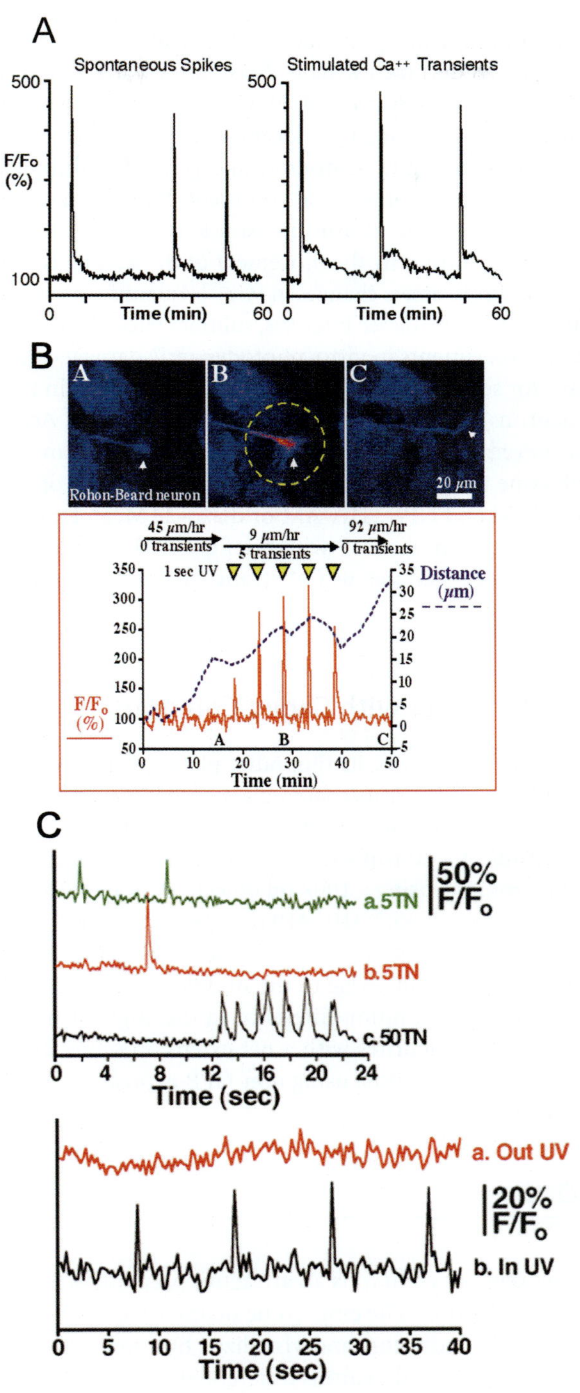

FIGURE 4. Reimposition of different classes of Ca^{2+} transients to evaluate their function. (*A*) Patterned stimulation of Ca^{2+} elevations in a neuron mimics spontaneous spikes (*left*). F/F_0 indicates fluorescence increase above baseline; 8 hours in culture. (*B*) stimulation of Ca^{2+} elevations in patterns of spontaneous growth cone Ca^{2+} transients. (*Top*) Sensory Rohon-Beard neuron growth cone (*arrowheads*) coloaded with fluo-3AM and NP-EGTA-AM in a stage-23 spinal cord; images at times indicated below. Yellow circle indicates the spot of ultraviolet light used to uncage Ca^{2+}. (*Bottom*) Fluorescence changes and distance migrated for this growth cone. Axon extension slowed when 5 UV flashes were applied at 5-min intervals (*yellow arrowheads*; 12/hr). (*C*) stimulation of Ca^{2+} elevations in patterns of spontaneous filopodial Ca^{2+} transients. (*Top*) Normalized fluorescence changes within regions of the filopodia captured separately at 125-msec intervals; the frequency of Ca^{2+} transients is higher in the filopodium on 50 μg/ml Tenascin than in those on 5 μg/ml Tenascin. (*Bottom*) Normalized fluorescence changes within regions of the filopodia captured sequentially at 125-msec intervals in response to uncaging Ca^{2+} from NP-EGTA at a frequency of 6 min^{-1}. (Adapted, with permission, from Gu and Spitzer 1995; Gomez and Spitzer 1999; Gomez et al. 2001.)

with fluo-4. Monitor baseline growth cone $[Ca^{2+}]_i$ and rate of axon outgrowth for at least 15 min prior to release of caged compounds with 1–2-sec flashes of UV light (340–380 nm) at different frequencies (Fig. 4B) (Gomez and Spitzer 1999). Ca^{2+} elevations mimicking spontaneous growth cone transients are generated by uncaging Ca^{2+} locally in growth cones. Suppress spontaneous growth cone transients by illumination of entire neurons to uncage BAPTA throughout the cells and prevent dilution by diffusion; this procedure suppresses Ca^{2+} transients for ~45 min. Determine the rate of axon extension in vivo at 2–5-min intervals by measuring axon length from the cell body to the leading margin of the growth cone, or by determining the movement of the leading edge of the growth cone relative to several stationary landmarks. To assess changes in $[Ca^{2+}]_i$, quantify the average pixel intensity within user-defined regions of growth cones at 15-sec intervals, subtract background, and normalize to baseline fluorescence.

For similar experiments in vitro, photodynamic damage appears to result from the stimulation of Ca^{2+} transients for several hours necessary to detect changes in the relatively slow rate of neurite extension, when neurons are grown on uncoated culture dishes. Accordingly, analyze more rapid neurite outgrowth achieved with neural tube explants grown on laminin. Laminin also suppresses spontaneous growth cone transients, facilitating temporal control of Ca^{2+} elevations via photorelease. Load cells with fluo-4AM and NP-EGTA-AM or diazo-2AM, and establish baseline rates of growth for each neuron. Subsequently, uncage Ca^{2+} with brief UV flashes delivered to the growth cone, generating Ca^{2+} transients that mimic spontaneous Ca^{2+} transients in their amplitude and duration (Lautermilch and Spitzer 2000).

Manipulating Filopodial Ca^{2+} Dynamics with Caged Compounds

To manipulate Ca^{2+} dynamics in filopodia, plate green fluorescent protein (GFP)-actin-expressing spinal cord explants on laminin-coated glass coverslips. After 6–18 hr in culture, load neurons with 2.5 μM NP-EGTA-AM for 1 hr, and then rinse in MR. Position motile growth cones with their leading edge 10 μm in front of a 20-μm spot and image at 15-sec intervals for ~30 min with a Bio-Rad MRC 600 confocal microscope using a 100x oil objective (or equivalent). Use a programmable shutter (e.g., Uniblitz) to pulse this spot with 310–410 nm UV light filtered from a 100-W mercury lamp for a duration of 200 msec every 10 sec (Fig. 4C) (Gomez et al. 2001). Measure the direction of neurite growth from the center of the leading edge to the middle of the neurite at the base of the growth cone using NIH Image. Score the maximum difference in the angle of neurite growth between 5 and 15 μm of growth. Do not include neurites with a net extension of <5 μm over a 30-min period. Test photorelease of Ca^{2+} in selected filopodia using non-GFP-expressing neurons coloaded with 2.5 μM fluo-4 and NP-EGTA, as a control.

ADVANTAGES AND LIMITATIONS

The developing *Xenopus* embryo is an attractive preparation in which to study aspects of early stages of neuronal differentiation. The embryonic neural tube can be dissected out, and neurons can be rapidly loaded with Ca^{2+} indicators and visualized by confocal microscopy. Dissociated cell cultures are relatively easy to prepare, and neurons differentiate as they do in vivo in several respects. Development is rapid, and features of neuronal differentiation are manifest during the first day following closure of the neural tube and during the equivalent time in culture (Ming et al. 2002).

There are limits to the success of the protocols we have outlined here, however. Use of AM Ca^{2+} indicator esters has the consequence of loading only the outer layers of cells in the intact neural tube; deep neurons are not loaded with indicator. This problem can be surmounted by the use of genetically encoded Ca^{2+} reporters. There are also limits to the duration of imaging in vivo due to movement of cells, as noted above. Finally, although it has been possible to study Ca^{2+} spikes and growth cone transients both in vitro and in vivo, it has been difficult to image filopodial transients in vivo because of the small dimensions of filopodia (~0.2 μm in diameter) and their three-dimensional rather than planar configuration.

It is worth remembering that the temporal and spatial parameters employed during imaging determine what can be observed. Brief events, like filopodial transients, will only be detected reliably by high-speed imaging. Longer events, such as growth cone transients, can be recorded at a lower image capture rate. Ca^{2+} transients that are propagated throughout neurons, such as Ca^{2+} spikes, do not require high spatial resolution for detection, whereas localized filopodial transients can be detected only at high magnification. It seems likely that there are additional classes of Ca^{2+} transients still to be discovered.

ACKNOWLEDGMENTS

We thank the members of our lab for helpful discussions. Supported by National Institutes of Health, NINDS NS15918 to N.C.S.

REFERENCES

Borodinsky L.N., Root C.M., Cronin J.A., Sann S.B., Gu X., and Spitzer N.C. 2004. Activity-dependent homeostatic specification of transmitter expression in embryonic neurons. *Nature* **429:** 523–530.

Cornell-Bell A.H., Finkbeiner S.M., Cooper M.S., and Smith S.J. 1990. Glutamate induces calcium waves in cultured astrocytes: Long-range glial signaling. *Science* **247:** 470–473.

Gomez T.M. and Spitzer N.C. 1999. *In vivo* regulation of axon extension and pathfinding by growth-cone calcium transients. *Nature* **397:** 350–355.

Gomez T.M., Robles E., Poo M.-M., and Spitzer N.C. 2001. Filopodial calcium transients promote substrate-dependent growth cone turning. *Science* **291:** 1983–1987.

Gorbunova Y.V. and Spitzer N.C. 2002. Dynamic interactions of cyclic AMP transients and spontaneous Ca^{2+} spikes. *Nature* **418:** 93–96.

Grynkiewicz G., Poenie M., and Tsien R.Y. 1985. A new generation of Ca^{2+} indicators with greatly improved fluorescence properties. *J. Biol. Chem.* **260:** 3440–3450.

Gu X. and Spitzer N.C. 1995. Distinct aspects of neuronal differentiation encoded by frequency of spontaneous Ca^{++} transients. *Nature* **375:** 784–787.

Gu X., Olson E.C., and Spitzer N.C. 1994. Spontaneous neuronal calcium spikes and waves during early differentiation. *J. Neurosci.* **14:** 6325–6335.

Holliday J. and Spitzer N.C. 1990. Spontaneous calcium influx and its roles in differentiation of spinal neurons in culture. *Dev. Biol.* **141:** 13–23.

Kao J.P., Harootunian A.T., and Tsien R.Y. 1989. Photochemically generated cytosolic calcium pulses and their detection by fluo-3. *J. Biol. Chem.* **264:** 8179–8184.

Lamborghini J.E. 1980. Rohon-Beard cells and other large neurons in *Xenopus* embryos originate during gastrulation. *J. Comp. Neurol.* **189:** 323–333.

Lautermilch N.J. and Spitzer N.C. 2000. Regulation of calcineurin by growth cone calcium waves controls neurite extension. *J. Neurosci.* **20:** 315–325.

Ming G., Scott T., Wong S.T., Henley J., Yuan X., Song H., Spitzer N., and Poo M. 2002. Adaptation in the chemotactic guidance of nerve growth cones. *Nature* **417:** 411–418.

Nieuwkoop P.D. and Faber J., eds. 1967. *Normal table of* Xenopus laevis *(Daudin): A systematic and chronological survey of the development of the fertilized egg till the end of metamorphosis*, 2nd edition. North-Holland, Amsterdam.

Ribera A.B. and Spitzer N.C. 1989. A critical period of transcription required for differentiation of the action potential of spinal neurons. *Neuron* **2:** 1055–1062.

Spitzer N.C. and Lamborghini J.E. 1976. The development of the action potential mechanism of amphibian neurons isolated in culture. *Proc. Natl. Acad. Sci.* **73:** 1641–1645.

Watt S.D., Gu X., Smith R.D., and Spitzer N.C. 2000. Specific frequencies of spontaneous Ca^{2+} transients upregulate GAD 67 transcripts in embryonic neurons. *Mol. Cell. Neurosci.* **16:** 376–387.

CHAPTER 46

Basics of Photoactivation

Graham C.R. Ellis-Davies

Department of Pharmacology and Physiology, Drexel University College of Medicine, Philadelphia, Pennsylvania 19102

Specific molecular interactions control cellular function. The photorelease of caged compounds (nucleotides, neurotransmitters, peptides, second messengers, proteins, etc.) can be used to control these interactions in living cells. Caged compounds are biological effector molecules whose active functionality has been chemically masked with a photoremovable protecting group (see Fig.1, top). Illumination produces a concentration jump from the caged molecule. This chapter discusses the basic principles underlying photoactivation, the properties of caging chromophores and commercially available caged compounds, and practical considerations for their effective use.

ADVANTAGES OF CAGED COMPOUNDS

Uncaging has many useful features and advantages compared to other methods (e.g., cell-flow techniques; see Hess 1993) for changing solute concentration inside or on living cells. These include:

1. *Intracellular release.* The intracellular compartment is completely accessible to photorelease technology, as cells are essentially transparent to uncaging light and the inert compound can be loaded into the cell: Release can be easily triggered inside cells.
2. *Speed of release.* Uncaging is fast (although it can vary from submicroseconds to milliseconds), so even channel-gating kinetics can be quantified. Conventional rapid mixing techniques (Hess 1993) produce concentration changes very slowly (several orders of magnitude more slowly than achieved using uncaging techniques) and thus can be kinetically limiting for some studies. Colquhoun et al. (1992) have developed a submillisecond method of solution change for excised patches.
3. *Timing of release.* By definition, the caged compound is inert until photoactivated. It can be loaded into the cell and be activated at any point during the experimental time course.
4. *Location of release.* Uncaging only occurs where light is incident; thus, uniform, global changes in concentration throughout the cytosol can be affected by even illumination of the whole cell. This circumvents all diffusional disequilibria, which can result from other techniques. In constrast, highly localized uncaging is also possible, by focusing excitation to a particular region of the cell (diffraction-limited release is possible with high-beam-integrity lasers). Two-photon photolysis (Denk 1994) is a specialized example of this as it produces both lateral and axial localization of excitation. Thus, substrate release occurs only where absorption of the uncaging light produces the electronic singlet state.
5. *Quantitation of release.* Repetitive, controlled release of the caged substrate is possible by calibration of uncaging efficiency as a function of incident light intensity or via intracellular imaging techniques.
6. *Noninvasive release.* Rapid flow techniques can cause physical perturbation of the sample. Photorelease provides a much less invasive means of producing concentration jumps.

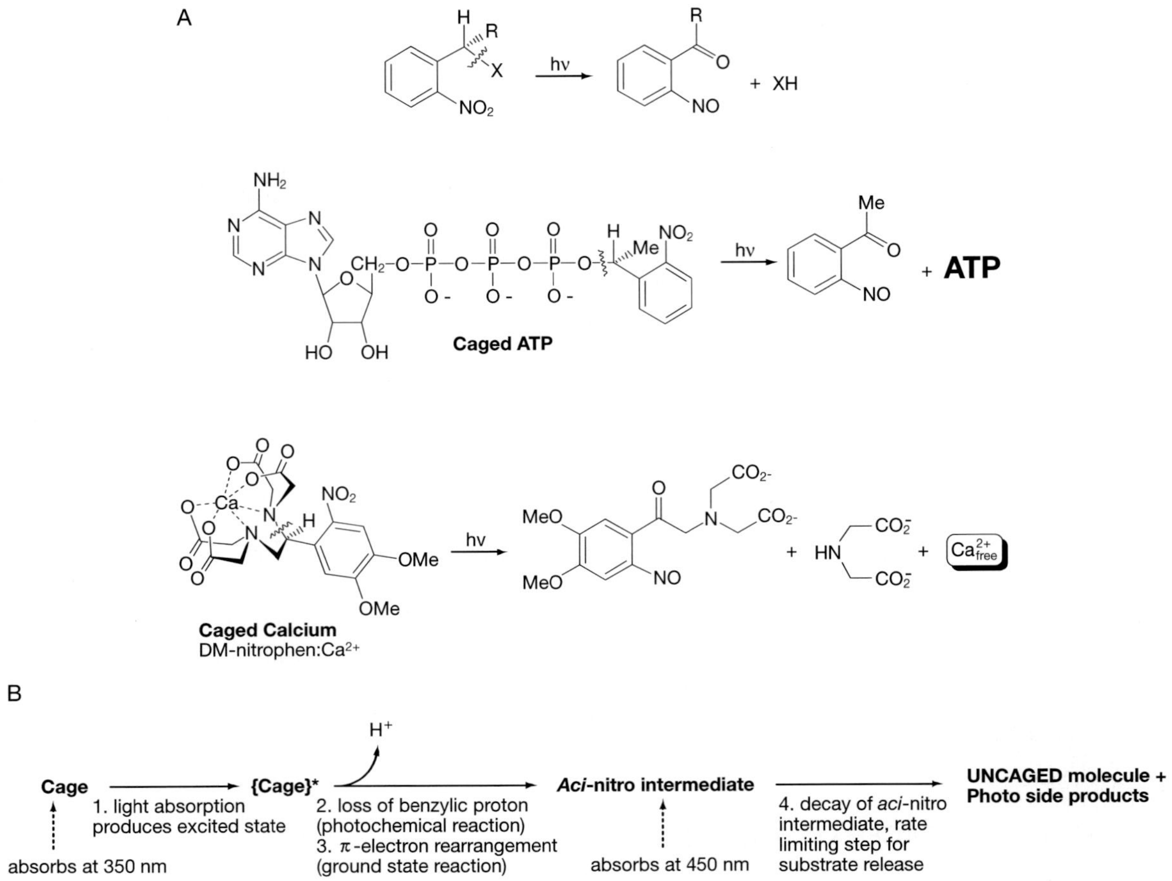

FIGURE 1. (*A*) Uncaging using nitrobenzyl photochemistry. Nitrobenzyl photochemistry can be used to cage and photorelease many different functional groups (i.e., X can be phosphate, thiophosphate, carboxylate, amine, phenol, alcohol, thiol, carbamate, or amide). Caged ATP and caged Ca^{2+} are examples of caged phosphate and caged amine, respectively. (*B*) General reaction mechanism for nitrobenzyl photochemistry. Uncaging of nitrobenzyl compounds takes place in four steps that are shown schematically. The rate-limiting step for substrate release is the decay of the *aci*-nitro intermediate, which absorbs at a much longer wavelength (450 nm) than the starting materials (350 nm).

CHARACTERISTICS OF CAGED COMPOUNDS

Several physiochemical characteristics of the photolabile molecule must be defined before it can be used experimentally. These include the purity, stability, inertness, and release rate of the caged compound. Each of these characteristics is discussed below.

Purity

Caged compounds must be extremely pure. Many chemicals are sold at a purity level of 95–99%, since this is quite adequate for most purposes. However, a 5% impurity of ATP in a 5.0 mM solution of caged ATP extrapolates to 0.25 mM ATP, which would be disastrous for the experiment. The tolerance for impurities may vary with the preparation being studied. If a caged compound purchased from a commercial source produces a response, prior to illumination, then the compound's chemical purity should be questioned. Techniques for rigorous purification of caged nucleotides have been published by Walker et al. (1988, 1989).

Stability

The caging of most molecules (e.g., ATP, glutamate, inositol triphosphate [IP3], and fluorescein) is accomplished by *covalent attachment* of the photoremovable protecting group. Irradiation breaks this bond, liberating the molecule that was caged. The caged compound must be hydrolytically and enzymatically stable for the duration of the experiment. Certain chemical bonds are much more stable than others. Ethers (e.g., caged fluorescein [Mitchison 1989], serine [Pirrung and Nunn 1992], and DMNPE-4 [Ellis-Davies 1998]), amines (e.g., DM-nitrophen [Kaplan and Ellis-Davies 1988] and caged epinephrine [Muralidharan and Nerbonne 1995]), and phenols (e.g., caged serotonin [Breitinger et al. 2000]) are all stable at pH 7. In contrast, *all* esters are somewhat hydrolytically unstable. Chemical instability of certain neurotransmitters could liberate enough of the caged species to produce desensitization before illumination. However, even if the originators of commercially available caged compounds were rigorous in their analysis of chemical stability, this would not guarantee that they have been as thorough in their purification of those products. Product shelf life is another thorny issue. Most developers of caged compounds probably do not study the long-term stability of their caged compounds. Most manufacturers make large batches of caged compounds and keep them until the last batch is almost sold out. This could be as long as 1–2 years. Few manufacturers are thought to have performed long-term stability studies on their compounds.

Inertness

A compound may be photolabile, but not caged (i.e., the compound is not biologically inert). For example, caged ATP cannot be hydrolyzed and is quite chemically stable, yet it is a mild competitive antagonist in certain tissues. Ideally, a caged compound should be neither an agonist nor an antagonist to any cellular receptor. A simple control for this is, of course, to apply working concentrations of the caged compound with activating concentrations of the effector to the tissue/cell.

Rate of Uncaging

By far the most widely used caging chromophores are the *ortho*-nitrobenzyl (NB) group, its derivative, the 4,5-dimethoxy-2-nitrobenzyl (DMNB) group, and their derivatives (see Fig. 2 and Table 1, entries 1 and 2). One of the clear advantages of the NB and DMNB groups, compared to all other caging

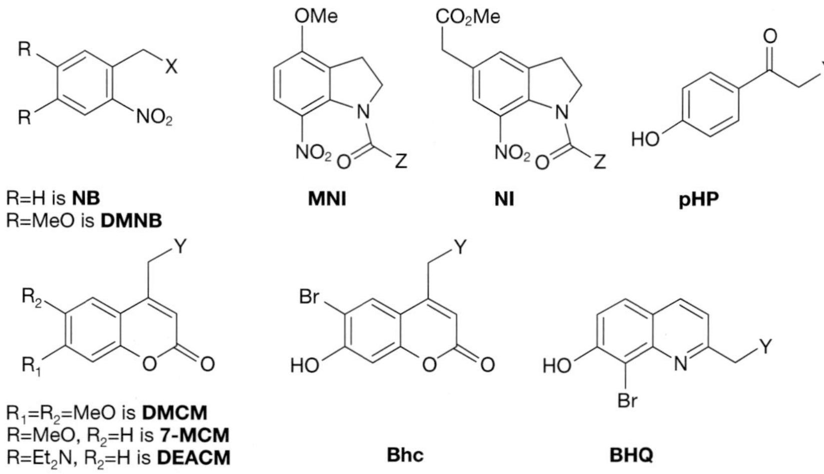

R=H is **NB**
R=MeO is **DMNB**
MNI
NI
pHP
R_1=R_2=MeO is **DMCM**
R=MeO, R_2=H is **7-MCM**
R=Et_2N, R_2=H is **DEACM**
Bhc
BHQ

FIGURE 2. Structures of caging chromophores. Several different chromophores can be used to make caged compounds. The NB and DMNB chromophores are generically useful (i.e., X is any functional group), whereas the other chromophores can only be used to cage acidic functionalities (i.e., phosphates or carboxylates; thus, Y is an acidic functionality and Z is alkyl).

TABLE 1. Summary of the physicochemical properties of caging chromophores

Cage	Absorption max (nm) (extinction/mM^{-1}cm^{-1})	Fluorescence	QY (chem)	QY (fluo)	2PCS (GM)	Rate (sec^{-1})
NB[a,b]	(~0.9 at 350 nm)	none	0.04–0.7	none	0.001	100–10^6
DMNB[b]	350 (5.2)	none	0.01–0.23	none	0.01	140–80,000
7-MCM	328 (13.3)	400	0.1	0.03	n.d.	10^8
DEACM	399 (19.5)	484	0.22	0.0058	n.d.	10^8
DMCM	349 (11.3)	444	0.04	0.022	n.d.	10^8
pHP[a,c]	very pH-dependent	n.d.	0.1–0.7	n.d.	n.d.	10^6-10^8
Bhc[c]	370 (15.0)	530	0.012–0.04	0.2–0.4	0.95	n.d.[d]
BHQ[c]	369 (2.6)	510	0.29	0.04	0.54	n.d.[d]
MNI	335 (3.6)	none	0.085	none	0.06	$>10^5$

Abbreviations and notes: QY = quantum yield (chemical or fluorescent); 2PCS = two-photon cross-section at 720–740 nm; n.d. = no data; NB = nitrobenzyl; DMNB = dimethoxynitrobenzyl; 7-MCM = 7-methoxycoumarinylmethyl; DEACM = 7-diethylaminocoumarinylmethyl; DMCM = 6,7-dimethoxycoumarinylmethyl; pHP = *para*-hydroxyphenacetyl; Bhc = 6-bromo-7-hydroxycoumarinylmethyl; BHQ = 8-bromo-7-hydroxyquinolino; MNI = 4-methoxy-7-nitroindolino. Values for the properties of 7-MCM, DEACM, and DMCM are for caged cAMP, Bhc, and MNI for caged glutamate, BHQ for "caged acetate."

[a] Absorption maxima are in the UV range (~280 nm) for both NB and pHP.

[b] The NB and DMNB are of such wide applicability as caging chromophores that it is impossible to make general statements as to their QY and rate of uncaging because these properties vary greatly depending on the type of functionality and compound caged. In contrast, the other cages summarized in the table have been developed for phosphate and carboxylic acids and are *only* useful for these functionalities.

[c] The absorption maxima of these cages are very pH-sensitive as they are phenols that deprotonate in the physiological range; e.g., pHP has a pK_a of 8.

[d] Probably uncage with a high rate, by analogy with other coumarin cages.

chromophores, is that they use a photochemistry that is applicable to every functional group tested so far; i.e., phosphate, thiophosphate, carboxylate, amine, phenol, alcohol, thiol, carbamate, and amide. These functional groups can all be protected and uncaged using nitrobenzyl photochemistry (see Corrie and Trentham 1993). The rate of the nitrobenzyl uncaging reaction varies greatly, depending on the functional group that has been caged and on the caging group itself. The excited-state reaction of the nitrobenzyl chromophore is removal of a proton from the benzylic carbon of the caging group. This photochemical reaction produces a reactive, ground-state intermediate (the *aci*-nitro species) which goes on to produce the liberated effector molecule, via a series of π-electron rearrangements. The rate of uncaging is dependent on these dark reactions and not on the photochemistry itself, which is very fast. Laser flash photolysis techniques have been used to make detailed studies of reaction rates in only a few caged compounds. Trentham and coworkers have shown that, for *ortho*-nitrobenzyl-caged ATP, the rate at which the disappearance of the *aci*-nitro species occurs is the same as the rate at which the photoproducts (ATP, H^+, and the caging group by-product; Walker et al. 1988) appear. Similar studies have been carried out for caged serine (Khan et al. 1993), glutamate (Corrie et al. 1993), and protons (Khan et al. 1993). We have shown that Ca^{2+} is released at the same rate as *aci*-nitro decay for the Ca^{2+} cages, NP-EGTA and DM-nitrophen (Ellis-Davies et al. 1996). It is on the basis of these studies that other researchers using the *o*-nitrobenzyl caging group *assume* that the uncaging rate corresponds to the disappearance of the *aci*-nitro species.

In the past few years, a range of other chromophores have been used to develop photolabile biological effector molecules. These are summarized in Figure 2 and Table 1. These caging chromophores are not generically applicable. (NB and DMNB cages have been used for all functional groups.) They can only be used to uncage acid-containing molecules (e.g., ATP, glutamate, and cAMP). The coumarin-based cages (DMCM, MCM, and DEACM; Eckardt et al. 2002) and the *para*-hydrophenacetyl (Park and Givens 1997) caged compounds all have fast (submicrosecond) rates of uncaging. By analogy with these determined rates, Bhc (Furuta et al. 1999) and BHQ (Fedoryak and Dore 2001) are also expected to uncage very quickly. Finally, inorganic chemical approaches based on transition metals are also possible (Bettache et al. 1996; Zayat et al. 2003).

The effective use of a particular caged compound can be predicated by a detailed assessment of the above physiochemical properties. Users are encouraged to read the original literature rather than reviews, and never to trust suppliers' catalogs! Users should be satisfied that the photolabile compound they wish to use really is caged. Table 2 provides a summary of the commercially available compounds that fall into this category.

EXPERIMENTAL USE OF CAGED COMPOUNDS

Cell Loading

Caged compounds are water-soluble and membrane-impermeant. Therefore, they must be loaded into the cytosol by (1) microinjection (Wilding et al. 1996), (2) permeabilization of the cell membrane (Zimmerman et al. 1995), or (3) whole-cell dialysis through a patch pipette (Neher and Zucker 1993). For quantitative studies, if the amount of uncaging cannot be measured in situ, it is vital that the intracellular concentration of the probe be known. Therefore, loading methods (2) and (3) are preferred, as they introduce a precisely defined concentration of cage into the cell.

The Illumination System

The light source for photolysis depends on the preparation and type of study. It should provide uniform release of the caged compound throughout the preparation (most caged compounds are photolyzed in the 300–380-nm near-UV range). If insufficient energy is delivered, and nonuniform uncaging occurs, much of the usefulness of the technique is lost. Frequency-doubled ruby lasers are most often used for contractile studies (Goldman 1986; Rapp et al. 1989). Flash lamps are adequate for secretory studies and are normally coupled to a microscope via the epifluorescence port. With the use of a pinhole, this setup can provide uncaging in restricted areas of the specimen. In some cases, illumination is effected from above the specimen plane, when a dichroic mirror is used to direct the light source to the sample. This can provide long-term concentration jumps in [Ca^{2+}], as the cell and patch pipette are illuminated by this route (Thomas et al. 1993). As an alternative to continuous light sources, pulsed sources, produced by placing a shutter in the path of a continuous source, can be used. Such systems using Xe arc lamps and CW Ar lasers have been described in detail (Parker 1992; Wang and Augustine 1995). The development of solid-state mode-locked Ti:sapphire lasers for two-photon imaging techniques provided a new light source for uncaging. Two-photon uncaging of neurotransmitters was first achieved by Denk in 1994 (the technique is discussed in Chapter 47).

All flash-photolysis studies use pulsed lasers as the light source. These lasers produce rapid pulses of intense, monochromatic light (e.g., the frequency-doubled ruby laser has a pulse width of 35 nsec and the Nd-YAG 4–6 nsec). Flash lamps, on the other hand, have much longer pulses of about 1 msec, which provides a number of advantages:

1. *Greater chemical conversion.* If cage lysis takes much longer than nonphotochemical de-excitation of the chromophore (e.g., 10 μsec vs. 1 nsec), then the chromophore can be re-excited many times during the long light pulse from a flash lamp, thus producing a better chemical yield for the same unit of energy.
2. *Fewer artifacts produced by the light pulse itself.*
3. *Cost.* A flash lamp costs about one-fifth of the price of an Nd-YAG laser, although an N_2 laser can be produced very cheaply (Engert et al. 1996).

Considerations When Measuring Photolysis

Quantitative physiological experiments with caged compounds require that the degree of photolysis be calibrated or measured for each experiment. In cases where a fluorescent indicator is available for the uncaged species (e.g., Ca^{2+}), this procedure is relatively straightforward. Nevertheless, care must be taken to control for the effects of cellular proteins on the dyes used (Konishi et al. 1991) and for effects of the cage/uncaging light itself upon the dye (see Neher and Zucker 1993; Thomas et al. 1993; Kasai et al. 1996; Parsons et al. 1996). An elegant method for the quantification of IP3 photolysis has

TABLE 2. Chemical properties of commercially available caged compounds

Cage	Quantum yield	Uncaging rate (sec^{-1})	Source
DM-nitrophen	0.18	3.8×10^4	CB, MP
NP-EGTA	0.23	6.8×10^4	MP
nitr-5	0.04	n.d.	CB, Sigma
CNB-glu	0.14	4.8×10^4	MP
CNB-GABA	0.16	3.6×10^4	MP
CNB-NMDA	0.43	32 and 1.7×10^3	MP
MNI-glu	0.085	$\sim 10^6$	Tocris
NPE-IP3	0.65	225 and 280	MP, CB
DMNB-fluorescein	n.d.	n.d.	MP
NPE-cADPribose	0.11	18	MP
NPE-ATP	0.48 or 0.63	90	CB, MP
diazo-2	0.03	2.3×10^3	MP
CNB-carbamoylcholine	0.8	1.7×10^4	MP

Abbreviations: CB = Calbiochem; MP = Molecular Probes; n.d. = no data.

been developed. The pH change produced upon uncaging in a quartz capillary placed in the focal plane of the microscope is measured with biscarboxyethylcarboxyfluorescein (BCECF) at various pulse strengths. A calibration curve can then be constructed for uncaging, and extrapolated to work in living cells (Khodakhah and Armstrong 1997).

COMMERCIALLY AVAILABLE CAGED COMPOUNDS

Table 2 presents a list of the commonly used commercially available caged compounds that have been used by many different research groups, suggesting that these compounds can work reliably. Most of these compounds have been well-characterized chemically, and their basic uncaging properties are summarized in Table 2.

CONCLUSION

Caged compounds are now widely used as chemical probes in neuroscience and many other fields of biology. They allow the downstream effects of a signaling molecule to be isolated from the upstream means of signal production. Furthermore, since rapid uncaging produces a single chemical species, uncaging establishes unambiguously the participation of the uncaged species in the effects initiated by the photoreleased molecule. These properties, as well as the others outlined above, make caged compounds useful tools for studying cellular signaling processes.

REFERENCES

Bettache N., Carter T., Corrie J.E., Ogden D., and Trentham D.R. 1996. Photolabile donors of nitric oxide: Ruthenium nitrosyl chlorides as caged nitric oxide. *Methods Enzymol.* **268:** 266–281.

Breitinger H.G., Wieboldt R., Ramesh D., Carpenter B.K., and Hess G.P. 2000. Synthesis and characterization of photolabile derivatives of serotonin for chemical kinetic investigations of the serotonin 5-HT(3) receptor. *Biochemistry* **39:** 5500–5508.

Colquhoun D., Jonas P., and Sakmann B. 1992. Action of brief pulses of glutamate on AMPA/kainate receptors in patches from different neurones of rat hippocampal slices. *J. Physiol.* **458:** 261–287.

Corrie J.E.T. and Trentham D.R. 1993. Caged nucleotides and neurotransmitters. In *Bioorganic photochemistry 2* (ed. H. Morrison), pp. 243–305. Wiley, New York.

Corrie J.E., DeSantis A., Katayama Y., Khodakhah K., Messenger J.B., Ogden D.C., and Trentham D.R. 1993. Postsynaptic activation at the squid giant synapse by photolytic release of L-glutamate from a "caged" L-glutamate. *J. Physiol.* **465:** 1–8.

Denk W. 1994. Two-photon scanning photochemical microscopy: Mapping ligand-gated ion channel distributions. *Proc. Natl. Acad. Sci.* **91:** 6629–6633.

Eckardt T., Hagen V., Schade B., Schmidt R., Schweitzer C., and Bendig J. 2002. Deactivation behavior and excited-state properties of (coumarin-4-yl)methyl derivatives. 2. Photocleavage of selected (coumarin-4-yl)methyl-caged adenosine cyclic 3′,5′-monophosphates with fluorescence enhancement. *J. Org. Chem.* **67:** 703–710.

Ellis-Davies G.C.R. 1998. Synthesis of photolabile EGTA derivatives *Tetrahedron Lett.* **39:** 953–957.

Ellis-Davies G.C.R., Kaplan J.H., and Barsotti R.J. 1996. Laser photolysis of caged calcium: Rates of calcium release by nitrophenyl-EGTA and DM-nitrophen. *Biophys. J.* **66:** 1006–1016.

Engert F., Paulus G.G., and Bonhoeffer T. 1996. A low cost UV laser for flash photolysis of caged compounds *J. Neurosci. Methods* **66:** 47–54.

Fedoryak O.D. and Dore T.M. 2002. Brominated hydroxyquinoline as a photolabile protecting group with sensitivity to multiphoton excitation. *Org. Lett.* **4:** 3419–3422.

Furuta T., Wang S.S.H., Dantzer J.L., Dore T.M., Bylee W.J., Callaway E.M., Denk W., and Tsien R.Y. 1999. Brominated 7-hydroxycoumarin-4-ylmethyls: Photolabile protecting groups with biologically useful cross-sections for two photon photolysis. *Proc. Natl. Acad. Sci.* **96:** 1193–1200.

Goldman Y.E. 1986. Laser pulsed release of ATP and other optical methods in the study of muscle contraction. In *Optical methods in cell physiology* (ed. P. De Weer and B.M. Salzberg), pp. 397–415. Wiley, Chichester, United Kingdom.

Hess G.P. 1993. Determination of the chemical mechanism of neurotransmitter receptor-mediated reactions by rapid chemical kinetic techniques. *Biochemistry* **32:** 989–1000.

Kaplan J.H. and Ellis-Davies G.C.R. 1988. Photolabile chelators for the rapid photorelease of divalent cations *Proc. Natl. Acad. Sci.* **85:** 6571–6575.

Kasai H., Takagi H., Ninomiya Y., Kishimoto T., Ito K., Yoshida A., Yosioka T., and Miyashita Y. 1996. Two components of exocytosis and endocytosis in phaeochromocytoma cells studied using caged Ca^{2+} compounds. *J. Physiol.* **494:** 53–65.

Khan S., Castellano F., Spudich J.L., McCray J.A., Goody R.S., Reid G.P., and Trentham D.R. 1993. Excitatory signaling in bacterial probed by caged chemoeffectors. *Biophys. J.* **65:** 2368–2382.

Konishi M., Hollingworth S., Harkins A.B., and Baylor S.M. 1991. Myoplasmic calcium transients in intact frog skeletal muscle fibers monitored with the fluorescent indicator furaptra. *J. Gen. Physiol.* **97:** 271–301.

Khodakhah K. and Armstrong C.M. 1997. Inositol trisphosphate and ryanodine receptors share a common functional Ca^{2+} pool in cerebellar purkinje neurons. *Biophys. J.* **73:** 3349–3357.

Mitchison T.J. 1989. Polewards microtubule flux in the mitotic spindle: Evidence from photoactivation of fluorescence *J. Cell Biol.* **109:** 637–652.

Muralidharan S. and Nerbonne J.M. 1995. Photolabile “caged” adrenergic receptor agonists and related model compounds. *J. Photochem. Photobiol.* **27:** 123–137.

Neher E. and Zucker R.S. 1993. Multiple calcium-dependent processes related to secretion in bovine chromaffin cells. *Neuron* **10:** 21–30.

Park C.H. and Givens R.S. 1997. New photoactivated protecting groups. 6. *p*-hydroxyphenacetyl: A phototrigger for chemical and biochemical probes. *J. Am. Chem. Soc.* **119:** 2453–2463.

Parker I. 1992. Caged intracellular messengers and the inositol phosphate signaling pathway. *Neuromethods* **20:** 369–396.

Parsons T.D., Ellis-Davies G.C.R., and Almers W. 1996. Photoactivation of nitrophenyl-EGTA triggers rapid exocytosis in pituitary melanotrophs. *Cell Calcium* **19:** 185–192.

Pirrung M.C. and Nunn D.S. 1992. Synthesis of photodeprotectable serine derivatives. Caged serine. *Bioorg. Med. Chem. Lett.* **2:** 1489–1492.

Rapp G., Poole K.J.V., Maeda Y., Ellis-Davies G.C.R., Kaplan J.H., McCray J., and Goody R.S. 1989. Lasers and flashlamps in research on the mechanism of muscle contraction. *Ber. Bunsen-Ges. Phys. Chem.* **93:** 410–415.

Thomas P., Wong J.G., Lee A.K., and Almers W. 1993. A low affinity Ca^{2+} receptor controls the final steps in peptide secretion from pituitary melanotrophs. *Neuron* **11:** 93–104.

Walker J.W., Feeney J., and Trentham D.R. 1989. Photolabile precursors of inositol phosphates. Preparation and properties of 1-(2-nitrophenyl)ethyl esters of *myo*-inositol 1,4,5-trisphosphate *Biochemistry* **28:** 3272–3280.

Walker J.W., Reid G.P., McCray J.A., and Trentham D.R. 1988. Photolabile 1-(2-nitrophenyl)ethyl phosphate esters of adenine nucleotide analogues. Synthesis and mechanism of photolysis. *J. Am. Chem. Soc.* **110:** 7170–7177.

Wang S.S.H. and Augustine G.J. 1995. Confocal imaging and local photolysis of caged compounds: Dual probes for synaptic function. *Neuron* **15:** 755–760.

Wilding M., Wright E.W., Patel R., Ellis-Davies G.C.R., and Whitaker M. 1996. Local perinuclear calcium signals associated with mitosis-entry in early sea urchin embryos. *J. Cell. Biol.* **135:** 1991–1999.

Zayat L., Calero, C., Albores P., Baraldo L., and Etchenique R. 2003. A new strategy for neurochemical photodelivery: Metal-ligand heterolytic cleavage. *J. Am. Chem. Soc.* **125:** 882–883.

Zimmermann B., Ellis-Davies G.C.R., Kaplan J.H., Somlyo A.V., and Somlyo A.P. 1995. Kinetics of prephosphorylation reactions and myosin light chain phosphorylation in smooth muscle. Flash photolysis studies with caged calcium and caged ATP. *J. Biol. Chem.* **270:** 23966–23974.

C H A P T E R 47

Two-Photon Uncaging Microscopy

Haruo Kasai,[1] Masanori Matsuzaki,[1] and Graham C.R. Ellis-Davies[2]

[1]*Department of Cell Physiology, National Institute for Physiological Sciences, Okazaki 444-8787, Japan;* [2]*Department of Pharmacology and Physiology, Drexel University College of Medicine, Philadelphia, Pennsylvania 19102*

The goal of two-photon uncaging microscopy is to exploit to the maximum the inherent optical sectioning power of two-photon excitation (2PE), but not in the usual way of imaging fluorescence emission from the focal volume. Rather, 2PE is used to produce highly localized concentration jumps of neurotransmitter (e.g., glutamate) in order to activate isolated clusters of receptors and so produce maps of receptor densities, in three-dimensional space, in complex structures such as the hippocampus. This chapter provides a detailed description of the development of a microscope designed to take advantage of all of the properties of caged neurotransmitters that undergo effective 2PE and rapid release (e.g., 4-methoxy-7-nitroinodinyl-glutamate [MNI-glu]). It also provides some examples of its effectiveness in cultured and acutely isolated hippocampal neurons.

Quantum theory says that absorption of light occurs when the transition moment between the ground state and the excited state is nonzero. This is only true when there is an inversion of orbital symmetry during the transition. (Electronic states have symmetries that are either even [*gerade* or "g" states] or uneven [*ungerade* or "u" states].) This principle is formalized in the parity selection rule for light absorption: Transitions from g to u or from u to g are allowed; transitions from g to g or from u to u are forbidden. 2PE gives rise to states in which the electronic wave function remains unchanged when the Cartesian coordinate system of the molecule is inverted through the center of symmetry, and so this process is strictly "geometrically forbidden" by the parity selection rule. Goeppert-Mayer (1931) realized that Paul Dirac's dispersion theory could apply to two-photon *excitation* as well as light transmission. She developed the idea of an "intermediate electronic excited state," which must have opposite symmetry of the ground and final excited state (so the parity rule for light absorption still applies), but overall, the selection rule for two-photon transitions is the exact opposite of one-photon: g to g is allowed but g to u is forbidden (Friedrich 1982).

BACKGROUND TO THE DEVELOPMENT OF TWO-PHOTON UNCAGING MICROSCOPY

The development of lasers revolutionized molecular spectroscopy, and, in the 1960s and 1970s, 2PE was used to study electronic excited states that had only been known theorectically (i.e., the forbidden g-to-g transitions, and especially the *gerade*-excited state energy levels) (for review, see Friedrich 1982). Birge and coworkers were the first to use 2PE to study biological chromophores such as rhodopsin (Birge 1986). Sheppard and Kampfner (1978) suggested that 2PE might be used for nonlinear scanning microscopy, but it was not until the pioneering work of Denk et al. (1990) that this idea was realized.

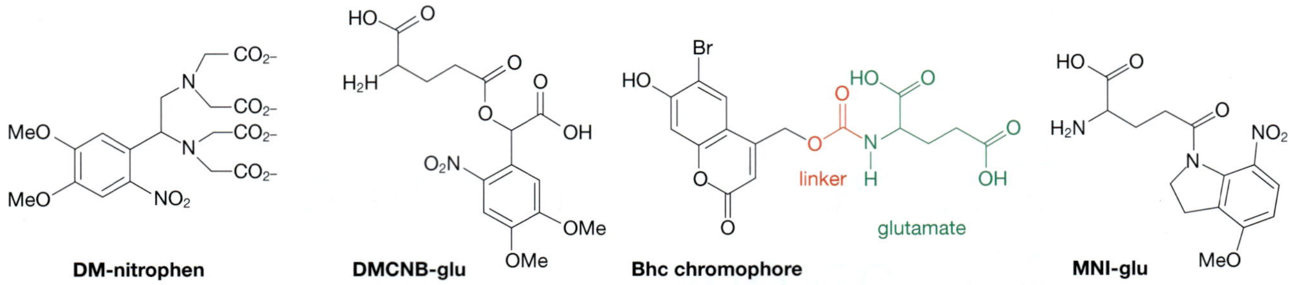

FIGURE 1. Structures of two-photon-sensitive caged compounds.

Photochemically initiated release of biochemically inert substrates (uncaging) is now used by many laboratories to study a very diverse range of cell types (plants, sea urchins, mammals, etc.). Except for a few examples (see, e.g., Denk 1994; Lipp and Niggli 1998), one-photon excitation has been used for uncaging. Denk (1994) found that the normal caging chromophore (*ortho*-nitrobenzyl) was fairly insensitive to 2PE (Denk had to use 640-nm light for 35 msec in each pixel, as the chromophore has a two-photon cross section of >0.0001 GM at 720 nm (1 GM = 10^{-50} $cm^4 s$ $photon^{-1}$) (cf. Brown and Webb 1998). Lipp and Niggli (1998) used uncaging of calcium by irradiation of DM-nitrophen (Fig. 1) (Ellis-Davies and Kaplan 1988) at 720 nm to mimic calcium sparks, and so initiate calcium waves in cardiac myocytes. Their work suggested that the electron-donating methoxy groups of the DM-nitrophen chromophore conferred sufficient absorptivity upon the *ortho*-nitrobenzyl cage to make it reasonably sensitive to 2PE and therefore useful for highly localized uncaging. This discovery led to the synthesis of DMNPE-4 (dimethoxynitrophenyl-EGTA-4) (Ellis-Davies 1998) and DMCNB-glutamate (dimethoxycarboxynitrobenzyl-glutamate) (Ellis-Davies 1999) (see Fig. 1 for structures), both of which are two-photon-sensitive (DelPrincipe et al. 1999; Ellis-Davies 1999). Simultaneous to the development of these new caged compounds, Furuta et al. (1999) introduced Bhc-glutamate as a caged neurotransmitter with a large two-photon cross section.

PROPERTIES OF CAGED GLUTAMATE FOR TWO-PHOTON PHOTOLYSIS

It is instructive to compare the properties of DMCNB-glu and Bhc-glu in order to define the essential chemical properties of caged glutamate for two-photon uncaging microscopy. First of all, the caged glutamate needs to be thermally very stable, so that slow aqueous decomposition does not take place in solution. Hydrolysis leads to the accumulation of free glutamate that could produce desensitization of receptors prior to uncaging. The Bhc chromophore is coupled to the amino group via a carbamate group, whereas the DMCNB cage is attached directly as an ester to the γ-carboxylate (Fig. 1). Carbamates are very stable at pH 7.4, whereas esters are somewhat hydrolytically sensitive, making Bhc-glu far superior to DMCNB-glu in this regard. Second, the two-photon cross section of the Bhc chromophore is ~100 times larger than the dimethoxynitrobenzyl chromophore (Brown and Webb 1998; Furuta et al. 1999). The third vital property of caged compounds for two-photon uncaging is that they are released very quickly. The 2PE volume has been approximated to a cylinder having dimensions of diameter of 0.3 μm and height of 0.9 μm (Brown et al. 1999). A small molecule at the center of this volume has a very brief residence time in the focal volume (~0.3 msec). Therefore, the half-time for uncaging must be much less than this period, otherwise caged molecules *in the excited state* would exit the initial focal volume, creating a diffuse "mist" of uncaged glutamate. Thus, slow uncaging does not take full advantage of the highly localized nature of 2PE (see Fig. 5d in Furuta et al. 1999), whereas very rapid uncaging does. DMCNB-glu is uncaged with a half-time of 5 μsec (Ellis-Davies 1999), whereas even though Bhc is probably uncaging quickly, *this still does not release glutamate*; rather, the released carbamate derivative of glutamate still must undergo a second thermal chemical reaction to give the desired final product. This process is known to take many milliseconds (Corrie et al. 1993).

Caged neurotransmitters for two-photon uncaging must have the following three chemical properties: (1) high chemical stability to aqueous hydrolysis at pH 7.4, (2) effective absorption of excitation light (i.e., good two-photon cross section), and (3) rapid uncaging of substrate (ideally, microsecond time constant). Since DMCNB-glu was somewhat susceptible to hydrolysis, we developed MNI-glu (Matsuzaki et al. 2000) as a caged glutamate that satisfies all three design criteria. Corrie and coworkers independently developed MNI-glu (Canepari et al. 2001; Papageorgiou and Corrie 2001). HPLC (high-performance liquid chromatography) detects no hydrolysis in solutions of MNI-glu at pH 7.4 after keeping a solution (1 mM) for 8 hr at room temperature. The same solution can also undergo numerous freeze-thaw cycles (−20°C) without any detectable deterioration of MNI-glu (limit of sensitivity is <1%). Wan and coworkers have recently shown that MNI-acetate is uncaged very quickly, probably with a time constant of <1 μsec (Morrison et al. 2002); therefore, by analogy, MNI-glu is also released quickly. Finally, MNI-glu has a two-photon cross section of about 0.06 GM (Matsuzaki et al. 2001).

MATERIALS

MNI-Glutamate

MNI-glu is purified with a reversed-phase HPLC (C18). The mobile phase is a linear gradient of acetonitrile in water containing 0.09% trifluoroacetic acid. The initial acetonitrile concentration is 10%, and this is increased linearly to 30% over 30 min. The fractions containing MNI-glu were pooled, lyophilized, and aliquoted into microcentrifuge tubes (0.3 μmole/tube), freeze-dried, and stored at −30°C. All of these procedures were carried out under yellow light illumination. MNI-glu is dissolved into a final recording solution at a concentration of 5–12 mM on the day of the mapping experiment.

Caution: *See Appendix 3 for appropriate handling of materials marked with <!>.*

PROCEDURE

Overview of General Procedure for Two-Photon Microscopy

The 2PE uncaging experiments follow six steps:

1. Align a laser beam<!> for 2PE excitation on the specimen.
2. Patch-clamp neurons in a brain slice for whole-cell perfusion of fluorescent tracers, such as calcein or fura-2, and for recording glutamate-induced current.
3. Acquire fluorescence images of dendrites, and search an appropriate region of interest (ROI) of x by y pixels (e.g., 32 by 16).
4. Apply MNI-glu around the ROI from a glass pipette that is placed just above the surface of the slice.
5. Record both two-photon-induced excitatory postsynaptic currents (2PEPSCs) and fluorescence values at the time of the irradiation at each pixel in the ROI. For three-dimensional mapping, systematically alter the z axis.
6. Wash out MNI-glu, and obtain high-quality fluorescence images.

Microscope and Electrophysiology Setup

This procedure uses an upright laser-scanning microscope (e.g., BX50WI and FluoView, Olympus) and a mode-locked Ti:sapphire laser (Tsunami, Spectra-Physics) fixed on an antivibration optical table (RS4000, New Port). A brain slice is placed on a microscope stage (30 × 40 cm), the position of which can be finely adjusted by a mechanical XY manipulator (Narishige) on the optical table (Fig. 2). Two mechanical manipulators on the stage control the headstage of a patch-clamp amplifier (Axon 200A) and a glass pipette for the application of caged compounds. The microscope system is shielded from electromagnetic waves by a Faraday cage covered with light-protected shielding cloth (Nihonkoden). MNI-glu is applied using a glass pipette with a 20-μm-diameter opening by positive

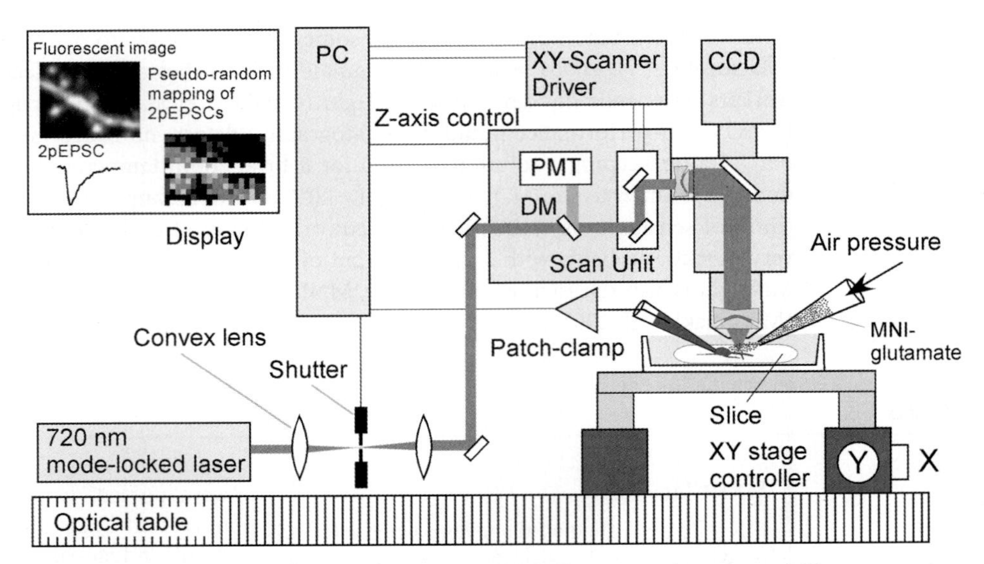

FIGURE 2. Schematic drawing of a two-photon uncaging microscope. A commercial laser-scanning microscope is modified to achieve full control of XY-scanner mirrors using an external scanner driver. A mode-locked laser beam is focused on a small aperture of a fast shutter using a pair of convex lens for rapid gating. MNI-glu is superfused locally on a slice preparation on which a whole-cell recording is performed. Automated mapping of glutamate currents is performed for a three-dimensional ROI.

air pressure (0.2–0.3 psi). Stability in the 2PE of MNI-glu was assessed by measuring two-photon fluorescence of the polar tracer Alexa 594, also applied to a brain slice using a glass micropipette: The coefficient of variation (CV) of such fluorescence values was 0.03.

Scan Unit and Photomultiplier

The two-photon uncaging microscope was constructed with minimal modifications to the commercial laser-scanning microscope. It is most important to achieve full control of the XY scanner; in this setup, external scanner drivers (CX6120, General Scanning) are connected to the driver ports of the scan unit (Fig. 2). The external scanner drivers are controlled by analog output of a digital-to-analog (D/A) converter (NI 6030E and PCI-6713, National Instruments). The output of the photomultiplier tube (PMT) is fed into an analog-to-digital (A/D) converter (NI 6110E). Instead of the detection of descanned fluorescence by the internal PMT (Fig. 2), the external PMT (R3896, Hamamatsu) is connected to a back port of the microscope. The advantages of such external PMTs are well-known (Mainen et al. 1999): They are more sensitive, but are also susceptible to background light interference, and so are used only when photon fluxes are too low for normal detection.

Mode-locked Laser Beam

A mode-locked Ti:sapphire laser (Tsunami, pumped by 5W Millenia, Spectra-Physics) is used at a wavelength of 720 nm for uncaging of MNI-glu. This is the shortest wavelength at which consistent, stable mode-lock can be achieved. The use of shorter wavelengths can cause more photodamage (Denk 1994). Peak wavelength and bandwidth of the laser are continuously monitored by a spectrometer (E201, Rees). A neutral-density filter wheel (VND-100, Sigma Koki) is placed at the input port of the microscope to attenuate the power at the specimen to a certain value (3–12 mW). Power loss within the microscope is estimated as 80% by using a power meter placed underneath the objective lens. The diameter of the beam must be reduced for rapid gating of the laser, since it is larger (3–5 mm) than the aperture of fast shutters (1.5 mm). A mechanical shutter (UHS1, Uniblitz) is placed at the focal point of two convex lenses (Fig. 2) for a pulse duration of 0.6–10 msec. A Pockels cell (350–50, Conoptics) is placed between two beam expanders for shorter pulses (50 µsec).

The group velocity dispersion at the specimen was minimized by chirp compensation optics using two SF-10 prisms (New Focus) that were separated by about 1 m. Without chirp compensation, pulse duration is expanded from 75 fsec to 330 fsec at the specimen, but only to about 100 fsec with compensation. Assuming that the majority of photodamage is due to two-photon absorption, however, the chirp compensation is not necessary if there is enough laser power.

The laser beam is aligned for optimal 2PE at the specimen after a warmup period of 1–2 hr. The beam is aligned using two mirrors in front of two irises that are set for the optimal laser alignment. First, a rough alignment is obtained using the two irises. Then the alignment is improved by maximizing a reflection image of an XY-scanner mirror that appears in the absence of infrared cut filter in front of the PMT. Third, two-photon fluorescence images are obtained for a fluorescent specimen, and the mirrors are finely readjusted to acquire homogeneous illumination that is brightest at the center of the image. Finally, the mirrors are readjusted, if necessary, so that the axis of the *x-z* image of a bead (diameter 3 μm or 0.1 μm; Molecular Probes) is precisely vertical.

Software

We have written custom software to control an XY scanner (DA channels 1 and 2), a shutter (DA channel 3), and a *z*-axis controller (DA channel 4). It also acquires current recording from the patch-clamp amplifier (AD channel 1) and output of the photomultiplier (AD channel 2). Custom software is essential because no commercial package is designed for this purpose. For image acquisition mode, the X-position is scanned by saw waves, and the Y-position is incremented at each cycle of the saw waves. Due to a delay in the movement of the X scanner during image acquisition, a small offset of the X-position occurs between the fluorescence image and the mapping experiment. Such offset is constant and is readily corrected for by precise positioning of the scanner during mapping to a corresponding position in the fluorescence image.

It is essential that the distribution of the positions of successive uncaging events in the mapping experiments be pseudorandomized to eliminate the effects of desensitization of receptor by previous uncaging at nearby pixels (Denk 1994), as illustrated in Figure 3D. If one image is composed of 2Nx*2Ny pixels, the position of the scanner (X(*i*),Y(*i*)) at *i*th uncaging (*i* = 0,…,4*Nx*Ny-1) is set as

X(*i*) = Nx*mod[*i*,2] + Px(int[mod[*i* ,4*Nx]/4])
Y(*i*) = Ny*int[mod[*i* , 4]/2] + int[*i* /(4*Nx)]

where Px is, for example, Px(*i*) = (0,4,2,6,1,5,3,7) when Nx = 8. The XY scanner is fixed for about 100 msec to a certain pixel, during which time the sample is irradiated with the laser beam for a short period, and the whole-cell current and the PMT output at the time of irradiation are recorded. The laser scan is then repositioned, in other pixels, and the process is repeated until all pixels are mapped. For example, it takes 52 sec to map an ROI consisting of 32 × 16 pixels. A lateral interval of pixel is 0.33 μm. For three-dimensional mapping, XY screens are performed sequentially for several *z*-axis positions with an interval of 0.8 μm (Fig. 4C). The error due to the sampling intervals is estimated as 0.1 (CV); the spatial resolution of glutamate mapping is 0.6 μm laterally and 1.4 μm axially. The simultaneously obtained fluorescence images are used to detect any structural alteration in the ROI during experiments. High-quality three-dimensional images, with a resolution of 0.083 μm laterally and 0.2 μm axially, are obtained before and after the mapping experiment for fine analysis of structures.

IMAGING AMPA RECEPTOR DENSITIES IN HIPPOCAMPAL NEURONS

The efficacy of 2PE of MNI-glu was demonstrated in cultured hippocampal neurons (Fig. 3). The caged compound did not activate any current at a concentration of 10 mM, nor did it affect the amplitude or the time course of miniature excitatory postsynaptic currents (mEPSCs), indicating that it does not exert any agonistic or antagonistic effect on AMPA (α-amino-3-hydroxy-5-methyl-4-isoxazole propionic acid) receptors (i.e., the caged compound is biologically inert). Two-photon uncaging of MNI-glu in a proximal region of the dendrites for a short period (50 μsec) elicited

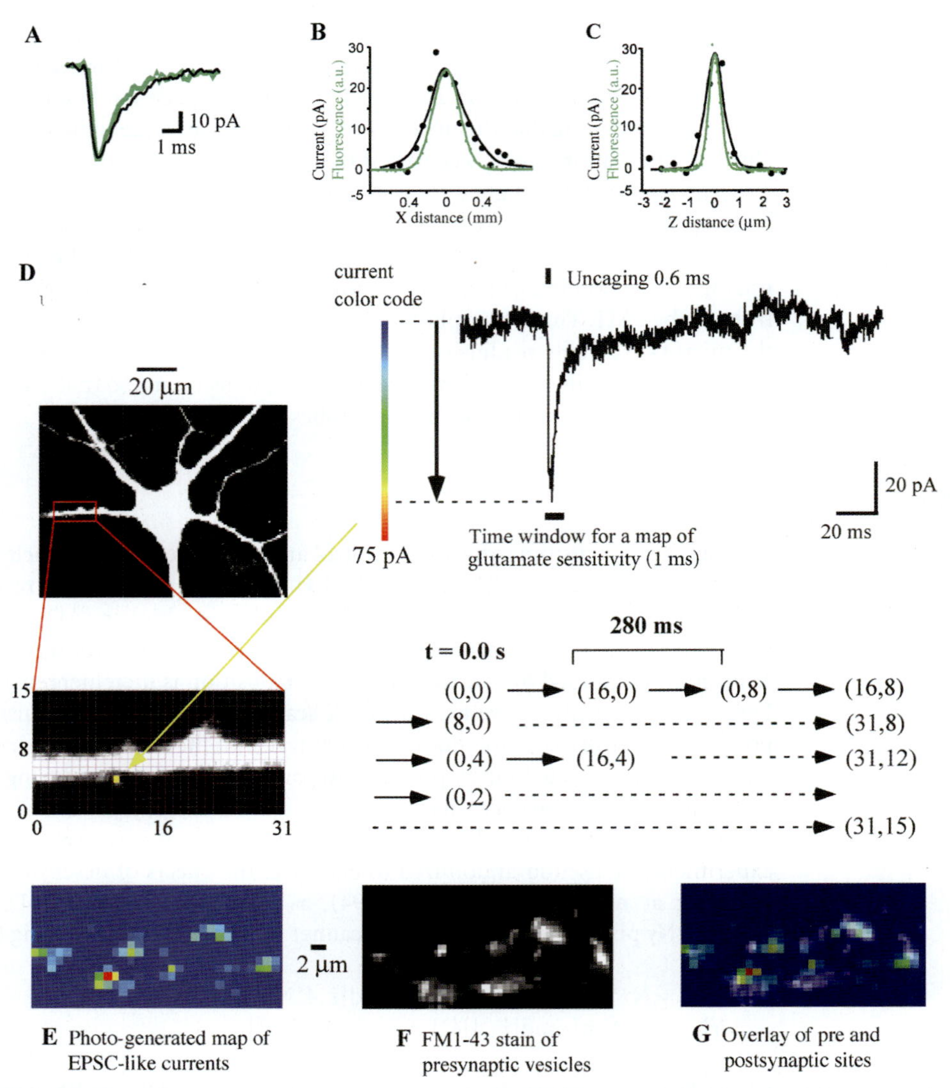

FIGURE 3. Functional mapping of glutamate sensitivities in hippocampal neurons. (*A*) Averaged current traces obtained from cultured hippocampal neurons in response to photolysis of MNI-glu for 50 μsec with 7 mW of 720-nm-wavelength light (*black*) and averaged mEPSCs (*green*) recorded from the same cell. (*B,C*) Amplitudes of 2pEPSCs (*black dots*) along horizontal (*B*) or vertical (*C*) lines crossing the center of an AMPA-receptor cluster, and fluorescence profiles of a 0.1-μm-diameter bead (*green dots*) along the horizontal (*B*) or vertical (*C*) axes. The smooth green lines (fluorescence) represent Gaussian fitting of the data, and the smooth black lines (current) are predicted from a theory as described by Matsuzaki et al. (2001). (*D*) Outline of the strategy for pseudorandom uncaging at individual pixels within a region of interest (ROI). Custom-written software permits full control of XY confocal mirrors and coordination of fluorescent imaging, uncaging period, and current measurement period by patch-clamp amplifier for a large number of pixels, in this case 512. The baseline noise of the current trace is relatively large, because the current was acquired in the absence of Mg in this particular experiment. Point uncaging of a 32 × 16-pixel ROI at each pixel in pseudorandom order generates a current map, as shown in part *E*. (*E*) Glutamate receptor density map produced upon the fura-2-filled postsynaptic space imaged in *D* presented in pseudocolor according to the scale in *D*. (*F*) In the same preparation, vesicles were stained with FM1-43 to image presynaptic terminals. (*G*) Overlay of images in *E* and *F*. (a.u) Arbitrary units.

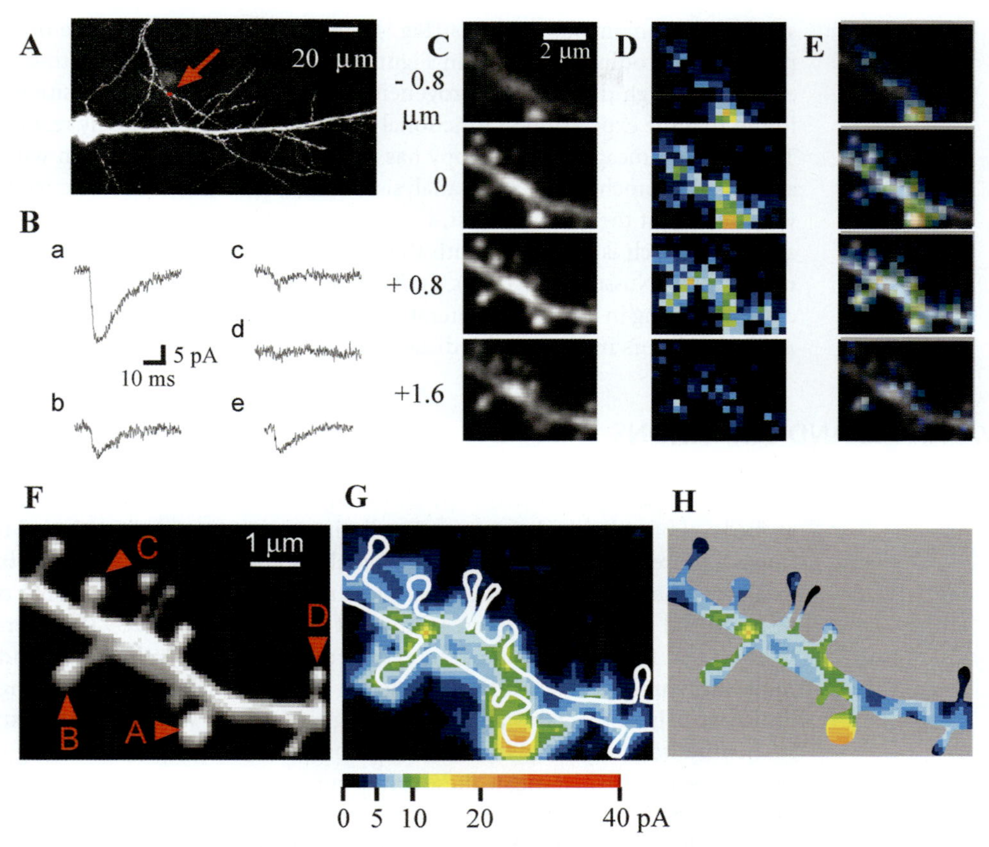

FIGURE 4. Three-dimensional mapping of glutamate receptor densities in brain slices. (*A*) Fluorescence image of a CA1 pyramidal neuron in slice. The arrow indicates the site of ROI. (*B*) 2pEPSCs induced at four different sites of the dendrite indicated as *A–D* in *F*. The traces were averages of data from four neighboring pixels. *e* shows an averaged trace of 10 mEPSCs recorded from the same neuron. (*C–D*) Fluorescence images at four *z*-axis sections as indicated (*C*), two-dimensional pseudocolor map of 2pEPSC amplitude (*D*), and overlays (*E*). (*F*) Three-dimensional reconstruction of the fluorescent images. (*G*) Maximal intensity plot of the glutamate sensitivity for the four sections shown in *D*. The map has been smoothed by linear interpolation. White lines indicate contours of the dendritic structures. (*H*) Glutamate sensitivity map within the dendritic contours.

2pEPSC (Fig. 3A, black), the time course of which was identical to those of spontaneous mEPSCs (Fig. 3A, green). Since the MNI-glu uncaging period is probably submicroseconds (Morrison et al. 2002), the extremely rapid onset (<100 μsec) of 2pEPSC indicates that the uncaged glutamate acts directly on glutamate receptors on the postsynaptic membrane. The current amplitude shows a supralinear (power exponent = 2.5 ± 0.6, mean ± S.D., $n = 7$) dependence on the radiation power (Matsuzaki et al. 2001). The time course of 2pEPSC can be predicted from the rapid photolysis (<50 μsec) of MNI-glu within a focal volume of 2PE with lateral and axial full-width-at-half-maximal (FWHM) diameter of 0.29 μm and 0.89 μm, respectively (green lines in Fig. 3B,C). The FWHM diameters of glutamate receptor activation were 0.45 μm laterally and 1.1 μm axially (black lines in Fig. 3B,C) using a high-NA (1.20) objective lens (Olympus, UplanApo60xW/IR). A slight loss of resolution was caused by the diffusion of glutamate prior to its binding to the receptors. Nevertheless, 2PE of MNI-glu permits the functional mapping of glutamate sensitivity with a spatiotemporal resolution similar to that achieved by activation of presynaptic terminals (~0.4 μm and ~0.1 msec) (Rusakov and Kullmann 1998).

An example of our methodology, when applied to hippocampal CA1 pyramidal neurons in acute slice preparations, is shown in Figure 4. The spatial resolution of AMPA-receptor mapping was slightly reduced, 0.6 μm laterally and 1.4 μm axially, as a result of the smaller NA (0.9) of the water-immersion lens (Olympus, LUMPlanFI60xW/IR2). The study therefore focused on those dendrites in which spines were relatively sparse (Fig. 4A,C, ~1 spine/μm). The time course of 2pEPSC (Fig. 4B, a–d) was similar to spontaneous mEPSCs (Fig. 4B, e). Maps of the amplitude of 2pEPSCs are represented by pseudocolor coding like those in Figure 3E, and are shown in Figure 4, D and G. Individual spines exhibited a high degree of heterogeneity in maximal glutamate sensitivity with a CV of 0.45. It was found that the expression of functional AMPA receptors was closely related to the geometry of spines. Two-photon uncaging microscopy has also been used in conjunction with nonstationary fluctuation analysis ("photochemical noise analysis") to measure the glutamate currents at a single spine head. It was found that there are, at most, 150 functional AMPA receptors in a mushroom spine (Matsuzaki et al. 2001), which is consistent with the estimated number of AMPA receptors using immunoelectron microscopy (Nusser et al. 1998). Smith et al. (2003) also used photochemical noise analysis to study distance scaling in Schaffer collateral synapses in CA1 pyramidal neurons, showing that the number of AMPA receptors increases with distance from the soma.

ADVANTAGES AND LIMITATIONS

Two-photon uncaging microscopy allows the optical stimulation of dendritic processes in brain slices at the level of individual synapses. The limitation of the technique is tied to its strength; namely, that it uses infrared light for stimulation. Spatial resolution for stimulation is 0.6 μm laterally and 1.4 μm axially for the mapping of AMPA receptors using an objective with an NA of 0.9. The resolution could be improved up to the optical resolution of the microscope (0.39 μm laterally and 1.28 μm axially) if light is used for observation of the results of two-photon stimulation and if the functional states of molecules can be measured by fluorescence imaging, e.g., by Ca^{2+} imaging. Even with these inherent limits, we believe that two-photon uncaging microscopy will provide an extremely useful tool for neuroscientists to study the dynamic, functional states of receptors in living cells. Electrical stimulation can also be used to investigate these states, but two-photon uncaging microscopy allows the same stimulus to be given repeatedly to a unique synapse (or many synapses in succession) in a complex neural network. It also provides the means to study the long-term structural and functional consequences of such stimulation (Noguchi et al. 2003; Matsuzaki et al. 2004).

The fundamental chemical event in learning and memory is hypothesized to be related to the secretion of the contents of synaptic vesicles and sensing of this signal by the postsynaptic receptors. By using two-photon uncaging from MNI-glu, it is possible to produce a very well-defined concentration jump of glutamate in a synaptic cleft, mimicking quantal release (Figs. 3A and 4B). This perhaps is the single most striking attribute of two-photon uncaging microscopy. It suggests that it would be useful to develop caged derivatives of all such secreted molecules. Although the two-photon microscope was developed by Watt Webb and coworkers in 1990, only now, with the development of new caged compounds that undergo two-photon photolysis, can the full potential of the invention to chemically probe neuronal membranes be fully realized.

ACKNOWLEDGMENTS

This work was supported by Grants-in-Aid from the Japanese Ministry of Education, Culture, Sports, Science, and Technology and from the Japan Society for the Promotion of Science (to H.K. and M.M.), and the National Institutes of Health/National Science Foundation and McKnight Fund for neuroscience (to G.E.D.), as well as by a research grant from the Human Frontier Science Program Organization (to G.E.D. and H.K.).

REFERENCES

Birge R. 1986. Two-photon spectroscopy of protein-bound chromophores. *Acc. Chem. Res.* **19:** 138–146.

Brown E.B. and Webb W.W. 1998. Two-photon activation of caged calcium with submicron, submillisecond resolution. *Methods Enzymol.* **201:** 356–380.

Brown E.B., Shear J.B., Adams S.R., Tsien R.Y., and Webb W.W. 1999. Photolysis of caged calcium in femtoliter volumes using two-photon excitation. *Biophys. J.* **76:** 489–499.

Canepari M., Nelson L., Papageorgiou G., Corrie J.E., and Ogden D. 2001. Photochemical and pharmacological evaluation of 7-nitroindolinyl and 4-methoxy-7-nitroindolinyl-amino acids as novel, fast caged neurotransmitters. *J. Neurosci. Methods* **112:** 29–42.

Corrie J.E.T., DeSantis A., Katayama Y., Khodakhah K., Messenger J.B., Ogden D.C., and Trentham D.R. 1993. Postsynaptic activation at the squid giant synapse by photolytic release of L-glutamate from a "caged" L-glutamate. *J. Physiol.* **465:** 1–8.

DelPrincipe F., Egger M., Ellis-Davies G.C.R., and Niggli E. 1999. Two-photon and UV-laser flash photolysis of the Ca^{2+} cage, dimethoxynitrophenyl-EGTA-4. *Cell Calcium* **23:** 85–91.

Denk W. 1994. Two-photon scanning photochemical microscopy: Mapping ligand-gated ion channel distributions. *Proc. Natl. Acad. Sci.* **91:** 6629–6633.

Denk W., Stricker J.H., and Webb W.W. 1990. Two-photon laser scanning microscopy. *Science* **248:** 73–76.

Ellis-Davies G.C.R. 1998. Synthesis of photolabile EGTA derivatives. *Tetrahedron Lett.* **39:** 953–957.

———. 1999. Localized photolysis of caged compounds. *J. Gen. Physiol.* **114:** 1a.

Ellis-Davies G.C.R. and Kaplan J.H. 1988. A new class of photolabile chelators for the rapid release of divalent cations: Generation of caged Ca and caged Mg. *J. Org. Chem.* **53:** 1966–1969.

Freidrich D.M. 1982. Two-photon molecular spectroscopy. *J. Chem. Educ.* **59:** 472–481.

Furata T., Wang S.S.H., Dantzer J.L., Dore T.M., Bylee W.J., Callaway E.M., Denk W., and Tsien R.Y. 1999. Bromonated 7-hydroxycoumarin-4-ylmethyls: Photolabile protecting groups with biologically useful cross-sections for two photon photolysis. *Proc. Natl. Acad. Sci.* **96:** 1193–1200.

Goeppert-Mayer M. 1931. Ueber elementarakte mit zwei quantenspruengen. *Ann. Phys.* **9:** 273–294.

Lipp P. and Niggli E. 1998. Fundamental calcium release vents revealed by two-photon photolysis of caged calcium in guinea-pig cardiac myocytes. *J. Physiol.* **508:** 801–809.

Mainen Z.F., Maletic-Savatic M., Shi S.H., Hayashi Y., Malinow R., and Svoboda K. 1999. Two-photon imaging in living brain slices. *Methods* **18:** 231–239.

Matsuzaki M., Honkura N., Ellis-Davies G.C.R., and Kasai H. 2004. Structural basis of long-term potentiation in single dendritic spines. *Nature* **429:** 761–766.

Matsuzaki M., Ellis-Davies G.C.R., Nemoto T., Miyashita Y., Iino M., and Kasai H. 2001. Dendritic spine geometry is critical for AMPA receptor expression in hippocampal CA1 pyramidal neurons. *Nat. Neurosci.* **4:** 1086–1092.

Matsuzaki M., Tachikawa A., Ellis-Davies G.C.R., Miyashita Y., Iino M., and Kasai H. 2000. Two-photon functional mapping of glutamate receptors in living hippocampal neurons. *Soc. Neurosci. Abstr.* **26:** 1131.

Morrison J., Wan P. Corrie J.E.T., and Papageorgiou G. 2002. Mechanisms of photorelease of carboxylic acids from 1-acyl-7-nitroindolines in solutions of varying water content. *Photochem. Photobiol. Sci.* **1:** 960–969.

Noguchi J., Matsuzaki M., Ellis-Davies G.C.R., and Kasai H. 2003. Functional distribution of NMDA and AMPA receptors in the dendrites of hippocampal CA1 pyramidal neurons studied with a dual-scanning two-photon microscope. *Soc. Neurosci. Abstr.* **33:** 476.5.

Nusser Z., Lujan R., Laube G., Roberts J.D., Molnar E., and Somogyi P. 1998. Cell type and pathway dependence of synaptic AMPA receptor number and variability in the hippocampus. *Neuron* **21:** 545–559.

Papageorgiou G. and Corrie J.E.T. 2001. Effects of aromatic substituents on the photocleavage of 1-acyl-7-nitroindolines. *Tetrahedron* **56:** 8197–8205.

Rusakov D.A. and Kullmann D.M. 1998. Extrasynaptic glutamate diffusion in the hippocampus: Ultrastructural constraints, uptake, and receptor activation. *J. Neurosci.* **18:** 3158–3170.

Shepherd C.J.R. and Kompfer R. 1978. Resonant scanning optical microscope. *Appl. Opt.* **17:** 2879–2882.

Smith M.A., Ellis-Davies G.C.R., and Magee J.C. 2003. Synaptic mechanisms of the distance-dependent scaling of Schaffer collateral synapses in CA1 pyramidal neurons. *J. Physiol.* **548:** 245–258.

CHAPTER 48

A Practical Guide: Chemical Two-Photon Uncaging

Diana L. Pettit and George J. Augustine

Department of Neurobiology, Duke University Medical Center, Durham, North Carolina 27710

Photolysis of caged compounds has been used to study the kinetics and spatial range of rate-limiting steps in biological processes (McCray and Trentham 1989; Parker and Yao 1991; Adams and Tsien 1993; Callaway and Katz 1993; Hess et al. 1995; Wang and Augustine 1995; Svoboda et al. 1996). However, restricting the spatial extent of photolysis is difficult when caged compounds are present throughout thick specimens, such as brain slices, because light on its way to and from the focal point will photolyze the caged compound in untargeted tissue (Fig. 1A). Out-of-focus light can be eliminated using "optical" two-photon excitation, in which two long-wavelength photons are needed to provide sufficient energy to photolyze a single caging group (Göppert-Mayer 1931; Denk 1994; Denk et al. 1995). Unfortunately, with few exceptions (Furuta et al. 1999; Matsuzaki et al. 2001; Smith et al. 2003; see Chapter 47), this approach yields little uncaging because long-wavelength photons generally do not efficiently photolyze caging groups (Denk 1994). This chapter describes a simple and economical way to eliminate out-of-focus uncaging during such experiments. The approach relies on attaching two caging groups to a single molecule, a concept proposed by Adams and Tsien (1993). The resulting requirement for two photons produces the free, uncaged molecule in proportion to the square of the light intensity (Fig. 1B). This phenomenon is called "chemical" two-photon uncaging (Pettit et al. 1997).

Chemical two-photon uncaging is useful for a wide range of applications, including both mapping of receptor location (see, e.g., Pettit et al. 1997; Pettit and Augustine 2000) and localized photostimulation of neurons via activation of excitatory glutamate receptors (Callaway and Katz 1993; Pettit et al. 1999; Dodt et al. 2002). Experimental preparations could include brain slices, cultured neurons, and, among other possibilities, whole brains in vivo. This chapter documents the utility of chemical two-photon uncaging in examining glutamate receptors of pyramidal neurons in hippocampal slices.

MATERIALS

Double-caged Glutamate Compound

Chemical two-photon uncaging requires a molecule that has two sites where inactivating cages can be attached. It is critical that each cage be sufficient to inactivate independently, so that the single-caged intermediate has no biological activity. This protocol uses a double-caged glutamate compound with cages at both the α and γ carboxyl groups (Wieboldt et al. 1994; Molecular Probes). Double-caged glutamate is hydrophobic, so it must be dissolved in dimethylsulfoxide (DMSO)<!>(to a final concentration of 0.1%) before adding it to the physiological saline solution.

Experimental Setup

The experimental setup must be capable of simultaneous whole-cell voltage-clamp recordings and confocal imaging (Noran, Odyssey), as described in Wang and Augustine (1995). Neurons are visualized using a 40× water-immersion objective (Olympus).

Light Source

Chemical two-photon uncaging requires a UV-light source. For this protocol, a continuous emission argon laser (Coherent, 305A) was used. It has been shown that 0.2–2 μJ of UV light is sufficient to obtain responses with the CNB double-caged glutamate (see Fig. 1B), corresponding to an energy density of 0.01–0.1 $\mu J/\mu m^2$ per flash. Thus, two-photon photolysis may be obtained with an inexpensive arc lamp, capacitor-discharge flash lamp, or an arc lamp connected to a pulse generator (Rapp and Guth 1988; Denk 1997). These and other continuous light sources require an electronic shutter to control the duration of illumination.

Caution: *See Appendix 3 for appropriate handling of materials marked with <!>.*

PROCEDURE

Chemical Two-Photon Uncaging

1. Prepare slices from hippocampus of rats 13–18 days old using standard techniques (Pettit et al. 1997).
2. Place the slices in a recording chamber and continuously perfuse with bicarbonate buffer (100–200 μM) containing the double-caged glutamate. Use a recirculating pump at a rate of 3–4 ml/min. The solution is stable for several hours.
3. Focus the output of a continuous emission argon laser <!> (Coherent, 305A) onto the specimen, through the objective, to form a spot ~3 μm in diameter.
4. Use conventional whole-cell patch-clamp recording methods (see, e.g., Pettit and Augustine 2000) to record the electrical currents that result from photolysis of the double-caged glutamate.

SHORT EXAMPLE OF APPLICATION

The axial resolution of photolysis using double-caged glutamate was tested by measuring currents elicited while varying the axial distance between the cell and the focal point of the UV-light beam. The relationship between the peak amplitude of glutamate-induced currents and axial position was described by Gaussian functions whose half-widths were used as a measure of axial resolution (Fig. 1C). For hippocampal pyramidal neurons, the mean half-width of this function was 43.5 μm ± 1.3 μm (S.E.M., $n = 21$) for single-caged glutamate and 18.7 ± 1.1 μm (S.E.M., $n = 18$) for double-caged glutamate. This represents a 57% improvement in axial resolution through the use of chemical two-photon uncaging. Similar results have been obtained in cerebellar Purkinje neurons (Wang et al. 2000). Limiting the duration of the photolytic pulse maximized the improvement in axial resolution. When light energy was doubled by increasing the duration of the light flashes, mean half-widths increased from 20.4 ± 1.6 μm to 31.1 ± 3.7 μm (S.E.M., $n = 7$); thus, the axial resolution depends on light intensity, just as is the case for optical two-photon uncaging.

At the energy levels required for photolysis, no signs of UV photodamage have been observed. Phototoxic effects may simply require much higher light levels than those required for photolysis. For example, Mendez and Penner (1998) demonstrated that a nonselective Ca^{2+}-permeable cation current is induced by UV irradiation of whole cells on the order of 1 $\mu J/\mu m^2$, whereas photolysis requires light levels on the order of 0.01–0.1 $\mu J/\mu m^2$.

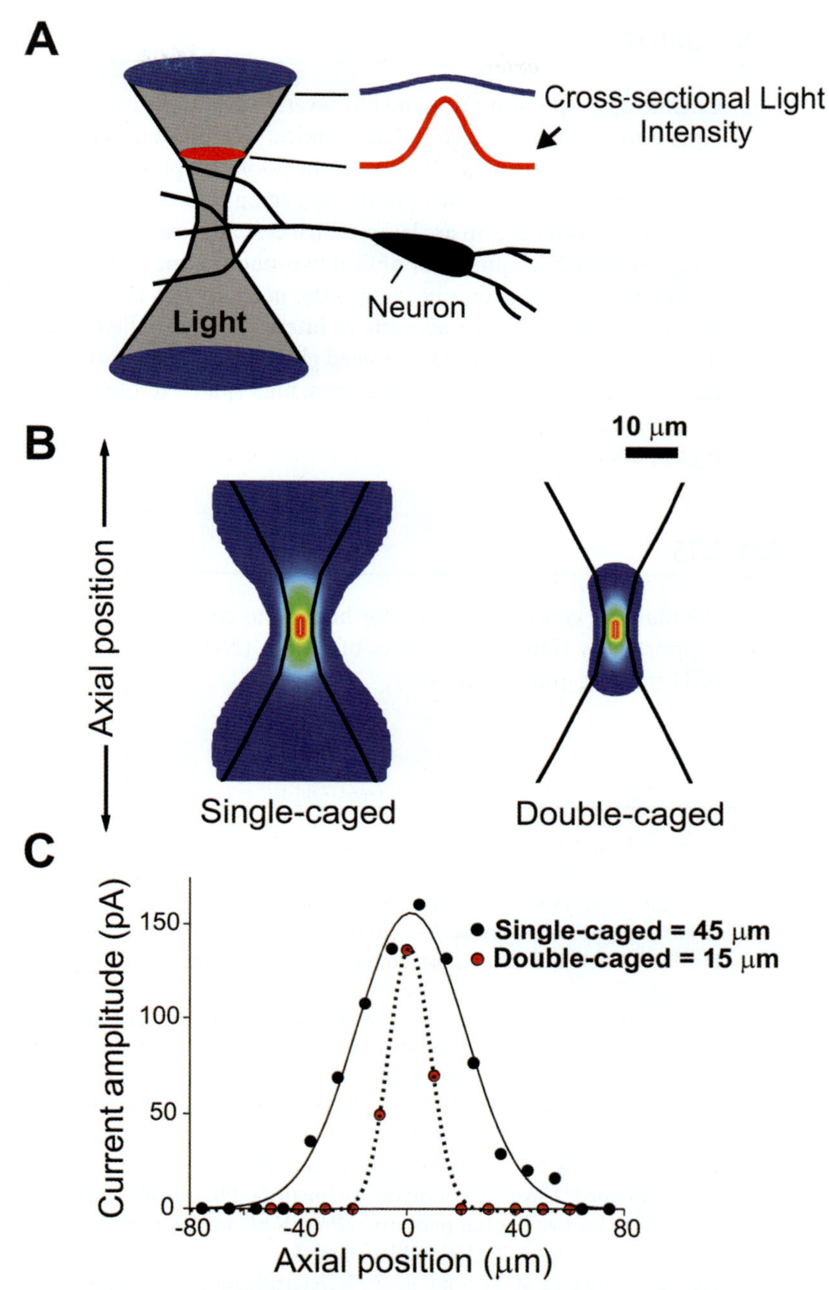

FIGURE 1. (*A, Left*) Schematic side view of a UV-light beam focused onto a neuron (*black*). At different horizontal planes (*red, blue*), the spot contains a constant total amount of light (*right*), although the light density increases as the beam comes into focus. (*B*) Side view of expected production of free glutamate following photolysis of single-caged and double-caged glutamate. (*Red*) Highest glutamate concentration; (*blue*) lowest glutamate concentration. (*C*) Relationship between axial position and the peak amplitude of currents evoked by single- and double-caged glutamate, obtained while varying the distance between the neuronal cell body and the focal plane of the UV light beam. The half-width of the Gaussian functions fit to these data is 40 μm for responses to single-caged glutamate and 15 μm for responses to double-caged glutamate. (Modified, with permission, from Pettit et al. 1997 [© Cell Press].)

ADVANTAGES AND LIMITATIONS

Chemical two-photon uncaging has several advantages over optical two-photon uncaging. Perhaps the biggest practical advantage is that chemical two-photon uncaging works more easily than optical two-photon uncaging, which is limited by the poor two-photon cross section of most existing caging groups (Denk 1994). Chemical two-photon uncaging is also more economical; effective light sources include an argon laser or even an arc lamp, which cost approximately $10,000 to $20,000. (A laser capable of the long wavelengths required for optical two-photon uncaging costs more than $100,000.) A disadvantage of chemical two-photon uncaging is the need for two independently effective caging sites, which may not be available on some agonists of interest. Because the effective spatial resolution of the method is limited by diffusion of singly uncaged glutamate from the site of photolysis, high spatial resolution also requires a high rate of photolysis. Thus, high spatial resolution is most readily achieved with a laser.

In summary, chemical two-photon uncaging is an effective and inexpensive procedure that offers high spatial resolution for local photolysis of caged compounds.

ACKNOWLEDGMENTS

We thank S. Wang and K. Gee for helping to develop chemical two-photon uncaging. This work was supported by National Institutes of Health (NIH) grants GM-65473, NS-17771, and NS-34045, and NIH training grant MH-15177.

REFERENCES

Adams S.R. and Tsien R.Y. 1993. Controlling cell chemistry with caged compounds. *Annu. Rev. Physiol.* **55:** 755–784.

Callaway E.M. and Katz L.C. 1993. Photostimulation using caged glutamate reveals functional circuitry in living brain slices. *Proc. Natl. Acad. Sci.* **88:** 7661–7665.

Crank J. 1975. *The mathematics of diffusion*, 2nd edition. Clarendon Press, Oxford, United Kingdom.

Denk W. 1994. Two-photon scanning photochemical microscopy: Mapping ligand-gated ion channel distributions. *Proc. Natl. Acad. Sci.* **91:** 6629–6633.

———. 1997. Pulsing mercury arc lamps for uncaging and fast imaging. *J. Neurosci. Methods* **72:** 39–42.

Denk W., Piston D.W., and Webb W.W. 1995. Two-photon molecular excitation in laser scanning microscopy. In *Handbook of bological confocal microscopy*, 2nd edition (ed. J.B. Pawley), pp. 445–458. Plenum Press, New York.

Dodt H.U., Eder M., Schierloh A., and Zieglgansberger W. 2002. Infrared-guided laser stimulation of neurons in brain slices. *Sci. STKE* **120:** PL2.

Furuta T., Wang S.S., Dantzker J.L., Dore T.M., Bybee W.J., Callaway E.M., Denk W., and Tsien R.Y. 1999. Brominated 7-hydroxycoumarin-4-ylmethyls: Photolabile protecting groups with biologically useful cross-sections for two photon photolysis. *Proc. Natl. Acad. Sci.* **96:** 1193–1200.

Göppert-Mayer M. 1931. Ueber Elementarakte mit zwei Quantenspruengen. *Ann. Phys.* **9:** 273.

Hess G.P., Niu L., and Wieboldt R. 1995. Determination of the chemical mechanism of neurotransmitter receptor-mediated reactions by rapid chemical kinetic methods. *Ann. N.Y. Acad. Sci.* **757:** 23–39.

Matsuzaki M., Ellis-Davies G.C., Nemoto T., Miyashita Y., Iino M., and Kasai H. 2001. Dendritic spine geometry is critical for AMPA receptor expression in hippocampal CA1 pyramidal neurons. *Nat. Neurosci.* **4:** 1086–1092.

McCray J.A. and Trentham D.R. 1989. Properties and uses of photoreactive caged compounds. *Annu. Rev. Biophys. Biophys. Chem.* **18:** 239–270.

Mendez F. and Penner R. 1998. Near-visible ultraviolet light induces a novel ubiquitous calcium-permeable cation current in mammalian cell lines. *J. Physiol.* **507:** 365–377.

Parker I. and Yao Y. 1991. Regenerative release of calcium from functionally discrete subcellular stores by inositol trisphosphate. *Proc. R. Soc. Lond. B Biol. Sci.* **246:** 269–274.

Pettit D.L. and Augustine G.J. 2000. Distribution of functional glutamate and GABA receptors on hippocampal pyramidal cells and interneurons. *J. Neurophysiol.* **84:** 28–38.

Pettit D.L., Wang S.S.-H., Gee K.R., and Augustine G.J. 1997. Chemical two-photon uncaging: A novel approach to optical mapping of glutamate receptors. *Neuron* **19:** 465–469.

Pettit D.L., Helms M.C., Lee P., Augustine G.J., and Hall W.C. 1999. Local excitatory circuits in the intermediate gray layer of the superior colliculus. *J. Neurophysiol.* **81:** 1424–1427.

Rapp G. and Guth K. 1988. A low cost high intensity flash device for photolysis experiments. *Pflugers Arch.* **411:** 200–203.

Smith M.A., Ellis-Davies G.C., and Magee J.C. 2003. Mechanism of the distance-dependent scaling of Schaffer collateral synapses in rat CA1 pyramidal neurons. *J. Physiol.* **548:** 245–258.

Svoboda K., Tank D.W., and Denk W. 1996. Direct measurement of coupling between dendritic spines and shafts. *Science* **272:** 716–719.

Wang S.S.-H. and Augustine G.J. 1995. Confocal imaging and local photolysis of caged compounds: Dual probes of synaptic function. *Neuron* **15:** 755–760.

Wang S.S., Khiroug L., and Augustine G.J. 2000. Quantification of spread of cerebellar long-term depression with chemical two-photon uncaging of glutamate. *Proc. Natl. Acad. Sci.* **97:** 8635–8640.

Wieboldt R., Gee K.R., Niu L., Ramesh D., Carpenter B.K., and Hess G.P. 1994. Photolabile precursors of glutamate: Synthesis, photochemical properties, and activation of glutamate receptors on a microsecond time scale. *Proc. Natl. Acad. Sci.* **91:** 8752–8756.

CHAPTER 49

A Practical Guide: Uncaging with Visible Light, Inorganic Caged Compounds

Leonardo Zayat, Luis Baraldo, and Roberto Etchenique

Departamento de Química Inorgánica, Analítica y Química Física, Inquimae, Facultad de Ciencias Exactas y Naturales, Universidad de Buenos Aires, Argentina

The chemistry of coordination compounds (i.e., inorganic substances centered on a metal atom surrounded by chemically bound nonmetallic atoms from ligand molecules) offers a wide range of photochemical reactions used in applications ranging from photography and lithography to light harvesting in solar cells. Despite this fact, they have not been exploited for uncaging biomolecules.

Of the 90 stable, or quasistable, elements in the periodic table, ~60 are metals, or present metal-like behavior. These elements have very rich chemistry. Most metals can form complexes with several ligands, which often present interesting photochemical properties. The properties of the coordination compounds can be tuned by carefully selecting the appropriated ligands for each coordination position available. Coordination compounds offer a number of advantages over organic substances as photoreleasers. A key advantage over other uncaging methods is the use of visible light. This chapter presents a brief panorama of metal-based caged compounds and their advantages, arising from chemical and physical properties.

The photochemistry of transition metal (d-group) compounds has been intensively studied over the last two centuries. The property of silver salts to darken in the presence of light was the basis of the widest application of photochemistry ever achieved: photography. The photochemistry of coordination compounds has also been exhaustively studied (Balzani and Carassitti 1970). The main difference between the photochemical properties of metal complexes and organic compounds is that although most organics photodecompose only in response to UV photons, many coordination complexes also respond to visible light.

A means of delivering nitric oxide to localized biological targets was proposed by Bourassa et al. (1997) in the form of photochemically active metal-nitrosyl complexes. Several such complexes have been found to release nitric oxide upon UV irradiation (Ford et al. 1998; Tfouni et al. 2003). Nitrite complexes, like the chromium-based *trans*$[Cr^{III}cyclam(ONO)_2]^+$, also undergo a photoredox reaction yielding nitric oxide as a major product (De Leo and Ford 1999). These inorganic caged compounds can release only nitric oxide, an inorganic-like biomolecule.

RUTHENIUM COMPLEXES

The use of Ru(II) complexes as photosensitizer dyes in solar cells was pioneered by Grätzel (Desilvestro et al. 1985; O'Regan and Grätzel 1991). Modified bipyridines (bpy) were used as ligands to produce

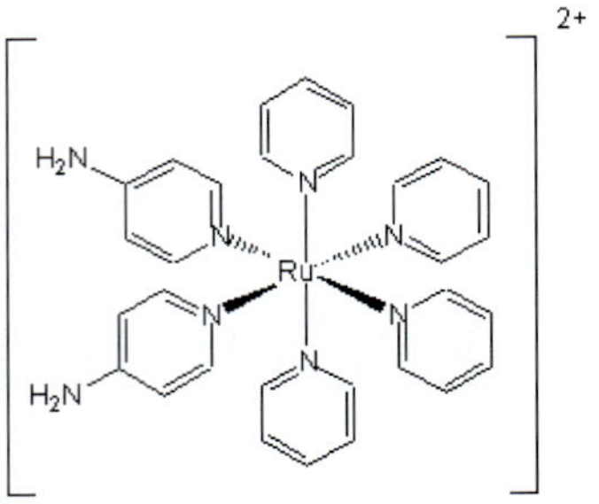

FIGURE 1. Schematic of $[Ru(bpy)_2(4AP)_2]^{2+}$.

highly efficient light-absorbing complexes. Upon irradiation, a charge transfer occurs between ruthenium (Ru) and bpy, generating an excited state that can lead to decomposition. One of the first reports of photodecomposition in a Ru-bpy complex was $[Ru(bpy)_2(py)_2]^{2+}$ (Bosnich and Dwyer 1966), which in aqueous solution releases one of the pyridines with quantum yield of $\phi = 0.4$ (Pinnick and Durham 1984).

The release of a ligand from a metal complex usually differs from the breaking of a covalent bond in the release from an organic molecule. Organic molecules tend to decompose through a series of steps, generating reactive intermediates or radicals, which decay into the final biomolecule through a dark process. Inorganic complexes usually release the ligand heterolitically, within a single photochemical step. The absence of subsequent dark reactions is the main reason that many inorganic photochemical processes are clean and very fast.

The fact that visible light ($\lambda > 400$ nm) can be used to release a ligand is a great advantage for the uncaging of compounds in living tissue, as it is much less harmful than UV light. Despite the lower energy of the coordination bond, it is strong enough to be stable at room temperature.

The existence on metals of multiple coordination positions (usually 6) allows the chemical derivatization of nonphotolabile ligands, and thus the tunability in absorption wavelength, hydrophylicity, redox potential, steric hindrance, and fluorescence properties of the compounds. Another advantage of metal caged compounds is that they can be imaged without using quartz optics. Good-quality objectives, suitable for normal light microscopy, can be used without the energy losses caused by glass absorption. Although most organic caged compounds require excitation in the 300–350-nm range, the absorption wavelength of two coordination caged compounds can be differentially tuned, providing two different visible irradiation ranges for the release of two different neurochemicals in different locations or at different times.

The first metal-based caged compound capable of releasing organic biomolecules was $[Ru(bpy)_2(4AP)_2]^{2+}$, where 4AP is 4-aminopyridine, a K^+ channel blocker, depicted in Figure 1 (Zayat

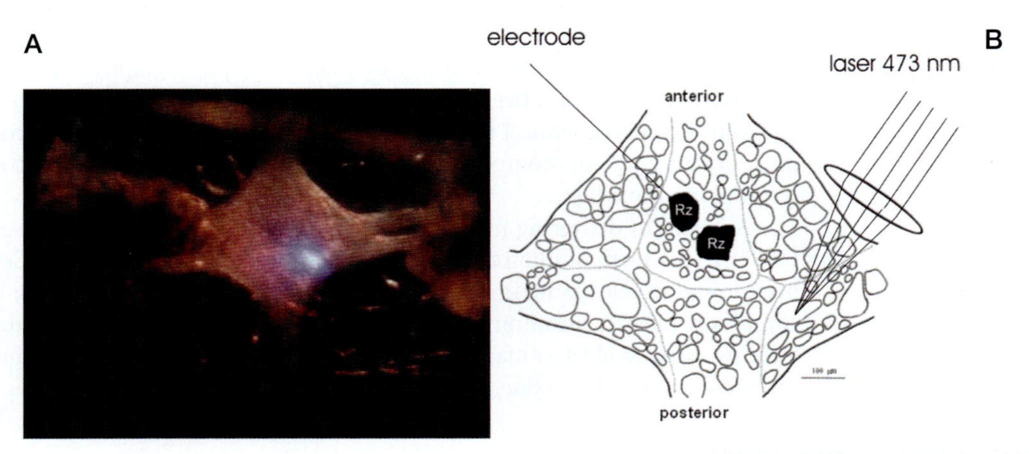

FIGURE 2. (*A*) Image of a leech ganglion during 473-nm irradiation over a posterolateral cell. (*B*) Schematic of the leech ganglion, showing the irradiated cell and the Retzius (Rz) neuron where the intracellular electrode is applied.

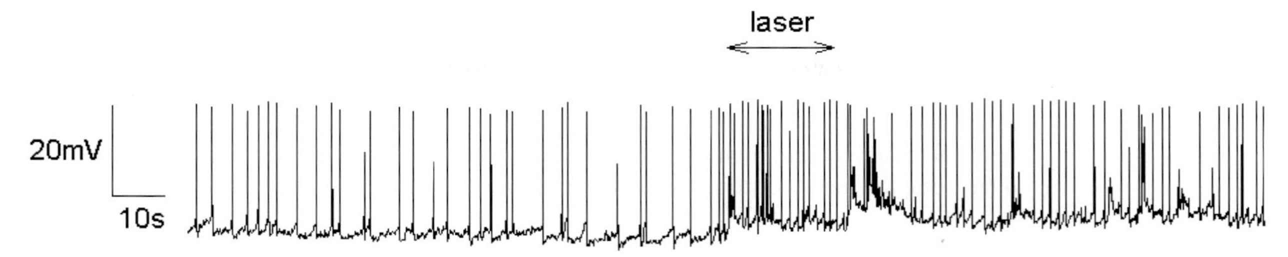

FIGURE 3. Spiking of the Rz neuron (see Fig. 2), showing the increase in frequency when a laser is applied on the lateral neuron. Direct light irradiation on the Rz neuron did not increase spiking frequency.

et al. 2003). The chloride salt of this complex is highly water-soluble, and its solutions absorb strongly at 490 nm. Irradiation at this wavelength results in decomposition, yielding free 4AP and $[Ru(bpy)_2(4AP)H_2O]^{2+}$ as the two unique photoproducts. The complex is remarkably stable (no dark decomposition is apparent in aqueous solution after 24 hr at ambient temperature), and it is nontoxic when applied in extracellular media. Quantum yield at 473 nm is ~0.03, which is significant for visible light decomposition. The caged complex fluoresces at about 590 nm, whereas the uncaged form does not fluoresce at all. This difference is useful for in situ quantification of photodecomposition. Biological assays using this complex have been made in *Hirudo medicinalis* leech ganglia (see Figs. 2 and 3) and mouse cortical slices. The complex also decomposes when subjected to focused and very high power instantaneous infrared light, probably through a two-photon absorption mechanism. Mouse cortical neurons show increased excitability when irradiated with light at 720–900 nm in the presence of Ru4AP (V. Nikolenko et al., in prep.).

Other complexes based in the same Ru-bpy fragment have been synthesized. Caged serotonin ($[Ru(bpy)_2(5HT)_2]^{2+}$) has properties similar to those of Ru4AP, with the absorption maxima shifted to slightly longer wavelengths (L. Zayat et al., in prep.). Ru complexes capable of photoreleasing nicotine, melatonin, and the NMDA (*N*-methyl-D-aspartate) agonist tetrazole-glycine are currently under investigation (R. Etchenique et al., in prep.).

PRACTICAL ISSUES

Although all of the general precautions regarding purity, stability, and biocompatibility associated with conventional caged compounds also apply, visible-light uncaging substances come with an additional requirement: All manipulations must be carried out either in the dark or using a high-quality filter that blocks the uncaging portion of the spectrum.

Most visible light sources can be used for uncaging. Xe-pulsed lamps are cheap and easy to use in the microsecond to millisecond range, but focusing is very difficult. This kind of source is suitable for kinetic investigation of fast channels. Several lasers are available in the 400–500-nm range. Argon (488 nm) lasers are suitable, but they are big and expensive, and can be replaced with the new blue DPSS Nd-YAG at 473 nm. Alternatively, Nd-YAG at 532 nm can be used for some Ru-caged compounds, but the efficiency of photorelease can be rather poor.

Ru-caged compounds also absorb strongly in the 330–380-nm range, making them compatible with the near-UV lasers used for organic caged compounds, although at the cost of their main convenience: the ability to provide mild irradiation.

Two-photon absorption of coordination caged compounds is achievable, and its potential is under investigation (R. Etchenique et al., in prep.). The visible or UV absorption bands could be excited using near infrared (720–950 nm) in a two-photon regime.

CONCLUSION

Metal-based caged compounds present several advantages over the organic phototriggers: Uncaging with visible light instead of UV, easy derivatization, wavelength tunability, intrinsic fluorescence, chemical tunability, and high stability. Although only a few caged neurochemicals have been prepared so far using the metal coordination strategy, the technique has a promising future.

REFERENCES

Balzani V. and Carassitti V. 1970. *Photochemistry of coordination compounds*. Academic Press, London.

Bosnich B. and Dwyer F.P. 1966. Bis-1,10-phenanthroline complexes of divalent ruthenium. *Aust. J. Chem.* **19:** 2229.

Bourassa J., DeGraff W., Kudo S., Wink D.A., Mitchell J.B., and Ford P.C. 1997. Photochemistry of Roussin's red salt, $Na_2[Fe_2S_2(NO)_4]$, and of Roussin's black salt, $NH_4[Fe_4S_3(NO)_7]$. In situ nitric oxide generation to sensitize gamma-radiation induced cell death. *J. Am. Chem. Soc.* **119:** 2853–2860.

De Leo M. and Ford P.C. 1999. Reversible photolabilization of NO from chromium(III)-coordinated nitrite. A new strategy for nitric oxide delivery. *J. Am. Chem. Soc.* **121:** 1980–1981.

Desilvestro J., Grätzel M., Kavan L., Moser J., and Augustynski J. 1985. Highly efficient sensitization of titanium-dioxide. *J. Am. Chem. Soc.* **107:** 2988–2990.

Ellis-Davies G.C.R. and Kaplan J.H. 1988. A new class of photolabile chelators for the rapid release of divalent cations: Generation of caged Ca and caged Mg. *J. Org. Chem.* **53:** 1966–1969.

Ford P.C., Bourassa J., Miranda K., Lee B., Lorkovic I., Boggs S., Kudo S., and Laverman L. 1998. Photochemistry of metal nitrosyl complexes. Delivery of nitric oxide to biological targets. *Coord. Chem. Rev.* **171:** 185–202.

O'Regan B. and Grätzel M. 1991. A low-cost, high-efficiency solar-cell based on dye-sensitized colloidal tio2 films. *Nature* **353:** 737–740.

Pinnick D.V. and Durham B. 1984. Photosubstitution reactions of $Ru(bpy)_2XY^{n+}$ complexes. *Inorg. Chem.* **23:** 1440–1445.

Tfouni E., Krieger M., McGarvey B.R., and Franco D.W. 2003. Structure, chemical and photochemical reactivity and biological activity of some ruthenium amine nitrosyl complexes. *Coord. Chem. Rev.* **236:** 57–69.

Zayat L., Calero, C., Albores P., Baraldo L., and Etchenique R. 2003. A new strategy for neurochemical photodelivery: Metal-ligand heterolytic cleavage. *J. Am. Chem. Soc.* **125:** 882–883.

CHAPTER 50

A Practical Guide: Infrared-guided Laser Stimulation of Neurons in Brain Slices

Hans-Ulrich Dodt, Matthias Eder, Anja Schierloh, and Walter Zieglgänsberger

Max-Planck-Institute of Psychiatry, 80804 Munich, Germany

This chapter describes an approach for easy and precise stimulation of single neurons in brain slices using caged neurotransmitters. The technique can be applied to neurobiological problems for which precise and rapid stimulation of neurons in brain slices is required. When a caged neurotransmitter is added to the superfusion medium, neurons in brain slices can be excited by shining light on them (Callaway and Katz 1993; Katz and Dalva 1994; Wieboldt et al. 1994). This very localized application allows dendrites to be scanned for the distribution of neurotransmitter receptors. Furthermore, by the stimulation of neighboring neurons, the connectivity of neuronal networks can be investigated. As laser stimulation can be performed quickly, this technique can also be used to search for synaptic connections between distant neurons with a low probability of connectivity.

MATERIALS

Setup for Infrared-guided Laser Stimulation

The setup shown in Figure 1A, manufactured by Luigs and Neumann, allows visualization of neurons in brain slices (Dodt and Zieglgänsberger 1994), with infrared-gradient contrast, and the visualization of the UV-light spot at the same time (Dodt et al. 1998). By directing the UV spot to the neuron of interest, "targeted photostimulation" can be achieved. The light from a UV source is introduced into the epifluorescence port of the microscope via a quartz fiber and reflected by a dichroic mirror onto the brain slice. As the light enters the objective from the back, the light-emitting end of the quartz fiber is demagnified, producing a small UV-light spot in the focus plane of the objective. By fine positioning of the quartz fiber, the light spot can be directed precisely to a point in the middle of the image plane.

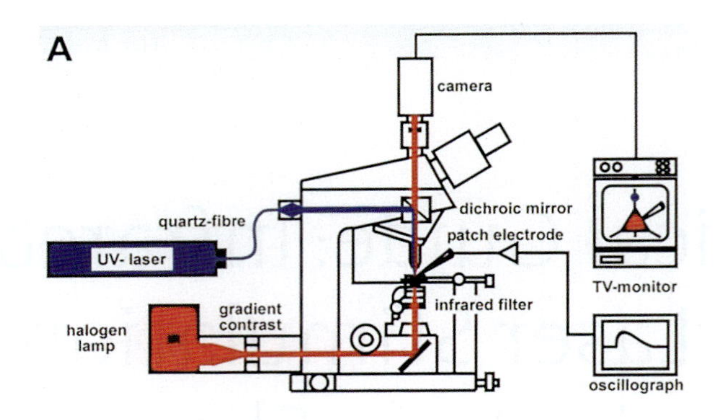

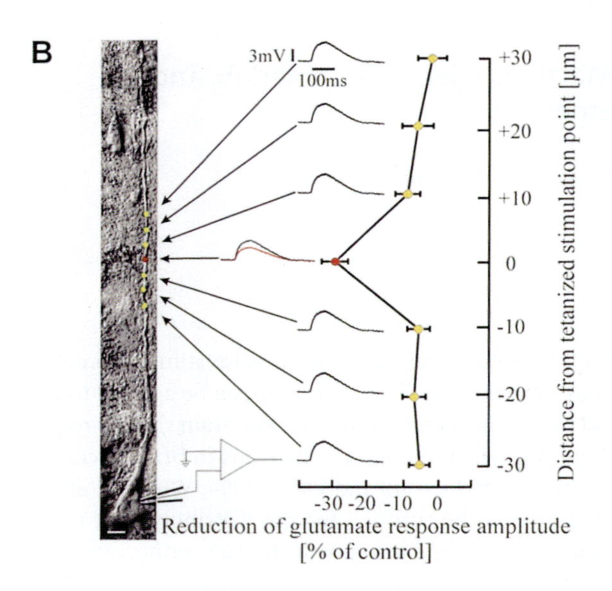

FIGURE 1. (*A*) Schematic drawing of the setup for infrared video microscopy and laser stimulation. The system is, in essence, a patch-clamp setup with infrared video microscopy using gradient contrast, combined with photostimulation by UV light, coupled by fiber-optics into the epifluorescence port. (*B*) LTD is spatially highly restricted. Example of a pyramidal neuron of neocortical layer V overlaid with seven UV-stimulation points on the dendrite (bar, 10 µm). The reduction of the glutamate responses after 5-Hz stimulation to the central point only is indicated in red. At all other stimulation sites, reductions were not significant compared with values before tetanic stimulation. Values were obtained by averaging five stimulations for each stimulation site on each neuron before and 10–20 min after the 5-Hz stimulation, and by calculating the relative reduction.

For light spots with diameters in the 10-µm range, a high-pressure mercury burner can be used as a light source. For light spots in the 1-µm range, and light flashes of a few milliseconds, a UV-laser (e.g., Enterprise II, Coherent) must be used to induce depolarizations of the neuron in the millivolt range (Dodt et al. 2002).

PROCEDURE

Infrared-guided Laser Stimulation of Neurons in Brain Slices

1. Prepare aliquots of the chosen caged neurotransmitter dissolved in standard Krebs-Ringer solution and store them at –20°C in the dark.

   Note: As caged glutamate, we use γ-(CNB-caged) glutamate (L-glutamic acid, γ-[α-carboxy-2-nitrobenzyl] ester, trifluoroacetic salt), and, as caged GABA, γ-(CNB-caged) GABA (γ-aminobutyric acid, α-carboxy-[2-nitrobenzyl] ester, trifluoroacetic salt) (both from Molecular Probes).

2. Prior to the experiment, and using the setup shown in Figure 1A, visualize the UV spot <!> by putting a small piece of white paper in the slice chamber. Then mark the position of the UV spot on the screen of the video monitor.

3. Remove the paper and replace it with a brain slice. Orient the slice such that the neuron of interest is positioned on the mark on the video monitor. The UV spot is visible in the brain slice due to autofluorescence of the tissue.

4. Submerge the slices in a chamber containing standard Krebs-Ringer solution, circulated with a peristaltic pump.

5. Just prior to the experiment, switch to a recirculation system to avoid wasting the expensive caged compound.
   a. Connect the slice chamber (via 0.3-mm inner diameter Teflon tubing) to a vial (a small standing cylinder) containing 1.5 ml of Krebs-Ringer solution (oxygenated continuously with carbogen [95% O_2/5% CO_2]).
   b. Circulate the medium in a closed loop system via a small peristaltic pump.
   c. Just before the recording is started, add a small aliquot (~30 μl) of highly concentrated caged neurotransmitter (25–100 mM) (prepared in step 1) to the vial to a final concentration of 0.25–1 mM.
6. After 20 min, scan the light spot along the dendrite of a patch-clamped pyramidal neuron and map the neurotransmitter sensitivity of the dendrite. This will reveal hot spots of neuronal excitability (Frick et al. 2001).

Caution: *See Appendix 3 for appropriate handling of materials marked with <!>.*

EXAMPLE OF APPLICATION

The above procedure showed a differential distribution of $GABA_A$ and $GABA_B$ receptors on neocortical neurons (Eder et al. 2001). Stimulation of other neurons allowed the identification of synaptic connections between the stimulated and the recorded neuron and was applied to the investigation of the neuronal network of the barrel cortex (Dodt et al. 2003).

The high spatial resolution of the laser stimulation can also be used to investigate the spatial extent of neurobiological phenomena related to neuronal plasticity (Dodt et al. 1999; Eder et al. 2002). To test for spatial specificity for, e.g., long-term depression (LTD), it is possible to stimulate sites, spaced 10 μm apart, along the neuronal dendrite (Fig. 1B). During the control and follow-up periods, 7 points along the dendrite are stimulated consecutively at low frequency (every 20 sec) by glutamate uncaging (laser spot diameter 1 μm, flashes of 3 msec). 5-Hz tetanic stimulation is applied to only one point. Figure 1B shows that a significant LTD is induced only at this point.

ADVANTAGES AND LIMITATIONS

Mapping the neurotransmitter sensitivity of the neuronal membrane with targeted photostimulation can be performed with all neurotransmitters synthesized as caged compounds, e.g., glutamate, GABA, aspartate, dopamine, noradrenaline, serotonine, and others. By adding specific receptor blockers, the distribution of many subtypes of neurotransmitter receptors can be studied, and, because only one pipette must be guided to the neuron, experiments of this type are much easier than those using microiontophoresis. As no leakage of neurotransmitter takes place prior to uncaging, the desensitization problems of iontophoresis are also circumvented.

One disadvantage of photostimulation, however, is its poor time resolution. After uncaging, the neurotransmitter diffuses over a large area of the neuronal membrane, as compared to a single synaptic site. Therefore, the time course of the induced membrane depolarization is about three times slower than that following synaptic stimulation. This makes spike timing unpredictable on the millisecond time scale. However, the experimental simplicity of laser stimulation now makes it feasible to investigate the neuropharmacology of cell compartments and to investigate the connectivity of neuronal networks.

REFERENCES

Callaway E.M., and Katz L.C. 1993. Photostimulation using caged glutamate reveals functional circuitry in living brain slices. *Proc. Natl. Acad. Sci.* **90:** 7661–7665.

Dodt H.U. and Zieglgänsberger W. 1994. Infrared videomicroscopy: A new look at neuronal structure and function. *Trends Neurosci.* **17:** 453–458.

Dodt H.U., Eder M., Frick A., and Zieglgänsberger W. 1999. Precisely localized LTD in the neocortex revealed by infrared-guided laser stimulation. *Science* **286:** 110–113.

Dodt H.U., Eder M., Schierloh A., and Zieglgänsberger W. 2002. Infrared-guided laser stimulation of neurons in brain slices. *Sci. STKE* **120:** PL2.

Dodt H.U., Frick A., Kampe K., and Zieglgänsberger W. 1998. NMDA and AMPA receptors on neocortical neurons are differentially distributed. *Eur. J. Neurosci.* **10:** 3351–3357.

Dodt H.U., Schierloh A., Eder M., and Zieglgänsberger W. 2003. Circuitry of rat barrel cortex investigated by infrared-guided laser stimulation. *Neuroreport* **14:** 623–627.

Eder M., Zieglgänsberger W., and Dodt H.U. 2002. Neocortical long-term potentiation and long-term depression: Site of expression investigated by infrared-guided laser stimulation. *J. Neurosci.* **22:** 7558–7568.

Eder M., Rammes G., Zieglgänsberger W., and Dodt H.U. 2001. GABA(A) and GABA(B) receptors on neocortical neurons are differentially distributed. *Eur. J. Neurosci.* **13:** 1065–1069.

Frick A., Zieglgänsberger W., and Dodt H.U. 2001. Glutamate receptors form hot spots on apical dendrites of neocortical pyramidal neurons. *J. Neurophysiol.* **86:** 1412–1421.

Katz L.C. and Dalva M.B. 1994. Scanning laser photostimulation: A new approach for analyzing brain circuits. *J. Neurosci. Methods* **54:** 205–218.

Wieboldt R., Gee K.R., Niu L., Ramesh D., Carpenter B.K., and Hess G.P. 1994. Photolabile precursors of glutamate: Synthesis, photochemical properties, and activation of glutamate receptors on a microsecond time scale. *Proc. Natl. Acad. Sci.* **91:** 8752–8756.

CHAPTER 51

Uncaging Calcium in Neurons

Kerry R. Delaney and Vahid Shahrezaei

Department of Biological Sciences, Simon Fraser University, Burnaby, British Columbia V5A 156, Canada

Photolabile or "caged" Ca^{2+} chelators can be used to change intracellular free-calcium concentrations $[Ca^{2+}]_i$ ($\Delta[Ca^{2+}]_i$) and, to a certain extent, control $[Ca^{2+}]$ in neurons with temporal and spatial resolution unmatched by microinjection or iontophoretic application. Although measurements of cytoplasmic $[Ca^{2+}]_i$ with fluorescent dyes are well suited to correlate $\Delta[Ca^{2+}]_i$ with physiological events such as neurotransmitter release or Ca^{2+}-dependent currents, photolabile Ca^{2+} chelators are particularly effective in demonstrating causal relationships between $\Delta[Ca^{2+}]_i$ and neuronal functions. This chapter discusses the properties of caged Ca^{2+} compounds and practical considerations for their use in neurons.

THE CHALLENGE OF MEASURING CHANGES IN INTRACELLULAR CALCIUM CONCENTRATIONS

Three main practical issues must be addressed for the effective use of caged Ca^{2+} compounds, including (1) loading the compound into the cell, (2) photoconverting the compound, and (3) estimating the spatiotemporal Ca^{2+} changes that result from photolysis. The last issue is by far the most difficult to resolve and, depending on the required accuracy of quantification, can be the main obstacle to their effective application to a particular research question.

Several factors contribute to the difficulty of determining $\Delta[Ca^{2+}]_i$ after photolysis. These can include the generation of one or more photoproducts with significant Ca^{2+} affinity, greater quantum yield for Ca^{2+}-bound versus free forms of the chelator, absorption of light across the depth of the cell, competition by endogenous buffers for released Ca^{2+}, and binding of protons and/or Mg^{2+}. Binding of photoreleased Ca^{2+} to cellular buffers and unphotolyzed caged chelator primarily determines $\Delta[Ca^{2+}]_i$ in the first few milliseconds after photolysis, whereas on a scale of 0.1 to several seconds, removal of Ca^{2+} from the cytoplasm by pumps must be considered.

Despite these difficulties, qualitative estimates of $\Delta[Ca^{2+}]_i$ often suffice to differentiate between hypotheses. In these cases, it is not necessary to solve multiple, simultaneous buffer equations for which good estimates of many of the critical parameters are usually unavailable. Precise quantification of $[Ca^{2+}]_i$ may not be required when Ca^{2+} is released in part of a neuron to test for localization of a Ca^{2+}-dependent process, or where $\Delta[Ca^{2+}]_i$ must be dissociated from transmembrane voltage changes. Quantification may also be unnecessary where control over the timing of Ca^{2+} elevation or decrease is the main objective.

Photolabile Ca^{2+} chelators have many applications. They have been used to rapidly release or buffer Ca^{2+} in order to characterize the time course and voltage dependence of Ca^{2+}-activated K^+ and nonspecific cation currents (Gurney and Lester 1987; Lando and Zucker 1989; Hall and Delaney 2002), and to study the inactivation kinetics of Ca^{2+} currents (Fryer and Zucker 1993; Johnson and Byerly 1993). Several groups have used caged Ca^{2+} to trigger neurotransmitter release to determine Ca^{2+} cooperativity, Ca^{2+} sensitivity, voltage dependence, and Ca^{2+}-mediated inhibition of transmitter release (Delaney and Zucker 1990; Mulkey and Zucker 1991; Heidelberger et al. 1994; Lando and Zucker 1994; Hsu et al. 1996; Bollmann et al. 2000; Schneggenburger and Neher 2000). Caged Ca^{2+} has also been used to estimate Ca^{2+} buffer capacity (al-Baldawi and Abercrombie 1995), to study Ca^{2+} sequestration by mitochondria and effects of pH on Ca^{2+} extrusion rates (Sidky and Baimbridge 1997). Finally, caged Ca^{2+} release (or buffering with Diazo-2) has been used to study the time course, and presynaptic versus postsynaptic dependence, of Ca^{2+} elevation for induction or maintenance of various activity-dependent synaptic-plasticity processes. Processes include facilitation, posttetanic, and long-term potentiation (Malenka et al. 1992; Mulkey and Zucker 1993; Kamiya and Zucker 1994; Fischer et al. 1997).

Effective use of caged Ca^{2+} compounds requires a thorough understanding of the principles and processes of neuronal Ca^{2+} buffering and homeostasis (see Sala and Hernandez-Cruz 1990; Neher and Augustine 1992; Roberts 1994; Neher 1995; Tank et al. 1995). A number of excellent reviews on the use of caged Ca^{2+} compounds are available (Adams and Tsien 1993; Zucker 1994; and Ellis-Davies 2003).

CHEMICAL PROPERTIES OF CAGED-CALCIUM CHELATORS

Although several cages have been developed and used during the last 15 years, only a few are currently commercially available. These are Nitr-5, DM-nitrophen (DMNP), nitrophenyl-EGTA (NP-EGTA), and Diazo-2. Properties of these and other caged compounds potentially available from the inventors are presented in Table 1.

Photolysis of a light-sensitive covalent bond, on the chelator, either decreases its affinity for Ca^{2+} (Nitr-5, DMNP, and NP-EGTA) or increases its affinity (Diazo-2; Adams et al. 1989). Therefore, it is possible to rapidly (~ 0.1 msec with a pulsed laser; ~1 msec with an arc-based flashlamp; 1 sec with an arc lamp) increase or decrease free $[Ca^{2+}]_i$ in neurons over a range from close to rest to several hundred micromolar. The photoproduct(s) has low, but not zero, affinity for Ca^{2+}; the affinity of both the photolyzed and the intact forms of the chelator, along with their relative proportions, determines the free $[Ca^{2+}]_i$ after photolysis. In practice, however, the useful range for caged Ca^{2+} compounds in neurons is often determined by the Ca^{2+} affinity of the intact cage, relative to physiological "resting" $[Ca^{2+}]_i$; this determines how much cage is bound (loaded) with Ca^{2+}, and thus how much Ca^{2+} is available for release. Figure 1 illustrates the change in free $[Ca^{2+}]_i$ produced as a function of percentage photolysis, assuming physiologically reasonable estimates of resting $[Ca^{2+}]_i$. Photolysis and release of Ca^{2+} are estimated to occur within a fraction of a millisecond for all the cages listed in Table 1 (McCray et al. 1992). Since a number of buffers are involved in setting the final equilibrium $[Ca^{2+}]_i$ after photolysis, and each of these buffers has a characteristic forward binding rate, there can be periods of time, after photolysis, when free $[Ca^{2+}]_i$ is much greater than predicted by the equilibrium solution of the relevant buffer equations (Fig. 2).

TABLE 1. Properties of photolyzable Ca^{2+} chelators

Cage	K_D for Ca (nM)	K_D Ca, products (mM)	ΔK_D for Ca	K_D for Mg (mM)	Quantum yield	Extinction coeff $M^{-1}cm^{-1}$
Azid-1	235	0.12	520	8	1	3.3×10^4
Diazo-2[a]	2200	0.0066	30	≈ 8		
DMNP	5	0.25[b]	50,000	0.0025	0.18	4,300
		3[c]	660,000			
DMNPE-4[d]	19	3.8	20,000	7	0.20	5,140
NP-EGTA	80	1	12,500	9	0.20 + Ca	975
					0.23 − Ca	
Nitr-5	145	0.006	54	8.5	0.012 + Ca	
					0.0035 − Ca	5,500

References: [a]Adams et al. (1989); [b]Ayer and Zucker (1999); [c]Kaplan and Ellis-Davies (1988); [d]DelPrincipe et al. (1999).

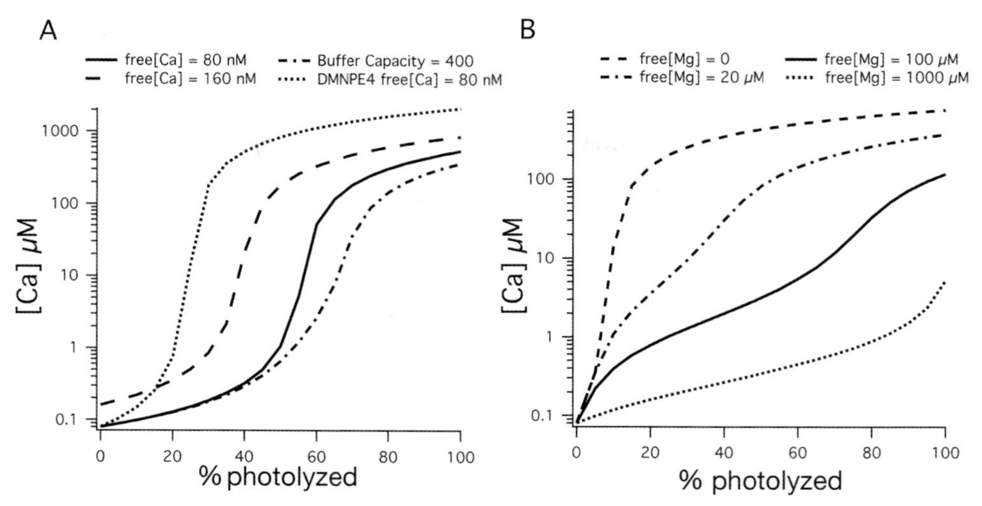

FIGURE 1. Equilibrium-free $[Ca^{2+}]_i$ as a function of photolysis for DM-nitrophen and NP-EGTA. (*A*) NP-EGTA or DMNPE-4 photolysis is simulated under conditions in which 5 mM of cage is loaded into a neuron with an endogenous buffer capacity of 100 (*solid and long dashed lines*; 200 μM of K_D 2 μM buffer) or 400 (*dotted line*). Simulations start with resting $[Ca^{2+}]_i$ of 80 nM (*solid and dotted lines*) or 160 nM (*dashed lines*) and resting $[Mg^{2+}]_i$ of 1 mM. (*B*) Free $[Ca^{2+}]$ as a function of DMNP photolysis for different levels of Mg^{2+}. For all simulations, 5 mM DMNP is added, free $[Ca^{2+}]$ prior to photolysis is 80 nM, and 200 μM of K_D 2 μM buffer is included. The free $[Mg^{2+}]$ is indicated next to each trace. Depending on the extent to which normal (i.e., ~1 mM) free Mg^{2+} levels are achieved after addition of the cage, actual changes in $[Ca^{2+}]$ will lie between the 0 Mg^{2+} (*dashed line*) and 1 mM Mg^{2+} (*dotted line*) cases. For these simulations, the K_D of photolyzed DMNP for Ca^{2+} was assumed to be 250 μM (Ayer and Zucker 1999).

Nitr-5 is a BAPTA derivative and has the advantages of low Mg^{2+} affinity, and pH independence. However, the dissociation constant (K_D) of unphotolyzed Nitr-5 is greater than typical estimates of resting $[Ca^{2+}]_i$, so most of the Nitr-5 in the cytoplasm is not Ca^{2+}-bound (~25% bound for 50 nM free $[Ca^{2+}]_i$). Other limitations of Nitr-5 are its low quantum yield and small change in affinity upon photolysis. Like BAPTA, the forward Ca^{2+}-binding rate of photolyzed Nitr-5 is fast, so Ca^{2+} released by photolysis rebinds to unphotolyzed Nitr-5 within a few milliseconds; Ca^{2+} binding to unphotolyzed Nitr-5 is probably the dominant factor in the recovery of $[Ca^{2+}]$ immediately after photolysis. Nitr-5 is therefore suitable for creating small (less than a few micromolar) flat "steps" in $[Ca^{2+}]_i$. The photoproducts of Nitr-5 are low-affinity forms of the chelator, and water. Control injections of prephotolyzed chelator in control experiments are recommended to check for effects not related to the rapid release of Ca^{2+}.

DMNP and NP-EGTA are derived from EDTA and EGTA, respectively. Their Ca^{2+}-binding affinities are sensitive to pH in the physiological range. The K_D of NP-EGTA is comparable to resting cytoplasmic $[Ca^{2+}]$, so it is better loaded than Nitr-5. Its main advantages over Nitr-5 are a much larger decrease in Ca^{2+} affinity upon photolysis and greater photolysis efficiency. As Figure 1A shows, NP-EGTA can be used to change $[Ca^{2+}]_i$ over a range from rest to several hundred micromolar. Starting from typical resting $[Ca^{2+}]_i$ and assuming moderate endogenous buffer capacity, about 40% photolysis will be sufficient to elevate steady-state $[Ca^{2+}]_i$ by several micromolar, and $[Ca^{2+}]_i$ elevation accelerates rapidly with more than 50% conversion (Fig. 1A). On the basis of the forward Ca^{2+}-binding rate, chemical equilibrium should be reached, between NP-EGTA, its photoproducts, and Ca^{2+}, within 10 msec after photolysis. Therefore, if significant photolysis is achieved within 1 msec, or less, a $[Ca^{2+}]$ transient several times larger than the equilibrium can occur, at least until a significant portion of the available cage is photolyzed (Fig. 2). DMNPE-4 is similar in most of its properties to NP-EGTA, with a significantly higher Ca^{2+} affinity, so it is more heavily loaded at resting $[Ca^{2+}]_i$, and it has a greater photolysis efficiency, so more Ca^{2+} can be released per flash. Unfortunately, however, DMNPE-4 is currently not commercially available (DelPrincipe et al. 1999).

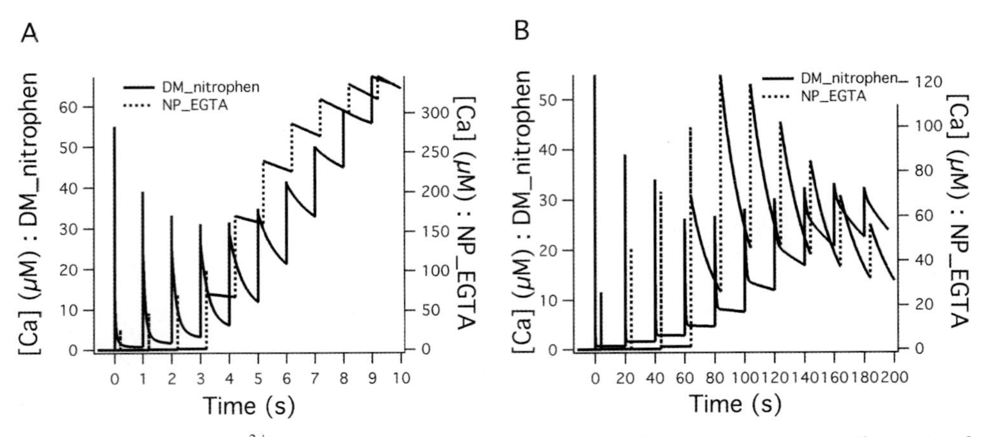

FIGURE 2. Changes in [Ca^{2+}] as a function of time after partial photolysis events. These simulations assume 20% photolysis of the remaining unphotolyzed cage per flash with 5 mM cage loaded into a neuron with an endogenous buffer capacity of 100 (200 μM of K_D 2 μM buffer). Simulations start with initial free [Ca^{2+}] of 80 nM and free [Mg^{2+}] of 100 μM. By the end of the tenth flash, about 90% of the cage is photolyzed. We approximated the effect of Ca^{2+} surface pumps and sequestration by a simple exponential mechanism that brings the free [Ca^{2+}] level back to the rest with a time constant of 5 sec in the absence of buffers and cage. Results for each cage are offset slightly for clarity. (*A*) For both cages, the earliest transient is due to rebinding to endogenous buffers and unphotolyzed cage. For DMNP, recovery of [Ca^{2+}] is initially dominated by the time required for equilibration between Ca^{2+}- and Mg^{2+}-bound, *unphotolyzed* DMNP when flashes are delivered at 1-sec intervals. After the initial rebinding to unphotolyzed cage and endogenous buffers is complete, NP-EGTA produces "steps" of $\Delta[Ca^{2+}]_i$ that recover little between flashes. (*B*) With 20 sec between photolysis events, the additional effects of Ca^{2+} removal by the cell can be seen. To simplify the simulation for NP-EGTA, the Mg^{2+} affinity was assumed to be zero which, even with a free [Mg^{2+}] of 1 mM, is reasonable given a 9 mM K_D for Mg^{2+}.

Upon photolysis, DMNP is estimated to change its affinity for Ca^{2+} by 50,000-fold (Ayer and Zucker 1999) to 660,000-fold (Kaplan and Ellis-Davies 1988). This property combined with a 5 nM dissociation constant for the unphotolyzed cage would appear to make it the ideal caged compound. However, it has a significant Mg^{2+} affinity that complicates its use in the presence of physiological levels of Mg^{2+}. In unique situations where Mg^{2+} can be eliminated or reduced, for example, by rapid exchange of low Mg^{2+} solution near the site of patching with a recording electrode, the full potential of DMNP to change $[Ca^{2+}]_i$ can be achieved (Fig. 1B) (Bollmann et al. 2000). Resting Mg^{2+} in cytoplasm is usually estimated to be in the 1 mM range. If free Mg^{2+} is present at 1 mM after loading the cage (e.g., because <1 mM of cage is present, or because the cage has been in the neuron long enough for Mg^{2+} equilibrium to be reestablished), then the Mg^{2+} significantly reduces Ca^{2+} loading, preventing exploitation of DMNP's high affinity for Ca^{2+}. The result is a reduction in the range over which $[Ca^{2+}]_i$ can be changed, and a dramatic increase in the amount of photolysis necessary to achieve micromolar $[Ca^{2+}]_i$ (Fig. 1B). Because it is unlikely that Mg^{2+} can equilibrate in neurons as rapidly as Ca^{2+} does, this presents a major unknown effect on the calculation of the degree of Ca^{2+} loading (and thus the amount of Ca^{2+} that will be released) when DMNP diffuses into distal dendritic compartments. In addition, photolysis will also cause release of Mg^{2+}, and thus, physiological effects of Mg^{2+} as a photoproduct may have to be considered in some experiments.

In the absence of Mg^{2+}, binding of released Ca^{2+} to unphotolyzed DMNP is similar to NP-EGTA occurring over a period of several (≥10) milliseconds (Zucker 1993). Thus, partial photolysis of both DMNP and NP-EGTA can produce a large $[Ca^{2+}]_i$ transient if the amount of Ca^{2+} released is capable of overwhelming the endogenous buffer capacity. When Mg^{2+} is present, it can further slow the equilibration of Ca^{2+} after photolysis of DMNP by delaying rebinding of Ca^{2+} to the unphotolyzed cage (Fig. 2A). Under some conditions, the presence of Mg^{2+} leads to a slower recovery of [Ca^{2+}] after a flash of photolysis light, and higher levels of $[Ca^{2+}]_i$ are reached sooner during continuous low-intensity photoconversion (Ayer and Zucker 1999).

Photoproducts of cages can potentially change cytosolic pH, so it is advisable to increase endogenous pH-buffer capacity when possible, e.g., by introducing at least equal amounts of pH buffer along with the chelator via injection or a patch pipette.

After release of Ca^{2+}, cellular processes such as surface pumping and internal sequestration operate to restore $[Ca^{2+}]_i$ to normal resting levels. With millimolar concentrations of caged compound, the total Ca^{2+} that must be removed after photolysis to restore resting $[Ca^{2+}]_i$ is large, so tens of seconds can be required. The temporally complex properties of Ca^{2+} (and Mg^{2+}) binding by photolyzed and unphotolyzed cages, endogenous buffers, and cellular Ca^{2+} removal processes are a problem that can sometimes be turned to experimental advantage, allowing $[Ca^{2+}]_i$ to be transiently elevated at intervals with repeated pulses of photolysis (Delaney and Zucker 1990; Zucker 1993). Some examples of the $\Delta[Ca^{2+}]_i$ transients that could result with intervals of one or several seconds between partial photolysis events for DMNP and NP-EGTA are illustrated in Figure 2.

PRACTICAL CONSIDERATIONS

Cell Loading

Cells must be loaded with enough Ca^{2+}-bound chelator for the released Ca^{2+} to be able to sufficiently alter $[Ca^{2+}]_i$ in the face of the endogenous buffering capacity of the cell. In addition, if a sustained rise is required, the release needs to oppose Ca^{2+} removal by pumps for some time. To enable released Ca^{2+} to effectively change free $[Ca^{2+}]_i$, the buffering capacity of the cage should be several times greater than that of the endogenous buffer. Chelator concentrations of 3–10 mM are attainable in practice and are sufficient for mammalian neurons. The chelators are generally soluble to ~100 mM in physiological solutions. Neurons can accommodate injections of 10% of their volume, so it is possible to achieve 2–10 mM chelator in cells, even using pressure injection.

Because caged chelators are Ca^{2+} buffers and must be introduced at high concentrations to effectively change $[Ca^{2+}]_i$, they can potentially interfere with physiologically generated Ca^{2+} changes that are required for processes like transmitter release and Ca^{2+}-dependent modulation of currents, prior to photolysis. The moderately slow forward binding rate of NP-EGTA and the slow unbinding of Mg^{2+} by DMNP reduce their effects on brief, cell-generated Ca^{2+} transients, e.g., those associated with action potentials. Nitr-5, on the other hand, binds Ca^{2+} significantly faster, which could cause problems by competing for incoming Ca^{2+}. Chelator can be loaded into neurons by

- iontophoresis through sharp electrodes,
- bath application of acetoxymethyl (AM) ester derivatives,
- pressure injection through beveled micropipettes (Lando and Zucker 1989; Delaney and Zucker 1990), and
- diffusion from whole-cell-recording pipettes (Neher and Augustine 1992; Schneggenburger and Neher 2000).

Usually, the loading method is determined by the size of the cell and the recording method that will be used to monitor the consequences of photolysis. Neurons have Ca^{2+} pumps and inward leaks that work to maintain a resting $[Ca^{2+}]_i$ of ~100 nM or less; changes in resting $[Ca^{2+}]_i$ resulting from addition of any exogenous Ca^{2+} buffer (free or preloaded with Ca^{2+}) will be transitory. Thus, following introduction of a fixed volume of caged chelator, the free $[Ca^{2+}]$ will eventually return to preinjection values and the Ca^{2+} loading of the chelator can be calculated from the free $[Ca^{2+}]$ (and free $[Mg^{2+}]$ in the case of DMNP) using the K_D(s) of the unphotolyzed chelator (Fig. 1). When loading chelators by extracellular application of AM ester derivatives, it is unlikely that influx into the cell and deesterification will occur fast enough to perturb normal resting levels of Ca^{2+} or Mg^{2+} even transiently. With whole-cell recording pipettes, the continuous exchange of cytoplasm with the solution in the tip makes it possible to achieve some level of $[Ca^{2+}]$ and/or $[Mg^{2+}]$ "clamping" in the soma or dendrite local to the pipette attachment site. This can be achieved by preloading the chelator in the pipette with a known $[Ca^{2+}]_i$ and/or reducing $[Mg^{2+}]$ to micromolar levels (Schneggenburger and Neher 2000); distal dendrites will, however, be largely at the mercy of Ca^{2+} homeostatic processes. A major unknown factor for DMNP is the extent to which [Mg^{2+}] will recover to resting concentrations by the time an experiment can be conducted after introduction of the cage.

Chelator can also transiently change the resting [Ca^{2+}]$_i$, when introduced by pressure injection. These changes occur local to the injection site, particularly if the injection proceeds rapidly into a large structure with a low surface-area-to-volume ratio, e.g., molluscan neural somata (Lando and Zucker 1989) or the squid giant presynaptic terminal (Delaney and Zucker 1990). Under these conditions, it is possible to inject the preloaded cage rapidly enough to overwhelm the endogenous buffers and the ability of surface pumps to maintain Ca^{2+} at resting levels for tens of seconds to minutes.

When structures of interest are small, a long-wavelength fluorescent indicator, such as Texas Red or fluorescein isothiocyanate (FITC), can be mixed at a known ratio with the chelator to estimate the concentration of pressure-injected chelator; fluorescence intensity of dye-chelator mixtures can be calibrated using short path-length microcuvettes. Determining the concentration of AM-ester-loaded or iontophoretically loaded chelators is more problematic and limits their use for quantitative studies.

Photolysis Efficiency

Because the change in [Ca^{2+}]$_i$ is nonlinearly related to the efficiency of cage photolysis (Fig. 1), it is important that sufficient photolysis be achieved to elevate [Ca^{2+}]$_i$ to effective levels before the cell removes the released Ca^{2+}. A twofold increase in photolysis can be the difference between the success and failure of a Ca^{2+} cage experiment. Less photolysis is required if more of the cage is loaded with Ca^{2+}. A period of activity producing a Ca^{2+} influx can be used to increase the loading of injected chelators immediately prior to photolysis. This will, of course, be associated with a rise in free [Ca^{2+}]$_i$, but it can also significantly increase the amount of Ca^{2+} that will be released (Fig. 1A). With several millimolars of cage, Ca^{2+} influx from many action potentials or repeated voltage clamp steps may be needed to load it significantly (Tank et al. 1995).

ILLUMINATION SYSTEMS

Light Sources

Photolysis can be achieved using any light source that has significant energy in the near-UV range ($\approx$350 nm), including conventional xenon and mercury arc lamps, UV lasers, and frequency-doubled or -tripled long-wavelength lasers (Parker et al. 1997). Two-photon excitation with long-wavelength light (Denk 1994) can be used and works very well for Azid-1 (Brown et al. 1999) and satisfactorily for DMNPE-4 (DelPrincipe et al. 1999). However, conversion efficiencies of commercially available cages are such that it is difficult to achieve high levels of [Ca^{2+}] without causing direct photodamage (Lipp and Niggli 1998; DelPrincipe et al. 1999). Wavelengths up to 390 nm will photolyze most caged Ca^{2+} molecules to some degree, but wavelengths in the range of 320–355 nm are particularly effective. When using arc lamps, glass lenses and filters should be avoided in the illumination path to minimize energy losses in this region.

Arc Lamps and Flash Lamps

Xenon and mercury arc lamps (75–150 W) operated in continuous mode can deliver sufficient light to photolyze ~20% of the chelator in about 1 sec, provided some effort is made to focus the arc optimally. Brief (~1-msec duration), high-intensity xenon arc discharges can be used to deliver light equivalent to 1 sec or more of continuous illumination (up to 300 mJ in the 330–380-nm band) (Rapp and Güth 1988). Mercury arc lamps can be briefly "pulsed" to deliver short-UV light with intensities equivalent to xenon arcs (Denk 1996), without the associated electromagnetic artifacts and acoustomechanical vibrations that must be dealt with when using xenon flash lamps (Zucker 1994). Xenon flash lamp systems are available from Rapp Optoelektronik, Chadwick-Helmuth, TILL Photonics, and Cairns. Chadwick-Helmuth only supplies a lamp socket, but suitable housings, including an elliptical or parabolic focusing mirror, are available from other sources. These can also be built relatively easily. TILL Photonics, Rapp Optoelektronik, and Cairns sell complete systems; Rapp Optoelektronik also offers customization of pulse width and high repetition rate on request. TILL, Rapp, and Cairns offer fiber-optic condenser couplings to standard microscopes. Charging of the capacitors limits the repetition rate of flash lamps to less than 1 Hz at full power, unless a second bank of capacitors is employed. Lower-power fast-repetition (e.g., 10–20 Hz) modes, essential for focusing and positioning the area of illumination, are provided by the commercial systems mentioned above.

Lasers

Although lasers provide a convenient collimated beam that can be focused easily to a small spot, powerful models still cost several times more than flash lamps. Frequency-doubled ruby lasers and XeF excimer lasers that have relatively long pulse durations (>10 nsec) can be pulsed at repetition rates of more than 1 Hz and can provide intensities comparable to flash lamps. Pulsed-nitrogen lasers (Spectra Physics) are less expensive and operate at 337 nm, an efficient wavelength for conversion. Their disadvantage is that they have pulse widths of 4 nsec or less, which may not be efficient for conversion of some cages. This pulse width could, however, be used to convert caged neurotransmitters and so should be tested prior to purchase. A moderately priced, frequency-tripled NdYAG laser with a 5-nsec pulse width (Minilite from Continuum) has been used for photolysis of caged inositol 1,4,5-triphosphate (Parker et al. 1997), but high photolysis rates were difficult to obtain without photodamage, possibly due to local heating. Compact, quasicontinuous pulsed UV lasers (355 nm, 100 kHz repetition rates, 100–500 mW) that are well-suited to uncaging Ca^{2+} are available from DPSS Lasers or Changchun New Industries Optoelectronics Tech. These can be gated on a pulse-by-pulse basis to provide temporal resolution of 10 μsec for the delivery of uncaging light (Shepherd et al. 2003). It is necessary to determine the level of illumination that, by itself, does not cause photodamage, especially with pulsed lasers that concentrate their power into a brief interval.

Setup and Accessories

For many purposes, a standard lamp housing attached to an epifluorescence microscope can be used, provided the illumination path is equipped with a shutter to control the duration of illumination. If visualization of fluorescence from Ca^{2+}-sensitive dyes is desired in combination with photolysis (Kao et al. 1989; Zucker 1993; Escobar et al. 1997), dichroic mirrors with high reflectance in the UV and one or more regions of the visible spectrum can be used to uncage chelators and then visualize fluorescent dyes. These are available from custom optical houses such as Chroma or Omega Optical. Caged FITC (Molecular Probes) suspended in a thin layer of agarose is very useful for initial alignment and to confirm the delivery of sufficient photolysis light prior to experimentation.

If optical fibers are used to deliver light to the preparation, then either quartz/fused silica or liquid light guides (e.g., Oriel Instruments) should be used. Liquid guides are less expensive and are available with a 3-mm diameter, which simplifies centering the arc on the input end of the fiber. Although light guides simplify light delivery, careful design of the beam coupling is required to avoid light losses. For laser coupling to microscopes, Oz Optics (Burlington, ON) and Prairie Technologies (Middleton, WI) offer several affordable custom solutions. Techniques for the fabrication and use of tapered quartz fibers are described by Eberius and Schild (2001). These can be positioned beside neurons to deliver photolysis light to areas as small as a few micrometers in diameter.

If working space permits, and focusing to a critical point is not required, the simplest way to deliver light for photolysis is to position a light source above the preparation and illuminate it directly. Elliptical reflectors make focusing the arc on the preparation simple because the UV and visible wavelengths are focused to the same point. Singlet quartz lenses, however, are not color-corrected and can require some empirical compensation for depth. Elliptical reflectors that wrap around the lamp are generally more efficient than lenses at gathering light. The large-diameter quartz lenses required to obtain a reasonable numerical aperture when distances between the arc and the specimen are greater than a few centimeters are also rather costly. High-quality, lens-based, or elliptical reflector housings can be obtained from Rapp Optoelektronik or Photon Technology International. Large elliptical mirrors (100–200 mm focal length) suitable for building a lamp housing can be obtained from Oriel Instruments, Melles Griolt, or other major optical supply companies for about $500 (2004 prices). The microbench series of opto-mechanical components from Linos Photonics or Thorlabs can be used to construct fiber-optic couplers and focusing lens assemblies without the need for custom machining.

When using arc lamps as photolysis light sources, removing infrared heat, without significant loss of UV wavelengths, requires some care; e.g., the transmission of inexpensive heat filters such as a KG1 or KG2 that are commonly used in microscope illuminators drops sharply below 350 nm. A UG11 glass filter has good transmission in the range of 300–370 nm and blocks wavelengths where fluorescent indicators such as fluo-4 or Ca-Green are excited. Such filters can be used to avoid photobleaching fluorescent indicators while photolyzing the cage with light from an external flashlamp.

If electrophysiological measurements are to be made alongside photolysis, it is important to avoid illuminating AgCl reference wires. These can generate photochemical artifacts that are easily misinterpreted as physiological responses to Ca^{2+} release.

Measurement of Photolysis Rates

Detailed methods for calibrating the rate of photolysis for DM-nitrophen are described by Zucker (1993). These methods are also suitable for other chelators. Either the change in the absorbance spectrum of the chelator following photolysis or the increase in free $[Ca^{2+}]_i$ resulting from photolysis is measured with a Ca^{2+} indicator dye using small volumes in cuvettes. Thin-walled, rectangular glass microcuvettes with path lengths of 20 or 50 μm can be purchased from VitroCom. Short lengths of these cuvettes can be filled by capillary action with solutions of chelator, and submerged in light mineral oil, or the ends sealed with grease, to prevent evaporation. Interactions between indicator dyes (e.g., fluo-3) and chelators have been reported (Zucker 1992); thus, for quantitative work, indicators must be recalibrated in the presence of the chelator. Detailed methods for in vitro calibration of photolysis using a variety of fluorescent Ca^{2+} indicators have been published by a number of groups (Neher and Zucker 1993; Thomas et al. 1993; Xu et al. 1997; Kishimoto et al. 2001). For methods that estimate photolysis rate by the rise in free $[Ca^{2+}]$ after photolysis, the concentration of chelator in the test solution must be determined accurately by absorption spectroscopy, with reference to published values of extinction coefficients, before addition of Ca^{2+}, to set free $[Ca^{2+}]_i$ near levels expected to occur in neurons.

Other Considerations

The purity of the chelator should be checked at the time of use. Commercial sources are generally about 90% pure, but the purity of each batch should be verified prior to use; batch variation is known to occur. Degradation of dissolved chelator is also known to occur, especially with repeated freeze-thaw cycles and time at −20°C. If solutions are kept for any length of time, the composition of the solution must be checked prior to use. Aliquots no larger than 50 μg should be purchased when possible. Alternatively, larger quantities can be dissolved, aliquoted, and dried, preferably using a stream of dry N_2, to provide single-use portions. Protect solutions both prior to and after loading cells from light less than 400 nm at all times, including light from microscope illuminators, using colored glass filters.

REFERENCES

Adams S.R. and Tsien R.Y. 1993. Controlling chemistry with caged compounds. *Annu. Rev. Physiol.* **55:** 755–784.

Adams S., Kao J., and Tsien R. 1989. Biologically useful chelators that take up Ca^{2+} upon illumination. *J. Am. Chem. Soc.* **111:** 7957–7968.

al-Baldawi N.F. and Abercrombie R.F. 1995. Cytoplasmic calcium buffer capacity determined with Nitr-5 and DM-nitrophen. *Cell Calcium* **17:** 409–421.

Ayer R.K., Jr. and Zucker R.S. 1999. Magnesium binding to DM-nitrophen and its effect on the photorelease of calcium. *Biophys J.* **77:** 3384–3393.

Bollmann J.H., Sakmann B., and Borst J.G. 2000. Calcium sensitivity of glutamate release in a calyx-type terminal. *Science* **289:** 953–957.

Brown E.B., Shear J.B., Adams S.R., Tsien R.Y., and Webb W.W. 1999. Photolysis of caged calcium in femtoliter volumes using two-photon excitation. *Biophys J.* **76:** 489–499.

Delaney K.R. and Zucker R.S. 1990. Calcium released by photolysis of DM-nitrophen stimulates transmitter release at squid giant synapse. *J. Physiol.* **426:** 473–498.

DelPrincipe F., Egger M., Ellis-Davies G.C., and Niggli E. 1999. Two-photon and UV-laser flash photolysis of the Ca^{2+} cage, dimethoxynitrophenyl-EGTA-4. *Cell Calcium* **25:** 85–91.

Denk W. 1994. Two-photon scanning photochemical microscopy: Mapping ligand-gated ion channel distributions. *Proc. Natl. Acad. Sci.* **91:** 6629–6633.

———. 1996. Pulsing mercury arc lamps for uncaging and fast imaging. *J. Neurosci. Methods* **72:** 39–42.

Eberius C. and Schild D. 2001. Local photolysis using tapered quartz fibres. *Pflugers Arch.* **443:** 323–330.

Ellis-Davies G.C. 2003. Development and application of caged calcium. *Methods Enzymol.* **360:** 226–238.

Escobar A.L., Rebez P., Kim A.M., Cifuentes E., Fill M., and Vergara J.L. 1997. Kinetic properties of DM-nitrophen and calcium indicators: Rapid transient response to flash photolysis. *Pfleugers Arch. Eur. J. Physiol.* **434:** 615–631.

Fischer T.M., Zucker R.S., and Carew T.J. 1997. Activity-dependent potentiation of synaptic transmission from L30 inhibitory interneurons of *Aplysia* depends on residual presynaptic Ca^{2+} but not on postsynaptic Ca^{2+}. *J. Neurophysiol.* **78:** 2061–2071.

Fryer M.W. and Zucker R.S. 1993. Ca^{2+}-dependent inactivation of Ca^{2+} current in *Aplysia* neurons: Kinetic studies using photolabile Ca^{2+} chelators. *J. Physiol.* **464:** 501–528.

Gurney A.M. and Lester H.A. 1987. Light-flash physiology with synthetic photosensitive compounds. *Physiol. Rev.* **67:** 583–617.

Hall B.J. and Delaney K.R. 2002. Contribution of a calcium-activated non-specific conductance of NMDAR mediated synaptic potentials in granule cells of the frog olfactory bulb. *J. Physiol.* **543:** 819–834.

Heidelberger R., Heinemann C., Neher E., and Matthews G. 1994. Calcium dependence of the rate of exocytosis in a synaptic terminal. *Nature* **371:** 513–515.

Hsu S.F., Augustine G.J., and Jackson M.B. 1996. Adaptation of Ca^{2+}-triggered exocytosis in presynaptic terminals. *Neuron* **17:** 501–512.

Johnson B.D. and Byerly L. 1993. Photo-released intracellular Ca^{2+} rapidly blocks Ba^{2+} current in *Lymnaea* neurons. *J. Physiol.* **462:** 321–347.

Kamiya H. and Zucker R.S. 1994. Residual Ca^{2+} and short-term synaptic plasticity. *Nature* **371:** 603–606.

Kao J.P.Y., Harootunian A.T., and Tsien R.Y. 1989. Photochemically generated cytosolic calcium pulses and their detection by Fluo-3. *J. Biol. Chem.* **264:** 8179–8184.

Kaplan J.H. and Ellis-Davies G.C. 1988. Photolabile chelators for the rapid photorelease of divalent cations. *Proc. Natl. Acad. Sci.* **85:** 6571–6575.

Kishimoto T., Liu T.T., Ninomiya Y., Takagi H., Yoshioka T., Ellis-Davies G.C., Miyashita Y., and Kasai H. 2001. Ion selectivities of the Ca^{2+} sensors for exocytosis in rat phaeochromocytoma cells. *J. Physiol.* **533:** 627–637.

Lando L. and Zucker R.S. 1989. "Caged calcium" in *Aplysia* pacemaker neurons. Characterization of calcium-activated potassium and nonspecific cation currents. *J. Gen. Physiol.* **93:** 1017–1060.

———. 1994. Ca^{2+} cooperativity in neurosecretion measured using photolabile Ca^{2+} chelators. *J. Neurophysiol.* **72:** 825–830.

Lipp P. and Niggli E. 1998. Fundamental calcium release events revealed by two-photon excitation photolysis of caged calcium in Guinea-pig cardiac myocytes. *J. Physiol.* **508:** 801–809.

Malenka R.C., Lancaster B., and Zucker R.S. 1992. Temporal limits on the rise in postsynaptic calcium required for the induction of long-term potentiation. *Neuron* **9:** 121–128.

McCray J.A., Fidler-Lim N., Ellis-Davies G.C., and Kaplan J.H. 1992. Rate of release of Ca^{2+} following laser photolysis of the DM-nitrophen-Ca^{2+} complex. *Biochemistry* **31:** 8856–8861.

Mulkey R.M. and Zucker R.S. 1991. Action potentials must admit calcium to evoke transmitter release (erratum *Nature* [1991] **351:** 419). *Nature* **350:** 153–155.

———. 1993. Calcium released by photolysis of DM-nitrophen triggers transmitter release at the crayfish neuromuscular junction. *J. Physiol.* **462:** 243–260.

Neher E. 1995. The use of Fura-2 for estimating Ca buffers and Ca fluxes. *Neuropharmacology* **34:** 1423–1442.

Neher E. and Augustine G.J. 1992. Calcium gradients and buffers in bovine chromaffin cells. *J. Physiol.* **450:** 273–301.

Neher E. and Zucker R.S. 1993. Multiple calcium-dependent processes related to secretion in bovine chromaffin cells. *Neuron* **10:** 21–30.

Parker I., Callamaras N., and Weir W.G. 1997. A high-resolution confocal laser-scanning microscope and flash photolysis system for physiological studies. *Cell Calcium* **21:** 441–452.

Rapp G. and Güth K. 1988. A low cost high intensity flash device for photolysis experiments. *Pflueg. Arch.* **411:** 200–203.

Roberts W.M. 1994. Localization of calcium signals by a mobile calcium buffer in frog saccular hair cells. *J. Neurosci.* **14:** 3246–3262.

Sala F. and Hernandez-Cruz A. 1990. Calcium diffusion modeling in a spherical neuron. Relevance of buffering properties. *Biophys. J.* **57:** 313–324.

Schneggenburger R. and Neher E. 2000. Intracellular calcium dependence of transmitter release rates at a fast central synapse. *Nature* **406:** 889–893.

Shepherd G.M., Pologruto T.A., and Svoboda K. 2003. Circuit analysis of experience-dependent plasticity in the developing rat barrel cortex. *Neuron* **38:** 277–289.

Sidky A.O. and Baimbridge K.G. 1997. Calcium homeostatic mechanisms operating in cultured postnatal rat hippocampal neurons following flash photolysis of nitrophenyl-EGTA. *J. Physiol.* **504:** 579–590.

Tank D.W., Regehr W.G., and Delaney K.R. 1995. A quantitative analysis of presynaptic calcium dynamics that contribute to short-term enhancement. *J. Neurosci.* **15:** 7940–7952.

Thomas P., Wong J.G., Lee A.K., and Almers W. 1993. A low affinity Ca^{2+} receptor controls the final steps in peptide secretion from pituitary melanotrophs. *Neuron* **11:** 93–104.

Xu T., Naraghi M., Kang H., and Neher E. 1997. Kinetic studies of Ca^{2+} binding and Ca^{2+} clearance in the cytosol of adrenal chromaffin cells. *Biophys. J.* **73:** 532–545.

Zucker R.S. 1992. Effects of photolabile calcium chelators on fluorescent calcium indicators. *Cell Calcium* **13:** 29–40.

———. 1993. The calcium concentration clamp: Spikes and reversible pulses using the photolabile chelator DM-nitrophen. *Cell Calcium* **14:** 87–100.

———. 1994. Photorelease techniques for raising or lowering intracellular Ca^{2+}. *Methods Cell Biol.* **40:** 31–63.

CHAPTER 52

A Practical Guide: Building a Simple Uncaging System

Karl Kandler,[1] Gunsoo Kim,[1] and Brigitte Schmidt[2]

[1]Department of Neurobiology, University of Pittsburgh, Pennsylvania 15261; [2]Center for Light Microscope Imaging and Biotechnology, Carnegie Mellon University, Pittsburgh, Pennsylvania 15213

This chapter describes an economical and easy-to-assemble ultraviolet (UV)-flash system suitable for rapid, focal photolysis of caged compounds. This system has been used to optically stimulate small dendritic segments of CA1 hippocampal neurons (Kandler et al. 1998) and map functional connections in auditory brain stem slices (Kim and Kandler 2003). It has also been combined with calcium imaging to selectively stimulate inhibitory presynaptic cell bodies (Kullmann et al. 2002). This approach should be applicable to a variety of cell types and systems in slices as well as dissociated neurons.

MATERIALS

Light Source

A 100-W mercury lamp burner<!> is used as the light source. In the system described here, a Series-Q arc lamp housing with a UV-grade fused silica condenser (Oriel) is used with a power supply from Opti Quip (Highland Mills). An electronic shutter, e.g., a Uniblitz Model LSG with $AlMgF_2$ coating (Vincent Associates), is used to control light-pulse length, and a pulse generator is used to time the shutter opening (e.g., Master 8, AMPI).

Optical Fiber

UV-transmitting, fused silica (high OH–) optical fibers are required. Fibers with an inner diameter of 5–50 µm can be obtained from CeramOptec Industries or Polymicro Technologies. A FITEL fiber cleaver from Fiber Instrument Sales is used to "cut" fibers to an appropriate length.

Manipulators

A fiber positioner (manual micromanipulator, see step 2 below), fiber chucks, some mounting hardware (base plates, mounting rods, screws, etc.), and breadboard (e.g., Newport; New Focus; Thorlabs) are required.

Microscope/Recording Setup

This protocol uses a fixed-stage microscope equipped with Nomarski/DIC optics (e.g., Olympus BX50WI or Zeiss Axioscope II) and a standard whole-cell patch-clamp recording setup.

Neurotransmitters

Caged substances, e.g., caged glutamate, can be obtained from Calbiochem, Dojindo Molecular Technologies, or Molecular Probes.

Caution: *See Appendix 3 for appropriate handling of materials marked with <!>.*

PROCEDURE

Assembly of the system requires no specialized knowledge and can be accomplished in 1–2 days. However, because the amount of free, uncaged compound depends on the intensity of the UV light that is delivered to the slice, it is important to maximize light capture into the fiber. This can be achieved by correctly aligning the optical fiber at the focal point of the condenser (see steps 2 and 3) and making sure that the "cut" of the fiber is straight and smooth. Examine all cuts in a stereomicroscope.

1. Set up the uncaging system as illustrated in Figure 1.
2. Attach the light-emitting end of the optical quartz fiber to an aluminum rod, mounted on a manual micromanipulator, and correctly align the optical fiber in the focal plane of the condenser; monitor the position of the fiber in relation to the brain slice through the microscope.
3. Check for optimal alignment by measuring the relative light output of the fiber with a UV-sensitive photodiode (e.g., Thorlabs) and a UV-band-pass filter (Oriel). Optimal alignment should also be tested while recording from a neuron using membrane currents to uncage glutamate as a sensitive indicator.
4. For off-line analysis, use a microscope-mounted video camera and a video frame grabber to document the exact location of uncaging sites in the slice.
5. If uncaging is unsuccessful, a good starting point for troubleshooting is to check the fiber alignment and the surface of the fiber cut.

EXAMPLES OF APPLICATIONS

In its first major application, this uncaging system was used to stimulate short segments of dendrites of hippocampal neurons with glutamate (Kandler et al. 1998). More recently, the system has been employed to map the spatial organization of glycinergic/GABAergic synaptic connections in the auditory brain stem (Kim and Kandler 2003). Figure 1b illustrates the experimental approach of a mapping experiment in a brain stem slice. A neuron in the lateral superior olive (LSO) was recorded in the whole-cell patch-clamp configuration using a pipette solution containing 60 mM chloride to increase the amplitude of chloride-mediated glycinergic and GABAergic synaptic currents. To map the location of presynaptic neurons in the medial nucleus of the trapezoid body (MNTB), glutamate was uncaged at >100 discrete locations, separated by ~50 μm, in and around the MNTB. The optical fiber (core diameter 20 μm) was moved manually, and the whole scanning procedure took ~1.5 to 2 hr. In the example shown, synaptic responses in the LSO neuron were elicited from only 12 uncaging locations (filled circles), which formed a sharp dorsoventrally oriented input map, mirroring the tonotopic organization of the MNTB-LSO pathway.

For the interpretation of input maps, information of the spatial resolution of the uncaging system is essential. This is determined by numerous factors, including the optical properties of the light delivery system, UV flash duration, the concentration of caged glutamate in the bath, the UV transmittance of the slice, and the geometry and physiological properties of the stimulated neuron. It is therefore necessary to empirically determine the spatial resolution of the uncaging system for each preparation. For the MNTB, the effective uncaging resolution was determined by recording from MNTB neurons while uncaging glutamate in their vicinity. Because only spiking MNTB neurons generate synaptic responses in the LSO, the distance from the cell body at which uncaging elicits action potentials was measured (Fig. 1c). In the example shown, spikes were elicited only if the center of the uncaging spot was less than 25 μm away from the cell body and a spatial resolution of 50 μm was assumed.

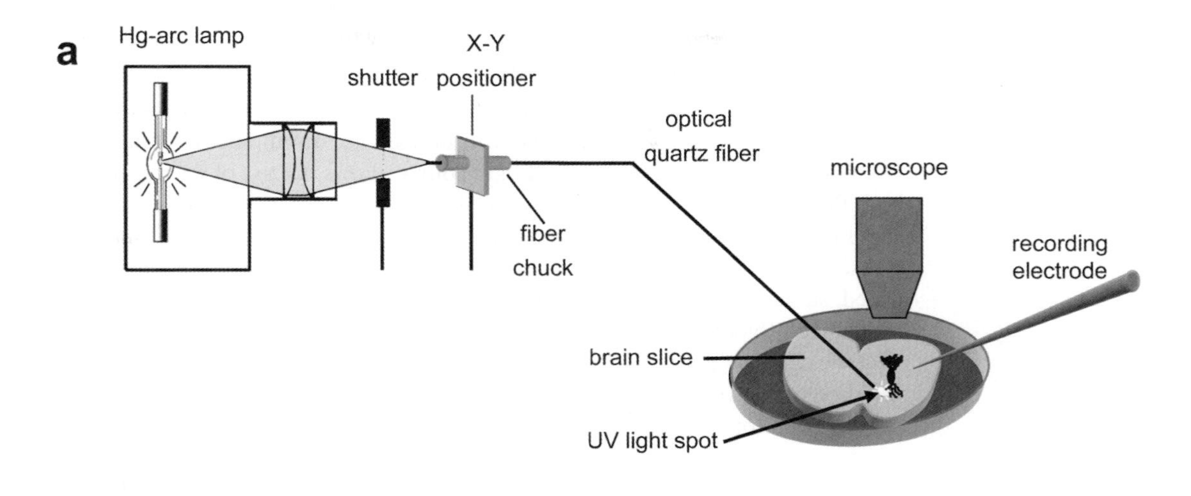

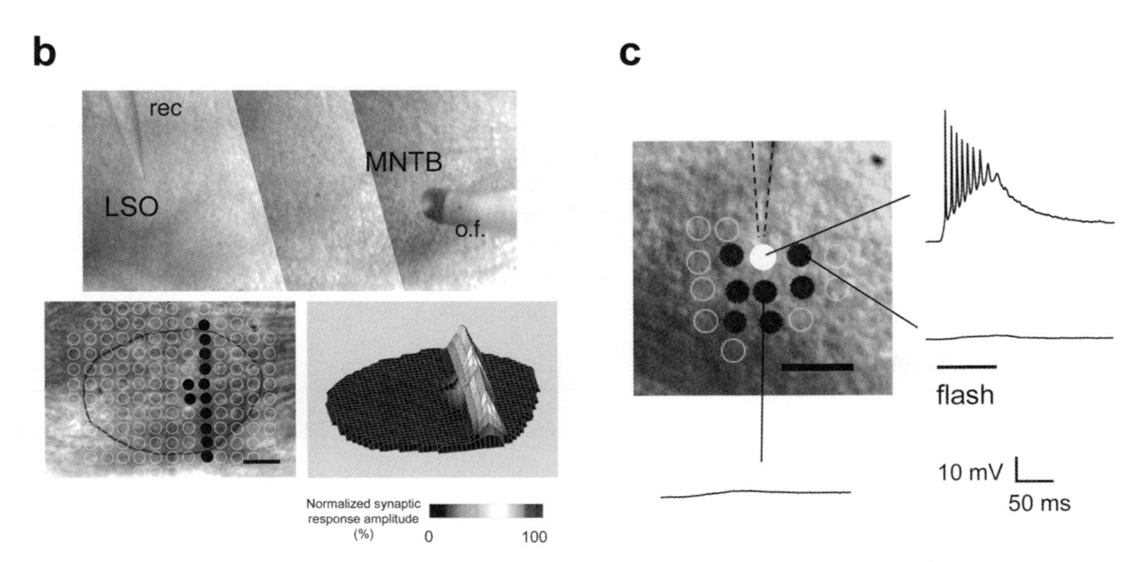

FIGURE 1. (*a*) Schematic illustration of the uncaging system. (*b*) Mapping functional connections in an auditory brain stem slice. The optical fiber (o.f.) is aimed at the medial nucleus of the trapezoid body (MNTB). A neuron in the lateral superior olive (LSO) is recorded in whole-cell patch-clamp mode (rec). In the panel below, each uncaging site in and around the MNTB (outlined by *black line*) is indicated by a circle. (*Filled circles*) Sites at which uncaging of glutamate elicited synaptic responses in the LSO neuron; (*open circles*) nonresponse sites. Successful uncaging sites form a contiguous narrow MNTB-LSO input map. Scale bar, 100 μm. In the 3D plot on the right, stimulation sites are encoded by the peak amplitude of corresponding synaptic responses. (*c*) Spatial resolution of uncaging glutamate. An MNTB neuron is recorded (electrode marked by *dashed line*) and glutamate is uncaged in its vicinity. Only the stimulation site (*white circle*) that is located directly over the cell body of the recorded MNTB neuron elicits action potentials and, thus, a synaptic response in the LSO. (*Black circles*) Uncaging sites from which only subthreshold depolarizations were elicted. (*Open circles*) Nonresponse sites. (*b*, *c*, Modified from Kim and Kandler 2003.)

ADVANTAGES AND LIMITATIONS

The major advantages of this uncaging system are its ease of use and its low costs, both of which make it easily available for most electrophysiological laboratories. This complete system comes at a fraction of the cost of a UV laser, the light source commonly used for uncaging (Katz and Dalva 1994; Wang and Augustine 1995; Pettit et al. 1997; Dantzker and Callaway 2000). Another advantage of the fiber-based system is that additional optical fibers can be added so that one setup can serve several electrophysiology rigs or that several fibers can be used for simultaneous multisite uncaging experiments (Kandler et al. 1998). The constant light source allows the use of a wide range of flash durations and high flash-repetition rates, limited only by the shutter speed.

In the configuration described here, the optical fiber is moved manually between uncaging positions, a rather time-consuming procedure. Therefore, if a large number of stimulation sites are to be tested, stable recordings over a considerable period (1–3 hr) of time are required. If such recordings are difficult to obtain, the use of a motorized micromanipulator with position-feedback ability will be necessary. Since both the fiber and the recording electrode approach the slice from above, a part of the slice is always obscured by the recording electrode. Although this can be problematic for some experiments, one usually can get around this limitation by adjusting the direction by which the recording electrode and fiber approach the slice.

Guiding the UV light to the slice with small optical fibers, instead of the optics of the microscope, minimizes the loss of UV light, allowing the use of shorter flash durations or lower concentrations of expensive caged compounds. However, use of optical fibers has the disadvantages that the UV light cannot be focused into the slice and that the diameter of the illuminated area cannot be changed during an experiment (this would require a change of fibers). These properties make the fiber system less suitable for experiments that require very restricted uncaging areas, such as is necessary for mapping the subcellular distribution of receptors (Pettit et al. 1997; Frick et al. 2001), or for stimulating single spines (Matsuzaki et al. 2001; Smith et al. 2003) which can be achieved by using high-magnification objectives combined with double-caged compounds (Wang et al. 2000) or by using two-photon uncaging (Brown et al. 1999; Furuta et al. 1999; Matsuzaki et al. 2001).

ACKNOWLEDGMENTS

Work in our laboratory has been supported by the NIDCD, the A.P. Sloan Foundation, and the Center for Neural Basis of Cognition.

REFERENCES

Brown E.B., Shear J.B., Adams S.R., Tsien R.Y., and Webb W.W. 1999. Photolysis of caged calcium in femtoliter volumes using two-photon excitation. *Biophys. J.* **76:** 489–499.

Dantzker J.L. and Callaway E.M. 2000. Laminar sources of synaptic input to cortical inhibitory interneurons and pyramidal neurons. *Nat. Neurosci.* **3:** 701–707.

Frick A., Zieglgansberger W., and Dodt H.U. 2001. Glutamate receptors form hot spots on apical dendrites of neocortical pyramidal neurons. *J. Neurophysiol.* **86:** 1412–1421.

Furuta T., Wang S.S., Dantzker J.L., Dore T.M., Bybee W.J., Callaway E.M., Denk W., and Tsien R.Y. 1999. Brominated 7-hydroxycoumarin-4-ylmethyls: Photolabile protecting groups with biologically useful cross-sections for two photon photolysis. *Proc. Natl. Acad. Sci.* **96:** 1193–1200.

Kandler K., Katz L.C., and Kauer J.A. 1998. Focal photolysis of caged glutamate produces long-term depression of hippocampal glutamate receptors. *Nat. Neurosci.* **1:** 119–123.

Katz L.C. and Dalva M.B. 1994. Scanning laser photostimulation: A new approach for analyzing brain circuits. *J. Neurosci. Methods* **54:** 205–218.

Kim G. and Kandler K. 2003. Elimination and strengthening of glycinergic/GABAergic connections during tonotopic map formation. *Nat. Neurosci.* **6:** 282–290.

Kullmann P.H., Ene F.A., and Kandler K. 2002. Glycinergic and GABAergic calcium responses in the developing lateral superior olive. *Eur. J. Neurosci.* **15:** 1093–1104.

Matsuzaki M., Ellis-Davies G.C., Nemoto T., Miyashita Y., Iino M., and Kasai H. 2001. Dendritic spine geometry is critical for AMPA receptor expression in hippocampal CA1 pyramidal neurons. *Nat. Neurosci.* **4:** 1086–1092.

Pettit D.L., Wang S.S., Gee K.R., and Augustine G.J. 1997. Chemical two-photon uncaging: A novel approach to mapping glutamate receptors. *Neuron* **19:** 465–471.

Smith M.A., Ellis-Davies G.C., and Magee J.C. 2003. Mechanism of the distance-dependent scaling of Schaffer collateral synapses in rat CA1 pyramidal neurons. *J. Physiol.* **548:** 245–258.

Wang S.S. and Augustine G.J. 1995. Confocal imaging and local photolysis of caged compounds: Dual probes of synaptic function. *Neuron* **15:** 755–760.

Wang S.S., Khiroug L., and Augustine G.J. 2000. Quantification of spread of cerebellar long-term depression with chemical two-photon uncaging of glutamate. *Proc. Natl. Acad. Sci.* **97:** 8635–8640.

CHAPTER 53

Ca^{2+} Uncaging in Nerve Terminals

Ralf Schneggenburger

AG Synaptische Dynamik und Modulation, Abteilung Membranbiophysik, Max-Planck-Institut für Biophysikalische Chemie, D-37077 Göttingen, Germany

This chapter describes a technique for Ca^{2+} uncaging in presynaptic nerve terminals, with emphasis on the calibration procedures necessary for quantitative Ca^{2+} imaging in combination with Ca^{2+} uncaging. The technique can, in principle, be applied to any nerve terminal accessible to whole-cell patch-clamp recordings. It has been successfully applied at the giant calyx of Held nerve terminals in brain-stem slice preparations (Schneggenburger and Neher 2000; Felmy et al. 2003; Wölfel and Schneggenburger 2003; see also Bollmann et al. 2000).

MATERIALS

Photolyzable Ca^{2+} Chelator

DM-nitrophen (DMN; Calbiochem) (for review, see Ellis-Davies 2003).

Fluorescent Ca^{2+} Indicator Dye

Fura-2FF pentapotassium salt (Teflabs, Austin, Texas).

Microscopic Setup

We use upright microscopes (ZEISS Axioskop FS2, or Olympus BX50WI) equipped with 60× water-immersion objectives (Olympus, LUMPlanFl). The light from a flashlamp (Rapp Optoelektronik) and the light from a monochromator used for fluorescence excitation (TILL-photonics) are fed via quartz light guides into a two-port epifluorescence condenser (TILL-photonics). A slow-scan CCD camera (480 × 640 pixel; TILL-photonics) is used for Ca^{2+} imaging with fura-2FF.

Stock Solutions

A 100 mM stock solution of DMN is made with double-distilled water. DMN must be protected from ambient light sources whenever possible and stored in aliquots at −80°C.

To examine the content of unphotolyzed DMN in the stock solution, a titration with $CaCl_2$ is made, with a solution containing 1 mM DMN and 0.1 mM of the high-affinity Ca^{2+} indicator, fura-2. A purity of approximately 80% is typically obtained. The purity of the DMN stock solution is taken into account when making the desired final solutions. The final solutions (Solutions 2–5; see Table 1) are based on a 2× stock solution (Solution 1), to which additional components are added using further stock solutions of appropriate concentration. The final volumes are made up by adding double-distilled water.

TABLE 1. Solutions

Solution no.	Purpose	Composition (in mM)
1	2X stock solution of basal components for Solutions 2–5	250 Cs-gluconate, 40 TEA-Cl<!>, 40 HEPES, 10 Na_2-ATP, 0.6 Na_2-GTP pH 7.2 (with CsOH<!>)
2	Determination of R_{min}	10 EGTA, 0.5 $MgCl_2$<!>, 1 DMN, 0.1 fura-2FF, in 1× Solution 1
3	Determination of $R_{10\mu M}$	10 DPTA<!>, 2.5 $CaCl_2$<!>, 0.5 $MgCl_2$, 1 DMN, 0.1 fura-2FF (free $[Ca^{2+}]$= 10.3 µM), in 1× Solution 1
4	Determination of R_{max}	10 $CaCl_2$, 0.5 $MgCl_2$, 1 DMN, 0.1 fura-2FF, in 1× Solution 1
5	Intracellular solution for Ca^{2+} uncaging	0.8 $CaCl_2$, 0.5 $MgCl_2$, 1 DMN, 0.1 fura-2FF, in 1× Solution 1
6	Intracellular solution for in-cell calibration of $R_{10\mu M}$	80 Cs-DPTA, 10 $CaCl_2$, 20 HEPES, 0.1 fura-2FF, 1 DMN (pH 7.2 with CsOH; free $[Ca^{2+}]$ = 10. 3 µM)
7	Extracellular solution for in-cell calibration of $R_{10\mu M}$	145 *N*-methyl-D-glucamine, 10 HEPES, 2.5 KCl<!>, 10 NaCl, 1 DPTA, 0.14 $CaCl_2$, 1 $MgCl_2$ (pH 7.4 with HCl<!>; free $[Ca^{2+}]$ = 10 µM)

Caution: *See Appendix 3 for appropriate handling of materials marked with <!>.*

CALIBRATION PROCEDURES

During experiments, the fluorescent ratio *R* of fura-2FF ($R = F_{350} / F_{380}$) is measured before and after flashes. The intracellular free-Ca^{2+} concentration, $[Ca^{2+}]_i$, is calculated from the measured ratio *R* according to Grynkiewicz et al. (1985):

$$[Ca^{2+}]_i = K_{eff} \cdot \frac{R - R_{min}}{R_{max} - R} \tag{1}$$

The calibration constants K_{eff}, R_{min}, and R_{max} must be determined experimentally. The calibration constants are first measured in vitro. Buffered Ca^{2+} solutions (Solutions 2, 3, and 4; see Table 1) are prepared using the low-affinity Ca^{2+} buffer, DPTA<!> (1,3-diamino-2-propanol-*N,N,N′,N′*,-tetraacetic acid, Fluka; K_d for Ca^{2+} = 80 µM at pH 7.2; see Heinemann et al. 1994). The solutions are filled into small glass cuvettes (50-µM path length; VitroCom), and their fluorescence is measured in the experimental setup (Fig. 1A1). These measurements also allow determination of the apparent half-maximal Ca^{2+} binding to fura-2FF (Fig. 1A2), by analyzing the fluorescence at 380 nm in the normalized spectra of Figure 1A1. Half-maximal Ca^{2+} binding of fura-2FF was found at 11.3 ± 2.5 µM $[Ca^{2+}]$ (n = 3). This value is consistent with previously published values (Xu-Friedman and Regehr 1999) when differences in ionic strength are considered, but it is lower than the value provided by the manufacturer of fura-2FF.

In-Cell Calibration

Because the properties of fluorescent dyes are likely different in cytoplasm as compared to pure aqueous solutions, an *in-cell* calibration of fura-2FF should also be made. Whole-cell voltage-clamp (holding potential, −80 mV) recordings of calyces of Held in brain-stem slices from 8–10-day-old rats are made with Ca^{2+} calibration solutions as (intracellular) pipette solutions, and the fluorescence ratio *R* is measured every 5 sec (Fig. 1B). The in-cell determination of R_{min} (using Solution 2) is usually not critical and gives values close to those measured in vitro (not shown).

For the in-cell calibration with a free $[Ca^{2+}]$ (10.3 µM) close to the K_d of fura-2FF (Fig. 1B2), a strongly Ca^{2+}-buffered intracellular solution with 80 mM DPTA, and without added ATP, was used

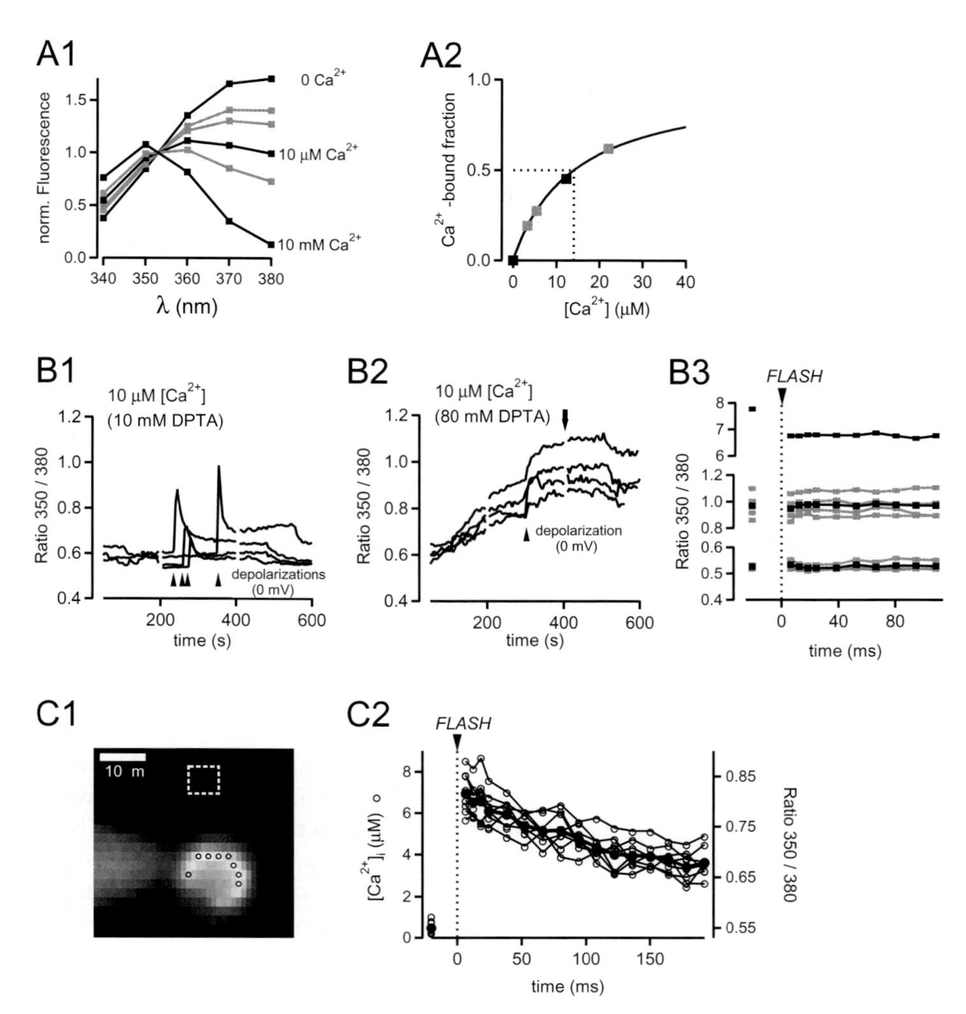

FIGURE 1. Calibration procedures for Ca^{2+} imaging in combination with Ca^{2+} uncaging. (*A1*) Spectra of fura-2FF at different free-Ca^{2+} concentrations, measured in thin glass cuvettes. (*A2*) Ca^{2+}-bound fraction of fura-2FF calculated from the measurements in *A1*. The fit with a Michaelis-Menten equation (*line*) indicates half-maximal binding at 14 μM [Ca^{2+}]. (*B1*) In-cell calibration measurements in $n = 4$ single calyx of Held nerve terminals, using a weakly Ca^{2+}-buffered intracellular solution with 10 μM free [Ca^{2+}] (Solution 3; see Table 1). Note the transient increase of the fluorescence ratio when the membrane potential was depolarized from −80 mV to 0 mV, indicating incomplete control of the cytoplasmic [Ca^{2+}] by the pipette solution. The extracellular solution also contained 10 μM [Ca^{2+}]. (*B2*) In-cell calibration measurements in $n =$ 4 calyces with a more strongly Ca^{2+}-buffered intracellular solution (Solution 6) and an extracellular solution with reduced Na^+ and Ca^{2+} concentrations (Solution 7). (*B3*) Determination of the change in the calibration ratios R_{min} (*lower panel*), $R_{10\mu M}$ (*middle*), and R_{max} (*upper panel*) after photolyzing DMN with a maximal flash. In the lower and middle panels, traces from single calyces are shown in gray, and black traces are averages across cells. (*C1*) Fluorescence image at 380-nm excitation of a calyx of Held during whole-cell recording. Pixel binning was 8 × 15. The pixels indicated by circles were analyzed to obtain the $[Ca^{2+}]_i$ traces shown in *C2*. (*Dashed box*) Region used for determining the background fluorescence. (*C2*) Fluorescence ratio change (*lines*) for the $n = 8$ individual pixels indicated in *C1*, and the corresponding change in $[Ca^{2+}]_i$ (*open circles*) before and after flash photolysis of Ca^{2+}-loaded DMN. (*Closed circles*) $[Ca^{2+}]_i$ averaged over all pixels.

(Solution 6). The extracellular solution had a low [Na^+] and a free [Ca^{2+}] of 10 μM, buffered by 1 mM DPTA (Solution 7). These solutions were designed to suppress the activity of cellular Ca^{2+}-extrusion mechanisms such as Ca^{2+}-ATPase and Na^+/Ca^{2+} exchanger. The fluorescence ratio R was observed to increase slowly during whole-cell recording (Fig. 1B2, $n = 4$ cells). After 300 sec of whole-cell recording, the holding potential was changed to 0 mV for 200 sec. This resulted in a further gradual increase in the fluorescence ratio *R*. It was assumed that at 0 mV, the intracellular and extracellular Ca^{2+} concentrations equilibrated by Ca^{2+} flux through open Ca^{2+} channels. Therefore, the calibration ratio for this intermediate [Ca^{2+}] of 10 μM ($R_{10μM}$) was taken as the value reached after 100 sec of depolarization (Fig. 1B2, arrow). This value ($R_{10\ μM}$ = 0.993 ± 0.11; mean ± S.D.; n = 4 cells) was similar to the corresponding value obtained in vitro at 10 μM free [Ca^{2+}] (R = 1.00 ± 0.1; n = 3), using Solution 3. The value of K_{eff} was calculated from $R_{10μM}$ after rearranging Equation 1.

In-cell determination of $R_{10μM}$ was also performed with an intracellular Ca^{2+} calibration solution with lower buffering strength (Solution 3). In these experiments (Fig. 1B1, n = 4 cells), depolarization to 0 mV resulted in a transient increase in the fluorescence ratio, despite the presence of nominally 10 μM [Ca^{2+}] on both sides of the membrane (the extracellular [Ca^{2+}] was 10 μM, buffered by DPTA). This indicates insufficient control of the cytoplasmic [Ca^{2+}] by the buffered Ca^{2+} solution in the patch pipette. Thus, care must be taken to design appropriate experimental conditions for the in-cell determination of fluorescence calibration constants similarly as described for the experiment shown in Figure 1B2.

Determination of Postflash Calibration Constants

Since the fluorescent properties of DMN change after photolysis (see Zucker 1992), the calibration constants used for calculating $[Ca^{2+}]_i$ following a flash must be corrected. This is done by eliciting a single flash with maximal intensity at the end of each in-cell calibration measurement. Figure 1B3 (lower panel) shows this procedure for n = 3 in-cell measurements of R_{min} (using Solution 2). The fluorescence ratio is unchanged after the flash. Figure 1B3 (middle panel) shows the same experiment for n = 4 in-cell determinations of $R_{10μM}$ (see Fig. 1B2; using Solutions 6 and 7). Although there was no apparent change in fluorescence ratio, a rise in the free [Ca^{2+}] from 10 μM to 12.3 μM is expected to occur after the flash, assuming a photolysis efficiency of 75%. Thus, for the calculation of K_{eff} after flashes, the corrected value for [Ca^{2+}] of 12.3 μM must be inserted into the rearranged Equation 1. Finally, Figure 1B3 (upper panel) shows the change of fluorescence ratio R in an in vitro determination of R_{max} (with Solution 4). From the series of calibration measurements of fura-2FF, the following average (n = 3–4) preflash and postflash (in brackets) calibration constants were obtained: R_{min} = 0.53 (0.53), K_{eff} = 150.4 μM (158.2 μM), R_{max} = 7.77 (6.79). Linearly interpolated postflash values can be used for experiments in which the flash intensity is reduced by neutral density filters.

SHORT EXAMPLE OF APPLICATION

A whole-cell patch-clamp recording from a calyx of Held nerve terminal is performed with Solution 5. After 3 min of whole-cell recording, a flash is given. The fluorescence of the cell and the surrounding tissue is measured at excitation wavelengths of 350 and 380 nm (exposure times = 5 msec; on-chip pixel binning of 8 × 15 pixel; see Fig. 1C1). The cellular fluorescence values are analyzed using the brightest eight superpixels selected in the area of the calyx under study, and the average fluorescence of an appropriate background region is subtracted. The background-corrected fluorescence ratios R are used to calculate the free $[Ca^{2+}]_i$ (Fig. 1C2). Note that $[Ca^{2+}]_i$ rises to slightly different peak values in the individual superpixels, but the variability is not large, indicating a relatively spatially homogeneous rise in $[Ca^{2+}]_i$.

ADVANTAGES AND LIMITATIONS

By stimulating nerve terminals using Ca^{2+} uncaging, spatial gradients of $[Ca^{2+}]_i$, which develop during the opening of voltage-gated Ca^{2+} channels, are avoided. By studying vesicle fusion and/or transmitter release in response to a spatially homogeneous $[Ca^{2+}]_i$ signal of measurable amplitude, it is possible to gain insights into the mechanism and Ca^{2+} requirements of vesicle fusion (Bollmann et al. 2000; Schneggenburger and Neher 2000; Felmy et al. 2003).

ACKNOWLEDGMENTS

I thank Erwin Neher for support, and for fruitful discussions on the use of photolyzable Ca^{2+} chelators.

REFERENCES

Bollmann J., Sakmann B., and Borst J. 2000. Calcium sensitivity of glutamate release in a calyx-type terminal. *Science* **289:** 953–957.

Ellis-Davies G.C.R. 2003. Development and application of caged calcium. *Methods Enzymol.* **360:** 226–238.

Felmy F., Neher E., and Schneggenburger R. 2003. Probing the intracellular calcium sensitivity of transmitter release during synaptic facilitation. *Neuron* **37:** 801–811.

Grynkiewicz G., Poenie M., and Tsien R. 1985. A new generation of Ca^{2+} indicators with greatly improved fluorescence properties. *J. Biol. Chem.* **260:** 3440–3450.

Heinemann C., Chow R.H., Neher E., and Zucker R.S. 1994. Kinetics of the secretory response in bovine chromaffin cells following flash photolysis of caged Ca^{2+}. *Biophys. J.* **67:** 2546–2557.

Schneggenburger R. and Neher E. 2000. Intracellular calcium dependence of transmitter release rates at a fast central synapse. *Nature* **406:** 889–893.

Wölfel M. and Schneggenburger R. 2003. Presynaptic capacitance measurements and Ca^{2+} uncaging reveal submillisecond exocytosis kinetics and characterize the Ca^{2+} sensitivity of vesicle pool depletion at a fast CNS synapse. *J. Neurosci.* **23:** 7059–7068.

Xu-Friedman M.A. and Regehr W.G. 1999. Presynaptic strontium dynamics and synaptic transmission. *Biophys. J.* **76:** 2029–2042.

Zucker R.S. 1992. Effects of photolabile calcium chelators on fluorescent calcium indicators. *Cell Calcium* **13:** 29–40.

CHAPTER 54

Direct Multiphoton Stimulation of Neurons and Spines

Hajime Hirase,[1,2*] Volodymyr Nikolenko,[2] and Rafael Yuste[2]

[1]Center for Molecular and Behavioral Neuroscience, Rutgers University, Newark, New Jersey 07102; [2]Department of Biological Science, Columbia University, New York, New York 10027

This chapter describes an optical method to directly stimulate a neuron (i.e., without using caged chemicals) using an infrared ultrafast mode-lock laser. This method can trigger action potentials in a targeted neuron when a laser beam is applied to the somatic membrane. Alternatively, it can mimic excitatory postsynaptic potentials (EPSPs) when applied to dendritic spines.

The protocol has been applied successfully to acute brain slices from the neocortex and hippocampus. It can be used in conjunction with slices bulk-loaded with calcium indicators, such as fura-2AM.

MATERIALS

Two-Photon Microscope

This protocol was optimized using an ultrafast mode-lock (repetition rate 76 Hz, pulse width ~100 fsec) laser beam, produced by a Ti:sapphire laser (Mira 800, Coherent), pumped by a solid-state Nd:YVO_4 laser (Verdi-5, Coherent). Similar results were obtained using a Chameleon laser (Coherent). The near-infrared laser beam was collimated and modulated by a Pockels cell (350-50BK, ConOptics), driven by a high-voltage DC power amplifier (Model 302, ConOptics). The output beam was then directed to a laser-scanning system (FluoView, Olympus) coupled to an upright microscope (BX50WI, Olympus). The Pockels cell was modulated by a signal generated using a square-wave generator (Master-8, A.M.P.I.). Transistor-to-transistor logic (TTL) signals marking the laser exposures (LINE ACTIVE and FRAME ACTIVE) were produced by the controller box (FV5-PSU) and used to trigger a square-signal-wave generator. The laser-scanning system and Z-positioning of the microscope objective were controlled by dedicated software supplied by the vendor (FluoView Software, Olympus). A more detailed description of the setup is available elsewhere (Majewska et al. 2000; Nikolenko et al. 2003).

Biological Tissue

The described protocol has been successfully applied using juvenile (postnatal day 7–14) C57 mouse neocortical and hippocampal acute slices (~300 µm thickness).

Fluorescent Indicator

To visualize the target cell, fluorescent indicator was applied either in bulk (fura-2AM) or with the whole-cell patch clamp (e.g., Alexa-488, fura-2, or Calcium Green I, Molecular Probes).

Caution: *See Appendix 3 for appropriate handling of materials marked with <!>.*

PROCEDURE

Direct Multiphoton Stimulation of Neurons and Spines

1. Prepare acute slices using a standard method. Bulk-load the slice with a fluorescent indicator (e.g., with fura-2AM ester), as described in Chapter 44. Alternatively, label a neuron by intracellular recording (e.g., whole-cell patch-clamp method) with a fluorescent indicator (Chapter 33).
2. Place the slice under the microscope objective (LUMPlanFL/IL 40x [NA = 0.8] or 60x [NA = 0.90], Olympus) and maintain in oxygenated artificial cerebrospinal fluid (125 mM NaCl, 3 mM KCl<!>, 10 mM glucose, 26 mM $NaHCO_3$, 1.1 mM NaH_2PO_4<!>, 2 mM $CaCl_2$<!>, 1 mM $MgSO_4$<!>; pH adjusted to 7.4, ~37°C) superfused through the recording chamber. Hold the slice in the recording chamber by placing a small weight ("harp") consisting of a C-shaped metal piece and a few thin nylon strings, making arcs over the metal piece. Alternatively, allow the slice to stick to the surface of the chamber by briefly (<1 min) draining its perfusate.
3. Scan the slice (xyt scan mode) with minimal laser intensity<!> to visualize fluorescence-loaded cells. Choose a cell to be stimulated (target) from the visualized cell population.
4. Once the target cell has been selected, locate the target cell at the soma level (or the corresponding dendritic spine) by adjusting the *z*-axis position of the objective using the control software. Mark the pixel coordinates of the cell's location and bring the focal point to the membrane surface by gradually lifting the *z*-axis position of the objective until the fluorescence signal diminishes.

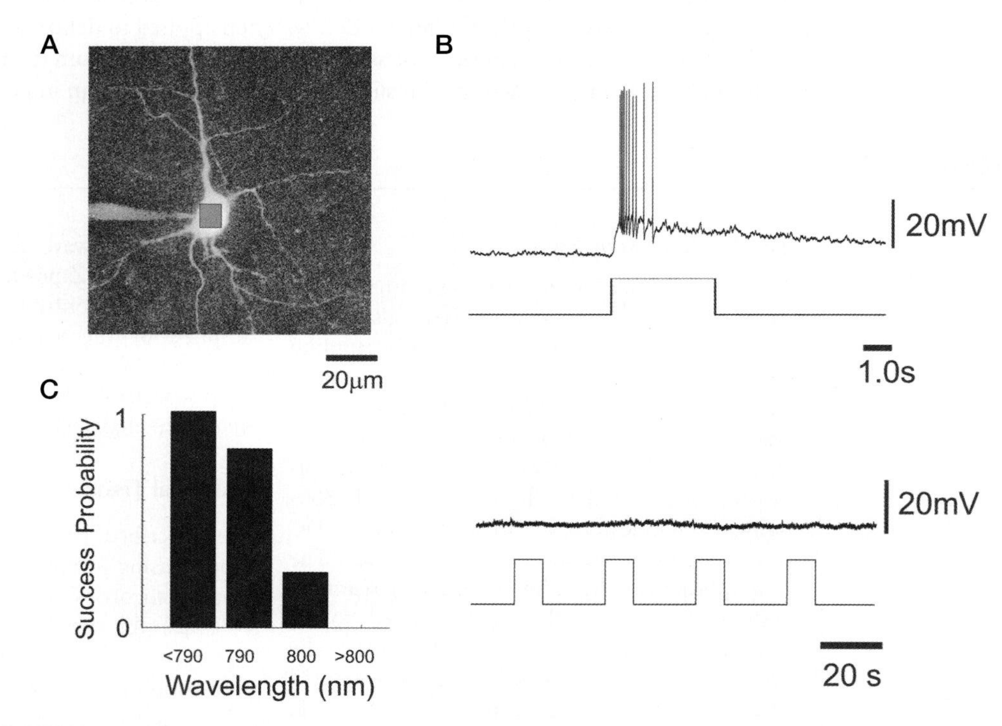

FIGURE 1. Optical stimulation of a cortical layer-5 pyramidal cell. (*A*) A layer-5 pyramidal cell was labeled with Alexa-488 through a whole-cell patch pipette. Laser irradiation was targeted at the dark square area. (*B, Top*) When the membrane surface was targeted, the cell was depolarized in response to laser irradiation and produced a burst of action potentials. (*Bottom*) When the soma was targeted with the laser, no change in membrane potential was recorded. (*C*) Bar chart indicating the relationship between probability of optical stimulation and excitation wavelength of the mode-lock laser beam. The probability of induction of optical stimulation is dependent on the excitation wavelength. (Modified, with permission, from Hirase et al. 2002 [©Wiley].)

5. Program the modulation signal to the Pockels cell in such a way that a high-intensity (>20 mW at the exit of the objective) excitation beam is applied only in the target cell area, keeping the rest of the scan with minimized laser intensity.
6. Repeated scanning (typically several times) of the membrane surface area of the target cell will lead to gradually increasing depolarizing responses of the target cell, eventually causing it to emit action potentials. Realization of the action potential generation can be monitored optically by calcium imaging of the soma, or electrically via whole-cell intracellular recording or by placing a glass pipette (~1 MΩ) close to the target cell to record the extracellular spikes.

EXAMPLE OF APPLICATION TO SOMATA AND SPINES

To confirm the effectiveness of the stimulation protocol, the membrane potential of the target cell was recorded using the whole-cell patch-clamp method. The cell was fluorescently labeled with Alexa-488 (Molecular Probes). Figure 1 shows an example of direct two-photon stimulation of a neuronal soma.

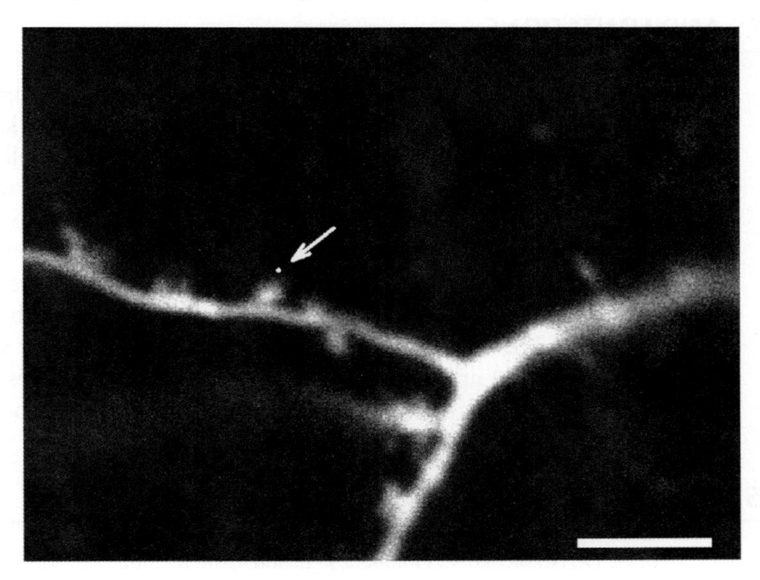

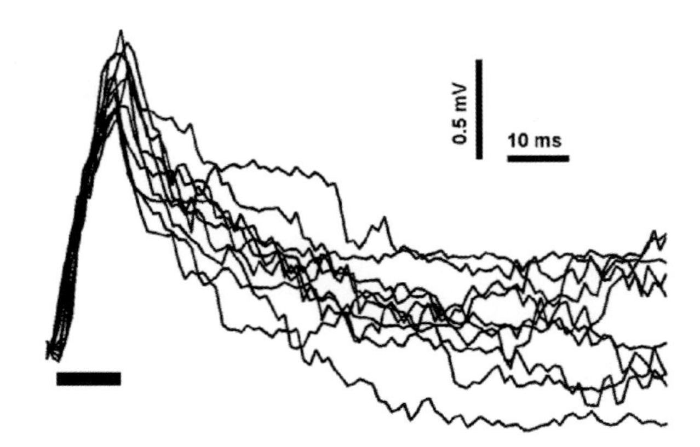

FIGURE 2. Optical stimulation of a spine. Experiment performed in a layer-V large pyramidal neuron from a P13 mouse visual cortex. The neuron was labeled with Alexa-488 (250 μM) using a patch pipette, and the spot above the spine was irradiated with 80–90 mW of average power on the sample (*top panel*). (*Bottom panel*) Simultaneous electrophysiological recording from the soma during the experiment. (*Black bar*) Duration of the irradiation. The laser stimulation produces a depolarizing response that resembles an EPSP. This effect is produced repeatedly.

A neuron was loaded with a fluorescent dye (Alexa 488) through a patch pipette (Fig. 1A). When the laser beam was targeted to the intra-soma area, there was a very small change in the membrane potential (Fig. 1B; bottom trace). In contrast, the membrane potential elicited a depolarizing response to the laser irradiation to the membrane and eventually emitted bursts of action potentials (Fig. 1B; top trace).

The probability of successful optical stimulation depends heavily on the wavelength of excitation (Fig. 1C). Generally, a wavelength of 790 nm or less produces reliable optical stimulation. Wavelengths greater than 800 nm do not produce reliable optical stimulation.

Figure 2 shows an example of a similar experiment, where the stimulated region is next to a dendritic spine (upper panel, spot). Laser irradiation triggers a small depolarization (bottom panel), which resembles the amplitude and kinetics of an EPSP. This optically triggered depolarization can be repeatedly elicited without any significant changes in its amplitude and kinetics. It is possible that the depolarization is due to the activation of the presynaptic terminal impinging on the spine. Regardless of its mechanisms, this method can be used to reliably stimulate any portion of the dendritic tree, without the use of caged compounds.

ADVANTAGES AND LIMITATIONS

An obvious advantage of this protocol is that it does not require the use of any extra chemicals (such as caged compounds) other than fluorescent markers used to label cell populations. This may be particularly important for in vivo experiments, where the application of exogenous chemicals is not straightforward. With recent advances in molecular biology, a wide variety of transgenic mice expressing fluorescent protein in various cell types have become available. These allow a combination of optical stimulation, as described here, and optical probing (Peterlin et al. 2000) to be used to map neuronal connectivity in various kinds of labeled interneurons without electrophysiology or immunocytochemistry. In addition, optical stimulation of spines can be a useful tool to investigate dendritic biophysics. The stimulation regime presented in this chapter works more efficiently for wavelengths shorter than 800 nm. When designing these experiments, it is important that the indicator be effectively excited at the wavelength of the beam used for direct multiphoton excitation (see Chapter 44).

REFERENCES

Hirase H., Nikolenko V., Goldberg J.H., and Yuste R. 2002. Multiphoton stimulation of neurons. *J. Neurobiol.* **51:** 237–247.

Majewska A., Yiu G., and Yuste R. 2000. A custom-made two-photon microscope and deconvolution system. *Pflugers Arch.* **441:** 398–408.

Nikolenko V., Nemet B., and Yuste R. 2003. A two-photon and second-harmonic microscope. *Methods* **30:** 3–15.

Peterlin Z.A., Kozloski J., Mao B.Q., Tsiola A., and Yuste R. 2000. Optical probing of neuronal circuits with calcium indicators. *Proc. Natl. Acad. Sci.* **97:** 3619–3624.

CHAPTER 55

A Practical Guide: Imaging Microglia in Live Brain Slices and Slice Cultures

Dana Kurpius and Michael E. Dailey

Department of Biological Sciences, University of Iowa, Iowa City, Iowa 52242–1324

This chapter describes methods for fluorescence labeling and time-lapse confocal imaging of microglia in neonatal or adult rodent brain tissue slices. The techniques used permit real-time analyses of microglial structure and dynamic remodeling in live rodent brain tissues. They are applicable to acutely prepared tissue slices from developing and adult animals, and to slice cultures derived from early postnatal day (<PND7) animals (see Dailey and Waite 1999; Stence et al. 2001; Petersen and Dailey 2004).

MATERIALS

Fluorescent Conjugates of Isolectin B_4 (IB_4) Derived from *Griffonia simplicifolia* Seeds

Several fluorescent IB_4 conjugates are commercially available (FITC-conjugated IB_4 from Sigma; and Alexa Fluor-488-, -568-, -or -647-labeled IB_4 from Molecular Probes). They all yield strong microglial staining and serve as excellent markers of parenchymal microglia in rodent brain tissues (Streit and Kreutzberg 1987). The Alexa-647 fluorophore is especially suitable for work in thick tissue slices because longer-wavelength light exhibits less tissue scatter.

Confocal Microscope

We use commercially available laser-scanning confocal microscopes (e.g., Leica) equipped with several lasers, computer-driven focus, and automated image capture.

Caution: *See Appendix 3 for appropriate handling of materials marked with <!>.*

PROCEDURE

Tissue Slice Preparation

1. Prepare acute tissue slices from developing and adult animals (a), or slice cultures from early <PND7 animals (b):
   a. Prepare acutely isolated hippocampal tissue slices from neonatal (PND5–7) or adult (6–8 weeks) Sprague Dawley rats or C57BL/6 mice (e.g., Harlan) as described by Dailey and Waite (1999).

Note: Briefly, rats or mice are decapitated, brains are removed in ice-cold Hanks' balanced salt solution (HBSS; GIBCO-BRL) supplemented with additional dextrose (6 mg/ml). Hippocampi are excised and sectioned transversely (400-μm-thick slices) using a manual tissue chopper (e.g., Stoelting).

b. For cultures, prepare slices from neonatal (PND5–7) animals following the static filter culture method described by Stoppini et al. (1991).

Fluorescent Labeling of Microglia

2. Label microglia in neonatal tissue slices and slice cultures by incubation for 1 hr at 37°C in culture medium (50% modified Eagle Medium [MEM], 25% HBSS, 25% horse serum, 2 mM glutamine, 0.044% $NaHCO_3$, and 10 units/ml penicillin-streptomycin<!>) containing fluorescent-conjugated IB_4 lectin (5 μg/ml). For mature hippocampal slices, label microglia by incubation in Sakmann's medium (125 mM NaCl, 2.5 mM KCl<!>, 2 mM $CaCl_2$<!>, 1.3 mM $MgSO_4$<!>, 1.25 mM

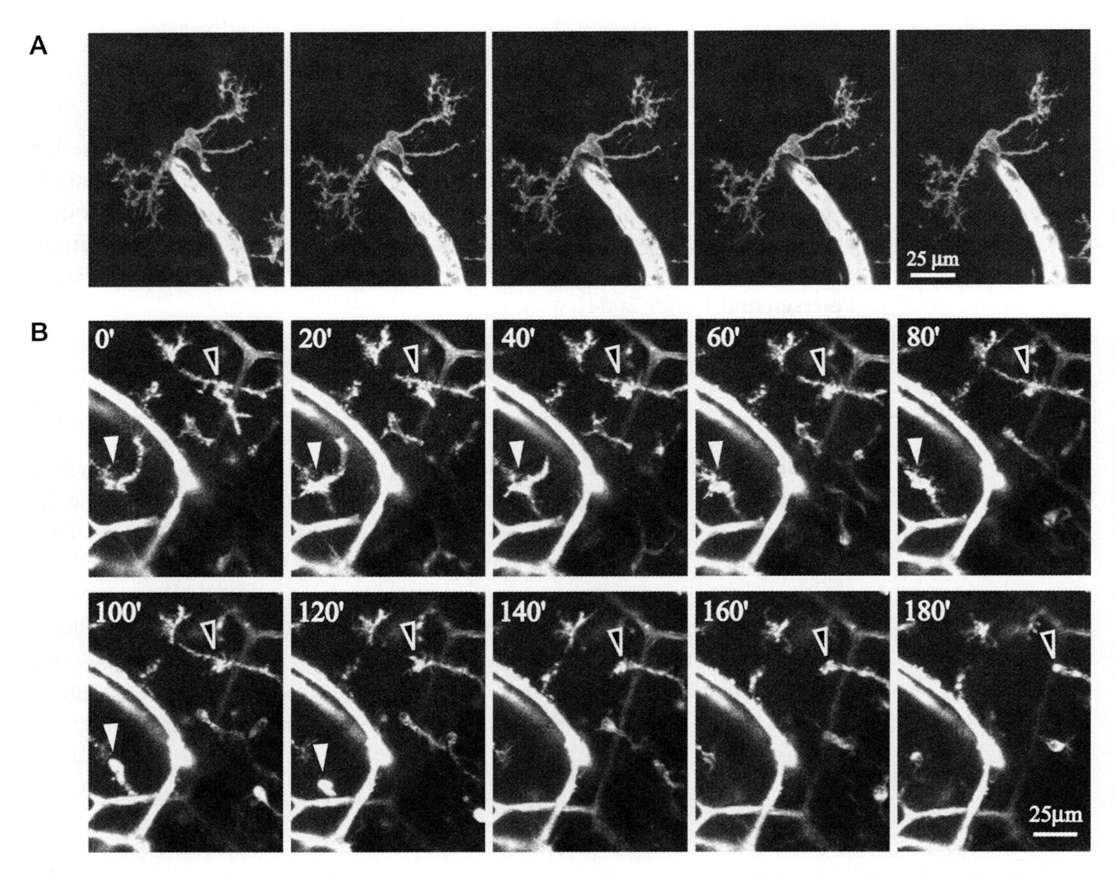

FIGURE 1. IB_4-labeled microglia in neonatal mouse brain tissue slices. (*A*) Tilt series of confocal images (15° rotation intervals) showing different perspectives of a single, resting microglial cell adjacent to a capillary in a fixed mouse hippocampal slice. The projection images were generated from a total of 102 optical sections collected at 0.2-μm *z*-step intervals using a 63×/1.2 water lens. The rotation series was created using Voxblast, a three-dimensional rendering program. Crossing eyes to fuse adjacent images gives perception of depth. (*B*) Time-lapse sequence in a live tissue slice shows activating microglia (*arrowheads*) transitioning from a ramified to locomotory state. Cells begin to retract extant branches prior to the extension of new, highly motile branches.

NaH_2PO_4<!>, 25 mM $NaHCO_3$, 10 mM glucose) containing Alexa 647-IB_4 (5 μg/ml)<!>. Incubate the slices in a 1.5-ml microcentrifuge tube and then rinse with dye-free medium for 5 min.

Confocal Imaging

3. To capture a large field of view ($\sim$500 $\times$ 500 μm) containing several (typically 10–30) cells, collect images using a 20$\times$/0.7 dry Plan Apo objective lens (e.g., Leica).

   Note: A typical time-lapse imaging experiment captures 15 confocal optical planes at 2-μm *z*-step intervals spanning a total depth of $\sim$30 μm in the axial (*z*) dimension. Imaging sessions typically last 8–15 hr, with no obvious signs of phototoxicity.

EXAMPLE OF APPLICATION

Figure 1A shows a tilt series of a confocal three-dimensional reconstruction of a single microglial cell situated near a capillary in a fixed neonatal mouse hippocampal tissue slice. Figure 1B shows microglial cell motility following activation in a time-lapse sequence from a neonatal mouse hippocampal slice.

ADVANTAGES AND LIMITATIONS

Tissue slices preserve the native tissue architecture and in vivo cellular milieu better than dissociated cell culture preparations. Using these methods, it has been shown that microglial cells undergo significant changes in cell morphology and motility over the first few hours after activation in live rodent brain tissue slices (Stence et al. 2001). Moreover, slice cultures have permitted studies of ongoing microglial behaviors over several days in vitro, including phagocytic uptake of dead neurons (Petersen and Dailey 2004). Thus, rodent tissue slices and slice cultures provide excellent models for studying microglial activation and motility in a central nervous system (CNS) tissue environment.

Fluorescent conjugates of IB_4 provide a simple, one-step method for labeling microglia in live or fixed brain tissues. IB_4 penetrates tissues efficiently and selectively stains microglia and endothelial cells lining blood vessels by binding to terminal α-D-galactosyl residues of glycoproteins on these cells. This selective surface staining provides excellent structural definition of microglia. The variety of fluorescent IB_4 conjugates allows flexibility for double- or triple-labeling with other fluorescent markers, including markers of live or dead cell nuclei (Dailey and Waite 1999).

Because IB_4 is nontoxic to live cells, it is amenable to time-lapse imaging in living tissues. However, there are some limitations that must be considered. First, it is essential to reduce illumination intensity to minimize phototoxic damage or bleaching. Fortunately, IB_4-labeled microglia are relatively hardy and can be subjected to confocal imaging conditions over many hours with few ill effects. Second, membrane turnover in activated microglia reduces surface labeling over the course of a lengthy time-lapse imaging session ($>$10 hr). This can result in a reduced fluorescent signal and "fading" of the microglia over the time course. Finally, imaging microglia in adult tissues is more challenging because, unlike neonatal tissues, the adult tissues require continuous perfusion with oxygenated media, and they are prone to showing signs of reduced movement under hypoxic conditions. These caveats aside, the present methods afford exciting opportunities to examine dynamic cell events as they occur in a brain tissue environment on a timescale of minutes, hours, and even days.

ACKNOWLEDGMENTS

This work was supported by the National Institutes of Health (NS-43468). Voxblast was developed at the University of Iowa.

REFERENCES

Dailey M.E. and Waite M. 1999. Confocal imaging of microglial cell dynamics in hippocampal slice cultures. *Methods* **18:** 222–230.

Petersen M. and Dailey M.E. 2004. Diverse microglial motility behaviors during clearance of dead cells in hippocampal slices. *Glia* **46:** 195–206.

Stence N., Waite M., and Dailey M.E. 2001. Dynamics of microglial activation: A confocal time-lapse analysis in hippocampal slices. *Glia* **33:** 256–266.

Stoppini L., Buchs P.-A., and Muller D. 1991. A simple method for organotypic cultures of nervous tissue. *J. Neurosci. Methods* **37:** 173–182.

Streit W.J. and Kreutzberg G.W. 1987. Lectin binding by resting and reactive microglia. *J. Neurocytol.* **16:** 249–260.

CHAPTER 56

Single and Multiphoton Fluorescence Recovery after Photobleaching

Edward Brown,[1] Ania Majewska,[2] and Rakesh K. Jain[1]

[1]*Edwin L. Steele Laboratory, Department of Radiation Oncology, Massachusetts General Hospital, Boston, Massachusetts 02114; Picower Center for Learning and Memory, Massachusetts Institute of Technology, Cambridge, Massachusetts 02114*

Fluorescence recovery after photobleaching (FRAP) is a technique for measuring kinetics of fluorescently labeled molecules. FRAP has been used to measure the diffusion coefficient (or analogous transport parameters) of labeled molecules in two-dimensional systems, such as cell membranes and lamellipodia (Feder et al. 1996), and in three-dimensional systems, such as tumor tissue or cell bodies (Chary and Jain 1989; Berk et al. 1997; Pluen et al. 2001). In addition to measuring diffusion coefficients, FRAP can be used to analyze binding kinetics (Kaufman and Jain 1991; Berk et al. 1997) and to quantify the connectivity of compartments, such as the dendrite/dendritic spine system (Majewska et al. 2000), plant plastids (Kohler et al. 1997), and the cell nucleus/cytoplasm system (Wei et al. 2003). The original "spot" FRAP technique (Peters et al. 1974; Axelrod et al. 1976) has undergone a variety of modifications to accommodate patterned photobleaching (Abney et al. 1992), continuous photobleaching (Wedekind et al. 1996), and more. This chapter discusses the three basic FRAP methods: traditional FRAP (Axelrod et al. 1976), multiphoton FRAP (MPFRAP) (Brown et al. 1999), and FRAP with spatial Fourier analysis (SFA-FRAP) (Berk et al. 1993). Each discussion is accompanied by a description of the appropriate mathematical analysis appropriate for situations in which the recovery kinetics are dictated by free diffusion. In some experiments, the recovery kinetics are dictated by the boundary conditions of the system, and FRAP is then used to quantify the connectivity of various compartments. Since the appropriate mathematical analysis is independent of the bleaching method, the analysis of compartmental connectivity is discussed last, in a separate section.

FRAP AND MPFRAP

In a FRAP experiment, a focused laser beam bleaches a region of fluorescently labeled molecules in a thin sample of tissue (Axelrod et al. 1976). The same laser beam, albeit greatly attenuated, is then used to generate a fluorescence signal from that region as the unbleached fluorophores diffuse into the sample. A photomultiplier tube, or similar detector, records the recovery in fluorescence signal. In a conventional (one-photon) FRAP experiment, simple analytical formulas can be fit to the fluorescence recovery curve in order to generate the two-dimensional diffusion coefficient of the fluorescent molecule, only if the sample is sufficiently thin (see FRAP Diffusion Analysis, p. 433). If the sample is not thin enough for the

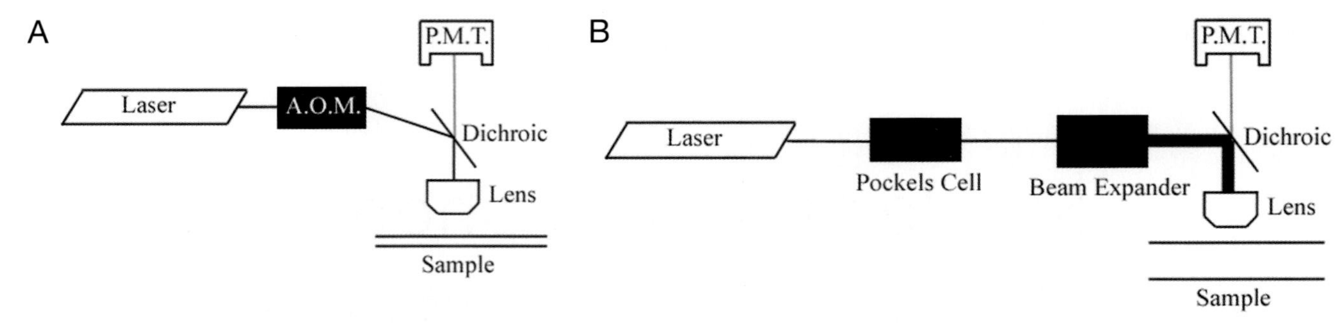

FIGURE 1. (*A*) Equipment for fluorescence recovery after photobleaching. (*B*) Equipment for multiphoton fluorescence recovery after photobleaching.

analytical solution to hold, the diffusion coefficient can be estimated by comparing the recovery time to that of molecules with known diffusion coefficients in samples of identical thickness.

In an MPFRAP experiment, a focused beam from a mode-locked laser provides both the bleaching and monitoring beam, generating fluorescence and photobleaching via multiphoton excitation (Brown et al. 1999). The intrinsic spatial confinement of multiphoton excitation means that the bleaching/monitoring volume is three-dimensionally resolved (Denk et al. 1990); consequently, there is no upper limit on the sample thickness. Simple analytical formulas can be applied to the fluorescence recovery curve to generate the three-dimensional diffusion coefficient of the fluorescent molecule.

FRAP Instrumentation

The primary instrumentation of one-photon FRAP consists of a laser source, an acousto-optic modulator (AOM), a dichroic mirror, an objective lens, a gated photomultiplier tube (PMT), and a data recording system such as an analog-to-digital (A/D) board or scaler (photon counting device) (Fig. 1A). The laser source is directed through the AOM to the dichroic mirror and objective lens and into the fluorescent sample.

The laser is typically an argon ion laser operating in TEM_{00} mode to produce a Gaussian transverse intensity profile, suitable for analysis of recovery curves (see FRAP Diffusion Analysis, p. 433). The laser must be modulated on a much faster timescale than the diffusive recovery time of the system, often requiring modulation times of fractions of a millisecond. This necessitates the use of an AOM as the beam modulation device because of its fast response time. To generate significant variation in transmitted intensity, the first diffraction maximum of the AOM should be used, not the primary transmitted beam.

MPFRAP Instrumentation

The primary instrumentation of MPFRAP consists of a laser source, Pockels cell, beam expander, dichroic mirror, objective lens, gated photomultiplier tube (PMT), and a data recording system (Fig. 1B). The laser source is directed through the Pockels cell to beam expander, dichroic mirror, and objective lens and into the fluorescent sample.

The laser is typically a mode-locked (~100-fsec pulses) Ti:sapphire laser. This beam is expanded to overfill the objective lens, thereby producing a uniformly illuminated back aperture, resulting in the formation of the highest resolution spot in the plane of the sample (Born and Wolf 1980). The intrinsic spatial confinement of multiphoton excitation produces a three-dimensionally defined bleach volume, whose size depends on the numerical aperture (NA) and wavelength of excitation light, and is typically $\sim 0.5 \times 0.5 \times 1$ µm. This extremely small bleached volume dissipates rapidly (hundreds of microseconds for smaller fluorescently labeled molecules such as FITC-bovine serum albumin [BSA] or green fluorescent protein [GFP]). Consequently, MPFRAP requires a beam-modulation system with response times as fast as 1 µsec. The AOM traditionally used in one-photon FRAP relies on diffraction of the laser beam to achieve intensity modulation, whereas 100-fsec pulses, typical of a mode-locked Ti:sapphire laser used in multiphoton FRAP, have a bandwidth of 15 nm. Different wavelengths

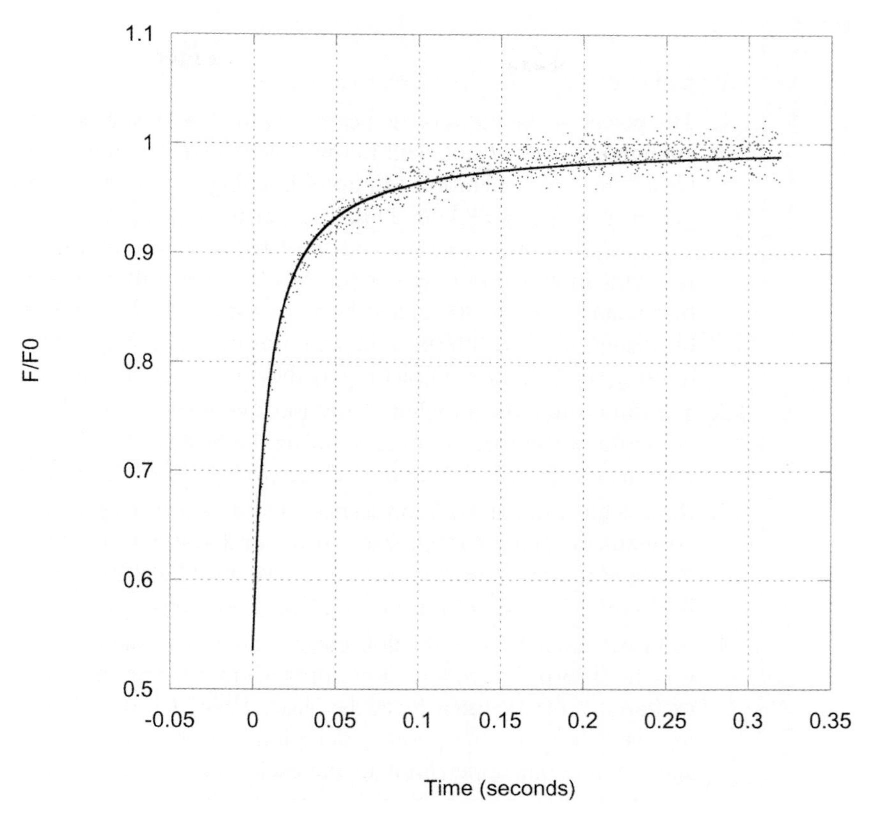

FIGURE 2. MPFRAP recovery curve. MPFRAP was performed on FITC-BSA in free solution. (*Solid line*) Least Chi-squared fit, producing a diffusion coefficient at 22°C of 5.9×10^{-7} $cm^2 sec^{-1}$.

of light will diffract in different directions; therefore, the nondiffractive Pockels cell is often used for MPFRAP beam modulation instead of an AOM.

PROCEDURE A

Fluorescence Recovery after Photobleaching

A FRAP or MPFRAP experiment consists of the following steps:

1. The laser modulator (AOM or Pockels cell) is set to a low state producing the "monitoring beam"<!>, generating fluorescence from the sample, which is collected by the objective lens and detected by the PMT. The scaler monitors the output of the PMT for a short duration (tens of microseconds to milliseconds), recording the "prebleach signal."
2. The PMT is then gated (i.e., the dynode voltage is set to zero) to avoid damage, and the amount of laser power sent to the objective lens by the laser modulator is switched to a high level, producing the "bleach beam" that photobleaches a fraction of the fluorophores at the sample. The modulator rapidly returns to the low state (after a total bleach time that depends on the sample dynamics; see Limitations below), and the PMT is subsequently ungated.
3. The fluorescence generated by the low intensity laser beam is then recorded continuously. As unbleached fluorophores diffuse into the region excited by the laser beam, the fluorescence signal recovers to equilibrium levels.
4. The fluorescence recovery curve is analyzed to yield diffusion coefficients or analogous parameters, as well as the fraction of immobile fluorophores (Fig. 2).

Caution: *See Appendix 3 for appropriate handling of materials marked with <!>.*

Limitations

A number of steps must be taken to ensure accurate determination of the diffusion parameters.

1. The power of the monitoring beam, i.e., the laser beam before and after the bleach, must be high enough to generate sufficient fluorescence signal, but not so high that it causes significant photobleaching. The generation of photobleaching by the monitoring beam can be detected easily by performing a test FRAP/MPFRAP experiment with no bleach pulse. If the fluorescence signal decays significantly over the course of the scan, the monitoring power must be lowered. If the resultant monitoring power is too low to allow sufficient signal-to-noise ratio, the monitoring beam can be repeatedly cycled between zero and a high power that causes some limited photobleaching, allowing intermittent recording of the fluorescence recovery at higher signal rates, while limiting the total photobleaching by the monitoring beam.
2. The duration of the bleaching flash must be short enough that no significant diffusion (i.e., recovery in fluorescent signal) occurs during the bleach pulse. A rule of thumb is that the bleach pulse must be less than 1/20th of the half-recovery time of the subsequent recovery curve.
3. If the acquisition of the fluorescence recovery curve does not occur for a long enough period, overestimation of the immobile fraction and underestimation of the diffusion recovery time can result. A rule of thumb is that the recovery curve must be visibly flat (i.e., any systematic change in signal is less than shot noise) for the latter half of the recording time.
4. In FRAP and MPFRAP, the fluorescence excitation rate is assumed to scale as Rate $\sim\sigma\langle I^b\rangle$, where σ is the absorption cross section (units of cm^2 for one-photon excitation and $cm^2 s$ for two-photon excitation), I is the intensity of the bleach beam, b is the number of photons absorbed in a bleaching event (one for one-photon excitation, two for two-photon, etc.) and $\langle\rangle$ denotes a time average. There is an upper limit to the excitation rate of a fluorescent molecule, however, because fluorescent molecules have excited-state lifetimes of $\tau_L = 10–15$ nsec and hence cannot be excited at a faster rate than $1/\tau_L$. Furthermore, the pulsed lasers used in MPFRAP have a duty cycle of $\tau_D =$ 12.5 nsec. Consequently, when the excitation rate of fluorophores during the bleaching pulse approaches a significant fraction of $1/\tau_L$ or $1/\tau_D$, the rate of excitation will deviate from $\sim\sigma\langle I^b\rangle$ and will asymptotically approach a limiting value, which depends on $1/\tau_L$ or $1/\tau_D$. This phenomenon is known as excitation saturation. FRAP or MPFRAP curves generated in the saturation regime, where the photobleaching rate does not scale as $\sim\sigma\langle I^b\rangle$, will produce erroneously low diffusion coefficients.
5. To avoid excitation saturation, a series of FRAP or MPFRAP curves must be generated at increasing bleach beam powers. The curves are then analyzed (see Diffusion Analysis below) using the bleach depth parameter and the diffusive recovery time as the fitting parameters. The bleach depth parameter of a FRAP or MPFRAP curve is proportional to the bleaching rate and will scale as $\sigma\langle I^b\rangle$ when the curve is not in the excitation saturation regime. When the bleach depth parameter measurably deviates from a $\sigma\langle I^b\rangle$ dependence, the bleaching is subject to excitation saturation and the diffusion coefficients will be erroneously low. A good rule of thumb is that the fractional deviation of the bleach depth parameter from $\sigma\langle I^b\rangle$ dependence should be less than 10%.
6. To convert a measured diffusive recovery time to a diffusion coefficient, the characteristic size of the bleached region must be known. In both FRAP and MPFRAP, the excitation probability as a function of position transverse to the beam axis at the focal spot (i.e., the transverse beam intensity profile to the bth power) can be well represented by a Gaussian function (see below), whose characteristic half-width at e^{-2} must therefore be determined in order to convert recovery times to diffusion coefficients. In the case of one-photon FRAP, the e^{-2} half-width of the excitation probability must be measured transverse to the beam axis only, whereas in MPFRAP, the $1/e$ half-width in both the transverse and axial directions must be measured. This is typically accomplished by scanning the focus of the laser beam across a subresolution (~10 nm or less) fluorescent bead and recording the fluorescent signal versus position of the bead. Unless the excitation beam of a one-photon FRAP system is provided by a confocal laser-scanning microscope, there is no mechanism for easily altering the position of the laser focus in the sample plane, so a simple method to measure the excitation probability is to scan a subresolution bead transversely across the stationary

beam focus with a stepper motor or piezoelectric motor (Schneider and Webb 1981). In an MPFRAP system, the laser position is usually governed by galvanometers and stepper motors as part of a multiphoton laser-scanning microscope system. Consequently, it is relatively easy to scan the laser across a stationary subresolution fluorescent bead to determine the transverse e^{-2} half-width, whereas the axial e^{-2} half-width can be determined by scanning the bead across the focus, using the focus stepper motor that accompanies most laser-scanning microscope systems.

DIFFUSION ANALYSIS: CONVENTIONAL (ONE-PHOTON) FRAP

If the sample thickness in a FRAP experiment is sufficiently thin, the complex three-dimensional hourglass shape of the focused bleaching beam (Born and Wolf 1980) can be ignored. This is because the bleaching only occurs in a thin slice at the focus of the beam, and the postbleach recovery kinetics occur laterally in a two-dimensional system. If the excitation laser is operating in TEM_{00} mode and significantly underfills the objective lens (i.e., the beam is significantly smaller than the back aperture of the objective lens), then the transverse intensity profile is a simple Gaussian, and an analytical formula for the fluorescence recovery curve can be derived. For a Gaussian laser beam, the fluorescence recovery curve describing free diffusion in a two-dimensional system is given by (Axelrod et al. 1976)

$$F(t) = F_{\infty} \sum_{n=0}^{\infty} \frac{(-\beta)^n}{n!} \frac{1}{(1 + n + 2nt/\tau_D)}$$

Where F_{∞} is the $t = \infty$ fluorescence signal, β is the bleach depth parameter, and τ_D is the two-dimensional diffusion recovery time. The fraction of immobile fluorophores in the sample is given by $(F_0 - F_{\infty})/F_{\infty}$, where F_0 is the prebleach fluorescent signal and "immobile" is defined as having a diffusive mobility significantly slower than the timescale of the experiment. The diffusion coefficient is given by $D = w^2/4\tau_D$, where w is the transverse e^{-2} half-width of the laser beam at the sample.

FRAP can be extended to thicker samples, but analytic derivations of the diffusion coefficient become problematic due to the complex nature of the hourglass-shaped focus laser distribution. Furthermore, the fluorescence recovery time becomes dependent on the thickness of the sample, which may or may not be known. In these cases, the FRAP technique is often limited to a simple comparison of recovery times between samples of unknown diffusion coefficients and samples with known diffusion coefficients that have been measured with analytical techniques such as those described above.

DIFFUSION ANALYSIS: MPFRAP

In an MPFRAP experiment, the highest spatial resolution is achieved by overfilling the objective lens, producing a diffraction-limited intensity distribution at the beam focus. The square (or higher power) of this intensity profile is well approximated by a Gaussian distribution, both transverse to and along the optical axis, although the half-width at e^{-2} is typically longer in the axial dimension than in the transverse dimension (Born and Wolf 1980). For an overfilled objective lens, inducing two-photon fluorescence and two-photon photobleaching, the fluorescence recovery curve, describing free diffusion in a three-dimensional system, is given by (Brown et al. 1999)

$$F(t) = F_{\infty} \sum_{n=0}^{\infty} \frac{(-\beta)^n}{n!} \frac{1}{(1 + n + 2nt/\tau_D)} \frac{1}{\sqrt{1 + n + 2nt/(R\tau_D)}}$$

Where F_{∞} is the $t = \infty$ fluorescence signal, β is the bleach depth parameter, and τ_D is the three-dimensional diffusion recovery time. The fraction of immobile fluorophores in the sample is given by $(F_0 - F_{\infty})/F_{\infty}$, where F_0 is the prebleach fluorescent signal and "immobile" is defined as having a diffusive mobility significantly slower than the timescale of the experiment. The diffusion coefficient is given by $D = w^2/8\tau_D$, where w is the transverse e^{-2} half-width of the laser beam at the sample and R is the square of the ratio of the axial e^{-2} half-width to the radial e^{-2} half-width (see Fig. 2).

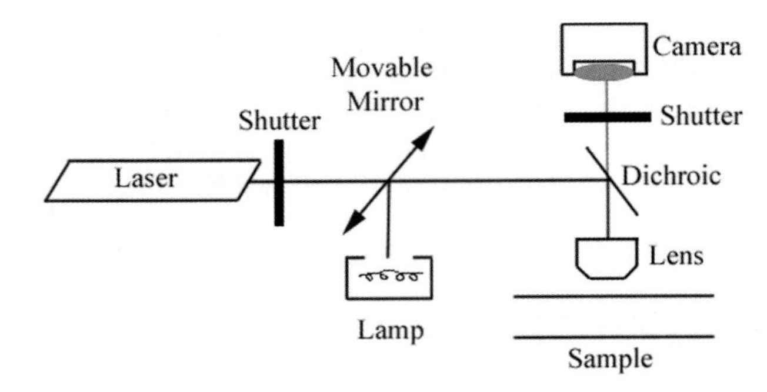

FIGURE 3. Equipment for FRAP with spatial Fourier analysis.

SFA-FRAP

Summary

In a FRAP experiment with spatial Fourier analysis, a focused laser beam is used to bleach out a region of fluorophore, as in the FRAP described above. Unlike conventional FRAP, however, the evolution of the bleached region is repeatedly imaged using a CCD camera with wide-field illumination provided by a mercury lamp. The sequential images of the recovery of the bleached spot are Fourier-transformed, and the decay of selected spatial frequency components produces the diffusion coefficient of the diffusing fluorophore, without requiring knowledge of the details of the bleaching distribution or sample thickness (Berk et al. 1993). Consequently, SFA-FRAP can be performed in thick samples, although its reliance upon epifluorescence means that the diffusion coefficients it measures are averages over the entire depth of view of the microscope and are not three-dimensionally resolved.

Instrumentation

The primary instrumentation of conventional SFA-FRAP consists of a laser source, dichroic mirror, two fast shutters, a galvanometer-driven movable mirror, an objective lens, a CCD camera and mercury arc lamp, and an image recording system such as a frame grabber card (Fig. 3). The laser source is directed through one shutter and dichroic mirror to the objective lens and into the fluorescent sample, whereas the lamp is directed by the movable mirror to be colinear with the laser.

The laser is typically an argon ion laser, as in FRAP. SFA-FRAP is traditionally used in thick (many hundreds of microns) tissues, and the imaged bleach spot is several tens of micrometers wide. Consequently, the diffusive recovery times can be relatively slow (several tens of milliseconds to many minutes), and the laser modulation rate does not have to be as rapid as in FRAP. Furthermore, the laser modulation is binary, with a bright bleaching flash and zero power being the only required states, and no intermediate power monitoring beam is needed. Therefore, a fast shutter is generally sufficient for SFA-FRAP, instead of a rapid (and more expensive) analog device such as the AOM or Pockels cell.

PROCEDURE B

A Typical SFA-FRAP Experiment

1. The laser is shuttered and the movable mirror directs the mercury lamp illumination into the sample. The CCD records a few prebleach images.
2. The CCD is shuttered to avoid damage, and the movable mirror deflects out of the laser path. The laser shutter opens briefly, allowing the bleaching flash through. The laser shutter closes again, ending the bleaching flash, and the CCD is unshuttered.

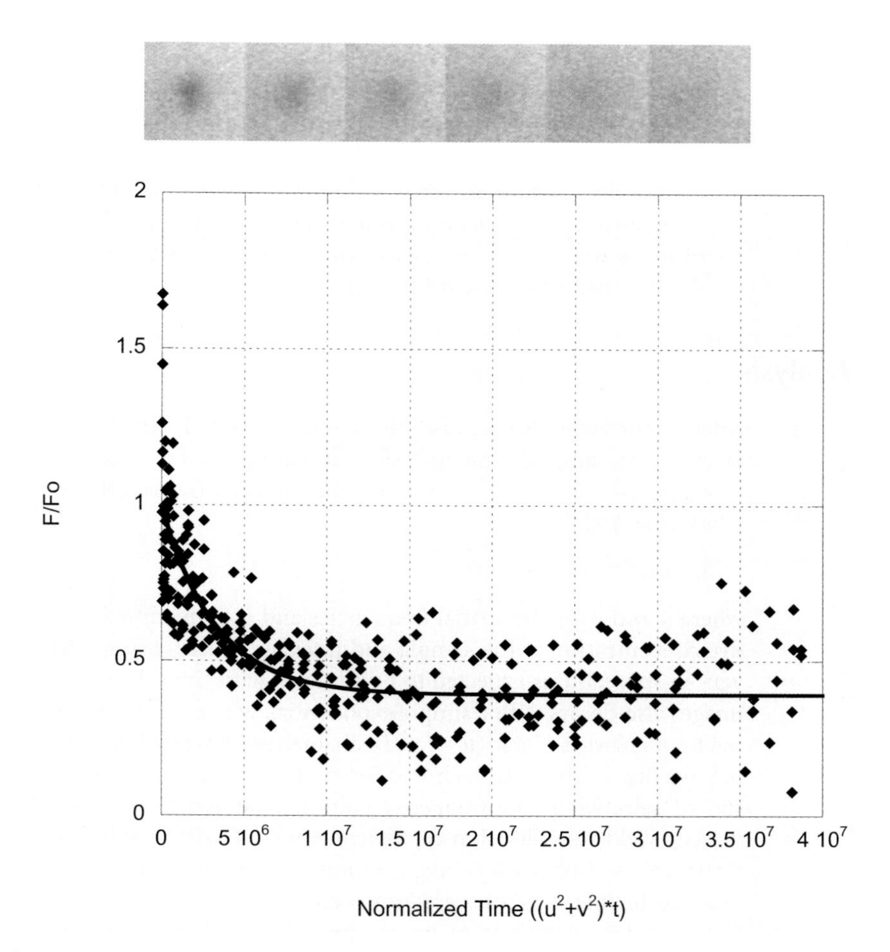

FIGURE 4. SFA-FRAP recovery images and recovery curve. SFA-FRAP was performed on FITC-IgG injected into the interstitium of an MCaIV tumor growing in the mammary fat pad of a SCID mouse. After the photobleaching flash, a series of 50 images of the bleached region was acquired, with representative images shown (*bottom*). The two-dimensional Fourier transform of each image was taken, and six spatial frequencies were tracked, all of which are plotted on normalized axes as shown (*top*). The exponential decay time was found by least Chi-squared fit, and produced a diffusion coefficient of 3.1×10^{-7} cm^{2}sec^{-1}. (Image courtesy of Trevor McKee.)

3. The movable mirror periodically shifts back into the light path, directing the mercury arc lamp illumination into the objective lens and allowing epifluorescence images to be captured as the bleached distribution recovers to equilibrium levels and the bleached spot disappears.
4. The series of images of the evolution of the bleached spot are Fourier-transformed, and six of the lowest spatial frequencies are plotted. The exponential decay of the spatial frequencies yields the diffusion coefficient of the labeled molecule, as well as the fraction of immobile fluorophores (see Fig. 4).

Limitations

A number of steps must be taken to ensure accurate determination of the diffusion parameters.

1. The epifluorescence lamp must be bright enough to generate sufficient signal, but not so bright as to cause significant photobleaching. Photobleaching due to the lamp can be easily detected by performing a test SFA-FRAP experiment with no bleach pulse. If the fluorescence signal decays significantly over the course of the image series, the epifluorescence lamp must be attenuated. Additionally, the duty cycle of the monitoring pulses can be altered, thereby reducing the total exposure time of the sample during the recovery period.

2. The duration of the bleaching flash must be short enough that no significant diffusion (i.e., recovery in fluorescent signal) occurs during the bleach pulse. A rule of thumb is that the bleach pulse must be less than 1/20th of the half-recovery time of the subsequent spatial frequency decay curve.
3. Duration of recording time/immobile fraction. If the acquisition of the fluorescence recovery images does not occur for a long enough period, overestimation of the immobile fraction and underestimation of the diffusion recovery time can result. A rule of thumb is that the decay curve of the spatial frequencies (see below) should be visibly flat (any change is less than shot noise) for the later half of the recording time.

Diffusion Analysis

After a photobleaching pulse, the concentration distribution of unbleached fluorophores $c(x,y,t)$ evolves according to the diffusion equation. If the concentration distribution is first Fourier-transformed with respect to x and y, the solution to this differential equation is a simple exponential (Berk et al. 1993)

$$C(u,v,t) = C(u,v,0)e^{-4\pi^2(u+v^2)Dt}$$

where u and v are the spatial frequencies and D is the diffusion coefficient. The relationship between dye concentration at the sample and intensity at the corresponding location in the CCD image may vary in space, but is expected to be constant with time. Therefore, the Fourier transform of the CCD images also decays with a simple exponential, where the exponential decay time of a given pair of spatial frequencies is $1/(4\pi^2(u^2 + v^2)D$. To analyze SFA-FRAP data, the series of CCD images of the evolving dye distribution are each Fourier-transformed, and the exponential decay of a number (typically six) of selected spatial frequency pairs is analyzed to produce the exponential decay time, which directly yields the diffusion coefficient. Unlike FRAP or MPFRAP, no knowledge of the initial spatial distribution of photobleaching is required. Note that this analysis ignores diffusion along the optical axis and furthermore uses epifluorescence images of the diffusing system, therefore generating a diffusion coefficient that is an average over the visible depth of the system. In other words, any depth-dependent differences in the diffusion coefficient will be averaged out to a single value (Berk et al. 1993).

COMPARTMENTALIZATION ANALYSIS

Summary

Single-photon and multiphoton FRAP can also be used to measure the diffusional coupling between two connected compartments. A characteristic time course for the diffusion or transport of different fluorescent molecules can be obtained by bleaching one compartment and monitoring the fluorescence recovery curve. This information can then be used to determine parameters such as resistivity and pore size of the separating barrier. MPFRAP has been used to examine diffusion of dyes between the excitatory synapse (dendritic spine) and its parent dendrite (Svoboda et al. 1996; Majewska 2000), and between plant plastids (Kohler et al. 1997). FRAP has examined the cell cytoplasm and the nucleus (Wei et al. 2003), as well as the turnover of fluorescently tagged actin filaments between the spine and dendrite (Star et al. 2002).

Instrumentation

A FRAP or MPFRAP instrument (described above) can be used for compartmentalization analysis. Since diffusion between compartments tends to be slower than that within a compartment, these experiments can often be performed in line-scan mode on a laser-scanning microscope. In line-scan mode, the excitation beam is scanned repeatedly along a single line that intersects an object of interest, and an x versus t image of the fluorescence is generated. The line-scan mode utilizes the acquisition electronics of the laser-scanning microscope and obviates the need to purchase a separate photon-counting device. This mode can, however, limit the acquisition speed, depending on the design of the microscope.

PROCEDURE C

FRAP and MPFRAP Compartmentalization Analysis

A FRAP and MPFRAP compartmentalization analysis experiment consists of the following steps:

1. The laser modulator (AOM or Pockels cell) is set to a low state producing the "monitoring beam," generating fluorescence from the sample, which is collected by the objective lens and detected by the PMT to measure the prebleach fluorescence. One of the compartments is chosen for the bleach/monitor paradigm (typically the smaller compartment).
2. The PMT is then gated (i.e., the dynode voltage is set to zero) to avoid damage, and the amount of laser power sent to the objective lens by the laser modulator is switched to a high level, producing the "bleach beam" which photobleaches a fraction of the fluorophores in one of the compartments. The modulator rapidly returns back to the low state (after a total bleach time that depends on the sample dynamics, as described above), and the PMT is subsequently ungated.
3. The fluorescence in the bleached spot is then recorded. As unbleached fluorophores diffuse from the unbleached compartment, the fluorescence signal recovers back to equilibrium levels.
4. The fluorescence recovery curve is analyzed to yield the characteristic coupling time, as well as the fraction of immobile fluorophores.

Limitations

1. Bleaching during the monitoring phase, bleach duration, and total acquisition time must be evaluated as in Points 1–3 in FRAP and MPFRAP "Limitations" described above.
2. Diffusional coupling is typically studied between two well-mixed compartments; i.e., where the timescale of diffusional equilibrium within the compartments is much faster than between compartments. This can be verified by spot bleaching within each of the compartments to determine the diffusion characteristics for fluorophores in each of the compartments.
3. Because the communicating compartments are well-mixed, the initial spatial profile of bleaching is essentially irrelevant because it is "washed out" as the bleached compartment undergoes diffusive mixing before significant communication with other compartments can occur. Consequently, measuring the bleach spot profile and avoiding saturation are not significant concerns. However, it is important to determine that bleaching occurs in only one compartment. This can be done by performing a line scan which intersects both compartments but restricting the bleaching pulse to a single compartment. In this case, both compartments can be monitored to ensure that bleaching is indeed spatially restricted.

Compartmentalization Analysis

The fluorescence recovery curve in a well-mixed, photobleached compartment diffusionally coupled to larger well-mixed compartments is given by the following equation

$$F(t) = F(\infty) - \Delta F_0 e^{-t/\tau}$$

where $F(\infty)$ is the fluorescence at $t = \infty$, ΔF_0 is the change in fluorescence level following the bleach pulse, and τ is the timescale of diffusion between the two compartments. The timescale t of recovery between compartments provides insight into the characteristic resistivity of the coupling pathway, the number of coupling pathways, the diffusion coefficient of the tracer, etc., depending on the geometry of the system (Majewska et al. 2000; Wei et al. 2003).

REFERENCES

Abney J.R., Scalettar B.A., and Thompson N.L. 1992. Evanescent interference patterns for fluorescence microscopy. *Biophys. J.* **61:** 542–552.

Axelrod D., Koppel D.E., Schlessinger J., Elson E., and Webb W.W. 1976. Mobility measurement by analysis of fluorescence photobleaching recovery kinetics. *Biophys. J.* **16:** 1055–1069.

Berk D.A., Yuan F., Leunig M., and Jain R.K. 1993. Fluorescence photobleaching with spatial Fourier analysis: Measurement of diffusion in light-scattering media. *Biophys. J.* **65:** 2428–2436.

———. 1997. Direct in vivo measurement of targeted binding in a human tumor xenograft. *Proc. Natl. Acad. Sci.* **94:** 1785–1790.

Born M. and Wolf E. 1980. *Principles of optics: Electromagnetic theory of propagation, interference and diffraction of light.* Pergamon Press, New York.

Brown E.B., Wu E.S., Zipfel W., and Webb W.W. 1999. Measurement of molecular diffusion in solution by multiphoton fluorescence photobleaching recovery. *Biophys. J.* **77:** 2837–2849.

Chary S.R. and Jain R.K. 1989. Direct measurement of interstitial convection and diffusion of albumin in normal and neoplastic tissues by fluorescence photobleaching. *Proc. Natl. Acad. Sci.* **86:** 5385–5389.

Denk W., Strickler J.H., and Webb W.W. 1990. Two-photon laser scanning fluorescence microscopy. *Science* **248:** 73–76.

Feder T.J., Brust-Mascher I., Slattery J.P., Baird B., and Webb W.W. 1996. Constrained diffusion or immobile fraction on cell surfaces: A new interpretation. *Biophys. J.* **70:** 2767–2773.

Kaufman E.N. and Jain R.K. 1991. Measurement of mass transport and reaction parameters in bulk solution using photobleaching. Reaction limited binding regime. *Biophys. J.* **60:** 596–610.

Kohler R.H., Cao J., Zipfel W.R., Webb W.W., and Hanson M.R. 1997. Exchange of protein molecules through connections between higher plant plastids. *Science* **276:** 2039–2042.

Majewska A., Brown E., Ross J., and Yuste R. 2000. Mechanisms of calcium decay kinetics in hippocampal spines: Role of spine calcium pumps and calcium diffusion through the spine neck in biochemical compartmentalization. *J. Neurosci.* **20:** 1722–1734.

Peters R., Peters J., Tews K., and Bahr W. 1974 Microfluorimetric study of translational diffusion of proteins in erythrocyte membranes. *Biochim. Biophys. Acta* **367:** 282–294.

Pluen A., Boucher Y., Ramanujan S., McKee T.D., Gohongi T., di Tomaso E., Brown E.B., Izumi Y., Campbell R.B., Berk D.A., and Jain R.K. 2001. Role of tumor-host interactions in interstitial diffusion of macromolecules: Cranial vs. subcutaneous tumors. *Proc. Natl. Acad. Sci.* **98:** 4628–4633.

Schneider M. and Webb W.W. 1981. Measurement of sub-micron laser beam radii. *Appl. Optics* **20:** 1382–1388.

Star E.N., Kwiatkowski D.J., and Murthy V.N. 2002. Rapid turnover of actin in dendritic spines and its regulation by activity. *Nat. Neurosci.* **5:** 239–246.

Svoboda K., Tank D.W., and Denk W. 1996. Direct measurement of coupling between dendritic spines and shafts. *Science* **272:** 716–719.

Wedekind P., Kubitscheck U., Heinrich O., and Peters R. 1996. Line-scanning microphotolysis for diffraction-limited measurements of lateral diffusion. *Biophys. J.* **71:** 1621–1632.

Wei X., Henke V.G., Strubing C., Brown E.B., and Clapham D.E. 2003. Real-time imaging of nuclear permeation by EGFP in single intact cells. *Biophys. J.* **84:** 1317–1327.

CHAPTER 57

All-Optical, In Situ Histology of Neuronal Tissue with Ultrashort Laser Pulses

Philbert S. Tsai,[1] Beth Friedman,[2] Chris B. Schaffer,[1] Jeffrey A. Squier,[3] and David Kleinfeld[1]

[1]Department of Physics and [2]Department of Neurosciences, University of California at San Diego, La Jolla, California 92093; [3]Department of Physics, Colorado School of Mines, Golden, Colorado 80401

This chapter describes the application of ultrashort laser pulses, ~100 fsec in duration, to image and ablate neuronal tissue for the purpose of automated histology. Histology is accomplished in situ by serial two-photon imaging of labeled tissue and removal of the imaged tissue with amplified pulses, as illustrated schematically in Figure 1A.

The ablation of tissue with ultrashort, infrared laser pulses requires large optical fluences, in excess of 1 J/cm^2. In the past, such high fluences had been achieved with commercially available ~100-MHz laser oscillators with ~100-fsec pulse duration, whose 1–10-nJ pulse energies could be focused to a less than 1-µm^2 spot size with high numerical aperture (NA) (i.e., tight focusing) objectives. Such beams have been used for fine-scale ablation of subcellular structures (Tirlapur and Konig 2002). More typically, ultrashort laser pulses have been amplified with an optical amplifier to obtain pulses with comparable duration and microjoule energies, albeit at repetition rates of only 1–10 kHz. These amplified pulses have been used to ablate a wide variety of tissues, including cornea (Loesel et al. 1996; Oraevsky et al. 1996; Juhasz et al. 1999; Lubatshowski et al. 2000; Maatz et al. 2000), dental tissue (Loesel et al. 1996; Neev et al. 1996), skin (Frederickson et al. 1993), and brain (Loesel et al. 1996, 1998; Suhm et al. 1996; Goetz et al. 1999; Tsai et al. 2003). This chapter describes the coupling of ultrashort pulse laser ablation with two-photon laser-scanning microscopy (TPLSM) for the purpose of serial histology of neuronal tissue (Tsai et al. 2003). This extends the histological use of TPLSM to imaging tissue throughout arbitrarily thick samples.

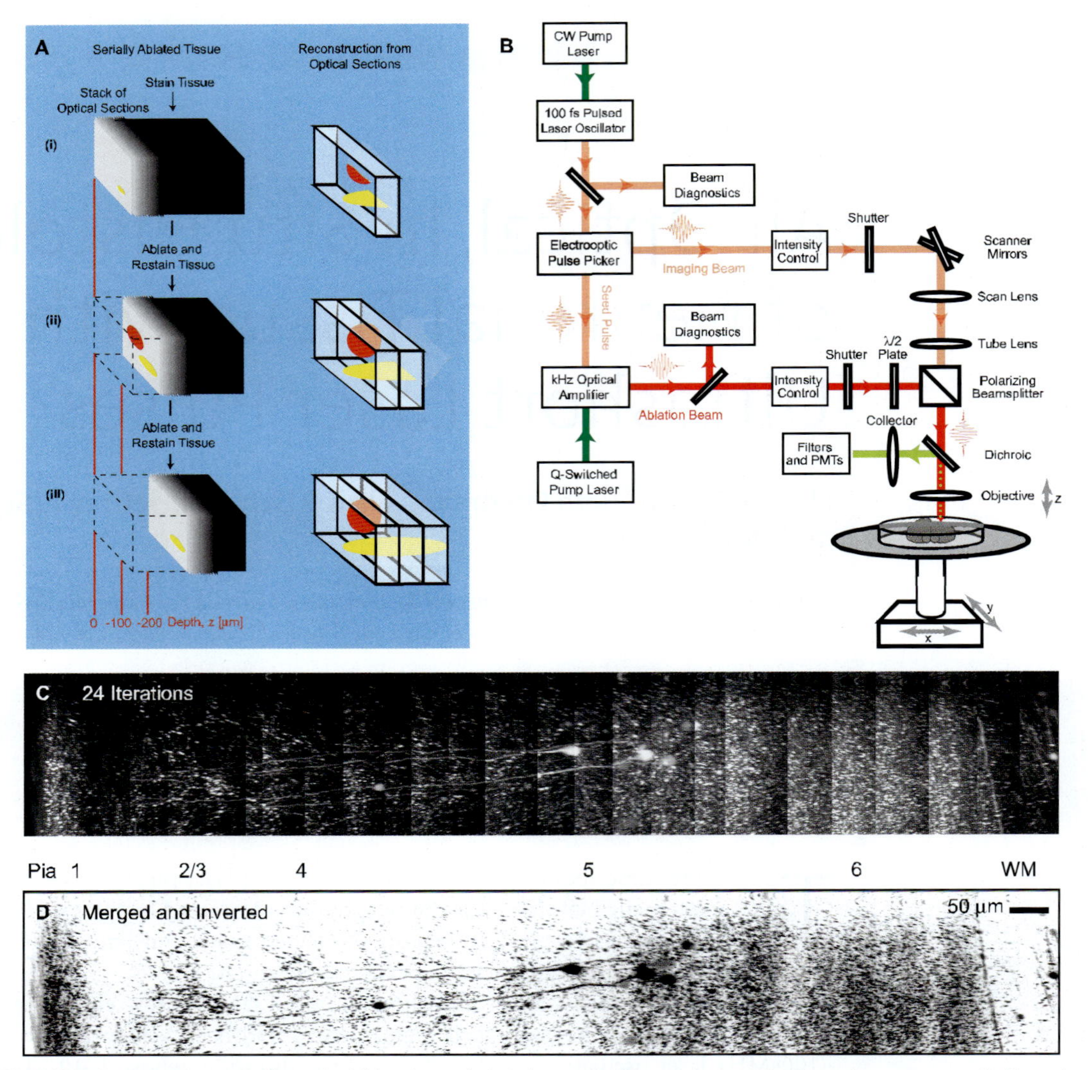

FIGURE 1. Overview of all-optical histology. (*A*) Schematic illustration of the iterative process of all-optical histology. (*i*) The tissue sample (*left column*) containing two fluorescently labeled structures is imaged by conventional two-photon laser-scanning microscopy to collect optical sections. Sections are collected until scattering of the incident light reduces the signal-to-noise ratio below a useful value; typically, this occurs at ~150 µm in fixed tissue. Labeled features in the resulting stack of optical sections are digitally reconstructed (*right column*). (*ii*) The top of the now-imaged region of tissue is ablated with amplified ultrashort laser pulses to expose a new surface for imaging. The sample is again imaged down to a maximal depth, and the new optical sections are added to the previously stored stack. (*iii*) The process of ablation and imaging is again repeated so that the structures of interest can be fully sectioned and reconstructed. (*B*) Schematic of laser and imaging systems. A train of 100-fsec pulses from an optically pumped Ti:sapphire oscillator are directed with scanning mirrors to image a preparation placed on a programmable translation stage. A portion of the imaging beam is picked off by an electro-optic pulse picker, e.g., a Pockels cell, for amplification in an optical amplifier before being directed to the sample for ablation. The tissue is moved laterally by the programmable translation stages to allow ablation over large areas, as well as imaging over multiple fields of view. (*C*) Iterative processing of a block of neocortex of a transgenic mouse with neurons labeled by the yellow-emitting variant of green fluorescent protein. Twenty-four successive cutting and imaging cycles are shown. The ablation laser was focused onto the cut face with a 20x magnification; 0.5-NA water objective and single passes, at a scan rate of 4 mm/sec, were made to optically ablate successive planes at a depth of 10 µm each with total thicknesses between 40 and 70 µm per cut. The energy per pulse was maintained at 8 µJ. Each stack of images represents a maximal side projection of all accumulated optical sections obtained using TPLSM at $\lambda = 920$ nm. The sharp breaks in the images shown in successive panels demarcate the cut boundaries. (*D*) A maximal projection through the complete stack with the breaks removed by smoothly merging overlapped regions. The contrast is inverted to emphasize the fine labeling. Anatomical regions are labeled above the figure, including the pia mater (Pia), the white matter (WM), and the cortical layers (1 to 6).

MATERIALS

The integration of the imaging and ablation optomechanical setup is shown schematically in Figure 1B. The individual components are as follows:

Two-Photon Laser-scanning Microscope

We used a custom-built system (Tsai et al. 2002) that incorporates a commercial Ti:sapphire laser oscillator (Mira-F900 pumped by a Verdi V-10, Coherent), commercially available computer interface boards (National Instruments), and software environment (LabVIEW, National Instruments). Alternatively, turnkey TPLSM systems are available from various commercial vendors.

Optical Amplifier

A subset of the nanojoule pulses from the Ti:sapphire oscillator are amplified in a multipass optical amplifier based on the system of Kapteyn and Murnane (Backus et al. 1998). This results in a 1-kHz train of 400-μJ pulses. Alternatively, appropriate amplified sources are now commercially available that produce up to 5-kHz trains of millijoule pulses (e.g., Libra from Positive Light). The beam of amplified pulses is integrated into the TPLSM with polarization optics to allow sequential two-photon imaging with 0.1–1-nJ pulses and ablation with 1–10-μJ pulses within the same apparatus. The amplified pulse energy is controlled with neutral density filters.

Fluorescently Labeled Tissue

A tissue preparation, typically fixed with 4% (w/v) paraformaldehyde<!> (PFA; Sigma) in phosphate-buffered saline (PBS; Sigma), is placed in the modified TPLSM and imaged using a water-immersion objective with high NA, i.e., 0.5–1.0. The structures of interest are tagged, either by surface application of a low-molecular-weight fluorescent dye or with the use of transgenic animals that selectively express a fluorescent protein.

Fast Programmable Translation Stages

The tissue is mounted on a set of fast translation stages, capable of speeds up to 8 mm/sec with micrometer accuracy (RCH22 series stages and 300 series controllers; New England Associated Technologies). These stages raster the tissue across the focus of the objective lens so that centimeter-sized areas can be ablated and allow for lateral displacement of the tissue so that multiple fields of view can be imaged using TPLSM.

Caution: *See Appendix 3 for appropriate handling of materials marked with <!>.*

PROCEDURE

Tissue Preparation and Laser Alignment

1. Adhere the fixed tissue to a petri dish by partially encasing the tissue with a solution of 2% (w/v) agarose (Sigma) in PBS, and allowing the agarose to cool and gel.

   Note: A thin layer of cyanoacrylate cement (Superbonder; Loctite) may be used to secure the tissue to the bottom of the petri dish.

2. Center and align the imaging and ablation beams. This can be accomplished with calibration samples that consist of a coverslip on top of a block of 100 μM fluorescein in 2% (w/v) agarose gel.

   a. Focus the imaging beam at the interface of the coverslip and the gel by locating the axial onset of fluorescence.

   b. Set the focus of the ablation beam using telescopes and adjustable mirror mounts in the amplified beam path to place the focus of the ablation beam at the center of the field of view.

   Note: Cavitation bubbles, visualized as dark spots in a sea of fluorescence, are generated when the focus of the ablation beam lies within the gel, but not when it lies within the coverslip.

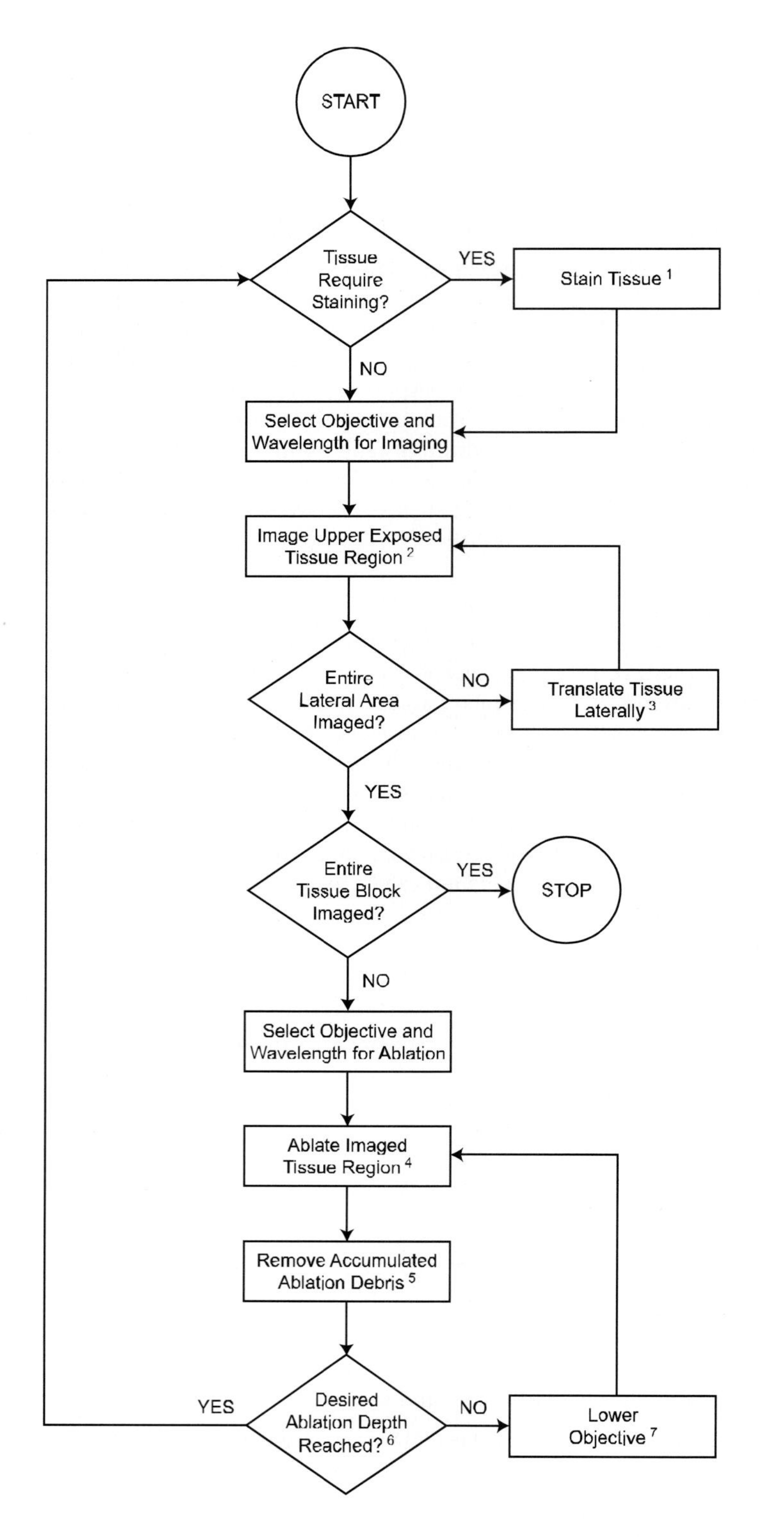

FIGURE 2. Flow chart of the all-optical histology. The following notes apply. (*1*) For example, to fluorescently label nucleic acids, apply 100 μM acridine orange (Molecular Probes) to the tissue surface for 3 min, followed by five washes with PBS for 1 min each. (*2*) Increase laser power approximately exponentially with depth to compensate for scattering and absorption losses. (*3*) Use the programmable translation stages to move to an adjoining lateral field of view for imaging. It is useful to allow ~10% spatial overlap with previously imaged fields of view to verify proper registration. (*4*) Program the translation stages to perform a raster scan across the entire lateral extent of the tissue. Parameters for ablation can be varied to trade off between speed and precision. Higher pulse energies can allow faster translation speed and larger step sizes, but they may increase the surface roughness of the tissue, thereby potentially reducing the quality of subsequent images. (*5*) Remove debris and hydrolysis bubbles resulting from the ablation process either with a continuous buffer flow across the tissue during the ablation or by manually washing the tissue with buffer between successive ablation planes. (*6*) The total depth of tissue to be removed per ablation round will vary with the tissue scattering properties and the strength of the fluorescent signal. It is important to ablate less depth than was imaged in the previous imaging round, as a means to verify registration. (*7*) To maximize spatial overlap, so as to minimize surface roughness, interlace the lines of ablation in successive planes by offsetting the start position in each plane by one half the lateral step size between lines.

Iterative Imaging and Ablation

3. The iterative processes of imaging and ablation are laid out as a flow chart in Figure 2 (technical details pertaining to particular steps are provided in the figure legend). Examples of cutting parameters used for sample ablation with a 1-kHz train of 100-fsec pulses are provided in Table 1. Note that these parameters represent cutting speeds that are overly conservative by a factor of between 2 and 10.

TABLE 1. Ablation parameters successively used with a 1-kHz train of amplified 100-fsec pulses

Objective NA	Energy (μJ)	Translation speed (μm/sec)	Lateral step size (μm)	Axial step size (μm)	Volume rate (mm^3/sec)
0.5 Water	2	4000	2.5	10	1×10^{-4}
0.5 Water	8	4000	10	10	4×10^{-4}
0.3 Water	20	4000	10	50	2×10^{-3}
0.2 Air	30	1500	10	20	3×10^{-3}

EXAMPLE OF APPLICATION

Layer-V pyramidal neurons from mouse neocortex were visualized via the iterative, optical histology described above (Fig. 1C,D). The tissue was obtained from the transgenic mouse strain B6.Cg-TgN(thy1-YFPH) 2Jrs (Feng et al. 2000), available through Jackson Labs. The animal was perfused with PBS followed by 4% (w/v) PFA in PBS. The perfusion volume was 5 ml/g body weight and the flow rate was 20 ml/min. The tissue was stored in 4% (w/v) PFA in PBS for postfixation before mounting in agarose.

Ablation was performed using a 20x magnification, 0.5-NA water-immersion objective. A beam of 8-μJ pulses, arriving at a rate of 1 kHz, was focused through the objective onto the sample. The sample was translated in a raster pattern at a speed of 4 mm/sec with a 10-μm lateral step between linear passes. The axial step between ablation planes was chosen to be 10 μm. Accumulated debris and hydrolysis bubbles under the water-immersion objective were removed by washing the tissue with buffer between successive ablation planes.

Imaging was performed with a 40x magnification, 0.8-NA water-immersion objective, which achieves 0.4-μm lateral resolution and 2-μm axial resolution (Williams et al. 1994). The 200 x 200-μm field was stored as an 800 x 800-pixel image. The axial step between images was 1 μm.

The result of 24 iterations of combined imaging and ablation along a radial axis of neocortex is shown in Figure 1C. The imaging in each iteration progressed in the dorsal-to-ventral direction (left to right in the figure), over a single 200 x 200-μm field of view through at least 100 μm of tissue depth. The ablation also progressed from the dorsal-to-ventral direction, removing an average of 60 μm per iteration. Each image stack was then projected along the coronal direction and overlaid to produce the composite in Figure 1C. The 24 individual stacks were then smoothly merged and contrast-inverted to produce the image in Figure 1D.

ADVANTAGES AND LIMITATIONS

The primary advantage of this all-optical histological technique is the ability to image large volumes of unfrozen and unembedded soft tissue in situ. Tissue distortion and misalignment due to the generation of thin physical sections are eliminated, allowing straightforward image registration and generation of three-dimensional maps and reconstructions. The technique is also conducive to automation for high-throughput applications.

The technique is limited to fluorescently labeled tissues whose optical properties are amenable to two-photon microscopy; i.e., the scattering depth must be large compared to the axial extent of the focal volume. The speed of TPLSM imaging is currently the limiting aspect of the technique (Tsai et al. 2003), by a factor of as much as 100, although faster imaging can be achieved with the use of resonant scanners.

REFERENCES

Backus S., Durfee C.G., III, Murnane M.M., and Kapteyn H.C. 1998. High power ultrafast lasers. *Rev. Sci. Instrum.* **69:** 1207–1223.

Feng G., Mellor R.H., Bernstein M., Keller-Peck C., Nguyen Q.T., Wallace M., Nerbonne J.M., Lichtman J.W., and Sanes J.R. 2000. Imaging neuronal subsets in transgenic mice expressing multiple spectral variants of GFP. *Neuron* **28:** 41–51.

Frederickson K.S., White W.E., Wheeland R.G., and Slaughter D.R. 1993. Precise ablation of skin with reduced collateral damage using the femtosecond-pulsed, terawatt titanium-sapphire laser. *Arch. Dermatol.* **129:** 989–993.

Goetz M.H., Fischer S.K., Velten A., Bille J.F., and Strum V. 1999. Computer-guided laser probe for ablation of brain tumours with ultrashort laser pulses. *Phys. Med. Biol.* **44:** N119–N127.

Juhasz T., Loesel H.L., Kurtz R.M., Horvath C., Bille J.F., and Mourou G. 1999. Corneal refractive surgery with femtosecond lasers. *IEEE J. Sel. Top. Quantum Electron.* **5:** 902–910.

Loesel F.H., Niemez M.H., Bille J.F., and Juhasz T. 1996. Laser-induced optical breakdown on hard and soft tissues and its dependence on the pulse duration: Experiment and model. *IEEE J. Quant. Electron.* **32:** 1717–1722.

Loesel F.H., Fischer J.P., Gotz M.H., Horvath C., Juhasz T., Noack F., Suhm N., and Bille J.F. 1998. Non-thermal ablation of neural tissue with femtosecond laser pulses. *Appl. Phys. B* **66:** 121–128.

Lubatschowski H., Maatz G., Heisterkamp A., Hetzel U., Drommer W., Welling H., and Ertmer W. 2000. Application of ultrashort laser pulses for intrastromal refractive surgery. *Graefe's Arch. Clini. Exp. Opthalmol.* **238:** 33–39.

Maatz G., Heisterkamp A., Lubatschowski H., Barcikowski S., Fallnich C., Welling H., and Ertmer W. 2000. Chemical and physical side effects at application of ultrashort laser pulses for intrastromal refractive surgery. *J. Optics A* **2:** 59–64.

Neev J., Da Silva L.B., Feit M.D., Perry M.D., Rubenchik A.M., and Stuart B.C. 1996. Ultrashort pulse lasers for hard tissue ablation. *IEEE J. Sel. Top. Quant. Electron.* **2:** 790–800.

Oraevsky A., Da Silva L., Rubenchik A., Feit M., Glinsky M., Perry M., Mammini B., Small W., and Stuart B. 1996. Plasma mediated ablation of biological tissues with nanosecond-to-femtosecond laser pulses: Relative role of linear and nonlinear absorption. *IEEE J. Sel. Top. Quant. Electron.* **2:** 801–809.

Suhm N., Gotz M.H., Fischer J.P., Loesel F., Schlegel W., Sturm V., Bille J.F., and Schroder R. 1996. Ablation of neural tissue by short-pulsed lasers—A technical report. *Acta Neurochir.* **138:** 346–349.

Tirlapur U.K. and Konig K. 2002. Targeted transfection by femtosecond laser light. *Nature* **418:** 290–291.

Tsai P.S., Nishimura N., Yoder E.J., Dolnick E.M., White G.A., and Kleinfeld D. 2002. Principles, design, and construction of a two photon laser scanning microscope for in vitro and in vivo brain imaging. In *In vivo optical imaging of brain function* (ed. R.D. Frostig), pp. 113–171. CRC Press, Boca Raton, Florida.

Tsai P.S., Friedman B., Ifarraguerri A.I., Thompson B.D., Lev-Ram V., Schaffer C.B., Xiong Q., Tsien R.Y., Squier J.A., and Kleinfeld D. 2003. All-optical histology using ultrashort laser pulses. *Neuron* **39:** 27–41.

Williams R.M., Piston D.W., and Webb W.W. 1994. Two-photon molecular excitation provides intrinsic 3-dimensional resolution for laser-based microscopy and microphotochemistry. *FASEB J.* **8:** 804–813.

CHAPTER 58

Imaging with Voltage-sensitive Dyes: Spike Signals, Population Signals, and Retrograde Transport

Efstratios K. Kosmidis,[1] Lawrence B. Cohen,[1,2] Chun X. Falk,[1,2] J.-Y. Wu,[3,4] and Bradley J. Baker[1]

[1]Department of Cellular and Molecular Physiology, Yale University School of Medicine, New Haven, Connecticut 06520; [2]RedShirtImaging, LLC, Fairfield, Connecticut 06485; [3]Department of Physiology, Georgetown University, Washington D.C. 20007; [4]WuTech Instruments, Inc., Gaithersburg, Maryland 20877

Optical measurements of membrane potential allow simultaneous measurements to be taken from many different locations. This is important in the studies of the nervous system where simultaneous activity can occur at the regional, cellular, and subcellular levels. A voltage-sensitive dye signal reflects the changes in transmembrane potential while ignoring the currents in the tissue caused by volume conductance, thus allowing a more accurate localization of the source of the activity (Zochowski et al. 2000).

We limit our discussion to "fast" signals (Cohen and Salzberg 1978). These signals occur only at the wavelengths related to the dye chromophore and are presumed to arise from membrane-bound dye; they follow changes in membrane potential with time courses that are rapid compared to the rise time of an action potential. Three types of signals will be discussed: (1) spike signals from individual neurons, (2) population signals recorded from thousands of neurons imaged onto single detectors, and (3) signals from hydrophobic dyes that have been retrogradely transported from axon to the cell body. A fourth kind of voltage-sensitive dye signal, from intracellularly injected dye, is described in Chapter 50.

MEASURING SMALL SIGNALS

Voltage-sensitive dye signals have been used in many different preparations, including individual neurons in tissue culture (Parsons et al. 1991), neurons in ganglia (London et al. 1987; Nakashima et al. 1992; Obaid et al. 1999), population signals from brain slices (Grinvald et al. 1982a; Komuro et al. 1993; Demir et al. 1994; Senseman 1996a,b; Sato et al. 1998; Wu et al. 2001; Bao and Wu 2003), and in vivo preparations (Orbach et al. 1985; Kauer et al. 1987; Prechtl et al. 1997; Lam et al. 2003), as well as signals from retrogradely transported dyes in embryonic chick spinal cord (Tsau et al. 1996; Wenner

A. Aplysia Abdominal Ganglion

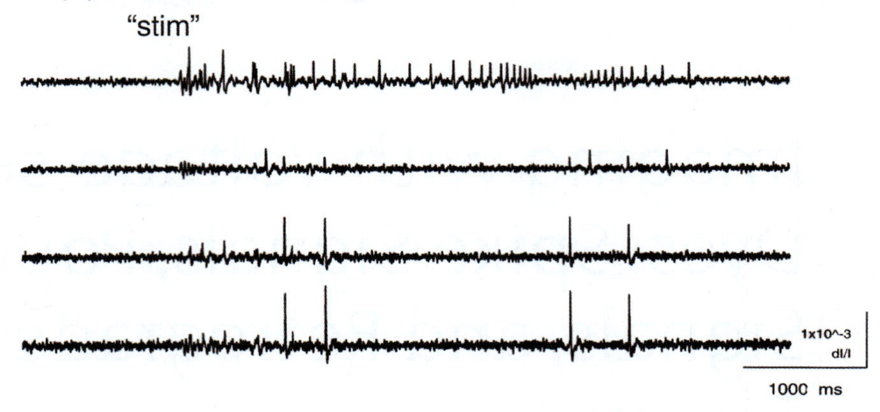

B. Cortical Slice

I. Electrical and optical signals

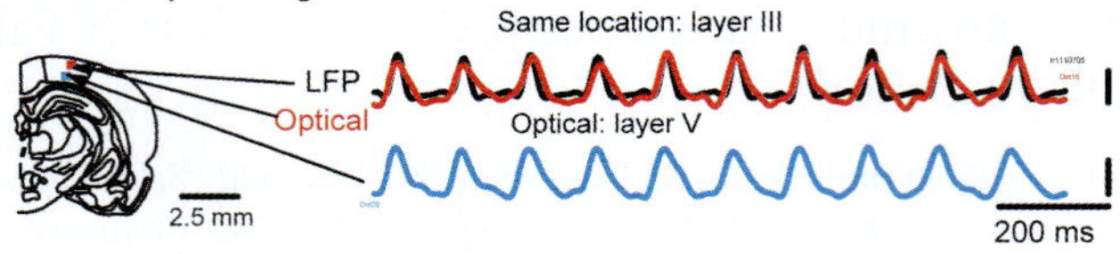

II. Propagating waves during the oscillation

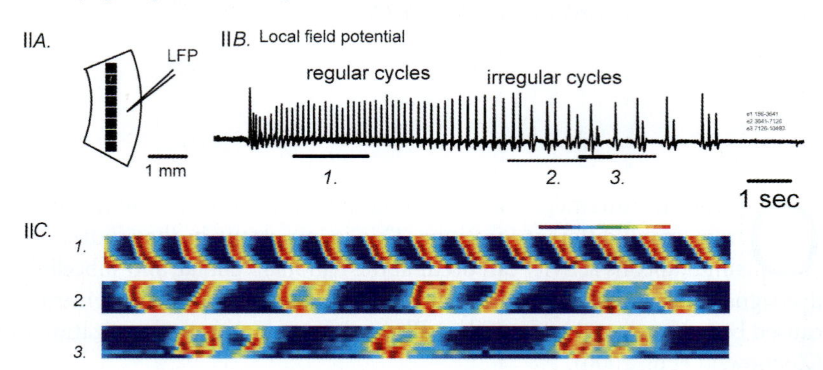

C. Turtle Olfactory Bulb

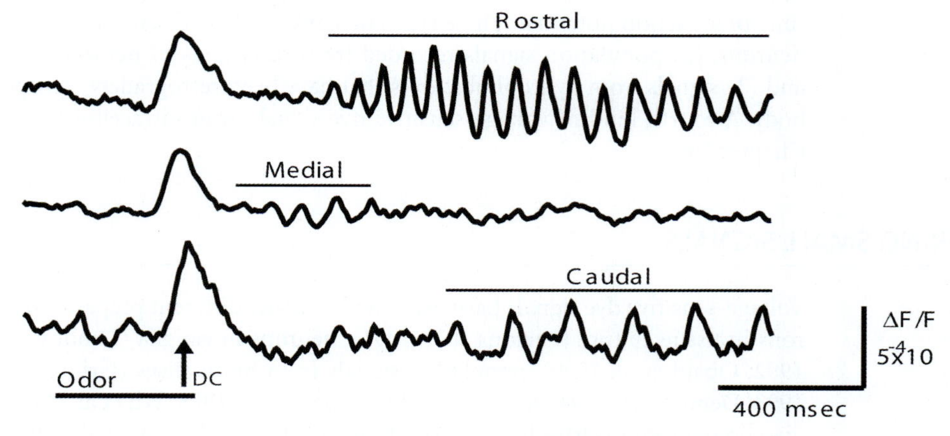

FIGURE 1. (*See facing page for legend.*)

et al. 1996) and lamprey cord (Hickie et al. 1996). As a result of these examples, it is possible to anticipate the fractional intensity changes that might be obtained from available dyes in a new experiment. Quite often these fractional changes are small; the fractional change of signal intensity can be 10^{-3} or less. Therefore, to record such signals, the noise must be in the range of 10^{-4} or less. Three examples of such signals are shown in Figure 1. Figure 1A shows the outputs of four photodiodes in a measurement of absorption signals from individual neurons of an *Aplysia* abdominal ganglion. Figure 1B shows population signals from a cortical slice illustrating spontaneous oscillations. Figure 1C illustrates the 5–15-Hz population signals from an in vivo turtle olfactory bulb in response to odor presentation to the olfactory epithelium. In these three instances, the fractional changes in the signal intensity are in the range 10^{-4} to 10^{-3}, and the noise is less than 10^{-4}. Because these signals are not tightly time-locked to a stimulus, signal averaging could not be used to reduce the noise.

Obviously, making a measurement where the noise is less than 10^{-4} of the total light requires that *all* sources of noise must be less than 10^{-4}. Five noise sources are briefly discussed. Four are relatively straightforward; the issue of shot noise is somewhat more complicated.

Noise in the Incident Light

The light emission from a tungsten filament light source is easily stabilized to one part in 10^{-5}. If a higher-intensity source is needed, a 150-W xenon arc lamp with a power supply from Opti-Quip has intensity fluctuations that are not too much larger than 10^{-4} of the incident light.

Vibrational Noise

Reducing the vibrational noise to less than 10^{-4} requires some effort. A number of precautions for reducing vibrational noise can be taken (Salzberg et al. 1977; London et al. 1987). The pneumatic isolation mounts of vibration isolation tables are more efficient in reducing vertical vibrations than in

FIGURE 1. Optical signals using voltage-sensitive dyes from three different preparations. In each case the signal represents a fractional intensity change of 10^{-3} or less, and the noise in the measurement is 10^{-4} or less. The staining must be sufficiently intense, the optics sufficiently efficient, and the array and amplifiers good enough to provide this kind of noise performance. (*A*) Spike signals from four photodiodes receiving light from four areas of an *Aplysia* abdominal ganglion. An absorption dye and transillumination were used. A light touch to the siphon skin was applied at the time labeled "stim"; vibrational noise from the stimulus can be seen in the second trace. Spike signals are seen on all four detectors. (Redrawn, from Falk et al. 1993.) (*B*) Population oscillations in a rat cortical slice. Absorption dye and transillumination were used. (*I, left*) The recording was done in a slice containing occipital cortex. A microelectrode recorded local field potential (LFP) from cortical layers II–III. Optical signals from the tissue surrounding the electrode (*red square*) and from layer V of the same column (*black square*) were simultaneously recorded with the LFP. (*I, right*) In layers II–III, simultaneous electrical (*black*) and optical (*red*) recordings had a similar waveform during the oscillations. In layer V, the optical signal (*blue*) has the same polarity. Bar, −50 µmV for LFP and 10^{-4} for optical recordings. (*IIA*) Eight optical detectors (*black squares*) were horizontally distributed in the deep cortical layers. Their signals were used in the images in IIC. A local field potential microelectrode was placed in layers II–III. (*IIB*) An oscillation epoch contains regular and irregular cycles. Optical signals during the periods 1, 2, and 3 (marked under the trace) are shown in IIC. (*IIC*) Pseudo-color images from optical signals recorded by the eight detectors. The optical signal from each detector was normalized to the maximum on that particular detector during that particular period, and normalized values were assigned to colors according to a color scale (*top right*). The X direction of the images is time (total time ~12 sec), and the Y direction is space in the horizontal direction in deep cortical layers, ~2.6 mm. (Redrawn, from Bao and Wu 2003.) (*C*) Population signals from three photodiodes receiving light from three regions of a turtle olfactory bulb during and following an odor application to the nose. A fluorescent dye was used with epi-illumination. Three different oscillations are illustrated. (Redrawn, from Lam et al. 2000.)

reducing horizontal movements. As a starting point, the use of air-filled soft rubber tubes (Newport) is recommended. For more severe vibration problems, an isolation table sold by Minus K Technology (www.minusk.com) may help. With its very low resonant frequency (~0.1 Hz), a large portion of vibration noise in the frequency range of biological signals can be eliminated.

Analog-to-Digital Converter Noise

To digitize a signal that is less than 10^{-4} of the resting intensity, the effective resolution of the analog-to-digital converter must be greater than 13 bits. Because noise in the converter degrades the nominal resolution by 1 or 2 bits, a nominal resolution of 16 bits is actually needed. The second-stage amplifier of a photodiode array camera provides a voltage gain of ~100× (Wu and Cohen 1993) after subtracting the DC light, which substantially increases the effective analog-to-digital resolution and eliminates the noises related to the multiplexing and digitization.

Dark Noise

Dark noise will be of concern in the situation where the light level is low; e.g, a single neuron is stained and only a small fraction of that neuron is imaged onto one pixel (Chapter 59). In other circumstances (e.g., the experiments in Fig. 1), the shot noise is much larger than the dark noise. If dark noise is a major contributor, it can be reduced by using a cooled CCD camera or, for very dim images, by using an image intensifier or a CCD chip (e.g., e2v CCD60) with a gain stage.

Shot Noise

The limit of accuracy with which light can be measured is set by the shot noise arising from the statistical nature of photon emission from presently available light sources. In a shot-noise-limited measurement, the signal-to-noise ratio (SNR) is proportional to the square root of the number of measured photons and the bandwidth of the photodetection system (Braddick 1960; Malmstadt et al. 1974). The basis for the square root dependence on intensity is illustrated in Figure 2. In 2A, the result of using a random-number table to distribute 20 photons into 20 time windows is shown. In 2B, the same procedure was used to distribute 200 photons into the same 20 bins. Clearly, relative to the average light level, there is more noise in the top trace (20 photons) than in the bottom trace (200 photons). In Figure 2, an SNR of about 3 was obtained with a photon count of 10 photons per msec. Following the square root relationship, an SNR of 10^{-4} will require a photon count of 10^8 photons/msec. This is an important result in the sense that it places two constraints on the apparatus. First, the optics must be efficient and the fluorescent dyes bright enough to generate 10^8 photons/msec/pixel at the imaging device. In epi-illumination measurements, both the excitation light and the emitted light pass through the objective; the intensity reaching the photodetector is proportional to the *fourth* power of numerical aperture (NA). Unfortunately, conventional microscope optics have low NAs at low magnifications. Salama (1988), Ratzlaff and Grinvald (1991), and Kleinfeld and Delaney (1996) took advantage of the high NA that can be achieved by using a camera or video lens in place of a microscope objective. At a magnification of 4×, a "Macroscope" based on a 25-mm focal length, 0.95 f, camera lens, had an image intensity 100 times greater than a conventional 4× microscope lens.

Second, the imaging apparatus must have a dynamic range (>80 db or >13 bits) large enough to measure a signal of 10^4 photons in a background of 10^8 photons. This dynamic range requirement cannot be met by the current generation of CCD cameras, which saturate at photon counts of less than 10^6 photons. At present, only photodiode array systems (NeuroPDA, RedShirtImaging, LLC) or a CMOS camera (SciMedia USA) has sufficient dynamic range. The measurements described here were made with the 464-pixel NeuroPDA. Larger fractional fluorescence changes (10^{-2}) are sometimes found in voltage-sensitive dye measurements (e.g., Chapter 59) and are often found in measurements using calcium indicator dyes. In these instances, CCD cameras provide adequate SNRs and better spatial resolution than photodiode arrays.

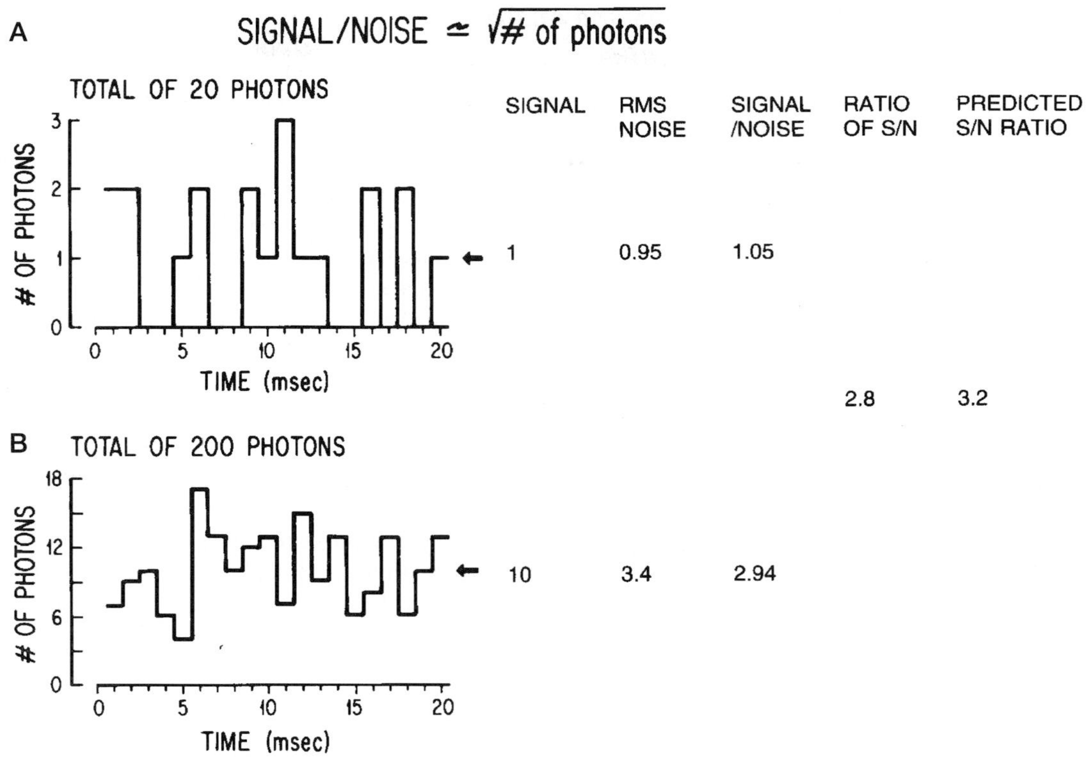

FIGURE 2. Plots of the results of using a table of random numbers to distribute 20 photons (*A*) or 200 photons (*B*) into 20 time bins. The result illustrates the fact that when more photons are measured, the SNR is improved. On the right, the SNR is measured for the two results. The ratio of the two SNRs was 0.43. This is close to the ratio predicted by the relationship that the SNR is proportional to the square root of the measured intensity. (Redrawn, from Cohen and Lesher 1986.)

EXAMPLES OF APPLICATIONS

Action Potentials from Individual Neurons in an *Aplysia* Ganglion

The original motivation for developing optical methods was to record all action potential activity of all neurons in a simple invertebrate ganglion during behaviors. In the first attempt to use voltage-sensitive dyes in ganglia (Salzberg et al. 1973), we were fortunate to be able to monitor activity in a single neuron because of severe photodynamic damage and a low SNR. Now, however, with better dyes and methods, the spike activity of hundreds of individual neurons can be recorded simultaneously (Zecevic et al. 1989; Nakashima et al. 1992; Obaid et al. 1999; Brown et al. 2004). In the experiments described below, the spike activity of about half of the 1000 cells in the *Aplysia* abdominal ganglion is measured.

Dye Staining

Considerable effort was made to optimize dye staining while causing minimal pharmacological effects on the gill-withdrawal behavior in *Aplysia*. Intact ganglia were incubated in a 0.15-mg/ml solution of the oxonol dye, RH155 (or its diethyl analog), using a protocol developed by Nakashima et al. (1992). Once stained, the dye was washed out of the bathing solution and signals were recorded for several hours. Oxonol dyes are moderately hydrophobic; a more hydrophobic dye would not penetrate through the connective tissue sheath, whereas a less hydrophobic dye would destain too rapidly.

Data Acquisition and Analysis

A photodiode array (NeuroPDA) was placed in the image plane formed by a microscope objective of 25× 0.4 NA. Because the very irregular image of a ganglion is formed on a rectilinear diode array, there is no simple correspondence between images of cells and photodetectors. The light from a larger cell body will fall on several detectors, and its activity will be recorded as simultaneous events on neighboring detectors (e.g., the signals on the bottom two detectors in Fig. 1A). In addition, because these preparations are multilayered, most detectors will receive light from several cells (e.g., the two different signals on the second detector in Fig. 1A). As a result, a sorting step is required to determine the activity in neurons from the spike signals on individual photodiodes. This sorting can be carried out using the human pattern recognition system or independent component analysis (ICA) (Brown et al. 2004).

The result of a complete analysis of one data set showed that 135 *Aplysia* neurons were activated by a light touch to the siphon. (Similar results were obtained by Nakashima et al. [1992] and Brown et al. [2004] in *Tritonia*.) This recording was estimated to be only about 50% complete. Thus, the actual number of activated neurons in the abdominal ganglion during the gill-withdrawal reflex was about 300. Thus, most neurons in the abdominal ganglion are modulated by the sensory stimulus. It is almost as if the *Aplysia* nervous system is designed such that every cell in the abdominal ganglion cares about this (and perhaps every) sensory stimulus. In addition, more than 1000 neurons in other ganglia are activated by this touch (Tsau et al. 1994). Clearly, information about this very mild and localized stimulus is propagated widely in the *Aplysia* nervous system.

Population Signals

In a measurement from a vertebrate brain stained by superfusing a solution of the dye over the surface, each of the 464 photodetectors will receive light from a substantial number (thousands) of neurons and neuronal processes. Because of scattering and out-of-focus light, this will be true even if the pixel size corresponds to a very small area of the brain. In this situation, the voltage-sensitive dye signal will be a population average of the change in membrane potential of all of these neurons and processes.

Brain Slices

This example shows wave patterns during theta oscillations in rat neocortical slices (Bao and Wu 2003).

Dye Staining

Absorption dyes were used because the absorption signals were larger than fluorescent signals in brain slices, as predicted and measured (Jin et al. 2002). The slices were stained with NK 3630 (aka RH482; see Momose-Sato et al. [1995] for molecular structure) and measured using transillumination. The signal on each detector (an area of $\sim 330 \times 330\ \mu m^2$) was recorded from ~5000 cortical neurons. The resting light intensity was $\sim 10^9$ photons/detector/msec and the oscillation signal was $\sim 10^{-4}$ of the resting light intensity.

Data Acquisition and Analysis

The oscillation was pharmacologically induced by adding carbachol and bicuculline to the perfusion media. In cortical layers II–III, the optical and local field potential signals had similar SNRs (Fig. 1B, I). In local field potential recordings, the oscillation had a phase reversal between superficial and deep layers, indicating a current sink/source pair located in the cortical column (Lukatch and MacIver 1997). This phase reversal is not seen in the optical signals (Fig. 1B, I), consistent with the assumption that optical signal represents transmembrane potential and is not affected by the local currents due to the geometrical arrangement of cortical neurons.

The oscillation was organized in epochs (Fig. 1B, IIA). Each epoch was composed of some "regular" cycles (relatively stable frequency and amplitude) followed by a period of "irregular" cycles (with variable frequency and amplitude, Fig. 1B, IIA). Each regular cycle was associated with one activation wave (Fig. 1B, IIC, image 1). This one-cycle one-wave pattern was seen during all regular and some irregular cycles. During irregular cycles, the initiation site of the wave and the propagating velocity varied in different cycles (Fig. 1B, IIC, images 2, 3). Complex patterns, such as collision and reflection of the waves, were also observed. In one of the cycles, the wave was apparently reflected at the edge of the tissue and the reflected waves propagated back and collided in the center, forming an "O" pattern (Fig. 1B, IIC, image 3).

This example shows that imaging the wave dynamics in cortical local networks is technically feasible. In brain slices, the SNR of absorption dyes is large, comparable to that of the local field potential recordings (Fig. 1B, I) (Jin et al. 2002), which allows an analysis of the wave dynamics without averaging.

Turtle Olfactory Bulb

Adrian (1950) discovered that odor stimuli induced an oscillation in the local field potential of the olfactory bulb. We were interested in making voltage-sensitive dye measurements during this oscillation for two reasons. First, we can measure the dye signal from many positions simultaneously and thus determine the spatial characteristics of the signal. Second, preliminary experiments suggested that the voltage-sensitive dye signals had about 25 times better spatial resolution than local field potential recordings (Zochowski et al. 2000).

Dye Staining

The turtle olfactory bulb was exposed by removing the overlying bone after the animal was anesthetized by immersion in ice water for 1 hour. The pial surface was incubated with a 0.02 to 0.1 mg/ml solution of a styryl dye, RH-414 (Grinvald et al. 1982b), RH795 (Grinvald et al. 1994), or JPW-1114 (Antic and Zecevic 1995) for 45 minutes and the dye solution was then washed out. After staining, the animal was warmed and then partially paralyzed with curare and its head held mechanically on the stage of a microscope. In some experiments, the turtle was anesthetized with urethane. The above dyes are relatively hydrophilic (di-methyl and di-ethyl); more hydrophobic dyes will not penetrate into the brain. Even if the dye staining is equal in all layers, deeper layers will receive less incident light, and less of the emitted light from deeper layers will reach the surface of the brain as a result of light scattering. Thus, the optical signal measured with epi-illumination will emphasize the signals from upper layers.

Data Acquisition and Analysis

In the measurements from the turtle olfactory bulb illustrated in Figure 1C, each detector received light from a 200-μm-square area (an approximate volume of 0.1 mm^3). In this in vivo preparation, some fraction of the photodiodes receive light from large blood vessels; on these detectors, erythrocyte movements result in excess noise. In different animals, this excess noise was present in 10–30% of the photodiodes receiving light from the bulb. The data illustrated in Figure 1C come from detectors that do not have this excess noise.

An olfactometer copied after Kauer and Moulton (1974) was used to apply a dilution of a saturated vapor of amyl acetate to the turtle's nose. The signals from three photodetectors from three different regions of the bulb are shown in Figure 1C. The rostral signal is often the largest and has a frequency in the range 10–18 Hz. The frequency of the caudal signal is exactly one half of the rostral signal frequency, and it is synchronized with a fixed phase shift. The medial signal has the shortest latency and briefest duration. Information about the spatial location, origin, or spread of the events cannot be obtained from the individual traces illustrated in Figure 1C. A series of pseudo-color frames (which can be displayed on a single page or as a movie) must be made using the data from all of the photodetectors. This pseudo-color display showed that both the rostral and the caudal oscillations were traveling waves. The rostral wave traveled in the rostral–caudal direction, whereas the caudal oscillation traveled in a lateral–caudal direction. Thus, the voltage-sensitive dye recording has provided a uniquely detailed picture of the spatial and temporal aspects of the oscillatory responses.

In Vivo Mammalian Brain

Measurements from in vivo mammalian preparations are more difficult than from turtles because the noises from the heartbeat and respiration are larger and because mammalian preparations are not as easily stained with voltage-sensitive dyes. However, three methods for reducing the movement artifacts from the heartbeat and respiration are, together, quite effective. First, a subtraction procedure can be used where two recordings are made, only one of which has a stimulus (Orbach et al. 1985). The second method reduces the movements of the brain by fixing a fluid-filled chamber onto the skull surrounding the craniotomy (e.g., Blasdel and Salama 1986). Third, the use of dyes that absorb and emit at longer wavelengths also reduces heartbeat noise (Shoham et al. 1999).

Retrograde Staining with Voltage-sensitive Dyes

One limitation of the above population signals is that all cell types in the region are stained and it is difficult to determine which cell type contributes which component to the overall signal. One method for obtaining cell-type-specific staining was recently explored in embryonic chick and in lamprey spinal cords. An identified cell class (motoneurons) was selectively stained by injecting a slurry of a very concentrated solution-cum-suspension of a hydrophobic styryl dye (di-octyl or di-dodecyl dye) into the ventral root. After 4–12 hours, some of this dye traveled to the motoneuron cell bodies in the chick cord. This dye was reasonably efficient in signaling changes in membrane potential, but no convincing evidence of signal detection from an individual motoneuron neuron was obtained (Tsau et al. 1996; Wenner et al. 1996). Later, in lamprey experiments, spike signals from individual neurons could be measured (Hickie et al. 1996) but not reliably. Further efforts to optimize this staining procedure are needed.

FUTURE DIRECTIONS

Because the voltage-sensitive dye-measuring apparatus is already near optimal, any improvement in the sensitivity of the voltage-sensitive dye measurements will now come from the development of better dyes. The vast majority of those previously synthesized are of the general class named polyenes (Hamer 1964), a group that is used to extend the wavelength response of photographic film. The best of these styryl and oxonol dyes have fluorescence changes of 10%/100 mV in a bilayer. Recently, Gonzalez and Tsien (1995) introduced a new scheme for generating voltage-sensitive signals using two chromophores and energy transfer. Although these fractional changes were also initially in the range of 10%/100 mV, more recent results are nearer 30% (T. Gonzalez and R. Tsien, pers. comm.). However, in order to obtain fast responses, one of the chromophores must be hydrophobic, and molecules of such hydrophobicity will not penetrate the brain tissue. Thus, it has not been possible to measure signals in intact tissues using this combination of dyes (T. Gonzalez and R. Tsien; A. Obaid and B.M. Salzberg, both pers. comm.). Finally, second harmonic generation provides an alternative to fluorescence-based voltage imaging (see Chapter 60).

Neuron-type-specific Staining

An important new direction is the development of methods for neuron-type-specific staining. Three quite different approaches have been tried. First, the use of retrograde staining procedures has recently been investigated in the embryonic chick and lamprey spinal cords (see above). Second, the use of cell-type-specific staining has been developed for fluorescein by Nirenberg and Cepko (1993). It might be possible to use similar techniques to selectively stain cells with voltage-sensitive or ion-sensitive dyes. Third, Siegel and Isacoff (1997) constructed a genetically encoded chimera of a potassium channel and green fluorescent protein (see Chapter 77). When introduced into a frog oocyte, this molecule had a (relatively slow) voltage-dependent signal with a fractional fluorescence change of 5%. More recently, Sakai et al. (2001) and Ataka and Pieribone (2002) developed similar constructs with very rapid kinetics. Neuron-type-specific staining would make it possible to determine the role of specific neuron types in generating the input–output function of a brain region. Unfortunately, in mammalian cells, these FP-voltage sensors are only minimally expressed in the extracellular membrane (Baker et al. 2004). They appear to be held in the endoplasmic reticulum. Solutions to this problem are under investigation.

Optical recording with voltage-sensitive dyes already provides unique insights into brain activity and organization. Clearly, improvements in sensitivity or selectivity would make these methods even more powerful.

ACKNOWLEDGMENTS

The authors are indebted to their collaborators Vicencio Davila, Amiram Grinvald, Kohtaro Kamino, David Kleinfeld, Ying-wan Lam, Les Loew, Jim Prechtl, Bill Ross, Brian Salzberg, Matt Wachowiak, Alan Waggoner, Joe Wuskell, and Michal Zochowski for numerous discussions about optical methods. The experiments carried out in our laboratories were supported by National Institutes of Health grants NS08437-DC05259 and NS36447.

REFERENCES

Adrian E.D. 1950. The electrical activity of the mammalian olfactory bulb. *Electroencephalogr. Clin. Neurophysiol.* **2:** 377–388.

Antic S., Major G., and Zecevic D. 1999. Fast optical recordings of membrane potential changes from dendrites of pyramidal neurons. *J. Neurophysiol.* **82:** 1615–1621.

Ataka K. and Pieribone V.A. 2002. A genetically-targetable fluorescent probe of channel gating with rapid kinetics. *Biophys. J.* **82:** 509–516.

Baker B.J., Lee H., Ataka K., Pieribone V., Cohen L.B., Isacoff E., and Kosmidis E. 2004. Plasma membrane expression of GFP-voltage sensors. *Soc. Neurosci. Abstr.* (in press).

Bao W. and Wu J.Y. 2003. Propagating wave and irregular dynamics: Spatiotemporal patterns of cholinergic theta oscillations in neocortex in vitro *J. Neurophysiol.* **90:** 333–341.

Blasdel G.G. and Salama G. 1986. Voltage-sensitive dyes reveal a modular organization in monkey striate cortex. *Nature* **321:** 579–585.

Braddick H.J.J. 1960. Photoelectric photometry. *Rep. Prog. Phys.* **23:** 154–175.

Brown G.D., Yamada S., and Sejnowski T.J. 2001. Independent component analysis at the neural cocktail party. *Trends Neurosci.* **24:** 54–63.

Cohen L.B. and Lesher S. 1986. Optical monitoring of membrane potential: Methods of multisite optical measurement. *Soc. Gen. Physiol. Ser.* **40:** 71–99.

Cohen L.B. and Salzberg B.M. 1978. Optical measurement of membrane potential. *Rev. Physiol. Biochem. Pharmacol.* **83:** 35–88.

Demir R., Haberly L.B., and Jackson M. 1994. Optical recordings of evoked and spontaneous epileptiform discharges and slices from piriform cortex. *Soc. Neurosci. Abstr.* **20:** 403.

Falk C.X., Wu J.-Y., Cohen L.B., and Tang A.C. 1993. Nonuniform expression of habituation in the activity of distinct classes of neurons in the *Aplysia* abdominal ganglion. *J. Neurosci.* **13:** 4072–4081.

Gonzalez J.E. and Tsien R.Y. 1995. Voltage sensing by fluorescence energy transfer in single cells. *Biophys. J.* **69:** 1272–1280.

Grinvald A., Manker A., and Segal M. 1982a. Visualization of the spread of electrical activity in rat hippocampal slices by voltage-sensitive optical probes. *J. Physiol.* **333:** 269–291.

Grinvald A., Hildesheim R., Farber I.C., and Anglister L. 1982b. Improved fluorescent probes for the measurement of rapid changes in membrane potential. *Biophys. J.* **39:** 301–308.

Grinvald A., Lieke E.E., Frostig R.D., and Hildesheim R. 1994. Cortical point-spread function and long-range lateral interactions revealed by real-time optical imaging of macaque monkey primary visual cortex. *J. Neurosci.* **14:** 2545–2568.

Hamer F.M. 1964. *The cyanine dyes and related compounds.* Wiley, New York.

Hickie C., Wenner P., O'Donovan M., Tsau Y., Fang J., and Cohen L.B. 1996. Optical monitoring of activity from individual and identified populations of neurons retrogradely labeled with voltage-sensitive dyes. *Soc. Neurosci. Abstr.* **22:** 321.

Jin W.J., Zhang R.J., and Wu J.Y. 2002. Voltage-sensitive dye imaging of population neuronal activity in cortical tissue. *J. Neurosci. Methods* **115:** 13–27.

Kauer J.S. and Moulton D.G. 1974. Responses of olfactory bulb neurones to odour stimulation of small nasal areas in the salamander. *J. Physiol.* **243:** 717–737.

Kauer J.S., Senseman D.M., and Cohen L.B. 1987. Odor elicited activity monitored simultaneously from 124 regions of the salamander olfactory bulb using a voltage sensitive dye. *Brain Res.* **418:** 255–261.

Kleinfeld D. and Delaney K. 1996. Distributed representation of vibrissa movement in the upper layers of somatosensory cortex revealed with voltage-sensitive dyes. *J. Comp. Neurol.* 375: **89**–108.

Komuro H., Momose-Sato Y., Sakai T., Hirota A., and Kamino K. 1993. Optical monitoring of early appearance of spontaneous membrane potential changes in the embryonic chick medulla oblongata using a voltage-sensitive dye. *Neuroscience* **52:** 55–62.

Lam Y.-W., Cohen L.B., and Zochowski M.R. 2003. Odorant specificity of three oscillations and the DC signal in the turtle olfactory bulb. *Eur. J. Neurosci.* **17:** 436–446.

Lam I.W., Cohen L.B., Wacnowiak M., and Zocnowsky M. R. 2000. Odors elicit three different oscillations in the turtle olfactory bulb. *J. Neurosci.* **20:** 749–762.

London J.A., Zecevic D., and Cohen L.B. 1987. Simultaneous optical recording of activity from many neurons during feeding in *Navanax. J. Neurosci.* **7:** 649–661.

Lukatch H.S. and MacIver M.B. 1997. Physiology, pharmacology, and topography of cholinergic neocortical oscillations in vitro. *J. Neurophysiol.* **77:** 2427–2445.

Malmstadt H.V., Enke C.G., Crouch S.R., and Harlick G. 1974. *Electronic measurements for scientists.* W.A. Benjamin, Menlo Park, California.

Momose-Sato Y., Sato K., Sakai T., Hirota A., Matsutani K., and Kamino K. 1995. Evaluation of optimal voltage-sensitive dyes for optical monitoring of embryonic neural activity. *J. Membr. Biol.* **144:** 167–176.

Nakashima M., Yamada S., Shiono S., Maeda M., and Sato F. 1992. 448-detector optical recording system: Development and application to *Aplysia* gill-withdrawal reflex. *IEEE Trans. Biomed. Eng.* **39:** 26–36.

Nirenberg S. and Cepko C. 1993. Targeted ablation of diverse cell classes in the nervous system *in viva. J. Neurosci.* **13:** 3238–3251.

Obaid A.L., Koyano T., Lindstrom J., Sakai T., and Salzberg B.M. 1999. Spatiotemporal patterns of activity in an intact mammalian network with single-cell resolution: Optical studies of nicotinic activity in an enteric plexus. *J. Neurosci.* **19:** 3073–3093.

Orbach H.S., Cohen L.B., and Grinvald A. 1985. Optical mapping of electrical activity in rat somatosensory and visual cortex. *J. Neurosci.* **5:** 1886–1895.

Parsons T.D., Salzberg B.M., Obaid A.L., Raccuia-Behling F., and Kleinfeld D. 1991. Long-term optical recording of patterns of electrical activity in ensembles of cultured *Aplysia* neurons. *J. Neurophysiol.* **66:** 316–333.

Prechtl J.C., Cohen L.B., Peseran B., Mitra P.P., and Kleinfeld D. 1997. Visual stimuli induce waves of electrical activity in turtle cortex. *Proc. Natl. Acad. Sci.* **94:** 7621–7626.

Ratzlaff E.H. and Grinvald A. 1991. A tandem-lens epifluorescence microscope: Hundred-fold brightness advantage for wide-field imaging. *J. Neurosci. Methods* **36:** 127–137.

Sakai R., Repunte-Canonigo V., Raj C.D., and Knopfel T. 2001 Design and characterization of a DNA-encoded, voltage-sensitive fluorescent protein. *Eur. J. Neurosci.* **13:** 2314–2318.

Salama G. 1988. Voltage-sensitive dyes and imaging techniques reveal new patterns of electrical activity in heart and cortex. *SPIE Proc.* **94:** 75–86.

Salzberg B.M., Davila H.V., and Cohen L.B. 1973. Optical recording of impulses in individual neurones of an invertebrate central nervous system. *Nature* **246:** 508–509.

Salzberg B.M., Grinvald A., Cohen L.B., Davila H.V., and Ross W.N. 1977. Optical recording of neuronal activity in an invertebrate central nervous system: Simultaneous monitoring of several neurons. *J. Neurophysiol.* **40:** 1281–1291.

Sato K., Momose-Sato Y., Hirota A., Sakai T., and Kamino K. 1998. Optical mapping of neural responses in the embryonic rat brainstem with reference to the early functional organization of vagal nuclei. *J. Neurosci.* **18:** 1345–1362.

Senseman D.M. 1996a. Correspondence between visually evoked voltage-sensitive dye signals and synaptic activity recorded in cortical pyramidal cells with intracellular microelectrodes. *Vis. Neurosci.* **13:** 963–977.

———. 1996b. High-speed optical imaging of afferent flow through rat olfactory bulb slices: Voltage-sensitive dye signals reveal periglomerular cell activity. *J. Neurosci.* **13:** 247–263.

Shorham D., Glaser D.E., Arieli A., Kenet T., Wijnbergen C., Toledo Y., Hildesheim R., and Grinvald A. 1999. Imaging cortical dynamics at high spatial and temporal resolution with novel blue voltage-sensitive dyes. *Neuron* **24:** 791–802.

Siegel M.S. and Isacoff E.Y. 1997. A genetically encoded optical probe of membrane voltage. *Neuron* **19:** 735–741.

Tsau Y., Wu J.Y., Hopp H.P., Cohen L.B., Schiminovich D., and Falk C.X. 1994. Distributed aspects of the response to siphon touch in *Aplysia:* Spread of stimulus information and cross-correlation analysis. *J. Neurosci.* **14:** 4167–4184.

Tsau Y., Wenner P., O'Donovan M.J., Cohen L.B., Loew L.M., and Wuskell J.P. 1996. Dye screening and signal-to-noise ratio for retrogradely transported voltage-sensitive dyes. *J. Neurosci. Methods* **70:** 121–129.

Wenner P., Tsau Y., Cohen L.B., O'Donovan M.J., and Dan Y. 1996. Voltage sensitive dye recording using retrogradely transported dye in the chicken spinal cord: Staining and signal characteristics. *J. Neurosci. Methods* **70:** 111–120.

Wu J.-Y. and Cohen L.B. 1993. Fast multisite optical measurements of membrane potential. In *Fluorescent and luminescent probes for biological activity* (ed. W.T. Mason), pp. 389–404. Academic Press, London.

Wu J.-Y., Guan L., Bai L., and Yang Q. 2001. An evoked all-or-none population activity in rat sensory cortical slices. *J. Neurophysiol.* **86:** 2461–2474.

Zecevic D., Wu J.Y., Cohen L.B., London J.A., Hopp H.P., and Falk C.X. 1989. Hundreds of neurons in the *Aplysia* abdominal ganglion are active during the gill-withdrawal reflex. *J. Neurosci.* **9:** 3681–3689.

Zochowski M., Wachowiak D.M., Falk C. X., Cohen L.B., Lam Y.-W., Antic S., and Zecevic D. 2000. Imaging membrane potential with voltage-sensitive dyes. *Biol. Bull.* **198:** 1–21.

CHAPTER 59

Dendritic Voltage Imaging

Maja Djurisic,[1] Srdjan Antic,[2] and Dejan Zecevic[1]

[1]Department of Cellular and Molecular Physiology; [2]Department of Neurobiology, Yale University School of Medicine, New Haven, Connecticut 06520

This chapter describes an optical technique that is used for imaging of electrical signals with voltage-sensitive dyes (VSD) from multiple sites on the processes of individual nerve cells, including thin terminal dendrites that are not accessible to electrical measurements. The goal of this approach is to understand biophysical and functional properties of individual neurons by monitoring integration of electrical signals simultaneously from many locations on neuronal processes.

Voltage imaging from individual nerve cells, selectively stained by intracellular injection of fluorescent dyes, has been used in the analysis of the electrical and functional structure of identified nerve cells in invertebrate ganglia and in experiments on vertebrate neocortical pyramidal neurons (Zecevic 1996; Antic et al. 1999, 2000; Antic 2003), as well as on mitral cells of the olfactory bulb (Fig. 1).

MATERIALS

Voltage imaging from individual vertebrate neurons in brain slices requires an optical monitoring apparatus integrated with patch pipette recordings and infrared (IR) video microscopy. The experimental setup used in this protocol is organized around a compound upright microscope (Olympus BX51WI) equipped with moveable head; stationary, rigid stage, dark-field, bright-field, IR-DIC, and fluorescence optics; two camera ports; and two micromanipulators mounted on the fixed stage. One camera port has a conventional CCD camera (CCD-300, Dage-MTI; www.dagemti.com) for visualizing neurons in slices under infrared-DIC illumination. The other camera port has a data acquisition CCD. Initial experiments used the NeuroCCD Imaging System (RedShirtImaging LLC), based on the Pixel Vision FastOne camera (80 x 80 pixel; back-illuminated, cooled CCD characterized by 2.7-KHz full-frame rate, 14-bit A-to-D resolution, well depth of 300,000 electrons/msec, frameshift time of 26 μsec, and a read noise of 35 electrons rms at 1-KHz frame rate). This system was recently replaced by a camera with lower read noise (9 electrons rms at 1-KHz frame rate), faster frameshift time (7 μsec), and the option to increase the frame rate from 2 KHz (full frame) to 3 KHz (2 × 2 binned), 5 KHz (3 × 3 binned), and 10 KHz (partial readout; 80 × 12 pixels) (NeuroCCD-SM, RedShirtImaging LLC). To reduce vibrations, the microscope must be placed on an isolation table. Model 350 BM-1 Minus-K vibration isolation systems (www.minusk.com) are recommended. The analysis and display of data were made using the NeuroPlex program (RedShirtImaging, LLC) written in IDL (Interactive Data Language, Research Systems). The fluorescent image of the preparation is projected via an optical coupler (0.1×) onto the CCD chip. Nikon 40×/0.9 NA or 60×/1 NA water-immersion objectives are most appropriate for measurements from vertebrate neurons in brain slices. A 250-W xenon, short-gap arc lamp (Osram, XBO 250 W/CR ORF) powered by a low-ripple-power supply (Model 1700XT/A, Opti-Quip, Highland Mills) was used as a source for excitation light. The best optical signals were obtained using an excitation interference filter of 520 ± 45 nm, a dichroic mirror with the central wavelength of 570 nm, and a 610-nm barrier filter (a Schott RG610).

PROCEDURE

Staining Individual Vertebrate Neurons

1. Prepare brain slices from 18–35-day-old Wistar rats, as described by Stuart and Sakmann (1994). Briefly, decapitate rats following halothane anesthesia and cut 300–400-μm-thick slices in ice-cold standard extracellular solution using a vibratome (e.g., Vibroslice VSLM1, WPI). (Standard extracellular solution for vertebrate neurons: 125 mM NaCl, 26 mM $NaHCO_3$, 20 mM glucose, 2.3 mM KCl, 1.26 mM KH_2PO_4, 2 mM $CaCl_2$, 1 mM $MgSO_4$, pH 7.4 when bubbled with 95% O_2, 5% CO_2.)
2. Incubate the slices for 30 min at 35°C and thereafter maintain slices at room temperature (24–25°C).
3. For somatic whole-cell recordings, use 2–5 MΩ patch pipettes. Fill the patch electrodes from the tip with standard intracellular solution by applying negative pressure for about 3 min. (Standard intracellular solution for vertebrate neurons: 115 mM potassium gluconate, 20 mM KCl, 4 mM Mg_2ATP, 10 mM phosphocreatine, 0.3 mM Na-GTP, 10 mM HEPES [adjust to pH 7.3 with KOH<!>]).
4. Backfill the electrodes with dye solution. (Voltage-sensitive dye solution for vertebrate neurons: Dissolve 3 mg/ml of the voltage-sensitive dye JPW1114 [Molecular Probes] in the standard intracellular solution.) Before filling the electrodes, filter the dye solution to eliminate microscopic particles. Use Millex-GV4 filters with 0.22-mm pore size (Millipore).
5. Patch individual neurons using DIC video microscopy. Do not apply pressure to the patch electrode after the whole-cell configuration is formed. Simple diffusion is sufficient to load the soma with an adequate amount of the dye.
6. Allow 30–60 min of free diffusion of the dye from the pipette into the cell body to complete the intracellular staining. After this time, remove the patch electrode by forming an outside-out patch. Incubate slices for an additional 2 hours to allow for the spread of the dye from the soma into distal dendritic processes before making optical recordings.

   Note: The major problem in staining vertebrate neurons from patch pipettes is leakage of dye from the electrode. Patching requires positive pressure to be applied to the patch pipette during electrode manipulation through the tissue while approaching healthy neurons. This pressure ejects the solution from the electrode. To avoid extracellular deposition of the dye and the resulting large background fluorescence, the tip of the electrode must be filled with dye-free solution. It is possible, using this approach, to load neurons without any leakage of the dye to the surrounding tissue. An example of the selective staining of a neuron is shown in Figure 1B.

Caution: *See Appendix 3 for appropriate handling of materials marked with <!>.*

EXAMPLE OF APPLICATION

A typical optical measurement from a mitral cell in the olfactory bulb slice is illustrated in Figure 1. The aim of the experiment was (1) to determine the pattern of spike initiation and propagation in the primary dendrite for the action potential evoked by an excitatory synaptic input and (2) to investigate the basic characteristics of the evoked excitatory postsynaptic potentials (EPSPs) at the site of origin, the glomerular dendritic tuft, including its attenuation along the primary dendrite. A mitral cell was loaded with a voltage-sensitive dye. A stimulating electrode was positioned in the olfactory nerve layer and the olfactory nerve was stimulated to produce an EPSP-evoked action potential in a postsynaptic mitral cell. Figure 1A shows a part of the composite DIC image of the stained mitral cell obtained under infrared light using a high-resolution CCD camera to document

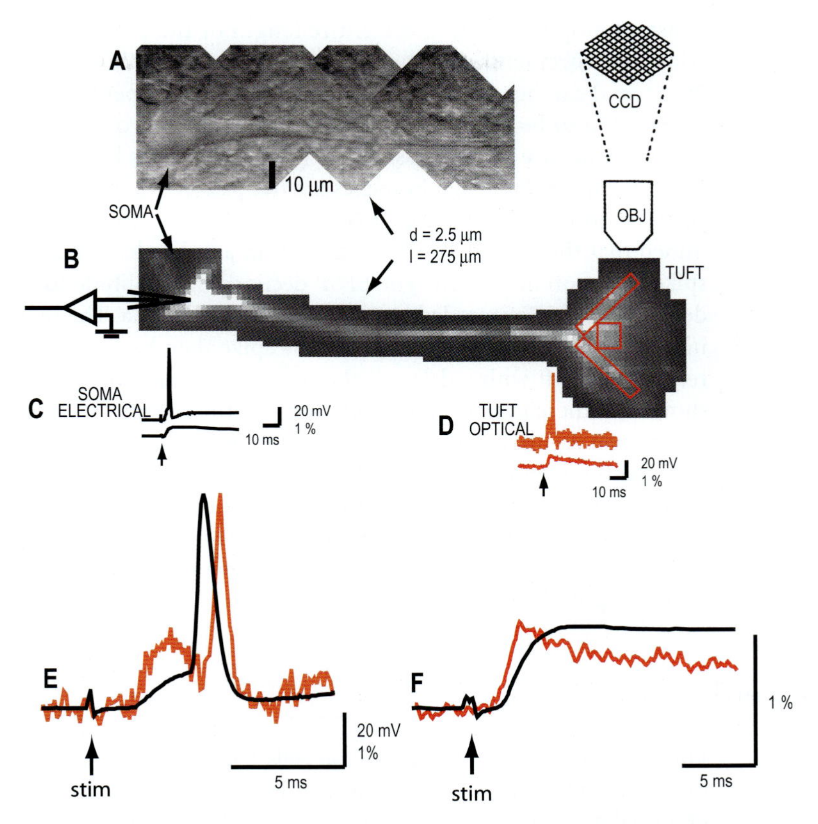

FIGURE 1. Voltage imaging of synaptic and action potentials from a mitral cell in the rat olfactory bulb slice. (*A*) A composite DIC image of the mitral cell soma and proximal part of the primary dendrite obtained under infrared illumination. (*B*) A composite fluorescence image of the same cell, stained with a voltage-sensitive dye, recorded with the fast, low-resolution (80 × 80 pixel) CCD camera used for voltage imaging. (*C, D*) Membrane potential was recorded from the soma with the patch electrode (*dark traces*) and simultaneously monitored optically from multiple locations on the primary dendrite, including terminal dendritic branches in the tuft (*red traces*). (*E*) Electrical signal from the soma (*dark trace*) and optical signal from the tuft (*red trace*; spatial average from selected area; *red rectangles*) compared on an expanded time scale. Action potential signals are scaled to the same height. The first spike was initiated in the soma-axon region and back-propagated into the tuft. (*F*) The shape of EPSP as a function of distance from the site of origin (tuft). The EPSP has a slower rate of rise in the soma (*dark trace*) compared to the tuft (*red trace*). Synaptic potential signals are scaled to the same height.

anatomical details of the neuron. Figure 1B shows a composite fluorescence image of the same cell recorded with the fast, low-resolution (80 × 80 pixel) CCD camera used for voltage imaging. Experiments were done at 35–36°C, at which the rising phase of the action potential is completed in about 0.4 msec. To avoid undersampling spike signals, voltage-imaging was done using partial CCD readout (12 × 80 pixels) and the frame rate of 10 KHz. Changes in light intensity from selected regions on the cell, corresponding to membrane potential changes, were displayed as a function of time and analyzed as illustrated in Figure 1C–F; panel C shows electrical recordings from the soma. The upper trace is a synaptically evoked action potential (80 mV in amplitude), following suprathreshold stimulation of the olfactory nerve. The lower trace is an EPSP evoked in another trial by subthreshold stimulation of the olfactory nerve. Simultaneous optical recordings from thin terminal dendritic branches in the tuft (not accessible to electrical measurements) are shown in Figure 1D. The comparison of action potential and synaptic potential signals recorded simultaneously from the two locations (soma and dendritic tuft), separated by a long primary dendrite, is shown in panels E and F. The temporal and spatial characteristics of the synaptic initiation of the spike were directly obtained from optical data. The amplitude calibration of optical data in terms of membrane potential is also possible in mitral cells. The action potential in these cells has constant amplitude along the entire length of the primary dendrite, as established by direct electrical measurements (Bischofberger and Jonas 1997; Chen et al. 1997; Christie and Westbrook 2003), and can be used as a calibration signal. The upper limit for the possible error in calibration was determined to be small (about 5%) by comparing two extreme cases in a computer simulation: (1) Tuft branches are totally passive and (2) tuft branches have full complement of v-activated channels. In addition, we know that branches are not passive (they can initiate spikes), so the error must be very small or nonexistent. By applying this calibration to the recording shown in Figure 1E, it was determined that the evoked EPSP was 22 mV in amplitude at the site of origin, but still subthreshold for spike initiation in the tuft. The EPSP declined in amplitude to 55% of the initial amplitude along the dendritic cable, as recorded by an electrode in the soma. The reduced EPSP resulted, however, in spike initiation in the soma-axon region. The spike then back-propagated along the primary dendrite and reached the tuft with a delay of about 1 msec, during the falling phase of the local EPSP. This result shows that there is a substantial difference in the voltage threshold between terminal dendritic tuft and soma-axon region of mitral cells. The biophysical basis for this difference is not well understood.

The broadening of the EPSP with distance from the site of origin was determined in the absence of the action potential, using subthreshold stimulation, as shown in Figure 1F. Signals from two locations are scaled to the same height to illustrate the change in the shape of the EPSP caused by the passive spread in the dendrite. In this measurement, the EPSP in the soma was attenuated to 57% of its initial amplitude in the tuft.

ADVANTAGES AND LIMITATIONS

Voltage imaging of individual nerve cells has two advantages: (1) The ability to make simultaneous measurements from many locations and (2) the ability to monitor membrane potential changes from neuronal processes that are not accessible to electrode recordings. Thin dendritic branches that receive synaptic inputs are likely to be the key compartment that determines the input–output transform of an individual neuron. Thus, ideally, one would like to be able to monitor, at multiple sites, subthreshold events as they travel from the synaptic sites of origin on neuronal processes and summate at particular locations to influence action potential initiation. This can now be done, at least in some preparations, using voltage imaging, as illustrated in Figure 1.

The major limitation of this technique is that the signal size is relatively small and its application requires attention to many technical details. Nevertheless, it is reasonable to predict that the technology best suited to the investigation of signal processing at the level of single neurons, and functional neuronal circuits, might ultimately be voltage imaging, which allows large-scale parallel recordings. It is clear that improving the sensitivity of the present technique by a factor of 10 would result in a more powerful method with broader application. Improvements in sensitivity can come from better staining protocols, better dyes, including genetically encoded probes, and from reduction of noise in the recording. Work is in progress in all of these areas.

The pharmacological effects and photodynamic damage caused by the dye represent a further potential limitation of this technique. Previous work, however, has shown that certain voltage-sensitive dyes have little or no pharmacological effect when applied within a limited concentration range from the outside or inside to some invertebrate neurons (Ross et al. 1977; Gupta et al. 1981; Zecevic et al. 1989; Antic and Zecevic 1995). Mammalian preparations have generally been found to be more sensitive to photodynamic injury than invertebrate preparations (Grinvald et al. 1982; Antic et al. 1999). With mitral cells, it was found that the resting membrane potential and action potentials recorded electrically from the soma were not affected during the staining period, indicating the absence of pharmacological effects. In addition, mitral cells were able to tolerate the high-intensity excitation light during successive recording trials, providing the exposures were kept relatively short (100–200 msec) and separated by dark intervals lasting for several minutes. In a stable experiment, it was routinely possible to acquire approximately 50 recording trials (illumination time: 200 msec/trial) before any effect of photodynamic damage was observed.

ACKNOWLEDGMENT

This work is supported by National Institutes of Health grant NS-42739.

REFERENCES

Antic S.D. 2003. Action potentials in basal and oblique dendrites of rat neocortical pyramidal neurons. *J Physiol.* **550:** 35–50.

Antic S.D. and Zecevic D. 1995. Optical signals from neurons with internally applied voltage-sensitive dyes. *J. Neurosci.* **15:** 1392–1405.

Antic S., Major G., and Zecevic D. 1999. Fast optical recordings of membrane potential changes from dendrites of pyramidal neurons. *J. Neurophysiol.* **82:** 1615–1621.

Antic S.D., Wuskell J.P., Loew L., and Zecevic D. 2000. Functional profile of the giant metacerebral neuron of helix aspersa: Temporal and spatial dynamics of electrical activity in situ. *J. Physiol.* **527:** 55–69.

Bischofberger J. and Jonas P. 1997. Action potential propagation into the presynaptic dendrites of rat mitral cells. *J. Physiol.* **504:** 359–365.

Chen W., Midtgaard J., and Shepherd G. 1997. Forward and backward propagation of dendritic impulses and their synaptic control in mitral cells. *Science* **278:** 463–467.

Christie J.M. and Westbrook G.L. 2003. Regulation of backpropagating action potentials in mitral cell lateral dendrites by A-type potassium currents. *J. Neurophysiol.* **89:** 2466–2472.

Grinvald A., Manker A., and Segal M. 1982. Visualization of the spread of electrical activity in rat hippocampal slices by voltage-sensitive optical probes. *J. Physiol.* **333:** 269–291.

Gupta R.K., Salzberg B.M., Grinvald A., Cohen L.B., Kamino K., Lesher S., Boyle M.B., Waggoner A.S., and Wang C.H. 1981. Improvements in optical methods for measuring rapid changes in membrane potential. *J. Membr. Biol.* **58:** 123–137.

Ross W.N., Salzberg B.M., Cohen L.B., Grinvald A., Davila H.V., Waggoner A.S., and Wang C.H. 1977. Changes in absorption, fluorescence, dichroism, and birefrigence in stained giant axons: Optical measurements of membrane potential. *J. Membr. Biol.* **33:** 141–183.

Stuart G.J. and Sakmann B. 1994. Active propagation of somatic action potentials into neocortical pyramidal cell dendrites. *Nature* **367:** 69–72.

Zecevic D. 1996. Multiple spike-initiation zones in single neurons revealed by voltage-sensitive dyes. *Nature* **381:** 322–325.

Zecevic D., Wu J.-Y., Cohen L.B., London J.A., Hopp H.-P., and Falk X.C. 1989. Hundreds of neurons in the *Aplysia* abdominal ganglion are active during the gill-withdrawal reflex. *J. Neurosci.* **9:** 3681–3689.

CHAPTER 60

Second Harmonic Imaging of Membrane Potential

Andrew C. Millard,[1] Aaron Lewis,[2] and Leslie M. Loew[1]

[1]Department of Cell Biology, Center for Cell Analysis and Modeling, University of Connecticut Health Center, Farmington, Connecticut 06030; [2]Division of Applied Physics, Hebrew University of Jerusalem, Jerusalem 91904, Israel

This chapter describes a relatively new approach for imaging membrane potential changes in single cells and multicellular preparations. It uses the nonlinear optical phenomenon known as second harmonic generation (SHG). Naphthylstyryl (ANEP) dyes, a class of electrochromic membrane-staining probes that have also been used to monitor membrane potential by fluorescence, produce SHG images of cell membranes with SHG intensities that are sensitive to voltage. The voltage sensitivity of this technique is significantly greater than the voltage sensitivity of fluorescence for these or any other dyes that have been tested. Thus, second harmonic imaging of membrane potential (SHIMP) has great promise for spatiotemporal mapping of electrical activity in neurons.

The work of Lawrence Cohen and colleagues in the mid 1970s led to the establishment of optical methods as a way to measure the electrical activity of cells in situations where traditional microelectrode methods are not possible or too limiting (Cohen et al. 1974). The authors' laboratory soon joined the effort to develop potentiometric dyes by applying rational design methods based on molecular orbital calculations of the dye chromophores and characterization of their binding and orientations in membranes (Loew et al. 1978, 1979a). Several important general-purpose dyes have emerged from this effort, including di-5-ASP (Loew et al. 1979b), di-4-ANEPPS (Fluhler et al. 1985; Loew et al. 1992), di-8-ANEPPS (Bedlack et al. 1992; Loew 1994), and TMRM and TMRE (Ehrenberg et al. 1988). The fluorescence signals from ANEP dyes have been particularly effective in studies aimed at mapping the activity of excitable cells in complex preparations (Wu et al. 1998; Obaid et al. 1999; Antic et al. 2000; Zochowski et al. 2000; Loew 2001).

In the mid 1980s, it was realized that the large charge redistribution that occurs upon absorption of a photon by the ANEP chromophores, which makes these dyes electrochromic, should also make them promising materials for SHG (Huang et al. 1988). SHG is a nonlinear optical process that can take place at the focus of an ultrafast near-infrared laser. As in the case of two-photon excitation of fluorescence (2PF), the probability of SHG is proportional to the square of the incident light intensity, so that 3D optical sectioning is a natural benefit of scanning microscopy with either of these nonlinear optical modalities. The physics behind these phenomena are, however, quite distinct. Whereas 2PF involves the near-simultaneous absorption of two photons to excite a fluorophore, followed by relaxation and noncoherent emission, SHG is a near-instantaneous process in which two photons are converted into a single photon of twice the energy. The SHG light propagates coherently in the forward direction. The intensity of the SHG signal depends on the molecular hyperpolarizability of the array of molecules that experience the intense laser field. The molecular hyperpolarizability, in turn, can be resonance-enhanced when the incident laser wavelength is close to twice the wavelength of an absorption band of the molecules; this resonance enhancement also depends on the sensitivity of this absorption band to electric fields; i.e., there is electrochromism. The full theory of SHG and its application to biological systems is beyond the scope of this chapter but has been thoroughly reviewed elsewhere (Moreaux et al. 2000; Loew et al. 2002; Campagnola and Loew 2003; Millard et al. 2003b; Pons et al. 2003). One key condition for SHG, in addition to a large molecular hyperpolarizability, is that the molecules producing an SHG signal be organized in a non-centrosymmetric array—a condition that is nicely met when the dye molecules stain one side of a cell membrane.

The relationship between SHG and electrochromism also prompted investigation of whether SHG from membranes stained with ANEP dyes could be sensitive to membrane potential (Bouevitch et al. 1993; Ben-Oren et al. 1996; Campagnola et al. 1999). Many of these earlier studies provided indications that the SHG sensitivity to membrane potential could be much larger than the typical best sensitivity of ~10% fluorescence change per 100 mV. However, only recently has SHG been measured in stained cells that are simultaneously voltage-clamped via whole-cell patch clamp (Millard et al. 2003a, 2004; Dombeck et al. 2004; Nemet et al. 2004). These experiments have allowed the precise characterization of the voltage sensitivity of SHG and the identification of the optimal wavelength for the incident laser fundamental light. The details of these SHIMP measurements are presented in this chapter along with a discussion of the prospects of this new technology for imaging membrane potential changes in neuronal preparations.

MATERIALS

Nonlinear Optical Imaging Setup

For nonlinear optical imaging of cell membranes, the authors have adapted a Fluoview scanning confocal imaging system (Olympus) for nonlinear optical imaging with an upright, stage-focusing Axioskop microscope (Carl Zeiss), as shown in Figure 1. A Mira 900 Ti:sapphire ultrafast laser (Coherent) is pumped by a 10-W Verdi doubled solid-state laser (Coherent) and purged with nitrogen gas from a liquid nitrogen dewar to make wavelengths above 930 nm accessible. BD2 Mirrors (Newport) are used to direct the beam through the various optical components, through the Fluoview scan-head, and, finally, into the microscope. The beam is first passed through a Faraday isolator (Electro-Optics Technology) to prevent back-propagating reflections from knocking the Mira out of mode-lock. A 700-nm long-pass filter (CVI Laser) then removes residual pump light as well as second-harmonic light produced within the laser system itself. A Pockels cell (Conoptics) is used to modulate the beam intensity, and then half- and quarter-waveplates (CVI Laser) are used to produce circularly polarized light at the sample, after which the beam enters the scan-head. The scanning beam enters the microscope from above, passing through a plano-concave, $f = -150$ mm, BK7 lens (Newport) to colocalize the bright-field focus and the focus for nonlinear excitation.

Simultaneous Imaging Mode

For nonlinear (simultaneous SHG and 2PF) imaging mode, an infinity-corrected 40× 0.8 NA water-immersion IR-Achroplan objective (Zeiss) focuses the ultrafast beam into the sample. The objective has a working distance of 3.6 mm, allowing access by patch pipettes for electrophysiological control and recording of the cell being imaged. 2PF is collected by the water-immersion objective, reflected by a dichroic mirror, and passed, through filters, to select emission wavelengths of interest. For imaging of styryl dyes, a 770-nm long-pass dichroic mirror (Chroma Technology, D in Fig. 1) is used with a sandwich (F_{2PF} in Fig. 1) of a 750-nm short-pass filter (CVI Laser) in combination with either a 540-nm band-pass filter (Chroma Technology) or a 675-nm band-pass filter (Chroma Technology). This setup covers wavelength ranges to either side of the emission spectrum crossover wavelength; i.e. ~615 nm for di-4-ANEPPS (Fluhler et al. 1985). The filtered 2PF is detected directly by a photomultiplier tube (Hamamatsu) that is connected to one of the Fluoview channel inputs via a PMT amplifier board (Olympus). SHG light is produced in the forward direction and is collected using a 0.9 NA condenser (Zeiss). The SHG light is then reflected from a band-reflecting mirror (M in Fig. 1) to focus through filters that are appropriate for the second harmonic wavelength, as detailed in Table 1, onto a photon-counting head (Hamamatsu H7421-40) that is connected to the second Fluoview channel input.

Cell System

As a model cellular system for membrane electrophysiology, the authors use N1E-115 mouse neuroblastoma cells. These are grown in Dulbecco's Modified Eagle Medium with 10% fetal bovine serum and 1% antibiotic-antimycotic (all from Invitrogen), maintained at 37°C with 5% CO_2. During patch-clamp experiments, an external buffer and an internal patch-pipette buffer are used. The external buffer is 20 mM HEPES (Merck Biosciences) in Earle's Balanced Salt Solution (EBSS, Sigma). The internal patch-pipette buffer is 140 mM potassium aspartate, 5 mM sodium chloride (both Sigma), and 10 mM HEPES, adjusted to pH 7.35 by stock potassium hydroxide and hydrochloric acid solutions.

Voltage-sensitive Dyes <!>

Voltage-sensitive dyes are produced using procedures adapted from Hassner et al. (1984). For routine research use, dyes may be purchased from Molecular Probes, and small samples may be obtained from the authors' laboratory upon request. The 1PF properties of the dyes are characterized using a hemispherical lipid bilayer (HLB) apparatus (Loew et al. 1979a; Loew and Simpson 1981). Aqueous solutions of the more hydrophobic dyes are complexed with cyclodextrin to facilitate and accelerate staining (Bullen and Loew 2001), using the following procedure.

First a 40× stock solution (4 mM) of the dye is prepared in 100% ethanol and then diluted to 2× in a 20 mM solution of carboxyethyl-γ-cyclodextrin (CE-γ-CD) (Cyclodextrin Technologies Development) in distilled water. Aliquots of this mixture (0.5 ml) are dehydrated in a rotary vacuum evaporator (Savant Instruments) and stored dry in a refrigerator. The solution is reconstituted at 1× when required to provide an aqueous dye solution at 100 μM. This complexed form of the dye is used because it provides more efficient staining than dyes used in combination with surfactants (Lojewska and Loew 1987), e.g., Pluronic F-127 (Molecular Probes).

Caution: *See Appendix 3 for appropriate handling of materials marked with <!>.*

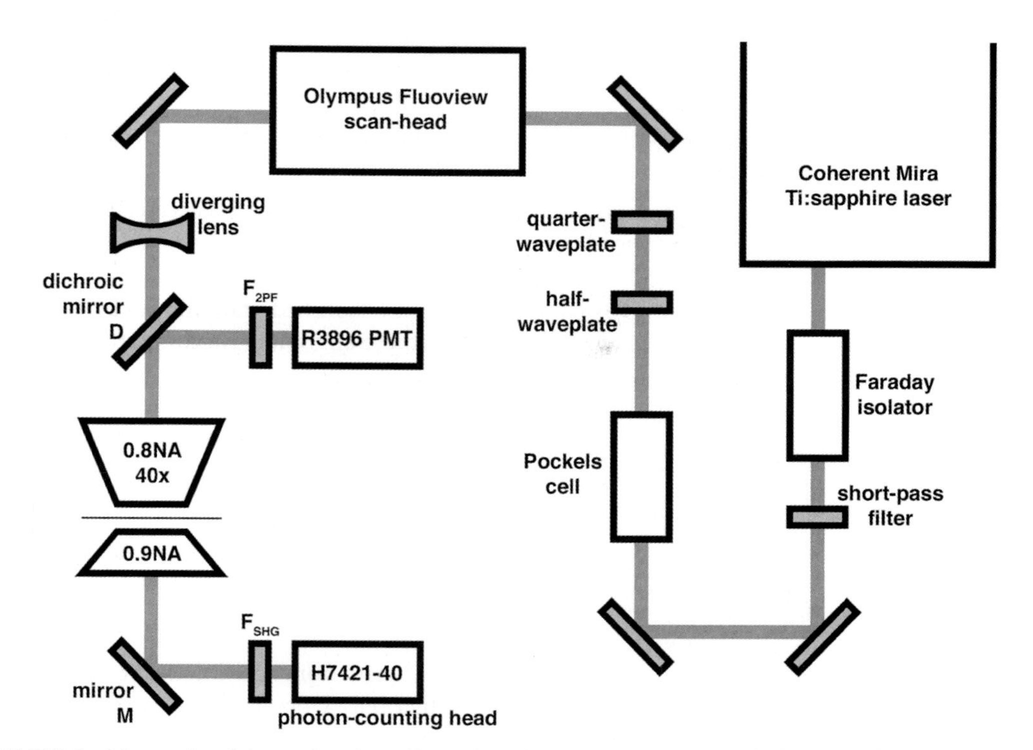

FIGURE 1. Schematic of the authors' nonlinear imaging system. Excitation light leaves the scan-head and is directed through a diverging lens into the microscope. The light passes through the dichroic (D) to enter the objective. 2PF is collected back through the objective and the dichroic reflects it through filters (F_{2PF}) into the photomultiplier tube. The condenser collects all transmitted light, directing it to the mirror (M), which selectively reflects SHG through filters (F_{SHG}) into the photon-counting head.

TABLE 1. SHG detection mirrors and filter sets, as shown in Figure 1, for particular excitation wavelengths

Wavelengths	Mirror, M	Filters, F_{SHG}
830–870 nm	CVI Laser TLM 1-425-45-UNP-0737 reflecting 425 ± 50 nm	Chroma Technology D425/40 and CVI Laser SPF-450, passing 425 ± 10 nm
890 nm and 910 nm	CVI Laser TLM1-450-45-UNP-0737 reflecting 450 ± 50 nm	Chroma Technology D460/50 and Oriel Instruments 57530, passing 450 ± 10 nm
930–970 nm	CVI Laser TLM1-475-45-UNP-0737 reflecting 475 ± 50 nm	Chroma Technology D470/40 and 475 ± 50 nm band-pass, passing 470 ± 15 nm

PROCEDURE

Second Harmonic Imaging of Membrane Potential

1. In preparation for patching and subsequent imaging, transfer the N1E-115 mouse neuroblastoma cells (see the Materials section for growth conditions) growing on glass coverslips (Fisher Scientific) to a 60-mm plastic dish (Becton Dickinson).

   Note: On an upright microscope, the sides of the plastic dish may need to be trimmed down to allow access by patch pipettes and to make it possible to switch between air and water objectives.

2. Hold the coverslips in place with small spots of vacuum grease and then add 3 ml of external buffer (see Materials list), sufficient to maintain a meniscus between the dish and the water-immersion objective.

3. Prepare patch pipettes on a micropipette puller (e.g., Brown-Flaming, Sutter Instruments) from 1.5-mm outer diameter, 0.86-mm inner diameter borosilicate glass (Sutter Instruments). The pipettes should have a resistance of ~6 MΩ when filled with internal buffer.

4. Set up electrophysiology.

   Note: The authors control a patch-clamp amplifier (Axon Instruments) by computer using LabVIEW (National Instruments) or Clampex (Axon Instruments). The software should provide diagnostics such as seal resistance during patching and synchronization of voltage- or current-clamping operations with non-linear imaging.

5. Using the air objective and bright-field imaging, select a cell for patching to a gigaohm seal (Penner 1995).

   Note: Intact patches are more stable than the whole-cell patches used for voltage- or current-clamping, so the configuration of the microscope should be switched from bright-field imaging to nonlinear imaging before forming the whole-cell patch.

6. When the whole-cell patch has been established, stain the cell by adding ~100 µl of the reconstituted dye/CE-γ-CD solution to achieve an overall dye concentration of ~3 µM for staining.

7. Perform imaging. The authors use Fluoview v.4.3, Olympus's imaging software designed for use with the Fluoview scanning confocal imaging system. Pin five (vertical sync) of the trigger (EXT-TRG) output of the Fluoview is used to trigger LabVIEW or Clampex as appropriate. Combined SHG and 2PF images are typically recorded in a time series and analyzed within Fluoview or using Wayne Rasband's ImageJ (http://rsb.info.nih.gov/ij/).

EXAMPLE OF APPLICATION

To determine the voltage-sensitivity of SHG from styryl dyes, the authors typically select single cells that have no physical contact with other cells. Once a cell is patched and stained, a series of 27 images is taken with the clamp voltage switched back and forth between 0 mV and a test voltage, *V*, after every 3 image frames. A total intensity value for each image is obtained by summing the intensity values of the pixels associated with the cell membrane; i.e., a total intensity value $I(t_i)$ is obtained for frame *i*. Fifteen total intensity values, $I_0(t_i)$, correspond to the 0 mV reference voltage and 12 values, $I_V(t_i)$, correspond to the test voltage, *V*. In addition to scatter, the $I_0(t_i)$ and $I_V(t_i)$ drift a little over time, due to a small but continuous incorporation of dye into the membrane during the course of the experiment. Thus, a normalization process that successfully corrects for the drift has been developed, as follows. The authors fit a second-order polynomial, $A_0(t_i)$, to the 15 $I_0(t_i)$ and then the normalized total intensity values, $N(t_i)$, are just $I(t_i) / A_0(t_i)$ for each i. The average relative signal change, C, for the test voltage is $\langle N_V(t_i)\rangle - \langle N_0(t_i)\rangle$. Each image series yields one value for C, which is considered a single experimental measurement, although, of course, each measurement obtained in this way actually represents data collected over an entire cell membrane and averaged over 15 0-mV images and 12 test-voltage images.

To illustrate SHIMP visually, an SHG image series can be processed to produce a montage such as that shown in Figure 2. Although on-the-fly averaging is not generally used while obtaining image series, each image in Figure 2 is a Kalman average of three acquisitions. Modulation of SHG intensity by membrane potential is apparent. The 81 acquisitions of the original image series, each corresponding to a ~2.5-sec acquisition time, took place without any major degradation in SHG intensity and with a stable response to the step changes in membrane potential. Such stability appears to be dependent on excitation wavelength, with greater degradation of both SHG and 2PF intensity at wavelengths shorter than 830 nm—attempts to image at 780 nm, for instance, typically result in rapid, readily visible damage (Campagnola et al. 1999)—along with gradual loss of cell viability with increased light exposure.

Figure 3 shows relative signal changes in SHG and 2PF versus clamp voltage for di-4-ANEPPS at 850 nm and at 910 nm. All four sets of data have been fit with straight lines going through the origin. The voltage sensitivity, either for SHG or 2PF (or 1PF), is defined as the slope of the linear fit for signal change relative to the 0 mV intensity, and a voltage sensitivity calculated in this way is expressed as a percentage change per 100 mV. From the linear fits, the voltage sensitivity for SHG is (−18.5 ± 0.5)% / 100 mV at 850 nm (on the basis of a total of 38 image sequences using ~30 cells) and (−26.3 ± 0.8)% / 100 mV at 910 nm (on the basis of 52 image sequences using ~40 cells). For 2PF detection at emission wavelengths between 490 nm and 560 nm, the voltage sensitivities are ~6.6%/100 mV and ~4.8%/ 100 mV, respectively. The best voltage sensitivity under 1PF is ~10%/100 mV at optimal excitation and emission wavelengths.

The data in Figure 3 show that the voltage sensitivity of SHG from di-4-ANEPPS is increased by ~42% between 850 nm and 910 nm, while the voltage sensitivity of 2PF is reduced by ~27%. Figure 4 shows the actual signal changes for −50 mV relative to 0 mV for excitation wavelengths between 830 nm and 970 nm. (Shorter wavelengths damage the cells too, whereas the authors' present laser system will not mode-lock beyond 985 nm.) The "fits" are meant only as a visual aid, but there is clearly an increase of the voltage sensitivity of SHG above 890 nm. This is good evidence for a two-photon resonance (Williams 1984) corresponding to the 465-nm one-photon absorption maximum of di-4-ANEPPS in lipid. Note that the quantum mechanical selection rules are different for two-photon processes compared with one-photon processes. This means that two-photon spectra rarely map exactly to one-photon spectra through a simple halving of excitation wavelength. Although it would be desirable to collect data at even longer excitation wavelengths, the data shown in Figure 4 suggest that the peak of the SHG voltage sensitivity lies in the range 960–980 nm, with the maximum value being ~38% / 100 mV. On two occasions, one at 950 nm and the other at 970 nm, signal changes of more than 20% were measured for a voltage change of −50 mV. Applying a voltage change of −100 mV to the same cells, signal changes of (43.4 ± 2.6)% and (41.6 ± 1.3)% were measured, respectively. For comparison, the voltage sensitivity at optimal excitation and emission wavelengths for 1PF from di-4-ANEPPS, as determined by the HLB apparatus, is ~10% / 100 mV. Over the course of these experiments, two dichroics (D in Fig. 1) were used, one for excitation wavelengths up to 910 nm and another for wavelengths longer than 910 nm.

The authors have also used their system for simultaneous electrophysiology and nonlinear imaging to determine the voltage sensitivity of SHG from other dyes in the same class of fluorophores as di-4-ANEPPS. These are di-4-ANEPMRF and di-4-ANEPMPOH, differing from di-4-ANEPPS only in the "aqueous anchor" attached to the quaternized nitrogen of the pyridinium moiety. Table 2 summarizes these results. The dyes' structural differences clearly affect their voltage sensitivities, although not in the same way for SHG versus 2PF versus 1PF. In all three cases, the SHG change is increased by going to the longer excitation wavelength, whereas the 2PF change is reduced.

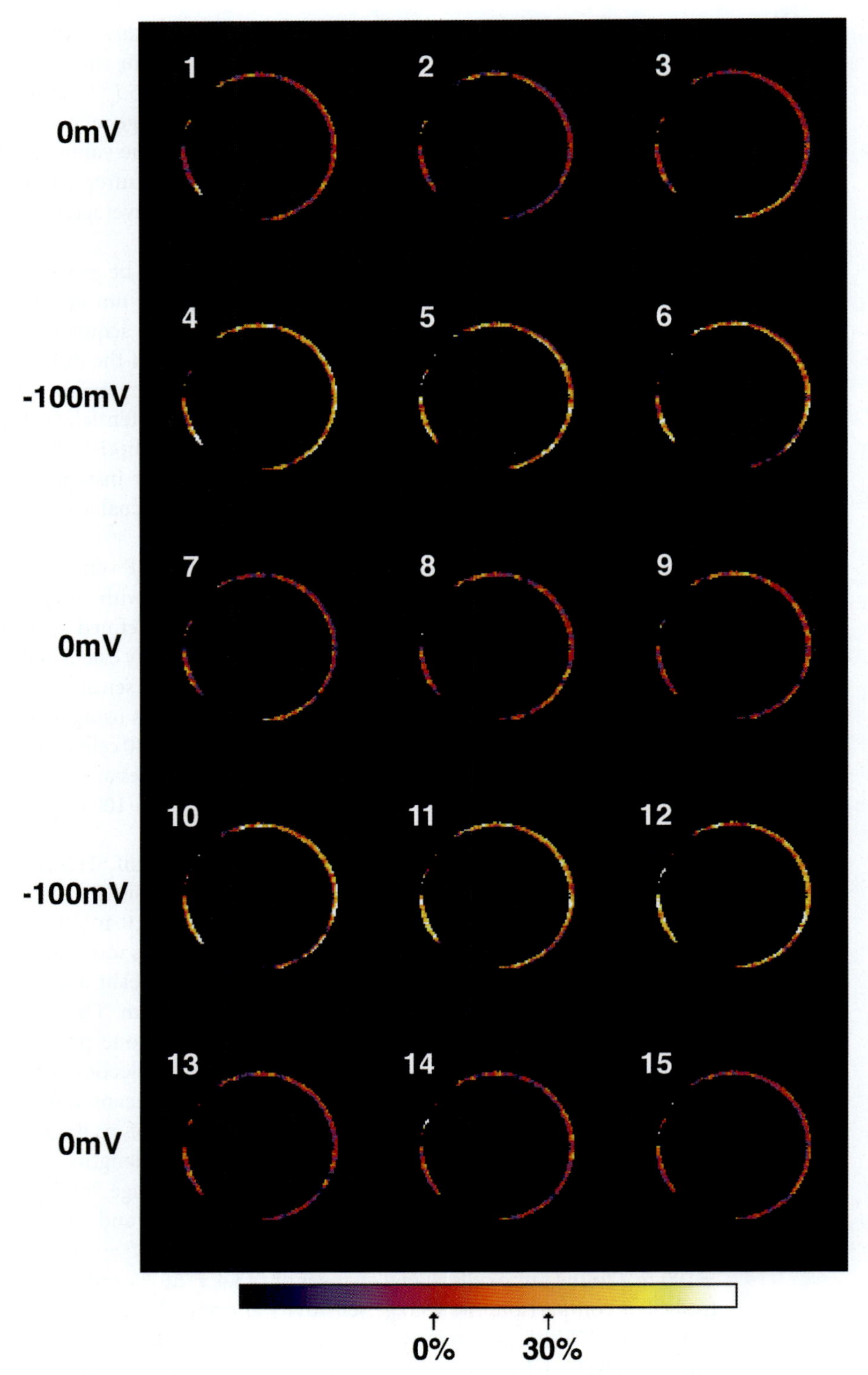

FIGURE 2. Montage of SHG images of an N1E-115 mouse neuroblastoma cell stained with ~3 μM di-4-ANEPPS and excited at 910 nm. Each image is a Kalman average of three acquisitions. This cell is patched, with the pipette entering the field of view from the lower left corner and resulting in a reduction in membrane in the lower left quadrant. A 2PF image of the cell (not shown) includes a significant number of filopodia in the upper left quadrant of the cell; they are sufficiently small that opposing membranes are within the optical coherence length of ~λ/10 (Campagnola et al. 2002), violating the non-centrosymmetric constraint, and hence the filopodia do not appear in the SHG images. The reduction in membrane SHG in the upper left quadrant of the cell may be related to the density of filopodia there. The

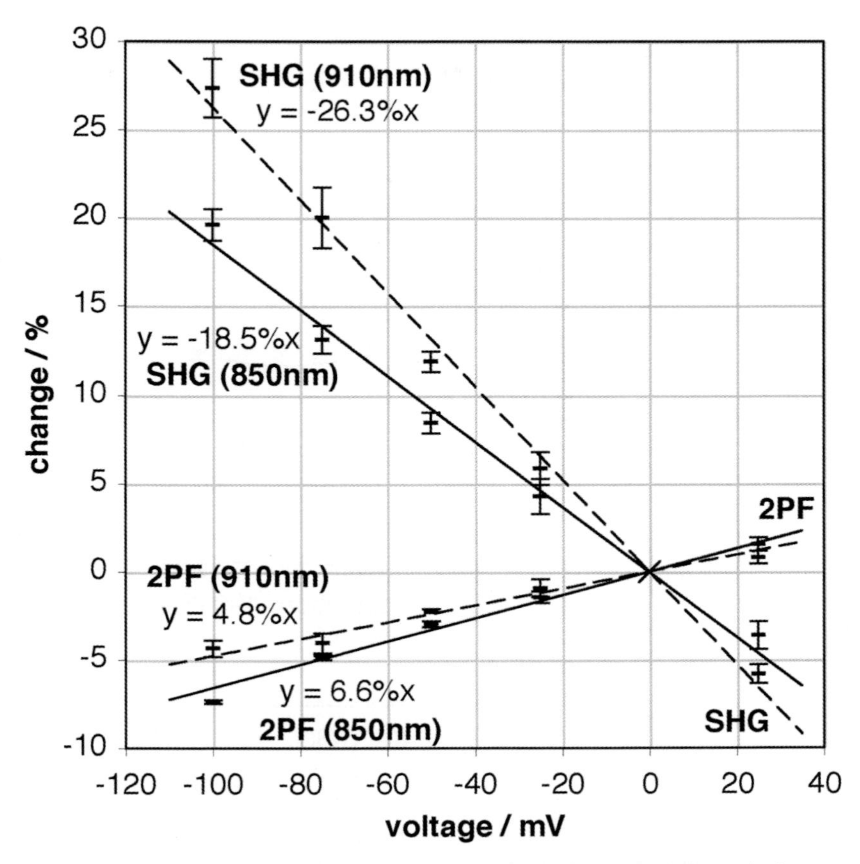

FIGURE 3. Signal changes versus voltage for SHG and 2PF from di-4-ANEPPS. Each data point indicates the mean change for a number of measurements, and each error bar shows the standard error in the mean. Changes are shown for two excitation wavelengths, 850 nm (linear fits in solid lines) and 910 nm (linear fits in dashed lines). In both cases, 2PF was collected for wavelengths between 490 nm and 560 nm. (Adapted, with permission, from Millard et al. 2004.)

images are from a time series, with time increasing from left to right and from top to bottom. The first, third, and fifth rows were all acquired at 0 mV, whereas the second and fourth rows were acquired at −100 mV. The color bar shows levels corresponding to the normalized reference intensity (marked 0%) and to a normalized intensity 30% above the reference. The average relative signal change for this image series is (27.0 ± 2.6)%. The images were processed using code written in Perl (http://www.perl.com/) and using the Netpbm package (http://netpbm.sourceforge.net/). The frames were first nudged, at most by a few pixels, to correct for a slight drift in the position of the cell during imaging. With the frames aligned, a mask corresponding to the cell membrane was generated. Within each frame, the mask was applied to select the bright membrane pixels, and a nearest-neighbor mean filter was applied, with the filter constrained to use only pixels within the mask. Just as the total intensity values were normalized as described in the Example of Application section, so were the intensity values of each pixel normalized across the frames by a running average interpolated from the values of that pixel in the frames corresponding to the 0 mV reference voltage. The images were displayed using ImageJ with a "Fire" look-up table. (Adapted, with permission, from Millard et al. 2004.)

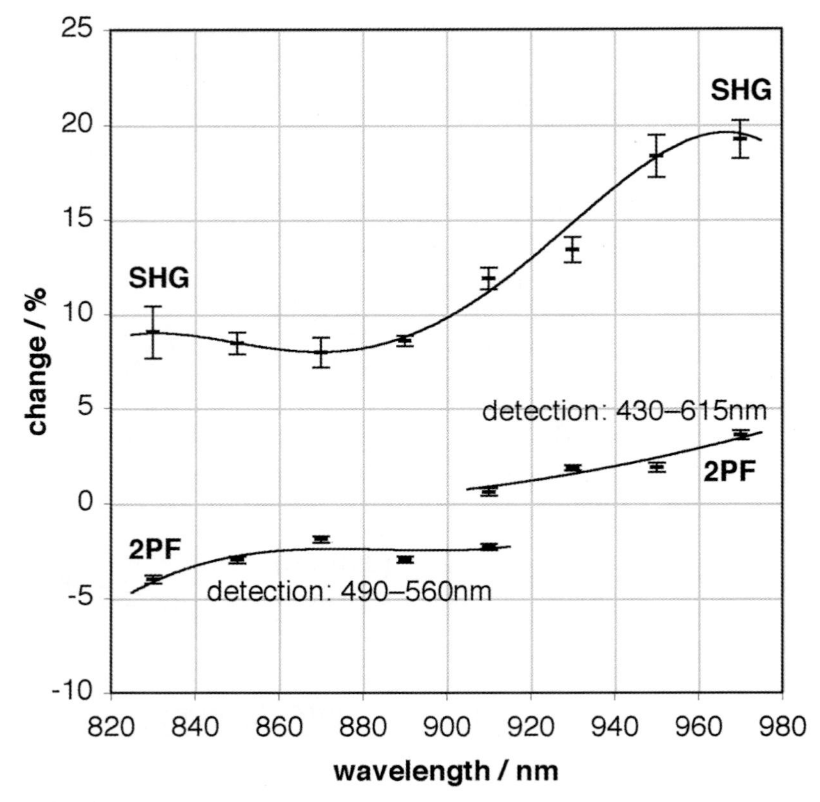

FIGURE 4. Signal changes for −50 mV change versus wavelength for SHG and 2PF from di-4-ANEPPS. The arbitrary fits are intended for trend visualization only. Each error bar shows the standard error of the mean. The 2PF data for excitation wavelengths of 830 nm to 910 nm were obtained from 2PF detection for wavelengths between 490 nm and 560 nm. The 2PF data for 910 nm to 970 nm were obtained from 2PF detection between 430 nm and 615 nm. (Although this wavelength range includes the SHG wavelength, there is essentially no backscattered SHG on the 2PF channel.) The difference in sign for 2PF changes from the two wavelength ranges indicates that the 2PF emission exhibits a biphasic response, similar to that of 1PF emission (Fluhler et al. 1985), a characteristic consistent with an electrochromic mechanism for voltage sensitivity. (Adapted, with permission, from Millard et al. 2004.)

ADVANTAGES AND LIMITATIONS

SHIMP appears to hold great promise as an alternative to fluorescence for the optical mapping of membrane potentials. Clearly, the fourfold increase in sensitivity to membrane potential of the SHG signal is a major advantage. Since SHG does not involve promotion of molecules to an excited state, the bleaching and photodynamic damage that accompany fluorescence may also be minimized; however, *resonance-enhanced* SHG is usually accompanied by collateral 2PF with its attendant

TABLE 2. SHG, 2PF, and 1PF signal changes for three dyes, produced by a change in transmembrane potential of −50 mV

	SHG	2PF	SHG	2PF	
Dye	excited at 850 nm		excited at 910 nm		1PF
di-4-ANEPPS	8.4 ± 0.6	−3.0 ± 0.2	11.9 ± 0.6	−2.3 ± 0.2	5
di-4-ANEPMRF	2.5 ± 0.5	−3.9 ± 0.4	6.1 ± 0.9	−3.3 ± 0.3	2
di-4-ANEPMPOH	7.7 ± 0.7	−4.0 ± 0.3	9.0 ± 0.5	−2.4 ± 0.3	0.8

SHG and 2PF data obtained using apparatus and methods described here; 1PF data obtained using HBL apparatus.

photodamage. Relatively minor enhancements to a 2PF microscope can enable second harmonic imaging, and the fact that these two modalities utilize separate light paths permits 2PF and SHG to be imaged simultaneously.

Ultimately, the practical test for whether SHIMP may complement, or perhaps even supplant, fluorescence as a method to image neuronal electrical activity will be whether it can genuinely image an action potential. Indeed, on the basis of its superior sensitivity, one might be tempted to conclude that SHIMP should already be better than fluorescence. However, a major challenge for the practical application of SHIMP is the size of the SHG signal. Great strides have been made in reducing the signal acquisition time using current SHIM apparatus, reduced from many minutes for a 100×100 pixel image in early experiments to just one second for a 512×512 image. Unfortunately, even these improved times are at least two orders of magnitude too slow to image action potentials. The key factor is that there are typically two orders of magnitude fewer SHG photons produced than there are fluorescent photons. Thus, the number of photons collected during an action potential is insufficient to overcome the shot noise inherent in the quantal nature of light. One solution is to limit the number of spatial points sampled in the laser-scanning microscope, most commonly by just scanning single lines (Dombeck et al. 2004), so that data can be acquired more rapidly. Of course, this will limit the ability to map electrical activity in a morphologically complex neuronal preparation—a 1D line scan is not a 2D image. Another approach would be to develop improved dyes with greater electrochromism and hyperpolarizabilities. Some preliminary promise has been shown by a radically different approach to dye development. In this approach, available electrochromic chromophores, such as ANEP, are linked to silver or gold nanoparticles (Peleg et al. 1996, 1999; Clark et al. 2000). The metal particles locally focus the laser electromagnetic field via plasmon resonance, thereby enhancing SHG from neighboring dye molecules. Naturally occurring pigments such as porphyrins and carotenoids have also been used as "harmonophores" (Millard et al. 1999), and retinal has been successfully used to image membrane potential in neurons (Nemet et al. 2004).

There are two other issues that must be considered if SHIMP is to be developed further: background SHG from intrinsic protein arrays, and tissue penetration. Highly ordered molecular arrays or biopolymers can produce extraordinarily large SHG signals. Indeed, these have been used for high-resolution 3D imaging of structures in cells and tissues such as collagen fibers, microtubules, and myofibrils (Campagnola et al. 2002; Dombeck et al. 2003; Mohler et al. 2003). The intrinsic SHG signals from collagen and from myofibrils in skeletal muscle can be so strong as to overwhelm signals from nearby ANEP-stained membranes. In practice, most common experimental neuronal preparations, including brain slices and in vivo imaging, do not have significant background signals from these sources, so the background is actually much lower than the typical contribution of autofluorescence to 2PF. However, because the SHG signal is most effectively collected in the transmitted light path, studies of brain slices would be limited and in vivo imaging precluded. The transmitted light path might become fully viable, at least for slice preparations, if high-powered amplified laser systems are used, as was demonstrated recently for 2PF imaging in thick specimens (Theer et al. 2003). Additionally, significant progress has been made in utilizing backscattered SHG to produce images of collagen from thick tissue specimens (Zoumi et al. 2002), and this technology should be adaptable to SHIMP.

Although some significant additional research will be required to make SHIMP a viable alternative to fluorescence for optical mapping of electrical activity in complex neuronal preparations, there are already clear directions that can be followed to overcome the obstacles. The high sensitivity to membrane potential of SHG from electrochromic membrane probes, as demonstrated in Figures 2, 3, and 4, motivates this pursuit.

ACKNOWLEDGMENTS

The authors thank Lei Jin, Joseph Wuskell, David Boudreau, and Mei-De Wei for their contributions to this work, and Paul Campagnola and Heather Clark for their advice. We gratefully acknowledge financial support under Office of Naval Research grant N0014-98-1-0703 and National Institutes of Health, National Institute of Biomedical Imaging and Bioengineering grant R01EB00196.

REFERENCES

Antic S., Wuskell J.P., Loew L., and Zecevic D. 2000. Functional profile of the giant metacerebral neuron of *Helix aspersa*: Temporal and spatial dynamics of electrical activity *in situ*. *J. Physiol.* **527:** 55–69.

Bedlack R.S., Wei M.-D., and Loew L.M. 1992. Localized membrane depolarizations and localized intracellular calcium influx during electric field-guided neurite growth. *Neuron* **9:** 393–403.

Ben-Oren I., Peleg G., Lewis A., Minke B., and Loew L.M. 1996. Infrared nonlinear optical measurements of membrane potential in photoreceptor cells. *Biophys. J.* **71:** 1616–1620.

Bouevitch O., Lewis A., Pinevsky I., Wuskell J.P., and Loew L.M. 1993. Probing membrane potential with nonlinear optics. *Biophys. J.* **65:** 672–679.

Bullen A. and Loew L.M. 2001. Solubility and intracellular delivery of hydrophobic voltage-sensitive dyes with chemically-modified cyclodextrins. *Biophys. J.* **80:**168a.

Campagnola P. and Loew L.M. 2003. Second-harmonic imaging microscopy for visualizing biomolecular arrays in cells, tissues and organisms. *Nat. Biotechnol.* **21:** 1356–1360.

Campagnola P.J., Wei M.-D., Lewis A., and Loew L.M. 1999. High-resolution nonlinear optical microscopy of living cells by second harmonic generation. *Biophys. J.* **77:** 3341–3349.

Campagnola P.J., Millard A.C., Terasaki M., Hoppe P.E., Malone C.J., and Mohler W.A. 2002. Three-dimensional high-resolution second-harmonic generation imaging of endogenous structural proteins in biological tissues. *Biophys. J.* **82:** 493–508.

Clark H.A., Campagnola P.J., Wuskell J.P., Lewis A., and Loew L.M. 2000. Second harmonic generation properties of fluorescent polymer encapsulated gold nanoparticles. *J. Am. Chem. Soc.* **122:** 10234–10235.

Cohen L.B., Salzberg B.M., Davila H.V., Ross W.N., Landowne D., Waggoner A.S., and Wang C.H. 1974. Changes in axon fluorescence during activity: Molecular probes of membrane potential. *J. Membr. Biol.* **19:** 1–36.

Dombeck D.A., Blanchard-Desce M., and Webb W.W. 2004. Optical recording of action potentials with second-harmonic generation microscopy. *J. Neurosci.* **24:** 999–1003.

Dombeck D.A., Kasischke K.A., Vishwasrao H.D., Ingelsson M., Hyman B.T., and Webb W.W. 2003. Uniform polarity microtubule assemblies imaged in native brain tissue by second-harmonic generation microscopy. *Proc. Natl. Acad. Sci.* **100:** 7081–7086.

Ehrenberg B., Montana V., Wei M.-D., Wuskell J.P., and Loew L.M. 1988. Membrane potential can be determined in individual cells from the nernstian distribution of cationic dyes. *Biophys. J.* **53:** 785–794.

Fluhler E., Burnham V.G., and Loew L.M. 1985. Spectra, membrane binding and potentiometric responses of new charge shift probes. *Biochemistry* **24:** 5749–5755.

Hassner A., Birnbaum D., and Loew L.M. 1984. Charge-shift probes of membrane potential: Synthesis. *J. Org. Chem.* **49:** 2546–2551.

Huang J.Y., Lewis A., and Loew L.M. 1988. Non-linear optical properties of potential sensitive styryl dyes. *Biophys. J.* **53:** 665–670.

Loew L.M. 1994. Voltage sensitive dyes and imaging neuronal activity. *Neuroprotocols* **5:** 72–79.

———. 2001. Mechanisms and principles of voltage sensitive fluorescence. In *Optical mapping of cardiac excitation and arrhythmias* (ed. D.S. Rosenbaum and J. Jalife), pp. 33–46. Futura Publishing, Armonk, New York.

Loew L.M. and Simpson L. 1981. Charge shift probes of membrane potential: A probable electrochromic mechanism for p-aminostyrylpyridinium probes on a hemispherical lipid bilayer. *Biophys. J.* **34:** 353–365.

Loew L.M., Bonneville G.W., and Surow J. 1978. Charge shift optical probes of membrane potential. Theory. *Biochemistry* **17:** 4065–4071.

Loew L.M., Campagnola P., Lewis A., and Wuskell J.P. 2002. Confocal and nonlinear optical imaging of potentiometric dyes. *Methods Cell Biol.* **70:** 429–452.

Loew L.M., Scully S., Simpson L., and Waggoner A.S. 1979a. Evidence for a charge-shift electrochromic mechanism in a probe of membrane potential. *Nature* **281:** 497–499.

Loew L.M., Simpson L., Hassner A., and Alexanian V. 1979b. An unexpected blue shift caused by differential solvation of a chromophore oriented in a lipid bilayer. *J. Am. Chem. Soc.* **101:** 5439–5440.

Loew L.M., Cohen L.B., Dix J., Fluhler E.N., Montana V., Salama G., and Wu J.-Y. 1992. A naphthyl analog of the aminostyryl pyridinium class of potentiometric membrane dyes shows consistent sensitivity in a variety of tissue, cell, and model membrane preparations. *J. Membr. Biol.* **130:** 1–10.

Lojewska Z. and Loew L.M. 1987. Insertion of amphiphilic molecules into membranes is catalyzed by a high molecular weight non-ionic surfactant. *Biochim. Biophys. Acta* **899:** 104–112.

Millard A.C., Jin L., Lewis A., and Loew L.M. 2003a. Direct measurement of the voltage sensitivity of second-harmonic generation from a membrane dye in patch-clamped cells. *Opt. Lett.* **28:** 1221–1223.

Millard A.C., Campagnola P.J., Mohler W., Lewis A., and Loew L.M. 2003b. Second harmonic imaging microscopy. *Methods Enzymol.* **361:** 47–69.

Millard A.C., Wiseman P.W., Fittinghoff D.N., Wilson K.R., Squier J.A., and Müller M. 1999. Third-harmonic generation microscopy by use of a compact, femtosecond fiber laser source. *Appl. Opt.* **38:** 7393–7397.

Millard A.C., Jin L., Wei M.-D., Wuskell J.P., Lewis A., and Loew L.M. 2004. Sensitivity of second harmonic generation from styryl dyes to trans-membrane potential. *Biophys. J.* **86:** 1169–1176.

Mohler W., Millard A.C., and Campagnola P.J. 2003. Second harmonic generation imaging of endogenous structural proteins. *Methods* **29:** 97–109.

Moreaux L., Sandre O., and Mertz J. 2000. Membrane imaging by second harmonic generation microscopy. *J. Opt. Soc. Am. B* **17:** 1685–1694.

Nemet B., Nikolenko V., and Yuste R. 2004. Second harmonic retinal imaging of membrane potential. *J. Biomed. Opt.* (in press).

Obaid A.L., Koyano T., Lindstrom J., Sakai T., and Salzberg B.M. 1999. Spatiotemporal patterns of activity in an intact mammalian network with single-cell resolution: Optical studies of nicotinic activity in an enteric plexus. *J. Neurosci.* **19:** 3073–3093.

Peleg G., Lewis A., Linial M., and Loew L.M. 1999. Nonlinear optical measurement of membrane potential around single molecules at selected cellular sites. *Proc. Natl. Acad. Sci.* **96:** 6700–6704.

Peleg G., Lewis A., Bouevitch O., Loew L.M., Parnas D., and Linial M. 1996. Gigantic optical non-linearities from nanoparticle enhanced molecular probes with potential for selectively imaging the structure and physiology of nanometric regions in cellular systems. *Bioimaging* **4:** 215–224.

Penner R. 1995. A practical guide to patch clamping. In *Single-channel recording*, 2nd edition (ed. E. Neher), pp. 3–30. Plenum Press, New York.

Pons T., Moreaux L., Mongin O., Blanchard-Desce M., and Mertz J. 2003. Mechanisms of membrane potential sensing with second-harmonic generation microscopy. *J. Biomed. Opt.* **8:** 428–431.

Theer P., Hasan M.T., and Denk W. 2003. Two-photon imaging to a depth of 1000 microm in living brains by use of a Ti:Al2O3 regenerative amplifier. *Opt. Lett.* **28:** 1022–1024.

Williams D.J. 1984. Organic polymeric and non-polymeric materials with large optical non-linearities. *Angew. Chem. Int. Ed. Engl.* **23:** 690–703.

Wu J.-Y., Lam Y.-W., Falk C., Cohen L.B., Fang J., Loew L., Prechtl J.C., Kleinfeld D., and Tsau Y. 1998. Voltage-sensitive dyes for monitoring mult-neuronal activity in the intact CNS. *Histochem. J.* **30:** 169–187.

Zochowski M., Wachowiak M., Falk C.X., Cohen L.B., Lam Y.W., Antic S., and Zecevic D. 2000. Imaging membrane potential with voltage-sensitive dyes. *Biol. Bull.* **198:** 1–21.

Zoumi A., Yeh A., and Tromberg B.J. 2002. Imaging cells and extracellular matrix in vivo by using second-harmonic generation and two-photon excited fluorescence. *Proc. Natl. Acad. Sci.* **99:** 11014–11019.

CHAPTER 61

Imaging Synaptic Vesicle Dynamics with Styryl Dyes

Silvio O. Rizzoli, Ute Becherer, Joseph Angleson, and William J. Betz

Department of Physiology and Biophysics, University of Colorado Medical School, Denver, Colorado 80262

The release of neurotransmitters from nerve terminals requires the fusion of synaptic vesicles with the plasma membrane (exocytosis). The vesicles are then retrieved from the membrane (endocytosis) and directed back to a pool of vesicles inside the terminal (vesicle recycling). This process has been studied for many years by recording postsynaptic cell responses to neurotransmitter release, and by electron microscopy of presynaptic terminals (investigating vesicular uptake of electron-dense markers). More recently, two new electrophysiological techniques have been developed—capacitance recording (which monitors changes in the surface area of cells) and amperometry (which investigates release of secretory granule products such as catecholamine or serotonin). Although these techniques have provided new information about secretory mechanisms, they are unsuitable for most types of synaptic preparations (Angleson and Betz 1997). Optical techniques can partially overcome this limitation. Various optical markers have been used, including sulforhodamine (Lichtman et al. 1985), fluorescently labeled antibodies (Kraszewski et al. 1996), GFP-labeled synaptic proteins (Chapter 81), and styryl dyes (FM; Betz and Bewick 1992). Among these, FM dyes have proved to be valuable in a number of different preparations, including neuromuscular junctions (NMJ) from the frog (Betz and Bewick 1992), snake (Teng et al. 1999), lizard (Lindgren et al. 1997), and larval *Drosophila* (Ramaswami et al. 1994; Kuromi and Kidokoro 1998); hippocampal neurons in culture (Ryan et al. 1993) and acute slice (Zakharenko et al. 2001); goldfish retinal bipolar cells (Lagnado et al. 1996; Zenisek et al. 2000); *Caenorhabditis elegans* neurons (Kay et al. 1999); and a variety of nonneuronal preparations (for review, see Cochilla et al. 1999).

THE DYE: PROPERTIES AND MECHANISM OF ACTION

Several FM dyes have been synthesized (Betz et al. 1996), with the most widely used being FM 1-43 (chemical structure shown in Fig. 1A). FM dyes ordinarily partition reversibly into lipid membranes, but are non-membrane-permeant. Consequently, dyes present in external media become trapped in endosomes that form when the cell is stimulated. Another useful feature of FM dyes is that their quantum yield increases several hundred times when partitioned in membranes versus aqueous solution.

All FM dyes have the same general structure: a positively charged hydrophilic head group and a lipophilic tail (usually formed by two aliphatic hydrocarbon chains) linked by a double-bond bridge (Fig. 1A). The tail ensures that the dye readily dissolves into the outer leaflet of membranes.

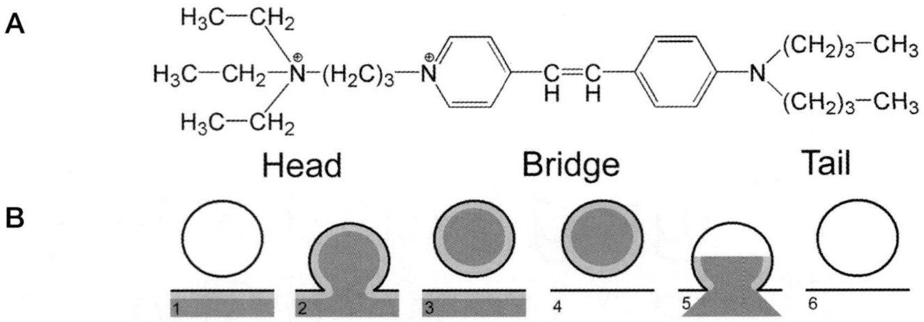

FIGURE 1. (*A*) Chemical structure of FM 1-43. The molecule is composed of a positively charged head group (which prevents the dye from permeating membranes) and a lipophilic tail (which ensures dye partition into membranes). These components are connected by a double-bond bridge; the number of double bonds in the bridge determines the spectral characteristics of the molecule. (*B*) FM dye staining and destaining. (*1*) The dye is added to the solution bathing the preparation. It does not have access to the intracellular space. (*2*) The preparation is stimulated. Vesicles exocytose and come in contact with the dye. (*3*) Vesicles endocytose and take up dye. (*4*) The dye is washed from the extracellular solution; it is removed from the plasma membrane, but remains trapped in internalized vesicles (*staining*). (*5*) The stained preparation is stimulated in absence of the dye. Vesicles fuse with the membrane and release dye (*destaining*). (*6*) Destained vesicles are internalized; at this point a new experiment of staining and destaining can be performed.

The length of the tail determines the dye's hydrophobicity and its ability to be washed out of membranes (for example, the time constants for dye wash-off are ~1 sec for the short-tailed FM 2-10, and 3–6 sec for FM 1-43). As a consequence, after exocytosis, FM 2-10 is released from synaptic vesicles more quickly than FM 1-43 (see, e.g., Pyle et al. 2000). The head group, which prevents the dyes from permeating membranes, carries two positive charges. The number of double bonds in the bridge connecting the aromatic rings of the head and tail groups determines the spectral properties of the dye. Dyes with one double bond (FM 1-43, FM 2-10) fluoresce at shorter wavelengths (yellow) than dyes with two or three (FM 4-64) double bonds (red). Differences in chemical structure between dyes have been useful in characterizing several biological processes. Different colored dyes were used to show synaptic vesicle mixing at the frog NMJ (Betz and Bewick 1992); dyes with different hydrophobicities have been used to identify different modes of recycling (Richards et al. 2000).

A typical FM dye experiment at a synapse is illustrated in Figure 1B. Preparations are bathed in a solution containing the dye, which labels the extracellular membranes (panel 1). Exocytosis is then induced (e.g., by electrical stimulation). Inner leaflets of vesicles come into contact with the dye-containing solution and become labeled (panel 2). Compensatory endocytosis results in internalization of loaded vesicles (panel 3). At this point, the dye is washed from the bath solution (panel 4), leaving the dye present only in endocytosed vesicles. Exocytosis can be monitored by stimulating these preparations in a dye-free bath; the fluorescent molecules wash out of the vesicles into the bath (which dramatically lowers their quantum yield) (panel 5). The decrease in vesicle fluorescence thus provides a direct measurement of vesicle release rate.

USING FM DYES: THE FROG NMJ PROTOCOL

For simple use of FM dyes, preparations must have little autofluorescence and must contain nerve terminals that do not overlap each other in different optical planes. In addition, constitutive uptake of dyes in neuronal or nonneuronal cells should be minimal. The frog cutaneous pectoris nerve–muscle preparation meets these criteria, being thin (only 5–6 fibers thick) and transparent, with relatively large nerve terminals that are well organized, linear, and in which nearly all exo- and endocytosis is mediated by recyclable synaptic vesicles.

MATERIALS

Solutions

Dissection and experiments are carried out in Normal Frog Ringer's Solution (NFR, containing 115 mM NaCl, 2 mM KCl <!>, 1.8 mM $CaCl_2$<!>, 5 mM HEPES; pH 7.2, osmolarity ~225 mOsm). High $[K^+]$ solution used for FM labeling contains 60 mM KCl and 57 mM NaCl. Dye concentrations are usually 2–10 μM for FM 1-43 <!> and for FM 4-64<!>, and 25-40 μM for FM 2-10 <!>.

Dissection

Remove the cutaneous pectoris nerve–muscle preparation from the frog, along with small portions of the skin and body wall to which it attaches (these are used later for mounting the preparation; dissection described by McMahan et al. 1972, 1980). Pin the preparation deep (dorsal) side-up (as that is the side that exhibits most surface nerve terminals) on a substrate such as Sylgard (Dow Corning). Carefully stretch it to ensure that the nerve terminals (that run along the muscle fibers) are straight, and remove the surface connective tissue. Avoid damaging muscle fibers or nerve cells; FM dyes stain the intracellular membranes of injured cells intensely, creating unwanted background fluorescence.

Instruments

Optimal visualization of FM dyes requires an upright epifluorescence microscope equipped with a 40–60X water-immersion objective. For FM 1-43, maximal fluorescence is observed at 465 nm excitation/560 nm emission (Henkel et al. 1996a); a conventional fluorescein filter set (435 nm excitation) can be used. Because phototoxicity is the major concern in FM imaging, the illumination should be kept at a minimum using neutral-density filters. For example, when using a 40X/0.75 NA water-immersion objective with a 100-W mercury lamp <!>, 5–15% transmission is sufficient. Note that phototoxicity usually appears before any clear signs of photobleaching and alters the ability of the preparations to cycle vesicles.

Image Capture

The preferred device for image capture is the CCD camera, which offers less spatial resolution than conventional film cameras, but has higher sensitivity and faster image acquisition rates. It also provides a live image display on a monitor.

Caution: *See Appendix 3 for appropriate handling of materials marked with <!>.*

PROCEDURE A

Staining

The general procedure for synaptic vesicle staining follows that presented in Figure 1. Nerve terminals are stimulated (electrically or by depolarization with high K^+) in the presence of the dye, allowing the vesicles to be recycled, and the dye is then washed from the chamber (Fig. 2A).

If imaging is to be carried out during stimulation, movements of the muscle must be inhibited. This can be achieved by application of curare <!> 10 μM), which blocks acetylcholine receptors, preventing muscles from twitching in response to electrical stimulation, but not in response to high-potassium stimulation, as the muscle fibers are directly depolarized beyond their mechanical threshold.

1. Introduce the preparation in dye-containing NFR <!> for ~1 min, to ensure diffusion of the dye into the synaptic cleft.

   Note: The dye stains the surface membranes of muscle cells and nerve terminals; a typical image is shown in the top panel of Figure 2A. The nerve terminal has a "railroad track" appearance.

2. Stimulate the preparation in presence of the dye (see text above for advice on dye concentrations).

   Note: Electrical tetanic stimulation of the nerve (delivered to the nerve via a suction electrode, for example, at a frequency of 30 Hz for 1–5 min) triggers cycling of 30–80% of all synaptic vesicles, ensuring adequate staining. High-potassium (60 mM) treatment of the preparation results in similar levels of dye uptake.

   The region between the "railroad tracks" fills with dye as vesicles are retrieved after undergoing exocytosis. The increased fluorescence intensity is proportional to the number of exocytosed vesicles.

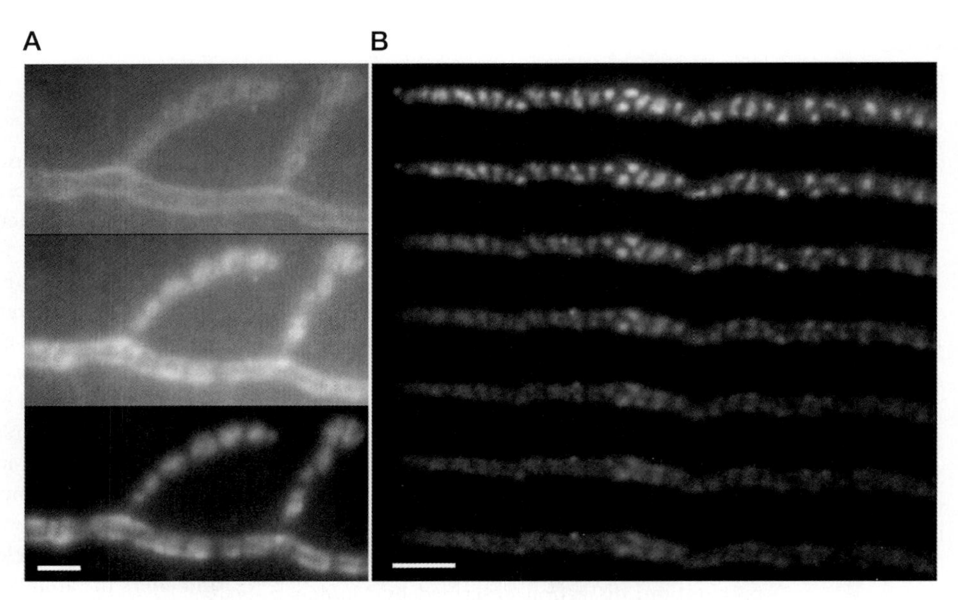

FIGURE 2. Staining and destaining of frog motor nerve terminals. (*A*) Nerve terminal shown during staining. Images were acquired and processed identically. (*Top panel*) The preparation had been bathed in a solution containing FM-dye for 1 min, which resulted in faint staining of the surface membrane. (*Middle panel*) The same preparation after stimulation (1-min tetanus, 30 Hz), followed by 15 min of rest in the same solution. The terminal is brightly stained and background fluorescence is high. (*Bottom panel*) After dye washout, the background staining is greatly reduced, and clusters of vesicles can be observed as distinct fluorescent spots. (*B*) Destaining of nerve terminals. A stained preparation was stimulated at 2 Hz for ~40 min, in the presence of 10 μM curare to prevent muscle movement. The images were taken approximately every 5 min. The nerve terminal loses most of its initial fluorescence. Bar, 5 μm. (Reprinted from Betz and Henkel 1994 [© The Rockefeller University Press].)

3. Allow the preparation to recover (also in presence of the dye), to ensure complete recycling of the vesicles. The length of the recovery period depends on the length of the stimulus; e.g., 1–3-min tetani require 10–20 min of rest.

   Note: For high-K^+ stimulation, recovery proceeds in dye-containing NFR. Figure 2A (middle panel) shows the same nerve terminal after stimulation for 1 min (30 Hz), followed by 15 min of recovery. The terminal took up dye, and it appears much brighter than in the top panel. However, since the dye is still present in the chamber, background fluorescence is also high.

4. Wash the dye from all surface membranes for 30–60 min using NFR at 4°C.

   Note: The low temperature ensures that little dye is lost through spontaneous vesicle release. To further minimize this factor, *Drosophila* preparations are usually washed in Ca^{2+}-free medium; this, however, is not necessary at the frog NMJ. Dye-scavenging compounds such as ADVASEP-7 (see below) can be used to shorten washing times.

5. Collect images.

   Note: (See Fig. 2A, bottom panel.) Fluorescent spots are now visible within the nerve terminal, and background fluorescence is low. The spots (usually 1–3 μm in diameter) represent clusters of synaptic vesicles containing several hundred to a few thousand vesicles.

PROCEDURE B

Destaining

Labeled preparations are stimulated in absence of the dye; upon synaptic vesicle exocytosis, the dye is released and diffuses away into the aqueous solution. Figure 2B shows images from a time-lapse sequence from an experiment in which a nerve terminal was stimulated for 40 min at 2 Hz. Over the course of the experiment, the terminal loses most of its FM dye.

Muscle twitching must be blocked to monitor dye loss during stimulation (see above). Tetanic stimulation causes faster dye release than low-frequency stimulation (stimulation at 30 Hz for 2–3 min releases similar amounts of dye as does 2-Hz stimulation for tens of minutes).

During destaining, the intensity of the excitation light should be minimized to avoid phototoxicity, without compromising image quality. A good rule of thumb is to adjust intensity (by use of neutral density filters) and exposure time such that the dynamic range of the initial (pre-destaining) images is 7–8 bits (i.e., the brightest pixel is only 250–300 counts above the dimmest). Additionally, the number of images acquired should be kept as low as possible.

EXAMPLES OF APPLICATIONS

Double labeling with FM 1-43 (yellow) and FM 4-64 (red) has been used to investigate vesicle mixing (Betz and Bewick 1992). Nerve terminals were fully loaded with FM 4-64, imaged, partially loaded with FM 1-43, and imaged again (with optics excluding the FM 4-64 signal). The conclusion was that complete mixing of synaptic vesicles occurs under strong stimulation.

FM dyes have been used to measure vesicle recycling in conjunction with electrophysiological measurements of neurotransmitter release (Betz and Bewick 1993). The minimum recycling time of vesicles was found to be ~1 min.

The mobility of vesicles inside terminals can be monitored by fluorescence recovery after photobleaching (FRAP). Lasers are used to bleach spots within FM-labeled vesicle clusters, and the recovery of the fluorescence is monitored. Vesicles appear to be immobile at rest, and no movement of labeled vesicles is seen inside the bleached spots during exocytosis (Henkel et al. 1996b).

At the frog NMJ, two vesicle-recycling pathways participate in endocytosis. One relies on the formation of membrane infoldings and endosomes, which are taken up and split into vesicles, whereas a small pool of vesicles cycles rapidly without such endocytic intermediates. The two pools were first identified by the ability of the more hydrophilic dye, FM 2-10, to selectively label the more rapidly cycling vesicles when washed off immediately after stimulation (Richards et al. 2000). Presumably, FM 2-10 can diffuse out of membrane infoldings that remain open to the outside solution, but not out of rapidly internalized vesicles.

FM DYE PHOTOCONVERSION

Another advantage of the FM dyes is that they can be used as endocytic markers, not only in fluorescence microscopy, but also in electron microscopy, through photoconversion, thereby increasing the spatial resolution of the studies.

This technique exploits the property of fluorescent dyes to generate free radicals (reactive oxygen species) when they emit photons. The general protocol is illustrated in Figure 3. Fluorescently labeled preparations are loaded with diaminobenzidine (DAB) and illuminated (Fig. 3A; nerve terminals fluoresce brightly). Free radicals, generated by the dye molecules, oxidize the DAB, which forms a stable, insoluble dark precipitate (Fig. 3B). The precipitate is identifiable by fluoresence and electron microscopy (Sandell and Masland 1988; Henkel et al. 1996a). Because DAB is oxidized only in the immediate vicinity of fluorescent molecules, the technique ensures that only labeled structures will contain the electron-dense precipitate. Figure 3C shows a cross-section image of a photoconverted frog motor nerve terminal; labeled vesicles are electron-dense (dark).

PROCEDURE C

Investigating the Frog NMJ Using Photoconversion

Dye photoconversion experiments are especially useful for describing the dynamics of endocytosis. At the frog NMJ, the majority of vesicles cycle through formation of endosomal intermediates, as identified by electron microscopy (Heuser and Reese 1973; Richards et al. 2000). At the snake NMJ, few such intermediates were observed; after short stimuli many labeled vesicles were observed, suggesting

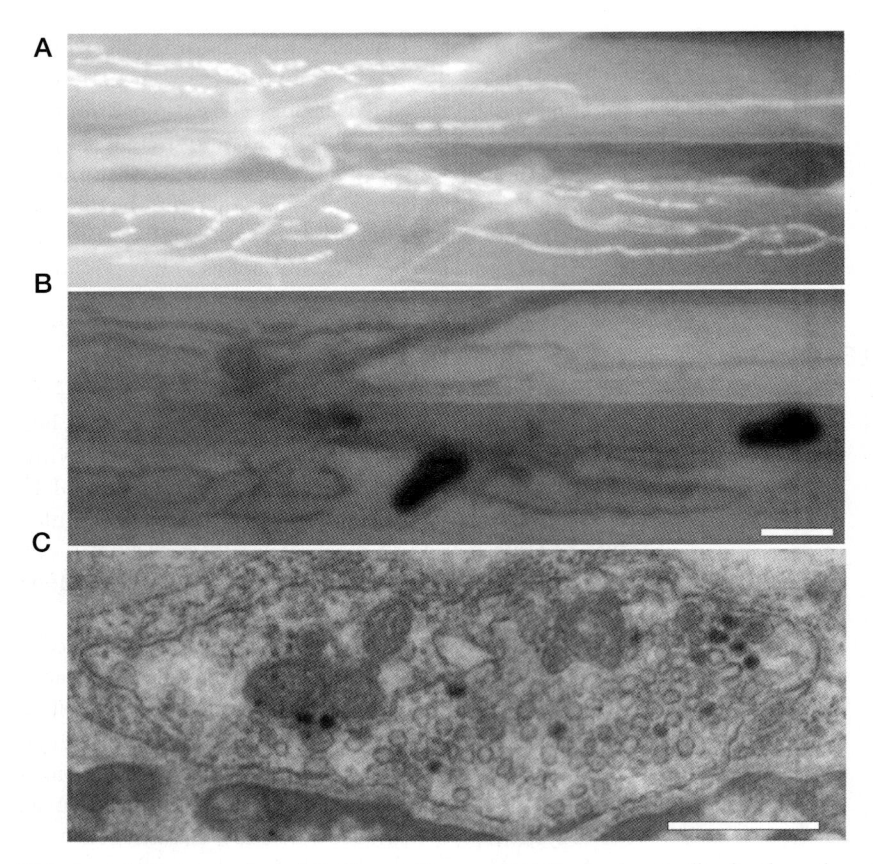

FIGURE 3. FM dye photoconversion. (*A*) Preparations are stained, chemically fixed, and soaked with diaminobenzidine (DAB). The dye within nerve terminals is brightly visible under the fluorescence microscope. The images in both *A* and *B* are taken with fluorescein optics; both FM fluorescence and glutaraldehyde autofluorescence are visible. (*B*) Photoconversion is complete. The DAB precipitate accumulates, and nerve terminals are seen as dark stripes on the muscle fibers, which still exhibit glutaraldehyde autofluorescence. The image was taken after ~35 min of illumination. Note also darkening of erythrocytes (almost invisible in *A*, but clearly shown in *B*), which indicates adequate photoconversion. Bar, 100 μm. (*C*) Typical image of a lightly stained, photoconverted nerve terminal in cross section. Labeled vesicles are electron-dense (dark). Bar, 0.5 μm.

direct endocytosis of vesicles from the surface membrane (Teng and Wilkinson 2000). The localization of different pools of vesicles within nerve terminals can also be investigated using photoconvertible dyes (Schikorski and Stevens 2001). The following procedure describes the use of photoconverting dyes to investigate aspects of vesicle metabolism in the frog NMJ.

1. Dissect the preparation (see Materials section above).
2. Stain the preparation (see Procedure A above).
3. Fix the preparation (still pinned as during staining) for 45–60 min, in ice-cold 1–2% glutaraldehyde<!> in NFR or in phosphate-buffered solution (PBS, 0.1 M, pH 7.2). (PBS: 28 mM NaH_2PO_4 <!>, 72 mM Na_2HPO_4 <!>, pH 7.2.)

   Note: Paraformaldehyde or glutaraldehyde is generally employed as a fixation agent. For good preservation of morphological features, use at least 0.5–1% (w/v) glutaraldehyde. Paraformaldehyde (1–2% w/v) can also be added to increase preservation, but it should be noted that it induces bursts of spontaneous release lasting for many minutes (Smith and Reese 1980). 1% glutaraldehyde/PBS has approximately the same osmolarity as the NFR, thus minimizing the negative effects of a change in osmolarity with fixation.

4. Cut the cutaneous pectoris muscle from the mounted preparation (rejecting the nerve and the portions connected to parts of skin and body wall) and wash it vigorously in PBS, at 4°C for 45–60 min.
5. Incubate the preparation for 10 min in 100 mM ammonium chloride <!> in PBS (to reduce autofluorescence of glutaraldehyde; Harata et al. 2001), and wash it again in PBS for 30 min.

Note: The intense washing eliminates free glutaraldehyde from the muscle; insufficient washing results in poor photoconversion, as unwashed fixative induces DAB to precipitate.

6. Re-mount the muscle on a Sylgard-coated surface and incubate it in 1.5 mg/ml DAB <!> (in PBS) for 30–60 min at 4°C.

   Note: If a transparent, crystalline precipitate forms upon DAB addition, the preparations should be rejected and longer wash times should be used on further experiments.

7. Illuminate the preparation for 30–60 min (as for imaging, see Materials section above) with a high-intensity light source <!>.

   Note: Intense illumination is necessary; for example, that provided by a 100-W mercury lamp. The objective used should provide a large enough photoconverted field for electronmicroscopic (EM) studies; we find 20x–25x objectives optimal. The nerve terminals appear bright (Fig. 3A). They rapidly lose fluorescence (20–30 min is usually sufficient for complete bleaching). Formation of the DAB precipitate follows, and the nerve terminals turn brown-black because of precipitate accumulation (Fig. 3B).

8. Remove the preparation, wash it in PBS (5–10 min), and process it for EM (as described in step 9).

   Note: Only dye in the nerve terminals on the top layer of muscle fibers undergoes efficient photoconversion. This is probably because of poor penetration of DAB deeper into the tissue; DAB conversion product is frequently seen in nerve terminals on both sides of the muscle, but not in those that lie deeper (Rizzoli and Betz, unpubl. observations).

9. Processing for EM: To avoid difficulty with identifying photoconversion spots after processing, cut them from the muscles and process them separately:

   a. Postfix photoconversion spots with 2% OsO_4<!> in PBS for 60 minutes.

   b. Dehydrate the tissue through an ascending series of ethanol solutions and propylene oxide <!> (10-min incubations in 50%, 70%, 90%, and 95% ethanol <!>, followed by 30-min incubations in 100% ethanol and 100% propylene oxide).

   c. Infiltrate the samples with epon resin<!> (50% epon resin solution) in propylene oxide for 12–16 hours.

   d. Replace the resin solution with 100% epon, and allow the propylene oxide to evaporate for 6–8 hours in open vials at room temperature.

   e. Embed the preparations in epon (12–16 hours at 60°C) and then section the epon blocks.

   f. Post-staining of thin sections should be light (e.g., 1-min incubations with 2% uranyl acetate <!> in 50% ethanol), ensuring that the DAB precipitate is clearly seen. A typical image of a photoconverted nerve terminal in cross section is shown in Figure 3C.

FM DYES IN OTHER SYNAPTIC PREPARATIONS

FM dyes have been used in a variety of systems. Figure 4 shows images typical of FM-loaded nerve terminals at the *Drosophila* larval NMJ (A), at the lizard NMJ (B), and at cultured hippocampal neurons

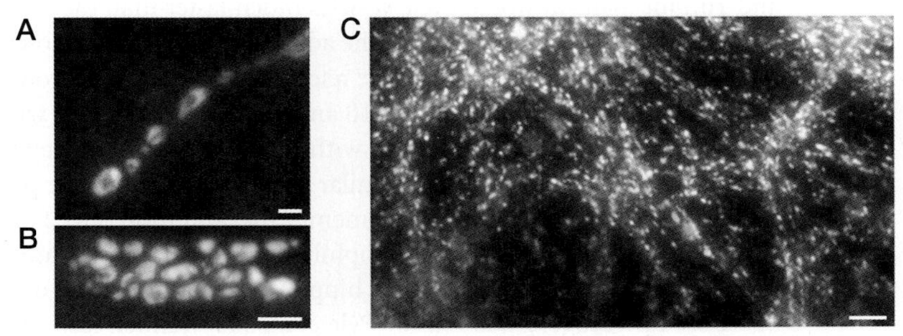

FIGURE 4. FM-labeled nerve terminals in different preparations. (*A*) *Drosophila* larval NMJ (bar, 3 μm). (*B*) Lizard (*Anolis*) NMJ (bar, 1 μm). (*C*) Cultured hippocampal neurons (fluorescent spots represent single presynaptic boutons; photo courtesy of Dr. V. Murthy; bar, 5 μm).

(C). Alhough the appearance of the labeling varies with the preparation, FM dyes still provide powerful tools in all these preparations. A brief discussion of FM use in these (and other) systems follows.

NEUROMUSCULAR JUNCTIONS

FM dyes can be used readily in NMJ preparations, in the frog (see above), rat (Ribchester et al. 1994), lizard (Lindgren et al. 1997), crayfish (Wang and Zucker 1998), and snake (Teng et al. 1999; Teng and Wilkinson 2000). In these preparations, the rate of FM uptake provides a reliable measure of presynaptic activity and can augment electrophysiological estimates of transmitter release. The destaining rate of the tissue indicates the rates of exocytosis and vesicle mobilization, which can be used, for example, to characterize the effects of various drugs and experimental conditions (Lindgren et al. 1997; Wang and Zucker 1998; Becherer et al. 2001). At the snake NMJ, where a sulfonic acid derivative of rhodamine (sulforhodamine 101; Lichtman et al. 1985) is typically used as endocytic tracer, FM dyes are used mainly for photoconversion.

FM dyes are also useful in the study of the larval *Drosophila* NMJ. As in vertebrate systems, uptake and release of the dye have been shown to be dependent on intact exo- and endocytosis (Ramaswami et al. 1994). A typical image of FM 1-43-stained *Drosophila* synaptic boutons is shown in Figure 4A. The dye is loaded only at the periphery of the boutons (compare with frog nerve terminals, Fig. 2). FM dye studies by Kuromi and Kidokoro (1998, 2000) have shown that *Drosophila* boutons contain two different vesicle pools, an exo/endo cycling pool (ECP), situated around the edges of boutons, and a reserve pool (RP), occupying the bouton centers (unlabeled in Fig. 4A). RP vesicles can be labeled only after prolonged tetanic stimulation.

HIPPOCAMPAL NERVE TERMINALS: THE "KISS-AND-RUN" HYPOTHESIS OF RECYCLING INVESTIGATED WITH FM DYES

Investigations into synaptic vesicle cycling in central nervous system preparations have also taken advantage of FM dyes. Presynaptic boutons of hippocampal neurons in culture are the most frequently used preparation for these purposes, with initial studies performed by Ryan et al. (1993). Unlike motor nerve terminals, hippocampal boutons have small numbers of synaptic vesicles (~200 vesicles/bouton; Schikorski and Stevens 1997), making them, at least theoretically, more sensitive to vesicle depletion. Interestingly, according to photoconversion experiments, hippocampal boutons also seem to cycle only a small proportion of their vesicles, as few as 10–18% (Harata et al. 2001). In contrast, 70–80% of vesicles at the frog NMJ are cycled. At the frog NMJ, most vesicles cycle through the slow endocytic pathway (see above; Richards et al. 2000), which relies on endosomal intermediates (Heuser and Reese 1973). Such a slow pathway would represent a significant bottleneck for vesicle recycling. An alternative means of vesicle recycling has been has been put forward in the kiss-and-run pathway (Fesce et al. 1994). In this proposed pathway, vesicles transiently fuse with the plasma membrane, through a relatively small pore, and endocytosis occurs simply in a reversal of fusion, by closure of the pore. A kiss-and-run mechanism would cycle vesicles much faster than the classic pathway. There was little evidence for fast cycling in this system until the advent of FM dyes, as shown in the next two paragraphs.

One experiment focused on the release of different dyes from vesicles during destaining (Klingauf et al. 1998; Pyle et al. 2000); FM 2-10 and FM 1-43 were used, with FM 1-43 being more hydrophobic than FM 2-10. If vesicles collapsed within the plasma membrane during exocytosis, both dyes would be expected to be released with similar kinetics. However, the process of endocytosis was faster than departitioning of the dye from the membrane, and more FM 2-10 than FM 1-43 was released; also, in subsequent studies, the more lipophilic FM 1-84 was released even more slowly than FM 1-43 (Klingauf et al. 1998), showing that hippocampal boutons rely on fast vesicle recovery. A time constant of ~1 sec for recovery was found (Pyle et al. 2000), much faster than predicted by the classic pathway.

Stevens and Williams (2000) combined imaging of FM 2-10 and FM 1-43 with electrophysiological recordings and used hyperosmolarity and action potential stimulation to release vesicles. This study showed that the two methods of stimulation cause different levels of dye release and uptake, but similar levels of neurotransmitter release. During hyperosmolarity-induced release, a proportion of the vesicles can release glutamate. They do not, however, lose or take up dye. It is possible that the vesicles connect to the plasma membrane via a pore through which glutamate, but not FM dyes, can pass; it is estimated that 20% of all vesicles behave in a similar manner during normal release.

SINGLE-VESICLE IMAGING WITH FM 1-43

FM dyes have also been used in conventional experiments, not unlike those performed on the frog NMJ, to study retinal bipolar cells from goldfish (Lagnado et al. 1996). The recent development of total internal reflection microscopy (TIRM, see Chapter 64) has made it possible to image single synaptic vesicles (Zenisek et al. 2000, 2002). Such experiments provide an accurate means of measuring exocytosis rates in single vesicles; they also allow direct measurement of less well understood aspects of vesicle cycling, such as vesicle trafficking (Zenisek et al. 2000). In this study, vesicles were observed approaching the plasma membrane and being captured at release sites, possibly via a mechanism referred to as "docking." Most vesicles were then seen to fuse with the release-site membrane, although some did occasionally detach. Partial release of dye, indicative of kiss-and-run behavior, was not seen in this system (Zenisek et al. 2002). Within a few milliseconds of vesicle/membrane fusion, the dye spread laterally into the membrane, indicative of full fusion (or a fusion pore that does not interfere with dye movement in the plane of the membrane). An important caveat is that this technique cannot monitor the behavior of unlabeled vesicles, and these represent ~99% of all vesicles.

SLICE PREPARATIONS

FM studies in delicate preparations such as acute brain slices are difficult because damaged membranes soak up large amounts of FM dye, which are hard to wash out. This problem can be alleviated by quenching the fluorescence of FM dye outside membrane-bound organelles. This can be achieved by applying sulforhodamine to labeled slices (Pyle et al. 1999). The sulforhodamine must be present throughout the imaging process, since, in these experiments, the FM dye is not washed from the slice. Another way to reduce background staining is to improve the washing process; the oligosaccharide ADVASEP-7 (a derivative of β-cyclodextrin), which has a high affinity for FM dyes, can be used in washing solutions to sequester dye molecules from biological membranes (Kay et al. 1999). The use of ADVASEP-7 has facilitated imaging in difficult systems such as rat brain and *Caenorhabditis elegans* slices. Zakharenko et al. (2001) used ADVASEP-7 to obtain high-quality images of acute hippocampal slices, demonstrating increases in presynaptic activity (as indicated by increased rates of FM destaining) during long-term potentiation—the first such description of the phenomenon in a noncultured preparation.

FM DYES: NOT JUST FOR VESICLE CYCLING

In mammalian cells, the plasma membrane has a distinct lipid composition; phosphatidylserine and phosphatidylethanolamine are found primarily on the internal leaflet of the membrane, and phosphatidylcholine is present on both leaflets. During apoptosis, the phospholipids in the plasma membrane are scrambled, resulting in the exposure of phosphatidylserine on the extracellular side of membranes. FM 1-43 fluorescence has been shown to increase upon phospholipid scrambling in Jurkat human leukemic T cells (Zweifach 2000). This has been interpreted as a response of the dye to scrambling-induced changes in the lipid-packing density, and not specifically to changes in phosphatidylserine density. FM 1-43 could therefore be used as an indicator for membrane scrambling and hence apoptosis.

A recent study suggested that FM 1-43 staining in hair cells from the mouse cochlea (Gale et al. 2001) was achieved via mechanotransducer channels (blocking mechanotransducer channel gating prevented dye uptake), raising the possibility that FM entry through ion channels may affect vesicle cycling studies. This conclusion was, however, contradicted by a similar study on inner hair cells from the guinea pig cochlea (Griesinger et al. 2002), in which dye uptake was unaffected by disruption of mechanotransduction.

CONCLUSION

Styryl dyes have been useful in the study of vesicle cycling because of three main characteristics: They reversibly stain membranes, they are impermeable to membranes, and their quantum yield is higher when bound to membranes than in aqueous solution. They can also be used as endocytic markers for electron microscopy. FM dye imaging was developed at the frog NMJ, but has been adopted in investigations of most synaptic systems, answering questions on aspects as diverse as vesicle mobility, endocytic mechanisms, refilling of docking sites, and localization of vesicle pools.

ACKNOWLEDGMENTS

We thank Steven Fadul and Dot Dill for excellent technical assistance.

REFERENCES

Angleson J.K. and Betz W.J. 1997. Monitoring secretion in real time: Capacitance, amperometry and fluorescence compared. *Trends Neurosci.* **20:** 281–287.

Becherer U., Guatimosim C. and Betz W. 2001. Effects of staurosporine on exocytosis and endocytosis at frog motor nerve terminals. *J. Neurosci.* **21:** 782–787.

Betz W.J. and Bewick G.S. 1992. Optical analysis of synaptic vesicle recycling at the frog neuromuscular junction. *Science* **255:** 200–203.

———. 1993. Optical monitoring of transmitter release and synaptic vesicle recycling at the frog neuromuscular junction. *J. Physiol.* **460:** 287–309.

Betz W.J. and Henkel A.W. 1994. Okadaic acid disrupts clusters of synaptic vesicles in frog motor nerve terminals. *J. Cell. Biol.* **124:** 843–854.

Betz W.J., Mao F., and Smith C.B. 1996. Imaging exocytosis and endocytosis. *Curr. Opin. Neurobiol.* **6:** 365–371.

Cochilla A.J., Angleson J.K., and Betz W.J. 1999. Monitoring secretory membrane with FM1-43 fluorescence. *Annu. Rev. Neurosci.* **22:** 1–10.

Fesce R., Grohovaz F., Valtorta F., and Meldolesi J. 1994. Neurotransmitter release: Fusion or 'kiss-and-run'? *Trends. Cell. Biol.* **4:** 1–4.

Gale J.E., Marcotti W., Kennedy H.J., Kros C.J., and Richardson G.P. 2001. FM1-43 dye behaves as a permeant blocker of the hair-cell mechanotransducer channel. *J. Neurosci.* **21:** 7013–7025.

Griesinger C.B., Richards C.D., and Ashmore JF. 2002. Fm1-43 reveals membrane recycling in adult inner hair cells of the mammalian cochlea. *J. Neurosci.* **22:** 3939–3952.

Harata N., Ryan T.A., Smith S.J., Buchanan J., and Tsien R.W. 2001. Visualizing recycling synaptic vesicles in hippocampal neurons by FM 1-43 photoconversion. *Proc. Natl. Acad. Sci.* **98:** 12748–12753.

Henkel A.W., Lubke J., and Betz W.J. 1996a. FM1-43 dye ultrastructural localization in and release from frog motor nerve terminals. *Proc. Natl. Acad. Sci.* **93:** 1918–1923.

Henkel A.W., Simpson L.L., Ridge R.M., and Betz W.J. 1996b. Synaptic vesicle movements monitored by fluorescence recovery after photobleaching in nerve terminals stained with FM1-43. *J. Neurosci.* **16:** 3960–3967.

Heuser J.E. and Reese T.S. 1973. Evidence for recycling of synaptic vesicle membrane during transmitter release at the frog neuromuscular junction. *J. Cell. Biol.* **57:** 315–344.

Kay A.R., Alfonso A., Alford S., Cline H.T., Holgado A.M., Sakmann B., Snitsarev V.A., Stricker T.P., Takahashi M., and Wu L.G. 1999. Imaging synaptic activity in intact brain and slices with FM1-43 in *C. elegans*, lamprey, and rat. *Neuron* **24:** 809–817.

Klingauf J., Kavalali E.T., and Tsien R.W. 1998. Kinetics and regulation of fast endocytosis at hippocampal synapses. *Nature* **394:** 581–585.

Kraszewski K., Daniell L., Mundigl O., and DeCamilli P. 1996. Mobility of synaptic vesicles in nerve endings monitored by recovery from photobleaching of synaptic vesicle-associated fluorescence. *J. Neurosci.* **16:** 5905–5913.

Kuromi H. and Kidokoro Y. 1998. Two distinct pools of synaptic vesicles in single presynaptic boutons in a temperature-sensitive *Drosophila* mutant, shibire. *Neuron* **20:** 917–925.

———. 2000. Tetanic stimulation recruits vesicles from reserve pool via a cAMP-mediated process in *Drosophila* synapses. *Neuron* **27:** 133–143.

Lagnado L., Gomis A., and Job C. 1996. Continuous vesicle cycling in the synaptic terminal of retinal bipolar cells. *Neuron* **17:** 957–967.

Lichtman J.W., Wilkinson R.S., and Rich M.M. 1985. Multiple innervation of tonic endplates revealed by activity-dependent uptake of fluorescent probes. *Nature* **314:** 357–359.

Lindgren C.A., Emery D.G., and Haydon P.G. 1997. Intracellular acidification reversibly reduces endocytosis at the neuromuscular junction. *J. Neurosci.* **17:** 3074–3084.

McMahan U.J., Edgington D.R., and Kuffler DP. 1980. Factors that influence regeneration of the neuromuscular junction. *J. Exp. Biol.* **89:** 31–42.

McMahan U.J., Spitzer N.C., and Peper K. 1972. Visual identification of nerve terminals in living skeletal muscle. *Proc. R. Soc. Lond. B Biol. Sci.* **181:** 421–430.

Pyle J.L., Kavalali E.T., Choi S. and Tsien R.W. 1999. Visualization of synaptic activity in hippocampal slices with FM1-43 enabled by fluorescence quenching. *Neuron* **24:** 803–808.

Pyle J.L., Kavalali E.T., Piedras-Renteria E.S., and Tsien R.W. 2000. Rapid reuse of readily releasable pool vesicles at hippocampal synapses. *Neuron* **28:** 221–231.

Ramaswami M., Krishnan K.S., and Kelly R.B. 1994. Intermediates in synaptic vesicle recycling revealed by optical imaging of *Drosophila* neuromuscular junctions. *Neuron* **13:** 363–375.

Ribchester R.R., Mao F., and Betz W.J. 1994. Optical measurements of activity-dependent membrane recycling in motor nerve terminals of mammalian skeletal muscle. *Proc. R. Soc. Lond. B Biol. Sci.* **255:** 61–66.

Richards D.A., Guatimosim C., and Betz W.J. 2000. Two endocytic recycling routes selectively fill two vesicle pools in frog motor nerve terminals. *Neuron* **27:** 551–559.

Ryan T.A., Reuter H., Wendland B., Schweizer F.E., Tsien R.W., and Smith S.J. 1993. The kinetics of synaptic vesicle recycling measured at single presynaptic boutons. *Neuron* **11:** 713–724.

Sandell J.H. and Masland R.H. 1988. Photoconversion of some fluorescent markers to a diaminobenzidine product. *J. Histochem. Cytochem.* **36:** 555–559.

Schikorski T. and Stevens C.F. 2001. Morphological correlates of functionally defined synaptic vesicle populations. *Nat. Neurosci.* **4:** 391–395.

———. 1997. Quantitative ultrastructural analysis of hippocampal excitatory synapses. *J. Neurosci.* **17:** 5858–5867.

Smith J.E. and Reese T.S. 1980. Use of aldehyde fixatives to determine the rate of synaptic transmitter release. *J. Exp. Biol.* **89:** 19–29.

Stevens C.F. and Williams J.H. 2000. "Kiss and run" exocytosis at hippocampal synapses. *Proc. Natl. Acad. Sci.* **97:** 12828–12833.

Teng H. and Wilkinson R.S. 2000. Clathrin-mediated endocytosis near active zones in snake motor boutons. *J. Neurosci.* **20:** 7986–7993.

Teng H., Cole J.C., Roberts R.L., and Wilkinson R.S. 1999. Endocytic active zones: Hot spots for endocytosis in vertebrate neuromuscular terminals. *J. Neurosci.* **19:** 4855–4866.

Wang C. and Zucker R.S. 1998. Regulation of synaptic vesicle recycling by calcium and serotonin. *Neuron* **21:** 155–67.

Zakharenko S.S., Zablow L., and Siegelbaum S.A. 2001. Visualization of changes in presynaptic function during long-term synaptic plasticity. *Nat. Neurosci.* **4:** 711–717.

Zenisek D., Steyer J.A., and Almers W. 2000. Transport, capture and exocytosis of single synaptic vesicles at active zones. *Nature* **406:** 849–854.

Zenisek D., Steyer J.A., Feldman M.E., and Almers W. 2002. A membrane marker leaves synaptic vesicles in milliseconds after exocytosis in retinal bipolar cells. *Neuron* **35:** 1085–1097.

Zweifach A. 2000. FM1-43 reports plasma membrane phospholipid scrambling in T-lymphocytes. *Biochem. J.* **349:** 255–260.

CHAPTER 62

A Practical Guide: Imaging FM Dyes in Brain Slices

Alan R. Kay

Department of Biological Sciences, University of Iowa, Iowa City, Iowa 52242

This chapter describes a method for monitoring the activity-dependent loading and unloading of synaptic vesicles with a fluorescent probe in intact or semi-intact neuronal systems. The method relies on the chemical properties of styryl pyridinium probes, such as FM1-43 (Cochilla et al. 1999), that are somewhat water-soluble but preferentially partition into lipid environments. The probe is taken up by endocytosis and only fluoresces once in the membrane. Stimulation of exocytosis releases the probe into the extracellular space, with a consequent decline in fluorescence, allowing the time course of release to be followed. In intact systems such as brain or brain slices, FM1-43 adsorbs to the extracellular surface of the plasma membrane and is resistant to removal by washing. This gives rise to an intense fluorescence signal associated with the extracellular membrane, which obscures the weaker signal associated with vesicular uptake. FM1-43 and its congeners can be removed by using a modified cyclodextrin (Advasep-7) that in effect serves as a soluble high-affinity scavenger for the probe, allowing it to be washed out of the preparation (Kay et al. 1999). Cyclodextrins (Szejtli 1998) are water-soluble, doughnut-shaped molecules with hydrophobic holes of the right size to accommodate FM1-43, forming what is called an "inclusion" complex.

This is a fairly general method that could be applied to a number of biological preparations, other than nervous systems. So far, the method has been applied to *Caenorhabditis elegans*, *Xenopus* tadpoles, lamprey spinal cord, fly brain, rat brain, and mouse brain.

MATERIALS

FM1-43 (Molecular Probes, Biotium) and Advasep-7 (Cydex, Biotium, Sigma).

A conventional or confocal microscope capable of imaging a fluophore (FM1-43) with an excitation maximum of 481 nm and an emission maximum of 582 nm.

PROCEDURE

Imaging FM Dyes in Brain Slices

1. Make up a stock solution of FM1-43<!> at a concentration of 150 μM in a stable buffered physiological solution; store at 2–8°C (buffered physiological solution 130 mM NaCl, 2.5 mM KCl<!>, 2 mM $CaCl_2$<!>, 2 mM $MgCl_2$<!>, 10 mM HEPES at pH 7.4).

2. Incubate the tissue in normal saline containing 2–10 μM of FM1-43 for 1–5 min, stirring and oxygenating the solution with a jet of 95/5 O_2/CO_2 (Normal saline: 125 mM NaCl, 2.5 mM KCl, 2 mM $CaCl_2$, 1.3 mM $MgSO_4$<!>, 25 mM $NaHCO_3$, 1.25 mM NaH_2PO_4<!>, and 25 mM glucose, bubbled with 95% O_2/5% CO_2).

   Note: Alternatively, because FM1-43 does not diffuse readily into the depths of tissue, it can be injected into the tissue using a patch pipette.

3. Stimulate the tissue electrically, or with high potassium chloride, to induce loading of the probe.
4. Wait for 1–2 min, to allow FM1-43 to be taken up by endocytosis. Wash the preparation with normal saline for 1 min. Replace the solution with normal saline containing 0.1–1 mM Advasep-7 and incubate for 1–2 min with stirring.

   Note: The inclusion complex is considerably more fluorescent than free FM1-43 and this leads to a boost in the overall fluorescence.

5. Wash the preparation with control saline for 5–20 min to remove the inclusion complex and free Advasep-7.
6. Monitor the fluorescence of the preparation under a microscope during the wash procedure.

   Note: If the loading process was satisfactory, fluorescent puncta should emerge and become evident. If washing steps were inadequate, repeat steps 4 and 5.

7. While focusing on fluorescently labeled synaptic terminals, image the fluorescence as a function of time, while stimulating release electrically or chemically.

   Note: Advasep-7 can be kept in the perfusing solution at a concentration of 10–100 μM to aid the removal of exocytosed probe. Stringent precautions should be taken to ensure that the preparation does not shift during stimulation.

Caution: *See Appendix 3 for appropriate handling of materials marked with <!>.*

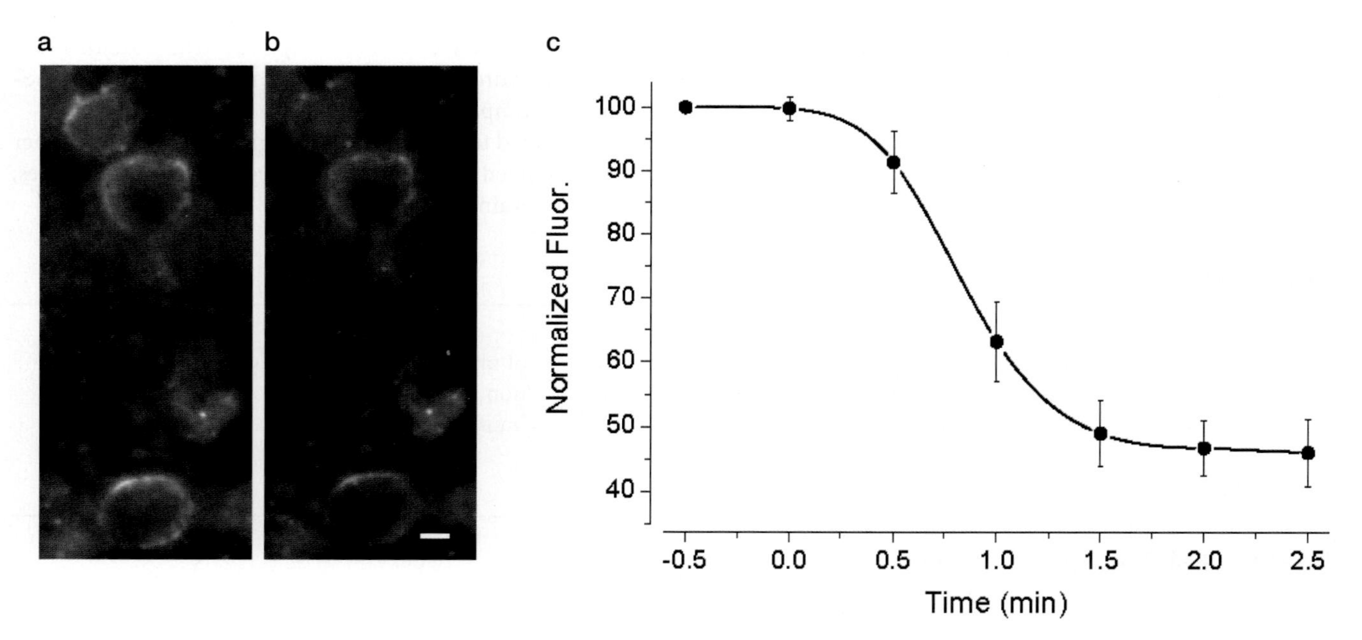

FIGURE 1. Synaptic terminals loaded with FM1-43 in the medial nucleus of the trapezoid body of the rat auditory brainstem. The nerve terminals in this region form the so-called calyces of Held that envelope the postsynaptic cell. (*a*) Prior to stimulation; (*b*) after stimulation with high potassium for 1.5 min. Bar, 10 μm. (*c*) Decline of fluorescence in the calyces of Held after high potassium stimulation. Average ± S.E.M. of ten terminals in four different slices. (Modified from Kay et al. 1999.)

EXAMPLE OF APPLICATION

Figure 1 shows a slice from the rat auditory brain stem that has been loaded with FM1-43 by potassium stimulation. Release was triggered by elevating the potassium concentration to 37.5 mM. The time course of destaining is plotted in Figure 1c.

ADVANTAGES AND LIMITATIONS

FM1-43 has been of great value in monitoring the spatiotemporal dynamics of synaptic transmission (Cochilla et al. 1999). It can also be used as a tool to locate synaptic terminations associated with single neurons or groups of neurons.

It is important to establish that the structures that load with FM1-43 are indeed synaptic terminations. This can be achieved by using antibodies against presynaptic marker proteins. FM1-43 itself is resistant to fixation; however, a fixable derivative, AM1-43, has recently become available (www.biotium.com). It is worth noting that, in rat brain slices, FM1-43 also labels capillaries and myelinated structures.

Endocytosis is not the only route for FM1-43 entry. It was recently found that FM1-43 can permeate through the mechanotransduction channel in hair cells, vanilloid receptors, and purinergic receptors (Meyers et al. 2003). In all cases, permeation could be reduced by channel blockers.

Stimulation, either electrical or through an increase in potassium, can lead to movement of a brain slice through the activation of capillary smooth muscle. Stringent measures should be taken to guard against any movement-induced artifacts by adequately securing the slice and by monitoring the location of labeled structures. Similar controls should be performed to guard against artifacts associated with photobleaching of the probe.

The rate of destaining is, to some extent, dependent on the rate of desorption of the probe from the membrane (Klingauf et al. 1998). FM probes with different chain lengths, and hence different rates of desorption, can be used to probe the precise nature of exocytosis. This has been used in an effort to understand the nature of the process of exocytosis, whether it occurs through a process of kiss-and-run or full-blown exocytosis (Zakharenko et al. 2002).

REFERENCES

Cochilla A.J., Angleson J.K., and Betz W.J. 1999. Monitoring secretory membrane with FM1-43 fluorescence. *Annu. Rev. Neurosci.* **22:** 1–10.

Kay A.R., Alfonso A., Alford S., Cline H.T., Haas K., Holgado A.M., Malinow R., Ryan T.A., Sakmann B., Snitsarev V.A., Stricker T.P., Takahashi M., and Wu L.-G. 1999. Imaging synaptic activity in intact brain and slices with FM1-43 in *C. elegans*, lamprey, *Xenopus* tadpole and rat. *Neuron* **24:** 809–817.

Klingauf J., Kavalali E.T., and Tsien R.W. 1998. Kinetics and regulation of fast endocytosis at hippocampal synapses. *Nature* **394:** 581–585.

Meyers J.R., MacDonald R.B., Duggan A., Lenzi D., Standaert D.G., Corwin J.T., and Corey D.P. 2003. Lighting up the senses: FM1-43 loading of sensory cells through nonselective ion channels. *J. Neurosci.* **23:** 4054–4065.

Szejtli J. 1998. Introduction and general overview of cyclodextrin chemistry. *Chem. Rev.* **98:** 1743–1754.

Zakharenko S.S., Zablow L., and Siegelbaum S.A. 2002. Altered presynaptic vesicle release and cycling during mGluR-dependent LTD. *Neuron* **35:** 1099–1100.

A Practical Guide: Imaging Zinc in Brain Slices

Alan R. Kay

Department of Biological Sciences, University of Iowa, Iowa City, Iowa 52242

This chapter provides a practical guide to imaging zinc within synaptic vesicles and the extracellular space of brain slices. Certain glutamatergic synaptic terminals in brain have high concentrations of exchangeable Zn^{2+} within their synaptic vesicles (Frederickson 1989). Classically, these terminals have been revealed by Timm's histochemical stain. More recently, a number of fluorimetric probes have become available that allow Zn^{2+} to be detected in live preparations. This chapter describes two methods, one for visualizing Zn^{2+} in synaptic vesicles, and another for detecting Zn^{2+} in the extracellular space.

These methods are applicable to any tissue with the high levels of loosely bound Zn^{2+} that are typically found in vesicles, and tissue that might have Zn^{2+} associated with macromolecules in the extracellular space.

MATERIALS

Reagents

Ca-EDTA (Sigma); DEDTC<!> (diethyldithiocarbamate, Sigma), a membrane-permeant transition metal chelator; EDPA<!> (ethylenediamine-*N,N′*-diacetic-*N,N′*-di-β-propionicacid, Aldrich), a transition metal chelator that does not suffer interference from calcium and magnesium; FluoZin-3 tetrapotassium salt (Molecular Probes); TPEN<!> [*N,N,N′,N′*-Tetrakis(2-pyridylmethyl)ethylenediamine, Aldrich, Sigma], a membrane-permeant transition metal chelator; Zinquin-free acid (Biotium, Dojindo, Toronto Research Chemicals, TefLabs), $ZnSO_4$ (0.051 M, Aldrich).

Microscope and Imaging System

This protocol requires a microscope equipped to image fluorophores with the following spectral characteristics: FluoZin-3 (ex 494, em 518) and Zinquin (ex 360, em 496). The author uses an Olympus BX50WI microscope fitted with a Princeton Instruments cooled CCD camera with illumination provided by a T.I.L.L photonics monochromator. A fluorimeter for detecting Zn^{2+} contamination is also required.

Caution: *See Appendix 3 for appropriate handling of materials marked with <!>.*

PROCEDURE A

Visualization of Intravesicular or Intracellular Zn^{2+}

1. Incubate the tissue in 10–30 μM Zinquin-free acid for 30 min to 1 hr.

   Note: This form of the probe crosses membranes passively, and if there is a high concentration of free or weakly bound Zn^{2+} in vesicles or in the cytoplasm, it will induce fluorescence.

2. To confirm the origin of any observed fluorescence, test whether the fluorescence is (a) resistant to the application of the membrane-impermeant Zn^{2+}-chelator Ca-EDTA 1 (mM) and (b) quenched by the application of the membrane-permeant chelators TPEN (0.1 mM) or DEDTC (1 mM).

   Note: Intracellular fluorescence will produce positive results for both tests.

PROCEDURE B

Testing Solutions and Labware for Zn^{2+} Contamination

1. Add 500 nM FluoZin-3 and 50 μM Ca-EDTA to a solution containing 140 mM NaCl, 2.5 mM KCl<!>, 2 mM $CaCl_2$<!>, 2 mM $MgSO_4$<!>, and 10 mM HEPES, pH 7.4, in a stirred methacrylate cuvette.

2. Allow the solution to equilibrate for 10 min, as the cuvette will introduce some Zn^{2+} contamination.

3. Calibrate the probe by adding $ZnSO_4$ at a concentration of 1 nM.

   Note: On addition of Zn^{2+} , the ion binds to FluoZin-3, giving rise to an immediate increase in fluorescence followed by a slow fall as the Zn^{2+} is chelated by Ca-EDTA. Repeated additions of Zn^{2+} can be made to the same solution as the Ca-EDTA chelates the ion. This technique can be used to detect Zn^{2+} elevations as low as 0.1 nM. Labware can be tested by simply introducing the item into the cuvette for a few seconds. FluoZin-3 has a K_d of 15 nM and is therefore sensitive to very low levels of Zn^{2+}. To detect Zn^{2+} release, it is important to ensure that nothing that comes in touch with the solution introduces Zn^{2+}. Solutions should not come into contact with glass or stainless steel, and certified "metal-free" pipette tips should be used (Fisher Scientific). All labware that comes into contact with solutions should be tested for Zn^{2+} contamination. All solutions should be stored in Teflon bottles. If the preparation is placed on a glass slide in the recording chamber, the slide should be covered with Saran Wrap as should a water immersion objective, as FluoZin-3 will leach Zn^{2+} from glass.

PROCEDURE C

Detecting the Release of Zn^{2+} into the Extracellular Space

1. Incubate tissue in 1–5 μM FluoZin-3 and 25–100 μM Ca-EDTA for 3–7 min, enough time for the probe to diffuse into the extracellular space. The tissue can be maintained for a short period (~20 min) in 1–3 ml of physiological saline by stirring and aerating the solution with a jet of 95% O_2/5% CO_2.

2. To determine whether there are steady levels of Zn^{2+} in the extracellular space, apply 100 μM membrane-impermeant Zn^{2+}-chelator EDPA.

   Note: A decrease in fluorescence indicates that there is free Zn^{2+} or Zn^{2+} weakly associated with macromolecules in the extracellular space.

3. To determine whether Zn^{2+} is released during synaptic activation of the slice, stimulate electrically, or chemically, while measuring fluorescence.

   Note: If the fluorescence increases in response to stimulation as a result of Zn^{2+} elevations, the addition of EDPA (to 100 μM), which can intercept Zn^{2+}, should prevent the increase in fluorescence or truncate elevations when added during stimulation (see Fig. 1B). It is important to confirm that the changes in fluorescence do not simply arise from changes in autofluorescence.

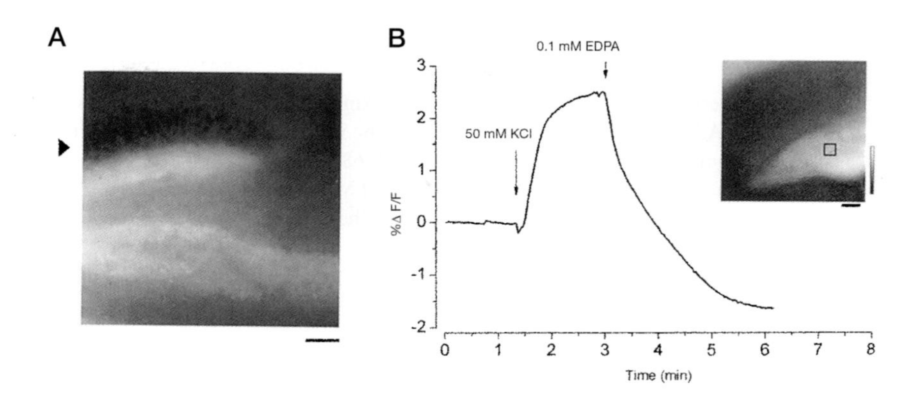

FIGURE 1. (*A*) Rat dentate gyrus loaded with 25 μM Zinquin free acid for 30 min. The arrow indicates the granule cell layer of the upper blade of the dentate gyrus. Bar, 100 μm. (*B*) The change in fluorescence in the hilus of a mouse hippocampal slice induced by 50 mM KCl in the presence of 2 μM FluoZin-3 and 50 μM Ca-EDTA. (*Inset*) Difference between fluorescence prior to stimulation and at the peak of the response. The hilus, triangular form in the lower right corner, exhibited the greatest change in fluorescence. Gray scale runs from 0 to 4.8% ΔF/F. Bar, 100 μm.

SHORT EXAMPLE OF APPLICATION

Figure 1A shows the dentate gyrus of a rat hippocampus loaded with Zinquin. Note that fluorescence is evident in the hilus and CA4, areas rich in vesicular Zn^{2+}, whereas cell bodies are not stained. Figure 1b shows the change in fluorescence induced in a mouse hippocampal slice by stimulation with 50 mM KCl and the truncation of the signal by the addition of EDPA.

ADVANTAGES AND LIMITATIONS

Zinquin is an unusual fluorimetric probe in that although it has a single carboxyl group, it is able to cross membranes passively (Snitsarev et al. 2001). Therefore, it is not necessary to use its acetyl ester form to load Zn^{2+}-rich vesicles. Zinquin can act as an ionophore, shuttling Zn^{2+} across membranes; however, the pH gradient across synaptic vesicles effectively prevents this from occurring (Snitsarev et al. 2001).

Because FluoZin-3 is so sensitive to Zn^{2+}, experiments using this dye are particularly susceptible to artifacts caused by contamination. Zn^{2+} can be introduced from rather unexpected sources, such as plastic tubing and polyethylene transfer pipettes (Fisher Scientific). It is important to scrutinize all phases of the experiment, and all components for their potential to introduce Zn^{2+} contamination. The inclusion of Ca-EDTA in the recording medium mops up contaminating Zn^{2+} , making it unnecessary to remove contaminating transition metals and Zn^{2+} from the bathing solution. However, if this approach is used, any additions that are made to the solution, e.g., high potassium, must also contain Ca-EDTA.

The electrodes used for electrical stimulation should be made of tungsten rather than stainless steel. The latter contains approximately 15% nickel, which can be released by electrolysis. Nickel interacts with FluoZin-3, increasing its fluorescence, potentially giving rise to artifactual signals.

It is important to recognize that a fluorimetric probe may signal Zn^{2+} release when Zn^{2+} is externalized but not actually released (Kay 2003). For example, if Zn^{2+} is weakly bound to a site within a synaptic vesicle that is presented to the extracellular space after exocytosis, the FluoZin-3, which under normal circumstances would remain bound, may chelate Zn^{2+}, giving rise to an artifactual elevation of Zn^{2+} in the extracellular space.

REFERENCES

Frederickson C.J. 1989. Neurobiology of zinc and zinc-containing neurons. *Int. Rev. Neurobiol.* **31:** 145–238.

Kay A.R. 2003. Evidence for chelatable zinc in the extracellular space of the hippocampus, but little evidence for synaptic release of Zn. *J. Neurosci.* **23:** 6847–6855.

Snitsarev V., Budde T., Stricker T.P., Cox J.M., Krupa D.J., Geng L., and Kay A.R. 2001. Fluorescent detection of Zn($^{2+}$)-rich vesicles with Zinquin: Mechanism of action in lipid environments. *Biophys. J.* **80:** 1538–1546.

CHAPTER 64

A Practical Guide: Imaging Exocytosis with Total Internal Reflection Microscopy

David Zenisek[1] and David Perrais[2]

[1]Yale University School of Medicine, Department of Cellular and Molecular Physiology, New Haven, Connecticut 06520; [2]Laboratoire de Physiologie Cellulaire de la Synapse, CNRS UMR 5091 et Université Bordeaux 2, Institut François Magendie, 33077 Bordeaux, France

Although electrophysiological techniques such as membrane capacitance measurements, electrochemical detection, and postsynaptic recordings are powerful for studying exocytosis, information concerning any steps prior to vesicle fusion must be inferred indirectly. We aim to investigate the steps leading up to vesicle fusion, by directly imaging synaptic vesicles and dense core granules prior to and including exocytosis.

A powerful technique for studying events near a cell surface is total internal reflection fluorescence microscopy (TIRFM). This technique allows selective imaging of fluorescent molecules that are closest to a high refractive index substance such as glass. TIRFM has been explored to study (1) exocytosis of single synaptic vesicles stained with FM1-43 in living goldfish retinal bipolar neurons and (2) exocytosis of single dense core granules stained with neuropeptide Y–enhanced green fluorescent protein (NPY-EGFP) in living bovine chromaffin cells. This chapter describes the basic theory behind TIRFM and provides a method for using TIRFM to image the release of vesicular markers after exocytosis.

TOTAL INTERNAL REFLECTION FLUORESCENCE MICROSCOPY

Principle

Light directed from a material of high refractive index (n_1) to one of lower refractive index (n_2) is totally reflected at angles equal to or greater than the critical angle, α_c, where

$$\alpha_c = \sin^{-1}(n_2/n_1) \tag{1}$$

At these angles, where light is totally reflected at the interface, an electromagnetic "evanescent field" is formed in the lower refractive index medium. This wave decays exponentially with distance from the interface with length constants that can be much shorter than the wavelength of illumination light. The length constant (d) of the exponential decay of this evanescent wave is governed by the function

$$d = (\lambda/4\pi)\,(n_1^2 \sin^2\alpha - n_2^2)^{-1/2} \tag{2}$$

where α is the angle at which the excitation beam hits the interface, and λ is the wavelength of the light. TIRFM (also known as evanescent field fluorescence microscopy) uses the thin evanescent wave to selectively excite fluorescent molecules in the portion of the lower refractive index medium nearest the interface of the two media. In cell biological or neurobiological studies, the low refractive index medium is usually the cytoplasm of a cell adherent to a higher refractive index coverslip. When a cell with fluorescently labeled secretory vesicles is firmly attached to a coverslip, the dye molecules will be excited only in those vesicles that lie within the thin evanescent field, thereby providing a high signal-to-noise ratio for vesicles within this region. In addition, the exponential drop in light intensity with distance allows small movements relative to the coverslip to be monitored by tracking object fluorescence. Using this method, single vesicles near (less than ~100 nm) the plasma membrane can be monitored as they travel to the membrane and undergo exocytosis.

Considerations for Optical Design

Several different microscope setups can be used for TIRFM (for review, see Axelrod 2001). A common setup uses a prism or hemicylinder, which is in optical contact with the coverslip via a layer of immersion oil to couple the laser beam to the interface. The fluorescence distribution near the interface is observed through the cell and through the layer of fluid surrounding it with a microscope objective lens. An alternative approach, first pioneered by Stout and Axelrod (1989), uses a high numerical aperture (NA) oil-immersion objective lens to guide the excitation light in an epi-illumination configuration onto the interface. This "prismless" or "objective-type" approach is preferable for our own experiments because it offers free access to the cell for recording electrodes and pipettes. In addition, it provides better image quality and efficient light collection, because high-NA objective lenses must be used for imaging. It also avoids the image distortion that may result when the plasma membrane is viewed through the cell. In the prismless configuration, even part of the emitted near-field fluorescence radiation can be collected through the glass substrate (Hellen and Axelrod 1987), further improving the fluorescence collection efficiency of the system.

Selection of Microscope Objective

The NA of an objective lens is equal to

$$\mathrm{NA} = n_1 \sin \alpha_{max} \tag{3}$$

where α_{max} is the maximum angle of incident light rays that can be achieved with the objective. Therefore, substituting Equation 3 into Equation 2 gives the equation for the shortest possible exponential length constant as a function of the objective:

$$d_{min} = \lambda/4\pi\ (\mathrm{NA}^2 - n_2^2)^{-1/2} \tag{4}$$

It is easy to see from Equation 4 that d becomes imaginary when NA < n_2, which results in the propagation of light rather than total internal reflection. Thus, to achieve total internal reflection, the objective must have an NA higher than the refractive index of the cell, typically about 1.37. With the growing popularity of TIRFM, Olympus, Zeiss, and Nikon now sell objectives with 1.45 NA, suitable for TIRFM. These objectives enable total internal reflection using standard glass coverslips and standard immersion oils. They are the most convenient objectives to use for TIRFM. For these objectives, the theoretical value of d_{min} is 82 nm, with $\lambda = 488$ nm. Although these objectives are suitable for most TIRFM applications, an Olympus 100X/1.65 NA (Caldwell 1997) is recommended for imaging vesicles in bipolar neurons, which are densely packed within the nerve terminal. The higher NA objective allows shallower ($d_{min} \approx 45$ nm) evanescent fields and, hence, better signal-to-noise ratios. Using the 1.65 NA objective does, however, have some drawbacks. In place of standard immersion oils, this objective requires di-iodomethane with sulfur. This oil is volatile and leaves a yellow sulfur residue, and it tends to be moderately fluorescent at 488 nm. The objective also requires expensive high refractive index coverslips ($n = 1.8$).

High refractive index coverslips can be reused up to ten times, but are too brittle to withstand standard washing procedures. Best results are obtained using soap and water followed by extensive rinsing. Because of these drawbacks, for most applications, the 1.45 NA objectives are recommended.

OPTICAL DESIGN OF THE TIRFM SETUP

Illumination

The setup is based on a commercial inverted microscope (IX70, Olympus) equipped with a side illumination port and side-facing filter cube (both available as options from Olympus), which allows for illumination through the side of the microscope. Figure 1 shows a diagram of the basic setup. Upon shutter opening, light from the 488 argon laser (MWK) is projected through a spatial filter (Newport) and expanded using a pair of apochromat lenses (focal distances 25 and 150 mm; Newport) mounted on an optical rail (Newport) separated by a distance of the sum of their focal distances. When aligned properly, a parallel beam should emerge from the two lenses with a beam width expanded by the ratio of the focal distance of the second lens the light travels through (f_2) to the focal distance of the first lens (f_1). Because the beam width from the laser is 1.5 mm and the focal distances in our beam expander are 25 mm and 150 mm, the beam is expanded to 9 mm.

The expanded beam is then reflected by a mirror and focused to the back focal plane by a focusing lens mounted on an optical rail just adjacent to the microscope (see Fig. 1). To achieve total internal reflection, the lens and mirror are mounted on translation stages, which can be moved to direct the beam to locations off-axis in the back focal plane of the high-NA objective lens (Apo 100x O HR, Olympus). To align the beam, the lens and mirror are adjusted so that the laser light is projected through the middle of the objective lens to the ceiling. At this point, the lens can be moved along the optical rail until the size of the spot on the ceiling is minimized, indicating that the laser light is focused to the back focal plane of the objective. Next, the laser beam is moved off-axis until total internal reflection is achieved. When using a focusing lens with a focal distance that is similar to the focal distance of the microscope, the beam will be reduced in size by the magnification factor of the objective. The focal distance of our lens is only slightly longer than the tube lens and thus illuminates a region of ~90 μm.

The angle of incidence light, α, is measured by centering on the objective a hemicylindrical prism of the same glass as the coverslips (Fig. 1, insert B). The laser light propagates through the hemicylinder and hits the interface with air normally and is therefore neither reflected nor deviated. The laser can thus be projected onto a screen, allowing α to be measured. Under the conditions described here, α typically ranges from 66° to 68° for bipolar cell experiments and from 57° to 59° for chromaffin cell experiments, which, using Equation 2, gives values for d of 41–43 nm and 60–68 nm, respectively.

Light Source

A 488-nm argon-ion laser is the light source of choice. In our setup, a multiline air-cooled argon-ion laser with an output power of about 100 mW is used. An acousto-optic tunable filter (Neos; see Chapter 98) is used to select specific laser lines. For exciting FM1-43 and EGFP fluorescence, the 488-nm line is used.

Image Capture and Analysis

The fluorescence light is collected by the same objective lens used for excitation and passes the dichroic mirror (dichroic mirror in Fig. 1) (Chroma) used for normal epifluorescence microscopy. A CCD camera (Hamamatsu Orca II ER 1394) mounted to an image intensifier (Gen IV, VideoScope International) on the side camera port of the microscope serves as imaging device. Images are collected and analyzed using a PC running MetaMorph (Universal Imaging). To synchronize illumination with data collection, the computer simultaneously sends a transistor-transistor logic (TTL) pulse to the shutter controller (Vincent Associates), which in turn opens the shutter (Vincent Associates).

Stabilization of the Microscope Objective Lens during Imaging

In the setup described here, the microscope stage drifted, relative to the objective lens, along the optical axis. This movement was compensated by a feedback-controlled focus-stabilization system. The

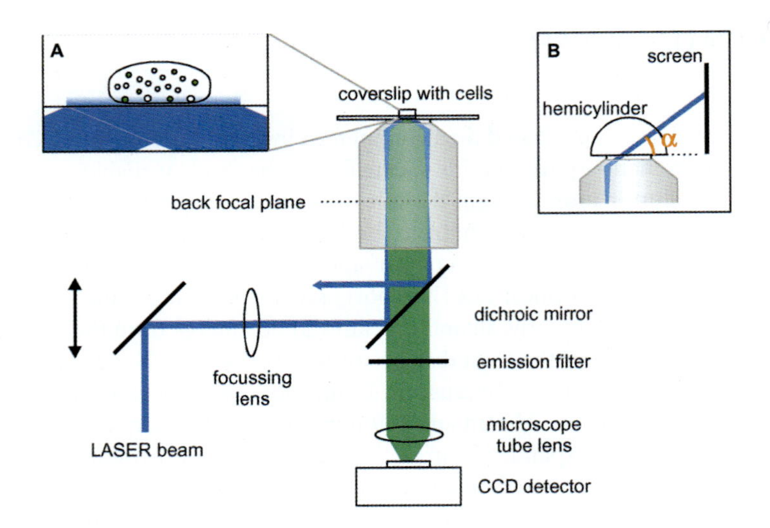

FIGURE 1. Schematic diagram of the prismless TIRFM setup. The expanded LASER beam is focused on the back focal plane of the high-NA objective. If a coverslip with stained cells is installed, an evanescent wave is generated and only the stained vesicles closest to the interface are visible (insert *A*). The angle of incidence α is measured by placing a hemicylinder on the objective, making contact with oil (insert *B*).

distance between the objective and the stage was continuously measured by a calibrated position sensor (PI-Foc, E-115.21, Physik Instrumente), and the objective was positioned by a piezo element (P-721.10, Physik-Instrumente).

MATERIALS

Buffers and Solutions

Depolarizing Buffer
95 mM NaCl, 25 mM KCl<!>, 2.5 mM $CaCl_2$<!>, 1 mM $MgCl_2$<!>, 10 mM HEPES (pH 7.2)

Standard Buffer
120 mM NaCl, 2.5 mM KCl, 2.5 mM $CaCl_2$, 1 mM $MgCl_2$, 10 mM HEPES (pH 7.2)

Culture Medium
45% DMEM<!>; 45% Ham's F-12; 10% fetal calf serum; 1 unit/ml penicillin/streptomycin, all from GIBCO-BRL

Low Calcium Buffer
120 mM NaCl, 2.5 mM KCl, 0.5 mM $CaCl_2$, 1.25 mM EGTA, 1 mM $MgCl_2$, 10 mM HEPES (pH 7.2)

Phosphate-Buffered Saline
155 mM NaCl, 1 mM KH_2PO_4, 6.5 mM NaH_2PO_4, pH 7.3, filter sterilized

Caution: *See Appendix 3 for appropriate handling of materials marked with <!>.*

PROTOCOLS AND PROCEDURE

Imaging Using TIRFM

On the day of the experiment, verify the alignment of the lens by viewing an undiluted aqueous suspension of 100-nm fluorescent beads (Molecular Probes) containing 2% solids on a cell-free coverslip. When aligned and focused properly, diffusing beads and beads adherent to the glass coverslip should be clearly resolved in the center of the field of view with very little fluorescent background. This is where total internal reflection occurs.

PROCEDURE A

Preparation of Dissociated Bipolar Cells

1. Triturate digested pieces of retina, prepared as described by Zenisek et al. (2002), on washed and rinsed high-refractive index coverslips ($n = 1.8$). After allowing the cells to settle, search the coverslip for bipolar cells with tightly adherent synaptic terminals.

   Note: Bipolar cells have distinctive morphology, containing a cell body with a single axon terminating in a single bulbous synaptic terminal.

2. Stain bipolar cell vesicles with 5 μm FM1-43 (Molecular Probes) in depolarizing buffer via local perfusion pipette aimed directly at the cell for 10 sec at room temperature. FM1-43 reversibly inserts into the outer leaflet of exposed plasma membrane and is taken up into vesicles via endocytosis. The dye is fluorescent in lipid environments, but nonfluorescent in aqueous solution.
3. Wash cells thoroughly with dye-free low-calcium buffer using a local perfusion pipette and bath exchange for more than 30 min to remove fluorescent dye from bipolar cell plasma membrane and from coverslip.

   Note: FM1-43 tends to adhere to glass.

4. Replace low-calcium buffer with standard buffer. Elicit exocytosis by depolarizing the bipolar cell with whole-cell patch pipette.
5. Image cells using TIRF illumination (see below).

PROCEDURE B

Preparation of Cultured Bovine Chromaffin Cells

1. Prepare dissociated bovine chromaffin cell cultures as described previously (Parsons et al. 1995), and plate them on plastic culture dishes in culture medium equilibrated at 37°C and 5% CO_2.
2. After 24 hours, resuspend the cells (the weakly adhering chromaffin cells are readily washed off while fibroblast remain adherent to the culture dish) in phosphate-buffered saline
3. Place the cell suspension (10^6 cells/ml) in an electroporation cuvette (0.4-mm gap size, BTX) with 20 μg of plasmid DNA (NPY-EGFP, Lang et al. 1997) and electroporate with one voltage pulse (20 msec, 220 V, T820 Electrosquare Porator, BTX).
4. Plate the cells on poly-D-lysine (0.1 mg/ml)-coated high-refractive index coverslips in preequilibrated culture medium.
5. Renew the medium after 1 day. Image the cells 2–4 days after transfection in standard buffer containing 5 mM $CaCl_2$ and 10 mM glucose.
6. View the cells under epifluorescence illumination (either by using a conventional epifluorescence illumination, such as an arc lamp<!>, or by moving the focusing lens of the TIRFM setup to obtain propagation of the laser light into the cell).

   Note: The transfection process is highly inefficient ($<1\%$ of cells treated are transfected). This step is neccesary to speed up the detection of transfected cells.

7. Image transfected cells using TIRF illumination (see below).

PROCEDURE C

Imaging Using TIRF Illumination

1. Run the CCD camera in focus mode to focus on the footprint of the cell. There should be only one focal plane in which objects are in focus.
2. Use camera-control software to open shutter and start recording TIRFM images.

EXAMPLES OF APPLICATIONS

Figure 2 shows a FM1-43-labeled bipolar cell, imaged with evanescent field microscopy. Single fluorescent synaptic vesicles appear as diffraction-limited solitary spots with TIRFM. Vesicles appear only within the footprint where the cell adheres to the coverslip, consistent with selective excitation of the cell surface by TIRFM. In sequences of images, the fluorescence of most vesicles fluctuates rapidly. Vertical motion of vesicles can be tracked by following the fluorescence intensity over time (Zenisek et al. 2000). When recording at video rate, a step depolarization applied through a patch-clamp electrode elicits a transient increase in fluorescent intensity, followed by the lateral diffusion of fluorescent signal indicating exocytosis of a vesicle (Fig. 2, B–G). This rise in fluorescence results from the released dye coming closer to the coverslip and changing orientation, thus being more strongly excited by the evanescent wave (Zenisek et al. 2002).

Figure 3 shows a chromaffin cell, transfected with NPY-EGFP, and imaged with TIRFM. Single granules are visible as fluorescent spots. Unlike most synaptic vesicles in bipolar cells, granules remained fairly stationary during 5-sec illumination periods. A step depolarization applied through a patch-clamp electrode caused a granule to release its fluorescent content, recorded at 20 Hz. The fluorescence rose rapidly, due to the released NPY-EGFP moving closer to the coverslip and to the dequenching of EGFP fluorescence following deacidification of the granule interior. The free NPY-EGFP evidently escaped in the cleft between the cell and the coverslip (D. Perrais et al., in prep.).

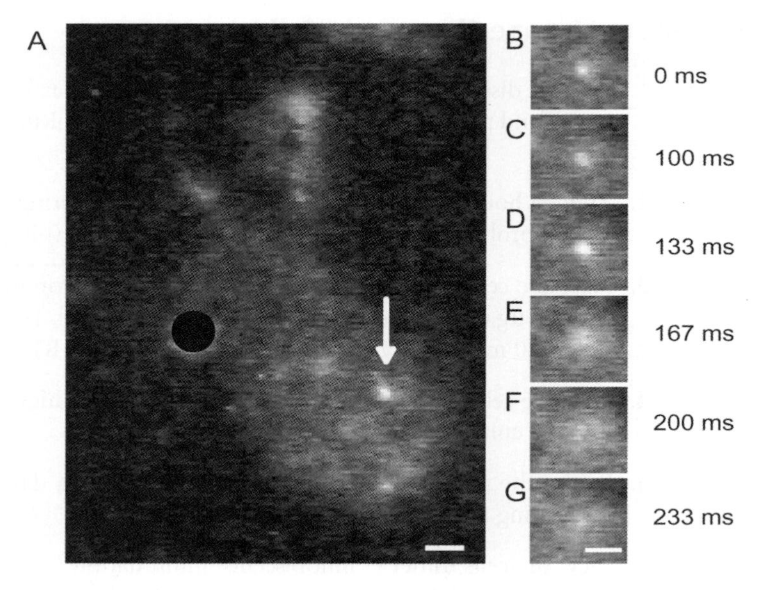

FIGURE 2. TIRFM imaging of bipolar cell vesicle exocytosis. (*A*) Image of a bipolar cell labeled with FM1-43 using TIRFM. The "footprint" of the terminal fluoresces faintly where it adheres to a glass coverslip. A more brightly fluorescent FM1-43-stained vesicle is indicated by an arrow. Bar, 1 μm. (*B–G*) Sequence of images of vesicle indicated by arrow in *A*. Images taken at the indicated times relative to the beginning of a 500-msec step depolarization. During the depolarization, the vesicle undergoes exocytosis. Note that the fluorescence of the vesicle increases and subsequently spreads over the surface of the terminal. Bar, 1 μm. For clarity, a piece of brightly fluorescent debris is masked out in *A*.

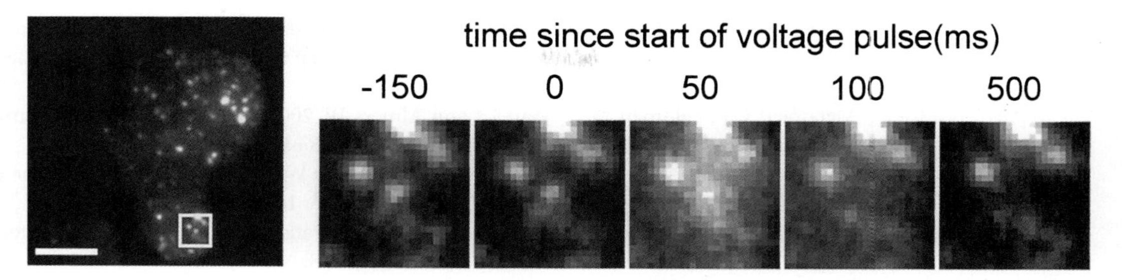

FIGURE 3. TIRFM imaging of a chromaffin cell transfected with NPY-EGFP and stimulated to exocytose. (*Left*) Image of a whole cell. Fluorescent spots correspond to single dense core granules stained with NPY-EGFP. (*Right*) Sequence of images corresponding to the square in the *left* image, at times indicated relative to the start of a 500-msec voltage pulse. The granule in the center releases NPY-EGFP that disappears completely in <500 msec. Bar, 1 μm.

ADVANTAGES AND LIMITATIONS

If fluorescent objects near the cell membrane are to be imaged, TIRFM offers a number of advantages over confocal or conventional epifluorescence microscopy: (1) TIRFM eliminates most background fluorescence since there is very little out-of-focus light; (2) TIRFM can provide improved *z*-discrimination of fluorescent objects; (3) TIRFM has a high collection efficiency of the emitted fluorescence, because no pinhole is needed; and (4) real-time imaging is readily achievable, since the whole-object plane is imaged at once, without scanning. Thus, with the development of digital cameras with high speeds and collection efficiencies, continuous image acquisition at a frequency >50 Hz is achievable, allowing release and diffusion kinetics of fluorescent molecules to be measured from single secretory vesicles (Zenisek et al. 2002). It is noteworthy, however, that the utility of TIRFM is limited to applications where the objects of interest reside within 200 nm of a flat, highly refractive substrate, usually a coverslip. In cell biology, this limitation restricts the practical uses of TIRFM to events at or near the plasma membrane in cells that adhere to glass.

CONCLUSION

This chapter presents a method for imaging single vesicles in neurons and chromaffin cells as they approach, and fuse with, the plasma membrane. Of the various TIRFM setups, the "prismless" approach is preferable because of its efficiency of high light collection, good image quality, and, most importantly, its free accessibility to the cell. TIRFM has proven useful in real-time studies of biological processes on or near the surface of single living cells, including exocytosis, endocytosis (Merrifield et al. 2002), and translocation of proteins into the plasma membrane (Teruel and Meyer 2002).

ACKNOWLEDGMENTS

D.Z. is supported by National Institutes of Health grant RO1-EY014990-01. D.P. was supported by a fellowship from the Human Frontier Science Program Organization.

REFERENCES

Axelrod D. 2001. Total internal reflection fluorescence microscopy in cell biology. *Traffic* **2:** 764–74.

Caldwell J.B. 1997. Ultra high NA microscope objective. *Opt. Photonics News* **11:** 44–46.

Hellen E.H. and Axelrod D. 1987. Fluorescence emission at dielectric and metal-film interfaces. *J. Opt. Soc. Am. B* **4:** 337–350.

Lang T., Wacker I., Steyer J., Kaether C., Wunderlich I., Soldati T., Gerdes H.-H., and Almers W. 1997. Ca2+-triggered peptide secretion in single cells imaged with green fluorescent protein and evanescent-wave microscopy. *Neuron* **18:** 857–863.

Merrifield C.J., Feldman M.E., Wan L., and Almers W. 2002. Imaging actin and dynamin recruitment during invagination of single clathrin-coated pits. *Nat. Cell Biol.* **4:** 691–698.

Parsons T.D., Coorssen J.R., Horstmann H., and Almers W. 1995. Docked granules, the exocytic burst, and the need for ATP hydrolysis in endocrine cells. *Neuron* **15:** 1085–1096.

Steyer J.A., Horstmann H., and Almers W. 1997. Transport, docking and exocytosis of single secretory granules in live chromaffin cells. *Nature* **388:** 474–478.

Stout A.L. and Axelrod D. 1989. Evanescent field excitation of fluorescence by epi-illumination microscopy. *Appl. Opt.* **28:** 5237–5242.

Teruel M.N. and Meyer T. 2002. Parallel single-cell monitoring of receptor-triggered membrane translocation of a calcium-sensing protein module. *Science* **295:** 1910–1912.

Zenisek D., Steyer J.A., and Almers W. 2000. Transport, capture and exocytosis of single synaptic vesicles at active zones. *Nature* **406:** 849–854.

Zenisek D., Steyer J.A., Feldman M.E., and Almers W. 2002. A membrane marker leaves synaptic vesicles in milliseconds after exocytosis in retinal bipolar cells. *Neuron* **35:** 1085–1097.

CHAPTER 65

A Practical Guide: Measuring Light-scattering Changes Associated with Secretion from Nerve Terminals

Brian M. Salzberg, Martin Muschol, and Ana Lia Obaid

Departments of Neuroscience and Physiology, University of Pennsylvania School of Medicine, Philadelphia, Pennsylvania 19104-6074

This chapter describes the use of light-scattering measurements to monitor directly events associated with the secretion of neuropeptides. In the neurosecretory terminals of the vertebrate neurohypophysis, the arrival of the action potential is coupled to the secretion of the neuropeptides, oxytocin and vasopressin. This excitation-secretion (E-S) coupling is mediated by a rise in intracellular calcium concentration (Douglas 1963, 1978), and, in mammals, the secretory event is accompanied by extremely rapid changes in light scattering, measured as transparency (Salzberg et al. 1985), that are relatively large ($\Delta I/I \sim 3 \times 10^{-4}$) compared with other intrinsic signals from neurons. These intrinsic optical signals provide a millisecond time-resolved monitor of events in the terminals that follow the entry of calcium, and they are intimately related to the release of neuropeptides. However, no one particular step in the sequence of events that couples excitation to secretion has been definitely implicated in the generation of the optical signals described in this chapter, and the identity of the physiological event or events remains the subject of active inquiry.

Secretion of peptides in the neurohypophysis results from bursts or trains of action potentials invading the terminals (Poulain and Wakerley 1982), rather than from the occurrence of individual impulses. The top trace in Figure 1 illustrates the changes in transmitted light intensity at 675 nm, when the terminals of an *unstained* CD-1 mouse neurohypophysis are stimulated at 16 Hz for 400 msec. The lower trace shows the same measurement, at a reduced gain, after the preparation was stained for 25 min in a 100 μg/ml solution of the potentiometric merocyanine-oxazolone dye, NK2367 (Salzberg et al. 1977). The intrinsic signal, at the top, consists of at least three separable components (Salzberg et al. 1985; Obaid et al. 1989). The rapid upstroke, which is an increase in large-angle light scattering, referred to as the E-wave for excitation, marks the arrival of the propagated action potential and action currents in the terminals, and probably has voltage- and current-dependent components (Cohen et al. 1972a,b). The large, long-lasting decrease in scattered-light intensity (an increase in transparency), the S-wave, for secretion, is considerably more interesting. (The S-wave is followed, in turn, by a variable, long-lasting recovery, the R-wave.) This S-wave has been studied for a

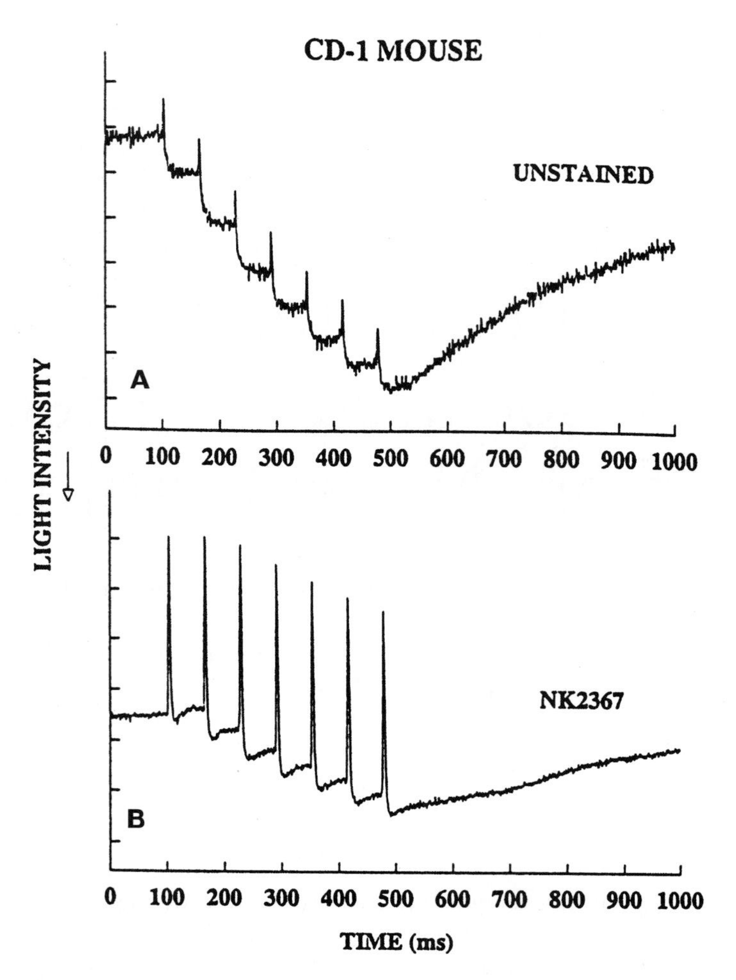

FIGURE 1. The optical responses to trains of stimuli (500 μsec at 16 Hz for 410 msec) of unstained and stained nerve terminals of the CD-1 mouse neurohypophysis. (*A*) The intrinsic optical signal recorded at 675 ± 35 nm from the nerve terminals of an unstained CD-1 mouse neurohypophysis during a stimulus train. The rapid upstrokes (E-waves) reflect the arrival of the action potentials (and action currents) in the terminals, and the sustained downward deflections (increases in transparency; S-waves) represent decreases in large-angle light scattering that are closely coupled to secretion of arginine vasopressin and oxytocin. (*B*) After staining the CD-1 mouse terminals with the potentiometric merocyanine-oxazolone dye NK2367 (100 μg/ml), extrinsic absorption signals that monitor the changes in membrane potential are superimposed on the light-scattering signals. All of the traces were recorded without averaging using a single photodiode (PV-444a from E.G.&G.) mounted behind a set of adjustable slits in the image plane of a compound microscope (Model UEM, Carl Zeiss, Inc.). 10x/0.4-NA objective (Wild Heerbrugg); response-time constant of the light-measuring system (10–90%) 330 μsec; AC coupling time constant 1.7 sec.

number of years (Salzberg et al. 1985, 1986; Obaid et al. 1989; Parsons et al. 1992; Salzberg and Obaid 1997) and, although it is intimately related to peptide release from these terminals, and it monitors some event that depends on Ca^{2+} entry, it does not necessarily reflect vesicle fusion, because it is absent in the frog (whose vesicles are indistinguishable from those of the mouse). This rapid-onset, but long-lasting, change in the intrinsic optical properties of the mammalian terminals shares many of the characteristics of the behavior of fast neurosecretory systems. These include dependence on stimulation frequency (Salzberg et al. 1985), paired-pulse facilitation, and exhaustion with high-frequency stimulation; dependence on extracellular Ca^{2+} (Salzberg et al. 1985); and sensitivity to Ca^{2+} channel blockers (Obaid et al. 1989) and to aminoglycoside antibiotics (Parsons et al. 1992), as well as to other interventions known to influence secretion, such as D_2O substitution for water (Salzberg et al. 1985). Indeed, changes in the amplitude of the S-wave are covariant with the effects on secretion of vasopressin and oxytocin produced by a wide variety of agents, and *all* of the manipulations that alter the light-scattering signal produce comparable effects on arginine vasopressin release, as determined by radioimmunoassay (Gainer et al. 1986) .

EFFECTS OF Ca^{2+} ON THE INTRINSIC OPTICAL SIGNAL COMPONENTS

Consider, for example, the effects of $[Ca^{2+}]_o$, illustrated in Figure 2. As has been known since the pioneering work of Douglas (1963; Douglas and Poisner 1964), calcium ion profoundly influences neurosecretory activity, and the extracellular concentration of Ca^{2+} would be expected to have large effects on the size of an intrinsic signal related to secretion from the terminals of the neurohypophysis. The records shown in Figure 2, A and B, are the averages of 16 sweeps, in 2.2 mM Ca^{2+} and 10 mM Ca^{2+}, respectively. The effects of Ca^{2+} are pretty clear, and the block of the S-wave by Cd^{2+} (not shown) is dramatic. Calcium, however, has other effects, including effects on threshold. Some of the effects of threshold variation can be eliminated by always using supramaximal stimulation, but notice that there remains some variation in the size of the E-waves that coincide in time with the terminal action potential. This very small light-scattering signal is related to the action potential, or the action current, or both (Cohen et al. 1972a,b), and is therefore roughly proportional to the number of terminals activated at a given time. On this admittedly crude assumption, the intrinsic signal is normalized to the size of the E-wave in order to compensate for changes in the invasibility of the tissue. This is demonstrated most clearly in traces C–F, in Figure 2, where the effect of Ca^{2+} on the light-scattering signal, per active terminal, is illustrated. The records labeled C (2.2 mM Ca^{2+}), D (0.1 mM Ca^{2+}), and E (5.0 mM Ca^{2+}) are here single sweeps, but, between each trial, 16 trials were averaged to improve the normalization to the initial upstroke, and these traces are shown on the right in panel F. Here, it is evident that $[Ca^{2+}]_o$ has a powerful effect on the amplitude of the S-wave, per active terminal. This is in contrast to the observation (Obaid and Salzberg 1996; data not shown) that micromolar concentrations of 4-aminopyridine enhance secretion by increasing the extent to which the action potential invades the highly ramified terminal arborization of the magnocellular neurons.

POSSIBLE MECHANISMS UNDERLYING SECRETION-COUPLED LIGHT SCATTERING

No evidence yet exists to implicate any particular step among the sequence of events that couples excitation to secretion in the generation of these very rapid intrinsic optical signals, and the identity of the physiological event responsible remains unclear. The fusion of secretory vesicles during exocytosis should result in the loss of relatively high-refractive-index particles and, thereby, reduce the refractive-index gradients in the terminals. However, the very early onset of the S-wave suggests that the optical signal could arise as a result of some calcium-dependent process prior to the fusion of the vesicles and the release of their contents. For example, it is possible that the light-scattering changes reflect a phase transition in the contents of the vesicles, or some rapid change in intracellular calcium stores, following calcium entry, but prior to secretion (Salzberg and Obaid 1997). In any event, these intrinsic optical signals are quite easily measured in a tissue like the mammalian neurohypophysis, where the relative homogeneity of the preparation, and the enormous proliferation of magnocellular neuron terminals, undoubtedly contribute to the generation of these relatively large optical changes. Magnocellular neurons, located in the hypothalamus, project their axons as bundles of fibers through the median eminence and infundibular stalk to terminate in the neurohypophysis, where neuropeptides and proteins are secreted into the local circulation. In the rat, for example, approximately 18,000 magnocellular neurons give rise to 40,000,000 terminals in the neurohypophysis, and nearly 99% of the excitable membrane belongs to terminals or secretory swellings (Nordmann 1977). At the same time, the precise control of the timing of the action potential in the terminals, the absence of any postsynaptic structure to further confound the interpretation of the optical signals, and the absence of any

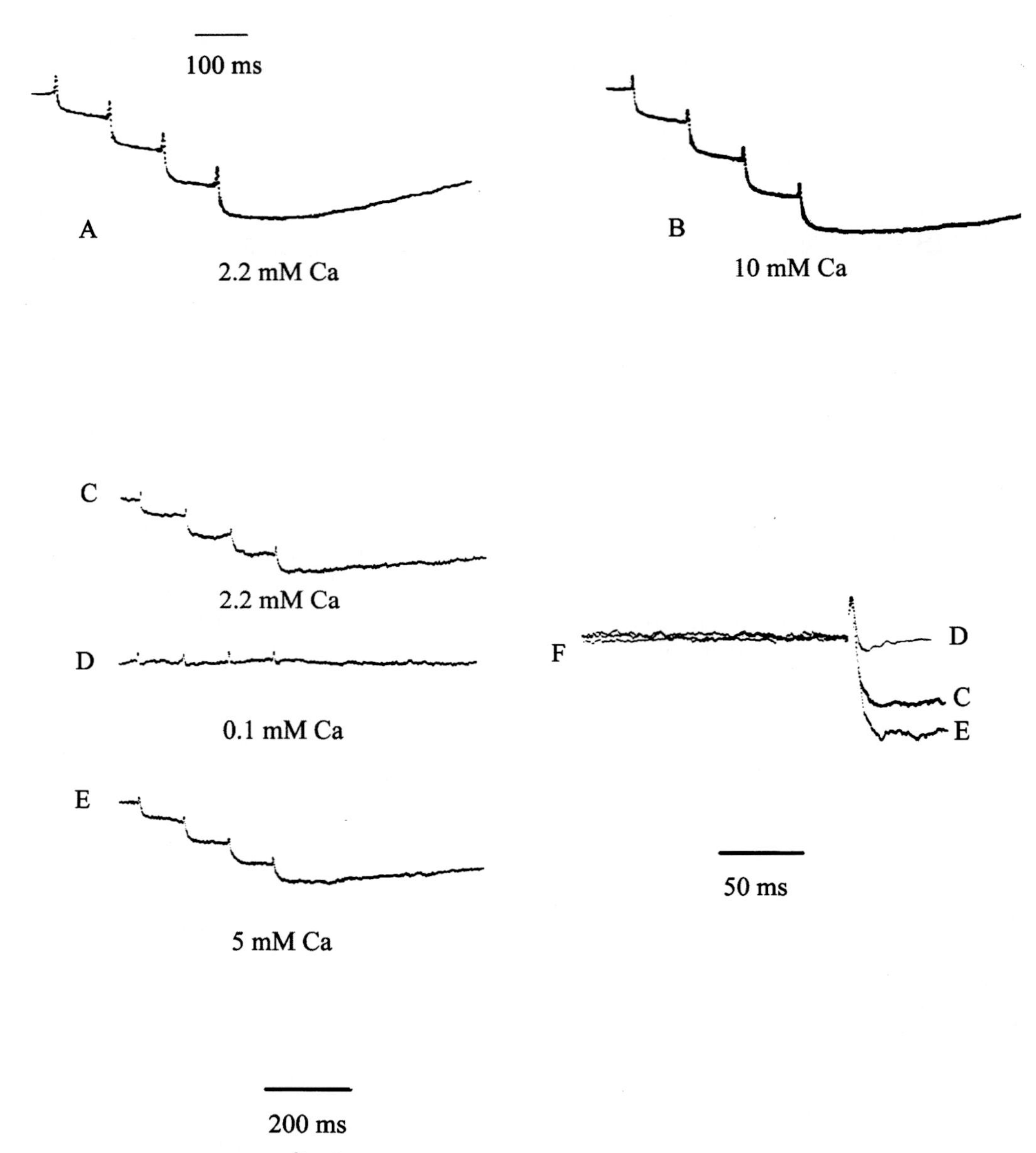

FIGURE 2. The effects of $[Ca^{2+}]_o$ on the amplitude of the S-wave of the light-scattering signal in the mouse neurohypophysis, when account is taken of the number of active terminals. (*A*) Light-scattering changes at 675 ± 26 nm accompanying stimulation at 10 Hz of an unstained mouse neurohypophysis in normal mouse Ringer's solution (2.2 mM Ca^{2+}). (*B*) Light-scattering changes in the same preparation 40 min after exposure to a Ringer's solution containing 10 mM Ca^{2+}. Traces *A* and *B* are each the average of 16 sweeps. (*C*) Light-scattering changes accompanying stimulation at 10 Hz in a different preparation, in normal mouse Ringer's solution (2.2 mM Ca^{2+}). (*D*) Light-scattering changes in the same preparation as *C* recorded 50 min after a reduction in $[Ca^{2++}]_o$ to 0.1 mM by Mg^{2+} substitution. (*E*) Light-scattering changes in the same preparation as *C* recorded 30 min after an increase in $[Ca^{2+}]_o$ to 5 mM. Records *C*, *D*, and *E* were each recorded without averaging. (*F*) Averages of 16 trials recorded under the same conditions as traces *C*, *D*, and *E*. These records were obtained immediately after the corresponding single trains, and are shown here normalized to the height of the E-wave (see text). The normalization factor for each trace is given in parentheses: trace *C* (1.00); trace *D* (0.72); trace *E* (1.16). Objective 10x, 0.4 NA; AC coupling time constant 3 sec; rise time of the light-measuring system (10–90%), 1.1 msec. Temperature 24–26°C.

barrier to the collection of the secretory products for biochemical assay make the neurohypophysis a particularly favorable system for understanding the intrinsic optical changes and how they are related to secretion of neuropeptides.

The aim of this procedure is to measure intrinsic optical signals (light-scattering changes) from mammalian nerve terminals. In practice, these signals are remarkably simple to record from any of the mammalian neurohypophyses (posterior pituitary; pars nervosa) that have been studied. To date, this approach has been used successfully in mouse, rat, and guinea pig. Instrumentation requirements and a protocol for preparation of the neurohypophysis are given below.

MATERIALS

The only requirements (Salzberg et al. 1977, 1985) are a light source that is stable to a few parts in 10^5, a vibration-isolation table, an imaging system such as a microscope, and a simple low-noise photodetector in the image plane of the objective. The latter can be a single silicon PIN photodiode, an avalanche photodiode, an array of photodiodes, a photomultiplier (although this is not recommended as the incident light intensities need to be quite high), or even a scientific CCD camera, having at least 14 bits of gray-scale resolution. (Because photodiode detectors are readily AC coupled, 8-bit resolution of the time-varying component of the signal is generally adequate when photodiodes are employed.)

Caution: *See Appendix 3 for appropriate handling of materials marked with <!>.*

PROCEDURE

Measuring Intrinsic Optical Signals from Mammalian Nerve Terminals

Isolation of the Neurointermediate Lobe (Comprising Neurohypophysis and Pars Intermedia)

1. Sacrifice female CD-1 mice, 30–40 days old, by guillotine and exsanguinate. Pin the head to the bottom of a Sylgard-lined dissection dish and remove the skin from the skull. Then open the skull along the dorsal midline and remove the dorsal cranium bilaterally.
2. Under a low-power dissecting microscope, cut the optic and olfactory nerves and reflect the brain caudally before removing it.

   Note: During this procedure, the infundibular stalk is automatically ruptured, leaving an infundibular stump and the entire pituitary gland in the base of the skull, held in place by thin connective tissue.
3. Circulate oxygenated mouse Ringer's solution (154 mM NaCl, 5.6 mM KCl<!>, 1 mM $MgCl_2$<!>, 2.2 mM $CaCl_2$<!>, 10 mM glucose, 20 mM HEPES, adjusted to pH 7.4 with NaOH<!>) over the pituitary. Use iridectomy scissors and fine (Dumont #5 INOX) forceps to remove the gland.
4. Once the pituitary has been isolated, separate the anterior pituitary (pars distalis) from the neurointermediate lobe (neurohypophysis, or pars nervosa, plus pars intermedia). The pars intermedia consists of a delicate lacework of tissue surrounding the neurohypophysis and provides a convenient site for pinning to the bottom of an experimental chamber.

Experimental Setup

5. Pin the neurohypophysis to the thin Sylgard bottom of a simple chamber so that the infundibular stalk lies clasped between a pair of Pt-Ir electrodes that are coated with Teflon, except where they contact the infundibulum.
6. Deliver brief shocks (100–200 V), lasting between 300 and 500 µsec, through a stimulus isolator.

   Note: The resulting changes in the transparency of the tissue are recorded by the photodetector, which is positioned in the image plane of the optical system behind an adjustable diaphragm, or mask, so that only light passing through the preparation is monitored. With adequate oxygenation, the resulting intrinsic optical signals can easily be recorded for several hours.

CONCLUSION

In the case of mammalian peptidergic neurosecretory terminals, the challenge lies not in detecting the changes in their optical properties, but in divining their meaning. The S-wave of the intrinsic optical signal has its onset within a few milliseconds of the arrival of the action potential, suggesting that it could arise as the result of some calcium-dependent process prior to the fusion of secretory granules and the release of their contents. The light-scattering changes described here may reflect a phase transition of the contents of the secretory granules, e.g., as the result of intragranular pH changes. Another possibility is that the S-wave monitors the contents or status of intraterminal calcium stores and that these reservoirs, located within a few hundred nanometers of the plasma membrane, have a direct role in exocytosis in mammalian neurosecretory terminals. In any event, the weak dependence on wavelength of the signal reflecting secretion (the S-wave), contrasted with the strong wavelength dependence of the extrinsic signals provided by calcium-indicator dyes, permits one to monitor simultaneously, in an AM-dye-loaded preparation, the Ca^{2+} transients in the nerve terminals and the time course of events intimately associated with the release of neuropeptides. The inherently fast responses of the two optical measurements should improve our ability to resolve early events in the coupling of excitation to secretion.

ACKNOWLEDGMENTS

This work was supported by U.S. Public Health Service grants NS-16824 and NS-40966 (B.M.S.).

REFERENCES

Cohen L.B., Keynes R.D., and Landowne D. 1972a. Changes in light scattering that accompany the action potential in squid giant axons; potential-dependent components. *J. Physiol.* **224:** 701–725.

———. 1972b. Changes in light scattering that accompany the action potential in squid giant axons: Current-dependent components. *J. Physiol.* **224:** 727–752.

Douglas W.W. 1963. A possible mechanism of neurosecretion-release of vasopressin by depolarization and its dependence on calcium. *Nature* **197:** 81–82.

———. 1978. Stimulus-secretion coupling: Variations on the theme of calcium activated exocytosis involving cellular and extracellular sources of calcium. *Ciba Found. Symp.* **54:** 61–90.

Douglas W.W. and Poisner A.M. 1964. Stimulus secretion coupling in a neurosecretory organ and the role of calcium in the release of vasopressin from the neurohypophysis. *J. Physiol.* **172:** 1–18.

Gainer H., Wolfe S.A., Jr., Obaid A.L., and Salzberg B.M. 1986. Action potentials and frequency-dependent secretion in the mouse neurohypophysis. *Neuroendocrinology* **43:** 557–563.

Nordmann J.J. 1977. Ultrastructural morphometry of the rat neurohypophysis. *J. Anat.* **123:** 213–218.

Obaid A.L. and Salzberg B.M. 1996. Micromolar 4-aminopyridine enhances invasion of a vertebrate neurosecretory terminal arborization: Optical recording of action potential propagation using an ultrafast photodiode-MOSFET camera and a photodiode array. *J. Gen. Physiol.* **107:** 353–368.

Obaid A.L., Flores R., and Salzberg B.M. 1989. Calcium channels that are required for secretion from intact nerve terminals of vertebrates are sensitive to ω-conotoxin and relatively insensitive to dihydropyridines. Optical studies with and without voltage-sensitive dyes. *J. Gen. Physiol.* **93:** 715–729.

Parsons T.D., Obaid A.L., and Salzberg B.M. 1992. Aminoglycoside antibiotics block voltage-dependent calcium channels in vertebrate nerve terminals. *J. Gen. Physiol.* **99:** 491–504.

Poulain D.A. and Wakerley J.B. 1982. Electrophysiology of hypothalamic magnocellular neurons secreting oxytocin and vasopressin. *Neuroscience* **7:** 771–808.

Salzberg B.M. and Obaid A.L. 1997. Triggered calcium release from intraterminal stores may play a direct role in neuropeptide secretion: Evidence from light scattering in mammalian nerve terminals. *Biophys. J.* **72:** A227.

Salzberg B.M., Obaid A.L., and Gainer H. 1985. Large and rapid changes in light scattering accompany secretion by nerve terminals in the mammalian neurohypophysis. *J. Gen. Physiol.* **86:** 395–411.

———. 1986. Optical studies of excitation and secretion at vertebrate nerve terminals. In *Optical methods in cell physiology* (ed. P. DeWeer and B.M. Salzberg), pp. 133–164. Society of General Physiologists and Wiley-Interscience, New York.

Salzberg B.M., Grinvald A., Cohen L.B., Davila H.V., and Ross W.N. 1977. Optical recording of neuronal activity in an invertebrate central nervous system: Simultaneous monitoring of several neurons. *J. Neurophysiol.* **40:** 1281–1291.

CHAPTER 66

Imaging with Quantum Dots

Jyoti K. Jaiswal,[1] Ellen R. Goldman,[2] Hedi Mattoussi,[3] and Sanford M. Simon[1]

[1]*The Rockefeller University, New York, New York 10021;* [2]*Center for Bio/Molecular Science and Engineering and* [3]*Division of Optical Sciences, U.S. Naval Research Laboratory, Washington, D.C. 20375*

Fluorescence microscopy is a commonly used approach for high-resolution biological imaging. However, existing fluorophores impose a number of limitations on what can be achieved by using this approach. Fluorescent quantum dots (QDs) are novel inorganic fluorophores that overcome many of the limitations of conventional fluorophores. This chapter describes their use for cellular and molecular tagging and discusses the utility of QDs for long-term and multispectral imaging in biology.

Monitoring signaling and interactions as cells grow and differentiate is the key to understanding organismal development. However, investigating these processes in multicellular organisms is a challenge; the intricacy of intercellular interactions is further complicated by the vast number of intracellular processes. As cellular interactions occur and change over time, it is necessary to use approaches that will allow not only simultaneous imaging of multiple biological interactions, but also imaging them over long periods. Fluorescence microscopy is among the most commonly used approaches for high-resolution, noninvasive imaging of live organisms (Emptage 2001; Stephens and Allan 2003). Organic fluorophores are the most commonly used tags for fluorescent imaging applications (Haugland 2002).

Despite their considerable advantages, organic fluorophores have several limitations in live cell imaging, including sensitivity to photobleaching, metabolic degradation, poor spectral versatility, and the need for customized approaches for their bioconjugation. Together, these limit the use of fluorescence imaging for long-term and simultaneous monitoring of multiple processes in live cells and tissues. Inorganic fluorescent nanocrystals, referred to as quantum dots, provide an alternative to organic fluorophores by circumventing these limitations (Jaiswal et al. 2003; Wu et al. 2003). This chapter focuses on the use of QDs for fluorescent labeling of cells and labeling of proteins in live cells.

The following features of QDs make long-term imaging and simultaneous in vivo imaging of multiple cells and molecules possible:

1. QDs are resistant to photodamage.
2. Many different QDs can be imaged simultaneously due to
   - narrow (<50 nm) emission profiles which reduce spectral overlap between QDs of different colors, and
   - the ability to use a single excitation line for several different QDs.
3. There is a common strategy for conjugating biomolecules to QDs of any color.
4. Enhanced brightness (for two-photon imaging, they are more than two orders of magnitude brighter than the best known organic fluorophores) (Larson et al. 2003).

MATERIALS

Fluorescence Microscope

QDs can be imaged using any type of fluorescence microscope, including epifluorescence, confocal, and multiphoton. However, unlike conventional fluorophores, a single wavelength of light can be used to excite QDs of several different colors. Since most commercially available QDs emit in the green-to-red region of the visible spectrum, a microscope capable of providing an excitation beam (from lamp or laser) in the UV to blue (~400-nm) region of the spectrum and capable of resolving multiple emission wavelengths could be used. As QDs are better excited by UV light, fixed cells can be imaged using a UV-light source. To minimize UV-induced photodamage, live cells should be imaged using a blue excitation light. For two-photon imaging, excitation at 800 nm is optimal, but any wavelength of light between 700 and 1000 nm could be also be used (Larson et al. 2003). The choice of emission filter would depend on the emission spectrum of the quantum dot in use.

Water-soluble Quantum Dots

There are at least two commercial suppliers of water-soluble QDs, namely, Quantum Dot Corporation and Evident Technology. QD Corporation provides QDs conjugated to avidin for use with biotinylated proteins and antibodies. Evident Technology offers QDs that can be conjugated with the amino or carboxyl terminus of a protein. QDs conjugated to specific antibodies can also be obtained from these suppliers. However, QDs are often synthesized and conjugated to specific biomolecules by researchers themselves. We use CdSe/ZnS QDs, which are rendered water-soluble by capping with DHLA (dihydroxylipoic acid) (Mattoussi et al. 2000). DHLA also causes the QD surface to be negatively charged, a property that has been used to conjugate QDs with biomolecules using the approach described in Procedure below.

Cells or Tissue for Labeling

QDs can be used to tag live cells and to label cell-surface proteins in live cells. They can also be used for labeling fixed cells and tissue sections. Depending on the application, the sample should be appropriately prepared before labeling with QDs.

PROCEDURE

Bioconjugation of the Quantum Dots

Depending on the surface chemistry of the QDs being used, the approach for bioconjugation can be varied. DHLA-capped QDs can be conjugated to proteins using positively charged adapters (Mattoussi et al. 2001). The adapter could be (1) naturally charged, e.g., avidin (Goldman et al. 2002); (2) a positively charged leucine zipper peptide (zb) (Mattoussi et al. 2001); or (3) a pentahistidine peptide (5x His) (Medintz et al. 2003). The commercially available QDs from QD Corporation are supplied in an avidin-conjugated form, and those from Evident Technologies are supplied with reagents for in vitro conjugation to avidin or other proteins. Both companies provide specific protocols for conjugating proteins to their QDs. The use of avidin permits stable conjugation of the QDs to ligands, antibodies, and other proteins, or other molecules that can be biotinylated. However, when using these QDs, it should be borne in mind that the presence of multiple avidin molecules on a single QD could lead to the production of a single fluorescent tag for multiple antibodies. For live cell staining, this could also cause aggregation of molecules, leading to altered activity, localization, or even inactivation. This problem can be overcome by using the mixed surface conjugation approach, as described by Mattoussi et al. (2001) (schematic in Fig. 1). This involves conjugating the QD to avidin in the presence of a molar excess of zb peptide-containing maltose-binding protein (MBP-zb). Due to their net positive charge, these proteins compete with each other to bind the negatively charged DHLA coat on the surface of the QD. Thus, by altering the ratio of these proteins in the mixture, their relative numbers on each QD can be regulated. This provides the potential to attach one biomolecule to each QD (Goldman et al. 2002).

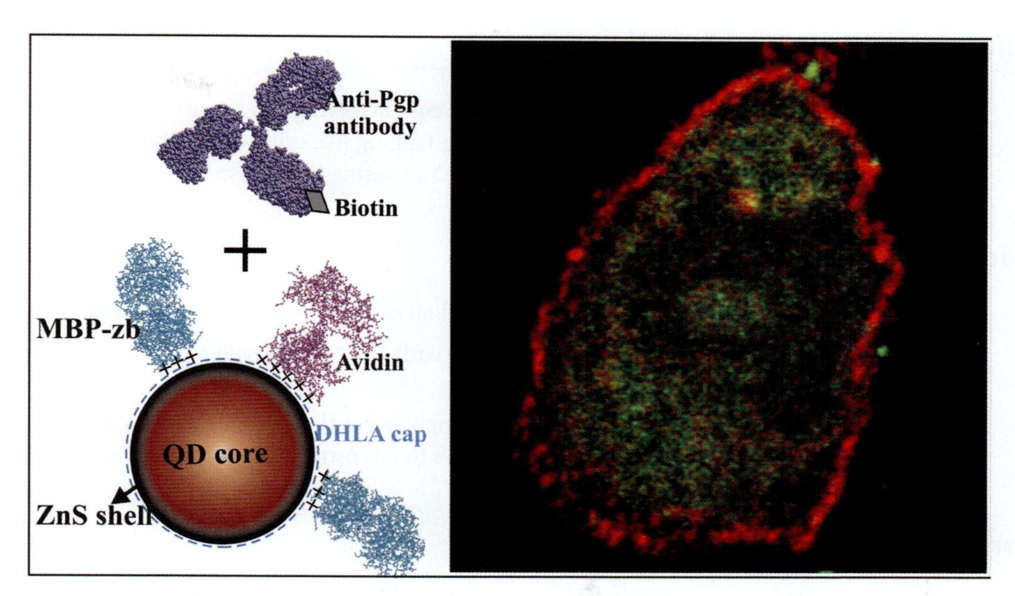

FIGURE 1. (*Left panel*) Mixed surface approach for conjugating avidin to DHLA-capped QDs. The MBP-zb molecules are used in molar excess to the avidin molecules, which allows regulation of the number of avidin molecules present on each QD. (*Right panel*) Optical section of a live cell expressing Pgp-GFP (*green*), with the cell surface labeled using QD-avidin (*red*). Imaging was performed using a Zeiss LSM510 microscope with a 63x water-immersion objective, 488-nm laser excitation, and 505–550-nm and 570–615-nm bandpass emission filters.

The protocol for bioconjugation varies depending on the surface properties of the hydrophilic QD used. This protocol details the mixed surface approach for bioconjugating DHLA-capped QDs.

This procedure requires the following reagents: DHLA-capped QD; Maltose-binding protein-zb (MBP-zb); ProteinG-zb (PG-zb). (Synthesis of QD, MBP-zb, and PG-zb proteins was carried out as described in Mattoussi et al. 2000 and references therein.)

1. Mix 100 pmoles of QD with 200 pmoles of avidin (Sigma) or PG-zb and 600 pmoles of MBP-zb, respectively, and adjust the final volume to 200 µl with 10 mM sodium tetraborate buffer (pH 9.0).
2. Allow the mixture to stand for 15 min at room temperature.
3. Add another 150 pmoles of MBP-zb to the mixture and stand at room temperature for a further 15 min.
4. Set up 500 µl of amylose (New England Biolabs) column, equilibrated using the sodium tetraborate buffer.
5. Load all of the QD bioconjugate mix onto the column and wash twice with the sodium tetraborate buffer.
6. Add 200 pmoles of the biotinylated molecule of interest (in case of avidin-conjugated QDs) or specific antibody (in case of PG-zb-conjugated QDs) to the column and allow it to stand for 30–60 min at room temperature.
7. Wash the column twice with sodium tetraborate buffer and elute using 10 mM maltose until all of the QDs are eluted from the column (QD elution can be easily monitored by placing the column in UV light and monitoring the resulting QD fluorescence).

   Note: This approach provides a pure population of conjugated QDs, free of unbound QDs, and the unbound biomolecules.

USING QDs TO LABEL LIVE CELLS

QDs are efficient tags for experiments involving long-term labeling of live cells (Jaiswal et al. 2003). Several approaches can be used for tagging live cells with QDs, including endocytosis, carrier peptide, binding to the cell surface, transfection using lipid-based reagent, and microinjection.

Endocytosis

This approach can be used to label all cells that are capable of endocytosis.

1. For labeling, incubate the cells with 1 μM DHLA-capped QDs in the appropriate growth media for 2–3 hours.
2. Remove excess QDs by washing several times with growth media or an appropriate buffer. This leads to localization of the QDs to endosomes (Fig. 2) (Jaiswal et al. 2003).

Carrier Peptide

This approach involves the use of a carrier peptide, which aids the uptake of QDs. It uses the Pep-1 (KETWWETWWTEWSQPKKKRKV) peptide, which has been shown to efficiently transport protein molecules into the cell (Morris et al. 2001).

1. For labeling, incubate the cells for 1 hour in serum-free media containing a preformed Pep1-QD complex (10 μM peptide and 100–500 nM QDs in 100 μl of serum-free media).
2. Remove excess complex by washing several times with growth media.

Although use of this approach allows cellular labeling using less QDs, it still appears to be dependent on the endocytic ability of cells, as labeling is abrogated in cells incubated at 4°C (J.K. Jaiswal et al., unpubl.).

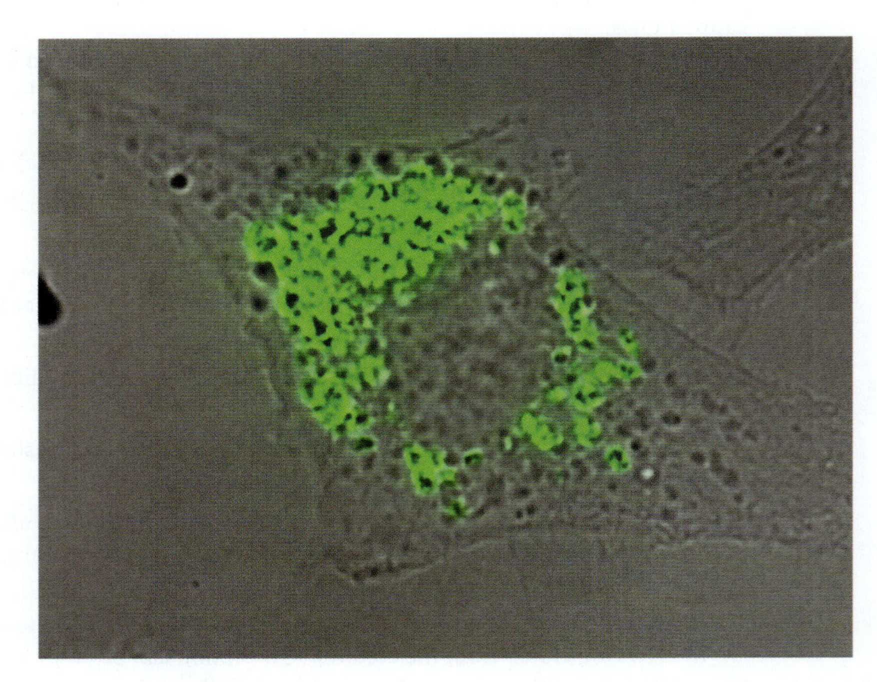

FIGURE 2. Overlay of brightfield and epifluorescence (*green*) image of HeLa cells, incubated for 2 hr at 37°C in complete medium (DMEM containing 10% fetal bovine serum [GIBCO-BRL]) with 1 μM DHLA-capped QDs (530-nm peak emission), followed by 15 min in complete medium without any QDs.

Cell-Surface Labeling

This approach can be used to label cells independent of endocytic activity (thus, it can also be used for prokaryotic cells).

1. Wash the cells free of growth media and incubate for 30 min at 4°C in 1 mg/ml solution of Sulfo-NHS-SS biotin (Pierce) in phosphate-buffered saline (PBS; Sigma).
2. Quench excess biotin by washing the preparation with Tris-buffered saline (TBS, 8 g of NaCl, 0.2 g of KCl<!>, and 3 g of Tris base in 1 liter of water, pH 7.4).
3. Incubate the cells in a serum-free medium containing 0.5–1 μM avidin-conjugated QDs for 10 min. The unbound QDs are removed by repeated washing.

Caution: *See Appendix 3 for appropriate handling of materials marked with <!>.*

Lipid-based Reagent

This approach also allows efficient and rapid labeling of cells irrespective of their ability to endocytose. Incubate 100 pmoles of DHLA-capped QDs in 100 μl of serum-free medium, containing a lipid-based transfection reagent appropriate for transfecting 3 μg of plasmid DNA. Lipofectamine 2000 (Invitrogen) and Fugene 6 (Roche) transfection reagents have been used to label tumor cells (Voura et al. 2004).

Microinjection

This approach is useful for localized labeling of neurons or other cells in situ. Carry out microinjection using 1–10 μM solution of QDs. This approach has been used to label isolated *Xenopus* eggs, which subsequently undergo normal development (Dubertret et al. 2002).

LABELING SPECIFIC PROTEINS WITH QDs

Bioconjugated QDs can be used for specific labeling of cellular proteins (Jaiswal et al. 2003; Wu et al. 2003). Specific proteins or cells can be tagged with QDs using one of two different approaches: One approach is to first conjugate the QD with either ligands or antibodies, and then to incubate the bioconjugate with the cells. A second approach is to first incubate the cells with a biotinylated primary antibody or ligand of interest and then, after washing, introduce an avidin-conjugated QD to bind these molecules (Fig. 1). Both of these approaches lead to specific labeling of intracellular and cell surface proteins (Jaiswal et al. 2003; Wu et al. 2003). However, in live cells, these approaches permit labeling of the cell surface proteins only (Fig. 1) (Jaiswal et al. 2003).

EXAMPLE OF APPLICATION

QDs can be used to tag cells or specific cellular components. The following example uses QDs for cell-specific labeling.

P-glycoprotein (Pgp) is a broad specificity drug efflux pump present on the surface of most cancer cells (Chen and Simon 2000). To specifically label Pgp proteins expressed on the cell surface using QDs, HeLa cells are transiently transfected with a plasmid encoding Pgp protein fused to the green fluorescent protein (Pgp-GFP). After 48 hours, the cells are incubated for 30 min in media containing the biotinylated monoclonal antibody (4E3), specific to the extracellular epitope of Pgp. Following this, cells are incubated for 15 min with QDs (peak emission 570 nm) conjugated with avidin, as described in Procedure A (see also Fig. 1). Similar results are obtained using non-biotinylated 4E3 antibodies and QDs conjugated to protein G-zb (Jaiswal et al. 2003).

REFERENCES

Chen Y. and Simon S.M. 2000. In situ biochemical demonstration that P-glycoprotein is a drug efflux pump with broad specificity. *J. Cell Biol.* **148:** 863–870.

Dubertret B., Skourides P., Norris D.J., Noireaux V., Brivanlou A.H., and Libchaber A. 2002. In vivo imaging of quantum dots encapsulated in phospholipid micelles. *Science* **298:** 1759–1762.

Emptage N.J. 2001. Fluorescent imaging in living systems. *Curr. Opin. Pharmacol.* **1:** 521–525.

Goldman E.R., Balighian E.D., Mattoussi H., Kuno M.K., Mauro J.M., Tran P.T., and Anderson G.P. 2002. Avidin: A natural bridge for quantum dot-antibody conjugates. *J. Am. Chem. Soc.* **124:** 6378–6382.

Haugland R.P. 2002. *Handbook of fluorescent probes and research products.* Molecular Probes, Eugene, Oregon.

Jaiswal J.K., Mattoussi H., Mauro J.M., and Simon S.M. 2003. Long-term multiple color imaging of live cells using quantum dot bioconjugates. *Nat. Biotechnol.* **21:** 47–51.

Larson D.R., Zipfel W.R., Williams R.M., Clark S.W., Bruchez M.P., Wise F.W., and Webb W.W. 2003. Water-soluble quantum dots for multiphoton fluorescence imaging in vivo. *Science* **300:** 1434–1436.

Mattoussi H., Mauro J.M., Goldman E.R., Anderson E.H., Sundar V.C., Mikulec F.V., and Bawendi M.G. 2000. Self-assembly of CdSe-ZnS quantum dot bioconjugates using an engineered recombinant protein. *J. Am. Chem. Soc.* **122:** 12142–12150.

Mattoussi H., Mauro J.M., Goldman E.R., Green T.M., Anderson G.P., Sundar V.C., and Bawendi M.G. 2001. Bioconjugation of highly luminescent colloidal CdSe-ZnS quantum dots with an engineered two-domain recombinant protein. *Phys. Stat. Sol.* **224:** 277–283.

Medintz I.L., Clapp A.R., Mattoussi H., Goldman E.R., Fisher B., and Mauro J.M. 2003. Self-assembled nanoscale biosensors based on quantum dot FRET donors. *Nat. Mater.* **2:** 630–638.

Morris M.C., Depollier J., Mery J., Heitz F., and Divita G. 2001. A peptide carrier for the delivery of biologically active proteins into mammalian cells. *Nat. Biotechnol.* **19:** 1173–1176.

Stephens D.J. and Allan V.J. 2003. Light microscopy techniques for live cell imaging. *Science* **300:** 82–86.

Voura E.B., Jaiswal J.K., Mattoussi H., and Simon S.M. 2004. Tracking metastatic tumour cell extravasation with quantum dot nanocrystals and fluorescence emission scanning microscopy. *Nat. Methods* (in press).

Wu X., Liu H., Liu J., Haley K.N., Treadway J.A., Larson J.P., Ge N., Peale F., and Bruchez M.P. 2003. Immunofluorescent labeling of cancer marker Her2 and other cellular targets with semiconductor quantum dots. Nat. Biotechnol. **21:** 41–46.

CHAPTER 67

Imaging Single Receptors with Quantum Dots

Sabine Lévi,[1] Maxime Dahan,[2] and Antoine Triller[1]

[1]*Biologie Cellulaire de la Synapse, Inserm U497, ENS, Paris, France;* [2]*Laboratoire Kastler Brossel, CNRS UMR8552, ENS, Paris, France*

This chapter describes a highly sensitive approach for tracking the motion of membrane molecules over extended time periods with single-molecule resolution. The technique uses nanometer-sized fluorescent ligands, covalently linked to the extracellular part of the object to be followed. Single-fluorophore epifluorescence imaging then informs on the membrane diffusion of the particle of interest. This approach has been used successfully in rat spinal cultured neurons to follow the membrane diffusion of the endogenous inhibitory glycine receptor (GlyR) (Dahan et al. 2003).

MATERIALS

This procedure used an inverted microscope (Olympus, IX70) equipped with a 60× objective (NA = 1.45, Olympus). QDot-605 and FM 4-64 were detected using a mercury arc lamp (excitation filter 525DF45) and appropriate emission filters (respectively, 595DF60 and 695AF55 from Omega Filters). For CY3 excitation, to achieve single fluorochrome detection, the sample was illuminated with a frequency-doubled YAG Crystal laser at 532 nm (~0.5 kW/cm^2). For detection of single QDot, the sample was illuminated with a mercury arc lamp. QDot and CY3 fluorescent images were obtained with an integration time of 75 and 100 msec, respectively, with a CCD camera (Micromax 512EBFT, Roper Scientific) with up to 1024 consecutive frames acquired with Metaview (Universal Imaging). Single-molecule trajectories were analyzed with custom-written program in MatLab (MathWorks).

Caution: *See Appendix 3 for appropriate handling of materials marked with <!>.*

PROCEDURE

Imaging Single Receptors with Quantum Dots

1. Use the following medium for incubations, washes, and imaging at 37°C: MEM Eagle medium (Eagle 1959) without phenol red, but containing 4 mM $NaHCO_3$, 10 mM HEPES, 6 g/liter glucose, 2 mM glutamine, 1 mM Na^+ pyruvate, and 1x B27 supplement.
2. Incubate the cells for 5 min with high dilution (~1 µg/ml) of primary antibody to label a small number of molecules.

3. After extensive washes, incubate the cells for 5 min in biotinylated secondary Fab antibody (10 µg/ml, Fab/Biotin ratio 1:1).
4. Following washes, incubate the neurons for 1 min in QDot-605-Streptavidin conjugate (~0.5 nM, Quantum Dot Corporation) in borate buffer (50 mM, pH 8–8.5) supplemented with 2% bovine serum albumin (BSA) and 200 mM sucrose.

   Note: The use of borate buffer is necessary to avoid nonspecific QDot labeling. If sufficient primary antibody is available, steps 3–4 can be avoided, and the primary antibody bound directly to QDots (IgG/QDot ratio 1:1).
5. Label presynaptic boutons for 30 sec with 1 µM FM4-64 in 40 mM KCl<!>. Wash and image cells.

EXAMPLE OF APPLICATION

GlyR lateral diffusion was studied in cultured neurons using a single-particle tracking (SPT) approach (Dahan et al. 2003). Data obtained with Cy3 fluorophore coupled directly to primary antibody (Fig. 1A) were compared with data obtained using QDot-605-Streptavidin conjugates (Fig. 1B,C). Cy3- and QDot-tagged receptors were detected in extrasynaptic and synaptic regions. Individual Cy3 molecules were identified by their single-step bleaching. Cy3-GlyR could be tracked only for short

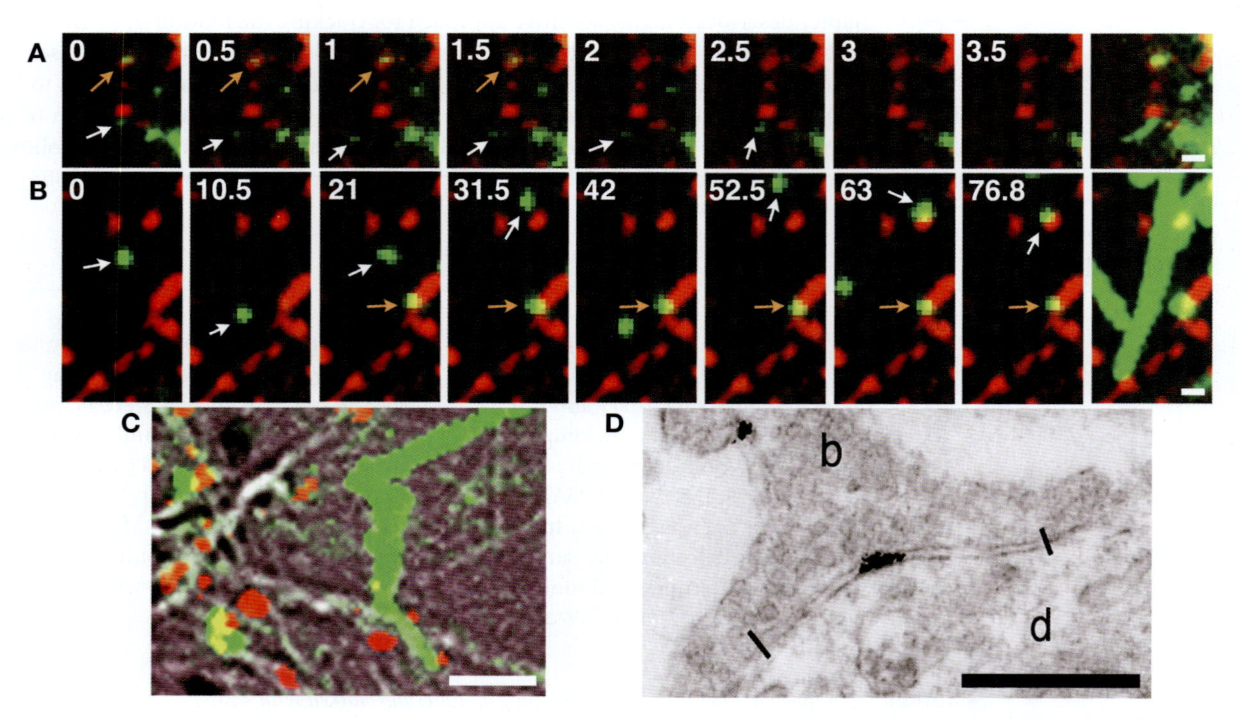

FIGURE 1. Comparison of lateral GlyR motion analyzed with Cy3 and QDot probes. GlyRs were detected in cultured neurons with Cy3 (*A*) or QDot-Streptavidin (*B–C*). (*Green*) GlyR spots; (*red*) FM4-64-labeled synaptic boutons. (*A–B*) Images were extracted from a sequence of 35 and 1024 images with an acquisition time of 100 msec and 75 msec for Cy3 and QDot, respectively. The time (sec) is indicated on each image. The last image is a maximum projection of the entire stack of images corresponding to GlyR trajectory. (*C*) Projection of time-lapse recording (1 Hz, 20 min) of QDot-GlyR trajectory (*green*) overlaid with FM 4-64 staining (*red*) and bright-field image. (*A–C*) Cy3 and QDot diffused rapidly in the extrasynaptic region (white arrows), whereas synaptic GlyRs were stable (*orange arrows*). Note the short membrane surface explored by Cy3-GlyR compared to QDot-GlyR. Long QDot imaging duration enabled the observation of a synaptic entry (*white arrow* in *B*). (*D*) EM detection of QDot-GlyR within the synaptic cleft. (d), dendrites; (b) synaptic boutons. The edges of the cleft are outlined. Bars: (*A–B*) 1 µm; (*C*) 5 µm; (*D*) 500 nm.

duration (~2.5 sec in Fig. 1A). In contrast, the photostability of QDots allowed QDot-GlyR trajectories to be visualized (Fig. 1B,C) for unprecedented durations (20 min, Fig. 1C). Single QDots were identified by the random intermittency of their fluorescence emission (Nirmal et al. 1996). For example, one QDot (white arrow in Fig. 1B) temporarily disappeared after 31.5 sec of recording. Long imaging duration enabled the observation of exchanges between extrasynaptic and synaptic domains, in which a GlyR alternated between free and confined diffusion states, respectively (white arrow in Fig. 1B). Silver-intensified and gold-toned QDots were detected using transmission electron microscopy (EM) with the same QDot immunolabeling protocol (Fig. 1D). EM analysis provided evidence that QDot-GlyR could access the core of the synapse (Fig. 1D).

ADVANTAGES AND LIMITATIONS

QDot staining is easy, fast, and nontoxic and gives low background fluorescence. QDots come in many colors and wide absorption spectra, yet narrow emission spectra, facilitating multicolor detection. Moreover, QDots, which are ~5–10 nm in diameter, intermediary between latex beads (500 nm) and conventional fluorophores (~1–4 nm), can still access confined cellular domains such as the synaptic cleft. As exemplified above, the photostability of QDots significantly extended the duration of recording periods. Another key feature of QDots is their strong fluorescence. With an integration time of 75 msec, the spots are detected with a signal-to-noise ratio of about 50 (an order of magnitude over traditional fluorophores). As a result, the lateral resolution with which individual QDot spots can be localized reaches 5–10 nm, well below the 40 nm achieved with Cy3 (Dahan et al. 2003). Furthermore, QDots can be excited with a mercury lamp instead of a laser beam, which is required for single Cy3 detection. Consequently, multiple individual fluorescent spots can be observed simultaneously. In addition, the same QDot probe can be used at both the optical and the electron microscope levels.

There are two main limitations to this approach. First, QDot blinking makes the analysis of long trajectories more complex, requiring sophisticated software. Second, QDots have not been coupled directly to a protein expressed from a transgene. However, it is now possible to couple QDots to free sulfhydryl groups of proteins (Quantum Dot Corporation coupling kit). Altogether, QDots are the tools of choice for SPT and many other ultrasensitive studies of cellular processes.

REFERENCES

Dahan M., Lévi S., Luccardini C., Rostaing P., Riveau B., and Triller A. 2003. Diffusion dynamics of glycine receptors revealed by single-quantum dot tracking. *Science* **302:** 442–445.

Eagle H. 1959. Amino acid metabolism in mammalian cell cultures. *Science* **130**: 432–437.

Nirmal M., Dabbousi B., Bawendi M.G., Macklin J.J., Trautman J.K., Harris T.D., and Brus L.E. 1996. Fluorescence intermittency in single cadmium selenide nanocrystals. *Nature* **383:** 802–804.

CHAPTER 68

A Practical Guide: Tracking Receptors by Imaging Single Molecules

Laurent Cognet,[1] Brahim Lounis,[1] and Daniel Choquet[2]

[1]*Centre de Physique Moléculaire Optique et Hertzienne, CNRS UMR 5798 et Université Bordeaux 1, 33405 Talence, France;* [2]*Laboratoire de Physiologie Cellulaire de la Synapse, CNRS UMR 5091 et Université Bordeaux 2, Institut François Magendie, 3077 Bordeaux, France*

This chapter describes imaging techniques using single optical labels, ranging from fluorescent dyes to scattering particles, for the study of the movement of individual or small assemblies of membrane proteins.These techniques have been used to track the movements of different types of plasma membrane proteins such as neurotransmitter receptors and adhesion proteins. They can be used to probe the degree of interaction between membrane proteins and cytoplasmic stabilizing elements in live cells. They have been used in a variety of cell culture preparations, such as cell lines, and primary hippocampal and spinal cord cultures (see Choquet and Triller 2003).

MATERIALS

Labels

For single-particle tracking (SPT) (Schnapp et al. 1988; Saxton and Jacobson 1997), particles are gold colloids at least 40 nm in diameter (nanogold) or latex, polystyrene, or silica particles, 200–1000 nm in diameter (Polyscience). These particles are visualized under video-enhanced DIC microscopy. Fluorescent labels such as semiconductor quantum dots (QDs, ~15 nm total diameter, including the biocompatible shell, QDot Corp.) and fluorescent latex particles (20–200 nm, Molecular Probes) are used for visualization under epifluorescence. For single-fluorophore detection (SFD) (Schmidt et al. 1996; Moerner and Orrit 1999) under epifluorescence, common labels are cyanine dyes<!> (Cy3 or Cy5, Amersham) or Alexa dyes<!> (Molecular Probes). For a summary of the available labels and examples of their applications, see Table 1.

Setup

This procedure uses an inverted microscope equipped with a 1.4-NA 100× oil-immersion objective (Olympus), DIC optics with an oil-immersion condenser, and epifluorescence. For SPT, high-resolution DIC images are acquired with fast digital cameras (e.g., Photometrics HQ). For SFD, a sensitive and rapid camera is used (e.g., Pentamax intensified CCD or Micromax back-illuminated with frame transfer, Roper Instruments). Fluorescence excitation sources are CW argon (Coherent), frequency-doubled Nd:Yag (Coherent), or He:Ne (JDS Uniphase) lasers<!>. Excitation laser beams enter through the fluorescence epiport and illuminate an area of 10 μm of the sample. Illumination intensities are of a few kW/cm^2. For each dye, an appropriate set of filters (Chroma) is required.

The total detection efficiency of a typical experimental SFD setup is in the range of 5–10%.

Caution: *See Appendix 3 for appropriate handling of materials marked with <!>.*

PROCEDURES

The general outline of a single fluorophore or particle-imaging experiment is first the attachment of the label to the receptor of interest through a specific high-affinity ligand (a natural ligand, an antibody, or a toxin) that recognizes the extracellular domain of the receptor in live cells. This must be performed at low enough label density to ensure that spots from single molecules or particles are resolved. A set of criteria (see below Signal Characterization and Sample Preparation) ensures that single labels are imaged. The use of monovalent labels (toxins or Fab fragments of antibodies) or low coating densities of ligands on particles helps ensure that labels are bound to single receptors. Imaging of the label, reflecting the behavior or properties of the underlying receptor, is then followed by image analysis. Bear in mind that it is usually impossible to know whether the label reveals the movement of individual or clustered receptors.

PROCEDURE A

Preparing the Label

For SPT, particles are coated with ligands via a sandwich of linkers to separate the ligand from the particle surface. This improves binding specificity. For example, sulfonated latex particles can be coated first by passive absorption of protein G (Sigma; 1 mg/ml for a 0.2% solution of 500-nm-diameter particles).

A primary anti-Fc antibody is then bound to protein G (0.1 mg/ml). This allows the coating of the bead with the anti-receptor antibody (1–100 µg/ml) oriented adequately to bind to the receptors on the cell surface. The surface density of antibodies on the bead is adjusted by titration with free Fc fragments. Prepared particles are then incubated for short times (5–10 min) with the live cells in the presence of 0.3% bovine serum albumin (BSA) to prevent nonspecific binding. Optical tweezers, formed by a focused Nd:Yag laser<!>, can be used to manipulate the particles in solution and place them in contact with the cells at known locations and times (Sterba and Sheetz 1998).

TABLE 1. Labels and areas of application for single particle tracking and single molecule detection

Labels	Latex	Gold	QDs	XFPs	Dyes
Typical applications	diffusion	diffusion	diffusion localization	localization stoichiometry interactions	diffusion localization interactions
Size	200–1000 nm	40 nm	10–15 nm	2 × 4 nm	<1 nm
Coupling	passive or chemical linkage	passive absorption	chemical linkage	genetic fusion	chemical linkage
Excitation	halogen lamp	halogen lamp	laser light or spectral lamps	laser light	laser light
Detection	DIC and video camera	DIC and video camera	low-light-level camera	low-light-level camera	low-light-level camera
Length of observation	5–10 min	5–10 min	few minutes	seconds	seconds
Precision of detection	1–10 nm	5–20 nm	10–40 nm	20–50 nm	20–50 nm

For SFD, dyes are linked to the ligand at a mean 1:1 (or lower) ratio by chemical reactions such as a NHS-ester reaction to free amines on an antibody. Receptors are then labeled by incubating live cells for a few minutes with the fluorinated antibody (0.1–1 μg/ml). Ligands must be sufficiently dilute that the signal arising from a molecule is optically resolved (typically, 1 molecule/μm^3, ~1 nM) and also that the signal from a single molecule is above that of all background sources (noise of the detector, autofluorescence from the sample). Use caution when using low auto-fluorescent immersion oils (FF type, Cargille Lab.) and acid cleaned coverslips (1 M HCl<!> for 15 min).

PROCEDURE B

Signal Characterization and Sample Preparation

In SFD experiments, it is important to thoroughly characterize the signal arising from the label-ligand couple immobilized on glass or embedded in a gel or polymers before performing experiments in cells. The signal arising from a single fluorophore should show digital photobleaching: It must be constant over time, and then drop instantaneously to background levels. All molecules should have comparable fluorescence levels. The signal arising from an immobile single molecule should display a diffraction-limited spot. In cells, the signal characteristics should be similar. For SPT or SFD, cells must be cultured on glass coverslips. Experiments are performed on labeled cells at 37°C in HEPES-buffered culture medium in a sealed chamber to prevent evaporation. Cells are used for no more than 30 min.

PROCEDURE C

Image Acquisition and Analysis

Images are acquired at fast rates (5–30 Hz). For SFD, a fast shutter in the excitation path prevents sample illumination in between frames to limit dye photobleaching.

For image analysis, particle or single dye identification is performed by custom-made or commercial softwares (e.g., Metamorph, Universal Imaging). Objects are fitted by comparison with reference signals. This allows for subwavelength localization of the labels, typically from below 50 nm for single dyes to 1 nm for SPT. Trajectories are then reconstructed on image stacks by connecting object positions from one image to the next. Low labeling densities are preferred to facilitate this process.

EXAMPLE OF APPLICATION

Figure 1 shows examples of traces recorded with both SPT and SFD used to detect α-amino-hydroxy-5-methyl-4-isoazole (AMPA)-subtype glutamate receptor movements in live cultured hippocampal neurons. AMPA receptors have been labeled with an antibody directed against the extracellular amino terminus of the GluR2 subunit (Chemicon). Long-term observation of receptor diffusion in the extrasynaptic membrane was observed by tracking 0.5-μm latex beads attached to the receptors (Fig. 1a,b), whereas receptor diffusion in and out of synapses could be observed by tracking Cy3–anti-GluR2-labeled receptors (Fig. 1c,d). These experiments reveal that AMPA receptors diffuse in both the synaptic and extrasynaptic membranes.

ADVANTAGES AND LIMITATIONS

The main advantage of particle and molecule tracking approaches is that they eliminate the implicit ensemble averages of conventional optical observations, gaining access to heterogeneity, dynamic fluctuations, diffusion, reorientation, colocalization, and conformational changes at the molecular level. Because labels can be localized at resolutions well below the diffraction limit, diffusion coefficients can be measured over a wide dynamic range (Saxton and Jacobson 1997; Cognet et al. 2002).

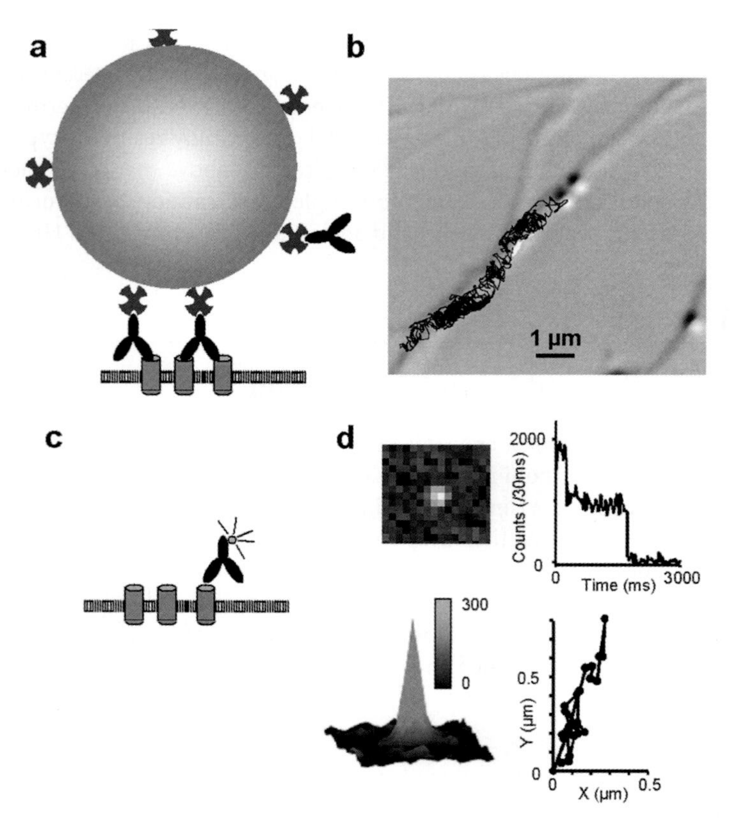

FIGURE 1. Imaging single AMPA receptor movements in neuronal membranes. (*a*) Schematic drawing of the attachment of a latex particle to receptors via antibodies directed against the extracellular amino terminus of the protein, as used in SPT experiments. (*b*) Superimposed images of cultured hippocampal neurons (8 days in vitro) and of the trajectory on a neurite of a 0.5 μm latex bead bound to surface AMPA receptors. The recording lasted 3 min, during which the bead-bound receptors explored an area of 5 μm^2. (*c*) Schematic drawing of the attachment of a single fluorescently labeled antibody to a receptor, as used in SFD experiments. (*d, top left*) Image of the fluorescent spot from a single molecule. (*d, top right*) Recording over time of the intensity of a fluorescence spot. The intensity displays characteristic sequential discrete photobleaching steps, indicating that the spot initially contained two individual fluorophores. (*d, bottom left*) Display of the intensity profile (*X,Y* are space, *Z* intensity) of a single fluorescent Cy5 molecule. The profile is diffraction-limited and matches the point spread function of the optical system. (*d, bottom right*) Example of the trajectory of a single antibody-labeled AMPA receptor recorded on cultured hippocampal neurons.

SPT and SFD have complementary properties. Although SPT allows for long observation times, the size of the label prevents access to specific neuronal areas such as the synaptic cleft (Borgdorff and Choquet 2002). In addition, it is difficult to precisely estimate the number of receptors bound to the particle. In contrast, fluorescent labels are short-lived, but their small size allows them to reveal receptor dynamics in constrained areas like synapses (Tardin et al. 2003).

An important limitation of SFD and SPT is the difficulty in distinguishing between the labels moving at the plasma membrane from those internalized in the cell. This means that rates of receptor endocytosis must be controlled during the experiments.

So far, experiments have only been performed in cell culture systems. Because of limited accessibility, SPT is unlikely to be suitable for use in tissue slices. However, it is likely that the small size of labels used in SFD will allow for their use in these preparations.

ACKNOWLEDGMENTS

This work has been supported by grants from the CNRS, the Council of the Région Aquitaine and the European community grant QLG3-CT-2001-02089.

REFERENCES

Borgdorff A. and Choquet D. 2002. Regulation of AMPA receptor lateral movement. *Nature* **417:** 649–653.

Choquet D. and Triller A. 2003. The role of receptor diffusion in the organization of the postsynaptic membrane. *Nat. Rev. Neurosci.* **4:** 251–265.

Cognet L., Coussen F., Choquet D., and Lounis B. 2002. Fluorescence microscopy of single autofluorescent proteins for cellular biology. *C.R. Physique* **3:** 645–656.

Moerner W.E. and Orrit M. 1999. Illuminating single molecules in condensed matter. *Science* **283:** 1670–1676.

Saxton M.J. and Jacobson K. 1997. Single-particle tracking: Applications to membrane dynamics. *Annu. Rev. Biophys. Biomol. Struct.* **26:** 373–399.

Schmidt T., Schutz G.J., Baumgartner W., Gruber H.J., and Schindler H. 1996. Imaging of single molecule diffusion. *Proc. Natl. Acad. Sci.* **93:** 2926–2929.

Schnapp B.J., Gelles J., and Sheetz M.P. 1988. Nanometer-scale measurements using video light microscopy. *Cell Motil. Cytoskel.* **10:** 47–53.

Sterba R.E. and Sheetz M.P. 1998. Basic laser tweezers. *Methods Cell Biol.* **55:** 29–41.

Tardin C., Cognet L., Bats C., Lounis B., and Choquet D. 2003. Direct imaging of lateral movements of AMPA receptors inside synapses. *EMBO J.* **22:** 4656–4665.

CHAPTER 69

A Practical Guide: Imaging Sodium in Dendrites

William Ross,[1] Joseph C. Callaway,[2] and Nechama Lasser-Ross[1]

[1]Department of Physiology, New York Medical College, Valhalla, New York 10595; [2]Department of Anatomy and Neurobiology, University of Tennessee Health Science Center, Memphis, Tennessee 38163

This chapter describes one-photon methods to detect the time course and spatial distribution of $[Na^+]_i$ changes in neurons in brain slices resulting from action potentials and synaptic activity. For two-photon sodium imaging, see Chapter 70. This technique has been applied to cerebellar Purkinje neurons (Lasser-Ross and Ross 1992; Callaway and Ross 1997), hippocampal pyramidal neurons (Jaffe et al. 1992), and presynaptic terminals (Regehr 1997). It is likely that it can be applied to other neurons in brain slices, particularly those that are relatively flat.

MATERIALS

Indicator

Membrane-impermeant, sodium-sensitive fluorescent indicator dye SBFI (sodium-binding benzofuran isophthalate) (Molecular Probes). SBFI binds Na^+ with a 1:1 stoichiometry with a K_D of ~17 mM (Minta and Tsien 1989).

Experimental Setup

A high-speed CCD camera and software are used to correlate electrophysiological and optical recordings (Lasser-Ross et al. 1991). The data shown in Figure 1 was obtained using a Photometrics (now Roper Scientific) AT200 cooled CCD. It was operated in the frame transfer mode. However, newer, faster, and more sensitive cameras are now available. The Roper Scientific Quantix 57, with a frame-transfer CCD chip, and the RedShirt Imaging NeuroCCD-SM256 are both recommended. These cameras have high quantum efficiency in the appropriate wavelength range and have sufficient spatial and temporal resolution for interesting measurements. The software used here was custom-written in the authors' laboratory.

Microscope

Olympus BX50WI equipped with epifluorescence port, 75-W Xenon arc lamp<!> and 40x and/or 60x water-immersion lenses. Equivalent microscopes from other manufacturers could be used.

Caution: *See Appendix 3 for appropriate handling of materials marked with <!>.*

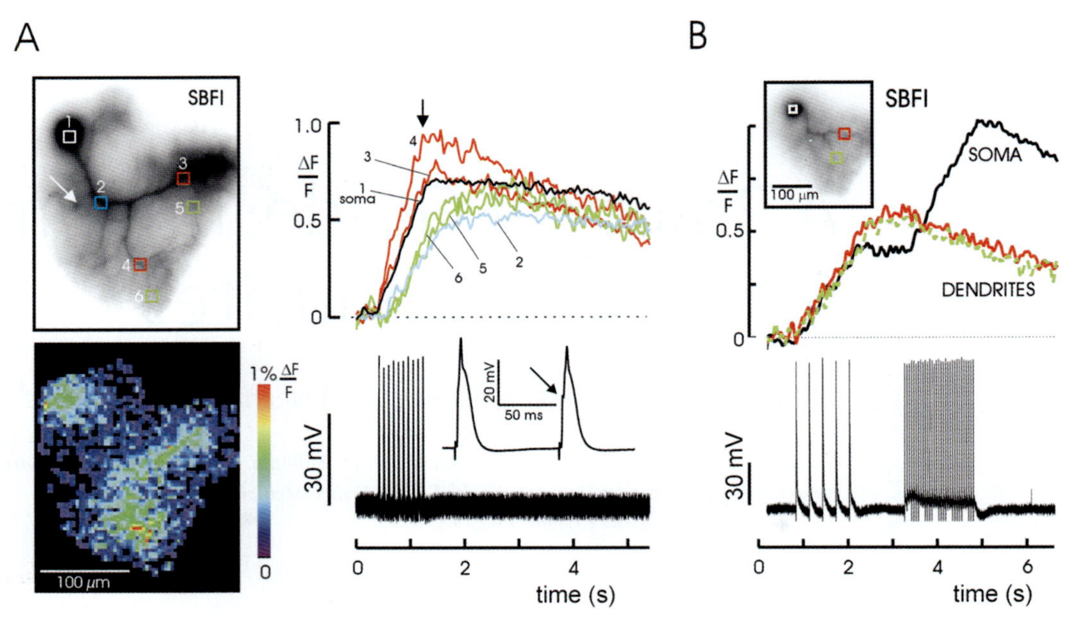

FIGURE 1. (*A*) Spatial distribution and time course of $[Na^+]_i$ changes in response to a train of ten climbing fiber (CF) responses. The fluorescence image in the upper left shows a guinea pig Purkinje cell filled with SBFI from a patch electrode on the dendrites (*white arrow*). The labeled boxes show the regions from which the individual traces on the right were measured. The pseudocolor image (*lower left*) shows the spatial distribution of the $[Na^+]_i$ changes measured at the time of the vertical black arrow over the traces. *Inset:* First two dendritically recorded CF electrical responses. Note the large Ca^{2+} spike at the peak of the response and the attenuated Na^+ spike (*small arrow*) on the rising edge of the potential change. Optical data are an average of ten sweeps. (*B*) $[Na^+]_i$ changes in response to a train of five CF stimuli followed by a train of 40 antidromic spikes. *Note:* Sodium changes were only detected in response to Na^+ spikes. Optical data are an average of 20 sweeps. (Adapted, with permission, from Callaway and Ross 1997.)

PROCEDURE

Imaging Voltage and Synaptically Activated Sodium Changes in Dendrites with a CCD Camera

Loading Cells with SBFI

1. Prepare a concentrated stock solution (10–20 mM) of SBFI in distilled water, and a stock of concentrated standard internal pipette solution (standard internal pipette solution: 140 mM K-gluconate<!>, 4 mM NaCl, 4 mM Mg-ATP, 0.3 mM Na-GTP, 10 mM HEPES, pH adjusted to 7.2–7.4 with KOH<!>).

2. Add an aliquot of the SBFI stock to an appropriate amount of the internal stock. Add water to achieve a final concentration of ~1–2 mM SBFI with proper osmotic balance.

   Note: No special procedures are required other than those that are normally used to prepare internal pipette solutions.

3. Use standard patch-pipette procedures (see Chapter 33) to load the chosen neuron with SBFI. Allow the pipette solution to diffuse throughout the cell, until all parts of the dendrites are visible.

   Note: Loading with sharp electrodes also works, although control over the final concentration is more difficult. The procedures are identical to those used to load neurons with membrane-impermeant calcium indicators (see Chapter 43).

Imaging Cells Loaded with SBFI

Fluorescence changes can be detected with the same filter set often used to detect fluorescence changes of the fura series of calcium indicators. However, to maximize the fluorescence signal, the authors recommend a wider excitation filter than usual (FW 25 nm), a dichroic mirror, centered at 420 nm, and a glass post filter (GG455). With these filters, the fluorescence of SBFI will decrease as $[Na^+]_i$ increases.

4. Stimulate the neuron either with the patch pipette used to load the cell or with an extracellular electrode. Record the $[Na^+]_i$ changes as changes from the baseline fluorescence levels recorded before stimulation.

   Note: The fluorescence changes of SBFI are small following electrical or synaptic stimulation. Therefore, it is important to arrange the fluorescence system to minimize noise. Recording a trial without stimulation and subtracting the drift in this trial from one with stimulation can usually correct for changes in fluorescence due to indicator or tissue bleaching.

Calibration of SBFI Fluorescence Changes

Changes in SBFI fluorescence are small, and usually proportional to $[Na^+]_i$ changes. However, the proportionality constant is difficult to determine. In the original papers describing the synthesis and use of SBFI, Tsien and his colleagues found that the properties of SBFI in the intracellular environment and in the cuvette were quite different (Harootunian et al. 1989). Other groups have calibrated SBFI in other preparations (e.g., see Rose and Ransom 1996), but it is not known whether this calibration is valid for neurons in slices. In addition, all measurements are a spatial average of $[Na^+]_i$ changes in a volume where the magnitude varies from point to point. Therefore, in most experiments, we found it more useful to present the results in terms of the fractional fluorescence change, $\Delta F/F$, without calibration. By doing this, information can be obtained about the time course, spatial distribution, and relative amplitude of the changes in $[Na^+]_i$ in different parts of the neuron. When comparing amplitudes, it is important to correct for background fluorescence in the slice and to consider only the indicator fluorescence. An approximate way to make this correction is to measure the fluorescence of the slice at a corresponding position away from the filled cell and to subtract this value from the total fluorescence of the slice at the position of the filled neuron.

EXAMPLE OF APPLICATION

Figure 1 shows two measurements of $[Na^+]_i$ changes in guinea pig cerebellar Purkinje cells in response to synaptic and electrical stimulation. Figure 1A shows the spatial distribution and time course of the $[Na^+]_i$ changes in response to climbing fiber (CF) activation. The spatial distribution corresponds closely to the known distribution of these synapses along the main dendrites of these cells, with weaker changes over the fine dendrites. There is also a strong signal in the soma due to voltage-gated Na^+ entry. Figure 1B compares the $[Na^+]_i$ changes from CF activation to the $[Na^+]_i$ changes due to a train of antidromic action potentials. The spike signals are only in the soma, corresponding to the known failure of Na^+ spikes to invade the dendrites of these cells (Stuart and Hausser 1994).

ADVANTAGES AND LIMITATIONS

Since the resting $[Na^+]_i$ in neurons is in the millimolar range, high concentrations of SBFI can be used to improve detection efficiency. SBFI is nontoxic, and we have not found any pharmacological effects of the indicator. SBFI responds to changes in $[K^+]_i$. However, changes in $[K^+]_i$ are not significant in most experiments. Two additional limitations are noted. First, the signals are small. The change in fluorescence was only about 1% in each example in Figure 1. In fact, signal averaging was needed to reduce the noise, and trains of stimuli were used to increase the signal. Therefore, this kind of measurement is very difficult to make if the responses cannot be repetitively activated. The current generation of cameras (see above) has better detection characteristics than the one used to generate the figure, but there is not much room for improvement. Therefore, better signals in the future are more likely to come from the use of two-photon excitation (see Chapter 70) and the synthesis of new optimally tuned and more fluorescent indicators.

Second, although the responses were time-locked to the stimuli, there was a significant delay in the responses at locations 5 and 6 in Figure 1A. This delay could reflect the diffusion of Na^+ from the site of entry along the main dendrite to these more peripheral locations. Na^+ diffusion from sites of entry is significant because Na^+ is not buffered in cells, unlike Ca^{2+}. In other experiments, we followed the diffusion of Na^+ in the axon and from the soma in these cells following electrical activity.

REFERENCES

Callaway J.C. and Ross W.N. 1997. Spatial distribution of synaptically activated sodium concentration changes in cerebellar Purkinje cells. *J. Neurophysiol.* **77:** 145–152.

Harootunian A.T., Kao J.P.Y., Eckert B.K., and Tsien R.Y. 1989. Fluorescence ratio imaging of cytosolic free Na^+ in individual fibroblasts and lymphocytes. *J. Biol. Chem.* **264:** 19458–19467.

Jaffe D.B., Johnston D., Lasser-Ross N., Lisman J.E., Miyakawa H., and Ross W.N. 1992. The spread of sodium spikes determines the pattern of dendritic Ca^{2+} entry into hippocampal neurons. *Nature* **357:** 244–246.

Lasser-Ross N. and Ross W.N. 1992. Imaging voltage and synaptically activated sodium transients in cerebellar Purkinje cells. *Proc. R. Soc. Lond. B Biol. Sci.* **247:** 35–39.

Lasser-Ross N., Miyakawa H., Lev-Ram V., Young S.R., and Ross W.N. 1991. High time resolution fluorescence imaging with a CCD camera. *J. Neurosci. Methods* **36:** 253–261.

Minta A. and Tsien R.Y. 1989. Fluorescent indicators for cytosolic sodium. *J. Biol. Chem.* **264:** 19449–19457.

Regehr W. 1997. Interplay between sodium and calcium dynamics in granule cell presynaptic terminals. *Biophys. J.* **73:** 2476–2488.

Rose C.R. and Ransom B.R. 1996. Intracellular sodium homeostasis in rat hippocampal astrocytes. *J. Physiol.* **491:** 291–305.

Stuart G.S. and Hausser M. 1994. Initiation and spread of sodium action potentials in cerebellar Purkinje cells. *Neuron* **13:** 703–712.

CHAPTER 70

A Practical Guide: Two-Photon Sodium Imaging in Dendritic Spines

Christine R. Rose

Physiologisches Institut, Ludwig-Maximilians-Universität München, 80336 München, Germany

Two-photon sodium imaging enables measurements of $[Na^+]_i$ transients in dendrites and spines in tissue slices (Rose et al. 1999; Rose and Konnerth 2001). This technique is also suitable for the determination of $[Na^+]_i$ transients in other cell types such as glial cells with high spatial resolution. It is also likely to be suited for measurement of Na^+ signals in vivo.

MATERIALS

A Na^+-sensitive fluorescent dye that is suitable for two-photon Na^+ imaging is SBFI (sodium-binding benzofuran isophtalate) (Minta and Tsien 1989; Rose et al. 1999; Rose and Konnerth 2001). SBFI is a ratiometric dye similar to calcium-sensitive dyes like fura-2 and has been employed for conventional, single-photon Na^+ imaging in many cell types (e.g., see Lasser-Ross and Ross 1992; Rose 1997; Chapter 69). The optimal Na^+-sensitive excitation wavelength of SBFI is between 380 and 400 nm, indicating that favorable excitation wavelengths for two-photon absorption are 760–800 nm (Denk et al. 1995).

SBFI's quantum efficiency is about six times lower than that of fura-2, and its fluorescence change when binding Na^+ is less pronounced than that of fura-2 with Ca^{2+} binding (Minta and Tsien 1989). This necessitates the use of relatively high dye concentrations (0.5–2 mM), and direct loading of the membrane-impermeable form of SBFI via a sharp microelectrode (Jaffe et al. 1992) or a patch pipette (Rose et al. 1999) is required for measurement in fine cellular processes in tissue slices. A possible concern is that SBFI does not discriminate sufficiently well between Na^+ and K^+ (Minta and Tsien 1989). However, SBFI's sensitivity to changes in intracellular $[K^+]$ is negligible under most experimental conditions (Rose 1997; Rose et al. 1999).

PROCEDURE

1. Load the chosen cell with intracellular saline containing 0.5–2 mM SBFI through a patch pipette.

   Note: For CA1 pyramidal neurons in mouse hippocampal slices, loading should last at least 45 min to ensure diffusion of the dye into the most distal dendrites.

2. After loading, excite between 760 and 810 nm.

   Note: Excitation results in bright fluorescence images, allowing the visualization of the entire neuron, including fine structures such as axons and spiny dendrites.

3. As required for all imaging measurements, choose the lowest possible excitation intensity to prevent dye bleaching or phototoxicity. However, because increases in Na^+ are reflected by decreases in the fluorescence emission of SBFI, do not set baseline fluorescence emission too low.

4. Calibrate the absolute $[Na^+]_i$ values in the cells (Rose et al. 1999).

   a. Fill the cells with SBFI using a patch electrode. Carefully withdraw the electrode to allow resealing of the cell membrane. Subsequently, perfuse the tissue slices with a Na^+-free calibration solution containing ionophores (3 μM gramicidin, 10 μM monensin<!>, 100 μM ouabain<!>). This step induces equilibration of Na^+ across the plasma membrane.

   b. Once a stable fluorescence emission in this solution is reached, which usually takes 20–30 min, perfuse the slices with calibration solutions of different Na^+ concentrations (e.g., 10, 20, 30, 50, 75, 100 mM Na^+). Stepwise changes in Na^+ concentrations result in stepwise changes in the SBFI fluorescence.

5. Plot normalized data from these calibration experiments. This should result in a graph that follows Michaelis-Menten kinetics and reveals an apparent K_d of ~26 mM at 22–23°C, which is close to the K_d value reported in situ with single-photon excitation (Donoso et al. 1992; Rose et al. 1999).

6. These in situ calibrations can then be used to express changes in SBFI fluorescence emission as $[Na^+]_i$ changes, assuming that the baseline $[Na^+]_i$ throughout the cell corresponds to the $[Na^+]$ of the pipette solution. The calibrations indicate that two-photon-induced SBFI fluorescence reliably reports $[Na^+]_i$ changes between 0 and ~75 mM $[Na^+]_i$ and is therefore well suited for the measurement of $[Na^+]_i$ alterations that occur during physiological conditions.

Caution: *See Appendix 3 for appropriate handling of materials marked with <!>.*

EXAMPLE OF APPLICATION

Two-photon Na^+ imaging combined with whole-cell patch-clamp recordings were used to measure activity-induced $[Na^+]_i$ transients in dendrites and spines of CA1 pyramidal neurons in rat hippocampal slices (Rose et al. 1999). A burst of back-propagating action potentials evoked $[Na^+]_i$ transients in the millimolar range throughout the proximal dendritic tree and in adjacent spines. The data suggested that the $[Na^+]_i$ accumulations in spines were caused by the opening of voltage-gated Na^+ channels on the spines themselves.

The same technique was employed to determine synaptically induced Na^+ transients in CA1 pyramidal cells (Rose and Konnerth 2001). A short burst of suprathreshold synaptic activity induced local dendritic $[Na^+]_i$ transients of about 10 mM. In presumed "active" spines (those that most likely received synaptic input), activity-induced Na^+ transients reached up to 35–40 mM (Fig. 1). It was found that the postsynaptic $[Na^+]_i$ transients were largely mediated by Na^+ entry through *N*-methyl-D-aspartate (NMDA) receptor channels and were detected during the coincident occurrence of synaptic potentials and back-propagating action potentials. The large amplitudes of the Na^+ transients and their location on dendritic spines suggested that this signal is an important determinant of electrical and biochemical spine characteristics.

ADVANTAGES AND LIMITATIONS

Generally, it is important to bear in mind that reporters of ion concentrations that bind a particular ion will act as a buffer and therefore distort ion transients. This problem, however, is almost negligible in the case of Na^+ imaging with SBFI because even dye concentrations of 2 mM are significantly below the K_d of SBFI (see above). However, binding of Na^+ to SBFI will alter the diffusion parameters of Na^+. Whereas Na^+ moves intracellularly with an apparent diffusion coefficient of about 13×10^{-6} cm^2sec^{-1} (Push and Neher 1988), SBFI-bound Na^+ diffuses much more slowly, with about 5×10^{-6} cm^2sec^{-1} (assuming that the diffusion coefficient of SBFI is similar to that of fura-2) (Push and Neher 1988).

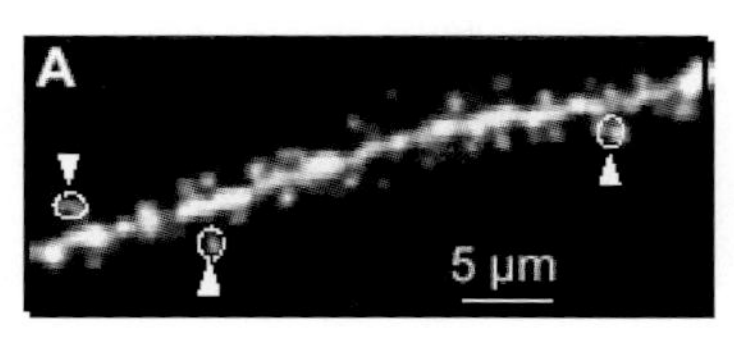

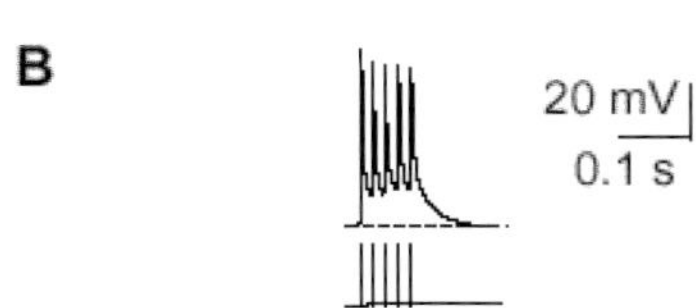

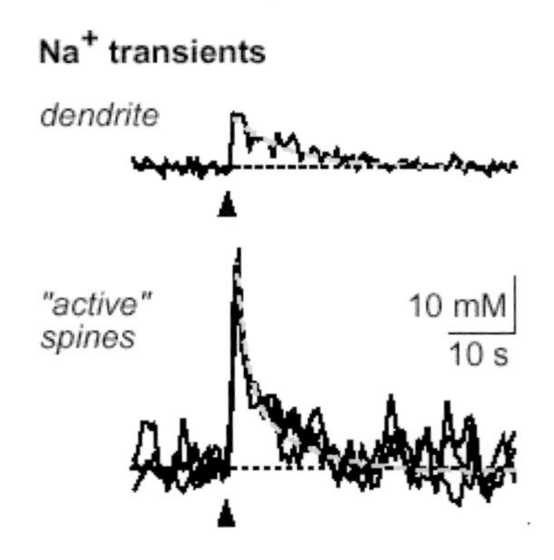

FIGURE 1. Synaptically induced Na^+ transients in dendrites and spines of CA1 pyramidal cells. (*A*) Image of the spiny dendrite of a CA1 pyramidal cell chosen for the experiment depicted in *B*. Arrowheads indicate spines from which measurements in *B* were taken. (*B*, *Upper trace*) Electrical response of the neuron to suprathreshold stimulation (five afferent stimuli at 50 Hz). (*Lower traces*) Average activity-induced Na^+ transients in the dendrite and in three single "active" spines (those spines that most likely received direct synaptic input). Data points of single spine traces were binned by a factor of 2 to improve the signal-to-noise ratio. (Modified, with permission, from Rose and Konnerth 2001 [©Society for Neuroscience].)

To date, SBFI is the only Na^+-sensitive fluorescent dye routinely used for Na^+ imaging experiments. However, its low quantum efficiency and small fluorescence change when binding Na^+ result in poor signal-to-noise ratios. This makes high-resolution measurement of Na^+ transients in fine processes a tedious task, and binning or averaging of several trials is often required to obtain satisfactory signals. But even this will not allow the detection of very small Na^+ changes (below the millimolar range). Clearly, the development of more sensitive fluorescent indicators for Na^+ is required to enable the detection of small Na^+ transients, e.g., during subthreshold synaptic activation of AMPA receptors in spines.

REFERENCES

Denk W., Piston D., and Webb W. 1995. Two-photon molecular excitation in laser-scanning microscopy. In *Handbook of biological confocal microscopy*, 2nd edition (ed. J.B. Pawley), pp 445–458. Plenum Press, New York.

Donoso P., Mill J., O'Neill S., and Eisner D. 1992. Fluorescence measurements of cytoplasmic and mitochondrial sodium concentration in rat vertricular myocytes. *J. Physiol.* **448:** 493–509.

Harootunian A., Kao J.P., Eckert B.K., and Tsien R.Y. 1989. Fluorescence ratio imaging of cytosolic free Na^+ in individual fibroblasts and lymphocytes. *J. Biol. Chem.* **264:** 19458–19467.

Jaffe D.B., Johnston D., Lasser-Ross N., Lisman J.E., Miyakawa H., and Ross W.N. 1992. The spread of Na^+ spikes determines the pattern of dendritic Ca^{2+} entry into hippocampal neurons. *Nature* **357:** 244–246.

Kafitz K.W., Rose C.R., Thoenen H., and Konnerth A. 1999. Neurotrophin-evoked rapid excitation through TrkB receptors. *Nature* **401:** 918–921.

Lasser-Ross N. and Ross W.N. 1992. Imaging voltage and synaptically activated sodium transients in cerebellar Purkinje cells. *Proc. R. Soc. Lond. B. Biol. Sci.* **247:** 35–39.

Minta A. and Tsien R.Y. 1989. Fluorescent indicators for cytosolic sodium. *J. Biol. Chem.* **264:** 19449–19457.

Push M. and Neher E. 1988. Rates of diffusional exchange between small cells and a measuring patch pipette. *Pflügers Arch.* **411:** 204–211.

Rose C. 1997. Intracellular Na^+ regulation in neurons and glia: Functional implications. *Neuroscientist* **3:** 85–88.

Rose C. and Konnerth A. 2001. NMDA receptor-mediated Na^+ signals in spines and dendrites. *J. Neurosci.* **21:** 4207–4214.

Rose C.R., Kovalchuk Y., Eilers J., and Konnerth A. 1999. Two-photon Na^+ imaging in spines and fine dendrites of central neurons. *Pflügers Arch. Eur. J. Physiol.* **439:** 201–207.

CHAPTER 71

Two-Photon Imaging of Chloride

Olga Garaschuk and Arthur Konnerth

Physiologisches Institut, Ludwig-Maximilians Universität München, 80336 München, Germany

This chapter describes an approach for two-photon Cl^- imaging in living cells. The approach has been tested successfully in various preparations, including cultured hippocampal neurons and cells in cerebellar, hippocampal, and cortical brain slices. It is also likely to be suitable for in vivo applications (see Stosiek et al. 2003).

MATERIALS

Membrane-permeant, Chloride-sensitive Fluorescent Indicator Dye

N-(ethoxycarbonylmethyl)-6-methoxyquinolinium bromide<!> (MQAE; Molecular Probes).

Two-Photon Laser-scanning Microscope

This procedure was developed using a custom-built microscope, based on a mode-locked laser system, operating at 740–800-nm wavelength, 80-MHz pulse repeat, and <130-fsec pulse width (Tsunami and Millenia; Spectra Physics). It also uses a laser-scanning system (either MRC 1024, Bio-Rad, or Olympus Fluoview, Olympus) coupled to an upright microscope (BX50WI, Olympus).

Caution: *See Appendix 3 for appropriate handling of materials marked with <!>.*

PROCEDURE

Staining Neurons with MQAE

1. Dissolve 6 mM MQAE<!> in the standard external saline (125 mM NaCl, 4.5 mM KCl<!>, 26 mM $NaHCO_3$, 1.25 mM NaH_2PO_4<!>, 2 mM $CaCl_2$<!>, 1 mM $MgCl_2$<!>, 20 mM glucose, pH 7.4, when bubbled with 95% O_2 and 5% CO_2).
2. Incubate cultured cells or brain slices with this solution for 10 minutes at 37°C and then rinse them with dye-free saline for 10–15 min.

   Note: This protocol allows high-quality staining of the upper 70–120 µm of a slice so that different types of neurons can be identified from their morphology (Marandi et al. 2002; see also Fig. 1).

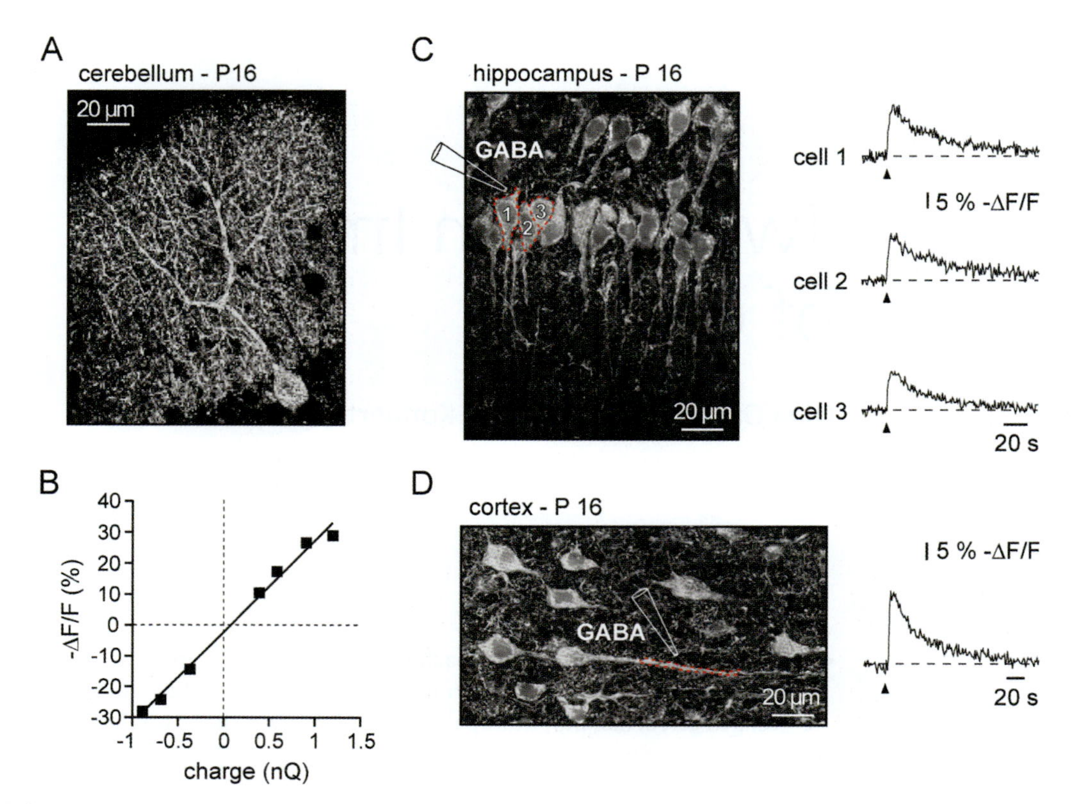

FIGURE 1. Two-photon images of MQAE-stained neurons in brain slices. (*A*) An image of a Purkinje cell from a 16-day-old rat (P16). Each image represents a maximal projection of seven to nine images taken with 1-μm steps. (*B*) Correlation between the amplitudes of GABA-mediated Cl^- transients and corresponding transmembrane anionic charges in an experiment using the gramicidin-based, perforated-patch technique. (*C*) GABA-mediated Cl^- transients in somata of CA1 pyramidal neurons (*right*) labeled with respective numbers on the image (*left*). (*D*) Dendritic GABA-mediated Cl^- transient (*right*) in a layer-5 pyramidal neuron of the cortex (*left*). (Modified from Marandi et al. 2002.)

3. Alternatively, cells in slices can be stained by a brief pressure ejection pulse (1 min, 70 kPa; Picospritzer II, General Valve) of ~400 fl of the pipette staining solution (25–100 mM MQAE, 2.5 mM KCl, 10 mM HEPES, osmolarity 310 mOsmol [with NaCl]) near the cells of interest.

   Note: In slices, this staining protocol yielded a stained area with a diameter of 200–300 μm. Individual neurons could be distinguished readily at depth up to 200–250 μm. Similar results were obtained in slices and in vivo using calcium indicator dyes (Stosiek et al. 2003).

   Caution: *See Appendix 3 for appropriate handling of materials marked with <!>.*

TWO-PHOTON IMAGING OF CELLS STAINED WITH MQAE

With one-photon imaging, MQAE is excited at wavelengths of 320–400 nm, and has an emission maximum at 460 nm (Verkman et al. 1989). With two-photon imaging, MQAE is excited efficiently at 740–760 nm. It is also possible to excite MQAE at longer wavelengths (up to 800 nm), but the intensity of the emitted light is lower (Marandi et al. 2002).

INTRACELLULAR CALIBRATION OF MQAE

The efficiency of quenching of quinolinium-based Cl^- indicators by Cl^- depends on the viscosity and/or polarity of the solvent (Jayaraman and Verkman 2000) and may, therefore, be different inside cells compared with in vitro tests. The calibration protocol introduced by Krapf et al. may be used for calibration of Cl^- levels in neurons in slices (Krapf et al. 1988; Marandi et al. 2002). Briefly, in situ calibration solutions containing different amounts of Cl^- (e.g. 0, 10, 20, 30, 40 mM) are prepared (basic solution: 0-40 mM KCl, 10 mM Hepes, 10 mM Na^+-gluconate, osmolarity 310 mOsmol (with K^+-gluconate), pH 7.4 [with KOH<!>])

The Cl^-/OH^- antiporter tributyltin chloride (10 μM) and the K^+/H^+ ionophore nigericin (10 μM) are added to each of these solutions. This treatment will break down the Cl^- gradient across the cell membrane and ascertain that the cytosolic Cl^- concentration ($[Cl^-]_i$) is equal to that of the corresponding calibration solution. The in situ calibration solutions (see below) are applied sequentially, and the intracellular steady-state fluorescence levels are measured. The mean fluorescence level in 0 Cl^--containing solution is defined as F_0. The F values for each calibration solution are plotted as F_0/F versus corresponding $[Cl^-]_i$ (so-called Stern-Volmer plot). The slope of the regression line is the reciprocal of an apparent dissociation constant (K_d). In our calibration experiments, the K_d of MQAE was 13 mM in cuvette and 40 mM inside neurons in brain slices (Marandi et al. 2002). (Solutions for cuvette calibration of MQAE<!> : 0–160 mM KCl, 30 mM HEPES, 5 mM MQAE, osmolarity 310 mOsmol [with sucrose], pH 7.4 [with KOH].)

EXAMPLE OF APPLICATION

Figure 1 shows examples of the two-photon Cl^- imaging experiments in different types of brain slices (modified from Marandi et al. 2002). Note that the low fluorescence of the extracellular space allows resolution of the secondary and tertiary dendritic branches of MQAE-stained neurons (Fig. 1A). As shown in Figure 1, C and D, the stained cells respond to brief (500-msec long) GABA applications with somatic and dendritic Cl^- transients. The combination of two-photon Cl^- imaging with gramicidin-based, perforated-patch-clamp recordings (Fig. 1B) reveals a clear correlation between the amplitudes of GABA-mediated transmembrane Cl^- fluxes and the amplitudes of MQAE-mediated fluorescence changes. The relation is linear and has the same slope for either direction of charge movement, indicating that MQAE is an accurate reporter of the transmembrane Cl^- charge transfer.

ADVANTAGES AND LIMITATIONS

MQAE provides easy and fast staining of neurons in brain slices with satisfactory fluorescence levels in cell bodies and dendrites. Advantages of MQAE-based two-photon Cl^- imaging include relatively high sensitivity and selectivity for Cl^-, insensitivity to changes in bicarbonate concentration and pH, low background fluorescence, and the possibility of prolonged continuous measurements. It is also important to mention that MQAE is quenched rapidly by Cl^- (<1 msec; Verkman et al. 1989) and is, thus, well suited for monitoring physiological changes in $[Cl^-]_i$, often occurring in the millisecond-to-second range. Furthermore, MQAE is quenched by a collisional quenching mechanism, which does not involve binding of Cl^- to the indicator dye molecule (Verkman 1990). MQAE, therefore, does not buffer Cl^-, and an increase in the intracellular dye concentration improves the signal-to-noise ratio without disturbing the time course of Cl^- transients. Its use in combination with two-photon microscopy minimizes bleaching and photochemical damage, which precluded the use of MQAE for fast continuous recordings in the past (Kaneko et al. 2001).

The limitations of the technique are due mainly to the fact that MQAE is a non-ratiometric dye, and is, thus, limited in terms of quantitative Cl^- measurements. Those interested in precise determination of $[Cl^-]_i$ may consider using the ratiometric Cl^--sensitive fluorescent protein Clomeleon (Kuner and Augustine 2000) (see Chapter 80). However, it should be borne in mind that Clomeleon has a slightly lower Cl^- sensitivity (K_d = 160 mM), that it is pH-sensitive, and that it has slower binding kinetics than MQAE (Verkman et al. 1989; Kuner and Augustine 2000). Another limitation of MQAE-based Cl^- imaging is dye loss through leakage. The leakage rate seems to be preparation-specific, ranging from 3% per hour in liposomes (Verkman et al. 1989) to 30% per hour in brain slices (Marandi et al. 2002). This limitation, however, is minor because it is easy to restain the preparation as necessary.

In summary, MQAE-based two-photon Cl^- imaging is a straightforward and simple approach for monitoring neuronal Cl^- dynamics.

REFERENCES

Jayaraman S. and Verkman A.S. 2000. Quenching mechanism of quinolinium-type chloride-sensitive fluorescent indicators. *Biophys. Chem.* **85:** 49–57.

Kaneko H., Nakamura T., and Lindemann B. 2001. Noninvasive measurement of chloride concentration in rat olfactory receptor cells with use of a fluorescent dye. *Am. J. Physiol. Cell Physiol.* **280:** C1387–1393.

Krapf R., Berry C.A., and Verkman A.S. 1988. Estimation of intracellular chloride activity in isolated perfused rabbit proximal convoluted tubules using a fluorescent indicator. *Biophys. J.* **53:** 955–962.

Kuner T. and Augustine G.J. 2000. A genetically encoded ratiometric indicator for chloride: Capturing chloride transients in cultured hippocampal neurons. *Neuron* **27:** 447–459.

Marandi N., Konnerth A., and Garaschuk O. 2002. Two-photon chloride imaging in neurons of brain slices. *Pflügers Arch.* **445:** 357–365.

Stosiek C., Garaschuk O., Holthoff K. and Konnerth A. 2003. In vivo two-photon calcium imaging using multicell bolus loading (MCBL). *Proc. Natl. Acad. Sci.* **100:** 7319–7324.

Verkman A.S. 1990. Development and biological applications of chloride-sensitive fluorescent indicators. *Am. J. Physiol.* **259:** C375–388.

Verkman A.S., Sellers M.C., Chao A.C., Leung T., and Ketcham R. 1989. Synthesis and characterization of improved chloride-sensitive fluorescent indicators for biological applications. *Anal. Biochem.* **178:** 355–361.

CHAPTER 72

A Practical Guide: Interferometric Detection of Action Potentials

Arthur LaPorta[1] and David Kleinfeld[2]

[1]*Laboratory of Atomic and Solid State Physics, Cornell University, Ithaca, New York 14853;* [2]*Department of Physics, University of California, La Jolla, California 92093-0319*

The technique described here is for the detection of individual action potentials in axons in vitro using intrinsic optical signals, and it has been demonstrated using axons dissected from lobster. The technique would also be applicable to axons and dendrites of cultured neurons with diameters somewhat larger than an optical wavelength. Using index matching techniques, it may be possible to extend the technique to smaller structures.

Early work showed that light scattering from a network of small neuronal processes is modulated by the propagation of action potentials (Cohen et al. 1968, 1972a,b). Studies of the scattering pattern obtained from a single axon are consistent with a change in the refractive index of the cell membrane. This change is proportional to the voltage across the membrane (Stepnoski et al. 1991) and presumably arises from alignment of dipoles in the membrane. The magnitude of the change was determined for isolated axons from *Aplysia* neurons, for which the relative change in refractive index was $\Delta n/n = 50 \times 10^{-6}$ per millivolt. This corresponds to an equivalent optical phase shift of 0.1 to 0.2 mradian per spike. These relatively small changes could be detected, in single trials, by measuring changes in light scattering with dark-field illumination (Stepnoski et al. 1991).

MATERIALS

The essential optical components include a 5-mW polarized laser (HeNe); single-mode polarization-preserving optical fiber with couplers; Wollaston prism, birefringent plate; water-immersion 20x microscope objective; electro-optic modulator, and polarization analyzer. The interferometric microscope must be compact and stiff, a combination that can be achieved with Microbench components (LINOS Photonics).

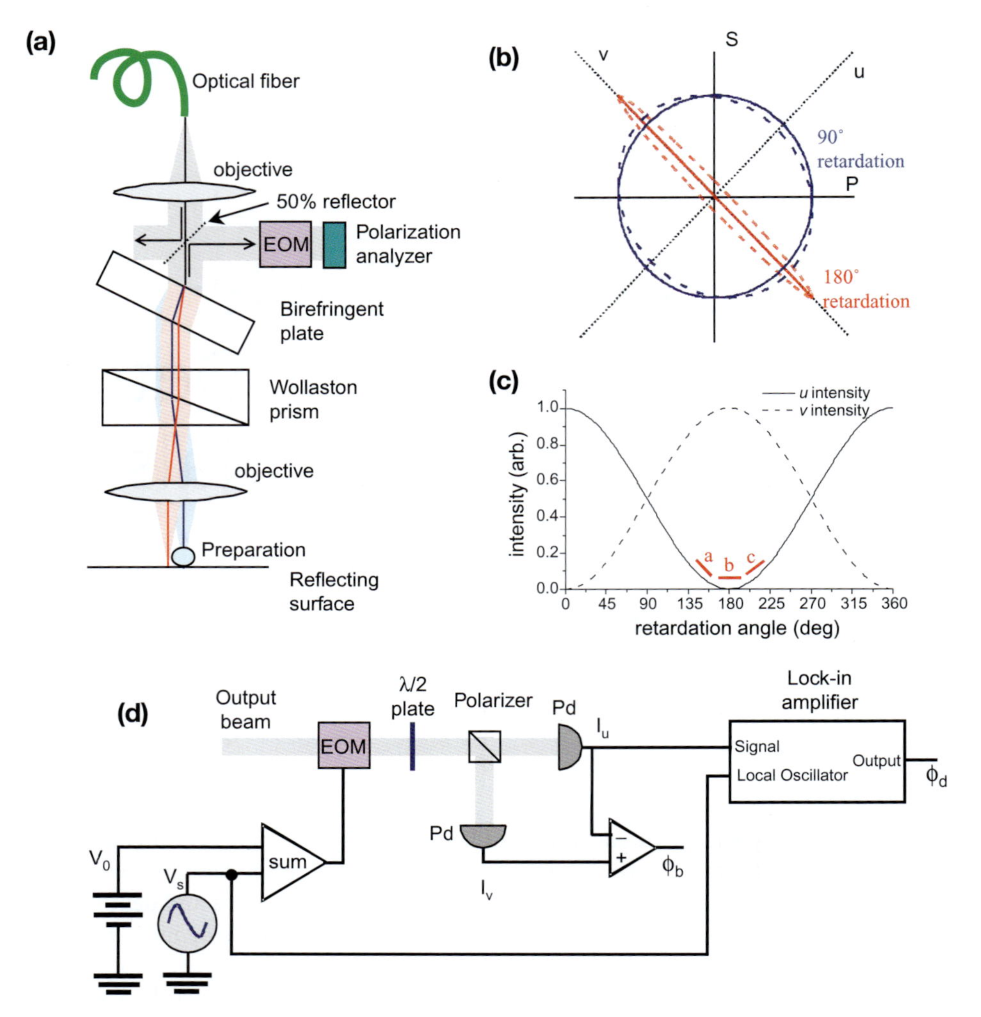

FIGURE 1. Schematic of the interferometer and polarization-based detection system for changes in optical path length. (*a*) Essential optics in the interferometer. The broad gray bar represents the total extent of the beam and the black line represents the central ray. After separation by polarization components, red and blue shading indicates S and P polarization components of the light. Laser light is introduced via a polarization-preserving optical fiber at the top and propagates down to the preparation plane through a microscope objective, polarization components, and a second water-immersion objective. The preparation plane is reflective at the excitation wavelength, and the light is retroreflected and ultimately coupled out of the microscope using a 50% reflector. The output beam is passed through an electro-optic modulator (EOM) and is detected with a polarization analyzer, shown in greater detail in *d*. See text for details. (*b*) The electric field as a function of optical phase shift where the S and P axes correspond to the polarization directions of the two rays (*red* and *blue*) in part *a* and the u and v axes correspond to the two channels of the polarization analyzer. The projection of the curves on the u and v axes gives the amplitude of that component that is transmitted to the detectors. (*c*) The intensity measured by the two channels of the analyzer as a function of optical phase shift. The short red lines are explained in the text. (*d*) Detailed schematic of the phase-sensitive detection scheme. The EOM introduces a phase shift of the S polarized field with respect to the P polarized field. A DC voltage, denoted V_0, is applied to correct for the phase shift of the resting axon. The λ/2 plate rotates the polarization state so that the analyzer detects the intensity along two diagonal directions u and v. For balanced detection the intensities I_u and I_v are nominally equal (see text) and are subtracted to give a phase signal that is denoted ϕ_b. For dark fringe detection, V_0 is set so that all light goes to the I_v detector. A phase modulation, denoted V_s with frequency f_s, is superimposed on the EOM drive signal. A deviation of the optical phase from the dark fringe (ϕ_d) causes a modulation at frequency f_s to appear in the photocurrent I_u (see text) which may be detected with a lock-in amplifier (LIA). The modulation signal V_s is connected to the reference input and the photocurrent I_u is connected to the signal input of the LIA, which is configured to measure the in-phase first harmonic component of the signal. The optical phase ϕ_d is obtained from the output of the LIA, which is equal to $(1/2)I_t(\phi_d)$ where I_t is the total intensity.

PROCEDURE

The goal of this procedure is to make a direct confirmation of the refractive index change by constructing a two-beam interferometer in which a beam that passes through an axon interferes with a beam that bypasses the axon. A change in the relative optical phase of the two beams is a measure of the change in total optical density of the preparation. To avoid randomly fluctuating phase shifts, the two beams must be transported by the same optical components. To accomplish this, the apparatus shown in Figure 1a is used. The apparatus uses polarization optics to create two distinct beams that propagate through the same optical system.

A polarization-preserving single-mode optical fiber that transports light from an HeNe laser acts as a point source for a microscope objective, producing a collimated Gaussian beam whose linear polarization angle is chosen to be at 45° with respect to the polarization optics that follow. The beam passes through an inclined quartz birefringent plate, which causes a displacement of the S and P polarization components with respect to each other. The beams then pass through a Wollaston prism, which introduces a relative angular deviation of the two polarization components, causing them to cross at a point beyond the rear face of the prism. The combined effect of the birefringent plate and Wollaston prism is similar to that of a Nomarski prism with an unusually large divergence angle. The S and P polarized beams are incident on the rear aperture of a 20X water-immersion microscope objective. The beam diameters are chosen to match the rear aperture of the objective, and the crossing point is placed at the rear focal plane of the objective. The microscope objective brings the collimated beams to diffraction-limited Gaussian focal spots in the image plane. The angular deviation of the two beams causes the focal spots to occur at distinct positions in the sample plane. The Wollaston angle was chosen to make this separation ~5 μm; the choice of separation distance depends on the application. Since the two beams cross at the rear focal plane of the objective, their central rays are parallel. The specimen chamber is positioned so that the two beams are normally incident, with their waist at the bottom reflective surface. As a result, the two beams are retro-reflected back into the microscope. The two beams are recombined by the Wollaston and birefringent plate and are coupled out of the apparatus by a 50% reflector.

The beam that emerges from the apparatus consists of a vertical polarization component that has passed through the axon and a horizontal polarization component that has bypassed the axon. The optical phase shift of the resting axon changes the polarization state of the output beam from linear to an arbitrary elliptical state. A change in the optical density or path length of the axon during an action potential would result in a perturbation to the elliptically polarized state.

To detect the small phase shifts expected in response to an action potential, the measurement must employ a differencing scheme to cancel technical fluctuations in the laser light intensity. Here, the output beam is passed through an electro-optic modulator (EOM), which is used to adjust the relative phase shift of S and P polarization components for the resting neuron. There are two ways that the EOM and polarization analyzer may be used (Fig. 1d). The nominal phase shift between the two polarizations may be set to 90°. In this case, the output beam has circular polarization (solid blue curve in Fig. 1b) and the analyzer, which is oriented to measure diagonal components I_u and I_v, sees a balanced signal. This corresponds to 90° in Figure 1c. A deviation in the phase causes a deformation of the circular state (dashed blue curve in Fig. 1b), which imbalances I_u and I_v. As a result, the difference signal $I_u - I_v$ is a measure in the phase in which the laser intensity cancels.

A more effective detection method (called the dark fringe method) is to set the nominal phase shift between the two polarizations to 180° so that the output state is linearly polarized along v and the intensity I_u is zero (solid red curve in Fig. 1b). An additional phase shift results in a small deviation from linear polarization that causes I_u to become non-zero. This signal cannot be used directly, since the intensity $I_u \propto 1 + \cos(180° + \phi)$ is second order in the phase ϕ and would have a vanishing signal-to-noise ratio. However, a first-order signal can be obtained by superimposing a sinusoidal modulation of frequency f_s on the EOM phase shift. For the resting axon, the modulation in the neighborhood of 180° produces no first-order signal in I_u because the slope of the I_u versus phase curve is zero (see line marked b in Fig. 1c). However, a positive phase shift in the axon would bias the net phase toward higher values (line marked c in Fig. 1c) where the slope of the curve is positive. In this case,

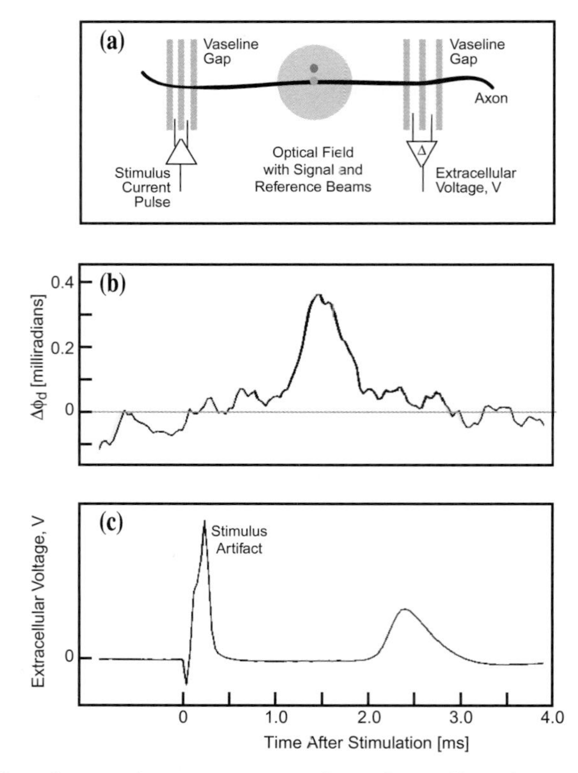

FIGURE 2. Results for interferometric measurements through a walking leg nerve from lobster, *Homarus americanus*. The nerve is maintained in artificial seawater. The nerve was microdissected down to a few, but more than one, axonal fiber. (*a*) The chamber that holds the preparation is a multicompartment structure that positions the nerve across a dielectric mirror that has a maximal reflection at 630 nm, close to the HeNe emission line. One beam (P polarization, *lower dot*) passes through the nerve and the second beam (S polarization, *upper dot*) passes about 5 μm away from the nerve. The nerve further rests on two platinum wires at one end, which serve as the voltage-stimulation source, and two at the other end, which serve as extracellular recording electrodes. Thin walls with slits that are filled with "Vaseline" provide electrical isolation between the chambers. (*b*) The optical signal, denoted by the change in output phase ϕ_d between the two beams, in response to a just suprathreshold stimulus. Shown is the stimulus triggered average of 100 samples. (*c*) The concomitantly measured and averaged electrical signal. It appears later than the optical signal since it is measured further along the axons.

the phase modulation produces a signal at frequency f_s in I_u with amplitude proportional to the axon phase shift. Similarly, a negative axon phase would bias the phase toward negative values (line marked a in Fig. 1c) and produce a modulation with negative amplitude in I_u. The amplitude of the signal at frequency f_s (and therefore the axon phase retardation) can be measured using a lock-in amplifier referenced to the EOM modulation signal as shown in Figure 1d. The primary advantage of this technique is that the photodetectors have much less noise near the carrier frequency (chosen to be >100 kHz) than near dc. In this configuration, both fluctuations in laser intensity and $1/f$ noise in the photodetectors are avoided, and the principal sources of noise are shot noise and fluctuations in the preparation itself.

SHORT EXAMPLE OF APPLICATION

Figure 2 shows the preparation chamber for the walking leg nerve dissected from the lobster and the optical phase shift recorded from a nerve. The nerve was microdissected so that only a small number of axonal fibers remained in the bundle. The second of the above detection schemes for phase detection was used. Action potentials were elicited by stimulating the nerve with a voltage pulse set just above threshold; this minimized the number of axons excited. The extracellular signals from the action potential(s) were detected using electrodes placed beyond the imaging point. An optical phase shift of approximately 0.4 mradian was measured after averaging 100 pulses. This signal may contain contributions from more than one axon.

ADVANTAGES AND LIMITATIONS

The advantage of this technique is that a direct measure of change in optical density is made at a specific point. The measurement is performed in such a way that noise in the measurement process itself (e.g., from fluctuations in laser intensity and detector noise) is minimized. The main limitations are that it requires an optically thin in vitro sample, since any distortion in the beams degrades the sensitivity of the polarization analyzer. The final signal-to-noise ratio seems to be dominated by fluctuation in the specimen itself. Although these fluctuations are hard to quantify, the main effects are probably Brownian motion of the physical structures and biochemical fluctuations in the axon cytoplasm.

REFERENCES

Cohen L.B., Keynes R.D., and Hille B. 1968. Light scattering and birefringence changes during nerve activity. *Nature* **218:** 438–441.

Cohen L.B., Keynes R.D., and Landowne D. 1972a. Changes in axon light scattering that accompany the action potential: Current-dependent components. *J. Physiol.* **224:** 727–752.

———. 1972b. Changes in light scattering that accompany the action potential in squid giant axons: Potential-dependent components. *J. Physiol.* **224:** 701–725.

Stepnoski R.A., LaPorta A., Raccuia-Behling F., Blonder G.E., Slusher R.E., and Kleinfeld D. 1991. Noninvasive detection of changes in membrane potential in cultured neurons by light scattering. *Proc. Natl. Acad. Sci.* **88:** 9382–9386.

CHAPTER 73

A Practical Guide: Intrinsic Optical Signal Imaging in Brain Slices

Brian A. MacVicar and Sean J. Mulligan

Brain Research Center, Department of Psychiatry, University of British Columbia, Vancouver, British Columbia V6T 2B5, Canada

This chapter describes the use of intrinsic optical imaging in brain slices. It outlines the procedures and the system equipment necessary for successful imaging acquisition and provides an introduction to the technique's applications.

Imaging of intrinsic optical signals (IOS) has been widely used to map patterns of activity in vivo; for example, visual cortex of the anesthetized cat (Malonek et al. 1997) and isolated perfused guinea pig brain (Federico et al. 1994; Federico and MacVicar 1996). Imaging IOS in brain slices (MacVicar and Hochman 1991; Joshi and Andrew 2001) are particularly useful for examining volume changes in neurons (Andrew and MacVicar 1994), swelling of astrocytes induced by K^+ uptake (MacVicar et al. 2002), spreading depression (Basarsky et al. 1998), and ischemic depolarization (Bahar et al. 2000). IOS imaging in vivo measures not only swelling alterations but also changes in blood flow and oxygenation of hemoglobin. (Malonek et al. 1997). However, IOS signals in both brain slices and in the intact brain measure alterations in the optical properties of tissue that are secondary results of neuronal activity. IOS signals do not directly measure neuronal activity.

MATERIALS

Imaging IOS requires a CCD camera, image acquisition hardware, and analysis software. It is suitable for use with an inverted or upright microscope with a perfusion bath and requires a tungsten or halogen light source with a stabilized power supply (<0.1% fluctuations).

PROCEDURES

The changes in light transmittance associated with neural activity are relatively small, and care must be taken to reduce sources of noise. In our experiments, the brain slices are submerged and superfused with ACSF (Basarsky et al. 1998). When an inverted or dissecting microscope is used, it is very important to maintain a stable depth of superfusion solution. Slight fluctuations in solution depth alter the light path and change the apparent light intensity. To prevent this, a coverslip should be placed in contact with the perfusion solution, directly above the slice in the light path. Alternatively, an upright microscope with a water immersion lens can provide imaging without an air/water interface. Brain slices are more transparent at the longer wavelengths, and the intrinsic signals are most readily observed with illumination in the range of 700–800 nm. A 750-nm 15-nm bandpass filter is typically used for illumination.

Signal Recovery

The video camera is the most important component in the IOS imaging system. The authors use a CCD camera (model 8912, Cohu). It is possible to use a cooled CCD with a 12-bit dynamic range. However, excellent results can be obtained with the 8-bit range of many scientific-grade CCD cameras. Because optical signals are small, high-gain settings are used, and the camera must have manual offset and gain controls. The black level is adjusted to ensure that the signal spans the input-voltage range of the A/D board of the frame grabber. If the gain is too high, the pixel values "wrap around" and the intensity measurements are misleading.

Data Acquisition and Analysis

The computer and frame grabber are important for signal averaging and analysis. IOS amplitudes are often close to the noise level of the standard video CCD cameras, and frame averaging is necessary to achieve sufficient noise reduction. The reduction in noise is a function of the square root of the number of frames. Averaging 8–16 frames should provide adequate noise reduction. The authors use the DT3155 frame grabber from Data Translation. Image acquisition and computation are controlled by Axon Imaging Workbench (AIW) software (INDEC) that was initially developed in the MacVicar lab. Frames are acquired and averaged at typical video rate frequencies of ~30 Hz. Intrinsic optical signals are measured with reference to control images that can be subtracted from subsequent images. Alternatively, intensities can be expressed as a percentage change with respect to control levels in specified regions of interest.

EXAMPLES OF APPLICATIONS

Intrinsic optical signals are generated by several cellular processes and can be used to analyze different dynamic properties of neural tissue. Cellular swelling, from neuronal activity or K^+ uptake into astrocytes, has been imaged by monitoring IOS in brain slices and in optic nerves (MacVicar and Hochman 1991; Andrew and MacVicar 1994; Holthoff and Witte 1996; MacVicar et al. 2002). Alterations in blood flow can be monitored by detecting the increased hemoglobin content in tissue (proportional to the absorption at 570 nm) or by detecting the changes in the state of hemoglobin from oxygenated to deoxygenated forms (proportional to the absorption at 600–630 nm) (Malonek et al. 1997). Cytochrome oxidase reduction due to increased oxygen demand can also be imaged in brain preparations by detecting absorption changes at 450 nm, 550 nm, and 600 nm (Federico et al. 1994). IOS imaging is very useful for monitoring the propagation of spreading depression (SD) or ischemic depolarization in brain slices (Federico et al. 1994; Basarsky et al. 1998; Muller and Somjen 1999; Obeidat et al. 2000; Joshi and Andrew 2001). An example of SD is shown in Figure 1. This example was induced by the perfusion of the Na^+, K^+, ATPase-inhibitor ouabain. Imaging IOS allowed the propagation of SD to be monitored through a cortical brain slice. The timing of the wave of increased light transmittance was correlated with the electrophysiological recording of the large negative shift in the extracellular field potential recording of the SD wave.

ADVANTAGES AND LIMITATIONS

Imaging IOS provides unparalleled views of neuronal activity patterns on a large scale. The principal advantage over other techniques, such as voltage-sensitive dyes, is that it is not necessary to stain or introduce an indicator dye to observe activity patterns in neuronal tissue. The main disadvantages of

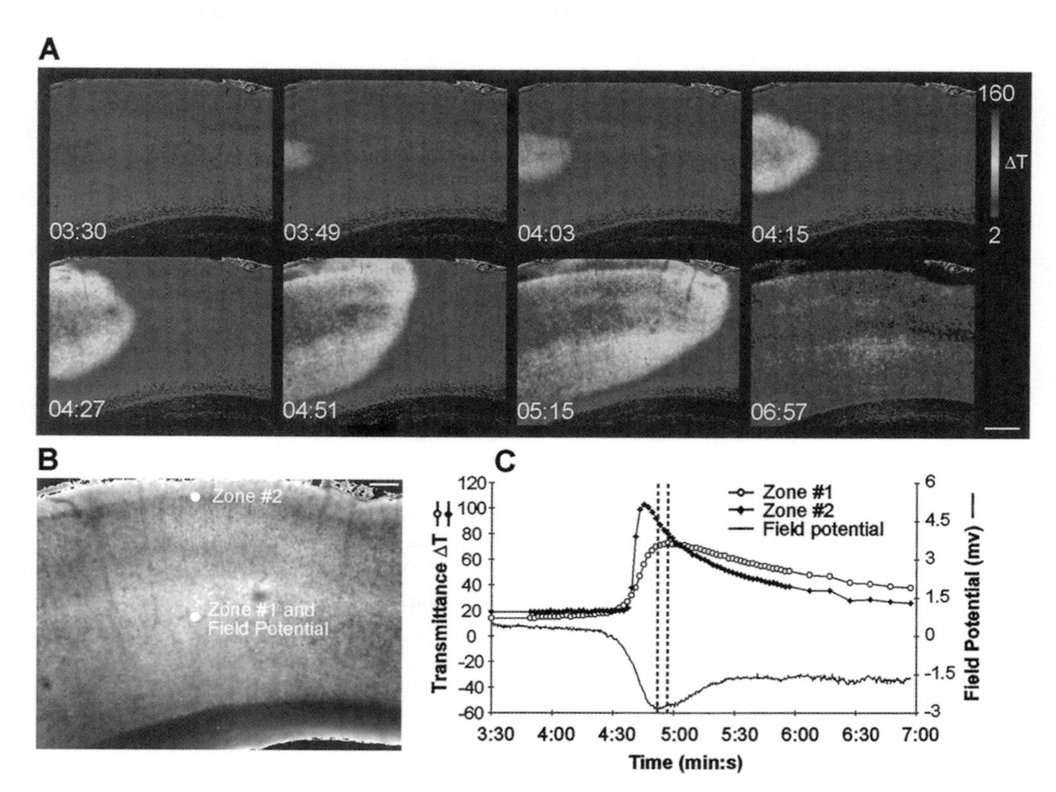

FIGURE 1. Imaging spreading depression (SD) in neocortical slices. (*A*) Intrinsic optical signals during superfusion of ACSF containing 100 μM ouabain. In this case, SD was initiated at the leftmost visible portion of the slice and propagated uniformly across the entire slice. (*B*) Description of the areas that were used for measurements. (*C*) Time course of the intrinsic optical signal and extracellular field potential for one region. Note the similar time course of the field potential and associated intrinsic optical signal. The dashed lines represent the peaks of the intrinsic optical signal and the field potential. Bars: *A*, 400 μm; *B* 200 μm. (Reprinted, with permission, from Basarsky et al. 1998 [© Society for Neuroscience].)

IOS imaging are the slower time course and the indirect nature of the signal. IOS are generated by secondary processes, not neuronal activity itself. The changes in cell volume, blood volume, and hemoglobin or cytochrome oxygenation occur in response to changes in neuronal activity; therefore, they follow with varying time delays. Because of this, the correlation with neuronal activity must be carefully established in each preparation.

ACKNOWLEDGMENTS

B.A.M. is a Canada Research Chair in Neuroscience and a Michael Smith Distinguished Scholar. S.J.M. is a fellow of the Heart and Stroke Foundation of Canada. The Canada Institutes of Medical Research supported work in the lab.

REFERENCES

Andrew R.D. and MacVicar B.A. 1994. Imaging cell volume changes and neuronal excitation in the hippocampal slice. *Neuroscience* **62:** 371–383.

Bahar S., Fayuk D., Somjen G.G., Aitken P.G., and Turner D.A. 2000. Mitochondrial and intrinsic optical signals imaged during hypoxia and spreading depression in rat hippocampal slices. *J. Neurophysiol.* **84:** 311–324.

Basarsky T.A., Duffy S.N., Andrew R.D., and MacVicar B.A. 1998. Imaging spreading depression and associated intracellular calcium waves in brain slices. *J. Neurosci.* **18:** 7189–7199.

Federico P. and MacVicar B.A. 1996. Imaging the induction and spread of seizure activity in the isolated brain of the guinea pig: The roles of GABA and glutamate receptors. *J. Neurophysiol.* **76:** 3471–3492.

Federico P., Borg S.G., Salkauskus A.G., and MacVicar B.A. 1994. Mapping patterns of neuronal activity and seizure propagation by imaging intrinsic optical signals in the isolated whole brain of the guinea-pig. *Neuroscience* **58:** 461–480.

Holthoff K. and Witte O.W. 1996. Intrinsic optical signals in rat neocortical slices measured with near-infrared dark-field microscopy reveal changes in extracellular space. *J. Neurosci.* **16:** 2740–2749.

Joshi I. and Andrew R.D. 2001. Imaging anoxic depolarization during ischemia-like conditions in the mouse hemi-brain slice. *J. Neurophysiol.* **85:** 414–424.

MacVicar B.A. and Hochman D. 1991. Imaging of synaptically evoked intrinsic optical signals in hippocampal slices. *J. Neurosci.* **11:** 1458–1469.

MacVicar B.A., Feighan D., Brown A., and Ransom B. 2002. Intrinsic optical signals in the rat optic nerve: Role for K(+) uptake via NKCC1 and swelling of astrocytes. *Glia* **37:** 114–123.

Malonek D., Dirnagl U., Lindauer U., Yamada K., Kanno I., and Grinvald A. 1997. Vascular imprints of neuronal activity: Relationships between the dynamics of cortical blood flow, oxygenation, and volume changes following sensory stimulation. *Proc. Natl. Acad. Sci.* **94:** 14826–14831.

Muller M. and Somjen G.G. 1999. Intrinsic optical signals in rat hippocampal slices during hypoxia-induced spreading depression-like depolarization. *J. Neurophysiol.* **82:** 1818–1831.

Obeidat A.S., Jarvis C.R., and Andrew R.D. 2000. Glutamate does not mediate acute neuronal damage after spreading depression induced by O_2/glucose deprivation in the hippocampal slice. *J. Cereb. Blood Flow Metab.* **20:** 412–422.

CHAPTER 74

Indicators Based on Fluorescence Resonance Energy Transfer

Roger Y. Tsien

Howard Hughes Medical Institute and Departments of Pharmacology and Chemistry and Biochemistry, University of California, San Diego, La Jolla, California 92093

One of the major new trends in the design of indicators for optically imaging biochemical and physiological functions of living cells has been the exploitation of fluorescence resonance energy transfer (FRET). FRET is a well-known spectroscopic technique for monitoring changes in the proximity and mutual orientation of pairs of chromophores. It has long been used in biochemistry and cell biology to assess distances and orientations between specific labeling sites within a single macromolecule or between two separate molecules (Stryer 1978; Lakowicz 1983; Uster and Pagano 1986; Herman 1989; Jovin and Arndt-Jovin 1989; Tsien et al. 1993). More recently, macromolecules or molecular pairs have been engineered to change their FRET in response to biochemical and physiological signals such as membrane potential, cyclic AMP (cAMP), protease activity, free Ca^{2+} and Ca^{2+}-CaM (calmodulin) concentrations, protein–protein heterodimerization, phosphorylation (Wouters et al. 2001) and reporter-gene expression (Table 1, Fig. 1). Because FRET is general, nondestructive, and easily imaged, it has proven to be one of the most versatile spectroscopic readouts available to the designer of new probes. FRET is particularly amenable to emission ratioing, which is more reliably quantifiable than single-wavelength monitoring and better suited than excitation ratioing to high-speed and laser-excited imaging.

PHOTOPHYSICAL PRINCIPLES OF FRET

FRET is the quantum-mechanical transfer of energy from the excited state of a donor fluorophore to the ground state of a neighboring acceptor chromophore or fluorophore. The acceptor must absorb light at roughly the same wavelengths as the donor emits. If the donor and acceptor are located within a few nanometers of each other and the mutual orientation of the chromophores is not too unfavorable, FRET becomes probable.

TABLE 1. Physiological indicators using fluorescence resonance energy transfer

Analyte or process	Donor and acceptor[a]	ΔFRET[b]	Wavelength (nm)[c] donor excitation	donor emission	acceptor emission	Maximum emission ratio change[d]	References
Depolarization	coumarin-phosphatidylethanolamine; bis(thiobarbiturate)trimethineoxonol	↓	414	450	560	1.8/(100 mV)	Gonzalez and Tsien (1997)
		↓	495	520	580	1.6–2.2	Adams et al. (1991, 1993)
cAMP	PKA catalytic subunit-FITC; PKA regulatory subunit-TROSu[e]						
Trypsin	BFP-(trypsin-sensitive linker)-GFP	↓	380	445	507	4.6	Heim and Tsien (1996)
Factor X_a	BFP-(factor X_a-sensitive linker)-GFP	↓	385	450	505	1.9	Mitra et al. (1996)
Caspase-3	BFP-(caspase-3-sensitive linker)-GFP	↓	380	440	511	?	Xu et al. (1998)
Ca^{2+}-CaM	GFP-CB_{SM}-BFP	↓	380	448	505	5.7	Romoser et al. (1997)
Ca^{2+}	GFP-CB_{SM}-BFP-CaMCN	↓	380	440	505	1.67	Persechini et al. (1997)
Ca^{2+}	ECFP-CaM-M13-EYFP	↑	433	476	527	2.1	Miyawaki et al. (1997, 1999)
Ca^{2+}	ECFP-CaM; M13-EYFP	↑	433	476	527	4	Miyawaki et al. (1997, 1999)
β-Lactamase expression	coumarin-cephalosporin-fluorescein (CCF2)	↓	409	447	520	70	Zlokarnik et al. (1998)

[a]Hyphens indicate covalent conjugation or fusion. Interacting donor and acceptor molecules are separated by semicolons.
[b]FRET indicates whether the efficiency of fluorescence resonance energy transfer is increased (↑) or decreased (↓) by the analyte or process.
[c]Donor excitation wavelength, donor emission wavelength, and acceptor emission wavelength, all in nanometers. Small differences (up to 8 nm) in the wavelengths cited by different laboratories for BFP and GFPs are probably not significant.
[d]Maximum factor by which emission ratio changes from zero to saturating levels of the analyte or process, except for depolarization of 100 mV amplitude.
[e]Tetramethylrhodamine N-hydroxysuccinimide.

Measurements of FRET

The efficiency of FRET in a population can be measured as a reduction in the fluorescence, excited-state lifetime, and susceptibility to photochemical reactions of the donor relative to the corresponding parameters when acceptors are remote or absent. If the acceptor is fluorescent, it can re-emit the transferred energy as its own fluorescence, which will be at longer wavelengths than those of the donor. The most robust and convenient way to image FRET is usually to ratio the acceptor and donor emission amplitudes. Such emission ratioing cancels out variations in the excitation intensity, the overall collection efficiency, and the number of donor–acceptor pairs in the sample-volume element, because these factors perturb the acceptor and donor signals equally. Measurements of donor excited-state lifetime require pico- to nanosecond time resolution in excitation and detection, which are expensive and rather rare capabilities in imaging systems. Measurements of donor photobleaching rate (Jovin and Arndt-Jovin 1989) can be accurate but are destructive and have poor time resolution.

Considerations for the Design of Probes

Space does not permit recapitulation of the physical basis and equations governing FRET, but some practical constraints and common misconceptions are worth mentioning. FRET should not be confused with donor emission of a real photon followed by reabsorption by an acceptor; the latter "trivial" process does not require molecular proximity, but rather, a sufficient optical density of acceptors anywhere within the path of the outgoing donor emission. FRET does not require that the fluorophores actually touch; in fact, such direct contact is usually undesirable because other very short range processes such as electron transfer or exciplex formation can then compete. Because both distance and mutual orientation affect FRET, it is hard to convert a FRET measurement into an absolute distance between fluorophores. However, if it is known that the probe exists in just two states, whose efficiencies of FRET have been empirically calibrated, measurement of FRET in the test specimen immediately indicates the relative occupancies of the two states. Even if one reference state has zero FRET, the corresponding emission ratio is usually non-zero and must be measured empirically. This undesired background arises because any wavelength that efficiently excites the donor has at least some slight ability to excite the acceptor as well, and any wavelength band that collects acceptor emission will also receive some donor emission as well. Minimizing the overlaps of the two excitation spectra and of the two emission spectra is equally or more important than maximizing the overlap of the donor emission and the acceptor excitation spectra.

Multiphoton excitation of the donor can initiate FRET just as one-photon excitation does. The only practical concern is that multiphoton excitation spectra are often unpredictable and not merely the one-photon spectra scaled to two or three times longer wavelengths. Therefore, the extent of undesirable overlap between the donor's and acceptor's multiphoton excitation spectra has to be checked empirically and is not reliably deducible from their one-photon spectra. Fortunately, the cyan and yellow mutants of green fluorescent protein (GFP), which are among the most promising donor–acceptor pairs for FRET, have proven well-suited to two-photon excitation. The effect of FRET on the ratio of 535-nm to 480-nm emissions was essentially the same whether the excitation was with single 440-nm photons or pairs of 770–810-nm photons (Fan et al. 1999).

TYPES OF INDICATORS USING FRET

Membrane Potential Indicators

Membrane voltage can be reported by its effect on the proximity of two fluorophores. One dye is a fluorescent ion, typically negative, that adsorbs strongly to the membrane and translocates from one side of the membrane to the other in response to membrane potential. The other dye is localized to one side of the membrane, typically the extracellular face of the plasma membrane. Hyperpolarization repels the mobile anion to that same face, encouraging FRET, whereas depolarization attracts the

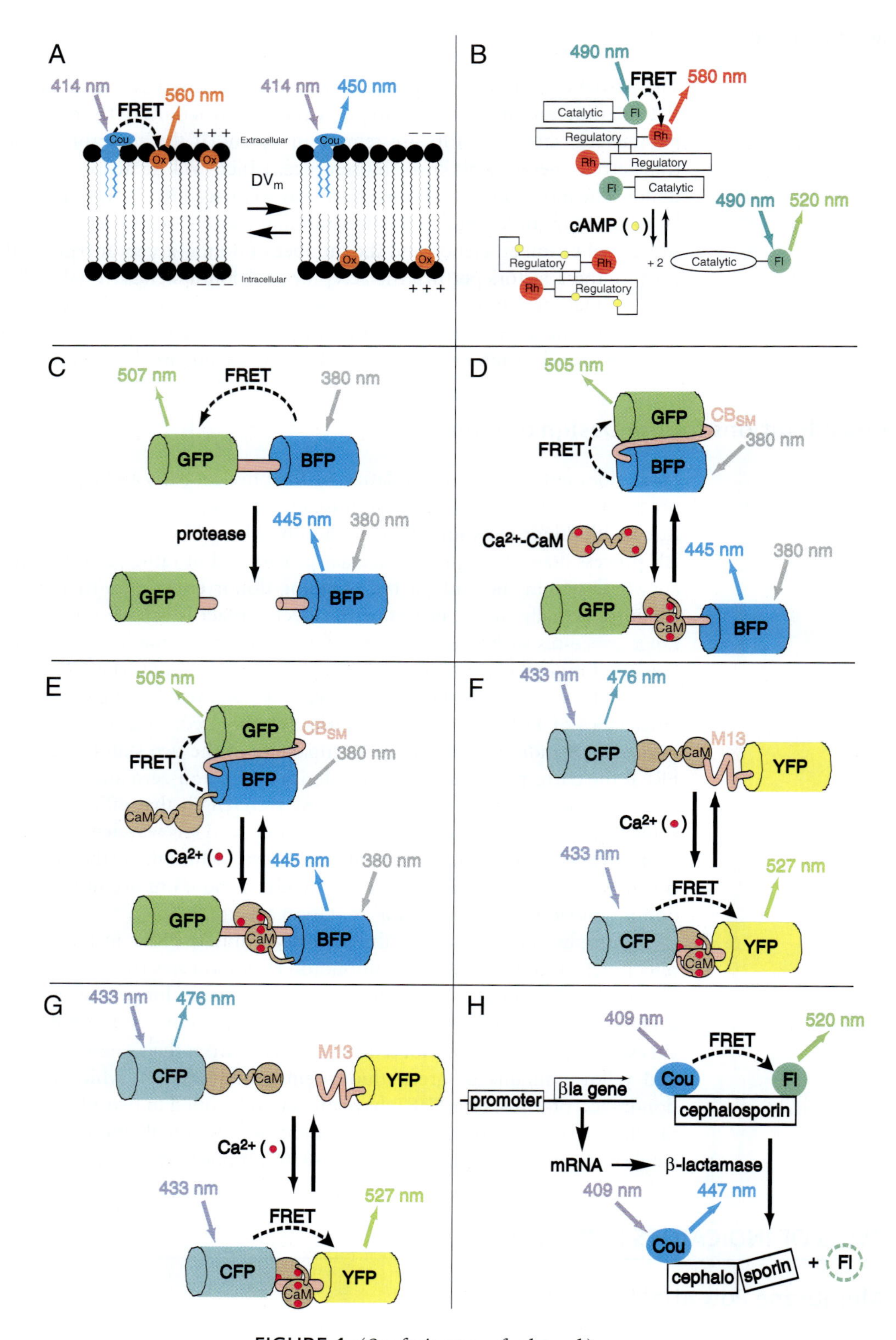

FIGURE 1. *(See facing page for legend.)*

anion to the cytoplasmic interface, increasing the average distance between the dyes and weakening FRET (Gonzalez and Tsien 1995, 1997). The major advantages of this two-fluorophore system are that it gives a much better combination of large signals and reasonably fast response kinetics than previous single-dye systems, especially when the readout is measured as the ratio of the donor and acceptor fluorescences. The ratio can change by as much as twofold for ~60 mV depolarization (J.E. Gonzalez, pers. comm.), and the response-time constant can be as little as 380 µsec at room temperature (Gonzalez and Tsien 1997), although not yet simultaneously. Such designed potentiometric indicators are used to image information processing in neural networks such as the swimming rhythm pattern generator in leech (Cacciatore et al. 1998). The main obstacle to wider application is that the two fluorophores are currently quite hydrophobic and can have difficulty reaching the desired target neurons in a complex tissue.

cAMP Indicators

cAMP can be imaged by monitoring the extent of FRET between the fluorescein-labeled catalytic subunit and rhodamine-labeled regulatory subunit of cAMP-dependent protein kinase (PKA). cAMP causes the dissociation of the two subunits and thereby disrupts FRET (Adams et al. 1991, 1993). This indicator is the only method currently available for imaging the dynamics of cAMP and has revealed subcellular diffusion gradients of cAMP in invertebrate neurons during synaptic stimulation and plasticity (Bacskai et al. 1993; Hempel et al. 1996). Now that it is commercially available, the major remaining deficiency is the dexterity required to introduce the 170-kD PKA holoenzyme complex into cells by pressure microinjection or diffusion from a patch-pipette in whole-cell mode.

GFP-based Systems

Proteins can be fluorescently tagged in vivo by fusion to GFP of the jellyfish *Aequorea victoria* (for a review of GFP-based systems, see Tsien 1998). Mutagenesis has made GFPs suitable for FRET by con-

---

FIGURE 1. Schematic mechanisms by which FRET-based indicators respond to biochemical signals. Drawings are not to scale. Incoming and outgoing arrows represent fluorescence excitation and emission, peaking at (but not confined to) the stated wavelengths. (*A*) Membrane potential indicators. The plasma membrane lipid bilayer (*black*) is loaded with a coumarin-labeled phospholipid (*blue*) and a membrane-permeant oxonol dye (*orange*). The illustration at left represents the normal negative resting potential where FRET from the donor coumarin to the acceptor oxonol is favored. The illustration at right shows how depolarization pulls the negatively charged oxonol to the inner leaflet of the membrane and decreases FRET. (*B*) cAMP indicators. The black outlines represent catalytic and regulatory subunits of cAMP-dependent protein kinase, labeled with fluorescein (Fl, *green circles*) and rhodamine (Rh, *red circles*), respectively. cAMP (*yellow circles*) binds to the regulatory subunits, changing their conformation and releasing the catalytic subunits. (*C*) Protease substrates. The green and blue cylinders, respectively, represent GFP and blue mutants (BFP). The protease-cleavable linker is shown as a pink tube. (*D*) Indicator for Ca^{2+}-CaM (*brown dumbbell*, with *red dots* representing the four Ca^{2+}) introduced by Romoser et al. (1997). The CaM-binding peptide from smooth muscle (CB_{SM}) connecting the GFP and BFP is shown in pink. Binding of Ca^{2+}-CaM to the CB_{SM} straightens the latter and disrupts intramolecular dimerization of the BFP and GFP. (*E*) Indicator for Ca^{2+} (Persechini et al. 1997) in which the CaM (with amino and carboxyl halves reversed) is fused via a short linker to the carboxyl terminus of the BFP. (Colors as in *D*.) (*F*) Indicator for Ca^{2+} (Miyawaki et al. 1997) in which a cyan mutant (CFP) of GFP, CaM (*brown*), a CaM-binding peptide ("M13," *pink*), and a yellowish mutant (YFP) of GFP are fused in order. (*G*) Indicator for Ca^{2+} and model for general intermolecular protein–protein interaction (Miyawaki et al. 1997), similar to *F* except that there is no covalent link between the CaM and the M13. (*H*) Indicator for gene expression using β-lactamase as a reporter. Transcription of the β-lactamase gene and expression of the enzyme cause cleavage of a cephalosporin conjugated to a coumarin (Cou, *blue*) and fluorescein (Fl, *green*) and disruption of intramolecular FRET. The hydrolysis of the cephalosporin β-lactam ring indirectly releases the fluorescein. The free thiol group left on the fluorescein (*dashed circle*) quenches the fluorescence of the latter.

centrating the excitation amplitude in a single, much amplified peak, and by generating new colors: blue, cyan, and yellowish green (Cubitt et al. 1995; Heim et al. 1995; Heim and Tsien 1996; Ormö et al. 1996; Tsien and Prasher 1998). The first pairs of GFPs suitable for FRET were blue mutants as donors and improved green mutants as acceptors. FRET was demonstrated by linking the two GFPs with a protease-sensitive linker and showing the loss of FRET upon addition of the protease, which was trypsin or Factor X_a (Heim and Tsien 1996; Mitra et al. 1996). Fortunately, the two GFPs themselves are quite resistant to most proteases. Such genetically encodable, tandem fusions with protease-specific cleavage sites offer the means to monitor protease activity inside transfected cells and organisms. For example, caspase-3 activity can be detected in apoptotic cells by flow cytometry (Xu et al. 1998).

Persechini Systems

Instead of using a protease to cleave a flexible linker, Romoser et al. (1997) made the linker a calmodulin (CaM)-binding peptide, CB_{SM}, derived from smooth muscle myosin–light-chain kinase. Binding of Ca^{2+}-CaM to the linker peptide should straighten the latter and increase the distance between the amino-terminal blue mutant and the carboxy-terminal GFP. Indeed, addition of Ca^{2+}-CaM markedly decreased the efficiency of FRET and increased the ratio of blue to green emissions by 5.7-fold. The GFP-CB_{SM}-BFP construct was expressed in bacteria, purified, and microinjected into mammalian cells, where it responded to cytosolic Ca^{2+} transients, albeit with signals much smaller than in vitro. The signals could be enhanced by coinjection of exogenous CaM, suggesting that the latter was limiting (Romoser et al. 1997). To create a self-contained Ca^{2+} sensor, the carboxyl terminus of the GFP-linker-BFP was fused to CaM, in which the amino- and carboxy-terminal halves were swapped to avoid certain steric constraints ("CaMCN"). The complete GFP-CB_{SM}-BFP-CaMCN fusion showed a ratio change of ~1.67-fold from zero to saturating Ca^{2+}, significantly less than that of GFP-CB_{SM}-BFP responding to unfused CaM (Persechini et al. 1997).

Cameleons

An independent approach to genetically encoded Ca^{2+} indicators was to include CaM within the fusion, consisting of cyan GFP-CaM-M13-yellow GFP, where M13 is another CaM-binding peptide domain from skeletal muscle myosin–light-chain kinase (see Chapter 78). Here the binding of Ca^{2+} to the CaM makes it wrap around the M13, which decreases the distance between the terminal GFP mutants and increases the efficiency of FRET. Cyan GFP (e.g., "ECFP") has become preferred because it is a much brighter and more photostable donor than blue GFP. The acceptor then has to be a yellowish GFP (e.g., "EYFP") to maintain enough separation between the donor and acceptor excitation spectra. The "E"s in these acronyms mean "enhanced" and refer to the inclusion of additional mutations that confer improved expression and folding in mammalian cells. These technical improvements made it possible to image these indicator molecules ("cameleons") biosynthesized in situ by transfected mammalian cells, thus obviating the need for microinjection. The Ca^{2+} affinity was easily tuned by mutation of the CaM, and the constructs were targetable to organelles or other privileged sites by appropriate trafficking signals (Miyawaki et al. 1997). More recently, the EYFP has been further mutated to reduce its sensitivity to acidification (Miyawaki et al. 1999).

Intermolecular Protein–Protein Interaction

Ca^{2+}-triggered intermolecular association of two separate fusions, ECFP-CaM and M13-EYFP, gives even larger FRET changes than the four-part chimera ECFP-CaM-M13-EYFP. The increased spectral signal presumably arises because in the absence of Ca^{2+}, the two GFPs can get much farther apart from one another than they can in the unimolecular cameleons. However, the cameleons are more reliable indicators of Ca^{2+} because their sensitivity to Ca^{2+} is independent of their absolute concentration, and because the CaM and M13 fused within a cameleon are almost completely indifferent to unlabeled bystander molecules of CaM and CaM-binding proteins (Miyawaki et al. 1999). Nevertheless, ECFP-

CaM and M13-EYFP show the feasibility of monitoring other dynamic protein–protein interactions in situ, by fusion with donor and acceptor GFPs (Miyawaki et al. 1997; Tsien and Miyawaki 1998).

Further examples of intermolecular protein–protein interactions detected in live single cells by FRET between GFP fusions include homodimerization of the nuclear transcription factor Pit-1 and heterodimer formation between Pit-1 and Ets-1, whereas interaction between Pit-1 and the estrogen receptor was undetectable (Day 1998). Similarly, association between the apoptosis-regulating proteins Bcl-2 and Bax was detectable in mitochondria by FRET, whereas Bcl-2 did not seem to interact with cytochrome *c* or with human papillomavirus E6 (Mahajan et al. 1998). In both these studies, the donor and acceptor were relatively primitive BFPs and GFPs, and no dynamic modulation of FRET was demonstrated. Nevertheless, they raise hopes for generalizing the use of FRET to image protein–protein interactions with spatial and temporal resolution in live cells.

β-Lactamase

Gene expression can be visualized in single living cells by using β-lactamase as a reporter enzyme to cleave novel membrane-permeant substrates and change their fluorescence from green to blue by disrupting FRET. This enzymatically amplified readout is several orders of magnitude more sensitive than GFP as a transcriptional reporter, although it does not track subcellular protein localization. It permits flow-cytometric selection and training of mammalian cell lines, and high-throughput screening of pharmaceutical candidate drugs (Zlokarnik et al. 1998). A cell line thus selected enabled demonstration that gene expression could be tuned to the frequency of inositol-1,4,5-trisphosphate and Ca^{2+} oscillations (Li et al. 1998). Novel cell clones and genetic elements responsive to acute stimuli can be found by enhancer trapping with β-lactamase as the reporter gene (Whitney et al. 1998).

CONCLUSION

It is safe to predict that applications of FRET will continue to expand. Detection of protein–protein interactions in live cells is increasingly important and popular. There is no more general way of designing detectors for physiological signals. Novel ways to attach the requisite donors and acceptors are being developed (Griffin et al. 1998; Tsien and Miyawaki 1998). Nevertheless, much trial and error is currently necessary to get FRET to work. Therefore, a major challenge will be to understand the empirical features and make FRET routinely and reliably transferable to new biological problems, circumstances, and laboratories.

REFERENCES

Adams S.R., Bacskai B.J., Taylor S.S., and Tsien R.Y. 1993. Optical probes for cyclic AMP. In *Fluorescent probes for biological activity of living cells—A practical guide* (ed. W.T. Mason), pp. 133–149. Academic Press, New York.

Adams S.R., Harootunian A.T., Buechler Y.J., Taylor S.S., and Tsien R.Y. 1991. Fluorescence ratio imaging of cyclic AMP in single cells. *Nature* **349:** 694–697.

Bacskai B.J., Hochner B., Mahaut-Smith M., Adams S.R., Kaang B.-K., Kandel E.R., and Tsien R.Y. 1993. Spatially resolved dynamics of cAMP and protein kinase A subunits in *Aplysia* sensory neurons. *Science* **260:** 222–226.

Cacciatore T.W., Brodfuehrer P., Gonzalez J.E., Tsien R.Y., Kristan W.B., Jr., and Kleinfeld D. 1998. Neurons that are active in phase with swimming in leech, and their connectivity, are revealed by optical techniques. *Soc. Neurosci. Abstr.* **24:** 1890.

Cubitt A.B., Heim R., Adams S.R., Boyd A.E., Gross L.A., and Tsien R.Y. 1995. Understanding, using and improving green fluorescent protein. *Trends Biochem. Sci.* **20:** 448–455.

Day R.N. 1998. Visualization of Pit-1 transcription factor interactions in the living cell nucleus by fluorescence resonance energy transfer microscopy. *Mol. Endocrinol.* **12:** 1410–1419.

Fan G.Y., Fujisaki H., Miyawaki A., Tsay R.-K., Tsien R.Y., and Ellisman M.H. 1999. Video-rate scanning two-photon excitation fluorescence microscopy and ratio imaging with yellow cameleons. *Biophys. J.* **76:** 2412–2420.

Gonzalez J.E. and Tsien R.Y. 1995. Voltage-sensing by fluorescence resonance energy transfer in single cells. *Biophys. J.* **69:** 1272–1280.

———. 1997. Improved indicators of cell membrane potential that use fluorescence resonance energy transfer. *Chem. Biol.* **4:** 269–277.

Heim R. and Tsien R.Y. 1996. Engineering green fluorescent protein for improved brightness, longer wavelengths and fluorescence energy transfer. *Curr. Biol.* **6:** 178–182.

Heim R., Cubitt A.B., and Tsien R.Y. 1995. Improved green fluorescence. *Nature* **373:** 663–664.

Hempel C.M., Vincent P., Adams S.R., Tsien R.Y., and Selverston A.I. 1996. Spatio-temporal dynamics of cAMP signals in an intact neural circuit. *Nature* **384:** 166–169.

Herman B. 1989. Resonance energy transfer microscopy. *Methods Cell Biol.* **30:** 219–243.

Jovin T.M. and Arndt-Jovin D.J. 1989. Luminescence digital imaging microscopy. *Annu. Rev. Biophys. Biophys. Chem.* **18:** 271–308.

Lakowicz J.R. 1983. *Principles of fluorescence spectroscopy*. Plenum Press, New York.

Li W., Llopis J., Whitney M., Zlokarnik G., and Tsien R.Y. 1998. Cell-permeant caged $InsP_3$ ester shows that Ca^{2+} spike frequency can optimize gene expression. *Nature* **392:** 936–941.

Mahajan N.P., Linder K., Berry G., Gordon G.W., Heim R., and Herman B. 1998. Bcl-2 and Bax interactions in mitochondria probed with green fluorescent protein and fluorescence resonance energy transfer. *Nat. Biotechnol.* **16:** 547–552.

Mitra R.D., Silva C.M., and Youvan D.C. 1996. Fluorescence resonance energy transfer between blue-emitting and red-shifted excitation derivatives of the green fluorescent protein. *Gene* **173:** 13–17.

Miyawaki A., Griesbeck O., Heim R., and Tsien R.Y. 1999. Dynamic and quantitative Ca^{2+} measurements using improved cameleons. *Proc. Natl. Acad. Sci.* **96:** 2135–2140.

Miyawaki A., Llopis J., Heim R., McCaffery J.M., Adams J.A., Ikura M., and Tsien R.Y. 1997. Fluorescent indicators for Ca^{2+} based on green fluorescent proteins and calmodulin. *Nature* **388:** 882–887.

Ormö M., Cubitt A.B., Kallio K., Gross L.A., Tsien R.Y., and Remington S.J. 1996. Crystal structure of the *Aequorea victoria* green fluorescent protein. *Science* **273:** 1392–1395.

Persechini A., Lynch J.A., and Romoser V.A. 1997. Novel fluorescent indicator proteins for monitoring free intracellular Ca^{2+}. *Cell Calcium* **22:** 209–216.

Romoser V.A., Hinkle P.M., and Persechini A. 1997. Detection in living cells of Ca^{2+}-dependent changes in the fluorescence emission of an indicator composed of two green fluorescent protein variants linked by a calmodulin-binding sequence. *J. Biol. Chem.* **272:** 13270–13274.

Stryer L. 1978. Fluorescence energy transfer as a spectroscopic ruler. *Annu. Rev. Biochem.* **47:** 819–846.

Tsien R.Y. 1998. The green fluorescent protein. *Annu. Rev. Biochem.* **67:** 509–544.

Tsien R.Y. and Miyawaki A. 1998. Seeing the machinery of live cells. *Science* **280:** 1954–1955.

Tsien R.Y. and Prasher D.C. 1998. Molecular biology and mutation of GFP. In *GFP: Green fluorescent protein: Properties, applications, and protocols* (ed. M. Chalfie and S. Kain), pp. 97–118. Wiley-Liss, New York.

Tsien R.Y., Bacskai B.J., and Adams S.R. 1993. FRET for studying intracellular signalling. *Trends Cell Biol.* **3:** 242–245.

Uster P.S. and Pagano R.E. 1986. Resonance energy transfer microscopy: Observations of membrane-bound fluorescent probes in model membranes and in living cells. *J. Cell Biol.* **103:** 1221–1234.

Whitney M., Rockenstein E., Cantin G., Knapp T., Zlokarnik G., Sanders P., Durick K., Craig F.F., and Negulescu P.A. 1998. A genome-wide functional assay of signal transduction in living mammalian cells. *Nat. Biotechnol.* **16:** 1329–1333.

Wouters F.S., Verveer P.J., and Bastiaens P.I. 2001. Imaging biochemistry inside cells. *Trends Cell Biol.* **11:** 203–211.

Xu X., Gerard A.L.V., Huang B.C.B., Anderson D.C., Payan D.G., and Luo Y. 1998. Detection of programmed cell death using fluorescence energy transfer. *Nucleic Acids Res.* **26:** 2034–2035.

Zlokarnik G., Negulescu P.A., Knapp T.E., Mere L., Burres N., Feng L., Whitney M., Roemer K., and Tsien R.Y. 1998. Quantitation of transcription and clonal selection of single living cells with β-lactamase as reporter. *Science* **279:** 84–88.

CHAPTER 75

Cellular Imaging of Bioluminescence

Jeffrey D. Plautz[1] and Steve A. Kay[2]

[1]*HighWire Press, Stanford University, Stanford, California 94305-6004;* [2]*The Scripps Research Institute, Department of Cell Biology, La Jolla, California 92037*

Bioluminescence results from a chemical reaction. Although there are numerous sources of bioluminescence in nature, including aequorin (Knight et al. 1991) and the bacterial *lux* genes (Liu et al. 1995), the most common bioluminescent reporter system used in research laboratories is based on luciferase, from the firefly (Wood 1995; Thompson et al. 1997; Welsh and Kay 1997). Luciferase binds luciferin; when the luciferin is oxidized and released, light is emitted. The reaction also requires energy (supplied in the cell as ATP) and oxygen. Several other factors can influence the rate of the reaction; however, these four components—enzyme, substrate, energy, and oxygen—are necessary and sufficient to complete the luciferase reaction (Ow et al. 1986).

Luciferase as a reporter gene has allowed a wide range of experiments that are difficult or impossible to carry out with any other current technology. Its quantitative reporting of transcriptional events and its short half-life of activity (3–5 hr; Thompson et al. 1991; Plautz et al. 1997b) have proven useful in a variety of experiments, ranging from in vitro measurements of promoter activation in transiently transfected culture cells (Day and Maurer 1990) to the recording of circadian rhythms in stably transformed bacteria (Liu et al. 1995), plants (Millar et al. 1992), and animals (Brandes et al. 1996; Geusz et al. 1997). Microscopic bioluminescent imaging of single glowing cells, however, has been difficult due to the relatively low, overall light output; it is not useful purely as a visual cell marker (Plautz et al. 1997a). This chapter presents techniques used in the authors' laboratories that have led to successful monitoring of single-cell bioluminescence over several days in cultured tissues.

PRACTICAL CONSIDERATIONS

Survival of the Sample

Live-tissue imaging presents many challenges. The most basic problem concerns keeping the sample alive and healthy. The sample must be alive to produce the ATP needed to supply energy to the luciferase reaction, and it must be healthy to synthesize the luciferase enzyme, in a manner relevant to the question addressed. In the case of long-term monitoring, this can be difficult.

Bacterial, plant, and insect cells are relatively easy to keep alive for days on the microscope stage. These cells generally do not require special atmospheric conditions and would even grow well in (sterile) dishes on the benchtop. Mammalian cells and tissues are more difficult. Although single-time-point or fast experiments may be carried out on suboptimally cultured tissue, special heated chambers with flowthrough media are generally required for good cell health (and therefore reliable data) during imaging. If the medium is not changed during a long-term (hours to days) experiment, then either large volumes of media must be used or the chamber must be sealed to the atmosphere to avoid medium depletion and sample drying due to evaporation.

Substrate Accessibility

Surviving samples must have access to oxygen and luciferin. Luciferin can generally be added directly to the medium and will remain effective for days or weeks at room temperature (Plautz et al. 1997a). The requirement for oxygen means that the sample cannot be placed directly under a coverslip if the imaging is to last more than a few minutes; in addition, the surface area and/or depth of the medium must be controlled so that sufficient gas can reach the tissue.

MATERIALS

Hardware Setup

Bioluminescence should be detected in a quantitative manner for the data to be most useful. Although bioluminescent imaging can be performed by exposing the sample to photographic film, modern microscopes, computers, and sensitive digital cameras make detection easier and more informative.

Microscope

The proper choice of microscope is important and will differ depending on each particular experimental setup. The main considerations are the sample type, and the sample's need for gas exchange. In general, inverted microscopes are more versatile than upright because the sample can sit on a coverslip on the bottom of a chamber. This allows the use of immersion objectives if desired. However, an upright microscope is useful for larger samples (where the end of the objective may fall below another part of the sample) or samples where the objective may be immersed directly in the culture medium.

Since bioluminescence is dim (often undetectable by eye), every emitted photon is important and should be captured. The light path should be as short and clear as possible. Internal mirrors and lenses should all be clean and aligned; every filter should be disengaged, and any light-splitting mirrors should be set so that 100% of the emitted bioluminescence goes to the detector.

Detector

The detector is extremely important. Successful imaging of bioluminescence necessitates an extremely sensitive camera head. Intensified and/or cooled CCD cameras meet this requirement. Hamamatsu model VIM intensified CCD and Photometrics liquid-nitrogen-cooled cameras are recommended; Princeton Instruments also offers cameras that appear to be suitable for single-cell bioluminescence imaging. The following are four critical performance characteristics to consider:

- Sensitivity
- Resolution
- Bit depth
- Dark current

Sensitivity is obviously important since every photon in bioluminescence imaging matters; photon loss should be minimized at the detector. Resolution determines the number of pixels in the chip, where a greater number of pixels offers greater detail recorded from the sample. Bit depth determines the number of gray levels detected at each pixel: 1-bit detectors show only the presence or absence of light at any particular pixel, whereas 12-bit detectors offer up to 4096 different levels of intensity at a single point. Finally, dark current directly affects the amount of time that the CCD may be exposed to the sample. High dark current allows noise to accumulate relatively quickly during signal capture. With the long exposure times typical for bioluminescence imaging, the accumulated noise

associated with relatively high dark current renders a long-exposure image useless.
Exposed CCD chips can be at least as sensitive as photographic film. Therefore, dark-room conditions in the imaging room are necessary to prevent bleaching of the image. The entire room should be light-tight; as an additional precaution, the microscope itself, or at least the sample, should also be protected. Optimally, the microscope should be contained in its own light-tight box with a sealed port for the computer cables; the practicality of this in the lab will depend on the experimental setup and available space.

Imaging Parameters

Three easily changeable parameters of the imaging system are the type of objective used, the exposure time, and the gain on the camera system. The interrelationships of these parameters are demonstrated in Figure 1.

Gain and Exposure Time

The overall signal increases as both gain and exposure time are increased. However, background also increases. By dividing signal by background, the curves change to show a point beyond which longer exposure times give diminishing returns with respect to image quality. The parameters must be set according to each particular experiment. For example, if a signal is bright and background can be minimized, a low gain may be appropriate. Likewise, very dim bioluminescent samples may need very long exposure times to accumulate sufficient signal to visualize much of anything. Although this also leads to higher background, these levels may be visually acceptable if the signal is sufficiently bright.

Objective Used

Objectives must be chosen according to the sample of interest. A 10× objective may have a higher NA than a 4× objective (allowing more light to be collected by it), but large samples may not fit in the field of the 10× objective. Therefore, to obtain an image of the whole sample, either use the 4× objective—its lower NA compensated for with increased gain and/or exposure time—or use the 10× objective and collect multiple images that can be electronically pieced together in postprocessing. The required level of detail (determined in large part by magnification) must also be considered in objective choice. Similarly, some samples are mounted in aqueous media; water-immersion objectives are most appropriate for these samples because the optics have been corrected for distortion caused by the different refractive indices of elements between the sample and the objective. Even though an oil-immersion objective with the same magnification may collect more photons than a water-immersion objective, the image itself may be somewhat distorted due to refractive-index mismatch. In addition, the immersion media may evaporate over long-term-imaging experiments.

The focal plane of the sample is also critical to maximum light throughput. Our experience has shown that light is best captured in the focal plane and that out-of-focus light contributes minimally to the overall signal. Some experimentation may be necessary to find the exact plane that provides the maximal signal.

The ideal policy is to have a wide array of objectives available for use, appropriate to a given imaging situation, along with a good working knowledge of their strengths and weaknesses. A more limited set of objectives may also be obtained with a specific set of samples or experiments in mind. Unfortunately, much of the equipment best suited for cellular bioluminescence imaging of live tissue (water-immersion objectives and sensitive CCD cameras) is still very expensive. In summary, the best choices of equipment and imaging parameters will be determined by the size and survival conditions of the sample, the overall light output of the sample, and the amount of data desired by the researcher.

DESCRIPTION OF A WORKING SETUP

In our laboratory, bioluminescence has been successfully imaged in whole animals and in intact living tissue with single-cell resolution (Fig. 2). These parameters have proven to balance exposure time and light throughput to allow quantitative imaging.

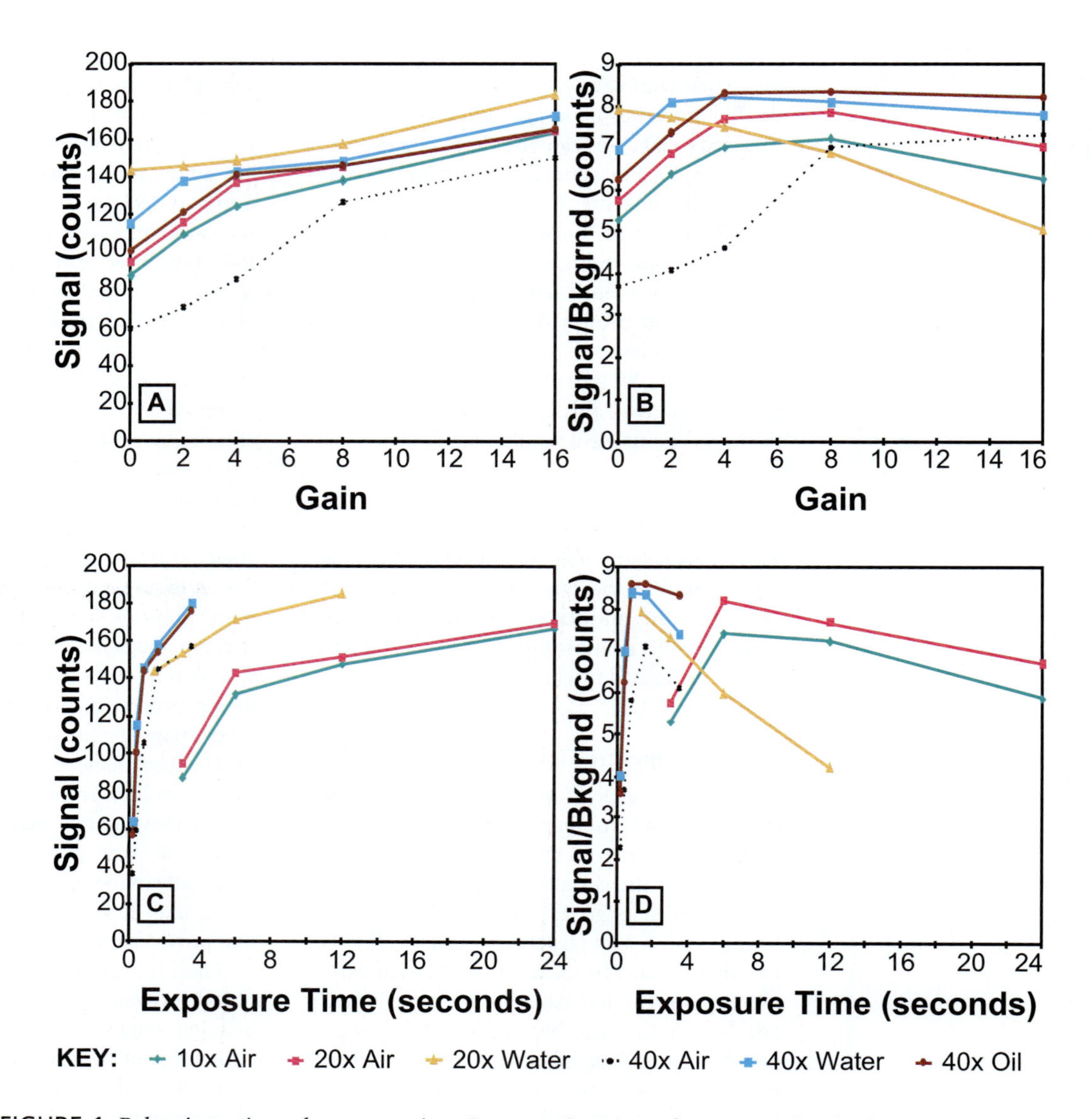

FIGURE 1. Balancing gain and exposure time. Increases in gain and exposure time both increase overall measured signal; however, gain and exposure time also affect background. When the signal-to-background ratio is calculated, a point is found beyond which higher gain and/or longer exposure times yield diminishing signal/background values. Although the data were taken from dim fluorescence rather than bioluminescence, the conclusions drawn can be extended to bioluminescence because both represent low-light detection. Data for these graphs were collected on an Olympus AX-70 upright microscope using a Hamamatsu C5810 color CCD camera. A 15-µm fluorescent bead was imaged in all cases. For image analysis, circles of equal area were placed either on the center of the bead or away from the bead in the background; the average brightness in each area was used to determine "signal" or "background," respectively. (*A*) Gain versus signal; (*B*) gain versus signal/background; (*C*) time versus signal; (*D*) time versus signal/background. The objectives in the key are listed in order of increasing NA. Gain values in *A* and *B* cover nearly the full range of the detector.

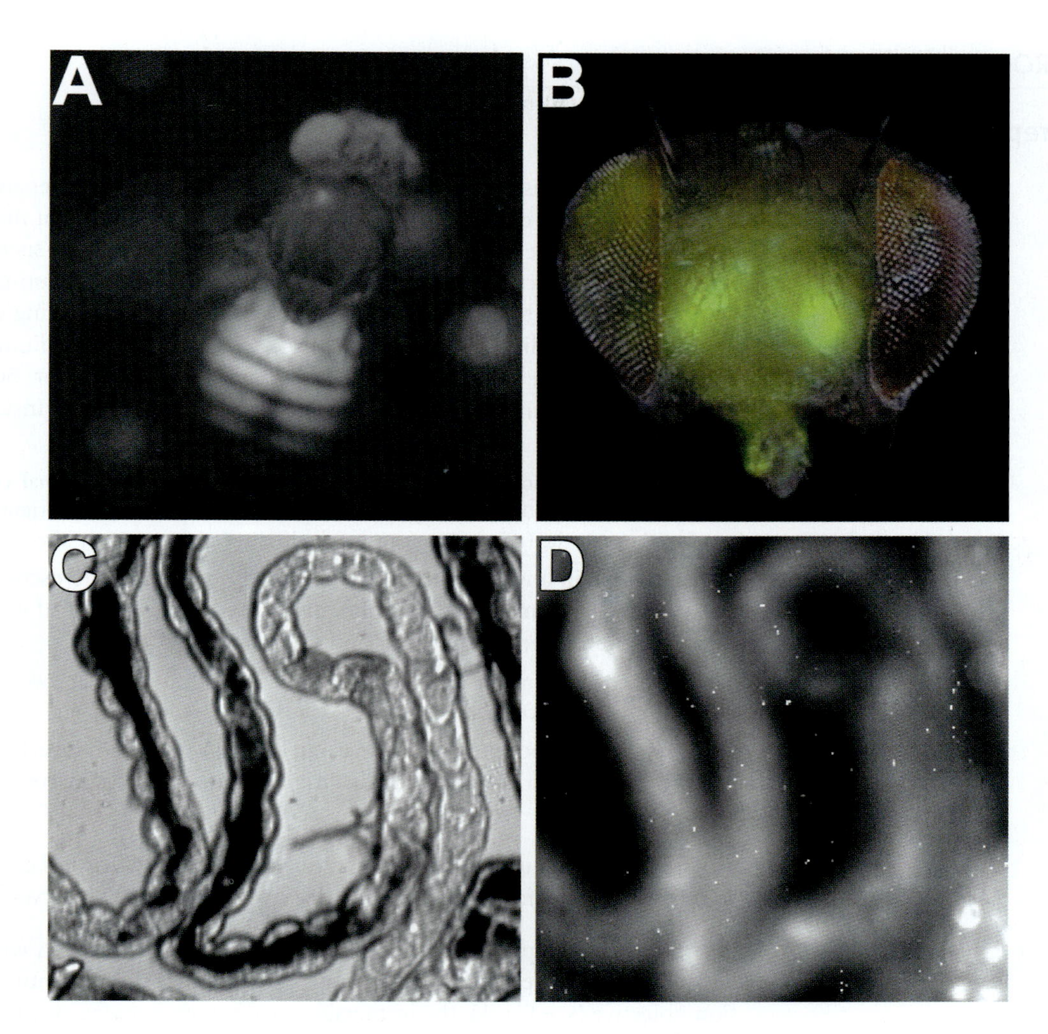

FIGURE 2. Whole-animal and single-cell imaging. (*A,B*) Bioluminescence can be detected from the whole animal. In *A*, the transgenic *Drosophila* (transformed with a construct that drives luciferase production from the digestive gene amylase; Plautz et al. 1997b) is expressing luciferase primarily from the anterior portion of the abdomen. Signal was captured for 1 hr with a 4X objective. Luciferase-mediated bioluminescence can be detected at the subanimal level. In the living period-luciferase transgenic head (Brandes et al. 1996) shown in *B*, bioluminescence was captured for 1 hr with a 10X objective. The tissue was then imaged separately with an external light source, and the two images were combined using Adobe Photoshop. Bioluminescence emanates primarily from the eyes, antennae, and proboscis. (*C,D*) Bioluminescence can also be detected from a single-period-luciferase *Drosophila* Malpighian tubule (a structure with kidney-like function). In *C*, the liquid-nitrogen-cooled CCD camera captured an image of the sample illuminated with ambient light; the sample was imaged for 0.5 sec with a 20X objective. This provides an outline of the entire sample and identifies single cells. Panel *D* shows the emitted bioluminescence from the same sample. The CCD chip was exposed for 1 hr.

PROCEDURE

Preparation of the Sample

In this example, cultured tissue from stably transformed luciferase-expressing *Drosophila* is imaged (Brandes et al. 1996; Plautz et al. 1997b). The procedure for preparation of the sample is given below.

1. Place the sample in the imaging environment. Select a large, healthy specimen that can be monitored whole or can act as a donor of dissected organs. For whole-animal imaging, anesthetize the animal according to local animal-handling guidelines; for imaging tissue, maintain the tissue in the culture medium (Plautz et al. 1997a). (Tissue Culture Medium: 85.9% S3 insect tissue culture medium, 12% fetal bovine serum [heat-inactivated for 30 min at 60°C]), 1% penicillin/streptomycin mixture<!>, 1% luciferin solution, 0.1% insulin solution [1 mg/ml stock].)

   Note: The concentration of the luciferin solution can vary, with a final concentration typically between 0.05 mM and 0.5 mM. Higher luciferin concentrations lead to brighter overall bioluminescence, but they use the (relatively expensive) luciferin reagent more rapidly.

2. Place the sample in a watertight chamber that can fit on the microscope stage. If the sample is not a whole living animal, fill the chamber with medium to ~ 20% of its volume. Seal the chamber with Parafilm to prevent evaporation of the medium.

   Note: The chamber should have an overall volume of 1–2 cc, with a flat coverslip bottom that is optically correct for the microscope.

3. Perform imaging at room temperature using an Olympus IX-70 inverted microscope. Cover the stage with black cloth during imaging in a darkened light-tight room. See below for typical imaging parameters.

   - *Exposure Time.* Bioluminescence is typically captured for 1 hr per exposure. Images are usully captured hourly for several consecutive days with no user intervention.

   - *Objective.* An Olympus 20x air objective balances a reasonable NA with single-cell resolution with an adequate field of view. Although better images can be captured with a water-immersion objective (see Fig. 1), the immersion medium evaporates over the course of a long-term experiment.

   - *Detector.* A Hamamatsu ORCA-ER cooled CCD camera is used for detection. The chip has 1324 x 1024-pixel resolution, but the signal in our experiments is intensified by binning the chip 8 x 8. Because the sample is so dim, the loss in resolution is an acceptable loss for the increase in signal. Gain is maximal (4x); the chip distinguishes among 4096 levels of intensity.

Caution: *See Appendix 3 for appropriate handling of materials marked with <!>.*

CONCLUSION

Bioluminescent imaging is emerging from its infancy. Its potential is great, since luciferase technology allows the visualization of real-time gene expression with single-cell, or even subcellular, resolution. Advances in imaging technology, as well as more widespread use of luciferase as a reported system, guarantee an increase in the number and range of systems that will surely benefit from this reporter. Still, experimental conditions vary from situation to situation. A proper understanding of the tradeoffs and components involved in single-photon detection will continue to be critical for successful cellular imaging of bioluminescence.

ACKNOWLEDGMENTS

The authors thank Dr. Reinhard Enders at Olympus microscopes for providing much of the equipment and technical expertise used in this research, and Butch Moomaw of Hamamatsu for help selecting low-light imaging cameras.

REFERENCES

Brandes C., Plautz J.D., Stanewsky R., Jamison C.F., Straume M., Wood K.V., Kay S.A., and Hall J.C. 1996. Novel features of *Drosophila period* transcription revealed by real-time luciferase reporting. *Neuron* **16:** 687–692.

Day R.N. and Maurer R.A. 1990. Pituitary calcium channel modulation and regulation of *prolactin* gene expression. *Mol. Endocrinol.* **4:** 736–742.

Geusz M.E., Fletcher C., Block G.D., Straume M., Copeland N.G., Jenkins N.A., Kay S.A., and Day R.N. 1997. Long-term monitoring of circadian rhythms in *c-fos* gene expression from suprachiasmatic nucleus cultures. *Curr. Biol.* **7:** 758–766.

Knight M.R., Campbell A.K., Smith S.M., and Trewavas A.J. 1991. Recombinant aequorin as a probe for cytosolic free Ca^{2+} in *Escherichia coli. FEBS Lett.* **282:** 405–408.

Liu Y., Golden S.S., Kondo T., Ishiura M., and Johnson C.H. 1995. Bacterial luciferase as a reporter of circadian gene expression in cyanobacteria. *J. Bacteriol.* **177:** 2080–2086.

Millar A.J., Short S.R., Hiratsuka K., Chua N.-H., and Kay S.A. 1992. Firefly luciferase as a reporter of regulated gene expression in higher plants. *Plant Mol. Biol. Rep.* **10:** 324–337.

Ow D.W., Wood K.V., DeLuca M., de Wit J.R., Helinski D.R., and Howell S.H. 1986. Transient and stable expression of the firefly luciferase gene in plant cells and transgenic plants. *Science* **234:** 856–859.

Plautz J.D., Kaneko M., Hall J.C., and Kay S.A. 1997a. Independent photoreceptive circadian clocks throughout *Drosophila. Science* **278:** 1632–1635.

Plautz J.D., Straume M., Stanewsky R., Jamison C.F., Brandes C., Dowse H.B., Hall J.C., and Kay S.A. 1997b. Quantitative analysis of *Drosophila period* gene transcription in living animals. *J. Biol. Rhythms* **12:** 204–217.

Thompson J.F., Hayes L.S., and Lloyd D.B. 1991. Modulation of firefly luciferase stability and impact on studies of gene regulation. *Gene* **103:** 171–177.

Thompson J.F., Geoghegan K.F., Lloyd D.B., Lanzetti A.J., Magyar R.A., Anderson S.M., and Branchini B.R. 1997. Mutation of a protease-sensitive region in firefly luciferase alters light emission properties. *J. Biol. Chem.* **272:** 18766–18771.

Welsh S. and Kay S.A. 1997. Reporter gene expression for monitoring gene transfer. *Curr. Opin. Biotechnol.* **8:** 617–622.

Wood K.V. 1995. Marker proteins for gene expression. *Curr. Opin. Biotechnol.* **6:** 50–58.

CHAPTER 76

Introduction of Green Fluorescent Protein into Hippocampal Neurons through Viral Infection

Roberto Malinow, Yasunori Hayashi, Mirjana Maletic-Savatic, Shahid H. Zaman, Jean-Christophe Poncer, Song-Hai Shi, José A. Esteban, Pavel Osten, and Ken Seidenman

Cold Spring Harbor Laboratory, Cold Spring Harbor, New York 11724

A number of methods have been used to express heterologous proteins in neurons. These include viral transfection (de Hoop et al. 1994; Pettit et al. 1995; Moriyoshi et al. 1996; Goins et al. 1997), lipofection (Holt et al. 1990), calcium phosphate (Watson and Latchman 1996), biolistic (Lo et al. 1994), and electroporation (see Chapters 12–14) approaches. Expression of green fluorescent protein (GFP), its more fluorescent mutant forms (e.g., EGFP [enhanced GFP]) (Cubitt et al. 1995; Heim et al. 1995), or their fusion protein derivatives, affords a number of informative possibilities in cellular neuroscience. EGFP is a soluble protein and appears to be homogeneously distributed within the cytosol of neurons when expressed. Thus, it reveals the structure of the neuron, including the cell body, and axonal and dendritic arbors. It is also sufficiently bright to reveal detailed structures such as axonal boutons and dendritic spines. When expressed as a fusion protein, EGFP can provide information about the distribution characteristics of the proteins within neurons. Furthermore, during single-cell electrophysiological studies, such expression can direct the investigator to record from a cell carrying a foreign gene. In this chapter, we describe the use of the Sindbis pseudovirus expression system to deliver GFP to neurons.

THE SINDBIS VIRUS EXPRESSION SYSTEM

Sindbis is a member of the alphaviruses, which are plus-stranded RNA viruses. Sindbis is related to the Semliki Forest virus that has previously been used for heterologous expression in neurons (de Hoop et al. 1994). The Sindbis virus infects a wide range of species including mammals, birds, reptiles, amphibians, and insects (Schlesinger 1993). Different strains of Sindbis virus can be used to selectively infect different cell types (Corsini et al. 1996). This chapter concentrates on one strain, DH(26S), that preferentially infects neurons over glia (50:1). To generate infective Sindbis virus particles that express a gene of interest, we have essentially followed the methods described by Bredenbeek et al. (1993). The methods have been used to express and analyze the expression of several proteins in various preparations (Shi et al. 1999; Hayashi et al. 2000; Zhu et al. 2000; Esteban et al. 2003).

The low level of pathogenicity of Sindbis virus in humans has allowed it to be classified as a Biosafety Level-2 (BL-2) agent by the NIH Recombinant DNA Advisory Committee. Nevertheless, before any constructs are synthesized, experiments should be cleared through an institutional biosafety committee. All personnel working with the Sindbis Expression System (Invitrogen) should be properly trained to work with BL-2 organisms. BL-2 precautions include the use of laminar flow hoods, laboratory coats, appropriate gloves and eye protection, and decontamination of infectious wastes. Sindbis virus can be inactivated by organic solvents, bleach, or autoclaving. In addition, the components of the Sindbis Expression System have been designed to guard against any potential health threats.

PROCEDURE A

Preparation of Sindbis Virus Expressing EGFP

MATERIALS

Recombinant Sindbis Plasmid Vector

pSinRep5 (Bredenbeek et al. 1993), containing the gene of interest, EGFP cDNA (derived from pEGFP-N1; Clontech). The EGFP gene is cloned between the subgenomic promoter and a poly(T) sequence. An alternative Sindbis expression vector, pSinRep(nsp2S^{726}), can be used for prolonged expression with decreased cytotoxicity in dissociated cultured neurons (see below). This vector is based on a mutant Sindbis virus containing a single-amino-residue change of P to S at position 726 (Dryga et al. 1997). The P726S mutation results in decreased viral replication and thus a lower-titer viral preparation.

Helper Virus Plasmid DH(26S)

The helper virus plasmid contains the gene for the structural proteins necessary for the production of virus particles (Schlesinger 1993). It is advantageous to use a highly expressing helper RNA containing mammalian tRNA sequence in its 5′ region (Bredenbeek et al. 1993). An alternative helper, termed DH-BB(tRNA/TE12), contains the 5′ tRNA sequence as well as neurotropic glycoprotein genes and can be used for efficient production of pSinRep(nsp2S^{726}) virus for infection of neurons (J. Kim and P. Osten, unpubl.).

Cells

Baby hamster kidney (BHK)-21 cells, plated in 10-cm (see step 3) and 35-mm (see step 10) dishes.

In Vitro Transcription

1. Linearize the recombinant Sindbis vector and the helper virus plasmid with appropriate restriction enzymes. Use 1 μg of each DNA.
2. Transcribe the linearized DNAs using a standard in vitro transcription protocol with an appropriate RNA polymerase (e.g., Ambion SP6 transcription kit).

   Note: The RNA thus produced is capped and polyadenylated. The Sindbis construct produces a recombinant transcript encoding the gene of interest and the native components essential for viral (and hence transgene) replication. This transcript does not encode the viral structural proteins necessary for the production of virus particles (Schlesinger 1993). These are encoded by the helper virus transcript (Bredenbeek et al. 1993).

Transfection

3. Use ~10 μg of each transcript to cotransfect BHK-21 cells (1×10^7 cells per sample, plated in a 10-cm dish), by electroporation.
4. Incubate the transfected cells at 37°C. After 24 hours, estimate the efficiency of transfection by examining the cells for cytopathic effects (elongation, detachment from substrate, etc.). By this time, most cells should be showing cytopathic effects.
5. Return the cells to the incubator for a further 12–24 hours.

   Note: For constructs driving EGFP expression, the efficiency of the electroporation step can easily be checked by examining the cells under an inverted microscope equipped for EGFP fluorescence imaging. Alternatively, if antibodies are available, immunocytochemistry can be performed on the cells after the viral supernatant has been removed. Using this method, more than 90% of cells should show the exogenous protein. This level of efficiency is necessary for the production of high-titer virus suspension. After 36–48 hours, the culture medium containing the viruses is collected (designated unpurified infective supernatant), and this can be used to infect other cells, such as neurons. At this time, almost all cells should exhibit cytopathic effects.

6. Collect the cell culture medium, containing the recombinant virus particles, and centrifuge in an ultracentrifuge using a swinging-bucket rotor (e.g., Beckman SW41) at 160,000*g* for 90 min at 4°C.
7. Store the resulting viral suspension at –80°C.

   Note: Essentially no helper virus RNA is packaged into the newly produced viral particles because the helper RNA either completely lacks a packaging signal, such as in the case of the DH-BB(tRNA/TE12) helper, or contains the signal, but packages very inefficiently due to other deletions in the Sindbis genome (Bredenbeek et al. 1993).This prevents any further replication of the virus and leads to the production of heterologous protein only when host cells are infected with the virus.

Titration of Sindbis Pseudovirus Suspension

8. Infect a known number of BHK-21 cells (10^5 cells/35-mm dish) with an unknown number of virus particles in a known volume of culture supernatant (e.g., 0.5–50 ml).

   Note: Because the Sindbis pseudovirus particles do not undergo a second round of infection and, thus, do not form plaques, the titer of the virus solution cannot be determined with a conventional plaque assay.

9. Incubate for 24 hours at 37°C and then determine the number of infected cells as a portion of the total number of cells.

   Note: The proportion of infected cells versus the total number of cells in several microscopic fields can be determined either by fluorescence (for EGFP or its fusion protein), by X-gal staining (for LacZ), or by immunostaining (for any protein for which an antibody is available).

10. Calculate the virus titer as follows:

    Virus titer = (Proportion of infected cells per visual field × 10^5)/(Total number of cells per visual field × Volume of virus solution)

    Notes: A typical yield from this protocol is 10^6 to 10^7 infective particles/ml, but in some cases, it can be as high as 10^8 infective particles/ml. Although the source of this variability is not known, it is due in part to the quality of RNA, viability of BHK-21 cells used for electroporation, and the length and species of cDNA used.

    Note that the amount of virus-containing solution required to generate an appropriate infection (i.e., injection in slices, as described below) is often determined empirically, without precise titration.

    For some experiments, in which a particularly efficient infection is required (e.g., infection of slices), it may be necessary to concentrate the virus suspension. The titer of virus can be increased by 50–100-fold as follows.

11. If necessary, concentrate the viral suspension as follows.
   a. Remove cell debris by centrifuging at 400*g* for 10 min and transfer the supernatant to an ultracentrifuge tube.
   b. Centrifuge the supernatant in an ultracentrifuge using a swinging-bucket rotor (e.g., Beckman SW41) at 160,000*g* for 90 min at 48°C.
   c. Aspirate the supernatant from top, leaving ~200 ml.
   d. Resuspend the pellet (which is usually invisible), aliquot into small amounts (~5 ml), and store at –80°C.

PROCEDURE B

Infecting Dissociated Hippocampal Cultured Neurons with Recombinant Sindbis Virus

In this procedure, dissociated hippocampal neurons are infected with Sindbis virus. A confluent monolayer of cortical astrocytes, derived from P1 rat pups (100,000 cells), is formed on a 25 x 25-mm coverslip, coated with poly-L-lysine. (Note: If neurons are plated without an astroglial feeder layer [i.e., directly on poly-L-lysine], Sindbis infection efficiency will be greatly reduced. Since Sindbis is a membrane-enveloped virus, it probably sticks to poly-L-lysine. Use an alternative substrate for plating neurons directly on cell culture plates [e.g., collagen].) Neurons are dissociated from E19 rat hippocampi by trypsin digestion followed by trituration and plated onto the astrocyte layer at 68,000 cells/coverslip (Banker and Goslin 1991; Maletic-Savatic and Malinow 1998).

MATERIALS

Cells and Culture Medium

Astrocytes and cortical neurons, as described in the introductory text, and serum-free medium for the cultivation of neurons and astrocytes.

MEM

(Earle's salt, GIBCO-BRL 11095-080), 0.001% ovalbumin, 25 ng/ml insulin, 1 mg/ml transferrin, 10 mM sodium pyruvate, 2 ppm progesterone<!>, 3 ppm SeO_2<!>, 0.5 nM putrescine (all from Sigma).

Caution: *See Appendix 3 for appropriate handling of materials marked with <!>.*

1. Culture cells as follows.
   a. Plate astrocytes and neurons on a 25-mm-square glass coverslip (Belco).
   b. Maintain cultures in serum-free medium (Banker and Goslin 1991) for 7–14 days at 35°C in an atmosphere of 5% CO_2.
2. Infect the cultured neurons by adding an aliquot (usually 5–50 ml) of the unpurified infective supernatant suspension (as prepared in Procedure A above).

   Note: Detectable expression is seen within 6 hours of infection. After 3 days, there are few obvious adverse effects on cell morphology. After 5 days, however, some cells show clear toxic effects. One study reports infection of cultured dorsal root ganglion cells with Sindbis virus, and expression of *lacZ* could be detected for more than 1 month (Corsini et al. 1996).

PROCEDURE C

Infecting Organotypic Hippocampal Slices

MATERIALS

Hippocampal Slices

Organotypic hippocampal slices are prepared as described previously by Stoppini et al. (1991). This protocol was optimized by taking animals of various ages and varying incubation and virus application times.

Virus

Sindbis-EGFP virus, containing the gene of interest, in suspension (as prepared in Procedure A above).

1. Prepare organotypic hippocampal slices from PND6 (postnatal day 6) to PND7 rat pups on Minicell-CM culture plate insert (Millipore), as described by Stoppini et al. (1991), in a medium described by Musleh et al. (1997).
2. Maintain the tissue slices at 35°C in an atmosphere of 5% CO_2. Change medium 2–3 times a week.
3. After 1–10 days, inject Sindbis-EGFP virus particles using a fine (tip ~20 µm) glass capillary by applying repeated pressure pulses (5 msec, 2 psi).

   Notes: Injection of virus into slices is conducted as described by Pettit et al. (1995).

   Visualization of virus solution with 0.1% Fast Green<!> helps handling.

   A minimum of 4–5 days after preparation is necessary for the slices to become adherent to the substrate filters.

3. Imaging and/or electrophysiological analysis can be carried out 12 hours to 5 days after infection.

IMAGING

Expression of EGFP in this manner produces very bright neurons, comparable with neurons loaded intracellularly with a high concentration of fluorescein. Neurons expressing EGFP can be imaged under epifluorescence and two-photon laser-scanning microscopy. For epifluorescence, both xenon 75-W and mercury 100-W light sources can be used.

For epifluorescence, filter sets from Chroma Technology, e.g., 31001 (Ex D480/30; Di 505DCLP; Em D535/40), or Omega, e.g., XF23 (Ex 485DF22; Di 505DRLP02; Em 535DF35), are suitable. These filter sets produce very little (<1%) overlap between EGFP and Texas Red signals. There can be significant overlap between the EGFP signal (which appears green to the eye) and background autofluorescence (which appears more yellow and is generally restricted to astrocytes).

ADVANTAGES AND LIMITATIONS

Several limitations to the use of Sindbis virus should be kept in mind. First, there is a limit to the size of the construct inserted in the viral genome. This is ~6 kb. Second, cell toxicity is associated with infection. Different tissues have different limits. Dissociated cultured neurons infected with pSinRep5 virus do not survive more than ~2 days following infection, whereas cells infected with the attenuated pSinRep(nsp2S^{726}) virus are viable for ~3–5 days. Neurons in cultured organotypic slices survive ~4–5 days following infection even with the pSinRep5 virus. We have found that neurons infected with GFP for less than 3 days have no difference in membrane or synaptic properties compared to noninfected neurons. Neurons in animals 12–21 days old infected in vivo (see Takahashi et al. 2003) do not survive more than ~72 hours after infection. Neurons in older animals are more poorly infected, and infection causes more toxicity earlier. In slice cultures, we have not been able to express two different proteins in individual neurons, by coinfection. There are reports, however, that dissociated neurons in culture can be coinfected with distinct Sindbis viruses (e.g., see Perez et al. 2001). These limitations can be overcome by using biolistic delivery, electroporation, or other viral vectors (e.g., HSV) (for details, see Carlezon et al. 2000; Chapter 6).

The delivery of recombinant proteins to neurons can be difficult, variable, and time-consuming. However, the Sindbis expression system provides a relatively easy and reproducible means of achieving such delivery. Generation of the infective particles can be rapid (within 1 week of having an appropriate subcloned construct). The strain used here is neurotropic, which is a particular advantage because it does not infect glia cells, and imaging provides better resolution. Infected cells remain viable for several days (at least) after infection and a large fraction of neurons can be infected simultaneously. Infection can be anatomically targeted to a small group of neurons, and gene expression is relatively rapid (detectable within hours). In conclusion, the Sindbis expression system has several features that make it preferable over other vector systems for some applications.

REFERENCES

Banker G.A. and Goslin K. 1990. *Culturing nerve cells.* MIT Press, Cambridge, Massachusetts.

Bredenbeek P.J., Frolov I., Rice C.M., and Schlesinger S. 1993. Sindbis virus expression vectors: Packaging of RNA replicons by using defective helper RNAs. *J. Virol.* **67:** 6439–6446.

Carlezon Jr., W.A., Nestler E.J., and Neve R.L. 2000. Herpes simplex virus-mediated gene transfer as a tool for neuropsychiatric research. *Crit. Rev. Neurobiol.* **14:** 47–67.

Corsini J., Traul D.L., Wilcox C.L., Gaines P., and Carlson J.O. 1996. Efficiency of transduction by recombinant Sindbis replicon virus varies among cell lines, including mosquito cells and rat sensory neurons. *BioTechniques* **21:** 492–497.

Cubitt A.B., Heim R., Adams S.R., Boyd A.E., Gross L.A., and Tsien R.Y. 1995. Understanding, improving and using green fluorescent proteins. *Trends Biochem. Sci.* **20:** 448–455.

de Hoop M.J., Olkkonen V.M., Ikonen E., Williamson E., von Poser C., Meyn L., and Dotti C.G. 1994. Semliki Forest virus as a tool for protein expression in cultured rat hippocampal neurons. *Gene Ther.* (suppl. 1) **1:** S28–S31.

Dryga S.A., Dryga O.A., and Schlesinger S. 1997. Identification of mutations in a Sindbis virus variant able to establish persistent infection in BHK cells: The importance of a mutation in the *nsP2* gene. *Virology* **228:** 74–83.

Esteban J.A., Shi S.H., Wilson C., Nuriya M., Huganir R.L., and Malinow R. 2003. PKA phosphorylation of AMPA receptor subunits controls synaptic trafficking underlying plasticity. *Nat. Neurosci.* **6:** 136–143.

Goins W.F., Krisky D., Marconi P., Oligino T., Ramakrishnan R., Poliani P.L., Fink D.J., and Glorioso J.C. 1997. Herpes simplex virus vectors for gene transfer to the nervous system. *J. Neurovirol.* (suppl. 1) **3:** S80–S88.

Haas K., Sin W.C., Javaherian A., Li Z., and Cline H.T. 2001. Single-cell electroporation for gene transfer in vivo. *Neuron* **29:** 583–591.

Hayashi Y., Shi S.H., Esteban J.A., Piccini A., Poncer J.C., and Malinow R. 2000. Driving AMPA receptors into synapses by LTP and CaMKII: Requirement for GluR1 and PDZ domain interaction. *Science* **287:** 2262–2267.

Heim R., Cubitt A.B., and Tsien R.Y. 1995. Improved green fluorescence. *Nature* **373:** 663–664.

Holt C.E., Garlick N., and Cornel E. 1990. Lipofection of cDNAs in the embryonic vertebrate central nervous system. *Neuron* **4:** 203–214.

Lo D.C., McAllister A.K., and Katz L.C. 1994. Neuronal transfection in brain slices using particle-mediated gene transfer. *Neuron* **13:** 1263–1268.

Maletic-Savatic M. and Malinow R. 1998. Calcium-evoked dendritic exocytosis in cultured hippocampal neurons. I. *Trans*-Golgi network-derived organelles undergo regulated exocytosis. *J. Neurosci.* **18:** 6803–6813.

Moriyoshi K., Richards L.J., Akazawa C., O'Leary D.D., and Nakanishi S. 1996. Labeling neural cells using adenoviral gene transfer of membrane-targeted GFP. *Neuron* **16:** 255–260.

Musleh W., Bi X., Tocco G., Yaghoubi S., and Baudry M. 1997. Glycine-induced long-term potentiation is associated with structural and functional modifications of alpha-amino-3-hydroxyl-5-methyl-4-isoxazolepropionic acid receptors. *Proc. Natl. Acad. Sci.* **94:** 9451–9456.

Perez J.L., Khatri L., Chang C., Srivastava S., Osten P., and Ziff E.B. 2001. PICK1 targets activated protein kinase C alpha to AMPA receptor clusters in spines of hippocampal neurons and reduces surface levels of the AMPA-type glutamate receptor subunit 2. *J. Neurosci.* **15:** 5417–5428.

Pettit D.L., Koothan T., Liao D., and Malinow R. 1995. Vaccinia virus transfection of hippocampal slice neurons. *Neuron* **14:** 685–688.

Schlesinger S. 1993. Alphaviruses—Vectors for the expression of heterologous genes (review). *Trends Biotechnol.* **11:** 18–22.

Shi S.H., Hayashi Y., Petralia R.S., Zaman S.H., Wenthold R.J., Svoboda K., and Malinow R. 1999. Rapid spine delivery and redistribution of AMPA receptors after synaptic NMDA receptor activation. *Science* **284:** 1811–1816.

Stoppini L., Buchs P.A., and Muller D. 1991. A simple method for organotypic cultures of nervous tissue. *J. Neurosci. Methods* **37:** 173–182.

Takahashi T., Svoboda K., and Malinow R. 2003. Experience strengthening transmission by driving AMPA receptors into synapses. *Science* **299:** 1584–1588.

Watson A. and Latchman D. 1996. Gene delivery into neuronal cells by calcium phosphate-mediated transfection. *Methods* **10:** 289–291.

Zhu J.J., Esteban J.A., Hayashi Y., and Malinow R. 2000. Postnatal synaptic potentiation: Delivery of GluR4-containing AMPA receptors by spontaneous activity. *Nat. Neurosci.* **3:** 1098–1106.

CHAPTER 77

Green Fluorescent Proteins for Measuring Voltage

Micah S. Siegel[1] and Ehud Y. Isacoff[2]

[1]*Computational and Neural Science Graduate Program, Howard Hughes Medical Institute, California Institute of Technology, Pasadena, California 91125;* [2]*Department of Molecular and Cell Biology, University of California, Berkeley, California 94720*

Measuring signal transduction, in large numbers of cells, with high spatial and temporal resolution, is a fundamental problem for studying information processing in the nervous system. To address this problem, a family of detectors that are chimeras between signal transduction proteins and fluorescent proteins has been designed. The prototype sensor is a genetically encoded probe that can be used to measure transmembrane voltage in single cells. This chapter describes a modified green fluorescent protein (GFP), fused to a voltage-sensitive K^+ channel, so that voltage-dependent rearrangements in the K^+ channel induce changes in the fluorescence of GFP. The probe has a maximal fractional fluorescence change of 5.1%, making it comparable to some of the best organic voltage-sensitive dyes. Moreover, the fluorescent signal is expanded in time in a manner that makes the signal 30-fold easier to detect than a traditional linear dye. DNA-encoded sensors have the advantage that they may be introduced into an organism noninvasively and targeted to specific brain regions, cell types, and subcellular compartments.

APPROACHES TO THE STUDY OF CELLULAR SIGNALING EVENTS

Use of Fluorescent Dyes as Sensors

Fluorescent indicator dyes have revolutionized our understanding of cellular physiology by providing continuous measurements in single cells and cell populations. There are two major impediments to progress using indicator dyes: (1) the lack of direct, noninvasive assays for most cellular communication events and (2) the difficulty of making observations on selected cell populations. Practically, these dyes must be synthesized chemically and introduced as hydrolyzable esters or by microinjection (Cohen and Lesher 1986; Gross and Loew 1989; Tsien 1989). Delivering indicator dyes to specific cell populations would be a significant advantage for many experiments, but this has proven to be a difficult problem. In the absence of such localization, optical measurements in neural tissue usually cannot distinguish whether a signal originates from activity in neurons or glia, nor can it determine which types of neurons are involved.

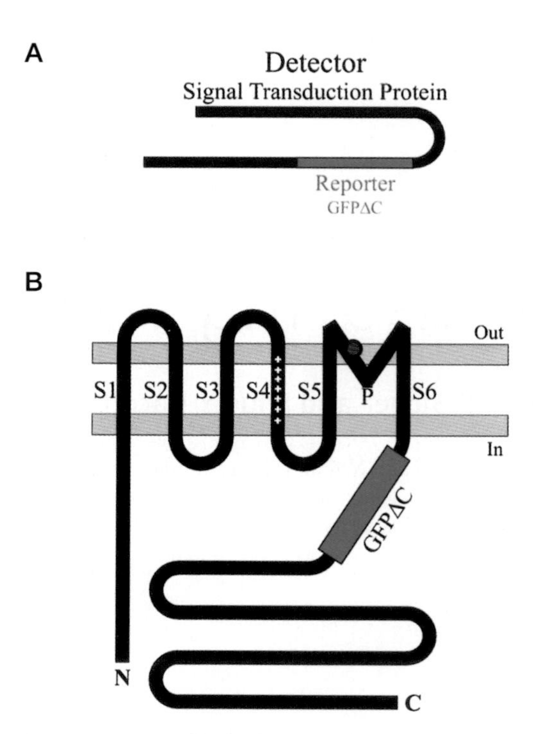

FIGURE 1. GFP illuminates signal transduction events. (*A*) Reporter/detector chimeras: GFP is fused to a signal transduction protein. (*B*) Putative orientation of the voltage-sensor prototype (FlaSh) in the cell membrane (*gray*). GFPΔC (*green*) was inserted in-frame into the Shaker K^+ channel (*black*). The fourth transmembrane segment S4 is positively charged. To make GFPΔC, amino acids 229–238 were deleted from the carboxyl terminus of wild-type GFP. A point mutation W434F (*red circle*) was made in the pore of the channel to eliminate ionic current through the sensor. Note that GFPΔC is intracellular.

Use of Protein-based Sensors

To increase our understanding of the signaling events that govern development, sensory transduction, learning, and memory, it is necessary to expand the range of signals that can be detected and to develop means of targeting sensors to specific cellular locations. One general approach to this problem is to construct sensors out of proteins—the biological molecules that transduce and transmit cellular signals (Siegel 1997). This approach would harness the high sensitivity and specificity of biological systems, which can detect an enormous range of signals with exquisite sensitivity. Moreover, it could permit the detection of signaling events throughout a signal-transduction cascade: from the earliest stage of membrane transduction, through the network of signaling relays and amplification steps, and finally to downstream events that occur anywhere from the nucleus back to the plasma membrane.

Because protein-based sensors are DNA-encoded, they can be placed under the control of cell-specific promoters, introduced in vivo or in vitro using gene-transfer techniques, and even targeted to specific subcellular compartments using protein signals recognized by protein-sorting machinery in the cell. In addition to solving the problem of sensor production and targeting, DNA-encoded sensors have the advantage that, unlike organic dyes, they can be rationally "tuned" by modification of their functional domains with mutations that are known to adjust their dynamic range of operation. Finally, by creating "biological spies" out of native proteins, the introduction of foreign substances that could interfere with cellular physiology is avoided.

A family of protein-based detectors for measuring signal transduction events in cells has been designed (Fig. 1A). These sensors are chimeras between a signal transduction protein fragment (detector) and a fluorescent protein (reporter). This chapter describes a prototypical example of this class of sensors: a GFP sensor that has been engineered to measure fast membrane-potential changes in single cells and in populations of cells.

FLUORESCENT SHAKER (FLASH) K⁺ CHANNEL: A SENSOR TO MEASURE CHANGES IN TRANSMEMBRANE POTENTIAL

In brief, a chimeric protein that is a modified GFP (Chalfie et al. 1994) was fused in-frame at a site just after the sixth transmembrane segment (S6; Fig. 1B) of the voltage-activated Shaker K^+ channel (Tempel et al. 1987; Baumann et al. 1988; Kamb et al. 1988). A detailed description of this sensor protein has been described previously (Siegel and Isacoff 1997 and references therein). (Henceforth, this chimera is called FlaSh, for *Fl*uorescent *Sh*aker.) To prevent FlaSh from loading down target cells with an additional potassium current, a point mutation was engineered into the pore of the channel. This mutation prevents ion conduction but preserves the channel's gating rearrangements in response to voltage changes.

The idea was that the voltage-dependent structural rearrangements in the channel would be transmitted to GFP, resulting in a measurable change in its spectral properties. The Shaker-GFP chimeric gene reports changes in membrane potential by a change in its fluorescence emission. In addition, the fluorescence response is amplified in time over the electrical event, drastically increasing the optical signal power per event. Temporal amplification in FlaSh is due to the response kinetics of the Shaker channel. Taken together, the properties of genetic encoding and temporal amplification allow the sensor to be delivered to selected cells in which action potentials may be detected with standard imaging equipment.

RESULTS

A Blueprint for FlaSh

The behavior of FlaSh was characterized in single *Xenopus laevis* oocytes by cRNA injection and voltage-clamp fluorimetry. Voltage steps from a holding potential of –80 mV evoked fluorescence emission changes and gating currents, but no ionic currents (Fig. 2A). The relation of the steady-state fluorescence change to voltage was sigmoidal and correlated closely with the steady-state gating-charge-to-voltage relation (Fig. 2B), indicating that in FlaSh, the fluorescent emission of GFP is coupled to the voltage-dependent rearrangements of the Shaker channel. The dynamic range of FlaSh is approximately –50 mV to –30 mV.

Kinetics of FlaSh

To determine whether FlaSh could respond to short-duration electrical activity, its fluorescence kinetics were investigated in response to brief voltage pulses. Surprisingly, short voltage transients of 3 msec and longer evoked long, stereotypical fluorescent responses (Fig. 2C). The magnitude of the fluorescence change was related to the duration of the step, but its kinetics of onset and recovery were constant for voltage spikes between 3 msec and 12.5 msec in duration. The entire collection of fluorescent responses was well fit by a single α-function with time constant of 23 msec for F_{on} and 105 msec for F_{off}.

The stereotypical fluorescence response was clearly visible in single-sweep recordings (data not shown). Subsequent events that occurred during the time course of the fluorescence change summated with the original response. For a train of identical brief pulses, the integral of the fluorescence response was constant at frequencies of 20 Hz and lower. Over this range, a linear filter model accounted well for the shape of the fluorescence response to a pulse train.

The response of FlaSh to physiologically realistic voltage traces was also examined. The voltage transient in response to light was measured for a variety of salamander retinal cell types (B. Rosko, pers. comm.), and these voltage transients were applied to oocytes expressing FlaSh. FlaSh reflected the dynamics of on-bipolar cell transients quite well (Fig. 2D, top), including the "sag" due to adaptation. Interestingly, FlaSh did not capture the response of wide-field amacrine cells (Fig. 2D, bottom). This makes sense because the response of the on-bipolar cell is within the dynamic range of FlaSh (–50 mV to –30 mV, highlighted in gray), whereas the response of wide-field amacrine cells falls outside this range.

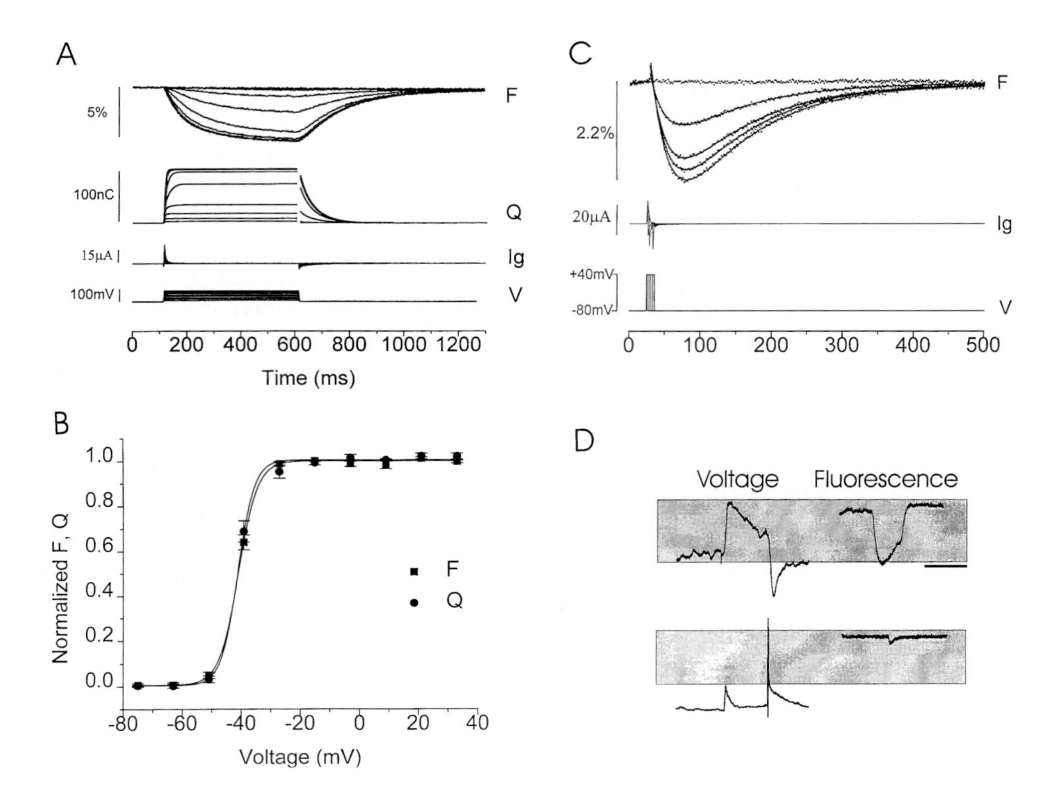

FIGURE 2. Membrane potential modulates fluorescence output in FlaSh. (*A*) Simultaneous two-electrode voltage-clamp recording and photometry show current and fluorescence changes in response to voltage steps (V) between –60 mV and 10 mV, in 10-mV increments. Holding potential was –80 mV. FlaSh exhibits on and off gating currents (I_g) but no ionic current. Integrating the gating current gives the total gating charge (Q) moved during the pulse. FlaSh fluorescence (F) decreases reversibly in response to membrane depolarizations. Traces are the average of 20 sweeps. Fluorescence scale, 5% ΔF/F. (*B*) FlaSh fluorescence is correlated to Shaker activation gating. The voltage dependence of the normalized steady-state gating-charge displacement (Q) overlaps the normalized steady-state fluorescence change (F). Both relations were fit by a single Boltzmann equation (*solid lines*). Values are mean ± S.E.M. from five oocytes. (*C*) The fluorescence response of FlaSh following short pulses from –80 mV to 40 mV (V) follows a characteristic trajectory that is well fit by a sum of two exponentials with time constants of 23 and 105 msec (F, *solid lines*). Pulses were 1.25, 3.75, 6.25, 8.75, and 11.25 msec in duration. Traces are the average of 25 sweeps. (*D*) FlaSh responses in oocytes to stereotypical voltage waveforms from salamander retinal cells. Waveforms from on-bipolar cell (*top*) and wide-field amacrine cell (*bottom*). The dynamic range of FlaSh (about –50 mV to –30 mV) is highlighted in gray. Voltage traces were recorded in salamander retinal preparation and played back into oocytes expressing FlaSh. Bar, 1 sec.

SETUP FOR IMAGING ON SINGLE CELLS

Voltage-clamp fluorimetry is performed on an inverted fluorescence microscope, with a 20X/0.7 NA or a 40X oil/1.3 NA objective that admits light from a large optical patch of membrane. Cells are controlled and measured electrophysiologically with a two-electrode voltage clamp. Photometry is done with a photomultiplier, which produces a voltage output that is proportional to the fluorescent emission of FlaSh. This fluorescence signal is digitized in parallel with current and voltage measurements of the voltage clamp so that the magnitude and kinetics of the channel functional transitions (measured electrically) can be related to the brightness changes of FlaSh.

DISCUSSION AND CONCLUSION

FlaSh is not a typical fluorescent voltage probe. Traditional "fast" voltage-sensitive dyes have been designed to respond quickly and linearly to membrane potential (Cohen and Lesher 1986; Gross and Loew 1989; Gonzalez and Tsien 1997; see also Chapter 58). In contrast, FlaSh provides a different solution to the underlying problem of detecting fast voltage transients: FlaSh gives long, stereotypical fluorescence pulses in response to brief voltage spikes. In response to a spike train, FlaSh behaves like a linear filter, with an "impulse response" that can be seen in Figure 2C. The temporally expanded response from FlaSh provides significant advantages for detecting individual electrical events, as the area under the response to single spikes is approximately 30-fold larger than the area under the input spike (converting units appropriately). A practical consequence of the FlaSh impulse response is that single action potentials might be visible across several frames of video-rate cameras.

One advantage of using the Shaker channel is that many mutations have been described that produce unique alterations in its voltage dependence and kinetics. This provides flexibility in tuning FlaSh to an operating range that best suits the signals of interest. For example, mutants with more negative operating range could be designed to detect activity in wide-field amacrine cells (Fig. 2D, bottom).

The success of FlaSh suggested that a modular approach could be used to produce optical sensors to detect other signaling events. This approach is being used to engineer signaling proteins so that changes in their biological activity are converted into changes in fluorescence emission. These sensor proteins have two domains: a "detector" that undergoes a conformational rearrangement during a cell-signaling event (e.g., a channel, receptor, enzyme, and G-protein) and a "reporter" fluorophore (e.g., GFP). Ideally, GFP is fused to the detector protein near a domain that undergoes a conformational rearrangement when the detector protein is activated. Movement in the detector domain alters the environment of GFP or places stress on the structure of GFP, thus altering its spectral properties. The constructs typically include a variant of GFP as a reporter, a signal transduction protein as a detector, and a subcellular targeting peptide.

Protein-based sensor proteins analogous to FlaSh may enable the noninvasive detection of activity in a variety of proteins, including receptors, G proteins, enzymes, and motor proteins. The developmental timing and cellular specificity of expression can be directed by placing the construct under the transcriptional control of a specific promoter. The combined ability to tune the sensor module via mutagenesis and to target the sensor to specific locations affords powerful advantages for the study of signal transduction events in intact tissues.

ACKNOWLEDGMENTS

We thank Botond Rosko for preparing voltage traces from the salamander retina; and Scott Fraser, Henry Lester, Carver Mead, Gilles Laurent, Norman Davidson, Sanjoy Mahajan, John Ngai, and members of the Isacoff lab for helpful discussions. Research was supported by the McKnight and Klingenstein foundations. M.S.S. is a Howard Hughes predoctoral fellow in the biological sciences.

REFERENCES

Baumann A., Grupe A., Ackermann A., and Pongs O. 1988. Structure of the voltage-dependent potassium channel is highly conserved from *Drosophila* to vertebrate central nervous system. *EMBO J.* **7:** 2457-2463.

Chalfie M., Tu Y., Euskirchen G., Ward W.W., and Prasher D.C. 1994. Green fluorescent protein as a marker for gene expression. *Science* **263:** 802–805.

Cohen L. and Lesher S. 1986. Optical monitoring of membrane potential: Methods of multisite optical measurement. *Soc. Gen. Physiol. Ser.* **40:** 71–99.

Gonzalez J.E. and Tsien R.Y. 1997. Improved indicators of cell membrane potential that use fluorescence resonance energy transfer. *Chem. Biol.* **4:** 269–277.

Gross D. and Loew L.M. 1989. Fluorescent indicators of membrane potential: Microspectrofluorometry and imaging. *Methods Cell Biol.* **30:** 193–218.

Kamb A., Tseng–Crank J., and Tanouye M.A. 1988. Multiple products of the *Drosophila Shaker* gene may contribute to potassium channel diversity. *Neuron* **1:** 421–430.
Siegel M.S. 1997. Genetic probes: New ways to watch cells in action. *Curr. Biol.* **7:** R556–R557.
Siegel M.S. and Isacoff E.Y. 1997. A genetically encoded optical probe of membrane voltage. *Neuron* **19:** 735–741.
Tempel B.L., Papazian D.M., Schwarz T.L., Jan Y.N., and Jan L.Y. 1987. Sequence of a probable potassium channel component encoded at *Shaker* locus of *Drosophila. Science* **237:** 770–775.
Tsien R.Y. 1989. Fluorescent probes of cell signaling. *Annu. Rev. Neurosci.* **12:** 227–253.

CHAPTER 78

Genetic Probes for Calcium Dynamics

Atsushi Miyawaki, Takeharu Nagai, and Hideaki Mizuno

Riken Brain Science Institute, 2-1 Hirosawa, Wako-shi, Saitama, 351-0198, Japan

To investigate how a neural circuit operates as an ensemble in the nervous system, or how contraction of a group of muscles is coordinated, Ca^{2+} dynamics are monitored as a signal that results from electrical activity. Over the past few years, several calcium probes have been generated using fluorescent proteins (Miyawaki 2003a,b), in particular, the green fluorescent protein (GFP) (Tsien 1998) from the bioluminescent jellyfish, *Aequorea victoria*. These probes employ simple GFP variants, circularly permuted GFP variants (Baird et al. 1999), or pairs of GFP variants that enable fluorescence resonance energy transfer (FRET) (Miyawaki et al. 1997). Because these probes can be expressed using gene transfer techniques, they have significant advantages over conventional organic fluorescent dyes. For instance, although the conventional optical imaging of brain tissue stained with voltage-sensitive dyes is a noninvasive technique for recording the activity of a number of neurons simultaneously, it collects signals from all cell types including glial cells, which represent a large fraction of the total membrane surface in the brain. In contrast, the selective introduction of genetically encoded probes into certain neurons will eliminate signals from glial cells. Moreover, it is possible to place the probes in specific subcellular compartments where the desired signals predominate. In this chapter, we focus on cameleons (Miyawaki et al. 1997) and ratiometric pericams (Nagai et al. 2001), both of which enable dual-ratio Ca^{2+} imaging. The molecular structures of these probes are shown in Figure 1.

CAMELEONS

FRET between two fluorophores is highly sensitive to the relative orientation and distance between the two fluorophores. This technique is amenable to emission ratioing, which is more quantitative than single-wavelength monitoring, and is also an ideal readout for fast imaging by laser-scanning confocal microscopy. Cameleons are chimeric proteins composed of a short-wavelength mutant of GFP, calmodulin (CaM), a glycylglycine linker, the CaM-binding peptide of myosin light-chain kinase (M13), and a long-wavelength mutant of GFP. Ca^{2+} binding to the CaM moiety of the cameleon initiates an intramolecular interaction between the CaM and M13 domains (Porumb et al. 1994). This interaction changes the chimeric protein from an extended conformation to a more compact one, thereby increasing the efficiency of FRET from the short- to the long-wavelength mutant of GFP (Miyawaki et al. 1997).

Since the prototype cameleon was released in 1997, several improvements have been made so that the cameleon (1) can be shifted to longer wavelengths, (2) is less sensitive to acidic pH, (3) exhibits more efficient maturity in mammalian cells at 37°C, and (4) exhibits a significantly wider dynamic range. First, whereas the original version has blue and green mutants of GFP as donor and acceptor, respectively, CFP and YFP have been substituted to make yellow cameleons (YCs). Subsequently, red cameleons have been constructed by incorporating RFP (DsRed) as an acceptor (Mizuno et al. 2001). Second, the original

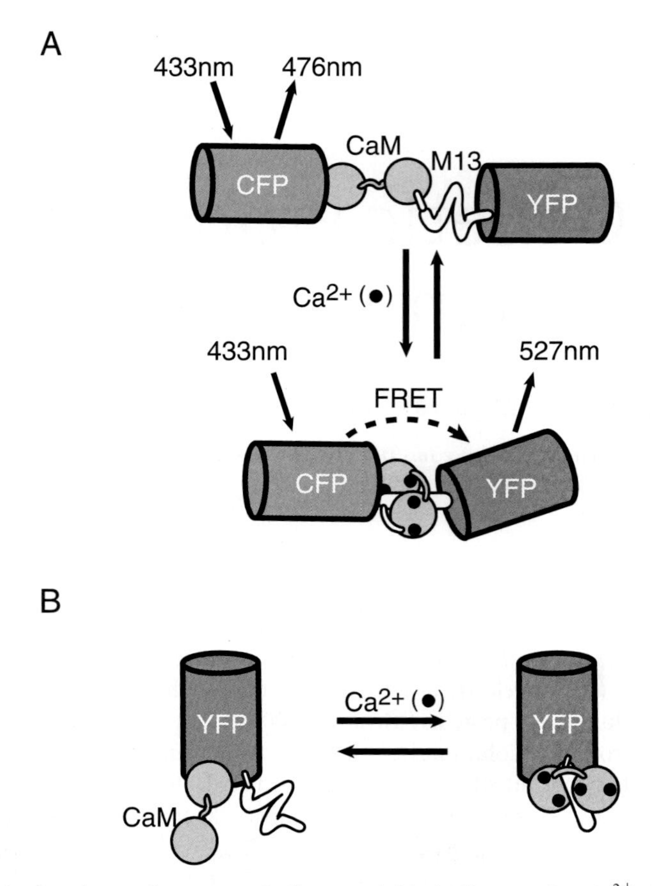

FIGURE 1. Schematic drawings of two genetically encodable indicators for Ca^{2+}. (*A*) Yellow cameleon. (*B*) Pericam. (Modified, with permission, from Tsien 2000.)

version of the YC (YC2.0) has high pH sensitivity because its acceptor, enhanced YFP (EYFP), is quenched by acidification with a pKa of 7. The pH sensitivity of YCs has been markedly reduced by introducing the V68L and Q69K mutations into EYFP (EYFP.1) (Miyawaki et al. 1999). The improved YCs, including YC2.1 and YC3.1, permit Ca^{2+} measurement without perturbation by pH changes between pH 6.5 and 8.0. Third, two bright versions of YFP, citrine (Griesbeck et al. 2001) and Venus (Nagai et al. 2002), which mature efficiently at 37°C, have recently been developed. The rapid maturation of Venus in YC2.12 or YC3.12, for example, allows the immediate detection of $[Ca^{2+}]$ transients after gene introduction in freshly prepared brain slices. Fourth, a new YC has been developed by substituting a circularly permuted variant of Venus in YC3.12. The resulting YC, YC3.60, shows five- to sixfold larger dynamic range than YC3.12. In this way, YCs have been improved mainly by optimizing the YFP component.

Responses of YC2.1 and YC3.1

YC2.1 yields an approximately twofold increase in emission ratio between zero and saturating Ca^{2+} (see Fig. 2A). YC2.1 shows a biphasic Ca^{2+} dependence (apparent dissociation constant K'_d, 100 nM and 4.3 μM; Hill coefficient *n*, 1.8 and 0.6) and is most responsive near basal cytosolic Ca^{2+} concentrations ($[Ca^{2+}]_c$s) (Fig. 2B). YC3.1 displays a monophasic Ca^{2+} response curve (K'_d, 1.5 μM; *n*, 1.1) and is expected to be helpful in quantifying relatively large $[Ca^{2+}]_c$ transients (Fig. 2B). Figures 2C and 2D compare the emission ratio responses of YC2.1 (C) and YC3.1 (D) to a supramaximal dose of histamine (0.1 mM) in HeLa cells. The initial peak and the subsequent plateau in $[Ca^{2+}]_c$ nearly saturated the YC2.1 response, although in one cell, a sustained oscillation with a mean frequency of 0.05 Hz was superimposed on the plateau. The application of cyproheptadine, a histamine antagonist, caused a large decrease in $[Ca^{2+}]_c$ to previous resting values. In contrast, YC3.1 showed apparently much

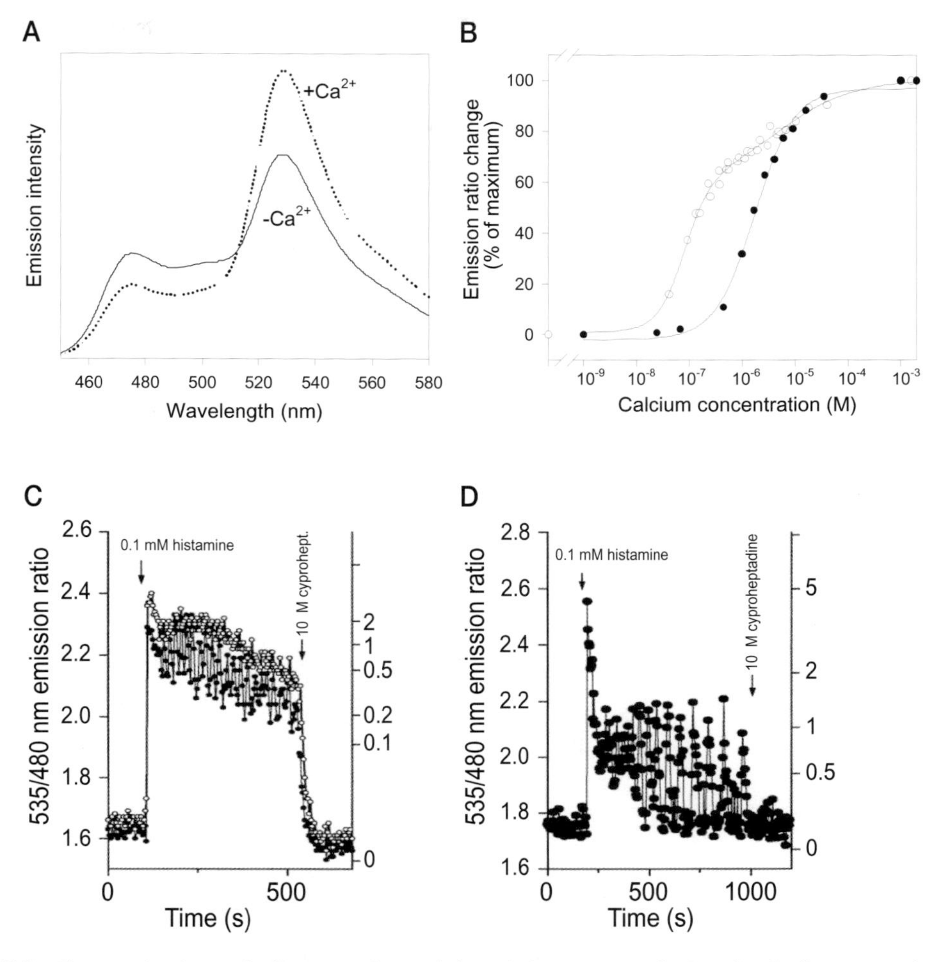

FIGURE 2. Characterizations of yellow cameleons. (*A*) Emission spectra of YC2.1 (excited at 432 nm) with 100 μM EGTA (*solid line*) and 100 μM $CaCl_2$ (*dotted line*) at pH 7.3. (*B*) Ca^{2+} titration curves of YC2.1 (*open circles*) and YC3.1 (*closed circles*). (*C, D*) Typical Ca^{2+} transients reported by YC2.1 (*C*) and YC3.1 (*D*) in HeLa cells induced by 0.1 mM histamine. (*B,C,D*, Reprinted, with permission, from Miyawaki et al. 1999 [© National Academy of Sciences].)

sharper $[Ca^{2+}]_c$ spikes. All spikes except the first reached only approximately halfway between R_{min} and R_{max}, indicating that their amplitudes were approximately equal to the K'_d of YC3.1 for Ca^{2+}, 1.5 μM. However, the relatively weak Ca^{2+} affinity also prevented the detection of the sustained plateau elevation between spikes. Therefore, the only apparent effect of cyproheptadine was to stop the oscillations.

Expansion of Ca^{2+} Responses of YCs

To achieve a Ca^{2+}-dependent large change in the relative orientation and distance between the fluorophores of CFP and YFP, a very recent work (Nagai et al. 2004) has used circularly permuted YFPs (cpYFPs), in which the amino and carboxyl portions were interchanged and reconnected by a short spacer between the original termini (Baird et al. 1999). Circular permutation was conducted on Venus, and new termini were introduced into surface-exposed loop regions of the β-can. One of the variants, cp173Venus, was given a new amino terminus at Asp-173, which is far removed at the opposite end of the β-can from the original amino terminus, Met1 (Fig. 3A). YC3.60 was generated by replacing Venus in YC3.12 with cp173Venus (Fig. 3B). Compared to YC3.12, YC3.60 is equally bright, but shows five- to sixfold larger dynamic range (Fig. 3C). Thus, YC3.60 gives a greatly enhanced signal-to-noise ratio (SNR), thereby enabling Ca^{2+} imaging experiments that were not possible with conventional YCs. YC3.60 shows a monophasic Ca^{2+} response curve (K'_d, 0.25 μM; *n*, 1.7).

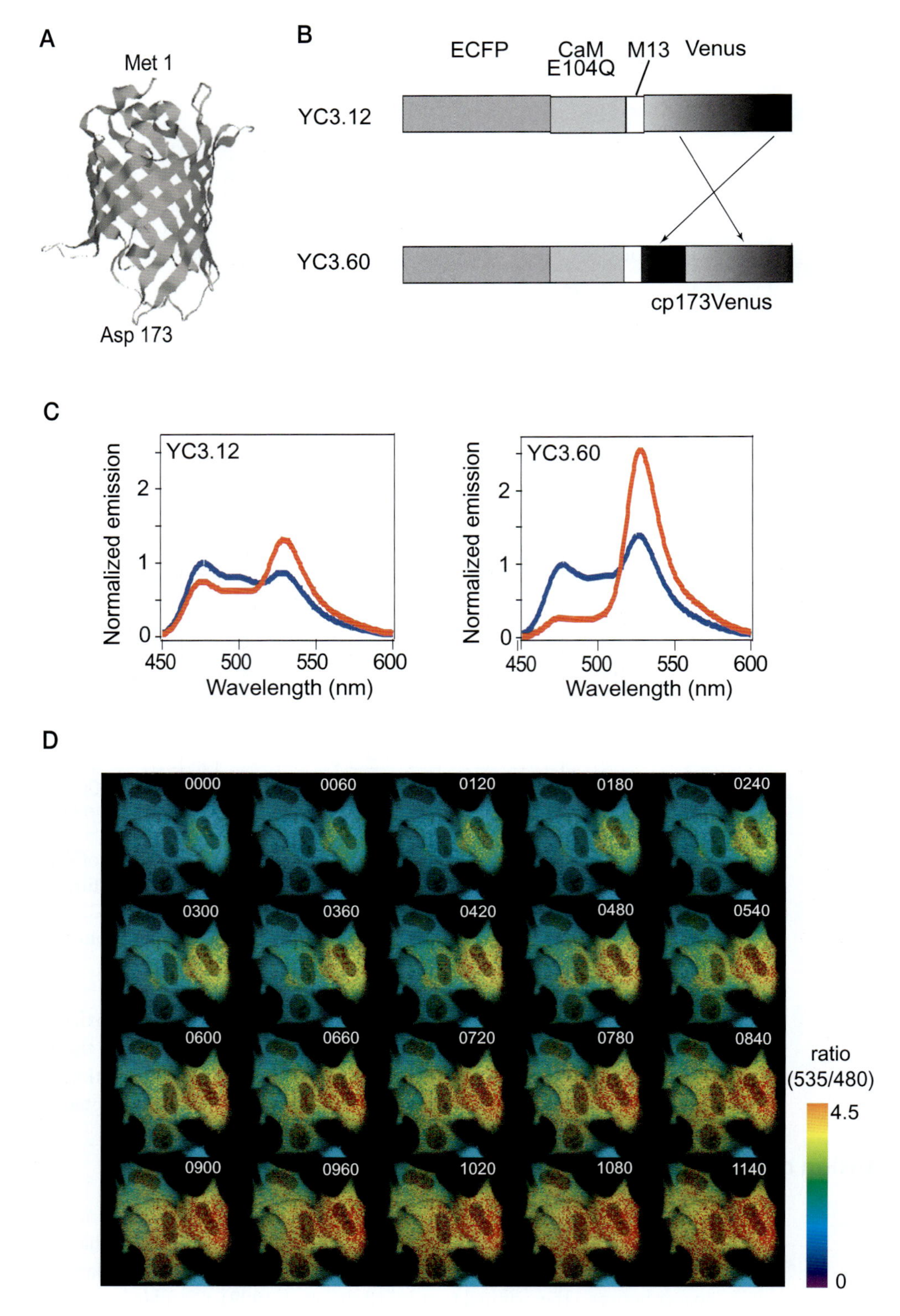

FIGURE 3. Development and performance of YC3.60. (*A*) The three-dimensional structure of GFP with the positions of the original (Met1) and the new (Asp173) amino termini indicated. (*B*) Domain structures of YC3.12 and YC3.60. (CaM) *Xenopus* calmodulin; (E104Q) mutation of the conserved bidentate glutamate (E104) at position 12 of the third Ca^{2+}-binding loop to glutamine. (*C*) Emission spectra of YC3.12 (*left*) and YC3.60 (*right*) (excitation at 435 nm) at zero (*blue line*) and saturated Ca^{2+} (*red line*). (*D*) A series of confocal pseudo-colored ratio images showing propagation of $[Ca^{2+}]_c$ in HeLa cells that expressed YC3.60.

PROCEDURE

Dual Emission Ratio Imaging of YCs

Briefly, the procedure for time-lapse $[Ca^{2+}]_c$ imaging in cultured cells is as follows (Miyawaki et al. 2000).

1. Attach HeLa cells to a coverslip in a petri dish. Transfect cells in the dish with 1 μg of cDNA using Lipofectin (Invitrogen).
2. Between 2 and 10 days after cDNA transfection, image HeLa cells on an inverted microscope (IX70) with a cooled CCD camera (Micromax, Roper Scientific). Expose cells to reagents in HBSS containing 1.26 mM $CaCl_2$<!>.

   Note: In the authors' setup, image acquisition and processing are controlled by a personal computer connected to a camera and filter wheels (Lambda 10–2, Sutter Instruments) using the program MetaFluor (Universal Imaging). The excitation filter wheel in front of the xenon lamp and the emission filter wheel (Lambda 10-2, Sutter Instruments) immediately below the CCD camera are also under computer control. Although the excitation filter wheel can be replaced with a fixed one, the emission filter wheel is required for imaging cameleons. Excitation light from a 75-W xenon lamp is passed through a 440DF20 (440 ± 10 nm) excitation filter. The light is reflected onto the sample using a 455-nm long-pass (455DRLP) dichroic mirror. The emitted light is collected with a 40× (numerical aperture [NA]: 1.35) objective and passed through a 480 ± 15 nm or 480 ± 25 nm band-pass filter (480DF30 or 480DF50, donor channel) for ECFP, and a 535 ± 12.5-nm band-pass filter (535DF25, FRET channel) for YFP. Interference filters are from Omega Optical or Chroma Technologies.
3. Define several factors for image acquisition, including (a) excitation power, which depends on the type of light source and neutral density filter, (b) NA of the objective, (c) time of exposure to the light, (d) image acquisition interval, and (e) binning. The last three factors should be considered in terms of whether temporal or spatial resolution is pursued.
4. Choose moderately bright cells (see below). In addition, the fluorescence should be uniformly distributed in the cytosolic compartment but excluded from the nucleus, as expected for a 74-kD protein without targeting signals. Select regions of interest so that pixel intensities are spatially averaged.
5. At the end of an experiment, convert fluorescence signals into values of $[Ca^{2+}]$. R_{max} and R_{min} can be obtained in the following way. To saturate intracellular indicator with Ca^{2+}, increase the extracellular $[Ca^{2+}]$ to 10–20 mM in the presence of 1–5 μM ionomycin. Wait until fluorescence intensity reaches a plateau. Then, to deplete the Ca^{2+} indicator, wash the cells with Ca^{2+}-free medium (1 μM ionomycin, 1 mM EGTA, and 5 mM $MgCl_2$<!> in nominally Ca^{2+}-free HBSS). The in situ calibration for $[Ca^{2+}]$ uses the equation $[Ca^{2+}] = K'_d[(R-R_{min})/(R_{max}-R)]^{(1/n)}$, where K'_d is the apparent dissociation constant corresponding to the Ca^{2+} concentration at which R is midway between R_{max} and R_{min}, and n is the Hill coefficient. The Ca^{2+} titration curve of YC2.1 can be fitted using a single K'_d of 0.2 μM and a single Hill coefficient of 0.62. Therefore, use $K'_d = 0.2$ μM and $n = 0.62$ for YC2.1. Use $K'_d = 1.5$ μM and $n = 1.1$ for YC3.1, and $K'_d = 0.25$ μM and $n = 1.7$ for YC3.60.

Caution: *See Appendix 3 for appropriate handling of materials marked with <!>.*

PRACTICAL CONSIDERATIONS

pH

pH-related artifacts were not an issue in the experiments that used HeLa cells, because agonist-induced $[Ca^{2+}]_c$ mobilization did not cause any intracellular pH changes detectable by the pH indicator, BCECF (data not shown). Correspondingly, comparisons of YC2 with YC2.1 and YC3 with YC3.1 showed no major differences in the reported $[Ca^{2+}]_c$ attributable to the difference in pH sensitivity. On the other hand, both YC2 and YC3 expressed in dissociated hippocampal neurons were perturbed by acidification following depolarization or glutamate stimulation. This problem was solved by using YC2.1 or YC3.1 (Miyawaki et al. 1999). YC2.12, YC3.12, and YC3.60 also give reliable Ca^{2+} responses over a physiological range of pH, from 6.5 to 6.2 (Nagai et al. 2004).

Concentration of Yellow Cameleons in Cells

The estimation of cameleon concentration in cells is essential for quantifying the tradeoff between optical detectability and Ca^{2+} buffering. It is important to consider the ability of various concentrations of YC3.1 to buffer Ca^{2+} transients in HeLa cells. During a 0.1-mM histamine challenge, sharp $[Ca^{2+}]_c$ transients followed by $[Ca^{2+}]_c$ oscillations can be observed in cells expressing < 200 μM YC3.1. In contrast, $[Ca^{2+}]_c$ recovers slowly to the baseline over a period of several hundred seconds, and oscillations are never observed when transfected HeLa cells expressed >300 μM YC3.1. Cameleons at cytosolic concentrations below 20 μM are too dim to give favorable SNRs with our current instruments. This limitation cannot be overcome by increasing the intensity of illumination because of YFP photochromism (see below).

Photochromism of Yellow Fluorescent Protein

If YFP is excited too strongly, its fluorescence will be reduced. This apparent bleaching is actually photochromism because the fluorescence recovers to some extent spontaneously and can be further restored by UV illumination (Dickson et al. 1997). Intense excitation of YCs also causes photochromism of the YFP moiety, which results in a decrease in the yellow:cyan emission ratio independent of Ca^{2+} change. The extent of photochromism is dependent on excitation power, numerical aperture of objective, and exposure time. Therefore, it is necessary to optimize these factors for each cell sample in order to minimize photochromism while still preserving a high SNR. Because the photochromism is partially reversible, the sampling interval is another factor that should be considered. Illumination at frequent intervals sometimes leads to a decrease in the resting ratio values of YCs. A better solution is to bin pixels at the cost of spatial resolution. The increased SNR permits the decrease in intensity of the excitation light with a neutral density filter and the observation of $[Ca^{2+}]_i$ oscillations without significant photochromism of the indicators.

Color CCD Camera

For dual-emission ratio imaging of YCs, consecutive data gathering has been described so far in this chapter; images are created by alternating filters in the emission path. For fast and simultaneous acquisition of YFP and CFP images, a color camera (Hamamatsu Photonics, C7780-22) composed of three CCD chips (RGB: red, green, and blue) and a prism may be used. The YFP and CFP images are captured by the G and B chips, respectively. In addition, to improve spatial resolution along the *z*-axis, a spinning-disk unit was placed in front of the camera. A confocal image of YC3.60-expressing HeLa cells is shown in Figure 3D. A series of ratio images in pseudo-color acquired at video rate (displayed at 16.7 Hz) shows how the increase in $[Ca^{2+}]_c$ appeared and propagated within the individual cells after stimulation with histamine.

PERICAMS

Circularly Permuted YFP

Wild-type GFP (WT-GFP) has a bimodal absorption spectrum with two peak maxima at 395 nm and 475 nm corresponding to the protonated (neutral) and deprotonated (anionic) states of the chromophore, respectively (Tsien 1998). The ionization state is modulated by a hydrogen-bond network, which is an intricate network of polar interactions between the chromophore and several surrounding amino acids. The chromophore of most GFP variants titrated with single pKa values, indicating that the internal proton equilibrium is disrupted by external pH.

Within the rigid "β-can" structure of GFP variants, Baird et al. (1999) found a site that can tolerate circular permutations, where two portions of the polypeptide are flipped around the central site. With obvious clefts in the β-can, the chromophore of circularly permuted GFPs (cpGFPs) seems to be more accessible to protons from outside the protein. The cpGFPs may be used to convert changes in the interaction between two protein domains into a change in the electrostatic potential of the chromophore; in other words, to transduce information about the interaction into a fluorescence signal. A less pH-sensitive YFP variant, EYFP.1 (Miyawaki et al. 1999), was subjected to circular permutation. The original amino and carboxyl termini were fused via a pentapeptide linker GGSGG, and Y145 and N144 became the new amino and carboxyl termini, respectively. The resulting chimeric protein is called cpEYFP.1 (Nagai et al. 2001).

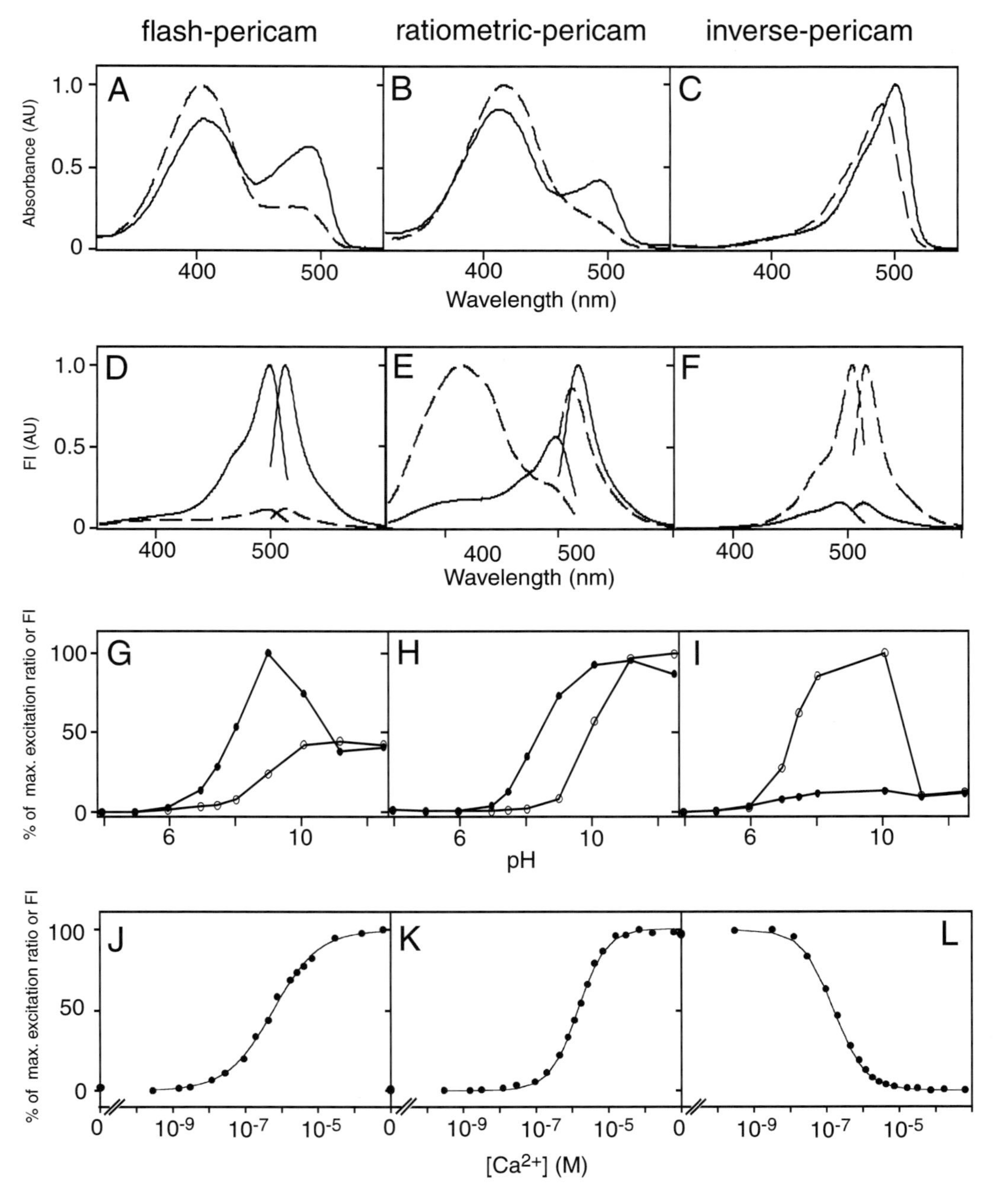

FIGURE 4. In vitro properties of flash pericam (*A*, *D*, *G*, and *J*), ratiometric pericam (*B*, *E*, *H*, and *K*), and inverse pericam (*C*, *F*, *I*, and *L*). Absorbance (*A–C*) and fluorescence excitation and emission (*D–F*) spectra of pericams. pH-dependence of normalized amplitudes in the 514-nm emission peak (*G*) and in the 516-nm emission peak (*I*) as well as the excitation ratio of 495/410 (*H*). (*A–I*) The spectra and data points were obtained in the presence (*solid circles*) or absence (*open circles*) *of* Ca^{2+}. (J–L) Ca^{2+} titration curves of pericams. (FI) Fluorescence intensity. (Reprinted, with permission, from Nagai et al. 2001 [© National Academy of Sciences].)

Construction of Pericams

We first made a construct in which cpEYFP.1 was fused to the carboxyl terminus of M13 through a tripeptide linker SAG, and through a GTG linker to the amino terminus of a CaM mutant (Fig. 1B), in which the conserved bidentate glutamate at position 104 in the third Ca^{2+}-binding loop had been changed to glutamine (Miyawaki et al. 1997). As the amino terminus of CaM and the carboxyl terminus of M13 were rather far apart (58 Å when the complex was formed), the β-can of cpEYFP.1 might be considerably twisted. However, this radically designed chimeric protein was fluorescent and, as we had hoped, showed Ca^{2+} sensitivity. The protein, having a circularly *per*muted EYFP.1 and a *CaM,* was named "pericam." The CaM and M13 domains projecting from cpEYFP.1 reminded us of the bill of a pelican. When excited at 485 nm, Ca^{2+}-bound pericam showed an emission peak at 520 nm, which was three times more intense than that of Ca^{2+}-free pericam (data not shown).

Three types of pericams have been generated by mutating several amino acids adjacent to the chromophore (Fig. 4) (Nagai et al. 2001). Of these, "flash pericam" becomes bright in the presence of Ca^{2+}, similar to G-CaMP (Nakai et al. 2000), another cpGFP-based Ca^{2+} probe, whereas "inverse pericam" dims. A third pericam, "ratiometric pericam" has an excitation wavelength that changes in a Ca^{2+}-dependent manner, and thereby enables dual excitation ratiometric Ca^{2+} imaging. Ratiometric pericam realizes quantitative Ca^{2+} measurement by minimizing the effects of several artifacts that are unrelated to changes in free intracellular Ca^{2+} concentration ($[Ca^{2+}]_i$). This pericam has been successfully used to monitor changes in $[Ca^{2+}]_i$ in cardiomyocyte mitochondria (Robert et al. 2001). That report has demonstrated that mitochondrial $[Ca^{2+}]$ ($[Ca^{2+}]_m$) oscillates synchronously with cytosolic $[Ca^{2+}]$ during beating.

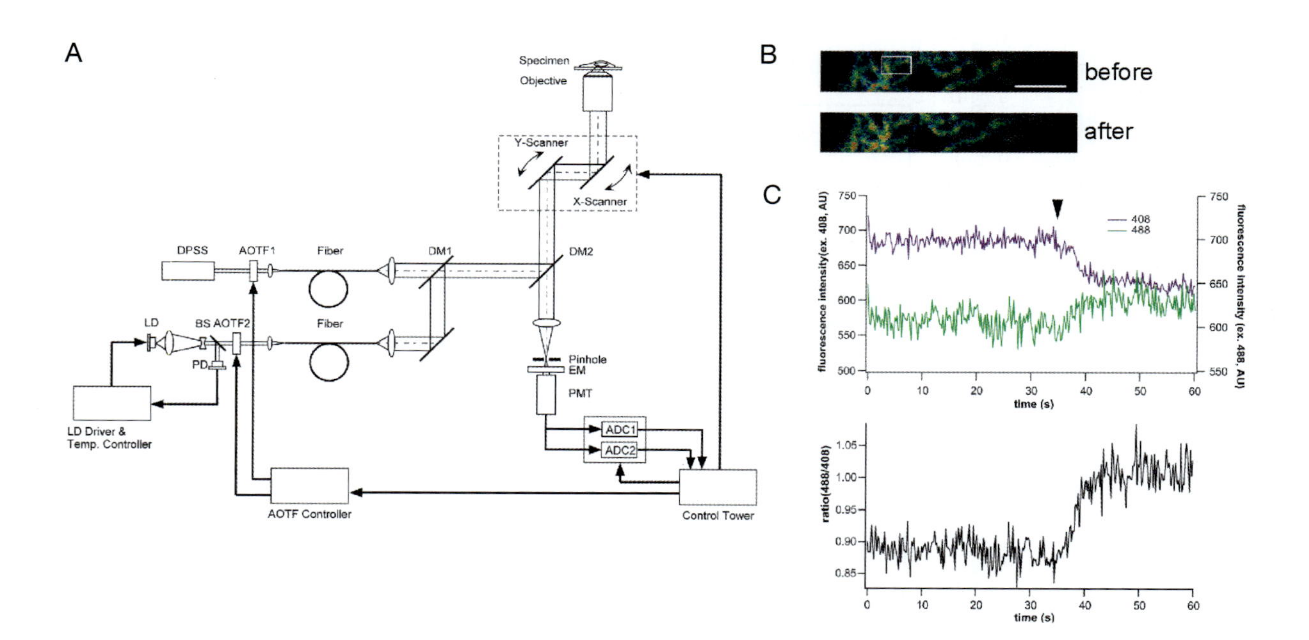

FIGURE 5. (*A*) Schematic diagram of the laser-scanning confocal microscopy system for fast dual-excitation ratiometric imaging. (DPSS) Diode-pumped solid-state laser; (LD) laser diode; (PD) photodiode; (BS) beam splitter; (DM) dichroic mirror; (EM) emission filter; (PMT) photomultiplier tube; (ADC) analog-to-digital converter. (*B*) Confocal and dual-excitation imaging of $[Ca^{2+}]_m$ using ratiometric pericam-mt. Ratio images before and after application of 1 μM histamine. Bar, 5 μm. (*C*) Time course of the averaged fluorescence signals from the white box in *B* with excitation at 488 (*green*) and 408 nm (*violet*) (*top*) and the ratio (*bottom*). The arrowhead indicates the time when histamine was applied. (Reprinted, with permission, from Shimozono et al. 2002 [©AAAS].)

Fast, Confocal Imaging of Calcium Using Ratiometric Pericam

In dual-excitation imaging, the excitation wavelength was alternated using a rotating wheel containing two band-pass filters or a high-speed grating monochromator. Use of the monochromator increased the rate at which the ratio measurement was conducted to ~10 Hz, allowing us to monitor the beat-to-beat changes in $[Ca^{2+}]_m$ of spontaneously contracting cardiac myocytes (Robert et al. 2001). These measurements were performed using conventional wide-field microscopy, which is suitable for producing the excitation peaks. However, monitoring of $[Ca^{2+}]$ change is often severely limited by the poor spatiotemporal resolution of the conventional wide-field microscopy. To obtain more reliable information on subcellular $[Ca^{2+}]$ change, it is necessary to increase the *z*-axis resolution and the rate of production and collection of the ratios of the excitation peaks. Figure 5A is a scheme for a modified laser-scanning confocal microscopy (LSCM) system for ratiometric pericam (Shimozono et al. 2002). Fast exchange between two laser beams was achieved using acousto-optic tunable filters (AOTFs). Samples were scanned on each line sequentially by a violet laser diode (408 nm) and a diode-pumped solid-state laser (488 nm). In this way, the ratios of the excitation peaks were obtained at frequencies of up to 200 Hz.

Calcium Transients in Motile Mitochondria

Although the cationic probe rhod2 has been widely used for measuring $[Ca^{2+}]_m$, the targeting specificity of this probe relies on the negative membrane potential of this organelle. On the other hand, the Ca^{2+}-sensitive photoprotein aequorin can be specifically targeted to the mitochondria and has been used for monitoring mitochondrial Ca^{2+} dynamics. However, aequorin requires the incorporation of coelenterazine, is irreversibly consumed by Ca^{2+}, and is very difficult to image because of its weak luminescence. To overcome these limitations, we have selectively targeted the ratiometric pericam to mitochondria in HeLa cells. First, changes in $[Ca^{2+}]_m$ were monitored by alternating the excitation wavelength automatically with conventional wide-field microscopy. In those studies, the excitation ratio acquisition rate was ~10 Hz, which was identical to the frame rate. Despite this high acquisition rate, the $[Ca^{2+}]_m$ measurements were often adversely affected by the rapid movement of the mitochondria, particularly at warmer temperatures. Therefore, using the modified LSCM technique, we increased the speed of excitation wavelength alteration so that it was faster than the movement of the mitochondria. With this method, the frame rate was 5 Hz and the excitation ratio-acquisition rate was 200 Hz. Although the frame rate did not allow us to fully monitor the rapid movement of the mitochondria, the high ratio-acquisition rate minimized the time lag between the two measurements used to produce each ratio signal. We believe that this $[Ca^{2+}]_m$ imaging method effectively corrects for the movement of mitochondria laterally or into and out of the optical section. After the application of histamine, spots of $[Ca^{2+}]_m$ increasing within a mitochondrion were identifiable (Fig. 5B), and the global increase in $[Ca^{2+}]_m$ was found to occur relatively slowly (Fig. 5C).

REFERENCES

Baird G.S., Zacharias D.A., and Tsien R.Y. 1999. Circular permutation and receptor insertion within green fluorescent proteins. *Proc. Natl. Acad. Sci.* **96:** 11241–11246.

Dickson R.M., Cubitt A.B., Tsien R.Y., and Moerner W.E. 1997. On/off blinking and switching behaviour of single molecules of green fluorescent protein. *Nature* **388:** 355–358.

Griesbeck O., Baird G.S., Campbell R.E., Zacharias D.A., and Tsien R.Y. 2001. Reducing the environmental sensitivity of yellow fluorescent protein. Mechanism and applications. *J. Biol. Chem.* **276:** 29188–29194.

Miyawaki A. 2003a. Fluorescence imaging of physiological activity in complex systems using GFP-based probes. *Curr. Opin. Neurobiol.* **13:** 591–596.

———. 2003b. Visualization of the spatial and temporal dynamics of intracellular signaling. *Developmental Cell* **4:** 295–305.

Miyawaki A., Griesbeck O., Heim R., and Tsien R.Y. 1999. Dynamic and quantitative Ca^{2+} measurements using improved cameleons. *Proc. Natl. Acad. Sci.* **96:** 2135–2140.

Miyawaki A., Mizuno H., Llopis J., Tsien R.Y., and Jalink K. 2000. Cameleons as cytosolic and intra-organellar calcium probes. In *Calcium Signalling: A practical approach* (ed. A.V. Tepikin), pp. 3–16. Oxford University Press, Oxford, United Kingdom.

Miyawaki A., Llopis J., Heim R., McCaffery J.M., Adams J.A., Ikura M., and Tsien R.Y. 1997. Fluorescent indicators for Ca^{2+} based on green fluorescent proteins and calmodulin. *Nature* **388:** 882–887.

Mizuno H., Sawano A., Eli P., Hama H., and Miyawaki A. 2001. Red fluorescent protein from *Discosoma* as a fusion tag and a partner for fluorescence resonance energy transfer. *Biochemistry* **40:** 2502–2510.

Nakai J., Ohkura M., and Imoto K. 2001. A high signal-to-noise Ca^{2+} probe composed of a single green fluorescent protein. *Nat. Biotechnol.* **19:** 137–141.

Nagai T., Sawano A., Park E.S., and Miyawaki A. 2001. Circularly permuted green fluorescent proteins engineered to sense Ca^{2+}. *Proc. Natl. Acad. Sci.* **98:** 3197–3202.

Nagai T., Yamada S., Tominaga T., Ichikawa M., and Miyawaki A. 2004. Expanded dynamic range of fluorescent indicators for Ca^{2+} by circularly permuted yellow fluorescent proteins. *Proc. Natl. Acad. Sci.* **101:** 10554–10559.

Nagai T., Ibata K., Park E.S., Kubota M., Mikoshiba K., and Miyawaki A. 2002. A variant of yellow fluorescent protein with fast and efficient maturation for cell-biological applications. *Nat Biotechnol.* **20:** 87–90.

Porumb T., Yau P., Harvey T.S., and Ikura M. 1994. A calmodulin-target peptide hybrid molecule with unique calcium-binding properties. *Protein Eng.* **7:** 109–115.

Robert V., Gurlini P., Tosello V., Nagai T., Miyawaki A., Di Lisa F., and Pozzan T. 2001. Beat-to-beat oscillations of mitochondrial Ca^{2+} in cardiac cells. *EMBO J.* **20:** 4998–5007.

Shimozono S., Fukano T., Nagai T., Kirino Y., Mizuno H., and Miyawaki A. 2002. Confocal imaging of subcellular Ca^{2+} concentrations using a dual-excitation ratiometric indicator based on green fluorescent protein. *Sci. STKE* **2002:** PL4. http://www.stke.org/cgi/content/full/OC_sigtrans;2002/125/pl4

Tsien R.Y. 1998. The green fluorescent protein. *Annu. Rev. Biochem.* **67:** 509–544.

———. 2000. Physiological indicators based on fluorescence resonance energy transfer. In *Imaging neurons: A laboratory manual* (ed. R. Yuste et al.), pp. 55.1–55.10. Cold Spring Harbor Laboratory Press, Cold Spring Harbor, New York.

CHAPTER 79

A Practical Guide: Targeted Recombinant Aequorins

Tullio Pozzan and Rosario Rizzuto

Department of Biomedical Sciences, University of Padova and CNR Unit for Study of Biomembranes, Padova, Italy

Aequorin is a small protein produced by the genus *Aequorea* that was widely used in the 1960s and 1970s as a probe to measure Ca^{2+} in living cells (Ridgway and Ashley 1967; Allen and Blinks 1978; Cobbold 1980). The invention of the carboxylate Ca^{2+} indicators (Tsien et al. 1982; Grynkiewicz et al. 1985), which are much simpler to load into intact living cells and to calibrate and image at the single-cell level, has led most groups to abandon aequorin. Yet, this latter Ca^{2+} indicator still offers some advantages over the fluorescent probes. In particular, the use of molecular biological techniques for expressing recombinant aequorin in mammalian cells, thus eliminating the need for microinjection, has opened new possibilities for this probe (Brini et al. 1995). Among the new uses of aequorin, one of the most interesting is the potential for targeting it specifically to different cellular locations (Rizzuto et al. 1992; Brini et al. 1993, 1997; Montero et al. 1995; Marsault et al. 1997), thus opening the possibility of monitoring selectively the dynamics of $[Ca^{2+}]$ with unprecedented spatial resolution. This chapter briefly discusses the problems concerned with targeting aequorin to different locations, the advantages and disadvantages offered by the steep dependence of luminescence on $[Ca^{2+}]$, and the instruments needed to obtain reliable measurements.

THE Ca^{2+} RESPONSE OF AEQUORIN

Aequorin, as produced by various *Aequorea* species, includes an apoprotein and a covalently bound prosthetic group (coelenterazine). When Ca^{2+} ions bind to three high-affinity sites (EF-hand type), aequorin undergoes an irreversible reaction, in which a photon is emitted and the oxidized coenzyme is released. For $[Ca^{2+}]$ between 10^{-7} M and 10^{-5} M, there is a relationship between the fractional rate of consumption (i.e., L/L_{max}, where L is the actual rate of photon emission and L_{max} is the maximal rate of aequorin discharge at saturating Ca^{2+} concentrations) and $[Ca^{2+}]$ (Allen et al. 1977; Blinks et al. 1978). Due to the cooperativity between the three binding sites, light emission is proportional to the second to third power of $[Ca^{2+}]$. It is important to stress that, unlike commonly used fluorescent indicators, aequorin is consumed upon photon emission and, thus, the probe content decreases with time. This decrease in probe concentration depends on the Ca^{2+} concentration to which aequorin is exposed. In addition, due to the steep relationship between $[Ca^{2+}]$ and photon emission, the average signal is biased toward the highest values, if, as often occurs within living cells, the concentration of the cation is inhomogeneous.

Wild-type aequorin is well suited for measuring $[Ca^{2+}]$ between 0.5 μM and 10 μM. However, in some intracellular compartments or regions, the $[Ca^{2+}]$ is much higher (e.g., the lumen of the endoplasmic reticulum [ER] and sarcoplasmic reticulum [SR], near Ca^{2+} channels and pumps). It is possible to generate aequorins with reduced Ca^{2+} affinities by mutating one (or possibly more) of the three EF-hand Ca^{2+}-binding sites (Kendall et al. 1992). The application of a modified aequorin to the study of [Ca^{2+}] microdomains in presynaptic terminals is discussed in Chapter 40.

The point mutation used by the authors ([D119A] Montero et al. 1995; Brini et al. 1997) affects the second EF-hand domain and produces a mutated aequorin whose affinity for Ca^{2+} is reduced by ~ 20-fold. The range of Ca^{2+} sensitivity can be expanded further by employing divalent cations other than Ca^{2+} or synthetic coelenterazine analogs (e.g., coelenterazine n; Montero et al. 1997).

The advantages of using recombinant aequorin include:

1. *Selective intracellular distribution.* Fluorescent probes are usually trapped mainly in the cytoplasm, but they are often in part sequestered, to a variable extent and depending on the cell type, into the lumen of organelles. In addition, slow leakage of the dyes into the extracellular medium also occurs. Recombinantly expressed aequorins, on the other hand, are not released by living cells, and their targeting is extremely selective. For example, less than 1% of aequorin targeted to the mitochondrial matrix is found in the cytoplasm.
2. *High signal-to-noise ratio (SNR).* The background chemiluminescence signal of living cells is intrinsically very low, unlike that of autofluorescence (particularly when low-excitation wavelengths are used). In addition, the ratio between the rate of photon emission by aequorin exposed to resting and saturating Ca^{2+} concentration is over 100,000. With fluorescent Ca^{2+} indicators this value is usually 3–10.
3. *Low Ca^{2+}-buffering effect.* The usual concentrations reached intracellularly with fluorescent probes are on the order of 10–100 μM, whereas those achieved with recombinantly expressed chimeric aequorins are in the range of 0.1–1 μM.
4. *Wide dynamic range.* The range of Ca^{2+} concentrations that can be measured with aequorins (wild-type and low-affinity mutants) is between 10^{-7} M and 10^{-3} M, whereas with most commonly used indicators, the usable range is from 10^{-8} M up to concentrations of a few micromolar.
5. *Possibility of coexpression with proteins of interest.* By transiently cotransfecting aequorin and a protein of interest, the Ca^{2+} signal comes only from the cell population expressing the protein investigated, thus bypassing the problem of variability of single-cell response. In the case of fluorescent indicators, this analysis is highly complex and time-consuming.

The disadvantages include:

1. *Overestimation of the average rise in cells (or compartments) with inhomogeneous behavior.* As discussed below, the aequorin signal, unlike that of fluorescent indicators, is biased toward the highest Ca^{2+} values. This may lead to substantial overestimations of the mean response of a cell population.
2. *Low-light emission.* The maximum number of photons in theory emittable by aequorin is 1 photon/molecule (in practice, less, about 0.6 photon/molecule). In the case of the fluorescent Ca^{2+} probes, the limit is set by the photobleaching of the dye, and it has been calculated to be more than 10,000 photons/molecule.
3. *Loading procedure.* Loading with fluorescent probes is extremely simple and requires a few tens of minutes. Aequorin must be introduced into cells via transfection or microinjection, and reconstitution with coelenterazine is necessary to generate the functional photoprotein.
4. *Necessity of reconstitution with a coenzyme.* The expressed polypeptide does not emit light if the cells are not supplemented externally with the coenzyme (see also below for the problems of reconstitution with coelenterazine).

STRATEGIES FOR TARGETING AEQUORINS TO SELECTIVE LOCATIONS

Selective targeting of recombinant polypeptides is a major issue of modern cell biology. The machinery and the signals that are responsible for the selective localization of endogenous proteins, in some cases, have been well characterized (e.g., mitochondria [Hartl et al. 1989] and ER [Pelham 1989]), but in others, are largely unknown. In the case of aequorin, we have adopted three different strategies for selectively targeting aequorins. Before describing them, it is worth mentioning that fusion of long polypeptides at the amino terminus of aequorin causes no appreciable alterations in the chemiluminescent properties and Ca^{2+} affinity of the probe. To the contrary, even small modifications at the carboxyl terminus result either in irreversible loss of chemiluminescence or in dramatic increases in the rate of Ca^{2+}-independent luminescence, which pose significant problems in quantitation of the signal (Nomura et al. 1991). The following three strategies have been used:

1. *Addition of a known targeting sequence at the amino terminus of aequorin.* This has been the case for aequorins targeted to the mitochondria (Rizzuto et al. 1992) and to the nucleus (Brini et al. 1993).
2. *Fusion of aequorin to the carboxyl terminus of a protein that contains its own targeting strategy.* In this case, it is important that the carboxy-terminal extension (due to the aequorin polypeptide) does not affect the targeting by itself. Using this approach, we have constructed the aequorins targeted to the SR (Brini et al. 1997), the plasma membrane (Marsault et al. 1997), and the intermembrane space of the mitochondria (Rizzuto et al. 1998).
3. *Fusion of aequorin to the carboxyl terminus of a polypeptide that binds firmly to an endogenous protein* (Sitia et al. 1990). This is the strategy adopted for the targeting to the ER lumen (Montero et al. 1995). In this latter case, the leader sequence, the VDJ, and the CH1 region of an IgG2 cDNA have been fused to the 3′ end of the aequorin cDNA. The CH1 domain binds with very high affinity to the endogenous ER lumenal protein Bip, and the chimeric aequorin is thus retained in the ER lumen (Montero et al. 1995).

It should be stressed that efficient and selective targeting is an essential part of the whole measurement, given that the spatial resolution of the method depends not on sophisticated hardware of subcellular imaging, but rather on the selective targeting of the probe. Thus, even a small fraction of missorted probe can profoundly affect the measurement, both qualitatively and quantitatively. It is therefore important that the subcellular localization is checked with extreme accuracy. For this purpose, both immunocytochemical approaches (at the optical and electron microscopy level) and functional approaches (subcellular fractionation, specific drugs) have been employed.

EXPERIMENTAL PROCEDURE

Transfection

In our laboratory, we currently employ three main transfection procedures as follows: calcium phosphate, electroporation, and particle gun. Other groups have used other standard procedures for loading aequorin cDNA into cells, including viral infection. Both transiently transfected and stable clones can be generated. Overall, recombinant expression of aequorin has, so far, proven to be efficient and totally innocuous for the cells, independent of the subcellular localization of the protein. Several different types of cultured cells have been transfected, from stable cell lines (HeLa, COS, 3T3, etc.) to primary cultures (skeletal muscle myoblast, neurons). Typically, the classic calcium phosphate procedure uses 40 μg of cDNA per tissue-culture dish.

Reconstitution

After expression, the recombinant apoprotein must be reconstituted into functional aequorin. This can be accomplished by incubating transfected cells with the chemically synthesized prosthetic group, coelenterazine (now commercially available from Molecular Probes). In our experience, coelenterazine is freely permeable across cell membranes, and reconstitution may occur within all intracellular compartments to which the photoprotein has been targeted. The reconstitution process is relatively slow (optimal reconstitution is usually observed after 1–2 hours of incubation with coelenterazine; Rizzuto et al. 1994). In compartments with low Ca^{2+}, the functional aequorin pool gradually increases during the reconstitution, as consumption is negligible. On the other hand, a major problem is posed by compartments with high Ca^{2+}. In fact, if the rate of discharge is the same as that of reconstitution, no active aequorin will be present at the end of the incubation with the prosthetic group. For this reason, we usually deplete these latter compartments of Ca^{2+} before carrying out the reconstitution (Montero et al. 1995; Brini et al. 1997).

The concentrations of coelenterazine usually employed vary between 1 μm and 10 μm, and the coenzyme is simply added to the medium from a stock solution in methanol. With most chimeric aequorins, coelenterazine is added to the complete growth medium containing 1–3% of serum. In the case of the aequorins trapped in the lumen of the ER or SR, it is necessary to first reduce the lumenal Ca^{2+} concentration. To achieve this, several protocols have been devised that take advantage of Ca^{2+} ionophores, inhibitors of the Ca^{2+} ATPases, or specific agents leading to opening of intracellular Ca^{2+} channels, e.g., caffeine (Montero et al. 1995; Brini et al. 1997).

Measurements

Instrumentation

For most of our experiments, we have employed a custom-built luminometer (Rizzuto et al. 1995). In this system, the perfusion chamber, which is on top of a hollow cylinder with a thermostat-controlled water jacket, is continuously perfused with buffer via a peristaltic pump. The cells are grown on glass coverslips that are then placed a few millimeters from the surface of a low-noise phototube. The photomultiplier (PMT) is kept in a dark, refrigerated box. An amplifier discriminator is built in the photomultiplier housing; the pulses generated by the discriminator are captured by a Thorn EMI photon-counting board, installed in a 486 IBM-compatibile computer. The board allows the storage of data in the computer memory for further analyses. A schematic of the setup is shown in Figure 1.

Calibration

To calibrate the crude luminescent signal in terms of $[Ca^{2+}]$, an algorithm has been developed (based on that of Blinks and coworkers [Allen et al. 1977; Blinks et al. 1978]) that takes into account the instant rate of photon emission and the total number of photons that can be emitted by the aequorin

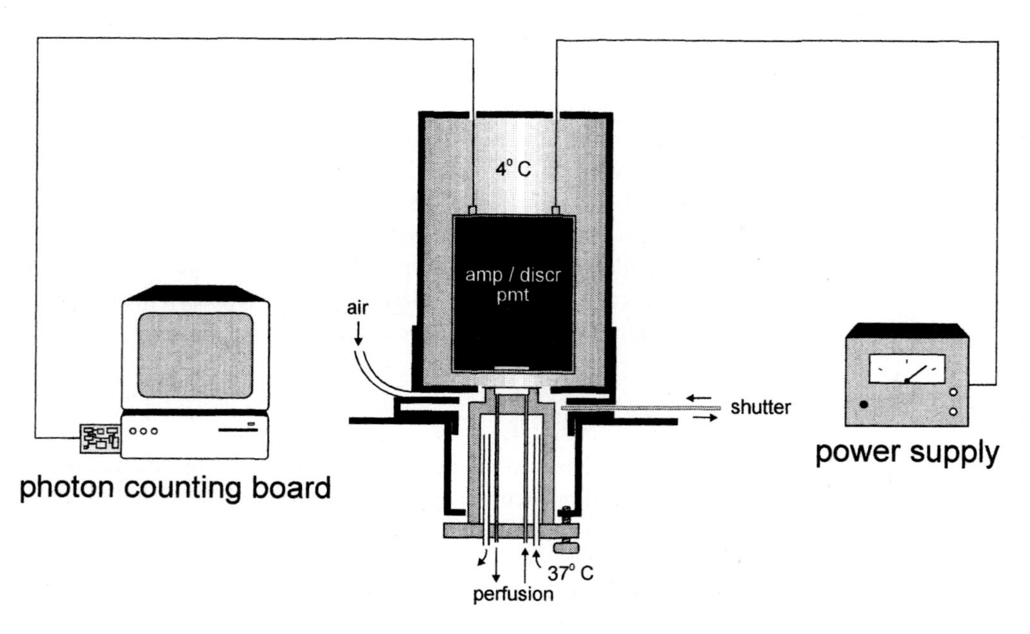

FIGURE 1. Schematic representation of the measuring apparatus. (amp/discr pmt) Amplifier/discriminator PMT.

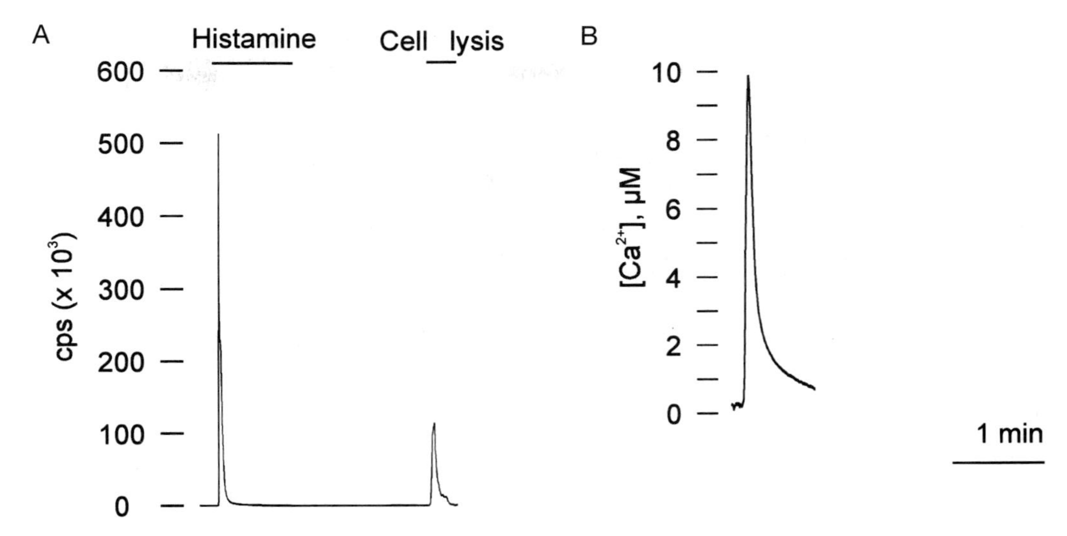

FIGURE 2. Typical experiments carried out with HeLa cells that were transiently transfected with the cDNA encoding aequorin targeted to the mitochondrial matrix. (*A*) The trace represents the kinetics of photon emission. (*B*) The trace represents the corresponding calibrated values. Where indicated, the cells were perfused with medium containing 200 mM histamine. "Cell lysis" refers to the perfusion with medium containing 10 mM $CaCl_2$ and 100 μM digitonin. For details, see the original papers by Rizzuto et al. (1994) and Brini et al. (1995).

of the sample. To obtain the latter parameter, at the end of each experiment the cells are lysed by perfusing them with a hyposmotic medium containing 10 μm $CaCl_2$ and a detergent (100 μm digitonin) to discharge all the aequorin that was not consumed during the experiment.

Result

Figure 2 shows a typical experiment carried out with HeLa cells transiently transfected with aequorin targeted to the mitochondrial matrix. The upper trace shows the kinetics of photon emission and the lower trace the calibrated values (using the algorithm described by Brini et al. 1997). As shown in the lower trace, the addition of histamine causes a very rapid and ample increase in the $[Ca^{2+}]$ of the mitochondrial matrix.

CONCLUSION

Recombinantly expressed aequorins represent a unique tool to investigate cellular Ca^{2+} homeostasis at the subcellular level. The main advantage of this approach with respect to the classic Ca^{2+} indicators resides in the highly selective targeting that can be achieved by appropriate modifications of the encoding cDNA. The techniques for expressing and reconstituting the functional photoproteins in various subcellular compartments are straightforward and can be applied to different model systems in vitro. The development of transgenic animals constitutively expressing the photoprotein will further expand the range of applications of this method.

ACKNOWLEDGMENTS

The experimental work described in this chapter was supported by grants to the authors from Telethon (project n 845, 850), the Human Frontier Science Program, the Biomed program of the European Union, the Armenise Foundation (Harvard), the Italian University Ministry, and the British Research Council.

REFERENCES

Allen D.G. and Blinks J.R. 1978. Calcium transients in aequorin-injected frog cardiac muscle. *Nature* **273:** 509–513.

Allen D.G., Blinks J.R., and Prendergast F.G. 1977. Aequorin luminescence: Relation of light emission to calcium concentration–A calcium-independent component. *Science* **195:** 996–998.

Blinks J.R., Allen D.G., Prendergast F.G., and Harrer G.C. 1978. Photoproteins as models of drug receptors. *Life Sci.* **22:** 1237–1244.

Brini M., Marsault R., Bastianutto C., Alvarez J., Pozzan T., and Rizzuto R. 1995. Transfected aequorin in the measurement of cytosolic Ca^{2+} concentration ($[Ca^{2+}]c$): A critical evaluation. *J. Biol. Chem.* **270:** 9896–9903.

Brini M., Murgia M., Pasti L., Picard D., Pozzan T., and Rizzuto R. 1993. Nuclear Ca^{2+} concentration measured with specifically targeted recombinant aequorin. *EMBO J.* **12:** 4813–4819.

Brini M., De Giorgi F., Murgia M., Marsault R., Massimino M.L., Cantini M., Rizzuto R., and Pozzan T. 1997. Subcellular analysis of Ca^{2+} homeostasis in primary cultures of skeletal muscle myotubes. *Mol. Biol. Cell* **8:** 129–143.

Cobbold P.H. 1980. Cytoplasmic free calcium and ameboid movement. *Nature* **285:** 441–446.

Grynkiewicz G., Poenie M., and Tsien R.Y. 1985. A new generation of Ca^{2+} indicators with greatly improved fluorescence properties. *J. Biol. Chem.* **260:** 3440–3450.

Hartl F.U., Pfanner N., Nicholson D.W., and Neupert W. 1989. Mitochondrial protein import. *Biochim. Biophys. Acta* **988:** 1–45.

Kendall J.M., Sala-Newby G., Ghalaut V., Dormer R.L., and Campbell A.K. 1992. Engineering the Ca^{2+}-activated photoprotein aequorin with reduced affinity for calcium. *Biochem. Biophys. Res. Commun.* **187:** 1091–1097.

Marsault R., Murgia M., Pozzan T., and Rizzuto R. 1997. Domains of high Ca^{2+} beneath the plasma membrane of living A7r5 cells. *EMBO J.* **16:** 1575–1581.

Montero M., Barrero M.J., and Alvarez J. 1997. $[Ca^{2+}]$ microdomains control agonist-induced Ca^{2+} release in intact HeLa cells. *FASEB J.* **11:** 881–885.

Montero M., Brini M., Marsault R., Alvarez J., Sitia R., Pozzan T., and Rizzuto R. 1995. Monitoring dynamic changes in free Ca^{2+} concentration in the endoplasmic reticulum of intact cells. *EMBO J.* **14:** 5467–5475.

Nomura M., Inouye S., Ohmiya Y., and Tsuji F.I. 1991. A C-terminal proline is required for bioluminescence of the Ca^{2+}-binding photoprotein, aequorin. *FEBS Lett.* **291:** 63–66.

Pelham H.R.B. 1989. Control of protein exit from the endoplasmic reticulum. *Annu. Rev. Cell Biol.* **5:** 1–23.

Ridgway E.B. and Ashley C.C. 1967. Calcium transients in single muscle fibers. *Biochem. Biophys. Res. Commun.* **29:** 229–234.

Rizzuto R., Simpson A.W.M., Brini M., and Pozzan T. 1992. Rapid changes of mitochondrial Ca^{2+} revealed by specifically targeted recombinant aequorin. *Nature* **358:** 325–328.

Rizzuto R., Bastianutto C., Brini M., Murgia M., and Pozzan T. 1994. Mitochondrial Ca^{2+} homeostasis in intact cells. *J. Cell Biol.* **126:** 1183–1194.

Rizzuto R., Brini M., Bastianutto C., Marsault R., and Pozzan T. 1995. Photoprotein-mediated measurement of calcium ion concentration in mitochondria of living cells. *Methods Enzymol.* **260:** 417–428.

Rizzuto R., Pinton P., Carrington W., Fay F.S., Fogarty K.E., Lifshitz L.M., Tuft R.A., and Pozzan T. 1998. Close contacts with the endoplasmic reticulum as determinants of mitochondrial Ca^{2+} responses. *Science* **280:** 1763–1766.

Sitia R., Neuberger M., Alberini C., Bet P., Fra A., Valetti C., Williams G., and Millstein C. 1990. Developmental regulation of IgM secretion: The role of the carboxy-terminal cysteine. *Cell* **60:** 781–790.

Tsien R.Y., Pozzan T., and Rink T.J. 1982. T-cell mitogens cause early changes in cytoplasmic free Ca^{2+} and membrane potential in lymphocytes. *Nature* **295:** 68–71.

CHAPTER 80

A Practical Guide: Imaging Synaptic Inhibition with Clomeleon, a Genetically Encoded Chloride Indicator

Ken Berglund, Robert L. Dunbar, Psyche Lee, Guoping Feng, and George J. Augustine

Department of Neurobiology, Duke University Medical Center, Durham, North Carolina 27710

Although several techniques are available to image excitatory processes in the brain, synaptic inhibition has remained largely invisible. Most synaptic inhibition in the brain arises from transmembrane fluxes of chloride ions (Cl^-), and thus, imaging intracellular Cl^- is, in principle, a good way to visualize the spatiotemporal dynamics of inhibition. This chapter discusses the suitability of Clomeleon, a genetically encoded indicator of Cl^-, as a tool for monitoring synaptic inhibition. Clomeleon is a fusion protein consisting of the cyan fluorescent protein (CFP) and yellow fluorescent protein (YFP), which are joined by a flexible 24-amino-acid linker (Kuner and Augustine 2000). Because of the close spatial proximity of the two fluorophores, fluorescence resonance energy transfer (FRET) occurs. Thus, exciting the CFP causes YFP to be excited and to emit yellow fluorescence. Binding of Cl^- to YFP (Jayaraman et al. 2000) quenches the YFP and decreases the degree of FRET. Thus, measuring the ratio of emission of YFP relative to that of CFP, while exciting the CFP, provides an absolute measure of the internal Cl^- concentration ($[Cl^-]_i$).

Clomeleon imaging should find a wide range of applications, both in vitro and in vivo. Such applications are by no means limited to brain; any cellular system that permits expression of exogenous DNAs or RNAs, or introduction of recombinant protein that has been expressed in some other way, should be an appropriate venue for Clomeleon technology. The specific goal of the work described here is to document our ability to target Clomeleon expression to neurons in the mouse brain and to use Clomeleon to measure $[Cl^-]_i$ in neurons of the superior colliculus, a region of the brain important for control of eye movements (Wurtz and Albano 1980).

MATERIALS

Transgenic mice expressing Clomeleon were created in collaboration with L.-S. Loo (in prep.). In brief, Clomeleon was expressed selectively under the control of a neuron-specific promoter, Thy1 (Feng et al. 2000), in the construct shown in Figure 1a. This construct was used to produce several lines of transgenic mice, each of which expresses Clomeleon in specific subsets of neurons. These mosaic patterns of expression are heritable and are therefore found in all individuals of a given line. One of the lines expresses Clomeleon within neurons in the superior colliculus, as well as in neurons of many other brain regions (Fig. 1b).

Caution: *See Appendix 3 for appropriate handling of materials marked with <!>.*

PROCEDURE A

Slice Preparation

Superior colliculus slices were prepared, using conventional methods (Lee et al. 1997), from 2–3-week-old mice. In brief, the brains were removed from anesthetized animals and placed in a cold artificial cerebrospinal fluid. A vibratome was used to make 250–300-μm-thick sagittal sections. The slices were then incubated for 30 minutes at 37°C prior to use.

PROCEDURE B

Optical Imaging

An upright epifluorescence microscope (E600FN, Nikon) was equipped with a mercury lamp, a dichroic mirror (515 nm), a filter wheel, excitation (440 ± 10 nm) and emission (485 ± 15 nm for CFP and 530 ± 15 nm for YFP) filters. Fluorescence excitation was produced by 50-msec-long light

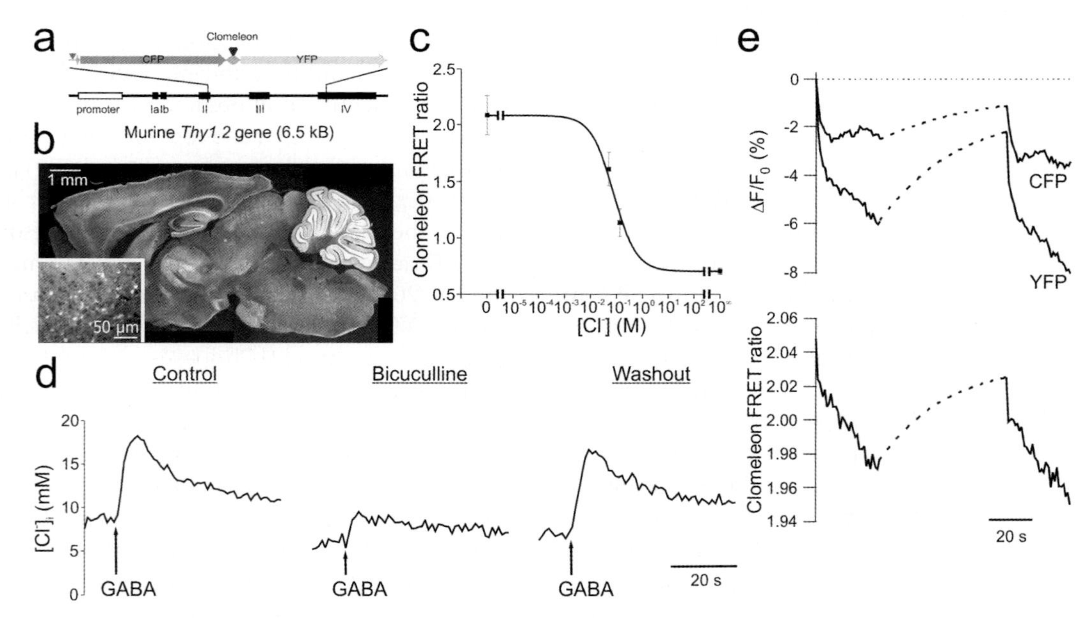

FIGURE 1. (*a*) Structure of the transgene of the construct used to produce Clomeleon transgenic mouse lines. (*b*) Expression pattern of Clomeleon in the brain of one mouse line. (*Inset*) Magnified view of the superior colliculus, showing expression in premotor neurons. (*c*) Chloride calibration curve of Clomeleon in the superior colliculusar neurons. (*d*) Response of a collicular neuron in the superior colliculus to the exogenously applied GABA. GABA (10 mM) was applied from a puffer pipette for 200 msec at the points indicated (*arrows*). (*e, top*) Reversible reduction of the fluorescence of the two Clomeleon fluorophores during intense light exposure. (*Dotted lines*) Presumed time courses of fluorescence recovery. (*Bottom*) Differential changes in donor and acceptor fluorescence cause a reversible reduction in the Clomeleon FRET ratio.

pulses (2 Hz) and fluorescence emission was alternately collected at each wavelength with a cooled CCD camera (CoolSNAP fx, Photometrics).

SHORT EXAMPLE OF APPLICATION

We first asked whether Clomeleon chronically expressed in transgenic mice was still capable of responding to changes in [Cl^-]. For this purpose, Clomeleon FRET signals were measured in individual superior colliculus cells in slices while manipulating [Cl^-] with a solution containing the K^+/H^+ ionophore, nigericin <!>, in combination with the OH^-/ Cl^- antiporter, tributyltin <!> (Inglefield and Schwartz-Bloom 1997). Nigericin clamps intracellular pH to the value of the extracellular pH. This is necessary to prevent [Cl^-] changes caused by ion movement through pH-dependent exchangers, such as the Cl^-/HCO_3^- exchanger (Krapf et al. 1988). Under such conditions, tributyltin equilibrates $[Cl^-]_i$ with $[Cl^-]_o$. Changes in Clomeleon FRET ratio were observed at [Cl^-] steps from 0 mM Cl^- to 135 mM Cl^- (Fig. 1c). A high concentration of F^- (125 mM) was used to saturate Clomeleon (Kuner and Augustine 2000). This caused a further reduction in the FRET emission ratio. Half-maximal quenching of Clomeleon occurred at 175 mM, with changes of [Cl^-] in the range of 10–100 mM being most sensitively reported by Clomeleon. Thus, chronically expressed Clomeleon does respond to changes in $[Cl^-]_i$.

Changes in Clomeleon signals resulting from the application of GABA (10 mM, 200 msec) from a puffer pipette (Fig. 1d) were also examined. GABA produced a transient increase in $[Cl^-]_i$ of approximately 10 mM, which decayed back to baseline within 30–50 seconds. The Cl^- transient was partially blocked by the GABA receptor antagonist, bicuculline (40 μM), indicating that it is due to activation of $GABA_A$ receptors. These results indicate that Clomeleon can report changes in neuronal $[Cl^-]_i$ associated with GABA receptors.

ADVANTAGES AND LIMITATIONS

Clomeleon offers two distinct advantages over existing chloride-sensitive dyes. The first is that the indicator can be targeted to specific cells via genetic targeting. Provided the appropriate promoter exists, it is possible to drive Clomeleon expression in any single cell type. This should also facilitate in vivo measurements of $[Cl^-]_i$. Second, Clomeleon is ratiometric and therefore permits the determination of absolute [Cl^-] within the cells of interest.

Two significant limitations of Clomeleon have thus far been identified. First, Clomeleon is sensitive to protons (Kuner and Augustine 2000). Although this interference is minimal under physiological conditions, it is worth measuring the internal pH of the cells under investigation so that Clomeleon responses can be accurately calibrated. A second limitation is that, under certain conditions, Clomeleon has complex time-dependent photochemical properties. Figure 1 shows an extreme example of such behavior. Clomeleon was excited by intense illumination, causing time-dependent drops in fluorescence at both emission wavelengths (upper panel). These changes could not be caused by fluorophore bleaching because they were partially reversible. The changes may instead reflect the peculiar photochromism properties of certain green fluorescent protein (GFP) derivatives (Dickson et al. 1997; Miyawaki et al. 1999). Whatever the mechanism, the more rapid loss of YFP fluorescence (as compared to the loss of CFP fluorescence) caused the FRET ratio to decrease (Fig.1, lower panel). This artifact would incorrectly indicate a decrease in $[Cl^-]_i$. In practice, such problems can be avoided by minimizing the intensity and duration of Clomeleon excitation.

In summary, this application demonstrates the utility of Clomeleon in the analysis of spatiotemporal dynamics of inhibitory networks in living brain.

ACKNOWLEDGMENTS

This work was supported by NIDA grant DA-15503. K.B. is an overseas research fellow of the Japan Society for the Promotion of Science, and R.L.D. was supported by an NRSA postdoctoral fellowship.

REFERENCES

Dickson R.M., Cubitt A.B., Tsien R.Y., and Moerner W.E. 1997. On/off blinking and switching behaviour of single molecules of green fluorescent protein. *Nature* **388:** 355–358.

Feng G., Mellor R.H., Bernstein M., Keller-Peck C., Nguyen Q.T., Wallace M., Nerbonne J.M., Lichtman J.W., and Sanes J.R. 2000. Imaging neuronal subsets in transgenic mice expressing multiple spectral variants of GFP. *Neuron* **28:** 41–51.

Inglefield J.R. and Schwartz-Bloom R.D. 1997. Confocal imaging of intracellular chloride in living brain slices: Measurement of $GABA_A$ receptor activity. *J. Neurosci. Methods* **75:** 127–135.

Jayaraman S., Haggie P., Wachter R.M., Remington S.J., and Verkman A.S. 2000. Mechanism and cellular applications of a green fluorescent protein-based halide sensor. *J. Biol. Chem.* **275:** 6047–6050.

Krapf R., Berry C.A., and Verkman A.S. 1988. Estimation of intracellular chloride activity in isolated perfused rabbit proximal convoluted tubules using a fluorescent indicator. *Biophys. J.* **53:** 955–962.

Kuner T. and Augustine G.J. 2000. A genetically encoded ratiometric indicator for chloride: Capturing chloride transients in cultured hippocampal neurons. *Neuron* **27:** 447–459.

Lee P.H., Helms M.C., Augustine G.J., and Hall W.C. 1997. Role of intrinsic synaptic circuitry in collicular sensorimotor integration. *Proc. Natl. Acad. Sci.* **94:** 13299–13304.

Miyawaki A., Griesbeck O., Heim R., and Tsien R.Y. 1999. Dynamic and quantitative Ca^{2+} measurements using improved cameleons. *Proc. Natl. Acad. Sci.* **96:** 2135–2140.

Wurtz R.H. and Albano J.E. 1980. Visual-motor function of the primate superior colliculus. *Annu. Rev. Neurosci.* **3:** 189–226.

CHAPTER 81

A Practical Guide: Synapto-pHluorins—Genetically Encoded Reporters of Synaptic Transmission

Gero Miesenböck

Department of Cell Biology, Yale University School of Medicine, New Haven, Connecticut 06510

PHluorins are pH-sensitive mutants of green fluorescent protein (GFP). Attached to proteins with defined cellular locations or itineraries, pHluorins report subcellular pH as well as protein transport between compartments of differing pH. A key application is the optical detection of neurotransmitter release ("synapto-pHluorins").

pHLUORINS

Since it is generally impossible to resolve the several dozen to several hundred vesicles at a typical synapse by optical means, vesicles undergoing exocytosis must be distinguished from resting vesicles spectrally. Synapto-pHluorins use pH-sensitive mutants of GFP, termed pHluorins, to afford this distinction (Miesenböck et al. 1998). Synaptic vesicles contain ATP-powered proton pumps that maintain a proton electrochemical gradient across the vesicle membrane; as a consequence, the vesicle interior is acidified to a pH of approximately 5.7. pHluorins located in these vesicles are in the "off" state. As a vesicle fuses with the plasma membrane, a proton-conducting pore opens to the extracellular space and the accumulated protons dissipate, switching the pHluorin "on."

Two classes of pHluorins, termed ecliptic and ratiometric, have been developed (Miesenböck et al. 1998). The chromophore of ecliptic pHluorin titrates, with a pK_a of about 7.1 (Sankaranarayanan et al. 2000), between a deprotonated, fluorescent, and a protonated, nonfluorescent, state (Miesenböck et al. 1998). The abundance of protons inside synaptic vesicles drives ecliptic pHluorin into the protonated form and quenches ("eclipses") its fluorescence. Exocytic events lead to chromophore deprotonation, which is detected as a sudden increase in fluorescence, excitable in the 450–490-nm band (Fig. 1). Both the original ecliptic pHluorin (GenBank accession number AF058695), and a newer "superecliptic" variant (GenBank accession number AY533296; this form carries the chromophore substitutions F64L and S65T, in addition to the ecliptic set of mutations), exist as mixtures of two conformers: the ecliptic pHluorin proper, whose fluorescence is excited at approximately 475 nm and quenched at pH <6.0, and a mildly pH-sensitive form that is excited at about 395 nm and remains fluorescent at pH <6.0 (Fig. 1a). The effect of the chromophore substitutions in superecliptic pHluorin

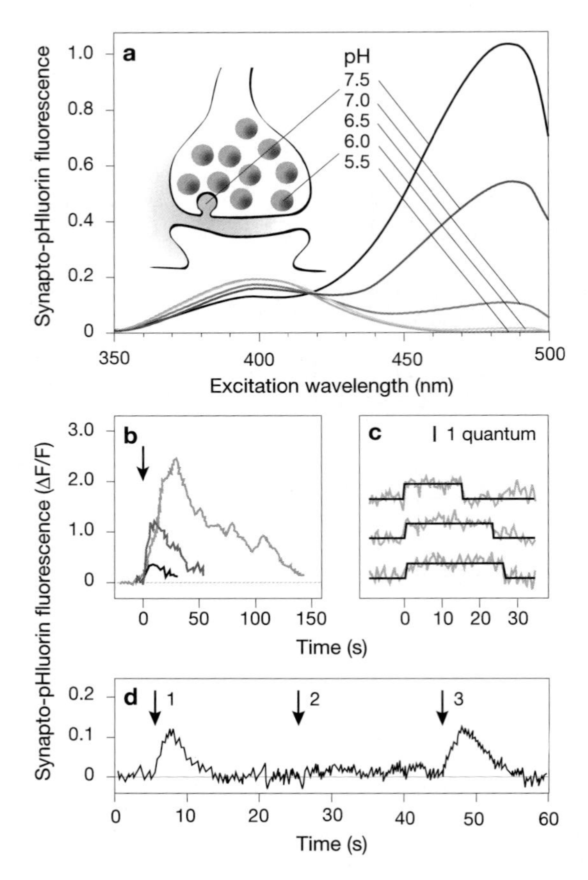

FIGURE 1. Imaging neurotransmission with synapto-pHluorin. (*a*) Fluorescence excitation spectra of superecliptic synapto-pHluorin as a function of pH, normalized to fluorescence excited at 491 nm at pH 7.5. In the context of a presynaptic terminal (*inset*), pHluorins located in resting vesicles (internal pH ~ 5.7) are nonfluorescent when illuminated at about 450–490 nm. Upon exocytosis, the vesicle interior becomes continuous with the extracellular space, pH rises to approximately 7.4, and the pHluorin complement of the fused vesicle becomes fluorescent. (*b*) Time course of synapto-pHluorin fluorescence of a single presynaptic terminal during 20 (*black trace*), 100 (*dark gray trace*), and 300 (*light gray trace*) action potentials. Action potentials were evoked by electrical field stimulation at 10 Hz; the arrow denotes the onset of the stimulus trains. (Modified, with permission, from Sankaranarayanan and Ryan 2000 [© Nature Publishing Group].) (*c*) Time course of synapto-pHluorin fluorescence of individual presynaptic terminals following a single action potential. Stepwise up (+ 1 quantum) and down (–1 quantum) transitions presumably correspond to the release and retrieval of single synaptic vesicles. The time elapsing between up and down transitions is variable, due to the stochastic nature of vesicle retrieval. (Modified, with permission, from Gandhi and Stevens 2003 [© Nature Publishing Group].) (*d*) Time course of synapto-pHluorin fluorescence of olfactory receptor neurons projecting to a single glomerulus in the antennal lobe of a fruit fly during the presentation of three different test odors (*arrows*). Odors 1 and 3 evoke neurotransmitter release, whereas odor 2 is ineffective. (Modified, with permission, from Ng et al. 2002 [© Elsevier].)

is to populate predominantly the ecliptic species, causing a 5.9-fold increase in its fluorescence, and to red-shift its excitation peak from 475 to 491 nm (Fig. 1a). At pH 7.5, superecliptic pHluorin emits at 22% the rate of enhanced GFP (EGFP).

Ratiometric pHluorin (GenBank accession number AF058694) possesses a bimodal excitation spectrum, similar to that of wild-type GFP. The excitation maximum lies at 395 nm at neutral pH but gradually shifts to 475 nm as pH is lowered (Miesenböck et al. 1998). Ratiometric comparison of the fluorescence intensities at the two excitation maxima provides a calibrated measure of pH (Miesenböck et al. 1998; Jankowski et al. 2001; Poschet et al. 2001, 2002; Ohara-Imaizumi et al. 2002; Olsen et al. 2002). Like all ratiometric indices, this measure is insensitive to variations in fluorophore concentration and optical path length. However, because ratiometric pHluorin lacks a nonfluorescent state at pH <6.0, exocytic events must be distinguished against a background of resting vesicles. This requires that individual secretory vesicles are resolved optically (see Chapter 61), or, if this is impossible, that the fraction of vesicles undergoing exocytosis is sufficiently large to alter the aggregate ratio detectably (Miesenböck et al. 1998).

TARGETING MODULE

The targeting module—the topogenic polypeptide sequence to which the pHluorin module is linked through gene fusion—determines the subcellular location or itinerary of the light emitter and thus the compartment or process to be monitored optically (Miesenböck et al. 1998; Jankowski et al. 2001; Poschet et al. 2001, 2002; Ohara-Imaizumi et al. 2002; Olsen et al. 2002). For imaging neurotransmitter release, superecliptic pHluorin is linked to the lumenally exposed carboxyl terminus of the v-SNARE VAMP/synaptobrevin-2, to create the membrane-bound "synapto-pHluorin" (Miesenböck et al. 1998).

GENE EXPRESSION

cDNAs encoding the pHluorin and the targeting module are fused via a short (typically 3–6 amino acids) flexible polypeptide linker. The construct is placed under the control of the appropriate constitutive, inducible, or cell-type-specific genetic control element and introduced into the cell, tissue, or organism of interest. The characteristics of the expression construct and the optimal DNA transfer method depend on the problem at hand. Approaches range from shotgun methods, for transient expression from viral promoters (Miesenböck et al. 1998; Sankaranarayanan and Ryan 2000; Gandhi and Stevens 2003), to cell-type-specific expression in transgenic or "knock-in" animals (Ng et al. 2002; Samuel et al. 2003).

OPTICAL IMAGING

The broad range of cellular processes, cell types, and organisms studied with pHluorins (Miesenböck et al. 1998; Jankowski et al. 2001; Poschet et al. 2001, 2002; Ng et al. 2002; Ohara-Imaizumi et al. 2002; Olsen et al. 2002; Kim et al. 2003; Samuel et al. 2003) is paralleled by an equally broad range of microscope configurations, light sources, and detectors (Yuste et al. 2000). Options range from wide-field illumination by arc lamps (Miesenböck et al. 1998; Gandhi and Stevens 2003) to lasers providing one- or two-photon excitation. The 488-nm argon ion laser line ideally matches the superecliptic excitation peak for single-photon excitation (Sankaranarayanan and Ryan 2000; Sankaranarayanan et al. 2000; Kim et al. 2003; Samuel et al. 2003); efficient two-photon excitation requires an ultrafast laser (pulse width 100–150 fsec, repetition rate ~ 80 MHz) operating in the 910–940-nm range (Ng et al. 2002). Emitted photons in the 500–560-nm band (peak at 508 nm) are separated from excitation light with suitable dielectric mirrors, interference filters, and/or absorptive optics (Yuste et al. 2000; Ng et al. 2002) and are detected by intensified (Miesenböck et al. 1998) or back-illuminated CCDs (Gandhi and Stevens 2003). Laser-scanning microscopy employs PMTs in confocal (Sankaranarayanan and Ryan 2000; Sankaranarayanan et al. 2000; Kim et al. 2003; Samuel et al. 2003) or whole-area mode (Ng et al. 2002). Instrumentation for excitation ratio imaging of ratiometric pHluorin is discussed elsewhere (Yuste et al. 2000).

EXAMPLES OF APPLICATIONS

Figure 1 illustrates the use of (super)ecliptic synapto-pHluorin for monitoring synaptic transmission in different types of neurons. In Figure 1b, the fluorescence of a single presynaptic terminal, formed by a rat hippocampal neuron in dissociated culture, was recorded while trains of 20, 100, and 300 action potentials were evoked by electrical field stimulation at 10 Hz (Sankaranarayanan and Ryan 2000). Stimulation caused transient increases in fluorescence (due to synaptic vesicle exocytosis) that decayed with time constants ranging from 4 to 50 sec (due to synaptic vesicle endocytosis and re-acidification). Under the assumptions that the synaptic terminal releases at most one vesicle per action potential, and that the probability of release is approximately 0.3, these fluorescence transients represent composites of approximately 6, 30, and 100 single-vesicle signals, respectively. Because individual vesicles add and subtract unitary quanta of fluorescence as they exocytose and recycle (Fig. 1c) (Gandhi and Stevens 2003), the time course of the composite signal reflects two physiological variables: the timing of individual release events and the distribution of waiting times for the retrieval process.

Figure 1d demonstrates the detection of synaptic transmission in an intact brain in response to a physiological stimulus. Synapto-pHluorin was expressed selectively in olfactory receptor neurons of a fruit fly, and fluorescence was measured by two-photon laser-scanning microscopy in a target region that receives synaptic input from neurons expressing one particular type of olfactory receptor (Ng et al. 2002). Odorants, acting as agonists on this particular receptor, evoke neurotransmitter release, which causes a rapid increase in synapto-pHluorin fluorescence, followed by a slower decay (time constant ~ 4 sec).

ADVANTAGES AND LIMITATIONS

Synapto-pHluorins possess five unique strengths as indicators of synaptic transmission. First, they are generated biosynthetically, without cofactor requirements. This permits the study of samples that are difficult to stain with exogenous molecules, such as intact tissues and organisms (Ng et al. 2002; Samuel et al. 2003). Second, indicator expression can be placed under the control of cell-type-specific genetic control elements (Ng et al. 2002; Samuel et al. 2003), ensuring a degree of specificity that synthetic indicator dyes lack (Zemelman and Miesenböck 2001). Third, in contrast to "single-shot" methods that image dye release (Betz and Bewick 1992; see also Chapters 61 and 62), the synapto-pHluorin signal is fully regenerative through multiple rounds of vesicle release and recycling. Fourth, the regenerative nature of the synapto-pHluorin signal allows it to provide direct information not only about vesicle release, but also about vesicle recycling (Sankaranarayanan and Ryan 2000, 2001; Gandhi and Stevens 2003). Finally, expression of synapto-pHluorin does not appear to interfere with neuronal function (Sankaranarayanan and Ryan 2000), which can be a concern when using chelating calcium indicators.

The limitations of synapto-pHluorins are tied to the biology of synaptic vesicles. The number of v-SNARE protein molecules (and thus synapto-pHluorins) present on the surface of a synaptic vesicle is smaller than the number of lipophilic dye molecules (Betz and Bewick 1992) that can be accommodated, leading to a smaller quantum of fluorescence per vesicle. At hippocampal synapses, about 10–20% of the total synapto-pHluorin pool is present at the cell surface under unstimulated conditions; this pool elevates background fluorescence and reduces the signal-to-noise ratio (Sankaranarayanan and Ryan 2000; Sankaranarayanan et al. 2000). If the synapse has a low basal release rate, and the vesicular and surface synapto-pHluorin pools do not intermix, prestimulus photobleaching (Gandhi and Stevens 2003; Samuel et al. 2003) can selectively eliminate the fluorescent surface pool and make even single-vesicle turnovers detectable (Gandhi and Stevens 2003). Once a vesicle exocytoses, its synapto-pHluorin complement remains fluorescent for an extended, but variable, period of time that is terminated by vesicle retrieval and acidification (Miesenböck et al. 1998; Sankaranarayanan and Ryan 2000, 2001; Gandhi and Stevens 2003). This period, which may last from a few hundred milliseconds to several dozens of seconds, constrains the temporal resolution of the method.

Synapto-pHluorin occupies an important position in neurobiology: it is the only probe available that can signal the activity of genetically specified synapses in intact neural tissue. Given the key roles of synaptic transmission in information processing and storage, knowledge of the dynamics of efferent synapses is an integral part of any functional description of neurons and circuits.

REFERENCES

Betz W.J. and Bewick G.S. 1992. Optical analysis of synaptic vesicle recycling at the frog neuromuscular junction. *Science* **255:** 200–203.

Gandhi S.P. and Stevens C.F. 2003. Three modes of synaptic vesicular recycling revealed by single-vesicle imaging. *Nature* **423:** 607–613.

Jankowski A., Kim J.H., Collins R.F., Daneman R., Walton P., and Grinstein S. 2001. In situ measurements of the pH of mammalian peroxisomes using the fluorescent protein pHluorin. *J. Biol. Chem.* **276:** 48748–48753.

Kim J.H., Udo H., Li H.L., Youn T.Y., Chen M., Kandel E.R., and Bailey C.H. 2003. Presynaptic activation of silent synapses and growth of new synapses contribute to intermediate and long-term facilitation in Aplysia. *Neuron* **40:** 151–165.

Miesenböck G., De Angelis D.A., and Rothman J.E. 1998. Visualizing secretion and synaptic transmission with pH-sensitive green fluorescent proteins. *Nature* **394:** 192–195.

Ng M., Roorda R.D., Lima S.Q., Zemelman B.V., Morcillo P., and Miesenböck G. 2002. Transmission of olfactory information between three populations of neurons in the antennal lobe of the fly. *Neuron* **36:** 463–474.

Ohara-Imaizumi M., Nakamichi Y., Tanaka T., Katsuta H., Ishida H., and Nagamatsu S. 2002. Monitoring of exocytosis and endocytosis of insulin secretory granules in the pancreatic beta-cell line MIN6 using pH-sensitive green fluorescent protein (pHluorin) and confocal laser microscopy. *Biochem. J.* **363:** 73–80.

Olsen K.N., Budde B.B., Siegumfeldt H., Rechinger K.B., Jakobsen M., and Ingmer H. 2002. Noninvasive measurement of bacterial intracellular pH on a single-cell level with green fluorescent protein and fluorescence ratio imaging microscopy. *Appl. Environ. Microbiol.* **68:** 4145–4147.

Poschet J.F., Boucher J.C., Tatterson L., Skidmore J., Van Dyke R.W., and Deretic V. 2001. Molecular basis for defective glycosylation and *Pseudomonas* pathogenesis in cystic fibrosis lung. *Proc. Natl. Acad. Sci.* **98:** 13972–13977.

Poschet J.F., Skidmore J., Boucher J.C., Firoved A.M., Van Dyke R.W., and Deretic V. 2002. Hyperacidification of cellubrevin endocytic compartments and defective endosomal recycling in cystic fibrosis respiratory epithelial cells. *J. Biol. Chem.* **277:** 13959–13965.

Samuel A.D., Silva R.A., and Murthy V.N. 2003. Synaptic activity of the AFD neuron in *Caenorhabditis elegans* correlates with thermotactic memory. *J. Neurosci.* **23:** 373–376.

Sankaranarayanan S. and Ryan T.A. 2000. Real-time measurements of vesicle-SNARE recycling in synapses of the central nervous system. *Nat. Cell Biol.* **2:** 197–204.

———. 2001. Calcium accelerates endocytosis of vSNAREs at hippocampal synapses. *Nat. Neurosci.* **4:** 129–136.

Sankaranarayanan S., De Angelis D., Rothman J.E., and Ryan T.A. 2000. The use of pHluorins for optical measurements of presynaptic activity. *Biophys. J.* **79:** 2199–2208.

Yuste R., Miller R.B., Holthoff K., Zhang S., and Miesenböck G. 2000. Synapto-pHluorins: Chimeras between pH-sensitive mutants of green fluorescent protein and synaptic vesicle membrane proteins as reporters of neurotransmitter release. *Methods Enzymol.* **327:** 522–546.

Zemelman B.V. and Miesenböck G. 2001. Genetic schemes and schemata in neurophysiology. *Curr. Opin. Neurobiol.* **11:** 409–414.

CHAPTER 82

Imaging Gene Expression in Live Cells and Tissues

Ricardo E. Dolmetsch, Natalia Gomez-Ospina, Eric Green, and Elizabeth A. Nigh

Department of Molecular Pharmacology, Stanford University School of Medicine, Stanford, California 94305

Regulated gene expression plays a critical role in the development of the nervous system as well as in plastic changes that occur in adult organisms. Monitoring the expression of genes has been essential for understanding the molecular mechanisms that underlie many complex biological processes. New advances in optical imaging technologies are producing powerful tools for monitoring changes in gene expression in single live cells and in intact organisms. These new technologies are particularly important for investigating the mechanisms that regulate the development of organisms and the plasticity of adult tissues like the brain. The goal of this chapter is to provide an overview of the available and emerging technologies to measure gene expression in live tissues and cells.

OVERVIEW: REGULATION OF GENE EXPRESSION

Gene expression is of central importance for many biological processes and is controlled by a complex network of regulatory mechanisms. At the most fundamental level, the expression of genes is regulated by the organization of chromosomes into transcriptionally active and transcriptionally silent domains, known, respectively, as euchromatin and heterochromatin. The organization of the chromatin into domains is regulated by enzymes that add methyl, hydroxyl, phosphate, ubiquitin, or acetyl groups to histones, the family of proteins that package DNA, and by enzymes that modify the DNA itself. Modified histones and DNA in turn recruit complexes of remodeling proteins that can either repress or activate the transcription of large groups of genes. The large-scale regulation of chromatin structure sets the stage for more detailed regulation of gene expression.

Within transcriptionally active chromatin domains, transcription factors control the expression of specific genes. Transcription factors (enhancers and repressors) are proteins that are recruited to binding sites on the DNA in the vicinity of genes, usually upstream of the initiation site, where they can both alter the local structure of the DNA–histone complexes and affect the recruitment of general transcription machinery including DNA-dependent RNA polymerases (Ptashne 1988; Zawel and Reinberg 1995). The binding of transcription factors to DNA and their ability to activate or repress transcription are tightly regulated, both by cell-intrinsic processes and by signaling pathways that are activated by extracellular stimuli. A large number of transcription factors have been identified, and a great deal is known about their regulation in response to signaling pathways. Comparatively little is known, however, about how transcription factors behave in intact cells and organisms during development and normal cellular function.

The activation of transcription leads to the production of messenger RNA (mRNA) that is itself subject to several mechanisms of regulation (Farina and Singer 2002; Black 2003). mRNAs are spliced by mRNA-remodeling enzymes to yield a variety of transcripts which encode proteins that may have distinct functions. The selection of splice isoforms is a regulated process. The stability of an mRNA and its localization within a cell are also subject to regulation and may have a profound effect on the expression of a specific protein. The concentration of specific cellular mRNAs, which is controlled by the activation of transcription factors, and the processing, localization, and degradation of an mRNA constitute another level at which gene expression is regulated.

A final level of gene regulation occurs during the process by which mRNAs are translated into proteins (Sonenberg and Dever 2003). Translation is a complex process composed of an initiation phase, an elongation phase, and a termination phase. All three phases can be regulated by signaling proteins that either are expressed in a developmentally regulated pattern or are activated by extracellular stimuli. In addition, the translation of a specific mRNA can be controlled by the binding of regulatory factors to the 3′ and 5′ untranslated region of an mRNA. By regulating the translation of a specific mRNA, a cell can rapidly alter the concentration of a protein in response to a signaling event. mRNA translation is therefore another key point of regulation in the process by which genes are converted into proteins.

OPTICAL METHODS FOR IMAGING GENE EXPRESSION

Measuring the Structure of Chromatin

The regulation of chromatin structure by histones and other DNA-associated proteins plays a central role in controlling gene expression in tissues during development and throughout the cell cycle (Narlikar et al. 2002; Grewal and Moazed 2003). The past few years have seen a surge of interest in the use of fluorescently labeled proteins to measure chromosome and chromatin structure in living cells. One approach has been to construct cell lines in which a multimerized binding site for a bacterial transcription factor is inserted into one or more chromosomal locations (for review, see Belmont 2001). GFP, fused to the bacterial transcription factor, is then introduced into the cell and used to localize the labeled chromosomal region in live cells by time-lapse microscopy (Robinett et al. 1996; Strukov et al. 2003). The most frequently used DNA motifs are the lac and tet operons, which bind the lac and tet repressor proteins, respectively. Because GFP-fusion proteins with the lac and tet repressor proteins bind to the lac and tet operons, but not to endogenous chromatin, they can be used to track chromatin reorganization in vivo. This approach has allowed the analysis of chromosomal condensation and relocalization during interphase, when the DNA cannot be detected by light microscopy. A related approach has been to use fluorescently tagged histones or other DNA-binding proteins to investigate chromosome reorganization (Kanda et al. 1998; Verschure et al. 1999; Lever et al. 2000). Because there are many histone-binding sites, it is difficult to measure histone motility over time. One group measured the motility of GFP-tagged histone H1 binding by fluorescence recovery after photobleaching (FRAP; see Chapter 56) and correlated changes in histone H1 diffusion with changes in the large-scale arrangement of the DNA (Lever et al. 2000). Cells were transfected with DNA constructs encoding a GFP-tagged histone; after expression, the histone-GFP was photobleached, and the recovery of fluorescence in a bleached area was measured in cells exposed to different kinds of stimuli. Changes in the rate of recovery after photobleaching were interpreted as changes in the regulated binding of histones to DNA and therefore as changes in the structure of chromosomes. This approach provides an indirect measurement of chromatin structure, but it has many potential drawbacks, including its poor spatial resolution, changes in chromatin dynamics associated with histone overexpression, and DNA damage associated with photobleaching.

A third approach for measuring the processes that modulate chromatin structure has been to study the interaction of transcription factors with chromatin-modulating enzymes. One example of this approach is the development of sensors for measuring the interaction of the transcription factor CREB and the histone modification enzyme CBP, which acetylates histones and activates transcription (Shaywitz and Greenberg 1999). The interaction of CREB with CBP has been imaged using both

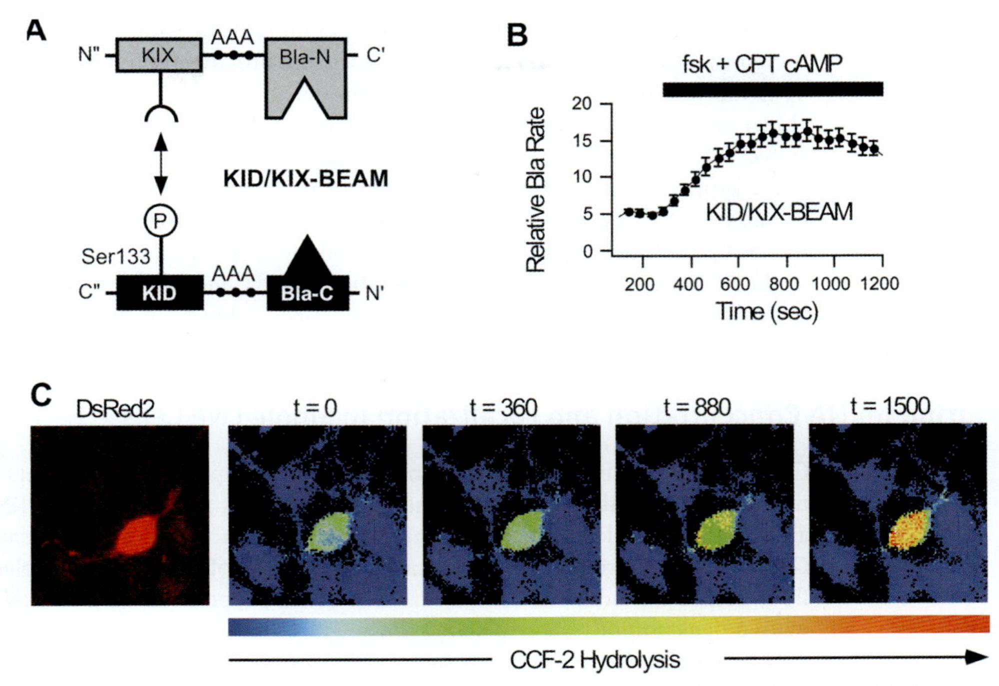

FIGURE 1. BEAM detection of CREB and CBP interaction in neurons. (*A*) Cartoon of BEAM strategy that uses β-lactamase (Bla) protein fragment complementation mediated by the phosphorylation-dependent interaction between the KID domain of CREB and the KIX domain of CBP. (*B*) Quantification of the time course of Bla activity in neurons transfected with the BEAM sensor and stimulated with forskolin (fsk) and CPT cAMP, agents that activate protein kinase A and phosphorylate the KID domain. (*C*) Time-lapse images of Bla activity using CCF-2 in neurons transfected with the BEAM sensor and stimulated with forskolin. The first image shows that the cell expresses red fluorescent protein as a transfection marker. (Modified from Spotts et al. 2002.)

protein fragment complementation and fluorescence energy transfer between GFP-tagged CREB and CBP. Protein fragment complementation assays are based on splitting enzymes or other proteins into complementary domains that are themselves inactive but that gain activity when brought together with their partner (Michnick 2001). The authors' group developed the BEAM system, which uses protein fragment complementation of the enzyme β-lactamase (Bla) to detect the interaction between any two proteins in intact cells. BEAM has been used to monitor the interaction between CREB and CBP in response to elevations of cAMP in neurons (see Fig. 1) (Spotts et al. 2002). In these experiments, a kinase-inducible binding domain of CREB, known as the KID, was fused to the initial 268 amino acids of Bla, whereas the KID-binding domain of CBP, called the KIX, was fused to the domain from amino acid 269 to the carboxyl terminus. The proteins were introduced into neurons, and the cells were loaded with an intracellular fluorescent sensor for Bla called CCF-2 that changes its fluorescence emission in response to Bla activity. Changes in Bla activity corresponding to the association of CREB and CBP could be detected within 5 sec of the stimulus, suggesting that this method is a rapid and efficient way of measuring protein dimerization in the nucleus. In principle, the BEAM system and other systems like it that use enzyme dimerization should be able to detect the interaction of a small number of protein molecules because the enzymatic activity will amplify the interaction of a few proteins to produce a large signal.

A second approach for detecting the interaction of transcription factors and enzymes that modify the chromatin structure is to use fluorescence resonance energy transfer (FRET; see Chapter 74) between proteins tagged with spectral variants of green fluorescent protein. The principle behind FRET is that when two fluorescent proteins with complementary emission and excitation spectra are brought into close proximity, there is transfer of energy from the fluorophore that is excited by the short-wavelength light to the fluorophore that is excited by the longer-wavelength light. The efficiency of energy transfer is nonlinearly dependent on the close proximity of the acceptor and the emitter, and therefore, FRET is a good method for measuring protein–protein interactions. A FRET-based probe of CREB and CBP has been made that can detect the recruitment of CBP to CREB in the nucleus (Mayr et al. 2001). This system is simpler than the BEAM system, as it does not require the introduction of an exogenous substrate into cells or tissues, but FRET is significantly less sensitive than BEAM and requires the interaction of thousands of GFP-tagged molecules for detection of a signal. In the existing sensor, when the KID domain was tagged with CFP and the KIX domain was tagged with YFP, the change in FRET upon dimerization was on the order of only 10% as compared with a 300% change in the BEAM system. Despite their limitations, FRET and BEAM may be generally useful for observing changes in chromatin activity if they are used to detect the binding of similar chromatin modification factors.

Measuring mRNA Concentration and Localization in Single Live Cells

One of the most direct methods for measuring gene expression is to monitor the concentration of an mRNA species within a cell. Current optical methods for measuring mRNA concentration rely on antisense oligonucleotide probes composed of either nucleotide analogs or artificial backbone analogs. Oligonucleotide probes complementary to an mRNA of interest are coupled to a fluorophore whose fluorescence is modulated by hybridization of the probe to the target mRNA. These methods for the in vivo detection of transcripts are known collectively as antisense imaging, and they are developing into powerful tools for the study of gene expression as well as for medical diagnostics and therapeutics. Antisense imaging promises to provide a means of noninvasively monitoring the expression of genes during disease as well as following pharmacological and gene therapies, although many obstacles still need to be overcome before the technology is ready for these applications. Here we discuss some of the basic principles of in vivo antisense imaging with emphasis on its use in living cells. Some aspects of the use of these technologies in animal models of disease and in human subjects are covered elsewhere (Lewis and Jia 2003).

All antisense imaging strategies use Watson–Crick pairing to generate hybrids between endogenous mRNAs and the fluorescent probes. Therefore, the cellular mRNA concentration can be monitored by measuring the amount of fluorescent probe bound to the native mRNA. Finding the best oligonucleotide backbone for fluorescent mRNA reporters has proved to be difficult, however. Unmodified DNA and RNA are rapidly degraded in vivo by various nucleases, making them poor candidates for probes. In addition, RNA duplexes activate a stress response in many cells and may cause degradation of endogenous messages via an RNA interference process. Nevertheless, in some cases DNA-based probes have been used successfully for imaging endogenous mRNAs (Dewanjee 1994; Cammilleri 1996; Sato 2001). To overcome the problems with degradation, DNA and RNA analogs have been synthesized that improve the stability, target affinity, and specificity of the probes. Several unnatural oligonucleotide backbones are resistant to nucleases that cause the degradation of the target mRNA. Phosphorothioate DNAs and 2′-O-methyl RNAs have been employed successfully to visualize mRNA in living cells (Dewanjee et al.1994; Kobori et al.1999), as have polynucleotides with non-ribose artificial backbones such as peptide nucleic acids (PNAs). PNAs seem to offer the greatest potential for use as in vivo imaging probes (Mardirossian et al. 1997; Lewis et al. 2002; Rao et al. 2003), but other artificial backbone analogs such as morpholinos (MORF) and trans-4-hydroxy-L-proline nucleic acid-phosphono nucleic acid (HypNA-pPNA) may also be useful. It is not clear which nucleic acid analog is the best for in vivo imaging of gene expression, but newly developed nucleic acid backbones will almost certainly offer many advantages relative to naturally occurring oligonucleotides.

Antisense imaging involves coupling an oligonucleotide to a label that can be detected optically. A variety of imaging labels and strategies have been used for visualization of the probe–mRNA interaction. Generally, antisense probes are coupled to fluorophores that become fluorescent upon probe–mRNA hybridization, and the activated fluorophore is detected by fluorescence microscopy (for review, see Paroo and Corey 2003). There are three basic strategies by which hybridization of a probe to a target mRNA changes the fluorescence of the mRNA probe: the interaction of the target and the probe disrupts the interaction between a quencher and a fluorophore unmasking fluorescence emission, the interaction generates new fluorophores by hydrolysis of a precursor or intercalation of a

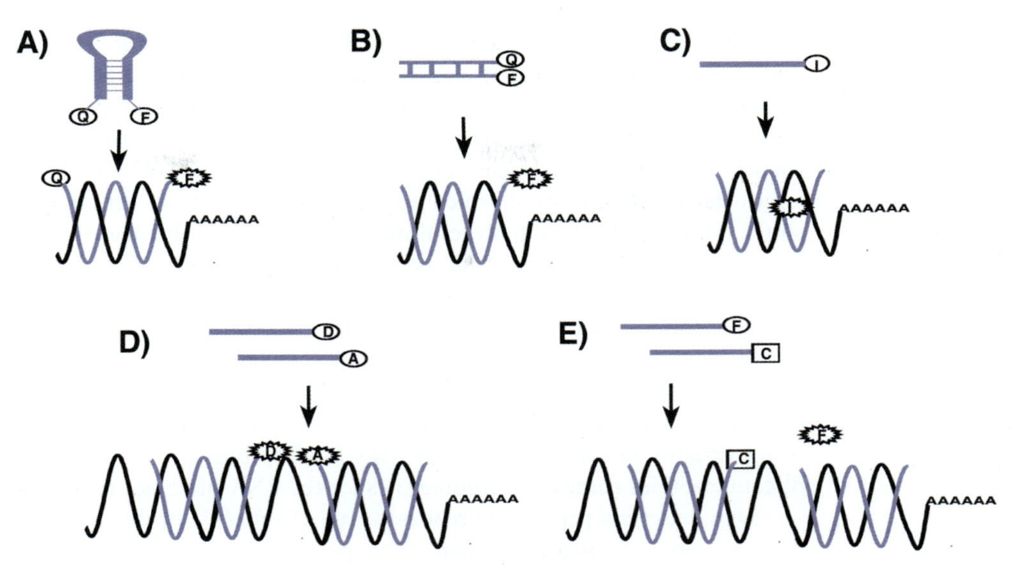

FIGURE 2. Antisense imaging strategies. (*A*) Molecular beacons become fluorescent when the hairpin is dissolved and the fluorophore and the quencher are separated (Q is the quencher; F is the fluorophore. (*B*) Duplex oligos use a similar strategy to detect a target mRNA but the quencher and the fluorophore are on opposite strands of the DNA probe. (*C*) Intercalating oligonucleotide probes increase their fluorescence in the presence of double-stranded nucleotides (I is the intercalator). (*D*) FRET relies on the nonradiative energy transfer between adjacent fluorophores (D is the donor and A is the acceptor fluorophore). (*E*) mRNA-triggered catalytic drug release uses a catalyst attached to a probe to liberate a fluorescent probe attached to a neighboring oligonucleotide (C is the catalyst).

dye into an RNA–DNA duplex, or the interaction brings donor and acceptor fluorophores together to achieve FRET. These strategies are shown in Figure 2 and are described in more detail below.

Molecular beacons and duplex oligonucleotide probes use quenchers to prevent the fluorophore from significantly emitting light until they bind to their target mRNA. Molecular beacons consist of hairpin oligonucleotides labeled with a fluorophore at one end and a quencher at the other. The hairpin brings the fluorophore close to the quencher when the probe is not hybridized to a target, but upon hybridization, the interaction between the quencher and the fluorophore is disrupted resulting in fluorescence emission. Duplex oligonucleotide probes are composed of two complementary strands; one strand has the fluorophore at the 5′ end and the complementary strand has a quencher at the 3′ end. The annealing of the antisense strand to the target mRNA disrupts the quencher–fluorophore interaction and results in increased fluorescence. Although most molecular beacons used to image mRNA have been DNA-based (Sokol et al. 1998; Perlette and Tan 2001), 2′-O′ methyl RNAs (Tsourkas et al. 2003), PNA-based beacons have been reported to be effective (Kuhn et al. 2002) and have the advantage of being more stable. Molecular beacons are widely used for the monitoring of real-time polymerase chain reactions (PCRs), semi-automated mutation analysis, and self-reporting oligonucleotide arrays. Design of both molecular beacons and duplex probes for use in live imaging, however, is complicated by the requirement that hairpin and probe melting and binding to the appropriate target must occur at physiological temperatures. This requirement makes the length and chemical composition of the probes, their denaturing kinetics, and their melting temperature important issues to consider in the design of beacons and duplex oligonucleotides.

A second approach for imaging mRNA involves probes that intercalate or hydrolyze upon hybridization with the target mRNA. Intercalating oligonucleotide probes consist of a single-stranded oligonucleotide probe with a covalently attached intercalator such as thiazole orange, which undergoes a fluorescent change following binding to a target and intercalation in the double-stranded oligonucleotide (Privat et al. 2002). This approach is promising but has the distinct disadvantage that most intercalating fluorophores are slightly fluorescent even when non-intercalated, thus raising the background of the system. Another interesting approach to generating hybridization-induced fluorescent probes is mRNA-triggered catalytic drug release (Ma and Taylor 2003). In this technology, a catalyst and a diacetyl fluorescein, a molecule that becomes fluorescent upon hydrolysis, are attached to two analogs that are complementary to adjacent sequences on the mRNA. Fluorescence is produced upon hydrolysis of the fluorescein by the nearby catalyst and is triggered specifically by the binding of both catalyst-bearing and fluorescein-bearing analogs to the mRNA. This approach offers the advantage of large catalytic amplification, which may make it possible to detect poorly expressed messages but may suffer from nonspecific signals resulting from catalysis of the diacetyl fluorescein in solution.

A final approach to detecting specific mRNAs is to use FRET between two mRNA probes to detect mRNA in living cells (Sei-Iida et al. 2000; Tsuji et al. 2000). In this case, when two fluorescently labeled oligonucleotides, each labeled with a donor or acceptor fluorophore, hybridize to adjacent locations on the transcript (<10 nm), there is energy transfer from the donor molecule to the acceptor fluorophore resulting in a change in the wavelength of light emission. One advantage of the FRET approach is that it may be possible to generate genetically encoded sensors for specific mRNAs by using variants of GFPs attached to RNA-binding proteins in conjunction with specially designed oligonucleotides containing binding sites for the RNA-binding proteins.

Despite several papers that report the use of antisense probes for imaging mRNA, this technology has not seen widespread acceptance. One problem is that most mRNAs are found at low concentrations inside cells, resulting in a small signal that is easily masked by the nonspecific autofluorescence of the cell. In addition, polynucleotides do not diffuse across cell membranes, so cells have to be loaded with the probes by transfection, electroporation, microinjection, or osmotic shock, all methods that are difficult to control and deleterious to cellular health. A further problem is that many of the probes have poor hybridization kinetics, so it is necessary to wait tens of minutes to hours for hybridization to occur and a signal to appear. Finally, short strands of nucleotides often have poor specificity at physiological temperatures, making it difficult to interpret the results of intracellular measurements.

Creative solutions to the problems of signal intensity and probe delivery are currently being developed. Brighter labels such as wave-shifting quantum dots promise to be useful for increasing the probe signals during mRNA imaging (Perez et al. 2002). Other technological advancements that could increase the signal-to-noise ratio are infrared fluorophores (Waddell et al. 2000), which would be insensitive to the endogenous autofluorescence of cells. Improvements are required in the delivery of antisense probes into cells and tissues, using better and more robust transfection methods or probes that are actively transported into cells. Naked oligonucleotides, viral vectors, cationic lipids, and new synthetic anionic carriers (Tavitian et al. 2002) have been used for the delivery of synthetic analogs. Other approaches have applied knowledge from cell and molecular biology to target probes to specific cells or tissues and enhance their intracellular uptake. For instance, PNAs have been conjugated to protein transduction domains such as *Drosophila* Antennapedia protein (Heckl et al. 2003), HIV tat peptide, and synthetic transduction peptides (Lewis et al. 2002), which leads to targeted internalization of the PNA. In addition, drug-targeting technology has produced some tools that are relevant for the use of antisense probes to image mRNA. Enhanced targeting can be achieved by conjugation of antisense imaging agents with a receptor ligand that is disease- or tissue-specific; e.g., somatostatin for tumor-associated somatostatin receptors (Mier et al. 2000) or IGF-1 for insulin receptors in breast cancer xenografts (Rao et al. 2003). Finally, a recent study reported the use of biotinylated ^{125}I-labeled PNAs and a streptavidin–monoclonal antibody OX26 conjugate to target the rat transfection receptor. The probe is able to efficiently traffic across the blood–brain barrier and the plasma membrane (Lee et al. 2002). These advances will lead to more flexibility in the use of mRNA detection for imaging gene expression.

Measuring the Activity of Promoters Using Reporter Genes

The most direct method of measuring the expression of a gene is to measure the levels of its mRNA, but the technical difficulties with this approach have given rise to other methods that are more experimentally tractable. Reporter genes are artificial genes in which an enzyme or a fluorescent protein is expressed downstream of the promoter for a gene of interest. Reporter genes are convenient because it is generally simpler to detect the expression of an enzyme or a fluorescent protein than it is to detect the mRNA itself.

There are two general approaches for generating a reporter gene. One approach is to ligate the putative regulatory regions from a gene into an expression plasmid containing a reporter molecule such as luciferase (discussed below). Most commonly, the DNA upstream of a gene's initiation site is amplified using PCR, and the amplified fragment is ligated into the promoter region of a plasmid, upstream of the initiation site for a reporter molecule. Several vectors are available from Promega, Invitrogen, and BD Biosciences that greatly facilitate the construction of a reporter plasmid. These plasmids typically contain a multiple cloning site (MCS) to ease insertion of the regulatory DNA of a gene. The MCS is followed by a TATAA transcription initiation site, a consensus ribosome-binding sequence, and the coding sequence for an enzyme or fluorescent protein. Under some circumstances it is useful to include a "minimal promoter," the regulatory DNA upstream of a promoter region that binds general transcription factors but does not by itself activate transcription. This approach increases the activity of a weak promoter, thus allowing analysis of the role of factors that have only small, modulatory effects on transcription. Because there is no general method for identifying what region of the chromosome contains the regulatory regions for a particular gene, it is generally necessary to make a series of plasmids that include different-length fragments of DNA from the region upstream of a gene.

A general limitation of plasmid-based reporter genes is that they may not contain all the regulatory elements of a gene and may not be regulated as the endogenous gene. An alternative method of generating a reporter gene is to use a bacterial artificial chromosome (BAC). BACs are bacterially propagated fragments of DNA that can contain up to 300 kilobases of genomic DNA. By inserting the DNA that encodes a reporter protein into the coding region of a gene within a BAC, a reporter gene that contains the majority of the regulatory sequences around a gene can be generated. BACs containing a specific gene of interest can be obtained from Invitrogen or from the Children's Hospital Oakland Research Institute (bacpac.chori.org/). To determine which BAC to obtain, it is necessary to map the gene to a specific chromosomal region. BACs can then be ordered that cover that region of the genome. More information on manipulating BACs is available in *Bacterial Artificial Chromosomes: Methods and Protocols from Methods in Molecular Biology* (Zhao and Stodolsky 2003).

Reporter analysis using BACs recreates more of the regulatory environment of the endogenous gene and is therefore thought to reflect endogenous regulation more faithfully. Using a BAC-based reporter gene is also useful for generating transgenic mice that express a reporter gene, because BAC reporter genes are less likely to suffer from the positional effects observed when inserting short fragments of DNA into the genome of mammals. The main disadvantage of a BAC reporter gene is that the molecular biology of BACs is more time-consuming than the molecular biology of conventional plasmids. In general terms, generating a BAC-based reporter gene involves cloning the reporter gene into the BAC using either bacterial or mammalian recombination of a PCR product containing homology arms and a resistance gene. The bacterial resistance gene is then excised from the final vector by using the CRE or FLP recombinases.

A second important consideration when generating a reporter gene is selecting the appropriate reporter protein. The most frequently used reporter genes encode either enzymes such as luciferase or β-lactamase, or fluorescent proteins such as GFP, YFP, and dsRED. The main advantage of fluorescent proteins is that they are easily detected by fluorescence microscopy in live cells and animals using widely available epifluorescence and confocal microscopes. Fluorescent proteins, however, are not particularly sensitive reporters of gene expression, as generally thousands or tens of thousands of molecules are required before expression is detectable. They are also slow to fold, and their fluorescence is sensitive to changes in the intracellular pH, making it difficult to quantify protein expression and imposing a lag on the rate at which it can be detected.

In contrast to fluorescent proteins, enzymes are capable of amplifying small changes in transcription into a large bioluminescent or fluorescent signal. Single-cell detection of enzymatic activity has several requirements. First, the enzyme has to be active in the intracellular environment. Second, the enzyme must have a luminescent or fluorescent substrate that changes its emission properties. Third, the substrate of the enzyme has to be easily loaded into cells and the product of the enzymatic reaction has to be retained by the cell long enough to obtain a measurement. Fourth, the enzyme has to be highly active to allow quantitation of enzymatic activity with reasonable time resolution. Fifth, the rate of substrate cleavage and fluorescent or luminescent emission has to be linearly related to the concentration of enzyme and substrate. For example, the product of the reaction should not inhibit the enzyme and the substrate should not allosterically modulate the enzyme.

There are numerous reporter enzymes including β-galactosidase, luciferase, β-lactamase, choline acetyl transferase, and secreted alkaline phosphatase. To date, only the first three have been used to detect gene expression in intact cells. β-Galactosidase (β-gal) was the first enzyme to be used as a single-cell reporter of gene expression in live cells (Fiering et al. 1991). Fluorescein β-digalactopyranoside (FDG) is a substrate for β-gal that yields free fluorescein upon cleavage. The product of FDG is readily detected by flow cytometry and by microscopy, and FDG is widely used for flow cytometric assays. However, FDG has several disadavantages when it comes to detecting gene expression in single cells in situ as a function of time. First, it is highly polar and therefore does not cross the cell membrane readily. It must therefore be loaded into cells using osmotic shock, which is inefficient, adversely affects cell health, and leads to variable loading among cells. The detection of FDG in cells is also highly problematic because the observed fluorescence is a function of both the amount of loading and the amount of β-gal activity, making it difficult to interpret differences in fluorescence intensity. An even more severe problem is that fluorescein, the cleavage product of FDG, only contains a single positive charge and therefore rapidly leaks out of cells at physiological temperatures. Thus, cells must be maintained on ice to allow accumulation of fluorescein in the cell in sufficient quantities for detection. Several attempts have been made to make FDG a better substrate for intracellular imaging; for instance, by adding carbon tails to make it more hydrophobic. These modifications have produced products that are weakly fluorescent and poorly cleaved by β-gal. These drawbacks limit the use of β-gal and FDG for use as an in vivo reporter gene, although the system is useful under some circumstances. Protocols for using the β-gal FDG system are available in Fiering et al. (1991) and Negulescu et al. (1994).

Currently, the best reporter system for detecting gene expression in single cells is composed of β-lactamase (Bla) and its substrate CCF-2. This system was originally developed in Roger Tsien's laboratory and is available commercially as the Gene Blazer system from Invitrogen (Raz et al. 1998; Zlokarnik et al. 1998). Bla has several advantages as a reporter gene over β-gal. First, Bla activity is entirely absent from mammalian cells, reducing any noise from endogenous enzymatic activity. Second, each molecule of Bla cleaves about 1000 molecules of the substrate per second, making it substantially more active than β-gal. The final advantage is that the substrate of Bla, CCF-2, is easily loaded into mammalian cells at room temperature and the product is retained by cells after cleavage. CCF-2 is available as an acetoxymethyl ester that readily diffuses into cells, where it is chemically modified by intracellular esterases which produce a fluorescent substrate that is trapped in the cell. This substrate is cleaved into a fluorescent product by Bla. The cellular retention of both the substrate and the product allows the signal to develop over time such that ~50 enzymes per cell can be detected with an incubation of 5 min. This permits the detection of poorly expressed genes.

In the intact CCF-2 molecule, excitation of coumarin with 440-nm light results in efficient energy transfer to a linked fluorescein that produces light of 530 nm. Cleavage of CCF-2 by Bla leads to separation of the two chromophores, interrupting energy transfer and producing light emission at 480 nm. Bla activity therefore causes the fluorescence emission of the substrate to change from green to blue. The change in the fluorescence spectrum of CCF-2 allows a ratiometric readout in which the fluorescence of the product can be divided by the fluorescence of the substrate to yield a ratio that increases in proportion to the amount of cleaved substrate. The ratiometric readout has several advantages. On the one hand, it neutralizes the effect of differences in CCF-2 loading among different cells because the fluorescences of both the substrate and the product are proportional to the amount of loaded dye. A more subtle advantage is that weakly expressing cells can be easily distinguished from cells that simply failed to load with the fluorescent substrate.

One main advantage of using the Bla–CCF-2 system is that it is possible to detect gene expression over time in single cells. Detecting a time course of gene expression in a Bla-expressing cell involves measuring the ratio of cleaved to intact substrate over time and calculating the change in the ratio. The change in the ratio is roughly proportional to the amount of Bla activity; however, a more accurate calculation also includes a term that compensates for the change in the concentration of intact CCF-2 during the experiment. The effective rate ($k_{\mathrm{Bla}}(\mathrm{t})$) of CCF-2 hydrolysis in a single cell is given by the following equation (for the full derivation, see Spotts et al. 2002):

$$k_{Bla}(t) = \alpha(t)\left(\frac{dR(t)_{460/530}}{dt}\right) \quad (1)$$

$$\alpha(t) = \frac{\left[F_{S530}\,F_{P460} - F_{S460}\,F_{P530}\right]}{\left[\left((F_{S530} - F_{P530})\,R(t)_{460/530}\right) + F_{P460} - F_{S460}\right]\left[F_{P460} - F_{P530}\,R(t)_{460/530}\right]} \quad (2)$$

where F_{S460} and F_{P460} are the 460-nm molar fluorescence (the fluorescence intensity divided by the concentration of the dye) of the intact CCF-2 substrate (S) and hydrolysis products (P), respectively, and F_{S530} and F_{P530} are the 530-nm molar fluorescences of the intact and hydrolyzed form of the CCF-2 substrate, respectively. Because F_{S460}, F_{P460}, F_{S530}, and F_{P530} are intrinsic properties of CCF-2 and are not significantly altered by the relative concentration of CCF-2 in a cell, Equations 1 and 2 provide a measure of the relative Bla activity in a neuron that is independent of the intracellular CCF-2 concentration.

Despite its advantages, real-time imaging of CCF-2 has some important limitations. First, both coumarin and fluorescein are highly susceptible to bleaching and therefore need to be illuminated with relatively weak light to prevent artifacts associated with differential bleaching of fluorescein and coumarin. Second, depletion of the substrate limits the amount of time that Bla activity can be measured to only a few hours, depending on the levels of Bla expression. Finally, CCF-2 is susceptible to changes in the pH of the intracellular solution, making it difficult to investigate signal transduction pathways that are accompanied by dramatic changes in pH.

In addition to β-gal and Bla, a third commonly used reporter enzyme is the photoprotein luciferase. Luciferase from the North American firefly *Photinus pyralis* is a catalyst for a reaction between luciferin and ATP that yields oxiluciferase, AMP, and one photon of light (Johnson and Shimomura 1972). Luciferase is widely used to measure gene expression in populations of cells in culture in a destructive assay that requires cell lysis. Luciferase, however, is not ideal for measuring gene expression in intact cells because the substrate, luciferin, permeates cell membranes inefficiently and the light emission from luciferase is generally weak. Typically, integration times on the orders of minutes are necessary to measure luciferase signals from single cells, even when using highly sensitive cooled CCD cameras and light-tight enclosures.

The main advantage of luciferase as a reporter molecule is the possibility of detecting photon emission noninvasively in live animals, since it offers some advantages relative to systems that depend on fluorescence excitation and emission. Mammalian tissues are relatively transparent to the red-shifted photons emitted by firefly luciferase, allowing the detection of active enzymes several centimeters into a tissue (Contag et al. 1998). Luminescence also has very good signal-to-noise properties because mammalian tissues do not emit light; the result is a very low background that allows easy detection of even quite weak light signals. In contrast, autofluorescence and light scattering limit fluorescence imaging to a few hundred micrometers into a tissue using multiphoton microscopes. Several recent studies have shown that tumor cells expressing firefly luciferase can be imaged in intact mice injected with luciferin. The detection system generally involves an intensified cooled camera and a light-tight enclosure. Animals carrying cells that express luciferase can be imaged about 5 min after intraperitoneal injection with luciferin (Edinger et al. 1999; Wu et al. 2002). Although this approach does not provide single-cell sensitivity, it does allow the measurement of gene expression in the tissues of an intact animal. Integration times on the order of 5 min are sufficient to image luciferase from about 400 highly expressing cells in the abdominal cavity of a mouse. Although the use of luciferase in intact animals is an exciting new tool, one potential problem with the technique is that luciferin distribution is not uniform in all tissues (Lee et al. 2003). It is therefore difficult to make quantitative conclusions about the expression levels of reporter genes in different tissues.

A wide variety of methods have been used to introduce reporter genes into cell lines and tissues. In general, the techniques are the same as those used to introduce plasmids into cells or to generate transgenic mice. Electroporation, calcium phosphate transfection, lipid-based transfection, and microinjection are all methods that work well for introducing reporters into cells in culture. Particle-mediated transfection (biolistics) is useful for introducing reporters into slices, and BAC transgenics are a good way of generating strains of reporter mice. Many viruses are not ideal for introducing reporter plasmids because they cause profound alterations in the regulation of gene expression in the recipient cells and often interfere with translation. Conventional transgenics using plasmid-based reporter genes are not a good method for producing reporter strains of mice because the regulation of the reporter gene is highly influenced by the location of the integration of the reporter gene in the genome and, in most cases, gives misleading results.

Reporter genes have a few important limitations that must be kept in mind when using them to report gene expression. Plasmid-based reporter genes, and to a lesser extent, BAC-based reporter genes lack regulation associated with chromatin remodeling complexes and therefore provide incomplete data about gene regulation. Reporter genes are also affected by interference from other exogenous genes that are overexpressed in the same cells. Plasmids that contain strong constitutive promoters such as cytomegalovirus (CMV) often reduce the expression of the reporter gene as a consequence of promoter competition for basal transcription factors. It is therefore important to consider which other plasmids are being introduced into the cell and the ratio of those plasmids to the reporter plasmid. An additional problem with reporter genes is that they report both the transcription and the translation of a gene. Therefore, changes in the expression of a reporter gene reflect both regulation of gene transcription and regulation of mRNA translation. The need for translation of the reporter gene also alters the time course of the apparent transcription event. Generally, a reporter protein can first be detected in about 3 hours, but increased mRNA levels can be detected within minutes. In addition, the degree of reporter gene enhancement is often much smaller than the increase in mRNA levels, possibly reflecting mRNA stabilization or limitations in the rate of mRNA translation into proteins. In summary, reporter genes are a useful method for imaging gene expression in intact cells and tissues but have significant limitations that must be taken into account when interpreting the data.

MEASURING PROTEIN TRANSLATION IN SINGLE CELLS IN REAL TIME

Reporter constructs measure the sum of transcriptional and translational control of a fusion protein. Because these constructs usually lack the endogenous upstream and downstream untranslated regions, they lack the mechanisms that commonly confer translational regulation for native mRNAs. This deficit is particularly significant in the study of gene expression in the nervous system and during development. It has long been known that many genes are regulated at a translational level in developing embryos, and it has also been suggested that neurons locally regulate the expression of particular genes far from the cell body.

Measuring translation in single cells requires an experimental setup that can monitor the accumulation of protein with time. The imaging of fluorescent proteins has proven to be a useful tool for this purpose, allowing experiments to be done without lysis of cells (as would be required to monitor the production of new proteins by incorporation of radioactivity). Researchers have used the untranslated regions from proteins (such as CaMKII) to confer this translational regulation onto a fluorescent protein. These constructs can be delivered by transfection or by viral infection to allow a larger population for study.

Although fluorescence imaging allows spatial discrimination of changing protein concentrations, these fluctuations may be due to either trafficking, diffusion, or local production. To distinguish between these possibilities has required strategies to isolate neurites from the potential supply of protein from the cell body. This can be done "mechanically" by severing neurites from their cell bodies. Using this preparation, Job and Eberwine (2001) transfected GFP mRNA directly into neurites and measured the accumulation of labeled protein. This experiment provides evidence that protein production in neurites does not require transcription. Alternatively, the use of fluorescently marked proteins allows for "optical" isolation by continuous photobleaching of the soma (Aakalu et al. 2001). The illumination of the cell body was proposed to eliminate the fluorescence of proteins translated in this region. This technique has the distinct advantage that cells remain intact and suitable for subsequent cell biological analysis.

Various mutations of the fluorescent tag have provided further insurance against the possibility that proteins enter neurites by trafficking or diffusion. A single point mutation of the standard GFP produces a destabilized GFP that has a shorter fluorescence lifetime (Li et al. 1998). Use of this tag has the dual advantages of eliminating background fluorescence in the cell and lessening the temporal window for proteins to move into distal regions. As well, the addition of a myristoylation site to the GFP produces a membrane-bound reporter that will not diffuse in the cytosol.

These experiments illustrate the potential of imaging experiments to uncouple the analysis of transcription and translation. Nevertheless, they have the disadvantage of monitoring the translation of a fluorescent protein rather than an endogenous protein, making the assumption that the coding sequence does not play a role in control of gene expression. They also overlook the maturation time of fluorescent proteins, which may allow some transit before a translated protein becomes fluorescent.

Overview and Summary

In summary, imaging methods for measuring gene expression have the potential to illuminate many problems in developmental biology and neuroscience. Current methods of measuring chromatin structure, mRNA abundance, and translation in single cells have recently been developed and can provide important experimental alternatives for attacking many problems. Further optimization of the methods of measuring gene expression, however, are still necessary to reduce many of the caveats associated with the methods that we have described here.

REFERENCES

Aakalu G., Smith W.B., Nguyen N., Jiang C., and Schuman E.M. 2001. Dynamic visualization of local protein synthesis in hippocampal neurons. *Neuron* **30:** 489–502.

Belmont A.S. 2001. Visualizing chromosome dynamics with GFP. *Trends Cell Biol.* **11:** 250–257.

Black D.L. 2003. Mechanisms of alternative pre-messenger RNA splicing. *Annu. Rev. Biochem.* **72:** 291–336.

Cammilleri S., Sangrajrang S., Perdereau B., Brixy F., Calvo F., Bazin H., and Magdelenat H. 1996. Biodistribution of iodine-125 tyramine transforming growth factor alpha antisense oligonucleotide in athymic mice with a human mammary tumour xenograft following intratumoral injection. *Eur. J. Nucl. Med.* **23:** 448–452.

Contag P.R., Olomu I.N., Stevenson D.K., and Contag C.H. 1998. Bioluminescent indicators in living mammals. *Nat. Med.* **4:** 245–247.

Dewanjee M.K., Ghafouripour A.K., Kapadvanjwala M., Dewanjee S., Serafini A.N., Lopez D.M., and Sfakianakis G.N. 1994. Noninvasive imaging of c-myc oncogene messenger RNA with indium-111-antisense probes in a mammary tumor-bearing mouse model. *J. Nucl. Med.* **35:** 1054–1053.

Edinger M., Sweeney T.J., Tucker A.A., Olomu A.B., Negrin R.S., and Contag C.H. 1999. Noninvasive assessment of tumor cell proliferation in animal models. *Neoplasia* **1:** 303–310.

Farina K.L. and Singer R.H. 2002. The nuclear connection in RNA transport and localization. *Trends Cell Biol.* **12:** 466–472.

Fiering S.N., Roederer M., Nolan G.P., Micklem D.R., Parks D.R., and Herzenberg L.A. 1991. Improved FACS-Gal: Flow cytometric analysis and sorting of viable eukaryotic cells expressing reporter gene constructs. *Cytometry* **12:** 291–301.

Grewal S.I. and Moazed D. 2003. Heterochromatin and epigenetic control of gene expression. *Science* **301:** 798–802.

Heckl S., Pipkorn R., Waldeck W., Spring H., Jenne J., von der Lieth C.W., Corban-Wilhelm H., Debus J., and Braun K. 2003. Intracellular visualization of prostate cancer using magnetic resonance imaging. *Cancer Res.* **63:** 4766–4772.

Job C. and Eberwine J. 2001. Identification of sites for exponential translation in living dendrites. *Proc. Natl. Acad. Sci.* **98:** 13037–13042.

Johnson F.H. and Shimomura O. 1972. Enzymatic and nonenzymatic bioluminescence. *Photophysiology* 275–334.

Kanda T., Sullivan K.F., and Wahl G.M. 1998. Histone-GFP fusion protein enables sensitive analysis of chromosome dynamics in living mammalian cells. *Curr. Biol.* **8:** 377–385.

Kobori N., Imahori Y., Mineura K., Ueda S., and Fujii R. 1999. Visualization of mRNA expression in CNS using 11C-labeled phosphorothioate oligodeoxynucleotide. *Neuroreport* **10:** 2971–2974.

Kuhn H., Demidov V.V., Coull J.M., Fiandaca M.J., Gildea B.D., and Frank-Kamenetskii M.D. 2002. Hybridization of DNA and PNA molecular beacons to single-stranded and double-stranded DNA targets. *J. Am. Chem. Soc.* **124:** 1097–1103.

Lee H.J., Boado R.J., Braasch D.A., Corey D.R., and Pardridge W.M. 2002. Imaging gene expression in the brain in vivo in a transgenic mouse model of Huntington's disease with an antisense radiopharmaceutical and drug-targeting technology. *J. Nucl. Med.* **43:** 948–956.

Lee K.H., Byun S.S., Paik J.Y., Lee S.Y., Song S.H., Choe Y.S., and Kim B.T. 2003. Cell uptake and tissue distribution of radioiodine labelled D-luciferin: Implications for luciferase based gene imaging. *Nucl. Med. Commun.* **24:** 1003–1009.

Lever M.A., Th'ng J.P., Sun X., and Hendzel M.J. 2000. Rapid exchange of histone H1.1 on chromatin in living human cells. *Nature* **408:** 873–876.

Lewis M., Jia F., Gallazzi F., Wang Y., Zhang J., Shenoy N., Lever S.Z., and Hannink M. 2002. Radiometal-labeled peptide-PNA conjugates for targeting *bcl-2* expression: Preparation, characterization, and in vitro mrna binding. *Bioconjug. Chem.* **13:** 1176–1180.

Lewis M.R. and Jia F. 2003. Antisense imaging: And miles to go before we sleep? *J. Cell. Biochem.* **90:** 464–472.

Li X., Zhao X., Fang Y., Jiang X., Duong T., Fan C., Huang C.C., and Kain S.R. 1998. Generation of destabilized green fluorescent protein as a transcription reporter. *J. Biol. Chem.* **273:** 34970–34975.

Ma Z. and Taylor J.S. 2003. PNA-based RNA-triggered drug-releasing system. *Bioconjug. Chem.* **14:** 679–683.

Mardirossian G., Lei K., Rusckowski M., Chang F., Qu T., Egholm M., and Hnatowich D.J. 1997. In vivo hybridization of technetium-99m-labeled peptide nucleic acid (PNA). *J. Nucl. Med.* **38:** 907–913.

Mayr B.M., Canettieri G., and Montminy M.R. 2001. Distinct effects of cAMP and mitogenic signals on CREB-binding protein recruitment impart specificity to target gene activation via CREB. *Proc. Natl. Acad. Sci.* **98:** 10936–10941.

Michnick S.W. 2001. Exploring protein interactions by interaction-induced folding of proteins from complementary peptide fragments. *Curr. Opin. Struct. Biol.* **11:** 472–477.

Mier W., Eritja R., Mohammed A., Haberkorn U., and Eisenhut M. 2000. Preparation and evaluation of tumor-targeting peptide-oligonucleotide conjugates. *Bioconjug. Chem.* **11:** 855–860.

Narlikar G.J., Fan H.Y., and Kingston R.E. 2002. Cooperation between complexes that regulate chromatin structure and transcription. *Cell* **108:** 475–487.

Negulescu P.A., Shastri N., and Cahalan M.D. 1994. Intracellular calcium dependence of gene expression in single T lymphocytes. *Proc. Natl. Acad. Sci.* **91:** 2873–2877.

Paroo Z. and Corey D.R. 2003. Imaging gene expression using oligonucleotides and peptide nucleic acids. *J. Cell. Biochem.* **90:** 437–442.

Perez J.M., Josephson L., O'Loughlin T., Hogemann D., and Weissleder R. 2002. Magnetic relaxation switches capable of sensing molecular interactions. *Nat. Biotechnol.* **20:** 816–820.

Perlette J. and Tan W. 2001. Real-time monitoring of intracellular mRNA hybridization inside single living cells. *Anal. Chem.* **73:** 5544–5550.

Privat E., Melvin T., Merola F., Schweizer G., Prodhomme S., Asseline U., and Vigny P. 2002. Fluorescent properties of oligonucleotide-conjugated thiazole orange probes. *Photochem. Photobiol.* **75:** 201–210.

Ptashne M. 1988. How eukaryotic transcriptional activators work. *Nature* **335:** 683–689.

Rao P.S., Tian X., Qin W., Aruva M.R., Sauter E.R., Thakur M.L., and Wickstrom E. 2003. ^{99m}Tc-peptide-peptide nucleic acid probes for imaging oncogene mRNAs in tumours. *Nucl. Med. Commun.* **24:** 857–863.

Raz E., Zlokarnik G., Tsien R.Y., and Driever W. 1998. β-lactamase as a marker for gene expression in live zebrafish embryos. *Dev. Biol.* **203:** 290–294.

Robinett C.C., Straight A., Li G., Willhelm C., Sudlow G., Murray A., and Belmont A.S. 1996. In vivo localization of DNA sequences and visualization of large-scale chromatin organization using lac operator/repressor recognition. *J. Cell Biol.* **135:** 1685–1700.

Sato N., Kobayashi H., Saga T., Nakamoto Y., Ishimori T., Togashi K., Fujibayashi Y., Konishi J., and Brechbiel M.W. 2001. Tumor targeting and imaging of intraperitoneal tumors by use of antisense oligo-DNA complexed with dendrimers and/or avidin in mice. *Clin. Cancer Res.* **7:** 3606–3612.

Sei-Iida Y., Koshimoto H., Kondo S., and Tsuji A. 2000. Real-time monitoring of in vitro transcriptional RNA synthesis using fluorescence resonance energy transfer. *Nucleic Acids Res.* **28:** E59.

Shaywitz A.J. and Greenberg M.E. 1999. CREB: A stimulus-induced transcription factor activated by a diverse array of extracellular signals. *Annu. Rev. Biochem.* **68:** 821–861.

Sokol D.L., Zhang X., Lu P., and Gewirtz A.M. 1998. Real time detection of DNA. RNA hybridization in living cells. *Proc. Natl. Acad. Sci.* **95:** 11538–11543.

Sonenberg N. and Dever T.E. 2003. Eukaryotic translation initiation factors and regulators. *Curr. Opin. Struct. Biol.* **13:** 56–63.

Spotts J.M., Dolmetsch R.E., and Greenberg M.E. 2002. Time-lapse imaging of a dynamic phosphorylation-dependent protein-protein interaction in mammalian cells. *Proc. Natl. Acad. Sci.* **99:** 15142–15147.

Strukov Y.G., Wang Y., and Belmont A.S. 2003. Engineered chromosome regions with altered sequence composition demonstrate hierarchical large-scale folding within metaphase chromosomes. *J. Cell Biol.* **162:** 23–35.

Tavitian B., Marzabal S., Boutet V., Kuhnast B., Terrazzino S., Moynier M., Dolle F., Deverre J.R., and Thierry A.R. 2002. Characterization of a synthetic anionic vector for oligonucleotide delivery using in vivo whole body dynamic imaging. *Pharm. Res.* **19:** 367–376.

Tsourkas A., Behlke M.A., and Bao G. 2003. Hybridization of 2′-o-methyl and 2′-deoxy molecular beacons to RNA and DNA targets. *Nucleic Acids Res.* **30:** 5168–5174.

Tsuji A., Koshimoto H., Sato Y., Hirano M., Sei-Iida Y., Kondo S., and Ishibashi K. 2000. Direct observation of specific messenger RNA in a single living cell under a fluorescence microscope. *Biophys. J.* **78:** 3260–3274.

Verschure P.J., van Der Kraan I., Manders E.M., and van Driel R. 1999. Spatial relationship between transcription sites and chromosome territories. *J. Cell Biol.* **147:** 13–24.

Waddell E., Wang Y., Stryjewski W., McWhorter S., Henry A.C., Evans D., McCarley R.L., and Soper S.A. 2000. High-resolution near-infrared imaging of DNA microarrays with time-resolved acquisition of fluorescence lifetimes. *Anal. Chem.* **72:** 5907–5917.

Wu J.C., Inubushi M., Sundaresan G., Schelbert H.R., and Gambhir S.S. 2002. Optical imaging of cardiac reporter gene expression in living rats. *Circulation* **105:** 1631–1634.

Zawel L. and Reinberg D. 1995. Common themes in assembly and function of eukaryotic transcription complexes. *Annu. Rev. Biochem.* **64:** 533–561.

Zhao S. and Stodolsky M. 2003. *Bacterial artificial chromosomes: Methods and protocols.* Humana Press, Totowa, New Jersey.

Zlokarnik G., Negulescu P.A., Knapp T.E., Mere L., Burres N., Feng L., Whitney M., Roemer K., and Tsien R.Y. 1998. Quantitation of transcription and clonal selection of single living cells with β-lactamase as reporter. *Science* **279:** 84–88.

CHAPTER 83

Imaging Olfactory Activity in *Drosophila* CNS with a Calcium-sensitive Green Fluorescent Protein

Allan M. Wong, Jorge Flores, and Jing W. Wang

Center for Neurobiology and Behavior, Columbia University, New York, New York 10032

Insects exhibit sophisticated odor-mediated behaviors controlled by an olfactory system that is genetically and anatomically simpler than that of vertebrates, providing an attractive system to investigate the mechanistic link between behavior and odor perception. Advances in neuroscience have been facilitated by modern optical imaging technologies—both in instrumentation and in probe design—that permit the visualization of functional neural circuits. We have developed an imaging system to monitor neural activity in the *Drosophila* antennal lobe that couples two-photon microscopy (see Denk et al. 1990; Chapter 7) with the specific expression of the calcium-sensitive green fluorescent protein, G-CaMP (Nakai et al. 2001; Wang et al. 2003). In this chapter, we discuss the practical considerations of this imaging system.

GENETIC CONSIDERATIONS OF THE CALCIUM PROBE

The yeast Gal4 transcriptional activator and the upstream activating sequence (UAS) have been exploited to create a binary gene expression system in *Drosophila* to express transgenes in a spatially restricted manner (Brand and Perrimon 1993). Many Gal4 lines exist that label defined subpopulations of the fly central nervous system (CNS), providing a platform upon which to genetically dissect the function of neural circuits. To study the olfactory response in the antennal lobe, we used *GH146-Gal4*, which expresses Gal4 in approximately 90 of the 200 projection neurons that connect the antennal lobe to the mushroom body and the protocerebrum (Stocker et al. 1997). The G-CaMP-coding sequence was fused to the UAS promoter to generate *UAS-GCaMP* transgenic flies (Wang et al. 2003). Flies harboring the *GH146-Gal4* and *UAS-GCaMP* transgenes express G-CaMP in projection neurons (Fig. 1B), such that changes in calcium concentration, an indicator of neural activity, can be monitored (Yuste and Katz 1991; Chapter 44).

The fluorescence intensity of labeled neurons is proportional to the concentration of the fluorescent probe. Low concentrations result in poor labeling, but high concentrations can alter the temporal dynamics of the calcium influx (Helmchen and Tank 2000). After some investigation, flies with four copies of the *UAS-GCaMP* transgene were chosen for further examination of the olfactory activity in the antennal lobe. Although fluorescence change $\Delta F/F$ of the same odor stimulation was similar among flies with different numbers of the transgenes (mean ± S.D. for the VM3 glomerulus in response

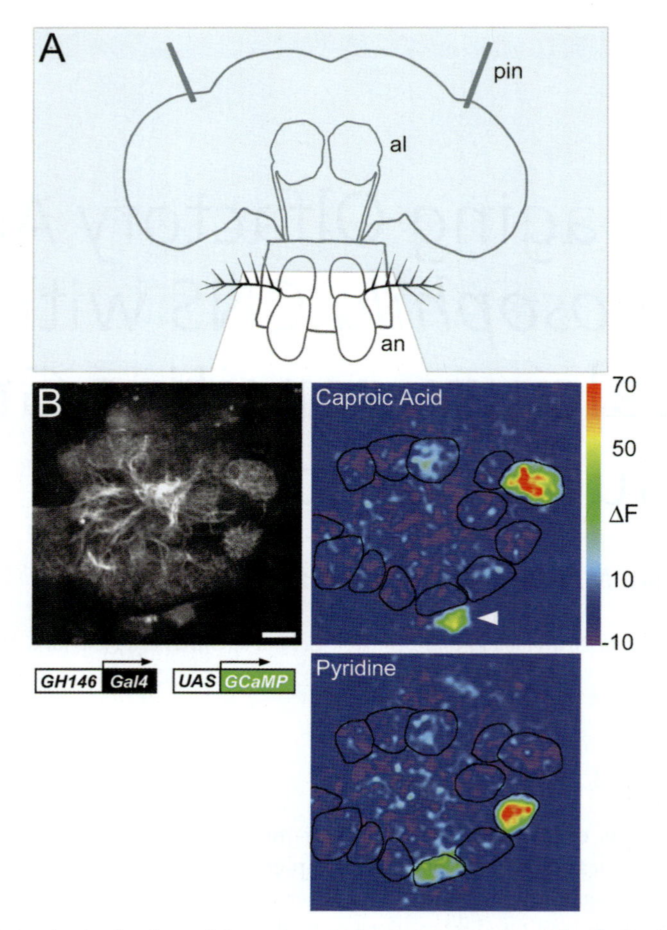

FIGURE 1. (*A*) A schematic view of the antennae–brain preparation. The fly brain was dissected in calcium-free saline, pinned to a thin Sylgard slab, and embedded in agarose. This preparation allows airborne odorants to reach the antennae (an) and imaging of the antennal lobe (al). (*B*) Flies harboring *GH146-Gal4* and *UAS-GCaMP* transgenes express G-CaMP in projection neurons and were analyzed by two-photon imaging in the antennal lobe. The gray-scale pre-stimulation image reveals glomerular structure (bar, 10 µm). The false color image shows peak glomerular response to caproic acid and pyridine. Images (128 x 128 pixels) were collected at 4 fps, and odorants at 16% saturated vapor were delivered for 2 sec. In this plane, caproic acid activates the DM3 glomerulus and a cell body (*arrowhead*) and pyridine activates the VM2 and VA3 glomeruli.

to isoamyl acetate at 33% saturated vapor: 122 ± 14, 114 ± 17, 122 ± 7 for 2, 4, and 6 copies of *UAS-GCaMP,* respectively, $n = 3$), flies with four or six copies of the *UAS-GCaMP* transgene showed better morphological structure in the antennal lobe than those with only two copies of *UAS-GCaMP*.

MATERIALS

Flies

Use adult flies (within 48 hours of eclosion) carrying four copies of the *UAS-GCaMP* transgene for this procedure. Older flies should be avoided because of their high background fluorescence.

***Drosophila* Adult Hemolymph-like Saline (AHL)**

108 mM NaCl, 5 mM KCl<!>, 2 mM $CaCl_2$<!>, 8.2 mM $MgCl_2$<!>, 4 mM $NaHCO_3$, 1 mM NaH_2PO_4<!>, 5 mM trehalose, 10 mM sucrose, 5 mM HEPES. Adjust to pH 7.5, filter, and store at 4°C.

Measure this solution using a freezing-point osmometer. The osmolarity of this saline is 265 mOsm, very close to the value of the adult *Drosophila* hemolymph (Singleton and Woodruff 1994). Use $CaCl_2$ and $MgCl_2$ solutions instead of powders to avoid precipitation in the saline.

Olfactometer

This technique uses a simple custom-made odor delivery system that permits quantitative control of odor stimulation by diluting saturated vapor up to 300-fold and delivering a precise amount of odorant to the brain preparation at a specific time during imaging. For more details about this instrument, see Wang et al. (2003).

Imaging

Calcium imaging was performed using a custom-made two-photon microscope (Denk et al. 1990). To select for the emission spectrum of G-CaMP (peak at 509 nm, Nakai et al. 2001) and block the infrared laser light and other background fluorescent light, a blocking filter (E700SP-special, Chroma Technology) in conjunction with a band-pass filter (HQ525/50m) was inserted between the focusing lens and the PMT light detector. The light source was a Mira 900 Ti:sapphire laser<!> at ~930 nm wavelength, pumped by a Verdi 10-W solid-state laser (Coherent). For more details, see Wang et al. (2003).

Caution: *See Appendix 3 for appropriate handling of materials marked with <!>.*

PROCEDURE

Antennae–Brain Preparation

The following procedure provides a *Drosophila* antennae–brain preparation that accommodates a water-immersion objective lens for imaging (Fig. 1A) and airborne odor stimulation.

1. Briefly anesthetize flies with CO_2 and then decapitate them.
2. Use an insect pin (0.1 mm diameter) to pin one of the heads through the proboscis onto a Sylgard-coated plate and flood with cold calcium-free AHL saline.

   Note: Cold calcium-free saline suppresses synaptic transmission during dissection, thereby maintaining the integrity of the preparation. Preparations dissected in normal AHL saline with 2 mM $CaCl_2$ are much less responsive to odor stimulation.
3. Use a #55 forceps to remove the head capsule and air sacs to expose the brain. Also remove the maxillary palps to leave the third antennal segment as the sole sensory organ that provides input to the antennal lobes.
4. Transfer the preparation onto a thin Sylgard slab on a microscope slide. Pin the antennae and brain in position with fine tungsten wires (California Fine Wire, 20 μm diameter).
5. Replace the calcium-free saline with normal AHL saline containing 2 mM $CaCl_2$. To minimize movement of the preparation during imaging, replace the saline with transparent agarose (Sigma) solution (2% in AHL saline) maintained at 38°C.
6. Prior to imaging, carefully remove the gel covering the antennae to allow odor stimulation and place a glass coverslip on top of the brain region for the water-immersion objective (Zeiss, IR-Achroplan, 40x, NA 0.80).
7. Carry out imaging within 4 hours of dissection.

   Note: At 2 and 4 hours after dissection, the odor responsivity was the same as that at 1 hour after dissection (mean ± S.D. of ΔF/F in the VM3 glomerulus in response to isoamyl acetate at 8% saturated vapor: 107 ± 10, 102 ± 14, 101 ± 17 for 1, 2, and 4 hours after dissection, respectively, $n = 4$).

Olfactory Stimulation/Imaging

For information on olfactory stimulation and imaging, please see Wang et al. 2003.

EXAMPLE OF APPLICATION

Figure 1B shows results from the experimental procedure described above. See figure legend for details.

ADVANTAGES AND LIMITATIONS

The optical imaging system described here is capable of simultaneously monitoring neural activity in a genetically defined population of neurons. It uses two-photon microscopy with the specific expression of the genetic calcium-sensitive probe, G-CaMP. The Gal4/UAS gene expression system targets G-CaMP expression in a population of sensory or projection neurons, and the power of two-photon microscopy to perform optical sections permits neural activity from individual glomeruli to be monitored in the three-dimensional antennal lobe. The dissociation constant (K_d) of G-CaMP is slightly larger than the resting calcium concentration, making it possible to detect small calcium influxes, evoked by odors at concentrations likely to be encountered by insects in their natural habitats. At present, this imaging system is limited by the availability of Gal4 lines that label neurons of interest. Although calcium influx is correlated with the number of action potentials (Wang et al. 2003; Chapter 44), the measurement of calcium influx is not an adequate reflection of fast neural dynamics in millisecond resolution, an intrinsic problem of all calcium probes, which can be mitigated by concurrent electrical recordings. This calcium imaging system should be useful to study higher-order neurons in the olfactory system to decipher the olfactory code and should be useful to study other sensory modalities in the fly.

REFERENCES

Brand A.H. and Perrimon N. 1993. Targeted gene expression as a means of altering cell fates and generating dominant phenotypes. *Development* **118:** 401–415.

Denk W., Strickler J.H., and Webb W.W. 1990. Two-photon laser scanning fluorescence microscopy. *Science* **248:** 73–76.

Helmchen F. and Tank D.W. 2000. A single-compartment model of calcium dynamics in nerve terminals and dendrites. In *Imaging Neurons* (ed. R. Yuste et al.), pp. 33.1–33.11. Cold Spring Harbor Laboratory Press, Cold Spring Harbor, New York.

Nakai J., Ohkura M., and Imoto K. 2001. A high signal-to-noise Ca^{2+} probe composed of a single green fluorescent protein. *Nat. Biotechnol.* **19:** 137–141.

Singleton K. and Woodruff R.I. 1994. The osmolarity of adult *Drosophila* hemolymph and its effect on oocyte-nurse cell electrical polarity. *Dev. Biol.* **161:** 154–167.

Stocker R.F., Heimbeck G., Gendre N., and de Belle J.S. 1997. Neuroblast ablation in *Drosophila* P [GAL4] lines reveals origins of olfactory interneurons. *J. Neurobiol.* **32:** 443–456.

Wang J.W., Wong A.M., Flores J., Vosshall L.B., and Axel R. 2003. Two-photon calcium imaging reveals an odor-evoked map of activity in the fly brain. *Cell* **112:** 271–282.

Yuste R. and Katz L. C. 1991. Control of postsynaptic calcium influx in developing neocortex by excitatory and inhibitory neurotransmitters. *Neuron* **6:** 333–344.

CHAPTER 84

Tracking Molecules in Intact Zebrafish

Ricardo Armisen,[1, 2] Michelle R. Gleason,[1, 2] Joseph R. Fetcho,[2] and Gail Mandel[1, 2]

[1]*Howard Hughes Medical Institute,* [2]*Department of Neurobiology and Behavior, State University of New York at Stony Brook, Stony Brook, New York 11794-5230*

This chapter describes an approach for monitoring the movement of tagged molecules in single neurons in intact embryonic and larval zebrafish. The technique has been used to acquire time-lapse images of the translocation of calcium/calmodulin-dependent protein kinase II (CaMKII) fused to the enhanced green fluorescent protein (EGFP) to synaptic sites in the spinal cord of a 2-day-old zebrafish in response to stimulation (Gleason et al. 2003).

MATERIALS

Microinjection Setup (Meng et al. 1999)

Injection tray (Westerfield 1995)
Fertilized zebrafish eggs
DNA encoding the fusion protein of interest under the control of a tissue-specific promoter
Dissecting stereomicroscope (20–125×), equipped with appropriate filter set and mercury arc lamp

Electrical Stimulation

Tungsten bipolar stimulating electrodes (0.010 inches, 5 MEG, 8 Degree; A-M Systems)
Three axes micromanipulator
Electrical stimulator
Zoom stereoscope (6.5–45×)

Microscope

An inverted microscope (Axiovert 100M) and a Zeiss laser-scanning confocal imaging system (LSM 510) or an upright microscope (Axioscop II FS) with a QLC100 Dual Nipkow Disk spinning confocal unit and an XR-10 CCD camera (Solamere Technology Group) and QED imaging software (QED Imaging)

Objectives

For the inverted microscope, we use a 63×/1.2-NA C-Apochromat, and for the upright microscope, we use a 63×/0.9-NA long-working-distance Achroplan (Zeiss).

PROCEDURE A

Screening, Paralysis, and Embedding

1. Microinject one-cell-stage zebrafish embryos (Higashijima et al. 1997) with ~1 nl of circular or linearized DNA dissolved in water to a final concentration of 25–50 ng/μl. Keep injected embryos in 10% Hanks' solution at 28°C.

   Note: To obtain a good individual, 100 embryos should be injected.
2. Two days after injection, screen the embryos for fluorescence in the desired cell type. Note the position of fluorescence relative to an obvious anatomical landmark.
3. To minimize movement artifacts, paralyze a positively labeled individual, by immersing it in a small volume of α-bungarotoxin<!> (1 mg/ml in distilled water) for 10–15 minutes.
4. Rinse the paralyzed embryo in 10% Hanks' solution (pH 7.4) (Westerfield 1995).
5. Embed embryo on its side in warm (43°C) 1.2% agar in 10% Hanks' solution in an "imaging chamber" (i.e., 35-mm petri dish with a round hole, covered from beneath by a 22-mm square coverslip, cut in the bottom).

Caution: *See Appendix 3 for appropriate handling of materials marked with <!>.*

PROCEDURE B

Electrical Skin Stimulation

1. For electrical stimulation of the skin, lay the embryo on its side such that the cell of interest is away from the coverslip and is accessible to the stimulation probes. If an upright microscope is used, carefully remove agar from the area that will be imaged once embedding is complete.
2. Under transmitted light and with the aid of a high-zoom stereoscope, carefully lower a pair of bipolar stimulating electrodes through the agar, onto the embryo, one or two segments away from the cell of interest. When the electrodes are near the embryo, use a 63× objective for finer positioning. Lower the electrodes until a slight but visible dimpling of the skin confirms contact.
3. Apply a single test pulse (0.1 msec in duration) to identify the stimulation threshold. Start at a low current strength of ~15 μA, and dial up the amplitude in successive single pulses while monitoring the area adjacent to the probe. A very small and local twitch in muscle signifies suitable stimulus strength.
4. Recenter the cell in the field (placement of stimulation probes may have moved the preparation), using fast scanning mode.
5. Adjust scanning parameters to the settings that will be used for the remainder of the experiment (see imaging procedures for details).
6. After obtaining baseline images, deliver a brief train of 30 pulses (0.1 msec in duration) at 100 Hz to the embryo at the threshold current amplitude. If one train is insufficient to cause an effect within the expected time interval, apply multiple trains and/or raise current amplitude.

PROCEDURE C

Bath Stimulation

1. For glutamate stimulation, lay the embryo on its side such that the cell of interest is closer to the objective and the disrupted side is farther from the objective. Use the previously identified anatomical landmark to center the cell of interest in the field of view.
2. Use a scalpel and fine forceps to carefully remove a small block of agar overlying the cell of interest.

3. Use a glass electrode that has been broken to a fine tip and short and gentle strokes to mechanically disrupt the embryo's skin on the side opposite to the cell of interest.
4. Initiate time-lapse experiment as above. After obtaining baseline images, add desired stimulant, such as glutamate.

PROCEDURE D

Imaging

For scanning confocal microscopy, due to the increased potential for bleaching and phototoxicity in the intact preparation, it is essential to use low laser strength and moderately fast scan time, particularly for prolonged experiments. The advantages of a spinning disk confocal microscope include real-time acquisition owing to fine temporal resolution, which allows kinetics studies on moving molecules. Additionally, reduced bleaching permits repeated experiments on the same preparation over longer time intervals.

1. Locate and center the cell using fast acquisition mode, with the maximum detector gain (photomultiplier tube for scanning and CCD camera for spinning confocal), an open pinhole (for scanning confocal), and minimum laser transmission.
2. Increase zoom step-wise to enlarge view of the desired region of the cell (e.g., dendrite).
3. Decrease detector gain and pinhole while taking moderately fast single images until acceptable signal-to-background levels are achieved.
4. Turn on line-by-line averaging. Decrease scan speed for the scanning confocal (typically to 3.1 sec/average) or increase acquisition time for the spinning confocal, to increase signal-to-noise ratio in the image.
6. Laser strength may be increased if the signal is dim. Keep the laser power as low as possible for obtaining an acceptable imaging, especially during prolonged experiments, to minimize photobleaching and photodamage.
7. Once these parameters are optimized, engage the time-series software.
8. After baseline images, apply stimulation (electrical pulses or addition of stimulant to the bath).

EXAMPLE OF APPLICATION

Figure 1 shows the translocation of CaMKII to postsynaptic sites in an identified interneuron in the intact zebrafish, in response to stimulation. Electrical stimulation of the skin or brief bath application of a 500 μM solution of glutamate resulted in the rapid, reproducible, and reversible appearance of punctate clusters distributed at synaptic sites on the dendrites and soma of the cell (Gleason et al. 2003)

ADVANTAGES AND LIMITATIONS

The intact preparation provides a meaningful context for the physiological event being studied. Other advantages offered by the young zebrafish include direct in vivo imaging, the ability to produce large numbers of labeled embryos easily using microinjection, and the existence of identified sensory circuits that can be exploited to activate a particular cell type. A limitation of the system is the fragility of 2–3-day-old embryos, which demands delicate physical manipulation of the fish during all stages preceding and during the experiment. In contrast to brain slices or isolated cells, nearly all original neural connections and sensory components are maintained in the intact preparation, so the occurrence of a downstream event may be precluded (or its manifestation enhanced) by some complex interplay of biological processes that are not fully understood.

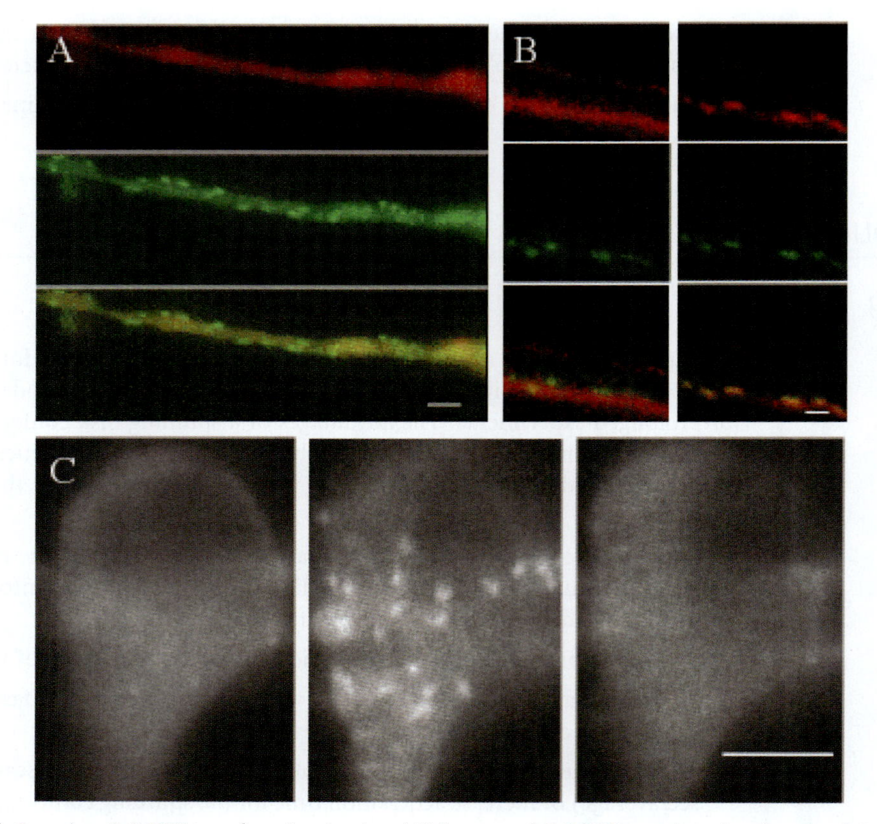

FIGURE 1. Imaging CaMKII translocation in vivo. (*A*) Images of CaMKII translocation in a dendrite of a commissural primary ascending (CoPA) spinal interneuron. (*Top*) Staining with DsRed; (*middle*) with GFP-CaMKII; (*bottom*) an overlay. The images were taken after stimulation of the skin (60 pulses, 0.2-msec duration, at 50 Hz and 32 μA) to activate sensory neurons that form glutamatergic synapses onto the dendrite. The stimulation led to the accumulation of the GFP-CaMKII (*middle panel*) in submembranous patches, while the DsRed staining remained diffuse. (*B*) CaMKII localizes to specific postsynaptic sites in dendrite from a CoPA spinal interneuron. (*Left*) Prestimulation images of a dendrite labeled with yellow fluorescent protein (YFP)-CaMKII (*red, top*) and PSD-95–CFP (cyan fluorescent protein) (*green, middle*), with an overlay (*bottom*). (*Right*) After stimulation (30 pulses, 0.2-msec duration, at 100 Hz and 30 μA), YFP-CaMKII forms clusters (*top*) that coincide (*bottom*) with the sites marked by the PSD-95–CFP. (*C*) Translocation of CaMKII following application of glutamate. The distribution of GFP-CaMKII in a CoPA cell soma is shown before (*left*) and ~ 1 min after (*middle*) application of a 500 μM glutamate solution containing 20 μM glycine. (*Right panel*) the recovery to a more uniform distribution ~ 3.5 min after washout of the glutamate solution. Bars: (*A*) 2 μm; (*B*) 1 μm; (*C*) 5 μm.

ACKNOWLEDGMENTS

This work was supported by the Howard Hughes Medical Institute (G.M.), National Institutes of Health grant NS-26539 (J.R.F.), and a National Science Foundation predoctoral fellowship (M.R.G.).

REFERENCES

Gleason M.R., Higashijima S., Dallman J., Liu K., Mandel G., and Fetcho J.R. 2003. Translocation of CaM kinase II to synaptic sites in vivo. *Nat. Neurosci.* **6:** 217–218.

Higashijima S., Okamoto H., Ueno N., Hotta Y., and Eguchi G. 1997. High-frequency generation of transgenic zebrafish which reliably express GFP in whole muscles or the whole body by using promoters of zebrafish origin. *Dev Biol.* **192:** 289–299.

Meng A., Jessen J.R., and Lin S. 1999. Transgenesis. *Methods Cell Biol.* **60:** 133–147.

Westerfield M., ed. 1995. *The zebrafish book: A guide for the laboratory use of zebrafish* (Brachydanio rerio), 3rd edition. University of Oregon Press, Eugene.

CHAPTER 85

Long-term, High-Resolution Imaging of Neurons in the Neocortex In Vivo

Anthony J.G.D. Holtmaat, Linda Wilbrecht, Alla Karpova, Carlos Portera-Cailliau, Barry Burbach, Josh T. Trachtenberg, and Karel Svoboda

Howard Hughes Medical Institute, Cold Spring Harbor Laboratory, Cold Spring Harbor, New York 11724

Neocortical tissue is dauntingly complex: 1 mm^3 contains nearly a million neurons, each of which connects to thousands of other neurons. The connections between neurons are synapses, tiny (~1 μm) junctions between neurons. The functional properties of the brain change in response to salient experiences during development and in the adult. The nature of these changes is poorly understood, but they are thought to take place, at least in part, at synapses. Developmental and use-dependent changes in the brain can occur within milliseconds, or unfold over hours, days, or weeks. To thoroughly understand adaptive plastic changes in the brain, it will be critical to monitor the structure and function of individual synapses in the intact brain over both these short and long timescales. Recent breakthroughs in technology allow living cells to be monitored in vivo over weeks to months, enabling the same neuron in the same animal to be studied as it develops, encounters different sensory experiences, or even as it learns. In this chapter, we discuss the details of procedures for long-term, high-resolution imaging of neurons in the neocortex in vivo. The techniques are illustrated with our own studies of structural plasticity in the adult brain.

TWO-PHOTON MICROSCOPY IN THE NEOCORTEX

High-resolution in vivo imaging of neurons in the cortex became possible with the invention of two-photon laser-scanning microscopy (2PLSM) (Fig. 1A) (Denk et al. 1990; Svoboda et al. 1997). This technique has key advantages over conventional, single-photon excitation techniques, such as confocal microscopy (Denk and Svoboda 1997). Two-photon excitation is the near-simultaneous (within femtoseconds) absorption of two photons coinciding on a fluorophore. The absorption rate depends quadratically on the illumination intensity, and therefore is confined to a small volume around the focal point. Scattered excitation light is too weak to generate fluorescence. Thus, signal is generated exclusively in a tiny focal volume, and all emitted fluorescence photons constitute useful signal. An additional, related advantage is that the longer wavelengths used to generate two-photon excitation penetrate scattering tissue more efficiently than the shorter wavelengths used to generate single-photon excitation of the same fluorophores. The scattering length increases approximately linearly

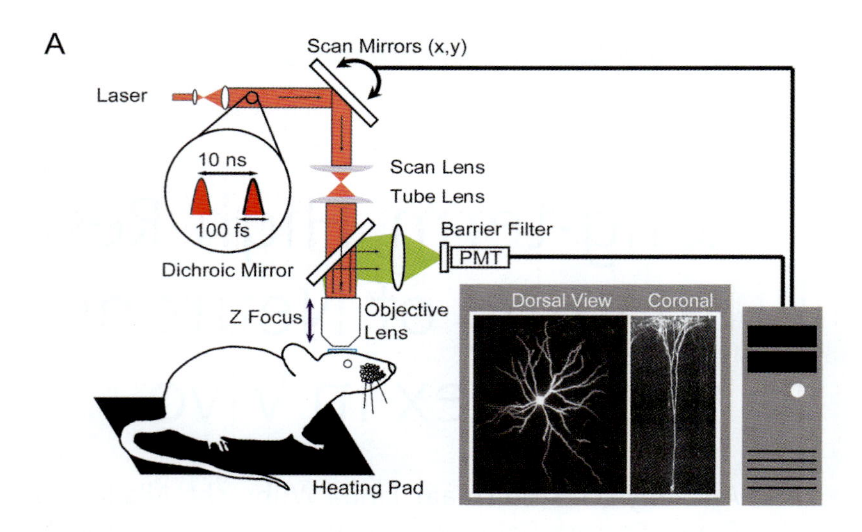

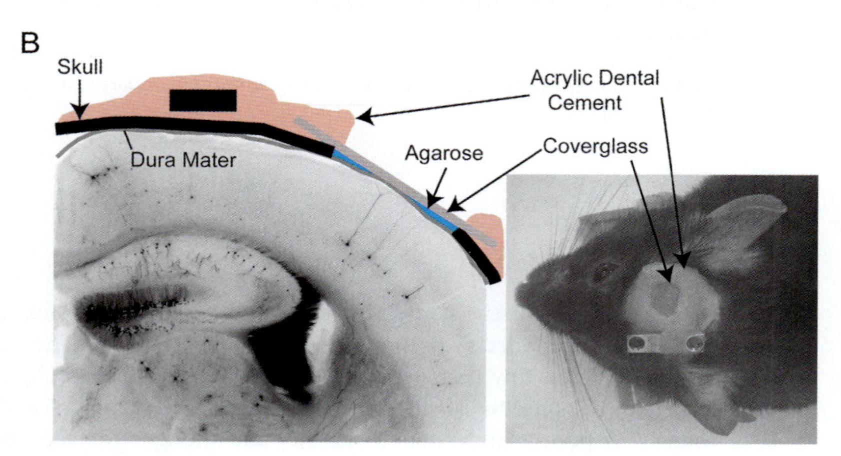

FIGURE 1. Two-photon laser-scanning microscopy in vivo. (*A*) Schematic of a two-photon laser-scanning microscope. The light source is a pulsed, mode-locked pulsed Ti:sapphire laser running at $\lambda \sim 910$ nm. The repetition rate is ~100 MHz and the pulse duration ~100 fsec. The power measured at the backfocal plane of the objective is < 200 mW. The laser beam is expanded and coupled into a pair of galvanometer scan mirrors. The mirrors are imaged onto the back focal plane of the objective using a scan lens and a microscope tube lens. The objective (typically Zeiss, 40x, 0.8 NA) focuses the excitation beam to a diffraction-limited spot. Fluorescence photons collected by the objective are reflected by a dichroic mirror and imaged onto a PMT detector. The image is constructed in a computer. (*B*) Experimental preparation for in vivo imaging. (*Left*) Schematic view of the imaging window superposed on a 100-µm section from a GFP-M mouse (Feng et al. 2000; Trachtenberg et al. 2002). (*Right*) Mouse with an implanted imaging window.

with increasing wavelengths (Oheim et al. 2001). In the cortex, the scattering length at 800 nm wavelength, $l_S^{(800)}$, measured in vivo is ~100 µm (Kleinfeld et al. 1998), and appears to decrease with increasing developmental age (Oheim et al. 2001). Typically, imaging is limited to the most superficial ~500 µm of the tissue, but this is sufficient for imaging the upper layers (1–4) of neocortex in the mouse.

FLUORESCENT MOLECULES

Imaging neurons requires the presence of fluorescent molecules in the cell of interest. Traditionally, synthetic small-molecule fluorophores have been delivered using bulk-labeling techniques or single-cell injection techniques. Bulk labeling gives weak and nonspecific labeling (Stosiek et al. 2003), whereas single-cell injections are technically more challenging and could damage the cell (Svoboda et al. 1997). In some cases, relatively noninvasive methods have been devised to label nonneuronal structures, including senile plaques in mouse models of Alzheimer's disease (Christie et al. 2001) and blood plasma (Kleinfeld et al. 1998).

For neurons, the fluorophore of choice for in vivo imaging is the green fluorescent protein (GFP) and related fluorescent proteins (XFPs) (Figs. 1B, 2, 3, 4). The power of GFP as a protein tag is well established. It is often not appreciated, however, that modern XFPs are excellent fluorophores: They have large extinction ratios and quantum efficiencies (comparable to some of the better synthetic fluorophores) over a large spectral range and are quite resistant to photobleaching (Tsien 1998; Campbell et al. 2002; Nagai et al. 2002). However, not all XFPs are equally useful. Red proteins, for example, tend to aggregate (but see Campbell et al. 2002) and are difficult to excite with the Ti:sapphire lasers used for two-photon imaging.

A high expression level of fluorophore is necessary to visualize small cellular compartments, such as dendritic spines. Using GFP strictly as a cytoplasmic marker, it is estimated that >1 μM of GFP is required to reliably detect dendritic spines, the smallest neuronal compartments. Thus, the protein concentrations needed for optimal imaging are higher than the vast majority of endogenous proteins expressed by neurons!

GFP has also been mutated to actively indicate subcellular activity (e.g., pH changes, Ca^{2+} influx) (Miyawaki et al. 1997; Miesenbock et al. 1998; Nagai et al. 2001; Nakai et al. 2001). Unfortunately, in the authors' experience, the fluorescence and/or signal-to-noise ratio of genetically encoded functional probes has not been sufficient to reliably image neuronal function at high resolution in the intact brain.

FLUORESCENT PROTEIN EXPRESSION IN VIVO

Delivery of a foreign gene(s) to neurons in vivo remains a challenge, in part because they are postmitotic cells and not easily accessible. Ideally, one would like to label neurons with cell-type specificity, noninvasively, without cytopathic side effects, and with controllable efficiency. Standard transfection methods, such as chemical transfection (calcium phosphate precipitation, liposomes, lipids, cationic polymers) and biolistic gene transfer, have not been successful in vivo (Washbourne and McAllister 2002).

Electroporation is an interesting exception. The permeabilization of cell membranes by the application of brief electrical pulses, known as electroporation, has been used extensively in vitro to introduce large molecules such as proteins or nucleic acids into cells (Neumann et al. 1999). More recently, however, in utero electroporation has been used to transfect large groups of neurons in developing chick and mouse embryos (Fukuchi-Shimogori and Grove 2001; Tabata and Nakajima 2001). The plasmid DNA is injected directly into the lateral ventricle of the embryo and then a brief electrical current is applied between two electrodes on either side of the embryo's head. Electroporation offers advantages over virus-mediated transfection because it lacks the biohazard risk and the potential toxicity of the virus itself. Moreover, a relatively pure population of neurons (e.g., pyramidal neurons) can be targeted depending on the embryonic age at which the electroporation is performed. Unfortunately, the method can be finicky, and low birth rates can occur due to the high voltages that are applied. The authors have used electroporation in E14–E16 mouse and rat embryos to label pyramidal neurons in neocortex (Fig. 2A, B) for later in vivo imaging. Under some conditions, electroporation can be used to transduce single neurons (Haas et al. 2001). Others have used electroporation to deliver RNAi constructs (Bai et al. 2003).

The use of recombinant viral vectors has been increasingly useful to express foreign genes in the adult brain. Recombinant viral vector technology exploits the ability of particular viruses to introduce transgenes into postmitotic neurons. Some recombinant viruses, such as Sindbis-, herpes simplex-, adeno-, adeno-associated (AAV)-, and lenti-based viral vectors, can transduce brain cells with a remarkable, but as yet poorly understood, efficiency for neurons (Hermens and Verhaagen 1998; Kay et al. 2001; Washbourne and McAllister 2002). Recombinant viruses can be produced relatively quickly and ultimately allow expression of different genes of interest in combinations under cell-type-specific promoters in a variety of species. Recent progress in this area has greatly diminished many of the cytopathic adverse effects of some viral vectors on their hosts (Kay et al. 2001). However, the expression of

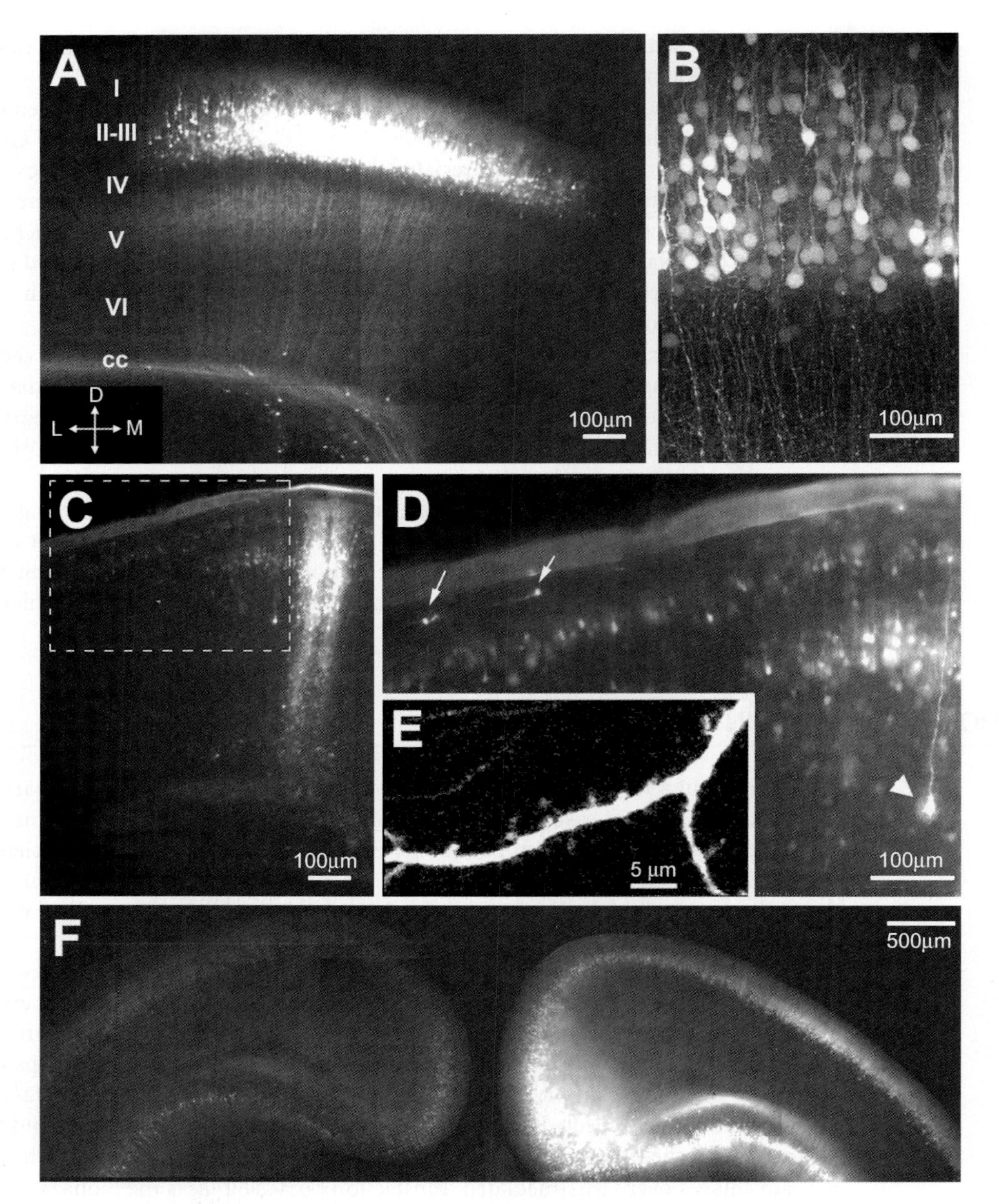

FIGURE 2. In utero electroporation and virus-mediated transfection. (*A*) Epifluorescence image of a fixed section through neocortex (rat, PND15) at the level of the hippocampus. In utero electroporation was performed at E15 with pCAG-Venus-YFP. (*B*) High-magnification image of the same section. (*C, D*) Images of a fixed section (rat, PND7). The pup was injected intracortically at PND2 with adenovirus-p-SYN-EGFP. The needle track can be seen as a vertically oriented column of brightly labeled neurons. (*D*) Blow-up from *C* (*dashed box*). Note that a variety of neurons are infected, including layer-5 pyramidal neurons (*large arrowhead* in *D*) and Cajal-Retzius cells in layer 1 (*small arrows* in *D*). (*E*) High-resolution two-photon image of a Cajal-Retzius cell dendrite in an anesthetized rat pup at PND5. (*F*) Montage of a section through the brain of a P7 mouse pup injected in utero at E16 with adenovirus-p-SYN-EGFP.

recombinant proteins, driven by some viral vectors (adenovirus, AAV, and lentivirus) can be slow, with delays ranging from days to weeks postinfection.

Viruses can be injected into embryos in utero, or into newborn or adult rodent brains. In utero injections of virus achieve similar results as electroporation (i.e., large numbers of cells transfected). One advantage of virus-mediated transfection in utero is that it can be less damaging and more rapid than electroporation. Additionally, the number of neurons transfected and the intensity of XFP signal can be titrated easily by diluting the virus. The authors have used adenovirus and lentivirus injections in utero and in newborn pups for live two-photon imaging of cortical neurons with excellent results (Fig. 2,C–E). Both lenti- and adenovirus can also be injected in utero into the ventricles of embryos to achieve widespread expression (Fig. 2F).

Germ-line transgenesis provides another powerful method for labeling neurons in vivo. In transgenic mice, the gene of interest is integrated into the genome of all body cells, but the temporal and spatial patterns of expression are in large part determined by cell-type-specific regulatory elements that are included in the transgene construct. The great advantages of using transgenic mice over transfection techniques include the non-immunogenic (non-cytopathic) expression of the recombinant protein and the stability and reproducibility of expression. Expression can be controlled spatially and temporally using tissue-specific and inducible promoters. Spatial control can be achieved by choosing from a limited variety of minimal promoters (Holtmaat et al. 1998; Wells and Carter 2001) or bacterial artificial chromosomes for transgenesis (Heintz 2001; Gong et al. 2003). Transgenic mice can be imaged after only a single surgical operation, and they can be crossed with other transgenic mice for differential labeling of neurons. Moreover, and unlike transfected animals, fluorescent expression in transgenic animals is long lived, usually persisting from development to adulthood. The main practical disadvantages of using transgenic mice are the long turnaround time between the design of a desired DNA construct and the generation of transgenic mouse lines, and the frequently high neuronal labeling density that makes optimal high-resolution microscopy difficult. Another problem is that fluorescent protein expression levels are often too low, and therefore too dim for in vivo imaging.

Long-term, high-resolution imaging of neurons in the neocortex in vivo was used recently to study the structural plasticity of neurons in the developing (A.J. Holtmaat et al., in prep.) and adult brain (Grutzendler et al. 2002; Trachtenberg et al. 2002). Using transgenic animals that express XFPs, it is possible to follow the movements and turnover of axons, dendrites, and spines over months. With the appropriate promoters, GFP (and other fluorescent molecules) can also be targeted to nonneuronal elements of the nervous system to study their dynamics in normal or pathologic tissues (see, e.g., Christie et al. 2001). With the advent of usable functional indicators, aspects of neuronal and synaptic function will also become accessible in vivo. For example, calcium-sensitive fluorophores in dendritic spines allow the direct measurement of glutamate release at excitatory synapses (Yuste and Denk 1995; Oertner et al. 2002).

MATERIALS

Imaging Equipment

For neurons expressing XFP, in vivo images of XFP-expressing neurons are acquired using a custom-built 2PLSM (Lendvai et al. 2000). A Ti:sapphire laser (Tsunami, Spectra Physics), running at $\lambda \sim 910$ nm, pumped by a 10-W solid-state laser (Millenia X, Spectra Physics) is used as a light source. The objective (40×, 0.8 NA) and scan lens were from Zeiss, the trinoc from Olympus, and the photomultiplier tube (PMT) from Hamamatsu. Detection optics with large apertures provide optimal fluorescence detection (Oheim et al. 2001). Image acquisition is achieved with custom software (MatLab) (Pologruto et al. 2003).

Caution: *See Appendix 3 for appropriate handling of materials marked with <!>.*

PROCEDURE

Long-term, High-Resolution Imaging of Neurons in the Neocortex In Vivo

The protocol described here was optimized for adult mice. With minor modifications, it is possible to image neurons in rats and mice as early as postnatal day (PND) 2. Mice (PND14 to PND511) in which a thy1 promoter drives the expression of XFP (GFP, line M; YFP, line H) (Feng et al. 2000; Trachtenberg et al. 2002) in a subset of cortical neurons are used for most of our studies.

Note: Humane treatment of animals must be observed at all times, and should follow the local facility guidelines for care and use of animals. Urethane can only be used for nonrecovery studies.

Surgery

1. Anesthetize transgenic mice deeply with an intraperitoneal injection of ketamine<!> (0.13 mg g^{-1} body weight) and xylazine<!> (0.01 mg g^{-1}) mixture, or an intraperitoneal injection of urethane<!> (1.5 mg g^{-1} body weight) for acute experiments. (ketamine/xylazine mixture = 12.5 mg/ml ketamine [2.6 ml of a 100 mg/ml stock] mixed with 0.65 mg/ml of xylazine [0.2 ml of a 100 mg/ml stock] in distilled water [20 ml]). ([urethane mixture = 20% urethane in saline (saline = 0.9% NaCl solution]).
2. Monitor the depth of anesthesia periodically by checking for lack of a toe-pinch reflex.
3. Lay the animal on a heating pad to maintain body temperature, and immobilize its head in a stereotaxic frame (e.g., Stoelting). Apply Vaseline directly to the eyes to maintain moisture and to protect from potential irritation by dental acrylic.
4. Administer dexamethasone<!> (0.02 ml at 4 mg ml^{-1}) by intramuscular injection to the quadriceps muscle.

   Note: This treatment minimizes the cortical stress response during chronic experiments.
5. Sterilize all surgical tools with 70% ethanol<!> and wash the scalp with ethanol or betadine. Using scissors, remove a flap of skin, ~1 cm square, covering the skull over the barrel cortex of both hemispheres.
6. Apply 1% xylocaine<!> (Lidocaine) to the skull and the exposed muscles using cotton swab applicators (no glue types; e.g., Pilgrim).
7. Remove all underlying fascia of the skull by scraping with a scalpel blade. Then separate the right lateral muscle (temporalis) from the bone and push it ventrally, avoiding bleeding.
8. Mark the cortical area of interest (e.g., barrel cortex at P35, is –2 mm Bregma, 4.25 mm lateral).
9. Apply a thin layer of cyanoacrylate adhesive (e.g., Vetbond, 3M) to the skull and the wound margins to stop bleeding and prevent the seepage of serosanguinous fluid. Avoid the area over the barrel cortex of the right hemisphere. This layer provides a better base for the dental acrylic to adhere to. Apply a thin layer of dental acrylic (Jet; Lang Dental) on top of the glue after it has dried.
10. Using a dental drill (1/4 bit) (Dental burrs, Henry Schein), slowly, to avoid heating and consequent damage to the dura, thin the circumference of a 5 × 5-mm region of the skull.

    Note: A circular shape is the simplest, with an island of skull left intact in the center. The cortical surface vasculature should become visible where the skull is thinned, and there should be minimal bleeding from the skull and none from the brain (Fig. 3A, B). Take care not to apply pressure to the skull and brain or to puncture the skull while drilling. (Sterile cortex buffer = 125 mM NaCl, 5 mM KCl <!>, 10 mM glucose, 10 mM HEPES, 2 mM $CaCl_2$, 2 mM $MgSO_4$<!> in distilled water, pH 7.4.)
11. Next, being careful not to exert any pressure on the brain, gently perforate the thinned skull superficially with the sharp point of a forceps and lift up the island of bone within the drilled circle to expose the dura.

    Note: This removal is most successful if performed under a bubble of cortex buffer. Gelfoam is used to control any bleeding from the dura and skull, and can be applied directly to the dura when wet. Prior to use, the Gelfoam (e.g., Pharmacia-Upjohn) should be cut into ~7-mm^2 blocks and soaked in cortex buffer.
12. Construct an optical window, ideally by covering the clear, unblemished, almost-dry dura with a thin layer of low-melting-point agarose (Sigma) (1.2%) in cortex buffer and a custom-made circular coverslip (5–7 mm diameter, No. 1 thickness) that should lie flush with the skull.

    Note: The glass should be as close to the brain as possible (~100 μm). The glass coverslip should be wider than the skull window and the agar should seal the brain from the air and dental cement. The 1.2% agarose must be made fresh and sterile for each experiment, and kept hot and mixed during the procedure. Cool the briefly hot agarose to less than ~37°C before applying it to the cortex—check the temperature by squirting a drop onto the back of your hand.

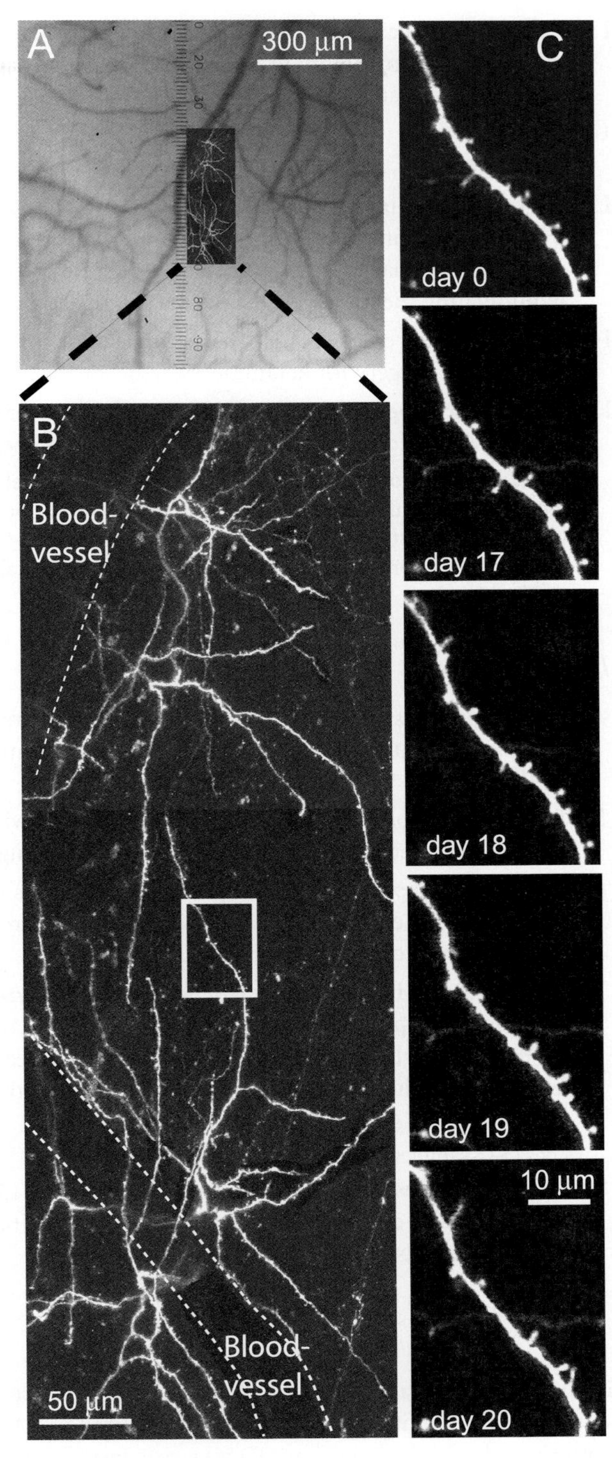

FIGURE 3. Long-term imaging of GFP-expressing pyramidal neurons in adult transgenic mice (GFP-M). (*A*) Bright-field image of surface vasculature. Superposed is the fluorescence image from *B*. The vasculature is used to find the same region of interest on subsequent imaging sessions. (*B*) Image of dendritic tufts from two layer-5 pyramidal neurons (projection of a stack of optical sections). Blood vessels and a region of interest (*white box*) are indicated. (*C*) Long-term chronic in vivo imaging of individual dendritic spines. The images are best projections (see Fig. 4).

13. Once the agar has set on the cortex, carefully remove any excess from the edges of the window and dry the remaining liquid on the skull. Do not allow any glue to come into contact with the agar—it will lose its transparency.
14. Seal the optical window to the skull with dental acrylic and cover all the exposed areas of skull, wound margins, and edges of the coverslip.
15. Clean a small titanium bar (home-made) with tapped screw holes and embed it into the acrylic over the intact (in this case, the left) hemisphere to stabilize the animal for subsequent imaging sessions. Ensure that the screw threads are kept clear of acrylic, and allow the acrylic to set before manipulating the mouse or the bar (30 min is typically sufficient).
16. Carry out acute experiments. These can be done for the duration of the anesthesia (typically up to 1 hour after surgery for ketamine/xylazine, or for 5 hours under urethane anesthesia).
17. For chronic experiments, allow the mouse to recover on the heating pad, then return it to its cage and allow an additional 7–10-day recovery period before imaging.

    Note: Windows often clear and improve during this delay. The cranial window can remain clear for months, without the use of antibiotics. To reduce stress for the mouse, minimize the size of the metal bar, which is used to stabilize the head, so that the mouse can freely move in its cage. A large cage with a high ceiling can help prevent the mice from knocking the dental acrylic and losing a bar or the entire window and headcap.

Imaging

4. For imaging sessions, anesthetize animals with ketamine/xylazine at ∼⅔ surgical dose (see above). Stabilize the animals under the microscope on a heating pad using screws placed through the bars on their heads.
5. Trail 1.5% agarose around the edge of the imaging window, making a small pool to hold water for the water-immersion objective to be used for imaging.
6. For neurons expressing XFP, acquire in vivo images of XFP-expressing neurons using a custom-built 2PLSM (Lendvai et al. 2000) with a Ti:sapphire laser light source<!>. See Materials section for more detail of setup.
7. In each animal, image the apical dendritic tufts of pyramidal neurons over a period of 3–150 days. Relocate the appropriate cells at each session using the unique vascular pattern in the vicinity of the cell. For high-magnification spine imaging, select 7–15 fields, each 50 × 50 μm, containing second and higher-order branches (Juraska 1982), for each cell.

    Note: Our image stacks typically consist of sections (512 × 512 pixels; ∼0.09 μm/pixel) collected in 1-μm steps (3–5 samples per resolution element).
8. Also collect low-magnification images (512 × 512 pixels; ∼0.3 μm/pixel; 3-μm steps) for overview. Take care to achieve close to identical fluorescence levels across imaged regions and imaging sessions by adjusting the excitation power.

Measurements

In our lab, spines are measured using custom software. Imaging tiny structures, such as dendritic spines, has certain limitations. Because of the limited NA of long-working-distance, water-immersion objectives, the resolution of our 3D images is insufficient to resolve spines reliably in the axial dimension. Therefore, structures that project mainly along the optical axis, below or above the dendrite, are not analyzed. All clear protrusions emanating laterally from the dendritic shaft, irrespective of apparent shape, are measured. Analysis is done blind, with the analyzer unaware of the experimental condition (i.e., cortical region, developmental age, manipulation). For each day of each region, images are aligned with each other using fiducial marks, such as dendritic branchpoints, that are typically stable across all imaging days. Images are analyzed in three dimensions rather than in projections. Figure 4A shows a standard maximum value projection. Note the out-of-focus haze, which can obscure small dendritic spines. Figure 4B shows a "best-value projection," a composite image spliced together from

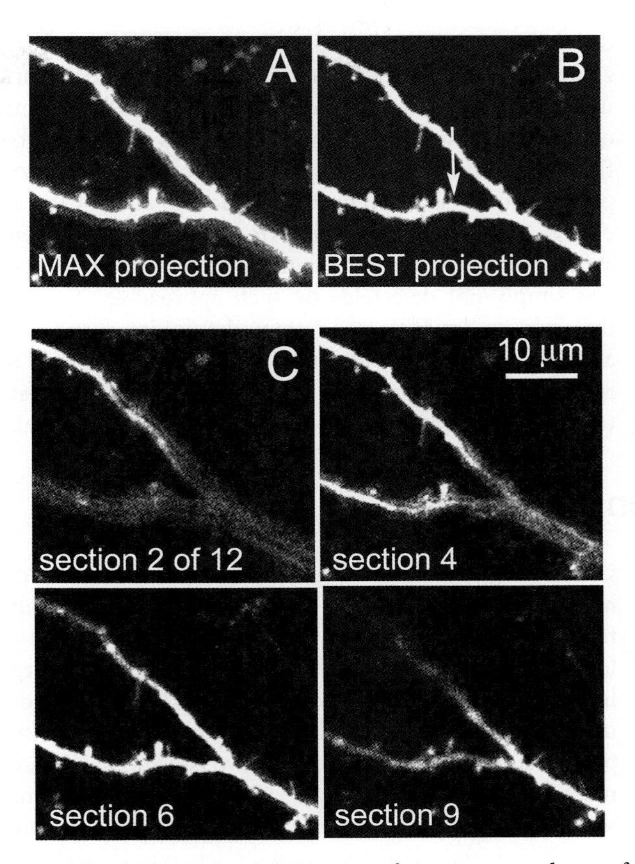

FIGURE 4. Morphometry of dendritic spines. (*A*) Because of movement and out-of-focus haze, maximum value projections are noisy. Small dendritic spines are often lost in the fuzz surrounding larger structures such as dendrites (*arrow*). (*B*) Best projection of the same image stack. For each spine, the best focal plane is identified and overlaid, preserving all structures in the reconstruction. (*C*) Raw data for *A* and *B*.

different sections containing the crispest view of dendritic segments. This image is representative of the signal-to-noise ratio in the three-dimensional stack and clearly shows the smallest spines. Control experiments, which include direct comparisons with electron microscopy, demonstrate that even the smallest dendritic spines can be reliably detected in vivo.

Troubleshooting

Assuming the transgenic mouse line chosen for in vivo imaging expresses the fluorescent protein at high enough concentrations in the cells of interest, the limiting step to high-resolution chronic imaging is typically in the surgery. Surgical operations must be clean, and performed consistently and quickly. The imaging window must be optically clear throughout the long-term imaging period. Blood from the surgery, regrowth of the skull after a few weeks or months, and infection can all obscure the imaging window. These events can be avoided by keeping the operations as sterile as possible, and by avoiding bleeding from the dura.

EXAMPLE OF APPLICATION

Neuronal function is determined to a large degree by the nature and organization of the synaptic connections it receives. Governing the organization of dendrites, axons, and the synaptic connections that link them are factors both intrinsic and extrinsic to the organism. During postnatal life, sensorimotor experience is the dominant extrinsic factor influencing the activity of neuronal networks. Experience-dependent activity sculpts the structure of axons and dendrites, as well as the efficacy and/or organization of impinging synapses. Elucidating the mechanisms of structural plasticity is essential to an understanding of the emergent network properties and memory formation. The predominant method of characterizing structural plasticity relies on postmortem examinations of neuronal morphology in fixed tissue. Only a single time point is collected, which requires group averaging and population comparisons to reveal significant differences in control and experimental conditions, where movement or change could really only be inferred (Volkmar and Greenough 1972; Zito and Svoboda 2002). Steady-state dynamics where, for example, the number of spines gained is equal to the number of spines lost, was simply not detectable by such approaches. Time-lapse imaging microscopy has unveiled rich dynamics of dendritic structure in developing cortical tissue in vitro (Dailey and Smith 1996; Engert and Bonhoeffer 1999; Maletic-Savatic et al. 1999) and in vivo (Fig. 2E) (Lendvai et al. 2000). More recently, long-term imaging experiments have revealed structural plasticity in the intact adult neocortex (Fig. 3C) (Grutzendler et al. 2002; Trachtenberg et al. 2002).

ADVANTAGES AND LIMITATIONS

Chronic high-resolution imaging has been achieved using the imaging window preparation presented here (Levasseur et al. 1975; Brown et al. 2001; Trachtenberg et al. 2002) as well as procedures involving skull thinning (Christie et al. 2001; Grutzendler et al. 2002). Each technique has distinct advantages. For the imaging window preparation, the only surgery required is the implantation of the imaging window, and therefore, imaging can be performed frequently. Skull-thinning is a difficult and invasive procedure that must be performed for every imaging session and is therefore best suited to protocols requiring no more than two time points. In addition, for high-resolution imaging, the skull must be thinned to less than ~50 μm, which is accompanied by a loss of vascularization and local skull degeneration, associated with inflammation and infection. A key advantage of the skull-thinning procedure is that time points can be separated by arbitrarily long periods.

Although high-resolution optical microscopy allows resolution and detection of the smallest neuronal compartments, including individual dendritic spines and axonal terminals, this technique is still fundamentally limited with respect to resolution. For example, it is difficult to determine whether two processes that appear to touch actually make a synaptic contact. In the authors' experience, a combination of optical microscopy with retrospective serial-section electron microscopy has proved especially powerful (Trachtenberg et al. 2002).

ACKNOWLEDGMENTS

We thank Brian Chen for help during early stages of this work, and Josh Sanes for transgenic mice. This work was supported by the Howard Hughes Medical Institute and the National Institutes of Health.

REFERENCES

Bai J., Ramos R.L., Ackman J.B., Thomas A.M., Lee R.V., and LoTurco J.J. 2003. RNAi reveals doublecortin is required for radial migration in rat neocortex. *Nat. Neurosci.* **6:** 1277–1283.

Brown E.B., Campbell R.B., Tsuzuki Y., Xu L., Carmeliet P., Fukumura D., and Jain R.K. 2001. In vivo measurement of gene expression, angiogenesis and physiological function in tumors using multiphoton laser scanning microscopy. *Nat. Med.* **7:** 864–868.

Campbell R.E., Tour O., Palmer A.E., Steinbach P.A., Baird G.S., Zacharias D.A., and Tsien R.Y. 2002. A monomeric red fluorescent protein. *Proc. Natl. Acad. Sci.* **99:** 7877–7882.

Christie R.H., Bacskai B.J., Zipfel W.R., Williams R.M., Kajdasz S.T., Webb W.W., and Hyman B.T. 2001. Growth arrest of individual senile plaques in a model of Alzheimer's disease observed by in vivo multiphoton microscopy. *J. Neurosci.* **21:** 858–864.

Dailey M.E. and Smith S.J. 1996. The dynamics of dendritic structure in developing hippocampal slices. *J. Neurosci.* **16:** 2983–2994.

Denk W. and Svoboda K. 1997. Photon upmanship: Why multiphoton imaging is more than a gimmick. *Neuron* **18:** 351–357.

Denk W., Strickler J.H., and Webb W.W. 1990. Two-photon laser scanning microscopy. *Science* **248:** 73–76.

Engert F. and Bonhoeffer T. 1999. Dendritic spine changes associated with hippocampal long-term synaptic plasticity. *Nature* **399:** 66–70.

Feng G., Mellor R.H., Bernstein M., Keller-Peck C., Nguyen Q.T., Wallace M., Nerbonne J.M., Lichtman J.W., and Sanes J.R. 2000. Imaging neuronal subsets in transgenic mice expressing multiple spectral variants of GFP. *Neuron* **28:** 41–51.

Fukuchi-Shimogori T. and Grove E.A. 2001. Neocortex patterning by the secreted signaling molecule FGF8. *Science* **294:** 1071–1074.

Gong S., Zheng C., Doughty M.L., Losos K., Didkovsky N., Schambra U. B., Nowak N.J., Joyner A., Leblanc G., Hatten M.E., and Heintz N. 2003. A gene expression atlas of the central nervous system based on bacterial artificial chromosomes. *Nature* **425:** 917–925.

Grutzendler J., Kasthuri N., and Gan W.B. 2002. Long-term dendritic spine stability in the adult cortex. *Nature* **420:** 812–816.

Haas K., Sin W.C., Javaherian A., Li Z., and Cline H.T. 2001. Single-cell electroporation for gene transfer in vivo. *Neuron* **29:** 583–591.

Heintz N. 2001. Bac to the future: The use of bac transgenic mice for neuroscience research. *Nat. Rev. Neurosci.* **2:** 861–870.

Hermens W.T. and Verhaagen J. 1998. Viral vectors, tools for gene transfer in the nervous system. *Prog. Neurobiol.* **55:** 399–432.

Holtmaat A.J., Oestreicher A.B., Gispen W.H., and Verhaagen J. 1998. Manipulation of gene expression in the mammalian nervous system: Application in the study of neurite outgrowth and neuroregeneration-related proteins. *Brain Res. Rev.* **26:** 43–71.

Juraska J.M. 1982. The development of pyramidal neurons after eye opening in the visual cortex of hooded rats: A quantitative study. *J. Comp. Neurol.* **212:** 208–213.

Kay M.A., Glorioso J.C., and Naldini L. 2001. Viral vectors for gene therapy: The art of turning infectious agents into vehicles of therapeutics. *Nat. Med.* **7:** 33–40.

Kleinfeld D., Mitra P.P., Helmchen F., and Denk W. 1998. Fluctuations and stimulus-induced changes in blood flow observed in individual capillaries in layers 2 through 4 of rat neocortex. *Proc. Natl. Acad. Sci.* **95:** 15741–15746.

Lendvai B., Stern E., Chen B., and Svoboda K. 2000. Experience-dependent plasticity of dendritic spines in the developing rat barrel cortex *in vivo*. *Nature* **404:** 876–881.

Levasseur J. E., Wei E.P., Raper A.J., Kontos A.A., and Patterson J.L. 1975. Detailed description of a cranial window technique for acute and chronic experiments. *Stroke* **6:** 308–317.

Maletic-Savatic M., Malinow R., and Svoboda K. 1999. Rapid dendritic morphogenesis in CA1 hippocampal dendrites induced by synaptic activity. *Science* **283:** 1923–1927.

Miesenbock G., Angelis D.A.D., and Rothman J.E. 1998. Visualizing secretion and synaptic transmission with pH-sensitive green fluorescent proteins. *Nature* **394:** 192–195.

Miyawaki A., Llopis J., Heim R., McCaffery J.M., Adams J.A., Ikura M., and Tsien R.Y. 1997. Fluorescence indicators for Ca^{2+} based on green fluorescent proteins and calmodulin. *Nature* **388:** 882–887.

Nagai T., Sawano A., Park E.S., and Miyawaki A. 2001. Circularly permuted green fluorescent proteins engineered to sense Ca2+. *Proc. Natl. Acad. Sci.* **98:** 3197–3202.

Nagai T., Ibata K., Park E.S., Kubota M., Mikoshiba K., and Miyawaki A. 2002. A variant of yellow fluorescent protein with fast and efficient maturation for cell-biological applications. *Nat. Biotechnol.* **20:** 87–90.

Nakai J., Ohkura M., and Imoto K. 2001. A high signal-to-noise Ca(2+) probe composed of a single green fluorescent protein. *Nat. Biotechnol.* **19:** 137–141.

Neumann E., Kakorin S., and Toensing K. 1999. Fundamentals of electroporative delivery of drugs and genes. *Bioelectrochem. Bioenerg.* **48:** 3–16.

Oertner T.G., Sabatini B.S., Nimchinsky E.A., and Svoboda K. 2002. Facilitation at single synapses probed with optical quantal analysis. *Nat. Neurosci.* **5:** 657–664.

Oheim M., Beaurepaire E., Chaigneau E., Mertz J., and Charpak S. 2001. Two-photon microscopy in brain tissue: Parameters influencing the imaging depth. *J. Neurosci. Methods* **111:** 29–37.

Pologruto T.A., Sabatini B.L., and Svoboda K. 2003. ScanImage: Flexible software for operating laser-scanning microscopes. *BioMed. Eng. OnLine* **2:** 13.

Stosiek C., Garaschuk O., Holthoff K., and Konnerth A. 2003. In vivo two-photon calcium imaging of neuronal networks. *Proc. Natl. Acad. Sci.* **100:** 7319–7324.

Svoboda K., Denk W., Kleinfeld D., and Tank D.W. 1997. In vivo dendritic calcium dynamics in neocortical pyramidal neurons. *Nature* **385:** 161–165.

Tabata H. and Nakajima K. 2001. Efficient in utero gene transfer system to the developing mouse brain using electroporation: Visualization of neuronal migration in the developing cortex. *Neuroscience* **103:** 865–872.

Trachtenberg J.T., Chen B.E., Knott G.W., Feng G., Sanes J.R., Welker E., and Svoboda K. 2002. Long-term in vivo imaging of experience-dependent synaptic plasticity in adult cortex. *Nature* **420:** 788–794.

Tsien R.Y. 1998. The green fluorescent protein. *Annu. Rev. Biochem.* **67:** 509–544.

Volkmar F.R. and Greenough W.T. 1972. Differential rearing effects on rat visual cortical plasticity. *Science* **176:** 1445–1447.

Washbourne P. and McAllister A.K. 2002. Techniques for gene transfer into neurons. *Curr. Opin. Neurobiol.* **12:** 566–573.

Wells T. and Carter D.A. 2001. Genetic engineering of neural function in transgenic rodents: towards a comprehensive strategy? *J. Neurosci. Methods* **108:** 111–130.

Yuste R. and Denk W. 1995. Dendritic spines as basic functional units of neuronal integration. *Nature* **375:**682–684.

Zito K. and Svoboda K. 2002. Activity-dependent synaptogenesis in the adult mammalian cortex. *Neuron* **35:** 1015–1017.

CHAPTER 86

A Two-Photon Fiberscope for Imaging in Freely Moving Animals

Fritjof Helmchen[1] and Winfried Denk[2]

[1]*Abteilung Zellphysiologie and* [2]*Abteilung Biomedizinische Optik, Max-Planck Institut für medizinische Forschung, 69120 Heidelberg, Germany*

This chapter describes a miniaturized two-photon-excited fluorescence microscope (in brief, two-photon fiberscope) designed for high-resolution imaging in freely behaving animals. The fiberscope is small and light enough to be carried by an adult rat. Mounted to the skull above a cranial window, it permits imaging of neurons with micrometer resolution to ~250 μm beneath the brain surface. Imaging of blood capillaries in the neocortex of awake, freely moving rats has been demonstrated (Helmchen et al. 2001). In combination with in vivo methods for labeling neurons with functional indicators, the two-photon fiberscope promises to enable optical recordings of neural activity during natural behavior.

FIBERSCOPE DESIGN

Mobilization of the microscope requires delivering the excitation light through an optical fiber (for reviews, see Delaney and Harris 1995; Helmchen 2002). For two-photon excitation, laser pulses of a Ti:sapphire laser (800–900-nm wavelength; 100-fsec pulse width at laser output; 80-MHz repetition rate) are propagated through a single-mode optical fiber. The output beam is collimated by a small lens and then focused through a water-immersion objective (Fig. 1a). Fluorescence can be detected either directly at the fiberscope headpiece using a small photomultiplier tube (PMT), or through a large-core fiber using a remote detector. The focus spot is scanned by inducing resonant flexural vibrations of the free fiber end in two dimensions (Fig. 1b). The fiberscope headpiece is mounted on a metal plate, which is glued to the skull. The original prototype was 7.5 cm long and weighed 25 g. Table 1 provides a list of parts used in our two-photon fiberscope prototypes.

FLUORESCENT DYE LABELING IN VIVO

Application of the two-photon fiberscope relies on fluorescence labeling in vivo. Tail-vein injection of a fluorescent dye (Chapter 93) was used for the demonstration of two-photon fiberscope imaging in freely moving rats (Helmchen et al. 2001). For imaging neural activity, individual neurons can be filled with organic indicator dyes via intracellular recording electrodes (high-resistance sharp electrodes

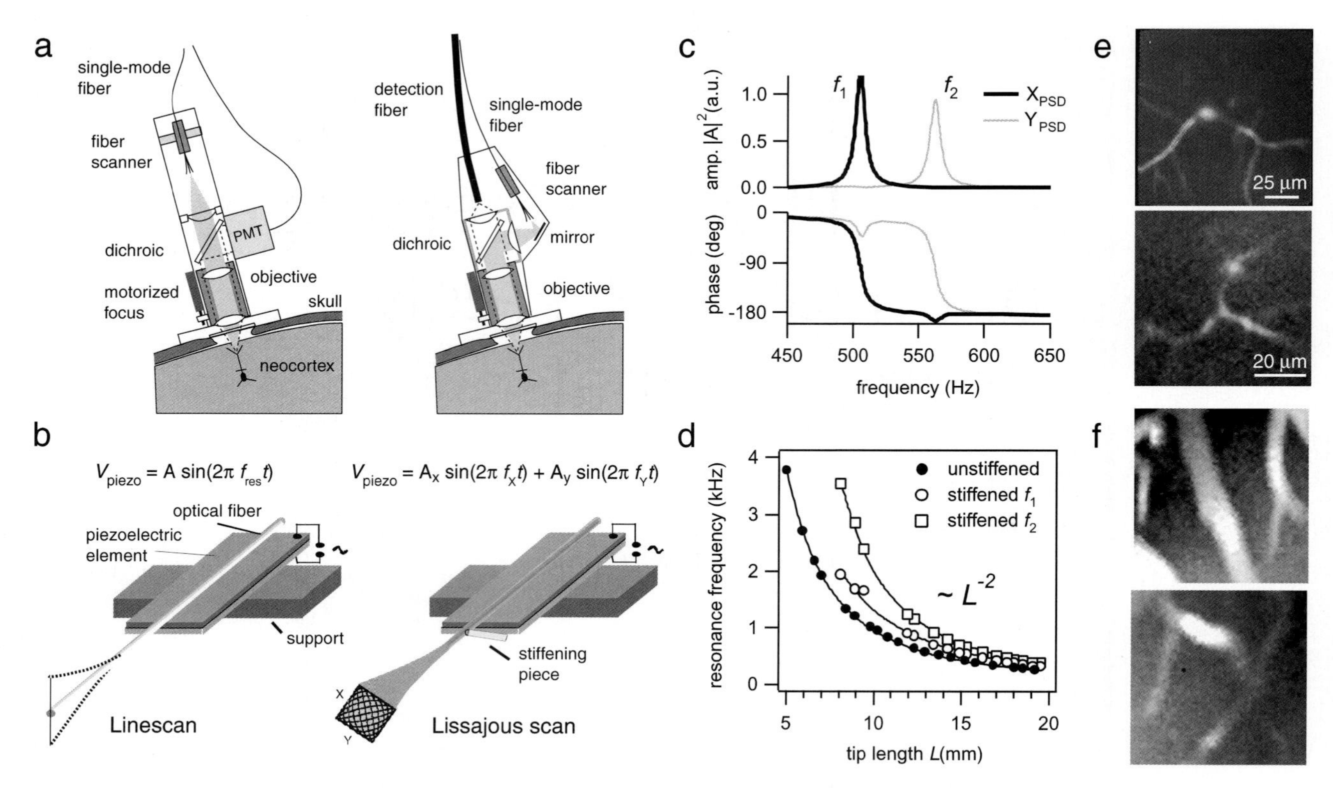

FIGURE 1. (*a*) Two-photon fiberscope designs. Fluorescence light (*dashed lines*) is detected either by a small headpiece-mounted PMT (version 1, *left*) or collected through a large-core multi-mode fiber (version 2, *right*). In these fluorescence collection modes, excitation light (*gray*) is either passed through a long-pass dichroic beamsplitter (*left*) or deflected by a short-pass dichroic (*right*). (*b*) Lissajous scanning in detail. Flexural vibration of the fiber end is excited with the piezoelectric bending element driven at the resonance frequency (f_{res}). Stiffening of the fiber end in one direction splits the resonance frequency in f_X and f_Y. Simultaneous excitation of vibrations in *X* and *Y* creates a Lissajous pattern, which permits area scanning (*right*). (*c*) Resonance curves measured for both directions using a position-sensitive detector (PSD). (*d*) Dependence of the resonance frequencies on the free length of the fiber end. (*e, f*) Two-photon fiberscope images; (*e*) dendrites of neurons in cortical layer 2/3 in head-fixed anesthetized rats; (*f*) fluorescently labeled blood vessels and capillaries in freely moving rats (parts *e* and *f*, Adapted, with permission, from Helmchen et al. 2001 [©Elsevier].)

[Svoboda et al. 1997; Helmchen et al. 2001] or whole-cell patch pipettes [Helmchen and Waters 2002]). This approach has been used for fiberscope measurements of dendritic calcium signals during anesthesia (Helmchen et al. 2001). However, single-cell labeling is laborious and—due to the restriction to at most a small number of neurons—reduces the utility of measurements in awake animals. For such experiments, population labeling techniques might be particularly useful, e.g., in vivo loading of populations of neocortical cells with calcium- or chloride-sensitive indicators by local application of membrane-permeant dyes (Stosiek et al. 2003; see Chapter 71). Alternatively, it should be possible to transfect neocortical neurons with a virus encoding activity-dependent fluorescent proteins (for review, see Guerrero and Isacoff 2001) using methods similar to those described for virally mediated induction of GFP expression in vivo (Lendvai et al. 2000; Kim et al. 2003).

TECHNICAL ASPECTS

Two crucial issues that are particular to the miniature two-photon fiberscope will be treated in more detail: (1) minimization of laser pulse length at the fiber output and (2) design of the compact scanning device.

TABLE 1. Components useful for designing a two-photon fiberscope

Component	Properties	Model (Vendor)
Single-mode fiber	core size 5–6 μm; NA 0.12–0.15; cladding diameter 125 μm	F-SF (Newport); FS-SN-4224 (3M); . . .
Piezoelectric element	custom; about 2 mm x 8 mm in size	(EDO, Polytec PI, Morgan Matroc, Piezo Systems, . . .)
Collimating lens	aspheric; NA 0.15	C280TM, C260TM (Geltech, Thorlabs)
Long-pass dichroic	cold mirror or dichroic beam splitter	M43-960 (Edmund Optics)
Short-pass dichroic	hot mirror, NIR blocking filter, or dichroic beam splitter	Calflex X (Linos Photonics)
Objective	water-immersion, NA>0.8; large fluorescence collection field of view	modified from Syncotec (now defunct); custom design (Throl Optical Systems)
Collection lens	small lens; NA matched to collection fiber	C230TM (Geltech)
Fluorescence collection fiber	large core >600 μm; high transmission in the visible	FT-600-URT (Thorlabs); MO2-534 (Edmund Optics)
Small PMT	high quantum efficiency in visible; weight 4 g	R7400U (Hamamatsu)
Small DC motor	weight 300 mg; gear ratio 1:125	smoovy motors (Micro Precision Systems)
Position-sensitive detector (PSD)		2L10SP (ON-TRAK Photonics); S5991 (Hamamatsu)

Two-Photon Excitation through Optical Fibers

Ultrafast laser pulses broaden in optical fibers because of material dispersion and—at the high laser powers required for deep imaging—because of nonlinear effects (Agrawal 1995). Such broadening limits the efficiency of two-photon excitation and must, therefore, be reduced to a minimum. Dispersion-induced broadening can be compensated by pre-chirping the laser pulses with a pair of diffraction gratings (Treacy 1969) or prism sequences (Walmsley et al. 2001) before coupling to the fiber. In the original publication (Helmchen et al. 2001), two reflection gratings (400 grooves/mm; 9.7° blaze angle; Richardson Gratings Lab) were used in double-pass configuration, providing around –60.000 fsec^2 at 17-cm spacing. The optimal grating or prism spacing can be found by measuring the pulse width with an autocorrelator and determining the spacing that yields the minimum pulse width. Note that, due to nonlinear optical effects in the fiber core (mainly self-phase modulation; Agrawal 1995), laser pulses progressively broaden at increasing power levels even with pre-chirping. Nonlinear pulse-broadening becomes significant at low average power (10 mW), and compensation becomes very poor at the high average power levels needed for imaging deep into tissue (Helmchen et al. 2002). Several approaches to alleviate the problem of nonlinear pulse-broadening, including the use of large-core specialty fibers (Helmchen et al. 2002; Ouzounov et al. 2002) and spectral, in addition to temporal, shaping of the input pulse (Clark et al. 2001), have been tested, albeit with limited success. A complete solution to the problem is offered by novel hollow-core optical fibers, for which distortion-free delivery of high-energy femtosecond pulses recently was demonstrated (Göbel et al. 2004).

Fiber End Resonance Scanning

A compact scanning device is essential for miniaturization. A simple mechanism is to move the tip of the illuminating fiber, which is in a plane conjugate to the focal plane (Giniunas et al. 1991; Delaney and Harris 1995). A piezoelectric bending element ("piezo") is used to induce resonant mechanical vibrations of the free fiber end (Fig. 1b). The bare fiber (coating stripped off) is glued to the piezo so that 1–2 cm of the fiber end remain free to vibrate. For a 125-μm (standard-)diameter fiber this results in a resonance frequency f_{res} between 200 Hz and 1000 Hz (Fig. 1d). In general, for transverse vibration of a cylindrical rod, f_{res} is given by (Bishop and Johnson 1960)

$$f_{res} = \frac{s_0^2}{4\pi}\frac{R}{L^2}\sqrt{\frac{E}{\rho}} \qquad (1)$$

where R is the radius, L the length of the rod, ρ the density (≈2.3 g/cm^3), E the elasticity module of the fiber (≈75 GPa), and $s_0 = 1.875$. Thus, f_{res} depends inversely on the square of the length L (Fig. 1d). Two-dimensional scanning is achieved by stiffening the fiber end in one direction with a short (2–4 mm) piece of bare fiber glued to the lower edge of the piezo and to the fiber at about 2–3 mm distance from the piezo (Fig. 1b). This results in two resonance frequencies of the fiber tip, which can be determined by driving the piezo with a sinusoidal waveform that is swept in frequency (Fig. 1c). Because the resonance widths are typically only 5–10 Hz, it is necessary to sweep the frequency slowly and finely so as not to miss the resonance. For accurate measurement of the resonance curve, the fiber tip can be imaged onto a 2D position-sensitive detector (Fig. 1c). This approach also permits a direct measurement of the Lissajous scan pattern that is excited when the piezo is driven with a superposition of two sine waves with the resonance frequencies for the orthogonal directions (Helmchen et al. 2001). With up to 100 V (peak-to-peak) drive amplitude, resonant vibrations of more than a millimeter peak-to-peak amplitude are reached. The Lissajous pattern depends on the ratio of the chosen frequencies. For details on how to choose the best frequencies within the resonance widths for optimizing the filling factor and the repeat frequency, see Helmchen et al. (2001). Because resonance is used, large phase shifts exist between the sine waves driving the piezo and the actual movement of the fiber tip (Fig. 1c). For image reconstruction it is, therefore, essential to correct these phase shifts, e.g., by using electronic adjustments of the position signals (Helmchen et al. 2001).

EXAMPLE OF APPLICATION

Pyramidal neurons in neocortical layer 2, loaded with calcium indicator via an intracellular electrode, were resolved in head-restrained, anesthetized rats (Fig. 1e). Dendritic calcium transients were measured using a line-scan mode, in which fiber-end vibration is induced in only one direction, by driving the fiber with only a single sine wave (Helmchen et al. 2001). Fluorescently labeled blood capillaries were imaged in awake, freely moving rats (Fig. 1f), demonstrating that stable images can be acquired in unrestrained animals, except during periods of strong head movement (Helmchen et al. 2001).

ADVANTAGES AND LIMITATIONS

In vivo imaging of neural activity in the mammalian brain using a standard two-photon microscope is limited to head-restrained animals. This usually requires experiments under anesthesia, which, of course, alter neural dynamics. Although imaging in awake, head-restrained animals might be possible, the immobilization of the head obviously places severe restrictions on an animal's behavior. The two-photon fiberscope, on the other hand, should, in combination with appropriate labeling techniques, permit measurement of single-cell as well as network activity in the brain of awake animals during natural behaviors. Such experiments may be essential to understand the cellular basis of neural function during behavior.

Currently, the two-photon fiberscope is limited in depth penetration and to a particular Lissajous scan pattern. Several improvements of the fiberscope are desirable and conceivable: (1) improvement of two-photon excitation through optical fibers (see above); (2) reduction in size and weight, e.g., by using special optics such as gradient-index (GRIN) lenses (Jung and Schnitzer 2003); and (3) remote control of the field of view. In addition, alternative approaches to the vibrating fiber scan may be employed, such as the use of micro-mirrors (Dickensheets and Kino 1996) or of coherent optical fiber bundles (Knittel et al. 2001), which could lead to even smaller headpieces and endoscopic imaging, but at the price of lower resolution.

REFERENCES

Agrawal G.P. 1995. *Nonlinear fiber optics*. Academic Press, San Diego.

Bishop R.E.D. and Johnson D.C. 1960. *Mechanics of vibration*. Cambridge University Press, Cambridge, United Kingdom.

Clark S.W., Ilday F.Ö., and Wise F.W. 2001. Fiber delivery of femtosecond pulses from a Ti:sapphire laser. *Opt. Lett.* **26:** 1320–1322.

Delaney P.M. and Harris M.R. 1995. Fiberoptics in confocal microscopy. In *Handbook of biological confocal microscopy* (ed. J.B. Pawley), pp. 515–523. Plenum Press, New York.

Dickensheets D.L. and Kino G.S. 1996. Micromachined scanning confocal optical microscope. *Opt. Lett.* **21:** 764–766.

Giniunas L., Juskaitis R., and Shatalin S.V. 1991. Scanning fibre-optic microscope. *Electron. Lett.* **27:** 724–726.

Göbel. W., Nimmerjahn A., and Helmchen F. 2004. Distortion-free delivery of nanojoule femtosecond pulses from a Ti:sapphire laser through a hollow-core photonic crystal fiber. *Opt. Lett.* **29:** 1285–1287.

Guerrero G. and Isacoff E.Y. 2001. Genetically encoded optical sensors of neuronal activity and cellular function. *Curr. Opin. Neurobiol.* **11:** 601–607.

Helmchen F. 2002. Miniaturization of fluorescence microscopes using fibre optics. *Exp. Physiol.* **87:** 737–745.

Helmchen F. and Waters J. 2002. Ca^{2+} imaging in the mammalian brain in vivo. *Eur. J. Pharmacol.* **447:** 119–129.

Helmchen F., Tank D.W., and Denk W. 2002. Enhanced two-photon excitation through optical fiber by single-mode propagation in a large core. *Appl. Opt.* **41:** 2930–2934.

Helmchen F., Fee M.S., Tank D.W., and Denk W. 2001. A miniature head-mounted two-photon microscope: High-resolution brain imaging in freely moving animals. *Neuron* **31:** 903–912.

Jung C.J. and Schnitzer M. 2003. Multiphoton endoscopy. *Opt. Lett.* **28:** 902–904.

Kim J., Dittgen T., Nimmerjahn A., Waters J., Pawlak V., Helmchen F., Schlesinger S., Seeburg P.H., and Osten P. 2003. Sindbis vector SINrep(nsP2S726): A tool for rapid heterologous expression with attenuated cytotoxicity in neurons. *J. Neurosci. Methods* **133:** 81–90.

Knittel J., Schnieder L., Buess G., Messerschmidt B., and Possner T. 2001. Endoscope-compatible confocal microscope using a gradient index-lens system. *Opt. Commun.* **188:** 267–273.

Lendvai B., Stern E.A., Chen B., and Svoboda K. 2000. Experience-dependent plasticity of dendritic spines in the developing rat barrel cortex in vivo. *Nature* **404:** 876–881.

Ouzounov D.G., Moll K.D., Foster M.A., Zipfel W.R., Webb W.W., and Gaeta A.L. 2002. Delivery of nanojoule femtosecond pulses through large-core microstructured fibers. *Opt. Lett.* **27:** 1513–1515.

Stosiek C., Garaschuk O., Holthoff K., and Konnerth A. 2003. In vivo two-photon calcium imaging of neuronal networks. *Proc. Natl. Acad. Sci.* **100:** 7319–7324.

Svoboda K., Denk W., Kleinfeld D., and Tank D.W. 1997. In vivo dendritic calcium dynamics in neocortical pyramidal neurons. *Nature* **385:** 161–165.

Treacy E.B. 1969. Optical pulse compression with diffraction gratings. *IEEE J. Quant. Electron.* **5:** 454–458.

Walmsley I., Waxer L., and Dorrer C. 2001. The role of dispersion in ultrafast optics. *Rev. Sci. Instrum.* **72:** 1–29.

CHAPTER 87

A Practical Guide: In Vivo Two-Photon Calcium Imaging Using Multicell Bolus Loading

Olga Garaschuk and Arthur Konnerth

Physiologisches Institut, Ludwig-Maximilians Universität München, 80336 München, Germany

This chapter describes an approach for in vivo two-photon Ca^{2+} imaging of large neuronal circuits with the resolution of individual cells. The approach was developed for in vivo imaging of the cortex, but it can be easily adapted for imaging other brain regions, including the cerebellum and olfactory bulb. It was recently used for in vivo recordings from individual spinal cord neurons in zebrafish larvae (Brustein et al. 2003). The staining technique can also be applied in brain slices of any developmental stage, from embryonic to adult.

MATERIALS

Anesthesia unit, including chamber for pre-anesthetic medication, a flow meter, and a vaporizer (latter items are for volatile anesthetic agents only). Consult the literature (e.g., Flecknell 2000) for the best choice of anesthesia for your species. Anesthetic procedures: For adult mice either ketamine/xylazine or urethane is recommended (0.1/0.01 mg/g and 1.5–1.9 mg/g body weight, respectively, intraperitoneally). Alternatively, mice can be anesthetized by inhalation of isoflurane (1–1.5% in pure O_2).

Surgical equipment, including a stereotaxic instrument, drill, warming plate to keep animal's body temperature constant (available from many providers; e.g., TSE-Systems)

Recording chamber, custom-made with central access opening (Stosiek et al. 2003)

Membrane-permeable calcium indicator dye (e.g., Calcium Green-1 AM, fura-2 AM, fluo-4 AM, indo-1 AM; Molecular Probes)

Manipulator and a pressure application device for injection of the staining solution into the brain. We used the LN-Mini manipulator from Luigs & Neumann, and Picospritzer II, from General Valve, respectively

Two-photon laser-scanning microscope (see Chapter 71 for details)

PROCEDURE

Staining Neurons with a Calcium Indicator Dye

1. Anesthetize the animal. Ensure that the surgical level of anesthesia has been reached (e.g., by testing the pinch withdrawal and the eyelid reflex). Inject ~ 50 μl of a local anesthetic agent (e.g., 2% lidocaine) subcutaneously at the location where the skin is to be removed (*optional*).
2. Remove the skin above the desired brain area. Thin the skull and polish it with a felt polisher (e.g., from Dr. Ihde Dental). Perform a small (~1 mm) craniotomy above an area devoid of large blood vessels. Use cyanoacrylic glue<!> to adhere the custom-made recording chamber to the skull, such that the middle of the chamber opening lies above the craniotomy.
3. Transfer the animal into the setup and place it onto a warming plate (38°C). Perfuse the recording chamber with a warm (37°C) standard external saline (125 mM NaCl, 2.5 mM KCl<!>, 26 mM $NaHCO_3$, 1.25 mM NaH_2PO_4<!>, 2 mM $CaCl_2$<!>, 1 mM $MgCl_2$<!>, 20 mM glucose, pH 7.4, when bubbled with 95% O_2 and 5% CO_2).
4. Dissolve acetoxymethyl (AM)-ester of the chosen indicator dye in dimethyl sulfoxide<!> plus 20% Pluronic F-127 (e.g., 2 g of Pluronic in 10 ml of DMSO<!>) to yield a dye concentration of 10 mM. Dilute this solution 1/10 or 1/20 with the standard pipette solution (150 mM NaCl, 2.5 mM KCl, 10 mM HEPES).
5. Fill a micropipette with this staining solution (pipette resistance 6–9 MΩ). Insert the pipette into the cortex and advance it along its axis until it reaches the desired depth (see Fig. 1A). Apply a pressure pulse (1 min, 70 kPa) to eject ~ 400 fl of the staining solution near the cells of interest. Remove the pipette. Wait for an hour to obtain a stable maximal fluorescence level in stained cells (Stosiek et al. 2003). This protocol yields a stained area with a diameter of 200–400 μm.

Caution: *See Appendix 3 for appropriate handling of materials marked with <!>.*

TWO-PHOTON IMAGING OF STAINED CELLS

Excitation light of 800 nm is used to image neurons stained with all indicator dyes tested (see above). The average power under the objective is <70 mW. With dye application pipette located 150–200 μm below the cortical surface, all cortical cells between the surface and 400 μm depth are stained (Stosiek et al. 2003). When imaging through the thinned skull (thickness of 8–10 μm), individual cells can be well resolved up to 200 μm below the cortical surface. Removing the skull above the imaging field further improves depth resolution, allowing the detection of individual cells up to 300 μm below the cortical surface. It should be stressed that the stability of recordings depends critically on the diameter of the craniotomy. Thus, openings larger than 1 mm in diameter are often accompanied by movement artifacts occurring at the heartbeat frequency.

EXAMPLE OF APPLICATION

Figure 1B–D shows examples of the in vivo two-photon Ca^{2+} imaging experiments in the barrel cortex of mice (modified from Stosiek et al. 2003). Figure 1B illustrates the quality of imaging data, obtained at different depths. Figure 1C shows Ca^{2+} transients in layer 2/3 neurons evoked by ionophoretic glutamate application in vivo. The glutamate-containing pipette was positioned less than 50 μm apart from the imaged cells. Ca^{2+} transients in Figure 1D were evoked by the deflection of the majority of whiskers on the contralateral side of the mouse's snout. Note that the signal-to-noise ratio is sufficient to allow individual, nonaveraged somatic Ca^{2+} transients to be distinguished clearly from the background noise.

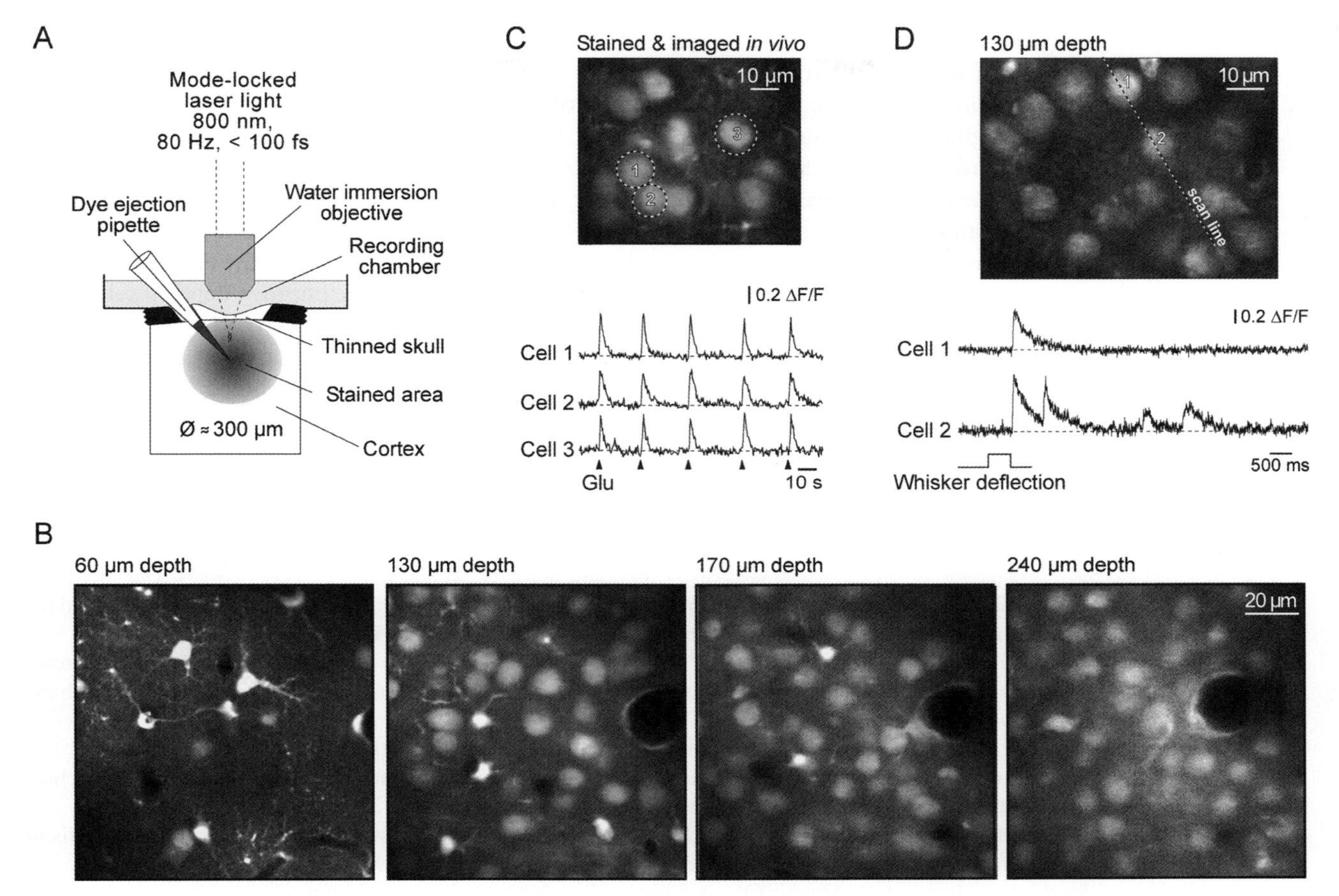

FIGURE 1. In vivo Ca^{2+} imaging of neuronal populations in the barrel cortex of mice. (*A*) Schematic drawing of the experimental arrangement. (*B*) High-magnification images of the barrel cortex of a 13-day-old mouse (P13) taken at increasing depth. (*C*) Ca^{2+} transients (*lower*) in 3 individual layer-2/3 neurons (as indicated in *upper*) of another P13 mouse evoked by five consecutive 500-msec ionophoretic glutamate applications. (*D*) Line-scan recordings of Ca^{2+} transients (*lower*) evoked in two layer-2/3 neurons by a deflection of the majority of whiskers on the contralateral side of the mouse's snout (P13 mouse). The position of the scanned line and the cells analyzed are indicated in *upper*. (*B–D*, Modified, with permission, from Stosiek et al. 2003.)

ADVANTAGES AND LIMITATIONS

The approach described, known as multicell bolus loading (MCBL; Stosiek et al. 2003), allows the simultaneous monitoring of Ca^{2+} levels in many individual neurons. The major difference between MCBL and other staining methods utilizing AM indicator dyes is that the indicators are delivered for a short period directly to the target cells. In particular, this approach improves the staining of neurons in the adult brain, which are, in general, not stained by AM indicator dyes bath-applied to brain slices. Additional advantages of MCBL include the need for only minor surgery and the possibility of restaining neurons and, thus, the ability to conduct long-lasting, perhaps even chronic (Christie et al. 2001), recordings.

Although MCBL allows many cells to be imaged simultaneously, the resolution of subcellular structures is lower, as compared to in vivo Ca^{2+} imaging of individual, microelectrode-loaded cells (Svoboda et al. 1997). This is for two obvious reasons. First, the image contrast is reduced due to the staining of many fine processes in the surrounding neuropil. Second, the dye concentration in MCBL-loaded cells is lower, on average 20 µM indicator dye, instead of <3–6 mM, when stained using a microelectrode (Svoboda et al. 1997; Stosiek et al. 2003). These limitations restrict the use of MCBL to analysis of somatic Ca^{2+} transients and make in vivo imaging of neuronal dendrites rather difficult. Furthermore, they reduce the depth resolution of recordings (200–300 μm compared with 500 μm when imaging cell dendrites of microelectrode-loaded cells). Future strategies for improving the quality of recordings include the use of longer wavelengths of the excitation light, larger numerical apertures of the objective lens, better transmittance of the optics, and higher photon sensitivity of the PMT. Because the proportion of scattered photons in the emitted fluorescence signal increases markedly with increasing imaging depth, a larger craniotomy and a larger effective angular acceptance of the detection optics (Oheim et al. 2001) should also significantly improve depth resolution by enabling the collection of a larger portion of scattered photons.

In conclusion, the approach described here is applicable for Ca^{2+} imaging of intact neurons both in vivo and in brain slices. It enables staining of adult neurons and, if combined with a miniature head-mounted two-photon microscope (Helmchen et al. 2001), it may allow in vivo two-photon imaging in freely moving animals.

REFERENCES

Brustein E., Marandi N., Kovalchuk Y., Drapeau P., and Konnerth A. 2003. 'In vivo' monitoring of neuronal network activity in zebrafish by two-photon Ca^{2+} imaging. *Pflugers Arch.* **446:** 766–773.

Christie R.H., Bacskai B.J., Zipfel W.R., Williams R.M., Kajdasz S.T., Webb W. W., and Hyman B.T. 2001. Growth arrest of individual senile plaques in a model of Alzheimer's disease observed by in vivo multiphoton microscopy. *J. Neurosci.* **21:** 858–864.

Flecknell P. 2000. *Laboratory animal anaesthesia.* Academic Press, New York.

Helmchen F., Fee M.S., Tank D.W., and Denk W. 2001. A miniature head-mounted two-photon microscope. high-resolution brain imaging in freely moving animals. *Neuron* **31:** 903–912.

Oheim M., Beaurepaire E., Chaigneau E., Mertz J., and Charpak S. 2001. Two-photon microscopy in brain tissue: parameters influencing the imaging depth. *J. Neurosci. Methods* **111:** 29–37.

Stosiek C., Garaschuk O., Holthoff K., and Konnerth A. 2003. In vivo two-photon calcium imaging using multicell bolus loading (MCBL). *Proc. Natl. Acad. Sci.* **100:** 7319–7324.

Svoboda K., Denk W., Kleinfeld D., and Tank D.W. 1997. In vivo dendritic calcium dynamics in neocortical pyramidal neurons. *Nature* **385:** 161–165.

CHAPTER 88

A Practical Guide: In Vivo Calcium Imaging in the Fly Visual System

Alexander Borst,[1] Winfried Denk,[2] and Juergen Haag[1]

[1]*Max-Planck-Institute for Neurobiology, Martinsried, Germany;* [2]*Max-Planck-Institute for Medical Research, Heidelberg, Germany*

The large motion-sensitive interneurons (lobula-plate tangential cells) of the fly represent a unique system for the in vivo study of dendritic integration and calcium signaling in individual neurons. In the blowfly, *Calliphora vicina*, there exist ~60 of these neurons per lobula plate. Each can be identified by its visual response characteristics, such as receptive field location and preferred direction of motion, as well as by its typical anatomy (Hausen 1984). The large dendrites of these neurons receive signals from thousands of retinotopically organized elements that are sensitive to the local direction of visual motion. Tangential neurons have been studied in detail with respect to the question of how their active and passive membrane properties (Borst and Haag 1996; Haag et al. 1997) give rise to their visual response properties (Haag and Borst 1996; Single et al. 1997; Haag et al. 1999). Lobula plate tangential cells are particularly well suited for optical recording in vivo (Borst and Egelhaaf 1992). These cells are located directly beneath the rear surface of the brain and are directly accessible upon opening the head capsule. They are visible in whole mount after fluorescent labeling, even in tissue that has not been cleared. In this chapter, we discuss the essential features of the lobula plate tangential cell experimental system and present practical guidelines, as well as recent technical improvements, for studying visually induced calcium dynamics.

MATERIALS

In this setup, a custom-built two-photon microscope was used with a novel stimulation device (Fig. 1). Briefly, the microscope consists of the following components: a 5-W-pumped Ti:sapphire laser (MaiTai, Spectra Physics), a Pockels cell (Conoptics), scan mirrors including drivers (Cambridge Technology), a scan lens (4401-302, Rodenstock), a tube lens (MXA 22018, Nikon), a dichroic mirror (DCSPR 25.5x36, AHF Tuebingen), and a 40x water immersion lens (Zeiss). The lens can be moved along all three axes by a stepper motor-driven micromanipulator (MP285-3Z, Sutter Instruments). Emitted light is filtered by a bandpass (HQ 535/50, Chroma) and collected by a multi-alkali photomultiplier (R6357, Hamamatsu). Visual stimuli are provided by a computer via a video projector (3M), filtered by a short pass (10SWF-450), and imaged by a total of three lenses in the path so that the pattern is focused into the reticule plane of the eyepiece. An adjustable aperture in the beam path is focused onto the center of eye-surface curvature. The image can be rotated in two axes around the fly by an articulated arm-like arrangement, which is driven by two stepper motors (Owis). The whole system is controlled by custom-written software.

PROCEDURE A

Whole-Mount Preparation

Female blowflies were briefly anesthetized with CO_2 and mounted ventral-side up with wax on a small preparation platform. The head capsule was opened from behind; the trachea and air sacs, which normally cover the lobula plate, were removed. To eliminate movements of the brain caused by peristaltic contractions of the esophagus, the proboscis of the animal was cut away and the gut was pulled out. This allowed intracellular recordings that were routinely stable for up to one hour.

PROCEDURE B

Cell Labeling

For electrophysiological recordings and intracellular injection of the calcium indicator Calcium Green (Ca-Green I hexapotassium salt, Molecular Probes), electrodes were pulled on a Brown-Flaming micropipette puller (P-97, Sutter Instruments) using thin-wall glass capillaries with an outer diameter of 1 mm (GC100TF-10, Clark). When filled with 2 M potassium acetate/0.5 M KCl/8.8 mM Ca-Green, they had resistances of about 30–40 MΩ. Cells were labeled by applying a hyperpolarizing current of –3 nA for about 5 minutes. A SEC-10L-amplifier (npi-electronics) was used throughout the experiments and was operated in the Bridge or DCC mode.

PROCEDURE C

Imaging

Despite the advantages that fly visual interneurons offer for optical recording in general, specific problems arise when trying to record visually evoked responses. This is because some of the intense light needed for fluorescence excitation almost inevitably reaches the photoreceptor and thus interferes with visual stimulation. This is true even if care is taken to illuminate only a small field of view, since the neuropil of the lobula plate strongly scatters light. Although with careful experimental design (Borst and Single 2000), visually evoked responses can be recorded, rather bright visual stimulation has to be used to compete successfully with photoreceptor activation by fluorescence excitation light. This effect can, however, be virtually eliminated by the use of two-photon microscopy (Denk et al. 1990). Here, pulsed infrared light is used to cause two-photon excitation, which, due to the quadratic intensity dependency, is limited to the focal region and thereby provides strong optical sectioning. This approach has been successfully employed for calcium recordings in the mammalian retina (Denk and Detwiler 1999; Euler et al. 2002). The lack of photoreceptor activation, even by intense infrared light shone into the back of the fly's head, was confirmed using electrophysiological recordings (no membrane voltage change is seen in response to opening the shutter of the two-photon excitation laser).

To avoid generating a false signal in the fluorescence channel, the stimulus light must be prevented from reaching the photomultiplier tube. This can be achieved by using appropriate nonoverlapping wavelength filters for fluorescence and stimulation light. In the system described here, short-pass filters were used, limiting the spectral range of the stimulus light to below 450 nm. The remaining short-wavelength light stimulates the photoreceptors strongly enough due to the UV sensitivity of the fly photo pigments (Hardie 1979). In order to avoid taxing the finite blocking ability of the filters, it is sometimes necessary to concentrate the stimulus light as much as possible. This can be achieved, for example, as follows (Fig. 1). Instead of using a screen onto which the visual pattern is projected, the fly is looking at the pattern through an eyepiece with the eyepoint (the illumination aperture) at the center of the eye curvature. This minimizes the amount of scattered photons and ensures that the light of the stimulus path is entering the fly visual system very efficiently.

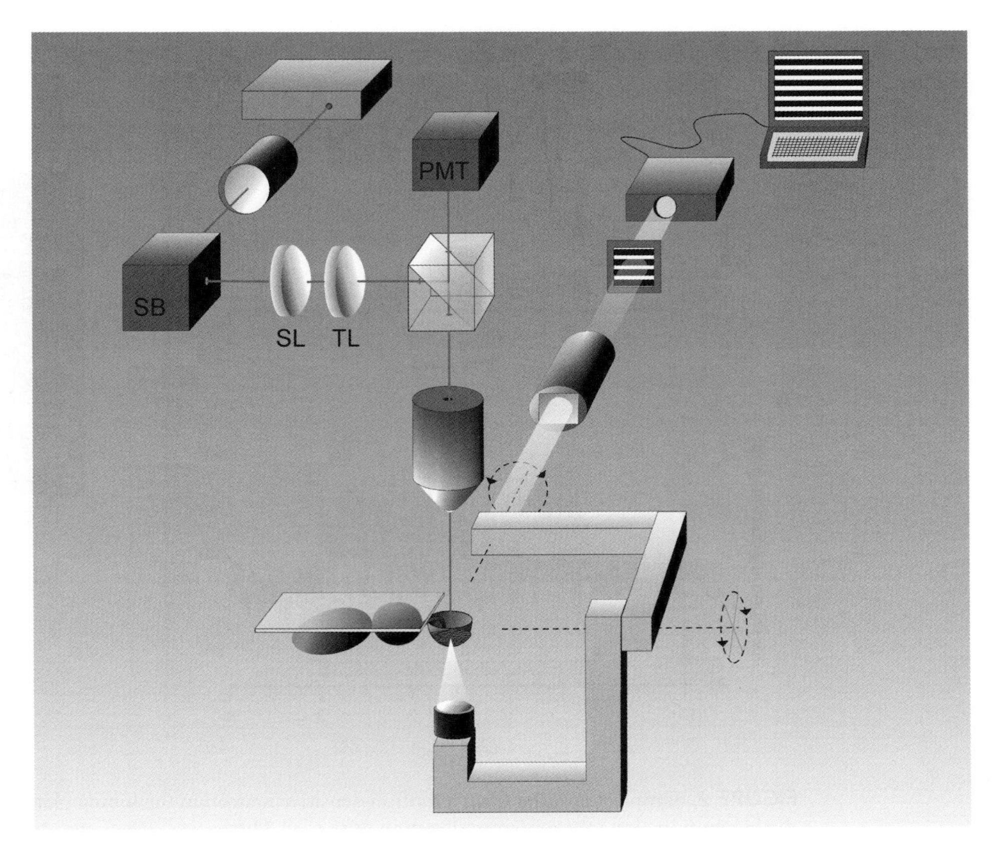

FIGURE 1. Schematic drawing of the setup used for two-photon calcium imaging and visual stimulation in vivo. The fly is shown under an objective lens, looking down onto a moving grating. From the top, a femtosecond laser pulse excites a fluorophore in the visual interneuron with the emitted photons being collected by a photomultiplier. (SB) Scan box, (SL) scan lens, (TL) tube lens, (PMT) photomultiplier tube.

[Ca^{2+}] MEASUREMENTS

Figure 2 gives an example of an in vivo experiment on a neuron that is sensitive to vertical image motion, a so-called VS2 cell. To the right, an image stack (50 images, 256 x 256 pixels, 1-μm steps, 200-μm image width) is combined into a single image ("extended focus"). The axon can be seen running from the center of the image to the left. The main dendrite extends from the upper right to the lower left corner of the image. In Figure 2a, the relative change in fluorescence, a measure of the cytosolic calcium concentration ([Ca^{2+}]), is shown during visual motion (bar below response traces) along the preferred direction of the cell. The optical signal is derived from a small dendritic branchlet of the cell. The [Ca^{2+}] rises steeply at the onset of motion, is modulated at the stimulus frequency in addition to the DC increase, and decays slowly after motion has stopped. Previous analyses of the mechanisms underlying such changes in free cytosolic calcium in lobula plate tangential cells have revealed that Ca^{2+} enters the cytosol from outside the cell, rather than being released from intracellular stores (Oertner et al. 2001). This influx is mainly voltage-dependent (Haag and Borst 2000) but has a minor component that enters through the nicotinic acetylcholine receptor (Single and Borst 2002). Acetylcholine is the main excitatory neurotransmitter in the insect central nervous system, as has also been demonstrated for the tangential cells (Brotz and Borst 1996). In Figure 2b, the axonal membrane potential is displayed, together with the simultaneously recorded optical signal. As shown, the neuron fires a barrage of spikes in addition to being slightly depolarized (Haag and Borst 1996). Unlike the

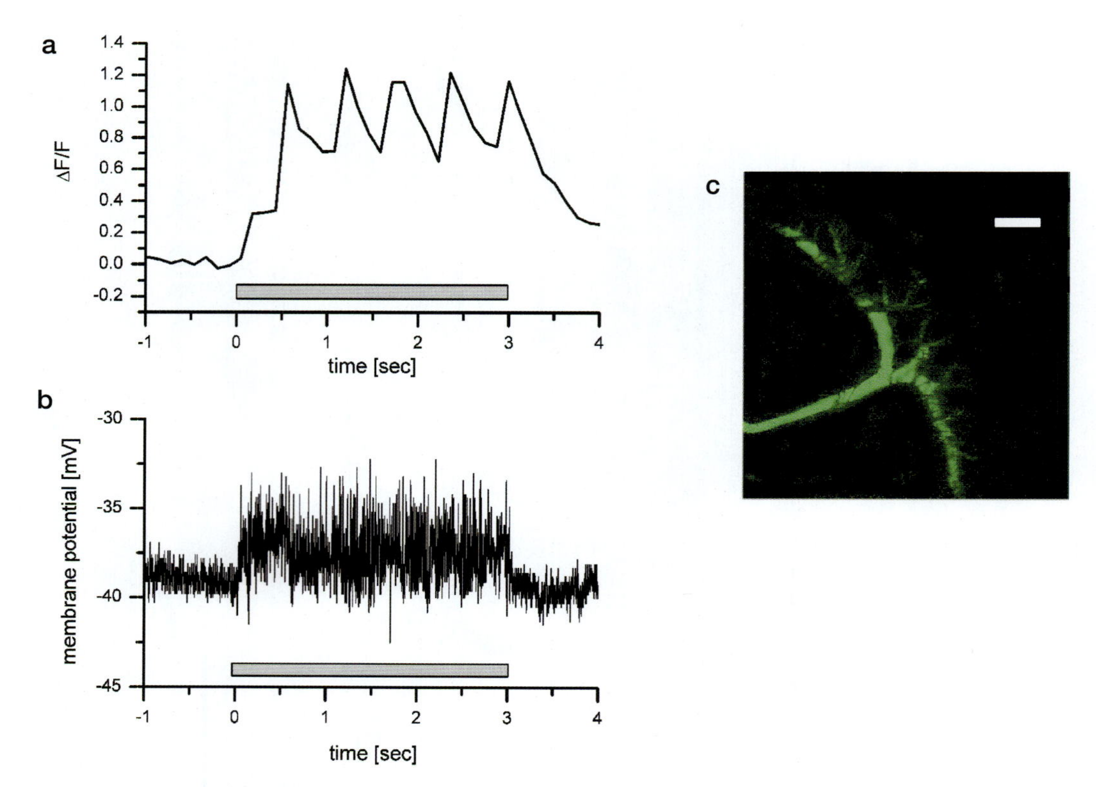

FIGURE 2. Sample recording from a motion-sensitive neuron in the lobula plate (VS2-cell). (*a*) Calcium signals during motion along the preferred direction of the cell. Fluorescence was stimulated at 850 nm. (*b*) Axonal membrane potential recorded simultaneously. (*c*) Projected two-photon image stack ("extended focus"). Bar, 20 μm. Stimulus parameters: mean luminance 500 cd/m^2, contrast 98%, spatial pattern wavelength 8.3°, stimulus velocity 10.8°/sec, and temporal frequency 1.3 Hz.

optical signal, the electrical signal is not modulated with the temporal stimulus frequency, i.e., the grating velocity divided by the spatial wavelength. This is exactly as predicted by the correlation-type of motion detector model (Reichardt 1961; Borst and Egelhaaf 1989): According to this model, local signals reflect both the direction of visual motion and pattern properties. Only after spatial integration do modulations cancel due to the phase shifts across the receptive field of the neuron, leaving a purely directionally selective signal in the axon of the cell (Egelhaaf et al. 1989; Single and Borst 1998). The great advantage of our current instrumentation over previous systems (Borst and Single 2000) lies in the fact that stimulus parameters can be chosen at will. In the experiment shown in Figure 2, the mean luminance level was ~500 cd/m^2. Furthermore, in contrast to 1P excitation, no reduction of stimulus contrast occurs due to stray excitation light.

REFERENCES

Borst A. and Egelhaaf M. 1989. Principles of visual motion detection. *Trends Neurosci.* **12:** 297–306.

———. 1992. In vivo imaging of calcium accumulation in fly interneurons as elicited by visual motion stimulation. *Proc. Natl. Acad. Sci.* **89:** 4139–4143.

Borst A. and Haag J. 1996. The intrinsic electrophysiological characteristics of fly lobula plate tangential cells. I. Passive membrane properties. *J. Comput. Neurosci.* **3:** 313–336.

Borst A. and Single S. 2000. In vivo calcium imaging in blowfly CNS. In *Imaging living cells: A laboratory manual* (ed. R.Yuste et al.), pp. 44.1–44.6. Cold Spring Harbor Laboratory Press, Cold Spring Harbor, New York.

Brotz T. and Borst A. 1996. Cholinergic and GABAergic receptors on fly tangential cells and their role in visual motion detection. *J. Neurophysiol.* **76:** 1786–1799.

Denk W. and Detwiler P.B. 1999. Optical recording of light-evoked calcium signals in the functionally intact retina. *Proc. Natl. Acad. Sci.* **96:** 7035–7040.

Denk W., Strickler J.H., and Webb W.W. 1990. Two-photon laser scanning fluorescence microscopy. *Science* **248:** 73–76.

Egelhaaf M., Borst A., and Reichardt W. 1989. Computational structure of a biological motion detection system as revealed by local detector analysis in the fly's nervous system. *J. Opt. Soc. Am. A* **6:** 1070–1087.

Euler T., Detwiler P.B., and Denk W. 2002. Directionally selective calcium signals in dendrites of starburst amacrine cells. *Nature* **418:** 845–852.

Haag J. and Borst A. 1996. Amplification of high-frequency synaptic inputs by active dendritic membrane processes. *Nature* **379:** 639–641.

———. 2000. Spatial distribution and characteristics of voltage-gated calcium currents within visual interneurons. *J. Neurophysiol.* **83:** 1039–1051.

Haag J., Theunissen F., and Borst A. 1997. The intrinsic electrophysiological characteristics of fly lobula plate tangential cells. II. Active membrane properties. *J. Comput. Neurosci.* **4:** 349–368.

Haag J., Vermeulen A., and Borst A. 1999. The intrinsic electrophysiological characteristics of fly lobula plate tangential cells. III. Visual response properties. *J. Comput. Neurosci.* **7:** 213–234.

Hardie R.C. 1979. Electrophysiological analysis of fly retina. I: Comparative properties of R1-6 and R7 and 8. *J. Comp. Physiol.* **129:** 19–33.

Hausen K. 1984. The lobula-complex of the fly: Structure, function and significance in visual behaviour. In *Photoreception and vision in invertebrates* (ed. M.A. Ali), pp. 523–559. Plenum Press, New York.

Oertner T.G., Brotz T., and Borst A. 2001. Mechanisms of dendritic calcium signaling in fly neurons. *J. Neurophysiol.* **85:** 439–447.

Reichardt W. 1961. Autocorrelation, a principle for the evaluation of sensory information by the central nervous system. In *Sensory communication* (ed W.A. Rosenblith), pp. 303–317. M.I.T. Press and Wiley, New York, London.

Single S. and Borst A. 1998. Dendritic integration and its role in computing image velocity. *Science* **281:** 1848–1850.

———. 2002. Different mechanisms of calcium entry within different dendritic compartments. *J. Neurophysiol.* **87:** 1616–1624.

Single S., Haag J., and Borst A. 1997. Dendritic computation of direction selectivity and gain control in visual interneurons. *J. Neurosci.* **17:** 6023–6030.

CHAPTER 89

Intrinsic Signal Imaging in the Neocortex: Implications for Hemodynamic-based Functional Imaging

Amiram Grinvald, Dahlia Sharon, Hamutal Slovin, and Ivo Vanzetta

Department of Neurobiology, The Weizmann Institute of Science, Rehovot, 76100 Israel

Optical imaging based on intrinsic signals (Grinvald et al. 1986) has provided a new level of understanding of the principles underlying cortical development, organization, and function. In the best cases, it provides a spatial resolution of 20 μm for mapping of cortical columns in vivo. This chapter provides a brief review of the development of this technique, the types of applications that have been pursued, and the general implications of some findings for other neuroimaging techniques based on hemodynamic responses, e.g., functional magnetic resonance imaging (f-MRI).

In the mammalian neocortex, cells that perform a given function or share common functional properties are often grouped together in cortical columns running from pia to white mater (Mountcastle 1957; Hubel and Wiesel 1962) (e.g., the orientation and ocular dominance columns of the visual cortex). Without knowing the function performed by each group of neurons, it is unlikely that the principles underlying the neural code and its implementation will be discovered. Therefore, attaining an understanding of the functional organization of a given cortical area is a key step toward revealing the fundamental mechanisms of its information processing. Thus, of special importance are experimental methods that allow the visualization of the functional organization of cortical columns, particularly methods providing high spatial and temporal resolution. To date, single- or multiunit extracellular recording techniques have provided the best tools for the study of the functional response properties of single cortical neurons and their synaptic interactions. Some studies have even used single-unit techniques for mapping the functional organization of larger cortical areas (Swindale et al. 1987), but since these techniques are not optimal for mapping, these studies required formidable efforts. Although multi-electrode techniques are more promising, the size and placement of these electrode arrays still pose severe problems.

Several imaging techniques that yield information about the spatial distribution of active neurons in the brain have been developed. Each of these techniques has significant advantages as well as limitations. For example, whereas the 2-deoxyglucose method (2-DG) permits postmortem visualization of active brain areas, or even of single cells, 2-DG is a one-shot approach: Only a single stimulus condition in a single animal can be assayed (the two-isotope 2-DG method permits the mapping of activity resulting from two stimulus conditions). Positron-emission tomography (PET) and f-MRI offer spectacular three-dimensional localization of active regions in the functioning human brain, but both have low spatial resolution. Other imaging techniques have also been applied in vivo with some success, but still suffer from either a limited spatial resolution, a limited temporal resolution, or both. Among these methods are radioactive imaging of changes in blood flow, electroencephalography, magnetoencephalography, and thermal imaging.

Application of optical imaging of intrinsic signals to studies of cortical development is reviewed in Chapter 73. Real-time optical imaging based on voltage-sensitive dyes in the neocortex is reviewed in Chapter 90. This review summarizes, in the main, work performed in our laboratory. More general reviews, describing earlier work carried out on simpler preparations, have been published elsewhere (Cohen 1973; Salzberg et al. 1986; Grinvald et al. 1988). More extensive reviews of optical imaging have also been published (Bonhoeffer and Grinvald 1996; Grinvald et al. 1988, 1999).

OVERVIEW

Currently, the best method for imaging the functional architecture of the cortex is based on slow changes in the intrinsic optical properties of active brain tissue. This method permits visualization of active cortical regions at a spatial resolution better than 50 μm and is immune to some of the problems associated with the use of extrinsic probes. The sources for these activity-dependent intrinsic signals include both changes in physical properties of the tissue itself that affect light scattering, and changes in the absorption, fluorescence, or other optical properties of intrinsic molecules. The fact that in many tissues small intrinsic optical changes are associated with metabolic activity has been known since the pioneering experiments of Kelin and Millikan on the absorption of cytochromes (Kelin 1925) and hemoglobin (Millikan 1937). The first optical recording of neuronal activity was made almost 50 years ago by Hill and Keynes (1949), who detected light-scattering changes in active nerves. Changes in absorption or fluorescence of intrinsic chromophores were extensively investigated by Chance and his colleagues (Chance et al. 1962) and by Jobsis and his colleagues (Jobsis 1977; Jobsis et al. 1977; for reviews, see Cohen 1973; Mayevsky and Chance 1982). Although intrinsic optical signals are usually very small or very noisy, it is still possible to use optical detection of intrinsic signals for the imaging of the functional architecture of the cortex (Grinvald et al. 1986; Frostig et al. 1990; Ts'o et al. 1990).

The basic experimental setup for optical imaging experiments is shown in Figure 1. The anesthetized animal is held in a stereotaxic frame (not shown). The brain is illuminated with flexible light guides, and digital pictures are acquired by the camera, which views the exposed brain through a cranial window. The data are analyzed either on the computer controlling the experiment or on a separate analysis computer, and the results are displayed on a color monitor. (For detailed practical guidelines, see Grinvald et al. 1999.) Significant methodological advances, which may reduce the data acquisition time for some types of maps, have been reported recently (Kalatsky and Stryker 2003).

The initial optical imaging studies investigated the well-known structural elements of the functional architecture such as ocular dominance in primary visual cortex and the "stripes" in V2 (Ts'o et al. 1990; Roe and Ts'o 1995), or the pinwheel-like organization of orientation preference (Bonhoeffer and Grinvald 1991, 1993; Bonhoeffer et al. 1995; Das and Gilbert 1995, 1997). Subsequently, methodological improvements made it possible to investigate more subtle features of cortical organization, such as direction-selective columns or spatial frequency columns (Malonek et al. 1994; Shmuel and Grinvald 1996; Weliky et al. 1996; Shoham et al. 1997). Similar progress has been obtained in exploring other visual areas. Ts'o and collaborators (1991, 1993) and Malach and his colleagues (1994) succeeded in imaging the separate pathways in thin, thick, and pale stripes in monkey V2. It has even been possible to demonstrate functional columns in visual areas farther up the processing stream, in areas V4 (Ghose and Ts'o 1997) and MT (Malonek et al. 1994; Malach et al. 1997). Tanaka and his colleagues used this method to image the functional organization in the inferotemporal (IT) area, one of the final stages of the visual pathway critical for object recognition (Wang et al. 1996). High-resolution maps were reported in the tree shrew (Bosking et al. 1997) and the ferret (see, e.g., Basole et al. 2003).

Although most optical imaging studies have been carried out in the visual cortex, this is by no means the only sensory system that can be studied using this method. Indeed, this method has lately proved to be a useful tool for investigating functional architecture in the somatosensory cortex of the

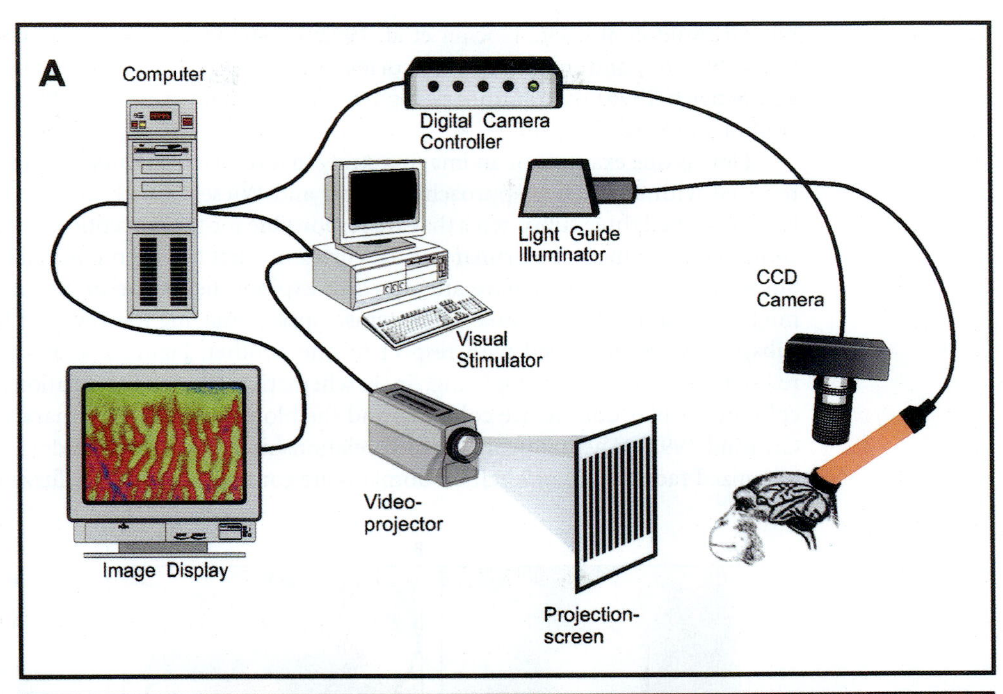

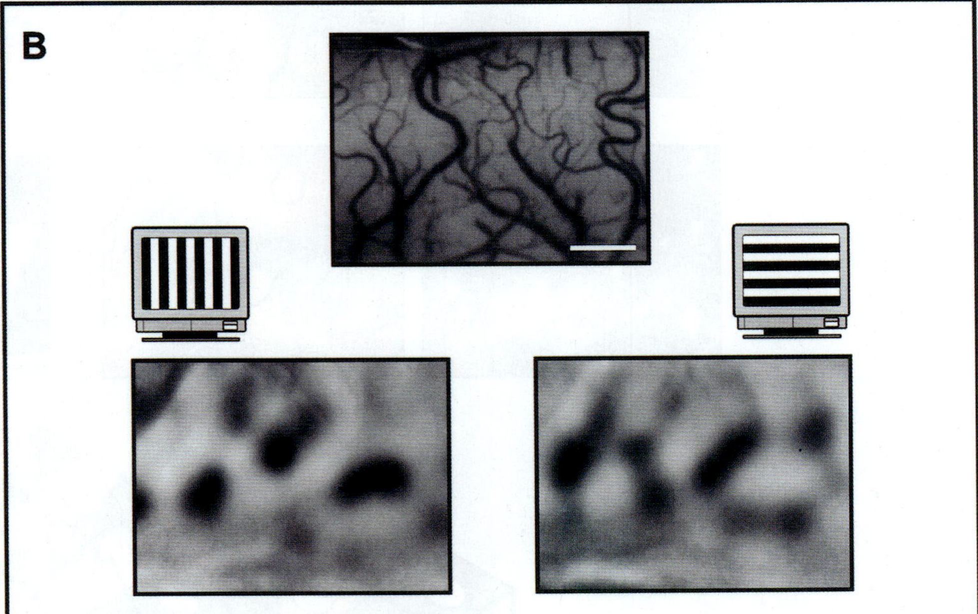

FIGURE 1. Optical imaging of functional maps in vivo. (*A*) The setup. Digital CCD images are taken of the animal's exposed cortex, which is sealed in a cranial window. The cortex is illuminated with light of 605 nm wavelength, and a camera is used to acquire digital images of the cortex. The acquired images are digitized in a camera controller, which transfers the data to a computer that controls the entire experiment. Functional maps are subsequently analyzed, and are displayed on a color monitor. To determine the quality of the maps during the imaging sessions, the data can be sent to a second computer for detailed, quasi on-line analysis. (*B*) Imaging functional architecture. An activity map for one orientation is obtained straightforwardly, by dividing the image captured during presentation of this orientation by the average of the images captured during presentation of all orientations. (*Top*) The image of the cortical surface, illuminated with green light to emphasize the vasculature. (*Bottom*) Activity maps evoked by visual stimulation with horizontal and vertical gratings. Bar, 1 mm. (*A*, Modified from Ts'o et al. 1990; *B*, modified from Bonhoeffer and Grinvald 1991.)

rat (Grinvald et al. 1986; Gochin et al. 1992; Frostig et al. 1994) and of the monkey (Shoham and Grinvald 2001) and in the auditory cortex of the guinea pig (Bakin et al. 1996) and the gerbil (Hess and Scheich 1996). Extraordinary high-resolution maps were also described in the olfactory bulb (Rubin and Katz 1999).

Here is one example of an imaging study that resolved an outstanding question that could not be resolved with alternative approaches. Hubel and Wiesel (1977) and Livingstone and Hubel (1984) have described three subsystems that are responsible for the perception of shape, depth, and color that combine in the macaque primary visual cortex in what has been referred to as the "ice-cube" model. Since then, many groups have attempted to explore the nature of any geometric relationships that might exist among these three subsystems. Optical imaging has revealed how the various columnar subsystems are organized with respect to one another. Figure 2 shows the experimental data and revised ice-cube model depicting, in a schematic fashion, the relationships between orientation columns, ocular dominance columns, and the blobs in monkey primary visual cortex (Bartfeld and Grinvald 1992). The following detailed relationships have been found: (1) Orientation-preference is organized radially; (2) orientation domains are continuous and have fuzzy boundaries; (3) iso-orien-

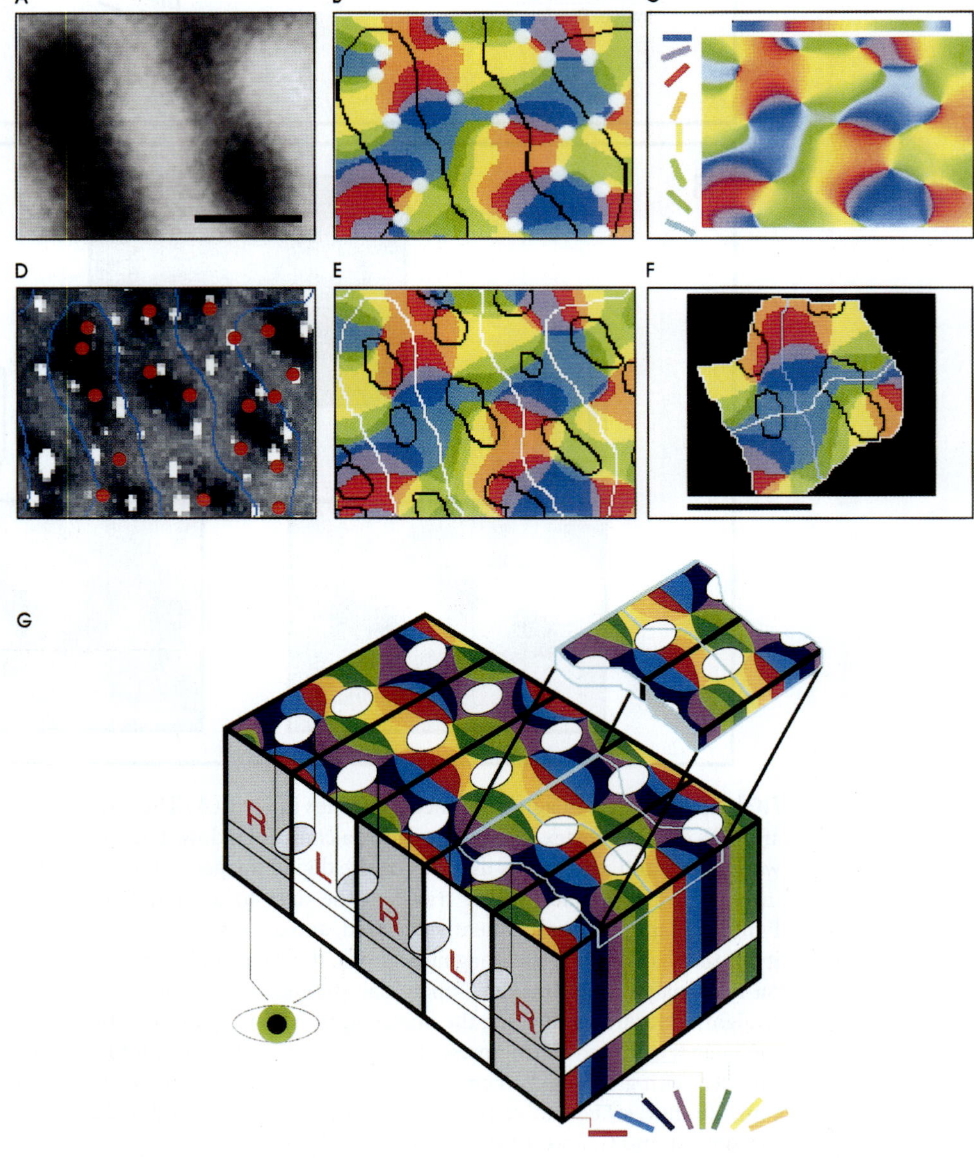

FIGURE 2. *(See facing page for legend.)*

tation lines tend to cross ocular dominance borders at 90°; (4) orientation pinwheels are centered on ocular dominance columns; (5) blobs are centered on ocular dominance columns; (6) the centers of the blobs and the centers of the orientation pinwheels are segregated; (7) there is a regular mosaic-like organization for each type of functional domain, without an overall pattern of repeating hypercolumns, each of which contains all types of functional domains. Similar results in the monkey have been independently reported by Blasdel (1992a,b) and by Obermayer and Blasdel (1993), using voltage-sensitive dyes, and in the cat by Hubener and his colleagues (1997), also imaging intrinsic signals.

Chronic Optical Imaging

Acute optical imaging utilizing anesthetized animals lasts from 1 to 6 days. Many interesting questions cannot be explored by performing acute experiments and require recording activity from behaving animals during performance of behavioral tasks. Particularly promising with regard to the feasibility of chronic optical imaging was the finding that cortical maps can be obtained through the intact dura or even through a thinned but closed skull (Frostig et al. 1990; Masino and Frostig 1996; Bosking et al. 1997; Schuett et al. 2002). These results were achieved using near-infrared light, which penetrates the tissue considerably better than does light of shorter wavelength.

Exploration of Behavior

It has also been demonstrated that optical imaging based on intrinsic signals can be used to investigate the functional architecture of the cortex in awake behaving animals (Grinvald et al. 1991). Such studies are of great interest, first because studies of many higher brain functions cannot be performed utilizing anesthetized animals. Second, studies of anesthetized preparations cannot determine whether the functional organization of a given cortical area is influenced by the behavior of the animal. Implantation of a transparent dura substitute has facilitated recoding from the cranial window for a long period of time (Arieli et al. 2002). Thus, ocular dominance and orientation columns were repeatedly imaged in the behaving macaque for a period of up to 1 year (Shtoyerman et al. 2000; Slovin et al. 2002). This progress indicates that it will be possible to successfully implement optical imaging techniques for the study of higher brain functions in awake behaving primates.

FIGURE 2. Relationship between pinwheels, ocular dominance columns, and blobs in primary visual cortex of the macaque monkey. (*A*) The optical map of ocular dominance from ~1.5 x 1 mm portion of striate cortex. The dark bands represent columns dominated by input from the right eye and the light bands by input from the left eye. Bar, 500 μm. (*B*) The borders of the ocular dominance columns (black lines taken from panel *A*) were overlaid onto the discrete "angle map" for orientation preference. The pinwheel centers were marked with circles (~67% were centered on ocular dominance columns). (*C*) The continuous angle map of orientation preference. The same unsmoothed digital data used in panel *B* are depicted using the color scale shown above the map. The oriented color bars at the right indicate the color code. (*D*) The cytochrome oxidase (CO)-rich blobs were marked on the histological photograph that corresponded exactly to the cortical area that was optically imaged. (*E*) Angle map from *B* and two overlays showing the relationships between blobs, iso-orientation domains, and ocular dominance columns. (*F*) Two adjacent fundamental modules, magnified from *E*. (*Bottom*) (*G*) The revised ice cube model, schematically illustrating the relationships found in the above maps. Black lines mark the borders between columns of neurons that receive signals from different eyes. This segregation is partially responsible for depth perception. White ovals represent groups of neurons responsible for color perception (blobs). The "pinwheels" are formed by neurons involved in the perception of shape, with each color marking a column of neurons responding selectively to a particular orientation in space. Note that both the blobs and the centers of the pinwheels lie at the center of the R or L columns. The iso-orientation lines (appearing as a border between two colors) tend to cross borders of ocular dominance columns (*black lines*) at right angles. The top "slice" above the "ice cube" model depicts two adjacent fundamental modules (400 μm x 800 μm). Each module contains a complete set of about 60,000 neurons, processing all three features of orientation, depth, and color. This scheme is oversimplified in that clockwise and counterclockwise pinwheels are perfectly interconnected. In reality, this relationship does not exist and fundamental modules (hypercolumns) do not form an orderly mosaic.

Developmental Studies

An area of investigation where chronic optical imaging has been most fruitfully applied is the study of postnatal experience-dependent plasticity and development in the neocortex (see Kim and Bonhoeffer 1994; Chapman et al. 1996; Godecke and Bonhoeffer 1996; Crair et al. 1997a,b, 1998; Chapman and Bonhoeffer 1998; Chapter 28). Such studies require longitudinal experiments, to determine changes in the cortical functional architecture over long periods of time. The technique of optical imaging is particularly useful in these studies, since it offers both the required spatial resolution and the ability to perform prolonged, comparative studies.

Human Clinical Studies

Another important application of optical imaging is mapping functional borders during human neurosurgery. It has been reported (MacVicar et al. 1990; Haglund et al. 1992) that optical imaging can be used to visualize activation of the human cortex in response to bipolar stimulation and during speech. More recently, Shoham and Grinvald performed optical imaging studies on humans to delineate borders of functional areas during neurosurgery. In a preliminary report (Shoham et al. 2001), they described the optical mapping of the human hand representation in the somatosensory cortex, which was subsequently confirmed during the surgery with differential EEG recording from a matrix of 16 surface electrodes. Godecke and colleagues applied the optical imaging method to map neocortical epileptic foci. In both these studies, it was found that the noise associated with the optical imaging of the human cortex was much larger than in animal experiments using the cranial window technique. Both groups also observed a large amount of activity-independent vascular noise. Despite these technical difficulties, it appears that optical imaging of functional borders in the human cerebral cortex is, in some cases, feasible (Cannestra et al. 1998). Thus, optical imaging may assist neurosurgeons in precisely locating the foci of epileptic events (Schwartz and Bonhoeffer 2001), or the borders of functional areas close to the site of surgical procedures, as well as obtaining high-resolution functional maps from that region.

Finally, can one hope to image human brain function noninvasively, using light, through the intact skull? The pioneering experiments by Jobsis (1977) on the cat, using transillumination with near-infrared light, and subsequent studies on human infants (Wyatt et al. 1986, 1990) have suggested that progress can be made in this direction. Furthermore, in a very innovative experiment, Chance and his colleagues (1993a,b) showed that, although light reflected from the cortex is attenuated by the thick skull, it can nevertheless be detected using photomultipliers. Kato and colleagues (1993) employed an array of illuminator-detector pairs (optodes) to obtain low-resolution optical images of human cerebral cortex. Evoked responses in the auditory and motor cortex were detected optically, using near-infrared light, and confirmed with electroencephalography. This report, and others (Hoshi and Tamura 1993; Gratton et al. 1995, 1997; Gratton 1997; Maclin et al. 2004), encourage the belief that it will be possible to design relatively inexpensive optical imagers for exploring cortical functional organization in human subjects, offering a spatial resolution of a few millimeters. Furthermore, a fast component of the intrinsic signal has been resolved by the Gratton group. Thus, it seems possible that this novel mode of optical imaging will offer the advantage of combining a spatial resolution comparable to that of PET and f-MRI, with a millisecond temporal resolution comparable to that of electroencephalography and magnetoencephalography (Wolf et al. 2003; Maclin et al. 2004). The simplicity and relatively low cost of such optical devices justify additional efforts in this direction.

SOURCES OF INTRINSIC SIGNALS

To optimally and faithfully image functional maps in the neocortex, it is crucial to understand the mechanisms underlying the intrinsic signals and their relation to the electrical activity of neurons. More than a century ago, Roy and Sherrington (1890) postulated that "the brain possesses an intrinsic mechanism by which its vascular supply can be varied locally in correspondence with local variations of functional activity." Indeed, in 1880, Angelo Mosso reported changes in cerebral blood flow produced by brain activity in patients with skull defects (Mosso et al. 1880). More modern imaging techniques have clearly demonstrated that there is a strong coupling between neuronal activity, local metabolic activity, and blood flow (Kety 1950; Lassen and Ingvar 1961; Sokoloff 1977; Raichle et al. 1983; Fox et al. 1986; Grinvald et al. 1986). More recently, Ogawa and his colleagues (1992) and Kwong et al. (1992) have found that the intrinsic signal is useful for functional brain mapping with MRI. However, only recently have the exact spatiotemporal characteristics of this coupling been revealed by optical imaging.

Although the intrinsic signal has different components, which originate from different sources, functional maps obtained at different wavelengths are very similar. Therefore, all of these components can be used for functional mapping (Frostig et al. 1990). Malonek and Grinvald (1996) used *Optical Imaging Spectroscopy*, a new technique providing simultaneous spectral information from many cortical locations in the form of a spatio-spectral image. The images obtained with imaging spectroscopy show the spectral changes at many wavelengths for each cortical point (Y- space versus wavelength λ). Using this technique, Malonek and Grinvald measured the spatial, temporal, and spectral characteristics of light reflected from the surface of the visual cortex following natural stimulation. Obtaining cortical spectra in this manner helps in (1) identifying signal sources by curve-fitting to known spectra; (2) determining the spatial precision of the signals; (3) evaluating the dynamics of the signals; (4) exploring the dynamics in different vascular compartments. Additional technical details were published elsewhere (Malonek and Grinvald 1996; Malonek et al. 1997; Vanzetta and Grinvald 1999, 2001; Vanzetta et al. 2004).

From the above studies, the sources of the intrinsic signals in reflection measurements appear to be the result of the following cellular or tissue changes.

- Activity-dependent oxygen consumption affecting hemoglobin saturation
- Changes in blood volume affecting the tissue light absorption
- Changes in blood volume affecting the tissue light scattering
- Changes in blood flow affecting hemoglobin saturation
- Activity-dependent changes in tissue light scattering

The first component of the intrinsic signal originates from activity-dependent changes in the oxygen saturation level of hemoglobin. This change in oxygenation itself contains two different components. The onset of the first component is an early one: an *increase* in the deoxyhemoglobin (Hbr) concentration, resulting from elevated oxygen consumption of the neurons due to their metabolic activity. This causes a darkening of the cortex due to increased absorption by Hbr molecules. This early phase was referred to as the "initial dip" in the f-MRI and related optical imaging literature. This interpretation (Grinvald et al. 1986; Frostig et al. 1990; Malonek and Grinvald 1996; Malonek et al. 1997; Vanzetta and Grinvald 2001; Vanzetta et al. 2004), however, remained controversial (Mayhew et al. 1998) for some time, as discussed below.

The second component of the intrinsic signal originates from an activity-dependent increase in cortical blood volume (CBV), increasing the amount of light absorbed by activated cortical regions in a wavelength-dependent fashion, following the absorption spectra of hemoglobin. Of great interest has been the level of spatial regulation of this component, especially since most current f-MRI studies utilize this component of the hemodynamic response. Previous optical imaging and imaging spectroscopy work in the visual cortex of cats and monkeys (Frostig et al. 1990; Malonek and Grinvald 1996; Shtoyerman et al. 2000; Vanzetta and Grinvald 2001) have reported a much weaker colocalization of CBV changes with the electrically active columns than the [Hbr] increase during the initial dip. This poorer colocalization is revealed by a lower signal-to-noise ratio in spatial maps of activation calculated from the CBV component relative to those from the initial dip. This is especially true for maps in which the response to a single stimulus condition is normalized by the response to a blank ("single-condition" maps). For "differential" imaging, where the response to a stimulus condition is normalized by or subtracted from the response to an "orthogonal" stimulus condition, the maps from the CBV are still somewhat poorer than from the initial dip. However, high-quality functional CBV maps have been obtained in cat and monkey visual cortex with differential imaging (Frostig et al. 1990; Vanzetta and Grinvald 2001). The differential procedure removes much of the CBV noise because it all but eliminates the contribution of the non-stimulus-specific part of the vascular response. These results obviously indicate that some of the CBV changes *do* colocalize with areas of increased neuronal activity.

The third component is delayed: It is an activity-related increase in blood flow, causing a *decrease* in the Hbr concentration. This happens because the blood rushing into the activated tissue contains higher levels of oxyhemoglobin (HbO_2). Another signal component originates from changes in blood volume, which are probably due to local capillary recruitment or a rapid filling of capillaries and dilation of venules in an area containing electrically active neurons. These blood-related components dominate the signal at wavelengths between 400 nm and 630 nm. However, blood volume changes in capillaries compress the cortical tissue, giving rise to a change in light scattering that exists at all wavelengths and exhibits a time course typical of blood volume changes rather than that of light-scattering changes that originated from electrical activity per se. This last significant component of the intrinsic signal arises from changes in light scattering that accompany cortical activation (Cohen et al. 1968; Tasaki et al. 1968; Grinvald et al. 1982). These changes are caused by ion and water movement, expansion and contraction of extracellular spaces, capillary expansion, or neurotransmitter release (for review, see Cohen 1973; see also Chapter 73). Light-scattering components become a significant source of intrinsic signals above 630 nm, and probably dominate the intrinsic signals in the near-infrared region above 800 nm.

The intrinsic signals that can be measured from the living brain are small. In optimal cases, the change in light intensity due to neuronal activity is no more than 0.1–6% of the total intensity of the reflected light, depending on wavelength. This means that intrinsic signals cannot be seen with the naked eye and that they have to be extracted from the cortical images with the appropriate data acquisition and analysis procedures. A major problem is that the biological noise associated with these measurements is in many cases larger than the signals themselves. Therefore, it is crucial to employ the proper procedures to extract the small signal of interest from the raw data. Such procedures have been developed, yielding high-resolution functional maps (for details, see Grinvald et al. 1999). To demonstrate the reliability of the data, the reproducibility of the optical maps obtained from the same area of cortex must be verified. The high degree of reproducibility that has been observed (see, e.g., Bonhoeffer and Grinvald 1993) gives confidence in the precision and reliability of cortical maps obtained in optical imaging. Note, however, that reproducibility of optical maps is a necessary, but not a sufficient, criterion. Blood vessels can give rise to signals that are reproducible but that do not correspond to neuronal activity. Therefore, electrophysiological verification is required to confirm new findings. Combining electrophysiology with optical imaging is described below (see Combining Intrinsic Optical Imaging with Other Techniques).

THE INITIAL DIP: EVIDENCE FOR EARLY DEOXYGENATION

Initially, the very existence of the initial dip in the intrinsic signal was controversial (Mayhew et al. 1998; Kohl et al. 2000; Lindauer et al. 2001; Martin et al 2002). In view of the substantial implications of this dip for mechanisms underlying the neurovascular coupling, cellular metabolism, and high-resolution functional imaging, it was constructive to verify the existence of the initial dip with raw data that did not require spectroscopic models. In this line, the raw action spectra, i.e., the amount of light reflected from active cortex as a function of wavelength, obtained from anesthetized cat and awake monkey in our laboratory proved useful (Vanzetta and Grinvald 2001). The late action spectra clearly displayed the two HbO_2 peaks, unambiguously showing an increased concentration of HbO_2 probably due to the delayed flow of more oxygenated blood. However, the double peak is evident only at late times of about >1 sec after the onset of electrical activity. At earlier times (e.g., ~0.6 sec), the action spectra exhibit only a rounded shape, which cannot be explained without assuming an early increase in the concentration of Hbr. Thus, these raw action spectra suggest that there is an early increase in Hbr concentration preceding the increase in HbO_2. This is in line with the observation of Mayhew et al. (1999, 2000) and Jones et al. (2001), who found an increase in Hbr concentration without a concomitant decrease in HbO_2, in this case in the rat. The initial dip has been confirmed with more direct measurement of intravascular oxygen concentration measurements, based both on phosphorescence quenching (Vanzetta and Grinvald 1999) and oxygen electrodes (Ances et al. 2001). It was also observed in awake subjects (Shtoyerman et al. 2000) and high-field blood-oxygenation level-dependent (BOLD)-f-MRI (Ernst and Henning 1994; Menon et al. 1995; Hu et al. 1997; Logothetis et al. 1999; Yacoub and Hu 1999; Grinvald et al. 2000; Kim et al. 2000ab). Recently, strong support for the interpretation of the origin of the initial dip has been provided by direct measurement of oxygen concentration changes in the cortical tissue itself, using oxygen electrodes (Thompson et al. 2003).

SINGLE-CONDITION MAPPING: IMPLICATIONS FOR FUNCTIONAL IMAGING BASED ON HEMODYNAMIC RESPONSES

The ultimate goal of high-resolution functional brain mapping, at the cortical column level, is single-condition rather than differential mapping, because appropriate orthogonal stimuli are rarely available. It is important to clarify the advantages and limitations of single-condition mapping compared with differential mapping. The spatial resolution of a functional map depends on the spread of a particular imaged signal beyond the site of electrical activity, the signal-to-noise ratio, and the instrument's spatial resolution. For example, the point spread function of the f-MRI signal is large. However, two distinct but adjacent sites, activated on the cortex, can be resolved at distances much smaller than the half-width of the signal spread. This can be done by differential imaging, provided that the signal-to-noise ratio is adequate—that is to say, by subtracting two maps that are known to be orthogonal to each other because they are evoked by stimuli that are known to activate complementary cortical areas. Ocular dominance columns and orientation columns belong to the rare examples where "orthogonal maps" really exist in nature. Using differential imaging, ocular dominance columns in human subjects have already been imaged with BOLD f-MRI (Menon et al. 1997; Menon and Goodyear 1999; Cheng et al. 2001; Goodyear and Menon 2001). The BOLD f-MRI signal is composed of (i) a "local" component, which is stimulus-specific, and probably originates from the capillaries, and (ii) a component referred to as the global signal, which originates from other sources, such as non-localized blood-volume and blood-flow changes that may not carry stimulus-specificity and hemodynamic activation of venules and large draining veins far away from the electrical activation site. For this reason, when two stimuli are known to be orthogonal, in the sense that they activate neurons in complementary cortical patches, comparing the activation patterns of the two stimuli enhances the signal's local component, while eliminating many, although not all, the global components.

There are two major limitations to differential imaging, however. First, differential imaging works perfectly well *only* in cases where the stimuli are orthogonal in the above sense. In the general case, orthogonal stimuli are impossible to define, particularly when it comes to domains dealing with higher cognitive functions. In such cases, subtracting responses from each other may lead to significant misinterpretation of the data. Second, by far the largest signals are from the draining veins, even veins smaller than 100 μm. Such large signals may spread up to 7 mm from retinotopic sites of activation (see, e.g., Shoham and Grinvald 2001). Importantly, the signals from such draining veins may be stimulus-specific and highly reproducible. Therefore, the use of differential imaging, high thresholds, signal averaging, and statistics does not guarantee removal of such artifacts. Single-condition mapping, in which the response to a single stimulus can be imaged directly, thus has great advantages over differential imaging and should be the ultimate goal of high-resolution functional brain imaging.

A significant advantage of optical imaging of the exposed cortical surface is that the spatial resolution of the optics is less than 1 μm, permitting unambiguous identification of blood vessel-derived responses and distinguishing them from cortical activation per se. In addition, because reflected photons are "cheap," they can be collected in large numbers to provide a signal-to-noise ratio that is higher than that accomplished with f-MRI. This allows visualization of the hemodynamic changes in the various cortical vascular compartments, without the need for elaborate image processing or statistical methods: In other words, viewing the results of optical imaging is direct and straightforward. Kim et al. (2000a) took advantage of the optical imaging protocol and have reported a striking technical advance in f-MRI: improving its spatial resolution for direct and selective visualization of individual cortical columns in the anesthetized cat. The authors used the early, negative phase of the BOLD f-MRI signal (the "initial dip"; the deoxygenation phase), to obtain single-condition maps of cortical columns. They further showed that the later, positive hyper-oxygenation phase of the BOLD signal, traditionally used for functional brain mapping, gives less accurate results than the early phase. These results, however, have remained controversial (Kim 2000b; Logothetis et al. 2000). Recently, it has been suggested that perfusion-based f-MRI, imaging presumably CBF changes that are largely limited to the arterioles and capillaries, as well as water exchange with the surrounding tissue, provides single-condition maps (Duong et al. 2001). Contrast-agent-based f-MRI and volume measurements also yield single-condition maps (Harrison et al. 2002). I. Vanzetta et al. (in prep.) confirmed, with optical imaging, that, after removal of vascular signals (e.g., by phase-contrast angiography), "cleaned" CBV-based maps are indeed rather precise (by comparing them with the well-established "gold standard" for mapping ocular dominance columns). Therefore, a component of blood volume change is regulated at the capillary level. Although optical imaging and f-MRI measure different mixtures of vascular parameters and each method exhibits different sensitivity to each component, it appears that single-condition mapping at the fundamental columnar level is feasible, not only with optical imaging, but also with f-MRI.

COMBINING INTRINSIC OPTICAL IMAGING WITH OTHER TECHNIQUES

It is clear that the study of cortical organization and function can greatly benefit from a combination of optical imaging with other techniques, such as tracer injections, electrical recording, and microstimulation. Since optical imaging of functional architecture can quickly and easily provide a picture of how certain functional parameters are represented on the cortical surface, it is an ideal tool for guiding electrophysiological recordings or tracer injections. Furthermore, histological data, such as the dendritic and axonal branching of single cells, can be directly correlated to the functional organization in the very same piece of tissue as depicted in Figure 3 (Malach et al. 1993, 1994, 1997; Kisvarday et al. 1994). Extensive work in this direction has been done by Fitzpatrick and his colleagues (Bosking et al. 1997).

In many cases, it is important to use metal or glass electrodes simultaneously with optical imaging. Furthermore, it is advantageous to be able to target the recording site to functionally identified cortical domains revealed by a previous session of optical imaging. For this purpose, it is useful to have a sealed cranial window coupled to a manipulable electrode. Arieli and Grinvald (2002) have designed such a device, which consists of a chamber with a square cover glass that is larger than the chamber diameter. The cover glass can be moved relative to the base of the cranial window. A hole in the glass covered with a rubber gasket allows the insertion of the electrode into the sealed chamber as illustrated in Figure 3, A and B.

This device has proved to be very useful for targeted tracer injections, electrical confirmation of optically obtained functional maps, and targeted microelectrode recordings. Shmuel and Grinvald (1996) and Shoham et al. (1997) have used it to perform both perpendicular and nearly tangential penetrations, studying the relationships between single-neuron responses and optically imaged functional domains for orientation and direction and for spatial frequency, respectively. The same device has proved equally useful for microstimulation during optical recording experiments. Slovin et al. (2003) and Strick et al. (2003) made use of it for evaluating the extent of cortical activation in response to microstimulation in the awake monkey.

CONCLUSIONS AND OUTLOOK

Optical imaging based on intrinsic signals is a method that allows investigators to map the spatial distribution of functional domains, offering unique advantages. No alternative noninvasive imaging technique for the visualization of functional organization in the living brain provides comparable spatial resolution. Optical imaging based on voltage-sensitive dyes (Chapter 90) does provide a comparable resolution and even offers a superior temporal resolution; however, extrinsic probes must be used. It is this level of resolution that can reveal *where* processing is performed—a necessary step for the understanding of the neural code at the population level. A key strength of intrinsic imaging is that the signals can be obtained in a relatively noninvasive manner over long periods of time. This is particularly important for chronic recordings, which can last up to a year. Optical imaging during chronic experiments allows the study of neocortical development in rodent (e.g., Kim and Bonhoeffer

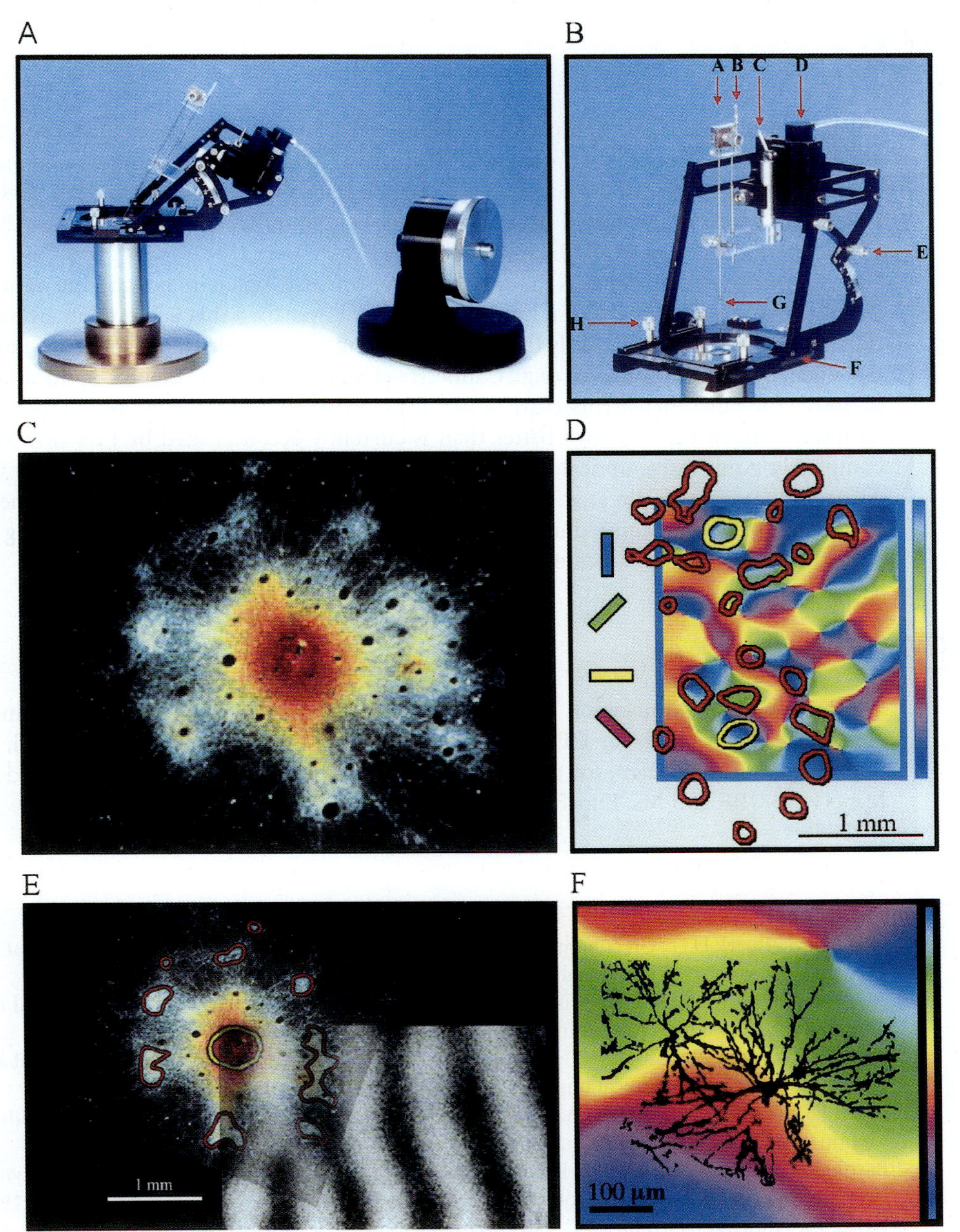

FIGURE 3. Combining optical imaging with other physiological techniques. Cranial window combined with an electrode manipulator for targeted electrical recordings or injections. (*A*) The X-Y and axial microdrive is shown on its stand at an angle of ~60°. The hydraulic microdrive itself is a Narishige hydraulic microdrive MO-11N, which can also be coarsely positioned with a manual manipulator. (*B*) Detail of *A*. (A) Tip of a tungsten microelectrode. (B) Pin for electrical connection with the electrode. (C) Bolt to lock coarse positioning of the manual manipulator. This enables the electrode guide to be advanced manually to penetrate the rubber gasket. Only then is the electrode pushed forward out of the protecting penetration tube and fine movement is achieved using the hydraulic microdrive. (D) Hydraulic microdrive connected to the black tiltable holder. (E) Two screws to lock the microdrive at various angles (range of 30°–90°). (F) Sliding glass. (G) Needle, which holds and protects the microelectrode, to penetrate the rubber gasket. (H) Four screws, which lock the upper part of the microdrive to its X-Y platform. (*Bottom*) (Panels *C–F*) Biocytin injection targeted at a monocular site in macaque monkey area V1. (*C*) Dark-field photomicrograph of a tangential cortical section showing a biocytin injection placed in the center of an ocular dominance column. Note extensive local halo around the injection site followed by clear axonal patches farther away. (*D*) Two injections superimposed on an orientation preference continuous angle map. The two injection sites (*yellow*) and their associated patches (*red*) were superimposed on this map. (*E*) Same injection site as in *C*, but superimposed on an optically imaged map of ocular dominance columns. The injection was targeted at a contralateral eye column (coded *black*) at the top left corner of the optically imaged area. Note the tendency of the biocytin patches to skip over the ipsilateral eye column (coded *white*). (*D*) Local connections and orientation domains. Dendritic arbors of two upper-layer neurons precisely superimposed on an orientation preference angle map taken from that same patch of cortex. Note that the dendritic arbors crossed freely through cortical regions having diverse orientation preferences. (From Arieli and Grinvald 2002 and Malach et al. 1993.)

1994; Crair et al. 1997a,b), and plasticity (Masino and Frostig 1996; Polley et al. 1999, 2004) and higher brain functions in behaving monkeys (Slovin et al. 2002). Whereas sophisticated methods of analysis have already been introduced (see, e.g., Sirovich et al. 1996; Everson et al. 1997; Gabbay et al. 2000) along with more sophisticated data acquisition techniques (Kalatsky and Stryker 2003), additional efforts in this direction would be rewarding.

Another major challenge is the application of intrinsic signal imaging to the human brain. So far, the quality of results obtained in human subjects is lower than that achieved in animal experimentation. Nevertheless, it seems likely that optical imaging will provide clinical benefits. Furthermore, during clinical use in neurological (e.g., Cannestra et al. 2001) and ophthalmic applications (Grinvald et al. 2004), the functional architecture of various sensory cortical areas can be mapped at unprecedented resolution, an order or two better than is currently accomplished by PET or f-MRI. Finally, the completely noninvasive optical imaging through the intact human skull devised by Gratton and his colleagues (1995) (Maclin et al. 2004) may provide an imaging tool offering both the spatial and the temporal resolutions required to expand our knowledge of the principles underlying the remarkable performance of the human cerebral cortex.

ACKNOWLEDGMENTS

We thank Chaipi Wijnbergen, Yoval Toledo, and Dov Etner for their assistance, and our previous coworkers: Frostig, Lieke, Gilbert, Wiesel, Ts'o, Bonhoeffer, Malonek, Shmuel, Glaser, Arieli, Shtoyerman, and Shoham, for their original contribution. This work was supported by grants from Minerva, BMBF, and Ms. Enoch.

REFERENCES

Ances B.M., Buerk D.G., Greenberg J.H., and Detre J.A. 2001. Temporal dynamics of the partial pressure of brain tissue oxygen during functional forepaw stimulation in rats. *Neurosci. Lett.* **306:** 106–110.

Arieli A. and Grinvald A. 2002. Combined optical imaging and targeted electrophysiological manipulations in anesthetized and behaving animals. *J. Neurosci. Methods* **116:** 15–28.

Arieli A., Grinvald A., and Slovin H. 2002. Dural substitute for long-term imaging of cortical activity in behaving monkeys and its clinical implications. *J. Neurosci. Methods* **114:** 119–133.

Bakin J.S., Kwon M.C., Masino S.A., Weinberger N.M., Frostig R.D. 1996. Suprathreshold auditory cortex activation visualized by intrinsic signal optical imaging. *Cereb. Cortex* **6:** 120–130.

Bartfeld E. and Grinvald A. 1992. Relationships between orientation preference pinwheels, cytochrome oxidase blobs and ocular dominance columns in primate striate cortex. *Proc. Natl. Acad. Sci.* **89:** 11905–11909.

Basole A., White L.E., Fitzpatrick D. 2003. Mapping multiple features in the population response of visual cortex. *Nature* **424:** 986–990.

Blasdel G.G. 1992a. Differential imaging of ocular dominance and orientation selectivity in monkey striate cortex. *J. Neurosci.* **12:** 3115–3138.

———. 1992b. Orientation selectivity, preference, and continuity in monkey striate cortex. *J. Neurosci.* **12:** 3139–3161.

Bonhoeffer T. and Grinvald A. 1991. Iso-orientation domains in cat visual cortex are arranged in pinwheel-like patterns. *Nature* **353:** 429–431.

———. 1993. The layout of iso-orientation domains in area 18 of cat visual cortex: Optical imaging reveals pinwheel-like organization. *J. Neurosci.* **13:** 4157–4180.

———. 1996. Optical imaging based on intrinsic signals. The *Methodology in brain mapping: The methods* (ed. A. Toga and J. Mazziotta), pp. 893–969. Academic Press, New York.

Bonhoeffer T., Kim A., Malonek D., Shoham D., and Grinvald A. 1995. The functional architecture of cat area 17. *Eur. J. Neurosci.* **7:** 1973–1988.

Bosking W.H., Zhang Y., Schofield B., and Fitzpatrick D. 1997. Orientation selectivity and the arrangement of horizontal connections in tree shrew striate cortex. *J. Neurosci.* **17:** 2112–2127.

Cannestra A.F., Pouratian N., Bookheimer S.Y., Martin N.A., Becker D.P., and Toga A.W. 2001. Temporal spatial differences observed by functional MRI and human intraoperative optical imaging. *Cereb. Cortex* **11:** 1047–3211.

Cannestra A.F., Black K.L., Martin. N.A., Cloughesy T., Burton J.S., Rubinstein E., Woods R.P., and Toga A.W. 1998. Topographical and temporal specificity of humann intraoperative optical intrinsic signals. *Neuroreport* **9:** 2557–2563.

Chance B., Cohen P., Jobsis F., and Schoener B. 1962. Intracellular oxidation-reduction states in vivo. *Science* **137:** 499–508.

Chance B., Kang K., He L., Weng J., and Sevick E. 1993a. Highly sensitive object location in tissue models with linear in-phase and anti-phase multi-element optical arrays in one and two dimensions. *Proc. Natl. Acad. Sci.* **90:** 3423–3427.

Chance B., Zhuang L., Unah C., Alter C., and Lipton L. 1993b. Cognition-activated low-frequency modulation of light absorption in human brain. *Proc. Natl. Acad. Sci.* **90:** 3770–3774.

Chapman B. and Bonhoeffer T. 1998. Overrepresentation of horizontal and vertical orientation preference in developing ferret area-17. *Proc. Natl. Acad. Sci.* **95:** 2609–2614.

Chapman B., Stryker M.P. and Bonhoeffer T. 1996. Development of orientation preference maps in ferret primary visual cortex. *J. Neurosci.* **16:** 6443–6453.

Cheng K., Waggoner R.A., and Tanaka K. 2001. Human ocular dominance columns as revealed by high-field functional magnetic resonance imaging. *Neuron* **32:** 359–374.

Cohen L.B. 1973. Changes in neuron structure during action potential propagation and synaptic transmission. *Physiol. Rev.* **53:** 373–418.

Cohen L.B., Keynes R.D., and Hille B. 1968. Light scattering and birefringence changes during nerve activity. *Nature* **218:** 438–441.

Crair M.C., Gillespie D.G., and Stryker M.P. 1998. The role of visual experience in the development of columns in cat visual cortex. *Science* **279:** 566–570.

Crair M.C., Ruthazer E.S., Gillespie D.C., and Stryker M.P. 1997a. Ocular dominance peaks at pinwheel center singularities of the orientation map in cat visual cortex. *J. Neurophysiol.* **77:** 3381–3385.

———. 1997b. Relationship between the ocular dominance and orientation maps in visual cortex of monocularly deprived cats. *Neuron* **19:** 307–318.

Das A. and Gilbert C.D. 1995. Long-range horizontal connections and their role in cortical reorganization revealed by optical recording of cat primary visual cortex *Nature* **375:** 780–784.

———. 1997. Distortions of visuotopic map match orientation singularities in primary visual cortex. *Nature* **387:** 594–598.

Duong T.Q., Kim D.-S., Ugurbil K., and Kim S.-G. 2001. Localized cerebral blood flow response at submillimeter columnar resolution. *Proc. Natl. Acad. Sci.* **98:** 10904–10909.

Ernst T. and Hennig J. 1994. Observation of a fast response in functional MR. *Magn. Reson. Med.* **32:** 146–149.

Everson R.M., Knight B.W., and Sirovich L. 1997. Separating spatially distributed response to stimulation from background. I. Optical imaging. *Biol. Cybern.* **77:** 407–417.

Fox P.T., Mintun M.A., Raichle M.E., Miezin F.M., Allman J.M., and van Essen D.C. 1986. Mapping human visual cortex with positron emission tomography. *Nature* **323:** 806–809.

Frostig R.D., Masino S.A., and Kwon M.C. 1994. Characterization of functional whisker representation in rat barrel cortex: Optical imaging of intrinsic signals vs. single-unit recordings. *Neurosci. Abstr.* **20:** 566.

Frostig R.D., Lieke E.E., Ts'o D.Y., and Grinvald A. 1990. Cortical functional architecture and local coupling between neuronal activity and the microcirculation revealed by *in vivo* high-resolution optical imaging of intrinsic signals. *Proc. Natl. Acad. Sci.* **87:** 6082–6086.

Gabbay M., Brennan C., Kaplan E., and Sirovich L. 2000. A principal components-based method for the detection of neuronal activity maps: Application to optical imaging. *Neuroimage* **11:** 313–325.

Ghose G.M. and Ts'o D.Y. 1997. Form processing modules in primate area V4. *J. Neurophysiol.***77:** 2191–2196.

Gochin P.M., Bedenbaugh P., Gelfand J.J., Gross C.G., and Gerstein G.L. 1992. Intrinsic signal optical imaging in the forepaw area of rat somatosensory cortex. *Proc. Natl. Acad. Sci.* **89:** 8381–8383.

Godecke I. and Bonhoeffer T. 1996. Development of identical orientation maps for two eyes without common visual experience. *Nature* **379:** 251–254.

Goodyear B.G. and Menon R.S. 2001. Brief visual stimulation allows mapping of ocular dominance in visual cortex using fMRI. *Hum. Brain Mapp.* **14:** 210–217.

Gratton G. 1997. Attention and probability effects in the human occipital cortex: An optical imaging study. *Neuroreport* **8:** 1749–1753.

Gratton G., Fabiani M., Corballis PM., and Gratton E. 1997. Noninvasive detection of fast signals from the cortex using frequency-domain optical methods. *Ann. N.Y. Acad. Sci.* **820:** 286–298.

Gratton G., Corballis P.M., Cho E., Fabiani M., and Hood D.C. 1995. Shades of gray matter: Noninvasive optical images of human brain responses during visual stimulation. *Psychophysiology* **32:** 505–509.

Grinvald A., Manker A., and Segal M. 1982. Visualization of the spread of electrical activity in rat hippocampal slices by voltage sensitive optical probes. *J. Physiol.* **333:** 269–291.

Grinvald A., Slovin H., and Vanzetta I. 2000. Non-invasive visualization of cortical columns by f-MRI. *Nat. Neurosci.* **3:** 105–107.

Grinvald A., Frostig R.D., Lieke E.E., and Hildesheim R. 1988. Optical imaging of neuronal activity. *Physiol Rev.* **68:** 1285–1366.

Grinvald A., Frostig R.D., Siegel R.M., and Bartfeld E. 1991. High resolution optical imaging of neuronal activity in awake monkey. *Proc. Natl. Acad. Sci.* **88:** 11559–11563.

Grinvald A., Lieke E., Frostig R.D., Gilbert C.D., and Wiesel T.N. 1986. Functional architecture of cortex revealed by optical imaging of intrinsic signals. *Nature* **324:** 361–364.

Grinvald A., Bonhoeffer T., Pollack A., Aloni E., Ofri R., and Nelson D. 2004. High resolution functional optical imaging: From the neocortex to the eye. *Ophthalmol. Clin. North Am.* **17:** 53–67.

Grinvald A., Shoham D., Shmuel A., Glaser E.D., Vanzetta I., Shtoyerman E., Slovin H., Wijnbergen C., Hildesheim R., Sterkin A., and Arieli A. 1999. In-vivo optical imaging of cortical architecture and dynamics. A. In *Modern techniques in neuroscience research* (ed. U. Windhorst and H. Johansson), pp. 893–969. Springer, Umea, Sweden.

Haglund M.M., Ojemann G.A., and Hochman D.W. 1992. Optical imaging of epileptiform and functional activity in human cerebral cortex. *Nature* **358:** 668–671.

Harrison R.V., Harel N., Panesar J., and Mount R.J. 2002. Blood capillary distribution correlates with hemodynamic-based functional imaging in cerebral cortex. *Cereb. Cortex* **12:** 225–233.

Hess A. and Scheich H. 1996. Optical and FDG mapping of frequency-specific activity in auditory cortex. *Neuroreport* **7:** 2643–2647.

Hill D.K. and Keynes R.D. 1949. Opacity changes in stimulated nerve. *J. Physiol.* **108:** 278–281.

Hoshi Y. and Tamura M. 1993. Dynamic multichannel near-infrared optical imaging of human brain activity. *J. Appl. Physiol.* **75:** 1842–1846.

Hu X., Le T.H., and Ugurbil K. 1997. Evaluation of the early response in fMRI in individual subjects using short stimulus duration. *Magn. Reson. Med.* **37:** 877–884.

Hubel D.H. and Wiesel T.N. 1962. Receptive fields, binocular interactions and functional architecture in the cat's visual cortex. *J. Physiol.* **160:** 106–154.

———. 1977. Ferrier lecture. Functional architecture of macaque monkey visual cortex. *Proc. R. Soc. Lond. B Biol. Sci.* **198:** 1–59.

Hubener M., Shoham D., Grinvald A., and Bonhoeffer T. 1997. Spatial relationships between three columnar systems in cat area 17. *J. Neurosci.* **17:** 9270–9284.

Jobsis F.F. 1977. Noninvasive, infrared monitoring of cerebral and myocardial oxygen sufficiency and circulatory parameters. *Science* **198:** 1264–1266.

Jobsis F.F., Keizer J.H., LaManna J.C., and Rosenthal M. 1977. Reflectance spectrophotometry of cytochrome aa3 in vivo. *J. Appl. Physiol.* **43:** 858–872.

Jones M., Berwick J., Johnston D., and Mayhew J. 2001. Concurrent optical imaging spectroscopy and laser-Doppler flowmetry: The relationship between blood flow, oxygenation, and volume in rodent barrel cortex. *Neuroimage* **13:** 1002–1015.

Kalatsky V.A. and Stryker M.P. 2003. New paradigm for optical imaging: temporally encoded maps of intrinsic signal. *Neuron* **22:** 529–545.

Kato T., Kamei A., Takashima S., and Ozaki T. 1993. Human visual cortical function during photic stimulation monitoring by means of near-infrared spectroscopy. *J. Cereb. Blood Flow Metab.* **13:** 516–520.

Kelin D. 1925. On cytochrome, a respiratory pigment, common to animals, yeast, and higher plants. *Proc. R. Soc. B* **98:** 312–339.

Kety S.S. 1950. Blood flow and metabolism of the human brain in health and disease. *Trans. Stud. Coll. Physicians Phila.* **18:** 103–108.

Kim D.S. and Bonhoeffer T. 1994. Reverse occlusion leads to a precise restoration of orientation preference maps in visual cortex. *Nature* **370:** 370–372.

Kim D.-S., Doung T.Q., and Kim S.-G. 2000a. High-resolution mapping of iso-orientation columns by fMRI. *Nat. Neurosci.* **3:** 164–199.

———. 2000b. Reply to "Can current fMRI techniques reveal the micro-architecture of cortex?" *Nat Neurosci.* **3:** 414.

Kisvarday Z.F., Kim D.S., Eysel U.T., Bonhoeffer T. 1994. Relationship between lateral inhibitory connections and the topography of the orientation map in cat visual cortex. *Eur. J. Neurosci.* **6:** 1619–1632.

Kohl M., Lindauer U., Royl G., Kühl M., Gold L., Villringer A., and Dirnagl U. 2000. Physical model for the spectroscopic analysis of cortical intrinsic optical signals. *Phys. Med. Biol.* **45:** 3749–3764.

Kwong K.K., Belliveau J.W., Chesler D.A., Goldberg I.E., Weisskoff R.M., Poncelet B.P., Kennedy D.N., Hoppel B.E., Cohen M.S., Turner R., Chang H.M., Brady T., and Rosen B.R. 1992. Dynamic magnetic resonance imaging of human brain activity during primary sensory stimulation. *Proc. Natl. Acad. Sci.* **89:** 5675–5679.

Lassen N.A. and Ingvar D.H. 1961. The blood flow of the cerebral cortex determined by radioactive krypton. *Exp. Basel* **17:** 42–43.

Lindauer U., Royl G,. Leithner C., Kuhl M., Gold L., Gethmann J., Kohl-Bareis M., Villringer A., and Dirnagl U. 2001. No evidence for early decrease in blood oxygenation in rat whisker cortex in response to functional activation. *Neuroimage.* **13:** 988–1001.

Livingstone M.S. and Hubel D.H. 1984. Anatomy and physiology of a color system in the primate visual cortex. *J. Neurosci.* **4:** 309–356.

Logothetis N.K. 2000. Can current fMRI techniques reveal the micro-architecture of cortex. *Nat. Neurosci.* **3:** 413.

Logothetis N.K., Guggenberger H., Peled S., and Pauls J. 1999. Functional imaging of the monkey brain. *Nat. Neurosci.* **2:** 555–562.

MacVicar B.A., Hochman D., LeBlanc F.E., and Watson T.W. 1990. Stimulation evoked changes in intrinsic optical signals in the human brain. *Soc. Neurosci. Abstr.* **16:** 309.

Maclin E.L., Low K.A., Sable J.J., Fabiani M. Gratton G. 2004. The event-related optical signal to electrical stimulation of the median nerve. *Neuroimage* **21:** 1798–804.

Malach R., Tootell R.B.H., and Malonek D. 1994. Relationship between orientation domains, cytochrome oxidase stipes and intrinsic horizontal connections in squirrel monkey area V2. *Cereb. Cortex* **4:** 151–165.

Malach R., Amir Y., Harel M., and Grinvald A. 1993. Novel aspects of columnar organization are revealed by optical imaging and in vivo targeted biocytin injections in primate striate cortex. *Proc. Natl. Acad. Sci.* **90:** 10469–10473.

Malach R., Schirman T.D., Harel M., Tootell R.B.H., and Malonek D. 1997. Organization of intrinsic connections in owl monkey area MT. *Cereb. Cortex* **7:** 386–393.

Malonek D. and Grinvald A. 1996. The imaging spectroscope reveals the interaction between electrical activity and cortical microcirculation; implication for functional brain imaging. *Science* **272:** 551–554.

Malonek D., Tootell R.B.H., and Grinvald A. 1994. Optical imaging reveals the functional architecture of neurons processing shape and motion in owl monkey area MT. *Proc. R. Soc. Lond. B Biol. Sci.* **258:** 109–119.

Malonek D., Dirnagl U., Lindauer U., Yamada K., Kanno I., and Grinvald A. 1997. Vascular imprints of neuronal activity. Relationships between dynamics of cortical blood flow, oxygenation and volume changes following sensory stimulation. *Proc. Natl Acad. Sci.* **94:** 4826–14831.

Martin C., Berwick J., Johnston D., Zheng Y., Martindale J., Port M., Redgrave P., and Mayhew J. 2002. Optical imaging spectroscopy in the unanaesthetised rat. *J. Neurosci. Methods* **120:** 25–34.

Masino S.A. and Frostig R.D. 1996. Quantitative long-term imaging of the functional representation of a whisker in rat barrel cortex. *Proc. Natl. Acad. Sci.* **93:** 4942–4947.

Mayevsky A. and Chance B. 1982. Intracellular oxidation-reduction state measured in situ by a multichannel fiber-optic surface fluorometer. *Science* **217:** 537–540.

Mayhew J., Johnston D., Berwick J., Jones M., Coffey P., and Zheng Y. 2000. Spectroscopic analysis of neural activity in brain: increased oxygen consumption following activation of barrel cortex. *Neuroimage* **12:** 664–675.

Mayhew J., Zheng Y., Hou Y., Vuksanovic B., Berwick J., Askew S., and Coffey P. 1999. Spectroscopic analysis of changes in remitted illumination: the response to increased neural activity in brain. *Neuroimage* **10:** 304–326.

Mayhew J., Hu D., Zheng Y., Askew S., Hou Y., Berwick J., Coffey P.J., and Brown N. 1998. An evaluation of linear model analysis techniques for processing images of microcirculation activity. *Neuroimage* 7: 49–71.

Menon R.S. and Goodyear B.G. 1999. Submillimeter functional localization in human striate cortex using BOLD contrast at 4 Tesla: Implications for the vascular point-spread function. *Magn. Reson. Med.* **4:** 230–235

Menon R.S., Ogawa S., Strupp J.P., and Ugurbil K. 1997. Ocular dominance in human V1 demonstrated by functional magnetic resonance imaging *J. Neurophysiol.* **77:** 2780–2787.

Menon RS, Ogawa S, Hu X, Strupp JP, Anderson P, Ugurbil K. 1995. BOLD based functional MRI at 4 Tesla includes a capillary bed contribution: Echo-planar imaging correlates with previous optical imaging using intrinsic signals. *Magn. Reson. Med.* **33:** 453–459.

Millikan G.A. 1937. Experiments on muscle hemoglobin in vivo; the instantaneous measurement of muscle metabolism. *Proc. R. Soc. B* **123:** 218–241.

Mosso A. Sulla circolazione del cervello dell'uomo. Atti. R. Accad. Lincei 5, 237–358 (1880) (See Swinnen S.P., 2002. Intermanual coordination: from behavioural principles to neural-network interactions. *Nat. Rev. Neurosci:* **3:** 348–359.)

Mountcastle V.B. 1957. Modality and topographic properties of single neurons of cat's somatic sensory cortex. *J. Neurophysiol.* **20:** 408–434.

Obermayer K. and Blasdel G.G. 1993. Geometry of orientation and ocular dimonance columns in primate striate cortex. *J. Neurosci.* **13:** 4114–4129.

Ogawa S., Tank D.W., Menon R., Ellermann J.M., Kim S.G., Merkle H., and Ugurbil K. 1992. Intrinsic signal changes accompanying sensory stimulation: Functional brain mapping with magnetic resonance imaging. *Proc. Natl. Acad. Sci.* **89:** 5951–5955.

Polley D.B., Kvasnak E., and Frostig R.D. 2004. Naturalistic experience transforms sensory maps in the adult cortex of caged animals. *Nature* **429:** 67–71.

Polley D.B., Chen-Bee C.H., and Frostig R.D. 1999. Two directions of plasticity in the sensory-deprived adult cortex. *Neuron* **24:** 623–637.

Roe A.W. and Ts'o D.Y. 1995. Visual topography in primate V2: Multiple representation across functional stripes.*J. Neurosci.* **15:** 3689–3715.

Raichle M.E., Martin W.R.W., Herscovitz P., Minton M.A., and Markham J.J. 1983. Brain blood flow measured with intravenous H2(15)0. II. Implementation and validation. *J. Nucl. Med.* **24:** 790–798.

Roy C. and Sherrington C. 1890. On the regulation of the blood supply of the brain. *J. Physiol.* **11:** 85–108.

Rubin B.D. and Katz L.C. 1999. Optical imaging of odorant representations in the mammalian olfactory bulb. *Neuron* **23:** 499–511.

Salzberg B.M., Obaid A.L., and Gainer H. 1986. Optical studies of excitation secretion at the vertebrate nerve terminal. *Soc. Gen. Physiol. Ser.* **40:** 133–164.

Schwartz T.H. and Bonhoeffer T. 2001. In vivo optical mapping of epileptic foci and surround inhibition in ferret cerebral cortex. *Nat. Med.* **7:** 1063–1067.

Schuett S., Bonhoeffer T., and Hubener M. 2002. Mapping retinotopic structure in mouse visual cortex with optical imaging. *J. Neurosci.* **22:** 6549–6559.

Shmuel A. and Grinvald A. 1996. Functional organization for direction of motion and its relationship to orientation maps in cat area 18. *J. Neurosci.* **16:** 6945–6964.

Shoham D. and Grinvald A. 2001. Visualization of the hand representation in the hand in macaque and Human area S-I using intrinsic signal optical imaging. *J. Neurosci.* **21:** 6820–6835.

Shoham D., Hubener M., Grinvald A., and Bonhoeffer T. 1997. Spatio-temporal frequency domains and their relationship to cytochrome oxidase staining in cat visual cortex. *Nature* **385:** 529–534.

Shtoyerman E., Arieli A., Slovin H., Vanzetta I., and Grinvald A. 2000. Long term optical imaging and spectroscopy reveal mechanisms underlying the intrinsic signal and stability of cortical maps in V1 OF behaving monkeys. *J. Neurosci.* **20:** 8111–21.

Sirovich L., Everson R.M., Kaplan E., Knight B.W., O'Brien E.V., and Orbach D. 1996. Modeling the functional organization of the visual cortex. *Physica D* **96:** 355–366.

Slovin, H., A. Arieli, R. Hildesheim, and A. Grinvald 2002. Long-term voltage-sensitive dye imaging of cortical dynamics in the behaving monkey. *J. Neurophys.* 88: 3421–3438.

Slovin H., Strick P.L., Hildesheim R., and Grinvald A. 2003. Voltage sensitive dye imaging in the motor cortex. I. Intra- and intercortical connectivity revealed by microstimulation in the awake monkey. *Soc. Neuroci. Abs.,* 554.8 .

Sokoloff L. 1977. Relation between physiological function and energy metabolism in the central nervous system. *J. Neurochem.* **19:** 13–26.

Swindale N.V., Matsubara J.A., and Cynader M.S. 1987. Surface organization of orientation and direction selectivity in cat area 18. *J. Neurosci.* **7:** 1414–1427.

Strick P., Grinvald A., Hildesheim R., and Slovin H. 2003. Voltage sensitive dye imaging in the motor cortex II. Cortical correlates of Go/No-Go delayed response task. *Soc. Neuroci. Abst.* 918.8.

Tasaki I., Watanabe A., Sandlin R., and Carnay L. 1968. Changes in fluorescence, turbidity and birefringence associated with nerve excitation. *Proc. Natl. Acad. Sci.* **61:** 883–888.

Thompson J.K., Peterson M.R., and Freeman R.D. 2003. Single-neuron activity and tissue oxygenation in the cerebral cortex. *Science* **299:** 1070–1072.

Ts'o D.Y., Gilbert C.D., and Wiesel T.N. 1991. Orientation selectivity of and interactions between color and disparity subcompartments in area V2 of Macaque monkey, *Neurosci. Abstr.* **17:** 431.

Ts'o D.Y., Roe A.W., and Shey J. 1993. Functional connectivity within V1 and V2: Patterns and dynamics. *Neurosci. Abstr.* **19:** 618.

Ts'o D.Y., Frostig R.D., Lieke E., and Grinvald A. 1990. Functional organization of primate visual cortex revealed by high resolution optical imaging, *Science* **249:** 417–420.

Vanzetta I. and Grinvald A. 1999. Cortical activity-dependent oxidative metabolism revealed by direct oxygen tension measurements; implications for functional brain imaging. *Science* **286:** 1555–1558.

———. 2001. Evidence and lack of evidence for the initial dip in the anesthetized rat: implications for human functional brain imaging. *Neuroimage* **13:** 959–967.

Vanzetta I. Slovin H. Omer D.B. Grinvald A. 2004. Columnar resolution of blood volume and oximetry functional maps in the behaving monkey; implications for FMRI. *Neuron* **42:** 843–54.

Wang G., Tanaka K., and Tanifuji M. 1996. Optical imaging of functional organization in the monkey inferotemporal cortex. *Science* **272:** 1665–1668.

Weliky M., Bosking W.H., and Fitzpatrick D. 1996. A systematic map of direction preference in primary visual cortex. *Nature* **379:** 725–728.

Wolf M., Wolf U., Choi J.H., Gupta R., Safonova L.P., Paunescu L.A., Michalos A., and Gratton E. 2003. Detection of the fast neuronal signal on the motor cortex using functional frequency domain near infrared spectroscopy. *Adv. Exp. Med. Biol.* **510:** 193–197.

Wyatt J.S., Cope M., Deply D.T., Wray S., and Reynolds E.O.R. 1986. Quantitation of cerebral oxygenation and haemodynamics in sick newborn infants by near–infrared spectrophotometry. *Lancet* **2:** 1063–1066.

Wyatt J.S., Cope D., Deply D.T., Richardson C.E., Edwards A.D., and Wray S. 1990. Reynolds EOR. Quantitation of cerebral blood volume in human infants by near-infrared spectroscopy. *J. Appl. Physiol.* **68:** 1086–1091.

Yacoub E. and Hu X. 1999. Detection of the early negative response in f-MRI at 1.5 Tesla. *Magn. Reson. Med.* **41:** 1088–1092.

CHAPTER 90

Voltage-sensitive Dye Imaging of Neocortical Activity

Amiram Grinvald, Dahlia Sharon, Alexander Sterkin, Hamutal Slovin, and Rina Hildesheim

Department of Neurobiology, The Weizmann Institute of Science, Rehovot, 76100 Israel

The processing of sensory information and the coordination of movement and higher brain functions are carried out by elaborate networks formed by millions of neurons. One primary question in brain research asks how single neurons and their intricate synaptic connections combine to form networks that are capable of performing such remarkable feats. Research has revealed that neocortical networks possess properties that are not evident in single neurons. Such properties can be investigated only by studying electrical activity in neuronal populations, an approach that requires imaging. Because the neocortex is organized in cortical columns (Mountcastle 1957; Hubel and Wiesel 1962), fundamental studies of cortical function require a spatial resolution that is better than 200 μm.

Most imaging techniques reflect indirect correlates of neuronal activity, such as the hemodynamic response, and do not record changes in membrane potentials per se. They provide high resolution in either the spatial or temporal domain, but not in both. Optical imaging utilizing voltage-sensitive dyes (VSDI), however, directly reflects membrane potential changes and has achieved high resolution in both the spatial and temporal domain, simultaneously (Grinvald et al. 1984, 1994; Orbach et al. 1985). Recent methodological modifications have led to a 30-fold improvement in the signal-to-noise ratio (SNR) of high-resolution VSDI, avoiding the need for signal averaging (Shoham et al. 1999; Kenet et al. 2003), including the behaving primate (Slovin et al. 2002). These advances have paved the way for a new era in functional cortical imaging for studying cortical dynamics.

Optical imaging based on intrinsic signals has the important advantage of measuring cortical population activity without using an extrinsic probe. However, because of the slow time course of intrinsic signals, which exhibit a rise time of seconds, intrinsic imaging cannot be used when the temporal aspect of neural coding is at issue. Although multi-electrode techniques are promising in this respect, the size and placement of the electrode arrays still pose problems, and, by nature, sample only a small fraction of a given cortical population. Furthermore, since it is not practical to use multi-electrode techniques for intracellular recordings, these techniques cannot be used to reveal the essential information contained in subthreshold synaptic potentials. VSDI does not suffer from these limitations and, thus, provides an attractive alternative technique. VSDI is an increasingly powerful tool for in vivo exploration of cortical population activity, providing sub-columnar spatial detail (50–100 μm) with millisecond precision.

This chapter reviews VSDI of in vivo population activity in the neocortex. VSDI of single cells and additional types of functional imaging using other probes are discussed elsewhere. Other chapters review applications of optical imaging based on intrinsic signals (see Chapter 89). This chapter begins by reviewing methodology and historical aspects of in vivo VSDI, and continues by discussing results obtained in a variety of mammalian species and paradigms, from sensory cortex of anesthetized rats and cats to motor cortex of behaving monkeys.

FROM IN VITRO SINGLE-CELL RECORDINGS TO IN VIVO POPULATION IMAGING

To perform optical imaging of electrical activity, the preparation under study is first stained with a suitable voltage-sensitive dye. The dye molecules bind to the external surface of excitable membranes and act as molecular transducers that transform any changes in membrane potential into optical signals. These optical signals are observed as changes in absorption or emitted fluorescence, and they respond to membrane potential changes in microseconds. The voltage-sensitive dye signals are linearly correlated with both the membrane potential changes and the membrane area of the stained neuronal elements. These optical changes are monitored with light-imaging devices positioned in a microscope image plane. Optical signals using voltage-sensitive dyes were first recorded by Tasaki et al. (1968) in the squid giant axon, and by Cohen and his colleagues in the squid giant axon and in individual leech neurons (Salzberg et al. 1973).

Voltage-sensitive dye *imaging* (multiple pixels) as opposed to *recording* (single pixel), began with photodiode arrays in invertebrate ganglia (Grinvald et al. 1981) and was subsequently employed in mammalian brain slices (Grinvald et al. 1982a) and in the isolated salamander olfactory bulb (Orbach and Cohen 1983). These initial results suggested that optical imaging could be a useful tool with which to study the mammalian brain in vivo as well. However, the initial in vivo experiments in rat visual cortex in 1982 revealed several complications that had to be overcome. One complication was the large amount of noise caused by respiratory and heartbeat pulsation. In addition, the relative opacity and packing density of the cortex limited the penetration of the excitation light and the ability of dyes to stain deep layers of the cortex. Subsequently, new, improved dyes that overcame these problems were developed (e.g., RH-414; Grinvald et al. 1982b), and an effective remedy for the heartbeat noise was found by synchronizing data acquisition with the electrocardiogram and subtracting a no-stimulus trial. These improvements facilitated in vivo imaging of several different sensory systems, including the retinotopic responses in the frog optic tectum (Grinvald et al. 1984), the whisker barrels in rat somatosensory cortex (Orbach et al. 1985), and experiments on the salamander olfactory bulb (Kauer et al. 1987; Kauer 1988; Cinelli and Kauer 1995; Cinelli et al. 1995). The development of more hydrophilic dyes improved the quality of the results obtained in cat and monkey visual cortex (e.g., RH-704 and RH-795; Grinvald et al. 1986, 1994).

The newest generation of voltage-sensitive dyes offers a 30-fold improvement in SNR over the early dyes. This was accomplished by designing dyes that are excited outside the absorption band of hemoglobin, thus minimizing pulsation and hemodynamic noise (Shoham et al. 1999). With this advance it has become possible to reveal the dynamics of cortical information processing and its underlying functional architecture at the necessary spatial and temporal resolution in both anesthetized and behaving animals. Additional advances related to the implantation of transparent artificial dura (Arieli et al. 2002) now allow chronic recordings to be taken over a long period of time. Optical imaging can be performed simultaneously with intracellular recording, extracellular recording, microstimulation, and tracer injection, thanks to the development of an electrode assembly attached to a cranial window (Arieli and Grinvald 2002).

It is important to comment on the relationship between the in vivo dye population signal and membrane potential changes. In simpler preparations, where single cells are distinctly visible, the dye signal looks just like an intracellular electrical recording (Salzberg et al. 1973, 1977; Grinvald et al. 1977, 1981, 1982b). In optical imaging from cortical tissue stained by topical application, however, the optical signal does not have single-cell resolution. Rather, it likely represents the sum of membrane potential changes in both pre- and postsynaptic neuronal elements, as well as a possible contribution from the depolarization of neighboring glial cells (Konnerth and Orkand 1986; Lev-Ram and

Grinvald 1986). Because of the much larger area of dendrites relative to that of cell somata (about 1000-fold), the voltage-sensitive dye signal in cortical tissue reflects, in the main, the postsynaptic potentials in the fine dendrites of cortical cells, rather than action potentials in cell somata. VSDI can therefore easily detect subthreshold synaptic potentials in the extensive dendritic arborization. Recent patch recordings from the rat somatosensory cortex, performed simultaneously with VSDI, indicated that the signals originate primarily from layer 2/3 and reflect dendritic postsynaptic potentials rather than somatic action potentials (Sterkin et al. 1998; Petersen et al. 2003a). Thus, dye signals are uniquely poised to supply data about subthreshold dendritic processing, providing information about aspects of neuronal processing that cannot usually be obtained from single-unit recordings or intracellular somatic recordings.

SPATIAL AND TEMPORAL RESOLUTION

Early VSDI utilized a cumbersome, and low-resolution, 12 x 12 "diode array camera" (Grinvald et al. 1981). Higher resolution has subsequently been achieved, due mostly to efforts by two groups in Japan: Kamino (Hirota et al. 1995) and Matsumoto (Iijima et al. 1992; Vranesic et al. 1994). Further increases in spatial resolution were achieved by Toyama and colleagues using a stroboscopic light (Toyama and Tanifuji 1991; Tanifuji et al. 1993). Today, commercial voltage-sensitive dye cameras, with up to 1,000,000 detectors (CCD or CMOS), are available from several vendors.

Detector technology, however, is not the factor that currently limits the spatial resolution achieved with VSDI. The properties of the dyes are a major limiting factor; in particular, the SNR that can be obtained with them, and the photodynamic damage that they cause. The development of suitable voltage-sensitive dyes is a key to the successful application of optical imaging for several reasons. First, different preparations often require dyes with different properties (Ross and Reichardt 1979; Cohen and Lesher 1986; Grinvald et al. 1988). Recently, it was shown that even within the same species, different cortical areas require different dyes (e.g., a given dye provided a high-quality signal in the rat somatosensory cortex but not in the rat olfactory bulb, although the bulb was well stained (Spors and Grinvald 2002). Second, the use of dyes is associated with difficulties that must be overcome. Under prolonged or intense illumination, dyes cause photodynamic damage. Additional difficulties include bleaching, limited depth of penetration into the cortex, and possible pharmacological side effects.

However, recent intracellular recordings in vivo have directly confirmed that stained cortical cells maintain their response properties (Sterkin et al. 1998; Grinvald et al. 1999, Fig. 23; Petersen et al. 2003a). Furthermore, long-term VSDI in awake monkeys indicated that, even after a year of imaging, monkey visual cortex continued to function normally: The animal maintained normal performance in tasks that required the cortical area being monitored. Thus, new dyes (Shoham et al. 1999) have largely alleviated the problems of pharmacological side effects and photodynamic damage. This means that extensive imaging sessions are now possible, allowing signal-averaging of many repeated presentations and improved SNRs, thus enhancing spatial resolution.

An additional factor that determines spatial resolution of VSDI is the structure of the cellular elements giving rise to the dye signal. Since the local dendritic trees of neurons have a diameter around 300–500 μm, a given pixel combines activity from cells with distant somas. Because dendritic trees of neurons, tuned to different functional properties, overlap only partially, sub-dendrite resolution is possible. The temporal resolution of VSDI is not limited by the response time of dye molecules, since these respond within microseconds to a change in the electrical field over the membrane (Cohen et al. 1974). The temporal resolution is, however, limited by the nature of population activity and the SNR. The power spectrum of other forms of population activity (EEG and LFP) suggests that most of the signal does not require a temporal resolution of more than 5–10 millisec. Achieving high SNRs depends largely, as explained above, on the dyes used and is somewhat more demanding in VSDI than in the sister technology of intrinsic imaging because the samples are shorter in time. This is due to the square-root relationship between the number of samples and the SNR. It is much easier to obtain a good SNR when signals are slower. For instance, the SNR in measurements of small signals is 33-fold higher when the signals have a rise time of 1 sec, as compared to a rise time of 1 millisec.

Empirically, we find that current dyes and equipment allow excellent SNR at a sampling rate of 5–10 msec with each pixel looking at a 64 × 64-μm area of cortex. Finer temporal and spatial sampling does not usually provide additional data, although exceptions do exist as noted in the study of the lateral spread beyond the retinotopic border, when a sampling rate of 0.6 msec was used (Derdikman et al. 2003; Petersen et al. 2003a,b). VSDI has been said to sacrifice spatial resolution to gain better temporal resolution. The above numbers actually show, however, that no such tradeoff exists with the present state of the art, relative to intrinsic imaging. Improvements in the dyes and in the spatial resolution of fast cameras have made it possible to obtain high-resolution functional maps of orientation columns, "lighting up" in milliseconds with an SNR even better than that obtained with the slow intrinsic signals (see Fig. 2 below). Additional developments will undoubtedly introduce further improvement.

In summary, VSDI of cortical activity is a particularly attractive technique for providing new insights into the temporal aspects of mammalian brain function. Among its advantages over other methodologies are (1) direct recording of the summed membrane potential changes of neuronal populations, including fine dendritic and axonal processes; (2) ability to measure these repeatedly from the same cortical region over an extended period of time, using different experimental or stimulus conditions; (3) imaging spatiotemporal patterns of activity of neuronal populations with a sub-millisecond temporal resolution; and (4) selective visualization of neuronal assemblies (see below). Several related reviews have been published elsewhere (Tasaki and Warashina 1976; Waggoner and Grinvald 1977; Cohen et al. 1978; Waggoner 1979; Grinvald 1984, 1985; Cohen and Lesher 1986; De Weer and Salzberg 1986; Salzberg et al. 1986; Loew 1987; Orbach 1988; Grinvald et al. 1991; Kamino 1991; Cinelli and Kauer 1992; for more detailed reviews, see Grinvald et al. 1988, 1999).

DISTRIBUTED PROCESSING: THE SPREAD OF ACTIVITY FAR BEYOND THE RETINOTOPIC/SOMATOTOPIC REPRESENTATIONS

One outstanding question that has benefited from real-time optical imaging is that of the distance in the cortical surface across which activation by a sensory point stimulus spreads. As mentioned above, the voltage-sensitive dye signal in cortical tissue reflects, in the main, the postsynaptic potentials in the fine dendrites of cortical cells rather than action potentials in cell somata. VSDI is thus ideally suited to answer this question.

The frog retinotectal connections offer a system that is topographically well organized: Each spot of light on the retina activates a small region in the optic tectum. The first optical imaging study investigating the spread of activation concentrated on visualizing the topographic distribution of sensory responses in the frog (Grinvald et al. 1984). The optical signals obtained from the tectum in response to discrete visual stimuli were found to correspond well to the known retinotopic map of the tectum. However, in addition to a focus of excitation, the spatial distribution of the signals showed smaller, delayed activity (3–20 msec) covering a much larger area than would be expected on the basis of classic single-unit mapping.

In rat primary somatosensory cortex, each whisker projects to a well-defined region termed the whisker barrel, and there is a simple somatotopic organization of the different whisker barrels side by side. This preparation thus offered a convenient opportunity to explore the question of activation spread in the mammalian brain. When the tip of a whisker was gently moved, optical signals were observed in the corresponding cortical barrel field (Derdikman et al. 2003). However, a discrepancy was noted between the size of an individual barrel, as recorded optically (1300 μm) and the histologically defined barrel (300–600 μm in layer IV of the cortex, showing neuronal somata rather than processes). The reason for this difference is probably that most of the optical signal originates from the superficial cortical layers in which neurons extend long processes to neighboring barrels (Orbach et al. 1985; Kleinfeld and Delaney 1996; Takashima et al 2001; Derdikman et al. 2003). Glutamate antagonists blocked the spread. Thus, postsynaptic activity in the dendritic processes could account for the detected spread (Petersen et al. 2003a,b). This lateral spread is both excitatory and inhibitory, and the balance probably depends on the stimulus parameters. Net surround inhibition was also documented at late times in the surround of an activated barrel (Orbach et al. 1985; Takashima et al. 2001; Derdikman et al. 2003).

In monkey striate cortex, retinotopic imaging experiments also showed activity over a cortical area much larger than predicted on the basis of standard retinotopic measurements in layer IV, but consistent with the anatomical finding of long-range horizontal connections in visual cortex (Gilbert and Wiesel 1983). The results of these experiments were used to calculate the *cortical point spread function*, which reflects the extent of cortical activation by retinal point stimuli. Figure 1 illustrates the cortical point spread function in macaque primary visual cortex. To show the relationship between the observed spread and individual cortical modules, the spread function is projected on a histological section of cytochrome oxidase blobs (Fig. 1B). The stimulus used here caused spiking, activated only in neurons residing in the marked small square, which contains just 4 blobs. However, more than

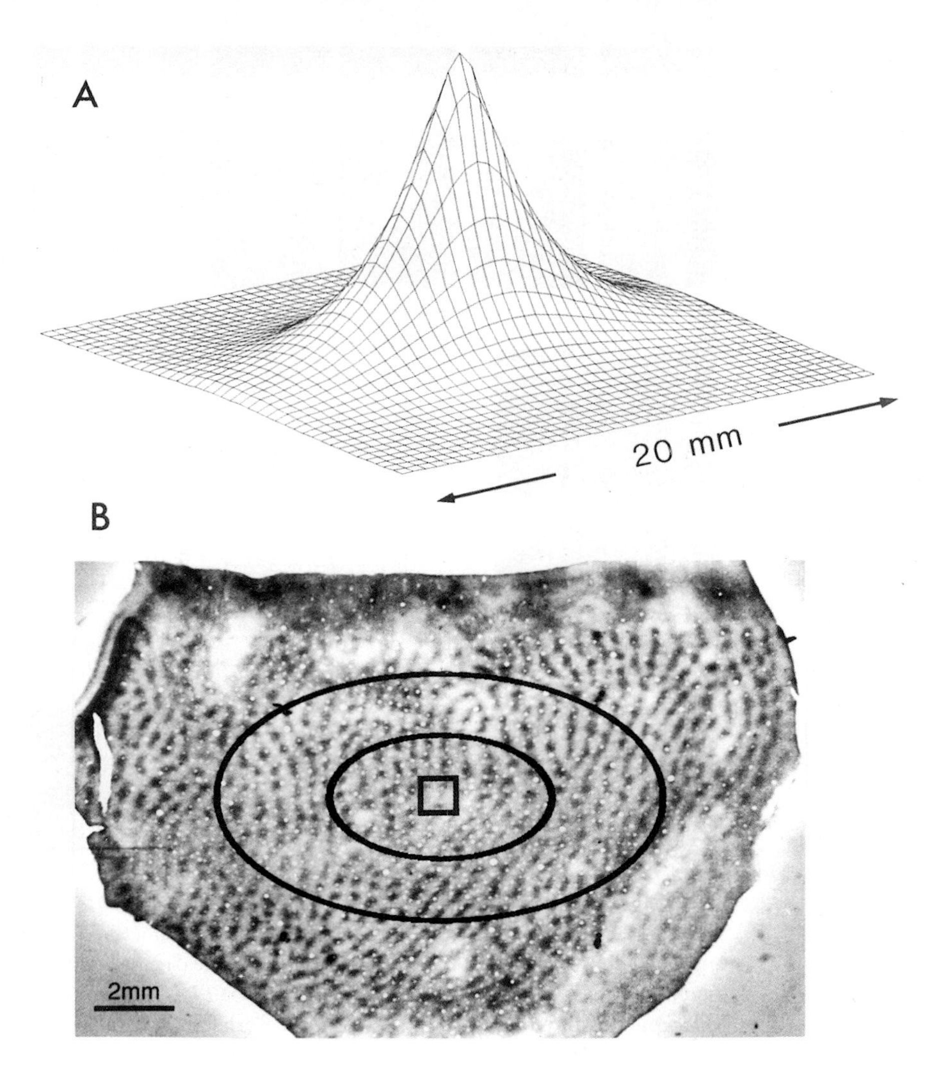

FIGURE 1. The number of functional domains involved in the processing of a small retinal image. (*A*) Calculation of the activity spread from a small patch in layer IV (the 1 x 1 mm square) within the upper cortical layers. At an eccentricity of ~ 6°, close to the V1/V2 border, such cortical activation would be produced by a retinal image of ~ 0.5 x 0.25° presented to both eyes. The "space constants" for the exponential spread measured with these dye experiments were 1.5 mm and 2.9 mm perpendicular and parallel to the vertical meridian, respectively. (*B*) The spread superimposed on a histological section showing the mosaics of cytochrome oxidase blobs, close to the V1/V2 border. The thin and thick stripes of V2 are also evident in the upper part. The center "ellipse" shows the contour where the amplitude of cortical activity drops to 1/e (37%) of its peak. The larger ellipse shows the contour where the spread amplitude drops to $1/e^2$ (14%). More than 10,000,000 neurons are included in the cortical area, bound by the large ellipse, containing a regular mosaic of about 250 blobs. (Modified, with permission, from Grinvald et al. 1994[© Society for Neuroscience].)

200 blobs had access to the information carried by the signal spread, albeit at lower amplitude. The apparent "space constant" for the spread was 1.5 mm along the cortical axis, parallel to the ocular dominance columns, and 3 mm along the perpendicular axis. The spread velocity was 0.1–0.2 m/sec

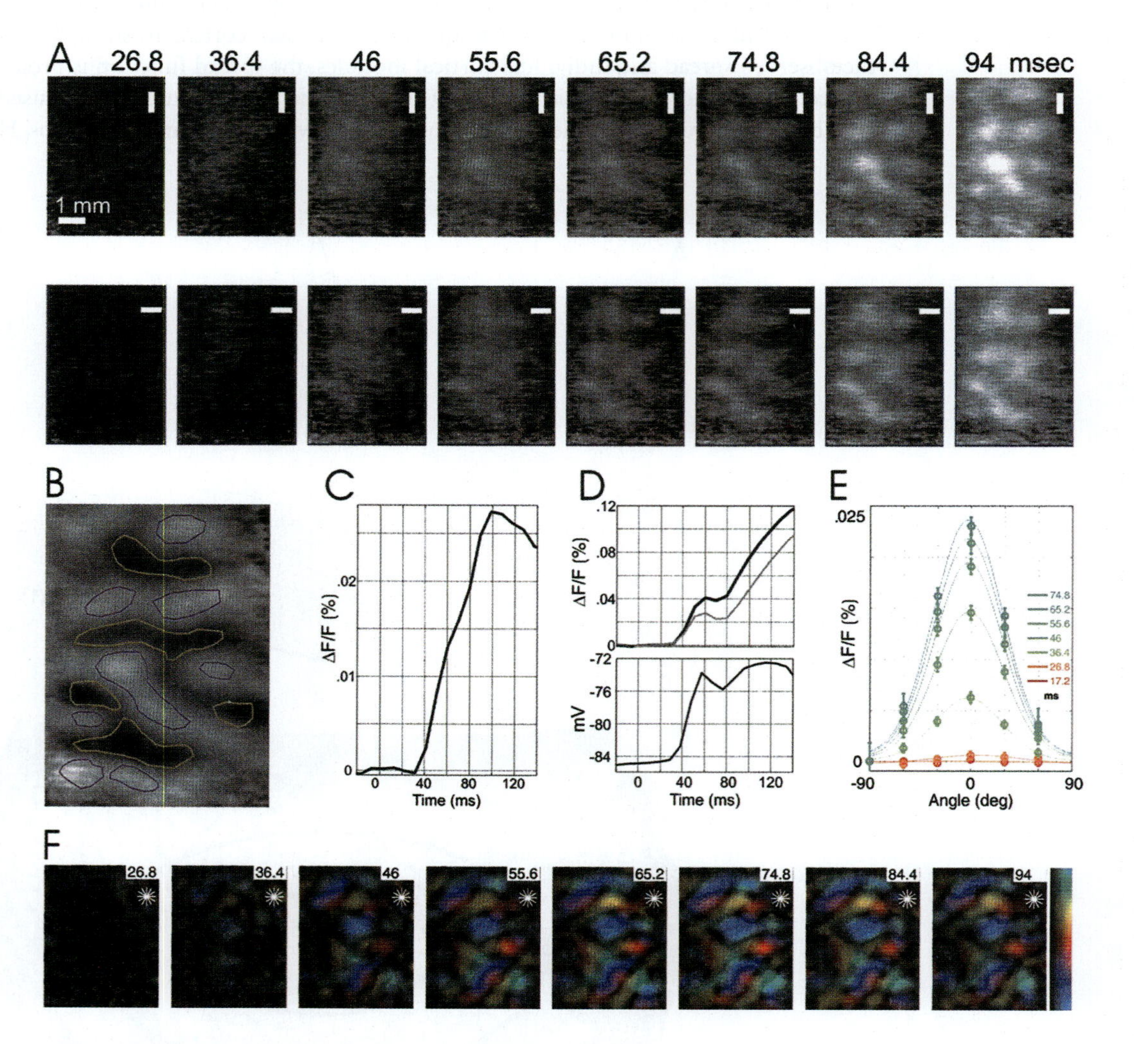

FIGURE 2. The dynamics of orientation tuning. (*A*) The two rows show the time series of single-condition orientation maps, in response to two orthogonal stimuli. (*B*) Spatial pattern revealed by differential imaging. (*C*) Temporal pattern of differential activity. The deceleration–acceleration (DA) notch is hardly visible here, thus it was overlooked in earlier studies. (*D*) (*Top*) The time course of the evoked response to preferred (*black*) and orthogonal (*gray*) orientations, calculated from the time series of single-condition maps shown in *A*. The two responses have the same onset latency (~ 30 msec); the response to the preferred orientation is larger right from the onset. The striking feature of the evoked responses is the DA notch. *C* and *D* were calculated for pixels in the regions marked in *B*. (*Bottom*) Confirmation of the DA notch with intracellular recordings performed in our lab (Sterkin et al. 1998). The notch in the intracellular data appears at the same time as that in the dye data, at around 50–80 msec. After the notch, the dye response continues to grow a lot more than the intracellular response; this could be due to the fact that the dye measures membrane potential changes over the entire dendritic arborization, and not just in the soma. (*E*) Orientation tuning curves at different times after stimulus onset, from another experiment. The curves after response onset (*green*) have the same shape but different amplitudes. The baseline of the curves was shifted to facilitate comparison. (*F*) Time series of orientation preference maps in polar depiction: Color represents preferred orientation (top to bottom of color scale on right is 0–180°), and brightness represents amplitude of differential response (left to right of color scale is 0–0.05%). Preferred orientation is steady from response onset. The VSDI data in *A–D* are raw, unfiltered. (Modified from Sharon and Grinvald 2002.)

(Grinvald et al. 1994). A recent study in anesthetized cat showed that this spreading subthreshold synaptic activity is priming the cortex and the cortex then responds differently to a subsequent stimulus; it has been reported that perception of motion when only stationary stimuli were displayed could be explained by this type of activity, which is totally elusive to unit recordings (Jancke et al. 2004).

The extensive lateral spread observed beyond the retinotopic borders in layer IV, both excitatory and inhibitory, indicates that the degree of distributed processing in primary visual cortex is much larger than often conceived—certainly a nontrivial challenge for theoreticians studying cortical networks. In these early studies of the visual response using VSDI, the spatial resolution did not enable exploration of such basic attributes of the visual response as orientation selectivity. Subsequent studies, described below, addressed this issue.

DYNAMICS OF SHAPE PROCESSING AT SUB-COLUMNAR RESOLUTION

A key goal for functional brain mapping is to obtain single-condition maps; that is, to acquire the pattern of activation evoked by a single stimulus condition, without the need to perform differential imaging of two orthogonal stimuli. This is important because for resolving most of the questions requiring imaging, truly orthogonal stimuli cannot be defined, and because analyzing single-condition data reveals processes that are difficult or impossible to see with differential data (as described within this section below; see also Chapter 89). With the technical advances in VSDI, we were able to obtain time series of single-condition maps millisecond by millisecond, at sub-columnar resolution. We used it to revisit fundamental questions regarding the emergence of orientation selectivity in visual cortex (Sharon and Grinvald 2002).

Two families of mechanisms have been proposed to play a role in the emergence of the highly orientation-selective responses of cortical visual neurons, a property not shared by their thalamic inputs. Feedforward-only models suggest appropriate alignment of thalamic input as the mechanism, whereas recurrent models suggest that intracortical interactions are more important. Measuring the dynamics of the response to the input by visual cortex, as enabled by VSDI, is relevant to this question because the feedforward explanation predicts that orientation selectivity should remain constant with time from stimulus onset, whereas if recurrent interactions are important, then selectivity should change as the cortical network performs its processing.

To answer the question of dynamics of orientation selectivity, we first acquire the crucial high-quality single-condition maps. A time series of the initial response is shown in Figure 2A for two orthogonal orientations. As soon as the response is observable, the two orthogonal stimuli are shown to preferentially activate complementary patches of cortex. This pattern is, of course, easily observed in the differential map, Figure 2B,C. To calculate orientation tuning curves, however, differential maps and time courses are of no use, and single-condition responses are needed (Fig. 2D).

In experiments where six different orientations were shown, the tuning curves at each point in time could be calculated directly from the single-condition responses (Fig. 2E). The immediate impression is that of tuning curves with a constant shape but changing amplitude. Indeed, the half-width at half-height of tuning curves was steady right from response onset (Sharon and Grinvald 2002). Sustained intracortical processing, therefore, does not seem to be needed to determine orientation tuning width, at least for the majority of the population.

There are additional aspects of the response that affect orientation selectivity, not just tuning width. As shown in Figure 2C, the modulation depth of the response (difference between preferred and orthogonal) does change, decreasing after a peak at 100 msec. An intriguing phenomenon is nearly undetectable in the differential time course, but very obvious when considering the evoked single-condition responses (Fig. 2D). There is a notch in the evoked response, equivalent to a deceleration followed by acceleration in the rise time—this is termed the evoked DA (deceleration–acceleration) notch. It is likely caused by a peaking at 50–80 msec of a suppressive mechanism. This is the kind of dynamics one would expect from intracortical processing. Furthermore, the evoked DA notch is more pronounced in response to the orthogonal stimulus than to the preferred (see Fig. 2D and Sharon and Grinvald 2002).

These results suggest that thalamic input may be the major determinant of orientation tuning *width* for most cortical neurons, but that intracortical processing is critical in *amplifying* the orientation-selective component of the response. They also hint that intracortical suppression contributes to this process by preventing the orthogonal response from increasing as rapidly as the response to the preferred orientation. The interaction between the evoked response and ongoing activity is described next.

SELECTIVE VISUALIZATION OF NEURONAL ASSEMBLIES

It has been suggested that neurons operate in assemblies (Hebb 1949), networks of neurons, that may or may not reside locally, which communicate coherently to perform the computations required for various tasks. Thus, in VSDI, neurons at the recorded site may belong to different neuronal assemblies. To explore cortical computations, it is often interesting to look at the dynamics of individual neuronal assemblies, rather than at the activity originating from heterogeneous populations containing many assemblies. A significant contribution of real-time optical imaging has been the visualization of the dynamics of coherent neuronal assemblies; i.e., neuronal assemblies in which the activity of cells is time-locked. The firing of a single neuron is used as a time reference to selectively visualize activity that is synchronized with it; i.e., only the activity in the assembly to which the reference neuron belongs (Grinvald et al. 1991; Arieli et al. 1995).

To isolate and study the spatiotemporal organization of neuronal assemblies, VSDI was combined with single-unit recordings and subsequent spike-triggered averaging of the optical recordings. With sufficient averaging (Fig. 3), the neuronal activity not time-locked to the reference neuron was averaged out, enabling the selective visualization of those cortical locations in which activity consistently occurred coherently with the firing of the reference neuron. The visual cortex (area 18) of anesthetized cats was stained with the dye RH-795, and either ongoing (spontaneous, in the absence of visual stimulation) or evoked activity was recorded continuously for 70 sec. We recorded simultaneously optical signals from 124 sites, together with electrical recordings of local field potentials (LFP) and single unit recordings (1–3 isolated units recorded with the same electrode). The spike-triggered averaging analysis showed that the averaged optical signal at the electrode site had a peak that temporally coincided with the occurrence of a peak in the LFP. The dye signal was similar to the local field potential recorded from the same site. This indicates the expected result that many neurons next to the electrode site had coherent firing patterns. Interestingly, however, the dye signals more distant from the electrophysiological recording site were heterogeneous, indicating that VSDI provides a better spatial resolution than field potential recordings.

In 88% of the neurons recorded during spontaneous activity, a significant correlation was found between the occurrence of a spike and the optical signal recorded in a large cortical region surrounding the recording site, including cortical sites up to 6 mm away. This result indicates that spontaneous activity of single neurons is not an independent process, but is time-locked to the firing or the synaptic inputs from numerous neurons, all activated in a coherent fashion, even without a sensory input. Surprisingly, it was found that the amplitude of this coherent ongoing activity, recorded optically, was often almost as large as the activity evoked by optimal visual stimulation. The amplitude of the ongoing activity, which was directly and reproducibly related to the spontaneous spikes of a single neuron, was, on average, as high as 54% of the amplitude of the visually evoked response by optimal sensory stimulation, recorded optically. Coherent activity was detected even at distant cortical sites up to 6 mm apart. On the other hand, in a few cases, it was found that the spontaneous activity of two adjacent neurons, isolated by the same electrode and sharing the same orientation preference, was correlated with two different spatiotemporal patterns of coherent activity, suggesting that adjacent neurons in the same orientation column can belong to different neuronal assemblies.

A subsequent study explored the relationship between the spatial pattern of population activity coherent with a single neuron's spikes during spontaneous and evoked activity. They were found to be very similar. Furthermore, it has been reported that the firing rate of a spontaneously active single neuron strongly depends on the instantaneous spatial pattern of ongoing population activity in a large cortical area: During spontaneous activity, whenever the instantaneous spatial population pattern correlated highly with the spatial population pattern evoked by the preferred stimulus of the recorded neuron, its firing rate increased. This was used to reconstruct the spontaneous activity of single neurons based on the global spontaneous activity (Tsodyks et al. 1999).

These results indicate that the spontaneous firing of single neurons is tightly linked to the cortical networks in which they are embedded. The idea of a neural network is a central concept in theo-

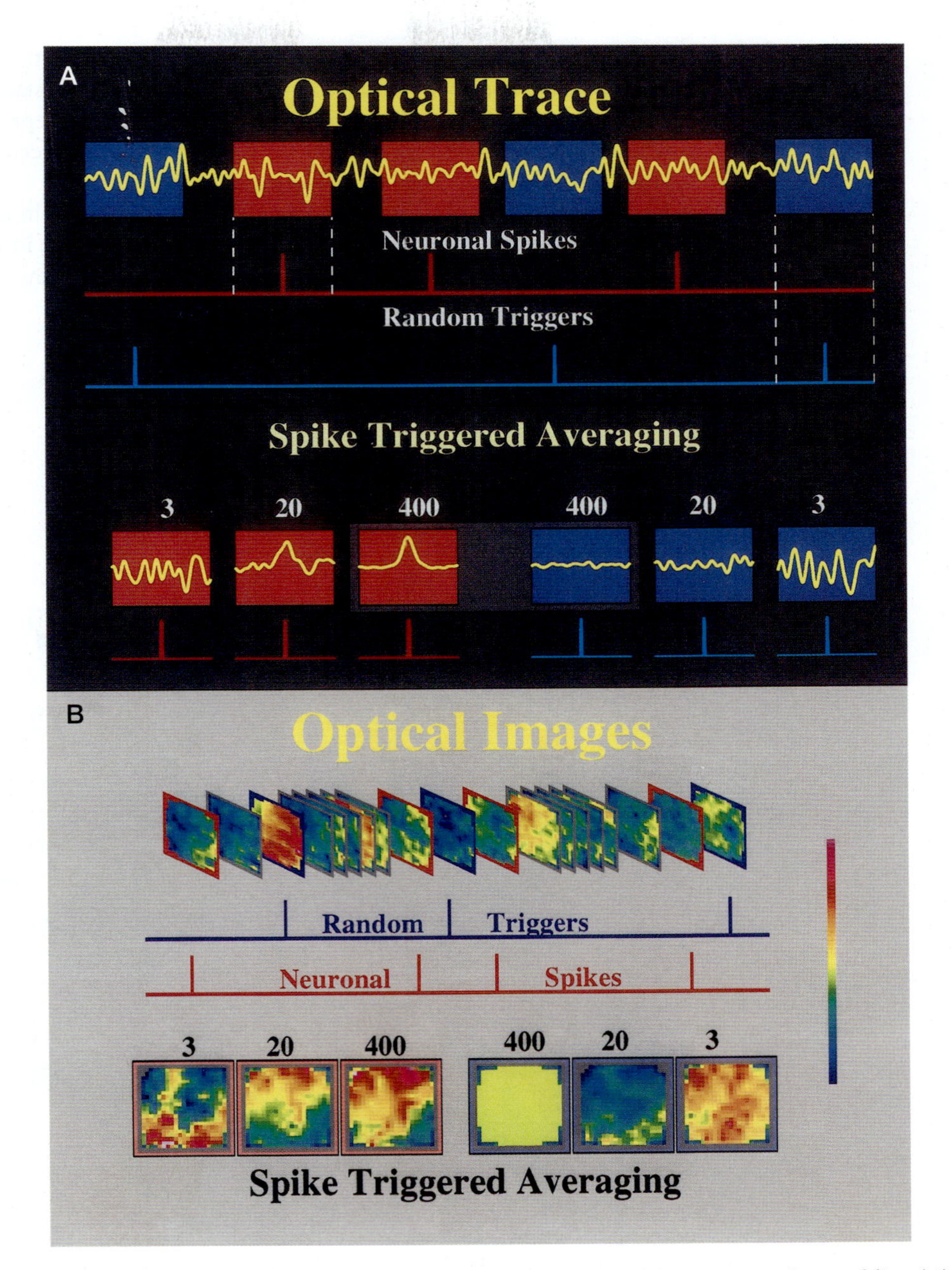

FIGURE 3. Procedure for selective visualization of the dynamics of coherent neuronal assemblies. (*A*) Time course of the optical signal obtained by spike-triggered averaging. The top yellow trace shows the amplitude of the optical signal reflecting compound electrical activity from a given cortical site, measured for 8 sec. The red trace below shows the simultaneously recorded action potentials from the reference neuron. The long recording session was subdivided into 1-sec time segments (*red windows* on the top trace) each centered on the timing of the action potential. The blue trace below shows random virtual spikes that are used as a control for the procedure (*blue windows*). The bottom traces in the red windows show the time course of the spike-triggered average after averaging 3, 20, and 400 time segments during which action potentials occurred. The traces in the blue windows show the results obtained from averaging the control virtual spikes. A clear coherent activity is detected already after averaging 20 events. (*B*) Spatial patterns of movies obtained by spike-triggered averaging. The top shows a series of images in the form of a movie instead of showing the activity in a single cortical site depicted in panel *A* above. The two traces below show the timing of simultaneously recorded action potentials and virtual action potentials that served as a control (*red* and *blue* traces, respectively). The bottom frame shows the spatial pattern observed at a given time resulting from spike-triggered averaging (three windows, *bottom left*). Note that the control patterns are rather flat already after averaging 20 random events without real action potentials (three *blue windows* at the *bottom right*). Color scale: From blue to pink indicates membrane potential changes from lowest to highest level.

retical brain research, and it is finally possible to directly visualize the cortical networks and their states in action, at high spatiotemporal resolution. Exploring cortical states is likely to reveal new fundamental principles about neural strategies for cortical processing, representations of objects, memories, context, expectations, and particularly about the interplay between internal cortical representations and the sensory input in primary sensory areas.

The finding that the amplitude of spontaneous ongoing activity in neuronal assemblies is nearly as large as evoked activity suggests that it may play an important role in shaping spatiotemporal patterns evoked by sensory stimuli (Arieli et al. 1996a). Therefore, it is important to be able to study the dynamics of ongoing and evoked activity without signal averaging.

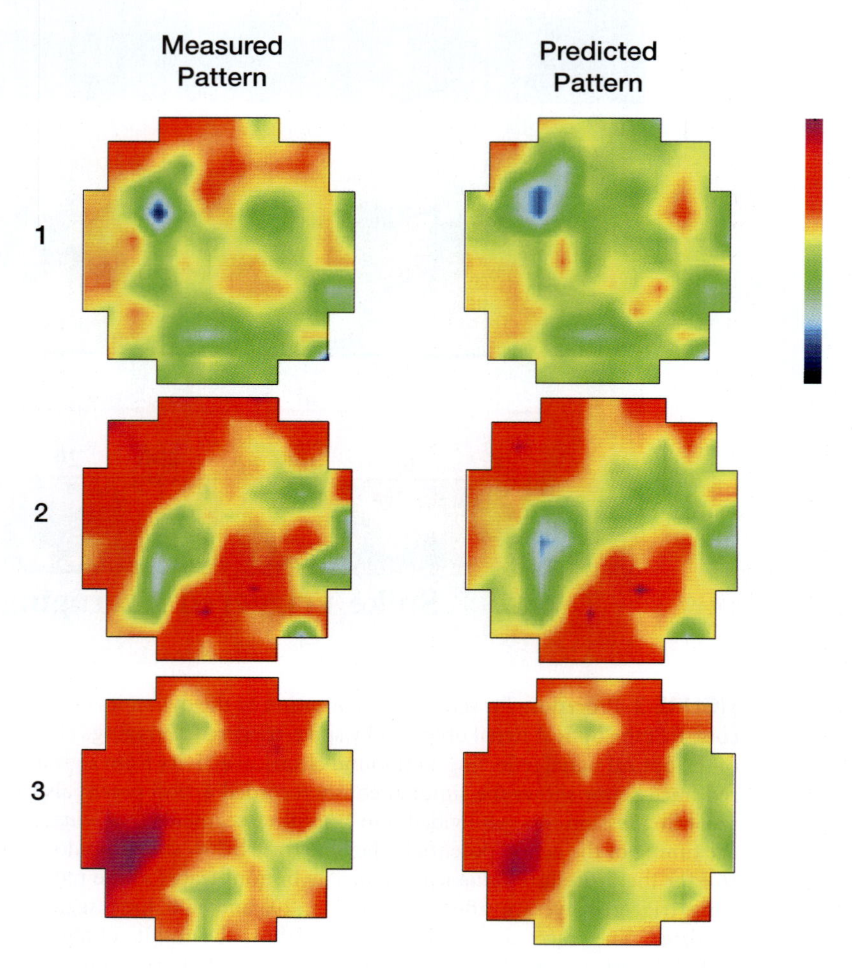

FIGURE 4. Predicting cortical evoked responses in spite of their large variability. Three examples of comparing predicted and measured responses are shown. A single trial response to a stimulus was predicted by summing the reproducible response component (estimated by averaging many trials of evoked responses) and the ongoing network dynamics (approximated by the initial state, i.e., the ongoing pattern just prior to the onset of the evoked response). (*Left columns*) Measured responses. (*Right columns*) Predicted responses, obtained by adding the initial state in each case. Color scale as in Fig. 3.

ONGOING ACTIVITY IN CORTICAL PROCESSING: IMAGING WITHOUT SIGNAL AVERAGING

Despite the large biological noise originating from respiration and heartbeat pulsations, reliable VSDI was accomplished even without signal averaging, using off-line correction procedures instead. This allowed, for the first time, exploration of the dynamics of spontaneous ongoing population activity and its interaction with evoked activity.

In the mammalian visual cortex, evoked responses to repeated presentations of the same stimulus exhibit a large variability. It has been found that this variability results from ongoing activity, reflecting the dynamic state of the cortical network. The central finding in the initial studies of ongoing activity using VSDI without signal averaging was that, despite this large variability, the evoked responses in single trials can be predicted by taking into account the preceding ongoing activity (Fig. 4). The evoked response in a given single-stimulus presentation can be predicted by summing the preceding ongoing activity with the evoked response, averaged over many presentations. This prediction is valid as long as the ongoing activity pattern, which presumably continues to change during the evoked response, is still similar to the initial state (Arieli et al. 1996a).

These findings indicate that the old notions of what "noise" in brain activity represents may have to be revised. Since the ongoing activity is often very large, it would be expected to play a major role in cortical function. It may provide the neuronal substrate for the dependence of sensory information processing on context, attention, behavioral and consciousness states, memory retrieval, and other aspects of cognitive function. Preliminary experiments have shown that behavior itself is also affected by ongoing activity (Arieli et al. 1996b).

The newly improved SNR of VSDI enabled exploration of the dynamics of ongoing activity and its relation to internal representations of sensory attributes. It was found that ongoing activity in the visual cortex of the anesthetized cat is composed of dynamically switching intrinsic cortical states, many of which closely correspond to orientation maps. When such an orientation state emerged spontaneously, it spanned several cortical hyper-columns and was usually followed by a state corresponding to a proximal orientation. Otherwise, it dissolved into an unrecognized state or noisy pattern after a few tens of milliseconds (Grinvald et al. 1991; Shoham et al. 1991; Kenet et al. 1997, 2003).

The hypothesis that in the awake animal ongoing activity is also composed of a set of states is currently being tested. Assuming such cortical states do exist also in the awake state poses a question: Why are switching orientations not usually perceived when we close our eyes? It is conceivable which states that are intrinsic to lower areas are constantly interacting with feedback from higher areas, as well as the incoming stimulation. They may not be as persistent in duration and as spatially coherent in the awake animal as in the anesthetized preparation. It has already been shown that spontaneous activity interacts with evoked responses and affects behavior. The above findings suggest that dynamically switching cortical states could express the brain's internal context, representing and influencing memory, perception, and behavior. Such studies obviously require the use of the awake behaving monkey rather than anesthetized subjects.

LONG-TERM VSDI OF CORTICAL DYNAMICS IN BEHAVING MONKEYS

Chronic optical imaging based on newer voltage-sensitive dyes was developed to facilitate exploration of the spatial and temporal patterns underlying higher cognitive functions in the neocortex of behaving monkeys. The technique was used to explore cortical dynamics, with high spatial and temporal resolution, over periods of up to 1 year from the same patch of cortex (Slovin et al. 2002). The visual cortices of trained macaques were stained one to three times a week and, immediately after each staining session, the monkey started to perform a simple behavioral task, while the primary visual cortex (V1) and the secondary area (V2) were imaged with a fast optical imaging system. Long-term, repeated VSDI from the same cortical area did not disrupt the normal cortical architecture, as confirmed repeatedly by optical imaging based on intrinsic signals. The spatial patterns of functional maps obtained by VSDI in response to full-field stimuli were essentially identical to those obtained from the same patch of cortex by imaging based on intrinsic signals. Upon comparing the relative amplitudes of the evoked signals and differential maps obtained using these two different imaging methods, it was found that the relative contribution of subthreshold versus spiking activity was higher in the voltage-sensitive dye signal than in the intrinsic signal. In the awake monkey, evoked activity could also be detected without signal averaging, similarly to the above-mentioned anesthetized cat experiments.

The same approach also proved useful in exploration of the motor cortex. Electrical recording studies during Go/No-Go tasks found that primary motor cortex (M1) neurons display changes in activity which begin primarily after a Move cue, whereas dorsal premotor (PMd) neurons, known to be engaged in motor planning, display changes in activity that can begin during the Instructed Delay periods. One monkey was trained to perform a Go/No-Go task. The monkey began a trial by placing his hand over a small photodiode in front of him. This caused a small white square to appear on a computer screen. After a variable pre-cue period (3000–4000 msec), the white square was changed to red (Go mode) or green (No-Go mode). Then, after a variable Instructed Delay period (300–1500 msec), a luminance change of the square told the monkey to follow the prior instruction: Uncover (Go) or continue to cover the photodiode (No-Go). In the PMd, the VSD signal changes began shortly after the onset of the Go or No-Go instructions (i.e., at the start of the Instructed Delay periods). Relatively large signals were observed in the PMd area during both Go and No-Go. The VSD signals were much smaller in M1 during the No-Go mode. These observations suggest that the preparation to perform or withhold a motor response is associated mainly with marked subthreshold activity in the PMd (Strick et al. 2003).

This work demonstrates a new way of recording and analyzing the dynamics of population activity in behaving monkeys, with high spatial and temporal resolution. The combination of VSDI with traditional electrical recordings can also be readily adapted to facilitate the selective visualization of coherent activity in neuronal assemblies involved in dynamic representations and processing of sensory input, as well as in the planning, control, and execution of motor output (Slovin et al. 2003; Strick et al. 2003). Another powerful combination is described below.

COMBINED VSDI AND MICROSTIMULATION IN FRONTAL AND MOTOR CORTEX

Despite extensive use of microstimulation in various brain regions to affect neuronal and behavioral responses (see, e.g., Newsome et al. 1989; Cohen and Newsome 2004), the spatiotemporal pattern of activation it evokes has not been characterized, leaving open many questions about the spread of activation, both direct, from the electric shock itself, and indirect, from synaptic activity. Microstimulation can be incorporated into VSDI studies of frontal and motor cortex of awake monkey.

The frontal eye field and neighboring area, 8Ar, of the primate cortex are involved in programming and execution of saccades. Electrical microstimulation in these regions elicits short-latency contralateral saccades. To determine how spatiotemporal dynamics of microstimulation-evoked activity are converted into saccade plans, the combination of VSDI and microstimulation in behaving monkeys was used. Short stimulation trains evoked a rapid and widespread wave of depolarization followed by an unexpectedly large and prolonged hyperpolarization. During this hyperpolarization, saccades are almost exclusively ipsilateral, suggesting an important role for hyperpolarization in determining saccade goal (Seidemann et al. 2002). This result demonstrates that neural activity, with composite spatiotemporal dynamics, can be elicited by microstimulation and emphasizes the importance of further characterization of microstimulation-evoked activity for the interpretation of the behavioral effects of microstimulation.

Intracortical stimulation has also been used as a tool to discover the pattern of muscle and/or movement representation in motor areas of the cortex. VSDI was used to examine the patterns of activation evoked by different stimulation parameters in the primary motor cortex (M1). These experiments used much finer stimulation conditions than the previous work. Single cathodal pulses in M1 (15–30 μA, 0.2-msec duration) evoked short-latency dye responses that continued for 40–50 msec after the onset of stimulation. These responses spread in an asymmetric manner from the stimulation site in M1; the width of the activated area was ~ 1.5–3 mm. Single pulses, larger than 15 μA, also evoked activity at spatially separate sites within M1. Short trains of pulses in M1 (2–20 pulses, 330–500 Hz, 30–100 μA) produced activation that spread from the electrode site to adjacent premotor areas and to the contralateral hemisphere. A distinct hyperpolarization followed the activation produced by a high-current (> 60 μA), short-stimulus train (2–20 pulses). Finally, long-stimulus trains (70 μA, 200 pulses, 330 Hz) evoked activation that spread over large portions of M1 and the adjacent premotor areas. The responses outlasted the stimulus train by several hundred milliseconds (Slovin et al. 2003). These results emphasize the importance of considering the spread of activation when evaluating the results of M1 stimulation.

CONCLUSIONS AND OUTLOOK

VSDI is based on the use of voltage-sensitive dyes, and the quality of results it produces is therefore limited by the properties of these dyes. The results presented here indicate that the technique has finally matured sufficiently to explore the neocortex in ways never feasible before. Furthermore, it is expected that the development of new voltage-sensitive dyes and future genetic engineering of suitable in vivo probes (see, e.g., Miyawaki et al. 1997; Siegel and Isacoff 1997; Chapter 77) will make the experiments much easier and will substantially improve the quality of results. Of particular importance are probes that will stain only specific cell types and/or specific cellular compartments (e.g., axons, somata, dendrites). Additional developments of other technologies, using second harmonic generation (Millard et al. 2003; see Chapter 60), multiphoton (Chapter 7), red-edge excitation (Kuhn and Fromherz 2003), and three-dimensional imaging, are highly desirable. Because VSDI excels at revealing spatiotemporal patterns of the synaptic potential—the input to a neocortical region, but lacking sensitivity to spiking its output—it should be highly fruitful to combine it with the new approach to calcium imaging in vivo that primarily reflects the output (Stosiek et al. 2003; Chapter 87). To increase the dimensionality of meaningful neurophysiological data obtained from the same patch of cortex, VSDI should be combined with targeted tracer injections, retrograde labeling, microstimulation, and intracellular and extracellular recordings. VSDI, combined with electrical recordings, allows investigators to obtain information about neuronal activity in coherent neuronal assemblies, within the neocortex, at subcolumnar resolution.

No alternative imaging technique for visualizing functional organization in the living brain provides a comparable spatial and temporal resolution. It is this level of resolution that makes it possible to address questions of both *where* and *when* processing is performed, promising that this technique will contribute to the study of *how*—revealing fundamental principles of neural coding and processing strategies at the population level.

ACKNOWLEDGMENTS

We thank Chaipi Wijnbergen, Yoval Toledo, and Dov Etner for their assistance, and our previous coworkers, Ron Frostig, Edmund Lieke, Doron Shoham, Daniel Glaser, Amos Arieli, Eyal Seideman, Tal Kenet, and Misha Tsodyks, for their original contributions. This work was supported by grants from the Grodetsky Center, Goldsmith, Glasberg, Heineman, and Korber foundations, BMBF, the ISF, and Ms. Enoch.

REFERENCES

Arieli A. and Grinvald A. 2002. Combined optical imaging and targeted electrophysiological manipulations in anesthetized and behaving animals. *J. Neurosci. Methods* **116:** 15–28.

Arieli A., Grinvald A., and Slovin H. 2002. Dural substitute for long-term imaging of cortical activity in behaving monkeys and its clinical implications. *J. Neurosci. Methods* **114:** 119–133.

Arieli A., Shoham D., Hildesheim R., and Grinvald. A. 1995. Coherent spatio-temporal pattern of ongoing activity revealed by real time optical imaging coupled with single unit recording in the cat visual cortex. *J. Neurophysiol.* **73:** 2072–2093.

Arieli A., Sterkin A., Grinvald A., and Aertsen A. 1996a. Dynamics of ongoing activity: Explanation of the large in variability in evoked cortical responses. *Science* **273:** 1868–1871.

Arieli A., Donchin O., Aertsen A., Bergman H., Gribova A., Grinvald A., and Vaadia E. 1996b. The impact of on going cortical activity on evoked potential and behavioral responses in the awake behaving monkey. *Neuroscience* **22:** 2022. (Abstr.)

Cinelli A.R. and Kauer J.S. 1992. Voltage sensitive dyes and functional-activity in the olfactory pathway. *Annu. Rev. Neurosci.* **15:** 321–352.

———. 1995. Salamander olfactory-bulb neuronal-activity observed by video-rate, voltage sensitive dyes imaging. II. Spatial and temporal properties of responses evoked by electrical stimulation. *J. Neurophysiol.* **73:** 2033–2052.

Cinelli A.R., Neff S.R., and Kauer J.S. 1995. Salamender olfactory-bulb neuronal-activity observed by video-rate, voltage sensitive dyes imaging. I. Characterization of the recording system. *J. Neurophysiol.* **73:** 2017–2032.

Cohen L.B. and Lesher S. 1986. Optical monitoring of membrane potential: Methods of multisite optical measurement. *Soc. Gen. Physiol. Ser.* **40:** 71–99.

Cohen L.B., Slazberg B.M., and Grinvald A. 1978. Optical methods for monitoring neurons activity. *Annu. Rev. Neurosci.* **1:** 171–182.

Cohen L.B., Salzberg B.M., Davila H.V., Ross W.N., Landowne D., Waggoner A.S., and Wang C.H. 1974. Changes in axon fluorescence during activity: Molecular probes of membrane potential. *J Membr Biol.* **19:** 1–36.

Cohen M.R. and Newsome W.T. 2004. What electrical microstimulation has revealed about the neural basis of cognition. *Curr. Opin. Neurobiol.* **14:** 169–177.

De Weer P. and Salzberg B.M., eds. 1986. Optical methods in cell physiology. *Soc. Gen. Physiol. Ser.*, vol. 40.

Derdikman D., Hildesheim R., Ahissar E., Arieli A., and Grinvald A. 2003. Imaging spatio-temporal dynamics of surround inhibition in the barrels somatosensory cortex. *J. Neurosci.* **23:** 3100–3105.

Gilbert C.D. and Wiesel T.N. 1983. Clustered intrinsic connections in cat visual cortex. *J. Neurosci.* **3:** 1116–1133.

Grinvald A. 1984. Real time optical imaging of neuronal activity: From single growth cones to the intact brain. *Trends Neurosci.* **7:** 143–150.

———. 1985. Real-time optical mapping of neuronal activity: From single growth cones to the intact mammalian brain. *Annu. Rev. Neurosci.* **8:** 263–305.

Grinvald A., Manker A., and Segal M. 1982a. Visualization of the spread of electrical activity in rat hippocampal slices by voltage sensitive optical probes. *J. Physiol.* **333:** 269–291.

Grinvald A., Salzberg B.M., and Cohen L.B. 1977. Simultaneous recordings from several neurons in an invertebrate central nervous system. *Nature* **268:** 140–142.

Grinvald A., Cohen L.B., Lesher S., and Boyle M.B. 1981. Simultaneous optical monitoring of activity of many neurons in invertebrate ganglia, using a 124 element 'Photodiode' array. *J. Neurophysiol.* **45:** 829–840.

Grinvald A., Frostig R.D., Lieke E., and Hildesheim R. 1988. Optical imaging of neuronal activity. *Physiol. Rev.* **68:** 1285–1366.

Grinvald A., Hildesheim R., Farber I.C., and Anglister L. 1982b. Improved fluorescent probes for the measurement of rapid changes in membrane potential. *Biophys. J.* **39:** 301–308.

Grinvald A., Lieke E.E., Frostig R.D., and Hildesheim R. 1994. Cortical point-spread function and long-range lateral interactions revealed by real-time optical imaging of macaque monkey primary visual cortex. *J. Neurosci.* **14:** 2545–2568.

Grinvald A., Anglister L., Freeman J.A., Hildesheim R., and Manker A. 1984. Real time optical imaging of naturally evoked electrical activity in the intact frog brain. *Nature* **308:** 848–850.

Grinvald A., Lieke E., Frostig R.D., Gilbert C.D., and Wiesel T.N. 1986. Functional architecture of cortex revealed by optical imaging of intrinsic signals. *Nature* **324:** 361–364.

Grinvald A., Bonhoeffer T., Malonek D., Shoham D., Bartfeld E., Arieli A., Hildesheim R., and Ratzlaff E. 1991. Optical imaging of architecture and function in the living brain. In *Memory: Organization and locus of change* (ed. L. Squire), pp. 49–85. Oxford University Press, New York.

Grinvald A., Shoham D., Shmuel A., Glaser E.D., Vanzetta I., Shtoyerman E., Slovin H., Wijnbergen C., Hildesheim R., Sterkin A., and Arieli A. 1999. In-vivo optical imaging of cortical architecture and dynamics. In *Modern techniques in neuroscience research* (ed. U. Windhorst and H. Johansson), pp. 893–969. Springer, New York.

Hebb D. 1949. The first stage of perception: Growth of the assembly. In *The organization of behavior: A neuropsychological theory*, pp. 60–78. Wiley, New York.

Hirota A., Sato K., Momosesato Y., Sakai T., and Kamino K. 1995. A new simultaneous 1020 site optical recording system for monitoring neuronal activity using voltage sensitive dyes. *J. Neurosci. Methods* **56:** 187–194.

Hubel D.H. and Wiesel T.N. 1962. Receptive fields, binocular interactions and functional architecture in the cat's visual cortex. *J. Physiol.* **160:** 106–154.

Iijima T., Matosomoto G., and Kisokoro Y. 1992. Synaptic activation of rat adrenal-medula examined with a large photodiode array in combination with voltage sensitive dyes. *Neuroscience* **51:** 211-219.

Jancke D., Chavane F., Naaman S., and Grinvald A. 2004. Imaging cortical correlates of a visual illusion. *Nature* **428:** 424–427.

Kamino K. 1991. Optical approaches to ontogeny of electrical activity and related functional-organization during early heart development. *Physiol. Rev.* **71:** 53–91.

Kauer J.S. 1988. Real-time imaging of evoked activity in local circuits of the salamander olfactory bulb. *Nature* **331:** 166–168.

Kauer J.S., Senseman D.M., and Cohen M.A. 1987. Odor-elicited activity monitored simultaneously from 124 regions of the salamander olfactory bulb using a voltage-sensitive dye. *Brain Res.* **25:** 255–61.

Kenet T., Arieli A., Grinvald A., and Tsodyks M. 1997. Cortical population activity predicts both spontaneous and evoked single neuron firing rates. *Neurosci. Lett.* **48:** S27.

Kenet T., Grinvald A., Tsodyks M., and Arieli A. 2003. Spontaneously occurring cortical representations of visual attributes. *Nature* **425**: 954–956.

Kleinfeld D. and Delaney K.R. 1996. Distributed representation of vibrissa movement in the upper layers of somatosensory cortex revealed with voltage-sensitive dyes. *J. Comp. Neurol.* **375:** 89–108.

Konnerth A. and Orkand R.K. 1986. Voltage sensitive dyes measure potential changes in axons and glia of frog optic nerve. *Neuroscience Lett.* **66:** 49–54.

Kuhn B. and Fromherz P. 2003. Anellated hemicyanine dyes in neuron membrane: Molecular stark effect and optical voltage recording. *J. of Phys. Chem. B* **107:** 7903–7913.

Lev-Ram R. and Grinvald A. 1986. K+ and Ca2+ dependent communication between myelinated axons and oligodendrocytes revealed by voltage-sensitive dyes. *Proc. Natl. Acad. Sci.* **83:** 6651–6655.

Loew L.M. 1987. *Optical measurement of electrical activity*. CRC Press, Boca Raton, Florida.

Millard A.C., Jin L. , Lewis A., and Loew L.M. 2003. Direct measurement of the voltage sensitivity of second-harmonic generation from a membrane dye in patch-clamped cells. *Opti. Lett.* **28:** 1221–1223.

Miyawaki A., Llopis J., Heim R., Caffery J.M., Adams J.A., Ikura M., and Tsien R.Y. 1997. Fluorescent indicators for Ca2+ based on green fluorescent proteins and calmodulin. *Nature* **388:** 882–887.

Mountcastle, V.B. 1957. Modality and topographic properties of single neurons of cat's somatic sensory cortex. *J. Neurophysiol.* **20:** 408–434.

Newsome W.T., Britten K.H., and Movshon J.A. 1989. Neuronal correlates of a perceptual decision. *Nature* **341:** 52–54.

Orbach H.S. 1988. Monitoring electrical activity in rat cerebral cortex. In *Spectroscopic membrane probes* (ed. L.M. Loew), vol. 3, pp. 115–135. CRC Press, Boca Raton, Florida.

Orbach H.S. and Cohen L.B. 1983. Simultaneous optical monitoring of activity from many areas of the salamander olfactory bulb. A new method for studying functional organization in the vertebrate CNS. *J. Neurosci.* **3:** 2251–2262.

Orbach H.S., Cohen L.B., and Grinvald A. 1985. Optical mapping of electrical activity in rat somatosensory and visual cortex. *J. Neurosci.* **5:** 1886–1895.

Petersen C.H., Grinvald A., and Sakmann B. 2003a. Spatio-temporal dynamics of sensory responses in layer 2/3 of rat barrel cortex measured in vivo by voltage-sensitive dye imaging combined with whole-cell voltage recordings and anatomical reconstructions. *J. Neurosci.* **23:** 1298–1309.

Petersen C.H., Hahn T., Mehta M., Grinvald A., and Sakmann B. 2003b. Interaction of sensory responses with spontaneous depolarization in layer 2/3 barrel cortex. *Proc. Natl. Acad. Sci.* **100:** 13638–13643.

Ross W.N. and Reichardt L.F. 1979. Species-specific effects on the optical signals of voltage sensitive dyes. *J. Membr. Biol.* **48:** 343–356.

Salzberg B.M., Davila H.V., and Cohen L.B. 1973. Optical: Recording of impulses in individual neurons of an invertebrate central nervous system. *Nature* **246:** 508–509.

Salzberg B.M., Obaid A.L., and Gainer H. 1986. Optical studies of excitation secretion at the vertebrate nerve terminal. *Soc. Gen. Physiol. Ser.* **40:** 133–164.

Salzberg B.M., Grinvald A., Cohen L.B., Davila H.V., and Ross W.N. 1977. Optical recording of neuronal activity in an invertebrate central nervous system; simultaneous recording from several neurons. *J. Neurophysiol.* **40:** 1281–1291.

Seidemann E., Arieli A., Grinvald A., and Slovin H. 2002. Dynamics of depolarization and hyperpolarization in the frontal cortex and saccade goal. *Science* **295:** 862–865.

Sharon D. and Grinvald A. 2002. Dynamics and constancy in cortical spatiotemporal patterns of orientation processing. *Science* **295:** 512–515.

Shoham D., Ullman S., and Grinvald A. 1991. Characterization of dynamic patterns of cortical activity by a small number of principal components. *Neuroscience* **17:** 1089. (Abstr.)

Shoham D., Glaser D.E., Arieli A., Kenet T., Wijnbergen C., Toledo Y., Hildesheim R., and Grinvald A. 1999. Imaging cortical architecture and dynamics at high spatial and temporal resolution with new voltage-sensitive dyes. *Neuron* **24:** 791–802.

Siegel M.S. and Isacoff E.Y. 1997. A genetically encoded optical probe of membrane voltage. *Neuron* **19:** 735–741.

Slovin H., Arieli A., Hildesheim R., and Grinvald A. 2002. Long-term voltage-sensitive dye imaging of cortical dynamics in the behaving monkey. *J. Neurophysiol.* **88:** 3421–3438.

Slovin H., Strick P.L., Hildesheim R., and Grinvald A. 2003. Voltage sensitive dye imaging in the motor cortex. I. Intra- and intercortical connectivity revealed by microstimulation in the awake monkey. *Soc. Neuroci. Abstr.* 554.8.

Spors H. and Grinvald A. 2002. Temporal dynamics of odor representations and coding by the mammalian olfactory bulb. *Neuron* **34:** 1–20.

Stosiek C., Garaschuk O., Holthoff K., and Konnerth A. 2003. In vivo two-photon calcium imaging of neuronal networks. *Proc. Natl. Acad. Sci.* **100:** 7319–7324.

Sterkin A., Lampl I., Ferster D., Grinvald A., and Arieli A. 1998. Real time optical imaging in cat visual cortex exhibits high similarity to intracellular activity. *Neurosci. Lett.* 51.S41.

Strick P., Grinvald A., Hildesheim R., and Slovin H. 2003. Voltage sensitive dye imaging in the motor cortex II. Cortical correlates of Go/No-Go delayed response task. *Soc. Neuroci. Abstr.* 918.8.

Takashima I. Kajiwara R., and Iijima T. 2001. Voltage-sensitive dye versus intrinsic signal optical imaging: Comparison of optically determined functional maps from rat barrel cortex. *Neuroreport* **12:** 2889–2894.

Tanifuji M., Yamanaka A., Sunaba R., and Toyama K. 1993. Propagation of excitation in the visual cortex studies by the optical recording. *Jpn. J. Physiol.* **43:** 57–59.

Tasaki I. and Warashina A. 1976. Dye membrane interaction and its changes during nerve excitation. *Photochem. Photobiol.* **24:** 191–207.

Tasaki I., Watanabe A., Sandlin R., and Carnay L. 1968. Changes in fluorescence, turbidity, and birefringence associated with nerve excitation. *Proc. Natl Acad. Sci.* **61:** 883–888.

Toyama K. and Tanifuji M. 1991. Seeing ecxcietation propagation in visual cortical slices *Biomed. Res.* **12:** 145–147.

Tsodyks M., Kenet T., Grinvald A., and Arieli A. 1999. The spontaneous activity of single cortical neuron depends the underlying global functional architecture. *Science* **286:** 1943–1946.

Vranesic I., Iijima T., Ichikawa M., Matsumoto G., and Knopfel T. 1994. Signal transmission in the parallel fiber Purkinje-cell system visualized by high resolution imaging. *Proc. Natl. Acad. Sci.* **91:** 13014–13017.

Waggoner A.S. 1979. Dye indicators of membrane potential. *Annu. Rev. Biophys. Bioeng.* **8:** 47-63.

Waggoner A.S. and Grinvald A. 1977. Mechanisms of rapid optical changes of potential sensitive dyes. *Ann. N.Y. Acad. Sci.* **303:** 217–242.

CHAPTER 91

A Practical Guide: Whole-Cell Recording and Voltage-sensitive Dye Imaging In Vivo

Carl Petersen

Laboratory of Sensory Processing, Brain and Mind Institute, EPFL, CH-1015 Lausanne, Switzerland

This chapter describes an in vivo technique for combining whole-cell recordings of membrane potentials of individual neurons with voltage-sensitive dye imaging (VSDI) of the ensemble neocortical network dynamics.

This approach has been tested on rodent barrel cortex, but it is likely to be applicable, with minor modifications, to other superficial brain areas (including other neocortical areas, cerebellum and olfactory bulb) and other species (such as cat or monkey).

MATERIALS

Voltage-sensitive Dye

RH1691 (Optical Imaging)

Fast Camera

NeuroCCD (Red Shirt Imaging, LLC) or Imager 3001/F (Optical Imaging) or MiCAM (Scimedia)

Optics

Stable DC power supply
Halogen light source
Band-pass filter 630 ± 10 nm
Shutter
Dichroic 650 nm
Long-pass filter 665 nm
Two large aperture video lenses mounted in tandem (Ratzlaff and Grinvald 1991)

Electrophysiology

Micromanipulator with step function and distance measurements
Patch-clamp amplifier

PROCEDURE A

Surgery and Staining

1. Deeply anesthetize the animal (see introductory text for suitable species) with either gas anesthesia (halothane<!> or isoflurane<!>); urethane<!>; or ketamine/xylazine<!>.

   Note: The dose of anesthetic must be adequate for surgical tolerance before proceeding and must conform to local animal care committee guidelines. Paw withdrawal, whisker movements, and eye-blink reflexes should be largely absent. The state of the animal must be carefully monitored during the surgery and staining procedures, which take ~3 hours to complete, and during the experimental recording session.

2. Keep the anesthetized animal on a heating blanket to maintain the rectal temperature at 37°C. Inject local anesthetic (lidocaine<!>) under the skin before making incisions.
3. After removing or reflecting the skin covering the skull, carefully scrape the bone clean.

   Note: Further cleaning of the bone can be performed, if necessary, with 1% hydrogen peroxide<!>.
4. Apply a thin layer of cyanoacrylate glue<!> to the surrounding region of the bone, away from where the craniotomy will be performed (this helps dental cement adhesion).
5. Glue a metal head plate to the skull with dental cement. The plate must be tangential to the region of cortex to be imaged, and the hole in the plate must be positioned over this region.
6. When the dental cement has hardened, immobilize the skull and minimize movement of the brain by fixing the head plate firmly between fixed metal posts.
7. Perform a craniotomy of 3 x 3 mm by drilling within the hole of the head plate. Be careful not to damage the underlying brain tissue.
8. Finally, remove the dura, leaving the pia of the underlying cortex exposed.
9. Dissolve the voltage-sensitive dye (Shoham et al. 1999) to 0.1–1 mg/ml in Ringer's solution (Ringer's solution: 135 mM NaCl, 5 mM KCl<!>, 5 mM HEPES, 1.8 mM $CaCl_2$<!>, 1 mM $MgCl_2$<!>).
10. Apply a small quantity (~250 µl) of this dye solution to the craniotomy. Seal the chamber with a glass coverslip, applying very slight pressure to prevent brain edema.
11. Leave the cortex to stain for 1–2 hours. During this period, the dye will diffuse into the superficial layers of the neocortex. At the end of the staining period, remove unbound dye by washing the cortex extensively with Ringer's solution.

    Note: The cortex should now have a pale blue color.
12. Cover the craniotomy with 1% agarose dissolved in Ringer's solution, and place a glass coverslip on top.

    Note: The glass coverslip should be large enough to cover the width of the craniotomy, but little more. This will allow access for whole-cell recordings to be made from both sides of the animal.

Caution: *See Appendix 3 for appropriate handling of materials marked with <!>.*

PROCEDURE B

Imaging

Illuminate with green light at ~530 nm and record the blood vessel pattern on the cortical surface using the camera (Fig. 1A). Move the focal plane 300 µm into the cortex and excite the voltage-sensitive dye with epifluorescent light at 630 nm.

Note: Emitted light is long-pass filtered (665 nm), forming the voltage-sensitive dye signal, which should be recorded at frame rates faster than 100 Hz. Heartbeat-related signals form the largest artifacts. The timing of these signals can be recorded via an electrocardiogram. The artifacts can then be removed by computer processing to improve the resolution of the collected signals.

PROCEDURE C

Whole-Cell Recording

1. Perform whole-cell recordings simultaneously with the voltage-sensitive dye imaging.

   Note: This allows the measurement of membrane potential of an individual cell. The ensemble-averaged membrane potentials of the neurons in the network are imaged with the voltage-sensitive dye.

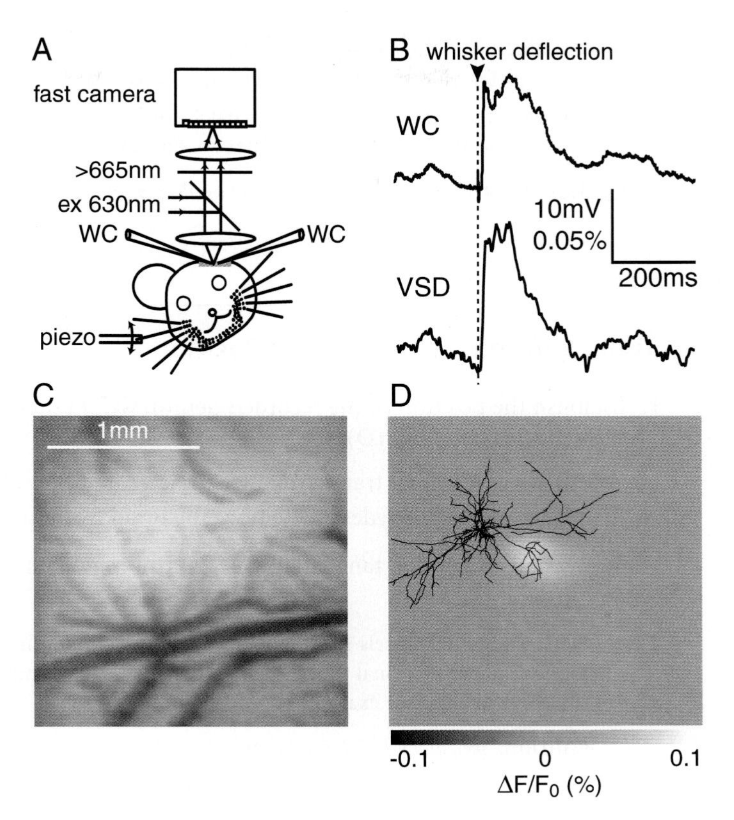

FIGURE 1. Whole-cell recording and voltage-sensitive dye imaging from barrel cortex. (*A*) Experimental setup for the whole-cell (WC) recording and voltage-sensitive dye (VSD) imaging. (*B*) Temporal changes in membrane potential of a neuron recorded in the whole-cell recording configuration (trace labeled WC) in a period during which a brief (2-msec) whisker stimulus was delivered. Voltage-sensitive dye imaging was performed simultaneously and quantified in a spatial region covering 200 x 200 µm immediately surrounding the somatic location of the neuron (trace labeled VSD). A subthreshold sensory response was recorded with a similar time course in both the whole-cell and the voltage-sensitive dye signals. (*C*) The blood vessel pattern on the cortical surface, imaged under green light illumination. (*D*) An image of the sensory response at 15 msec following whisker stimulus, recorded by the voltage-sensitive dye. The sensory depolarizing response at this early time point following a single whisker deflection is localized to a barrel column in somatosensory cortex. In the next milliseconds, the sensory response can spread to cover several square millimeters of cortex. Superimposed on the functional VSD image are the reconstructed axon (*thin lines*) and dendrites (*thick lines*) of the layer 2/3 pyramidal neuron.

2. Fill whole-cell pipettes with intra-pipette solution containing 3 mg/ml biocytin (to allow staining of the recorded neurons). (Intra-pipette solution: 135 mM potassium gluconate<!>, 4 mM KCl<!>, 10 mM HEPES, 10 mM phosphocreatine; 4 mM MgATP, and 0.3 mM Na3GTP; pH 7.2 adjusted with KOH<!>.)

   Note: Pipettes should have a resistance of ~5 MΩ. Monitor the tip resistance by applying brief voltage steps of 5 mV in the voltage-clamp mode while measuring the current flow on an oscilloscope. Apply a positive pressure of ~200 mmHg on the pipette.

3. Slowly advance the electrode through the agarose under the coverslip and into the cortex. When the tip is close to the chosen recording site, reduce the positive pressure to ~30 mmHg.
4. Advance the pipette in 2-µm steps until the tip resistance suddenly increases (indicating contact with a cell membrane). Release the pressure in the pipette and apply light suction until a gigaohm seal is formed.
5. Establish whole-cell recording configuration by rupturing the membrane inside the pipette. This can be achieved by either applying brief suction pulses or slowly increasing the suction pressure.

6. After collecting data, slowly retract the whole-cell recording pipette, while monitoring whole-cell capacitance transients with 5-mV voltage steps.

   Note: During the retraction, the excised patch configuration should be established. This ensures that the neuron remains intact and viable for later anatomical staining.

   Caution: *See Appendix 3 for appropriate handling of materials marked with <!>.*

PROCEDURE D

Anatomical Analysis of Neuronal Structure and Position

1. Establish the position of the recorded neuron within the functional map obtained using the voltage-sensitive dye (Fig. 1D):

   a. Perfuse the animal transcardially with ice-cold 0.1 M phosphate buffer, pH 7.2, and then with 4% paraformaldehyde<!>. Remove the brain from the skull and fix overnight at 4°C.

   b. Section the cortex tangentially at 100 μm and stain for biocytin using ABC kit (Vectastain Laboratories).

      Note: The blood vessels initially imaged with green illumination now provide the link between the location of the neuronal processes visualized by the anatomical stain and the functional voltage-sensitive dye images (see example of application below).

      Caution: *See Appendix 3for appropriate handling of materials marked with <!>.*

EXAMPLE OF APPLICATION

Applied to the barrel cortex, these techniques have provided information concerning the relationship between individual neurons and the ensemble spatiotemporal dynamics of sensory responses (Petersen et al. 2003a,b). Figure 1A shows the experimental setup for the whole-cell (WC) recording and voltage-sensitive dye (VSD) imaging. Figure 1B shows the temporal changes in membrane potential of a neuron recorded in the whole-cell recording configuration (trace labeled WC) in a period during which a brief whisker stimulus was delivered. Voltage-sensitive dye imaging was performed simultaneously and quantified in a spatial region covering an area of 200 x 200 μm immediately surrounding the somatic location of the neuron. A subthreshold sensory response was recorded with a similar time course in both membrane potential and voltage-sensitive dye signal. Figure 1C illustrates the blood vessel pattern on the cortical surface imaged under green light illumination. Figure 1D shows an image of the spatial extent of an early time point during the sensory response recorded by the voltage-sensitive dye. After this experiment, the brain was fixed and stained for biocytin, which had been introduced into the neuron during the whole-cell recording. By aligning the neuronal structure and the blood vessel pattern, the reconstructed dendritic (thick lines) and axonal (thin lines) arbors can be superimposed on the voltage-sensitive dye image.

ADVANTAGES AND LIMITATIONS

Both whole-cell (Hamill et al. 1981) and voltage-sensitive dye (Salzberg et al. 1973) recordings can measure rapid changes in membrane potential. Whereas the whole-cell recording provides information about membrane potential at only one point in space, the voltage-sensitive dye signal can be imaged with a fast camera to resolve the spatiotemporal dynamics. However, voltage-sensitive dyes give small signals—often on the order of 10%/100 mV (Zhang et al. 1998). Under experimental conditions, where the dye is nonselectively applied to brain tissue, as described here, it is not possible to resolve voltage-sensitive dye signals from individual neurons or their processes (although this has been achieved by intracellular application of voltage-sensitive dye [Antic et al. 1999]). The technique of applying voltage-sensitive dyes will thus only be useful for measuring the ensemble activity of neuronal networks. The combination of whole-cell recordings with voltage-sensitive dye imaging allows the membrane potential of individual neurons to be placed in the context of the ensemble spatiotemporal dynamics of the network. This is a powerful tool for extending our knowledge of brain function.

It is important to note several limitations of the voltage-sensitive dye technique, which largely stem from the small signal amplitudes and the nonselective staining technique. First, although a close correlation between layer 2/3 pyramidal neuron subthreshold membrane potential changes and the local RH1691 voltage-sensitive dye signal has been reported (Petersen et al. 2003a,b), this may not be the case under all experimental conditions. A number of factors, including the details of the staining procedure, anesthesia, and brain region, might alter the nature of the voltage-sensitive dye signal. Second, there are activity-dependent changes in the intrinsic optical absorption and reflection properties of the brain (Grinvald et al. 1986). These intrinsic optical signals can interfere with voltage-sensitive dye measurements, since, in some cases, they have similar amplitudes. However, they can be distinguished by slower time courses, wavelength-dependence, and, of course, the presence/absence of voltage-sensitive dye.

In summary, if careful experimental measurements and controls are performed, voltage-sensitive dye imaging can be used to define the ensemble spatiotemporal subthreshold membrane potential dynamics within which the membrane potential changes of individual neurons are embedded. Data from these individual neurons can be recorded simultaneously with the ensemble dynamics using the whole-cell technique.

REFERENCES

Antic S., Major G., and Zecevic D. 1999. Fast optical recordings of membrane potential changes from dendrites of pyramidal neurons. *J. Neurophysiol.* **82:** 1615–1621.

Grinvald A., Lieke E., Frostig R.D., Gilbert C.D., and Wiesel T.N. 1986. Functional architecture of cortex revealed by optical imaging of intrinsic signals. *Nature* **324:** 361–364.

Hamill O.P., Marty A., Neher E., Sakmann B., and Sigworth F.J. 1981. Improved patch-clamp techniques for high-resolution current recording from cells and cell-free membrane patches. *Pflügers Arch.* **391:** 85–100.

Petersen C.C.H., Grinvald A., and Sakmann B. 2003a. Spatiotemporal dynamics of sensory responses in layer 2/3 of rat barrel cortex measured in vivo by voltage-sensitive dye imaging combined with whole-cell voltage recordings and neuron reconstructions. *J. Neurosci.* **23:** 1298–1309.

Petersen C.C.H., Hahn T.T.G., Mehta M., Grinvald A., and Sakmann B. 2003b. Interaction of sensory responses with spontaneous depolarisation in layer 2/3 barrel cortex. *Proc. Natl. Acad. Sci.* **100:** 13638–13643.

Ratzlaff E.H. and Grinvald A. 1991. A tandem-lens epifluorescence macroscope: Hundred-fold brightness advantage for wide-field imaging. *J. Neurosci. Methods* **36:** 127–137.

Salzberg B.M., Davila H.V., and Cohen L.B. 1973. Optical recording of impulses in individual neurones of an invertebrate central nervous system. *Nature* **246:** 508–509.

Shoham D., Glaser D.E., Arieli, A., Kenet, T., Wijnbergen, C., Toledo, Y., Hildesheim R., and Grinvald A. 1999. Imaging cortical dynamics at high spatial and temporal resolution with novel blue voltage-sensitive dyes. *Neuron* **24:** 791–802.

Zhang J., Davidson R.M., Wei M.D., and Loew, L.M. 1998. Membrane electric properties by combined patch clamp and fluorescence ratio imaging in single neurons. *Biophys. J.* **74:** 48–53.

CHAPTER 92

A Practical Guide: In Vivo Imaging of Tumors

Edward Brown, Lance L. Munn, Dai Fukumura, and Rakesh K. Jain

Edwin L. Steele Laboratory, Department of Radiation Oncology, Massachusetts General Hospital, Boston, Massachusetts 02114

Light microscopy of tumors, as for other thick, scattering tissues such as the brain or the developing embryo, is limited by light penetration and optical access. Because of problems with light penetration (i.e., scattering and absorption) in tumors, epifluorescence and confocal microscopy are typically limited to the outer 50–100 μm of the accessible tumor tissue (Jain 1998). Recently, a marked improvement in light penetration was made using the multiphoton laser-scanning microscope, illuminating tumors to depths of up to 400 μm (Brown et al. 2001). However, even this achievement is insufficient to provide images of tumors growing within laboratory mice, and the advances have done nothing to improve optical access, i.e., the ability of the imaging apparatus to reach the tumor surface, which remains a significant limitation. Most mouse tumors must be exteriorized for examination under the light microscope, a procedure that limits the duration and repeatability of imaging. This practical guide describes the generation of chronic window preparations in the mouse. These preparations allow an implanted tumor to grow for several weeks in an optically accessible location in vivo, making it possible to examine the living tumor with high-resolution light microscopy in a repetitive manner.

Two chronic window preparations are discussed: (1) the dorsal skinfold chamber, which allows in vivo imaging of tumors growing in the subcutaneous space (Leunig et al. 1992), and (2) the cranial window, which allows in vivo imaging of tumors growing on the brain surface (Yuan et al. 1994). Both procedures are suitable for use in mice and rats. The size of incisions, etc., specified in this chapter are for mice: Incisions, etc., in rats are proportionally larger.

MATERIALS

These procedures require general surgical supplies and equipment, including ketamine/xylazine anesthesia<!>. Ketamine (Abbot Laboratories) is used at 100 mg/kg body weight and xylazine (Phoenix Pharmaceuticals) at 10 mg/kg body weight. The custom-made titanium frames used in the procedures were produced in-house (see Fig. 1).

Caution: *See Appendix 3 for appropriate handling of materials marked with <!>.*

FIGURE 1. Dorsal skinfold chamber.

PROCEDURE A

Dorsal Skinfold Chamber

1. Anesthetize the animal and remove the hair from the animal's back using hair clippers (e.g., Oster) and depilatory cream. Use two custom-made titanium frames (weight 3.2 g) (see Fig. 2) to bracket a fold of skin, which is gently stretched from the back of the mouse and held in place with sutures.

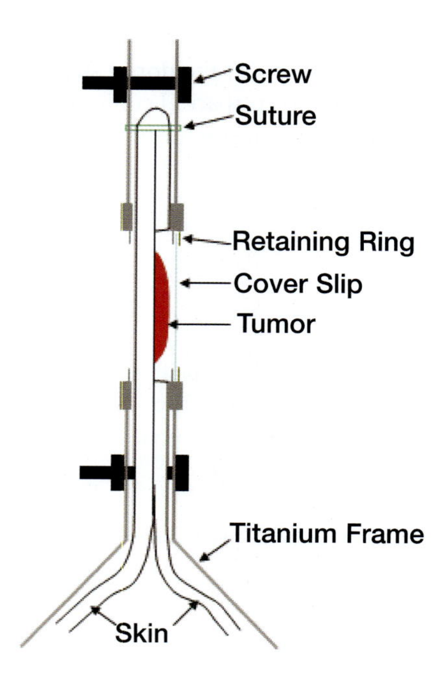

FIGURE 2. Cross-section of the dorsal skinfold chamber.

Note: The frames are fitted with coverslip-retaining rings and are held together with three small screws, spaced around the window area: The two lower screws penetrate the skinfold through small incisions.

2. Use a scalpel to cut a 15-mm-diameter circle of one layer of skin (including epidermis, subcutaneous tissue, and striated muscle), and cover the remaining layer with a glass coverslip laid in one of the titanium frames and held in place with a retaining ring. Allow the animals to recover from surgery for 48 hours before implanting the tumor.
3. To implant the tumor, anesthetize the animal and carefully remove and discard the coverslip. Implant a 2-μl tumor cell suspension (~10^5 cells) or an ~1 mm^3 piece of tumor tissue at the center of the chamber. Reseal the chamber with a new coverslip and wait 1–2 weeks (depending on tumor type) for the tumor to grow.
4. To image the chamber, anesthetize the animal, lie it on its side on a heated stage, and gently fasten the chamber to the microscope stage with a custom-built metal frame.

PROCEDURE B

The Cranial Window

1. Anesthetize the animal. Fasten the head of the animal within a stereotactic frame (Fig. 3).
2. Make a longitudinal incision of the skin between the occipet and the forehead.
3. Remove a disk of skin from the top of the skull, and scrape off the periosteum underneath the temporal crests.

FIGURE 3. Cranial window.

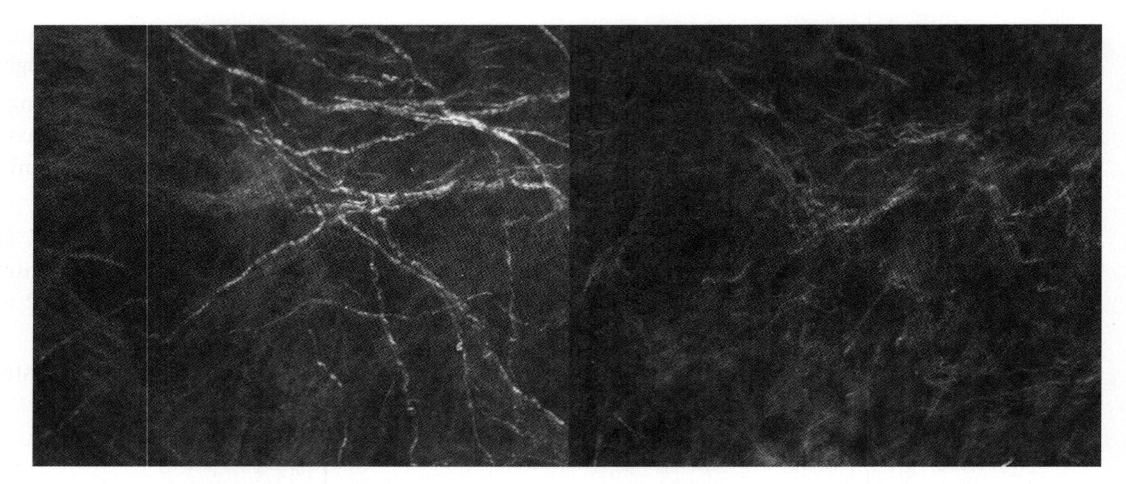

FIGURE 4. Application of the dorsal skinfold chamber to study the collagen matrix in tumors undergoing treatment with the hormone relaxin. Images were taken 2 weeks apart. Each image is a maximum intensity projection of 20 images taken in 5-μm steps, and is 500 μm across.

4. Draw a 6-mm circle over the frontal and parietal regions of the skull. Use a high-speed dental drill with a 0.5-mm burr tip (e.g., Fine Science Tools) to make a groove around the margins of the circle. Gradually deepen the groove by gentle repetitive drilling until the bone flap becomes loose.
5. Use a Malis dissector (e.g., Codman) to separate the bone flap from the dura mater underneath and remove it. Place Gelfoam (e.g., Pharmacia) on the cut edge to absorb blood and saline. Continuously rinse the dura mater with sterile saline.
6. Make a small incision close to the sagittal sinus and insert iris microscissors. Completely cut the dura and arachnoid membranes from the surface of both hemispheres, avoiding any damage to the sagittal sinus.
7. Place a 1-mm^3 piece of tumor tissue in the center of the exposed region, either on the pial surface or in the brain parenchyma. Then fix a 7-mm diameter round cover glass to the bone with cyanoacrylate glue.
8. To image the window, anesthetize the mouse and fasten it in a stereotactic apparatus to minimize motion artifacts.

SHORT EXAMPLE OF APPLICATION

An HSTS human sarcoma was implanted in a dorsal skinfold chamber in a *s*evere *c*ombined *i*mmuno*d*eficient (SCID) mouse and allowed to grow for 2 weeks. The mouse was then treated with the hormone relaxin for another 2 weeks via an implanted osmotic pump. During the treatment, the tumor was imaged repeatedly using second harmonic generation to observe fibrillar collagen (Brown et al. 2003). Images from the day 1 and day 14 of treatment are shown in Figure 4. Extensive applications of these models are discussed elsewhere (Jain et al. 2002).

ADVANTAGES AND LIMITATIONS

The surgical preparations described here allow repetitive high-resolution light microscopy of tumors growing in vivo in the subcutaneous space or in the brain. However, the lifetime of the preparations is not infinite. The dorsal skinfold chamber typically lasts for up to 5 weeks before the animal must be sacrificed, whereas the cranial window lasts for the lifetime of the animal.

ACKNOWLEDGMENT

This chapter is based on two related reviews by Jain et al. (2001, 2004).

REFERENCES

Brown E., Campbell R., Tsuzuki Y., Xu L., Carmeliet P., Fukumura D., and Jain R.K. 2001. In vivo measurement of gene expression, angiogenesis, and physiological function in tumors using multiphoton laser scanning microscopy. *Nat. Med.* **7:** 864–868.

Brown E., McKee T.D., di Tomaso E., Pluen A., Seed B., Boucher Y., and Jain R.K. 2003. Dynamic imaging of collagen and its modulation in tumors in vivo using second harmonic generation. *Nat. Med.* **9:** 796–801.

———. 1998. The next frontier of molecular medicine: Delivery of therapeutics. *Nat. Med.* **4:** 655–657.

Jain R.K., Munn L.L., and Fukumura D. 2001. Transparent window models and intravital microscopy. In *Tumor models in cancer research* (ed. B.A. Teicher), pp. 647–671. Humana Press, Totowa, New Jersey.

———. 2002. Dissecting tumour pathophysiology using intravital microscopy. *Nat. Rev. Cancer* **2:** 266–276.

Jain R.K., Brown E., Munn L.L., and Fukumura D. 2004. Intravital microscopy of normal and diseased tissues in the mouse. In *Live cell imaging: A laboratory manual* (ed. D.L. Spector and R.D. Goldman). Cold Spring Harbor Laboratory Press, Cold Spring Harbor, New York. (In press.)

Leunig M., Yuan F., Menger M.D., Boucher Y., Goetz A.E., Messmer K., and Jain R.K. 1992. Angiogenesis, microvascular architecture, microhemodynamics, and interstitial fluid pressure during early growth of human adenocarcinoma LS174T in SCID mice. *Cancer Res.* **52:** 6553–6560.

Yuan F., Salehi H.A., Boucher Y., Vasthare U.S., Tuma R.F., and Jain R.K. 1994. Vascular permeability and microcirculation of gliomas and mammary carcinomas transplanted in rat and mouse cranial windows. *Cancer Res.* **54:** 4564–4568.

CHAPTER 93

Two-Photon Imaging of Cortical Microcirculation

David Kleinfeld[1] and Winfried Denk[2]

[1]*Department of Physics, University of California, La Jolla, California 92093;*
[2]*Max-Planck Institute for Medical Research, Department of Biomedical Optics, D-69120 Heidelberg, Germany*

Brain homeostasis depends on adequate levels of blood flow to ensure the delivery of nutrients and to facilitate the removal of catabolic products. The exchange of material between constituents in the blood and neurons and glia occurs at the level of individual capillaries, vessels that are 5–8 μm in caliber. An increase in the electrical activity within populations of neurons may result in transient increases and decreases of blood flow in activated and nearby quiescent regions, respectively (Woolsey et al. 1996). This chapter describes how cortical blood flow at the level of individual capillaries that lie deep to the cortical surface can be observed with two-photon laser-scanning microscopy (TPLSM) techniques (Kleinfeld et al. 1998; Kleinfeld and Denk 2000; Chaigneau et al. 2003).

Redistributions of blood flow, together with changes in the blood oxygenation levels, form the basis of functional magnetic resonance imaging (MRI) (Ogawa et al. 1990) and intrinsic optical imaging (Blasdel and Salama 1986; Grinvald et al. 1986). The study of capillary flow, in parallel with measures of cell physiology, forms the basis of understanding these imaging modalities (Devor et al. 2003). A second area is the study of blood flow in conjunction with rodent models of stroke (Pinard et al. 2002; Friedman et al. 2004).

MATERIALS

The most common species used in previous studies have been Wistar or Sprague-Dawley rats and Swiss-Webster mice (Woolsey et al. 1996; Hudetz 1997). Whole-animal imaging experiments usually require the use of an anesthetic to immobilize the animal in a calm and pain-free state (Flecknell 1987; Short 1987), although the use of conscious but immobilized preparations (Ono et al. 1986; Kleinfeld et al. 2002) or conscious animals with a fiber microscope (Helmchen et al. 2001) is an alternative. Anesthetics, by their nature, have a strong effect on the activity in the nervous system and therefore on the hemodynamics coupled to neural activity (Table 1). Beyond these issues, the microscopy described here follows standard in vivo TPLSM (Denk et al. 1994; Svoboda et al. 1997; Kleinfeld et al. 1998; Lendvai et al. 2000; Tsai et al. 2002).

TABLE 1. Anesthetics<!> and doses used in studies of cerebral blood flow in rat

Agent (delivered i.p.)	Initial dose (per kg rat)	Supplement (per kg rat)	Representative reference
α-Chloralose	50 mg	40 mg/hr	Lindauer et al. (1993)
Ketamine plus	100 mg	30 mg	Wei et al. (1995)
xylazine	50 mg	15 mg as required	
Sodium pentabarbitol	60 mg	10 mg as required	Rovainen et al. (1993)
Sodium pentabarbitol plus	25 mg		Hudetz et al. (1992)
ketamine	30 mg		
Urethane	1000 mg	100 mg as required	Kleinfeld et al. (1998)
Urethane plus	600 mg		Ngai et al. (1988)
α-chloralose	50 mg		

Initial dosing is approximate and depends on the emotional state of the animal.

PROCEDURE

Two-Photon Imaging of Cortical Microcirculation

1. Anesthetize animals with urethane<!> at an initial level of 1.2 g/kg, delivered intraperitoneally as two half doses over a period of 10 min. Supplement this initial dose at 0.2 g/kg as necessary during the procedure.
2. Treat the animals prophylactically with atropine (0.1 mg subcutaneously (s.c.). every 2 hours), to inhibit respiratory distress. Maintain body fluids by s.c. injection of physiological saline (0.9% [w/v] NaCl), supplemented with 5% (w/v) dextrose (10 ml/kg/hour). Protect the eyes against desiccation by applying drops of mineral oil, and maintain body temperature near 37°C using a heating blanket (Harvard Instruments).
3. Once the animal is anesthetized, secure it in a stereotaxic holder (no. 960, David Kopf Instruments), modified to rotate about the anterior–posterior axis, and make a succession of glancing cuts along an outline of the desired craniotomy with a high-speed handpiece (Patterson Dental Supply) equipped with a no. 1/2 round bur. Stop cutting just as the bone begins to craze.
4. Remove the resulting bone flap with forceps, and protect the now exposed surface of the dura from desiccation with a sponge (Gelfoam, Upjohn) soaked with physiological saline.
5. To hold the animal's head rigidly to the optical apparatus, fix a metal frame to the skull surrounding the craniotomy with dental acrylic cement (Kleinfeld and Denk 2000).

   Note: To achieve a reliable connection between acrylic and bone, the contact regions on the bone must be carefully cleaned of soft tissue and rubbed with a cotton applicator, prior to application of a thin layer of low-viscosity cyanoacrylate cement (Superbonder, Loctite).
6. Once the frame is in place, remove the dura and fill the interior of the chamber with 1 to 2% (w/v) low-melting-point agarose (Sigma), dissolved in artificial cerebral spinal fluid (ACSF) (125 mM NaCl, 5 mM KCl<!>, 10 mM glucose, 10 mM HEPES, 3.1 mM $CaCl_2$<!>, 1.3 mM $MgCl_2$<!>; pH 7.4). Note that this ACSF contains neither carbonate nor phosphate, which precipitate when the solution is boiled to dissolve the agarose (Kleinfeld and Delaney 1996).
7. Seal the chamber using a cover glass (no. 1, cut to size).
8. Label blood plasma by tail-vein injection:
   a. Submerge the tail in warm water (37°C) for ~2 min (to make the veins dilate) and then place a soft clamp at the base of the tail.
   b. Starting as close to the tip of the tail as possible, insert a 24-gauge catheter and, for 250-g rats, inject a 0.5-ml bolus of 5% (w/v) fluorescein isothiocynate dextran (70–2000 kD, Sigma) solution.

Caution: *See Appendix 3 for appropriate handling of materials marked with <!>.*

SHORT EXAMPLE OF APPLICATION

Figure 1a shows the angioarchitecture of a cube of cortex, ~150 μm on edge, that contains microvessels. Successive, rapidly acquired planar images of such microvessels revealed a succession of dark objects that moved across a sea of fluorescently labeled serum (Fig. 1b). The dark objects are red blood cells (RBCs), which exclude the dye and are therefore not fluorescent. The change in position of the spots between successive images is proportional to the velocity (arrows; Fig. 1b) of the flow in the

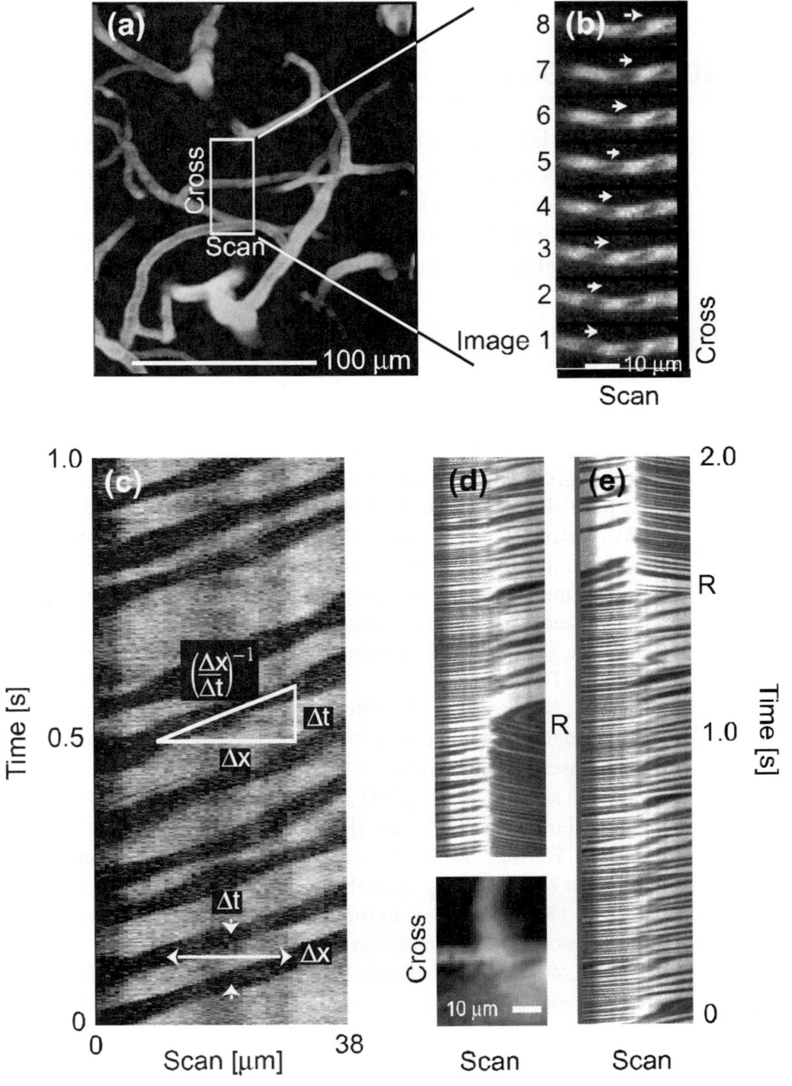

FIGURE 1. Parameterization of blood flow in a capillary. (*a*) Horizontal view in the vicinity of a capillary. The image is the maximal projection from a contiguous set of 100 planar scans acquired every 1 μm between 310 and 410 μm below the pial surface. (*b*) Successive planar images, acquired every 16 msec, through a capillary at a depth of 450 μm. The change in position of a particular unstained object, interpreted as a RBC, is indicated by the series of arrows (→); the speed of the RBC is 0.11 mm/sec. (*c*) Line scan through a capillary at a depth of 600 μm. In this imaging modality, the RBCs appear as dark bands. The annotations illustrate how RBC flow can be parameterized as a function of space and time. The instantaneous flux is $1/\Delta t$, the linear density is $1/\Delta x$, and the velocity is $\Delta x/\Delta t$, as indicated. (*d, e*) Examples of reversals in the direction of flow, labeled *R*, observed by simultaneously monitoring RBC motion in two collinear arms of a T-junction at a depth of 260 μm. The lower panel in part *d* is a planar scan of the junction. The two data sets correspond to ~2-sec intervals of line-scan measurements that were performed 5 min apart.

capillary. To achieve high time resolution, repetitive scans along the central axis of a capillary were acquired. This allowed the flow of RBCs to be characterized. The motion of RBCs leads to dark bands in the data set (Fig. 1c). The average time between bands at a fixed position, denoted Δt, is inversely proportional to the flux, the average distance between bands at a fixed time, denoted Δx, is inversely proportional to the density of RBCs, and the average slope of the band, $\Delta t/\Delta x$, is inversely proportional to the velocity of the RBCs (Fig. 1c). These three quantities are related by flux = density · velocity. A particularly interesting case involves flow at a "T"-junction, for which 2 of the 3 arms may be simultaneously imaged. For the examples of Figures 1d and 1e, the flow is seen to spontaneously reverse direction in one of the arms.

ADVANTAGES AND LIMITATIONS

Two-photon laser-scanning microscopy can be used to image the flow of RBCs more than 600 μm below the pial surface in rat. This range encompasses layer 2/3 of neocortex, as well as the superficial part of layer 4—the level of the dominant thalamic input to neocortex. It is likely that future studies will be most valuable when blood flow measurements are performed simultaneously with measurements of neighboring neuronal or glial electrical and ionic activity.

REFERENCES

Blasdel G.G. and Salama G. 1986. Voltage-sensitive dyes reveal a modular organization in monkey striate cortex. *Nature* **321:** 579–585.

Chaigneau E., Oheim M., Audinat E., and Charpak S. 2003. Two-photon imaging of capillary blood flow in olfactory bulb glomeruli. *Proc. Natl. Acad. Sci.* **100:** 13081–13086.

Denk W., Delaney K.R., Kleinfeld D., Strowbridge B., Tank D.W., and Yuste R. 1994. Anatomical and functional imaging of neurons and circuits using two photon laser scanning microscopy. *J. Neurosci. Methods* **54:** 151–162.

Devor A., Dunn A.K., Andermann M.L., Ulbert I., Boas D.A., and Dale A.M. 2003. Coupling of total hemoglobin concentration, oxygenation, and neural activity in rat somatosensory cortex. *Neuron* **39:** 353–359.

Flecknell P.A. 1987. *Laboratory animal anesthesia: An introduction for research workers and technicians.* Academic Press, San Diego.

Friedman B., Nishimura N., Schaffer C.B., Kleinfeld D., and Lyden L.D. 2004. Heterogeneous changes in blood flow in response to single vessel blockages and MCA occlusions in rat parietal cortex as revealed by in vivo two-photon laser imaging. *Stroke* **35:** 81.

Grinvald A., Lieke E.E., Frostig R.D., Gilbert C.D., and Wiesel T.N. 1986. Functional architecture of cortex revealed by optical imaging of intrinsic signals. *Nature* **324:** 361–364.

Helmchen F., Fee M.S., Tank D.W., and Denk W. 2001. A miniature head-mounted two-photon microscope: High-resolution brain imaging in freely moving animals. *Neuron* **31:** 903–912.

Hudetz A.G. 1997. Blood flow in the cerebral capillary network: A review emphasizing observations with intravital microscopy. *Microcirculation* **4:** 233–252.

Hudetz A.G., Weigle C.G.M., Fendy F.J., and Roman R.J. 1992. Use of fluorescently labeled erythrocytes and digital cross-correlation for the measurement of flow velocity in the cerebral microcirculation. *Microvasc. Res.* **43:** 334–341.

Kleinfeld D. and Delaney K.R. 1996. Distributed representation of vibrissa movement in the upper layers of somatosensory cortex revealed with voltage sensitive dyes (erratum *J. Comp. Neurol.* [1997] **378:** 594). *J. Comp. Neurol.* **375:** 89–108.

Kleinfeld D. and Denk W. 2000. Two-photon imaging of neocortical microcirculation. In *Imaging neurons: A laboratory manual* (ed. R. Yuste et al.), pp. 23.1–23.15. Cold Spring Harbor Laboratory Press, Cold Spring Harbor, New York.

Kleinfeld D., Mitra P.P., Helmchen F., and Denk W. 1998. Fluctuations and stimulus-induced changes in blood flow observed in individual capillaries in layers 2 through 4 of rat neocortex (erratum *Proc. Natl. Acad. Sci.* [1999] **96:** 8307c). *Proc. Natl. Acad. Sci.* **95:** 15741–15746.

Kleinfeld D., Sachdev R.N.S., Merchant L.M., Jarvis M.R., and Ebner F.F. 2002. Adaptive filtering of vibrissa input in motor cortex of rat. *Neuron* **34:** 1021–1034.

Lendvai B., Stern E.A., Chen B., and Svoboda K. 2000. Experience-dependent plasticity of dendritic spines in the developing rat barrel cortex in vivo. *Nature* **404:** 876–881.

Lindauer U., Villringer A., and Dirnagl U. 1993. Characterization of CBF response to somatosensory stimulation: Model and influence of anesthetics. *Am J. Physiol.* **264:** H1223–H1228.

Ngai A.C., Ko K.R., Morii S., and Winn H.R. 1988. Effect of sciatic nerve stimulation on pial arterioles in rats. *Am. J. Physiol.* **254:** H133–H139.

Ogawa S., Lee T.-M., Nayak A.S., and Glynn P. 1990. Oxygenation-sensitive contrast in magnetic resonance image of rodent brain at high fields. *Magn. Reson. Med.* **14:** 68–78.

Ono T., Nakamura K., Nishijo H., and Fukuda M. 1986. Hypothalamic neuron involvement in integration of reward, aversion and cue signals. *J. Neurophysiol.* **56:** 63–79.

Pinard E., Nallet MacKenzie E.T., Seylaz J., and Roussel S. 2002. Penumbral microcirculatory changes associated with peri-infarct depolarizations in the rat. *Stroke* **33:** 606–612.

Rovainen C.M., Woolsey T.A., Blocher N.C., Wang D.-B., and Robinson O.F. 1993. Blood flow in single surface arterioles and venules on the mouse somatosensory cortex measured with videomicroscopy, fluorescent dextrans, nonoccluding fluorescent beads, and computer-assisted image analysis. *J. Cereb. Blood Flow Metab.* **13:** 359–371.

Short C.E. 1987. *Principles and practice of veterinary anesthesia*. Williams and Wilkins, Baltimore, Maryland.

Svoboda K., Denk W., Kleinfeld D., and Tank D.W. 1997. In vivo dendritic calcium dynamics in neocortical pyramidal neurons. *Nature* **385:** 161–165.

Tsai P.S., Nishimura N., Yoder E.J., White A., Dolnick E., and Kleinfeld D. 2002. Principles, design and construction of a two photon scanning microscope for in vitro and in vivo studies. In *Methods for in vivo optical imaging* (ed. R. Frostig), pp. 113–171. CRC Press, Boca Raton, Florida.

Wei L., Rovainen C.M., and Woolsey T.A. 1995. Ministrokes in rat barrel cortex. *Stroke* **26:** 1459–1462.

Woolsey T.A., Rovainen C.M., Cox S.B., Henger M.H., Liange G.E., Liu D., Moskalenko Y.E., Sui J., and Wei L. 1996. Neuronal units linked to microvascular modules in cerebral cortex: Response elements for imaging the brain. *Cereb. Cortex* **6:** 647–660.

CHAPTER 94

Imaging Neuronal Activity with Calcium Indicators in Larval Zebrafish

Joseph R. Fetcho

Department of Neurobiology and Behavior, SUNY at Stony Brook, Stony Brook, New York 11794

This chapter describes a method for using calcium indicators to monitor neuronal activity in transparent larval zebrafish. The approach has been used to determine which neurons in the brain and spinal cord are active during swimming and escape behaviors in normal and mutant lines of zebrafish (Fetcho and O'Malley 1995; O'Malley et al. 1996; Ritter et al. 2001; Gahtan et al. 2002; Liu et al. 2003).

MATERIALS

This procedure is carried out on 4–5-day-old larval zebrafish using Calcium Green dextran (10,000 MW, Molecular Probes). The author uses a Zeiss laser-scanning confocal imaging system (LSM 510) on an inverted microscope (although an upright would work, too). The recommended objectives are a Zeiss 25×/0.8 Plan-neofluar multi-immersion (used with water) or a Zeiss 63x/1.2 NA C-Apochromat water-immersion lens.

PROCEDURE

Imaging Neuronal Activity with Calcium Indicators in Larval Zebrafish

Method for Backfilling

1. Pull the injection pipettes from 1.5-mm microelectrode glass (A-M Systems) on a pipette puller (e.g., Sutter Instruments model P 87).
2. Break off the tip of the pulled pipette so that the diameter is about 10–20 μm, roughly 1/4 to 1/3 the width of the spinal cord in a 5-day-old larval fish.
3. Backfill the electrode with about 0.5 μl of a 50% solution of Calcium Green dextran (10,000 MW, Molecular Probes) dissolved in 10% Hanks' solution (Westerfield 1995).
4. Anesthetize the appropriate number of 4–5-day-old larval zebrafish in MS 222 (Sigma) (0.02 g of MS222 per 100 ml of 10% Hanks' solution) and place them in a plastic petri dish containing a few

Present address: Department of Neurobiology and Behavior, Cornell University, Ithaca, New York 14853.

millimeters' depth of solid agar (1.2% Bacto-Agar, DIFCO) or low-melting-point agarose to prevent damage to the electrode and the fish during injection.

5. Pressure-inject the indicator solution into the spinal cord (or muscle) of the fish using a picospritzer device (Parker Instrumentation) and a micromanipulator. A pressure of 10–20 psi is typically applied in brief 3-msec pulses while observing the orange/red colored indicator entering the tissue.
6. After injection, place the fish in 10% Hanks' solution in different wells of a multiwell culture dish for individual tracking.
7. Store the fish overnight at room temperature and image the next day.

   Note: The delay between labeling and imaging allows ample time for transport of indicator and recovery of the fish from the injection.

Preparation of the Fish for Imaging

8. Restrain the fish by embedding part or all of the larvae (typically we study fish about 5 days old) in agar.

   Notes: To image neuronal activity, it is important to keep the preparation as still as possible. Zebrafish neurons are small (most <10 μm in diameter) and if the fish shifts position by only a few micrometers, the cell or cells being imaged will be out of focus. The fish can use cutaneous respiration and will survive in the agar if it is kept moist. The agar does allow some movement, but brings the fish back very close to its original position after the movement, so that the neurons remain in focus.

   Embedding the larval fish in agar without compromising its health (which usually means killing it) requires practice because the fish are very delicate.

   In the author's lab, imaging is usually done with inverted microscopes, so the objectives come from below. The fish are held in a small glass-bottomed petri dish (3.5 cm) with a coverslip glued to a hole in the bottom. The fish is embedded in agar on the surface of the coverslip. The embedding is done by first coating the coverslip with a thin (about one-fish thickness) layer of agar. The anesthetized fish is then placed in a groove cut in the agar, and a tiny amount of agar (kept just above its melting point in a dry block heater [VWR]) is added to fill in the groove and cover the fish. For optimal imaging, the fish should be against the coverslip—it may float up when the agar is added. Once the fish has been covered with a thin layer of agar, and the agar has been allowed to set, 10% Hanks' solution is added on top of the agar to keep the fish moist. The fish is healthy if the blood cells are moving rapidly in the vessels running through the spinal cord.

Image Acquisition and Analysis

The imaging process involves collecting a confocal time series from the cells of interest. The behavior of interest is elicited during the time series to determine whether the cells increase in intensity in association with the behavior. The cells must not be exposed to too much light because the indicators used for the imaging are phototoxic, and exposing them to intense light for too long kills the cells.

When using the confocal microscope, the strategy is to fully open the confocal aperture and increase the gain of the photomultiplier to collect and amplify as much light as possible from the specimen. Under these conditions, the intensity of the laser light used for imaging (15 mW argon 488 line in Zeiss 510 at 75% power) can be attenuated to 0.1–1% of its maximal intensity while still allowing imaging of the responses of neurons. Practice in identifying cell types is important, as it is harder to identify cells under low light levels because the details of processes are not as obvious as under brighter illumination.

The conditions useful for imaging responses are, unfortunately, not ideal for recording cell morphology. For morphology, the confocal aperture must be reduced to allow good optical sectioning of the tissue. This typically requires increased illumination of the specimen to acquire an adequate signal, which risks damaging the cells. Consequently, it is advisable to collect the physiological data first and then increase light intensity and close the confocal aperture to obtain high-quality images of morphology.

Once a series of images of the neuron(s) before and after the behavior has been acquired, the question is whether the cell was brighter after the behavior than before. This can be assessed by using the confocal software. This will determine the magnitude and time course of any increase in fluorescence intensity in the imaged neurons. In the author's experience, roughly a 10% or more increase in fluorescence ($\Delta F/F$) is observed in the soma for a single action potential.

EXAMPLE OF APPLICATION

Figure 1 shows an example in which calcium imaging was used to explore the pattern of activation of neurons in a mutant line of zebrafish. This mutant produces extra reticulospinal Mauthner cells (Liu et al. 2003). There is normally only one Mauthner cell on each side of the brain; the cells are involved in generating an escape behavior that fish use to avoid predators. In the mutant line, there are extra Mauthner

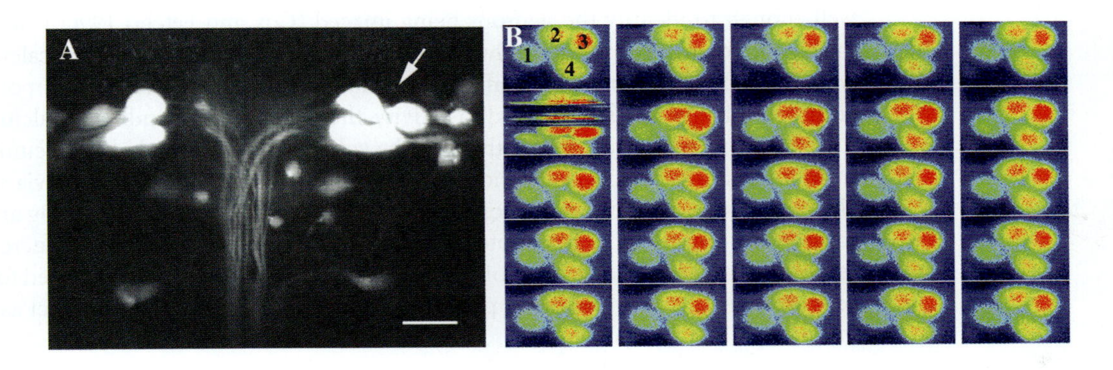

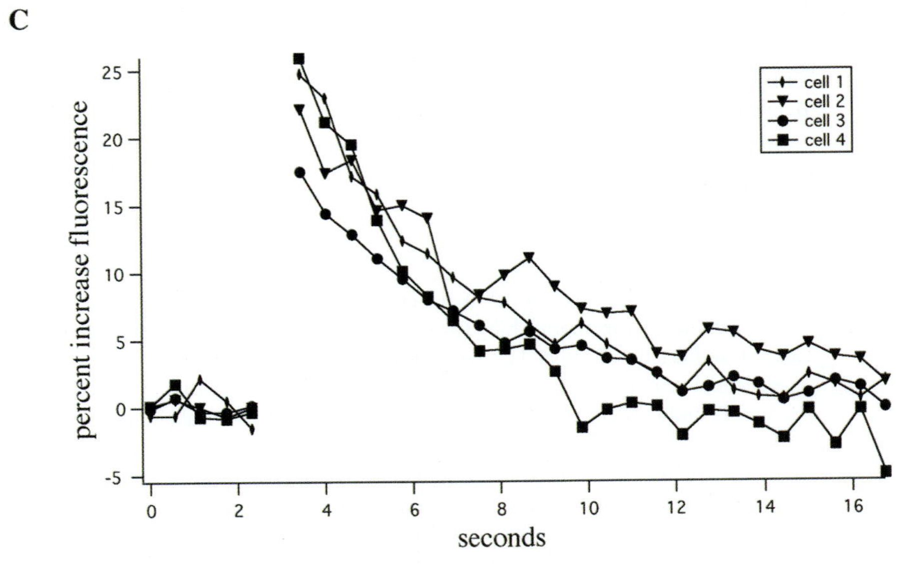

FIGURE 1 . Calcium imaging of extra Mauthner cells in the mutant line *deadly seven*. (*A*) Confocal images of the hindbrain of an intact mutant fish in which the Mauthner cells were backfilled by injection of Calcium Green into caudal spinal cord. There is a cluster of Mauthner cells on both the right (*arrow*) and left sides of the hindbrain, where a normal fish would only have one cell. (*B*) Calcium imaging of the cells during an escape, shown in pseudocolor with red being the brightest, reveals that all of the cells increase in brightness when an escape occurs. A movement artifact from the escape is evident in panel 6. Frames are ordered from left to right with 578 msec between frames. (*C*) Quantification of the intensities in the cells labeled 1–4 in the trial in *B*. The normalized intensities ($\Delta F/F$) are plotted versus time and show that all cells responded during the escape. Bar in *A*, 30 μm. (Reprinted, with permission, from Liu et al. 2003 [© Society for Neuroscience].)

cells, visible in the confocal images in panel A. Calcium imaging of these neurons during escapes, shown in B and quantified in C, reveals that all of the cells are active, demonstrating that the extra cells are integrated into the escape circuit. The performance of the mutant line in escapes is close to normal.

ADVANTAGES AND LIMITATIONS

The key advantage of calcium imaging is that it allows easy identification of active neurons in the intact fish while it is generating behavior. This is not possible with electrophysiology, which requires paralyzed, dissected preparations. The ability to image neuronal activity with single-cell resolution in fish is a powerful tool for investigating the neuronal basis of behavior in normal and mutant lines of fish.

There are two major drawbacks to calcium imaging in this context. First, the injections used to label the cells may damage the circuits of interest. This can be circumvented by using other labeling approaches, such as blastomere injections or injections of AM dyes, but these do not typically label well enough to identify the neurons being imaged (Cox and Fetcho 1996; Brustein et al. 2003). The problem of damage is most effectively avoided by using genetically encoded calcium indicators, which we have applied recently in zebrafish (Higashijima et al. 2003). Another important problem is that although calcium levels are related to activity, the relationship is indirect. Calcium levels can rise due to an influx from outside the cell through the voltage-gated channels, but calcium can also be released from internal stores upon activation of a neuron. It can also enter the cell via synaptically activated channels. These different sources of calcium, along with effects of spike timing and saturation of indicator, make it more difficult to determine the relationship between a given increase in calcium (fluorescence) and the level of activity of the cell. In vivo, the indicators are best used for determining which cells are active and for looking at patterns of activity, rather than for an exact assessment of the level of activity of individual neurons.

ACKNOWLEDGMENTS

This work was supported by National Institutes of Health grant NS-26539. Many individuals participated in the development of the techniques and in the experiments using them, with major contributions to the approaches described here coming from Drs. Don O'Malley and Katharine Liu.

REFERENCES

Brustein E., Marandi N., Kovalchuk Y., Drapeau P., and Konnerth A. 2003. "In vivo" monitoring of neuronal network activity in zebrafish by two-photon Ca(2+) imaging. *Pflugers Arch.* **446:** 766–773.

Cox K.J. and Fetcho J.R. 1996. Labeling blastomeres with a calcium indicator: A non-invasive method of visualizing neuronal activity in zebrafish. *J. Neurosci. Methods* **68:** 185–191.

Fetcho J.R. and O'Malley D.M. 1995. Visualization of active neural circuitry in the spinal-cord of intact zebrafish. *J. Neurophysiol.* **73:** 399–406.

Gahtan E., Sankrithi N., Campos J.B., and O'Malley D.M. 2002. Evidence for a widespread brain stem escape network in larval zebrafish. *J. Neurophysiol.* **87:** 608–614.

Higashijima S., Masino M.A., Mandel G., and Fetcho J.R. 2003. Imaging neuronal activity during zebrafish behavior with a genetically encoded calcium indicator. *J. Neurophysiol.* **90:** 3986–3997.

Liu K.S., Gray M., Otto S.J., Fetcho J.R., and Beattie C.E. 2003. Mutations in deadly seven/notch1a reveal developmental plasticity in the escape response circuit. *J. Neurosci.* **23:** 8159–8166.

O'Malley D.M., Kao Y.-H., and Fetcho J.R. 1996. Imaging the functional organization of zebrafish hindbrain segments during escape behaviors. *Neuron* **17:** 1145–1155.

Ritter D.A., Bhatt D.H., and Fetcho J.R. 2001. *In vivo* imaging of zebrafish reveals differences in the spinal networks for escape and swimming movements. *J. Neurosci.* **21:** 8956–8965.

Westerfield M. 1995. *The zebrafish book: A guide for the laboratory use of zebrafish (*Brachydanio rerio*)*, 3rd edition. University of Oregon Press, Eugene.

CHAPTER 95

Microscopy and Microscope Optical Systems

Frederick Lanni[1] and H. Ernst Keller[2]

[1]*Department of Biological Sciences, Carnegie Mellon University, Pittsburgh, Pennsylvania 15213;* [2]*Microscope Division, Carl Zeiss, Inc., Thornwood, New York 10594*

Given the 300-year history of the light microscope since Leeuwenhoek and Hooke, one may expect this invention to have matured. Rather, the past decades have seen a remarkable set of advances in microscope optical systems and the uses to which the microscope has been put. Advances in optical materials and fabrication, electronic detectors and cameras, lasers, and computers have led to enormous advances in microscopy and the fields of science and technology in which the light microscope is a major tool. Nevertheless, the optics that constitute the core of a microscope embody a set of principles, understood for more than a century, that continue to spark the imagination and give rise to new imaging technologies and discovery.

USE AND CARE OF MICROSCOPES

It is our experience that most users of microscopes learn by trial and error long before reading a book (or chapter!) on the subject. Therefore, a discussion of use and care of the instrument has been put "up front" in this chapter.

The three essential components of any microscope are the objective, the condenser, and a precision stage system to position the specimen. In a transmitted-light microscope, objective and condenser are separate components, whereas in an incident-light microscope, the objective serves as its own condenser. Use and care of the instrument are defined by the characteristics and limitations of these components; for example, objectives differ in an important way from lenses in other types of image-forming instruments, such as cameras. To minimize aberration and maximize resolution, objectives are designed with restricted use conditions. A high-quality camera lens can be adjusted to focus over a large subject range without obvious change in performance over the corresponding range of demagnification. In contrast, microscope objectives are designed to image at fixed magnification a particular plane in object space onto a detector located at a particular position in image space. The gain is that the image formed by a microscope is much closer to the diffraction limit of resolution than a camera photograph. Alteration of a microscope optical system in a way that causes the objective to be focused at a distance from the specimen that differs from the "design" working distance will result in degraded image quality. Likewise, when the refractive index of the specimen differs from the value upon which the objective design was based, large losses in contrast and resolution can result when focusing deeply. Sensitivity to deviation from design conditions is generally much greater for lenses of higher numerical aperture (NA). In an ideal case, the objective would be designed for use with an immersion fluid in which the refractive index was the same as the index of the specimen. A practical solution is to design a series of objectives for use under specific optical conditions, using standard immersion fluids (Table 1). Condensers, likewise, are designed for specific uses. Large working distance (2–5 cm) is a common requirement for transilluminating cells in flasks or dishes on an inverted microscope, but this limits the NA to moderate values. Dry, high-resolution condensers are designed to operate with 3–4 mm clearance between the lens and a standard microscope slide (1 mm thickness). For the highest-resolution work in transmitted light, oil-immersion condensers are used with 1 mm clearance to the slide.

The most important use rule in microscopy is to avoid mechanical damage to the objective and condenser. Clearly, optical surfaces are most susceptible to damage by abrasion, but permanent misalignment is an equally serious problem commonly caused by impact or deformation of the metal or ceramic barrel of the lens. For two reasons, microscope lenses and condensers are generally positioned very close to the specimen. As with any image-forming instrument, high magnification is obtained by placing the object close to the front focal plane of a short focal-length lens or system of lenses. Second, high resolution depends on capture by the objective of light rays that exit the specimen at high angle; i.e., the imaging system must have high *aperture*. Even though condensers are not usually corrected for image formation, the same rules apply; high demagnification (concentration of light) and high aperture are required. For low-power optics, the working distance can range from millimeters to >1 cm, but for widely used high-NA immersion objectives, the range is 0.06–0.25 mm. Therefore, microscope users must take great care to avoid collision between objective and specimen, condenser and specimen, objective and condenser, objective and stage, etc., caused by movement of the focus drive, turret, stage, or condenser focus drive.

Of equal importance is cleanliness of optical surfaces. Settled dust particles or fingerprints anywhere in the lens system, particularly between specimen and final image, will cause degraded image quality and can even create artifactual image features. The most common problem is oil residue or a fingerprint on the front lens surface of a "dry" objective, or dried residue on the top surface of a cover glass. In a fluorescence microscope, a single particle of fluorescent dust that settles onto the final lens surface of the objective can cause a large increase in background. Oil-immersion optics present a different set of problems; after use, the oiled lens surfaces should be gently wiped with lens tissue and the objective or condenser stored in a dust-free container. It is generally not necessary to remove all traces of oil between uses. For the same reason, it is very important not to carelessly change oil types; differences in refractive index or actual immiscibility can cause serious image degradation. Microscope manufacturers supply oil of the appropriate index for their lenses. If blending of immersion fluids is required, compatible standard oils can be obtained from Cargille (Cedar Grove, NJ). Changeover to an oil of different index requires complete removal of the residue of the previous oil. If the oils are miscible, this is most easily accomplished simply by repeated application of the new oil along with gentle removal using lens tissue. The other case where complete removal of oil is necessary is in changing from oil immersion to glycerine or water immersion with a multi-immersion objective. In this case, use of a droplet of xylene on lens tissue will remove oil.

Spillover of specimen materials poses both a cleanliness and damage hazard. In microscopy of living specimens, a common hazard is spillage of culture medium into the condenser on an upright microscope, or into everything but the condenser on an inverted microscope. Dried culture medium leaves a heavy residue of salt crystals and protein that can permanently damage lens coatings and surfaces and quickly corrode metal components, iris leaves, bearings, and more. Especially in perfusion setups, it is well worth using a sheet rubber dam or shield over at-risk components. Optical and precision mechanical components that get wet with culture medium should be carefully cleaned with distilled water and carefully dried. This may entail disassembly and reassembly of filter sets, condensers, or other components. After use in saline, seawater, pond water, or culture medium, the fluid contact surfaces of a direct immersion objective should always be gently rinsed with distilled water and dried with lens tissue. In some cases, direct immersion objectives will contact medium that contains bioactive compounds, fluorescent vital stains, or microorganisms. In this case, special cleaning procedures may be required to prevent contamination of subsequent specimens.

TABLE 1. Standard objective types for biological microscopy

Immersion	n	NA range	Design condition
Dry	1.00	0.13–0.95	cover glass; std thickness = 0.17 mm, "#1-1/2"cell culture dishes or flasks (0–2.0 mm plastic or glass); adjustable corrector or interchangeable window optics
		0.8–0.95	adjustable corrector for cover glass thickness
Water	1.33	0.3–1.0	direct; no cover glass
		0.5–1.2	indirect; adjustable correction for cover glass thickness
Glycerol	1.47	0.6–1.35	fused silica cover glass—0.2 mm
Oil	1.52	0.5–1.4	(cover glass)
Multi-imm.	var.	0.5–0.9	adjustable correction for n and cover glass thickness

PRINCIPLES OF LIGHT MICROSCOPY

Paraxial Image Formation

Many of the basic properties of image-forming instruments (telescopes, cameras, microscopes) can be understood in terms of ray optics. In particular, an optic consisting of nothing more than a hemispherical boundary between two materials such as glass and air can be shown to have the properties of a lens. This means that, for a set of rays that all propagate close to and nearly parallel to a specific axis, the optic will form a focused image of a small object located close to the axis at some distance from the spherical surface. This result also holds for thin lenses and for more complicated optics composed of sets of lenses. The restriction to low-angle, near-axial rays is known as the paraxial limit, in which the Law of Refraction can be linearized and lens action can be summarized in two rules:

1. The lens has both a front and back **focal length** defined on a common axis through its center.
2. Rays entering the lens parallel to the axis are refracted so as to exit on a path that intersects the focal point on the opposite side.

Using only these two rules, familiar equations can be derived for magnification and **conjugate distances** (object and image location) (Fig. 1). Multi-element lens systems can be represented by serial application of the rules, where the image from the "nth" lens surface is the real or virtual object for the following surface. In the paraxial limit, multi-element lens systems act like single lenses, possessing front and back focal lengths measured out from mathematically defined front and back **principal planes**, respectively. An important result, known as the Helmholtz relation, is that the front and back focal lengths are in the ratio of the **refractive indices** of the object and image spaces; i.e., $f'/f = n'/n$, regardless of the specific sequence of lens elements between the entrance and exit surfaces.

A very concise derivation of useful results can be made by reference to the ray diagram of Fig. 1. In general, object-space quantities are represented by unmarked symbols; n, f, s. Image-space variables are "primed"; n', f', s' (Table 2). The relation between object and image locations can be derived by considering the similar triangles in the ray diagram:

$$h/(s-f) = h'/f \quad (1a)$$

$$h'/(s'-f') = h/f \quad (1b)$$

Solving for s and s', and using the definition of magnification ($h'/h = M$)

$$s = (1 + 1/M)f \quad (2a)$$

which shows that the object must be located a small distance outside the front focal length, and

$$s' = (M + 1)f' \quad (2b)$$

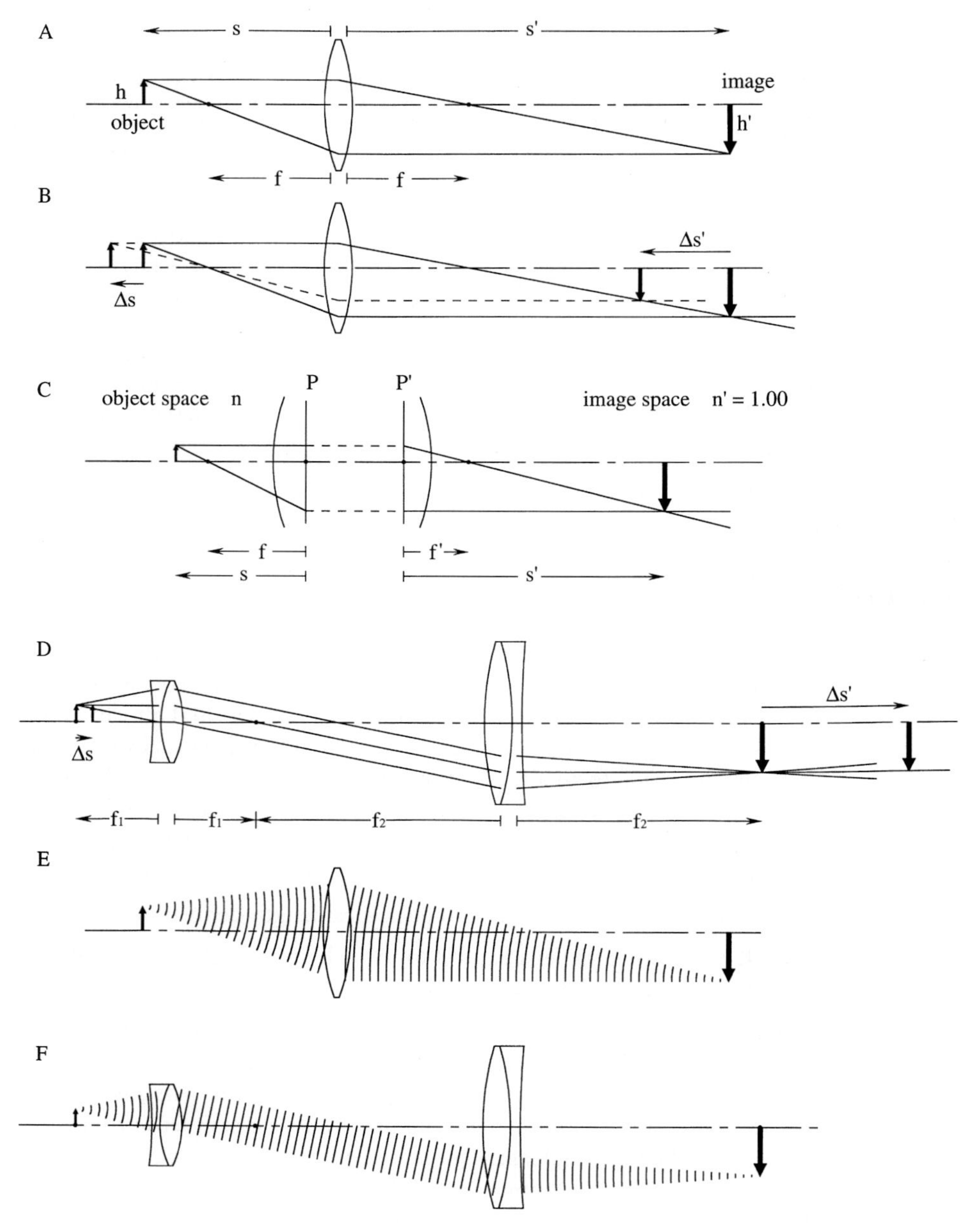

FIGURE 1. Ray and wave diagrams for lens systems. (*A*) Ray trace for a thin lens; the principal ray from the object arrowhead enters the lens parallel to the axis and emerges on a path intersecting the back focal point. Likewise, the ray from the arrowhead passing through the near focal point emerges parallel to the axis. The intersection of the two rays defines the location and magnification of the image (Eqs. 4a and 5). (*B*) Focusing by axial displacement of the object changes both the image location and the magnification. (*C*) In the paraxial limit, a lens system composed of an arbitrary number of spherical surfaces on a common axis can be represented by front and rear focal points measured out from a pair of principal planes (P and P′). Principal rays arriving at the input plane are transferred directly to the output plane as shown. The locations of the principal planes and the focal points are functions of the curvature, spacing, and refractive indices of all of the lens elements in the assembly. The ratio of the system focal lengths is determined only by the object space and image space indices; $f/f' = n/n'$ (Helmholtz relation). (*D*) Two lenses separated by the sum of the focal lengths form an afocal combination in which the magnification is constant. When the object is located in the front focal plane of lens 1, the image is located in the back focal plane of lens 2. (*E* and *F*) Wave-optical diagrams corresponding to *A* and *D*, respectively. The action of a lens is to change the radius of curvature of incident spherical waves. In *E*, a diverging spherical wave is converted to a converging wave. In *F*, the diverging wave is first collimated (infinite radius of curvature), then converged to form an image at a finite conjugate distance. In both *E* and *F*, the wavelength is greatly exaggerated.

which shows that the image will be located a large distance beyond the back focal length. Approximately, $(M + 1)f'$ is equal to the body tube length (BT) in finite-tube systems, or the focal length of the "tube lens" (f_{TL}) in infinite-conjugate (IC) systems (see below). In ratio

$$s'/s = M\,f'/f$$

$$= n'M/n \qquad \text{(using the Helmholtz relation)} \qquad (3)$$

Solving for the relation between conjugate distances s and s'

$$1 = f/s + f'/s' \qquad \text{(general form)} \qquad (4a)$$

For non-immersed optical systems ($n' = n$ and $f' = f$), Equation 4a can be put in the familiar form

$$1/f = 1/s + 1/s' \qquad (4b)$$

For computation of image distance, the Gaussian optics form of 4a is used

$$s' = f's/(s - f) \qquad (4c)$$

Transverse magnification clearly varies with object distance, growing without bound as the object is brought up to the front focal point. Using Equations 3 and 4c

$$M = f/(s - f) \qquad (5)$$

The axial magnification is found by differentiation of Equation 4c

$$ds'/ds = -\,(f/f')\,(s'/s)^2$$

$$= -\,(n/n')\,(n'M/n)^2 \qquad (6a)$$

Using the definition of axial magnification (Table 2)

$$M_{ax} = (n'/n)M^2 \qquad (6b)$$

showing that it is proportional to the square of the transverse magnification.

Afocal combinations, in which an objective and a converging lens (or tube lens) are separated by the sum of their respective focal lengths, are a special case because the system focal length is infinite (Fig. 1D). Therefore, a different ray construct is used for derivation of the paraxial relations. In an afocal system, magnification is constant (independent of s and s') and equal to the ratio of the objective and tube lens focal lengths

$$M = h'/h = f_{TL}/f'_{obj} \qquad (7)$$

Likewise, the conjugate distances are related by a linear equation (cf. Eq. 4c)

$$s' = f_{TL} - (n'/n)M^2\,(s - f) \qquad (8)$$

TABLE 2. Paraxial optic definitions and formulas

f, f'		object-side and image-side focal lengths of lens system
s, s'		object and image conjugate distances
h, h'		object and image offsets from axis
f'/f	$= n'/n$	Helmholtz relation for any coaxial system of spherical lenses n = refractive index of object immersion medium n' = refractive index of the image space (usually air; $n' = 1.00$) for nonimmersed optics, $n' = n$, and $f' = f$
M	$= h'/h$	definition of transverse magnification for a microscope, $M >> 1$ M is not constant, but is a function of s in all but afocal systems
M_{ax}	$= \vert ds'/ds \vert$	definition of axial magnification as the derivative between object and image conjugate distances

When the object is located in the front focal plane of the objective, the image is formed in the rear focal plane of the tube lens. This differs significantly from the Gaussian optics result (Eq. 2a), in which the object is located at a slightly greater distance. In the afocal case also, axial magnification is proportional to the square of transverse magnification. If s and s' are measured from the front and rear focal planes rather than from the principal planes (i.e., $s - f = \Delta s$ and $s' - f_{TL} = \Delta s'$), the conjugate-distance relation can be put into the particularly simple form

$$\Delta s' = -(n'/n)M^2 \Delta s \tag{9}$$

When an afocal system is set up so that the object is in focus when it is in the front focal plane of the objective (i.e., when the camera is located in the back focal plane of the tube lens), rays from each point in the specimen are collimated on the path between the objective and the tube lens. This is an idealized model for infinite-conjugate (IC) or "infinity-corrected" microscopes (see below). Real IC systems are not necessarily exactly afocal; generally, the objective and tube lens are separated by a design distance other than the sum of the focal lengths to optimize compensation of aberration.

Measurement of Depth When There Is an Index Mismatch

Because axial position (depth) within a specimen is encoded through sharpness of focus, calibrated movement of the microscope focus drive provides the primary means for quantitative analysis of 3D structure. The paraxial optics formulas provide the basic relation between stage increment and focus shift (Galbraith 1955). If, for example, a low-power non-immersed objective is focused on a particular point within a mounted specimen, light diverging from that point is more strongly diverged by refraction upon exiting the cover glass (Fig. 2). In the paraxial limit, this has the effect of creating a virtual object at a lesser depth. If the actual source point is at depth z_P, the virtual source appears at z_V, with the two coordinates in proportion to the ratio of the immersion refractive index to specimen index

$$z_V = (n_{imm}/n)z_P \tag{10}$$

When the focus drive is adjusted to move the stage closer to the objective by an increment Δz_S, the virtual source remains fixed. The paraxial convergence point of the actual rays shifts away from the objective, so that the increment in focus depth exceeds the stage increment

$$\Delta z_F = -(n/n_{imm})\,\Delta z_S \tag{11}$$

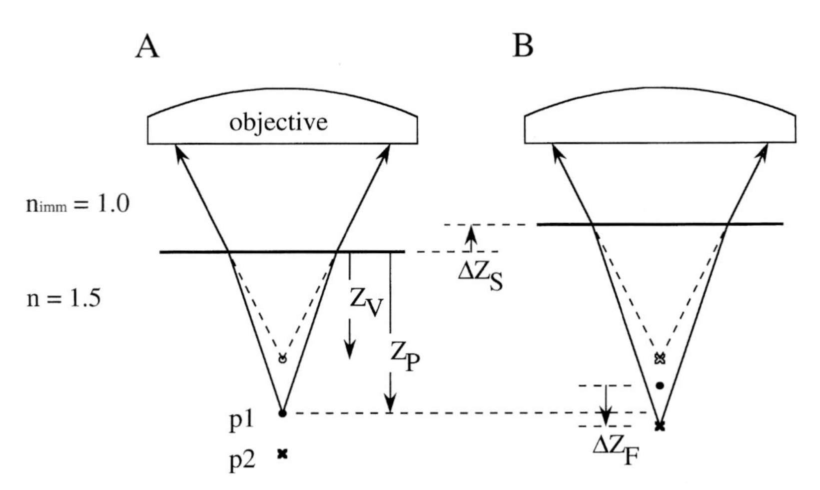

FIGURE 2. Correction for refraction in depth measurement by focus. In this example, a dry objective is used to view a specimen mounted in high-index medium below a cover glass ($n > n_{imm}$). (*A*) When the microscope is focused on particle p1, the rays exiting the specimen appear to diverge from a virtual object "o" in the paraxial limit. (*B*) Adjustment of the stage to bring p2 into focus requires a smaller displacement than the actual axial separation of p1 and p2. In the opposite circumstance ($n < n_{imm}$), the stage increment exceeds the focus increment. In either case, the optical focus shift is related to the stage displacement by the ratio of the immersion and specimen indices: $\Delta Z_F = -(n/n_{imm})\,\Delta Z_S$.

Intervening layers of different index, such as a cover glass, do not affect this result. The correction can be significant; for an embedded specimen, n/n_{imm} = 1.52. When the immersion index exceeds the specimen index (such as in the use of an oil-immersion lens to look into living specimens), the virtual source is deeper into the specimen than the actual source point, and the stage increment exceeds the increment in focus depth. Equations 10 and 11 apply without modification. Only in the case where $n = n_{imm}$ are the two increments identical. The difference between stage and focus increment must be corrected to eliminate axial distortion when 3D reconstructions are computed from serial-focus image sets. With high-NA lenses, index mismatch not only causes focus shift, but also introduces focus-dependent spherical aberration. Therefore, choice of an immersion system to minimize mismatch is always the best situation.

Light as a Wave

For discussion of most of the principles and practice of light microscopy, it is sufficient to consider light as a scalar wave field. Clearly, the vector nature of the electromagnetic field cannot be ignored in discussion of focusing of light by a high-aperture lens, or the measurement of polarized components of an image field. Likewise, lasers, absorption, fluorescence, and detection of light are most easily discussed in a quantum picture. There are many matching concepts between wave optics and quantum mechanics, so there is intrinsic value in understanding image-forming instruments in a wave picture.

A light wave has several characteristics that can be represented as a graph of a periodic function such as

$$A(x,t) = A_0 \sin[k(x - \mathrm{v}t)] \tag{12}$$

A_0 scales the sine function, which oscillates between the values ±1. Therefore, A_0 represents the peak **amplitude** of the wave. The argument of the sine function is linear in both position (x) and time (t), to represent both the spatial pattern of the wave field and the fact that it propagates in the x direction steadily at speed v. The direction of propagation would be in the $+x$ direction for positive values of v, and in the $-x$ direction for negative values of v. $A(x,t)$ can be graphed on either a spatial axis (at a specified time) or a time axis (at a specified position), and, in either case, it is a graph of amplitude versus the independent coordinate. It is important to realize that $A(x,t)$ can be graphed in three-dimensional space, where A is then a scalar amplitude value, and where Equation 12 is seen to represent plane waves because $A(x,t)$ is independent of y and z and is therefore of constant value on any surface where the argument of the sine function is constant. In this case, these surfaces are planes normal to the x axis. In a more general representation

$$A(\underline{r},t) = A_0 \sin[\underline{k} \cdot \underline{r} - \omega t + \phi] \tag{13}$$

represents a scalar plane wave field oriented normal to the **wave vector** $\underline{k}$, and propagating in the direction defined by $\underline{k}$.

In addition to amplitude and velocity, the periodic wave field $A(x,t)$ has an exactly defined wavelength and frequency, and therefore represents an idealized monochromatic wave. Because sin(x) has a natural wavelength of 2π, the constant k is the scale factor that sets the wavelength, λ. Effectively, k is the **spatial frequency** of the field in radians/unit length λ

$$\mathrm{k} = \| \underline{k} \| = 2\pi/\lambda \tag{14}$$

The **frequency** of the wave ν is simply the number of cycles that occur per unit time, so that the product of frequency and wavelength must equal the **wave speed**

$$\nu\lambda = \mathrm{v} \tag{15a}$$

or, with frequency expressed in radians/sec (ω)

$$\omega/\mathrm{k} = \mathrm{v} \tag{15b}$$

For light in vacuum, v = c. When a wave enters a material medium in which its speed is altered, the frequency must remain constant—otherwise, there would be nonconservation of "cycles." Therefore, both λ and v must change proportionally. The **phase** of $A(x,t)$ at a particular location and time is the shift of the wave relative to some standard graph such as $\sin(kx)$. Phase shift is usually expressed in angular units of degrees or radians, or in fractions of a wavelength. For example, $\cos[k(x - \nu t)]$ lags $\sin[k(x - \nu t)]$ by a phase shift of 90°. In Equation 13, ϕ is a phase angle that represents an offset in the argument of the sine function. It causes a shift in the graph by a distance $(\phi/2\pi)\lambda$. To represent linear and elliptical states of polarization, the amplitude of a wave field can be written as a complex vector quantity, $\underline{A}(\underline{r},\underline{t})$. Because light is a transverse wave in isotropic materials, the vector representing the amplitude must be oriented normal to the direction of propagation. For example,

$$\underline{A}(x,t) = \underline{e}_y A_0 \sin[k(x - \nu t)] \tag{16}$$

represents plane waves propagating in the $+x$ direction, linearly polarized in the y direction. Finally, **intensity** is a measure of the energy carried by a wave field. In analogy with the AC delivery of power to a resistor, which is proportional to the mean square voltage or current, the intensity of a wave field is proportional to the time-average of the square of the amplitude. In reduced units

$$I = \langle A^2 \rangle = A_0^2 \langle \sin^2[\omega t] \rangle = (^1/_2)\, A_0^2 \tag{17}$$

In a quantum description, the energy carried by the field is proportional to the number density of photons and to the energy carried per photon (e_ν)

$$e_\nu = h\nu = hc/\lambda \tag{18}$$

where h is Planck's constant. A light quantum carries energy far greater than the average thermal energy per mode in a molecule. Therefore, absorption of light by a molecule can initiate many processes that do not occur otherwise.

Huygens' principle of wave propagation is a fundamental and powerful concept that is most dramatically illustrated by attempting to isolate a single ray of light by use of a collimator and a pinhole in a distant opaque screen. In such an experiment, the observation is that collimated light (plane waves) incident on the pinhole emerges as expanding spherical waves rather than as a thread-like collimated beam, showing that the concept of a "ray" has meaning only as the local director of an extended wavefront. Huygens formulated this experimental result (easily observable in water waves) into a picture in which every differential patch on a wavefront acts as a source of an expanding spherical wavelet. At a time increment Δt, each of these wavelets will have expanded to a radius $c\Delta t$, and the "propagated" wavefront is the net result of the superposition and interference of all of the wavelets (Fig. 3). Using Huygens' principle, it is straightforward to show that an unbounded plane wavefront continues to propagate as such, and that a complete spherical wave grows as an expanding sphere (of diminishing amplitude). The principle is rigorously expressed in Kirchoff's diffraction integral, which is of great use in computation of scalar fields for real optical systems. For the purposes of the present discussion, Huygens' picture can be used to describe the propagation of plane waves incident on a planar boundary between two materials where the wave speed is altered from ν_1 to ν_2 (Fig. 4). The construction shows geometrically that the transmitted wave is refracted according to the rule $\nu_1/\sin\theta_1 = \nu_2/\sin\theta_2$. This result corresponds directly to the Law of Refraction

$$n_1 \sin\theta_1 = n_2 \sin\theta_2 \tag{19}$$

and shows that the **refractive index** (n) is the ratio of the speed of light to the speed of the wave in each material (Table 3)

$$n_i = c/\nu_i \tag{20}$$

Equations 15 and 20 lead to the corresponding result for wavelength, which must also change proportionally with ν

$$\lambda = \lambda_0/n \tag{21}$$

where λ_0 is the wavelength in vacuum. For light propagating through a sequence of materials that differ in refractive index, the total time of travel for a wavefront on a particular ray path is simply the sum of the path length divided by the wave speed in each material

$$\Delta T = L_1/\nu_1 + L_2/\nu_2 + L_3/\nu_3 + \ldots \tag{22}$$

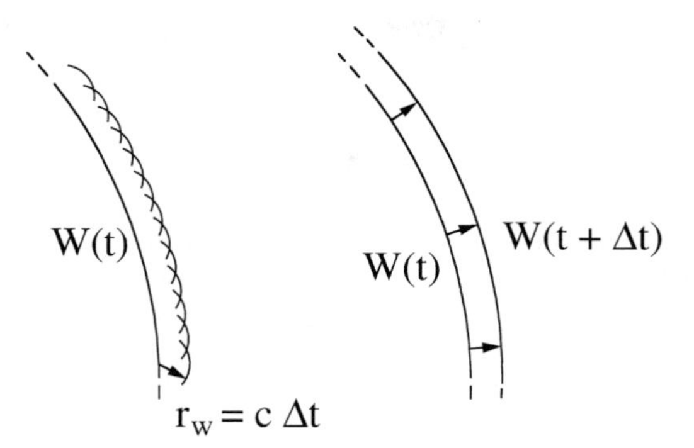

FIGURE 3. Huygens' construction. Each infinitesimal area element on wavefront $W(t)$ acts as a source of an expanding hemispherical wavelet proportional in amplitude to the local value of W. After a time interval Δt, each wavelet will have expanded to a radius vt, where v is the wave speed (here shown as c, the speed of light). The constructive interference of the wavelets in the forward direction defines a continuous wave front $W(t + \Delta t)$ that is advanced and generally altered in shape and amplitude distribution. An unbounded plane wave maintains its planarity and uniform amplitude. An expanding or contracting spherical wave remains spherical, but with decreasing or growing amplitude, respectively. In general, an arbitrary wave front, including partial plane or spherical waves, will undergo diffractive modification as it is propagated.

The optical path, OP, is defined as $c\Delta T$, the total path length that would be traveled by the wave in vacuum. Through Equation 20

$$OP_{1,N} = n_1L_1 + n_2L_2 + n_3L_3 + \ldots + n_NL_N \tag{23}$$

Clearly, OP can be computed easily on any known ray path. However, in a fundamental sense, OP is a constant of integration resulting from Fermat's principle.

Resolution in Transmitted-light Microscopes: The Abbe Model

The most fundamental difference between microscopes and other types of imaging instruments is the high aperture of the optics. By this it is meant that the specimen can be illuminated by light rays enter-

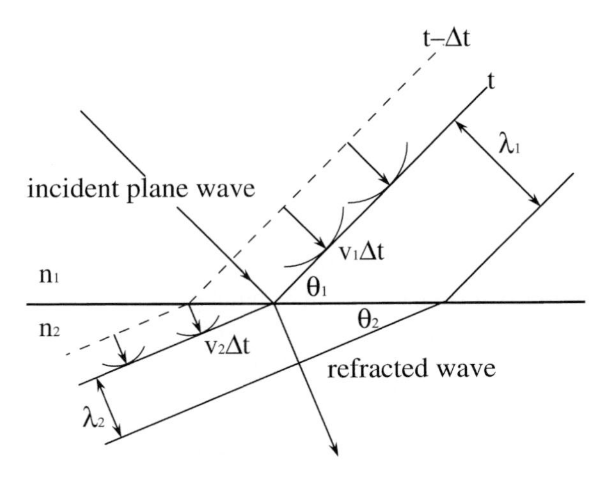

FIGURE 4. The relation between refraction and wave speed. Huygens' construction shows that the direction of propagation of a wave front is altered at a boundary where wavelets differ in speed. In a time interval Δt, wavelets expand to a radius of $v_1\Delta t$ in medium 1, and to $v_2\Delta t$ in medium 2. A comparison of triangles shows that $(1/v_1)\sin\theta_1 = (1/v_2)\sin\theta_2$. Because the frequency of the wave must be identical in the two media, the wavelength (λ) must change in proportion to v. With the refractive index (n) defined as c/v, Snell's Law is obtained directly: $n_1 \sin\theta_1 = n_2 \sin\theta_2$. Also, $\lambda_i = \lambda_0/n_i$ in each medium. The reflected wave in medium 1 is not shown.

TABLE 3. Optical constants relevant to microscopy

Visible light wavelength band (λ)		390–750 nm
Speed of light in free space (c)		3.0×10^8 m/sec (30 cm/nsec)
Frequency band of visible light		$6.7 - 4.3 \times 10^{14}$ Hz
Important wavelengths		
Peak human scotopic vision	556 nm	(yellow-green)
Mercury green lamp	546.1 nm	standard optical design wavelength
Low-pressure sodium lamp	589.0 nm	standard yellow wavelength for refractometry
(Doublet)	589.6 nm	
Red helium-neon laser	632.8 nm	
Important refractive indices (n)		
Air	1.00028	
Water	1.333	(20°C)
Culture medium or saline	1.335	
Animal cells	1.36	(average, by refractometry)
Glycerol	1.47	
Silica	1.46	
Crown glass	1.52	
Average refractive index increment for proteins and nucleic acids:		0.0018 per gm/deciliter

ing over a wide range of angles, and that the objective can form an image from light rays that exit the specimen over a similar angular range. The relationship between aperture and resolution in a microscope was described by E. Abbe (1840–1905) in a model in which the specimen is considered to be a planar diffraction grating mask on a slide (Fig. 5). The mask is presumed to be immersed in a medium of refractive index n, so that the wavelength is λ_0/n. This picture is the appropriate model for transmitted-light microscopy, in which there is spatial coherence across the light waves impinging upon the specimen from a particular direction. If the grating is illuminated by a beam of monochromatic plane waves (also called a light pencil) from the condenser, the transmitted light exits as a set of plane-wave diffraction orders that travel at well-defined angles relative to the incident light direction. For a grating composed of parallel slits of period p, and light of wavelength λ_0/n, diffraction orders occur at angles (θ_N) for which the optical path increment per slit is an integer multiple of the wavelength. This is the condition that guarantees purely constructive interference between wavelets expanding from each slit. In the case where the plane waves arrive at the grating at normal incidence

$$p \sin \theta_N = N \lambda_0/n, \qquad N = 0, \pm 1, \pm 2, \pm 3, \ldots, \pm N_{max} \qquad (24)$$

where N_{max} is defined by the requirement that $|\sin \theta_N| \leq 1$. Each diffraction order is a set of featureless plane waves, showing clearly that the information specifying the structure of the grating must be car-

FIGURE 5. The Abbe model: coherent image formation. (*A*) The condition for constructive interference in a diffraction grating is that the optical path difference between neighboring "slits" or sources is an integer number of wavelengths; $s_1 + s_2 = N\lambda$. This is clearly dependent on both the direction from which light is incident on the grating and the direction from which the grating is viewed, and leads directly to Bragg-type relations (Eqs. 24 and 26). (*B*) A grating is composed of a large number of equi-spaced sources, in which case there is nearly complete destructive interference for light exiting the grating in directions for which the sources are not all in phase, and completely constructive interference in the specific directions from which the sources all appear in phase. (*C*) Plane waves normally incident on a grating are scattered as a set of diffraction orders, in which the neighboring slits differ in phase by 0, 1, 2, ... or more cycles of the wave. Each diffraction order is a set of plane waves exiting the grating at a specific angle (Eq. 25). A finer grating produces more widely divergent orders (compare *C* and *E*). (*D*) Wave-optical origin of diffraction orders. Each slit produces cylindrical waves that act as Huygens' wavelets. These interfere to produce plane waves at specific angles to the grating. For the wavelength and grating period shown, three such angles occur. These are shown in *E* as five diffraction orders. (*F*) When light is incident obliquely, the orders are shifted in angle, with the zero order propagating in the forward direction. (*Continued on facing page.*)

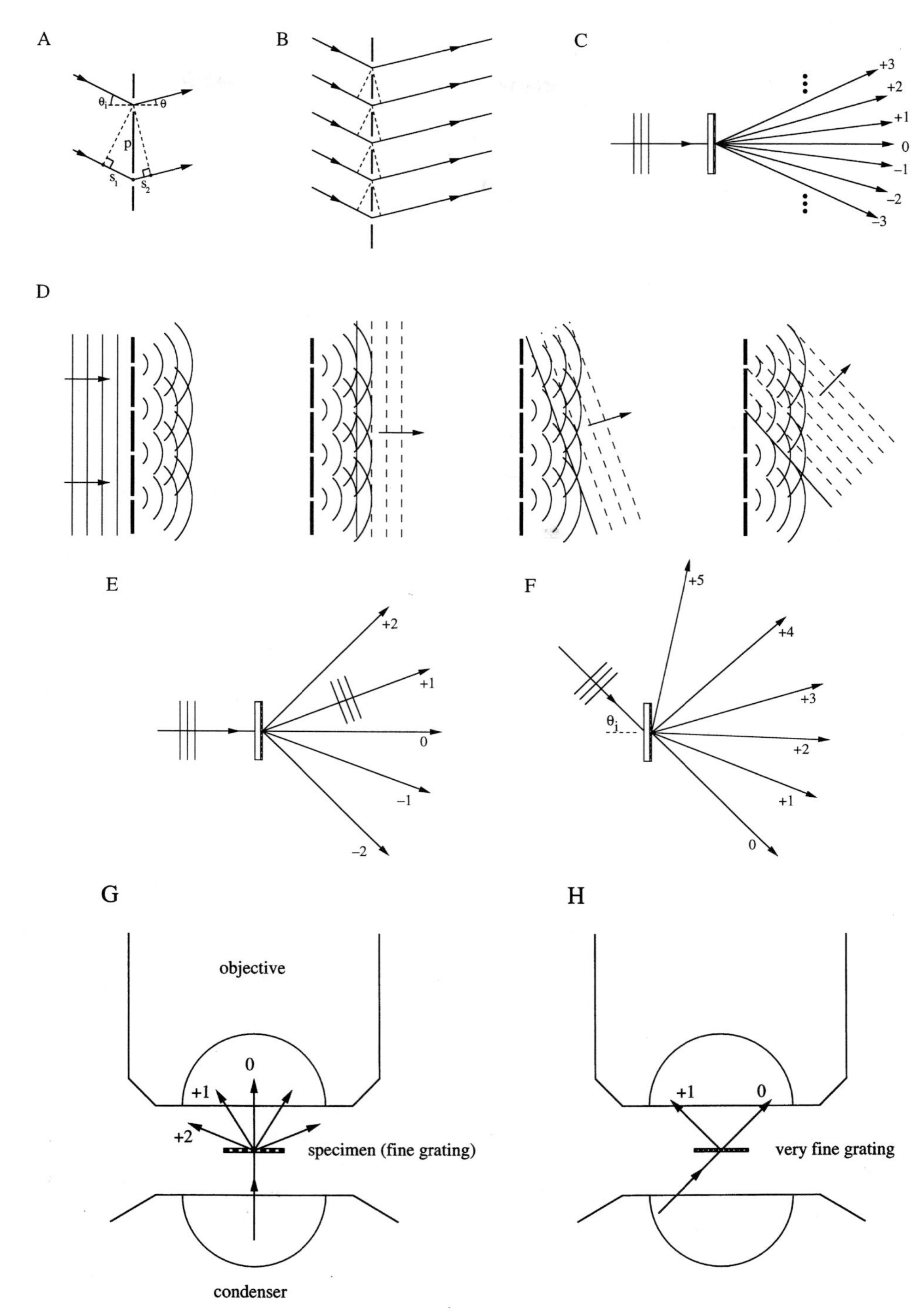

FIGURE 5. (*Continued from facing page.*) (*G*) A fine diffraction grating as an elementary object in a transmitted-light microscope. With normally incident illumination, the zero-order and both first-order waves are captured by the objective and interfere to form the image. In *G*, the second-order waves exit the grating at an angle exceeding the aperture angle of the objective. A very fine grating would produce orders so widely divergent that only the zero order would enter the microscope. No interference pattern would be formed in the image plane and the grating would be unresolved. (*H*) With oblique illumination, the widely divergent first order of a very fine grating is captured by the objective along with the zero order. Interference between these waves produces a fundamental image of the grating. The Abbe resolution limit is defined by the maximally oblique condition where the incidence angle is the aperture angle of the objective and the grating period is such that the first order exits at the opposite of this angle (Eq. 27). In the usual circumstance, the condenser produces waves incident on the grating from a range of directions.

ried by the angle, amplitude, and phase of each order *relative to the others*. In the Abbe picture, the objective functions to capture and redirect the diffraction orders so that they interfere to produce a magnified image of the grating. Equation 24 shows that the orders become more widely divergent (and fewer in number) for finer gratings

$$\theta_N = \text{Arcsin}[N\,\lambda_0/np] \tag{25}$$

In this simple case, the finest grating that would produce an image is the one for which only the three central orders ($N = -1, 0, 1$) could enter the objective (Fig. 5G). In other words, the maximum aperture angle for the objective ($\pm\theta_{max}$) sets the limit on θ_1. Abbe's insight was that the fundamental periodicity in the grating image was produced by interference between the zero order ($N = 0$) and either one of the first orders ($N = \pm 1$). When the incident plane waves are oblique (incidence angle = θ_i), the condition for constructive interference is a slight modification of Equation 24

$$p(\sin\theta_N - \sin\theta_i) = N\,\lambda_0/n \tag{26}$$

In particular, when $\theta_i = -\theta_{max}$ the zero order will exit the specimen and enter the objective at $-\theta_{max}$, and the first order can then enter at an angle as large as $+\theta_{max}$ (Fig. 5H). In this case, Equation 26 defines the transverse resolution limit of the microscope, p_{min}, as the finest grating period that can be resolved

$$p_{min}(\sin\theta_{max} - \sin(-\theta_{max})) = p_{min}(2\sin\theta_{max}) = \lambda_0/n$$

or

$$p_{min} = \lambda_0/(2n\sin\theta_{max}) \tag{27}$$

The quantity $n\sin\theta_{max}$ is of fundamental importance in microscopy, and defines the **numerical aperture (NA)** of the optical system when n is the index of the immersion medium. The Abbe resolution limit is therefore most often put in the form

$$p_{min} = \lambda_0/(2\ \text{NA}) \tag{28}$$

For mid-visible wavelengths, the Abbe resolution can be as high as 0.2 μm (Table 4). In cases where the condenser NA differs from the objective NA, Equation 26 reduces instead to

$$p_{min} = \lambda_0/(\text{NA}_{cond} + \text{NA}_{obj}) \tag{29}$$

For an idealized objective in which $\theta_{max} = 90°$, the NA would equal the immersion fluid refractive index, n. Practical constraints on lens design limit θ_{max} to <90°, and NA to values close to but less than n (Table 1).

The Abbe picture makes evident the essential function of both the condenser and objective in transmitted light microscopy. The condenser not only serves to concentrate light into the field of view, but also provides incident light over a wide range of angles. This ensures that the objective aperture is fully utilized in the capture of diffraction orders from all possible "elementary gratings" constituting the specimen—up to the transverse resolution limit, and regardless of grating orientation in the plane of the slide. Less obvious, but just as important, the NA of the condenser and objective also determine the axial resolution, or depth of field, of the microscope. Again, considering the specimen to be a diffraction grating illuminated by a single set of plane waves, it is the superposition of diffraction orders that produces an interference pattern in the image plane of the microscope. This pattern

TABLE 4. Transverse resolution and depth of field versus NA ($\lambda_o = 0.55$ μm)

NA	Immersion	p_{min} (μm)	δ (μm)
0.3	dry	0.92	6.1
0.5	dry	0.55	2.2
0.75	water	0.37	1.3
1.2	water	0.23	0.51
1.25	oil	0.22	0.53
1.4	oil	0.20	0.43

not only exists in the image plane, but is extended in image space; i.e., it has a large depth of focus, which is equivalent to poor axial resolution (Fig. 6). This situation is changed when a condenser is used to illuminate the grating from many directions simultaneously. Because the light incident from different directions is mutually incoherent, the interference patterns are formed independently in the image space. In the image plane, all of the patterns sum in register to give a sharp image of the grating. Out of the image plane, the individual patterns are progressively shifted out of register by an increment depending on the actual angles of the incoming diffraction orders. The net result is that the grating image loses sharpness over a finite axial range in image space, which defines the **depth of focus.** In a microscope, where the camera is in a fixed position with respect to the optical system, this must be translated into an equivalent focus range. Normalizing the depth of focus by the axial magnification (Eq. 6b) defines the **depth of field** (δ)

$$\delta = n\,\lambda_0/\mathrm{NA}^2 \tag{30}$$

This equation will vary by a numerical factor depending on the measure of image sharpness used in its derivation, but the important point is that depth of field in a transmitted-light microscope depends inversely on the square of the system NA (Table 4). A similar result holds in fluorescence microscopy (see Eq. 39). Reduced depth of field is equivalent to enhanced axial resolution. Increased NA improves both the transverse and axial resolution, and therefore strongly improves the 3D resolution of the microscope.

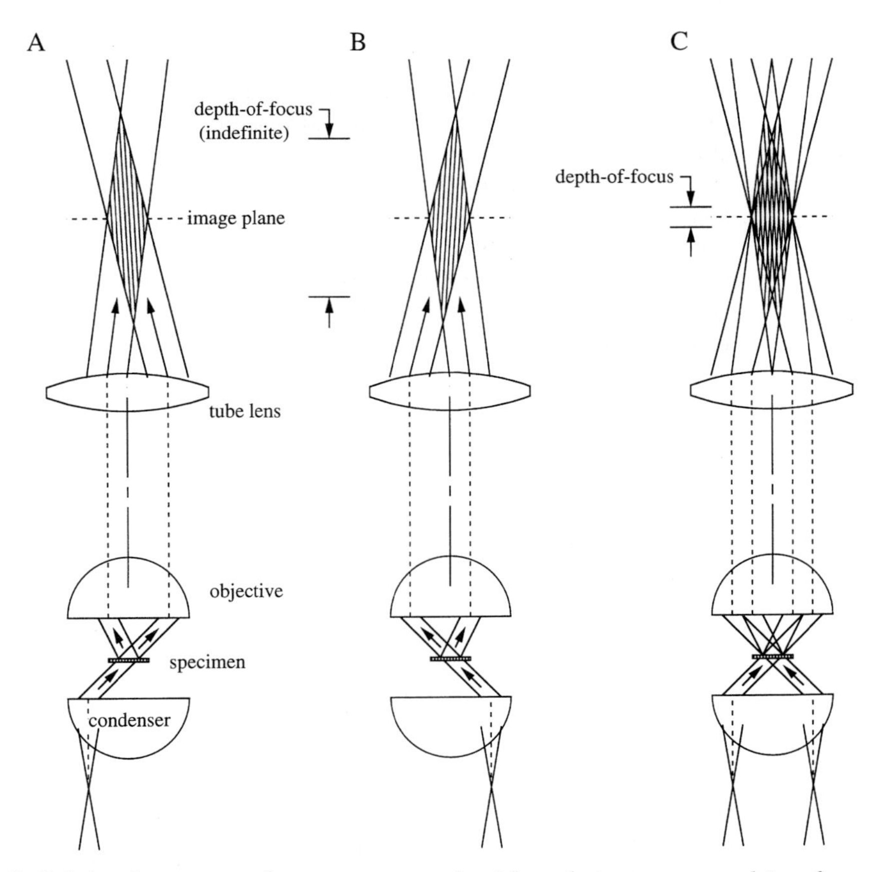

FIGURE 6. Relation between condenser aperture and axial resolution; superposed interference patterns define the depth of field. (*A*) A fine grating object illuminated obliquely from one direction produces two diffraction orders that interfere in the image plane to produce a fundamental image. The interference pattern has indefinite depth of focus; therefore, the microscope has no axial resolution. (*B*) Light incident from a different part of the condenser pupil produces a similar image. (*C*) The superposition of these two fields reinforces the image only in the neighborhood of the image plane. Use of the full condenser aperture (not shown) superposes many such interference patterns and minimizes the range over which the grating is sharply focused.

The strong variation of the depth of field with NA is of very practical and routine use. It is often difficult to initially bring a specimen into focus in a high-NA system because of the absence of visible features in the defocused field. Reduction of the condenser aperture by closing the condenser iris greatly increases the depth of field, and therefore the contrast of out-of-focus features. They can then be discriminated against background and brought into focus. Once focus is set, the condenser iris is opened to increase resolution. In general, visibility (a combination of contrast and resolution) is optimized at an intermediate setting of the iris. However, maximum resolution is obtained with maximum aperture. In video-camera-based imaging systems, the "black level" and gain of the video signal amplifier can be adjusted to greatly increase contrast at full aperture.

Immersion Optics

High NA as prerequisite for obtaining high resolution (both transverse and axial) was the driving force behind development of immersion optics. Light exiting the specimen into the objective is refracted most strongly at the cover glass-to-air interface, where there is a large difference in refractive index (Fig. 7). Not only is there increased reflective loss for rays exiting the cover glass at high angle, but also total internal reflection (TIR) of rays incident above the critical angle, which, for air–glass refraction, is 41.8°. This limitation is overcome by filling the air space between the cover glass and objective with an immersion fluid having a higher index. In oil-immersion systems, the cover glass, oil, and objective front lens are all nearly index-matched, so that rays propagate without deviation into the objective aperture. Reduction of the severity of refraction between the specimen and the objective is important more generally because it makes possible the design of lenses having very low aberration at high NA, the key to high-resolution imaging. In general, the ideal circumstance is a lens designed for use with an immersion fluid having the same refractive index as the *specimen*, although this is not always possible in practice.

The optical structure of the specimen affects ray paths into the microscope. Even if the actual biological object is only weakly refractive or absorptive, the index of the mounting medium and the thickness of the cover glass both have optical effects that can lead to focus-dependent spherical aberration in the image (Frisken Gibson 1990; Frisken Gibson and Lanni 1991; McNally et al. 1994; Scalettar et al. 1996). The situation is most severe in the most common case; "**dry**" or **air-immersion objectives** for biological use. These lenses are designed to compensate for the refraction that occurs as

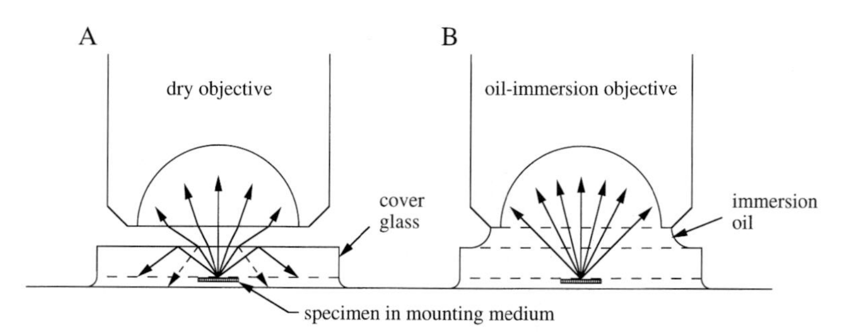

FIGURE 7. Ray diagram for immersion optics. (*A*) Light waves exiting a specimen are refracted at the cover glass/air interface, and again at the air/lens interface. This introduces spherical aberration and causes reflective loss (*dashed arrows*), deviation of high-angle rays out of the aperture range of the objective, and total internal reflection of rays exiting above the air/glass critical angle. Dry objectives are designed to compensate for the spherical aberration introduced by a cover glass of standard thickness (#1-1/2). (*B*) In an oil-immersion system, the specimen mounting medium, cover glass, oil, and objective are all nearly index-matched. In this case, waves propagate without refraction, minimizing aberration and reflective loss. Additionally, diffraction orders that would be lost due to refraction in a dry system enter the immersion objective and contribute to the sharpness of the image.

rays exit the specimen through the cover glass–air interface, where there is a relatively large index difference of 0.5. The correction depends on use of a cover glass of standard thickness, by convention 0.17 mm, or "#1-1/2." Furthermore, the specimen must be mounted in high-index medium immediately below the cover glass. If the specimen is not in contact with the cover glass, the intervening layer of mounting medium will essentially act as additional cover glass thickness, throwing off the correction. Likewise, focusing into a thick specimen below a cover glass also adds to the ray paths in the high-index medium and causes aberration that grows with depth. In the ideal case, the lens would be designed for use with an immersion fluid that is isorefractive with the specimen, with or without a cover glass. In living cells, the average index is 1.36, only slightly greater than for water, 1.33. Therefore, the ideal case can be approached with **water-immersion objectives**. Direct-immersion objectives, of most use on upright microscope stands, are of fixed correction. They are designed for contact with culture medium and require no cover glass between cells and lens. Indirect-immersion objectives are designed for viewing living specimens through a cover glass, with distilled water as the coupling fluid. These lenses incorporate a variable correction that must be set depending on cover glass thickness over the range 0.15–0.20 mm (#1–#2). For the best results, cover glass thickness should be measured with a micrometer caliper prior to cell culture or specimen assembly. **Oil-immersion objectives** (n = 1.51–1.52) are designed for use with specimens mounted in high-index medium. Cover glass thickness is less important in this case, because of the near-index match between immersion oil and glass. Other types of corrections exist, such as for viewing cells on inverted microscopes through plasticware of various thickness. Glycerol-immersion objectives and variable-correction multi-immersion objectives (water/direct–water/cover glass–glycerol/direct–glycerol/cover glass–oil) are less common, but useful, optical systems (Table 1).

The Sine Relation, Body Tube Length, and Infinite-Conjugate Systems

Given the importance of NA in the performance of a microscope, it is clear that paraxial optics cannot provide a basis for deriving resolution limits or for minimizing aberration. The diffraction-limited performance provided by a microscope over a flat field is a result of design to satisfy a wave-optical restriction known as the Sine Relation. In essence, the condition requires not only that the optical path (or propagation delay) between a source point (A) and its corresponding geometric image point (A') be constant for all possible rays connecting A and A', but also that the path be stationary for any choice of A in the design object plane and within the design field of view. The mathematical expression of the Sine Relation maps the angle (U) of a ray entering the objective, to the angle (U') of that ray as it arrives at the image plane

$$n \sin U = M \sin U' \tag{31}$$

It is straightforward to show that a microscope which satisfies the sine relation for a specimen located exactly at the design "working distance" will not satisfy the condition for a specimen shifted toward or away from the lens—even if the camera, film, or eyepiece is correspondingly shifted to the position of best focus. To accommodate this restriction in a practical way, microscopes are designed to a standard **tube length**, traditionally defined as the distance behind the objective at which the primary image is formed + 10 mm (Fig. 8). For biological microscopes, the standard body tube length has been 160 mm. In the past 20 years, there has been a major design change to **infinite-conjugate** (**IC**) systems (Keller 1995), in which the objective is followed by a tube lens, usually a high-quality singlet, doublet, or triplet (Table 5). In IC systems, the objective is designed so that light from a point source located exactly in the front focal plane emerges from the back pupil of the objective as collimated light. This is equivalent to formation of the image of the point source at "infinity." The function of the tube lens is to converge each collimated set of rays to form the primary image at a distance of one focal length (Fig. 8). The primary advantage of an IC system is that a space is provided between objective and tube lens in which flat optics (filters, waveplates, dichroic reflectors, crystal prisms, polarizers, electro-optic devices) can be inserted without causing significant image shift, focus shift, or aberration. A secondary advantage is that the focus drive can move the objective over a limited range relative to the microscope (and fixed stage) without causing magnification error or aberration.

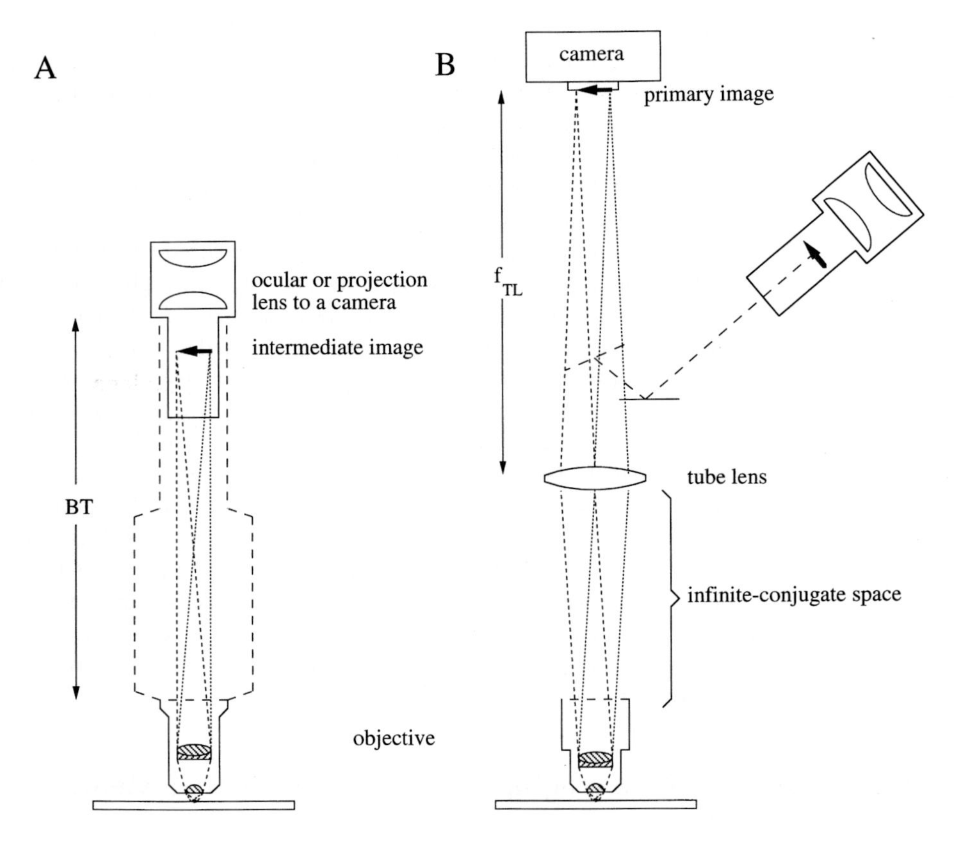

FIGURE 8. Standard layout for 160-mm and infinite-conjugate (IC) microscopes. (*A*) In a finite tube-length system, the objective converges light from an in-focus object, forming an intermediate image, by design, 10 mm below the rim of the body tube. An ocular or projection lens is used to form a secondary image in a camera. For biological microscopes, the standard body tube length (BT) has traditionally been 160 mm. (*B*) In an infinite-conjugate system, the objective collimates light from point sources in an in-focus object. The primary image is then formed in the back focal plane of a separate tube lens. Flat optics can be inserted into the infinite-conjugate space with minimal aberrating effect.

The Sine Relation applies to both finite- and infinite-conjugate objectives. When U' is a **marginal ray** (i.e., when $U = \theta_{max}$ and $U' = U'_{max}$), then Equation 31 can be put in a particularly useful form

$$n \sin \theta_{max} = \text{NA} = M \sin U'_{max} \tag{32}$$

Because M>>1 in microscope systems, U'_{max} is generally a small angle. Therefore, to a good approximation, $\sin U'_{max}$ is simply the ratio of half the diameter of the objective pupil ($d_p/2$) to the tube length or the focal length of the tube lens (f_{TL}). Therefore,

$$d_p = 2\,(\text{NA}/M)\,f_{TL} \tag{33}$$

This simple relation shows that the diameter of the clear aperture of different objectives, as seen from the back side, increases with NA and decreases with *M*. It is important in fluorescence microscopy where the specimen is illuminated through the objective, and where there should be a good match between the projected size of the light source and the diameter of the objective back pupil. Or, in laser-scanning microscopes, there will be an optimum laser beam expansion ratio to match the Gaussian beam diameter to each objective back pupil.

The most important practical implication of the Sine Relation is that aberration will be minimized when the use conditions of the microscope match the design conditions. A problem such as loss of parfocality between eyepieces and camera often results from misadjustment of the camera tube length. If the microscope is simply refocused to compensate for this problem, the specimen will be offset from

TABLE 5. Tube lens focal length in IC systems

Leica	200 mm
Nikon	200 mm
Olympus	180 mm
Zeiss	160 mm

Use of an IC objective on a microscope stand that incorporates a tube lens of focal length (f_2) different from the design value (f_1) will have the effect of altering the magnification by the factor f_2/f_1.

its design position, and the image will show increased aberration. With the growth in use of microscopes for optical sectioning or imaging at depth, a common problem is spherical aberration caused by refractive index mismatch between immersion fluid and specimen. The index range in living cells (1.34–1.45, average 1.36) is closer to that of water (1.33) than to that of immersion oil (1.52). Therefore, although high-NA oil-immersion objectives generally give outstanding performance imaging adherent living cells on a cover glass, focusing into the specimen causes spherical aberration to accumulate at a rate of 1/3 wave per micrometer (Frisken Gibson and Lanni 1991). At a depth of 10 μm, the loss in image sharpness is easily noticeable. This problem can be minimized by use of water-immersion objectives (direct or indirect), where there is better match between specimen and immersion index. Design parameters and use conditions are permanently marked on objective barrels (Table 6).

Principal Optical Components and Köhler Illumination

Best utilization of the full aperture and field of a microscope is usually achieved in Köhler Illumination, the condition in which (1) the condenser and specimen are set in their correct axial position relative to the objective and (2) the light source is imaged into the back focal plane of the condenser. This involves basic setup steps such as centering of the light source and focusing the collector lens, and routine adjustments such as approximately focusing the specimen, then centering and focusing the condenser. The importance of Köhler illumination can be seen in a ray diagram showing light propagation through the principal components of the microscope. In the simplest transmitted-light system, these would be, in order: (1) light source, (2) collector lens, (3) field iris, (4) condenser aperture iris, (5) condenser lens, (6) specimen, (7) objective, (8) body tube lens (infinite-conjugate system), and (9) oculars or (10) camera primary image plane (Fig. 9). The collector lens is usually of moderately high aperture to obtain high brightness and forms a magnified image of the light source in the back focal plane of the condenser. This guarantees that every point in the image of the light source is completely defocused over the entire field of view by the condenser, so that the illumination of the specimen is very even. As shown in Figure 9, the image of the light source at **4** fills the entire condenser pupil, so that the full **illuminating numerical aperture** (INA) can be utilized. The variable iris at **4** masks the light source image that is allowed to enter the condenser and sets the maximum angle at which light enters the specimen. The condenser iris is therefore an **aperture stop**, which sets the INA. In contrast, the diameter of the variable iris at **3** sets the cone angle of each converging ray bundle that forms the source image, and therefore sets the diameter of the illuminated field in the specimen. Iris **3** is therefore a **field stop**. In practice, the condenser is moved axially to put the image of the iris **3** on the specimen. Setup of the microscope can be quickly accomplished by the following sequence of steps:

1. Center and focus light source image projected into the condenser.
2. Focus microscope on an easily visible feature in the specimen. This properly sets the position of the specimen relative to the objective.
3. Close field stop enough to put its image at least partially in the field of view. If the condenser is far from focus or far off-center, the image of the field stop may be highly blurred or completely out of view, initially.
4. Center and focus condenser to get a sharp, centered image of the field stop on the specimen.

TABLE 6. Objective identifier markings

Identifier	Symbol	Typical value/explanation
Degree of correction		
Achromat	—	basic correction for chromatic aberration (CA)
Fluar	—	moderate correction for CA
Apochromat	—	high correction for CA
Plan	—	flat field correction
Magnification	(M, x)	0.5–40x (dry), 10–100x (water), 40–100x (oil)
Numerical aperture	(NA)	0.02–0.75 (dry), 0.3–1.2 (water), 0.5–1.4 (oil)
Immersion type		
dry	(no marking)	
water (direct)	W, WI	
water (cover glass)	W Korr	adjustable correction for cover glass thickness
glycerol	G, Glyc	fused silica cover glass—0.2 mm
oil	Oil, Oel	
multi-	Imm	adjustable correction for water, glycerol, oil
Tube length/cover glass thickness		
160-mm body tube	160/0.17	standard cover glass
	160/—	unspecified, or none
infinite-conjugate (IC)	∞/0.17	standard cover glass
	∞/—	unspecified, or none
Specialized use		
phase contrast	Ph1, Ph2, Ph3	standard phase annuli/phase plates
polarized light	Pol, DIC	strain-free lens elements
UV fluorescence	U-, U340/380	UV transmissive lens elements
dark field	iris	internal iris for variable NA

5. Open the field and aperture irises. For visual use of the microscope, the aperture is often not opened fully because this improves contrast at the expense of some resolution. The best setting is specimen-dependent. For maximum resolution, the aperture iris must be fully opened.
6. Focus on the object of interest. When this is done, check to see whether the image of the field iris is still sharp, and adjust the condenser focus, if necessary. In most microscopes, the main focus drive moves the stage and condenser as a unit relative to the objective, or moves the objective relative to the stage and condenser. In either case, the set position of the condenser relative to the objective is not maintained as focus is changed. Ideally, the focus drive would move the stage relative to fixed condenser and objective.

It can now be seen that, in Köhler illumination, light is sent into the specimen over a wide angular range. Each collimated pencil of light passing through the specimen represents an essentially independent set of plane waves, because each pencil originates in a different part of the image of the lamp filament. In other words, the pencils are mutually incoherent. In the Abbe model, the image formed by a single light pencil was considered. In Köhler illumination, roughly 10^6–10^7 independent light pencils pass through the specimen when a high-NA condenser is used at full aperture. For most biological specimens such as living cells, the zero-order diffraction is much stronger than higher orders, resulting in a bright, even background. The net result is that the highest-resolution images formed by a microscope in transmitted light also have low contrast and therefore low visibility. The development of video-enhanced contrast (VEC) has revolutionized transmitted-light microscopy by enabling instruments to be used at full aperture (Allen et al. 1981; Inoue 1981; Inoue and Spring 1997).

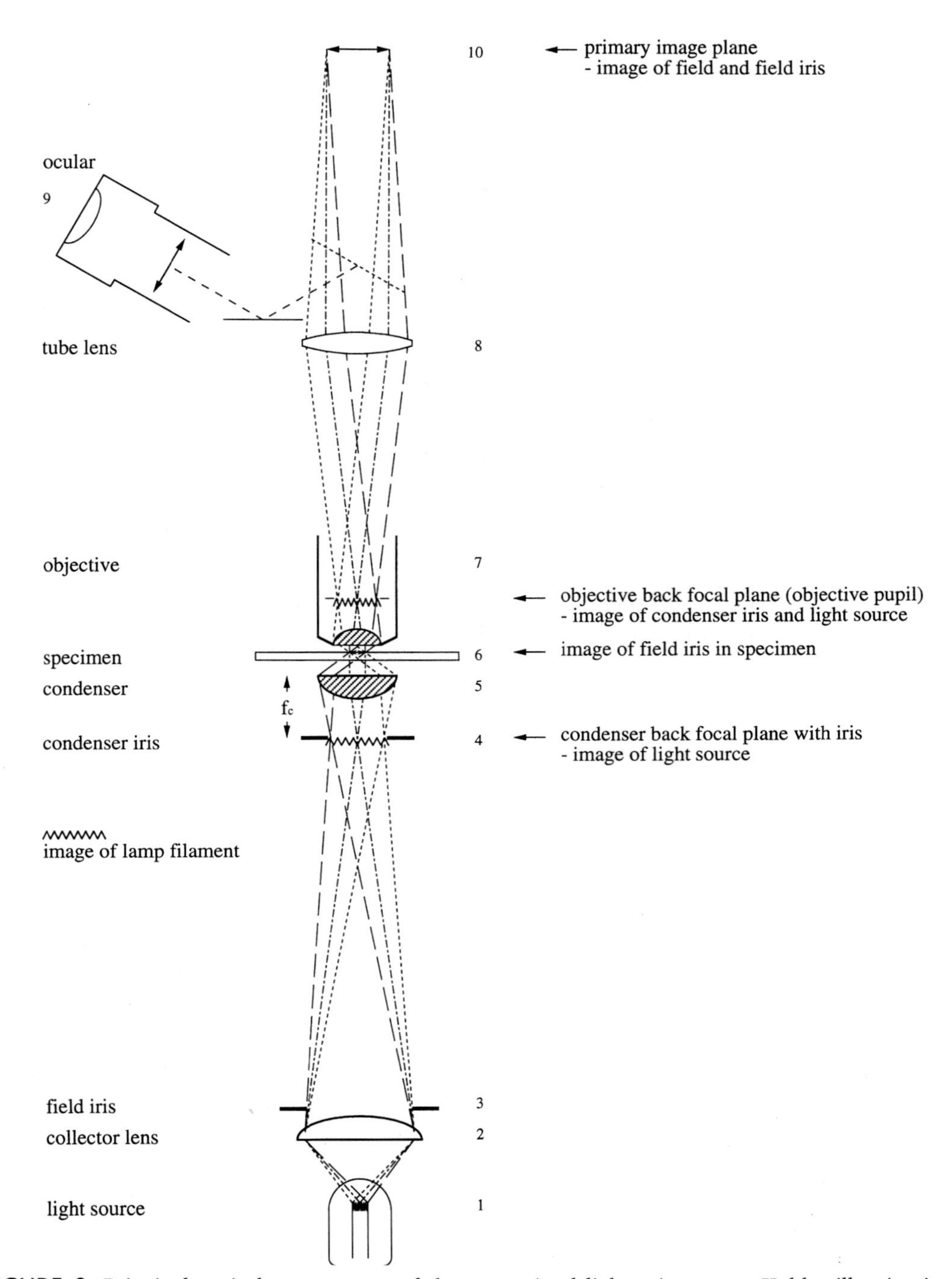

FIGURE 9. Principal optical components of the transmitted-light microscope. Köhler illumination is shown in an IC system. The collector optics (2) form an image of the light source in the back focal plane of the condenser (4). This guarantees that the lamp will be completely defocused and the specimen evenly illuminated. The source image is masked by the condenser iris, which controls the illumination aperture. Additionally, the field iris (3) is located so that its demagnified image is projected into the specimen plane. The collimated components of the transmitted illumination are converged by the objective (7) to form an image of the light source in its back focal plane. Spherical waves originating in the specimen are collimated by the objective and converged by the tube lens to form an image in its focal plane. The primary magnification is the ratio of tube lens to objective focal lengths.

In the Köhler setup (Fig. 9), the optical conjugate planes can be easily identified. An image of the lamp filament is formed in the back focal plane of the condenser and, along with an image of the condenser iris, in the back focal plane (pupil) of the objective. The filament is completely defocused in the planes of the field stop, specimen, and primary image. An image of the field stop is formed in the specimen and in the primary image plane. Microscope stands are designed with this setup in mind, so it is important to note that large alterations in the system, such as would be caused by moving the light source away from its design position, could result in reduced INA and/or uneven illumination of the field of view. It can also be seen that irregularities in the image of the light source, such as the bright filament wires in an incandescent lamp, will not cause unevenness in the field of illumination, but instead cause unevenness in the angular distribution of the incoming light. Because this represents under-utilization of the available NA, both transverse and axial resolution are affected. Use of a ground-glass light diffuser helps, but is optically inefficient. Ideally, the image of the light source in the condenser pupil should be uniformly bright and matched to the pupil diameter. In a **light scrambler**, this is accomplished by focusing the source into a 1-mm fiber optic. A lens is then used to magnify the uniformly bright output end of the fiber so that its image fills the pupil. Use of a light scrambler produces high field and angular uniformity, and maximum resolution (Inoue and Spring 1997; G.W. Ellis light scrambler/www.technicalvideo.com). For certain applications where a laser is used in place of a conventional light source, active scrambling is required to eliminate speckle due to coherence. This is most easily accomplished by use of a rotating diffuser (Hard et al. 1977), or by vibrating the fiber optic in a scrambler.

In fluorescence or other incident-light microscopes, the objective also acts as the condenser. In this case, it is not practical to place the aperture iris in the objective pupil, because it would needlessly reduce the NA, and therefore the resolution and brightness of the image. In this case, the aperture iris is placed between the light source and field iris along with a lens that images the aperture iris into the objective pupil (see Fig. 17).

OPTICAL SYSTEMS FOR IMAGING OF LIVING CELLS

The natural or experimentally induced optical properties of living cells provide the basis for contrast generation in a wide variety of special-purpose optical systems for the microscope. These optical systems can be placed in two main groups, those dependent on the refractive properties of cells, and those dependent on the absorptive and luminescence properties (Table 7).

Phase Contrast

The study of living cells by light microscopy was revolutionized by Zernike's development of phase-contrast optics, which make visible the refractive structure of nearly transparent objects. With a few notable exceptions such as the erythrocyte and chloroplast, most cells or organelles present a very low optical absorbance to visible light. Additionally, the average refractive index (1.36–1.38) of most cells in culture is not much greater than that of the surrounding medium (typ. 1.335), although refractive heterogeneity within cells and in the extracellular matrix adds up to a formidable light-scattering cross section in dense tissue. Therefore, the passage of light waves through a single cell results mainly in the distortion of the wave fronts due to all of the structures within the cell that differ in refractive index relative to the immediate surround. No significant change in amplitude or intensity (brightness) would be observed. More-refractive organelles slightly retard the passage of each wave front. This phase delay, usually expressed as a difference in optical path ($\Delta n \times L$; see Eq. 23), is a result of the lower speed of the wave in the region of greater index. For single cells, this difference is usually small; the average phase delay for a cell of 5 µm thickness and average index 1.36 in culture medium of index 1.335 is $(1.36 - 1.335) \times 5 = 0.125$ µm, or approximately one-quarter wavelength. A subcellular structure will cause a much smaller retardation; the lamellipod of fibroblasts, which contains mainly actin at a concentration of 40 mg/ml (4 gm/dl) (Abraham et al. 1999), causes a wave front delay of only 1.3 nm, or $\lambda/430$. Because essentially no amplitude reduction of the light waves occurs to create contrast, and because the phase delays are usually very small, the generation of sufficient contrast for visual discrimination requires converting phase shifts into brightness variation.

TABLE 7. Microscope optical systems and modes used in biological imaging

Contrast due to refractive structure within the object:	
Phase contrast	(Zernike phase contrast)
Differential interference contrast	(DIC, Nomarski phase contrast)
Reflection interference	
Polarization	
Confocal scanning reflectance and interference	
Hoffman modulation contrast, Varel	
Contrast due to absorption of light within the object:	
Amplitude contrast/bright field	
Luminescence	
fluorescence	
multiband fluorescence	
confocal scanning fluorescence	
fluorescence polarization	
total internal reflection fluorescence (TIRF)/Evanescent wave excitation	
fluorescence lifetime	
fluorescence resonance energy transfer (FRET)	
multiphoton scanning fluorescence	
phosphorescence	
delayed fluorescence	
near-field scanning fluorescence	

Zernike phase contrast is based on the general principle that refracted or locally phase-shifted wave fronts can be represented as the summed or superposed amplitudes of **scattered waves** and the unperturbed waves (or **direct waves**) that illuminate the specimen. In other words, the phase-distorted wave fronts exiting the specimen into the microscope consist of (1) the plane-wave field that would be present with no specimen in place (or with a perfectly index-matched specimen in place) and (2) the light scattered as spherical waves by the refractive features in the specimen. In general, the scattered waves will be weak relative to the direct light. Additionally, Zernike recognized that the scattered waves must lag in phase by 90° the direct light, to account for the absence of absorption (180° lag) or stimulated emission (no lag). Generation of a phase-contrast image occurs by optically processing the direct and scattered light differently, then allowing the two fields to interfere. In a phase-contrast microscope, two optical elements effect this transformation: a mask, or **phase annulus**, inserted into the back focal plane of the condenser, and a **phase plate** located in the back focal plane of the objective (Fig. 10). Because this plane is usually located within the multi-lens structure, phase plates are integral parts of phase-contrast objectives. In the Köhler setup, the annulus masks the image of the light source so that illumination is directed into the specimen from a restricted set of directions forming a hollow cone, rather than from all of the directions subtended by the full condenser aperture. An image of the annulus is formed in the objective coincident with the phase plate, so that 100% of the direct light passes through a special annular zone on the plate. On the other hand, the scattered light, being in the form of spherical waves expanding out of the specimen, passes mainly through the complementary zones of the plate, which constitute most of its area. The annular zone in the phase plate affects the direct light in two ways; it is semitransparent to attenuate the direct light to an amplitude comparable to the strongest expected scattered light, and it phase-shifts the direct light to change the phase difference from 90° to 180°. This creates the condition under which phase delay can be seen as brightness variation. In practice, the direct light is phase-advanced by 90°, by making the phase plate thinner in the attenuation zone. In the image plane of the microscope, the phase-advanced direct light provides uniform reference waves over the entire field of view. The scattered-light amplitude field, which would produce a **dark-field image** if the direct light were blocked in the objective, interferes with the direct light to produce the **phase-contrast image**. Because of the 180° phase difference between the two fields, the interference process results in attenuation or gain in brightness, and therefore an image with significant contrast. In most phase-contrast microscopes, the phase-shifting is

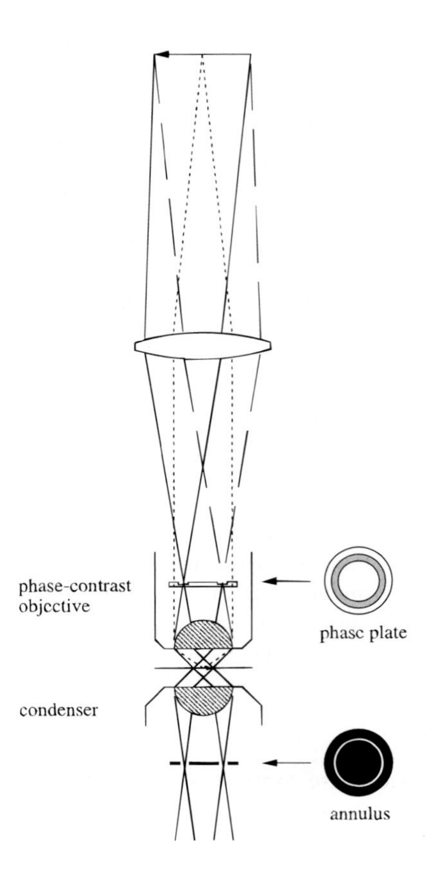

FIGURE 10. Optical components of a phase-contrast microscope. The principal components of the transmitted-light microscope (Fig. 9) are augmented by a mask in the condenser back focal plane and a phase plate in the objective. The mask is usually in the form of an annulus, and the phase plate in the form of a semitransparent ring. In use, the annulus restricts the illumination to be incident on the specimen from a set of directions that form a "hollow cone." Because the annulus is completely defocused by the condenser, illumination of the specimen is uniform. In the objective, the phase plate is coincident with the image of the annulus. This allows for attenuation and phase-shifting of the unscattered light. Most of the scattered light (*dashed lines*) passes through the transparent zones of the phase plate. The phase-shifted direct light and the scattered light interfere in the image plane to produce a high-contrast image.

designed so that more-refractive structures show up as darker features in the image (Fig. 11). This is known as positive phase contrast. In high-NA phase-contrast systems, cellular structures as thin as lamellipods can be discriminated under typical conditions.

The beauty of phase contrast is its simplicity, ease of alignment, insensitivity to polarization and birefringence effects (a major advantage for viewing cells through plasticware), and isotropy of response to different in-plane orientations of a particular feature in the specimen. Its difficulties include the need for a phase-contrast objective, the fact that the full INA is not utilized, the optical effect of the phase plate on other modes of microscopy, and a well-known "halo" artifact in the image around sharp boundaries of the object. The halo is due to the passage of some of the scattered light through the direct-light zone of the phase plate. This is an unavoidable consequence of the need for an annulus of finite width to allow for sufficient illumination. The halo often significantly increases the dynamic range of the image, making it difficult to apply VEC methods to this mode of microscopy. Experimental phase-contrast systems exist that suppress the halo and remove the restriction on VEC (Ellis 1988). Given the limitations of phase contrast, it remains a remarkably reliable and useful optical mode.

In practice, alignment of phase-contrast optics is a simple addendum to the Köhler setup procedure. After focusing and centration steps 1–6:

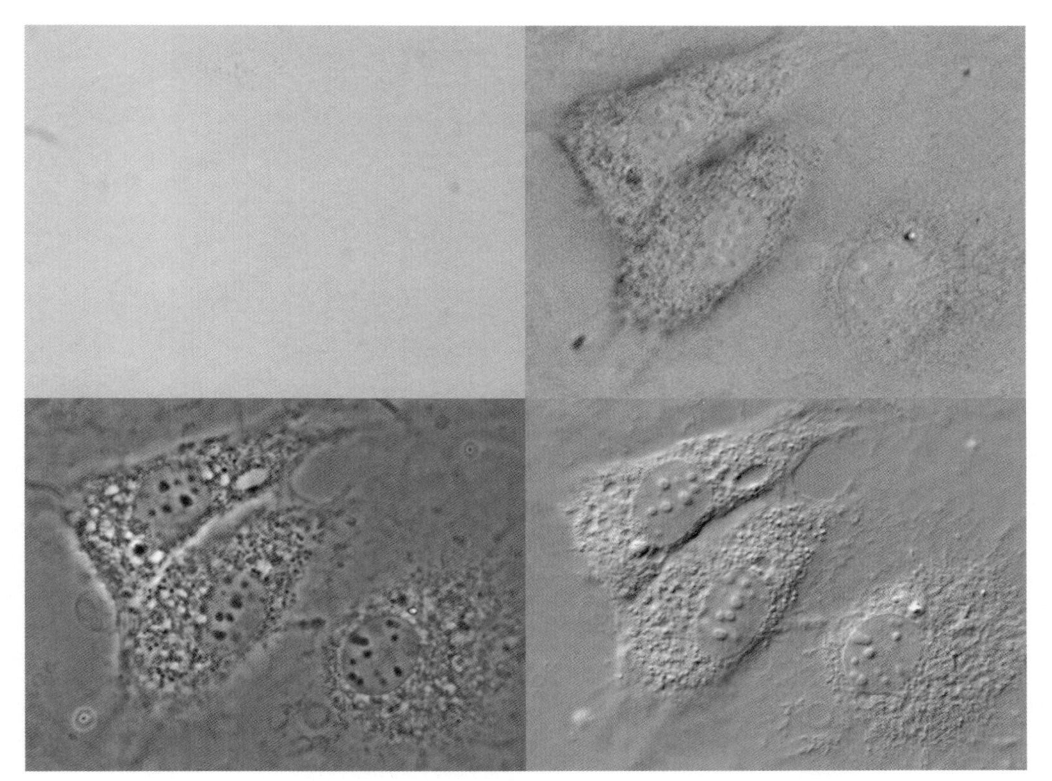

FIGURE 11. Fibroblasts in bright field, phase contrast, and VEC-DIC. (*Upper left panel*) Bright-field image of adherent fibroblasts in physiological medium, 100x 1.3 NA. No features produce sufficient contrast for visibility. (*Upper right*) Digitally contrast-stretched bright-field image shows weak amplitude effects of scatter in the object. (*Lower left*) Zernike phase-contrast image, no digital enhancement. (*Lower right*) Nomarski phase-contrast image with digital enhancement (VEC-DIC). Field of view width, 120 μm.

7. Switch condenser turret or slider to the phase annulus corresponding to the objective in use.
8. Remove one ocular and insert a **phase telescope** or **Bertrand lens**, or switch to an internal telescope setting. Focus the telescope (not the microscope!) to obtain a sharp image of the phase annulus and phase plate in the back focal plane of the objective.
9. Adjust the centration of the annulus (not the centration of the condenser!) relative to the phase plate, so that the bright image of the annulus falls exactly within the annular zone of the phase plate.

Mismatched diameters of the two annular components signify that the incorrect phase annulus is in place in the condenser. Vignetting of the image of the light source within the annular mask will occur if the condenser is out of focus.

Differential Interference Contrast

For many live-cell imaging applications, Nomarski differential interference contrast (DIC) has supplanted the Zernike system. The main advantage of DIC over phase contrast is that the optical elements required for DIC do not mask or obstruct the condenser or objective pupils, so that the instrument can be used at full NA. This improves resolution (particularly axial resolution), obviates halo artifacts, and produces an image well-suited for electronic enhancement of contrast. However, DIC is essentially a phase-contrast imaging mode. In this case, phase shifts caused by the refractive structure of the specimen are encoded in a field of **polarized light**, the two superposed components of which are then mutually offset and analyzed to show refractive index gradients.

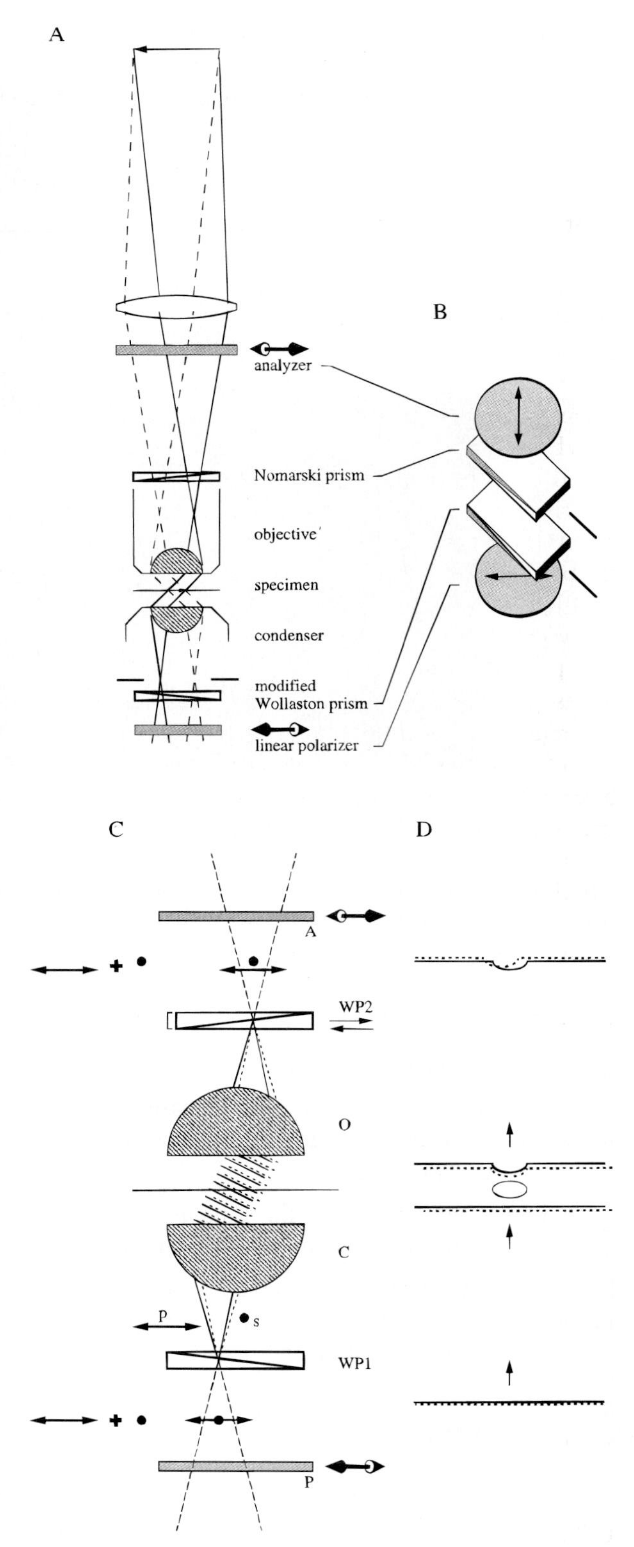

FIGURE 12. Optical components of a differential interference contrast microscope. (*A*) The principal layout of Fig. 9 is augmented with a polarizer, modified Wollaston prism, Nomarski prism, and analyzer. (*B*) The polarizer and analyzer are set nominally to extinction (crossed), and the birefringent prisms are inserted at 45° to the polarizer axes. (*C*) Mode of action of the principal DIC components. For simplicity, Wollaston prisms (WP1, WP2) located in the condenser and objective back focal planes are shown. In practice, Nomarski-type prisms are used. The prism shear axes are parallel to the page. (*Continued on facing page.*)

The essential DIC optical components consist of (1) a **linear polarizer** inserted between the light source and the condenser, (2) a **modified Wollaston prism** mounted close to the iris in the condenser back focal plane, (3) a **Nomarski prism** inserted immediately behind the objective, and (4) a **linear polarizer (analyzer)** mounted before the tube lens and image plane (Fig. 12). The polarizer and analyzer are crossed; i.e., if the polarizer is oriented at 0° (east–west), the analyzer is oriented at 90° (north–south). With the Wollaston and Nomarski prisms removed, this optical configuration is equivalent to a polarizing microscope set for maximum extinction of transmitted light. In this state, birefringent features in the specimen would show up bright against a dark background. The prisms are inserted at 45° (NW–SE), and can greatly change the optical response, depending on their relative adjustment. The prisms are each composed of two thin crystal optic wedges, in which the optical axis in one wedge is normal to the optical axis in the other. The complete prism is formed by cementing the wedges to form a thin plate, which is anisotropic. One particular direction in the plane of the plate is the **prism shear axis**; a beam of light entering normal to one face of the prism is split into two emergent beams that are orthogonally polarized and deviated by an angle (Fig. 13). The prism shear axis and the normal axis define the shear plane in which the emergent beams separate, and define the polarization axes of the two beams; one beam will be polarized in the shear plane (p), and the other beam will be polarized across the shear plane (s). The angle of beam deviation is set by the design of the prism. For microscopes, this angle is so small that there is no observable separation of the emergent light. In terms of wavefronts, Figure 13 shows that the angular deviation of the two beams is equivalent to a constant phase shift per unit length introduced by the prism across its face, in a direction parallel to the shear axis. The phase shift per unit length is equal but opposite for the p and s waves. The relative intensities of the emergent p and s beams depend on the polarization axis of the incoming light relative to the prism shear axis; if the incoming beam is p-polarized (polarized parallel to the shear axis), only a p-polarized beam will emerge. In the opposite case, only an s-polarized beam will emerge. A balanced case occurs when light that is linearly polarized at 45° to the shear axis enters the prism; 45° linear polarization is equivalent to the in-phase vector summation of two fields, one polarized parallel to the shear axis (p), and the other across the shear axis (s), each with 70.7% amplitude. The emergent p and s beams are then of equal brightness. Circularly polarized incident light would also produce equally bright emergent linearly polarized beams, as would 45° elliptically polarized light. In addition to being anisotropic, Wollaston and Nomarski prisms are directional; i.e., collimated light sent in from the opposite side of the prism will also split into p and s beams, but with exchanged polarizations.

Basic DIC image formation can be visualized apart from the microscope. Consider first a collimated beam of light into which is placed a polarizer, followed by a thin specimen, a lens, a Wollaston prism with its shear axis at 45° to the polarization, an analyzer crossed to the polarizer, and a screen on which to project the image (Fig. 14). The prism is placed in the back focal plane of the lens. The

---

FIGURE 12. (*Continued from facing page.*) Considering only light diverging from a single point in the condenser back focal plane, WP1 slightly deviates the p- and s-polarized waves. This angular deviation (shear) is converted into a small spatial offset when the condenser collimates the light. The offset p- and s-polarized plane waves traversing the specimen are modified in phase by its refractive structure. The objective reverts the differential offset of the plane waves to angular convergence, which is then compensated by an inverted Wollaston prism (WP2). Transverse offset of the inverted prism has the effect of adding a uniform phase bias across the objective pupil. In the figure, this is shown as phase lag of p relative to s. The analyzer (A) extinguishes the original component of linear polarization. (*D*) DIC wave optics. The uniform, linearly polarized plane wave below WP1 is sheared by the prism and condenser and phase-distorted by the specimen. When the shear is removed by the objective and WP2, a phase shift is produced between the p and s components where the waves interacted with refractive index gradients in the specimen. The phase shift produces elliptically polarized light, which is not completely extinguished by the analyzer. Therefore, the DIC image highlights refractive index boundaries in the specimen oriented across the shear axis. With zero phase bias (WP2 compensates WP1 exactly), the background appears dark, index boundaries appear bright. With sufficient phase bias, the background is mid-range gray, with index boundaries appearing bright or dark depending on the sign of the index gradient.

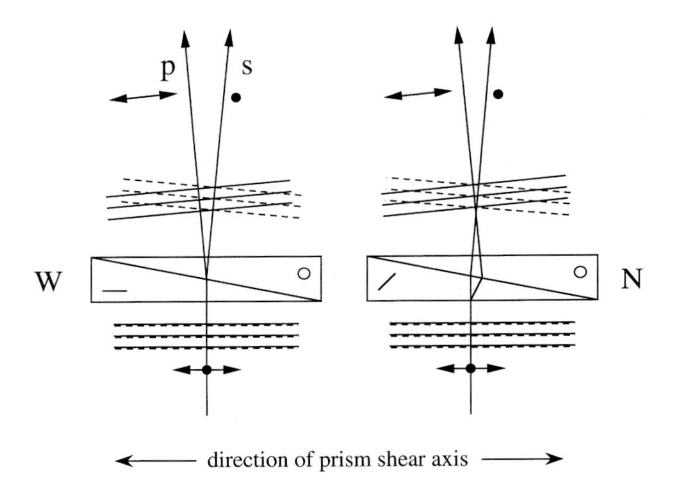

FIGURE 13. Ray and wave diagrams for Wollaston and Nomarski prisms. Prisms of this type are usually made from wedges of quartz, a positively birefringent crystal. The angular deviation between *p*- and *s*-polarized light produced by double refraction in the prism is a function of its wedge angle. For DIC microscopy, this angle is so small that separation of the *p* and *s* waves is not seen. In both prism types, the optic axis of the quartz in one wedge is normal to the axis in the other. The main difference between the prisms is that the convergence point for deviated rays is within the Wollaston prism, but outside the Nomarski prism. This allows the Nomarski prism to be used with conventional condensers and objectives in which the back focal plane is not easily accessible.

polarized plane wave field entering the specimen gets phase-delayed in regions of higher refractive index, and emerges with wave front distortion but with essentially unchanged polarization. The lens brings the wave fronts to an approximate focus at the location of the Wollaston prism, after which the wave fronts then expand toward the analyzer and screen. In the absence of the Wollaston prism, the entire field would be extinguished by the analyzer, and the image would be dark. With the prism in place, the *p* and *s* components of the field are deviated slightly in angle and arrive at the analyzer with a small spatial offset along the direction of the prism shear axis. Across all parts of the field where the wave fronts are undistorted, the offset causes no change in the net state of polarization; the *p* and *s* components add in-phase to produce the original linear polarization, which is then extinguished. However, across parts of the field where the wave fronts show a **phase gradient**, the offset can produce a **phase difference** between the *p* and *s* components. This changes the local polarization of the field from linear to elliptical, so that the analyzer passes some light. The net result is that the image will show features in the specimen where there is a change in optical path due to a difference in refractive index or thickness. The image is "differential" because the offset is very small; i.e., phase differences from neighboring points are compared by interference. More specifically, the idealized DIC image will show **optical path gradients** in the specimen, with maximum contrast occurring when the gradient is along the direction of the shear axis of the prisms. Mathematically, the gradient of optical path in the specimen is a vector field, and the DIC image shows, at each point in the image, the (scalar) component of the gradient parallel to the direction of the prism shear axis, and within the depth of field. For a thorough technical discussion of DIC optical systems, see Pluta (1989a,b).

In a real DIC microscope, the full set of optics—polarizer, Wollaston prism, condenser, specimen, objective, Nomarski prism, and analyzer—enable the DIC image to be formed in a high-NA system. First, the polarizer and Wollaston prism are nominally set to balance the *p* and *s* fields; i.e., the polarizer defines 0°, and the prism is set at 45°. The location of the modified Wollaston prism close to the back focal plane of the condenser is fundamentally important because the lens then converts the angular deviation of the *p*- and *s*-polarized light into a fixed transverse spatial offset of the fields that exit the condenser and propagate through the specimen (Fig. 15). The offset distance is "differential" in that it is, by design, approximately equal to the resolution limit. After passage of the light through the specimen, where phase delays are caused by its refractive structure, the objective re-converts the spatial offset of the *p* and *s* fields to angular convergence. Ideally, an inverted Wollaston prism locat-

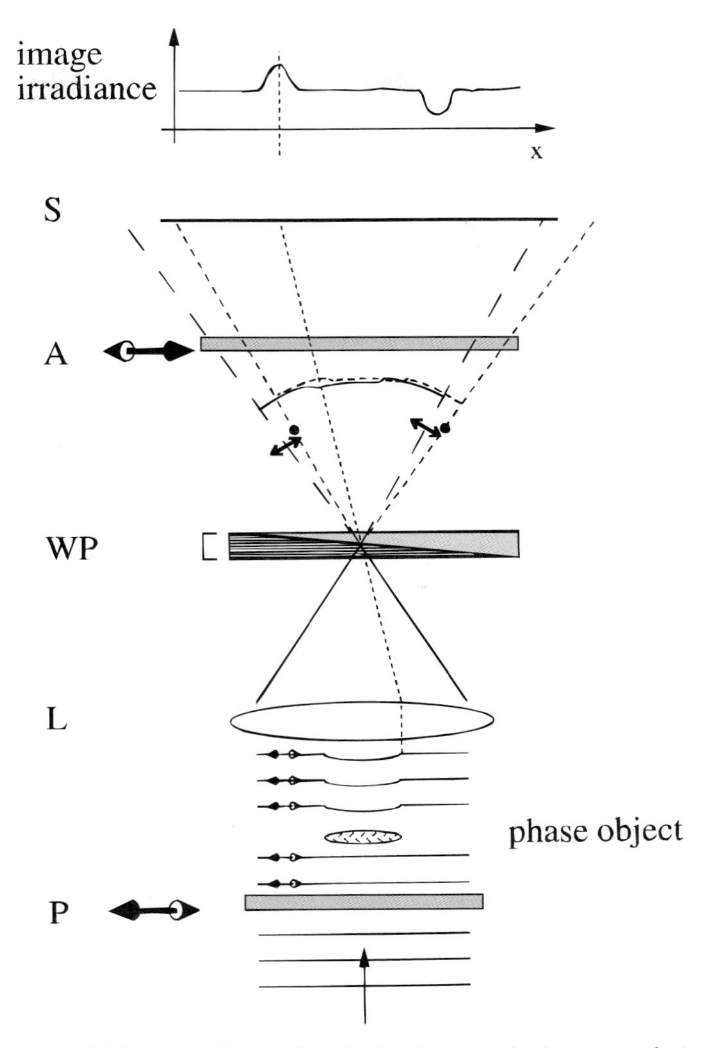

FIGURE 14. Elementary DIC optics. DIC can be obtained by simply shearing polarized collimated light that traverses the specimen. Components: linear polarizer (P), lens (L), Wollaston prism (WP), analyzer (A), image receiving screen or film (S). As shown, the refractive structure of the object causes a phase delay in the wave fronts. The lens and Wollaston prism convert the delay into an angular shift between *p*- and *s*-polarized components, giving the linearly polarized light a small elliptical component. If the prism is offset, a uniform phase bias is also introduced, here shown as *p* lagging *s*. At every point in the field, the analyzer passes the linear component normal to the original polarization. In the schematic, this creates an image in which one edge of the phase object appears brighter than background, and the other edge darker than background. The illumination numerical aperture (INA) in this elementary system is zero (only collimated light is sent into the specimen from the condenser), therefore no axial resolution and reduced transverse resolution would result. Finite INA would produce a different bias for every set of plane waves traversing the specimen, therefore, contrast would be lost.

ed in the back focal plane (pupil) of the objective, and also oriented at 45°, would recombine the converging *p* and *s* fields. In most objectives, however, the pupil is located inaccessibly within the lens groups. Nomarski's solution to this problem was to modify the prism so that it could be located immediately behind the objective (see Fig. 13). In this location, the Nomarski prism recombines the *p* and *s* fields, effectively removing the offset. The phase gradients impressed on the offset *p* and *s* fields in passage through the specimen are therefore converted into phase differences, creating elliptical polarization in the recombined field. When the Nomarski prism exactly compensates for the effect of the Wollaston prism, the action of the analyzer is to extinguish the light at all locations where there

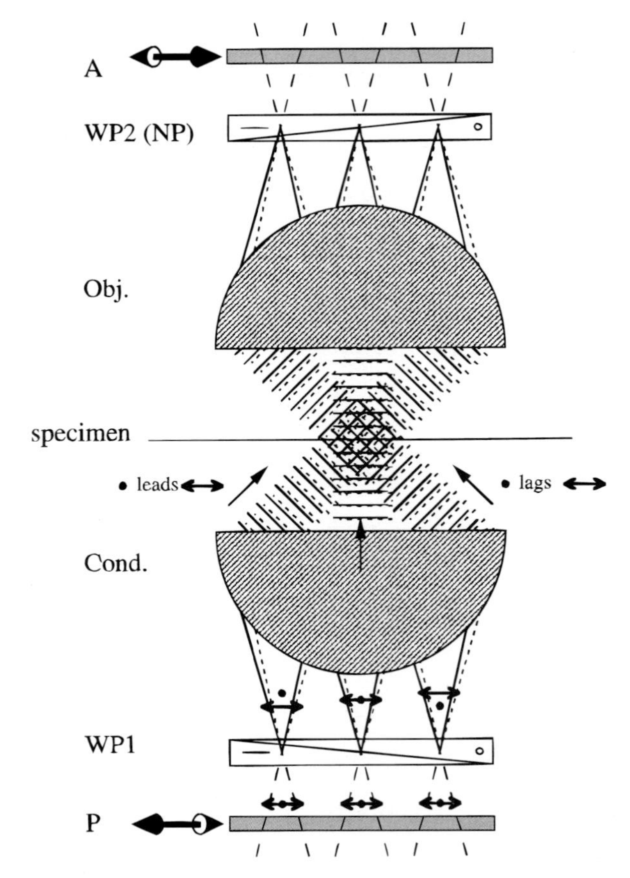

FIGURE 15. High-NA DIC microscope: ray and wave optics. The essential feature of a high-NA DIC system is the use of two polarization-shearing prisms to compensate for linear variation in bias across the condenser pupil. To simplify the diagram, Wollaston prisms are shown in the back focal planes of the condenser and objective. Three source points are shown in the condenser pupil (coincident with WP1), radiating into the condenser. Each gives rise to a pencil of plane waves that traverses the specimen at a specific angle. In addition to an angular deviation between *p*- and *s*-polarized light, the prism introduces a bias phase shift that is different for each location along the prism shear axis. In the left-most pencil, the *s*-polarized field leads *p*. In the on-axis pencil, *s* and *p* are in phase. In the right-most pencil, *s* lags *p*. The spatially variant bias is exactly compensated by the inverted prism (WP2) that follows the objective. Essentially, WP1 is imaged onto WP2 so that the bias is matched at each point. A transverse shift in either prism along its shear axis causes a bias mismatch that is uniform across the pupil. The design of the prisms is specific to the focal lengths, pupil diameters, and pupil locations of the condenser and objective.

was no phase gradient. The resulting image is dark in background with bright features showing refractive edges in the specimen, similar to dark-field optics. Like a dark-field image, both "increasing" and "decreasing" index boundaries are brighter than background. To resolve this ambiguity in the sign of the gradient, the Nomarski prism serves a second important function; offsetting the prism along its shear axis (by turning a fine screw to slide the prism) uniformly shifts the relative phase of the recombining beams, so that the polarization of the emergent light arriving at the analyzer can be varied from linear through elliptical to circular, etc. The phase shift of the *p* wave relative to the *s* wave is referred to as the **bias**. With zero bias, the analyzer causes the image background to appear dark, and phase gradients bright. As the bias is increased, the background becomes uniformly brighter, and phase gradients appear brighter or darker than background, depending on the direction of the phase gradient with respect to the direction of the prism offset. Changing the sign of the bias reverses the directional response. For weakly refractive objects such as cells, the signal-to-noise ratio (SNR) of the DIC image peaks at a bias setting slightly greater than the largest phase shift caused by the specimen. However, the SNR is generally not much less for bias settings as high as 90°, where the linearity of DIC is highest, and the brighter image can be detected and processed as a video-rate signal. This is the basis of the widely used method of **video-enhanced contrast** (VEC) developed for DIC systems by R.D. Allen, and for polarization microscopy by S. Inoué. Additionally, the total light dose to the specimen is lower than would be needed to get a usable image at zero bias. In microscopes where the Nomarski prism is fixed, the bias can be set by use of a **compensator** (polarizer and waveplate) to produce elliptically polarized light before the Wollaston prism in the condenser. Recent work on use of a compensator in place of the analyzer in a DIC system along with algebraic image processing (Geometric Phase Shift DIC; Cogswell et al. 1997) has demonstrated precision linearization of the imaging response.

The use of two prisms in the Nomarski DIC system is the key feature that enables sharp images to be formed at high NA. As shown in the Köhler setup in Figure 9, the condenser and the objective together transfer an image of the light source and first shearing prism onto the second prism (which is inverted). Across the image of the light source, the second prism introduces a phase shift per unit length that exactly compensates the linear phase shift between *p* and *s* polarizations introduced by the first prism. Offsetting the second prism along its shear axis does not change the phase shift per unit length but simply adds or subtracts a constant phase difference across the pupil, hence changing the bias. Therefore, the use of paired prisms allows DIC image formation to occur with the same bias for every light pencil in the condenser aperture, regardless of the direction from which it traverses the specimen. In this condition, maximum contrast and resolution are attained. The sharp depth of field that results from use of the full condenser and objective NA gives DIC systems a strong "optical sectioning" characteristic.

In addition to operation at full aperture and formation of a high signal-to-noise image, DIC systems provide a number of other advantages. Although the lenses must be free of strain birefringence, objectives used in DIC systems do not have to contain a special element such as a phase plate. This is an advantage not only in manufacture and cost, but also because the phase plate complicates the point spread function in fluorescence microscopy and increases back-reflection of light in the incident-light illuminators used in both fluorescence and reflection-interference systems. In general, switchover from DIC to fluorescence requires only the removal of the analyzer to avoid attenuation of the fluorescence image. For high-resolution fluorescence microscopy, the Nomarski prism must also be removed, because it shears the fluorescence image into two very slightly offset polarized components. This can significantly lower the peak brightness of sharply defined features (Kontoyannis et al. 1996). The chief disadvantage of DIC is that the contrast transfer function is directional; i.e., phase gradients oriented along the shear axis show up with much greater contrast than gradients that are nearly normal to the shear axis. In some cases this can be used to advantage; for example, to reduce the contrast of a highly anisotropic cell process, such as an axon, to make visible internal organelles of much lower contrast (P. Forscher, pers. comm.). Image pairs with orthogonal orientations of the shear axis can be obtained with a centerable rotation stage, but this is often not practical under experimental conditions. A simple method for optical rotation of the shear axis relative to the microscope has yet to be developed.

The striking optical sectioning characteristic of DIC systems is currently driving both theory and experiment on 3D DIC object estimation (Feineigle 1996; Preza et al. 1996; Preza 1998; Kagalwala and Kanade 1998). The aim is to obtain an estimate of the refractive structure of the object, $\Delta n(x,y,z)$. Pioneering work has resulted in the formulation of computational models for DIC optical components, and for partially coherent transmitted-light image formation, and computational methods for refining an object estimate.

Fluorescence and Fluorescence Optical Systems

At the present time, fluorescence microscopy has developed into the most widely used method for study of both fixed and living specimens. This is due to the high specificity possible with a vast array of fluorescent molecules ranging from stains and labeled antibodies to molecular analogs, physiological indicators, and expressible markers such as green fluorescent protein (GFP) and phytofluors. It is also due to the high sensitivity of fluorescence detection. Under practical conditions, very low background is possible in fluorescence microscopy. It is therefore not surprising that four of the five demonstrated single-molecule tracking/imaging techniques are fluorescence-based (Betzig and Chichester 1993; Nie et al. 1994; Funatsu et al. 1995; Sase et al. 1995; Vale et al. 1996; Smith et al. 1999; Moerner and Orrit 1999; Weiss 1999).

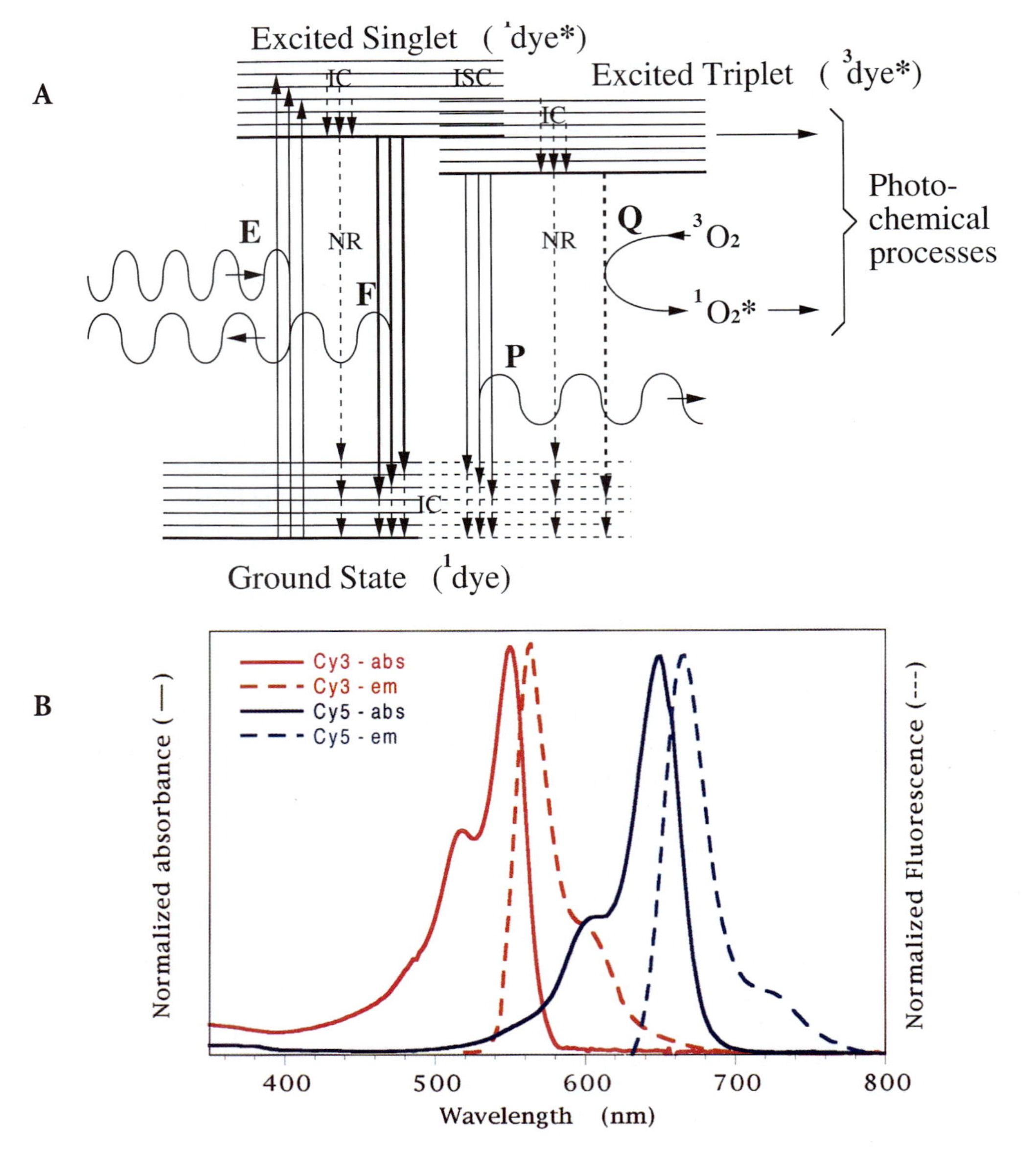

FIGURE 16. (*A*) Jablonski energy-level diagram of photophysical processes in organic dyes. Each set of horizontal lines represents vibrational levels in a single electronic stationary state of the molecule. The electronic ground state (lowest energy) is shown along with two excited electronic states. Vertical solid lines represent radiative transitions between states, whereas vertical dotted lines represent nonradiative transitions: (E) excitation, (IC) internal conversion, (F) fluorescence, (NR) nonradiative transition, (ISC) intersystem crossing, (P) phosphorescence, (Q) quenching by energy transfer. The ground state is usually an electronic singlet (all electrons spin-paired). At ambient temperature, very few molecules would be in any state other than the lowest vibrational level of the ground state; therefore, excitation processes originate in this level. Absorption of a photon results in excitation of the dye to a stationary state of greater energy. In general, the energy of the photon must match the energy difference between the initial and final states (multiphoton absorption is not shown here). Closely spaced vibrational levels plus thermal motion in the molecule allow for a range of photon energies to match a transition, and cause the excitation to be seen spectroscopically as an absorption band rather than as a sharp absorption line. Excitation by absorption normally occurs with no change in spin-pairing; therefore, the excited state is also a singlet. Fast relaxation processes convert vibrational energy into thermal motion and bring the molecule to the lowest vibrational level of the excited state. Under normal conditions, the lifetime of this state is in the nanosecond range. Fluorescence originates from this level, and can be the main relaxation process in a bright dye. (*Continued on facing page.*)

Unlike phase-contrast methods, fluorescence follows the absorption of light by one or more molecules in the specimen. (For brevity, all fluorescent markers are referred to here as "dyes." The full range includes synthetic dyes, modified and native biochemical fluors, and semiconductor "quantum dots.") As shown in a Jablonski diagram (Fig. 16), this excitation process usually brings the dye from its singlet electronic ground state to its lowest excited singlet state. Each of these electronic states is composed of a manifold of vibrational substates, which broadens the range of photon energies (wavelengths) that can cause the transition. At normal temperatures, dye molecules are virtually always in the lowest vibrational level of the ground state, so excitation occurs from this level into the vibrational manifold of the lowest excited state. This excitation band is usually graphed as an **absorption spectrum** (extinction coefficient versus wavelength). For organic dyes, extinction coefficients range from 50,000 to 250,000/molar/cm. Vibrational relaxation of electronic states (internal conversion) occurs with great rapidity, so that an excited-state dye quickly ends up in the lowest vibrational level of the excited electronic state. Excited-state lifetimes are usually in the nanosecond range, but can be affected by the relative rates of nonradiative relaxation, spontaneous emission, stimulated emission, intersystem crossing, quenching, or direct photochemistry. Fluorescence is observed when a photon is emitted by the dye as it returns directly to the ground state. Because of the energy loss that accompanies the initial internal conversion, fluorescence occurs at a lower photon energy (longer wavelength) than excitation. The return to the electronic ground state may leave the dye transiently in a vibrationally excited state prior to rapid internal conversion to the lowest ground-state level. This allows for a range of emitted photon energies, observed as a **fluorescence emission spectrum** (Fig. 16B). In most cases, the absorption spectrum is closely related to the **fluorescence excitation spectrum**, which is determined by measuring emission flux as a function of excitation wavelength. When expressed on a quantum basis of photons fluoresced versus photons absorbed, the quantum yield (QY) of the dye is the fractional probability with which fluorescence will occur relative to other relaxation modes. For "bright" dyes, QY usually exceeds 0.1, and may range as high as 0.9 (Tsien and Waggoner 1995). Spontaneous spin-unpairing of two electrons in the excited state can occur due to molecular collisions. This process, known as intersystem crossing, is enhanced by spin-orbit coupling when the dye contains one or more heavy atoms (Cl, Br, I). The resulting triplet state is reactive and highly susceptible to quenching due to its long intrinsic lifetime. If not quenched, a triplet can relax by phosphorescence, emission of a photon at a longer wavelength than the fluorescence. Triplets are very effectively quenched by dissolved molecular oxygen (O_2), which is prevalent, highly diffusible, and has a triplet ground state. Oxygen quenching of dye triplets can result in the production of singlet oxygen, a long-lived excited state of O_2, or several forms of oxygen radicals. Dye triplets can also react directly with other organic molecules, especially intracellular redox intermediates. Two results of this photochemistry relevant to microscopy of living cells are photobleaching (or fading) and phototoxicity.

FIGURE 16. (*Continued from facing page.*) The closely spaced vibrational levels of the ground state, along with thermal motion, allow for a range of emitted photon energies. Fluorescence (and phosphorescence) is therefore normally seen as an emission band. A number of processes or conditions can cause spin-unpairing and conversion (intersystem crossing) of the excited singlet to a triplet state. The radiative transition from the triplet state to the ground state (phosphorescence) occurs with low probability. Therefore, triplet states generally persist until quenched or until the dye is involved directly in a chemical reaction. In biological specimens, dissolved oxygen (O_2) is a highly effective quencher of dye triplet states. In this photochemical reaction, the ground-state oxygen molecule (which is a triplet) can be excited to a reactive singlet state. Singlet oxygen can initiate reactions leading to bleaching of the dye and to phototoxicity. (*B*) Absorption and fluorescence spectra of two cyanine dyes. The excitation spectrum is similar to the absorption spectrum but rolls off more sharply in the overlap region. For cyanine 3, peak extinction occurs at 550 nm (*green*), peak emission at 570 nm (*orange*). For cyanine 5, peak extinction occurs at 649 nm (*red*), peak emission at 670 nm (*far-red*). In both cases, Stokes' shift, which represents the difference in excitation and emission energy, is ~20 nm. Both dyes show very high extinction (150,000 and 250,000/molar/cm, respectively) and good quantum yield when conjugated to protein (>15% and >28%, respectively). Spectra provided courtesy of R. Mujumdar, Carnegie Mellon University, Center for Light Microscope Imaging and Biotechnology.

Evolution has put the intrinsic chromophore of GFP in the core of a β-barrel structure (Ormo et al. 1996). In that environment, the "dye" is protected from collisions with water molecules and is isolated from frequent encounters with dioxygen. GFP is one of the most stable fluorescent labels known. Synthetic dyes held as cryptands in soluble macromolecules such as cyclodextrins have also been shown to be significantly stabilized against photochemical degradation (Guether and Reddington 1997). In fixed specimens, antioxidants can be used to slow fading. In some living preparations, oxygen scavengers can be used to remove dissolved oxygen for limited periods of time. Glucose oxidase, along with glucose and catalase, is a highly effective scavenger that can be added to culture medium at the time of infusion.

Because fluorescence is the result of absorption of a photon by a molecule, followed by emission of a photon of longer wavelength, the essential feature of any fluorescence microscope is a means to excite the specimen with color-filtered light, and to separate the excitation from the emission wavelength band by use of a second filter. This allows the fluorescence image to be formed in a dark background to give maximum sensitivity. Normally, the degree of labeling in a biological specimen is so low that only a small fraction of the illumination traversing the specimen is absorbed by the fluorescent marker molecules, and a fraction of this is re-emitted. In short, the fluorescence image will be weak relative to the brightness of the illumination. Therefore, the central problem in fluorescence microscopy is obtaining high-efficiency illumination of the specimen, but very high rejection of the illumination band from the image along with efficient capture of the emission. The most common optical configuration giving the required high performance is **incident-light illumination** combined with a filter set that includes (1) an **excitation filter**, (2) a **dichroic or wavelength-selective reflector**, and (3) an **emission filter** (Fig. 17). Broadband (white) light from an intense lamp is directed into the microscope through the excitation filter, which is chosen to match the excitation band of the dye. The resulting filtered light is deflected by a matched dichroic reflector into the back pupil of the objective. Therefore, in this "epi-fluorescence" setup, the objective functions as a condenser in addition to its usual role. High-NA objectives therefore function as more powerful condensers, as well as collecting a greater fraction of the fluoresced light. High performance also results from the fact that the illumination exits the objective into the specimen and is therefore directed away from the microscope. Back-scattered and back-reflected light that re-enters the objective is of unchanged wavelength, and therefore is mostly deflected out of the microscope by the dichroic reflector. Fluorescence emission captured by the objective is efficiently transmitted by both the dichroic reflector and the matched emission filter and is focused to form an image. The small fraction of the back-scattered illumination that passes the dichroic reflector is yet many times brighter than the fluorescence, so the emission filter must effectively block this light, which would otherwise appear as a background in the image. The second function of the emission filter is to block out the fluorescence of other dyes that may interfere with the tracer being imaged. Overall, stringent conditions are placed on the optical quality of the excitation, dichroic, and emission filters. The dichroic reflector is usually a multilayer thin film interference filter formed on the surface of a thin glass substrate, and designed for 45° angle of incidence. In general, these are not overcoated and hence are very easily damaged by careless handling. A small scratch will cause significant transmission of unwanted light. Excitation and emission filters are designed for normal incidence, and many are composite, consisting of a colored glass substrate on which one or more multilayer interference filters are deposited. The interference films can provide sharp-cut band-pass characteristics, back-reflecting out-of-band light. However, simple interference filters transmit weak side bands. Colored glass, on the other hand, usually shows long-pass transmission or relatively broad band-pass characteristics, but can have extremely high absorbance outside the transmission range. The composite filter can therefore be designed as a band pass with extremely high blocking power at other wavelengths. Dichroics and band-pass filters usually are designed to be set in a particular orientation relative to the light source (and are usually marked as such by the manufacturer). In the simplest cases, this is because the colored glass substrate is absorptive and luminescent, whereas the interference films are reflective. Excitation filters placed with the reflective side toward the light source eliminate most out-of-band light by reflection, whereas the opposite orientation would require that the colored glass substrate absorb most of the out-of-band light and dissipate the resulting heat. Likewise, in emission filters, the reflective interference films eliminate most excitation-band light in the first step, minimizing the excitation of luminescence in the colored glass substrate.

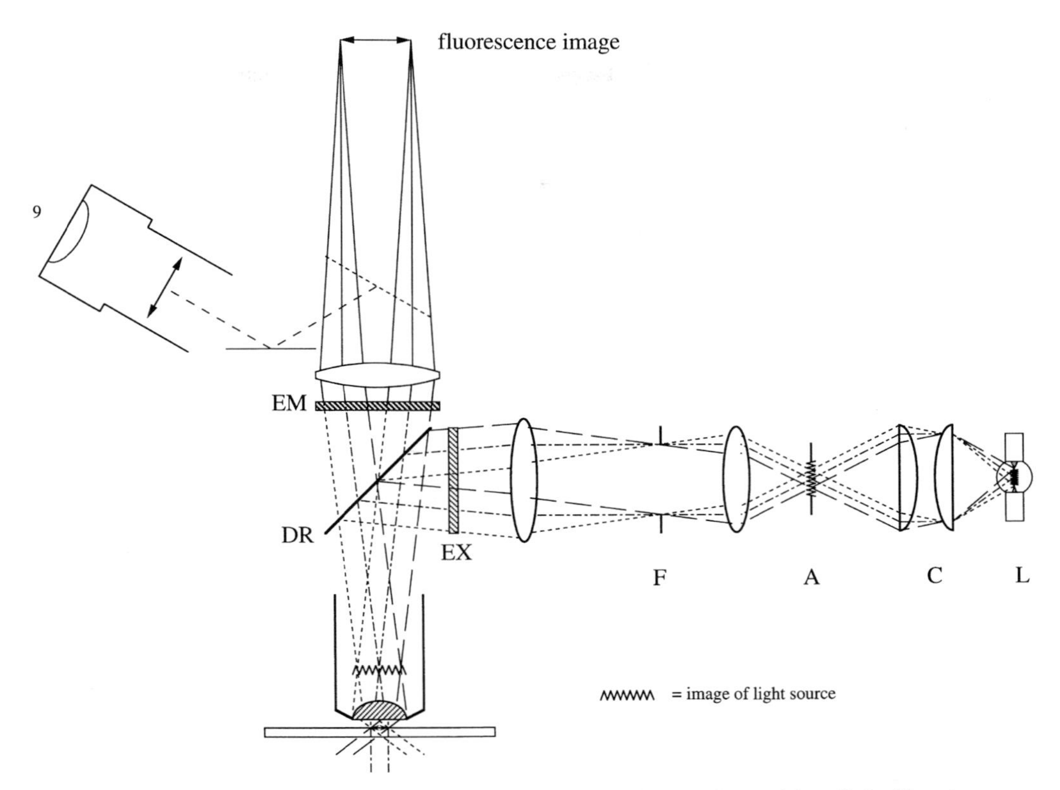

FIGURE 17. Principal optical components of a fluorescence microscope. Incident-light illuminator components (L) arc lamp, (C) collector lens assembly, (A) illumination aperture iris (coincident with image of light source), (F) field iris. Fluorescence filter set: (EX) exciter filter, (DR) dichroic reflector, (EM) emission filter. In this optical system, the objective functions also as the condenser. Unlike a transmitted-light microscope, the aperture iris is imaged into the condenser (objective) back pupil, rather than being physically located at this position. In this way, there is no obstruction of the image-forming light path. The fluorescence microscope differs from a reflectance microscope only in the wavelength selectivity of the filters and dichroic reflector. Use of a green band-pass excitation filter (such as for rhodamine) along with a blue-band dichroic reflector (such as for fluorescein) suffices to generate a reflection-interference image of moderate to good quality. In this case, the band-pass filter functions to increase longitudinal coherence.

Fluorescence filter sets of three basic types are widely used in microscopy (Fig. 18). For specimens labeled with a single fluorochrome, the highest optical throughput can be obtained with a long-pass emission filter. For example, the universal fluorescein (FITC) set consists of a blue bandpass "exciter," a blue-reflecting dichroic, and a yellow glass long-pass "emitter." Likewise, a rhodamine filter set consists of a green band-pass exciter, green-reflecting dichroic, and red-orange glass long-pass emitter. However, when fluorescein is used along with rhodamine in doubly labeled specimens, the yellow long-pass filter is replaced by a green band-pass filter. This is necessary to block the orange emission from rhodamine that is weakly excited by blue light. Without this change, both dyes would contribute to the FITC image. No change is required in the rhodamine filter set because fluorescein is not excited by green light. Five to seven sequential band-pass sets can be overlapped to span the spectrum from UV excitation/violet emission to far-red excitation/near-IR emission (Waggoner et al. 1989). In some applications, a panel of two or more spectrally distinct dyes is used along with a multiband filter set; for direct fluorescence color imaging with color film or a color electronic camera, a multiband exciter is used along with a complementary multiband dichroic and emitter pair. In this type of filter set, the dichroic/emitter pass bands interleave on the long-wavelength side of the exciter bands. For high-spatial precision multicolor imaging, a single multiband dichroic and emitter pair can be used along with a set of single-band exciters in an external filter changer. In this way, the optics in the imaging light path remain fixed, eliminating a source of alignment error in superposition of images.

A

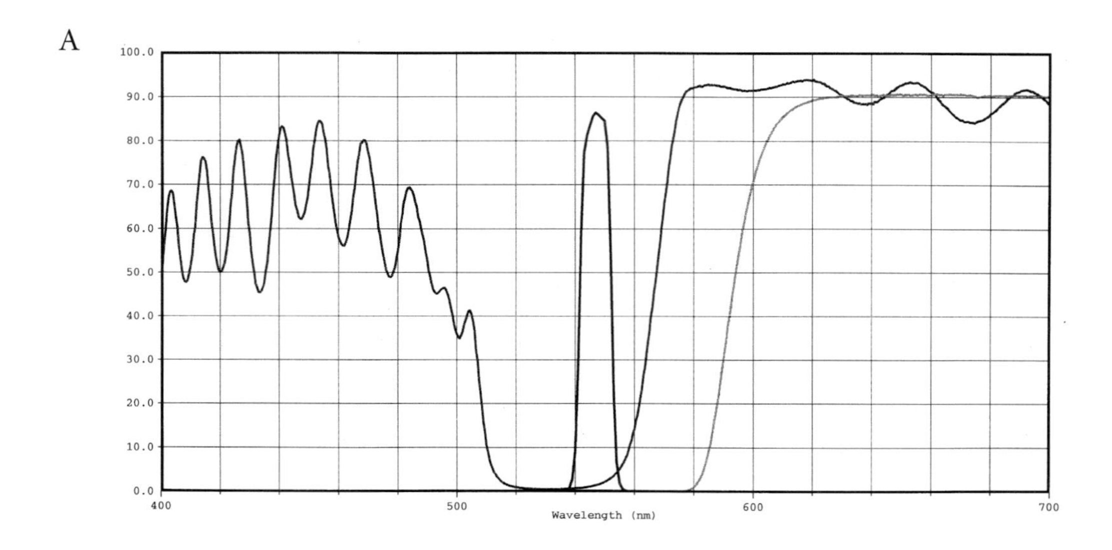

B

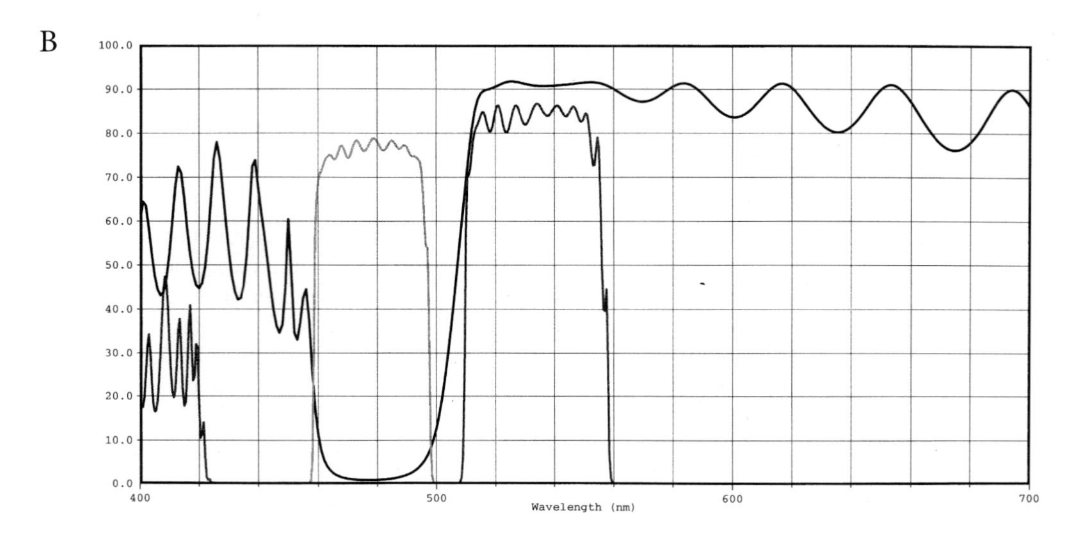

C

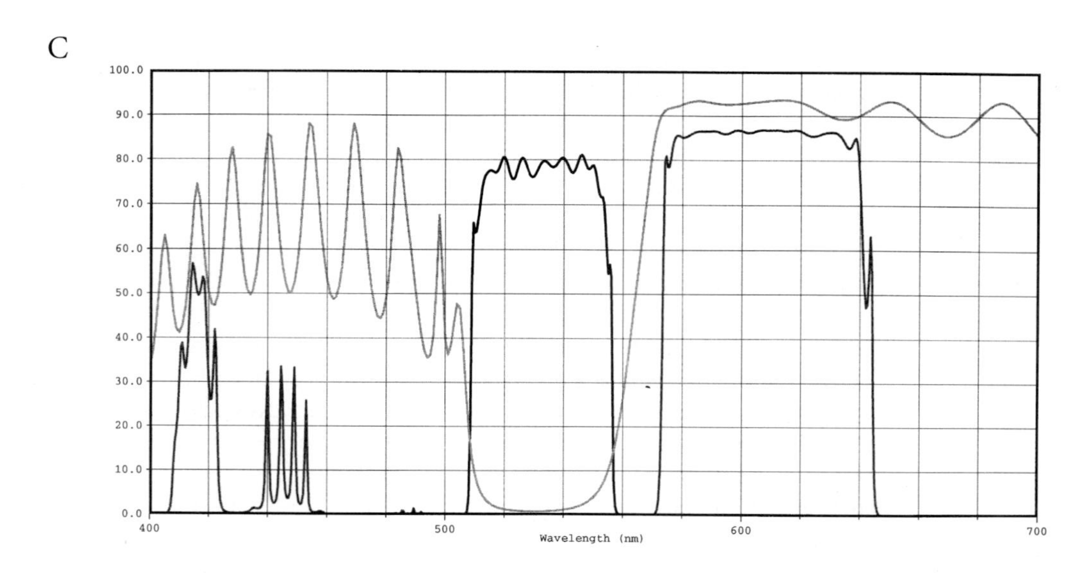

FIGURE 18. (*Figure continues on facing page.*)

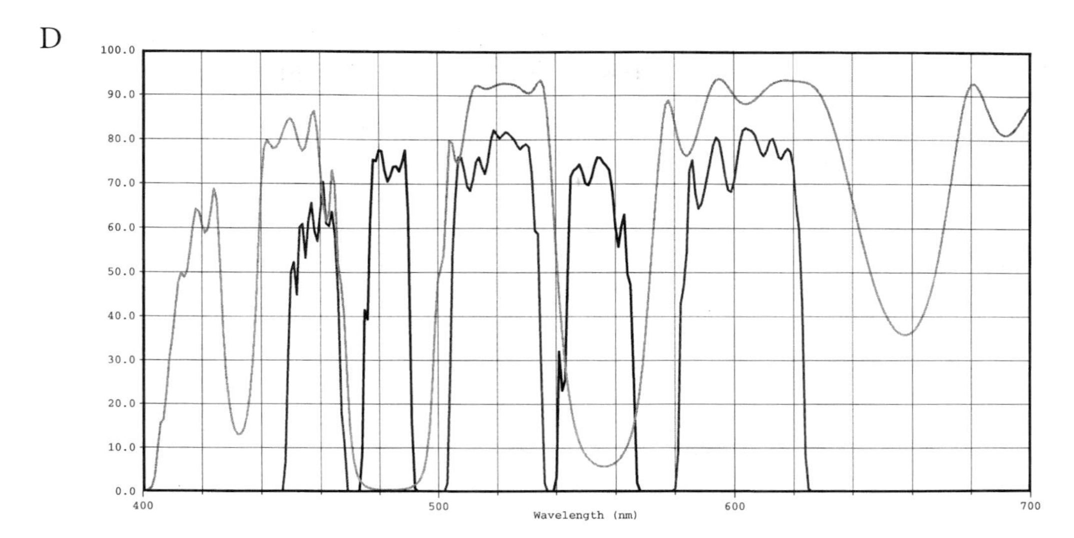

FIGURE 18. Spectra of representative fluorescence filter sets. Spectra are plotted as percent transmittance for the filters (normal incidence) and for the dichroic reflector (45° incidence). Each set consists of three elements: an excitation filter, a dichroic reflector, and an emission filter. The essential feature of each set is the complete separation of the excitation and emission pass bands, and the intervening transition from high reflectivity to high transmissivity in the dichroic reflector. Because of its multilayer structure, the dichroic reflector generally has complex transmission bands at short wavelengths that are of no consequence in filter set performance. In contrast, the exciter and emitter are generally strongly blocked, out of band. In the simplest type of set (*A*), a band-pass excitation filter is used along with a long-pass emission filter. This maximizes the detection of fluorescence in the image, but would give mixed-color images for a doubly labeled specimen in which both dyes were excited to some degree within the single illumination band. The set shown, green excitation/orange-red emission, would be suitable for rhodamine- and cyanine3-type dyes. In sets *B* and *C*, sharp-cutoff band-pass filters are used for both excitation and emission in the fluorescein and rhodamine bands, respectively. Even though blue light, which excites fluorescein efficiently, also excites rhodamine weakly, rhodamine emission is blocked from the fluorescein-band image by the band-pass emission filter in *B*. In *C*, the green exciter filter cuts off above the fluorescein excitation band. Filter sets of this type therefore give outstanding performance with doubly labeled specimens. More complex multiband filters are used with color cameras, or in cases where movement of optical elements is disadvantageous. In *D*, a dual-band exciter is used along with a dual-band emitter; the pass bands interdigitate to eliminate cross talk. The intervening reflector is designed to be highly reflective in the two excitation bands and highly transmissive in the emission bands. The filter set can also be used with separate excitation filters in a motorized wheel or with another type of external wavelength-selecting device. Filter curves provided courtesy of P. Millman and M. Stanley, Chroma Technology. Key to Chroma filter sets: (*A*) #11002 Basic set for green-excitation dyes; (*B*) #41001 Hi-Q set for fluorescein (FITC), red-shifted GFP, BodiPy, fluo-3, and diO; (*C*) #41002 Hi-Q set for rhodamine, tetramethylrhodamine (TRITC), and diI; (*D*) #51004v2 Multiband set for FITC and TRITC.

The most common light sources for fluorescence microscopy are high-pressure mercury and xenon arc lamps. These sources both appear to output intense white light, but they actually differ greatly in spectral distribution. Mercury arcs produce light concentrated in bands centered around the atomic emission lines of the gas-phase mercury atom, most notably at the following wavelengths: 254 nm (UV), 265 nm (UV), 365 nm (UV), 405 nm (violet), 436 nm (deep blue), 546 nm (green), and 578 nm (yellow). Therefore, this source is ideal if there is good coincidence between one of these bands and the excitation band of a specific dye. Out-of-band power is not insignificant, as evidenced by the fact that mercury arcs are used routinely with fluorescein-like dyes (480–490 nm peak excitation), as well as for in-band rhodamine-like dyes. In contrast, xenon arcs emit a less-intense continuum over the range 250–800 nm, with superimposed emission lines in the 450–500 nm and 800–1050 nm ranges. Therefore, good output can be obtained at any wavelength by use of the appropriate band-pass filter. For certain important applications, the continuum has advantages. For example, in ratio imaging of the calcium indicator dye fura-2 (Grynkiewicz et al. 1985), the largest response is obtained by UV excitation at 340 and 380 nm. High power is available at these wavelengths in a mercury arc, but both excitation filters must be able to block the much more intense emission at the 365-nm peak of the band. This is much less a problem with the xenon spectrum. In quantitative imaging applications, chronic spatial instability of the arc (particularly in mercury lamps) can cause image-to-image brightness variation. For both types of arcs, good ventilation is required for removal of heat, for removal of ozone, and for safety in the event of a mercury arc explosion.

The physical size of an arc affects not only its brightness, but also the efficiency with which it can deliver light to the illuminated field of a microscope. The brightest lamps are the widely used DC "short-arcs"—usually 100-W mercury or 75-W xenon—whereas larger lamps produce more light overall, but lower brightness within the arc projected area. In a Köhler-illuminated system, the image of the light source should be matched exactly to the back pupil of the objective; if the image diameter is less, the full INA will be underutilized; if it is greater, light will be wasted, and stray light will be needlessly introduced into the microscope. Because the NA of the lamp **collector lens** (NA_{coll}) effectively sets the maximum diameter of the field of view (FOV) in the specimen, and the diameter of the light source (d_s) sets the INA, a simple relation exists between the four quantities when the full aperture is used (i.e., when INA = NA_{obj})

$$d_s = (NA_{obj}/NA_{coll})\ (FOV) \tag{34}$$

an example of the "optical invariant" for image-forming systems. With $NA_{coll} = 0.5$ and with FOV = 150 μm for a 100x 1.3-NA objective, the optimal source diameter is 0.4 mm. In such a case, short-arc lamps are very efficient. For low-magnification objectives (particularly those of high NA), the large back pupil can be most effectively filled by a larger arc. Optical scramblers can be used to produce a circular, homogeneous source matched to the diameter of the pupil. A number of other light sources used in fluorescence microscopes include tungsten-halogen incandescent lamps, xenon flashlamps, pulsed mercury arcs, and actively scrambled lasers.

The combination of a 75–100-W short arc lamp and high-NA optics can produce intense irradiation of the specimen, a situation requiring attention particularly for live-cell imaging. After collimation and filtering, 1–10 mW can be delivered to the back pupil of a 100x objective. Focused into a 150-μm field of view, a 5-mW flux produces an irradiance of 30 W/cm^2, 300-fold greater than bright sunlight integrated over all wavelengths. Therefore, it is a general rule in fluorescence microscopy to minimize light exposure in any wavelength band in which the specimen or a dye within the specimen has a significant extinction coefficient. As a comparison, this level of irradiance is still far less than in a laser-scanning confocal microscope, where 10 μW focused into the objective diffraction limit (0.5 μm) produces 50,000 times the solar flux.

Many variations exist on the basic epifluorescence microscope. Use of laser excitation, such as is common in scanning confocal microscopes, obviates the need for an excitation filter. Use of acousto-optic tunable filters (AOTFs) (see Chapter 98), liquid-crystal tunable filters (LCTFs) (see Chapter 100), interferometers, diffraction gratings, and other wavelength-selective devices can replace one or more of the basic elements of the fluorescence filter set.

Principles of Fluorescence Microscopy

Image formation in a fluorescence optical system differs in a fundamental way from transmitted-light and reflected-light microscopes. The molecular excited states that lead to fluorescence have a finite average lifetime in the nanosecond range, and thus normally do not maintain any degree of coherence with the excitation field. Essentially, dye molecules in the specimen emit fluorescence independently, and therefore are mutually incoherent. In place of the Abbe picture, in which the specimen is represented as a collection of diffraction gratings that interact with plane wave illumination, it is now more appropriate to consider point sources of spherical waves representing the **spontaneous emission** of light from individual molecules. This model was first developed by Rayleigh (J.W. Strutt) in the analysis of telescope performance in imaging of stars (equivalent to point sources because of their great distance). The second important result of the mutual incoherence of emission from dye molecules is that

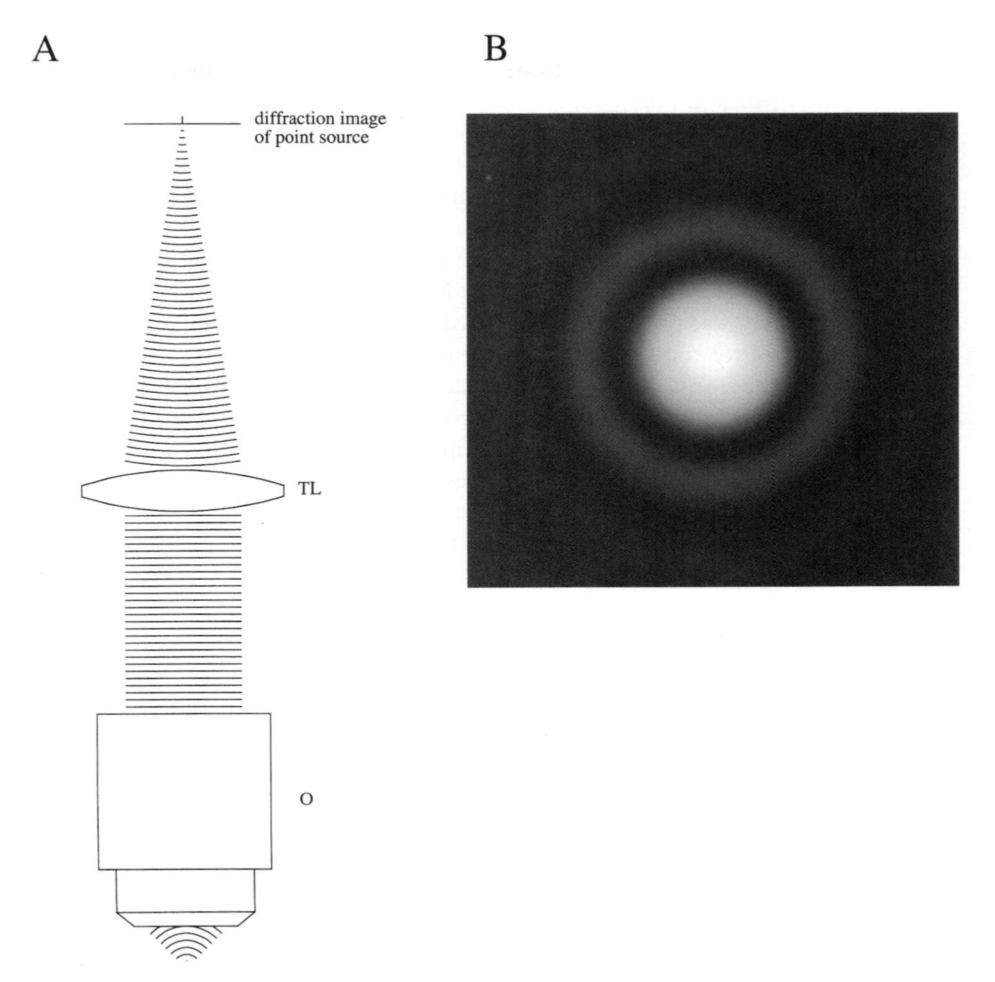

FIGURE 19. Wave front capture by an objective. The Airy pattern—diffraction image of a point source. Resolution in a fluorescence microscope is set by the effect of diffraction on the image of a point source. (*A*) Because of the finite aperture of an objective lens, a microscope captures only part of the spherical wave front emitted by a dye molecule in the specimen. The finite transverse extent of the wave front, and its finite wavelength, cause phase and amplitude variation to occur as it converges to a focus in the image plane. The result of this fundamental process is that the light is not concentrated to a point, but is distributed in a characteristic pattern consisting of a central bright disk surrounded by faint rings. Ideally, this is Airy's pattern, in which 84% of the incoming power is encircled by the first dark ring. Aberration (including defocus) causes the pattern to become more diffuse and less bright. In the image plane, the diameter of the central disk is (Magnification) x $1.22\lambda/NA$, usually several tens of micrometers. In unmagnified coordinates, the radius of the Airy disk defines Rayleigh's resolution limit, $0.61\lambda/NA$. For a high-NA lens this can be as small as 0.22 μm. (*B*) Highly magnified view of Airy pattern, shown in a logarithmic visibility gray scale.

the **fluorescence image** is simply a superposition or sum of the individual brightness patterns contributed by every fluorescent molecule in the specimen. This situation can be changed by **stimulated emission**, which has been utilized in fluorescence microscopy (Dong et al. 1995), but spontaneous emission predominates under usual conditions. However, the independence of emission from dye molecules does not obliterate all traces of coherence; the finite bandwidth ($\Delta\lambda$) of fluorescence emission guarantees temporal or "longitudinal" correlation, usually expressed as a propagation distance over which a wave train holds a single frequency; $L_{\|} = (\lambda_{avg})^2/(\Delta\lambda)$, typically 5–10 μm for dyes with 50-nm emission bandwidth. Additionally, the waves emitted by each dye molecule are transversely coherent. This self-coherence has been utilized to gain axial resolution in interferometric 4Pi microscopes (Hell and Stelzer 1992) and by direct fluorescence image interferometry in I^2M or I^5M microscopes (Gustafsson et al. 1995).

In the Rayleigh picture, a dye molecule emitting light can be pictured as a point source of scalar spherical waves. It is more accurate to treat the molecule as a dipole antenna radiating polarized light (Axelrod 1989), but it is not necessary for the present discussion. Only a circular section of the diverging wave fronts is captured by the objective (Fig. 19), and the "cone angle" of this set of waves is defined by the NA of the lens. Essentially, the objective (or objective plus tube lens in modern infinite-conjugate microscopes) captures a circular section of each expanding wave front and transforms it into a converging wave front headed for a focus in the image plane—or in a neighboring plane, if the object is not in focus. Because the wave front has been limited by the circular pupil of the objective and is no longer a complete spherical wave, diffraction modifies the converging wave in both amplitude and phase. The result is that convergence to a point does not occur; rather, the image of the point source consists of a bright central spot surrounded by a set of progressively attenuated rings (Fig. 19B). This diffraction image, derived (for telescopes) by G.B. Airy in 1834, sets the transverse resolution limit in fluorescence microscopy. Rayleigh defined the limiting resolution condition by the offset superposition of the diffraction images of two stars, where the shift (s_{min}) was equal to the radius of the first dark ring of the Airy pattern. In a fluorescence microscope this would correspond to viewing a specimen in which two point-like fluorescent particles were close enough so that their in-focus diffraction images overlapped to that same degree. For microscopes, the mathematical form of the normalized Airy pattern intensity

$$i_A = [2\, J_1(u)/u]^2 \tag{35}$$

contains the argument $u = 2\pi$ NA r/λ_0, where r is a demagnified radial coordinate. The first dark ring in the Airy pattern occurs where $u = 3.831706$, the first nonzero root of the first-order Bessel function, J_1. The corresponding value of r is, therefore, $(3.83/2\pi)\lambda_0/\mathrm{NA}$, which gives directly Rayleigh's transverse resolution formula

$$s_{min} = 0.61\, \lambda_0/\mathrm{NA} \tag{36}$$

For high-NA objectives, this limit can approach 0.2 µm. Because it is based on a subjective criterion, the Rayleigh formula differs from Abbe's diffraction result (Eq. 28). Fourier transformation of the Airy pattern shows that it is, in fact, a band-limited function with the same spectral range as the Abbe limit. In that sense, there is no difference in the resolution limits for transmitted-light and fluorescence microscopes, even though the optical models differ greatly. The actual size of an Airy pattern in the image plane of the microscope will be enlarged by the magnification. In terms of the diameter of the central spot, or Airy disk

$$D_{min} = 2\, M\, s_{min} = 1.22\, M\, \lambda_0/\mathrm{NA} \tag{37}$$

For a 100x 1.3-NA objective, the Airy disk diameter would be 50 µm; for a 10x 0.3-NA lens, 20 µm. It is this dimension relative to the detector element size in the imaging device (e.g., CCD element or film grain size) that determines how accurately the recorded image matches the true image field (see Nyquist sampling, below).

Clearly, the Airy pattern represents only the in-focus image of a fluorescent point object. If the point source is defocused, i.e., axially displaced from the design focus plane, the geometric image point will also be axially displaced. The converging wave front in the image space will either (1) impinge on the detector before contracting to an Airy pattern or (2) fully contract and partially diverge before reaching the detector. In either case, a less-bright and more-diffuse image will be seen. The Airy pattern represents the idealized **in-focus 2D point spread function** (**PSF**) for a fluorescence microscope. The entire through-focus series of stacked images, in which the Airy pattern is simply the central plane, constitutes the **3D PSF** for the microscope. The depth of field in fluorescence (δ_F) is determined by the loss in brightness of the Airy disk with defocus. This is conservatively defined as the stage increment that causes the Airy disk to become a dark spot surrounded by concentric rings. Use of **Kirchhoff's diffraction integral** to derive the on-axis variation of the 3D PSF shows that its normalized form is

$$i_{ax} = [\sin(\nu)/\nu]^2 \tag{38}$$

in which the argument, $v = \pi\, NA^2\, \Delta z/2n\lambda_0$, is linear in the focus shift Δz. The first axial zero of the 3D PSF occurs where $v = \pi$, resulting in an axial Rayleigh resolution formula

$$\delta_F = 2\, n\, \lambda_0/NA^2 \tag{39}$$

As in the transmitted-light case (Eq. 30), the depth of field depends strongly on NA. For the highest-NA objectives, δ_F is in the range 0.7–0.9 µm. Equation 39 represents a conservative criterion: distinguishing two collinear axial point sources by serial adjustment of focus. When observing distinct point sources, much smaller differences in axial location can be discriminated. In this case, **Rayleigh's quarter-wave criterion** (**RQWC**) defines the depth of field by the effect of focus adjustment on the radius of curvature of the spherical wave front that converges to form the point-source image. When the curvature change is such that there is one-quarter wavelength of phase difference at the margin of the objective pupil (relative to zero difference at the center), there will be significant destructive interference at the center of the image. Formulated in terms of object-space ray angles and coordinates (Taylor and Salmon 1989)

$$\delta_{RQWC} = \lambda_0/(8n \sin^2[{}^1/_2 \sin^{-1}(NA/n)]) \tag{40}$$

For low- or moderate-NA systems, the RQWC formula gives a depth of field equal to ${}^1/_4$ of the axial Rayleigh criterion (Eq. 39). For high-NA systems, the RQWC focus range is as small as 0.13 µm. This limit is generally not attained in fluorescence microscopy, most likely because of signal-to-noise limitations, but it has been approached in transmitted-light systems (Inoué 1989). Because transverse resolution sharpens as 1/NA in both dimensions, and axial resolution sharpens as $1/NA^2$ (or better), the "volume resolution" of the microscope sharpens as $1/NA^4$.

This particular example in fluorescence microscopy illustrates an important point: As the Airy pattern is blurred with defocus, there is little change in the total flux of light arriving at the image plane from the point source. The detected photons are simply distributed over a larger region in the camera. In phase-contrast or DIC microscopy, defocused features blend into the relatively bright, incoherent background. In fluorescence, the background is dark, so that out-of-focus features are noticeable. The out-of-focus features in fluorescence have a more serious effect because of the much lower total photon flux and lower signal-to-noise ratio (SNR). The need for quantitative accuracy in fluorescence imaging of cells has driven the development of systems for fluorescence optical sectioning microscopy, most notably through confocal or multiphoton scanning, but also through computational deblurring, and encoding methods (see below; Chapters 6 and 101).

The Importance of Optical Efficiency and NA

Unlike transmitted-light imaging where the light flux is high, fluorescence microscopy is usually photon-limited. This is particularly true for imaging of living cells, where photobleaching of dyes and photochemical toxicity almost always place severe limits on the allowable light dose to the specimen. Therefore, in fluorescence, it is always best to maximize optical efficiency through careful selection of light source, filter sets, and objective. Because both light-condensing and light-collecting efficiencies of an objective increase with NA, it is always best to use the highest-NA objective compatible with the application.

The relation between NA and light collection efficiency can be derived by considering a small source located at the focus of a lens. The fraction of radiated power that enters the lens within the cone angle defined by θ_{max} is proportional to the solid angle taken up by the lens pupil as seen from the source

$$P = P_0\, (1 - \cos\theta_{max})/2 \tag{41a}$$

$$= (P_0/2)\, (1 - [1 - \sin^2\theta_{max}]^{1/2}) \tag{41b}$$

$$= (P_0/2)\, (1 - [1 - (NA/n)^2]^{1/2}) \tag{41c}$$

For lenses of low to moderate NA, the square root term can be accurately simplified

$$P = 0.32\ P_0\ (\mathrm{NA}/n)^2 \tag{41d}$$

showing that the efficiency grows as the square of the NA or better.

The overall effect of NA on optical efficiency and image brightness is a result of three factors: efficiency of concentration of illumination, efficiency of collection of emission, and sharpness of image formation. In the usual Köhler setup, the incident-light illuminator forms an image of the light source in the back pupil of the objective, which also functions as the condenser. Therefore, the total power focused into the field of view will be equal to the brightness of the source image multiplied by the area of the pupil. Using the result (Eq. 33) that the pupil radius is proportional to NA/M, the total power will be proportional to the square of this ratio. The excitation intensity is proportional to the total power normalized by the field-of-view area, which is proportional to the square of $1/M$. Therefore

$$I_{ex} = B\ (\mathrm{NA}/M)^2/(1/M)^2 \tag{42a}$$

$$= B\ \mathrm{NA}^2 \tag{42b}$$

where B is an effective source brightness. A more detailed analysis shows that, ideally

$$I_{ex} = (1/4)\ B_0\ \mathrm{NA}^2 \tag{42c}$$

where B_0 is the brightness of the filtered source (the factor 1/4 is derived from a thermodynamic restriction on image brightness). The fluorescence emitted by a point source in the specimen will be proportional to I_{ex}. The total collected emission power (P_{em}) reaching the camera will be proportional to both I_{ex} and the light collection efficiency of the objective. Using the previous results 41a and 42c

$$P_{em} = k_F I_{ex}\ (1 - \cos\theta_{max})/2$$

$$= (k_F/4)\ B_0\ \mathrm{NA}^2\ (1 - [1 - (\mathrm{NA}/n)^2]^{1/2})/2 \tag{43a}$$

With a quadratic approximation to the square root for low to moderate NA

$$P_{em} = 0.32\ (k_F B_0/4n^2)\ \mathrm{NA}^4 \tag{43b}$$

showing that the total detected fluorescence is independent of magnification and increases as the fourth power of NA or better. Of greater significance is the **peak brightness** of the Airy pattern image of a point source. This can be estimated by normalizing P_{em} to the area of the Airy disk in the image plane

$$B_F = P_{em}/\pi R_A^{\ 2}$$

$$= 0.32\ (k_F B_0/4n^2)\ \mathrm{NA}^4/\pi(0.61\ M\ \lambda_0/\mathrm{NA})^2$$

$$= B_{em}\ \mathrm{NA}^6/M^2\ \lambda_0^{\ 2} \tag{44}$$

This remarkable result, sixth-order dependence on NA, would be seen only in the limit where the camera pixel density was better than or equal to the optical bandwidth, and the object was much smaller than the resolution limit (Fig. 20). As pointed out above, alteration of M or NA changes the size of the Airy disk relative to the detector pixel elements and may lead to low signal per pixel due to oversampling, or loss of recorded resolution due to undersampling. If an auxiliary magnifier is adjusted to maintain the detector pixel density equal to the optical bandwidth as the NA is changed, the peak brightness increases simply as NA^4 (Eq. 43b). For a diffraction-limited line object, such as a labeled microtubule or actin filament, the image brightness should increase as NA^5/M^2. When the source of fluorescence is spatially extended (such as when imaging an indicator dye in a cell) and the image features are not diffraction-limited, the total detected light flux will depend on NA^4 and the brightness will depend on NA^4/M^2. Finally, if the fluorescence is collected over the entire field of view (such as from a labeled cell monolayer), both the total signal and the brightness will depend roughly on NA^4/M^2, because the number of illuminated cells in the field of view will decrease as $1/M^2$. This strong dependence in all situations is the main reason behind use of high-NA lenses for fluorescence microscopy.

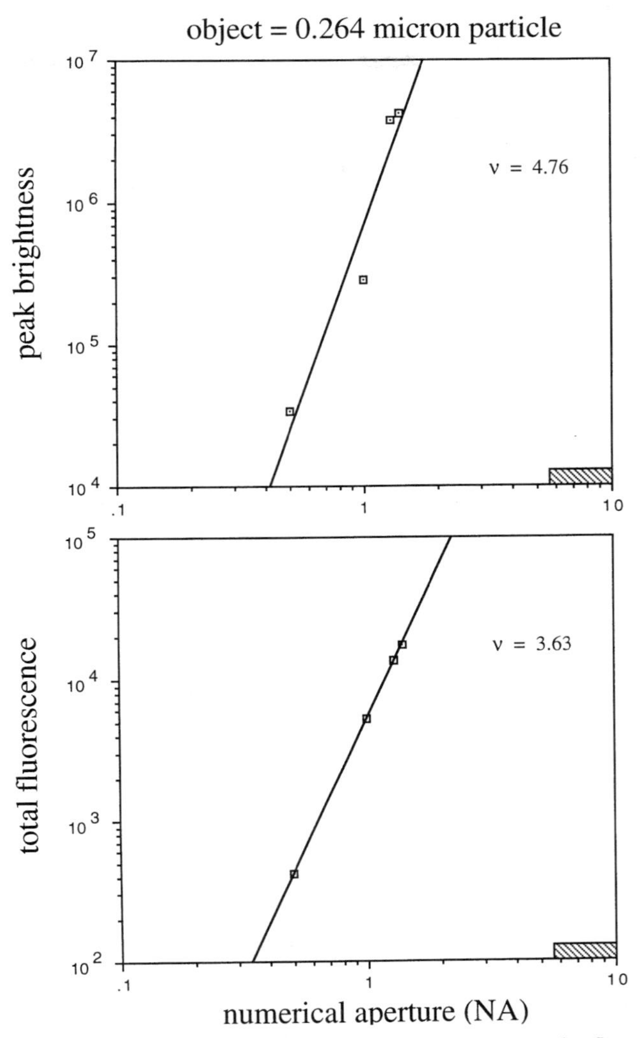

FIGURE 20. Total flux and peak brightness versus NA–point source. In the fluorescence microscope, the objective captures emitted light but also serves as the condenser. The efficiency of the lens in each function increases approximately as the square of the NA (Eqs. 41–44). Therefore, the total light flux in the image of a point source should increase with the fourth power of NA. Additionally, the size of the Airy disk formed by the microscope when imaging a point source decreases as 1/NA. Therefore, at constant magnification, the brightness of a point source image should increase as the total flux normalized by the area of the Airy disk; i.e., as the sixth power of NA. In practice, this will be reduced by the finite spatial extent of the test particle. Measurements made using fluorescent 0.26-μm particles embedded in optical cement show 3.63-power dependence for the total flux, and 4.76-power dependence for the brightness, clearly demonstrating the importance of NA in high-resolution fluorescence imaging.

OPTICAL SECTIONING MICROSCOPY

The Optical Transfer Function in Fluorescence Microscopy

In fluorescence microscopy, the brightness of the image is usually directly related to a measurement of interest: local density of labeled molecules, calcium ion concentration, rate of energy transfer, etc. Generally, quantitative interpretation of pixel values is confounded by the simple fact that a specimen is three-dimensional (and irregular) and its image is two-dimensional. In other words, the image is only a projection of the specimen, and that projection is usually composed of superimposed in-focus and out-of-focus contributions. Several options exist for partially or fully correcting this situation: (1) Reduction of the system NA to increase depth of field so that the entire specimen is in focus. This alternative is equivalent to reducing the specimen to a 2D model and is usually unattractive due to loss of image brightness and resolution. (2) Ratio imaging of the fluorescence signal of interest against a reference dye distribution. The ratio method is widely used to partially correct the images of soluble physiological indicator dyes for the irregular shape of a cell, but is much less useful in structural studies. (3) Removal of out-of-focus contributions from one or more images by optical spatial filtering or computational methods. This approach is known as **optical sectioning microscopy** (**OSM**). The most widely used methods providing direct optical sectioning are confocal scanning fluorescence microscopy (White et al. 1987; see Chapter 6) and multiphoton scanning fluorescence microscopy (Denk et al. 1990; see Chapters 7 and 8). Optical sections can also be obtained by computational deconvolution, also known as computational optical sectioning microscopy (COSM) applied to a serial-focus set of fluorescence images, or by recently developed and adapted optical spatial encoding methods utilizing a grating mask (see Chapter 101). When optical sections can be obtained throughout a serial-focus image set, the result is effectively a 3D reconstruction of the object. In many cases, however, it is the elimination of the out-of-focus component—which mixes contributions from different volume elements of the object—more than getting a 3D reconstruction that is important.

Not surprisingly, the degree to which a microscope can produce optical sections (directly or indirectly) is related to the sharpness of its depth of field, which is a strong function of NA and of the absence of aberration. This is clearly related to both the transverse and axial sharpness of the point spread function of the microscope. The characteristics of the PSF, and its effect on image formation are most apparent in the **optical transfer function** (OTF), a mathematical expression of the spectrum and weighting of spatial frequency information that can be captured by the microscope in the form of images of the object. In general, there is a Fourier transform relationship between images and spectra, although the exact form of the relation will depend on the optical system in use; transmitted-light, fluorescence, reflected-light, or interference. The OTF is related to the Ewald sphere in X-ray crystallography, which sets a limit on the number of Fourier coefficients that can be determined from the X-ray diffraction pattern and used to reconstruct the unit cell of the crystal. Both functions are defined in **reciprocal space**, where the k_x, k_y, and k_z coordinate axes are in units of spatial frequency. However, in crystallography, there is no imaging step; the diffraction order angles and intensities are measured individually and directly as the crystal is rotated, phases are determined indirectly, and the resulting set of Fourier coefficients are inverse-transformed to produce a model of the unit cell. In light microscopy, the situation is slightly different: First, light exits the specimen into the microscope simultaneously over the full range of angles defined by the NA. Second, the fixed cone of ray directions defined by the NA and the optic axis selectively transmits, or *filters*, the diffracted or fluoresced light on its way to the image plane. The filtering or weighting function constitutes the OTF, and its effect is to limit resolution and alter contrast. In a formal sense, the chief aim in optical sectioning microscopy is to expand the OTF, partially or fully reverse its filtering effect, and recover an accurate model of the true object.

In fluorescence, the relationship between true object and image is particularly straightforward. As described above, individual dye molecules in the specimen fluoresce independently and, therefore, lack mutual coherence. As a result, the fluorescence image is a linear superposition of the intensity patterns due to every point source constituting the object. Sources in the plane of focus contribute Airy patterns to the image, each centered on the corresponding geometric image point. A source displaced from the plane of focus by a distance z will contribute a blurred pattern that is a slice cut from the 3D PSF at a distance z from the Airy pattern plane; mathematically, the contribution to the image at location (x, y) can be expressed differentially

$$\delta i(x, y; \Delta z) = \mathrm{obj}(x', y', z')\ \mathrm{psf}(x-x', y-y'; \Delta z-z')\ \delta V' \quad (45)$$

where $\mathrm{obj}(\underline{r}')$ is the fluorescent label density in the object at (x', y', z'), and the PSF is centered on that source point and evaluated at coordinates that are both transversely and axially offset to the image point (x,y). The PSF is weighted by the object source density because fluorescence is linearly proportional to the concentration of the dye label. The excitation field strength is treated as a constant over the region of interest (this would be different in the case of confocal, multiphoton, or other spatial encoding systems). The image is then the integrated contribution of all source points in the object, and has the form of a **convolution** of the true object with the PSF

$$i(x, y; \Delta z) = \iiint \text{obj}(x', y', z')\ \text{psf}(x-x', y-y'; \Delta z-z')\ dV' \quad (46a)$$

In this expression, the variables x and y represent demagnified image-plane coordinates, whereas Δz is the focus drive increment. With that definition, the convolution can be written compactly as

$$i(\underline{r}) = \iiint \text{obj}(\underline{r}')\ \text{psf}(\underline{r}-\underline{r}')\ dV' \quad (46b)$$

where $i(\underline{r})$ can represent a 3D data set consisting of a stack of serial-focus images. The Fourier transform (FT) is generally a **deconvolver**. In a formal sense, 3D transformation into the spatial frequency coordinate system (k_x, k_y, k_z) deconvolves the object from the PSF

$$I(\underline{k}) = \text{FT}[i(\underline{r})] \quad (47a)$$

$$= \text{FT}[\iiint \text{obj}(\underline{r}')\ \text{psf}(\underline{r}-\underline{r}')\ dV'] \quad (47b)$$

$$= \text{FT}[\text{obj}(\underline{r})] \cdot \text{FT}[\text{psf}(\underline{r})] \quad (47c)$$

and defines the fluorescence microscopy OTF as the Fourier transform of the fluorescence PSF

$$\text{OTF}(\underline{k}) = \text{FT}[\text{psf}(\underline{r})] \quad (48)$$

Therefore, the transformed image set is formally equivalent to the transform of the true object weighted (multiplied) by the OTF

$$I(\underline{k}) = \text{OBJ}(\underline{k}) \cdot \text{OTF}(\underline{k}) \quad (49)$$

In this relation, the filtering effect of the OTF can be clearly seen; spatial frequency components (Fourier coefficients) of the object for which the OTF is small-valued will be attenuated in the data. Where the OTF is zero, information on object Fourier coefficients is not at all present in the data.

The transform of the true object, $\text{OBJ}(\underline{k})$, is a Fourier spectrum, each value of which is a Fourier coefficient that represents a fixed plane-wave grating that contributes to the structure of the object. For example, the coefficient $\text{OBJ}(\underline{k}'')$ represents the differential contribution

$$\delta\ \text{obj}(\underline{r}) = \text{OBJ}(\underline{k}'') \cdot \{\cos[\underline{k}''\cdot\underline{r}] - i \sin[\underline{k}''\cdot\underline{r}]\} \cdot \delta k_x\ \delta k_y\ \delta k_z/(2\pi)^3 \quad (50)$$

which is a fixed sinusoidal plane-wave field oriented with wave fronts normal to the wave vector $\underline{k}''$, with spatial frequency $k''/2\pi$ cycles/mm and period $2\pi/k''$ mm (where $k'' = \|\underline{k}''\|$). In the Abbe model of transmitted-light microscopy (see above), the object is pictured as a superposition of 2D gratings composed of periodic refractive index variations. This model can be directly extended to 3D (Streibl 1985), in which case it becomes equivalent to Bragg diffraction from crystal planes. In the case of fluorescence, actual gratings do not exist as such, but only represent virtual periodic distributions of "label density" in the specimen. The main point is that, when formulated in terms of Fourier spectra, the physical meaning of the transfer function is readily apparent.

The PSF (or its Fourier transform, the OTF) clearly determines the resolution and optical sectioning capability of the fluorescence microscope. Fourier transformation of the PSF shows that it is a **band-limited function**; i.e., the OTF turns out to be non-zero only within a sharply defined volume in reciprocal space (Fig. 21). For direct image formation, that volume has the shape of a chord torus, rotationally symmetric about the k_z axis. The most notable feature of the OTF band limit is the axial "missing cone" region, a result of the limited range of angles encompassed by the objective aperture. The geometry of the torus in this reciprocal space is significant. The simple fact that it is oblate signifies that axial resolution is inferior to transverse. Object spatial frequency components that are purely transverse (plane-wave analogs of the flat diffraction gratings considered in the Abbe theory) map into the (k_x, k_y) plane, which cuts through the OTF in its largest diameter. The radius of the OTF therefore corresponds to the highest transverse spatial frequency (k_{tr}) admitted by the objective, a grating pattern for which the period equals Abbe's resolution limit (Eq. 28)

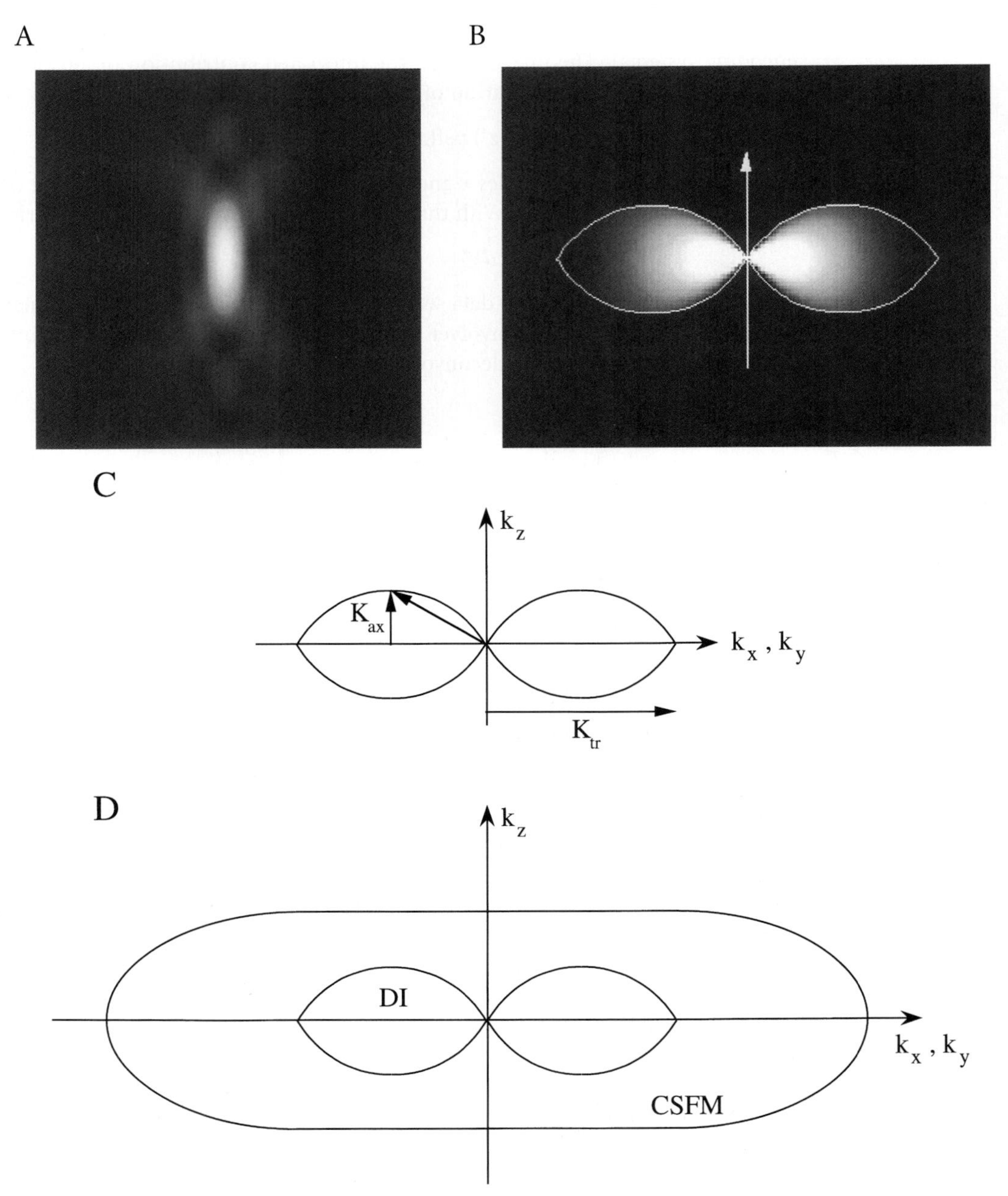

FIGURE 21. The point-spread function and the 3D optical transfer function—direct image formation. (*A*) Central axial section through a computed PSF for a fluorescence microscope. The PSF is rotationally symmetric about the vertical (*z*) axis. A horizontal slice through the center of the PSF would show the Airy pattern as the focused image. A horizontal slice through any plane offset from the center would show a defocused image consisting of diffuse rings, but representing the same total light flux. (*B*) Fourier transformation of the PSF shows that it is a band-limited function: The OTF is nonzero only within a toroidal region of reciprocal space symmetric around the k_z-axis. In this figure, the OTF was drawn for an objective of NA 1.25. Within the passband, the OTF is a strong lowpass filter. The central "missing-cone region" causes loss of spatial frequency information that defines the axial structure of the object. The OTF "inflates" with increasing NA, reducing the relative volume of the missing cone. (*C,D*) Band limits of the OTF for direct image formation (DI) and for confocal scanning (CSFM). Both figures are rotationally symmetric around the k_z axis. The radius of the toroid (K_{tr}) defines the Abbe resolution limit. The oblique wave vector in *C* has the highest spatial frequency axial component within the pass band (K_{ax}). (*D*) The confocal OTF is also a low-pass band-limited filter, but without a missing-cone region. Its volume is ~8X that of the DI OTF.

$$k_{tr} = (4\pi n/\lambda_0) \sin \theta_{max} \quad (51a)$$

$$= 2\pi/(\lambda_0/2NA) \quad (51b)$$

Quite a different situation exists for the Fourier coefficients that fall on the k_z axis. These correspond to purely axial spatial frequency components in the object: fixed plane-wave gratings oriented along the axis of the microscope. Because of the missing cone region in the OTF, every one of these axial components is zeroed out except for the "DC" Fourier coefficient that maps to the origin of reciprocal space. The same is true for the off-axis coefficients that fall into the missing cone region, and which correspond to near-axial tilted gratings. As a result of this severe loss of axial information, true optical sectioning does not occur directly. Within the OTF pass band, the coefficients with the highest frequency axial projection (k_{ax}) map to the apex of the torus

$$k_{ax} = (2\pi n/\lambda_0)\ (1 - \cos \theta_{max}) \quad (52a)$$

$$= (2\pi n/\lambda_0)\ (1 - [1 - (NA/n)^2]^{1/2}) \quad (52b)$$

For lenses of low to moderate NA, the square root term can be accurately simplified to show that the corresponding axial period is similar to the classical depth of field (Eqs. 30 and 39)

$$k_{ax} = 2\pi/(1.56n\ \lambda_0/NA^2) \quad (52c)$$

Within the toroidal volume the OTF is a low-pass filter, much more strongly weighting the coefficients close to the origin ($k = 0$). Therefore, regardless of the 3D range of reciprocal space over which object Fourier coefficients are distributed, Equation 49 restricts transfer to only those coefficients that fall within the OTF pass band, progressively attenuating those of higher spatial frequency within the band. Not surprisingly, the volume enclosed by the OTF pass band increases as NA^4 or better, and the volume lost in the missing cone region decreases. In confocal scanning microscopes, optical spatial filtering greatly modifies the PSF and OTF. The confocal OTF is also band-limited, but has no missing cone region and encloses approximately an eightfold greater volume in reciprocal space (Fig. 21). In multiphoton scanning microscopes, nonlinear excitation of fluorescence leads to a similar result: eightfold expansion of the OTF and elimination of the missing cone region. In both scanning instruments, direct optical sectioning is obtained.

Deconvolution and Estimation

The transfer relation for direct fluorescence imaging (Eq. 49) suggests that a refined estimate of the true object can be obtained by compensating for the filtering effect of the OTF. The simplest computational scheme is some form of inverse filter applied to the data. In an idealized, noiseless case, the inverse filter would have the form $1/\mathrm{OTF}(\underline{k})$ in reciprocal space, and would be applied to $I(\underline{k})$. However, because of the band limitation, the best possible result of inverse filtering would be perfect recovery of all object Fourier coefficients within the OTF volume. This still leaves the severe effect of the missing cone. In practice, the result is far worse: In peripheral zones of the OTF where its transmittance is low, 1/OTF will be very large. Because real image data will always have spatially distributed noise in it, inverse filters cause high-frequency noise to dominate the object estimate. Therefore, any practical computational method of deconvolution must be stable when operating on noisy data (Preza et al. 1992). An alternative to the simple inverse is the Wiener filter or **Wiener inverse** (Agard et al. 1989), which is the linear filter that minimizes the integrated square error when averaged over all possible square-integrable objects in an unbiased way. The mathematical form of the Wiener inverse in reciprocal space depends on both the OTF and the noise **power spectrum**, $\langle |N(\underline{k})|^2 \rangle$

$$W(\underline{k}) = \begin{cases} \mathrm{OTF}^*(\underline{k})/[|\mathrm{OTF}(\underline{k})|^2 + \langle|N(\underline{k})|^2\rangle] & \text{within band limit,} \\ 0 & \text{outside band limit.} \end{cases} \quad (53)$$

When the noise level is very low, $W(\underline{k})$ is similar to 1/OTF. In zones where the OTF has very low amplitude relative to the noise power spectral density, $W(\underline{k})$ rolls off to balance accuracy against stability. In practice, the **object estimate** ($E(\underline{r})$) is computed as an inverse Fourier transform of the filtered data

$$E(\underline{r}) = \mathrm{FT}^{-1}[W(\underline{k})\ I(\underline{k})] \tag{54}$$

However, even in the best of circumstances, the Wiener inverse can only give good estimates of in-band object Fourier coefficients. Therefore, it is not particularly useful for dealing with the OTF in conventional direct imaging systems, although it should perform well with confocal or multiphoton scanning data.

As pointed out by pioneering groups (Harris 1964; Agard et al. 1989; Fay et al. 1989; Carrington 1990), computational deconvolution of direct-image data requires both **regularization** for stability and **extrapolation** in reciprocal space for estimation of out-of-band Fourier coefficients. Extrapolation provides mainly the axial spatial frequency information that is cut out by the missing-cone region of the OTF. Mathematically, extrapolation is possible when the image set contains a sufficiently large number of independent measurements (pixel values) of sufficient precision to overdetermine the in-band coefficients. A number of elegant computational deconvolution methods have been devised that iteratively converge on an optimum estimate of obj($\underline{r}$), given a serial-focus image set as data, along with a measured or computed system PSF or OTF. All of the methods make essential use of constraints or a priori knowledge for reliable extrapolation. The most important of these is nonnegativity of the object label density function that is being estimated. Another is the finite field in which the object is located. For typical image sets that may consist of 8–128 focus planes, each 512 × 512 pixels × 10–16 bits, iteration to convergence requires minutes to hours of processing time on a powerful computer equipped with hundreds of megabytes of RAM and fast disks. A number of deconvolution algorithms have undergone commercial development with or without associated computerized digital microscope systems (for Internet sites, see References). One set of algorithms known as XCOSM (X-windows computational optical sectioning microscopy) was developed and placed in the Internet public domain by the NIH Resource in Biomedical Computing formerly located at Washington University, St. Louis, Missouri. (See http://www.essrl.wustl.edu/~preza/xcosm and http://www.omrfcosm.omrf.org.) XCOSM includes a regularized linear least-squares (RLLS) algorithm, which is an advanced Wiener filter (Preza et al. 1992), and a maximum-likelihood estimator based on the iterative expectation-maximization (E-M) algorithm. Although E-M is computationally very demanding, it readily incorporates the important constraints of nonnegativity and finite field. Additionally, it can be adapted to situations where the data sets are incomplete, or where the PSF is shift-variant; i.e., varies from location to location within the field of view. The form of the E-M algorithm used in fluorescence microscopy was specifically derived for Poisson processes, which is a realistic model for the independent fluorescence of dye molecules in an illuminated specimen (Vardi et al. 1985; Holmes 1988; Krishnamurthi et al. 1995).

In practice, E-M takes the form of an array of correction factors used to modify an existing estimate of the object

$$E_{(n+1)}(\underline{r}) = E_{(n)}(\underline{r})\ \{\textstyle\iiint\{\ i(\underline{r}')\ [\iiint E_{(n)}(\underline{r}'')\ \mathrm{psf}(\underline{r}'-\underline{r}'')\ dV''\ ]^{-1}\}\ \mathrm{psf}(\underline{r}'-\underline{r})\ dV'\} \tag{55}$$

In this iteration formula, $E_{(n)}(\underline{r})$ is the current estimate of the 3D object, $i(\underline{r})$ is the 3D image data, psf($\underline{r}$) is the 3D PSF, and $E_{(n+1)}(\underline{r})$ is the new estimate. In each iteration cycle, the data are weighted by the inverse of the convolution of the current estimate with the PSF, then blurred by convolution with the coordinate-inverted PSF. This generates a 3D array of weighting coefficients, which is used to update the current estimate by multiplication. Therefore, each iteration requires four 3D operations: (1) convolution of the current estimate with the PSF, (2) division of the data by the blurred estimate, (3) convolution of the weighted data with the coordinate-inverted PSF, and (4) multiplication of the current estimate by the resulting coefficient array. The initial estimate can be a smoothed version of the data set or can be as simple as a non-zero constant field. Because the correction is multiplicative, any pixels in the initial estimate initially set to zero will retain that value. This provides a means for directly incorporating the object field constraint. Additionally, because the fluorescence PSF and data are nonnegative everywhere, negative values cannot be generated by the composed multiplication,

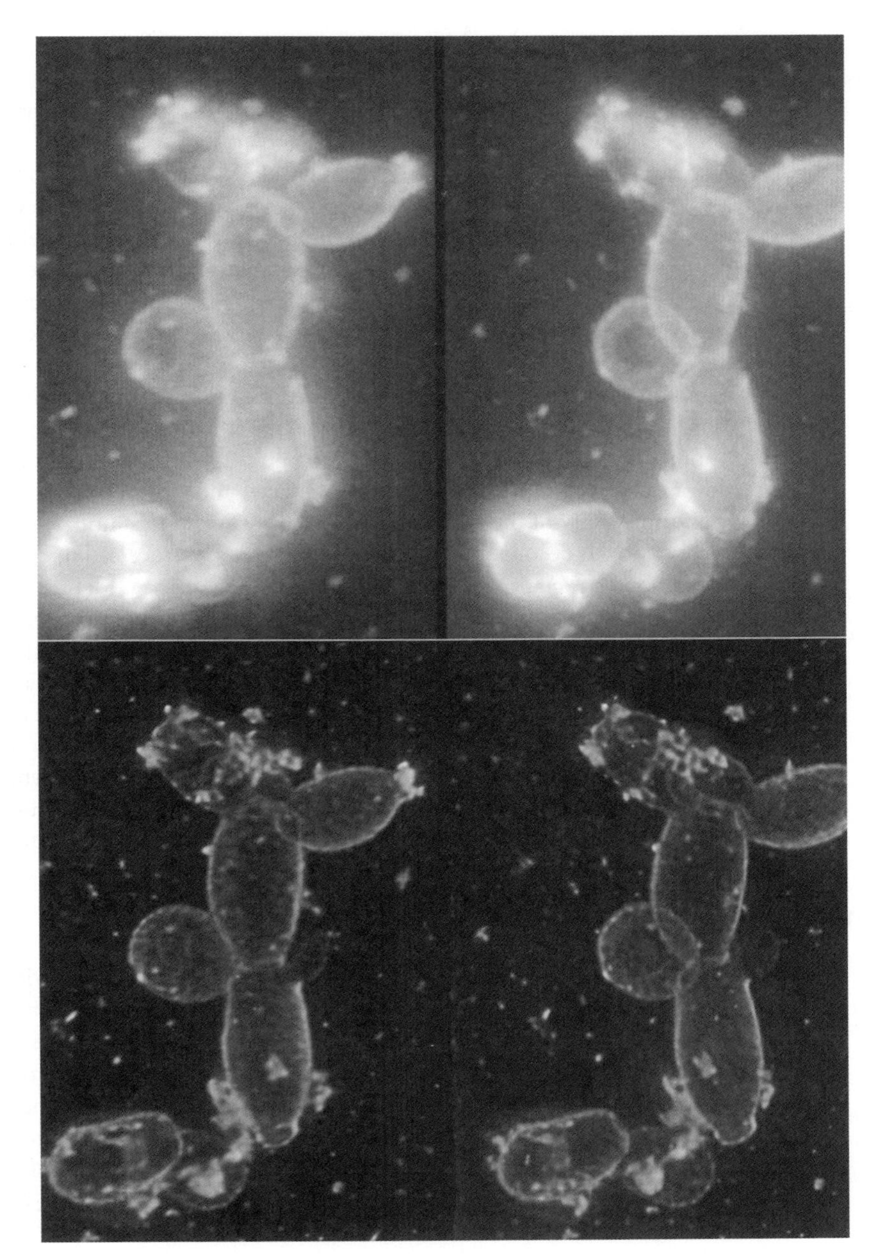

FIGURE 22. Computational deconvolution of serial-focus fluorescence image data. Image stacks showing yeast cells labeled with a fluorescent marker of the plasma membrane. The stacks are displayed as stereo pairs. An automated microscope system equipped with a 100x 1.35-NA objective was used, with a focus increment of 0.1 μm, and with the magnification at the camera set for Nyquist sampling (10 pixels/μm). (*Upper panel*) Image data prior to deconvolution. (*Lower panel*) Object estimate after 1000 rounds of expectation-maximization (Eq. 55). (Images provided courtesy of J.-A. Conchello and W. Goldman, Washington University, Institute for Biomedical Computing and Department of Molecular Biology.)

division, and convolution operations, provided that the initial estimate is also nonnegative. E-M iteration, as well as the other rigorous deconvolution algorithms, requires computation of discrete convolutions in every cycle. Practically, these are implemented by utilizing the equivalence between convolution and multiplication of Fourier transforms; the **discrete Fourier transform** (**DFT**) is computed for each of the two functions to be convolved, the transforms are multiplied, and the inverse DFT is computed (Fig. 22).

Reliable deconvolution in fluorescence microscopy requires attention to detail in every stage of data acquisition and processing. The microscope must be free from stage drift, and the focus drive must be computer-controlled and free from backlash. Feedback stabilization of "focus" (more correctly, of axial setting) can null drift to nanometer levels (Lanni 1993). The high-NA objective, dichroic reflector, and emission filter must be of the highest quality, and the camera must be parfocal with the design ocular settings to ensure that no spherical aberration is introduced by body tube length error. High geometric precision, linearity, and dynamic range are all essential camera characteristics. Cooled CCD cameras providing 10–16 bits of dynamic range and very low dark count rates have become the standard detector in this application (Hiraoka et al. 1988). Additional magnification must be incorporated based on the objective and the size of the CCD elements in the camera array (see Nyquist sampling). Depending on the "chip"used in the camera, elements range from 6.45 μm to 23.0 μm. Division of this number by the total system magnification gives the effective pixel size in object-space coordinates. The fluorescence illuminator should be IR-blocked and adjusted to match the objective INA and to maximize uniformity of the excitation field. This provides best utilization of the available camera dynamic range. Prior to deconvolution, images should be preprocessed to compensate for (1) gain and baseline differences between CCD elements in the camera array, (2) light source instability, and (3) systematic effects such as photobleaching and focus-dependent background due to luminescence from immersion oil. For a given objective, background light in a fluorescence microscope increases with illuminated field diameter. This wastes dynamic range, and is a source of spatial noise. Reduction of the field iris diameter to encircle the object of interest, or at least to encircle the camera field of view (which is usually smaller than the ocular field of view) minimizes this contribution. For small objects, a reduced illumination field also produces partial confocality in the excitation, which is advantageous in deconvolution (Hiraoka et al. 1990). If the PSF of the microscope is measured for use in deconvolution of reduced-field data, it is important that it be measured under the same field condition.

Determination of PSFs and OTFs for Deconvolution

The PSF can be derived from basic principles for model systems with or without aberrations (Frisken Gibson and Lanni 1991; Kontoyannis and Lanni 1996) or can be measured by imaging of subresolution fluorescent particles. Neither route is a trivial exercise, because precision and accuracy are needed. Computation has the advantage of absence of noise and unlimited dynamic range, but the disadvantage of being only as accurate as the mathematical model of the optical system. Measurement has the advantage of accurately including the aberrations of a real system, but the disadvantages of measurement error and spatial noise. Because the optical properties of the specimen itself can introduce aberration, the highest-precision work in deconvolution utilizes PSFs that are computationally refined from measurements in situ (Carter et al. 1993; Carrington et al. 1995; Scalettar et al. 1996). In this procedure, 0.1–0.2-μm fluorescent polystyrene latex microspheres are added sparsely to the specimen to serve as pseudo-point sources. PSF image stacks are registered on-axis and averaged to reduce noise; processing may include rotational averaging around the centroid, which improves precision but obliterates any actual transverse asymmetry that may be intrinsic to the microscope or caused by the specimen. Because the test particles have a small but finite diameter, the measured PSF is a convolution of the true PSF with the source density within the microsphere. The true PSF can be estimated by deconvolving a uniform sphere from the measured and averaged PSF data.

Alternatively, the OTF can be obtained through measurement or computation. Whereas the PSF is computed via numerical evaluation of a modified Kirchhoff diffraction integral, OTF formulas derived from first principles are generally expressible in terms of analytic functions within the OTF band limit (Frieden 1967; Wang and Frieden 1990). In principle, a computed OTF can be converted to its corresponding PSF via the inverse Fourier transform. In the more usual case, a computed or measured PSF is Fourier-transformed to obtain a discretized OTF, which is then used directly in an iterative deconvolution algorithm.

Nyquist Sampling and Restoration of the Image Field

When viewed by eye, the image field of a microscope is perceived as a continuous distribution of irradiance. This is true even though the retina is essentially a discrete array of photodetector elements in which light is detected as discrete photocounts. When the image field is detected with an electronic camera, the output is sampled and digitized (true also for film). Under these conditions, pixellation and quantization become apparent. Overly coarse pixellation degrades resolution; overly fine pixellation ("empty magnification") needlessly reduces the field of view and increases variance by reducing the number of photocounts per pixel. Clearly, optimum pixellation should be determined by the resolution limit of the optical system that forms the image.

Pixellation and resolution are related by the **Sampling Theorem** (Goldman 1968), which sets an important restriction on the information content of images. The theorem relates the frequency content of a signal to the density of measured values that is required for *exact* reconstruction of the signal in the absence of noise. The signal can be a time series, as in communications, or it can be a spatial pattern such as a spectrum or image. Specifically, if an image, $f(x)$, is limited to spatial frequencies lower than a band limit value F_0 (cycles/mm, for example), then $f(x)$ can be reconstructed exactly at all values of x if the signal is known at sample points spaced at an increment $\Delta x = 1/\text{bandwidth}$; in this case, $\Delta x = 1/(2F_0)$ mm. In principle, the sample points must extend throughout the spatial domain, but a finite set is usually sufficient to determine $f(x)$ accurately for an object well within a finite field. The basis of the Sampling Theorem is in representation of a finite frequency spectrum as a Fourier series truncated to a single period by a window function of width $2F_0$; inverse Fourier transformation of this representation shows that the nth coefficient in the series is actually a sample value of the image, $f(n\Delta x)$, at location $x = n\Delta x$. Furthermore, each sample value is the weighting coefficient of a continuous sampling function centered at $x = n\Delta x$

$$f(x) = \Sigma_n f(n\Delta x) \sin[2\pi F_0(x - n\Delta x)]/2\pi F_0(x - n\Delta x) \quad (56a)$$

$$= \Sigma_n f(n\Delta x) \operatorname{sinc}[2\pi F_0(x - n\Delta x)] \quad (56b)$$

Together, the set of terms constitute an exact interpolation series for $f(x)$. The quantity "1/bandwidth" represents a maximum interval between samples known as the **Nyquist limit.** Undersampling results in loss of resolution and can result in aliasing—the generation of artifactual spatial frequency components.

For microscopes, it is the Abbe resolution limit (p_{min}; see Eq. 28) that sets the information density. Specifically, the Abbe limit is the period of the highest spatial frequency transferred to the image field by the optical system ($k_{tr}/2\pi$; see Eqs. 51). The bandwidth of a microscope is therefore $2(k_{tr}/2\pi)$, which is also the diameter of the OTF. A noiseless 2D image field is therefore completely defined by point samples in a square array spaced at 1/bandwidth

$$\Delta s_{Nyq} = 1/[2(k_{tr}/2\pi)] \quad (57a)$$

$$= \lambda_0/4\ \text{NA} \quad (57b)$$

i.e., the Nyquist sample interval equals one-half the Abbe limit (Lanni and Baxter 1992). For high-NA objectives, Δs_{Nyq} is therefore close to 0.1 μm (Nyquist density = 10 pixels/μm). In real terms, the sampling interval is the camera pixel size divided by the total system magnification. For Nyquist sampling, this number must not exceed λ_0/4NA. A CCD camera with 19 μm square elements used with a 100x 1.3 NA objective, for example, requires a total system magnification of 190x for Nyquist sampling in the mid-visible range. Therefore, additional magnification ≥1.9x must be utilized. Needless to say, most biological imaging with electronic cameras is carried out "sub-Nyquist," where the recorded image is of significantly lower resolution than the actual optical field of the microscope. This is usually dictated by the need for a sufficient field of view (an advantage still held by photographic film!). However, there is no getting around meeting or exceeding the Nyquist density to properly match camera to microscope. The fact that the OTF is circular rather than square means that there is some redundancy in image data sampled at the Nyquist density over a square array as defined above. However, there has yet been no practical use made of this latter fact in microscopy.

With Nyquist sampling, it is possible to computationally convert pixellated data into a "continuous" image field by use of a 2D interpolation series analogous to Equation 56

$$f(x,y) = \Sigma_m \Sigma_n f(m\Delta s, n\Delta s)\ \mathrm{sinc}[2\pi F_0(x - m\Delta s)]\ \mathrm{sinc}[2\pi F_0(y - n\Delta s)] \quad (58)$$

Although rarely done, this has the advantage of producing a field that is independent of arbitrary offsets between the sampling array (or **Nyquist lattice**) and the image features of interest. As the simplest possible example, the Airy pattern image of a point source could fall exactly centered on a pixel, centered at the adjoining corners of four pixels, or at an arbitrary offset in between. In each case, the pixellated image will appear differently, with 16–21 pixels located within the Airy disk (4–5 pixels across). The band-limited reconstruction, on the other hand, should produce the same Airy pattern in every case.

Real cameras strictly do not produce image data in the form of point samples as required for direct application of the Sampling Theorem. At the present time, use of cooled CCD cameras provides the greatest optical efficiency, dynamic range, and geometric precision in image capture. Pixel values in the digitized image represent summed photocounts in the contiguous-square "bins" that constitute the CCD array. In contrast, the interpolation series (Eqs. 56 and 58) are based on point-sampled values of the image field. The difference is illustrated by the example of the Airy pattern image; even though the Airy disk is defined by a prominent circular node at which the brightness drops to zero, no pixel in the CCD image will ever read "zero" because of the finite area over which each CCD element captures photons. Direct use of CCD camera output in the interpolation formula results in a blurred reconstruction of the Airy pattern. The effect of square-pixel summing can be reversed, in the absence of noise, by use of an inverse filter to estimate point samples from real square-pixel data. The directly usable form of this filter (Lanni and Baxter 1992) is a discrete 2D **convolution kernel**, centered at K_{00}

$$K_{00} = (16/\pi^2)\ (\lambda_c)^2$$

$$K_{m'n'} = (16/\pi^2)\ (\lambda_c + \Sigma_m (-1)^m/(2m-1)^2)\ (\lambda_c + \Sigma_n (-1)^n/(2n-1)^2) \quad (59)$$

where λ_c is Catalan's constant (0.91596559), and where the summations run 1 to $|m'|$ or 1 to $|n'|$, or are zero when $m' = 0$ or $n' = 0$, respectively. This is applied directly to square-pixel image data to estimate point-samples for use in an interpolation series (Eq. 58). The interpolation series can be computed to any degree of fineness, but the main advantage is that the result is, in principle, independent of the original pixellation.

Not surprisingly, in optical sectioning microscopy, the Nyquist criterion applies to axial sampling, or focus increment, as well as to pixellation. Using Equation 52 for the axial spatial frequency limit of the OTF, the **Nyquist focus increment** equals 1/(axial bandwidth)

$$\Delta z_{\mathrm{Nyq}} = 1/2(k_{ax}/2\pi) \quad (60a)$$

$$= \lambda_0/2n\ (1 - [1 - (\mathrm{NA}/n)^2]^{1/2}) \quad (60b)$$

$$= 2\ \delta_{\mathrm{RQWC}}\ \text{(see Eq. 40)} \quad (60c)$$

For a high-NA objective, Δz_{Nyq} is approximately 0.3 µm. However, for direct image formation, this situation is not strictly analogous to transverse sampling requirements because the axial range in a serial-focus image set is usually much less than needed to completely blur out-of-focus features. Therefore, the sample points do not run out to infinity or to where the signal drops to arbitrarily small values as required by the Sampling Theorem. In a confocal or multiphoton scanning microscope, the object is strongly attenuated with defocus, so that a finite number of sample points (focus planes) can accurately represent the axial structure of the object in an interpolation series representation.

An additional point regarding confocal and multiphoton scanning microscopes is that the transverse and axial cutoffs of the OTF are doubled relative to direct image formation. Nyquist densities are therefore doubled in each dimension; in other words, resolution is increased, but so is the sampling requirement that must be satisfied to utilize the additional information.

PRACTICAL LIMITS ON IMAGE QUALITY

Aberration

Both optical aberration and background light cause loss of contrast and resolution in all modes of microscopy and can cause a microscope to perform far below the fundamental limits to which it is designed. Aberration can originate in the optical system of the microscope or in the optical properties of the mounted specimen. For example, mixing of dissimilar immersion oil residues can cause irregular refraction of light rays exiting the specimen. Likewise, because glycerol is highly hygroscopic, its uniformity as an immersion fluid is affected by ambient humidity. As discussed above, a difference between the immersion and specimen refractive indices leads to focus-dependent spherical aberration (SA), most noticeable as axial asymmetry in the PSF. When SA is severe, the fluorescence image of a point particle will show prominent concentric rings on one side of focus and featureless blur on the other. The central maximum of the in-focus image will be attenuated, and the brightness of the surrounding rings increased in comparison to the Airy pattern. This reduces contrast in any image. SA also causes elongation of the central maximum of the PSF, which dramatically reduces axial resolution. In fluorescence systems, dichroic reflectors and emission filters must be selected to minimize wave-front distortion. To begin with, a high-quality filter with perfectly flat surfaces can cause image offset due to **wedge**—the condition where the flat surfaces are not exactly parallel. This condition makes the "flat optic" into a weak prism (flat optics made with precisely parallel surfaces are known as etalons) and is not considered an aberration because wedge does not degrade the image. Much more serious is surface curvature, a common condition in general-purpose filters composed of two or more cemented glass substrata. In this case, the filter acts as a weak, irregular lens, causing astigmatism, spherical aberration, or other distortions in the image. Manufacturers of fluorescence filter sets know that emission filters for microscopy must be of **imaging grade**, preferably selected for flatness by examination of the filter in a transmission interferometer. Clearly, the same high quality is needed in any optic inserted between the objective and image detector: prisms, analyzers, wave plates, or electro-optic devices. In fluorescence optical systems used simultaneously with DIC, the Nomarski prism will shear the fluorescence image into two slightly offset, orthogonally polarized components (Kontoyannis et al. 1996). This is not an aberration in the usual sense and is of little consequence when imaging extended fluorescent features in an object. However, the shear can significantly reduce the brightness over background of punctate and line objects at the diffraction limit.

Background Light

Sources of background in fluorescence microscopy can be found in both the microscope and specimen. Emission filters must with great efficiency block light at the excitation wavelength that is back-reflected into the detector path of the instrument. Even with incident-light illumination, where most of the excitation propagates away from the microscope, the back-reflected light flux can be orders of magnitude greater than the fluorescence. Paired excitation and emission filters for fluorescence microscopy characteristically attenuate by factors $>>10^4$ in complementary pass bands. A heat filter should be used in the illuminator to block the intense IR light produced by incandescent and arc lamps. Stray IR light in the 750–1000-nm band is not directly visible, but may be well within the spectral range of an electronic camera. Luminescence can also be excited within objectives and other glass optical components, particularly by UV light. Over the past two decades, objectives have become greatly improved by the use of low-absorbance, low-luminescence glasses and optical cements. Some of the best colored glasses used in emission filters are also luminescent. Used alone, these glasses make excellent long-pass filters, but will cause a background glow when there is a significant flux of back-reflected excitation to be blocked. This can be minimized by use of immersion objectives to reduce reflection. Colored glasses are often used in combination with interference multilayers to obtain the best band-pass filter characteristics. In this case, luminescence is minimized when the light is first incident on the reflective multilayer, followed by the colored glass.

The most common sources of background light in fluorescence microscopy are in the assembled specimen, including immersion fluid, cover glass, and slide. Only immersion oil of fluorescence grade should be used. Contamination of the oil by marker inks, stains, and dust should be avoided (lint is often fluorescent because of the brighteners used in laundry detergents!). Background due to the immersion fluid depends on the amount of fluid in the illuminated volume between cover glass and objective. This is clearly less for high-NA lenses having short working distances, and is also dependent on the depth to which the objective is focused. Cover glasses are rarely a significant source of luminescence, but they should be carefully cleaned of dried culture medium before immersion fluid is applied. Dried medium causes severe light scatter and is intensely fluorescent. Slides are six times thicker than cover glasses and, if not of good quality, can be a significant source of luminescence. For most live-cell microscopy, culture dishes or chambers with cover glass windows are used, eliminating the slide as a factor. By far, the living specimen and culture medium are the major contributors of background light. In culture medium, Phenol Red used as a visual pH indicator is moderately fluorescent and should not be used in microscopy. Base medium and serum also contain many weakly fluorescent biochemicals and proteins. Luminescence from this source can be reduced by use of a "thin" culture chamber that minimizes the volume of medium that is illuminated. In general, background light increases with the diameter of the illuminated field; therefore, the field iris should be reduced when the object of interest is much smaller than the design FOV. Because background sources are less strongly excited outside of the blue end of the spectrum, use of long-wavelength fluorescent dyes has grown rapidly. A fundamental limitation on background reduction is set by **Raman emission**, an inelastic scattering process that converts a small fraction of the incident light to isotropic emission at longer wavelengths. Raman emission generally originates in water within the specimen, or in the immersion fluid. Although it is well beyond the scope of this discussion, Raman scatter, back-scatter, and phosphorescence can all be gated out through time-resolved fluorescence imaging (Herman 1998). Because of its noise content, background light is one of the factors that ultimately limits sensitivity in fluorescence microscopy (photobleaching and excited-state lifetime are the other factors). Nevertheless, the fluorescence microscope has emerged as an unmatched instrument for molecular level imaging, including single-molecule tracking and spectroscopy (Gross and Webb 1986; Kron and Spudich 1986; Betzig and Chichester 1993; Nie et al. 1994; Funatsu et al. 1995; Sase et al. 1995; Vale et al. 1996; Dickson et al. 1997; Femino et al. 1998; Waterman-Storer and Salmon 1998; Waterman-Storer et al. 1998; Moerner and Orrit 1999; Smith et al. 1999; Weiss 1999).

CONCLUSION

In this chapter, we have laid out principles and practical considerations on basic light microscopy as applied to biological specimens. Much of the discussion centers on the effect of high aperture in the image formation process and a description of the main phase- and fluorescence-contrast systems. Not covered are the many interesting and useful special-purpose instruments and methods, such as confocal scanning fluorescence, multiphoton scanning fluorescence, reflection-interference, total internal reflection fluorescence, near-field scanning, polarization, fluorescence polarization, resonance energy transfer, and interferometric fluorescence microscopies. Nor have we covered microscope automation, which is revolutionizing the data-gathering power of the core instrument. In the bibliography accompanying this chapter are references to texts and articles that cover more deeply special topics in optics and applications. We have every reason to expect that these remarkable developments will continue.

ACKNOWLEDGMENT

F. Lanni thanks the National Science Foundation for past support through the Science and Technology Centers program.

GENERAL REFERENCES ON MICROSCOPY AND OPTICS

Born M. and Wolf E. 1980. *Principles of optics*, 6th edition. Pergamon, New York.

Haugland R.D. 1996. *Handbook of fluorescent probes and research chemicals*, 6th edition. Molecular Probes, Inc., Eugene, Oregon.

Herman B. 1998. *Fluorescence microscopy*, 2nd edition. Springer/BIOS Scientific, Oxford, United Kingdom.

Herman B. and Jacobson K., Eds. 1990. *Optical microscopy for biology*. J. Wiley & Sons, New York.

Inoué S. and Spring K.R. 1997. *Video microscopy—The fundamentals*, 2nd edition. Plenum Press, New York.

Longhurst R.S. 1973. *Geometrical and physical optics*, 3rd edition. Longman, London.

Matsumoto B., Ed. 1993. Cell biological applications of confocal microscopy. *Methods Cell Biol.*, vol. 38.

Pawley J.B., Ed. 1995. *Handbook of confocal fluorescence microscopy*, 2nd edition. Plenum Press, New York.

Piller H. 1977. *Microscope photometry*. Springer-Verlag, Heidelberg.

Pluta M. 1989a. *Advanced light microscopy*, vol. 1: *Principles and basic properties*. Elsevier, New York.

———. 1989b. *Advanced light microscopy*, vol. 2: *Specialized methods*. Elsevier, New York.

———. 1989c. *Advanced light microscopy*, vol. 3: *Measuring techniques*. Elsevier, New York.

Rebhun L.I., Taylor D.L., and Condeelis J.S., eds. 1988. Optical approaches to the dynamics of cellular motility. *Cell Motil. Cytoskel.*, vol. 10. A.R. Liss, New York.

Slayter E.M. 1976. *Optical methods in biology*. Krieger Publishing, Huntington, New York.

Spector D., Goldman R.D., and Leinwand L.A. 1998. *Cells: A laboratory manual*, vol. 2: *Light microscopy and cell structure*. Cold Spring Harbor Laboratory Press, Cold Spring Harbor, New York.

Taylor D.L. and Wang Y.-L., eds. 1989. Fluorescence microscopy of living cells in culture. Part B. Quantitative fluorescence microscopy—Imaging and spectroscopy. *Methods Cell Biol.*, vol. 30.

Wang Y.-L. and Taylor D.L., eds. 1989. Fluorescence microscopy of living cells in culture. Part A. Fluorescent analogs, labeling cells, and basic microscopy. *Methods Cell Biol.*, vol. 29.

Wilson T. and Sheppard C. 1984. *Theory and practice of scanning optical microscopy*. Academic Press, London.

REFERENCES

Abraham V.A., Krishnamurthi V., Taylor D.L., and Lanni F. 1999. The actin-based nanomachine at the leading edge of migrating cells. *Biophys. J.* **77:** 1721–1732.

Agard D.A., Hiraoka Y., Shaw P., and Sedat J.W. 1989. Fluorescence microscopy in three dimensions. *Methods Cell Biol.* **30:** 353–377.

Allen R.D., Allen N.S., and Travis J.L. 1981. Video-enhanced contrast, differential interference contrast (AVEC-DIC) microscopy: A new method capable of analyzing microtubule-related motility in the reticulopodial network of *Allogromia laticollaris*. *Cell Motil. Cytoskel.* **1:** 291–302.

Axelrod D. 1989. Fluorescence polarization microscopy. *Methods Cell Biol.* **30:** 333–352.

Betzig E. and Chichester R.J. 1993. Single molecules observed by near-field scanning optical microscopy. *Science* **262:** 1422–1425.

Carrington W.A. 1990. Image restoration in 3D microscopy with limited data. *Proc. SPIE* **1205:** 72–83.

Carrington W.A., Lynch R.M., Moore E.D.W., Isenberg G., Fogarty K.E., and Fay F.S. 1995. Superresolution three-dimensional images of fluorescence in cells with minimal light exposure. *Science* **268:** 1483–1487.

Carter K.C., Bowman D., Carrington W., Fogarty K., McNeil J.A., Fay F.S., and Lawrence J.B. 1993. A three-dimensional view of precursor messenger RNA metabolism within the mammalian nucleus. *Science* **259:** 1330–1335.

Cogswell C., Smith N.I., Larkin K.G., and Hariharan P. 1997. Quantitative DIC microscopy using a geometric phase shifter. *Proc. SPIE* **2984:** 72–81.

Denk W., Strickler J.H., and Webb W.W. 1990. Two-photon laser scanning fluorescence microscopy. *Science* **248:** 73–76.

Dickson R.M., Cubitt A.B., Tsien R.Y., and Moerner W.E. 1997. On/off blinking and switching behaviour of single molecules of green fluorescent protein. *Nature* **388:** 355–358.

Dong C.Y., So P.T., French T., and Gratton E. 1995. Fluorescence lifetime imaging by asynchronous pump-probe microscopy. *Biophys. J.* **69:** 2234–2242.

Ellis G.W. 1988. Scanned aperture light microscopy. *Proc. Electron Microsc. Soc. Am.* **46:** 48–49.

Fay F.S., Carrington W., and Fogarty K.E. 1989. Three-dimensional molecular distributions in single cells analysed using the digital imaging microscope. *J. Microsc.* **153:** 133–149.

Feineigle P. 1996. "Motion analysis and visualization of biological structures imaged via the Nomarski differential interference contrast microscope." Ph.D. thesis, Carnegie Mellon University, Pittsburgh, Pennsylvania.

Femino A.M., Fay F.S., Fogarty K., and Singer R.H. 1998. Visualization of single RNA transcripts in situ. *Science* **280:** 585–590.

Frieden B.R. 1967. Optical transfer of the three-dimensional object. *J. Opt. Soc. Am.* **57:** 56–66.

Frisken Gibson S.F. 1990. "Modeling the 3D imaging properties of the fluorescence light microscope." Ph.D. thesis, Carnegie Mellon University, Pittsburgh, Pennsylvania.

Frisken Gibson S. and Lanni F. 1991. Experimental test of an analytical model of aberration in an oil-immersion objective lens used in three-dimensional light microscopy (reprinted in **9:** 154–166 [1992]). *J. Opt. Soc. Am. A* **8:** 1601–1613.

Funatsu T., Harada Y., Tokunaga M., Saito K., and Yanagida T. 1995. Imaging of single fluorescent molecules and individual ATP turnovers by single myosin molecules in aqueous solution. *Nature* **374:** 555–559.

Galbraith W. 1955. The optical measurement of depth. *Q. J. Microsc. Sci.* **96:** 285–288.

Goldman S. 1968. *Information theory*. Dover, New York.

Gross D. and Webb W.W. 1986. Molecular counting of low-density lipoprotein particles as individuals and small clusters on cell surfaces. *Biophys. J.* **49:** 901–911.

Grynkiewicz G., Poenie M., and Tsien R.Y. 1985. A new generation of Ca2+ indicators with greatly improved fluorescence properties. *J. Biol. Chem.* **260:** 3440–3450.

Guether R. and Reddington M.V. 1997. Photostable cyanine dye beta-cyclodextrin conjugates. *Tetrahedron Lett.* **38:** 6167–6170.

Gustafsson M.G.L., Agard D.A., and Sedat J.W. 1995. Sevenfold improvement of axial resolution in 3D widefield microscopy using two objective lenses. *Proc. SPIE* **2412:** 147–156.

Hard R., Zeh R., and Allen R.D. 1977. Phase-randomized laser illumination for microscopy. *J. Cell Sci.* **23:** 335–343.

Harris J.L. 1964. Diffraction and resolving power. *J. Opt. Soc. Am.* **54:** 931–936.

Hell S. and Stelzer E.H.K. 1992. Properties of a 4pi confocal fluorescence microscope. *J. Opt. Soc. Am. A* **9:** 2159–2166.

Herman B. 1998. *Fluorescence microscopy*, 2nd edition. Springer/BIOS Scientific, Oxford, United Kingdom.

Hiraoka Y., Sedat J.W., and Agard D.A. 1988. The use of a charge-coupled device for quantitative optical microscopy of biological structures. *Science* **238:** 36–41.

———. 1990. Determination of three-dimensional imaging properties of a light microscope system. Partial confocal behavior in epifluorescence microscopy. *Biophys. J.* **57:** 325–333.

Holmes T.J. 1988. Maximum-likelihood image restoration adapted for noncoherent optical imaging. *J. Opt. Soc. Am. A* **5:** 666–673.

Inoué S. 1981. Video image processing greatly enhances contrast, quality, and speed in polarization-based microscopy. *J. Cell Biol.* **89:** 346–356.

———. 1989. Imaging of unresolved objects, superresolution, and precision of distance measurement with video microscopy. *Methods Cell Biol.* **30:** 85–112.

Inoué S. and Spring K.R. 1997. *Video microscopy—The fundamentals*, 2nd edition. Plenum Press, New York.

Kagalwala F. and Kanade T. 1998. Computational model of image formation process in DIC microscopy. *Proc. SPIE* **3261:** 193–204.

Keller H.E. 1995. Objective lenses for confocal microscopy. In *Handbook of biological confocal microscopy*, 2nd edition (ed. J.B. Pawley), pp. 111–126. Plenum Press, New York.

Kontoyannis N.S. and Lanni F. 1996. Measured and computed point spread functions for an indirect water-immersion objective used in three-dimensional fluorescence microscopy. *Proc. SPIE* **2655:** 34–42.

Kontoyannis N.S., Krishnamurthi V., Bailey B., and Lanni F. 1996. Three-dimensional fluorescence microscopy of cells. *Proc. SPIE* **2678:** 6–14.

Krishnamurthi V., Liu Y.-H., Bhattacharyya S., Turner J.N., and Holmes T.J. 1995. Blind deconvolution of fluorescence micrographs by maximum-likelihood estimation. *Appl. Opt.* **34:** 6633–6647.

Kron S.J. and Spudich J.A. 1986. Fluorescent actin filaments move on myosin fixed to a glass surface. *Proc. Natl. Acad. Sci.* **83:** 6272–6276.

Lanni F. 1993. Feedback-stabilized focal plane control for light microscopes. *Rev. Sci. Instrum.* **64:** 1474–1477.

Lanni F. and Baxter G.J. 1992. Sampling theorem for square-pixel image data. *Proc. SPIE* **1660:** 140–147.

McNally J.G., Preza C., Conchello J.-A., and Thomas L.J. 1994. Artifacts in computational optical sectioning microscopy. *J. Opt. Soc. Am. A* **11:** 1056–1067.

Moerner W.E. and Orrit M. 1999. Illuminating single molecules in condensed matter. *Science* **283:** 1670–1676.

Nie S., Chiu D.T., and Zare R.N. 1994. Probing individual molecules with confocal fluorescence microscopy. *Science* **266:** 1018–1021.

Ormo M., Cubitt A.B., Kallio K., Gross L.A., Tsien R.Y., and Remington S.J. 1996. Crystal structure of the Aequorea victoria green fluorescent protein. *Science* **273:** 1392–1395.

Pluta M. 1989a. *Advanced light microscopy*, vol. 1: *Principles and basic properties*. Elsevier, New York.

———. 1989b. *Advanced light microscopy*, vol. 2: *Specialized methods*. Elsevier, New York.

Preza C. 1998. "Phase estimation using rotational diversity for differential interference contrast microscopy." Ph.D. thesis, Washington University, St. Louis, Missouri.

Preza C., Snyder D.L., and Conchello J.-A. 1996. Imaging models for three-dimensional transmitted-light DIC microscopy. *Proc. SPIE* **2655:** 245–256.

Preza C., Miller M.I., Thomas L.J. , and McNally J.G. 1992. Regularized linear method for reconstruction of three-dimensional microscopic objects from optical sections. *J. Opt. Soc. Am. A* **9:** 219–228.

Sase I., Miyata H., Corrie J.E.T., Craik J.S., and Kinosita K. 1995. Real time imaging of single fluorophores on moving actin with an epifluorescence microscope. *Biophys. J.* **69:** 323–328.

Scalettar B.A., Swedlow J.R., Sedat J.W., and Agard D.A. 1996. Dispersion, aberration, and deconvolution in multi-wavelength fluorescence images. *J. Microsc.* **182:** 50–60.

Smith D.E., Babcock H.P., and Chu S. 1999. Single-polymer dynamics in steady shear flow. *Science* **283:** 1724–1727.

Streibl N. 1985. Three-dimensional imaging by a microscope. *J. Opt. Soc. Am. A* **2:** 121–127.

Taylor D.L. and Salmon E.D. 1989. Basic fluorescence microscopy. *Methods Cell Biol.* **29:** 207–237.

Tsien R.Y. and Waggoner A.S. 1995. Fluorophores for confocal microscopy. In *Handbook of biological confocal microscopy*, 2nd edition (ed. J.B. Pawley), pp. 267–279. Plenum Press, New York.

Vale R.D., Funatsu T., Pierce D.W., Romberg L., Harada Y., and Yanagida T. 1996. Direct observation of single kinesin molecules moving along microtubules. *Nature* **380:** 451–453.

Vardi Y., Shepp L.A., and Kaufman L. 1985. A statistical model for positron emission tomography. *J. Am. Stat. Assoc.* **80:** 8–20.

Waggoner A.S., DeBiasio R., Conrad P., Bright G.R., Ernst L., Ryan K., Nederlof M., and Taylor D. 1989. Multiple spectral parameter imaging. *Methods Cell Biol.* **30:** 449–478.

Wang S.-I. and Frieden B.R. 1990. Effects of third-order spherical aberration on the 3-D incoherent optical transfer function. *Appl. Opt.* **29:** 2424–2432.

Waterman-Storer C.M. and Salmon E.D. 1998. How microtubules get fluorescent speckles. *Biophys. J.* **75:** 2059–2069.

Waterman-Storer C.M., Desai A., Bulinski J.C., and Salmon E.D. 1998. Fluorescent speckle microscopy: A method to visualize the dynamics of protein assemblies in living cells. *Curr. Biol.* **8:** 1227–1230.

Weiss S. 1999. Fluorescence spectroscopy of single biomolecules. *Science* **283:** 1676–1683.

White J.G., Amos W.B., and Fordham M. 1987. An evaluation of confocal versus conventional imaging of biological structures by fluorescence light microscopy. *J. Cell Biol.* **105:** 41–48.

INTERNET SITES

G.W. Ellis fiber optic light scrambler (TM):
Technical Video Ltd., P.O. Box 693, Woods Hole, Massachusetts 02543
http://www.technicalvideo.com

Video Microscopy Solutions-Digital Microscopy Internet Magazine
Ted Inoue
http://www.videomicroscopy.com/

Public-access algorithms for deconvolution (XCOSM) developed through the NIH Resource in Biomedical Computing, Institute for Biomedical Computing, Washington University:
http://www.essrl.wustl.edu/~preza/xcosm/ and http://www.omrfcosm.omrf.org

Commercial sites for deconvolution:
http://aqi1.aqi.com/ (AutoQuant Imaging Inc.)
http://www.api.com/dvsystems.html (Applied Precision, Inc.)
http://www.lsr.co.uk/ (Life Science Resources Ltd.)
http://www.scanalytics.com/Catalog/deconvolution2.html (Scanalytics, Incorporated)
http://www.svi.nl/ (Scientific Volume Imaging b.v.)
http://www.vaytek.com/decon.htm (VayTek, Inc.)
http://www.zeiss.de/axiovision (Carl Zeiss GmbH)

CHAPTER 96

Practical Limits to Resolution in Fluorescence Light Microscopy

Ernst H. K. Stelzer

Light Microscopy Group, Cell Biology and Biophysics Programme, European Molecular Biology Laboratory, Heidelberg, Germany

In a perfect optical system, the numerical aperture (NA) and the wavelength (λ) determine the resolution. In a real fluorescence microscope, the number of photons collected from a specimen determines the contrast and, thus, the resolution. The number of picture elements per resolvable unit area, the number of photons, and the aberrations affect the contrast (Castleman 1993). The theory of confocal fluorescence microscopy claims that the resolution in all three dimensions is improved over that of wide-field fluorescence microscopes. This implies that laser scanners should be used instead of video cameras and that the improved lateral resolution can be used to visualize otherwise unidentifiable features. Such claims are backed up by theoretical comparisons of confocal and wide-field microscopy. In transmission and reflection contrasts, the resolution in terms of the cutoff frequency is improved by a factor of two (Sheppard and Choudhury 1977), whereas the full width at half-maximum of a point object is improved by a factor $1/\sqrt{2} \approx 1/1.4 \approx 0.7$ (Brakenhoff et al. 1979). On the other hand, every user of a confocal fluorescence microscope knows that faint features in an object which provides a low signal are not easily distinguished. In fact, it is common experience that human visual perception becomes much worse in the dark.

This chapter summarizes the influence of noise and the number of picture elements per unit area on the lateral resolution of wide-field and confocal fluorescence microscopes. The main purpose is to understand how an image of two closely spaced small objects is affected by noise. The chapter outlines how appropriate numbers can be estimated. It does not attempt to establish a strict mathematical framework. The emphasis is placed on understanding how resolution, contrast, dynamic range, and signal-to-noise ratio (SNR) are related and what affects them in fluorescence microscopy (Stelzer 1998).

AIRY DISKS DESCRIBE IMAGES OF POINT-LIKE LIGHT SOURCES

In microscopy, the spatial distribution of a point light source in the image plane is equivalent to a system response. It is referred to as the amplitude point-spread function (amplitude PSF) and is used to describe the properties of the optical components (Born and Wolf 1980). Although the distribution of the amplitudes cannot be seen directly, the intensity PSF (i.e., an image) can be directly visualized by placing a piece of paper into the optical path or by indirect observation through a camera (Fig. 1). The intensity PSF extends in all three dimensions. Because of the cylindrical symmetry of microscope

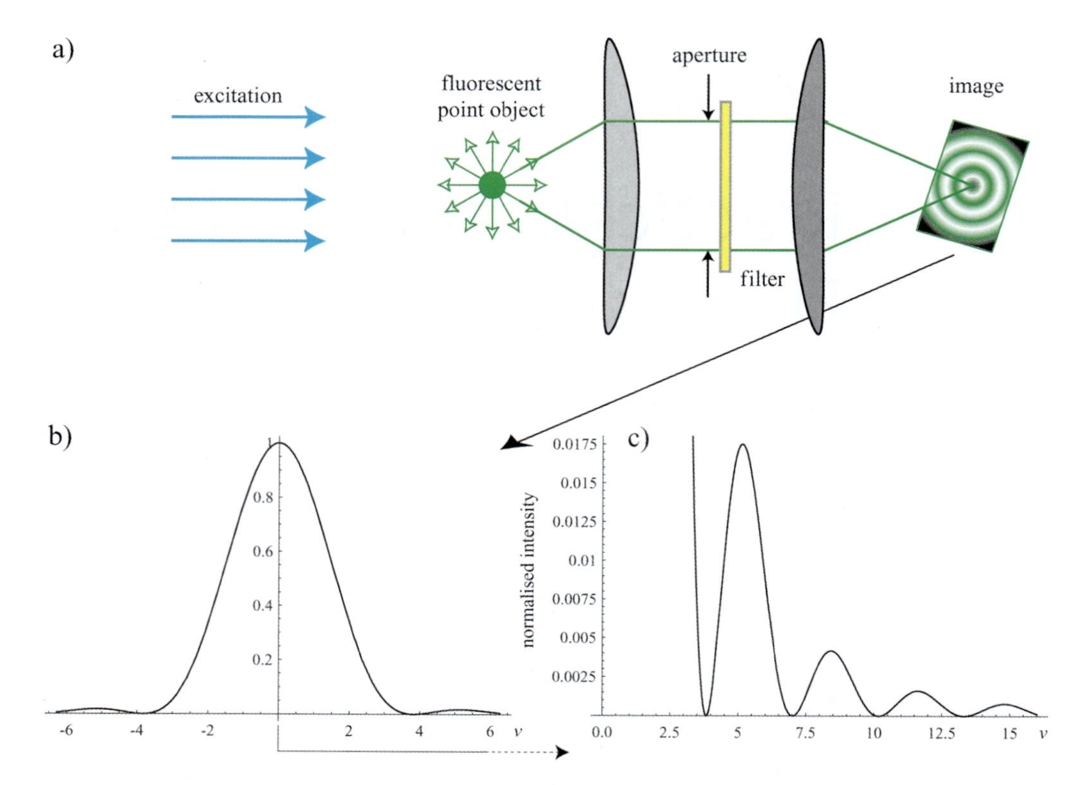

FIGURE 1. Airy pattern and Airy disk. (*a*) In a microscope, a beam of light excites the fluorophores of a point-like object. The fluorescence emission occurs with an even probability in all directions. A small fraction of the emitted light is collected by a set of lenses, spectrally filtered, and focused into an image plane where it appears as an Airy disk. (*b*) The intensity distribution of a point-like light source in the focal plane of a perfect optical system is described by the Airy pattern. The variable v is the distance from the optical axis normalized by the wavelength and the NA. (*c*) The Airy pattern has a main maximum and higher-order maxima with positions in between where the intensity becomes zero.

objective lenses, the two lateral components can be regarded as equal. The rotational symmetric Airy pattern describes the intensity distribution in the plane of focus as a function of the distance from the optical axis (Hopkins 1943).

The lateral Airy pattern tells us nothing about the component of the PSF along the optical axis. For reasons of simplicity, we purposely ignore the axial component and restrict all analyses to influences on the lateral resolution.

CONTRAST AND RESOLUTION ARE RELATED TERMS

In fluorescence microscopy, the imaging process is described using intensity PSFs. The resultant image is a sum of Airy disks (Fig. 2). Let us assume we form an image of two point objects. If the two objects are very far apart in object space, the images are very far apart, too, and easily separable. If the two objects are very close, their individual images overlap and the combined image may appear as a single image resulting from a brighter and/or larger single object. In general, the two images will overlap to a certain extent and will consist of two peaks with a gap between. The deeper the gap, the easier it is to distinguish, i.e., to resolve, the two objects in the image.

To quantify how well the two objects are separable, i.e., how well they are resolved, we introduce the term "contrast." It is defined as the difference between the lowest intensity found between the images of the two point objects and the highest intensity in the images. Contrast can be calculated and plotted as a function of the center–center distance s between two objects (Fig. 3). Because the height

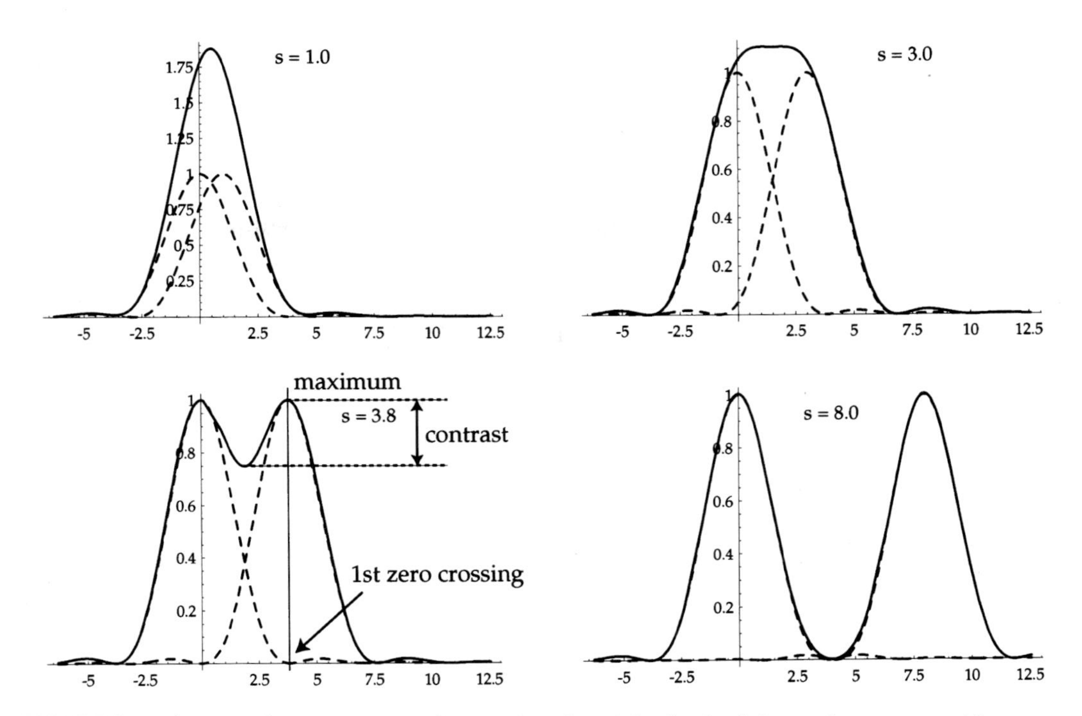

FIGURE 2. Definition of contrast versus distance functions. The depth of the gap between two Airy patterns depends on their center–center distance or separation s. If the two Airy patterns are very close to each other, the gap does not exist ($s = 1$ and $s = 3$). As the distance increases, the depth of the gap increases and hence the contrast improves. Note that the scale is different in the upper two graphs. In the case $s = 3.8$ (the Rayleigh criterion), the maximum of the right Airy pattern overlaps with the first zero crossing of the left Airy pattern.

of a single Airy disk is one, and the minimum between two Airy disks is at best zero, the highest achievable contrast is one. This is achieved only when the distance between the two objects is large. If two objects come closer to each other, the contrast in the image decreases. At a certain distance, the contrast becomes zero. The two maxima are no longer distinguishable, and hence a contrast ceases to exist.

The distance s, at which the contrast becomes zero, is referred to as the contrast cutoff distance and defines the smallest distance between two objects that can be resolved in the image. The relationship between contrast and distance is called the contrast-versus-distance function, which is similar in concept to a contrast transfer function.

How is resolution defined? Resolution is the distance of two objects at which their image provides a certain contrast. The ambiguity associated with the term resolution stems from the fact that several definitions make perfect sense. Some examples include:

- The cutoff distance is the distance beyond which two objects cannot be distinguished. The Sparrow criterion defines the resolution of an optical system through its cutoff distance.
- The Rayleigh criterion uses the distance at which the contrast is 26.4%. At this distance, the maximum of one Airy pattern coincides with the first minimum of the other Airy pattern.

Any contrast between 0% and 99.99% can be used to define a resolution. No definition is fundamentally better. The important message of this discussion is that it does not make any sense to discuss contrast and resolution as if they were independent.

SAMPLING REDUCES THE CONTRAST

Until now, we assumed that the Airy pattern was smooth and described by an infinite number of samples. In a real optical system, however, we use a camera or a laser scanner to generate a finite number of picture elements (pixels); i.e., we sample the image (Fig. 4). Each pixel summarizes the response of

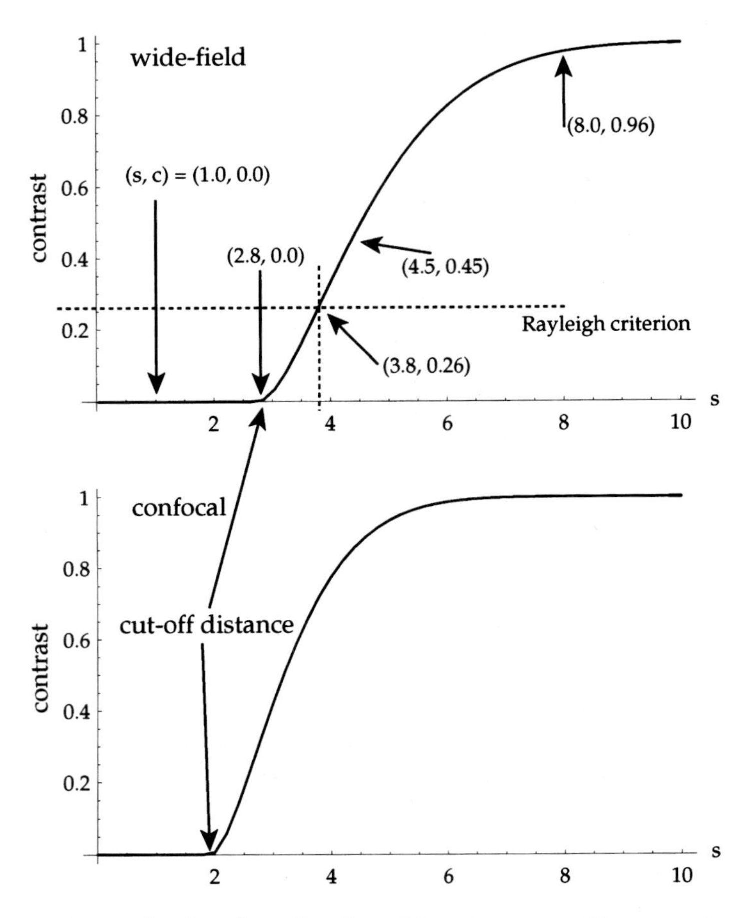

FIGURE 3. The contrast c can be plotted as a function of the center–center distance s of two Airy disks. Only beyond a certain distance does the contrast start to exist. The upper figure refers to wide-field fluorescence microscopy and the lower figure refers to confocal fluorescence microscopy. The cutoff distance and the slope depend on the width of the Airy disk. The cutoff distance in a wide-field microscope is larger than in a confocal microscope. The traditional way is to plot contrast over the inverse of the distance. This measure describes how many objects are present per unit length and is called spatial frequency. The smaller the width of the Airy disk, the higher is the cutoff frequency. The dashed lines indicate the Rayleigh criterion.

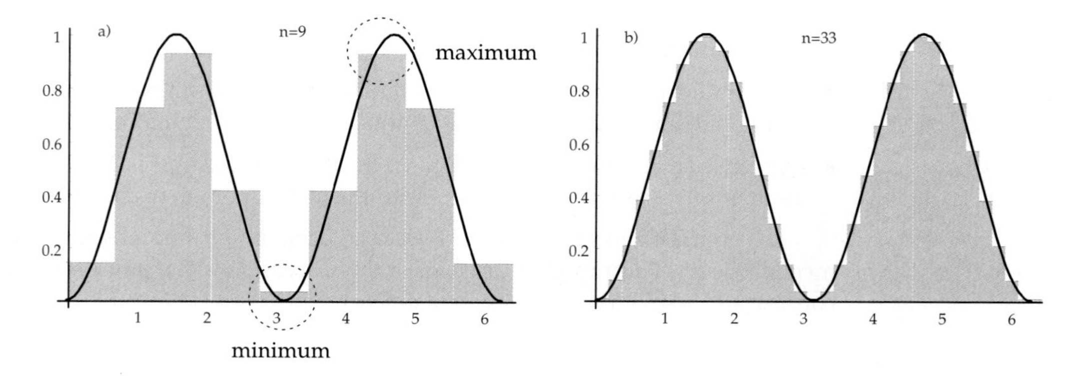

FIGURE 4. Influence of sampling on contrast. The finite number of pixels in a camera or a scanner breaks the continuous function down into a series of intensity values. Each of these values represents the intensity integrated over a small area. This procedure causes maxima to become smaller and minima to become larger. The main effects are a decrease of the contrast and an ambiguity concerning the positions of minima and maxima.

TABLE 1. Contrast ranges in percent achievable for various numbers of picture elements per Airy pattern, per Airy disk, and per Airy object consisting of two Airy disks at the Rayleigh distance (s=3.8)

Pixels per Airy pattern	Disk	Object	Contrast (%)
64	4096	5888	26.4–26.2
32	1024	1536	26.2–25.8
16	256	384	25.8–24.1
8	64	96	24.0–17.6
4	16	24	17.0–0.0
2	4	6	0.0
1	1	2	0.0

the optical system in a certain area. When looking at the Airy pattern, note that the zero crossings occur at points. However, because the pixels have a finite extent, one never looks at points, but always at areas. The areas integrate the intensity. Therefore, it is impossible to detect zero intensity. Since this effect increases the minimum value as well as it decreases the maximum value, the consequence is that the sampling process reduces the contrast. The cutoff distance is increased, and consequently the resolution (independent of the definition) is decreased. Thus, whenever one uses a camera in an optical system, one will never be able to achieve the theoretical maximal resolution.

The effect of sampling can be estimated by relating the extent of a pixel to the diameter of the Airy disk. The question is, How many pixels are required to cover the area below two adjacent Airy disks to achieve a certain contrast? The more nonoverlapping pixels are used, the smaller becomes the effect due to sampling (see Table 1). Analysis shows that four samples per Airy pattern along the main axis of two objects, which under ideal conditions provide a contrast of 26.4%, result in contrasts between 0% and 17%. At least eight samples per Airy pattern are required to guarantee a contrast of at least 17.6% in an Airy object under all circumstances. Using a water-immersion objective lens with a NA of 1.2 at a wavelength of 488 nm and assuming the typical field to consist of 512 samples per axis, the pixel size, i.e., the pixel–pixel distance has to be on the order of 60 nm.

$$1.22 \bullet \frac{488 \text{ nm}}{1.2} \Big/ 8 = 496 \text{ nm} \Big/ 8 = 62 \text{ nm}$$

The horizontal field size should be at most 32 μm.

$$512 \frac{\text{samples}}{\text{line}} \bullet \frac{62 \text{ nm}}{\text{sample}} = 31.8 \frac{\mu\text{m}}{\text{line}}$$

Otherwise, objects that are about (496 nm/2 =) 248 nm apart have a contrast that is less than 17.6%. A camera that observes a field size of 60 μm with 512 samples per line or less will already be blind to such small object distances.

NOISE REDUCES THE CONTRAST

Whatever intensity value is measured, it is only an approximation of the number of photons that have been emitted or scattered by an object. The variation of the signal during repeated observations is called noise. Noise induces an uncertainty in the quantification (see, e.g., Carlsson 1991) of intensity and, hence, an uncertainty in the contrast that has been measured. Because an error has to be accepted in the estimation of the intensity, it must be accepted that the contrast is in general underestimated. However, every reduction of the contrast causes an increase of the cutoff distance and, hence, a decrease of the resolution.

It is obvious that the effects of noise and sampling have to be combined (Fig. 5) and will always increase the cutoff distance. The effects of sampling can be estimated, but it is not as easy to estimate the noise in a single pixel. In transmission or reflection contrasts, the number of photons will be very

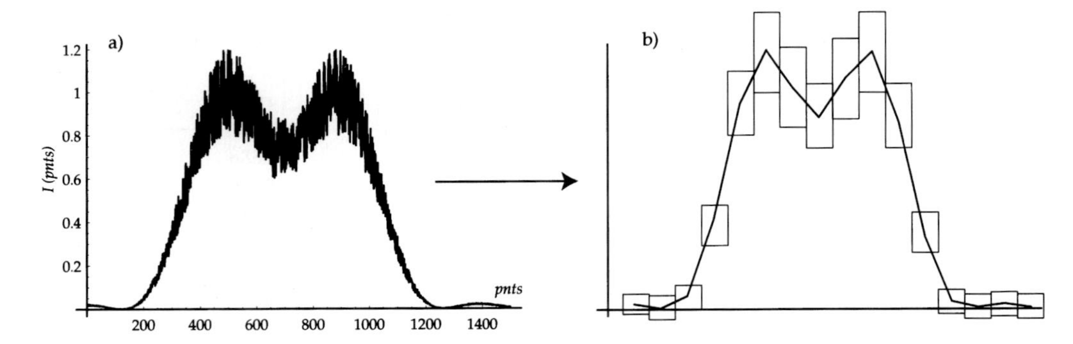

FIGURE 5. Influence of noise on contrast. (*a*) About 20% noise has been added to the image of two point-like objects at the Rayleigh distance. (*b*) Twelve points are used to represent the sum (i.e., eight points per Airy pattern). The heights of the boxes indicate the uncertainties associated with each intensity value in the Airy pattern.

high and the most likely sources of noise will be due to the electronic equipment and variations in the illumination intensity. In fluorescence microscopy, the number of fluorescence photons is very small and Poisson noise will be dominating.

DYNAMIC RANGE INFLUENCES THE CUTOFF FREQUENCY

Once the noise relative to the signal has been estimated, one can calculate the number of distinguishable gray levels; i.e., the dynamic range in a pixel or in an image. The dynamic range is essentially the ratio of the signal over the noise and, in the case of a fluorescence microscope, the square root of the number of photons.

$$\text{Dynamic range (in a fluorescence microscope)} = \frac{signal}{noise} = \frac{signal}{\sqrt{signal}} = \sqrt{signal}$$

The noise provides us with a lower limit below which different signals cannot be distinguished. Coming back to the contrast-versus-distance functions defined earlier, their cutoff distances were determined by assuming an infinitely high SNR and, hence, an infinitely high dynamic range. The cutoff is the distance at which the function crosses the zero contrast line. In the simplest case, noise is

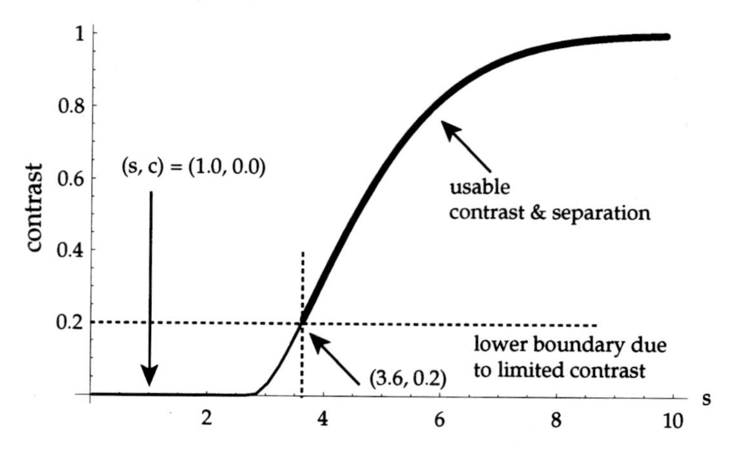

FIGURE 6. Influence of noise on cutoff distance. Noise is introduced by imposing a lower limit to the contrast. The simplest case is a constant distance-independent noise level. A noise level of 20% is assumed. This reduces the usable portion of the contrast versus distance function. Objects whose separation is smaller than the cutoff due to the lower boundary cannot be distinguished.

introduced by raising this level (Fig. 6). Only those distances that provide at least a certain contrast can be distinguished. In estimating the resolution in a noise-limited situation, this level must be used as a reference. Obviously, the cutoff distance is directly affected. It should be noted that the digital number which describes the intensity of a pixel is only proportional to the number of photons but not identical to it. The microscope manufacturer should be able to provide a function that calculates it.

IMPROVING THE RESOLUTION INCREASES NOISE

Improving the resolution means that the observed area or volume is decreased (Jonkman et al. 2003). If object and object preparation procedures remain the same, this means that the number of observable fluorophores and, in consequence, the number of collectable photons, is decreased. Because the dynamic range ultimately depends on the number of photons that can be collected from a given number of fluorescent molecules, any improvement that decreases the observable region has to increase the noise. This imposes a limit to any technique that intends to improve resolution by decreasing the size of a PSF. If, for example, the observable volume is decreased by a factor of two, the signal is reduced by at least a factor of two and the observation period has to be increased by a factor of four simply to maintain the SNR.

REFERENCES

Born M. and Wolf E. 1980. Elements of the theory of diffraction. In *Principles of optics*, pp. 370–458. Pergamon Press, Oxford, United Kingdom.

Brakenhoff G.J., Blom P., and Barends P. 1979. Confocal scanning light microscopy with high aperture immersion lenses. *J. Microsc.* **117:** 219–232.

Carlsson K. 1991. The influence of specimen refractive index, detector signal, integration and non-uniform scan speed on the imaging properties in confocal microscopy. *J. Microsc.* **163:** 167–178.

Castleman K.R. 1993. Resolution and sampling requirements for digital image processing, analysis and display. In *Electronic light microscopy* (ed. D. Shotton), pp. 71–93. Wiley-Liss, New York.

Hopkins H.H. 1943. The Airy disk formula for systems of high relative aperture. *Proc. Phys. Soc.* **55:** 116–120.

Jonkman J.E.N., Swoger J., Kress H., Rohrbach A., and Stelzer E.H.K. 2003. Resolution in optical microscopy. *Methods Enzymol.* **360:** 416–446.

Sheppard C.J.R. and Choudhury A. 1977. Image formation in the scanning microscope. *Opt. Acta* **24:** 1051–1073.

Stelzer E.H.K. 1998. Contrast, resolution, pixelation, dynamic range and signal-to-noise ratio. *J. Microsc.* **189:** 15–24.

Lasers for Multiphoton Microscopy

Frank W. Wise

Department of Applied Physics, Cornell University, Ithaca, New York 14853

Multiphoton microscopy (Denk et al. 1990) has been shown to be a powerful new technique for the visualization and quantitative investigation of dynamic cell processes. Microanalytical chemistry and micropharmacology also become possible with nonlinear optical excitation. Despite these advanced capabilities, further instrumentation development is needed before these techniques find widespread acceptance. The proliferation of multiphoton microscopies is currently limited, partially by a lack of suitable illumination sources. Extremely stable, user-friendly sources of femtosecond-duration optical pulses are required, with the output wavelength tunable in the visible or near-infrared (IR) regions of the spectrum.

Fortunately, a revolution in short-pulse sources also occurred in 1990, with the discovery of Kerr-lens mode locking in titanium-doped sapphire (Ti:sapphire; Spence et al. 1991). Mode-locked Ti:sapphire lasers are extremely stable, solid-state sources of high-power 100-fsec pulses at wavelengths in the near IR, and commercial versions have been available since 1991. With a Ti:sapphire source, it is possible to construct microscope workstations that can be operated by reasonably knowledgeable users.

This chapter discusses the features of existing laser sources used in multiphoton microscopy and considerations for developing lasers for use with a broader spectrum of fluorescent probes. A brief introduction to lasers is given below.

INTRODUCTION TO LASERS: HOW THEY WORK

As indicated in Figure 1, there are two key elements of a laser: the gain medium and the optical resonator. The gain medium is an amplifier for light, and the optical resonator (two mirrors in the simplest case) feeds the light back into the gain medium for continued amplification of the laser beam.

Ordinarily, light is not amplified when it passes through matter; it is either transmitted with no gain or loss, or absorbed. This is a property of matter in thermal equilibrium and is the reason that optical amplifiers do not occur in nature. If the energy of an incident light photon (a packet of electromagnetic energy) matches the difference between allowed energy levels of an atom, the light is absorbed because more atoms occupy the lower of the two energy levels than the higher level. For amplification to occur, the population of the upper level must be greater than that of the lower level. This never occurs in equilibrium and is referred to as a population inversion. An inversion can be generated by exciting some of the atoms from their ground state up to an excited state. This is referred to as the pumping process and obviously requires energy.

If the excited state has a reasonably long lifetime, the atoms accumulate in that state while it is being pumped, thus creating an inversion. Once an inversion is created in an optical resonator, laser

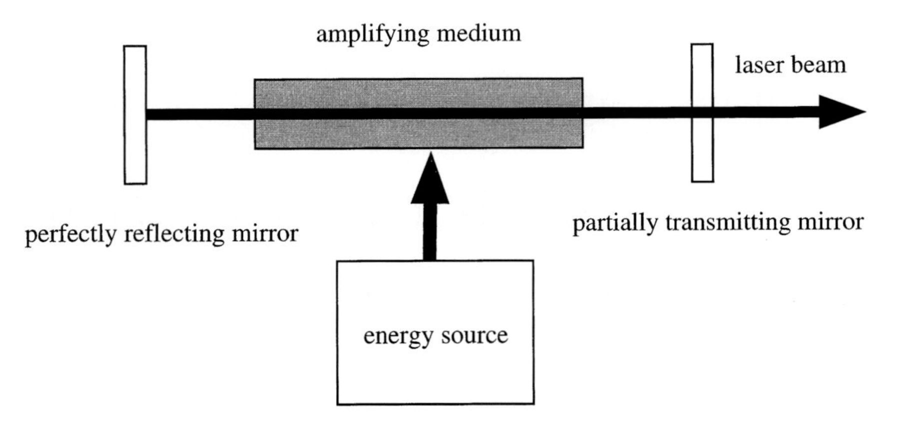

FIGURE 1. Principle of operation for a simple laser.

action starts. A single photon of light propagating along the axis of the laser is amplified by the process of stimulated emission—the presence of that photon in the gain medium stimulates an atom in the excited state to make a transition to the lower state. At that point, there are two photons (the original one and the stimulated one), so the light has been amplified. The initially weak light is reflected from the mirrors of the resonator so that it passes repeatedly through the gain medium, getting amplified each time. Eventually, the amplification must decrease (referred to as saturation) because a finite amount of energy is available to create the population inversion. Thus, the beam builds up to a steady high intensity. One of the mirrors of the resonator (the output coupler) allows a fraction of the light to leak out, and that produces the output beam.

Because the emission wavelength of an amplifying medium depends on its energy levels, each material tends to produce a unique wavelength. Thus, lasers are specialized tools, with each one limited to a narrow range of emission wavelengths. Lasers have been made of thousands of different materials, in solid, liquid, and gas phases. However, only a small fraction of these have found common use. These range from semiconductor lasers with dimensions much less than 1 mm and producing approximately 1 mW of power, to gas and solid-state lasers producing thousands of watts.

FEMTOSECOND-PULSE LASERS

The operation of a continuous-wave laser, which produces a steady output beam of constant power, is described above. It is often more useful to produce the output power in the form of pulses, because a pulse can produce very high instantaneous power. In particular, ultrashort pulses allow us to expose atoms, molecules, and materials to high instantaneous power (thousands [kW] to millions [MW] of watts) while the average power incident on the sample is low, only milliwatts. Another benefit of short pulses is that they can be used to initiate and record dynamic processes as fast as the time scale of the pulse.

Pulse generation in the most common femtosecond lasers can be qualitatively understood in terms of a simple picture. Imagine that we take a continuous-wave laser and replace the high-reflecting mirror with a special "nonlinear" mirror that has a higher reflectance for intense light than for weak light. Weak light is absorbed by the mirror, and so never builds up in the laser cavity. A brief fluctuation of intense light will be preferentially reflected and will be amplified many times. As the pulse builds in intensity, the peak of the pulse is reflected more than the low-intensity wings. The nonlinear mirror thus sharpens and narrows the pulse with each reflection. A nanosecond-duration fluctuation can evolve to a 100-fsec pulse this way.

One particular implementation of the nonlinear mirror is a semiconductor saturable-absorber mirror or SESAM. Kerr-lens mode-locked (KLM) lasers, which include Ti:sapphire lasers and dominate the commercial femtosecond-laser market, do not have a nonlinear mirror. However, these lasers do have an effective saturable absorber that works to promote pulse formation in the manner described above.

Interested readers should consult one of the many books on lasers for further information; a well-written textbook on lasers is Silvfast (1996). For facts about many kinds of lasers, see Hecht (1992).

PERFORMANCE CHARACTERISTICS OF THE LASER SOURCE

Table 1 summarizes the features of lasers used in multiphoton microscopy. The great majority of publications in multiphoton microscopy to date have used Ti:sapphire lasers. A significant exception is the work of White and coworkers, performed with an additive-pulse mode-locked Nd:YLF laser (Wokosin et al. 1996). This laser generates 1-psec pulses at a wavelength of 1.05 μm, which are compressed to approximately 150 fsec, external to the laser. Although the availability of Ti:sapphire lasers has been a crucial factor in the development of multiphoton microscopy, it does not completely solve the illumination problem. The use of Ti:sapphire lasers is ultimately limited by several factors. First, Ti:sapphire lasers are tunable from 700 to 1000 nm, and so provide two-photon illumination for dyes with single-photon absorption between approximately 350 and 500 nm (in fact, it is known that the two-photon cross sections of many dyes peak at wavelengths somewhat shorter than twice that of the linear absorption peak [Xu and Webb 1996]). New sources will be needed to excite the numerous useful chromophores with linear absorptions below 350 nm and above 500 nm. Second, a Ti:sapphire laser is pumped by an argon-ion laser, which is large, inefficient, and very expensive to purchase, operate, and maintain. Widespread use of multiphoton excitation is unlikely if the cost of the illumination source is a significant fraction of the cost of the microscope, as is the case with current Ti:sapphire laser systems. Finally, although commercial Ti:sapphire lasers are stable enough and acceptably user-friendly for specialized research laboratories, turnkey instruments will be needed if multiphoton microscopy is to become a routine tool.

It is straightforward to list the critical performance characteristics of a source for multiphoton-excitation laser-scanning microscopy. The following capabilities and characteristics are desired:

- The pulse duration should be approximately 100 fsec. Longer pulses (up to 10 psec) may be useful for two-photon microscopy, although with a reduction in performance. Excitation with three or more photons places a premium on peak power and, thus, minimum pulse duration.
- The pulse repetition rate should be between 100 and 800 MHz for microscopy, to achieve reasonable scan rates. The upper limit of 800 MHz is determined by typical chromophore lifetimes of a few nanoseconds. The repetition rate could be reduced to approximately 1 MHz in nonimaging applications, e.g., uncaging of bioeffector molecules.
- The peak power of each pulse must be approximately 10 kW to achieve good signal-to-noise ratios (SNRs) at fast scan rates. This implies an emitted-pulse energy of 1 nJ for a 100-fsec pulse and an average power of 100 mW for a 100-MHz repetition rate.
- Ideally, the wavelength should be tunable from 500 to 1000 nm. It is highly unlikely that a single source will be able to cover this entire range. On the other hand, many applications require only one or two loosely specified wavelengths. Because Ti:sapphire lasers cover the range from 700 to 1000 nm, new capabilities will be provided by sources emitting short pulses between 500 and 700 nm.
- Fluctuations in the power must be low (<1%), and the source must supply a high-quality Gaussian beam with good pointing stability.
- The source must be compact and robust and require little maintenance. Together with the need for low fluctuations, these imply an all-solid-state design. Fiber coupling of the source to the microscope would eliminate the need for some alignment of the source. Turnkey operation is important for most groups, and low-utility consumption would be an advantage in some installations.
- For maximum impact, the source should be commercially available, ideally at a cost of less than $50,000. A price much higher than this will put the source beyond the reach of many research groups.

No existing laser technology meets the needs of multiphoton microscopy as listed here.

TABLE 1. Laser sources for multiphoton microscopy

Laser	Full name	Mode-locking method[a]	Pump laser	Average power	Spectrum range (nm)	Pulse duration	Pulse repetition rate (MHz)
Ti:sapphire	titanium:sapphire	KLM	argon-ion laser or frequency-doubled Nd:YVO_4	1 W	700–1000	100 fsec	100
Nd:YLF	neodymium:yttrium lithium fluoride	APM	800-nm diode laser	1 W	1050	1 psec	100
Nd:YVO_4	neodymium:yttrium vanadate	not mode-locked	800-nm diodes	10 W	1050		
Cr:LiSAF, Cr:LiSGAF	chromium:lithium strontium aluminum fluoride, chromium:lithium strontium gallium fluoride	KLM, SESAM	670-nm diodes	100 mW	800–900	100 fsec	100
Cr:forsterite	chromium:forsterite	KLM	Nd:YAG laser, Nd: fiber laser, Nd:YVO_4 laser, Yb:fiber laser	500 mW	1200–1300	100 fsec	100
Nd:glass	neodymium:glass	SESAM	800-nm diodes	100 mW	1050	100 fsec	100
Yb:glass	ytterbium:glass	SESAM	915-nm diodes	100 mW	1000–1100	500 fsec	100

[a]KLM refers to Kerr-lens mode-locking, APM refers to additive-pulse mode-locking, SESAM refers to a semiconductor saturable-absorber mirror.

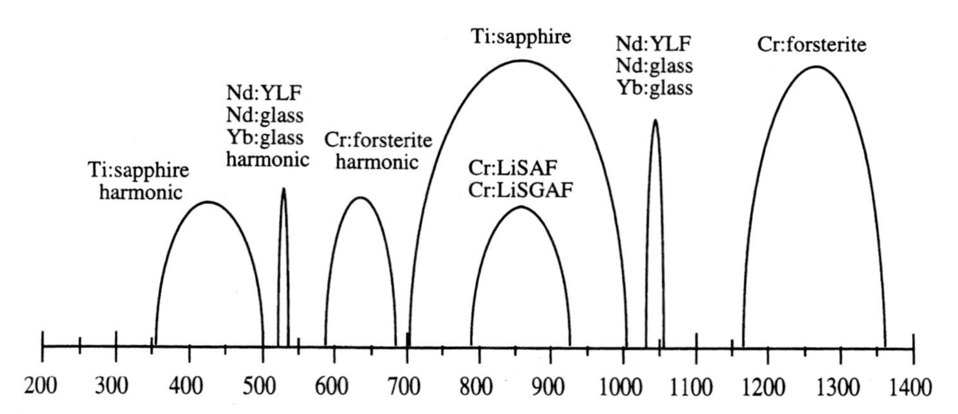

FIGURE 2. Wavelengths available from mode-locked lasers and their second harmonics. The heights of the different arcs roughly indicate the power available at the corresponding wavelengths.

EXISTING LASER SOURCES

Ti:Sapphire Lasers

The ion-laser-pumped Ti:sapphire laser still offers the greatest performance in terms of power and wavelength tunability. Femtosecond Ti:sapphire lasers are available from Spectra-Physics; Coherent; and Clark/MXR. This is likely to be the choice of research laboratories desiring the greatest flexibility for at least the next few years. The Ti:sapphire can also be pumped by the second-harmonic frequency of a diode-pumped Nd:YVO_4 laser. A laser based on this approach and designed for industrial applications is just becoming available commercially from Coherent, Laser Group. The hardened design sacrifices some performance: The laser emits a single wavelength about 800 nm, and the average power is approximately 300 mW, compared to the 1–2 W obtained with ion-laser pumping. Second-harmonic generation extends the range of a high-power Ti:sapphire laser to include wavelengths between 350 and 500 nm.

Cr:LiSAF Lasers

Much interest has accompanied the development of femtosecond-pulse lasers pumped directly by laser diodes. The first diode-pumped 100-fsec lasers were based on Cr:LiSAF (Kopf et al. 1994) and related materials such as Cr:LiSGAF (Yanovsky et al. 1995). These lasers also emit in the range 800–900 nm, but do not offer performance comparable to that of the Ti:sapphire owing to the limited pump power available from the diodes and inferior thermal properties of the laser materials. Average powers up to approximately 100 mW are obtained with 100-fsec pulses, and the wavelength tunability is restricted to approximately 50 nm. Because the performance is below that of the Ti:sapphire in all parameters, these lasers are only likely to find application where compactness, stability, reduced cost, and perhaps the lack of a cooling-water requirement are at a premium. Two-photon-excited images of the mitochondria of muscle cells stained with rhodamine 123 were produced using a diode-pumped Cr:LiSGAF laser, demonstrating the utility of these sources (R. Williams et al., unpubl.). Diode-pumped Cr:LiSAF lasers are commercially available from Time-Bandwidth Products.

Nd:YLF Lasers

For intense excitation at 1 μm, the Nd:YLF source developed by Ferguson and coworkers and referred to above is the basis of a commercial unit produced by Microlase Optical Systems. Mode locking of this laser is accomplished using an interferometric technique that requires feedback control of the resonator, and a pulse-compressor is needed to obtain femtosecond pulses. Thus, this is a complicated system, and it is reasonable to expect that it will be out-performed in the future by simpler and cheaper sources. Diode-pumped Nd:glass lasers directly produce 60-fsec pulses (Aus der Au et al. 1997), although the power is a factor of 10 lower than that of the compressed output from Nd:YLF. It seems to be only a matter of time before mode locking of diode-pumped Yb:glass lasers is accomplished, and this should produce average powers of hundreds of milliwatts. Commercial products should follow soon after.

LASER SOURCE DEVELOPMENT IN THE 500–700-NM SPECTRUM RANGE

Although the optimal wavelength for two-photon excitation of a specific chromophore is often still unknown, there are numerous instances where two-photon excitation is effective near twice the wavelength of the single-photon absorption peak. These imply that wavelengths between 500 and 700 nm, in the gap between the fundamental and second-harmonic wavelengths of Ti:sapphire, are important. For example, wavelengths between 250 and 300 nm are needed for (one-photon) excitation of intrinsic fluorophores, such as certain proteins, amino acids, and nucleic acids. Studies of tryptophans also employ wavelengths about 300 nm. Two-photon excitation of these fluorophores is accomplished with wavelengths between 500 and 600 nm. Other important applications are likely to require longer wavelengths. The two-photon absorption cross section of the calcium indicator indo-1 is increasing as the exciting wavelength reaches 700 nm, the shortest wavelength available from a Ti:sapphire laser (Xu 1996). NADH, DAPI, and Dansyl are other important fluorophores exhibiting two-photon cross sections that peak below 700 nm and, thus, out of the reach of the Ti:sapphire (Xu 1996). Finally, preliminary experiments indicate that the 500–700-nm band will be necessary for optimal cage photolysis (W. Zipfel et al., unpubl.). This is consistent with the observation that single-photon cage photolysis is optimized by excitation between 300 and 350 nm (McCray and Trentham 1989). The three-photon-excitation cross sections of tryptophan, serotonin, and dopamine were also found to be increasing at the short-wavelength limit of Ti:sapphire (Maiti et al. 1997).

These examples illustrate the likely importance of the 500- to 700-nm wavelength region in multiphoton microscopy. There are no known solid-state laser media that emit in this range and are capable of supporting femtosecond-duration pulses. Thus, it is currently not possible to design and construct a femtosecond-pulse solid-state laser in this spectral region. Even if laser-gain media could be identified, visible sources for optical pumping are quite limited and, in particular, high-power laser diodes do not exist below 670 nm. The most promising approach to the generation of short pulses at these wavelengths is to produce high-power femtosecond pulses at longer wavelengths in the IR (i.e., 1–2 μm), where solid-state laser media and pump sources exist. The intense IR pulses can then be converted efficiently to desired wavelengths in the visible region of the spectrum using nonlinear optics.

Lasers based on the Cr:forsterite crystal appear to be promising for this approach. The gain bandwidth of Cr:forsterite extends from 1150 to 1360 nm, so second-harmonic wavelengths between 575 and 680 nm are possible with this material. This is illustrated in Figure 2, along with the corresponding ranges for other laser crystals. Cr:forsterite has been mode-locked using the same mechanism operative in Ti:sapphire (Pang et al. 1993), and pulses as short as 25 fsec are produced (Yanovsky et al. 1993). A diode-pumped and frequency-doubled Cr:forsterite laser could therefore satisfy the requirements of a practical source for two-photon-excited laser-scanning microscopy.

Liu and coworkers demonstrated diode-pumping of a femtosecond Cr:forsterite laser using a double-clad Yb-doped fiber (Liu et al. 1998). This is a compact, stable source of the required IR pulses; the same workers have also shown that these pulses can be converted to wavelengths of about 620 nm with nearly 50% efficiency (Liu et al. 1997). It appears that Cr:forsterite will meet some of the need for a source in the 500–700-nm region. Yb:glass lasers will provide short pulses near 530 nm if efficient frequency-doubling of femtosecond pulses at 1 μm can be demonstrated.

ACKNOWLEDGMENTS

This work was supported by the National Institutes of Health under award RR-10075. The author thanks Professor W. Webb and the members of his research group for numerous stimulating discussions.

REFERENCES

Aus der Au J., Kopf D., Morier-Genoud F., and Keller U. 1997. 60-fs pulses from a diode-pumped Nd:glass laser. *Opt. Lett.* **22:** 307–309.

Denk W., Strickler J.H., and Webb W.W. 1990. Two-photon laser scanning fluorescence microscopy. *Science* **248:** 73–76.

Hecht J. 1992. *The laser guidebook*, 2nd edition. McGraw-Hill, Blue Ridge Summit, Pennsylvania.

Kopf D., Weingarten K., Brovelli L., Kamp M., and Keller U. 1994. Diode-pumped 100-fs passively modelocked Cr:LiSAF laser using an anti-resonant Fabry-Perot saturable absorber. *Opt. Lett.* **19:** 2143–2145.

Liu X., Qian L.J., and Wise F.W. 1997. Efficient generation of 50-fs red pulses by frequency-doubling in lithium triborate. *Opt. Commun.* **144:** 265–268.

Liu X., Qian L.J., Wise F.W., Zhang Z., Itatani T., Sugaya T., Nakagawa T., and Torizuka K. 1998. Femtosecond Cr:forsterite laser diode pumped by a double-clad fiber. *Opt. Lett.* **23:** 129–131.

Maiti S., Shear J.B., Williams R.M., Zipfel W.R., and Webb W.W. 1997. Measuring serotonin distribution in live cells with three-photon excitation. *Science* **275:** 530–532.

McCray J.A. and Trentham D.R. 1989. Properties and uses of photoreactive caged compounds. *Annu. Rev. Biophys. Chem.* **18:** 239–270.

Pang Y., Yanovsky V., Wise F., and Minkov B. 1993. Self-modelocked Cr:forsterite laser. *Opt. Lett.* **18:** 1168–1170.

Silvfast W.T. 1996. *Laser fundamentals*. Cambridge University Press, Cambridge, United Kingdom.

Spence D., Kean P., and Sibbett W. 1991. 60-femtosecond pulse generation from a self mode-locked Ti:sapphire laser. *Opt. Lett.* **16:** 42–44.

Wokosin D.L., Centonze V., White J.G., Armstrong D., Robertson G., and Ferguson A. 1996. All-solid-state lasers facilitate multiphoton excitation fluorescence imaging. *IEEE J. Select. Top. Quantum Electron* **2:** 1051–1065.

Xu C. 1996. "Multiphoton excitation of molecular fluorophores in nonlinear laser scanning microscopy." Ph.D. thesis, Cornell University, Ithaca, New York.

Xu C. and Webb W.W. 1996. Measurement of two-photon excitation cross-sections of molecular fluorophores with data from 690 to 1050 nm. *J. Opt. Soc. Am. B* **13:** 481–489.

Yanovsky V., Pang Y., and Wise F. 1993. Generation of 25-fs pulses from a Kerr-lens modelocked Cr:forsterite laser with optimized group-delay dispersion. *Opt. Lett.* **18:** 1541–1543.

Yanovsky V., Wise F.W., Cassanho A., and Jenssen H.P. 1995. Kerr-lens modelocked diode-pumped Cr:LiSGAF laser. *Opt. Lett.* **20:** 1304–1306.

CHAPTER 98

Acousto-optic Tunable Filters for Microscopy

Elliot S. Wachman

ChromoDynamics, Inc., Lakewood, New Jersey 08701

As the number of fluorescent probes for biological research has proliferated, a pressing need to improve the spectral versatility of fluorescence microscopes has arisen. At the same time, dramatic improvements in camera and computing technology have engendered increasing interest in the study of sub-millisecond timescale live-cell dynamics, an area that requires sub-millisecond light shuttering and wavelength-switching. The need for these two features—spectral versatility and speed—has led ChromoDynamics to develop a new instrument based on the technology of acousto-optic tunable filters (AOTFs), the AOTF microscope. This instrument includes both an AOTF-based illumination source and an AOTF-based imaging system.

SPECTRAL ILLUMINATION SOURCES

Illumination wavelength is most commonly controlled by an interference wheel placed in front of the light source. In this approach, wavelength and pass bands are fixed, and switching times are typically 50 msec. A monochromator-based source is an alternative technique, which, if properly designed, can provide 2–3-msec switching times between wavelengths. Monochromator systems have a fixed grating-dependent spectral resolution.

ChromoDynamics has developed an AOTF illumination system that can be fiber-coupled into either the trans- or epi-illumination port of a microscope. It can be tuned from 420 nm to 750 nm, and it provides four different bandwidth settings at each wavelength and switching times of less than 50 µsec.

SPECTRAL IMAGING

Spectral imaging technologies can be loosely divided into those that are band sequential (where the individual wavelength images are taken one after another) and those that are non-band sequential (where multiple "nonspectral" image data are acquired and then subsequently transformed to provide spectral image information). Examples of the former include interference filter wheels, liquid crystal tunable filters (LCTFs), and AOTFs; examples of the latter include Fourier transform imaging spectroscopy (FTIS) and tomographic imaging.

Interference filter wheels have fixed spectral parameters and relatively slow switching times, whereas LCTFs have broad tuning capabilities, fixed bandwidth, and similar 50-msec switching times. In FTIS, a Sagnac interferometer is used to acquire interferometric imaging data at a variety of interferometer settings, and the resulting data are Fourier-transformed to provide a spectral image set with upward of 50 bands, ranging between 400 nm and 700 nm. In tomographic imaging, light is bounced off a diffraction grating and the various multiple diffraction orders are separated and captured on a single CCD chip for subsequent processing and extraction of the contained spectral information. In both these techniques, individual wavelength ranges cannot be selected, spectral resolution cannot be changed, and variations in signal to noise across the tuning range cannot be compensated for (as is possible with band sequential techniques). Additionally, FTIS-typical data sets take at least 30 sec to acquire, whereas tomographic imaging has prodigious computational processing requirements.

Unlike these techniques, AOTFs have the capability for broad tunability together with variable bandwidth operation and sub-millisecond switching times. ChromoDynamics has developed an AOTF-based imaging system, which docks easily to the camera port of any fluorescence microscope. Image fidelity is excellent, and the device can be tuned from 465 nm to 800 nm. Bandwidths can be varied to one of eight different settings at every wavelength, and settings can be switched in ~50 μsec.

These AOTF devices, protected by two U.S. patents (Wachman et al. 1998a,b), have been used for experiments ranging from fluorescence and multispectral reflectance imaging to multispectral imaging of PAP smears. The following chapter briefly describes the operation of these devices and gives examples of some of their applications.

TECHNICAL BACKGROUND

Acousto-optic Tunable Filters

An acousto-optic crystal is one whose optical properties vary in the presence of an acoustic wave. In particular, when an (ultrasonic) acoustic wave propagates in such a material, the periodic regions of compression and rarefaction it sets up act as a diffraction grating for incident light. Such an acoustic wave may be efficiently generated by applying a radio-frequency (RF) electronic signal to an acoustic transducer bonded to the crystal surface. Varying the frequency of this RF signal modifies the frequency of the acoustic wave and the resulting grating spacing in the crystal, thereby causing different wavelengths of the incident light to be diffracted. By proper selection of the RF drive frequency, therefore, the crystal may be "tuned" to select a particular color among those incident on its input face; this light then exits the crystal at an angle distinct from that of the incident light. This is the principle of operation of an AOTF.

A single AOTF crystal can be tuned continuously over one octave in frequency anywhere in the ultraviolet to near-infrared regions of the spectrum. In addition, by using multiple RF inputs simultaneously, the bandwidth of the filtered light can be broadened or even custom-tailored to the sample being studied. Finally, changing the amplitude of the RF input(s) alters the intensity of the AOTF-filtered light. Since the RF field inside the crystal can be changed in tens of microseconds, the time it takes the acoustic wave to travel across a typical crystal, an acousto-optic crystal can serve as a fast light-shutter as well as a variable-band-pass, variable-throughput tunable filter.

Why then doesn't everyone use acousto-optic filters on his microscope? The answer is threefold. First, typical interference filters used for fluorescence microscopy have an out-of-band rejection of 10^{-6} relative to their peak transmission. For AOTFs this figure is ordinarily around 10^{-2}. Second, although the efficiency of an AOTF for polarized laser light can be nearly 100%, for broadband, incoherent, unpolarized light, as is typical for both fluorescence excitation and imaging, the throughput of an AOTF can be much lower than that of a conventional filter. This is due both to the relatively narrow natural bandwidths of illumination and imaging AOTFs relative to standard interference filters, and to the polarization-selectivity of the acousto-optic process (the *s*- and *p*-polarized light filtered by the AOTF exit the crystal in different directions, and only one of these is ordinarily used). Finally, AOTFs used for imaging applications have historically had quite poor image quality. Prior to our work, the best spatial resolution achieved through an imaging AOTF was 1.5–2.0 μm. This is clearly inadequate for many applications.

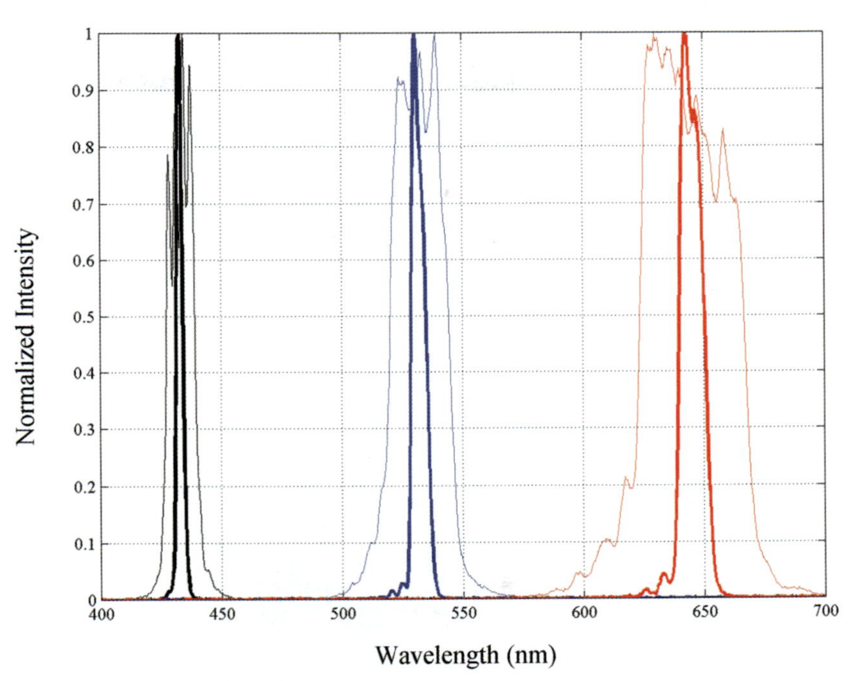

FIGURE 1. Output of the AOTF illumination system at various wavelengths. Thicker curves represent lowest bandwidth settings, thinner curves represent highest bandwidth settings.

Technological Improvements

The difficulties experienced in the development of the ChromoDynamics illumination and imaging systems have been addressed using a variety of techniques.

Illumination System

The ChromoDynamics system uses a 300-W short-arc xenon lamp with two AOTF crystals, placed back-to-back to provide a cumulative out-of-band rejection of 10^{-4}:1. (For comparison, it should be noted that monochromators typically provide 10^{-3}:1 rejection.) This setup also allows combination of both polarizations of the AOTF-filtered output. The crystals used have a natural bandwidth of ~8 nm in the center of their tuning range, so that when used in conjunction with a multiple RF driver, variable bandwidths are produced that can be as broad as those usually used for fluorescence excitation. The result is output powers between 0.5 mW and 1.0 mW. With further improvements in the optical design, this figure should increase significantly. Representative spectra produced by the AOTF illumination system are shown in Figure 1.

Imaging System

For imaging, the use of two AOTF crystals is not practical. Instead, an apodized transducer is used to provide improved out-of-band rejection. This technique allows the amplitude of the acoustic field to be shaped inside the crystal and, due to the physics of the acousto-optic process, leads to a corresponding shaping of the AOTF spectral output. The result is a rejection ratio of 10^{-3}:1 away from peak, about an order of magnitude lower than is possible without apodization. This decrease in side-lobe magnitude is critical for obtaining high-quality images, and provides rejection adequate for bright-field and some fluorescence work. As with nearly all other spectral technologies, however, the rejection of the AOTF, even with this improvement, is not sufficient for most fluorescence work, and a supplementary long-pass filter is frequently used to help remove residual excitation light. A typical imaging AOTF output spectrum is shown in Figure 2.

a

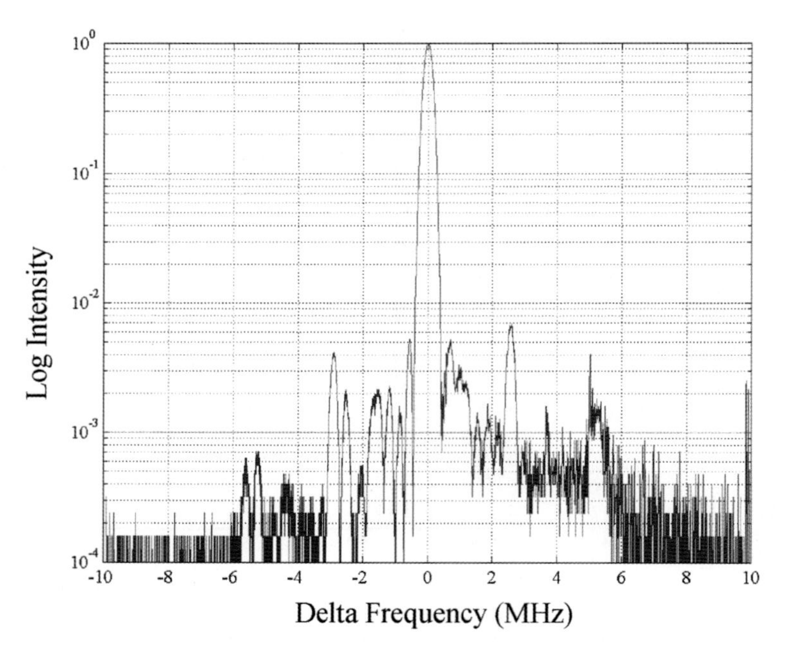

b

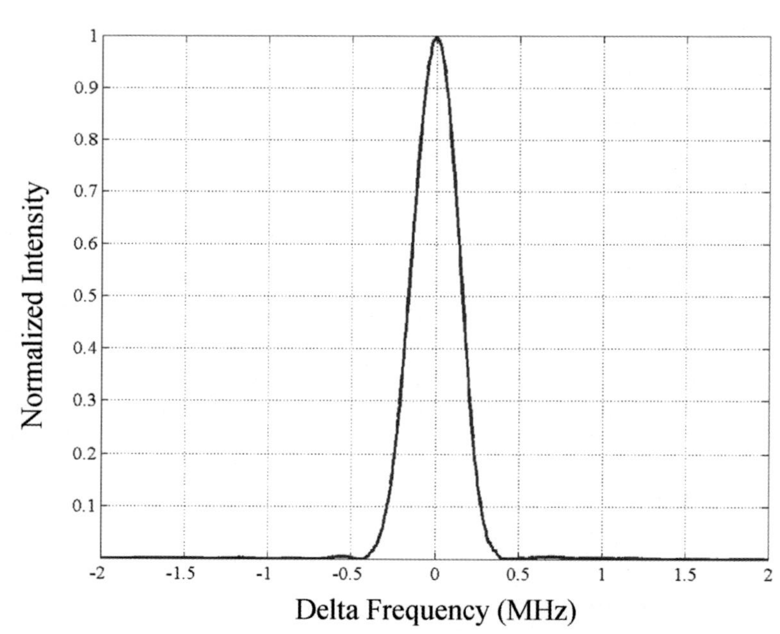

FIGURE 2. Spectral response of the AOTF imaging system. Curve is shown at 633 nm on (*a*) log and (*b*) linear scales.

For improved throughput, optimized coatings are used on all surfaces, and the AOTF is driven up to 8 radio frequencies simultaneously. (Although both filtered polarizations have been used, as described in connection with the illumination source, only one of these polarizations provides the image quality required for microscopy applications.) A total device efficiency of ~25% is obtained for unpolarized light at all wavelengths. In other words, if 10 mW of unpolarized light exits the microscope and enters the AOTF imaging system, ~2.5 mW will arrive at the CCD camera at its output. For

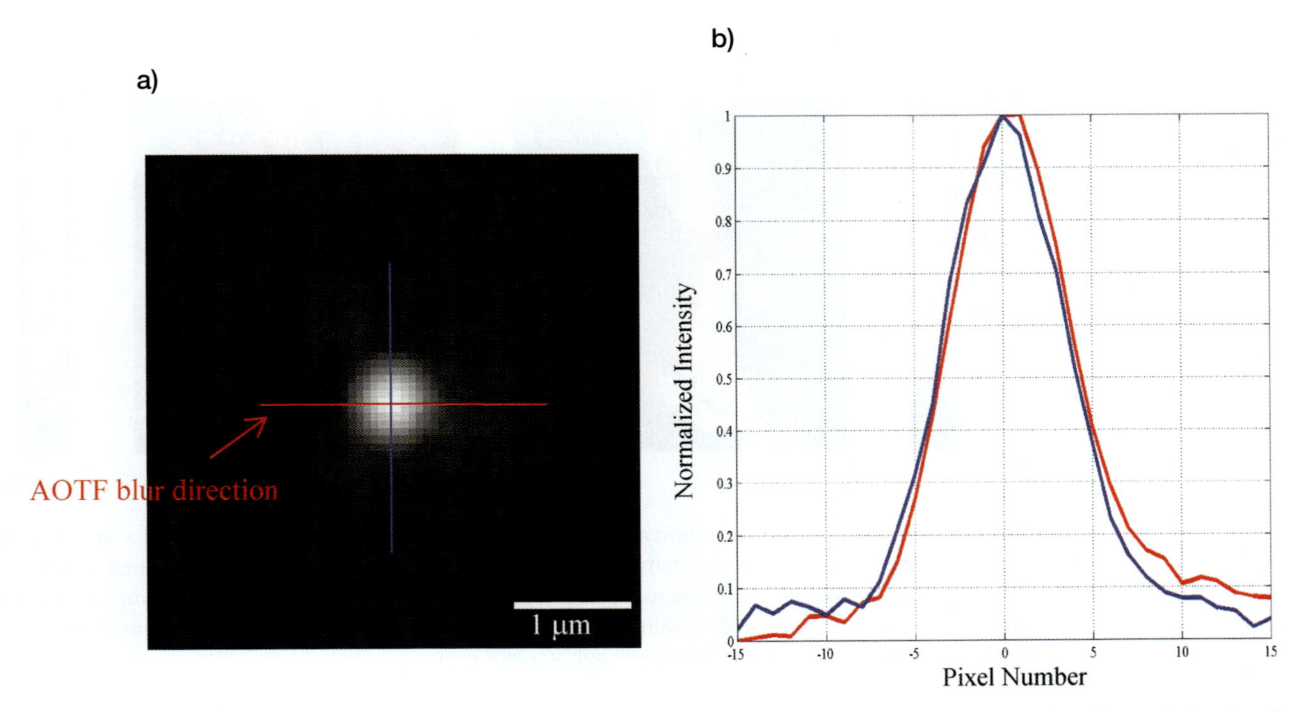

FIGURE 3. Blur-free images with the AOTF imaging system. A fluorescence image of a sub-resolution bead (0.1 µm diameter) imaged by the AOTF module is displayed in *a*. Image distortion due to the acousto-optic interaction occurs only in the horizontal plane of this figure. In *b* is an intensity profile comparing the bead intensity profiles in the vertical and horizontal directions, illustrating that there is no detectable blur produced by the AOTF.

many experiments, this loss can be adequately compensated for by choosing an appropriately high quantum-efficiency camera.

Image blur in an AOTF is caused by the spreading of the acoustic wave as it travels through the crystal. In the ChromoDynamics system, light coming from a particular point in the sample plane enters the crystal in a well-defined direction. However, interaction with this spreading acoustic wave causes the filtered light to leave the crystal in a cone of different directions. When this crystal output is focused on the image plane, a blurred image results. AOTF image blur can be virtually eliminated by (1) apodizing the transducer (as discussed above); (2) using as long a crystal as can be currently fabricated (Wachman et al. 1996, 1997); and (3) placing the AOTF in the nominally collimated portion of a 1:1 relay assembly. The result is diffraction-limited imaging, as illustrated in Figure 3. Finally, it should be noted that the exit face of the AOTF crystal is cut in such a manner that chromatic image shift—the change in position of images of different wavelengths—is negligible.

SOME APPLICATIONS OF THE AOTF MICROSCOPE

In Vivo Measurements of Oxygen Tension and Oxygen Saturation in a Mouse

The spectral tunability and speed of the AOTF microscope are well-suited for measuring oxygen dynamics in living tissue. In particular, our instrument has been used to generate steady-state oxygen saturation (SO_2) and oxygen tension (PO_2) maps in the cerebral cortex of mice as the inspiratory O_2 was varied from hypoxia (10%) through normoxia (21%) to hyperoxia (60%) (Shonat et al. 1997a). Oxygen saturation maps were obtained from a spectral analysis of reflected absorbance images collected at 11 different wavelengths in the 500–600-nm region through the imaging AOTF. Oxygen tension maps were obtained by frequency-domain measurements of phosphorescence lifetimes of an injected palladium–porphyrin compound; this measurement required synchronized modulation of

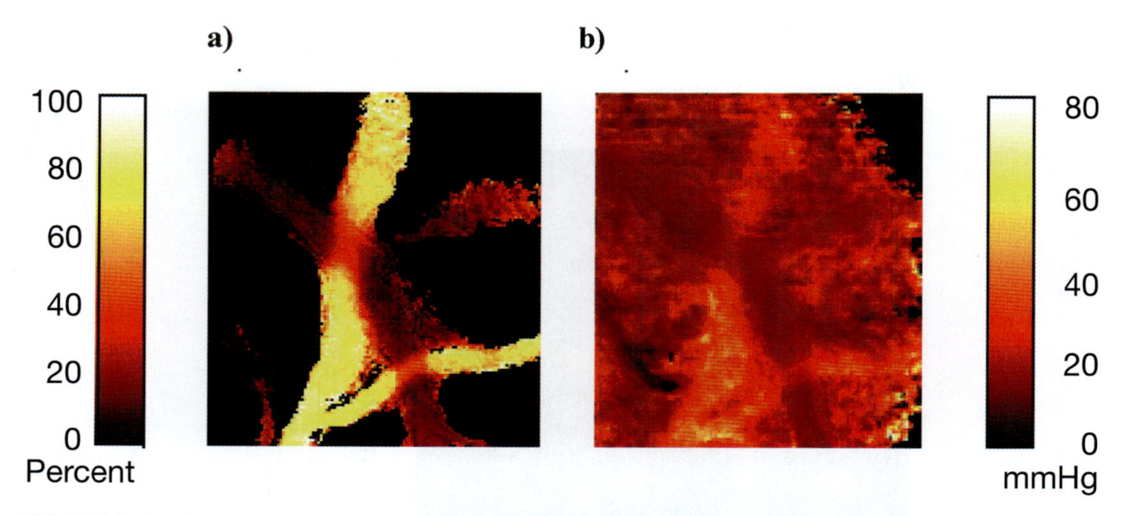

FIGURE 4. In vivo oxygen saturation and oxygen tension maps in the cerebral cortex of a mouse. Maps of oxygen saturation (*a*) and oxygen tension (*b*) were obtained near-simultaneously through a cranial window in a mouse using the AOTF microscope. Oxygen saturation data were obtained by imaging reflectance spectroscopy; oxygen tension data were determined by frequency-domain lifetime measurements of palladium–porphyrin, an oxygen-sensitive phosphorescent probe.

both excitation and imaging AOTFs on a sub-millisecond timescale. The two sets were taken sequentially, each set requiring approximately one minute of data acquisition.

Figure 4 shows representative oxygen saturation (a) and oxygen tension (b) maps taken in this manner. This work represents the first time 2D SO_2 maps have been taken in the cortex, as well as the first time SO_2 and PO_2 maps have been taken in succession with the same instrument, clearly demonstrating the potential of the AOTF microscope for in vivo physiological measurements. Preliminary results have also been reported of investigations of cortical oxygen response in a mouse following amphetamine injection (Shonat et al. 1997b).

Multiparameter Fluorescence

In many application fields, it is of great interest to label as many intracellular components as possible. Although scores of fluorescent probes are available, the breadth of their excitation and emission profiles makes it difficult to distinguish more than four or five distinct probes in a single sample using conventional interference filters.

Using an AOTF imaging system, however, dyes with closely spaced emission profiles that would be indistinguishable using interference filters can be clearly resolved by taking a series of images at different wavelengths within the band pass of a single filter. Choice of band center and width for these images can be varied to obtain cleanest separation of the different dyes. It is possible to distinguish in solution four different dyes whose emission peaks are all within 15 nm of each other (data not shown). Preliminary results using this technique in cells are shown in Figure 5, where seven proteins are simultaneously labeled in SKBR-3 cells using four different interference filter sets.

Multicolor Sub-millisecond Neuroscience

Because nerve cells often have a very complex cytoarchitecture and their functionality is highly compartmentalized, direct study of specific cellular functions is difficult using invasive approaches with low spatial resolution such as electrophysiology. This has led many researchers to use optical imaging techniques in order to elucidate function in the nervous system. Many of the most basic events of neuron function, however, occur on a timescale of a millisecond or less, making optical studies of neural cell function a challenging task. A number of areas in neuroscience research could be advanced tremendously

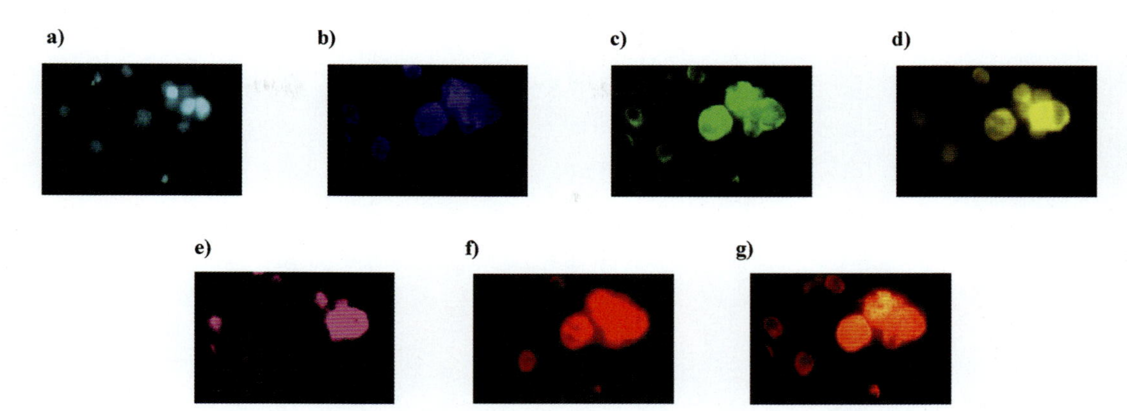

FIGURE 5. Seven-color cellular imaging with acousto-optic tunable filters. (*a*) Nucleus labeled with DAPI; (*b*) *her-2/neu* labeled with FITC; (*c*) *vegf* labeled with Cy3; (*d*) *cyclin D1* labeled with Cy3.5, (*e*) *ras* labeled with Cy5; (*f*) *c-myc* labeled with Cy5.5, and (*g*) *p53* labeled with Cy7.

by the use of a high-resolution imaging instrument, capable of multicolor detection with millisecond time resolution. The impact would be most obvious in the area of synaptic function. Simultaneous imaging of pre- and postsynaptic calcium and/or voltage together with transmitter release, for example, would significantly increase our understanding of the mechanisms of synaptic function.

There are several detectors available that can provide the required temporal and spatial resolution, and multiple colors can be imaged on a single chip with suitable splitting optics. However, for three or more colors, this approach is difficult to implement without risking channel overlap and compromising spatial resolution. A more elegant approach is to acquire images of the various probes successively using a fast wavelength-switching device in conjunction with a fast readout camera. AOTFs are ideal for this application.

In a preliminary study, an AOTF-based multiline laser source was used to study the spatial distribution of presynaptic calcium transients within the active zones of the neuromuscular junction of adult frogs following single action potential stimulations (Wachman et al. 2004). The study was extended to include simultaneous imaging of postsynaptic calcium transients as well. A system based on our AOTF imaging system that will enable studies of three or more probes is currently under development.

ACKNOWLEDGMENTS

Portions of the research described in this section were supported by grants from the National Science Foundation and the National Institutes of Health.

REFERENCES

Shonat R.D., Wachman E.S., Niu W., Koretsky A.P., and Farkas D.L. 1997a. Near-simultaneous hemoglobin saturation and oxygen tension maps in mouse brain using an AOTF microscope. *Biophys. J.* **73:** 1223–1231.

———. 1997b. Near-simultaneous hemoglobin saturation and oxygen tension maps in the mouse cortex during amphetamine stimulation. In *Oxygen transport to tissue XX* (ed. A.G. Hudetz and D.F. Bruley), pp. 149–158. Plenum Press, New York.

Wachman E.S., Farkas D.L., and Niu W. 1998a. Submicron imaging system having an acousto-optic tunable filter. U.S. Patent 5,796,512.

———. 1998b. Light Microscope having acousto-optic tunable filters. U.S. Patent 5,841,577.

Wachman E.S., Niu W., and Farkas D.L. 1996. Imaging acousto-optic tunable filter with 0.35-micrometer spatial resolution. *Appl. Optics* **35:** 5220–5226.

———. 1997. AOTF microscope for imaging with increased speed and spectral versatility. *Biophys. J.* **73:** 1215–1222.

Wachman E.S., Poage R.E., Farkas D.L., and Meriney S.D. 2004. Spatial distribution of calcium entry evoked by single action potentials within the presynaptic active zone. *J. Neurosci.* **24:** 2877–2885.

CHAPTER 99

A Practical Guide: Arc Lamps and Monochromators for Fluorescence Microscopy

Rainer Uhl

BioImaging Zentrum (BIZ), Ludwig Maximilians University, München, Germany

This chapter discusses the design and performance of an arc lamp-based illumination system for fluorescence microscopy. Although both lasers and arc lamps can be used as light sources for fluorescence microscopy, arc lamps are preferred for wide-field applications because of their lower price, wavelength selectability, and lack of interference fringes from reflections inside the optical system.

MATERIALS

The Design Principle

The system must deliver the highest flux of monochromatic light to the specimen plane, while maintaining a high degree of spatial homogeneity. Any such system is subject to a fundamental optical limitation; the quantity $n \times NA \times d$ is a constant (Helmholtz-Lagrange invariability), where n is the index of refraction, NA the numerical aperture, and d the diameter of an optical sender or receiver system. Hence, concentrating the light onto a small spot by forming a demagnified image of a light source will increase the NA of the cone of light in the image plane. If a dry objective with NA = 0.75 demagnifies a 1-mm source to a spot of 0.25-mm diameter, the usable cone of light from the source has a NA no larger than 0.1875 (0.75 x 0.25/1). Sizes and NA of source and target therefore require careful matching.

Choosing the Right Arc Lamp

Arc lamps<!> are the smallest non-laser light sources available. Mercury arc lamps have particularly compact arcs, and hence very high light flux densities. However, their use is restricted to the exact wavelengths of the Hg lines (Fig. 1), or to spectral regions that are well separated from these lines, otherwise severe spectral distortions may occur. Between lines, mercury arc lamps are no brighter than xenon short-arc lamps, which cover a broad spectral range with only a few lines in the blue and infrared regions (Fig. 1).

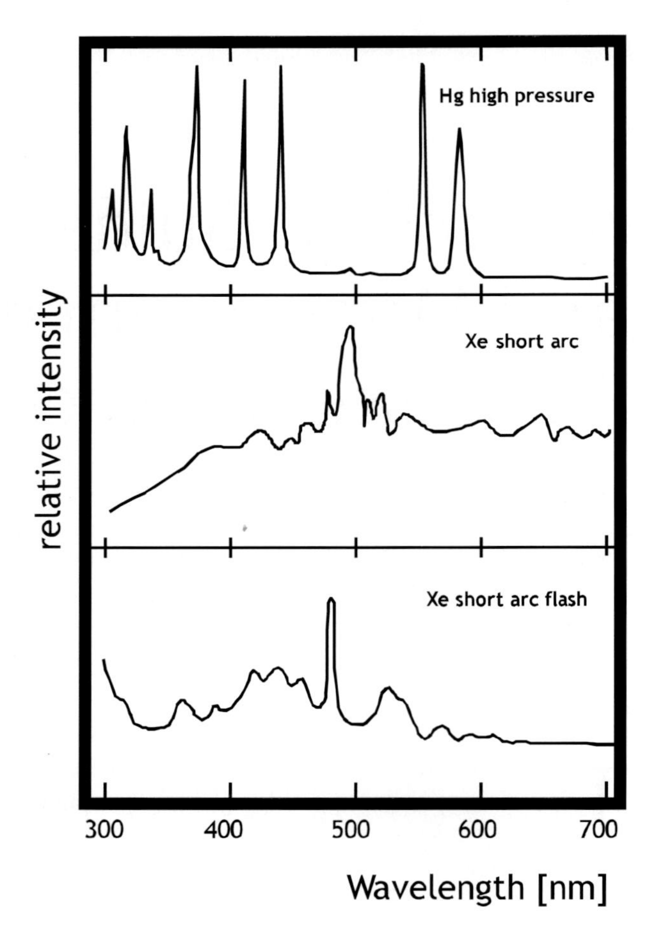

FIGURE 1. Spectral distribution curve for Hg arc lamps, continuous Xe short-arc lamps, and pulsed-Xe arc lamps.

The highest photon flux density of commonly available Xe short-arc lamps is achieved with 75-W models, in particular the Ushio model UXL-75-XO, which has a usable hot spot of 0.3 mm × 0.5 mm. However, as with Hg lamps, high photon fluxes reduce the life expectancy of the bulb. The brightest Xe lamps (75 W) outlast their Hg counterparts by a factor of two (400 vs. 200 hours), but so-called high-stability versions may live up to 15× longer. Unfortunately, their electrode geometry produces larger hot spots and hence does not allow equally high photon flux densities for illuminated fields <300 µm.

Xe flashlamps cannot provide the same brightness as DC lamps; the flashlamp's efficiency to transform electrical input into light output is only 5–10%, compared to 50% for DC lamps. Moreover, their arcs are generally much larger (1.5–6 mm) than those of a DC lamp. Consequently, 10–100× greater photon flux densities may be obtained from DC lamps. If image brightness is not of the highest importance, or if extended exposure times are tolerable, pulsed-arc sources offer some advantages: They only radiate when needed and so produce virtually no heat and last for the lifetime of the instrument.

Fiber Coupling

Coupling the light from the light source into the microscope using a flexible light guide has several advantages: It keeps the hot lamp away from the sample, it allows mechanical uncoupling from the microscope, and it scrambles the light, thereby improving spatial homogeneity. Single-core quartz fibers lose less than 1% light per meter of length; for moderately long fibers, most loss is due to reflections at the entrance and exit windows. It is crucial to match the fiber NA and diameter to the illumination task. For instance, a maximally bright spot of 270 μm under the microscope (when illuminated with a 40x oil-immersion objective of NA 1.3, corresponding to the diagonal length of a 2/3" CCD chip) requires a fiber of 1.25-mm diameter and NA 0.22. Liquid light guides, with their higher NA and larger diameters, will never allow a comparable concentration of the transmitted energy onto a similarly small spot.

Wavelength Selection

Interference filters or a monochromator can be used to select a small spectral portion of the light from an arc source. Several wavelength changers, which allow switching between different interference filters, are available. Filter wheels can do this very quickly and with only minor mechanical vibrations, yet they do not allow flexible light-exposure protocols. Other mechanical devices that move filters in and out of the beam are slower and cause vibration, which necessitates mechanical uncoupling through a fiber. The Sutter Instrument DG-4 wavelength switcher controls two galvanometer-driven mirrors that direct the beam through one of several filters. However, with the steadily increasing number of dyes available (see Appendix 2), a more flexible system such as a monochromator has several advantages (see below).

PROCEDURE

A Monochromator-based Illumination System: Polychrome V

Combining a standard microscope arc lamp with a commercial monochromator does not yield very satisfactory performance: The stepping motors commonly employed are slow, the intensity throughput is low because the optical geometry of the monochromator does not match the requirements of epifluorescence illumination, and the lamp collector optics are seldom achromatic. The following is a description of the basic features of a matched combination of light source, monochromator, and epifluorescence condenser, designed and optimized for fluorescence microscopy. This unit, called Polychrome V, was developed in collaboration with TILL Photonics. The name "Polychrome" indicates that it can provide many different colors of light for excitation, and the model number "V" indicates that it has undergone several rounds of evolution. Similar systems are now available from PTI, Kinetic Imaging, and Cairn Research. The considerations described here apply to any monochromator whose performance approaches theoretical limits.

SHORT EXAMPLE OF APPLICATION

Lamp and Monochromator Optics

The Polychrome consists of two separate optical modules: a lamp module illuminating the entrance slit of a monochromator with maximal brightness, and a monochromator module selecting the desired wavelength and coupling the monochromatic light into an optical fiber.

The lamp module employs a toroidal mirror that, if used by itself, would form a virtually aberration-free, achromatic 1:1 image of the arc. Two matched and mutually chromatically correcting aplanatic lenses introduce a magnification and thus reduce the NA. In the Polychrome V a NA of 0.57 is collected from the arc lamp and reduced to 0.15 before it enters the spectrograph through an adjustable entrance slit.

The slit in the monochromator module is positioned at the focal point of a 20° off-axis parabolic mirror (f = 85 mm, diameter = 25.4 mm; LT-Ultra GmbH). The resulting parallel white light is aimed at a plane diffraction grating (Carl Zeiss) that has 1302 lines/mm, a diameter of 28 mm, and is blazed at 400 nm where its diffraction efficiency exceeds 80%. The diffracted light is directed, at an angle of 20° with respect to the incoming light, onto an achromatic UV-transmitting lens doublet that focuses it onto a variable exit slit. The light passing this slit is collected by an achromatic lens combination and guided to the microscope by a single fiber quartz light guide of 1.25 mm diameter with NA 0.22.

Wavelength selection is achieved by rotating the grating, which is mounted onto a galvanometric scanner (Model 6230, CTI). An analog command voltage or a USB-command-string is used to determine the scanner excursion; the relationship between command voltage and wavelength involves two parameters, determined by a calibration procedure similar to that used with a pH meter.

Epifluorescence Condenser

Standard epifluorescence condensers do not have optimal throughput, they cannot guarantee optimal homogeneity in the specimen plane, and they are not suitable for fiber illumination, all because they are optimized for sources of much smaller diameter and higher NA. Moreover, they employ Köhler illumination rather than critical illumination that images the source into the specimen plane. The latter is preferred for fiber-coupled systems because the cone of light leaving the fiber is spatially inhomogeneous, whereas the intensity at the fiber exit face is homogeneous.

A suitable achromatic, aplanatic condenser provides the highest possible coupling efficiency. To achieve maximal image brightness, only a small spot is illuminated. This corresponds to the area seen by a 2/3" CCD camera. Both an aperture stop and a field stop may be placed in the beam; the former to reduce the overall brightness, the latter to allow selection of the illuminated spot. All lenses are multilayer antireflection-coated and exhibit less than 0.5% reflection per surface over a wavelength range of 330–640 nm.

ADVANTAGES AND LIMITATIONS

Optical Throughput

The output power drops no more than 50% between 320 nm and 500 nm. Photon fluxes of 2×10^{23} to 5×10^{23} $m^{-2}sec^{-1}$ can be achieved in the specimen plane, implying a photon-transfer efficiency close to the theoretical limits for the lamp. A single fluorophore of FITC or GFP then emits up to 10^4 photons per second, enough for single-molecule detection. Reducing the width of the entrance slit reduces the output intensity without affecting the bandwidth.

Spectral Purity

Entrance and exit slit adjustments limit the spectral bandwidth to between 1 and 14 nm; larger bandwidths would require a grating with fewer lines per millimeter. There should be zero output at wavelengths more than 14 nm from the peak, but a small plateau extends some 50 nm on either side of the peak due to stray light caused by mechanical imperfections of the parabolic mirror and grating. Monochromators cannot reject the same level of stray light as a suitable filter combination can, but with a suitable short-pass filter in the excitation beam, matching the dichroic beam splitter's reflectivity range, monochromators can achieve similarly dark backgrounds as interference filters.

Dynamic Performance

The time to reach the excitation wavelength and the exposure duration are limited by the dynamic performance of the galvanometric scanner carrying the diffraction grating. The wavelength is changed from the rest value (260 nm, not transmitted by the microscope optics) to the desired wavelength and back. The largest wavelength jump, to 680 nm, requires the scanner to move by 17.5° of its 20° range. Any wavelength is reached within 2 nm in less than 1.5 msec. The accuracy of the scanner is maintained to within 0.1 nm over the lifetime of the system.

The fast scanner allows the exposure time to be set with millisecond precision. Unlike rotating filter wheels, the fast scanner exposure time may be set independently for each wavelength, and complex wavelength protocols may be executed under computer control. Thus, as with most modern single-lens reflex cameras, the image brightness may be determined by the exposure time. Since photodamage depends on the integral photon dosage and not on the peak intensity, at least with photon fluxes achieved with arc lamps, neutral density filters are unnecessary.

One drawback is the non-zero transition time of the scanner, during which illumination with undesired wavelengths occurs. The relative error increases with decreasing exposure time. In cases where the dye absorbs more strongly during the transition than at the final wavelength (Grynkiewicz et al. 1985), correct calcium concentrations may still be derived from ratio experiments if the calibration values R_{min} and R_{max} are determined using the same exposure times as used in the experiment (see Messler et al. 1996). Another cure is to "blindfold" the detector during wavelength transitions. This is feasible with photomultipliers, gateable image intensifiers, or CCD cameras employing interline-transfer sensors that can shift the charge from the image to the storage zone in ~ 2 μsec. Frame transfer cameras, on the other hand, require up to a millisecond for this shift, causing smear if pixels are illuminated during charge shifting.

The output power of the system is stable both on a millisecond and second timescale, with a relative root mean square noise of $< 3 \times 10^{-3}$ in both cases.

REFERENCES

Grynkiewicz G., Poenie M., and Tsien R.Y. 1985. A new generation of calcium indicators with greatly improved fluorescence properties. *J. Biol. Chem.* **260:** 3440–3450.

Messler P., Harz H., and Uhl R. 1996. Instrumentation for multiwavelength excitation imaging. *J. Neurosci. Methods* **69:** 137–147.

CHAPTER 100

The Use of Liquid-Crystal Tunable Filters for Fluorescence Imaging

Kenneth R. Spring

Laboratory of Kidney and Electrolyte Metabolism, National Heart, Lung and Blood Institute, National Institutes of Health, Bethesda, Maryland

Fluorescence microscopy allows the detection of very few molecules, often with great specificity. Fluorescent dyes are sensitive to their environment and may undergo alterations in their absorption or emission spectra as well as changes in their polarization sensitivity (i.e., the extinction of the fluorophore as a function of orientation to a polarized excitation beam). It is, therefore, very useful to be able to scan the emission spectrum of fluorophores and to examine the dependence of the fluorescence emission on the polarization of the excitation beam (Bentley et al. 1985; Checovich et al. 1995; Chen et al. 1997; Nir 1997). Multispectral imaging of fluorescently labeled microscopic specimens has increased in popularity as new dyes and better detectors have become available (Garin et al. 1996; Speicher et al. 1996; see also Chapters 3 and 74). Liquid-crystal tunable filters (LCTFs) are well suited to emission-spectrum scanning and the determination of the polarization sensitivity of the emitted fluorescence (Hoyt and Benson 1992; Wright et al. 1996). This chapter reviews the operating and design principles of LCTFs, briefly comparing them to alternative available devices, and giving some examples of their present and future utility for fluorescence microscopy.

TUNABLE FILTER OPERATING PRINCIPLES

Two general classes of tunable filters are described—those utilizing birefringent elements and those using dichromatic dye-absorption elements (Wright et al. 1996).

Birefringent Filters

The requirements of astronomers for narrow-band filters led to the development of polarization-interference filters by Bernard Lyot in 1933. The basic building block of each stage of the Lyot filter is a birefringent crystal, oriented at 45° between two linear polarizers. Although the input polarizer delivers equal light intensities into the ordinary and extraordinary axes of the birefringent crystal, the propagation speeds of the two rays differ, and they emerge with a phase delay or retardation. When the two components are recombined, the polarization state is altered, and transmission through the second polarizer varies according to wavelength.

The maximum retardance (R) of the birefringent crystal depends on its thickness (d) and the difference in the refractive indices for the ordinary and extraordinary rays (n_o and n_e, respectively) at the wavelength of the incident illumination

$$R = d\,(n_e - n_o) \tag{1}$$

The phase delay (Γ) caused by the crystal is given by

$$\Gamma = \frac{2\pi R}{\lambda} \tag{2}$$

where λ is the wavelength of the incident light. The transmission (T) of the crystal depends on Γ, and hence, as follows

$$T = \frac{1}{2}\cos^2\frac{\Gamma}{2} \tag{3}$$

A Lyot filter consists of several birefringent crystal stages in series. Generally, the crystals are selected for a binary sequence of retardance so that the transmission is maximal at the wavelength determined by the thickest crystal retarder. The other stages serve to block the transmission of unwanted wavelengths. Lyot filters have been constructed for solar astronomy with a bandwidth as narrow as 0.025 nm (Beckers et al. 1975).

Liquid Crystals

Liquid crystals are rod-like molecules that self-assemble to produce an oriented birefringent layer that has retardance and alters the state of polarized light. When a liquid crystal is placed between crossed polars, the fraction of light that is transmitted then depends on the retardation caused by the liquid-crystal layer. The retardance of the liquid-crystal layer can be altered by the imposition of a low-voltage AC field. A liquid-crystal retarder usually consists of two high-throughput film polarizers sandwiching a glass cell containing the thin liquid-crystal layer (5–25 μm thick). The interior glass walls of the liquid-crystal cell are coated with transparent indium–tin oxide electrodes, and a thin polymer layer is applied over the indium–tin oxide film. The polymer is brushed by buffing so that many microscopic ridges are produced in its surface. The liquid crystals tend to orient themselves parallel to the polymer surface and the ridges. The buffing direction in the opposing polymer layers may be oriented in parallel to produce a nematic liquid crystal or at right angles to produce a twisted-nematic liquid crystal. Imposition of a voltage causes the oriented liquid-crystal molecules to tilt away from the polymer surfaces along the lines of the electric field and thereby reduce or lose their optical activity.

Liquid-Crystal Tunable Filters

A wavelength-selective liquid-crystal filter may be constructed from a stack of fixed birefringent-crystal and variable liquid-crystal retarders as shown in Figure 1. In this simplified example of a Lyot filter, 4 polarizers are separated by 3 layers of birefringent-crystal/liquid-crystal combinations with retardances in binary steps denoted as R, 2R, and 4R. Each stage produces a series of transmission bands, described by Equation 3 and depicted in Figure 1. The product of the transmission bands is the single wavelength curve at the bottom of the figure. In practice, a Lyot LCTF may have as many as 11 polarizers and 10 liquid-crystal layers and use a built-in microprocessor to tune all the stages (e.g., the Varispec from Cambridge Research and Instrumentation).

Dichromatic Dye-absorption Filters

Dye-absorption filters were proposed by Edwin Land in 1950 to produce color television. The filter uses dichromatic color polarizers combined with a rapidly switchable retarder to alter the transmission of incident polychromatic light and produce broadband red, green, and blue (RGB) outputs. Land's design only became a practical reality in the 1970s when fast liquid-crystal and ferroelectric

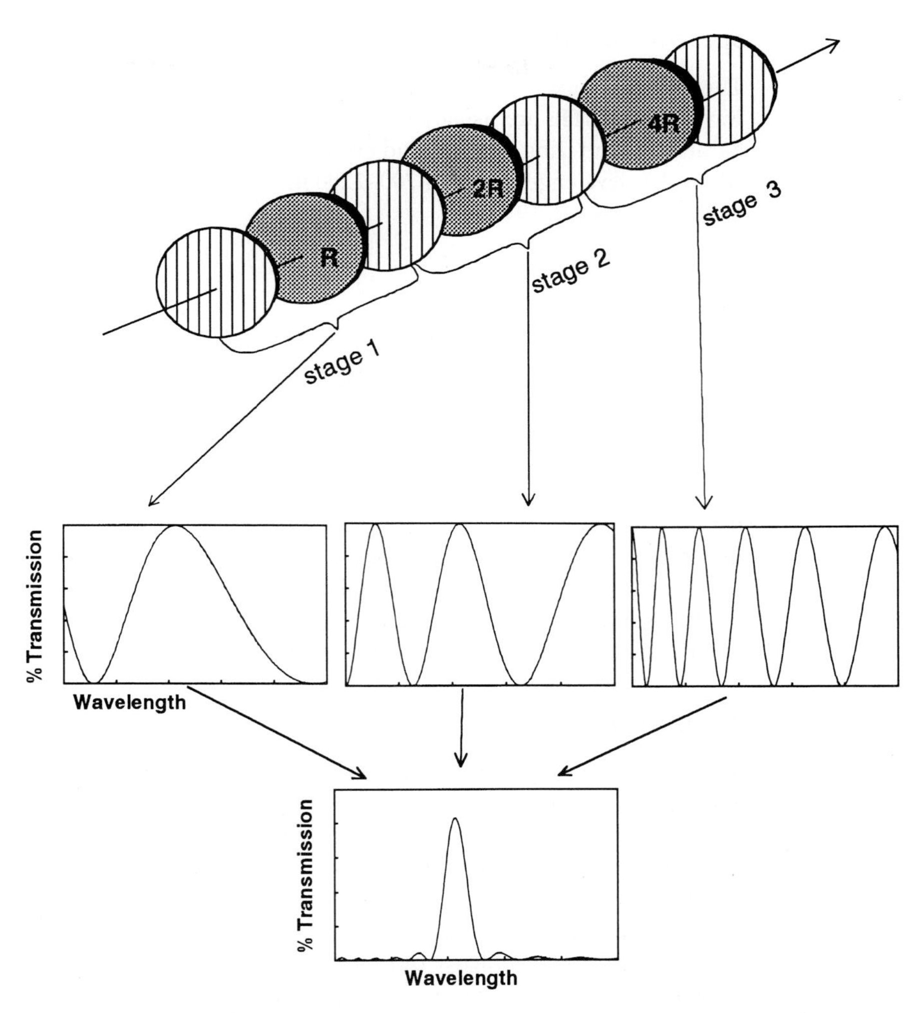

FIGURE 1. Design and principle of operation of a Lyot LTCF. Linear polarizers are indicated by disks with vertical lines, and the variable retarders (disks denoted by R) are oriented at 45° to the plane of polarization. The retardance is indicated on each stage, with the resultant transmission spectra shown below (figure from Cambridge Research and Instrumentation; reprinted with permission).

polarization switches became available. In these devices, the wavelength and bandwidth of the transmitted light are fixed, as determined by the broadband dyes used in the dichromatic filters.

LCTF PERFORMANCE CHARACTERISTICS

Critical Parameters

LCTFs of the Lyot type have very high spatial resolution and a large, clear aperture (up to 35 mm) and do not produce image shifts as the transmission wavelength is changed. They are best utilized on the detection side of the fluorescence microscope because of their somewhat limited spectral range and susceptibility to damage from the output of arc lamps. Typical LCTFs have an angle of acceptance of approximately ±7° from normal, generally adequate for use of the device in the microscope emission path. Wavelength changes can be completed in 25 msec at 25°C (i.e., time constant of 15 msec). The critical parameters in LCTF performance for fluorescence microscopy are spectral range, bandwidth, and transmission. Each is considered in turn.

Spectral Range

The spectral range of LCTFs depends on the choice of polarizers, coatings, and liquid crystals. Visible-wavelength LCTFs are usually limited to a range of 400–700 nm. The UV-spectral limitation arises because of scattering by the liquid crystals at short wavelengths. Near-IR LCTFs have been constructed for the wavelength range of 700–2200 nm and are useful for Raman microscopy (Hoyt and Benson 1992; Wright et al. 1996).

Bandwidth

The bandwidth of an LCTF is determined during its design by choice of the number of stages and their retardance. The narrower the bandwidth, the more stages are needed; LCTFs with bandwidths of 0.25–50 nm have been produced (Wright et al. 1996). LCTFs designed for use as RGB filters for true color imaging have far fewer stages than high-resolution, narrow-band LCTFs designed for spectral scanning. Since LCTF bandwidth is fixed in wave numbers, it is not constant as a function of wavelength but increases across the spectrum. For example, bandwidth increases from 10 nm to 40 nm as the wavelength changes from 400 nm to 700 nm. Proper correction of spectra requires that the increase in bandwidth as a function of wavelength be taken into account.

Rejection of unwanted wavelengths is also affected by the number of stages in the LCTF, with typical leakage values of less than 1 part in 10^4. The bandwidth and rejection properties of LCTFs are generally acceptable for fluorescence microscopy.

Transmission

Transmission by LCTFs is limited both by the requirement for polarized light and by the wavelength dependence of the liquid crystals (see above). Fluorescence emission is generally unpolarized, so at least 50% of the emitted fluorescence must be sacrificed when an ideal polarizer is introduced. Further losses occur because of the need for multiple polarizers and fixed retarders and the limited transmissive properties of the liquid crystals. Typical transmissions of polarized light by a 10-nm-bandwidth LCTF range from a minimum of 6% at 450 nm to a maximum of 30% at 650 nm. Spectra must be corrected for the dependence of LCTF transmission on wavelength. This is a relatively simple matter, because the transmission is a monotonic function of wavelength (see Eq. 3). Although such low-percent transmissions are generally acceptable for transmitted light microscopy, they are not acceptable for fluorescence microscopy, particularly in comparison to the 60–80% achieved by interference filters for unpolarized light.

LCTF Designs for Improved Transmission

Two design changes have been used to increase the throughput of LCTFs: (1) reduction of the number of stages and (2) recovery of the lost polarization component.

Reduction of the Number of Stages

Reduction of the number of stages of an LCTF can be achieved by restricting the free spectral range, increasing the bandwidth, or employing a different design (e.g., the Land or Solc filters; see Wright et al. 1996). In the limiting case, exemplified by the RGB LCTF, peak transmission is still only ~25–35% for randomly polarized light, although bandwidth is increased to cover the visible spectrum with the three bands. Out-of-band rejection is usually relatively poor, and attenuation ratios lower than 100:1 are common. These devices cannot be used for spectral scanning, but they can effectively convert a monochrome camera to a frame-sequential color camera (e.g., ferroelectric devices from Displaytech and liquid-crystal devices from Cambridge Research and Instrumentation).

Recovery of the Lost Component of Polarization

Recovery of the lost component of polarization has been realized in the Lyot LCTF design from Cambridge Research and Instrumentation, illustrated in Figure 2. The randomly polarized fluorescence is first passed through a polarizing beam splitter that is oriented so that the *p* polarization component is on the optical axis and the *s* component is orthogonal. The LCTF has its polarizers oriented to accept only *p* polarization, so the *p*-oriented image passes through the device. The *s* component is directed to a twisted-nematic half-wave retarder, where it is converted to *p* polarization. The converted *s* component image then passes through the LCTF adjacent to the unmodified *p* component image. The emerging images are combined by another polarizing beam splitter after the unmodified *p* component is converted to *s* polarization by a second twisted-nematic half-wave retarder. Alignment of the two images is readily achieved with integral gimbal-mounted mirrors. The dual image LCTF requires a physically larger device, both in diameter (as it must accommodate two images side by side) and in optical path length, with the additional prisms, mirrors, and retarders.

Transmission is virtually doubled in this design with peak values of 45–50% at 600–700 nm and a minimum value of 35% at 500 nm for a 20-nm bandwidth LCTF used in our laboratory (Nitschke and Spring 1995). An additional benefit of the dual-image LCTF arises from the use of twisted-nematic half-wave retarders rather than fixed-retarder plates. For proper compensation, the retardance must be adjusted as a function of wavelength to maintain a half-wave throughout the spectral range. Control of the retardance of the *p* and *s* components of the fluorescence image passing through the LCTF means that the polarization of the fluorescence signal can be determined, e.g., by turning off either the *p* or *s* components totally or by varying the relative retardance in either path.

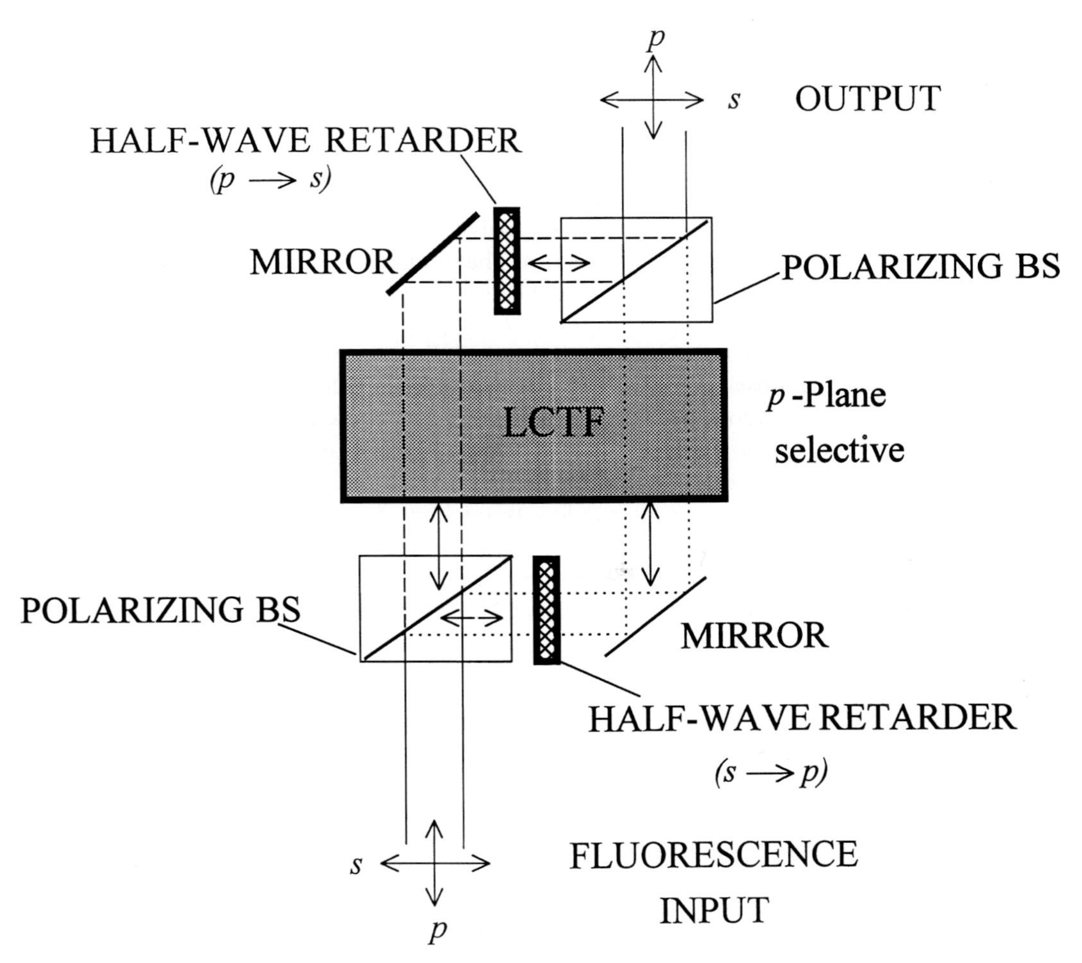

FIGURE 2. Optical paths in a polarization-insensitive LCTF. Design of a high-throughput device that processes both *p* and *s* components of randomly polarized light. The LCTF is *p*-plane-selective, so *s* polarized light is converted to *p* by a beam-splitting prism (BS) and liquid-crystal half-wave retarder ($s \rightarrow p$) at the bottom of the figure. After passing through the LCTF, both beams are recombined by the beam-splitting prism–half-wave retarder assembly shown at the top of the figure.

COMPARISON OF LCTFs TO OTHER METHODS FOR SPECTRAL SCANNING OF FLUORESCENCE EMISSION

There are three competing approaches for determination of the spectral content of entire images of fluorescent microscopic objects: filter wheels, acousto-optical tunable filters (AOTFs), and interferometers. Another approach that is not considered here is the recording of the spectrum of small portions of the microscope field of view with a rapid-scan spectrometer (Nir 1997). This spectroscopic method is limited to selected subregions and is not comparable to wide-field imaging with LCTFs. It is noteworthy that accurate spectral scans by any method require characterization of the spectral sensitivity of the entire system, including both the wavelength-selection device and the detector. The relative strengths and weaknesses of each of these alternatives are considered in comparison to LCTFs.

Filter Wheels

Filter wheels in the emission path have high transmission, excellent rejection of unwanted wavelengths, reasonably high speed, and relatively low cost. Limitations are the requirement to select filters in advance, mechanical difficulties in changing filters, and image shift due to lack of filter-surface flatness (the "wedge"), or imperfections in the filter mount. The mechanical component of the filter wheel can introduce vibrations, misregistration, or timing limitations into the detection system. Filter wheels have been most frequently employed on the excitation side of the fluorescence microscope, where the above factors are less of a concern; however, they have been used with success on the emission side of systems for combinatorial karyotyping (Speicher et al. 1996).

Acousto-optical Tunable Filters

AOTFs are becoming increasingly popular in microscopy (Chatton and Spring 1993) as excitation-wavelength selection devices. They require a collimated, polarized beam that has been shaped by cylindrical optics and focused accurately into the device. They have the virtues of extraordinary speed (switching rates up to the MHz range), very high transmission of polarized laser beams (85–90%), narrow bandwidth, tolerance to high light fluxes, and a lack of moving parts (e.g., see AOTFs from the Brimrose Corp. of America). The requirement for a well-focused and collimated beam has limited the application of AOTFs in microscopy to laser-based illumination systems (Nitschke and Spring 1995). The bandwidth of AOTFs is often too narrow to permit adequate detection of fluorescence emission. AOTFs exhibit dispersion that results in a positional change as a function of wavelength and image smear. Up to now, the low throughput due to the narrow bandwidth and the beam-collimation requirements, as well as the dispersion of light, have precluded the practical application of these devices on the detection side of a fluorescence microscope. For further details on AOTFs in fluorescence microscopy, see Chapter 98.

Interferometry

An interferometric-based scanner has found application in spectral scanning of the emission of a fluorescence microscope (Garin et al. 1996). The design employed, a Sagnac interferometer, utilizes a stepping-motor-controlled mirror to alter the optical path length in the device. A large number of interference images must be obtained, as the entire spectrum is scanned in small increments (~1 nm). The resultant stack of several hundred interferograms is processed by Fourier methods to generate a multispectral image in which the spectrum of each pixel may be obtained. The apparatus is relatively bulky, slow (several minutes per spectrum), and expensive, but has proven valuable in multispectral imaging of chromosomes.

USE OF LCTFs IN MICROSCOPY

At present, applications of LCTFs are limited to a few spectral-scanning studies (Nitschke and Spring 1995; Timbs and Spring 1996; Wright et al. 1996) and to the use of the RGB devices for multispectral imaging (Wright et al. 1996). The virtues of LCTFs—high resolution, large aperture, lack of image shift, and reasonable speed—have to be balanced against their limitations—limited spectral range, nonconstant bandwidth, lower transmission, and increased optical path length. Although new designs for LCTFs are under development (Wright et al. 1996), the existing devices provide the microscopist with the unique capability to determine, with high fidelity, the spectral and polarization properties of emitted fluorescence in the entire field of view.

REFERENCES

Beckers J., Dickson L., and Joyce R. 1975. Observing the sun with a fully tunable Lyot-Ohman filter. *Appl. Opt.* **14:** 2061–2066.

Bentley K.L., Thompson L.K., Klebe R.J., and Horowitz P.M. 1985. Fluorescence polarization: A general method for measuring ligand binding and membrane microviscosity. *BioTechniques* **3:** 356–366.

Chatton J.-Y. and Spring K.R. 1993. Light sources and wavelength selection for widefield fluorescence microscopy. *Bull. Microsc. Soc. Am.* **23:** 324–333.

Checovich W.J., Bolger R.E., and Burke T. 1995. Fluorescence polarization—A new tool for cell and molecular biology. *Nature* **375:** 254–256.

Chen C.-S., Martin O.C., and Pagano R.E. 1997. Changes in the spectral properties of a plasma lipid analog during the first seconds of endocytosis in living cells. *Biophys. J.* **72:** 37–50.

Garin Y., Macville M., du Manoir S., Buckwald R.A., Schröck E., Cabib D., and Ried T. 1996. Spectral karyotyping. *Bioimaging* **4:** 65–72.

Hoyt C.C. and Benson D.M. 1992. Merging spectroscopy and digital imaging enhances cell research. *Photonics Spectra* (**November**): 92–96.

Nir I. 1997. Multispectral imaging completes the picture. *Laser Focus World* (**October**): 133–137.

Nitschke R. and Spring K.R. 1995. Electro-optical wavelength selection enables confocal ratio imaging at low light levels. *J. Microsc. Soc. Am.* **1:** 1–11.

Speicher M.R., Ballard S.G., and Ward D.C. 1996. Karyotyping human chromosomes by combinatorial multi-fluor FISH. *Nat. Genet.* **1:** 368–410.

Timbs M.M. and Spring K.R. 1996. Hydraulic properties of MDCK cell epithelium. *J. Membr. Biol.* **153:** 1–11.

Wright H., Crandall C.M., and Miller P. 1996. Active filters enable color imaging. *Laser Focus World* (**May**): 85–90.

Fluorescence Grating Imager Systems for Optical-sectioning Microscopy

Frederick Lanni

Department of Biological Sciences, Carnegie Mellon University, Pittsburgh, Pennsylvania 15213

In fluorescence microscopy, optical sectioning is defined as the attenuation or removal of out-of-focus features from an image, and is a prerequisite for quantitative analysis of three-dimensional structure or function within the specimen. A single optical section (OS) is a view of a slice within the specimen centered axially on the plane of focus. A through-focus stack of optical sections therefore constitutes a 3D representation of the specimen. For the past 20 years, optical sectioning has been driven by the development of the laser confocal scanning fluorescence microscope (CSFM), the two-photon scanning fluorescence microscope (TPSFM), and deconvolution algorithms. In 1997, Wilson and colleagues described both a principle and an optical system to achieve nearly direct optical sectioning in a conventionally illuminated reflectance microscope (Neil et al. 1997; Wilson et al. 1997, 1998a; Juskaitis et al. 1998). The essential idea was the use of "structured light" or striped illumination to achieve focus-dependent selectivity. The extension to fluorescence was straightforward (Neil et al. 1998, 2000; Wilson et al. 1998b,c; Ben-Levy and Peleg 1999; Lanni and Wilson 1999; Lagerholm et al. 2003; Vanni et al. 2003), and there are now instruments available from Carl Zeiss (ApoTome; www.zeiss.com) and from Optem-Avimo (Optigrid; www.thales-optem.com) that incorporate the concept. In addition, much more complex coherent structured-light systems for optical sectioning have been developed, but these are beyond the scope of this chapter. For the purposes of the present discussion, the basic instrument is referred to as a fluorescence grating imager (FGI). The attractive characteristics of the FGI are its versatility and performance, given its simplicity and relatively low cost.

OPERATING PRINCIPLE

In a grating imager, only a single modification is made to the microscope. A movable Ronchi grating mask is placed in the field iris plane of the incident-light illuminator, thus producing striped illumination in the specimen within the depth of focus of the objective (Fig. 1). The actual focus range of the stripes can be made very sharp if a fine grating is used with an objective of high numerical aperture (NA). Therefore, the image formed by the microscope will consist of striped in-focus features superposed with uniformly illuminated out-of-focus features (background). The period of the projected stripes is the actual grating period divided by the demagnification ratio between the illumina-

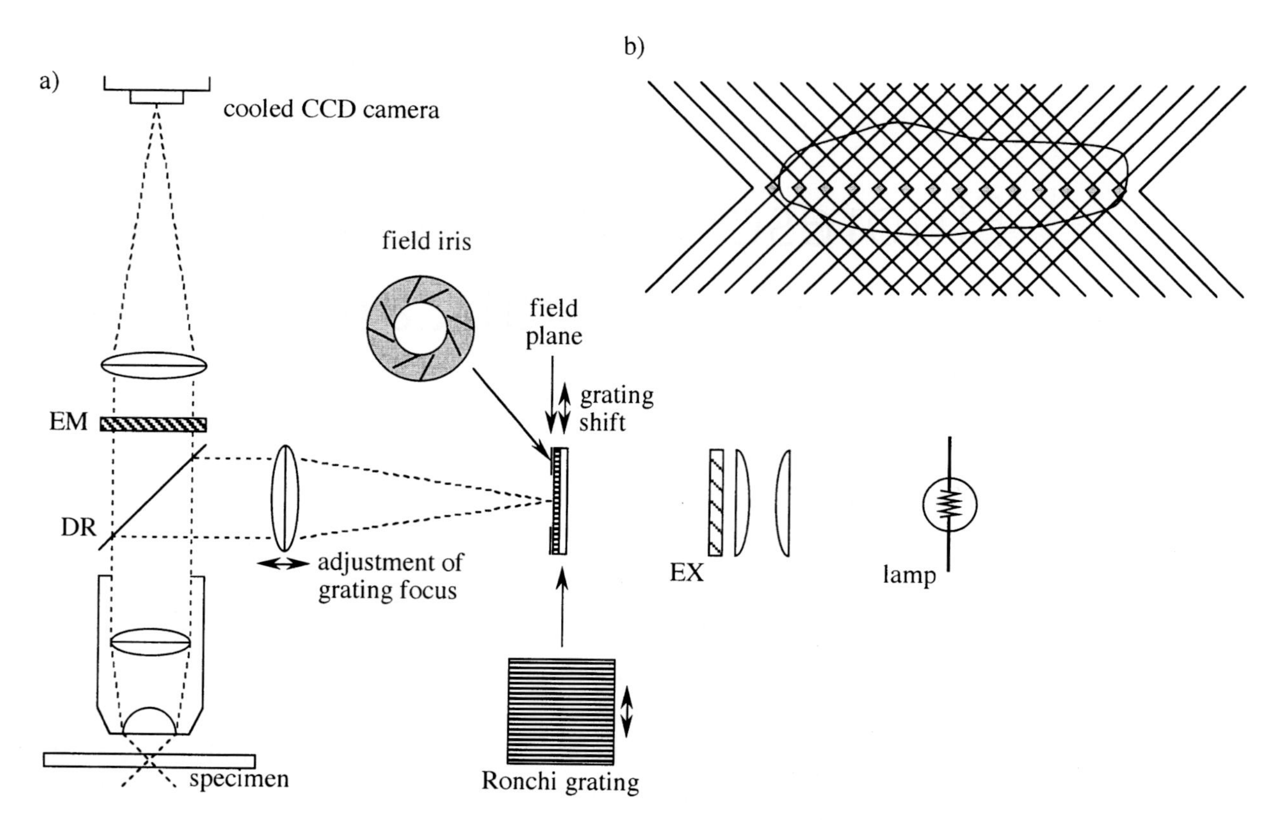

FIGURE 1. Schematic of fluorescence grating imager. (*a*) Optics layout showing movable grating mask inserted at location of illuminator field iris. Ronchi gratings (equal-width square-wave pattern) in the range 8–50 line pairs per mm (LP/mm) are used, depending on the demagnification ratio determined by the objective and the focal length of the illuminator tube lens. A means for bringing the grating to a sharp focus in the plane of focus of the objective is necessary, here shown as an axial adjustment of the illuminator tube lens. To minimize aberration in the grating image, the excitation filter is shown removed from its usual position to precede the grating. This is conveniently done by use of an external filter wheel between lamp collector lens and microscope stand. Additionally, the dichroic reflector must be flat and strain-free. Actuators for grating movement can be piezoelectric or electromagnetic, or the grating may be fixed but refractively shifted by a separate optic. Shift increments are generally 1/3 or 1/4 of the grating period, so will be in the range 5.00–42.0 μm. Computer control synchronizes discrete grating movement and CCD readout between periods of light exposure. A fast light shutter (not shown) is essential for precision exposure timing and minimization of unnecessary exposure. (*b*) Schematic showing finite grating depth of focus within a 3D specimen. Geometrically, it can be seen that both grating spatial frequency and objective NA will affect the depth of focus of the pattern.

tor field iris plane and the specimen, which is objective-dependent (and generally not equal to the specified objective magnification). When the grating is shifted so as to move the projected stripes transversely across the object, the fluorescence detected in any pixel element in the image will consist of (1) a steady or DC component, due to the out-of-focus background, plus (2) an oscillating (AC) component due to the shifting of the stripes across in-focus structures, and (3) shot noise due to both the DC and AC signal components. In brief, the amplitude of the AC component is the in-focus signal level for that pixel. By shifting the grating between a minimum of three defined positions to obtain three images, a very simple digital image processing operation can subtract the background and demodulate the AC component to produce an optical section.

COMPUTATION OF THE OPTICAL SECTION

The simplest projected grating image is a sine-wave pattern, which, when translated across the specimen, produces a sinusoidal modulation of the fluorescence. Translation of the grating is denoted by the phase shift (ϕ), where, for example, a shift of 90° is equivalent to movement equal to 1/4 of the grating period. In any pixel, the fluorescence signal can be expressed as a periodic function of ϕ, plus a constant term due to out-of-focus background

$$f(\phi) = \text{DC term} + \text{AC term} = a_0/2 + (a_1 \cos \phi + b_1 \sin \phi) \quad (1)$$

The in-focus part of the total fluorescence signal in that pixel is the AC amplitude, which is given by the Pythagorean sum of the sine and cosine coefficients $\sqrt{(a_1^2 + b_1^2)}$. It can be shown generally that the AC amplitude can be computed on a pixel-by-pixel basis from three images made with the grating shifted between three distinct positions. Three images are necessary because there are three unknown coefficients in Equation 1. The most commonly used shift sequences are given in Table 1, along with corresponding formulas for computing the optical section. The out-of-focus image component, which is the same in all images in the set, is removed by computing differences between image pairs (Fig. 2). The resulting bipolar, striped images are then demodulated by algebraic operations, which effectively form the Pythagorean sum.

A real projected grating will include spatial harmonics. For a perfect Ronchi (square wave) grating, only odd harmonics occur: $f(\phi) = dc + (a_1 \cos \phi + b_1 \sin \phi) + (a_3 \cos 3\phi + b_3 \sin 3\phi) + (a_5 \cos 5\phi + b_5 \sin 5\phi) + \cdots$. The amplitude of the AC term is still the in-focus part of the total fluorescence signal in that pixel. However, now it can be seen that the simple image processing formulas will not give $\sqrt{(a_1^2 + b_1^2)}$, exactly. The harmonics occur as error terms and may appear as fine stripes in the optical sections. Two factors can minimize or eliminate this problem: (1) The shift sequence and algorithm used can suppress one or more harmonics. As originally pointed out by Wilson et al. (1997; Neil et al. 1997), the 1/3-period shift sequence exactly compensates the 3rd harmonic, therefore, the first error term is due to the 5th harmonic. Since, for a Ronchi grating, the amplitudes drop off as 1/n, this error is small. (2) Because the incoherent modulation transfer function (MTF) of the microscope decreases with spatial frequency (to zero at the inverse of Abbe's resolution limit), the microscope attenuates the higher harmonics that are projected into the specimen, further reducing the error terms. In principle, harmonic error can be eliminated altogether by choosing the Ronchi so that its 5th harmonic matches or exceeds Abbe's resolution limit for the objective. This is not a severe restriction. With the period of the 5th harmonic set equal to Abbe's resolution limit (λ/2NA), the fundamental period will equal 5 x (λ/2NA). This is 2.5 times more coarse than the optimal period (λ/NA, see Eq. 3 below), but is projected with high contrast. Alternatively, the grating could be chosen so that the 3rd harmonic would be equal to Abbe's limit, in which case the fundamental period would be 3 x (λ/2NA). This is close to the optimal period (λ/NA), but will have lower contrast than with 5th harmonic suppression.

In general, a wide variety of grating shift sequences can be formulated, all of which, in principle, provide the basis for computation of an optical section. Sequences of more than three images per focus plane can be used to compute the contribution of higher harmonics, or, for example, to eliminate 2nd harmonic error. However, this is a tradeoff in terms of speed and light exposure. The particular shift sequence used, the number of images recorded, and the processing formula all affect var-

TABLE 1. FGI image-processing formulas

Grating shift sequence	Computation of optical section	Uniform exposure
1/3-period, 3 images 0°–120°–240°	$(2^{1/2}/3)\ [\ (i_0 - i_{120})^2 + (i_{120} - i_{240})^2 + (i_{240} - i_0)^2]^{1/2}$	yes
1/4-period, 4 images 0°–90°–180°–270°	$(1/2)\ [\ (i_0 - i_{180})^2 + (i_{90} - i_{270})^2]^{1/2}$	yes
1/4-period, 3 images 0°–90°–180°	$(2^{-1/2})\ [\ (i_0 - i_{90})^2 + (i_{90} - i_{180})^2]^{1/2}$	no

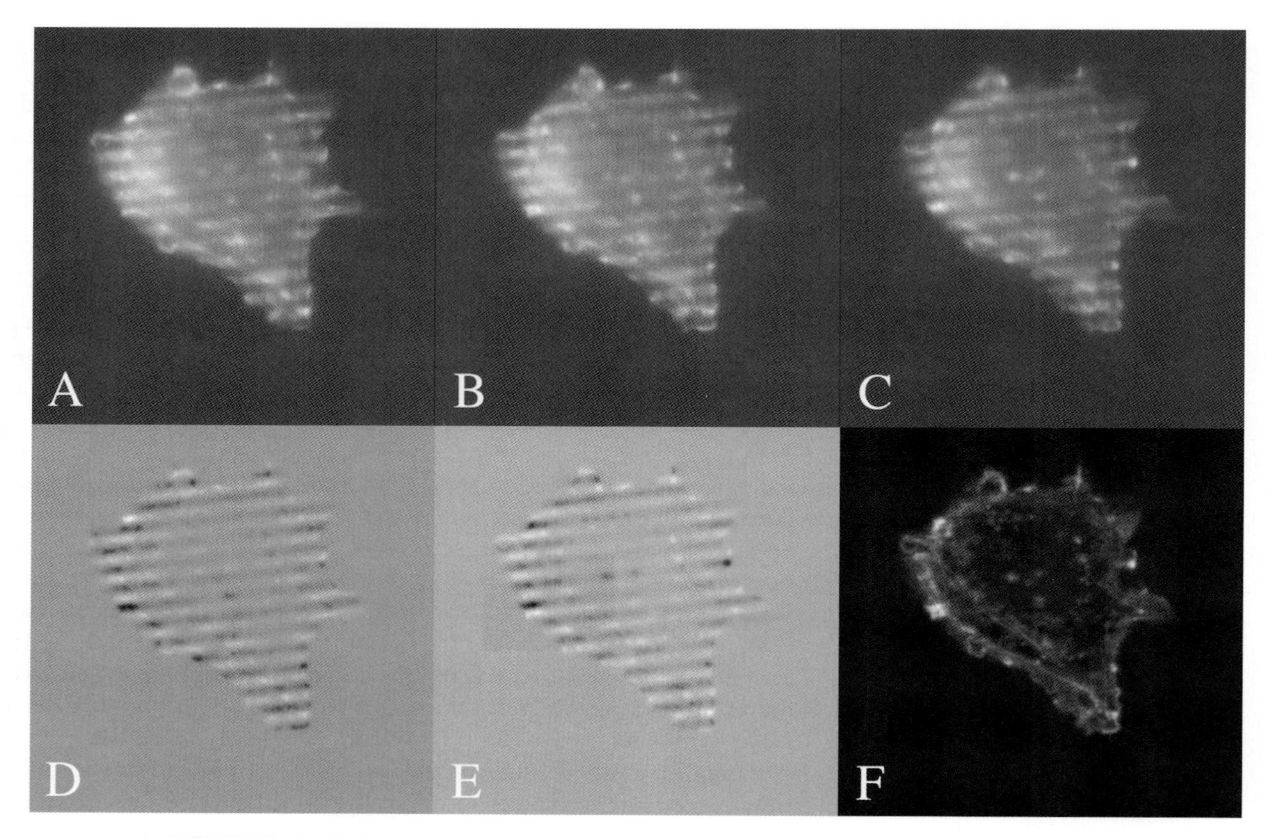

FIGURE 2. Optical section computation from image data showing the actin cytoskeleton in a 3T3 fibroblast. Cells were grown on a cover glass under standard incubator conditions, fixed, permeabilized, and stained with rhodamine-phalloidin to show F-actin. (*A,B,C*) i_0, i_{90}, i_{180} with focus set close to the adherent basal region of the cell. (*D,E*) Difference images (*A,B*) and (*B,C*) as in Table 1, formula 3. In *D* and *E*, the gray scale is bipolar with zero at mid-range gray. (*F*) Optical section computed by Pythagorean summation of *D* and *E*. Field of view: 39 μm.

ious performance characteristics of the instrument: For example, as noted, the 0°–120°–240° shift sequence eliminates 3rd harmonic error. It also uniformly exposes the plane of focus, which minimizes patterned photobleaching (Table 1). In general, light exposure of the specimen in the grating imager is comparable to conventional fluorescence microscopy. Because the Ronchi passes 50% of the incident light, the three exposures required for one optical section deliver 1.5 times the light dose needed for a conventional image, in which the in-focus features are equally bright (but superposed with out-of-focus features).

SHARPNESS OF THE OPTICAL SECTION

On the basis of an analysis of the optical transfer function of the FGI (Lagerholm et al. 2003), it can be shown that the optical section thickness, δ, is related to the NA and projected grating period (L) by the formula

$$\delta = (\lambda/2) / \{[n^2 - (NA - \lambda/L)^2]^{1/2} - [n^2 - NA^2]^{1/2} \} \qquad L > \lambda/ 2NA \tag{2}$$

The combination of a high-NA objective and a fine grating can produce sharp sections that, in principle, match the optical-sectioning performance of a confocal scanner. Ideally, the sharpest sectioning is obtained when the projected grating period is twice Abbe's resolution limit;

$$\delta_{min} = (\lambda/2) / \{ n - [n^2 - NA^2]^{1/2} \} \qquad \text{for } L = \lambda/ NA \tag{3}$$

In practice, a two- to fourfold coarser grating generally gives better performance due to increased stripe contrast. As in any form of fluorescence microscopy, FGI signal per pixel is reduced as sectioning is sharpened.

SIGNAL-TO-NOISE RATIO

Relative to a confocal scanner, performance of the FGI is limited ultimately by photon counting noise due to out-of-focus (background) features in the object. Unlike a confocal scanner, in which the pinhole blocks most background light prior to detection, background light in an FGI is detected in each image, then removed by digital subtraction (Table 1 formulas). Both signal (S) and background (B) photocount levels can be treated as Poisson-distributed variables, with means N_S and N_B. The subtraction steps remove the background mean, N_B, but leave a residual noise with a level roughly equal to the Poisson root-mean-square (RMS) value, $\sqrt{N_B}$. For example, if the background amounts to 10^4 counts, and the in-focus signal 10^3, $\sqrt{N_B} = \sqrt{10^4} = 100$, therefore a signal-to-noise ratio (SNR) of roughly $10^3/100 = 10$ is expected. If the in-focus signal were 100 in a background of 10^4, it could barely be distinguished from the noise. A more detailed analysis shows the RMS additivity of noise in the image processing steps leading to the optical section. The intrinsic FGI noise level can be estimated from the ideal case of removal of a perfectly defocused background in a "0°–120°–240°" image set with S = 0 (blank specimen). By definition, N_B is the same for all three images, and the RMS noise level in the resulting blank optical section is $1.155\sqrt{N_B}$, only 15% greater than the noise level in a single image. Therefore, a simple SNR approximation is the ratio of S to the RMS sum of the Poisson noise levels in the signal ($\sqrt{(S+B)}$) and the background ($\sqrt{B}$): $SNR \approx S/\sqrt{(S + 2B)}$.

SAMPLING IN THE IMAGE PLANE

What sampling is required for a camera to pick up all of the detail in the optical image field formed by a microscope? This is set by Nyquist's sampling theorem, when noise is not a limiting factor: The density of sample points must exceed the inverse of the signal bandwidth. In the case of a microscope, Abbe's resolution limit, $\lambda/2NA$, is the period of the finest grating that can be resolved, and sets the signal bandwidth. The spatial frequency of that grating is $2NA/\lambda$ (cycles/μm), and the bandwidth is defined as the frequency interval from $-2NA/\lambda$ to $2NA/\lambda$, or $4NA/\lambda$. Therefore, the Nyquist sampling interval in nonmagnified coordinates is 1/bandwidth, or $\lambda/4NA$; i.e., the camera pixel spacing divided by total magnification must be $\leq \lambda/4NA$ (Lanni and Baxter 1992). For typical high-NA microscopes, $\lambda/4NA$ is generally close to 0.1 μm. When using a camera having, for example, 6.45 μm CCD elements, a total magnification ≥ 64.5x is required to achieve Nyquist sampling. In real cameras, finite pixel size also causes spatial averaging of the fine details in the image field (Lanni and Baxter 1992). The effect of finite CCD pixel size in the grating imager is a reduction in percentage modulation in an image of the grating projected upon a uniformly fluorescent specimen. Clearly, if the grating period matched the pixel size, no grating would be seen. A graph of percentage modulation versus number of camera pixels across one grating line pair (Fig. 3) shows that modulation is preserved to a high degree when at least 10 pixels span a projected line pair. As an example, a grating chosen to set $L = 5\lambda/2NA$ would also make $L/10$ exactly equal to $\lambda/4NA$, the Nyquist interval.

PERFORMANCE

A very basic measure of performance for any optical-sectioning fluorescence microscope is its axial response, defined as the graph of pixel brightness versus focus increment for a uniform planar specimen. In practice, a stable fluorescent film that is thin compared to the smaller of (1) the wave-optical depth of field ($\sim 2n\lambda/NA^2$) or (2) the optical-sectioning limit, must be used to obtain a meaningful measurement. Figure 4 shows the typical axial response of a grating imager based on a Zeiss Axiovert 200M operated with a 1.30 NA objective (Zeiss 100x Plan-Neofluar) and 1.33 μm projected grating period, as measured using a 0.05-μm rhodamine-labeled thin film specimen. The optical section thickness, defined by the graphical full-width at half-maximum (FWHM), was 0.65 μm. In compari-

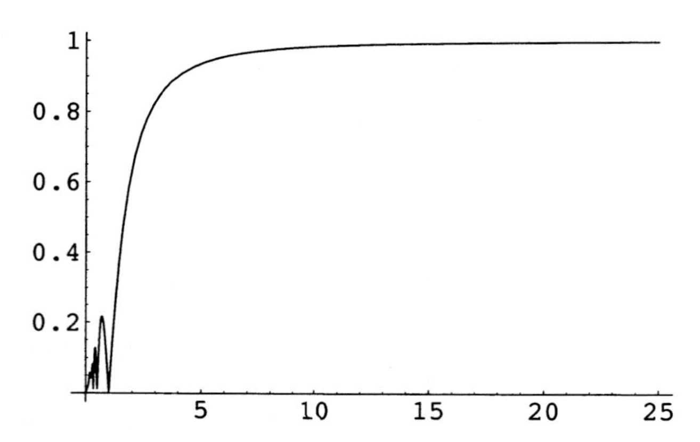

FIGURE 3. Effect of finite camera CCD element size on image contrast. Graph shows the contrast of a projected sine-wave grating versus number of contiguous pixels per cycle (N). The plotted function is $| \sin(\pi/N)/(\pi/N) |$.

son, Equation 2 gives a value $\delta = 0.623$ µm for NA = 1.3, L = 1.33 µm, and $\lambda = 570$ nm (strictly, Eq. 2 should distinguish the excitation and emission wavelengths, but these are usually close, in this example 540 nm and 605 nm, respectively). The experimental FWHM is comparable to a confocal scanner

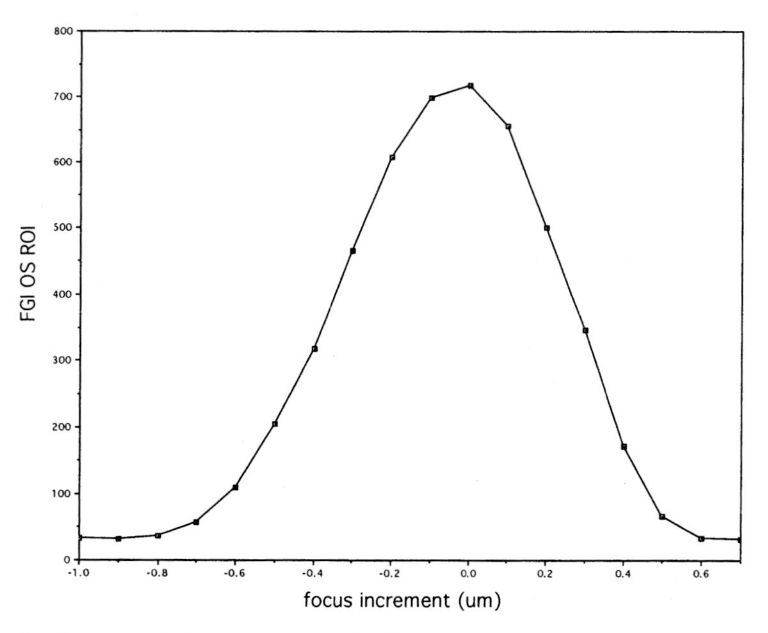

FIGURE 4. Grating imager axial response. Graph shows optical section signal level versus focus drive movement in a small region of interest (ROI) in the center of an image of a thin film. For the test reported here, poly-[methyl methacrylate] covalently labeled with rhodamine (Rh-PMMA) was spin-cast from chlorobenzene onto a cover glass to form a thin film. A simple reflectance interferometer was used to determine the film thickness (50 nm) based on fringe shift and a refractive index equal to 1.49 for PMMA. The cover glass was then mounted with the film in contact with optical cement (Norland Optical Adhesive #60, index = 1.56) on a slide, and UV light was used to polymerize the cement. It was previously determined that Rh-PMMA is insoluble in NOA#60. The advantages of this type of standard specimen are its simple refractive structure and similar refractive indexes (glass 1.52, PMMA 1.49, NOA 1.56). The full-width at half-maximum (FWHM) defines the sharpness of the optical section, here 0.65 µm. Instrument: Zeiss Axiovert 200M with Queensgate piezoelectric grating shifter, 100x 1.30 NA oil-immersion objective, 50 LP/mm Ronchi grating projected to 1.33-µm period, Hamamatsu Orca-ER CCD camera with 6.45-µm pixels, and QED Imaging InVivo instrument control software.

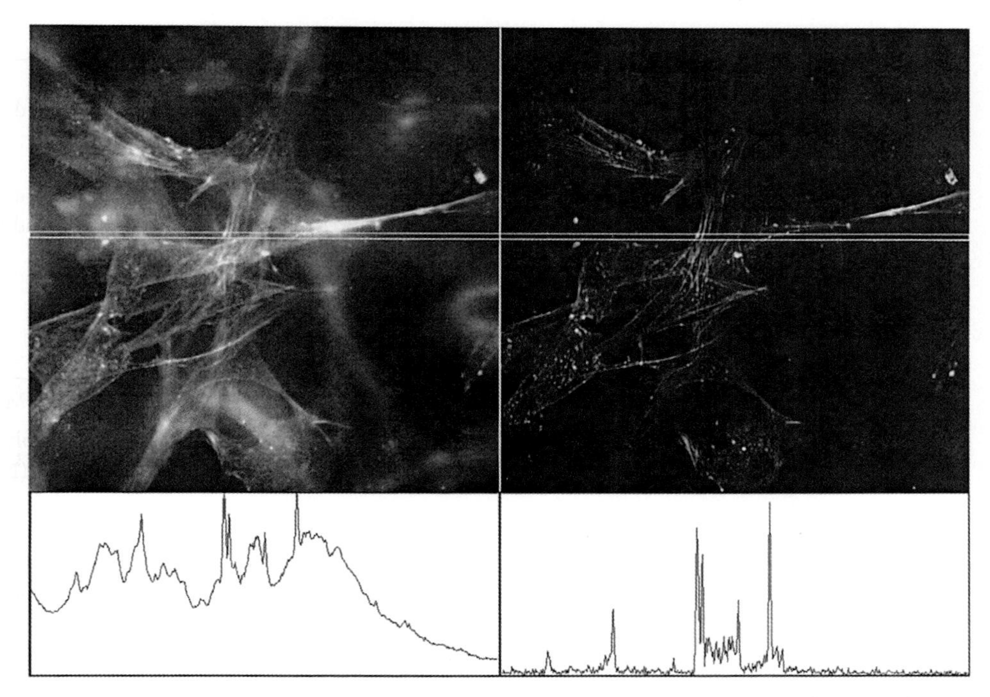

FIGURE 5. Quantification of actin cytoskeletal structures in a grating imager optical section (*upper right*) versus the corresponding conventional image (*upper left*). (*Lower panels*) Pixel-row profile between guidelines shows fluorescence from out-of-focus features mixed with in-focus features, and how attenuation of the background in the optical section makes quantification possible. Gray scale in optical section profile (*lower right*) is boosted 3x.

operated with the same objective. Furthermore, the FGI axial response could be further sharpened by use of a finer grating because, in this example, L exceeded λ/NA by a factor of three (see Eq. 3).

In cell biological applications, where out-of-focus image features are generally less than an order of magnitude greater in brightness than in-focus features, grating imagers generally produce fluorescence optical sections comparable to high-quality confocal scans. Figure 5 shows the quantitative effect of optical sectioning in a specimen of chick embryo fibroblasts grown in a 3D collagen gel, then

TABLE 2. Use factors affecting FGI performance

Minimization of aberration
proper immersion / dial-in spherical aberration correction
proper parfocalization of CCD (at ± 0 diopter ocular setting)
imaging-grade excitation filters (or use external filter wheel)
strain-free dichroic reflector
Alignment of light source
Use of full illumination numerical aperture (INA)
Stability of light source
Precision of exposure timing
Shift sequence: cancellation of harmonics and uniformity of summed exposure in plane of focus
Grating parfocus
Axial color correction
Speed vs. movement in specimen
Nyquist sampling of image field in camera (pixel spacing ≤ λ/4NA)
Oversampling of grating image (N > 10 pixels/line pair)
Camera linearity / >10-bit digitization
Minimization of photobleaching

TABLE 3. Comparison of FGI to other optical-sectioning methods

Advantages	
	Simple accessory to microscope; all other modes unaffected
	Fast compared to point-scan CSFM
	Low light exposure (3 x 50%)
	Get optical section from single focus plane; image stack not required
	Deconvolution not required; processing is noniterative and fast
	Conventional light source
	Standard filter sets
	Axial bandwidth matches CSFM
	Low cost
Disadvantages	
	Slower than Yokogawa-type spinning-disk multipoint CSFM
	Background rejection less than CSFM
	Photobleaching or specimen movement gives stripe artifact

fixed and stained for F-actin with rhodamine phalloidin. The left panel in the figure shows a fluorescence image in the absence of optical sectioning, the right panel shows the FGI optical section at the same plane of focus. A pixel-row profile across the images between the guidelines shows the large effect that out-of focus light would have on quantification of actin density in the cytoskeleton, and the degree of improvement obtained with the grating imager.

As in any form of high-resolution microscopy, a number of use-related factors have been identified that have a significant effect on performance. Table 2 lists those known to date. These can be classified generally as reduction in aberration and grating defocus (which reduce projected grating contrast and image contrast), optimization of the light source and minimization of light exposure, and optimization of sampling. In comparison to confocal scanning, the grating imager has a number of advantages and disadvantages, as listed in Table 3. An outstanding advantage is that the grating imager is compatible with use of a large number of standard high-quality fluorescence filter sets. The only wavelength range limitation so far encountered is longitudinal chromatic aberration when using UV excitation. Finally, although the grating imager has been developed as a fast and compact means of obtaining optical sections, it is possible to apply 3D deconvolution methods to the unprocessed serial-focus image sets. Although this is computationally much more demanding, such processing can be done off-line while still obtaining fast optical sections through one of the image-processing formulas in Table 1.

ACKNOWLEDGMENTS

The author gratefully acknowledges past support from the National Science Foundation IID and STC Programs, and from Carl Zeiss Inc., Thornwood, New York. The author also thanks Winfried Denk for communicating his deep insight into the problem of optical-sectioning microscopy, which inspired much of the past and current work.

REFERENCES

Ben-Levy M. and Peleg E. 1999. Imaging measurement system. U.S. Patent #5,867,604.

Juskaitis R., Neil M.A.A., and Wilson T. 1998. Microscopy using an interference pattern as illumination source. *Proc. SPIE* **3261:** 27–28.

Lagerholm B.C., Vanni S., Taylor D.L., and Lanni F. 2003. Cytomechanics applications of optical sectioning microscopy. *Methods Enzymol.* **361:** 175–197.

Lanni F. and Baxter G.J. 1992. Sampling theorem for square-pixel image data. *Proc. SPIE* **1660:** 140–147.

Lanni F. and Wilson T. 1999. Grating image systems for optical sectioning fluorescence microscopy of cells, tissues, and small organisms. In *Imaging neurons: A laboratory manual* (ed. R. Yuste et al.), 8.1–8.9. Cold Spring Harbor Laboratory Press, Cold Spring Harbor, New York.

Neil M.A.A., Juskaitis R., and Wilson T. 1997. Method of obtaining optical sectioning by using structured light in a conventional microscope. *Opt. Lett.* **22:** 1905–1907.

———. 1998. Real time 3D fluorescence microscopy by two beam interference illumination. *Opt. Commun.* **153:** 1–4.

Neil M.A.A., Squire A., Juskaitis R., Bastiaens P.I.H., and Wilson T. 2000. Wide-field optically sectioning fluorescence microscopy with laser illumination. *J. Microsc.* **197:** 1–4.

Vanni S., Lagerholm B.C., Otey C., Taylor D.L., and Lanni F. 2003. Internet-based image analysis quantifies contractile behavior of individual fibroblasts inside model tissue. *Biophys. J.* **84:** 2715–2727.

Wilson T., Juskaitis R., and Neil M.A.A. 1997. A new approach to three-dimensional imaging in microscopy. *J. Anal. Morphol.* **4:** 231.

Wilson T., Neil M.A.A., and Juskaitis R. 1998a. Real-time three-dimensional imaging of macroscopic structures. *J. Microsc.* **191:** 116–118.

———. 1998b. Optically sectioned images in widefield fluorescence microscopy. *Proc. SPIE* **3261:** 4–6.

———. 1998c. Microscopy imaging apparatus and method. International Patent #WO 98/45745.

CHAPTER 102

Analysis of Dynamic Optical Imaging Data

Bijan Pesaran,[1] Andrew T. Sornborger,[2] Nozomi Nishimura,[3] David Kleinfeld,[3] and Partha P. Mitra[4]

[1]*Department of Biology, Caltech, Pasadena, California 91125;* [2]*Departments of Mathematics and Engineering, University of Georgia, Athens, Georgia 30602;* [3]*Department of Physics, University of California at San Diego, La Jolla, California 92093-0319;* [4]*Cold Spring Harbor Laboratory, Cold Spring Harbor, New York 11724*

Dynamic optical imaging experiments generate large, multivariate data sets that contain signal and noise components of considerable spatiotemporal complexity. Advances in available computational power now make it possible to identify and remove noise components and characterize signal structure in a timely manner through the use of modern signal and image processing techniques. The goal of this chapter is to present these techniques and illustrate their application to the analysis of optical imaging data.

Noise in imaging data arises from two broad categories of sources, biological and technical. Biological sources, such as cardiac and respiratory cycles, are routinely present, as is noise arising from motion of the experimental subject and slow vasomotor oscillations (Mayhew et al. 1996; Mitra et al. 1997; Kalatsky and Stryker 2003). In studies of evoked activity, ongoing brain activity, unrelated to the stimulus, interferes with the signals of interest. Technical noise sources in imaging experiments include photon counting statistics, electronic instrumentation, 50/60 Hz electrical activity, CCD camera refresh functions, and building vibrations, to name but a few. All of this activity combines to mask the neuronal signals of interest.

This chapter presents signal and image processing techniques that have proven useful in the analysis of optical imaging data. The main techniques are drawn from multitaper spectral analysis, harmonic analysis, and the singular value decomposition (SVD). The tools are illustrated on two sets of optical imaging data derived from (1) rat primary somatosensory cortex and (2) cat primary visual cortex.

The spectral methods presented here play a central role in separating physiological artifacts from stimulus responses with trial averaging. Although the procedures overlap to a certain degree, the identification and removal of physiological artifacts make use of the harmonic analysis method, whereas stimulus response characterization uses the periodic stacking method.

Application area	Method
Removal of physiological artifacts	Harmonic analysis
Stimulus response with trial averaging	Periodic stacking method

MATHEMATICAL METHODS

Four tools for the analysis of optical imaging data are presented below: multitaper spectral estimation, harmonic analysis, the SVD (in two different forms), and the periodic stacking method. This presentation is focused on the use of the tools rather than on their derivation. Further information on the technical aspects of the discussion is available in Thomson (1982); Percival and Walden (1993); Mitra and Pesaran (1999); and Sornborger et al. (2003b). With regard to the choice of software, calculations are typically performed using MATLAB (Mathworks), a general purpose language for numerical analysis and visualization. MATLAB has a routine for the calculation of the Slepian functions used in a spectral analysis called **dpss** (discrete prolate spheroidal sequences [Slepian sequences]). MATLAB also has routines to calculate fast Fourier transforms (**fft**) and singular value decompositions (**svd**). MATLAB's statistics toolbox contains a routine to calculate the cumulative and inverse F-distribution functions (**fcdf** and **finv**, respectively). Its signal processing toolbox contains a routine for calculating a multitaper power spectrum estimate (**pmtm**). All of these routines are helpful for coding the methods outlined below.

Multitaper Spectral Estimation

Spectral estimation is based on the premise that the frequency domain is the appropriate basis upon which to examine dynamic activity. This assumes that the activity is stationary. Although this is not usually true of neural activity on long timescales, say hours, it is not unreasonable to suggest that on a sub-second timescale, neural processes change very little. The approach is to then repeat the calculation on neighboring windows overlapping in time, usually displaced by a fixed amount. The result, called the spectrogram, is a time–frequency representation of the function being calculated.

Conventional spectral analysis involves multiplying time series data by a single time series of the same length known as a taper, or more conventionally, as a window function. Examples of such single tapers are Hamming, Hanning, and Cosine tapers. We use multitaper methods in which many tapers are used to operate on a single window in time of the data. The tapers used are the Slepian functions, or discrete prolate spheroidal sequences (DPSS), which form a set of orthogonal functions. The Slepian functions are characterized by a single parameter, W, also called the bandwidth parameter. This parameter specifies the frequency and bandwidth of the Slepian functions. For a given frequency half-bandwidth W and length N, there are approximately $2NW$ Slepian functions $w_t(k)$ ($k = 1,\ldots,\lfloor 2NW \rfloor$, $t = 1,\ldots N$) that have their power concentrated in the frequency range $[-W,W]$.

Step 1: Computing the Slepian Functions

The Slepian functions are characterized by their length, N, and bandwidth parameter, W. The parameters N and W determine the maximum number of usable functions, $K = \lfloor 2NW{-}1 \rfloor$, and they are selected by the investigator using a knowledge of the dynamics of the processes under investigation. Selection is best made iteratively, visual inspection, and some degree of trial and error. $2NW{-}1$ gives the number of effectively independent frequencies over which the spectral estimate is effectively smoothed, so that the variance in the estimate is typically reduced by a factor $2NW{-}1$. Thus, the value of $2W$ is chosen according to the degree of smoothing required. As a rule of thumb, fixing the time bandwidth product NW at a small number (typically 3 or 4) and then varying the window length in time until sufficient spectral resolution is obtained is a reasonable strategy.

Step 2: Computing the Tapered Fourier Transforms

The next step is the computation of the tapered Fourier transforms of the data x_t ($t = 1,\ldots,N$), for each taper $w_t(k)$ ($k = 1,\ldots,K$)

$$\tilde{x}_t(f) = \sum_{t=1}^{N} w_t(k)x_t \exp(-2\pi ift) \tag{1}$$

It is important to note, before taking the tapered Fourier transform, that the data are typically padded with zeros (Mitra and Pesaran 1999) to the nearest power of 2 greater than γN where γ is an integer greater than 2. The zeros are added to one end of the time series, after they have been multiplied by the tapers.

Step 3: Direct Spectral Estimate

The simplest example of the multitaper method is the direct multitaper spectral estimate, $S_{MT}(f)$, which is simply the average over individual tapered spectral estimates

$$S_{MT}(f) = \frac{1}{K}\sum_{k=1}^{K} |\tilde{x}_k(f)|^2 \tag{2}$$

The spectrum may be computed with a moving window to obtain a spectrogram that provides a time–frequency representation of the data.

Harmonic Analysis

Multitaper methods provide a robust and efficient way to carry out harmonic analysis: the analysis of discrete sinusoidal components of activity present in a continuous background. This allows the detection, estimation of parameters, and extraction of the sinusoidal activity on a short moving window.

An optimally sized analysis window is needed. It must be sufficiently small to capture the variations in the amplitude, frequency, and phase, but long enough to have the frequency resolution to separate the relevant peaks in the spectrum, both artifactual and originating in the desired signal.

Step 1: Detection and Estimation of a Sinusoid in a Colored Background

The presence of a sinusoidal component in colored noise background may be detected by a test based on a goodness-of-fit F-statistic (Thomson 1982). The activity is modeled as sinusoid of frequency, f, with a certain amplitude and phase added to a random noise process that is locally white on a scale given by the bandwidth parameter W of the tapers.

$$a(t) = \sum_n A_n \cos[f_n t + \phi_n] + \delta a(t) \tag{3}$$

The amplitude, A_n, and phase, ϕ_n, are given by the complex amplitude, $\mu_n(f_n)$, of the cosine wave. This complex amplitude can be estimated using the tapered Fourier transforms of the data, $x_k(f)$, and the Fourier transform of the tapers themselves at zero frequency, $U_k(0)$. For this application, we find that the data should be padded by a factor of at least 25, and we usually pad by a factor of 100.

$$\mu_n(f_n) = \frac{A_n}{2}\exp(i\phi_n) \tag{4}$$

$$\hat{\mu}_n(f_n) = \frac{\sum_k \tilde{x}_k(f_n) U_k(0)}{\sum_k |U_k(0)|^2}$$

The goodness-of-fit F-statistic, which allows us to test the hypothesis that the sine wave is present at that frequency, is given by

$$F(f_n) = \frac{(K-1)|\hat{\mu}_n(f_n)|^2 \sum_k |U_k(0)|^2}{\sum_k |\tilde{x}_k(f_n) - \hat{\mu}_n(f_n) U_k(0)|^2} \tag{5}$$

The quantity in Equation 5 is F-distributed with $(2,2K-2)$ degrees of freedom. The significance level is chosen to be $1-1/N$ so that, on average, there will be one false detection of a sinusoid across all frequencies.

If the cause of the estimated sinusoid is considered to be noise, as may occur from regular breathing, heartbeat, or electrical noise, it can be subtracted from the data, and the spectrum of the residual time series may be obtained as before.

Step 2: Removal of Periodic Components

Sometimes the parameters of periodic components vary slowly in time, drifting in center frequency, amplitude, or phase. The parameters A_n, f_n, and ϕ_n can be estimated as a function of time by using a moving time window. The goal is to estimate the smooth functions $A_n(t)$, $f_n(t)$, and $\phi_n(t)$ to give the component to be subtracted from the original time series.

The frequency F-test described above is used to determine the fundamental frequency tracks $f_n(t)$ in Equation 3. The time series used for this purpose may be either a single time series in the data or an independently monitored physiological time series. The sequence of the fundamental frequency over time is then used to construct the frequencies for the harmonics and sums and differences of individual oscillations, usually respiration and cardiac rhythms, generated by interactions between them. The final set $f_n(t)$ contains all these frequencies.

The estimated sinusoids are reconstructed for each analysis window, and the successive estimates are overlap-added to provide the final model waveform for the artifacts. If more precision is required, the estimates for the amplitude and phase for each window can be interpolated to each digitization point to allow for nonlinear phase changes over the shift between each window. This is akin to using a shift in time between two successive analysis windows of the sampling rate but achieved at far less computational cost.

More details on implementation of this procedure are given below in the Data Analysis Strategy and Example of Applications sections.

Multivariate Time Series Methods

To this point in the presentation, the operations have been described on univariate data, but optical imaging experiments record many pixels of activity simultaneously, which leads to multivariate time series.

The SVD is a general matrix decomposition of fundamental importance that is equivalent to principal component analysis in multivariate statistics, and generates low-dimensional representations for complex multidimensional time series. These low-dimensional representations are formed from distinct modes that are orthogonal to each other. Consequently, the SVD is a powerful tool to reduce the number of interesting dimensions of the data and to characterize coherent states of activity. Here, we present two applications of the SVD, one to imaging data in its more usual space and time dimensions (the space–time SVD), and one in which the time dimension has been Fourier-transformed into frequency to give the space–frequency SVD. Another application of the SVD, for extracting responses to periodically presented stimuli, is presented below.

Space–Time SVD

The space–time SVD is a one-step operation on the space–time data $I(x,t)$. The SVD of such data is given by

$$I(x,t) = \sum_n \lambda_n I_n(x) a_n(t) \tag{6}$$

where $I_n(x)$ are the eigenmodes of the "spatial correlation matrix" $\int_{-\infty}^{\infty} I(x,t)I(x',t)dt$. Similarly, $a_n(t)$ are the eigenmodes of the "temporal correlation matrix" $\int_{-\infty}^{\infty} I(x,t)I(x,t')dx$. The eigenvalues, λ_n, give the amount of power or variance in each of the ordered space and time eigenmodes. Their relative values give an indication of how large the signal is compared to the noise. Discarding modes with small eigenvalues allows a data dimension reduction for the purposes of visual inspection.

Applications of the SVD on space–time imaging and receptive field data are abundant in the literature: see Golomb et al. (1994) for a didactic presentation. However, in our experience, the SVD, when applied to the space–time data, suffers from a severe drawback because there is no reason why the neurobiologically distinct modes in the data should be orthogonal to each other. In practice, this means segregation of the activity may be prevented because different sources of fluctuations may appear in a single mode of the decomposition, or a single activity pattern may appear across different modes.

In the next section, we give a more effective way of separating distinct components in the data using a decomposition analogous to the space–time SVD, but in the space–frequency domain. The success of the method stems from the fact that the data in question are better characterized by a frequency-based representation.

Space–Frequency SVD

The basic idea is to project the space–time data to a frequency interval, and then perform an SVD on these space–frequency data (Thomson and Chave 1991; Mann and Park 1994; Mitra et al. 1997). Projecting the data on a frequency interval can be performed effectively by using DPSS with the appropriate bandwidth parameter.

Step 1: Constructing the Space–Frequency Matrix

Given the $N_x \times N$ space–time data matrix $I = I(x,t)$, the space–frequency data corresponding to the frequency band $[f - W, f + W]$ are given by the $N_x \times K$ matrix of complex numbers.

$$\tilde{I}(x,k;f) = \sum_{t=1}^{N} I(x,t) w_k(t,W) \exp(2\pi i f t) \tag{7}$$

Step 2: SVD of the Space–Frequency Matrix

We are considering the SVD of the $N_x \times K$ complex matrix with entries $\tilde{I}_n(x;f)$ for fixed f.

$$\tilde{I}_n(x;f) = \sum_n \lambda_n(f) \tilde{I}_n(x;f) \tilde{a}_n(k;f) \tag{8}$$

This SVD can be carried out as a function of the center frequency f, using an appropriate choice of W. At each frequency f one obtains a singular value spectrum $\lambda_n(f)$ $(n = 1,2,...,K)$, the corresponding (in general complex) spatial mode $\tilde{I}_n(x;f)$, and the corresponding local frequency modes $\tilde{a}_n(k;f)$. The interval W separates independent values of frequency in this analysis. The frequency modes can then be projected back into the time domain to give (narrow-band) time-varying amplitudes of the complex eigenimage (Mann and Park 1994).

Step 3: A Measure of Spatial Coherence

In the space–frequency SVD computation, an overall coherence $\bar{C}(f)$ may be defined as (it is assumed that $K \leq N_x$)

$$\bar{C}(f) = \frac{\lambda_1^2}{\sum_{n=1}^{K} \lambda_n^2(f)} \tag{9}$$

The overall coherence spectrum then reflects how much of the fluctuation in the frequency band $[f - W, f + W]$ is captured by the dominant spatial mode. The value ranges between 0 and 1 and, for random data, $\bar{C}(f) \approx 1/K$. This sets a threshold for significance.

Periodic Signals, Trial Averaging, and High-Resolution Methods

Experimental data on the dynamical response of a noisy system to a stimulus typically consist of repeated measurements of the response of the system to one or more stimuli. The most common method for increasing the signal-to-noise ratio (SNR) for stimulus–response data is to average repeated measurements of the response, a method called trial averaging. The methods described above can all be used with simple trial-averaging of responses. Simple trial averaging can be improved, however, by using high-resolution sinusoid detection methods in the frequency domain that were presented above.

One approach that makes use of sinusoid detection methods in the frequency domain is called the *periodic stacking method* (Sornborger et al. 2003b). This method was developed to denoise and characterize the response to stimulus in optical imaging data of the intrinsic signal in cats and macaques. A related method has been used to characterize periodic electrical activity in the heart (Sornborger et al. 2003a). The technique is similar in spirit to that presented by Kalatsky and Stryker (2003), but involves extraction of stimulus information at all harmonics of the stimulus frequency, not just the fundamental. When the signal lies in a distinct band of frequencies, the SNR of periodic stacking estimates significantly increases relative to the trial-averaged estimate. Further improvements can be made in multivariate estimates, in which a subspace of the vector space within which the images lie can be identified as containing statistically significant signal.

The Periodic Stacking Method

We begin by considering univariate stimulus–response data. During an experiment, multiple responses to a stimulus are measured. We assume all response measurements, $x_m(t)$ defined for $0 < t < T$, are of equal duration and concatenate all the M responses to a given stimulus. The resulting function, of duration MT, we denote by $X(t)$. We define $X(mT + t) \approx x_m(t)$ where $x_m(t)$ is the measured response to the mth repetition of the stimulus, of duration T. Since we are measuring M responses to the same stimulus, the signal $X(t)$ is a combination of a T-periodic piece and measurement noise, ε.

$$X(t) = \sum_f \alpha_f \exp(2\pi i f t/T) + \varepsilon(t) \tag{10}$$

To understand the structure of the signal in Fourier space, we perform a Fourier transform on the signal

$$X(t) = \sum_q \tilde{x}_q \exp(2\pi i q t/MT) \tag{11}$$

The coefficients $\tilde{x}_q$ are then given by the expression

$$\tilde{x}_q = \sum_f \alpha_f \delta(q - fM) + \tilde{\varepsilon}(q) \tag{12}$$

where $\tilde{\varepsilon}(q)$ indicates the Fourier transform of the noise, ε. From this expression, we see that the signal is a sequence of harmonics, α_f, at frequencies that are integral multiples of the base, stimulus frequency $2\pi/T$ (i.e., $q/M = f$).

We can estimate the signal from the frequency and complex amplitude of the harmonics that carry the response. We use multitaper harmonic analysis described above to accurately determine the amplitude and phase of the periodic response. This also gives an estimate of the noise and the statistical significance of deterministic sinusoids in a signal using Equations 3, 4, and 5. Since we know the periodicity of the repeated stimulus, we identify and extract the sinusoids in the data that lie at multiples of that base frequency. Response contributions that are not located at frequencies commensurate with the base frequency are discarded as noise. We then recombine the estimated sinusoids to form an estimate of the response as before. This procedure is equivalent to demodulating the data at the stimulus frequency and harmonics of the stimulus frequency and summing the demodulates.

The periodic stacking method can be extended to the case of a multivariate data set using the SVD. It is impractical to analyze a typical set of images pixel by pixel, due to the fact that there are often $p = 10{,}000$ pixels or more per image. So we first perform a singular value decomposition on the data, $I(x,t)$. Usually, most of the variance in the data is captured in the first 100 or 200 eigenfunctions. We can therefore consider the compressed data set

$$I(x,t) = \sum_{n=1}^{Q} \lambda_n I_n(x) a_n(t) \qquad (13)$$

where Q = 200, for example. With this step, we have thrown away P-200 eigenfunctions. However, we hope not to have thrown away the signal. One should always investigate as many eigenfunctions as possible for signal features, especially with unfamiliar data, so as to minimize the risk of throwing away signal in high-index eigenfunctions.

The next step in the analysis is to use multitaper harmonic analysis, as described above, to estimate the amplitudes A_n for the time courses $a_n(t)$ at each harmonic of the stimulus frequency. For this application, the bandwidth of the estimate should be chosen to be slightly less than the stimulus frequency. This choice avoids any significant overlap that might introduce correlations between harmonic estimates.

Following the multitaper harmonic analysis procedure, we can obtain the estimates for sinusoidal components at each harmonic of the stimulus frequency. Then we check, harmonic by harmonic, to see whether there are any statistically significant sinusoids in any of the first Q time series $a_n(t)$. As described above, statistical significance is determined by checking the value of the cumulative distribution function of the F-statistic for a given harmonic component is larger than $1 - 1/N$. We then assemble an estimate of the statistically significant periodic content

$$\hat{I}(x,t) = \sum_{n=1}^{Q} \lambda_n I_n(x) \hat{a}_n(t) \qquad (14)$$

where $\hat{a}_n(t)$ is the sum over statistically significant estimates for each harmonic of the stimulus frequency.

Estimates obtained using the multivariate signal estimation method, outlined above, have higher SNRs than trial-averaged estimates, largely because this method makes use of a measure of the statistical significance of the harmonics across all Q eigenfunctions $a_n(t)$. Therefore, eigenfunctions with no statistically significant content are discarded. Noise associated with these eigenfunctions is thereby eliminated, in contrast to trial-averaged estimates.

The above discussion is simplified, as we only consider the case where a single repeated stimulus is presented to the animal. The intrinsic optical signal measures changes in reflectance of the cortex due to subtle changes in blood oxygenation. Responses to multiple stimuli in optical imaging measurements of the intrinsic signal typically consist of a change in the global blood oxygenation and blood volume that is not related to any particular stimulus (the *nonspecific* response), accompanied by relatively smaller local changes in deoxy- and oxyhemoglobin concentrations that change depending on the stimulus (the *stimulus-specific* response). This approach can also be extended to distinguish between these two aspects of the signal (A.T. Sornborger et al., in prep.).

DATA ANALYSIS STRATEGY

Step 1: Visualization of the Raw Data

Direct visualization of the raw data is the first step to check the quality of the experiment and direct further analysis. Individual time series from the images and movies of the images should be examined. If the images are noisy, for example, due to large shot noise, truncation of a space–time SVD, with possibly some additional smoothing, provides a simple noise reduction step for the visualization (see Fig. 1).

A space–time SVD of the data is computed and the leading principal component time series modeled as a sum of sinusoids. This is useful for two reasons: (1) The images in question typically have many pixels, and it is impractical to perform the analysis separately on all pixels. (2) The leading SVD

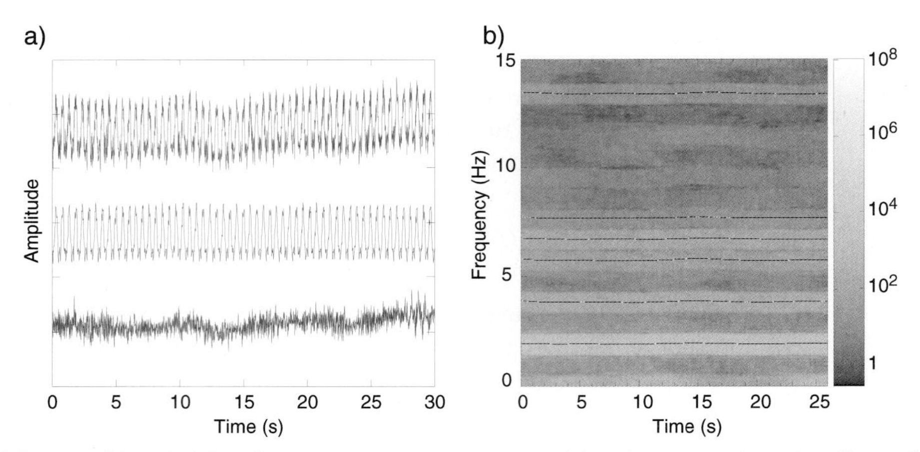

FIGURE 1. Optical imaging data from rat somatosensory cortex. (*a*) Noise suppression of cardiac and respiratory rhythms. The figure shows the results of filtering the respiratory and cardiac rhythms from a single principal component mode in the time domain. The top curve is the raw mode. The middle curve shows the reconstructed noise signal using the overlap-add technique described in the text. The bottom curve shows the residual signal after noise suppression. (*b*) Time–frequency representation of the raw mode from the top of *a*. The black lines show the estimated frequency tracks. The fundamental of the respiratory rhythm was at 1.5 Hz and that of the cardiac rhythm at 7 Hz.

modes capture a large degree of global coherence in the oscillations. However, the procedure may as well be applied to individual image time series.

Step 2: Preliminary Characterization

The next stage aims to identify the various artifacts and determine a preliminary characterization of the signal. A time–frequency spectral estimate described above should be calculated. This can be done on individual pixels, or a space–time SVD can be calculated first, followed by operations on the leading principal components (Fig. 2).

A more powerful characterization is obtained by the space–frequency SVD. There is sufficient frequency resolution in optical data so that, as a practical matter, the oscillatory artifacts tend to segregate well. Studying the overall coherence spectrum reveals the degree to which the images are dominated by the respective artifacts at the relevant frequencies, and the corresponding leading eigenimages show the spatial distribution of these artifacts more cleanly compared to the space–time SVD. Moreover, provided the stimulus response does not completely overlap the artifact frequencies, a characterization is also obtained of the spatiotemporal distribution of the stimulus response.

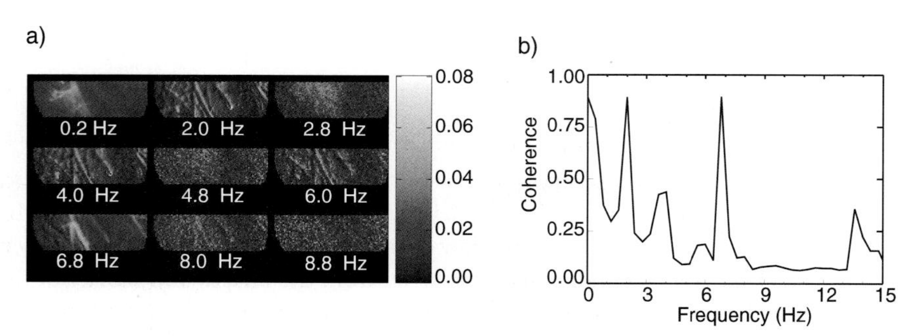

FIGURE 2. Space–frequency SVD. (*a*) The amplitude of the dominant eigenimages as a function of center frequency. The respiratory rhythm is present as a derivative highlighting the blood vessels at 2 Hz and harmonics. The cardiac rhythm is present as an increase in luminescence on the vessel at 6.8 Hz. At intermediate frequencies, spatial structure is diminished. (*b*) Overall coherence for the eigenimages in *a*.

Step 3: Artifact Removal

Based on the preliminary inspection stage, one can proceed to remove the various artifacts to the extent possible. The techniques described in this chapter are most relevant to artifacts that are sufficiently periodic, such as cardiac/respiratory artifacts, 50/60 Hz and other frequency-localized noise such as building or fan vibrations. An example of artifact removal from data from rat primary somatosensory cortex is presented below.

Step 4: Stimulus-Response Characterization

This may be the most delicate step, since the goal of the experiment is usually to find the stimulus response that is not known a priori. If the stimulus is presented periodically and/or repeatedly, as is usually the case, the characterization of stimulus response is fairly straightforward using the periodic stacking method. An example of this technique on data from cat visual cortex is presented below. Alternatively, coherent activity related to a single presentation of the stimulus may also be efficiently extracted by the space–frequency SVD technique if the stimulus response is known a priori to inhabit a particular frequency band (including a low-frequency band). An example of the use of this technique may be found in Prechtl et al. (1997).

EXAMPLE OF APPLICATIONS

In this section, we present an application of the tools described above to intrinsic signal optical imaging data. There are two stages to the analysis, exploratory and confirmatory. Exploratory analysis determines parameters of interest and the structure of any noise present. Following this, the noise is filtered and the signal structure characterized. The first application presents techniques for denoising intrinsic optical imaging data from rat primary somatosensory cortex. The second application presents techniques for characterizing signal structure from intrinsic optical imaging data from cat primary visual cortex.

Application 1

Experimental Method

Wide-field fluorescent imaging was used to record optical signals from the surface of rat somatosensory cortex. The subjects were Sprague-Dawley rats prepared and maintained as described previously (Kleinfeld and Delaney 1996). In brief, bone and dura were removed from a 4 x 4-mm region of the primary vibrissa areas of parietal cortex. The exposed cortex was then stained with the dye RH-795 (Molecular Probes). A metal frame was fixed to the skull and surrounded the craniotomy as a means to rigidly hold the head of the animal to the optical apparatus. With the addition of agarose gel and a cover glass window, this frame further served as an optically clear chamber that sealed and protected the cortex; resealing the craniotomy was crucial for the mechanical suppression of excessive motion that would otherwise result from changes in cranial pressure with each heartbeat and breath.

The fluorescent yield from the cortical surface was recorded with a CCD camera (no. PXL 37; Photometrics). The signal was calculated as the change in fluorescence relative to the background level; i.e., $\Delta F = (I - I_{avg})/I_{avg}$ where I_{avg} is the average intensity in the record. For the sample data shown in Figure 1, the pixel field was 30 x 90, the sampling rate was 95.4 Hz, and records were 3000 frames (286 sec) in length. Each digitized CCD pixel collected up to an estimated 300,000 electrons, so that the sensitivity per pixel per sample (or per frame) was limited to a fractional change of ~0.002. The electrocardiogram, as an indicator of heart rate, and chest expansion, as an indicator of respiration, were further recorded. For the sample data shown in Figure 1, no stimuli were applied to the rat during data acquisition.

We now consider practical issues involved in implementing harmonic analysis on our wide-field imaging data. These considerations hold for both voltage-sensitive dye images (Fig. 1a) and for intrinsic optical imaging of rat somatosensory cortex. Prior to the harmonic analysis, the impact of artifacts associated with large blood vessels was reduced by excluding pixels inside such blood vessels from further analysis. The main harmonic artifacts in the data sets were due to breathing, heartbeat, mixing of the breathing and heartbeat, and a vibrational mode in the building.

As a first means to decrease processing time, the dominant frequency of each harmonic artifact was assumed to be independent of the spatial extent of the images. To identify the artifacts, all pixels in each image frame were averaged to generate a univariate time series and calculate the spectra from the entire time window. In practice, a range of frequencies for each artifact was selected by hand, since the relevant frequencies may drift during the acquisition of the image series. In typical data sets, heartbeat, breathing, and vibration artifact frequencies remained stable enough over several hours to use a fixed set of frequency ranges. The selected frequency ranges were chosen to be wide enough that they covered the variations of the artifact, but narrow enough that only one harmonic artifact was present in each range.

Sliding windows were used to detect and track changes in the amplitude of the artifact over time. These windows were typically 1.5 sec long and were shifted by 0.1 sec. The frequency for each artifact was fit at each point in time by calculating the direct multitaper spectral estimate in each sliding time window (Eqs. 1–4). The frequency with the largest F-statistic within the user-selected artifact band was selected to represent the dominant artifact frequency at that point in time (Fig. 1b). It is possible for several frequencies for one artifact in a time window to be significant according to the F-statistic test, either because the frequency of the artifact may shift, or because the frequency resolution of the spectral estimate is less than the Rayleigh range.

When the time course of artifact frequencies has been identified, the phase and amplitude of the harmonic artifacts can be calculated for each pixel in the series of images. A direct approach is to treat each pixel in the image data as a time series that is used to calculate the phase and amplitude of each artifact at that point in space. This approach can be computationally intensive due to a large number of pixels. Alternatively, as a second procedure to decrease processing time, the space–time SVD was used to reduce the number of components that require harmonic analysis. In practice, a typical number of significant modes was determined empirically to be 0.1 of eigenmodes; for the present example with 3000 modes (Fig. 1a), about 300 independent time series would be analyzed. The harmonic analysis to determine phase and amplitude at the frequency of the artifacts was performed only on the SVD time eigenmodes associated with significant eigenvalues.

In both the pixel-by-pixel and the space–time SVD analysis, each harmonic artifact was modeled by using a sliding window in time. Finally, the reconstructed artifact was subtracted from the original time series (Percival and Walden 1993). For the space–time SVD analysis, the artifact-free time eigenmodes and their corresponding eigenvalues and space eigenmodes were used to calculate a new series of images.

Application 2

Experimental Method

The experiments were carried out on adult (2–5 kg) cats (*Felis domestica*). Anesthesia was induced with intramuscular injections of xylazine [Rompun (Miles), 2 $mg \cdot kg^{1}$] and ketamine [Ketaset (Fort Dodge Laboratories), 10 $mg \cdot kg^{1}$] and was maintained with intravenous (i.v.) infusion of pentothal (Astra) (1–3 $mg \cdot kg^{1} \cdot hr^{1}$). Muscular paralysis was induced by i.v. infusion of pancuronium bromide (Abbott) (1.3 $mg \cdot kg^{1} \cdot hr^{1}$). The state of anesthesia was monitored and maintained carefully in accordance with the National Institutes of Health guidelines. The animals were respired mechanically and the end-expiratory concentration of CO_2 was kept at 3.5–4%. Blood pressure, electroencephalogram, electrocardiogram, and core body temperature were monitored and maintained within normal physiological ranges. A craniotomy and durotomy exposed a region of V1 cortex corresponding to 2–8° eccentricity in the visuotopic representation. A cylindrical, stainless steel, glass-topped chamber was attached to the skull with screws and plumber's epoxy (Propoxy 20, Hercules), and was filled with inert silicone oil. The cortex was illuminated uniformly with 600-nm light and imaged through a tandem-lens configuration by using a cooled 12-bit charge-coupled device (PXL, Photometrics, 536–389 pixels) that was synchronized to the cardiac and respiratory cycles.

An example of results using the periodic stacking method is shown in Figure 3. In panels a and c, we plot the first two eigenfunctions resulting from an SVD of the estimated signal $\hat{I}(x, t)$. In panels b and d, we plot estimates of their time courses plotted with one sigma (i.e., one standard deviation of the standard error of the mean) error bars. The vertical lines denote changing stimulus. The first segment is the response to a 0° oriented drifting grating, the second is the response to a 30° oriented drifting grating, etc. These two eigenfunctions make up 95% of the estimated signal. The eigenfunction in a is responsible for most of the signal at 0° and 90°, and the eigenfunction in c is responsible for most of the signal at 45° and 135°. The envelopes of the responses of these two eigenfunctions form a cosine and sine. This is due to the periodic nature of the stimulus (rotate the orientation of a drifting grating by 180° and the grating is back at 0°). The spatial dappling of the eigenfunctions gives rise to the classic result (Blasdel 1992a,b; Everson et al. 1998) that singularities or pinwheels exist in the response to oriented drifting grating stimuli. As the orientation of the drifting grating changes, the maximum response rotates about the pinwheels. Note that the sharp changes in dynamics in the signal at the boundaries between the changing stimuli are accurately captured. These changes were introduced artificially when the data were concatenated.

ADVANTAGES AND LIMITATIONS

Spectral methods, with averaging of selected frequency bands, rejection of physiological artifacts, and statistical tests of significance, provide a robust means by which to analyze imaging data with significant correlations in space and time. For example, the raw data of Figure 3 had a SNR of 0.0002, which was increased to 21 after processing. The ability to compute confidence limits on the results of the analysis provides a means to compare and contrast features across time and across different data sets. The main limitation of this approach is that, in general, when no model for signal is proposed, the resolution of the estimates is limited because the time–bandwidth product must be greater than 1.

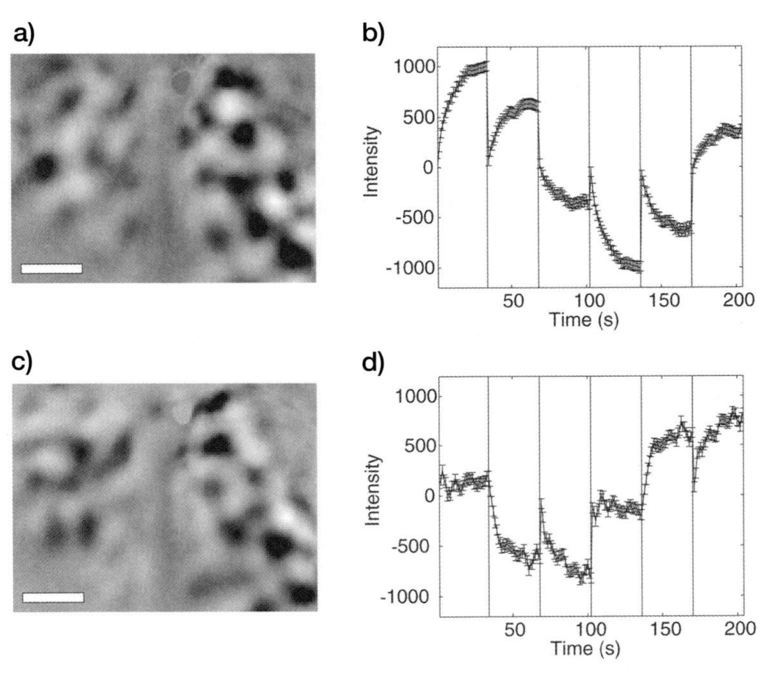

FIGURE 3. Results from a periodic stacking estimate of the stimulus-specific response of cat primary visual cortex to oriented drifting grating stimuli at 0°, 30°, ..., 150°. (*a,c*) The first two eigenfunctions of a singular value decomposition of the estimated signal $\hat{I}(x,t)$. (*b,d*) Estimates of the time courses plotted with one-sigma (i.e., one standard deviation of the standard error of the mean) error bars. See text for discussion.

ACKNOWLEDGMENTS

The authors thank L. Sirovich and T. Yokoo for their collaboration and C. Sailstad for the use of her optical imaging data. This work was supported by grants from the NCR (RR-13419 to D.K.), NINDS (NS-41096 to D.K.), and a National Science Foundation predoctoral fellowship to N.N. Many of the techniques described here were devised or refined as part of the NIMH sponsored "Workshop on the Analysis of Neural Data" at the Marine Biological Laboratories.

REFERENCES

Blasdel G.G. 1992a. Differential imaging of ocular dominance and orientation selectivity in monkey striate cortex. *J. Neurosci.* **12:** 3115–3138.

———. 1992b. Orientation selectivity, preference, and continuity in monkey striate cortex. *J. Neurosci.* **12:** 3139–3161.

Everson R.M., Prashanth A.K., Gabbay M., Knight B.W., Sirovich L., and Kaplan E. 1998. Representation of spatial frequency and orientation in the visual cortex. *Proc. Natl. Acad. Sci.* **95:** 8334–8338.

Golomb D., Kleinfeld D., Reid R.C., Shapley R.M., and Shariman B.I. 1994. On temporal codes and the spatiotemporal response of neurons in the lateral geniculate nucleus. *J. Neurophysiol.* **72:** 2990–3003.

Kalatsky V.A. and Stryker M.P. 2003. New paradigm for optical imaging: Temporally encoded maps of intrinsic signal. *Neuron* **38:** 529–545.

Kleinfeld D. and Delaney K. 1996. Distributed representation of vibrissa movement in the upper layers of somatosensory cortex revealed with voltage sensitive dyes. *J. Comp. Neurol.* **375:** 89–108.

Mann M.E. and Park J. 1994. Global-scale modes of surface temperature. *J. Geophys. Res.* **99:** 25819–25833.

Mayhew J.E., Askew S., Zheng Y., Porrill J., Westby G., Redgrave P., Rector D., and Harper R. 1996. Cerebral vasomotion: 0.1 Hz oscillation in reflected light imaging of neural activity. *Neuroimage* **4:** 183–193.

Mitra P.P. and Pesaran B. 1999. Analysis of dynamic brain imaging data. *Biophys. J.* **76:** 691–708.

Mitra P.P., Ogawa S., Hu X.P., and Ugurbil K. 1997. The nature of spatiotemporal changes in cerebral hemodynamics as manifested in functional magnetic resonance imaging. *Magn. Reson. Med.* **37:** 511–518.

Percival D.B. and Walden A.T. 1993. *Spectral analysis for physical applications.* Cambridge University Press, New York.

Prechtl J., Cohen L.B., Pesaran B., Mitra P.P., and Kleinfeld D. 1997. Visual stimuli induce waves of electrical activity in turtle cortex. *Proc. Natl. Acad. Sci.* **91:** 12467–12471.

Sornborger A., Sirovich L., and Morley G. 2003a. Extraction of periodic multivariate signals: Mapping of voltage-dependent dye fluorescence in mouse heart. *IEEE Trans. Med. Imag.* (in press).

Sornborger A., Sailstad C., Kaplan E., Knight B., and Sirovich L. 2003b. Spatio-temporal analysis of optical imaging data. *Neuroimage* **18:** 610–621.

Thomson D.J. 1982. Spectrum estimation and harmonic analysis. *Proc. IEEE* **70:** 1055–1996.

Thomson D.J. and Chave A.D. 1991. Jackknifed error estimates for spectra, coherences, and transfer functions. In *Advances in spectrum analysis and array processing* (ed. S. Haykin), pp. 58–113. Prentice Hall, Englewood Cliffs, New Jersey.

Electromagnetic Spectrum

Marilu Hoeppner

The *electromagnetic spectrum.* The different forms of electromagnetic radiation include radio waves, IR rays, visible light, UV light, X-rays, and gamma rays. All electromagnetic rays exhibit the properties of waves, each having a characteristic wavelength (λ) and frequency (ν). Electromagnetic waves travel with the speed of light c, which in vacuum is 2.99792458×10^8 m/sec ($\sim 3 \times 1^8$ m/sec). The relationship between the wavelength, the frequency, and the speed of light is given by $c = \lambda\nu$. Thus, it is possible to refer to electromagnetic radiation in terms of wavelength or frequency. Electromagnetic radiation may also be considered in terms of quanta or photons. The relationship between the energy of the electromagnetic wave E and the frequency is given by $E = h\nu$, where h is Planck's constant ($h = 4.14 \times 10^{-15}$ eV sec or $h = 6.63 \times 10^{-34}$ J sec). The higher the frequency of the electromagnetic wave (shorter the wavelength), the greater the amount of energy carried by a photon. Abbreviations: (*Photon energy*) eV–electron volt (1 eV = 1.60219×10^{-19} joule); keV–kiloelectron volt (10^3 eV); MeV–megaelectron volt (10^6 eV). (*Frequency*) Hz–hertz (cycles per second); kHz–kilohertz (10^3 Hz); MHz–megahertz (10^6 Hz); GHz–gigahertz (10^9 Hz); THz–terahertz (10^{12} Hz); PHz–petahertz (10^{15} Hz); EHz–exahertz (10^{18} Hz). (*Wavelength*) m–meter; km–kilometer (10^3 m); Mm–megameter (10^6 m); mm–millimeter (10^{-3} m); µm–micrometer (10^{-6} m); nm–nanometer (10^{-9} m); pm–picometer (10^{-12} m); fm–femtometer (10^{-15} m). Figure provided by Marilu Hoeppner.

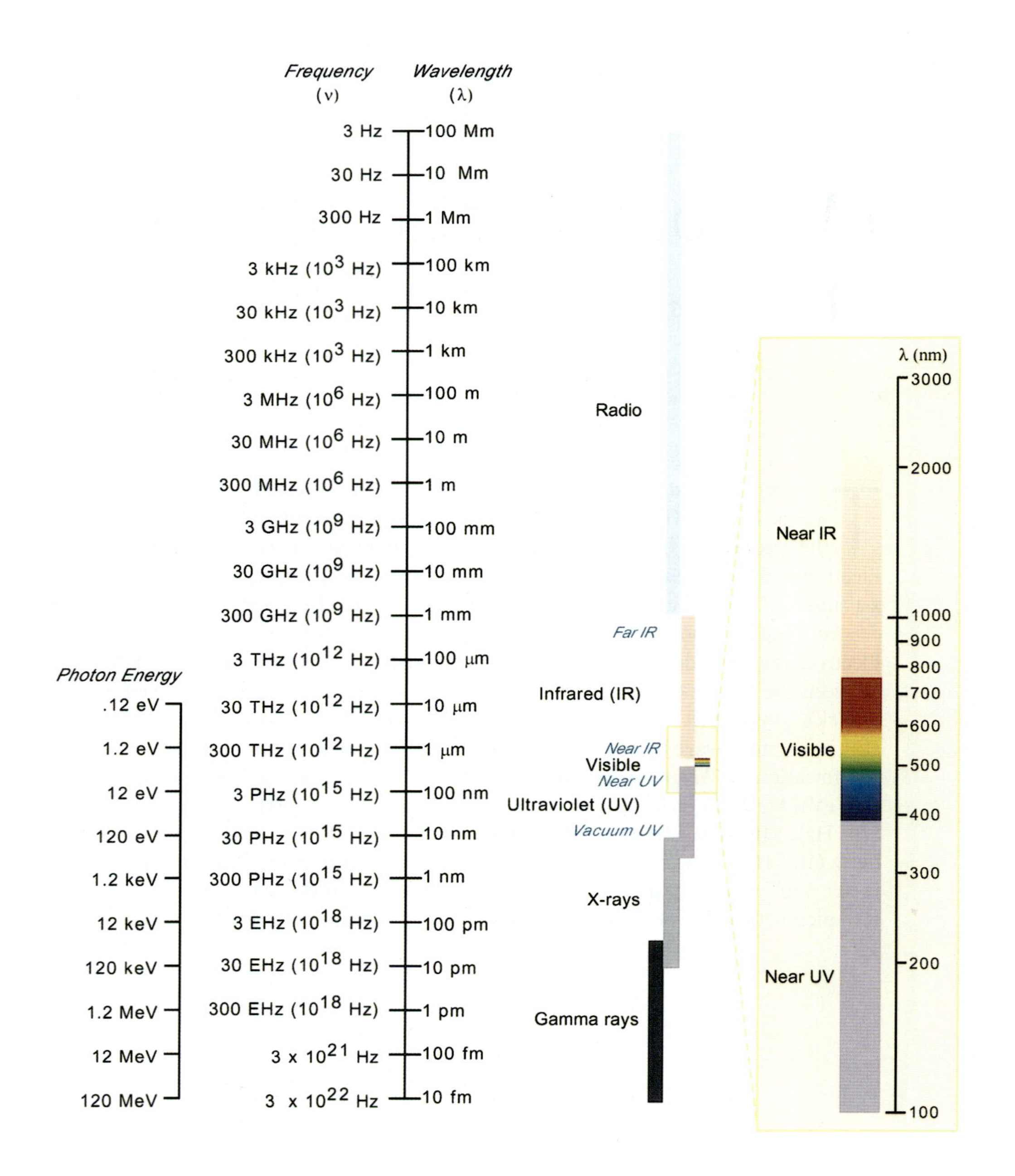
Frequency (ν)
Wavelength (λ)
3 Hz
100 Mm
30 Hz
10 Mm
300 Hz
1 Mm
3 kHz (10^3 Hz)
100 km
30 kHz (10^3 Hz)
10 km
300 kHz (10^3 Hz)
1 km
3 MHz (10^6 Hz)
100 m
30 MHz (10^6 Hz)
10 m
300 MHz (10^6 Hz)
1 m
3 GHz (10^9 Hz)
100 mm
30 GHz (10^9 Hz)
10 mm
300 GHz (10^9 Hz)
1 mm
3 THz (10^12 Hz)
100 µm
30 THz (10^12 Hz)
10 µm
300 THz (10^12 Hz)
1 µm
3 PHz (10^15 Hz)
100 nm
30 PHz (10^15 Hz)
10 nm
300 PHz (10^15 Hz)
1 nm
3 EHz (10^18 Hz)
100 pm
30 EHz (10^18 Hz)
10 pm
300 EHz (10^18 Hz)
1 pm
3 x 10^21 Hz
100 fm
3 x 10^22 Hz
10 fm
Photon Energy
.12 eV
1.2 eV
12 eV
120 eV
1.2 keV
12 keV
120 keV
1.2 MeV
12 MeV
120 MeV
Radio
Far IR
Infrared (IR)
Near IR
Visible
Near UV
Ultraviolet (UV)
Vacuum UV
X-rays
Gamma rays
λ (nm)
3000
2000
1000
900
800
700
600
500
400
300
200
100
Near IR
Visible
Near UV

APPENDIX 2

Microscopy: Lenses, Filters, and Emission/Excitation Spectra

TABLE 1. Excitation and emission maxima for typical fluorochromes

	Exc (nm)	Em (nm)		Exc (nm)	Em (nm)		Exc (nm)	Em (nm)
Acridingelb	470	550	DIDS Thiocyanatostilbene	342	418	Pararosanilin-Feulgen	560	625
Acridinorange + DNA	502	526	DiIC$_{3/4/12/22}$ Indocarbocyanine	540–560	556–575	PBFI	340/380	420
Acridinorange + RNA	460	650	DiOC$_{7/16/2}$ Carbocyanine	550	580	Phosphin	465	565
Acriflavin-Feulgen	480	550–600	DiSC$_{1/2/3/6}$ Thiacarbocyanine	559	585	C-Phycocyanin	605	645
Alizarinkomplexon	595		Diaminonaphtylsuäfonsäure	340	525	B-Phycoerythrin	546/565	575
Allphycocyanin	630	660	DTAF	495	528	R-Phycoerythrin	480–550	578
AMCA	345	425	Ethidiumbromid + DNA	510	595	POPO-1	434	456
7-Amino-Actinomycin D	555	655	Euchrysin	430	540	POPO-3	534	570
Amino-Methylcumarin	354	441	Evans Blue	550	611	Primulin	410	550
6-Amino-Quinolin	360	443	Feulgen	480	560	Propidiumjodid	536	617
Auramin	460	550	Fluo-3	480	520	Pyren	343	380–400
BAO (Ruch)	280	460	Fluoresceindiacetat FDA	499		Pyronin	490–580	530–610
BCECF	430/480	520	Fluoresceinisothiocyanat FITC	490	525	QUIN 2	340/365	490
Berberinsulfat	430	550	Fluoro-Gold	350–395	530–600	Quinacrine mustard	440	510
BFP (Blue Fluoresc. Protein)	382	448	Blue FluoSpheres	360	415	Resorufin	571	585
BOBO-1	462	481	Crimson FluoSpheres	625	645	Rhodamin	540–560	580
BOBO-3	570	602	Red FluoSpheres	580	605	Rhodamine 123	540–560	580
BODIPY	503	512	Dark Red FluoSpheres	650	690	Rhodamine phalloidin	550	575
BODIPY 581/591 phalloidin	584	592	Yellow-Green FluoSpheres	490	515	Rhodol green	500	525
Calcein	495	500–550	FURA-2	340/380	500/530	Säurefuchsin	540	630
Calcein blue	375	420–450	GFP green	471	503	SBFI	340/380	420
Calcium Crimson	588	611	GFP Mutant W	458	480	SNARF	480	600/650
Calcium Green	554	575	GFP Wild Type	396	508	Stilben SITS, SITA	365	460
Calcium Orange	506	531	Hoechst 33258 (Bisbenzimid)	365	480	Sulfaflavin	380–470	470–580
Cascade blue	376,399	423	Hoechst 33342 (Bisbenzimid)	355	465	Tetracyclin	390	560
Catecholamine	410	470	Indo-1	360	410/480	Tetrametylrhodamin	540	566
Chromomycin/Mitramycin	436–460	470	Life/Dead Viability/ Cytotoxicity Kit			Texas red	595	620
Coriphosphin	460	575	Lissamin-Rodamin B	535	580	Thioflavin S	430	550
CPM	385	471	Lucifer Yellow	428	540	Thiazinrot	510	580
CTC 5-cyano-2,3-ditolyl-	602		Magdalrot	540	570	TOTO-1 + DNA	509	533
Tetrazolium-Chlorid			Merocyanin	555	578	TOTO-3 + DNA	642	661
Cyanine Cy2	489	505	4-Metylumbelliferon	360	450	TRITC	540	580
Cyanine Cy3	575	605	Mithramycin	420	575	XRITC	560	620
Cyanine Cy5	640	705	NBD - Amine	460–485	534–542	YOYO-1	491	509
Dansylchlorid	380	475	NBD - Chlorid	480	510–545	YOYO-3	642	660
DAPI + DNA	359	461	Nil Rot	485	525	Xylenolorange	377	610
DiBAC4 Dibutylbarbitursäure Trimethinoxonol	439	516	Olivomycin	350–480	470–630			

Data from E. Keller (Carl Zeiss, Inc.)

TABLE 2. Filter specifications

Fluorochrome[a]	Filter manufacturer	Filter set
Single dye sets		
DAPI	Zeiss	01 or 02
	Nikon	UV-lA, UV-2A, or UV-2B
	Olympus	U-Excitation
	Leitz	A or A2
	Chroma	31000
FITC	Zeiss	10
	Nikon	B-2H, B-1H
	Olympus	B + G520 or IB + G520
	Leitz	L3 or L3.1
	Chroma	31001
Propidium iodide	Zeiss	15 or 14
Rhodamine	Nikon	G-1B or G-2A
Texas red	Olympus	G-Excitation
	Nikon	M2, N2, or N2.1
	Chroma	31002, 31004, 31005
SpectrumGreen	Zeiss	17 or 10
	Nikon	B-2E, B-1E, B-2H, or B-1H
	Olympus	B/G520 or IB/G520
	Leitz	U or L3.1
SpectrumOrange	Zeiss	15 or 14
	Nikon	G-1B or G-2A
	Olympus	G-Excitation
	Leitz	M2, N2, or N2.1
Dual dye sets		
DAPI + FITC	Chroma	51000
	Omega	XP50
DAPI + propidium iodide	Chroma	51002
DAPI + Texas red	Chroma	51003
FITC + Texas red	Chroma	51006
	Omega	XP53
FITC + propidium iodide	Chroma	51005
SpectrumGreen + SpectrumOrange	Zeiss	23
SpectrurnGreen + propidium iodide	Zeiss	23, 19, 16, 11, or 09
	Nikon	B-3A, B-2A, or B-1A
	Olympus	B or IB
	Leitz	H3, I2/3, or K3
SpectrumOrange + DAPI	Imagenetics	DAPI/IO4, IO2/IO4, or DAPI/IO4c/O10c
Triple dye sets		
DAPI + FITC + TRITC	Chroma	61000
DAPI + FITC + propidium iodide	Chroma	61001
DAPI + FITC + Texas red	Chroma	61002
	Omega	XP56
SpectrumGreen SpectrumOrange + DAPI	Imagenetics	DAPI/IO4c/IO10c

Reprinted, with permission, from Bieber (1994, ©John Wiley & Sons).

[a]SpectrumGreen and SpectrumOrange are available from GIBCO/BRL. A variety of other filters are available from the respective manufacturers.

TABLE 3. Properties of objective lenses

Objective lens	Magnification	NA	WD (mm)	Comments	Manufacturer[a]
Achromat	4x	0.1	16		N
	10x	0.25	6.1		O
	20x	0.40	3.0		O
	40x	0.65	0.45		O
	60x	0.80	0.15		O
	100x	1.25	0.13	oil immersion	O
Plan Achromat	0.5x	0.02	7.0		N
	1x	0.04	3.2		N
	2x	0.05–0.06	5–7.5		N,O
	4x	0.10	22–30		N,O
	10x	0.25	10.5		N,O
	20x	0.40	1.30		N,O
	40x	0.65	0.56		N,O
	50x	0.90–0.50	0.20–0.40	oil immersion	N,O
	100x	1.25	0.15–0.17	oil immersion	N,O
Plan Fluor	1.25x	0.04	3.5		Z
(Neo Fluor)	2.5x	0.075	9.3		Z
	4x	0.16	13.0		N,O
	5x	0.15	13.6		Z
	10x	0.30	5.6–10.0		N,O,Z
	20x	0.50	1.3–1.6		N,O,Z
	40x	0.75	0.33–0.51		N,O,Z
	60x	1.25	0.10		N,O
	63x	1.25	0.10	oil immersion	Z
	100x	1.30	0.05–0.10	oil immersion	N,O,Z
Epiplan-Neofluor	1.25x	0.035	3.0		Z
	5x	0.15	13.7	darkfield WD = 5.5	Z
	10x	0.30	5.7	darkfield WD = 3.5	Z
	20x	0.50	1.4	darkfield WD = 1.2	Z
	20x	0.65	0.29	oil, polarizing	Z
	50x	0.80	0.58		Z
	50x	1.0	0.29	oil, polarizing	Z
	100x	0.90	0.24	darkfield WD = 0.27	Z
	100x	1.3	0.13	oil, polarizing	Z
Plan Apochromat	1.25x	0.04	5.10		O
	2x	0.08	6.20		N,O
	4x	0.16	12.2		N,Z
	10x	0.32	1.9		N,Z
	10x	0.45	2.8	optimum resolution	N,Z
	20x	0.60	0.45		N,Z
	20x	0.75	0.61	optimum resolution	N,Z
	40x	0.95	0.13–0.16	correction	N,O,Z
	40x	1.00	0.31	oil immersion	N,Z
	60x	1.40	0.10	oil immersion	N,O
	63x	1.40	0.09	oil immersion	Z
	100x	1.40	0.09–0.10	oil immersion	N,O,Z

[a]This information was collected from Zeiss (Z), Olympus (O), and Nikon (N). NA indicates the numerical aperture of the lens; WD indicates the working distance of the lens. For specialty objectives, contact the respective manufacturers.

FILTERS USED IN FLUORESCENCE MICROSCOPY

The primary filtering element in the epifluorescence microscope is the set of three filters housed in the fluorescence *filter cube*, also called the *filter block*: the *excitation* filter, the *emission filter*, and the *dichroic beamsplitter*. A typical filter cube is illustrated schematically in Figure A3.1.

- The **excitation** filter (also called the *exciter*) is a color filter that transmits only those wavelengths of the illumination light that efficiently excite a specific dye. Common filter blocks are named after the type of excitation filter: UV or U (ultraviolet) for exciting DAPI, Indo-1, etc.; B (blue) for exciting FITC; and G (green) for exciting TRITC, Texas Red, etc. Although shortpass filter designs were used in the past, bandpass filter designs are now used.
- The **emission** filter (also called the *barrier filter* or *emitter*) is a color filter that attenuates all of the light transmitted by the excitation filter and very efficiently transmits any fluorescence emitted by the specimen. This light is always of longer wavelength (more to the red) than the excitation color. These can be either bandpass filters or longpass filters. Common barrier filter colors are blue or pale-yellow in the U-block; green or deep-yellow in the B-block; and orange or red in the G-block.
- The **dichroic beamsplitter** (also called the *dichroic mirror* or *dichromatic beamsplitter*[1]) is a thin piece of specially coated glass (the substrate) set at an angle 45° to the optical path of the microscope. This coating has the unique ability to reflect one color (the excitation light) but transmit another color (the emitted fluorescence). Current dichroic beamsplitters achieve this with great efficiency, i.e., with greater than 90% reflectivity of the excitation along with approximately 90% transmission of the emission. This is a great improvement over the traditional gray half-silvered mirror, which reflects only 50% and transmits only 50%, giving only about 25% efficiency.

Most microscopes have a slider or turret that can hold from two to four individual filter cubes. It must be noted that the filters in each cube are a matched set, and mixing filters and beamsplitters

[1]The term "dichroic" is also used to describe a type of crystal or other material that selectively absorbs light depending on the polarization state of the light. (Polaroid plastic film polarizer is the most common example.) To avoid confusion, the term "dichromatic" is sometimes used.

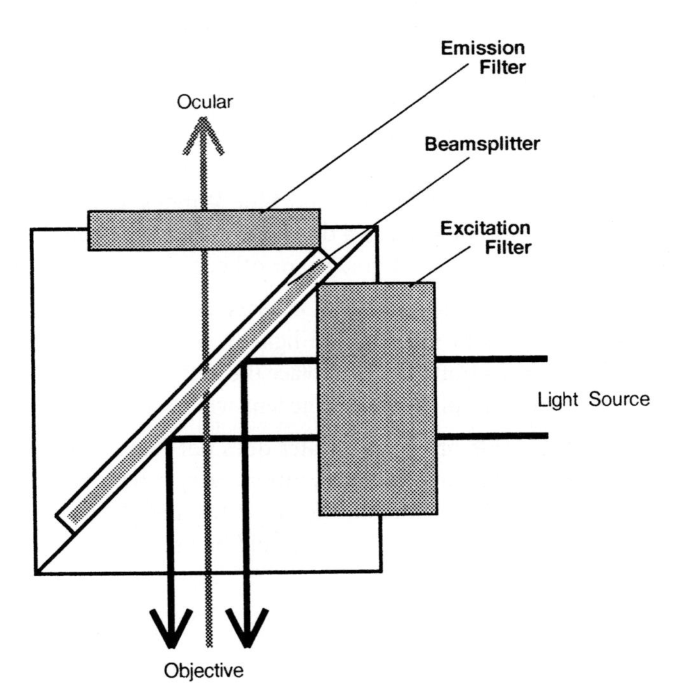

FIGURE 1. Schematic of a fluorescence filter cube.

should be avoided unless the complete spectral characteristics of each filter component are known. Other optical filters can also be found in fluorescence microscopes:

- A heat filter (also called a hot mirror) is incorporated into the illuminator collector optics of most but not all microscopes. It attenuates infrared light (typically wavelengths longer than 800 nm) but transmits most of the visible light.
- Neutral-density filters, usually housed in a filter slider or filter wheel between the collector and the aperture diaphragm, are used to control the intensity of illumination.
- Filters used for techniques other than fluorescence, such as color filters for transmitted-light microscopy and linear polarizing filters for polarized light microscopy, are sometimes installed. (Information courtesy of Chroma Technology Corp.)

CARE AND CLEANING OF OPTICAL EQUIPMENT

Purpose of Cleanliness

Microscopes provide optimal images only if lens surfaces are kept clean. Cleanliness requirements are most stringent in polarizing and interference microscopes, but a fingerprint on any lens of a microscope can spoil the image. Similarly, it is important to use clean, unscratched, high-quality slides and coverslips and to keep fingerprints and all other dirt from their surfaces.

How to Clean a Lens (Adapted from Inoué 1986)

1. Never clean any optic dry.
2. Use only high-quality lens paper.
3. Do not touch anything (even lens paper) to a lens surface except as a last resort. Never rub a lens with lens paper or other items such as Q-Tips, commercial facial tissue, or Kimwipes. Small dust particles are always present in the air and if rubbed into a microscope lens, these dust particles can cause microscopic scratches in the lenses, and the image will be degraded. Commercial facial tissue or bathroom tissue contains diatom frustules (glass) as a filler. These must be avoided for lens cleaning. Lens coatings must be treated with respect.
4. Only use oil on lenses designed to be used with oil. It can be time-consuming to clean a low-power lens that has accidentally been immersed in oil. Dry lenses may become soiled. Soak an unused Q-Tip with a solvent and very gently roll over the surface of the lens once. Pass a second Q-Tip with solvent over the lens gently—barely touch the surface. Various solvents are used. The glass cleaners Sparkle and Blue Windex seem to work well and are inexpensive. If the dirt is water-soluble, place a drop of distilled water on the lens surface and then gently blot it up with lens paper (good quality) or other tissue; however, the lens should not be contacted directly.
5. Inspect the lens using an inverted ocular from the microscope as a magnifier. Observe the objective lens in the reflected light from a lamp (generally in the ceiling). To observe microscopic imperfections and dirt, place the lens to be inspected and cleaned under a dissecting microscope and focus on the lens. The lens will need to be illuminated from above.
6. If Sparkle or distilled water does not remove the dirt, then try ether. Xylene and benzene have been used, but these solvents are very carcinogenic and might also harm the lens cements. Use a 1:1:1 mixture of chloroform, water, and alcohol (well-shaken). Ether also works quite well, but it is very flammable and should be used in a well-ventilated area with no flames nearby.

Keeping Lenses Clean

1. Always clean oil off lens after each session.
2. Keep objectives that are not in use in the screw-top containers in which they came. Always inspect lens surfaces with an inverted ocular before putting them away.

3. Some microscopists find it useful not to clean the lens between slides if they are continually using an oil immersion lens with a series of slides. Instead, wipe off excess oil (especially important in inverted microscopes) with lens paper and place new slide with a bit of new oil onto the old oil, being careful to avoid air bubbles.
4. If using seawater, be especially careful to remove it from all metallic and lens surfaces as it will corrode metals and etch away some lens antireflex coatings.
5. Keep condensers, compensators, etc. in boxes, plastic "baggies," or petri dishes until used.
6. Prohibit all smoking in microscopy laboratories because tobacco tar accumulates on lens surfaces and reduces their efficiency.

Precautions That Help Avoid Damage

1. Keep seawater and other corrosive fluids away from microscopes.
2. Avoid placing microscopes in vulnerable locations, e.g., near the edge of a table. Place high-quality microscopes on air-cushioned optical tables to avoid image deterioration due to microscope movement.
3. When removing objectives, etc., use two hands, and hold the objective over the table so that if an accident occurs, the distance the lens has to fall is minimized.

Personal Safety

1. Never open an ether bottle if there is a flame in the room.
2. Observe the no smoking regulation.
3. Avoid damage to the retinas of your eyes by protecting them from unfiltered arc sources (mercury, xenon, etc.), strobe lamps, and especially lasers.
4. If a high-pressure mercury arc lamp should explode or even stop working, ***evacuate the room without delay***. If possible, open a window as you exit and then notify your institution's safety office.

REFERENCES

Bieber F. 1994. Microscope and image analysis. In *Current protocols in human genetics* (ed. N.C. Dracopoli et al.), p. 4.4.5. John Wiley, New York

Inoué S.1986. *Video microscopy*. Plenum Press, New York.

Spector D.L., Goldman R.D., and Leinwand L.A., Eds. 1998. Microscopy: Lenses, filters, and emission/excitation spectra. In *Cells: A laboratory manual,* vol. 3. *Subcellular localization of genes and their products*, pp. A3.1–A3.10. Cold Spring Harbor Laboratory Press, Cold Spring Harbor, New York.

APPENDIX 3

Cautions

GENERAL CAUTIONS

Please note that the Cautions Appendix in this manual is not exhaustive. Readers should always consult individual manufacturers and other resources for current and specific product information. Chemicals and other materials discussed in text sections are not identified by the icon <!> used to indicate hazardous materials in the protocols. However, they may be hazardous to the user without special handling. Please consult your local safety office or the manufacturer's safety guidelines for further information.

The following general cautions should always be observed.

- **Become completely familiar with the properties of substances used before** beginning the procedure.
- **The absence of a warning** does not necessarily mean that the material is safe, since information may not always be complete or available.
- **If exposed to toxic substances,** contact your local safety office immediately for instructions.
- **Use proper disposal procedures** for all chemical, biological, and radioactive waste.
- **For specific guidelines on appropriate gloves**, consult your local safety office.
- **Handle concentrated acids and bases with great care.** Wear goggles and appropriate gloves. A face shield should be worn when handling large quantities. Do not mix strong acids with organic solvents because they may react. Sulfuric acid and nitric acid especially may react highly exothermically and cause fires and explosions. Do not mix strong bases with halogenated solvent because they may form reactive carbenes that can lead to explosions.
- **Handle and store pressurized gas containers** with caution because they may contain flammable, toxic, or corrosive gases; asphyxiants; or oxidizers. For proper procedures, consult the Material Safety Data Sheet that must be provided by your vendor.
- **Never pipette solutions using mouth suction.** This method is not sterile and can be dangerous. Always use a pipette aid or bulb.
- **Keep halogenated and nonhalogenated solvents separately** (e.g., mixing chloroform and acetone can cause unexpected reactions in the presence of bases). Halogenated solvents are organic solvents such as chloroform, dichloromethane, trichlorotrifluoroethane, and dichloroethane. Some nonhalogenated solvents are pentane, heptane, ethanol, methanol, benzene, toluene, *N,N*-dimethylformamide (DMF), dimethyl sulfoxide (DMSO), and acetonitrile.
- **Laser radiation**, visible or invisible, can cause severe damage to the eyes and skin. Take proper precautions to prevent exposure to direct and reflected beams. Always follow manufacturer's safety guidelines and consult your local safety office. See caution below for more detailed information.
- **Flash lamps**, due to their light intensity, can be harmful to the eyes. They also may explode on occasion. Wear appropriate eye protection and follow the manufacturer's guidelines.
- **Photographic fixatives and developers** also contain chemicals that can be harmful. Handle them with care and follow manufacturer's directions.
- **Power supplies and electrophoresis equipment** pose serious fire hazard and electrical shock hazards if not used properly.
- **Microwave ovens and autoclaves** in the lab require certain precautions. Accidents have occurred involving their use (e.g., to melt agar or bacto-agar stored in bottles or to sterilize). If the screw top is not completely removed and there is not enough space for the steam to vent, the bottles can explode and cause severe injury when the containers are removed from the microwave or autoclave. Always completely remove bottle caps before microwaving or autoclaving. An alternative method for routine agarose gels that do not require sterile agar is to weigh out the agar and place the solution in a flask.
- **Ultra-sonicators** use high-frequency sound waves (16–100 kHz) for cell disruption and other purposes. This "ultrasound," conducted through air, does not pose a direct hazard to humans, but the associated high volumes of audible sound can cause a variety of effects, including headache, nausea, and tinnitus. Direct contact of the body with high-intensity ultrasound (not medical imaging equipment) should be avoided. Use appropriate ear protection and display signs on the door(s) of laboratories where the units are used.
- **Use extreme caution when handling cutting devices** such as microtome blades, scalpels, razor blades, or needles. Microtome blades are extremely sharp! Use care when sec-

tioning. If unfamiliar with their use, have someone demonstrate proper procedures. For proper disposal, use the "sharps" disposal container in your lab. Discard used needles *unshielded*, with the syringe still attached. This prevents injuries (and possible infections; see Biological Safety) while manipulating used needles, since many accidents occur while trying to replace the needle shield. Injuries may also be caused by broken Pasteur pipettes, coverslips, or slides.

GENERAL PROPERTIES OF COMMON CHEMICALS

The hazardous materials list can be summarized in the following categories:

- Inorganic acids, such as hydrochloric, sulfuric, nitric, or phosphoric, are colorless liquids with stinging vapors. Avoid spills on skin or clothing. Spills should be diluted with large amounts of water. The concentrated forms of these acids can destroy paper, textiles, and skin, as well as cause serious injury to the eyes.
- Inorganic bases such as sodium hydroxide are white solids that dissolve in water and under heat development. Concentrated solutions will slowly dissolve skin and even fingernails.
- Salts of heavy metals are usually colored powdered solids that dissolve in water. Many of them are potent enzyme inhibitors and therefore toxic to humans and to the environment (e.g., to fish and algae).
- Most organic solvents are flammable volatile liquids. Avoid breathing the vapors, which can cause nausea or dizziness. Also avoid skin contact.
- Other organic compounds, including organosulfur compounds such as mercaptoethanol or organic amines, can have very unpleasant odors. Others are highly reactive and should be handled with appropriate care.
- If improperly handled, dyes and their solutions can stain not only your sample, but also your skin and clothing. Some of them are also mutagenic (e.g., ethidium bromide), carcinogenic, and toxic.
- All names ending with "ase" (e.g., catalase, β-glucuronidase, or zymolase) refer to enzymes. There are also other enzymes with nonsystematic names like pepsin. Many of them are provided by manufacturers in preparations containing buffering substances, etc. Be aware of the individual properties of materials contained in these substances.
- Toxic compounds are often used to manipulate cells. They can be dangerous and should be handled appropriately.
- Be aware that several of the compounds listed have not been thoroughly studied with respect to their toxicological properties. Handle each chemical with the appropriate respect. Although the toxic effects of a compound can be quantified (e.g., LD_{50} values), this is not possible for carcinogens or mutagens where one single exposure can have an effect. Realize that dangers related to a given compound may also depend on its physical state (fine powder vs. large crystals/diethylether vs. glycerol/dry ice vs. carbon dioxide under pressure in a gas bomb). Anticipate under which circumstances during an experiment exposure is most likely to occur and how best to protect yourself and your environment.

HAZARDOUS MATERIALS

Note: In general, proprietary materials are not listed here. Kits and other commercial items as well as most anesthetics, dyes, fixatives, and stains are also not included. Follow the manufacturer's safety guidelines that accompany these products.

Acridine orange may be a mutagen and may be harmful by inhalation, ingestion, or skin absorption. Wear appropriate gloves and safety glasses. Do not breathe the dust.

Aminobenzoic acid may be harmful by inhalation, ingestion, or skin absorption. Wear appropriate gloves and safety glasses.

Ammonium chloride, NH_4Cl, may be harmful by inhalation, ingestion, or skin absorption. Wear appropriate gloves and safety glasses and use in a chemical fume hood.

Anesthetics: Follow manufacturer's safety guidelines.

Benzyl alcohol is an irritant and may be harmful by inhalation, ingestion, or skin absorption. Wear appropriate gloves and safety glasses. Keep away from heat, sparks, and open flame.

Benzyl benzoate is an irritant and may be harmful by inhalation, ingestion, or skin absorption. Avoid contact with the eyes. Wear appropriate gloves and safety glasses.

Blood (human) and blood products and Epstein-Barr virus. Human blood, blood products, and tissues may contain occult infectious materials such as hepatitis B virus and HIV, which may result in laboratory-acquired infections. Investigators working with EBV-transformed lymphoblast cell lines are also at risk of EBV infection. Any human blood, blood products, or tissues should be considered a biohazard and should be handled accordingly. Wear disposable appropriate gloves, use mechanical pipetting devices, work in a biological safety cabinet, protect against the possibility of aerosol generation, and disinfect all waste materials before disposal. Autoclave contaminated plasticware before disposal; autoclave contaminated liquids or treat with bleach (10% [v/v] final concentration) for at least 30 minutes before disposal. Consult the local institutional safety officer for specific handling and disposal procedures.

α-Bungarotoxin is a potent neurotoxin and is harmful by inhalation, ingestion, or skin absorption. Wear appropriate gloves and safety glasses and always use in a chemical fume hood. Do not breathe the dust.

$CaCl_2$, *see* **Calcium chloride**

Calcium chloride, $CaCl_2$, is hygroscopic and may cause cardiac disturbances. It may be harmful by inhalation, ingestion, or skin absorption. Do not breathe the dust. Wear appropriate gloves and safety goggles.

Calcium nitrate, $Ca(NO_3)_2$, is a strong oxidizer and reacts violently upon contact with many organic substances. Handle with great care. It may be harmful by inhalation, ingestion, or skin absorption. Wear appropriate gloves and safety glasses. Keep away from heat, sparks, and open flame.

$Ca(NO_3)_2$, *see* **Calcium nitrate**

Carbonyl cyanide-M-chlorophenyl-hydrazone (CCCP) may be harmful by inhalation or ingestion. Thermal decomposition will produce cyanide (poison). Wear appropriate gloves and use in a chemical fume hood.

CCCP, *see* **Carbonyl cyanide-M-chlorophenyl-hydrazone**

Cesium hydroxide, CsOH, may be harmful by inhalation, ingestion, or skin absorption. It is extremely destructive to the mucous membranes and upper respiratory tract; inhalation may be fatal. Do not breathe the dust. Wear appropriate gloves and safety glasses and always use in a chemical fume hood.

$C_6H_5CH_3$, *see* **Toluene**

CH_3CH_2OH, *see* **Ethanol**

CsOH, *see* **Cesium hydroxide**

Cyanoacrylate adhesives are harmful by inhalation, ingestion, and skin absorption. Immediate bonding of tissues can occur. Do not pull apart. Inhalation may cause lightheadedness. Wear appropriate gloves and safety glasses. Follow manufacturer's safety guidelines.

DAB, *see* **3,3′-Diaminobenzidine tetrahydrochloride**

Dexamethasone may be harmful by inhalation, ingestion, or skin absorption. Overexposure may cause reproductive disorders and possible risks of harm to the unborn child. Wear appropriate gloves and safety glasses. Do not breathe the dust.

Diaminobenzidine tetrahydrochloride (DAB) is a carcinogen. Handle with extreme care. Avoid breathing vapors. Wear appropriate gloves and safety glasses and use in a chemical fume hood.

Diethyl ether, Et_2O or $(C_2H_5)_2O$, is extremely volatile and flammable. It is irritating to the eyes, mucous membranes, and skin. It is also a CNS depressant with anesthetic effects. It may be harmful by inhalation, ingestion, or skin absorption. Avoid breathing the vapors. Wear appropriate gloves and safety glasses and always use in a chemical fume hood. Explosive peroxides can form during storage or on exposure to air or direct sunlight. Keep away from heat, sparks, and open flame.

***N,N*-Dimethylformamide** (DMF), $HCON(CH_3)_2$, is a possible carcinogen and is irritating to the eyes, skin, and mucous membranes. It can exert its toxic effects through inhalation, ingestion, or skin absorption. Chronic inhalation can cause liver and kidney damage. Wear appropriate gloves and safety glasses and use in a chemical fume hood.

Dimethyl sulfoxide (DMSO) may be harmful by inhalation or skin absorption. Wear appropriate gloves and safety glasses and use in a chemical fume hood. DMSO is also combustible. Store in a tightly closed container. Keep away from heat, sparks, and open flame.

DMSO easily penetrates the skin and is added as a transport mediator in some creams. This property makes it a carrier for other substances, thus potentiating their risks.

DMF, *see* ***N,N*-Dimethylformamide**

DMSO, *see* **Dimethyl sulfoxide**

Epon resin, *see* **Resins**

Ethanol, CH_3CH_2OH, may be harmful by inhalation, ingestion, or skin absorption. Wear appropriate gloves and safety glasses.

Ether, *see* **Diethyl ether**

Et_2O or $(C_2H_5)_2O$, *see* **Diethyl ether**

Fast Green is a carcinogen and may be harmful by inhalation, ingestion, or skin absorption. Wear appropriate gloves and safety glasses and always use in a chemical fume hood.

Formaldehyde, HCHO, is highly toxic and volatile. It is also a possible carcinogen. It is readily absorbed through the skin and is irritating or destructive to the skin, eyes, mucous membranes, and upper respiratory tract. Avoid breathing the vapors. Wear appropriate gloves and safety glasses and always use in a chemical fume hood. Keep away from heat, sparks, and open flame.

Formamide is teratogenic. The vapor is irritating to the eyes, skin, mucous membranes, and upper respiratory tract. It may be harmful by inhalation, ingestion, or skin absorption. Wear appropriate gloves and safety glasses and always use a chemical fume hood when working with concentrated solutions of formamide. Keep working solutions covered as much as possible.

Gluconic acid may be harmful by inhalation, ingestion, or skin absorption. Wear appropriate gloves and safety glasses.

Glutaraldehyde is toxic. It is readily absorbed through the skin and is irritating or destructive to the skin, eyes, mucous membranes, and upper respiratory tract. Wear appropriate gloves and safety glasses and always use in a chemical fume hood.

HCHO, *see* **Formaldehyde**

HCl, *see* **Hydrochloric acid**

H_3COH, *see* **Methanol**

$HCON(CH_3)_2$, *see* **Dimethylformamide**

H_2O_2, *see* **Hydrogen peroxide**

Hydrochloric acid, HCl, is volatile and may be fatal if inhaled, ingested, or absorbed through the skin. It is extremely destructive to mucous membranes, upper respiratory tract, eyes, and skin. Wear appropriate gloves and safety glasses and use with great care in a chemical fume hood. Wear goggles when handling large quantities.

Hydrogen peroxide, H_2O_2, is corrosive, toxic, and extremely damaging to the skin. It may be harmful by inhalation, ingestion, or skin absorption. Wear appropriate gloves and safety glasses and use only in a chemical fume hood.

Isoamyl acetate is flammable and is an irritant. It may be harmful by inhalation, ingestion, or skin absorption. Wear appropriate gloves and safety goggles. Do not breathe the vapors or mist. Keep away from heat, sparks, and open flame.

KCl, *see* **Potassium chloride**

K-gluconate, *see* **Potassium gluconate**

KOH, *see* **Potassium hydroxide**

Laser radiation, both visible and invisible, can be seriously harmful to the eyes and skin and may generate airborne contaminants, depending on the class of laser used. High-power lasers cause permanent eye damage and can burn exposed skin, ignite flammable materials, and activate toxic chemicals that release hazardous by-products. Avoid eye or skin exposure to direct or scattered radiation. Do not stare at the laser and do not point the laser at anyone. Wear appropriate eye protection and use suitable shields that are designed to offer protection for the specific type of wavelength, mode of operation (continuous wave or pulsed), and power output (watts) of the laser being used. Avoid wearing jewelry or other objects that may reflect or scatter the beam. Some non-beam hazards include electrocution, fire, and asphyxiation. Entry to the area in which the laser is being used must be controlled and posted with warning signs that indicate when the laser is in use. Always follow suggested safety guidelines that accompany the equipment and contact your local safety office for further information.

- **Ion lasers** present a hazard due to high-voltage, high-current power supplies. Always follow manufacturer's suggested safety guidelines.
- **Ultraviolet lasers** present a hazard due to invisible beam, high-energy radiation. Always use beam traps, scattered-light shields, and fluorescent beamfinder cards.
- **Blue-green lasers** present a hazard due to photothermal coagulation. Blue and green wavelengths are readily absorbed by blood hemoglobin.

Magnesium chloride, $MgCl_2$, may be harmful by inhalation, ingestion, or skin absorption. Wear appropriate gloves and safety glasses and use in a chemical fume hood.

Magnesium sulfate, $MgSO_4$, may be harmful by inhalation, ingestion, or skin absorption. Wear appropriate gloves and safety glasses and use in a chemical fume hood.

MeOH or **H_3COH**, *see* **Methanol**

Mercury lamps, due to their light intensity, can be harmful to the eyes. They also may explode on occasion. Wear appropriate eye protection and follow manufacturer's guidelines.

Methanol, MeOH or **H_3COH**, is poisonous and can cause blindness. It may be harmful by inhalation, ingestion, or skin absorption. Adequate ventilation is necessary to limit exposure to vapors. Avoid inhaling these vapors. Wear appropriate gloves and goggles and use only in a chemical fume hood.

Methylene blue is irritating to the eyes and skin. It may be harmful by inhalation, ingestion, or skin absorption. Wear appropriate gloves and safety glasses.

$MgCl_2$, *see* **Magnesium chloride**

$MgSO_4$, *see* **Magnesium sulfate**

Monensin may be fatal by inhalation, ingestion, or skin absorption. Wear appropriate gloves and safety goggles and use only in a chemical fume hood.

MOPS, *see* **3-(*N*-Morpholino)-propanesulfonic acid**

3-(*N*-Morpholino)-propanesulfonic acid (MOPS) may be harmful by inhalation, ingestion, or skin absorption. It is irritating to mucous membranes and upper respiratory tract. Wear appropriate gloves and safety glasses and use in a chemical fume hood.

MQAE, *see* ***N*-(Ethoxycarbonylmethyl)-6-methoxyquinolinium bromide**

Na_2HPO_4, *see* **Sodium hydrogen phosphate**

$NaH_2PO_4/Na_2HPO_4/Na_3PO_4$, *see* **Sodium phosphate**

NaOH, *see* **Sodium hydroxide**

***N*-(Ethoxycarbonylmethyl)-6-methoxyquinolinium bromide (MQAE)** is potentially harmful by inhalation, ingestion, or skin absorption. Wear appropriate gloves and safety glasses.

NH_4Cl, *see* **Ammonium chloride**

Nigericin is highly toxic and may be fatal by inhalation, ingestion, or skin absorption. It is irritating to the eyes, skin, mucous membranes, and upper respiratory tract. Wear appropriate gloves and safety goggles and use in a chemical fume hood. Do not breathe the dust.

Nitrogen (gaseous or liquid) may be harmful by inhalation, ingestion, or skin absorption. Wear appropriate gloves and safety glasses. Consult your local safety office for proper precautions.

Osmium tetroxide, OsO_4 (osmic acid), is highly toxic if inhaled, ingested, or absorbed through the skin. Vapors can react with corneal tissues and cause blindness. There is a possible risk of irreversible effects. Wear appropriate gloves and safety goggles and always use in a chemical fume hood. Do not breathe the vapors.

OsO_4, *see* **Osmium tetroxide**

Ouabain is toxic to human cells by inhalation, ingestion, or skin absorption. Wear appropriate gloves and safety goggles and use in a chemical fume hood. Do not breathe the dust.

Paraformaldehyde is highly toxic. It is readily absorbed through the skin and is extremely destructive to the skin, eyes, mucous membranes, and upper respiratory tract. Avoid breathing the dust. Wear appropriate gloves and safety glasses and use in a chemical fume hood. Paraformaldehyde is the undissolved form of formaldehyde.

Permount, *see* **Toluene**

Phenol Red may be harmful by inhalation, ingestion, or skin absorption. Wear appropriate gloves and safety glasses and use in a chemical fume hood.

Phenylthiocarbamide is highly toxic and may be fatal if swallowed. It is harmful by inhalation, ingestion, or skin absorption. Wear appropriate gloves and safety goggles and use only in a chemical fume hood. Do not breathe the dust.

Picric acid powder (Trinitrophenol) is caustic and potentially explosive if it is dissolved and then allowed to dry out. Care must be taken to ensure that stored solutions do not dry out. Handle all concentrated acids with great care. It is also highly toxic and may be harmful by inhalation, ingestion, or skin absorption. Wear appropriate gloves and goggles.

Pluronic acid may be harmful by inhalation, ingestion, or skin absorption. Wear appropriate gloves and safety glasses.

Potassium chloride, KCl, may be harmful by inhalation, ingestion, or skin absorption. Wear appropriate gloves and safety glasses.

Potassium gluconate may be harmful by inhalation, ingestion, or skin absorption. Wear appropriate gloves and safety glasses.

Potassium hydroxide (KOH) and KOH/methanol, are highly toxic and may be fatal if swallowed. They may be harmful by inhalation, ingestion, or skin absorption. Solutions are corrosive and can cause severe burns. They should be handled with great care. Wear appropriate gloves and safety goggles.

Progesterone may be harmful by inhalation, ingestion, or skin absorption. Do not breathe the dust. Wear appropriate gloves and safety glasses and use in a chemical fume hood.

Putrescine is flammable and corrosive and may be harmful by inhalation, ingestion, or skin absorption. Wear appropriate gloves and safety glasses. Keep away from heat, sparks, and open flame.

Resins are suspected carcinogens. The unpolymerized components and dusts may cause toxic reactions, including contact allergies with long-term exposure. Avoid breathing the vapors and dusts. Wear appropriate gloves and safety goggles and always use in a chemical fume hood. Sensitivity to these chemicals may develop with repeated contact.

Selenium dioxide, SeO_2, may be fatal by inhalation, ingestion, or skin absorption. It is highly toxic by inhalation of the dust or vapor. Wear appropriate gloves and safety glasses and always use in a chemical fume hood.

SeO_2, *see* **Selenium dioxide**

Sodium hydrogen phosphate, Na_2HPO_4 (sodium phosphate, dibasic), may be harmful by inhalation, ingestion, or skin absorption. Wear appropriate gloves and safety glasses and use in a chemical fume hood.

Sodium hydroxide (NaOH) and solutions containing NaOH are highly toxic and caustic and should be handled with great care. Wear appropriate gloves and a face mask. All other concentrated bases should be handled in a similar manner.

Sodium phosphate, $NaH_2PO_4/Na_2HPO_4/Na_3PO_4$, is an irritant to the eyes and skin. It may be harmful by inhalation, ingestion, or skin absorption. Wear appropriate gloves and safety goggles. Do not breathe the dust.

Spermidine may be corrosive and harmful by inhalation, ingestion, or skin absorption. Wear appropriate gloves and safety glasses and use in a chemical fume hood.

Streptomycin is toxic and a suspected carcinogen and mutagen. It may cause allergic reactions. It may be harmful by inhalation, ingestion, or skin absorption. Wear appropriate gloves and safety glasses.

TCA, *see* **Trichloroacetic acid**

***N,N,N′,N′*-Tetrakis(2-pyridlmethyl)ethylenediamine (TPEN)** may be harmful by inhalation, ingestion, or skin absorption. Wear appropriate gloves and safety glasses.

Toluene, $C_6H_5CH_3$, vapors are irritating to the eyes, skin, mucous membranes, and upper respiratory tract. Toluene can exert harmful effects by inhalation, ingestion, or skin absorption. Do not inhale the vapors. Wear appropriate gloves and safety glasses and use in a chemical fume hood. Toluene is extremely flammable. Keep away from heat, sparks, and open flame.

TPEN, *see* **(*N,N,N′,N′*-Tetrakis(2-pyridlmethyl)ethylenediamine**

Tricaine is an irritant and may be harmful by inhalation, ingestion, or skin absorption. Wear appropriate gloves and safety glasses.

Tricaine methane sulfonate is an irritant and may be harmful by inhalation, ingestion, or skin absorption. Wear appropriate gloves and safety glasses.

Trichloroacetic acid (TCA) is highly caustic. Wear appropriate gloves and safety goggles.

Trinitrophenol, *see* **Picric acid**

Trypsin may cause an allergic respiratory reaction. It may be harmful by inhalation, ingestion, or skin absorption. Do not breathe the dust. Wear appropriate gloves and safety goggles. Use with adequate ventilation.

Uranyl acetate is toxic if inhaled, ingested, or absorbed through the skin. Wear appropriate gloves and safety glasses and use in a chemical fume hood.

Urethane is a mutagen and suspected carcinogen. It is also highly toxic and is readily absorbed through the skin. It is harmful by inhalation, ingestion, or skin absorption. Wear appropriate gloves and safety glasses. Do not breathe the dust and use only in a chemical fume hood.

UV light and/or **UV radiation** is dangerous and can damage the retina. Never look at an unshielded UV light source with naked eyes. Examples of UV light sources that are common in the laboratory include hand-held lamps and transilluminators. View only through a filter or safety glasses that absorb harmful wavelengths. UV radiation is also mutagenic and carcinogenic. To minimize exposure, make sure that the UV light source is adequately shielded. Wear protective appropriate gloves when holding materials under the UV light source.

Xylazine may be harmful by inhalation, ingestion, or skin absorption. Wear appropriate gloves and safety glasses.

Zinc sulfate, $ZnSO_4$, may be harmful by inhalation, ingestion, or skin absorption. Wear appropriate gloves and safety glasses.

$ZnSO_4$, *see* **Zinc sulfate**

Index